NATURAL ARSENIC IN GROUNDWATERS OF LATIN AMERICA

Arsenic in the Environment

Series Editors

Jochen Bundschuh
International Technical Cooperation Program, CIM (GTZ/BA),
Frankfurt, Germany
Instituto Costarricense de Electricidad (ICE), San José, Costa Rica
Royal Institute of Technology (KTH), Stockholm, Sweden

Prosun Bhattacharya
KTH-International Groundwater Arsenic Research Group, Department of Land and
Water Resources Engineering, Royal Institute of Technology (KTH), Stockholm,
Sweden

ISSN: 1876-6218

Volume 1

ISGSD

International Society of
Groundwater for
Sustainable Development

Cover photo

The cover photo of volume 1 shows a surface water course in Atacama desert, northern Chile, which is naturally contaminated by arsenic deriving from the volcanic mountain chain of the Andes, which can be seen in the background.

The photo is courtesy of Professor Roger Thunvik, KTH-International Groundwater Arsenic Research Group, Department of Land and Water Resources Engineering, Royal Institute of Technology (KTH), Stockholm, Sweden.

Natural Arsenic in Groundwaters of Latin America

Editors

J. Bundschuh

*International Technical Cooperation Program, CIM (GTZ/BA),
Frankfurt, Germany
Instituto Costarricense de Electricidad (ICE), San José, Costa Rica
Royal Institute of Technology (KTH), Stockholm, Sweden*

M.A. Armienta

*Instituto de Geofísica, Universidad Nacional Autónoma de México (UNAM),
Mexico City, Mexico*

P. Birkle

*Instituto de Investigaciones Eléctricas, Gerencia de Geotermia,
Cuernavaca, Morelos, Mexico*

P. Bhattacharya

*KTH-International Groundwater Arsenic Research Group, Department of Land
and Water Resources Engineering, Royal Institute of Technology (KTH), Stockholm,
Sweden*

J. Matschullat

Technical University, Bergakademie, Freiberg, Germany

A.B. Mukherjee

*Environmental Sciences, Department of Biological and Environmental Sciences,
University of Helsinki, Finland*

CRC Press
Taylor & Francis Group
Boca Raton   London   New York   Leiden

CRC Press is an imprint of the
Taylor & Francis Group, an **informa** business

A BALKEMA BOOK

The digital elevation models PIA3395 (World: Figure 74.1), PIA3377 (North America; Figures 13.1; 14.1; 15.1; 17.1; 24.1; 29.1; 72.2), PIA3364 (Central America: Figures 10.1; 12.1) and PIA3388 (South America; Figures 1.1; 5.1; 7.1; 8.1; 33.1; 41.1; 43.1; 74.5) (http://photojournal.jpl.nasa.gov) [Courtesy NASA/JPL-Caltech].

The satellite image of Spain is by Angela King of Geology.com using Landsat data from NASA (Figures 19.1; 30.1; 37.1; 39.1) (http://geology.com/world/satellite-image-of-spain.jpg), it is property of Geology.com and reproduced with permission.

CRC Press/Balkema is an imprint of the Taylor & Francis Group, an informa business

© 2009 Taylor & Francis Group, London, UK

Typeset by Vikatan Publishing Solutions (P) Ltd., Chennai, India
Printed and bound in Great Britain by Cromwell Press Ltd, Trowbridge, Wiltshire

Published by: CRC Press/Balkema
       P.O. Box 447, 2300 AK Leiden, The Netherlands
       e-mail: Pub.NL@taylorandfrancis.com
       www.crcpress.com – www.taylorandfrancis.co.uk – www.balkema.nl

Library of Congress Cataloging-in-Publication Data
Natural arsenic in groundwaters of Latin America / editors, J. Bundschuh ... [et al.].
    p. cm. -- (Arsenic in the environment)
  Includes bibliographical references and index.
  ISBN 978-0-415-40771-7 (hardback : alk. paper) -- ISBN 978-0-203-88623-6 (ebook : alk. paper)
1. Groundwater -- Arsenic -- Latin America. 2. Arsenic -- Toxicology -- Latin America. 3. Hydrogeology -- Latin America. I. Bundschuh, Jochen. II. Title. III. Series.

TD427.A77N385 2008
551.49098--dc22

2008025441

ISBN: 978-0-415-40771-7 (Hbk)
ISBN: 978-0-203-88623-6 (eBook)

About the book series

Although arsenic has been known as the 'silent toxin' since ancient times, and the hazard of arsenic of geogenic origin in drinking water resources was described from different places around the world long ago—e.g. in Argentina in 1917—it is only in the last two decades, that it has received overwhelming public attention worldwide. As a consequence of the biggest arsenic calamity in the world detected more than two decades ago in Southern Asia, there has been an exponential rise in scientific interest that has triggered high quality research. Since then arsenic hazard in drinking water resources, soils, plants and air of predominantly geogenic origin, the propagation of arsenic in the food chain, the chronic affects of arsenic ingestion by humans, and their toxicological and related public health consequences, have been described in many parts of the world, and every year arsenic hazard is discovered in new countries or regions. These discoveries together with the recent lowering of the regulatory limits for arsenic concentration in drinking water to 10 μg/l, means that number of people known to be exposed to drinking water concentrations of arsenic higher than the regulatory limit are increasing, and the estimate of 200 million people exposed worldwide to arsenic through drinking water alone, may now be an underestimate.

The book series "Arsenic in the Environment" is an inter- and multidisciplinary source of information, making an effort to link the occurrence of geogenic arsenic in different environments and media including ground- and surface water, soil and air, and its effect on human society. The series fulfills the growing interest on the arsenic issue worldwide which is being accompanied by stronger regulations on arsenic contents in drinking water and food, which are adapted not only by the industrialized countries, but are also increasingly adapted by the developing countries.

The book series covers all fields of research concerning arsenic in the environment with an aim to present an integrated approach from its occurrence in rocks and their mobilization into the ground- and surface water, soil and air, its transport therein, the pathways of arsenic and their introduction into the food chain and finally the uptake by humans. Human arsenic exposure, bioavailability, metabolism and toxicology are treated together with related public health effects and risk assessments in order to better manage the arsenic-bearing terrestrial and aquatic environments to reduce human arsenic exposure. Arsenic removal technologies and other methodologies to mitigate the arsenic problem are addressed not only from the technological, but also from economic and social point of view considering legislative and political issues and international cooperation, as e.g. international agreements or programs for mitigating the arsenic problem. Only such inter- and multidisciplinary approaches, would allow case-specific selection of optimal mitigation measures of to provide arsenic safe drinking water, food, and air.

We have an ambition to make this book series an international, multi- and interdisciplinary source of knowledge and a platform for arsenic research oriented to the direct solution of problems with considerable social impact and relevance rather than only focusing on cutting edge and breakthrough research in physical, chemical, toxicological, medical and other specific issues on arsenic on a broader environmental realm. The book series shall also form a consolidated source with information on worldwide occurrences of arsenic, which otherwise is dispersed and often hard to access, is bundled, and therefore anybody can easily find and access information about arsenic related topics. It shall also have a role to increase awareness and knowledge among administrators, policy makers and company executives, on the problem and to improve the international and bilateral cooperation on geogenic arsenic hazard and its effects globally.

As consequence we presume this book series as comprehensive information base, which includes authored or edited books from world-leading scientists on their specific field of arsenic research, but will also contain volumes with selected papers from international or regional congresses or other scientific events. Further on, the abstract books of the homonymous international congress series, which we organize biannually in different parts of the world shall appear in this book series. The series shall be open for anybody, in person or as scientific associations, society and scientific networks, for bringing in new book projects. Supported by a strong multi-disciplinary editorial board, book proposals and manuscripts will be peer reviewed and evaluated.

Jochen Bundschuh
Prosun Bhattacharya
(Series Editors)

Editorial board

Table of Contents

About the book series	VII
Editorial board	IX
List of contributors	XIX
Acknowledgements	XXXVII
Editors' foreword	XXXIX

Section I: Regional introduction and overview

1 Occurrence, health effects and remediation of arsenic in groundwaters
of Latin America — 3
*J. Bundschuh, M.E. García, P. Birkle, L.H. Cumbal,
P. Bhattacharya & J. Matschullat*

2 The presence of arsenic in drinking water in Latin America and its effect
on public health — 17
M.L. Castro de Esparza

Section II: Arsenic occurrence and genesis in sedimentary
and hard-rock aquifers

South America

3 Arsenic in groundwater and sediments from La Pampa province, Argentina — 35
P.L. Smedley, H.B. Nicolli, D.M.J. Macdonald & D.G. Kinniburgh

4 Arsenic hydrogeochemistry in groundwater from the Burruyacú basin,
Tucumán province, Argentina — 47
*H.B. Nicolli, A. Tineo, C.M. Falcón, J.W. García, M.H. Merino,
M.C. Etchichury, M.S. Alonso & O.R. Tofalo*

5 Mineralogical study of arsenic-enriched aquifer sediments at Santiago del
Estero, Northwest Argentina — 61
O. Sracek, M. Novák, P. Sulovský, R. Martin, J. Bundschuh & P. Bhattacharya

6 Intermediate to high levels of arsenic and fluoride in deep geothermal
aquifers from the northwestern Chaco-Pampean plain, Argentina — 69
*M.G. García, C. Moreno, M.C. Galindo, M. del V. Hidalgo,
D.S. Fernández & O. Sracek*

7 The origin of arsenic in waters and sediments from Papallacta lake area
in Ecuador — 81
L.H. Cumbal, J. Bundschuh, V. Aguirre, E. Murgueitio, I. Tipán & C. Chávez

8   Arsenic contamination, speciation and environmental consequences
     in the Bolivian plateau   91
     J. Quintanilla, O. Ramos, M. Ormachea, M.E. García, H. Medina, R. Thunvik,
     P. Bhattacharya & J. Bundschuh

9   Using GIS to define arsenic-anomalous catchment basins considering
     drainage sinuosity   101
     A.B. Silva

Central America and Mexico

10   Natural arsenic groundwater contamination of the sedimentary aquifers
     of southwestern Sébaco valley, Nicaragua   109
     M. Altamirano Espinoza & J. Bundschuh

11   Arsenic and water quality of rural community wells in San Juan de Limay,
     Nicaragua   123
     L. Morales, C. Puigdomènech, A. Puntí, E. Torres, C. Canyellas,
     J.L. Cortina & A.M. Sancha

12   Volcanic arsenic and boron pollution of Ilopango lake, El Salvador   129
     D.L. López, L. Ransom, J. Monterrosa, T. Soriano, F. Barahona,
     R. Olmos & J. Bundschuh

13   The abundance of natural arsenic in deep thermal fluids of geothermal
     and petroleum reservoirs in Mexico   145
     P. Birkle & J. Bundschuh

14   Development of a geographic information system for Zimapán
     municipality in Hidalgo, Mexico   155
     E. Ortiz, R. Reséndiz, E. Ramírez, V. Mugica & M.A. Armienta

15   Determining the origin of arsenic in the Lagunera region aquifer,
     Mexico using geochemical modeling   163
     C. Gutiérrez-Ojeda

16   Arsenic mobilization in aquatic sediments of an impacted mining area,
     north-central Mexico   171
     N.A. Pelallo-Martinez, M.C. Alfaro-De la Torre,
     R.H. Lara-Castro & J. Castro-Larragoitia

17   Contamination of drinking water supply with geothermal arsenic
     in Los Altos de Jalisco, Mexico   179
     R. Hurtado-Jiménez & J.L. Gardea-Torresdey

Other countries and general processes

18   Geogenic arsenic in an Australian sedimentary aquifer: Risk awareness
     for aquifers in Latin American countries   193
     B. O'Shea & J. Jankowski

19   Arsenic in a Triassic sandstone aquifer, Castellón, Spain   205
     M.V. Esteller, E. Giménez & I. Morell

20   Arsenic distribution in the groundwater of the Central Gangetic plains
     of Uttar Pradesh, India   215
     Parijat Tripathi, AL. Ramanathan, Pankaj Kumar, Anshumali Singh, P. Bhattacharya,
     R. Thunvik & J. Bundschuh

21 Temporal variations of groundwater arsenic concentrations
    in southwest Bangladesh   225
    M. Jakariya, P. Bhattacharya, M. Manzurul Hassan, K. Matin Ahmed,
    M. Aziz Hasan, Sabiqun Nahar & J. Bundschuh

Section III: Analytical methods for arsenic and laboratory studies

22 Rapid, clean and low-cost assessment of inorganic and total arsenic
    in food by visible and near-infrared spectroscopy   235
    R. Font, A. De Haro-Bailón, D. Vélez, R. Montoro & M. Del Río-Celestino

23 Infield detection of arsenic using a portable digital voltameter, PDV6000 245
    M. Wajrak

24 The use of synchrotron micro-X-ray techniques to determine arsenic
    speciation in contaminated soils   255
    J.L. López-Zepeda, M. Villalobos, M. Gutiérrez-Ruiz, F. Romero,
    M. Marcus & G. Sposito

25 Arsenic speciation study using X-ray fluorescence and cathodic
    stripping voltammetry   265
    L.A. Valcárcel, A. Montero, J.R. Estévez & I. Pupo

26 Dissolution kinetics of arsenopyrite and its implication on arsenic
    speciation in the environment   273
    J. Cama, M.P. Asta, P. Acero & G. De Giudicci

Section IV: Arsenic in soil, plants and food chain issues

Arsenic in soils

27 Geogenic enrichment of arsenic in histosols   285
    T.R. Rüde & H. Königskötter

28 Sorption and desorption behavior of arsenic in the soil   295
    S. Tokunaga & M.G.M. Alam

29 Effect of wastewater irrigation on arsenic concentration in soils and selected
    crops in the state of Hidalgo, Mexico   303
    C.A. Lucho-Constantino, H.M. Poggi-Varaldo, L.M. Del Razo, M.E. Cebrian,
    I. Sastre-Conde, R.I. Beltrán-Hernández & F.R. Prieto-García

30 Arsenic determination in soils from a mining zone in the eastern Pyrenees,
    Catalonia (Spain)   311
    M.J. Ruiz-Chancho, J.F. López-Sánchez & R. Rubio

Arsenic in plants and food

31 Bioavailability of arsenic species in food: Practical aspects for human
    health risk assessments   319
    J.M. Laparra, D. Vélez, R. Montoro, R. Barberá & R. Farré

32 Determination of arsenic content in seafood products in the school meals
    distribution program, *Junta Nacional de Auxilio Escolar y Becas,*
    region VII, Chile   327
    S. Vilches, G. Andrade, O. Muñoz & J.M. Bastías

33   Arsenic contamination from geological sources in environmental
       compartments in a pre-Andean area of Northern Chile   335
       O. Díaz, R. Pastene, N. Núñez, E. Recabarren G., D. Vélez & R. Montoro

34   Total arsenic content in vegetables cultivated in different zones in Chile   345
       A.M. Sancha & N. Marchetti

35   Assimilation of arsenic into edible plants grown in soil irrigated
       with contaminated groundwater   351
       I.M.M. Rahman, M. Nazim Uddin, M.T. Hasan & M.M. Hossain

36   Investigation of arsenic accumulation by vegetables and ferns
       from As-contaminated areas in Minas Gerais, Brazil   359
       H.E.L. Palmieri, M.A.B.C. Menezes, O.R. Vasconcelos,
       E. Deschamps & H.A. Nalini, Jr.

37   Arsenic in plant samples from a contaminated mining area in the
       eastern Pyrenees, Catalonia (Spain)   365
       M.J. Ruiz-Chancho, J.F. López-Sánchez & R. Rubio

38   Soil-to-leaf transfer factor for arsenic in peach (*Prunus persica* L.)   371
       D.L. Orihuela, J.C. Hernández, R.J. López-Bellido, S. Pérez-Mohedano,
       L. Marijuán & N.R. Furet

39   Arsenic uptake and distribution in broccoli, cauliflower and radish plants
       grown on contaminated soil   379
       M. Del Río-Celestino, M.M. Villatoro-Pulido, M.I. De Haro-Bravo,
       R. Font & A. De Haro-Bailón

40   Arsenic mobility in the rhizosphere of the tolerant plant *Viguiera dentata*   387
       R. Briones-Gallardo, G. Vázquez-Rodríguez & M.G. Monroy-Fernández

Section V: Toxicology and metabolism

41   Survey of arsenic in drinking water and assessment of the intake of arsenic
       from water in Argentine Puna   397
       S.S. Farías, G. Bianco de Salas, R.E. Servant, G. Bovi Mitre, J. Escalante,
       R.I. Ponce & M.E. Ávila Carrera

42   Chronic arseniasis in El Zapote, Nicaragua   409
       A. Gómez

43   Transfer of arsenic from contaminated dairy cattle drinking water to milk
       (Córdoba, Argentina)   419
       A. Pérez-Carrera, C. Moscuzza & A. Fernández-Cirelli

44   Molecular mechanisms of arsenic-induced carcinogenesis   427
       K.K. Singh, M. Vujcic & M. Shroff

45   Early signs of immunodepression induced by arsenic in children   435
       L. Vega, G. Soto, A. Luna, L. Acosta, P. Conde, M. Cebrián, E. Calderón,
       L. López & M. Bastida

46   Evaluation of human arsenic contamination in the district of Santa Bárbara,
       Minas Gerais, Brazil   447
       N.O.C. Silva, C.A. Rocha, T.V. Alves, E. Deschamps,
       S.M. Oberdá & J. Matschullat

47 Effects of fluoride and arsenic on the central nervous system 453
   D.O. Rocha-Amador, L. Carrizales, J. Calderón, R. Morales & M.E. Navarro

48 Neurotoxicity of arsenic 459
   M.E. Gonsebatt, J. Limón-Pacheco, E. Uribe-Querol, G. Gutiérrez-Ospina,
   V.M. Rodríguez, M. Giordano, L.M. Del Razo & L.C. Sánchez-Peña

49 Mouse liver cytokeratin 18 (CK18) modulation by sodium arsenite 467
   P. Ramírez, L.M. Del Razo & M.E. Gonsebatt

50 Effects of selenium deficiency on diabetogenic action of arsenite in rats 473
   J.A. Izquierdo-Vega, L.C. Sánchez-Peña, L.M. Del Razo & C. Soto

51 Histological characteristics of sural nerves in rats exposed to arsenite 481
   E. García-Chávez, L.M. Del Razo, B. Segura, H. Merchant & I. Jiménez

52 Arsenic-induced p53-DNA binding activity in epithelial cells 489
   M. Sandoval, M. Morales, A. Ortega, E. López-Bayghen & P. Ostrosky-Wegman

53 Microbial volatilization of arsenic 495
   S. Čerňanský, M. Urík & J. Ševc

Section VI: Treatment and remediation of arsenic-rich groundwater

Natural geological materials—available locally and regionally

54 Feasibility of arsenic removal from contaminated water using indigenous limestone 505
   M.A. Armienta, S. Micete & E. Flores-Valverde

55 Characterization of Fe-treated clays as effective As sorbents 511
   B. Doušová, A. Martaus, D. Koloušek, L. Fuitová, V. Machovič & T. Grygar

56 Natural red earth: An effective sorbent for arsenic removal from Sri Lanka 521
   M. Vithanage, K. Mahatantila, R. Chandrajith & R. Weerasooriya

57 Adsorption of As(V) onto goethite: Experimental statistical optimization 527
   M. Alvarez-Silva, A. Uribe-Salas, F. Nava-Alonso & R. Pérez-Garibay

Chemical methods

58 Subsurface treatment of arsenic in groundwater—experiments at laboratory scale 537
   H.M. Holländer, P.-W. Boochs, M. Billib, T. Krüger, J. Stummeyer & B. Harazim

59 Two-step *in situ* decontamination of mine water enriched with arsenic and iron 547
   B. Doušová, T. Brůha, A. Martaus, D. Koloušek, R. Pažout & V. Machovič

60 Arsenic removal from groundwater using ferric chloride and direct filtration 555
   R.G. Fernández, B. Petrusevski, J. Schippers & S. Sharma

61 The use of iron-coated LECA for arsenic removal from aqueous solutions
   under batch and flow conditions 565
   I. Cano-Aguilera, A.F. Aguilera-Alvarado, G. de la Rosa, R. Fuentes-Ramírez,
   G. Cruz-Jiménez, M. Gutiérrez-Valtierra, M.L. Ramírez-Ramírez & N. Haque

62 Polymer-supported Fe(III) oxide particles: An arsenic-selective sorbent 571
   L.H. Cumbal & A.K. SenGupta

63 Application of coagulation-filtration processes to remove arsenic
   from low-turbidity waters 581
   A.M. Sancha & C. Fuentealba

64   Arsenic removal from groundwater by coagulation with polyaluminum
      chloride and double filtration   589
      R.G. Fernández, A.M. Ingallinella & L.M. Stecca

65   A simple electrocoagulation set up for arsenite removal from water   595
      P.D. Nemade, S. Chaudhari & K.C. Khilar

Other technologies

66   Arsenic in the environment and its remediation by a novel filtration method 605
      T.R. Roth & K.J. Reddy

67   Arsenic removal by solar oxidation in groundwater of Los Pereyra,
      Tucumán province, Argentina   615
      J. d'Hiriart, M. del V. Hidalgo, M.G. García, M.I. Litter & M.A. Blesa

68   Removal of arsenic from groundwater using environmentally reactive
      iron nanoparticles   625
      S.R. Kanel & H. Choi

69   Phytoremediation of arsenic by sorghum (*Sorghum biocolor*) under
      hydroponics   643
      N. Haque, N.S. Mokgalaka, J.R. Peralta-Videa & J.L. Gardea-Torresdey

70   Potential use of sedges (Cyperaceae) in arsenic phytoremediation   649
      M.T. Alarcón-Herrera, O.G. Núñez-Montoya, A. Melgoza-Castillo,
      M.H. Royo-Márquez & F.A. Rodriguez Almeida

71   Filter development from low cost materials for arsenic removal from water   657
      G. Muñiz, L.A. Manjarrez-Nevárez, J. Pardo-Rueda, A. Rueda-Ramírez,
      V. Torres-Muñoz, M.L. Ballinas-Casarrubias & G. González-Sánchez

72   Arsenic removal from water of Huautla, Morelos, Mexico using
      capacitive deionization   665
      S. Garrido, M. Aviles, A. Ramirez, C. Calderon, A. Ramirez-Orozco, A. Nieto,
      G. Shelp, L. Seed, M.E. Cebrian & E. Vera

73   Low-cost technologies for arsenic removal in the Chaco-Pampean plain,
      Argentina   677
      M.E. Morgada de Boggio, I.K. Levy, M. Mateu, M.I. Litter,
      P. Bhattacharya & J. Bundschuh

Section VII: Innovative and sustainable options for arsenic mitigation: Some experiences

74   Arsenic-safe aquifers as a socially acceptable source of safe drinking
      water—What can rural Latin America learn from Bangladesh experiences?   687
      J. Bundschuh, P. Bhattacharya, M. von Brömssen, M. Jakariya, G. Jacks,
      R. Thunvik & M.I. Litter

75   Mitigation actions respond to arsenic exposure in Brazil   699
      E. Deschamps, S.M. Oberdá, J. Matschullat, N.O.C. Silva & O.R. Vasconcelos

Subject index   705

Locality index   735

Author index   741

List of contributors

P. Acero
Institut de Ciències de la Terra "Jaume Almera" (ICTJA-ESP), Barcelona, Catalonia, Spain

L. Acosta,
Sección Externa de Toxicología, Centro de Investigación y de Estudios Avanzados del Instituto Politécnico Nacional (CINVESTAV-IPN), Mexico City, Mexico

A.F. Aguilera-Alvarado
Facultad de Química, Universidad de Guanajuato, Guanajuato, Gto., Mexico

V. Aguirre
Scientific Research Center of Escuela Politecnica del Ejercito (ESPE), Sangolqui, Ecuador

K.M. Ahmed
Department of Geology, University of Dhaka, Dhaka, Bangladesh

M.G.M. Alam
National Institute of Advanced Industrial Science and Technology, Tsukuba, Ibaraki, Japan

M.T. Alarcón-Herrera
Centro de Investigación en Materiales Avanzados (CIMAV), Chihuahua, Chih., Mexico

M.C. Alfaro-De la Torre
Facultad de Ciencias Químicas, Universidad Autónoma de San Luis Potosí (UASLP), San Luis Potosí, S.L.P., Mexico

M.S. Alonso
Departamento de Ciencias Geológicas, Facultad de Ciencias Exactas y Naturales, Universidad de Buenos Aires, Buenos Aires, Argentina

M. Altamirano Espinoza
Centro para la Investigación en Recursos Acuáticos de Nicaragua (CIRA/UNAN), Universidad Nacional Autónoma de Nicaragua, Managua, Nicaragua

M. Alvarez-Silva
Unidad Saltillo, Centro de Investigación y de Estudios Avanzados del Instituto Politécnico Nacional (CINVESTAV-IPN), Ramos Arizpe, Coah., Mexico

T.V. Alves
Laboratório de Contaminantes Metálicos, Divisão de Vigilância Sanitária, Instituto Octávio Magalhães, Fundação Ezequiel Dias (FUNED), Belo Horizonte, Brazil

G. Andrade
Escuela de Ingeniería en Alimentos, Universidad del Bío-Bío, Chillán, Chile

M.A. Armienta
Instituto de Geofísica, Universidad Nacional Autónoma de México (UNAM), Mexico City, Mexico

M.P. Asta
Institut de Ciències de la Terra "Jaume Almera" (ICTJA-ESP), Barcelona, Catalonia, Spain

M.E. Ávila Carrera
Grupo InQA-Investigación Química Aplicada, Facultad de Ingeniería, Universidad Nacional de Jujuy, S.S. de Jujuy, Prov. de Jujuy, Argentina

M. Aviles
Instituto Mexicano de Tecnologia del Agua (IMTA), Jiutepec, Mor., Mexico

M.L. Ballinas-Casarrubias
Facultad de Ciencias Químicas, Universidad Autónoma de Chihuahua, Chih., Mexico

F. Barahona
Departamento de Fisica, Facultad de Ciencias y Humanidades, Universidad de El Salvador, San Salvador, El Salvador

R. Barberá
Facultad de Farmàcia, Universitat de València, Burjassot, Valencia, Spain

J.M. Bastías
Departamento de Ingeniería en Alimentos, Universidad del Bío-Bío, Chillán, Chile

M. Bastida
Jurisdicción Sanitaria, Secretaria de Salubridad y Asistencia, Zimapán, Hgo., Mexico

R.I. Beltrán-Hernández
Centro de Investigaciones Químicas (CIQ), Universidad Autónoma del Estado de Hidalgo (UAEH), Pachuca, Hgo., Mexico

P. Bhattacharya
KTH-International Groundwater Arsenic Research Group, Department of Land and Water Resources Engineering, Royal Institute of Technology (KTH), Stockholm, Sweden

G. Bianco de Salas
Gerencia Química, Comisión Nacional de Energía Atómica, San Martín, Prov. de Buenos Aires, Argentina

M. Billib
Institute of Water Resources Management, Hydrology and Agricultural Hydraulic Engineering, University of Hanover, Hanover, Germany. Now at: Chair of Hydrology and Water Resources Management, Brandenburg University of Technology, Cottbus, Germany

P. Birkle
Instituto de Investigaciones Eléctricas, Gerencia de Geotermia, Cuernavaca, Morelos, Mexico

M.A. Blesa
Gerencia Química, Comisión Nacional de Energía Atómica, and Escuela de Posgrado, Universidad de Gral. San Martín, San Martín, Prov. de Buenos Aires, Argentina

P.-W. Boochs
Institute of Water Resources Management, Hydrology and Agricultural Hydraulic Engineering, University of Hanover, Hanover, Germany and Now at: Chair of Hydrology and Water Resources Management, Brandenburg University of Technology, Cottbus, Germany

G. Bovi Mitre
Grupo InQA-Investigación Química Aplicada, Facultad de Ingeniería, Universidad Nacional de Jujuy, S.S. de Jujuy, Prov. de Jujuy, Argentina

R. Briones-Gallardo
Facultad de Ingeniería-Instituto de Metalurgia, Universidad Autónoma de San Luis Potosí (UASLP), San Luis Potosí, S.L.P., Mexico

T. Brůha
Institute of Chemical Technology in Prague, Prague, Czech Republic

J. Bundschuh
International Technical Cooperation Program, CIM (GTZ/BA), Frankfurt, Germany and Instituto
Costarricense de Electricidad (ICE), San José, Costa Rica

C. Calderón
Instituto Mexicano de Tecnología del Agua (IMTA), Jiutepec, Mor., Mexico

E. Calderón
Sección Externa de Toxicología, Centro de Investigación y de Estudios Avanzados del Instituto
Politécnico Nacional (CINVESTAV-IPN), Mexico City, Mexico

J. Calderón
Facultad de Medicina, Universidad Autónoma de San Luis Potosí (UASLP), San Luis Potosí, S.L.P.,
Mexico

J. Cama
Institut de Ciències de la Terra "Jaume Almera" (ICTJA-ESP), Barcelona, Catalonia, Spain

I. Cano-Aguilera
Facultad de Química, Universidad de Guanajuato, Guanajuato, Gto., Mexico

C. Canyellas
Departament d'Enginyeria Química, Universitat Politècnica de Catalunya, Barcelona, Spain

L. Carrizales
Facultad de Medicina, Universidad Autónoma de San Luis Potosí (UASLP), San Luis Potosí, S.L.P.,
Mexico

M.L. Castro de Esparza
Centro Panamericano de Ingeniería Sanitaria y Ciencias del Ambiente (CEPIS/SDE/OPS-PERU),
Lima, Peru

J. Castro-Larragoitia
Facultad de Ingeniería/Instituto de Geología, Universidad Autónoma San Luis de Potosí (UASLP),
San Luis Potosí, S.L.P., Mexico

M. Cebrián
Sección Externa de Toxicología, Centro de Investigación y de Estudios Avanzados del Instituto
Politécnico Nacional (CINVESTAV-IPN), Mexico City, Mexico

S. Čerňanský
Department of Ecosozology and Physiotactics, Faculty of Natural Sciences, University in
Bratislava, Bratislava, Slovakia

R. Chandrajith
Department of Geology, University of Peradeniya, Peradeniya, Sri Lanka

S. Chaudhari
Centre for Environmental Science and Engineering, Indian Institute of Technology IIT Bombay,
Powai, Mumbai, India

C. Chávez
Scientific Research Center of Escuela Politecnica del Ejercito (ESPE), Sangolqui, Ecuador

H. Choi
Department of Environmental Science and Engineering, Gwangju Institute of Science and
Technology (GIST), Gwangju, Korea

P. Conde
Sección Externa de Toxicología, Centro de Investigación y de Estudios Avanzados del Instituto Politécnico Nacional (CINVESTAV-IPN), Mexico City, Mexico

J.L. Cortina
Departament d'Enginyeria Química, Universitat Politècnica de Catalunya, Barcelona, Spain

G. Cruz-Jiménez
Facultad de Química, Universidad de Guanajuato, Guanajuato, Gto., Mexico

L.H. Cumbal
Centro de Investigación Científica, Escuela Politecnica del Ejercito (ESPE), Sangolqui, Ecuador

J. d'Hiriart
Facultad de Ciencias Naturales e Instituto Miguel Lillo, Universidad Nacional de Tucumán, San Miguel de Tucumán, Prov. de Tucumán, Argentina

G. De Giudicci
Department of Earth Science, University of Cagliari, Cagliari, Italy

A. De Haro-Bailón
Instituto de Agricultura Sostenible (CSIC), Córdoba, Spain

M.I. De Haro-Bravo
Centro IFAPA Alameda del Obispo, Córdoba, Spain

G. de la Rosa
Facultad de Química, Universidad de Guanajuato, Guanajuato, Gto., Mexico

L.M. Del Razo
Sección Externa de Toxicología, Centro de Investigación y de Estudios Avanzados del Instituto Politécnico Nacional (CINVESTAV-IPN), Mexico City, Mexico

M. Del Río-Celestino
Centro IFAPA Alameda del Obispo, Córdoba, Spain

E. Deschamps
Fundação Estadual do Meio Ambiente (FEAM), Belo Horizonte, Minas Gerais, Brazil

O. Díaz
Facultad de Química y Biología, Universidad de Santiago de Chile, Santiago de Chile, Chile

B. Doušová
Institute of Chemical Technology in Prague, Prague, Czech Republic

J. Escalante
Grupo InQA-Investigación Química Aplicada, Facultad de Ingeniería, Universidad Nacional de Jujuy, S.S. de Jujuy, Prov. de Jujuy, Argentina

M.V. Esteller
Centro Interamericano de Recursos del Agua (CIRA), Universidad Autónoma del Estado de México, Toluca, Mexico

J.R. Estévez
Centro de Aplicaciones Tecnológicas y Desarrollo Nuclear (CEADEN), Havana, Cuba

M.C. Etchichury
Museo Argentino de Ciencias Naturales Bernardino Rivadavia, Buenos Aires, Argentina

C.M. Falcón
Cátedra de Hidrogeología, Facultad de Ciencias Naturales e Instituto Miguel Lillo, Universidad Nacional de Tucumán, Tucumán, Prov. de Tucumán, Argentina

S.S. Farías
Gerencia Química, Comisión Nacional de Energía Atómica, San Martín, Prov. de Buenos Aires, Argentina

R. Farré
Facultad de Farmàcia, Universitat de València, Burjassot, Valencia, Spain

D.S. Fernández
Servicio Geológico Minero Argentino, Delegación Tucumán, Tucumán, Prov. de Tucumán, Argentina

R.G. Fernández
Centro de Ingeniería Sanitaria (CIS), Facultad de Ciencias Exactas, Ingeniería y Agrimensura, Universidad Nacional de Rosario, Rosario, Prov. de Santa Fe, Argentina

A. Fernández-Cirelli
Centro de Estudios Transdisciplinarios del Agua, Facultad de Ciencias Veterinarias, Universidad de Buenos Aires, Buenos Aires, Argentina

E. Flores-Valverde
Posgrado en Ciencias e Ingeniería Ambientales, Universidad Nacional Autónoma de México (UNAM), Mexico City, Mexico

R. Font
Instituto de Agricultura Sostenible (CSIC), Córdoba, Spain

C. Fuentealba
División de Recursos Hídricos y Medio Ambiente, Facultad de Ciencias Físicas y Matemáticas, Universidad de Chile, Santiago de Chile, Chile

R. Fuentes-Ramírez
Facultad de Química, Universidad de Guanajuato, Guanajuato, Gto., Mexico

L. Fuitová
Institute of Chemical Technology in Prague, Prague, Czech Republic

N.R. Furet
Instituto Superior de Ciencias y Tecnologías Superiores, La Habana, Cuba

M.C. Galindo
Facultad de Ciencias Naturales and Inst. Miguel Lillo, Universidad Nacional de Tucumán, Tucumán, Prov. de Tucumán, Argentina

J.W. García
Cátedra de Hidrogeología, Facultad de Ciencias Naturales e Instituto Miguel Lillo, Universidad Nacional de Tucumán, Tucumán, Prov. de Tucumán, Argentina

M.E. García
Instituto de Investigaciones Químicas, Universidad Mayor de San Andrés, La Paz, Bolivia

M.G. García
CIGeS, Facultad de Ciencias Exactas, Físicas y Naturales (FCEFyN), Universidad Nacional de Córdoba, Córdoba, Prov. de Córdoba, Argentina

E. García-Chávez
Sección Externa de Toxicología, Centro de Investigación y de Estudios Avanzados del Instituto Politécnico Nacional (CINVESTAV-IPN), Mexico City, Mexico

J.L. Gardea-Torresdey
Department of Chemistry, The University of Texas at El Paso, El Paso, TX, USA

S. Garrido
Instituto Mexicano de Tecnologia del Agua (IMTA), Jiutepec, Mor., Mexico

E. Giménez
Área Dptal. Ciencia y Tecnología Agroforestal y Ambiental, Facultad de Ciencias y Artes, Universidad Católica De Ávila (UCAV), Ávila, Spain

M. Giordano
Instituto de Neurobiología, Universidad Nacional Autónoma de Mexico (UNAM), Querétaro, Mexico

A. Gómez
UNICEF-Nicaragua, Managua, Nicaragua

M.E. Gonsebatt
Instituto de Investigaciones Biomédicas, Universidad Nacional Autónoma de México (UNAM), Mexico City, Mexico

G. González-Sánchez
Depto. de Medio Ambiente y Energía, Centro de Investigación en Materiales Avanzados, S.C., Chih., Mexico

T. Grygar
Institute of Inorganic Chemistry, Rež, Czech Republic

C. Gutiérrez-Ojeda
Instituto Mexicano de Tecnología del Agua (IMTA), Jiutepec, , Mexico

G. Gutiérrez-Ospina
Instituto de Investigaciones Biomédicas, Universidad Nacional Autónoma de México (UNAM), Mexico City, Mexico

M. Gutiérrez-Ruiz
LAFQA, Instituto de Geografía, Universidad Nacional Autónoma de México (UNAM), Mexico City, Mexico

M. Gutiérrez-Valtierra
Facultad de Química, Universidad de Guanajuato, Guanajuato, Gto., Mexico

N. Haque
Environmental Science and Engineering PhD Program, The University of Texas at El Paso, El Paso, TX, USA

B. Harazim
Federal Institute of Geosciences and Natural Resources of Germany (BGR), Hanover, Germany

M. Aziz Hasan
Department of Geology, Dhaka University, Dhaka, Bangladesh

M.T. Hasan
Applied Research Laboratory, Department of Chemistry, University of Chittagong, Chittagong, Bangladesh

M. Manzurul Hassan
Department of Geography and Environment, Jahangirnagar University, Savar, Dhaka, Bangladesh

J.C. Hernández
Escuela Politécnica Superior, Universidad de Huelva, Huelva, Spain

Hidalgo, M. del V.
Facultad de Ciencias Naturales e Instituto Miguel Lillo, Universidad Nacional de Tucumán, San Miguel de Tucumán, Prov. de Tucumán, Argentina

H.M. Holländer
Institute of Water Resources Management, Hydrology and Agricultural Hydraulic Engineering, University of Hanover, Hanover, Germany. Now at: Chair of Hydrology and Water Resources Management, Brandenburg University of Technology, Cottbus, Germany

M.M. Hossain
Institute of Forestry and Environmental Sciences, University of Chittagong, Chittagong, Bangladesh

R. Hurtado-Jiménez
El Colegio de la Frontera Norte, A.C., Ciudad Juárez, Chih., Mexico

A.M. Ingallinella
Centro de Ingeniería Sanitaria (CIS), Facultad de Ciencias Exactas, Ingeniería y Agrimensura, Universidad Nacional de Rosario, Rosario, Prov. de Santa Fe, Argentina

J.A. Izquierdo-Vega
Sección Externa de Toxicología, Centro de Investigación y de Estudios Avanzados del Instituto Politécnico Nacional (CINVESTAV-IPN), Mexico City, Mexico

G. Jacks
KTH-International Groundwater Arsenic Research Group, Department of Land and Water Resources Engineering, Royal Institute of Technology (KTH), Stockholm, Sweden

M. Jakariya
KTH-International Groundwater Arsenic Research Group, Department of Land and Water Resources Engineering, Royal Institute of Technology (KTH), Stockholm, Sweden and NGO Forum for Drinking Water and Sanitation, Lalmatia, Dhaka, Bangladesh

J. Jankowski
School of Biological, Earth and Environmental Sciences, The University of New South Wales, Sydney, NSW, Australia

Í. Jiménez
Depto. de Fisiología, Biofísica y Neurociencias, Centro de Investigación y de Estudios Avanzados del Instituto Politécnico Nacional (CINVESTAV-IPN), Mexico City, Mexico

S.R. Kanel
Department of Civil Engineering, Auburn University, Auburn University, Auburn, CA, USA

K.C. Khilar
Department of Chemical Engineering, Indian Institute of Technology IIT Bombay, Powai, Mumbai, India

D.G. Kinniburgh
British Geological Survey, Wallingford, OX, UK

D. Koloušek
Institute of Chemical Technology in Prague, Prague, Czech Republic

H. Königskötter
Institute of Geography, Ludwig-Maximilians-University, Munich, Germany

T. Krüger
Institute of Water Resources Management, Hydrology and Agricultural Hydraulic Engineering, University of Hanover, Hanover, Germany and Now at: Chair of Hydrology and Water Resources Management, Brandenburg University of Technology, Cottbus, Germany

P. Kumar
School of Environmental Sciences, Jawaharlal Nehru University, New Delhi, India

J.M. Laparra
Instituto de Agroquímica y Tecnología de Alimentos (CSIC), Burjassot, Valencia, Spain

R.H. Lara-Castro
Instituto de Metalurgia, Universidad Autónoma de San Luis Potosí (UASLP), San Luis Potosí, S.L.P., Mexico

I.K. Levy
Gerencia Química, Comisión Nacional de Energía Atómica, San Martín, Prov. de Buenos Aires, Argentina

J. Limón-Pacheco
Instituto de Investigaciones Biomédicas, Universidad Nacional Autónoma de México (UNAM), Mexico City, Mexico

M.I. Litter
Gerencia Química, Comisión Nacional de Energía Atómica, and Escuela de Posgrado, Universidad de Gral. San Martín, San Martín, Prov. de Buenos Aires, Argentina

D.L. López
Department of Geological Sciences, Ohio University, Athens, OH, USA

E. López-Bayghen
Depto. de Genética y Biología Molecular, Centro de Investigación y de Estudios Avanzados del Instituto Politécnico Nacional (CINVESTAV-IPN), Mexico City, Mexico

L. López
Instituto Nacional de Salud Pública, Cuernavaca, Mor., Mexico

R.J. López-Bellido
Escuela Politécnica Superior, Universidad de Huelva, Huelva, Spain

J.F. López-Sánchez
Departament de Química Analítica, Universitat de Barcelona, Barcelona, Spain

J.L. López-Zepeda
LAFQA, Instituto de Geografía, Universidad Nacional Autónoma de México (UNAM), Mexico City, Mexico

A. Luna,
Sección Externa de Toxicología, Centro de Investigación y de Estudios Avanzados del Instituto Politécnico Nacional (CINVESTAV-IPN), Mexico City, Mexico

C.A. Lucho-Constantino
Depto. de Biotecnología y Bioingeniería, Centro de Investigación y de Estudios Avanzados del Instituto Politécnico Nacional (CINVESTAV-IPN), Mexico City, Mexico and Depto. de Ingeniería en Biotecnología, Universidad Politécnica de Pachuca, Zempoala, Hgo., Mexico

D.M.J. Macdonald
British Geological Survey, Wallingford, OX, UK

V. Machovič
Institute of Chemical Technology in Prague, Prague, Czech Republic

K. Mahatantila
Department of Geology, University of Peradeniya, Peradeniya, Sri Lanka

L.A. Manjarrez-Nevárez
Facultad de Ciencias Químicas, Universidad Autónoma de Chihuahua, Chih., Mexico

N. Marchetti
División de Recursos Hídricos y Medio Ambiente, Facultad de Ciencias Físicas y Matemáticas, Universidad de Chile, Santiago de Chile, Chile

M. Marcus
Advanced Light Source, Lawrence Berkeley National Laboratory, Berkeley, CA, USA

L. Marijuán
Huntsman-Tioxide, Madrid, Spain

A. Martaus
Institute of Chemical Technology in Prague, Prague, Czech Republic

R. Martin
Facultad de Ciencias Exactas y Tecnologias, Universidad Nacional de Santiago del Estero (UNSE), Prov. de Santiago del Estero, Santiago del Estero, Argentina

M. Mateu
Gerencia Química, Comisión Nacional de Energía Atómica, San Martín, Prov. de Buenos Aires, Argentina

J. Matschullat
Interdisciplinary Environmental Research Center, TU Bergakademie Freiberg, Freiberg, Germany

H. Medina
Instituto Nacional de Salud Ocupacional (INSO-SNS), La Paz, Bolivia

A. Melgoza-Castillo
Instituto Nacional de Investigaciones Forestales, Agrícolas y Pecuarias (INIFAP), Universidad Autónoma de Chihuahua (UACH), Chihuahua, Chih., Mexico

M.A.B.C. Menezes
Centro de Desenvolvimento da Tecnologia Nuclear, Comissão Nacional de Energia Nuclear, Belo Horizonte, Minas Gerais, Brazil

H. Merchant
Instituto de Investigaciones Biomédicas (IIB), Universidad Nacional Autónoma de México (UNAM), Mexico City, Mexico

M.H. Merino
Instituto de Geoquímica (INGEOQUI), San Miguel, Prov. de Buenos Aires, Argentina

S. Micete
Posgrado en Ciencias e Ingeniería Ambientales, Universidad Nacional Autónoma de México (UNAM), Mexico City, Mexico

N.S. Mokgalaka
Department of Chemistry, The University of Texas at El Paso, El Paso, TX, USA

M.G. Monroy-Fernández
Facultad de Ingeniería-Instituto de Metalurgia, Universidad Autónoma de San Luis Potosí (UASLP), San Luis Potosí, S.L.P., Mexico

A. Montero
Centro de Aplicaciones Tecnológicas y Desarrollo Nuclear (CEADEN), Havana, Cuba

J. Monterrosa
Fundacion de Amigos del Lago de Ilopango, San Salvador, El Salvador

R. Montoro
Instituto de Agroquímica y Tecnología de Alimentos (CSIC), Burjassot, Valencia, Spain

L. Morales
Departament d'Enginyeria Química, Universitat Politècnica de Catalunya, Barcelona, Spain

M. Morales
Depto. de Genética y Biología Molecular, Centro de Investigación y de Estudios Avanzados del Instituto Politécnico Nacional (CINVESTAV-IPN), Mexico City, Mexico

R. Morales
Facultad de Psicología, Universidad Autónoma de San Luis Potosí (UASLP), San Luis Potosí, S.L.P., Mexico

I. Morell
Instituto Universitario de Plaguicidas y Aguas, Universitat Jaume I, Castellón, Spain

C. Moreno
Facultad de Ciencias Naturales and Inst. Miguel Lillo, Universidad Nacional de Tucumán, Tucumán, Prov. de Tucumán, Argentina

M.E. Morgada de Boggio
Gerencia Química, Comisión Nacional de Energía Atómica, San Martín, Prov. de Buenos Aires, Argentina

C. Moscuzza
Centro de Estudios Transdisciplinarios del Agua, Facultad de Ciencias Veterinarias, Universidad de Buenos Aires, Buenos Aires, Argentina

V. Mugica
Universidad Autónoma Metropolitana-Azcapotzalco, Col Reynosa, Tamaulipas, Mexico City, Mexico

G. Muñiz
Facultad de Ciencias Químicas, Universidad Autónoma de Chihuahua, Chih., Mexico

O. Muñoz
Instituto de Ciencia y Tecnología de los Alimentos, Universidad Austral de Chile, Valdivia, Chile

E. Murgueitio
Scientific Research Center of Escuela Politecnica del Ejercito (ESPE), Sangolqui, Ecuador

S. Nahar
Department of Geology, Dhaka University, Dhaka, Bangladesh

H.A. Nalini, Jr.
Departamento de Geologia, Universidade Federal de Ouro Preto, Ouro Preto, Minas Gerais, Brazil

F. Nava-Alonso
Unidad Saltillo, Centro de Investigación y de Estudios Avanzados del Instituto Politécnico Nacional (CINVESTAV-IPN), Ramos Arizpe, Coah., Mexico

M.E. Navarro
Facultad de Psicología, Universidad Autónoma de San Luis Potosí (UASLP), San Luis Potosí, S.L.P., Mexico

M. Nazim Uddin
Applied Research Laboratory, Department of Chemistry, University of Chittagong, Chittagong, Bangladesh

P.D. Nemade
Centre for Environmental Science and Engineering, Indian Institute of Technology IIT Bombay, Powai, Mumbai, India

H.B. Nicolli
Instituto de Geoquímica (INGEOQUI), San Miguel, Prov. de Buenos Aires, Argentina and Consejo
Nacional de Investigaciones Científicas y Técnicas (CONICET), Argentina

A. Nieto
Enpar Technologies Inc., Guelph, ON, Canada

M. Novák
Institute of Geological Sciences, Faculty of Science, Masaryk University, Brno, Czech Republic

N. Núñez
Programa Agrícola CODELCO Chile, Calama, Chile

O.G. Núñez-Montoya
Facultad de Zootecnia, Universidad Autónoma de Chihuahua (UACH), Chihuahua, Chih., Mexico

S.M. Oberdá
Fundação Estadual do Meio Ambiente (FEAM), Belo Horizonte, Minas Gerais, Brazil

R. Olmos
Departamento de Fisica, Facultad de Ciencias y Humanidades, Universidad de El Salvador, San
Salvador, El Salvador

D.L. Orihuela
Escuela Politécnica Superior, Universidad de Huelva, Huelva, Spain

M. Ormachea
Instituto de Investigaciones Químicas, Universidad Mayor de San Andrés, La Paz, Bolivia

A. Ortega
Depto. de Genética y Biología Molecular, Centro de Investigación y de Estudios Avanzados del
Instituto Politécnico Nacional (CINVESTAV-IPN), Mexico City, Mexico

E. Ortiz
Universidad Autónoma Metropolitana-Azcapotzalco, Col Reynosa, Tamaulipas, Mexico City,
Mexico

B. O'Shca
Department of Geology, Dickinson College, Carlisle, PA, USA

P. Ostrosky-Wegman
Depto. de Genética y Toxicología Ambiental, Instituto de Investigaciones Biomédicas, Universidad
Nacional Autónoma de México (UNAM), Mexico City, Mexico

H.E.L. Palmieri
Centro de Desenvolvimento da Tecnologia Nuclear, Comissão Nacional de Energia Nuclear, Belo
Horizonte, Minas Gerais, Brazil

J. Pardo-Rueda
Facultad de Ciencias Químicas, Universidad Autónoma de Chihuahua, Chih., Mexico

R. Pastene
Facultad de Química y Biología, Universidad de Santiago de Chile, Santiago de Chile, Chile

R. Pažout
Institute of Chemical Technology in Prague, Prague, Czech Republic

N.A. Pelallo-Martinez
Facultad de Ciencias Químicas, Universidad Autónoma de San Luis Potosí (UASLP), San Luis
Potosí, S.L.P., Mexico

J.R. Peralta-Videa
Department of Chemistry, The University of Texas at El Paso, El Paso, TX, USA

A. Pérez-Carrera
Centro de Estudios Transdisciplinarios del Agua, Facultad de Ciencias Veterinarias, Universidad de Buenos Aires, Buenos Aires, Argentina

R. Pérez-Garibay
Unidad Saltillo, Centro de Investigación y de Estudios Avanzados del Instituto Politécnico Nacional (CINVESTAV-IPN), Ramos Arizpe, Coah., Mexico

S. Pérez-Mohedano
Huntsman-Tioxide, Madrid, Spain

B. Petrusevski
UNESCO-IHE, International Institute for Infrastructural, Hydraulic and Environmental Engineering, Delft, The Netherlands

H.M. Poggi-Varaldo
Depto. de Biotecnología y Bioingeniería, Centro de Investigación y de Estudios Avanzados del Instituto Politécnico Nacional (CINVESTAV-IPN), Mexico City, Mexico

R.I. Ponce
Grupo InQA-Investigación Química Aplicada, Facultad de Ingeniería, Universidad Nacional de Jujuy, S.S. de Jujuy, Prov. de Jujuy, Argentina

F.R. Prieto-García
Centro de Investigaciones Químicas (CIQ), Universidad Autónoma del Estado de Hidalgo (UAEH), Pachuca, Hgo., Mexico

C. Puigdomènech
Departament d'Enginyeria Química, Universitat Politècnica de Catalunya, Barcelona, Spain

A. Puntí
Departament d'Enginyeria Química, Universitat Politècnica de Catalunya, Barcelona, Spain

I. Pupo
Centro de Aplicaciones Tecnológicas y Desarrollo Nuclear (CEADEN), Havana, Cuba

J. Quintanilla
Instituto de Investigaciones Químicas, Universidad Mayor de San Andrés, La Paz, Bolivia

I.M.M. Rahman
Applied Research Laboratory, Department of Chemistry, University of Chittagong, Chittagong, Bangladesh

AL. Ramanathan
School of Environmental Sciences, Jawaharlal Nehru University, New Delhi, India

A. Ramirez
Instituto Mexicano de Tecnologia del Agua (IMTA), Jiutepec, Mor., Mexico

E. Ramírez
Universidad Autónoma Metropolitana-Azcapotzalco, Col Reynosa, Tamaulipas, Mexico City, Mexico

P. Ramírez
Facultad de Estudios Superiores Cuautitlán, Universidad Nacional Autónoma de México (UNAM), Mexico City, Mexico

A. Ramirez-Orozco
Instituto Mexicano de Tecnologia del Agua (IMTA), Jiutepec, Mor., Mexico

M.L. Ramírez-Ramírez
Facultad de Química, Universidad de Guanajuato, Guanajuato, Gto., Mexico

O. Ramos
Instituto de Investigaciones Químicas, Universidad Mayor de San Andrés, La Paz, Bolivia

L. Ransom
Department of Geological Sciences, Ohio University, Athens, OH, USA

E. Recabarren G.
Facultad de Ingeniería, Universidad de Santiago de Chile, Santiago de Chile, Chile

K.J. Reddy
Department of Renewable Resources and School of Energy Resources, University of Wyoming, Laramie, WY, USA

R. Reséndiz
Universidad Autónoma Metropolitana-Azcapotzalco, Col Reynosa, Tamaulipas, Mexico City, Mexico

C.A. Rocha
Laboratório de Contaminantes Metálicos, Divisão de Vigilância Sanitária, Instituto Octávio Magalhães, Fundação Ezequiel Dias (FUNED), Belo Horizonte, Brazil

D.O. Rocha-Amador
Facultad de Medicina, Universidad Autónoma de San Luis Potosí (UASLP), San Luis Potosí, S.L.P., Mexico

F.A. Rodriguez Almeida
Facultad de Zootecnia, Universidad Autónoma de Chihuahua (UACH), Chihuahua, Chih., Mexico

V.M. Rodríguez
Environmental and Community Medicine, The University of Medicine and Dentistry of New Jersey and Rutgers, Piscataway, NJ, USA

F. Romero
LAFQA, Instituto de Geografía, Universidad Nacional Autónoma de México (UNAM), Mexico City, Mexico

T.R. Roth
Department of Renewable Resources and School of Energy Resources, University of Wyoming, Laramie, WY, USA

M.H. Royo-Márquez
Instituto Nacional de Investigaciones Forestales, Agrícolas y Pecuarias (INIFAP), Universidad Autónoma de Chihuahua (UACH), Chihuahua, Chih., Mexico

R. Rubio
Departament de Química Analítica, Universitat de Barcelona, Barcelona, Spain

T.R. Rüde
Institute of Hydrogeology, RWTH Aachen University, Aachen, Germany

A. Rueda-Ramírez
Facultad de Ciencias Químicas, Universidad Autónoma de Chihuahua, Chih., Mexico

M.J. Ruiz-Chancho
Departament de Química Analítica, Universitat de Barcelona, Barcelona, Spain

A.M. Sancha
División de Recursos Hídricos y Medio Ambiente, Facultad de Ciencias Físicas y Matemáticas, Universidad de Chile, Santiago de Chile, Chile

L.C. Sánchez-Peña
Sección Externa de Toxicología, Centro de Investigación y de Estudios Avanzados del Instituto Politécnico Nacional (CINVESTAV-IPN), Mexico City, Mexico

M. Sandoval
Depto. de Genética y Biología Molecular, Centro de Investigación y de Estudios Avanzados del Instituto Politécnico Nacional (CINVESTAV-IPN), Mexico City, Mexico

I. Sastre-Conde
Conselleria d'Agricultura i Pesca, Gov. of Islas Baleares, Palma de Mallorca, Islas Baleares, Spain

J. Schippers
UNESCO-IHE, International Institute for Infrastructural, Hydraulic and Environmental Engineering, Delft, The Netherlands

L. Seed
Enpar Technologies Inc., Guelph, ON, Canada

B. Segura
Facultad de Estudios Superiores Iztacala (FES), Universidad Nacional Autónoma de México (UNAM), Mexico City, Mexico

A.K. SenGupta
Department of Civil and Environmental Engineering at Lehigh University, Bethlehem, PA, USA

R.E. Servant
Gerencia Química, Comisión Nacional de Energía Atómica, San Martín, Prov. de Buenos Aires, Argentina

J. Ševc
Institute of Geology, Faculty of Natural Sciences, Comenius University in Bratislava, Bratislava, Slovakia

S. Sharma
UNESCO-IHE, International Institute for Infrastructural, Hydraulic and Environmental Engineering, Delft, The Netherlands

G. Shelp
Enpar Technologies Inc., Guelph, ON, Canada

M. Shroff
Department of Cancer Genetics, Roswell Park Cancer Institute, Buffalo, NY, USA

A.B. Silva
Universidade Estadual de Feira de Santana, (UEFS), Bahia, Brazil

N.O.C. Silva
Laboratório de Contaminantes Metálicos, Divisão de Vigilância Sanitária, Instituto Octávio Magalhães, Fundação Ezequiel Dias (FUNED), Belo Horizonte, Brazil

Anshumali Singh
School of Environmental Sciences, Jawaharlal Nehru University, New Delhi, India

K.K. Singh
Department of Cancer Genetics, Roswell Park Cancer Institute, Buffalo, NY, USA

P.L. Smedley
British Geological Survey, Wallingford, OX, UK

T. Soriano
Departamento de Fisica, Facultad de Ciencias y Humanidades, Universidad de El Salvador, San Salvador, El Salvador

C. Soto
Depto. de Sistemas Biológicos, Universidad Autónoma Metropolitana, Xochimilco (UAM-X), Mexico City, Mexico

G. Soto
Sección Externa de Toxicología, Centro de Investigación y de Estudios Avanzados del Instituto Politécnico Nacional (CINVESTAV-IPN), Mexico City, Mexico

G. Sposito
Ecosystem Sciences Division, University of California at Berkeley, Berkeley, CA, USA

O. Sracek
Institute of Geological Sciences, Faculty of Science, Masaryk University, Brno, Czech Republic

L.M. Stecca
Centro de Ingeniería Sanitaria (CIS), Facultad de Ciencias Exactas, Ingeniería y Agrimensura, Universidad Nacional de Rosario, Rosario, Prov. de Santa Fe, Argentina

J. Stummeyer
Federal Institute of Geosciences and Natural Resources of Germany (BGR), Hanover, Germany

P. Sulovský
Institute of Geological Sciences, Faculty of Science, Masaryk University, Brno, Czech Republic

R. Thunvik
KTH-International Groundwater Arsenic Research Group, Department of Land and Water Resources Engineering, Royal Institute of Technology (KTH), Stockholm, Sweden

A. Tineo
Cátedra de Hidrogeología, Facultad de Ciencias Naturales e Instituto Miguel Lillo, Universidad Nacional de Tucumán, Tucumán, Prov. de Tucumán, Argentina

I. Tipán
Scientific Research Center of Escuela Politecnica del Ejercito (ESPE), Sangolqui, Ecuador

O.R. Tofalo
Departamento de Ciencias Geológicas, Facultad de Ciencias Exactas y Naturales, Universidad de Buenos Aires, Buenos Aires, Argentina

S. Tokunaga
National Institute of Advanced Industrial Science and Technology, Tsukuba, Ibaraki, Japan

E. Torres
Departament d'Enginyeria Química, Universitat Politècnica de Catalunya, Barcelona, Spain

V. Torres-Muñoz
Facultad de Ciencias Químicas, Universidad Autónoma de Chihuahua, Chih., Mexico

Parijat Tripathi
School of Environmental Sciences, Jawaharlal Nehru University, New Delhi, India

A. Uribe-Salas
Unidad Saltillo, Centro de Investigación y de Estudios Avanzados del Instituto Politécnico Nacional (CINVESTAV-IPN), Ramos Arizpe, Coah., Mexico

E. Uribe-Querol
Instituto de Investigaciones Biomédicas, Universidad Nacional Autónoma de México (UNAM), Mexico City, Mexico

M. Urík
Institute of Geology, Faculty of Natural Sciences, Comenius University in Bratislava, Bratislava, Slovakia

L.A. Valcárcel
Centro de Aplicaciones Tecnológicas y Desarrollo Nuclear (CEADEN), Havana, Cuba

O.R. Vasconcelos
Departamento de Engenharia Sanitária e Ambiental, Universidade Federal de Minas Gerais (UFMG), Belo Horizonte, Minas Gerais, Brazil

G. Vázquez-Rodríguez
Facultad de Ingeniería-Instituto de Metalurgia, Universidad Autónoma de San Luis Potosí (UASLP), San Luis Potosí, S.L.P., Mexico

L. Vega
Sección Externa de Toxicología, Centro de Investigación y de Estudios Avanzados del Instituto Politécnico Nacional (CINVESTAV-IPN), Mexico City, Mexico

D. Vélez
Instituto de Agroquímica y Tecnología de Alimentos (CSIC), Burjassot, Valencia, Spain

E. Vera
Centro de Investigación y de Estudios Avanzados del Instituto Politécnico Nacional (CINVESTAV-IPN), Mexico City, Mexico

S. Vilches
Escuela de Ingeniería en Alimentos, Universidad del Bío-Bío, Chillán, Chile

M. Villalobos
LAFQA, Instituto de Geografía, Universidad Nacional Autónoma de México (UNAM), Mexico City, Mexico

M.M. Villatoro-Pulido
Centro IFAPA Alameda del Obispo, Córdoba, Spain

M. Vithanage
Department of Geology and Geography, University of Copenhagen, Denmark International Water Management Institute (IWMI), Battaramulla, Sri Lanka

M. von Brömssen
KTH-International Groundwater Arsenic Research Group, Department of Land and Water Resources Engineering, Royal Institute of Technology (KTH), Stockholm, Sweden and Ramböll Sweden, Stockholm, Sweden

M. Vujcic
Department of Cancer Genetics, Roswell Park Cancer Institute, Buffalo, NY, USA

M. Wajrak
Edith Cowan University, Joondalup, WA, Australia

R. Weerasooriya
Institute of Fundamental Studies, Kandy, Sri Lanka

Acknowledgements

The editors thankfully acknowledge the enormous input and dedication of the following colleagues for their time-consuming efforts to review contributions, for editorial handling and for English language corrections where authors were no native speakers. We particularly thank the editorial board members of the book series "Arsenic in the Environment" for editorial handling, review and editing of the manuscripts (in alphabetical order): Nazmul Haque (66, 67, 70), Wolfgang H. Höll (55, 59, 61), Jan Hoinkis (54, 62, 63, 73), Douglas B. Kent (2, 15, 29), Walter Klimecki (49), Dina Lopéz (1, 7, 10, 22, 24, 25), Jack Ng (35, 36, 40), Andrew Meharg (32, 34, 46, 48, 75), Britta Planer-Friedrich (9, 26), Peter Ravenscroft (8, 27, 42, 56), Olle Selinus (44, 47), Kenneth G. Stollenwerk (14, 16, 19, 33), Miroslav Stýblo (41, 43, 45, 50, 51), Jenny Webster Brown (6, 12, 13, 17), Walter Wenzel (60, 64, 68), and Yan Zheng (11, 74).

We would like to thank Miroslav Stýblo and Wolfgang H. Höll, for their help during preparation of the subject index in the areas of arsenic toxicology/health effects and arsenic remediation, respectively.

For further revision and language editing we thank: María Eugenia Gonsebatt (42), Douglas B. Kent (7, 21, 41, 73), Andrew Meharg (43–45, 47, 51, 52, 75), Jack Ng (22, 25, 30, 31, 39, 65, 71, 72), Hugo Nicolli (41), and Kenneth G. Stollenwerk (1, 3–5, 7, 9, 10, 11, 18, 24, 26, 28, 37, 38, 50, 54, 55, 57–64, 66–70).

Should we have forgotten to mention another referee we humbly apologize—this was certainly unintentional.

The editors also thank the group of scientists who reviewed the book manuscripts in a first stage (in alphabetical order): Jordi Cama (Institut de Ciències de la Terra "Jaume Almera", Barcelona, Spain), Deoraj Harry Caussy (Word Health Organization, New Delhi, India), D. Chandrasekharam (Indian Institute of Technology, IIT Bombay, India), Jose Luis Cortina (Universitat Politècnica de Catalunya, Barcelona, Spain), Luis H. Cumbal (Escuela Politecnica del Ejercito, Sangolqui, Ecuador), Luz María Del Razo (CINVESTAV-IPN, Mexico City, Mexico), Mercedes Del Río-Celestino (IFAPA-Centro Alameda del Obispo, Córdoba, Spain), Eleonora Deschamps (Fundação Estadual do Meio Ambiente, Belo Horizonte, Brazil), Oscar Díaz Schulz (Universidad de Santiago de Chile, Santiago de Chile, Chile), Wolfgang Driehaus (GEH Wasserchemie GmbH & Co. KG, Osnabrueck, Germany), Barbora Doušová (Institute of Chemical Technology in Prague, Prague, Czech Republic), María Eugenia Gonsebatt (UNAM, Mexico City, Mexico), Hartmut M. Holländer (University of Hanover, Hanover, Germany), Ana Maria Ingallinella (Universidad Nacional de Rosario, Rosario, Argentina), Marta I. Litter (Comisión Nacional de Energía Atómica, San Martín, Argentina), Hugo B. Nicolli (Instituto de Geoquímica, San Miguel, Argentina), Darrel Kirk Nordström (USGS, Denver, USA), Lars-Åke Persson (International Maternal and Child Health (IMCH), Uppsala University, Sweden), Roser Rubio (Universitat de Barcelona, Barcelona, Spain), Ondra Sracek (Masaryk University, Brno, Czech Republic), Thomas R. Rüde (RWTH Aachen University, Aachen, Germany), Doris Stüben [University of Karlsruhe (TH), Germany], and Deni Vélez (Instituto de Agroquímica y Tecnología de Alimentos, Valencia, Spain).

Prosun Bhattacharya would like to thank the Swedish Development Cooperation Agency (Sida-SAREC and Sida-Natur), Swedish Research Council for Environment, Agricultural Sciences and Spatial Planning (Formas), the Strategic Environmental Research Foundation (MISTRA) Sweden, for support to the participants in this conference as well as for the editorial time for this volume.

Editors' foreword

Groundwater resources for human consumption are naturally contaminated by elevated levels of arsenic (As) in many parts of the world. Severe health effects have been associated with high As-concentrations in groundwater used for drinking purposes. These As-concentrations often exceed the World Health Organization (WHO) provisional guideline value of 10 µg/l, which is being adapted by an increasing number of countries around the globe.

During the last three to four decades, As-rich groundwaters in Southern and South-Eastern Asia have received much attention. However, the situation seems to be equally important in Latin America. There, the number of studies is still relatively low, and the extent and severity of As-exposure in the population has only marginally been evaluated. Arsenic occurrence in groundwater in Argentina, Bolivia, Brazil, Chile, Mexico, Nicaragua, Peru, and other Latin American countries deserve more attention. Recently, elevated groundwater As-concentrations as well as As-related health effects were detected in Nicaragua, Ecuador and El Salvador—countries where the groundwater arsenic problem was not assumed to exist. The true extent of risk associated with chronic As-toxicity still has to be determined.

Sustainable land-use and agricultural practices in Latin American countries are regionally threatened by the use of high As irrigation water. Elevated levels of natural As in groundwater from geogenic sources is therefore an issue of primary environmental concern, and limits the use of these resources, and hinders socio-economic growth. Hence there is a need to improve our understanding of the genesis of As-rich groundwaters, constraints on the mobility of As in groundwater and other environmental compartments, As-uptake from soil and water by plants, As-propagation through the food chain, health impacts on human beings, life stock, and other animals, assessment of environmental health risks and impacts, and As-removal technologies, to improve the socio-economic status of the affected regions.

These facts have inspired us to edit this inter- and multidisciplinary volume, making it a first attempt to compile as much information as possible on the arsenic problem in Latin America, a continent from which not much is internationally known on this topic (with a few notable exceptions, but only from just a few countries). Our interest was to produce a high quality book, but at the same time try to include as much information as practible on the arsenic issues in Latin America. Thus, readers will encounter quiet a variety of contributions, including some that "only" present new data, e.g., from countries, where the arsenic problem was not know so far, since we consider them as important source of information, as well as comprehensive and complex interpretations and experimental results.

This volume intends to bring together geo-scientists, specialists from public health, from chemical and engineering sciences involved in arsenic-related issues with a focus on Latin America. The book impressively demonstrates that arsenic-related problems in Latin America are of the same order of magnitude as, e.g., in Southern Asia, where the problem has obtained much more interest from the general public and the international scientific world. The volume is designed to (1) create interest within the Latin American countries, affected by the presence of arseniferous aquifers, (2) address the international scientific community in general, (3) update the current status of knowledge on the dynamics of natural arsenic from bedrock and soils via aquifers and groundwater to the food chain, (4) continue the important worldwide forum on improved and efficient techniques for As-removal in regions with elevated arsenic levels in groundwater, (5) increase awareness among administrators, policy makers and company executives, and (6) improve the international cooperation on arsenic-related studies in Latin America.

Jochen Bundschuh
María Aurora Armienta
Peter Birkle
Prosun Bhattacharya
Jörg Matschullat
Arun B. Mukherjee
(editors)

Section I
Regional introduction and overview

CHAPTER 1

Occurrence, health effects and remediation of arsenic in groundwaters of Latin America

J. Bundschuh
International Technical Cooperation Program, CIM (GTZ/BA), Frankfurt, Germany
Instituto Costarricense de Electricidad (ICE), San José, Costa Rica

M.E. García
Instituto de Investigaciones Químicas, Universidad Mayor de San Andrés, La Paz, Bolivia

P. Birkle
Instituto de Investigaciones Eléctricas, Gerencia de Geotermia, Cuernavaca, Morelos, Mexico

L.H. Cumbal
Centro de Investigación Científica, Escuela Politecnica del Ejercito (ESPE), Sangolqui, Ecuador

P. Bhattacharya
KTH-International Groundwater Arsenic Research Group, Department of Land and Water Resources Engineering, Royal Institute of Technology (KTH), Stockholm, Sweden

J. Matschullat
Interdisciplinary Environmental Research Center, TU Bergakademie Freiberg, Freiberg, Germany

ABSTRACT: At least 4 million people depend on drinking water with toxic arsenic (As) concentrations in Argentina, Bolivia, Chile, Mexico and Peru, which primarily originate from geogenic sources. In other Latin-American countries, the occurrence of the problem and/or the number of exposed people is still unknown. This chronic As exposure is associated with neurological and dermatological problems as well as carcinogenic effects. In contrast to urban areas, practically no action has been taken by the authorities to mitigate the As problem for the rural population, which often depends on As-contaminated water as their only available drinking water resource. This lack of interest has slowed the development of low-cost remediation methods for small communities or single houses. However, various suitable remediation techniques have been developed at the laboratory scale. In a limited number of cases these techniques have been tested and proven in the field and have helped to mitigate As problems. Examples of remediation techniques include solar oxidation methods, phytoremediation, and the use of natural materials as adsorbents for As removal from drinking water. Therefore, the problem is not a technological one, since viable solutions are already available. The problem lies in convincing the responsible authorities to consider the As occurrence as a natural health risk and support the development and the application of remediation methods for rural areas.

1.1 INTRODUCTION TO THE ARSENIC PROBLEM IN LATIN AMERICA

At least four million people rely upon water that has been contaminated by arsenic (As) ($>$50 µg/l) in Latin American countries such as Argentina, Chile, Mexico, and Peru. The magnitude of As contamination in some countries, such as Argentina and Mexico, makes this issue a primary public health concern. For example, in Argentina (and also in Chile up until 1970) more than 1% of the population is exposed to the contaminated water. In Bolivia, Brazil, Costa Rica, Ecuador, El Salvador, and Guatemala, As in drinking water has been detected, but the numbers of affected individuals is still unknown. In other Latin American countries, the possible existence of a groundwater As problem has not yet been assessed. For example, in Nicaragua, As exposure and the related severe health effects were only detected four years ago. With advances in modern analytical methods used for As detection at low concentrations, and the introduction of new national As limits for drinking water (10 µg/l), several countries that had until now "safe" As levels, will be classified as having unsafe concentrations and the numbers of people exposed will significantly increase in the near future. This limit for As concentrations in drinking water was introduced recently in Nicaragua and Argentina, and there are plans to implement it in Chile and Mexico in the near future.

In most of these countries, water contamination by As is caused by the occurrence of geogenic As, mostly related to volcanism in the Andes (Argentina, Bolivia, Chile, Peru) and their continuation in Middle America (Nicaragua, Mexico, El Salvador). From these sources, As is released into the environment (ground- and surface waters, soils, etc.) by natural dissolution, rock weathering (Argentina, Chile, Bolivia, Peru, Nicaragua, El Salvador, Mexico), and/or by mining activities (Chile, Bolivia, Peru, Mexico). Other sources of As release, which are of minor and very local importance, are artificial (e.g., due to electrolytic metal producing processes (Brazil)), and agricultural activities (e.g., the use of As-containing plaguicides).

This chapter is comprised of three parts: First is a country-by-country state of the art overview of the occurrence and the respective sources of As in groundwater and surface water used for drinking purposes in Latin America. Second, it addresses the numbers of people exposed and affected in terms of effects that have been already observed, and possible health effects that may arise in the future. The third part of this chapter briefly discusses remediation methods that have been applied in both urban and rural areas, and the measures that must be taken in the future to mitigate As contamination of drinking water in "Rural Latin America". In this discussion low-cost remediation techniques suitable for small communities and for single households are included as possible solutions.

1.2   SOURCES AND PRESENT FORMS OF ARSENIC IN LATIN AMERICA

Arsenic is present in water, mainly groundwater, due to natural process and also due to many labor activities such as mining where As can be found at dangerously high concentrations. Mine workers and people living in mining areas are exposed to As and can develop illnesses related to this element. This is a problem for several Latin American countries where intense mining activity and natural contamination have generated a high health risk for these populations.

Arsenic is typically released into the environment in an inorganic form and tends to adsorb strongly to soils. Leaching into subsurface soils is generally not significant, except under reducing conditions. Physical soil characteristics, such as pH, dissolved oxygen, organic carbon content, cation exchange capacity, and iron oxide content tend to govern the leaching potential. Soluble forms of As in soil may either run off into surface water bodies or leach into shallow groundwater. Arsenic may also be introduced into aqueous systems through natural weathering of soil and rock. Arsenic transport in water depends not only on the form of As, but also on the interaction of As with other materials. Arsenate [As(V)] is the predominant form of As found in groundwater, although arsenite [As(III)] may be present in significant proportions depending on local geology and water pH and redox characteristics.

1.3   REGIONAL DISTRIBUTION OF ARSENIC IN LATIN AMERICA

1.3.1   *General*

The presence of As in ground- and surface waters from geogenic sources (including those released through mining activities) has been documented in Argentina, Chile, Peru and Mexico (Fig. 1.1)

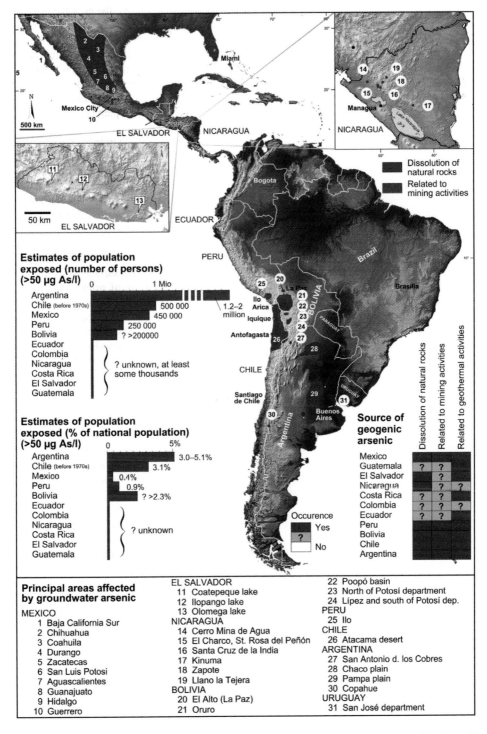

Figure 1.1.   Map of Latin America showing hotspot locations of ground- and surface waters with elevated As from geogenic sources (>50 µg/l), the principal origin and pathway of its release, by leaching from rocks of from mining activities, and the country-by-county data of the exposed population (>50 µg/l). See text for references.

over the last several decades. In these countries, As concentrations in the range of 0.01 to several mg/l are common. In other countries, such as Bolivia, Brazil, Ecuador, Nicaragua and El Salvador, this problem has only been detected recently and investigated over the last few years (Fig. 1.1). In Nicaragua—a country where the contamination of groundwater with As was not considered a problem—elevated As concentrations in groundwater as well as As-related health effects have recently been documented.

1.3.2   *Argentina*

Arsenic contamination in groundwater is of primary environmental concern in extended parts of the Chaco-Pampean plain region in Argentina. This area consists of Tertiary aeolian loess-type deposits in the Pampean plain and of predominantly fluvial sediments of Tertiary and Quaternary age in the plains of the Chaco region, all covering a surface area of about 978,634 km^2 (Fig. 1.1).

The As source in this region is geogenic and associated with a Holocene volcanic ash bed (about 6 mg/kg As); the highly soluble rhyolitic volcanic glass component (5–20%) of the sandy sediments; and sediments originating from metamorphic and acid magmatic rocks (Bundschuh *et al.* 2004).

A high variability in the groundwater As concentration is caused by several hydrogeological and hydrogeochemical factors. Arsenic may be leached from the volcanic glass and transported into the aquifer. The spatial distribution and variability of As concentrations, even in the same (mostly) shallow phreatic aquifer, is high. Areas with high groundwater As concentrations have been found to correspond to areas with prevalent Na-HCO$_3$-waters, elevated pH, and elevated groundwater residence times; although a more or less homogenous aquifer matrix at the shallow levels characterize the wells exhibiting both low and high concentrations of As in groundwater (Bhattacharya *et al.* 2006, Bundschuh *et al.* 2004).

1.3.3   *Uruguay*

The presence of geogenic As in groundwater was recently and for the first time reported from Raigón aquifer, located in the south of San José department in Uruguay (Fig. 1.1; Guérèquiz *et al.* 2006), where it spreads over an agricultural area of 1800 km^2 with a population of 47,000. Ongoing investigations by these authors have found an average As content between 25 and 50 µg/l in 22 groundwater samples with some exceeding 50 µg/l. Guérèquiz *et al.* (2006) assume that the occurrences of As is related to the same source (volcanic ash) as in the nearby province of Santa Fe (Argentina) and the Puelche aquifer (Buenos Aires province).

1.3.4   *Chile*

Several areas of northern Chile have a long and well-documented history of high As concentrations in drinking water (Sancha *et al.* 2004). In this region, the water that supplies most of the towns and villages in the Atacama desert is obtained from rivers that originate in the Andean mountain range, which contains numerous sources of geogenic As (Fig. 1.1). Typical As concentrations of these river waters are 200 to 900 µg/l. The provincial capital Antofagasta (250,000 inhabitants) has had a unique pattern of As exposure. The As concentration in Antofagasta's public drinking water supply rose sharply in 1958 (Hopenhayn-Rich *et al.* 2000) when a new water supply was used (containing 800 µg As/l). The As concentrations in this water supply system remained elevated until 1970 when the first As removal plant was inaugurated, thereby reducing As concentrations in drinking water to an average value of 40 µg/l.

1.3.5   *Bolivia*

The central and southern parts of Bolivia's Andean highland are the areas with the worst environmental problems related to geogenic As. The release of As into the environment occurs either by natural leaching/weathering of volcanic rocks (and the sediments derived from them) or from mining activities. In Poopó basin (Fig. 1.1), surface water samples from local rivers and Poopó lake itself have As concentrations between 90 and 140 μg As/l in areas not affected by mining activities, and up to 2.0 mg As/l in rivers affected by mining activities. Arsenic concentrations in groundwater range between 10 and 90 μg/l (Bundschuh and García 2008, García *et al.* 2006, García and Bundschuh 2006). In Poopó lake, the highest As concentrations were found in the dry season (210–220 μg/l versus 20 μg/l in the wet season), whereas in most tributary rivers maximum As values correspond to the wet season (10–380 μg/l versus 10–50 μg/l in the dry season), indicating for both areas different principal As mobility controls (García and Bundschuh 2006). In contrast, sediments from the rivers and lakes contain similar As concentrations in both the wet and the dry season.

1.3.6 *Peru*

In Peru, geogenic As contaminants are present in Aricota lake, which is used for the water supply of the city of Ilo (Fig. 1.1). This lake is fed by the Collazas and Salado rivers, which pass through the area of Yucamane volcano. The rocks in this volcanic area are the principal As source. Recently, high groundwater As concentrations were also reported from the area of Puno city.

1.3.7 *Ecuador*

Since Ecuador is within the Pacific ring of fire, the As release to waters, sediments, and soils is volcanic in origin. Arsenic in natural waters has been determined in the northern Andean region of Ecuador. Geothermal waters from El Carchi, Imbabura, Pichincha, Cotopaxi, and Tungurahua provinces show As levels from 113 to 844 μg As/l (Cumbal *et al. unpubl. data*). Surface waters and sediments are also As-contaminated in this region. El Angel river in the Carchi province receives thermal waters and shows As in the range of 64 to 113 μg As/l (Cumbal *et al. unpubl. data*).

In the central part of the Andean region, natural As is present in the Papallacta lake. This lagoon is fed by the Tambo river and geothermal residual waters and shows As concentrations that vary from 104 to 360 μg As/l (Cumbal *et al.* 2006). In addition, As above the Ecuadorian standard (10 μg/l) was also found in springs that are used as drinking water sources in the towns of Tumbaco, Guayllabamba, Cumbaya, Yaruqui, El Quinche, Pifo, and Puembo in, Pichincha province.

Although As contamination has not been totally identified and evaluated in Ecuador, it is believed that around 200,000 people in rural areas may be exposed to As ingestion by water and food consumption.

1.3.8 *Nicaragua*

Groundwater As was reported in Nicaragua for the first time in 1996 (ECO/OPS 1997). The As source is geogenic and due to the weathering of volcanic rocks. However detailed hydrogeochemical studies of the oxidizing aquifers in volcanic rocks and derived sediments, and the mobilization mechanisms are still missing. The highest recorded As concentrations (about 1320 μg/l) were documented in a tube well in El Zapote village (Fig. 1.1; Barragne-Bigot 2004, Gomez 2004 and 2006).

In 2003, UNICEF studied the As concentrations of groundwater from 77 wells in five areas (Barragne-Bigot 2004). This study delimited four areas (Fig. 1.1) with groundwater As concentrations exceeding the national limit of 10 μg/l: (1) Cerro Mina de Agua (municipality Villanueva), (2) communities El Charco and Santa Rosa del Peñón (municipality Santa Rosa del Peñón), (3) community Cruz de la India (municipality Santa Rosa del Peñón), and (4) community Kinuma (municipality La Libertad) (Fig. 1.1). The waters have a roughly neutral pH (6 to 8) and As is predominantly present as As(V). Total As concentrations range from 10 to 107 μg/l. A total of 1270 people in these areas are exposed to toxic levels of As in drinking water (Barragne-Bigot 2004).

In 2005, UNICEF sampled 54 wells in the community of Llano La Tejera (municipality Jinotega) with 714 inhabitants. 88% of the houses have their own wells. As concentrations vary from below detection limit up to 1200 µg As/l (average 100.4 µg/l) making 87% of the wells unsuitable for drinking (Nicaraguan limit 10 µg As/l) (Bundschuh *et al.* 2006, Larios and Bundschuh *unpubl. data*).

A study of Altamirano Espinoza and Bundschuh (2009) of the alluvial aquifer of the southwestern part of Sébaco valley composed by sediments from volcanic rocks and groundwater under oxidizing conditions (200 < Eh < 400 mV) related high groundwater As concentrations with tectonic faults having hydrothermal alteration. In 21 of 57 water samples, total As concentrations ranging from 10 to 122 µg/l, exceeded the national limit of 10 µg As/l.

1.3.9   *El Salvador*

At least three lakes have high As concentrations in El Salvador: Ilopango, Coatepeque, and Olomega (Fig. 1.1). Only Ilopango and Coatepeque have been investigated. More than 200,000 people inhabit the Ilopango lake drainage basin (184.9 km^2), many of whom use the lake water. High As concentrations between 150 and 770 µg/l make this water unsuitable for human consumption. Two As sources have been identified for Ilopango waters: (1) the internal sediments of the lake, which contain As-rich volcanic products deposited by the last eruptions of this caldera, and (2) the material transported to the lake by the Chaguite river, whose As load is produced by the leaching and erosion of volcanic deposits within the Illopango lake basin (López *et al.* 2009). The same authors assume that the ash of the last calderic eruption of Ilopango (about 2000 years ago) covers all of El Salvador and could be the source of As contamination for other surface and subsurface water bodies and could affect other environments. Hot springs located along the southwestern shore of Coatepeque lake and the leaching of sediments and volcanic products are the likely As sources. Numerous hot springs and fumaroles along the volcanic chain of El Salvador are discharging As-rich fluids. Its fate in the environment and its impact on the population has not been assessed.

1.3.10   *Mexico*

Elevated As concentrations are encountered in groundwater in different regions of Mexico, that cover large parts of the states of Baja California Sur, Chihuahua, Durango, Coahuila, Guanajuato, Hidalgo, Zacatecas, San Luis Potosi, Aguascalientes, but the problem is also described from the states of Guerrero, Jalisco, Michoacán, Morelos, Nuevo León, Puebla and Sonora. Principal anthropogenic sources are two ore melters located in San Luis Potosí and Torreón, Coahuila state (Fig. 1.1). The source of As is predominantly geogenic although a substantial As contamination is also caused by the mining activities.

In the mining area of Zimapán (Hidalgo state; Fig. 1.1), the main As-bearing rocks are massive sulfide ores, with pyrite, arsenopyrite, pyrrholite, and other As-bearing minerals. Arsenic release into the environment occurs through both the natural dissolution/weathering of the As-rich rocks, and through mining activities (e.g., through tailings with up to 22,000 mg As/kg; Armienta *et al.* 2005). Consequently, the groundwater in the area of Zimapán has high As concentrations (190–650 µg As/l; average 380 µg/l; Armienta *et al.* 2005).

The Salamanca aquifer system (1900 wells), located in Guanajuato state (Fig. 1.1), is naturally affected by As from geogenic sources, but also from anthropogenic sources, such as the solid wastes of several chemical industries. The highest documented As concentration (280 µg/l) was documented in the groundwater of a well (Rodriguez *et al.* 2005).

In the village of Tlamacazapa (Guerrero state; Fig. 1.1), As release into the environment (groundwater, soil) is a natural process produced by the dissolution/weathering of As-bearing rocks. Soil and rock analyses show that As is present in concentrations of up to 56 mg/kg and 26 mg/kg, respectively. Arsenic and the concentration of other contaminants typical of sewage water are strongly correlated. The presence of sewage apparently promotes the As release from aquifer materials. It is likely that As mobilization is the result of a desorption associated with arsenate-phosphate competition for sorption sites (Cole *et al.* 2005).

Recently, As-contaminated groundwater was reported from the southern part of the state of Baja California Sur (Fig. 1.1). Here, groundwater is pumped from about 500 wells, which are used for: (1) providing drinking water for cities and towns (122 wells), (2) irrigation (82 wells), and (3) providing water supplies in rural areas (296 open wells). The results of the groundwater analyses reveal that 20% of the sampled wells have concentrations exceeding the WHO limit of 10 µg/l and that 13% are exceeding the Mexican limit of 25 µg/l. Thereby the mining area San Antonio-El Triunfo was found to be the area with the highest As concentrations in groundwater (100–500 µg/l) followed by the Sierra El Mechudo mountain range (30–400 µg/l), whose As source is related to volcanic rocks (Semarnat 2005).

1.3.11   *Other countries*

In other countries, with geological setting related to Andean or Middle American volcanism, sparse information on the occurrence of As from geogenic sources in surface and groundwater exists. Further investigations are needed to fully assess the magnitude of As contamination in Latin America.

1.4   TOXICOLOGICAL PROBLEMS RELATED TO ARSENIC CONTAMINATION

1.4.1   *General*

Long-term exposure to As in drinking water has been associated with an increased risk of cancer and hyperkeratosis, skin pigmentation, degenerative effects on the circulatory system, neurotoxicity, and hepatotoxicity. In Latin America, the problem of As poisoning has not been addressed with the necessary attention, and the actual number of people at risk of chronic As toxicity is not yet known. This lack of knowledge regarding As occurrence and its related health risks deserves immediate attention. Another problem is that national estimates of populations exposed to high As levels are not based on any universal protocol. Different countries use different concentration limits and different analytical techniques to quantify As concentrations, making it impossible to directly compare data from one country to the next. For example, in Argentina the population exposed to As in drinking water is estimated to be 2 million. This population estimate is based on an As concentration range of 2–2900 µg/l (Sancha and Esparza 2000). In contrast, Bundschuh *et al.* (2000, 2004) used an As concentration of 50 µg/l, and estimated the exposed population to be 1.2 million. In Northern Chile, predominantly in Antofagasta, about half a million people were exposed between 1955 and 1970 to drinking water with average As concentrations of 600 µg/l. The installation of treatment plants has reduced these values to an average of 40 µg/l (Sancha *et al.* 1998). About 250,000 people are exposed to drinking water exceeding the national limit of 0.05 mg As/l in Peru (Esparza 2004), and about 450,000 people are exposed to drinking water also exceeding the national limit of 50 µg As/l in Mexico (Avilés and Pardón 2000, Finkelman *et al.* 1993). In countries such as El Salvador, Ecuador, and Nicaragua, no estimates are available for populations at-risk, but several hundred thousand people are probably exposed. Both WHO and the USEPA (US Environmental Protection Agency) have recently (2005) adopted a 10 µg As/l limit for drinking water. Nicaragua followed as the first country in Latin America to adopt this value. Chile and Argentina recently followed and lowered their national drinking water standards to 10 µg/l, while other Latin American countries continue to debate whether to introduce it. Based on this new As concentration limit of 10 µg/l, the estimated number of persons exposed to excessive As concentrations will increase significantly.

1.4.2   *Pathways of human exposure to geogenic arsenic*

Populations may be As-exposed from geogenic sources in several ways. The most common, and the most studied in Latin America, is the uptake of As through drinking water. In terms of the numbers of exposed people, Argentina and Mexico are the most affected countries in Latin America (and are the third most affected countries in the world), followed by Peru and Bolivia. There is a similar regional distribution of As uptake through food. As concentrations in food are affected by two primary factors: (1) the uptake of As by crops from As-rich irrigation water or from As-rich soils, and (2) the use of As-rich water during food preparation. In contrast, human exposure to solids containing geogenic As (solids, dusts, atmosphere) is mostly restricted to active mining areas (e.g., Andean highland of Argentina, Chile, Bolivia, Peru, and some areas of Brazil and Mexico), but can also be expected in areas where winds transport As-rich sediments, such as the Pampa-Chaco plain of Argentina. Both the human uptake of geogenic As through the food chain and through exposure to airborne dust have been very poorly studied in Latin America and deserve attention in the near future.

1.4.3   *Case studies from Latin America*

The groundwater As problem in the Córdoba province (Argentina) has been documented since 1913 (up to 800 μg As/l). Astolfi (1981) found skin lesions, keratosis and epithelioms in the population of this region. In this study, 100 patients presented different grades of diverse types of lesions related to As ingestion (hiperhydrosis, melanodermia, hyperkeratosis, infection, ulceration, and cancer). Astolfi found a clear association between As exposure time and severity of the lesions.

In a study of Taco Pozo (Argentina; Chaco province), Concha (2001) found an average ingestion of 210 μg As/l in drinking water/person. When combined with a food source of corn porridge containing 300–440 μg As/l (Concha 2001) the average resident of Taco Pozo had a daily As uptake of approximately 1–2 mg As/person, which explains the high occurrence of skin cancer observed in that area (M. Biaggini, *pers. com.* 2005).

In Antofagasta (Chile) water from the Toconce river was introduced in 1958 as the new primary source of drinking water. The fact that this river water contains around 800 μg As/l was not considered. Consequently, in 1962 the first cases of chronic As-related health effects were reported in Antofagasta. A total of 500,000 people were exposed to high As levels for 12 years until an As removal plant was installed in 1970 by the public water supply company. The installation of this removal plant resulted in a decrease in As concentrations of drinking water to an average of 40 μg/l.

Cortes *et al.* (2004) studied As concentrations in the urine of 8th grade students in Antofagasta. They found As concentrations of 55 μg/l for the year 2000, whereas in 1977 the urine concentrations in students were significantly higher (71 to 152 μg/l). In an area not affected by the As problem (Santiago de Chile) corresponding values were much lower (13 to 20 μg/l).

Also in Chile, Hopenhayn-Rich *et al.* (2000) investigated and compared the mortality of two regions, those of Antofagasta, with its well-documented history of high As exposure, and those of Valparaíso (Fig. 1.1) located to the South, which can be considered as a low-exposure city. The results indicate an increase in the late fetal, neonatal, and postneonatal mortality rates for Antofagasta, compared to Valparaíso, during those time periods. The highest As-exposure from drinking water was reported for Antofagasta.

The first Central American cases of arsenicosis were reported in 1996 in El Zapote, a rural community of Nicaragua (Gomez 2004, and Gomez 2009) where the people were exposed to contaminated drinking water from a public tube well for two years (1994–96). As concentrations in the tube well were as high as 1320 μg/l. As contamination was also detected in private hand wells used before 1994 and after 1996 (45–66 μg/l). Between July and October of 2002, a complete medical examination was given to 111 inhabitants of El Zapote proper, and those who lived in the surrounding community between 1994 and 1996. Participants were divided into two groups according to the average As concentration ingested during the last 8 years: High As ingestion (80–380 μg/l) and low As ingestion (<80 μg/l). Keratosis and hyper-pigmentation characteristics of chronic arsenicosis were strongly associated with the high As ingestion group (Gomez, this volume). The typical cutaneous manifestations evident on these patients confirmed the diagnosis of the first reported collective case of chronic arsenicosis in Nicaragua and Central America (Gomez, this volume). The significant statistical relationship between high As ingestion and respiratory damage further supports the presumption that chronic ingestion of inorganic As could be associated with damage to the respiratory system (Gomez 2009).

The population potentially exposed to As ingestion from groundwater in Mexico is estimated to be ∼2.0 million (CNA 1999), with concentrations varying from 30 to about 1200 μg/l (Cebrian *et al.* 1994).

In Zimapán (Mexico), a low-income community in a mining area that has been active over the last 400 years and has nearly 15,000 inhabitants, groundwater is the only available source of drinking water. For more than 12 years the inhabitants have consumed As-rich water (190 to 650 μg/l; average 380 μg/l), causing a variety of health impacts (Armienta *et al.* 2005).

In the Lagunera region, which extends over large parts of Durango and Coahuila states (Mexico), the chronic exposure to groundwater As is a endemic problem, which was described for the first time already in the year 1958.

In the Salamanca aquifer system (Mexico), which is exploited by 1900 wells, high As concentrations of up to 280 μg/l (from geogenic and anthropogenic sources) pose a risk to the population (Rodriguez *et al.* 2004). The people of Salamanca are not only exposed to As through the consumption of As-rich drinking water, but are exposed in other ways since they live in the vicinity of several As-producing industries (Rodriguez *et al.* 2004). In addition, the As risk for this area is further increased by the presence of various other contaminants. For example, two carcinogens of different origin, As and organic compounds, acting on the same organ or tissue may increase the damage. There is no local epidemiological information that may allow for the determination of the degree of correlation between As exposure and cancer, although this relationship is expected to exist.

The people from Tlamacazapa village (Guerrero state) display toxic health effects related to exposure to As and other trace metals. This exposure has come primarily from the uptake of drinking water from shallow wells (Cole *et al.* 2005). However, detailed studies of this exposure have not yet been undertaken. Studies are also lacking for other regions of Mexico and many sites in Latin America where high As exposures are likely to be found.

1.5   PREVIOUS EXPERIENCES AND NEEDS IN ARSENIC REMEDIATION IN LATIN AMERICA

The As problem in drinking water has already been solved in most of Latin America's larger urban areas with the installation of water treatment plants. However, many of them are not working properly or are proving to be too expensive to operate. For example, the provincial capital Antofagasta (since the 1970s), Calama, San Pedro, and other cities and large towns in northern Chile treat their water successfully predominantly through flocculation with FeCl3 and subsequent filtration. In Peru, a treatment plant (using flocculation by ferric chloride) was constructed in the city of Ilo in 1982, but it has high operation costs and is not working properly (Esparza 2004). In the cities and some large towns of Argentina, e.g., in the provinces of La Pampa, Santa Fe, and Santiago del Estero, coagulation methods and inverse osmosis have been applied to remove As from drinking water. All of these treatment methods are expensive and are typically inefficient. Therefore, the concerned countries, mainly Chile and Argentina, are presently developing new methods and improving existing systems of As remediation. The purpose of producing new and improved methods is the reduction of treatment costs and maintenance needs with improved automation.

In contrast to urban areas, practically no action has been taken by the authorities, or international and bilateral cooperation agencies, to mitigate the As problem affecting rural populations. This makes the dispersed rural population the most disadvantaged group because they often depend on As-contaminated water as their only available drinking water resource. Typically, these communities are not aware of As toxicity. For these reasons, the reduction of As exposure in rural communities should be a primary goal for both governmental and international organizations in the very near future.

The remediation methods mentioned above are, in most cases, not suitable for small communities, and especially not for the dispersed rural population, typical of the continent. This is, for example, the case in the Chaco-Pampean plane (Argentine), where about 12% of the population are living in dispersed settlements consisting of less than 50 inhabitants, which belong mostly to the poorest members of the regional population. Low-cost remediation methods are needed for small communities of less than 50 inhabitants. These methods should require very little handling and maintenance.

During the last few years some new As removal techniques have been developed at the laboratory scale, but only a few of them have been applied or proven in the field. Examples include: (1) solar oxidation methods (García *et. al* 2004, Nieto *et al.* 2004), (2) phytoremediation (e.g., using algae *Lessonia nigrescens* Hansen (2004), or lacustrine algae (Bundschuh *et al.* 2007), or the use of (3) biomass (González-Acevedo *et al.* 2005, Haque *et al.* 2005), or (4) natural or activated clay and lime as adsorbents for As removal from drinking water.

One of the low-cost technologies is the so-called 'solar oxidation and removal of As' (SORAS). Countries like Bangladesh have used this technology to reduce As pollution in drinking water at the rural community level, and some others like Chile and Argentina are trying to test and implement this method. It is based on the oxidation and precipitation of As, assisted by light in the presence of citric acid. This procedure consists of filling plastic bottles with the contaminated water, then adding some drops of lemon juice to the bottles, and then leaving them to rest under sunlight for a few hours. The capacity of As removal using this method at the household level has been well proven (García *et al.* 2004).

The application of the SORAS technology to a rural area of Chile was implemented (Cornejo 2004) for a simple and low-cost decontamination of water from the Camarones river. The water of this river is used for human consumption and irrigation. Physicochemical analyses were carried out to characterize the water. Variables such as ion concentrations, lemon juice volume, and exposure time were determined in synthetic as well as real samples.

The photocatalytic method using TiO_2 as a catalyst is another recently developed, low-cost technology for As removal, suitable for small communities and at the household-scale. Nieto (2004) developed a technique to cover the inside of borosilicate bottles with a film of TiO_2. His laboratory experiments have confirmed that when these bottles are filled with the As-contaminated water, and exposed to sunlight, As is removed from the liquid, demonstrating that this technique is a suitable low-cost, small scale method of As removal.

1.6   CONCLUSIONS

The As contamination of the environment (ground- and surface water, soils, air) of geogenic origin is an important environmental problem in Latin American perspective for sustainable development.

1. Arsenic is released predominantly from host rocks by dissolution/weathering or by mining activities. The As release impacts the water supply and produces toxicological effects in the population.
2. In the affected areas of Latin America, the chronic As exposure could be associated with neurological and dermatological problems and carcinogenic effects. The population with the highest risk of As exposure is the poor population of rural areas, especially those living in dispersed communities of less than 50 inhabitants.
3. The current methods available for As removal are, in most cases, not suitable for small communities, and especially not for dispersed rural populations. These communities require low-cost remediation methods, which must be very simple in terms of handling and maintenance procedures. Some methods that meet these requirements have already been developed, such as solar oxidation and phytoremediation.

4. Therefore, As removal from drinking water in rural areas is not a technological problem. The problem is convincing the responsible authorities to become interested in the problem. As remediation of rural drinking water will not be possible if the local and national authorities of the affected countries, along with bilateral or international cooperation agencies, do not recognize that the As contamination of (ground)water in the rural areas of Latin America is one of the most important natural health risks of the present century, and that it is their responsibility to solve it. They must recognize that the As contamination of groundwater is an issue and a problem that will challenge the UN Millennium Development Goals for sustainable development on a global scale. Therefore, all the responsible parties should consider doing their best to equip people for a better life in those parts of the world where As in water affects the wellbeing of the population and their potential for sustainable development. It should be stressed here that before using any technology for the As removal, sociological surveys are necessary to evaluate the educational level, as well as the skills and attitude of the people to accept new water remediation ideas and techniques that responsible parties of the world should bring among the village people of the Latin American countries.

REFERENCES

Altamirano Espinoza, M. and Bundschuh, J.: Natural arsenic enrichment in the southwestern basin of the Sébaco Valley, Nicaragua. In: J. Bundschuh, M.A. Armienta, P. Birkle, P. Bhattacharya, J. Matschullat and A.B. Mukherjee (eds): *Natural arsenic in groundwater of Latin America.* Taylor and Francis/Balkema, Leiden, The Netherlands, 2009 (this volume).

Armienta, M.A., Rodriguez, R., Cruz, O., Aguayo, A., Ceniceros, N, Villaseñor, G., Ongley, L.K. and Mango, H.: Environmental behavior of arsenic in a mining zone: Zimapán, México. In: J. Bundschuh, P. Bhattacharya and D. Chandrasekharam (eds): *Natural arsenic in groundwater: occurrence, remediation and management.* Taylor and Francis/Balkema, The Netherlands, 2005, pp. 125–130.

Astolfi, E.A.N., Maccagno, A., García Fernández, J.C., Vaccaro, R. and Stimola, R.: Relation between As in drinking water and skin cancer. *Biolog. Trace Element Res.* 3 (1981), pp. 133–143.

Avilés, M. and Pardón, M.: Remoción de arsénico de agua mediante coagulación-floculación a nivel domiciliario. Federación Mexicana de Ingeniería Sanitaria y Ciencias del Ambiente, FEMISCA, Mexico City, Mexico, 2000, pp. 1–10.

Barragne-Bigot, P.: Contribución al estudio de cinco zonas contaminadas naturalmente por arsénico en Nicaragua. UNICEF, Managua, Nicaragua, 2004.

Bhattacharya, P., Claesson, M., Bundschuh, J., Sracek, O., Fagerberg, J., Jacks, G., Martin, R.A., Storniolo, A. and Thir, J.M.: Distribution and mobility of arsenic in the Rio Dulce alluvial aquifers in Santiago del Estero Province, Argentina. *Sci. Total Environ.* 358 (2006), pp. 97–120.

Bundschuh, J., Farías, B., Martin, R., Storniolo, A., Bhattacharya, P., Cortes, J., Bonorino, G. and Albouy, R.: Groundwater arsenic in the Chaco-Pampean Plain, Argentina: case study from Robles county, Santiago del Estero Province. *Appl. Geochem.* 19 (2004), pp. 231–243.

Bundschuh, J. and García, M.E.: Rural Latin America—A forgotten part of the global groundwater arsenic problem? In: P. Bhattacharya, A.L. Ramanathan, J. Bundschuh, D. Chandrasekharam and A.B. Mukherjee (eds): *Groundwater for sustainable development: problems, perspectives and challenges.* Taylor and Francis/Balkema, Leiden, The Netherlands, 2008.

Bundschuh, J., García, M.E. and Bhattacharya, P.: Control mechanisms of seasonal variation of dissolved arsenic and heavy metal concentrations in surface waters of Lake Poopó basin, Bolivia. Geological Society of America Annual Meeting, Philadelphia, 22–25 Oct. 2006, *Geological Society of America Abstracts with Programs* 38:7 (2006), p. 320.

Bundschuh, J., García, M.E. and Alvarez, M.T.: Arsenic and heavy metal removal by phytofiltration and biogenic sulfide precipitation—a comparative study from Poopó Lake basin, Bolivia. *Abstract volume 3rd International Groundwater Conference IGC-2007, Water, Environment and Agriculture—Present Problems and Future Challenges*, February 7–10, 2007, Tamil Nadu Agricultural University, Coimbatore, India, 2007, p. 152.

Cebrian, M.E., Albores, A., García-Vargas, G., Del Razo, L.M. and Ostrosky, P.: Chronic arsenic poisoning in humans. The case of Mexico. In: J.O. Nriagu (ed): *Arsenic in the environment*, Part II: *Human health and ecosystem effects.* Wiley, New York, 1994, pp. 93–107.

CNA: *Normas oficiales Mexicanas*. Comisión Nacional del Agua, Mexico City, Mexico, 1999.

Cole, J.M, Smith, S. and Bethune, D.: Arsenic source and fate at a village drinking water supply in Mexico and its relationship to sewage contamination. In: J. Bundschuh, P. Bhattacharya and D. Chandrasekharam (eds): *Natural arsenic in groundwater: occurrence, remediation and management.* Taylor and Francis/Balkema, Leiden, The Netherlands, 2005, pp. 67–75.

Concha, G.: *Metabolism of inorganic arsenic and biomarkers of exposure*. PhD Thesis, Institute of Environmental Medicine, Karolinska Institute, Stockholm, Sweden, 2001.

Cornejo, L., Mansilla, H.D., Arenas, M.J., Flores, M., Flores, V., Figueroa, L. and Yáñez, J.: Removal of arsenic from waters of the Camarones river, Arica, Chile, using the modified SORAS technology. In: M.I. Litter and A. Jiménez González (eds): *Advances in low-cost technologies for disinfection, decontamination and arsenic removal in waters from rural communities of Latin America (HP and SORAS methods)*. AOS Project AE/141, Digital Grafic, La Plata, Argentina, 2004, pp. 85–97.

Cortez, S., Pino, P., Atalah, E., Silva, C. and Jara, M.: Exposición a arsénico ambiental en niños de Antofagasta, II región, Chile. In: A.M. Sancha (ed): *Tercer Seminario Internacional sobre Evaluación y Manejo de las Fuentes de Agua de Bebida contaminadas con Arsénico* (proceedings available as CD), Universidad de Chile, November 08–11, 2004, Santiago de Chile, Chile, 2004.

Cumbal, L., Aguirre, V., Murgueitio, E., Tipán I. and Chavez, C.: El origen del arsénico en aguas y sedimentos de la Laguna de Papallacta. Taller de distribución del As en Iberoamérica, 27–30 November 2006, Red Temática 406RT0282 Iberoarsen, Centro Atómico Constituyentes, San Martín, Buenos Aires, Argentina, 2006, pp. 87–90.

ECO/OPS: Epidemiología ambiental: un proyecto para América Latina y El Caribe. Centro Panamericano de Ecología Humana y Salud División de Salud y Ambiente, Agencia de Protección Ambiental, Programa Internacional sobre Seguridad Química, Red de Epidemiología Ambiental, Lima, Peru, 1997.

Esparza, M.L.: Presencia de arsénico en el agua de bebida en América Latina y su efecto el la salud pública. In: A.M. Sancha (ed): *Tercer Seminario Internacional sobre Evaluación y Manejo de las Fuentes de Agua de Bebida contaminadas con Arsénico* (proceedings available as CD), Universidad de Chile, November 08–11, 2004, Santiago de Chile, Chile, 2004.

Finkelman, J., Corey, G. and Calderon, R.: Environmental epidemiology: A project for Latin America and the Caribbean. ECO, Metepec, Mx, Mexico, 1993.

García, M.G., Lin, H.J., Custo, G., d'Hiriart, J., Hidalgo, M. del V., Litter, M.I. and Blesa, M.A.: Advances in solar oxidation removal of arsenic in waters of Tucumán, Argentina. In: M.I. Litter and A. Jiménez González (eds): *Advances in low-cost technologies for disinfection, decontamination and arsenic removal in waters from rural communities of Latin America (HP and SORAS methods)*. AOS Project AE/141, Digital Grafic, La Plata, Argentina, 2004, pp. 43–63.

García, M.E., Bundschuh, J., Ramos, O., Quintanilla, J., Persson, K.M., Bengtsson, L. and Berndtsson, R.: Heavy metals in aquatic plants and their relationship to concentrations in surface water, groundwater and sediments—A case study of Poopó basin, Bolivia. *Rev. Quim. Boliv.* (2006).

García, M.E. and Bundschuh, J.: Control mechanisms of seasonal variation of dissolved arsenic and heavy metal concentrations in surface waters of Lake Poopó basin, Bolivia. Geological Society of America Annual Meeting, Philadelphia, 22–25 Oct. 2006, *Geological Society of America Abstracts with Programs* 38:7 (2006), p. 320.

Gómez, A.: Chronic arsenicosis in El Zapote, Nicaragua, 1994–2002. In: A.M. Sancha (ed): *Tercer Seminario Internacional sobre Evaluación y Manejo de las Fuentes de Agua de Bebida contaminadas con Arsénico* (proceedings available as CD), Universidad de Chile, November 08–11, 2004, Santiago de Chile. Chile.

Gómez, A.: Chronic arsenicosis and respiratory effects in El Zapote, Nicaragua: In: J. Bundschuh, M.A. Armienta, P. Birkle, P. Bhattacharya, J. Matschullat and A.B. Mukherjee (eds): *Natural arsenic in groundwater of Latin America*. Taylor and Francis/Balkema, Leiden, The Netherlands, 2009 (this volume).

González-Acevedo, Z.I., Cano-Aguilera, I. and Aguilera-Alvarado, A.F. : Removal and recovery of arsenic from aqueous solutions by sorghum biomass. In: J. Bundschuh, P. Bhattacharya and D. Chandrasekharam (eds): *Natural arsenic in groundwater: occurrence, remediation and management.* Taylor and Francis/Balkema, Leiden, The Netherlands, 2005, pp. 255–261.

Guérèquiz, A.R., Mañay, N., Goso Aguilar, C. and Bundschuh, J.: Assessment of the environmental risk caused by the presence of arsenic in the western area of the Raigón aquifer, department of San José, Uruguay. Taller de distribución del As en Iberoamérica, 27–30 November 2006, Red Temática 406RT0282 Iberoarsen, Centro Atómico Constituyentes, San Martín, Buenos Aires, Argentina, 2006, pp. 111–112.

Hansen, H.K., Rojo, A., Oyarzun, C., Ottosen, A.R. and Mateus, E.: Biosorption of arsenic by *Lessonia nigrescens* in wastewater from cooper smelting. In: A.M. Sancha (ed): *Tercer Seminario Internacional sobre Evaluación y Manejo de las Fuentes de Agua de Bebida contaminadas con Arsénico* (proceedings available as CD), Universidad de Chile, November 08–11, Santiago de Chile, Chile, 2004.

Haque, N., Morrison, G., Perrusquía, G., Cano-Aguilera, I., Aguilera-Alvarado, A.F. and Gutiérrez-Valtierra, M.: Sorption of arsenic on sorghum biomass: A case study. In: J. Bundschuh, P. Bhattacharya and D. Chandrasekharam (eds): *Natural arsenic in groundwater: occurrence, remediation and management.* Taylor and Francis/Balkema, Leiden, The Netherlands, 2005, pp. 247–253.

Hopenhayn-Rich, C., Browning, S., Hertz-Picciotto, I., Ferreccio, C., Peralta, C. and Gibb, H.: Chronic arsenic exposure and risk of infant mortality in two areas of Chile. *Environ. Health Perspect.* 108 (7): 2000, pp. 667–673.

Lopez, D., Ramson, L., Monterrosa, J., Soriano, T., Barahona, J. and Bundschuh, J.: Volcanic pollution of arsenic and boron at Ilopango lake, El Salvador. In: J. Bundschuh, M.A. Armienta, P. Birkle, P. Bhattacharya, J. Matschullat and A.B. Mukherjee (eds): *Natural arsenic in groundwater of Latin America.* Taylor and Francis/Balkema. Leiden, The Netherlands, 2009 (this volume).

Nieto, J., Aguilar, J., Ponce, S., Rodriguez, J., Solis, J. and Estrada, W.: TiO$_2$ thin films deposited by spray pyrolysis inside glass tubes for use in the HP technology for water potabilization. In: M.I. Litter and A. Jiménez González (eds): *Advances in low-cost technologies for disinfection, decontamination and arsenic removal in waters from rural communities of Latin America (HP and SORAS methods).* AOS Project AE/141, Digital Grafic, La Plata, Argentina, 2004, pp. 153–155.

Rodriguez, R., Armienta, M.A. and Mejia, J.A.: Arsenic contamination of the Salamanca aquifer system in Mexico: A risk analysis. In: J. Bundschuh, P. Bhattacharya and D. Chandrasekharam (eds): *Natural arsenic in groundwater.* Taylor and Francis/Balkema, Leiden, The Netherlands, 2005, pp. 77–83.

Sancha, A.M. and C. de Esparza, M.L.: Arsenic status and handling in Latin America. Univ. Chile, Grupo As de AIDIS/DIAGUA, CEPIS/OPS, Lima, Peru, 2000.

Sancha, A.M., Fuentealba, C. and Campos, C.: Assessing arsenic removal technologies for drinking water production: the Chilean experience. In: A.M. Sancha (ed): *Tercer Seminario Internacional sobre Evaluación y Manejo de las Fuentes de Agua de Bebida contaminadas con Arsénico* (proceedings available as CD), Universidad de Chile, November 08–11, 2004, Santiago de Chile, Chile, 2004.

Sancha, A.M., O'Ryan, R., Marchetti, N. and Ferreccio, C.: Análisis de riesgo en la regulación ambiental de tóxicos: caso del arsénico en Chile. *XXVI Congreso Interamericano de Ingeniería Sanitaria y Ambiental,* Lima, Peru, 1998.

SEMARNAT: Secretaría de Medio Ambiente y Recursos Naturales del Gobierno del Estado de Baja California Sur, (2005) http://www.semarnat.gob.mx/estados/bajacaliforniasur/.

CHAPTER 2

The presence of arsenic in drinking water in Latin America and its effect on public health

M.L. Castro de Esparza
Centro Panamericano de Ingeniería Sanitaria y Ciencias del Ambiente (CEPIS/SDE/OPS-PERU), Lima, Peru

ABSTRACT: In Latin American countries at least four and a half million people drink water containing arsenic (As) at levels which pose a risk to their health to such an extent that in certain areas this has become a public health problem. Those most at risk are people living in scattered rural areas, who drink untreated water and are unaware of the risk to which they are exposed. The presence of As in the environment and in water sources for human consumption is due to natural geological factors, anthropogenic activities, including mining and metal smelting, electrolytic processes for the production of high quality metals and, to a lesser extent, the use of organic As-based pesticides in agriculture. Toxicological and epidemiological studies in Latin American countries confirm the information and indicate that chronic ingestion of As in drinking water results in skin lesions, hyperpigmentation and hyperkeratosis of the palms of the hands and soles of the feet; it also causes nervous system disorders, diabetes mellitus, anemia, liver disorders, vascular illnesses and skin, lung and bladder cancer. Recent studies have also shown the effects caused by the ingestion of As on the intellectual development of children.

2.1 INTRODUCTION

In various Latin American countries, including Argentina, Chile, Mexico, El Salvador, Nicaragua, Peru, Bolivia and Brazil, at least four and a half million people drink water containing arsenic (As) at concentrations that pose a risk to their health. The As content of water, especially groundwater, in some cases reaches 1000 μg/l. In other regions of the world, such as India, Bangladesh, China and Taiwan the problem is even greater (Chowdhury *et al.* 2000, Smith 2000b, Bhattacharya *et al.* 2002, Chakraborti *et al.* 2002). According to our information, around 35 million people in Bangladesh are exposed to As; several million of them are children. In the United States, more than 350,000 people drink water that contains more than 50 μg/l of As, and more than 2.5 million are supplied with water having an As content in excess of 25 μg/l (Science 1987, Marcus and Rispin 1988).

The purpose of this work is to provide information from Latin America on this environmental and public health problem, which must be addressed in order to minimize its effects and to reduce arsenicism in the affected areas. This chapter provides a bibliographic record of the presence of As in drinking water and its effects on the health of people thus exposed.

2.2 THE ORIGIN OF ARSENIC IN DRINKING WATER

In general in Latin America, the presence of As in drinking water sources used for human consumption results from natural geological factors (Mexico, Argentina, Chile, Peru) (Sancha *et al.* 1998, Bundschuh *et al.* 2004, Bhattacharya *et al.* 2006), mining and refining of metals (Chile, Bolivia, Peru), electrolytic processes producing metals of high quality such as cadmium and zinc (Brazil) and, to a lesser extent, the use of organic As-containing pesticides (Mexico) (Cebrián *et al.* 1994).

Natural As in surface water and groundwater in Latin America is associated with Tertiary and Quaternary volcanism in the Andes mountains, a continually active process as shown by lava flows, fumaroles, thermal springs and geothermal phenomena related to volcanism in the so-called "Pacific Ring of Fire". This volcanism can also influence critical water properties such as high pH, alkalinity, hardness, salinity and the presence of boron, fluorine, silica and vanadium.

These geological processes produced the important deposits of copper, principally in Chile, Peru and Bolivia, the exploitation and smelting of which has helped to increase already high levels of environmental As. The geographical conditions in these areas, characterized by high altitude, a scarcity of water and adverse weather conditions, have largely limited the development of large cities. Exposure is higher in Chile, Argentina and Mexico than in other countries in the region (Sancha and Esparza 2000).

Arsenic in surface and groundwater derives from mineral dissolution, erosion, disintegration of rocks and from atmospheric deposition. It is also present in aerosols and it may be found in water in the trivalent and pentavalent forms, depending on the environmental conditions. Oxidized forms are found more often in surface water, while the reduced forms are more common in groundwater, particularly at greater depths.

In Argentina, the higher As content in water is natural. The As concentration in groundwater in the affected areas is variable, from lower values of 100 μg/l up to values in excess of 1000 μg/l. The source of As in groundwater in central and northern Argentina is volcanic ash, although activities associated with farming may be responsible to a lesser extent (Benitez *et al.* 2000, Nicolli *et al.* 1989, Bundschuh *et al.* 2004, Bhattacharya *et al.* 2006).

In Mexico, As is present within the volcanic belt where As-rich rocks and soils contaminate groundwater. Another possible source of contamination is organic As-containing pesticides, which have been used since before 1945.

In Chile, As is present in all ecosystems in the north of the country because of the widespread occurrence of Quaternary volcanism. Between 1955 and 1970 in Antofagasta, the average As concentration in the water was 598 μg/l. Current values indicate an average of 40 μg/l (Sancha *et al.* 1998).

In Bolivia, the largest water source for La Paz receives runoff water from the mining area of Milluni before it reaches the treatment plant. In the smelting town, where three small companies operate (Calbol, Hormet and Bustos), 0.7% of the As contained in burned wastes is released into the environment. The soil and drinking water have been found to contain As, with the highest concentrations occurring in a recreation area near the Bustos smelter (ECO/PAHO 1997). There is also information on the presence of As in rivers, lakes and ponds located in the southwest of Bolivia (Quintanilla 1992).

In Peru, the city of Ilo draws water from the Ilo river for domestic and industrial use, which originates in Aricota lake. Aricota lake is fed by two rivers: the river Callazas and the river Salado, which have As concentrations of 640 and 1680 μg/l, respectively. These rivers pass close to the Yucamane volcano, which appears to be the source of As (Siveroni 1989). Recent studies show the presence of As in the area around Puno and in the departments of Tacna and Moquegua.

Brazil uses 1500 tons of As per year for the electrolytic production of zinc and cadmium. Industrial waste and atmospheric emissions of arsine are the main sources of environmental contamination by As. High As concentrations were found close to the wastewater outfall near Enseño Cove (Barcellos 1992, Finkelman *et al.* 1993), a situation that contributes to the contamination of drinking water sources. Furthermore, in the Minas Gerais mining districts of Nova Lima and Santa Barbara, mining activities associated with the extraction and refinement of gold causes the contamination of water with As. Arsenic concentrations in water were from 0.4–350 μg/l (Matschullat *et al.* 2000).

2.3   TOXICOLOGY OF ARSENIC

The main pathways for human exposure to As are ingestion and inhalation. Arsenic can accumulate in the body as a result of chronic exposure and, at certain concentrations, can cause problems such as changes to the skin (relaxation and dilatation of the pores) with secondary effects on the nervous system; irritation of the respiratory organs and gastrointestinal tract and altered hematopoiesis. It can also accumulate in the bones, muscles and skin and, to a lesser degree, in the liver and kidneys. People who have suffered prolonged exposure to inorganic As in drinking water exhibit hyperkeratosis of the hands and feet, the principal manifestation of which is skin pigmentation and calluses on the palms and soles of the feet. The presence of As in water, its degree of contamination and the incidence of skin diseases in Argentina and Mexico are described in another section of this chapter.

Experiments with laboratory animals indicate that inorganic trivalent As is more toxic than the pentavalent form because pentavalent compounds have less effect on enzyme action. However, *in vivo*, these can be reduced to trivalent compounds. The toxicity of As depends on its oxidation state, chemical structure and solubility in the biological medium. The scale of As toxicity declines in the following order: arsine > inorganic As(III) > organic As(III) > inorganic As(V) > organic As(V) > As compounds and elemental As. The toxicity of inorganic As(III) is 10 times higher than that of inorganic As(V) and the lethal dose for adults is 1–4 mg As/kg. For the more common forms such as AsH_3, As_2O_3, As_2O_5, this dosage varies between 1.5 mg/kg and 500 mg/kg of body mass (NAS 1999).

In drinking water, As is generally found in the form of arsenate and 40 to 100% can be absorbed easily (Frederick *et al.* 1994) in the gastrointestinal tract. Ingested inorganic As passes into the blood stream where it links with hemoglobin and in 24 hours can be found in the liver, kidneys, lungs, spleen and skin. It is found in greater concentrations in the skin, bones and muscles than other tissues. Accumulation in the skin stems from the ease of reaction with proteins (with sulphydryl groups) (Environmental Health Directorate of Health Canada 1992).

Metabolic changes in As speciation occur essentially in the liver, where endogenous thiols play a critical role in the conversion of As(III) and As(V). It appears that glutathione (GSH) acts as a reducing agent. The inorganic As(III) forms can be methylated [oxidation and formation of methylarsenic As(V)] if the functional group S-adenosylmethionine (SAM) is accepted. The probable end-product of continuous methylation is dimethylarsenate (DMA). The methylarsenic As(III) and intermediate As(III) forms may be toxic and inhibit glutathione reductase (GR), a key enzyme in the metabolism of GSH and whose action (GR) is critical in maintaining the redox reactions of cells (NAS 1999, Albores *et al.* 1997).

In the human body inorganic As(III) and inorganic As(V) act through different mechanisms. The behavior of inorganic As(V) is similar to that of phosphate, but differs in the stability of its esters. The esters of phosphoric acid are stable, which enables the existence of deoxyribonu-cleic acid (DNA) and adenosine 5-triphosphate (ATP). In contrast, the acidic esters of As(V) can be hydrolyzed. The enzymes can accept arsenate and incorporate it into compounds such as ATP, but the resulting compounds are hydrolyzed immediately. Therefore, arsenate can deactivate the oxidizing metabolism of ATP synthesis. In contrast, inorganic As(III) has a high affinity for thiol groups in proteins and can deactivate a variety of enzymes, such as pyruvate dehydrogenase and 2-oxoglutarate dehydrogenase (Frederick *et al.* 1994, National Academy of Sciences 1999). However, monomethylarsenate (MMA) and dimethylarsenate (DMA) do not form strong bonds with human biological molecules. This explains why it is less toxic than inorganic As.

The kinetics relating to the toxicity of inorganic As, including cancer has not yet been established. The most acceptable explanation is that a chromosome abnormality is induced without acting directly on DNA.

Ingested inorganic As is absorbed by the tissues and then progressively eliminated by methylation. It is excreted in urine through the kidneys. When ingestion is greater than excretion it tends to accumulate in the hair and nails. The normal concentrations of As in urine, hair and nails are 5–40 µg/kg, 80–250 µg/kg and 430–1080 µg/kg, respectively (National Academy of Sciences 1999). Human sensitivity to the toxic effects of As varies, depending on genetics, metabolism, diet, state of health, sex, and many other factors. These factors should be taken into account in any risk assessment of As exposure. Those at greater risk have a poor ability to methylate As and, therefore, retain more, the most vulnerable being children and people who are undernourished (Acosta *et al.* 2006).

In some species of mammals it has been shown that inorganic and organic As are teratogenic and oral ingestion affects fetal growth and prenatal viability (Andreson *et al.* 1999). A supplement with a high As content in the diet (e.g., 350–4500 ng/g) affects growth and reproduction in animals (National Academy of Sciences 1999). Studies have shown that urine is the best biomarker for measuring absorbed inorganic As, as blood, hair and nails are less sensitive to exposure (National Academy of Sciences 1999). It has been shown that children are more sensitive than adults to As poisoning and are the most affected by arsenicism, because of malnutrition and lack of sanitation in scattered (poor) rural areas (Acosta *et al.* 2006).

Chronic ingestion of As in drinking water causes nervous system disorders, diabetes mellitus (Del Razo *et al.* 2005), anemia, liver disorders, vascular illnesses and skin, lung and bladder cancer (Del Razo *et al.* 2000, Albores *et al.* 2001, Endo *et al.* 2003, Kirk and Sarfaraz 2003, Rossman 2003). Other studies have shown that As alters the heme biosynthesis pathway enzymes activity and As induced cell signal transduction in an individual chronically exposed (Aguilar *et al.*1999, Qian *et al.* 2003). Recent studies have also shown the effects caused by As ingestion on the intellectual development of children (Borja *et al.* 2001, Wasserman and Zhongqi 2004).

2.3.1 *Arsenicism*

Drinking As-rich water over a long period is unsafe, and chronic ingestion of inorganic As at low doses (as sometimes observed in natural water) is a recognized cause of arsenicism (Beck *et al.* 2003).

In Bangladesh four stages of arsenicism are recognized:

- *Preclinical*: the patient shows no symptoms but As can be found in tissue and urine samples.
- *Clinical*: at this stage it affects the skin. Darkening of the skin (melanosis) is observed, frequently of the palms of the hands, with dark patches appearing on the chest, back, limbs and gums. A more serious symptom is keratosis or hardening of the skin to form nodules on the palms of the hands and soles of the feet. WHO calculates that this stage requires 5 to 10 years of exposure to As.
- *Complications*: more pronounced clinical symptoms and effects on the internal organs are observed. Studies have reported enlargement of the liver, kidneys and spleen. There is also information linking this stage with conjunctivitis, bronchitis and diabetes.
- *Malignity*: development of tumors or cancer of the skin or other organs. In this stage, the person affected may develop gangrene or cancer of the skin, lungs or bladder.

In the first two stages, if the patient drinks water that is free from As, recovery will be almost complete. The third stage can be reversed but not the fourth stage (Bangladesh Centre for Advanced Studies 1997).

2.3.2 *Treatment*

Treatment generally involves providing the patient with As-free drinking water. The next step is to monitor the patient and ensure that he is no longer exposed to the element.

Other treatments proposed are chelation and improved nutrition. Chelation has been used in West Bengal (India) and Bangladesh, but it is not known whether it can remove As from the skin and it is not effective if the patient continues to drink contaminated water. Evidence from Taiwan shows that nutritional factors can modify the risk of cancer associated with As. Taking vitamins (A and multi-vitamin supplements) and improved nutrition (protein) can improve patient's condition, particularly when the skin is affected.

The United States Environmental Protection Agency (USEPA) classifies As as a group A carcinogen because of evidence of its adverse effects on health. Exposure to 50 µg/l can cause 31.33 cases of skin cancer per 1000 inhabitants and a proposal has been made to reduce the acceptance

Table 2.1. Guideline values for arsenic in drinking water established by various regulatory agencies.

Country/Organization	Maximum level of contamination (MLC)[1], µg/l
Argentina	50/10 [2]
Brazil	10
Canada	25
Mexico	25
Chile	10
Peru	50
USA	10
World Health Organization (WHO)	10
Germany and other EU countries	10
India	50
China	50
Taiwan	20

[1]Compiled from different sources; [2] lowered in the 2007 to 10 µg/l

limit from 50 µg/l to 10–20 µg/l. The International Agency for Research on Cancer has classified it in group 1 because of evidence on carcinogenicity in human beings. Natural elimination from the human body is through urine, feces, perspiration and skin epithelium (peeling). There is evidence linking As with cancer. However, the carcinogenic dose is not known.

Some studies on the toxicity of As indicate that many current standards based on WHO guidelines are too high and suggest that limit values should be re-evaluated on the basis of epidemiological studies. For example, in Taiwan it is calculated that the limit should be reduced from 20 to as low as 5 µg/l. In other cases it would appear that these values should be increased, in accordance with regional conditions. In Latin America, it has been shown that similar levels of As under different conditions (climate, nutrition and others) result in different effects.

2.4   EXPOSED POPULATION AND EPIDEMIOLOGICAL STUDIES

The following is a summary of epidemiological studies carried out in Latin American countries, where part of the population is exposed to As in drinking water.

2.4.1   *Argentina*

The problem has been known about for nearly 100 years when epidemiologists from Córdoba and other provinces found and associated skin damage to As in the drinking water. The first pathological manifestations were known as Bell Ville disease and then as endemic regional chronic hydroarcenicism (ERCH).

It is estimated that 2,000,000 inhabitants are exposed to As levels in the range 2–2900 µg/l (Sancha and Castro de Esparza 2000). The most seriously affected provinces are Salta, La Pampa, Córdoba, San Luis, Santa Fe, Buenos Aires, Santiago del Estero, Chaco, and Tucumán (Pinedo and Ligarán 1998, Álvarez *et al.* 2000).

One of the most seriously affected areas is in the province of Chaco. The most evident effects were darkening of the skin, lesions, hyperkeratosis, warts, melanosis, leukodermia, basal cell carcinoma and senile keratomiasis with a high incidence of cancer of the urethra, urinary tract and bladder. It has been shown that, with As poisoning, keratosis predominates over hyperpigmentation. The types of cancer found were skin cancer and internal cancers (66% of which were in the lungs). Levels of As in the water in this area, particularly in the neighborhood of San Martin, are in excess of 700 µg/l (Finkelman *et al.* 1993, Benitez *et al.* 2000).

In the province of Santiago del Estero, deaths associated with arsenicism have been reported since 1983. According to data provided by the Secretary for Epidemiology, serious cases in children and women have been found. Reported effects include lesions on the soles of the feet, arms and trunk, melanoderma on the arms and trunk, lesions on the palms of the hands, leucoderma on the back and chest.

In 1997 in Mili, 71 groundwater samples were analyzed, 52% of which had high concentrations of As, up to 2400 μg/l (Herrera *et al.* 2002). At Gran Porvenir, in the department of Banda, Santiago del Estero (474 inhabitants), where samples were taken from 36 households out of a total of 103, the water was found to contain As concentrations between 2 and 143 μg/l (19 samples were collected at random), of which 58% contained critical concentrations. If we consider that all the samples contained higher than permitted levels of As and that more than 50% of inhabitants drink this water, it is clear that there is a public health problem (De Paredes 1997). Later studies on shallow groundwater in the counties of La Banda and Robles in Santiago del Estero province revealed total (t-As) concentration in the range of 7.0 to 14,969 μg/l with mean and median concentrations of 743 μg/l and 53.6 μg/l respectively (Bejarano and Nordberg 2003, Claesson and Fagerberg 2003, Bhattacharya *et al.* 2006). The aqueous speciation of As was dominated by As(V) with a mean concentration of 617 μg/l, while the concentrations of As(III) was low, in the range from 1.9 to 45% (mean 125 μg/l, see Claesson and Fagerberg 2003 for details). However, dominance of As(III) was noted in the samples with high DOC concentrations where the abundance of trivalent species ranged from 46 to 66% of the total As in groundwater (Bhattacharya *et al.* 2006).

A study was conducted of the metabolism of inorganic As in children in three villages in northern Argentina. Arsenic concentrations in drinking water from San Antonio de los Cobres and Taco Pozo was 200 μg/l, and that from Rosario de Lerma was 650 μg/l. Arsenic concentrations in the blood and urine of children from the first two villages were 10 to 30 times higher (9 and 380 μg/l, respectively) than in the blood and urine of children from Rosario de Lerma. The fact that higher percentages of inorganic As were found in children's urine than in that of adults shows that children are more sensitive than adults (Concha *et al.* 1998).

In the province of Córdoba, located in the centre of the country, As concentrations in water are in excess of 100 μg/l, increasing the risk of cancer and skin changes. Between 1986 and 1991 studies of deaths caused by cancer of the bladder, lungs and kidneys were carried out in 26 districts, which had previously been classified based on the As concentration in drinking water. The As concentration in the area where exposure was greatest (San Justo and Union) was 178 μg/l. The studies showed a clear dose-response relation between As in the water and the risk of cancer. Mortality rates from bladder cancer were 2.14 for men and 1.82 for women (95% confidence) in the two districts where exposure was highest. The mortality rates for lung cancer, expressed as low, medium and high rates, were 0.92, 1.54 and 1.77, respectively for men and 1.24, 1.34 and 2.16 for women. Mortality rates from kidney cancer were similar: 0.87, 1.33 and 1.57 for men and 1.00, 1.36 and 1.81 for women ($p < 0.001$). These studies did not find a clear relation between the As and mortality due to skin and liver cancer (Hopenhayn-Rich *et al.* 1996, Hopenhayn-Rich *et al.* 1998).

In order to update the map of the risk of arsenicism and to identify critical areas, the province of Córdoba was divided up based on geography: western or mountains and eastern or plains. During the study, 100 samples of water (groundwater) were taken from forty locations. The departments identified as critical were San Justo, M. Juarez, Union, Río Cuarto and Río Primero where lung cancer mortality rate was higher as compared to other cancers: in the western or mountain region, 20.1%; in the eastern or plains region, 37.4%; and to a lesser extent skin cancer, 1.9–2.1%. Other types of cancer affected the prostate, colon, bladder, kidney and larynx. It was found that 57% of the areas studied had concentrations in excess of the values permitted by WHO guidelines (Pinedo and Zigarán 1998).

At present, the water supply regulator of the city of Santa Fe is carrying out three studies with support from PAHO: (1) a risk map of 213 water supplies; (2) an epidemiological study of endemic regional chronic hydroarcenicism; and (3) a correlation between As in drinking water and deaths from 5 associated cancers (Corey 2002).

2.4.2   *Chile*

The most serious contamination occurs between latitudes $17°30'$ and $26°05'$ south and between longitude $67°00'$ and the Pacific Ocean. The cities having the highest exposure to As are Antofagasta, Calama, Santiago, Rancagua, Taltal, Tocopilla, La Serena and San Pedro de Atacama. Approximately 500,000 inhabitants were exposed to As contamination until treatment plants were constructed. In Antofagasta, between 1955 and 1977 the average As concentration was found to be 598 μg/l while the present values give an average of 40 μg/l (Sancha *et al.* 1998).

Studies in Region II between 1950 and 1993 showed the risk of death caused by different cancers associated with As were, principally, from bronchopulmonary, bladder and renal cancers (Rivera and Corey 1995). In 1970 the annual rate of dermatosis associated with chronic arsenicism was 20 per 100,000. Concentrations of As in the hair and urine of those exposed were above normal values and a study that included 100 children with dermatosis associated with arsenicism detected 19 cases of Raynaud's phenomenon (Finkelman *et al.* 1993).

Between 1994 and 1996 a study was carried out in the northern region of Chile (Arica, Iquique, Copiapó and Antofagasta) aimed at relating As exposure to the risk of contracting lung cancer. Cases of lung cancer and two hospital controls were evaluated. One control was a patient with cancer and the other was a patient without cancer; neither diagnosis was related to As. Over 20 months, 151 cases of lung cancer were admitted together with 409 controls (167 with cancer and 242 without cancer). There was a clear dosage-response relation between the average concentration of As and the risk of cancer (adjusted against a linear regression model that included sex, age and whether the patients smoked tobacco) at 95% confidence, of 1, 1.7 (0.5–5.1), 3.9 (1.2–13.4), 5.5 (2.2–13.5) and 9.0 (3.6–22) for As concentrations from below detection level to as high as 400 μg/l (Ferreccio *et al.* 1998).

In 1998 a risk analysis study of As exposure was carried out as required by the environmental toxins regulations. The project consisted of developing a water, food and air baseline. Samples of water and food were taken all over the country while air was sampled in the north and central zones; these figures were compared with information from the national ingestion and inhalation baseline. The heath impact assessment of exposure to As made use of an ecological study and case and control studies, by comparing with death rates caused by As-associated cancers (lung, bladder, liver, skin and kidney). The results indicate that the greatest contribution of As to total exposure in the northern zone comes from drinking water (41.7 to 85.3%). This also applied to Santiago and Rancagua (72.7 to 69.3%). In the south, diet acquired greater importance and, in general, the contribution made by the air was around 1%, with the exception of Copiapó where it was 12.2%. In the north, death rates from lung, skin and bladder cancers were higher. The study shows that the higher death rate in the north is largely attributable to exposure to As. The relative risks for different cancers (bladder, urinary tract, lung, liver, kidney and larynx) were reported together with heart diseases, ischemia, chronic respiratory illnesses and chronic As-related dermatosis.

Smith *et al.* (2000a) carried out a study on skin lesions suffered by the inhabitants of Atacama. This was a study of eleven families in the Chiu-Chiu area, whose drinking water contained 750 to 800 μg/l of As, and eight families in the village of Caspana who were used as a control group. The results showed that four of the six men who drank water contaminated with As for more than 20 years had skin lesions; no cases were recorded in adult women, but two additional cases were found among adolescents. The prevalence of skin lesions in these small groups is comparable to Taiwan and India where the population is more susceptible, probably because of their nutritional level. The inhabitants of the Atacama region showed no alarming cases despite prolonged exposure; perhaps their resistance was due to food rich in vitamin A.

Studies have been carried out on trends in infant mortality associated with exposure to As in two Chilean cities: Antofagasta and Valparaíso. The study was retrospective and aimed to evaluate time and location patterns for infant mortality between 1950 and 1996, using univariant statistical graphing techniques and Poisson's linear regression analysis. The results showed high rates of fetal, neonatal and postneonatal mortality in Antofagasta, though not in Valparaíso. The linear regression analysis indicated an association between exposure to As and fetal, neonatal and postneonatal mortality. These findings indicate, but not definitively, an association between As and the increase in infant mortality in Antofagasta (Hopenhayn-Rich *et al.* 2000).

2.4.3   *Mexico*

The first information on As contamination dates from 1962, when 40 serious cases and one death were reported in the urban area of Torreón, Coahuila.

The presence of As in drinking water is a problem found in the Durango, Coahuila, Zacatecas, Morelos, Aguas Calientes, Chihuahua, Puebla, Nuevo León, Guanajuato, San Luis Potosi and Sonora aquifers, Zimapán valley and the Lagunera region, where concentrations have been found that exceed the values given in NOM-127SSA1 (50 μg As/l) (Finkelman *et al.* 1993, Avilés and Pardón 2000). It is estimated that around 450,000 people are exposed.

Studies of As began in the Lagunera region in the states of Durango and Coahuila. Chronic endemic As poisoning was found in this area with extreme outbreaks affecting both animals and humans. A study of As contamination in 128 water wells in 11 districts found a range of 8 to 624 μg/l; more than 50% of the samples contained more than 50 μg/l. It is thought that around 400,000 people were exposed to drinking water with As concentrations greater than 50 μg/l. This evidence suggests the existence of a medium-term public health problem. Of these groups, a total of 489,634 people face an individual maximum cancer risk of between 4.5×10^{-2} and 5.7×10^{-2}; a total of 609,253 people face a risk of between 5.2×10^{-3} and 4.1×10^{-2} (Vega Gleeson 2001).

In the Lagunera region other signs and symptoms of arsenicism have been found, including a 0.7% prevalence of black-foot disease. In order to establish an epidemiological surveillance system in this region, a project entitled "Identification of Early Health Risk for Exposure to Arsenic" was carried out. It involved an evaluation of the genotoxic risks and excretion in urine of porphyrins and methylated As derivatives. The objective was to inform decisions regarding measures to control arsenicism.

At present, Mexico and Argentina, with the support of PAHO, are carrying out a socio-economic study of the impact of As on public health and the viability of alternatives for removing As from the drinking water supply.

In Mexico a study has been carried out to examine the possible relationships between chronic exposure to inorganic As and the risk of cancer and other, non-cancerous, diseases. An important factor is the average MA(III) concentration, which is significantly higher in the urine of exposed individuals who had skin lesions as compared to those who drank water contaminated with inorganic As but had no skin lesions. These results suggest that urinary concentrations of MA(III), (the most toxic species among identified metabolites of inorganic As), may serve as an indicator to identify individuals with increased susceptibility to the toxic and cancer-promoting effects of arsenicism (Valenzuela *et al.* 2004).

2.4.4   *Peru*

The south of Peru contains semi-desert areas, in which populations are served by drinking water supplies derived from rivers that flow from the Andes towards the Pacific Ocean. Elevated concentrations of As have been found in some of these rivers. For example, the river Locumba (500 μg As/l), which flows through Puno and Moquegua (the Ilo valley), constitutes the drinking water supply for approximately 250,000 inhabitants, who thus become exposed to As (Esparza 2000, Esparza and Sancha 2000).

In 1994, As concentrations were determined in 53 samples of drinking water, river, well and spring water in the Rimac river basin. It was found that 84.9% of the samples exceeded the limit recommended by the WHO (Infante and Palomino 1994). Nevertheless, no cases of As poisoning have been recorded. In 1999 another study of drinking water in the province of Huaytara, Huancavelica, was carried out. Arsenic concentrations in 31 samples yielded an average of 246 μg/l. The highest concentration was found in Pachac, probably because of the presence of a warehouse used to store fertilizers and As-based pesticides (Flores 1999).

In 2002 an evaluation of the river Locumba found As concentrations between 200 and 400 μg/l. The inhabitants of the valley have been drinking this water for many years and no cases of arsenicism have been reported. In Puno, As concentrations up to 180 μg/l have been found in recently drilled wells. A study will be carried out to evaluate alternative means of removing As (Esparza 2005).

2.4.5 *Nicaragua*

In 2001, UNICEF invited different national and international institutions to a workshop to discuss the problem of As contamination in drinking water faced by the inhabitants of El Zapote and surrounding districts in the Sébaco valley. A form was provided at this meeting for local health personnel to record "Care for patients with arsenicism". One hundred and eleven people who had drunk water contaminated with As attended and it was found that those who had ingested higher levels of As were suffering from paresthesia, edema of the lower limbs, burning sensation in the eyes, skin lesions and respiratory problems. Keratosis and hyperpigmentation typical of chronic arsenicism were found. Two patients had splenomegalia and hypertension and there were a few cases of hepatomegalia and anemia.

It was recommended that a program for prevention, treatment and control should be implemented to guarantee continued care and attention for the patients suffering from arsenicism. In addition, an education campaign in the affected communities and surrounding areas was recommended (Gómez 2002).

Another study carried out by UNICEF identified eight areas where As concentrations were higher than normal (Santa Rosa del Peñon, La Cruz de la India, Cerro Mina de Agua, Kinuma, El Mojon and Las Pilas). More detailed and systematic studies were recommended, together with the use of less costly analytical methodologies (PIDMA/UNICEF 2002).

2.4.6 *Bolivia*

The communities at risk are located in Alto Lima II, to the north of the city of El Alto, in the province of Murillo in the department of La Paz, and the community of Vinto, in the city of Oruro. The total population in these communities is approximately 20,000 people (ECO/PAHO 1997).

A study in El Alto indicated that 70% of children between 5 and 7 years of age had excessive concentrations of As in their urine. The effects of oral ingestion on children were skin lesions and neurological symptoms. A similar evaluation in the mining area of Vinto, located 7 km from the city of Oruro, found that the most important means of contamination are from water, air and dust and that children are at greater risk of exposure to As than adults (ECO/PAHO 1997).

Arsenic concentrations in drinking water of 210 to 12,600 μg/l were found in regions where the regional endemic chronic arsenicism was identified. These concentrations are sufficient to cause long term toxic effects in the population (Quintanilla 1992).

Vahter *et al.* (1995) have conducted research on health hazards caused by As in drinking water from tube wells in Andean highlands. Aspects including adverse reproductive outcome, interactions with nutritional status, and wide range exposure to As among pregnant women consuming contaminated water were studied. The research indicated, that As easily passes the placenta and, therefore, the fetuses are heavily exposed to As. A follow-up project aims at elucidating the exposure to As via breast milk, drinking water and food in early childhood to determine the extent to which such exposure alters child development.

2.4.7 *Brazil*

There has been no reports on the natural occurrences of As in drinking water in Brazil. Instead, As contamination results from mining and smelting activities. The Minas Gerais mining districts of Nova Lima and Santa Barbara (located in southeastern Brazil) are situated around Au mining areas. In April 1998, urine samples were taken from 126 schoolchildren aged 9.8 ± 1.1.

The results showed toxicologically high As concentrations, with a range of 2.2–106 µg/l and a mean of 25.7 µg/l). Twenty per cent of the individuals that participated showed elevated As concentrations such that adverse health effects cannot be excluded on a long-term basis. The As concentrations in water were 0.4–350 µg/l (Matschullat *et al.* 2000).

2.5   CONCLUSIONS AND RECOMMENDATIONS

In Latin America approximately 4,800,000 inhabitants are exposed to As in their drinking water. Their health is being affected to such a degree that in countries such as Mexico and Argentina it is considered a public health problem. More locally, at-risk groups include workers in the mining and metallurgical industries.

The most affected individuals are scattered rural dwellers who drink untreated water and are unaware of the risks to which they are exposed. These people require assistance from health, hygiene and environmental authorities to plan the provision of water and to undertake other activities as part of a program of prevention and control of the risks associated with drinking water containing higher than recommended As concentrations. Such programs should involve the authorities, the community and the local health systems.

Toxicological and epidemiological studies confirm findings from earlier scientific studies, which indicated that chronic ingestion of As in drinking water causes a variety of health problems. These include skin lesions, such as hyperpigmentation and hyperkeratosis of the hands and feet; nervous system disorders; diabetes mellitus; anemia; liver disorders; vascular diseases (peripheral vascular illnesses such as myocardial infarctions and thickening of the arteries); and skin, lung and bladder cancer, particularly in children.

The effects of arsenicism on human reproductive health have not been demonstrated. Nevertheless, it is known that As can pass into the placenta. More research is required to understand its genotoxic effects.

Children are more susceptible than adults to As poisoning. Children who are the most susceptible to As poisoning are those who suffer the effects of malnutrition and inadequate hygiene in poor and rural areas.

Research is required into the effects of As on the health of people exposed to low concentrations in water and other exposure routes. A variety of steps must be taken in order to address the problem of As in at-risk areas. Cases of arsenicism must be identified. Sources of As-free water must be found or suitable technology must be used to remove As from drinking water. Once alternative sources of drinking water are developed for long term use, progress should be monitored and care provided for patients suffering the effects of arsenicism (including vitamin supplements, lotions for those with keratosis, etc.). Arsenic concentrations in drinking water supplies and the effectiveness of treatment, if applicable, should be monitored periodically.

Epidemiological and As removal studies as well as national policies, strategies and standards should be developed within an integrated conceptual framework. This should be accomplished with participation of the different social stakeholders. It is recommended that a reliable and consistent analytical capability be developed so that the results of laboratory and field studies can be compared. Standard validated analytical methods and procedures should be available, together with control samples to ensure the quality of the results obtained.

International bodies and health authorities should publish the fact that the number of people suffering from arsenicism will increase if methods to mitigate the problem are not introduced in time. It is important to demonstrate the social and economic effects of arsenicism on families and how mitigation methods can reduce these effects.

One method of analysis and prediction would be to apply a simulation methodology that would allow testing of real models in an affected area. The methodology must contain an epidemiology component as well as a component to evaluate the socio-economic effects on families and the community. The cost of doing no intervention must also be evaluated.

REFERENCES

Acosta, L., Bastida, M., Calderón, A., Cebrián, M., Conde, M., López, C., Luna, A., Soto, P. and Vega, L.: Assessment of lymphocyte subpopulations and cytokine secretion in children exposed to arsenic. *The FASEB Journal* (2006), express article 10.1096 and fj. Published online.

Aguilar, C., Albores, A., Borja, H., Cebrián, M., Del Razo, L., Hernández, Z. and García, V.: Altered activity of heme biosynthesis pathway enzymes in individual chronically exposed to arsenic in Mexico. *Arch. Toxicol.* 73 (1999), pp. 90–95.

Albores, A., Cebrián, M., Del Razo, L., García-Gonsebatt, M., Kelsh, M., Otrosky, W., Montero, R. and Vargas, H.: Altered profile of urinary arsenic metabolites in adults with chronic arsenicism. *Arch Toxicol.* 71 (1997), pp. 211–217.

Albores, A., Brambila, C., Calderon, A., Del Razo, L., Quintanilla, V. and Manno, M.: Stress proteins induced by arsenic. *Toxicol. Appl. Pharmacol.* 177 (2001), pp. 132–148.

Álvarez, J., Esparza, M.L., Rivero, S. and Liberal, V.: Community participation in reducing risk by exposure to arsenic in drinking water. 22nd International Water Services Congress and Exhibition. Buenos Aires, Sep. 1999; *Water Supply* 16 (2000), pp. 618–619.

Andreson, W., Brown, J., Del Razo, L., Kenyon, E. and Kitchlin, K.: An integrated pharmacokinetic and pharmacodynamic study of arsenite action. 1. Heme oxygenase induction in rats. *Teratog. Carcinog. Mutagen.* 19 (1999), pp. 385–402.

Avilés, M. and Pardón, M.: *Remoción de arsénico de agua mediante coagulación-floculación a nivel domiciliario*. Federación Mexicana de Ingeniería Sanitaria y Ciencias del Ambiente (FEMISCA), Mexico City, Mexico, 2000.

Bangladesh Centre for Advanced Studies: Arsenic special issue. *BCAS Newsletter Jan.–Mar. 1997*, 8:1 (1997), pp. 1–8.

Beck, B., Dubé, E., Schoen, A. and Sharma, R.: Arsenic toxicity at low doses: epidemiological and mode of action considerations. *Toxicol. Appl. Pharmacol.* 198 (2003), pp. 253–267.

Bejarano, G. and Nordberg, E.: *Mobilisation of arsenic in the Río Dulce alluvial cone, Santiago del Estero Province, Argentina*. MSc Thesis, Dept. of Land and Wat. Res. Eng., KTH, Stockholm, Sweden, 2003.

Benítez, M., Osicka, R., Gimenez, M. and Garro, O.: *Arsénico total en aguas subterráneas en el centro-oeste de la provincia de Chaco*. Comunicaciones Científicas y Tecnológicas, Chaco, Argentina, 2000.

Bhattacharya, P., Frisbie, S.H., Smith, E., Naidu, R., Jacks, G. and Sarkar B.: Arsenic in the environment: a global perspective. In: B. Sarkar (ed): *Handbook of heavy metals in the environment* (Chapter 6). Marcell Dekker Inc., New York, 2002, pp. 145–215.

Bhattacharya, P., Classon, M., Bundschuh, J., Sracek, O., Fagerberg, J., Jacks, G., Martin, R.A., Storniolo, A. del S. and Thir, J.M.: Distribution and mobility of arsenic in the río Dulce alluvial aquifers in Santiago del Estero Province, Argentina. *Sci. Total Environ.* 358 (2006b), pp. 97–120.

Borja, A., Calderón, J., Díaz Barriga, F., Goleen, A., Jiménez, C., Navarro, M., Rodríguez, L. and Santos, D.: Exposure to arsenic and lead and neuropsychological development in Mexican children. *Enviroment Research Section* A85 (2001), pp. 69–76.

Bundschuh J., Farías, B., Martin, R., Storniolo, A., Bhattacharya, P., Cortes, J., Bonorino, G. and Albouy, R.: Groundwater arsenic in the Chaco-Pampean Plain, Argentina: case study from Robles county, Santiago del Estero Province. *Appl. Geochem.* 19 (2004), pp. 231–243.

Calderon, R., Hudgens, E., Le, C., Schreinemachers, D. and Thomas, D.: Excretion of arsenic in urine as a function of exposure to arsenic in drinking water. *Environ. Health Perspect.* 107:8 (1999), pp. 663–667.

Cebrián, M.E., Albores, A., García-Vargas, G. and Del Razo, L.M.: Chronic arsenic poisoning in humans: the case of Mexico. In: J.O. Nriagu (ed): *Arsenic in the environment. Part II: Human health and ecosystem effects*. John Wiley and Sons, Inc., New York, 1994, pp. 94–100.

Cebrián, M., Córdova, E., Del Razo, L., Garrido, E. and Hernández, Z.: Effects of arsenite on cell cycle progression in a human bladder cancer cell line. *Toxicology* 207 (2004), pp. 49–57.

Chakraborti, D., Rahman, M.M., Paul, K., Chowdhury, U.K., Sengupta, M.K., Lodh, D., Chanda, C.R., Saha, K.C. and Mukherjee, S.C.: Arsenic calamity in the Indian subcontinent: What lessons have been learned. *Talanta* 58 (2002), pp. 3–22.

Chowdhury, U., Badal, K., Biswas, B., Chowdhury, T., Samanta, G., Mandal, B., Basu, G., Chanda, Ch., Lodh, D., Saha, K., Mukherjee, S., Roy, S., Kabir, S., Quamruzzaman, Q. and Chak, D.: Groundwater arsenic contamination in Bangladesh and West Bengal, India. *Environ. Health Perspect.* 108:5 (2000), pp. 393–397.

Claesson, M. and Fagerberg, J.: *Arsenic in groundwater of Santiago del Estero—Sources, mobility patterns and remediation with natural materials*. MSc Thesis, Dept. of Land and Wat. Res. Eng., KTH, Stockholm, Sweden, 2003.

Concha, G., Nermeli, B. and Vathter, M.: Metabolism of inorganic arsenic in children with chronic high arsenic exposure in northern Argentina, *Environ. Health Perspect.* 106:6 (1998), pp. 355–359.

Corey, G.: *Epidemiological study of endemic regional chronic hydroarcenicism.* Enress, Santa Fe, Argentina, 2002, http://www.enress.gov.ar/docs/Congreso%20FENCAP%20parte%201.pdf.

De Paredes, G.: *Trabajo de investigación sobre hidroarsenicismo (HACRE).* Santiago del Estero, Argentina, 1997.

Del Razo, L., De Vizcaya, R., Izquierdo, V., Sanchez, P. and Soto, C.: Diabetogenic effects and pancreatic oxidative damage in rats subcronically exposed to arsenite. *Toxicol. Lett.* 160 (2005), pp. 135–142.

Del Razo, L., Gonsebatt, M., Gutiérrez, M. and Ramírez, P.: Arsenite induces DNA-protein crosslink and cytokeratin expression in the WRL-68 human hepatic cell line. *Press. Carcinogenesis* 21:4 (2000), pp. 701–706.

ECO/PAHO: *Evaluación de riesgos para la salud en la población expuesta a metales en Bolivia.* Pan American Center for Human Ecology and Health, Division of Health and Environment, Pan American Health Organization, World Health Organization, Mexico City, Mexico, 1997.

Environmental Health Directorate of Health Canada: *Arsenic guidelines for Canadian drinking water quality.* Ottawa, Ontario, Canada, 1992.

Endo, G., Fukushima, S., Kinoshita, A., Kuroda, K., Morimura, K., Salim, E., Shen, J., Wanibuchi, H., Wei, M. and Yoshida, K.: Understanding arsenic carcinogenicity by the use of animal models. *Toxicol. Appl. Pharmacol.* 198 (2003), pp. 366–376.

Esparza, M.L.: El problema del arsénico en el agua de bebida en América Latina. *Prevención de Riesgos* 55, pp. 29–34, Santiago de Chile, Chile.

Esparza, M.L.: *Estudio para el mejoramiento de la calidad del agua de pozos en zonas rurales de Puno.* Lima, Peru, 2005.

Ferreccio, C., Gonzales, C., Milosavjlevic, V., Marshall, G., Sancha, A. and Smith, A.: Lung cancer and arsenic concentrations in drinking water in Chile. *Epidemiol.* 11:6 (2000), pp. 673–679.

Finkelman, J., Corey, G. and Calderon, R.: *Environmental epidemiology: A project for Latin America and the Caribbean.* Metepec, ECO/PAHO, Lima, Peru, 1993.

Flores, Y.: *Análisis químico toxicológico y determinación del arsénico en aguas de consumo directo en la provincia de Huaytará, Departamento de Huancavelica.* Thesis, School of Pharmacy and Biochemistry, UNMSM, Lima, 1999.

Frederick, P., Kenneth, B. and Chien-Jen, C.: Health implications of arsenic in drinking water. *J. AWWA* 86:9 (1994), pp. 52–63.

Gómez C.A.: *Monitoreo y atención de intoxicados con arsénico en el Zapote, Municipio de San Isidro Departamento de Matagalpa, Nicaragua 1994–2002.* In UNICEF: El arsénico y metales pesados en aguas de Nicaragua (CD), Managua, Nicaragua, 2002.

Herrera, H., Farías, B., Martín, R., Cortés, J., Storniolo, A. and Thir, J.: *Origen y dinámica del arsénico en el agua subterránea del Departamento de Robles—provincia de Santiago del Estero.* Universidad Nacional de Santiago del Estero, Argentina, http://www.unesco.org.uy/phi/libros/congreso/28herrera.pdf, 2002.

Hopenhayn-Rich, C., Biggs, M., Fuchs, A., Bergoglio, R., Tello, E., Nicolli, H. and Smith, H.: Bladder cancer mortality associated with arsenic in drinking water in Argentina. *Epidemiol.* 7 (1996), pp. 117–124.

Hopenhayn-Rich, C., Biggs, M. and Smith, H.: Lung and kidney cancer mortality associated with arsenic in drinking water in Córdoba, Argentina. *Int. J. Epidemiol.* 27 (1998), pp. 561–569.

Hopenhayn-Rich, C., Browning, S., Hertz-Picciotto, I., Ferreccio, C., Peralta, C. and Gibb, G.: Chronic arsenic and risk of infant mortality in two areas of Chile. *Environ. Health Perspect.* 108:7 (2000), pp. 667–673.

Iglesias, S. and González, M.: Situación de la contaminación atmosférica en Lima metropolitana y Callao. *J. Research Institute of the School of Geology, Mines and Geographical Sciences* 4:7 (2001), Lima, Peru, http://sisbib.unmsm.edu.pe/bibvirtual/publicaciones/geologia/v04_n7/situa_contam.htm

Infante, L. and Palomino, S.: *Cuantificación espectrofotométrica de arsénico en aguas de consumo humano en la vertiente del río Rímac.* Thesis, School of Pharmacy and Biochemistry, UNMSM, Lima, Peru, 1994.

Kirk, T. and Sarfaraz, A.: Oxidative stress as a possible mode of action for arsenic carcinogenesis. *Toxicol. Lett.* 137 (2003), pp. 3–13.

Marcus, W. and Rispin, A.: Threshold carcinogenicity using arsenic as an example. In: R.C. Cothern, M.A. Mehlamn and E. Marcus: Risk assessment and risk management of industrial and environmental chemicals. *Advances in Modern Environmental Toxicology* No. 15, Princeton Publishing, Princeton, New Jersey, 1988.

Matschullat, J., Perobelli, B., Deschamps, E., Ribeiro Figueiredo, B., Gabrio, T. and Schwenk, M.: Human and environmental contamination in the Iron Quadrangle, Brazil. *Appl. Geochem.* 15:2 (2000), pp. 181–190.

NAS (National Academy of Sciences): *Arsenic in drinking water.* National Academy Press, Washington, DC, 1999.

Nicolli, H.B., Suriano, J.M., Gómez, M.A., Ferpozzi, L.H. and Baleani, O.: Groundwater contamination with arsenic and other trace elements in an area of the Pampa Province of Córdoba, Argentina. *Environ. Geol. Water Sci.* 14 (1989), pp. 3–16.

PIDMA/UNICEF: Puntos de abastecimiento de agua contaminada por arsénico y plomo identificados en Nicaragua en julio del 2002. In UNICEF: *El arsénico y metales pesados en aguas de Nicaragua* (CD). Managua, Nicaragua, 2002.

Pinedo, M. and Zigarán, A.: Hidroarsenicismo en la Provincia de Córdoba, actualización del mapa de riesgo e incidencia. *XXVI Inter-American Congress of Sanitary and Environmental Engineering*, Lima, Peru, 1998.

Qian, Y., Castranova, V. and Shi, X.: New perspectives in arsenic-induced cell signal transduction. *J. Inorg. Biochem.* 96 (2003), pp. 271–278.

Quintanilla, J.A.: Evaluation of arsenic in bodies of superficial water of the south Lipez of Bolivia (southwest). *Proceedings International Seminar: Arsenic in the Environment and its Incidence on Health*, Universidad de Chile, Santiago de Chile, Chile, 1999, pp. 109–121.

Rivera, M. and Corey, G.: Tendencia del riesgo a morir por cánceres asociados a la exposición crónica al arsénico, II Región Antofagasta, 1950–1993. *Cuad. Med. Soc. XXXVI* 4 (1995), pp. 39–51.

Rossman, T.: Mechanism of arsenic carcinogenesis: an integrated approach: fundamental and molecular mechanisms of mutagenesis. *Mutation Res.* 533 (2003), pp. 37–65.

Sancha A.M., O'Ryan R., Marchetti, N. and Ferreccio, C.: Análisis de riesgo en la regulación ambiental de tóxicos: caso del arsénico en Chile. *XXVI Inter-American Congreso of Sanitary and Environmental Engineering,* Lima, Peru, 1998.

Sancha A.M.C. and de Esparza M.L.: *Arsenic status and handling in Latin America.* Universidad de Chile, Grupo As de AIDIS/DIAGUA, CEPIS/PAHO, Lima, Peru, 2000.

Siveroni, M.: *Eliminación de arsénico del Río Locumba, para uso doméstico e industrial de la ciudad de Locumba.* IV Congreso de Ingeniería Sanitaria y Ambiental. 1989, Lima, Perú, 1989.

Smith, A., Arroyo, A., Mazumder, G., Kosnett, M., Hernandez, A., Beeris, M., Smith, M. and Moore, L.: Arsenic-induced skin lesions among Atacameño people in northern Chile despite good nutrition and centuries of exposure. *Environ. Health Perspect.* 108:7 (2000a), pp. 617–620.

Smith, A., Lingas, E. and Rahman, M.: Contamination of drinking-water by arsenic in Bangladesh: a public health emergency. *Bulletin of the World Health Organization* 78:9 (2000b), pp. 1093–1103.

Valenzuela, O., Borja, A., García, V., Cruz, G., García, M., Calderón, A. and Del Razo, L.: *Urinary trivalent arsenic species in a population chronically exposed to inorganic arsenic.* Cinvestav-IPN, Instituto Mexicano del Seguro Social, Facultad de Medicina/Universidad Juárez del Estado de Durango, Servicios de Salud del Estado de Hidalgo, Mexico, 2000.

Vahter, M., Concha, G., Nermell, B., Nilsson, R., Dulout, F. and Nataranjan, A.: A unique metabolism of inorganic arsenic in native Andean women. *Eur. J. Pharmacol. Environ. Toxicol. Pharmacol. Sect.* 293 (1995), pp. 455–462.

Vega G.S.: *Riesgo sanitario ambiental por la presencia de arsénico y fluoruros en los acuíferos de México.* Comisión Nacional del Agua, Gerencia del Saneamiento y Calidad del Agua, Mexico City, Mexico, 2001.

WHO: *Towards an assessment of the socioeconomic impact of arsenic poisoning in Bangladesh.* Water Sanitation and Health, Dhaka, Bangladesh, 2000.

Wasserman, G., Liu, X., Parvez, F., Ahsan, H., Factor-Litvak, P., van Geen, A., Slavkobich, V., Lolacono, N., Cheng, Z., Hussain, I., Momotaj, H. and Graziano, J.: Water arsenic exposure and children's intellectual function in Araihazar Bangladesh. *Environ. Health Perspect.* 112:13 (2004), pp. 1329–1333.

Section II
Arsenic occurrence and genesis in sedimentary
and hard-rock aquifers

South America

CHAPTER 3

Arsenic in groundwater and sediments from La Pampa province, Argentina

P.L. Smedley
British Geological Survey, Wallingford, OX, UK

H.B. Nicolli
Instituto de Geoquímica (INGEOQUI), San Miguel, Prov. de Buenos Aires, Argentina
Consejo Nacional de Investigaciones Científicas y Técnicas (CONICET), Argentina

D.M.J. Macdonald & D.G. Kinniburgh
British Geological Survey, Wallingford, OX, UK

ABSTRACT: Arsenic (As) in pumped groundwaters from the Quaternary loess aquifer of northern La Pampa, Argentina, has concentrations in the range <4–5300 µg/l, most being present as the oxidized arsenate form. Other anions and oxyanions (B, F, Mo, V, U) also often have high concentrations. These trace elements show positive correlations with both pH and alkalinity. Arsenic concentrations are particularly high in pumped groundwaters and porewaters beneath small topographic depressions which act as zones of seasonal discharge and restricted groundwater flow. Evaporation of water in these small internal drainage systems can be significant but is not responsible for the observed high As concentrations. Accumulation of As (dissolved and sorbed) through flow towards the depression and lack of flushing are likely controls. The As may be derived from a number of minerals but sorption/desorption reactions involving Fe oxides and possibly Mn oxides are considered important controls on the mobility of As, sorption being weakest at high pH. Modeling suggests that competition from other anions, especially vanadate, for binding sites on Fe oxides can further enhance the concentrations of As in the groundwater.

3.1 INTRODUCTION

The Chaco-Pampean plain of Argentina represents one of the largest identified high-As groundwater provinces in the world, extending over an area of around 1 million km^2. Several studies over the last few years have highlighted problems with high groundwater As concentrations in the region (Nicolli *et al.* 1989, 2000, 2001, 2004, Uriarte *et al.* 2001, 2002, Farías *et al.* 2003, Bundschuh *et al.* 2004). In a number of cases, documented average groundwater As concentrations are in excess of 100 µg/l, several samples containing in excess of 1000 µg/l (Table 3.1). The concentrations are among the highest observed in low-temperature natural groundwaters. As large areas of Argentina are semi-arid, surface water resources are scarce and hence groundwater is an important resource for both public and private supply.

Health problems linked to chronic exposure to As from drinking water have been documented in the region since the early 20th century. These include the condition locally known as Bell Ville disease, a skin disorder manifested by pigmentation changes and keratosis, first documented in the town of Bell Ville in Córdoba province. Recognized health problems also include skin, bladder and lung cancer (Hopenhayn-Rich *et al.* 1996, Cabrera and Gómez 2003, Bates *et al.* 2003, 2004).

The affected groundwaters are present in Quaternary loess deposits, occasionally reworked by fluvial processes. These form a superficial cover over the Chaco-Pampean plain, in places reaching several hundreds of meters thick. The loess deposits are dominantly silts and often contain

Table 3.1.   Summary statistical data for As (μg/l) in groundwater from the Chaco-Pampean plain.

Province (Argentina)	Min	Median	Mean	Max	n	Reference
Córdoba	18.7	255	418	3810	60	Nicolli et al. (1989)
Tucumán, Los Pereyra (shallow)	19.7	185	272	758	31	Warren (2001)
Tucumán, Los Pereyra (deep)	0.25	5.9	12	70	25	Warren (2001)
Tucumán, Salí river (shallow)	12.2	45.8	159	1660	42	Nicolli et al. (2000, 2004)
Tucumán, Salí river (deep)	11.4	33.7	37.9	107	26	Nicolli et al. (2000, 2004)
Córdoba/B. Aires/Santa Fe/San Luis	10	67	108	593	66	Farías et al. (2003)
Santiago del Estero (1998)	2	–	170	2400	65	Bundschuh et al. (2004)
La Pampa	<4	150	414	5300	108	Smedley et al. (2002)

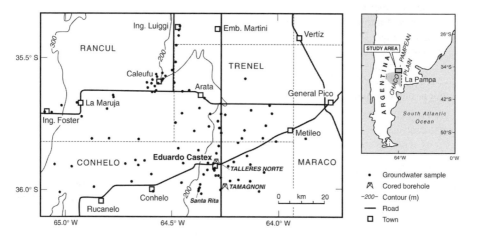

Figure 3.1.   Map of northern La Pampa province, showing groundwater sampling locations and locations of cored boreholes (Talleres Norte and Tamagnoni) (from Smedley et al. 2002).

a significant component of intermixed volcanic glass shards. Occasional discrete volcanic ash horizons also occur. The loess sediments typically have silica and alkali contents comparable to dacite and the ash deposits are rhyolitic. Calcretes are also well-developed, occurring throughout the sedimentary sequence as components of the matrix or as nodules, veins or extensive sheets. These, together with palaeosols, have developed during the Quaternary history of the region under the semi-arid climatic conditions (Zárate and Fasano 1989).

This chapter summarizes the results of hydrogeochemical investigations of groundwaters, sediments and sediment extracts carried out in the northern part of the province of La Pampa (Fig. 3.1). The study area consists of a gently undulating plain with a slight easterly slope of around 0.0014. The area covers 110 × 70 km and is dominantly rural but includes several towns, the largest of which is Eduardo Castex. The Quaternary loess aquifer is the only source of water supply in the area and the groundwater generally occurs under unconfined conditions.

3.2   GROUNDWATER CHEMISTRY

Groundwater chemistry investigations in the study area have been described by Smedley et al. (2002). Pumped groundwater samples have been collected from 108 existing wells and boreholes which vary in depth from 6–140 m. The groundwaters are mainly of Na-HCO$_3$ or Na-mixed-anion type, although salinity is often high under the ambient semi-arid conditions, and some more saline samples are of Na-Cl composition. Total-dissolved-solids concentrations have a large range from

730–11,500 mg/l (Smedley *et al.* 2002). Groundwater pH values are neutral to alkaline (7.0–8.7 in pumped groundwater) and alkalinity is often high (HCO_3 195–1440 mg/l; Fig. 3.2), resulting largely from silicate hydrolysis reactions. The groundwaters are also overwhelmingly oxic, with detectable dissolved oxygen, often high nitrate concentrations (NO_3-N < 0.2–140 mg/l) and low concentrations of dissolved Fe and Mn.

Arsenic concentrations in the 108 pumped groundwater samples lie in the range <4–5300 µg/l (median 150 µg/l; Table 3.1; Fig. 3.2). 95% of these samples have concentrations exceeding the WHO guideline value for As in drinking water of 10 µg/l; 73% exceed the old Argentine national standard of 50 µg/l (in 2007 a new standard of 10 µg/l was introduced). Arsenic is dominantly present in solution as the oxidized form, As(V) (Smedley *et al.* 2002). Most of the groundwaters with concentrations of As greater than 1000 µg/l are from wells or boreholes with depths less than 40 m (Fig. 3.3).

A number of other solutes which form anions and oxyanions in oxic alkaline conditions also have high concentrations in the Pampean groundwaters. These include B, F, Mo, V and U, which reach up to 13.8 mg/l, 29.2 mg/l, 990 µg/l, 5.4 mg/l and 250 µg/l, respectively (Fig. 3.2). Although fewer samples were analyzed for Se, concentrations are also often high, reaching up to 40 µg/l.

Positive correlations are apparent between As and both pH and alkalinity (HCO_3; Fig. 3.4). Good positive correlations are also found with F and V, as well as weaker correlations with a number of other anion- and oxyanion-forming elements (Fig. 3.4). In contrast, Se does not correlate with As,

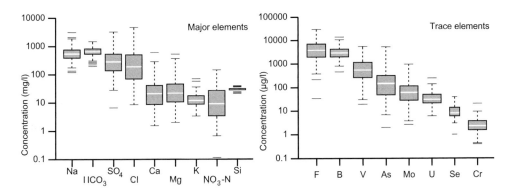

Figure 3.2.   Box plots of selected major ions and trace elements in the groundwaters of northern La Pampa. Boxes represent the interquartile range, whiskers are 1.5 times the interquartile range and white horizontal lines are the median values. 'Outliers' are shown as isolated horizontal lines. Box widths are proportional to the square root of the numbers of samples.

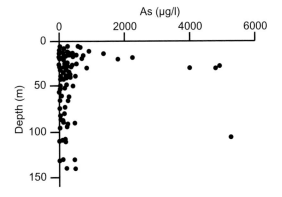

Figure 3.3.   Arsenic variation with borehole or well depth in the pumped groundwaters of the study area.

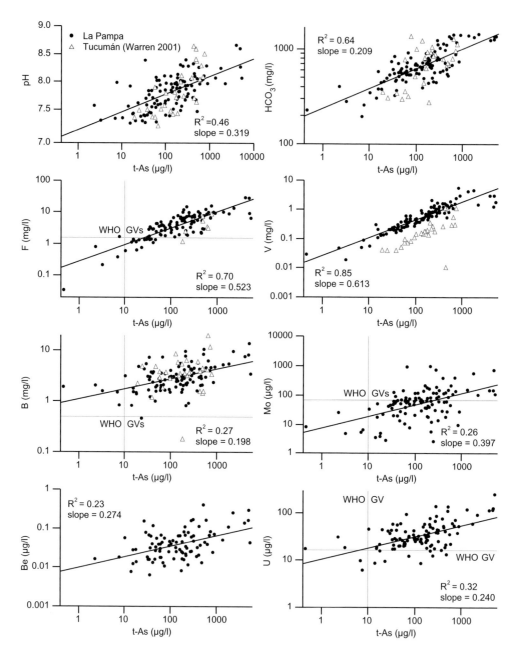

Figure 3.4.   Variation of As with other anion and oxyanion-forming species in groundwaters from La Pampa (after Smedley *et al.* 2002) and Tucumán (Warren 2001). Trace-element concentrations have not been left-censored. Correlation coefficients refer to La Pampa data only. Where applicable, WHO guideline values (GVs) are also given (WHO 2004).

concentrations indicating instead a broad positive correlation with salinity. The correlation with dissolved P is also poor.

Associations in alkaline, oxidizing groundwaters between many dissolved anions and oxyanion species have been reported by a number of workers from elsewhere in Argentina (Nicolli *et al.* 1989,

2001, 2004, Warren 2001) and the USA (Robertson 1989). Warren (2001) described the chemistry of groundwater in the analogous Quaternary loess aquifer of Los Pereyra area, Tucumán. Data for groundwaters from the Tucumán shallow aquifer (borehole depths <40 m) are plotted alongside those from La Pampa in Figure 3.4. The dataset shows comparable concentration ranges for pH, HCO_3, F and B, although the highest concentrations of As (up to 758 μg/l) are less extreme than those found in La Pampa. The correlation between As and V is equally strong in the Tucumán samples, although V concentrations are consistently lower. The similarities between the data from the two regions point to similar aquifer conditions and geochemical processes. Differences in V concentration ranges most likely relate to variations in source mineralogy.

Porewaters extracted by centrifugation from two cored boreholes, named as Talleres Norte and Tamagnoni (Fig. 3.5), also show broad positive correlations between As and pH, HCO_3, B, F, V and U (Fig. 3.5), although concentrations of some diminish at extremely high As concentrations. The absolute concentrations of the oxyanions can vary significantly between the two sites. The boreholes were drilled to 26.5 m and 30 m respectively, Talleres Norte located in a flat-lying area in Eduardo Castex and likely to be a recharge area, Tamagnoni on the edge of a small topographic depression, 10 km to the south. This depression has been water-filled over the last few years although water levels fluctuate as a result of seasonal evaporation. Like many other topographic depressions in the region, it is likely to be a zone of seasonal groundwater discharge resulting in

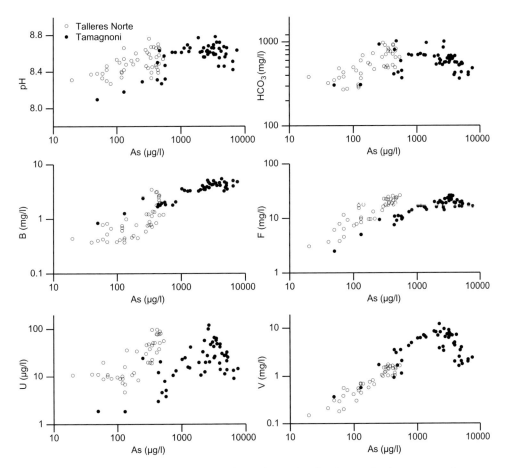

Figure 3.5.   Variation of As with some other anion and oxyanion-forming species in porewaters from two cored boreholes (Talleres Norte and Tamagnoni) in the study area.

local upward flow. Porewaters from the Tamagnoni borehole reveal very high Cl concentrations in the topmost 5 m or so of the saturated aquifer as a result of evaporative concentration in the shallow subsurface (although they are strongly undersaturated with halite and not likely to have derived Cl from dissolution of this mineral). Chloride concentrations fall to around 100 mg/l or less at depths greater than 10 m below surface (Fig. 3.6).

In contrast, As concentrations show an overall increase with depth in the porewaters from the Tamagnoni borehole (Fig. 3.6). This cannot be related to simple evaporative concentration because of the inverse relationship with Cl. The concentrations of As (and other oxyanions) increase in higher-pH Na-HCO$_3$ groundwaters at greater depth in the porewaters. Arsenic concentrations in the Tamagnoni porewaters reach up to 7500 μg/l, an order of magnitude higher than is found in the porewaters of Talleres Norte (up to 530 μg/l). Concentrations of V are also an order of magnitude higher in the Tamagnoni porewaters. These much higher oxyanion accumulations in the topographic low area are believed to be the result of groundwater flow directions (upwards towards the depression) resulting in an accumulation of solutes locally due to poor flushing and increased residence time.

The alkaline, high-As and associated high-oxyanion groundwaters found in La Pampa appear to be a feature largely specific to the shallow Quaternary loess aquifers of Argentina. Warren (2001) found that deeper groundwaters (>70 m) in a Tertiary aquifer in Tucumán province had generally much lower concentrations of As (up to 70 μg/l) and V (up to 130 μg/l). Nicolli *et al.* (2001) also found much lower concentrations of As and V in deep groundwaters from a Tertiary aquifer in the Salí river basin of Tucumán. Arsenic concentrations were 11.4–107 μg/l and V were 48–113 μg/l which, although sometimes high, were less extreme than those seen in the overlying shallow Quaternary aquifer (As 12.2–1660 μg/l; V 30.7–300 μg/l). Nicolli *et al.* (2001) also found relatively low concentrations in deep groundwaters from the Burruyacú basin of Tucumán (As of 13.8–37 μg/l; V of 13.3–61 μg/l), compared to those in the shallow aquifer above (As of 15.8–1610 μg/l; V of 17–1090 μg/l).

These distinct groundwater compositions with depth imply significant differences in sediment chemistry and/or groundwater flow in the different aquifers. Controlling factors could include stratigraphic variations in mineralogy, content of volcanic ash and amount of groundwater flushing since sediment deposition. Determination of the relative significance of these processes is difficult without more information on the mineralogy, geochemistry and groundwater flow in these deeper aquifers.

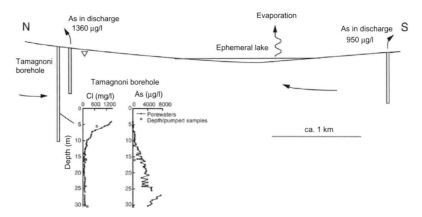

Figure 3.6.   Sketch cross section of a topographic depression in the southern part of the study area, showing As concentrations in samples from abstraction boreholes and porewater profiles for Cl and As in the cored Tamagnoni borehole [35°59.55′ S 64°15.57′ W].

3.3   SEDIMENT CHEMISTRY

Investigations of the chemistry of sediments in the loess aquifer of La Pampa have been described by Smedley *et al.* (2005). Some of the salient features are summarized here.

The loess sediments contain abundant plagioclase, with variable amounts of quartz, alkali feldspar, altered ferromagnesian minerals and heavy minerals (especially ilmenite and magnetite) within a smectite-illite clay matrix. Volcanic glass is abundant in some samples, as are secondary Mn oxides.

Summary statistics for up to 50 sediment samples from the study area (including 5 samples of rhyolitic ash) are given in Table 3.2. The silicic nature of most of the loess silts reflects the composition of the Andean volcanic rocks from which they are dominantly derived.

Concentrations of total As in the sediments lie in the range 3–18 mg/kg (median 7 mg/kg). These are broadly comparable with the concentrations of As found in average sediments, which are typically around 5–10 mg/kg (Smedley and Kinniburgh 2002). Vanadium concentrations in the Pampean sediments are in the range 22–174 mg/kg (median 90 mg/kg), lowest concentrations being from the rhyolitic ashes. These compare to concentrations of 39–99 mg/kg in loess deposits from other parts of the world and an estimated average upper crustal V concentration of 60 mg/kg (Taylor *et al.* 1983). Likewise, Taylor *et al.* (1983) found U concentrations of 1.8–3.0 mg/kg in loess deposits elsewhere and quoted an average upper crustal U concentration of 2.5 mg/kg. These values compare to U concentrations of 0.9–5.1 mg/kg (median value 2.5 mg/kg) in the Pampean loess samples. The concentrations of these oxyanion-forming elements are therefore at the high end of the ranges of concentrations found in loess and other sediments elsewhere, but do not appear to be exceptional.

3.4   CONTROLS ON ARSENIC MOBILITY

The sources of As in the sediments are difficult to define because only very small amounts of the element need to be released from the solid phase to produce the concentrations of dissolved As observed in the groundwaters. Hence, many minerals may be potential candidates, including silicates (volcanic glass, pyroxene, amphibole, biotite, clays) and oxides (magnetite, ilmenite or secondary Fe, Al and Mn oxides). Sulfide minerals are rare and not considered significant. The

Table 3.2.   Summary statistical data for samples of loess sediment (including rhyolitic ash samples) from the study area (major oxides, LOI and TOC in weight %, trace elements in mg/kg).

Parameter	Min	Median	Max	*n*	Parameter	Min	Median	Max	*n*
SiO_2	39.25	61.69	69.41	50	As	3	7	18	50
TiO_2	0.37	0.71	0.85	50	Ba	276	529	1430	50
Al_2O_3	10.18	15.14	16.72	50	Cd	<1	<1	2	36
Fe_2O_{3T}	2.1	4.83	5.97	50	Co	5	12.5	19	50
Mn_3O_4	0.06	0.095	0.15	50	Cr	<2	23.5	33	48
MgO	0.48	1.86	2.8	50	Cu	6	26	36	50
CaO	1.51	3.35	21.97	50	F	<500	<500	984	50
Na_2O	1.68	2.69	4.79	50	Mo	<1	<1	6	50
K_2O	1.32	2.36	4.04	50	Ni	<1	12	18.5	50
P_2O_5	0.08	0.15	0.24	50	Rb	38	80	118	50
ZrO_2	0.02	0.02	0.04	34	Sb	<1	<1	1	50
BaO	0.04	0.07	0.18	34	Se	<1	<1	1	50
LOI	1.83	4.79	18.28	50	Sr	142	335	752	50
TOC	0.02	0.08	0.94	41	Th	5	7.6	12	36
					U	0.9	2.5	5.1	50
					V	22	90	174	36
					Zn	36	56	105	50

volcanic components of the loess deposits are likely to be important primary sources of As and other oxyanions. These are fine-grained, recently erupted and unstable in the weathering environment. The As associated with them may therefore be in a relatively labile form. However, the sediments also contain abundant Fe, Mn and Al oxides which are known to play an important role in the cycling of As between the solid and dissolved phase. The metal oxides are therefore likely to be a significant control on the As concentrations in the groundwaters.

Although the sediments do not appear to have extremely high As concentrations when compared to sediments elsewhere, profiles from the two cored boreholes at Talleres Norte and Tamagnoni show some notable correlations between porewater As concentrations and total sediment As concentrations (Fig. 3.7).

Correlations are also apparent between porewater As and both oxalate-extractable and hydroxylamine-extractable As concentrations, measured to assess the contribution of As from amorphous or poorly-structured Fe oxides and Mn oxides, respectively. The highest observed total and extractable As concentrations are found in the deepest part of the Tamagnoni borehole, where the

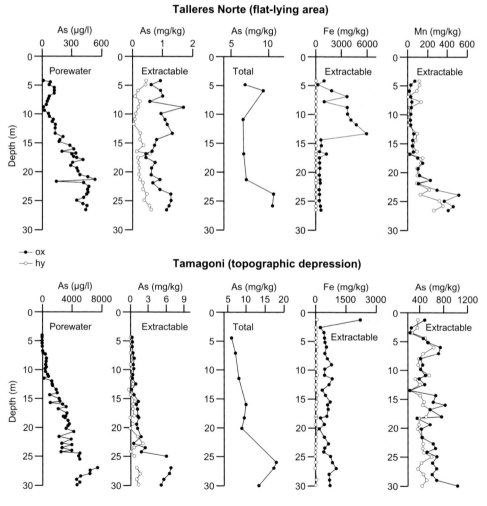

Figure 3.7.   Porewater, sediment and sediment extract (oxalate- (ox) and hydroxylamine- (hy) extractable) profiles for As, Fe and Mn in the Talleres Norte and Tamagnoni boreholes (after Smedley *et al.* 2005).

porewaters contain up to 7500 μg/l As. Here, up to 50% of the As appears to be oxalate-extractable (Fig. 3.7) and is therefore relatively labile.

Relationships with extractable Fe and Mn concentrations are less clear, but there appears some correlation at least in the Talleres Norte profile with Fe_{ox} at shallow depths and Mn_{ox} at greater depth. This suggests that As is associated at least in part with poorly-structured Fe and Mn oxides. No correlation was found between dissolved As and Al_{ox}, although a control by Al oxide cannot be discounted. Magnetite can potentially be extracted in oxalate solutions and the Fe_{ox}-As_{ox} relationships between mean that magnetite may be an additional source and sink of As. As the sediments and groundwaters of the area are overwhelmingly oxic, controls on As mobility by these oxides must be achieved by sorption/desorption reactions involving dominantly arsenate [As(V)]. The pH-dependence of arsenate sorption on Fe and Al oxides is well-known (e.g., Rietra *et al.* 1999, Goldberg and Johnston 2001, Dixit and Hering 2003). Binding of arsenate to these metal oxides should therefore be at its weakest in the high-pH waters.

Exploratory modeling using the diffuse double-layer model and the Dzombak and Morel (1990) default database for hydrous ferric oxide (HFO) in PHREEQC (Parkhurst and Appelo 1999) indicates that competitive effects from other anionic solutes may also have a significant impact on dissolved As concentrations. Smedley *et al.* (2005) modeled the sorption of As by a 1 g/l suspension of HFO by equilibrating with solutions having the chemistry of the porewaters in the Talleres Norte and Tamagnoni boreholes. Modeling was carried out by inclusion of As alone and then by inclusion of potentially competing anions. The models suggested that the log K_d values (ratio of sorbed to solution As) were up to an order of magnitude lower when the competitors (especially V and P) were included compared to when As was considered as the only sorbing anion. This suggests that high concentrations of elements such as V can have a strong competitive effect and can further enhance As mobility. A more detailed analysis requires better models and databases (Rietra *et al.* 1999).

3.5   CONCLUSIONS

Groundwater from the Quaternary loess aquifer of northern La Pampa has spatially variable but often high concentrations of As. Pumped groundwater samples from boreholes and wells have an observed range of <4–5300 μg/l. The dissolved As correlates positively with several other trace elements which also reach unusually high concentrations, V concentrations reaching up to 5.4 mg/l, F up to 29 mg/l, B up to 14 mg/l, Mo up to 990 μg/l and U up to 250 μg/l. The groundwater is universally oxic with a neutral to alkaline pH (7.0–8.7) and the dissolved As is present predominantly as As(V). Highest concentrations of these anion and oxyanion species tend to be present in the high-pH, Na-HCO_3 groundwaters.

The mineral sources of the As in the groundwater are difficult to identify unequivocally. The aquifer sediments have unremarkable concentrations of As (3–18 mg/kg) although porewater As concentrations correlate broadly with host sediment As concentrations. The sediment As data indicate that weak binding of As to the sediments rather than exceptionally high As concentrations in the source are responsible for the high groundwater As concentrations observed. Oxalate and hydroxylamine sediment extracts from cored boreholes show some association between As and either Fe or Mn oxides in different parts of the profiles and suggest that sorption/desorption reactions involving these minerals may be important in controlling As mobility. The evidence for an association with Al oxides is less clear but these may also be involved.

Some of the highest concentrations of As and associated trace-elements are found beneath topographic depressions. Porewater from a cored borehole in one investigated topographic depression (Tamagnoni borehole) had As concentrations up to 7500 μg/l. The depressions are likely to be zones of periodic groundwater discharge and increased residence time. This increased residence time is believed an important factor controlling the accumulation of such high concentrations of As and other oxyanions.

Modeling of As sorption to HFO suggests that competition from other anions can have a significant effect on the amounts of sorbed As, the biggest effect being from vanadate at the concentrations of solutes observed. Although hydrous ferric oxide is probably not the dominant Fe(III) oxide present in the Pampean aquifer, it serves as a useful indicator of the likely reactions involved in As release to groundwater in the region.

As permanent surface-water courses are absent in La Pampa, groundwater is the only available source of water for public and private supply. Options for mitigation are therefore very limited. Water treatment (reverse osmosis) is carried out in many urban areas in the region. This is effective but expensive and currently less used in domestic settings. Abstraction from deeper boreholes may offer some benefits in terms of water quality as our data suggest that the groundwaters with the most extreme As concentrations (greater than 1000 μg/l) are generally found in boreholes or wells of less than 40 m depth. However, since 95% of the groundwaters sampled have As concentrations above the WHO and the new (since 2007) Argentine guideline value for drinking water, while only 73% exceed the old Argentine national As standard, drilling deeper wells within the loess deposits is still unlikely to provide groundwater with acceptably low As concentrations. Groundwaters from deeper pre-Quaternary aquifers in other parts of the Chaco-Pampean plain have been found to contain lower concentrations of As and associated anionic species, often below drinking-water limits. These offer some potential as alternative sources of water where they occur, but their capital cost implications as well as long-term viability and lateral extent across the Chaco-Pampean plain need to be assessed more thoroughly.

REFERENCES

Bates, M.N., Rey, O., Biggs, M.L., Hopenhayn, C., Moore, L.F., Kalman, D., Steinmaus, C. and Smith, A.: Arsenic in drinking water and bladder cancer in Argentina: Results of a case-control study. *Epidemiology* 14:5 (2003), pp. S123–S123.

Bates, M.N., Rey, O.A., Biggs, M.L., Hopenhayn, C., Moore, L.E., Kalman, D., Steinmaus, C. and Smith, A.H.: Case-control study of bladder cancer and exposure to arsenic in Argentina. *Am. J. Epidemiol*, 159:4 (2004), pp. 381–389.

Bundschuh, J., Farías, B., Martin, R., Storniolo, A., Bhattacharya, P., Cortes, J., Bonorino, G. and Albouy, R.: Groundwater arsenic in the Chaco-Pampean plain, Argentina: Case study from Robles county, Santiago del Estero province. *Appl. Geochem.* 19:2 (2004), pp. 231–243.

Cabrera, H.N. and Gómez, M.L.: Skin cancer induced by arsenic in the water. *J. Cutan. Med. Surg.* 7:2 (2003), pp. 106–111.

Dixit, S. and Hering, J.G.: Comparison of arsenic (V) and arsenic (III) sorption onto iron oxide minerals: Implications for arsenic mobility. *Environ. Sci. Technol.* 37:18 (2003), pp. 4182–4189.

Dzombak, D.A. and Morel, F.M.M.: *Surface complexation modelling—hydrous ferric oxide*. John Wiley and Sons, New York, 1990.

Farías, S.S., Casa, V.A., Vazquez, C., Ferpozzi, L., Pucci, G.N. and Cohen, I.M.: Natural contamination with arsenic and other trace elements in ground waters of Argentine Pampean plain. *Sci. Total Environ.* 309:1–3 (2003), pp. 187–199.

Goldberg, S. and Johnston, C.T.: Mechanisms of arsenic adsorption on amorphous oxides evaluated using macroscopic measurements, vibrational spectroscopy, and surface complexation modeling. *J. Colloid Interface Sci.* 234:1 (2001), pp. 204–216.

Hopenhayn-Rich, C., Biggs, M.L., Fuchs, A., Bergoglio, R., Tello, E.E., Nicolli, H. and Smith, A.H.: Bladder-cancer mortality associated with arsenic in drinking water in Argentina. *Epidemiology* 7 (1996), pp. 117–124.

Nicolli, H.B., Suriano, J.M., Gómez Peral, M.A.G., Ferpozzi, L.H. and Baleani, O.A.: Groundwater contamination with arsenic and other trace-elements in an area of the Pampa, province of Córdoba, Argentina. *Environ. Geology Water Sci.* 14:1 (1989), pp. 3–16.

Nicolli, H.B., Tineo, A., García, J., Falcón, C. and Merino, M.: Trace-element quality problems in groundwater from Tucumán, Argentina. *Water-Rock Interaction 2001*: Balkema, Leiden, 2001, pp. 993–996.

Nicolli, H.B., Tineo, A. and García, J.W.: Estudio hidrogeológico y de calidad del agua en la cuenca del Río Salí, provincia de Tucumán, *Revista de Geología Aplicada a la Ingeniería y al Ambiente*, Buenos Aires, Argentina, (2000), pp. 82–100.

Nicolli, H.B., Tineo, A., García, J.W., Falcón, C.M., Merino, M.H., Etchichury, M.C., Alonso, M.S. and Tofalo, O.R.: The role of loess in groundwater pollution at Salí River Basin, Argentina. *Water-Rock Interaction* 11, Balkema, Leiden, The Netherlands, 2004, pp. 1591–1595.

Parkhurst, D.L. and Appelo, C.A.J.: User's guide to PHREEQC (Version 2)—A computer program for speciation, batch-reaction, one-dimensional transport, and inverse geochemical calculations. Water-Resources Investigations Report 99-4259, USGS, 1999.

Rietra, R.P.J.J., Hiemstra, T. and van Riemsdijk, W.H.: The relationship between molecular structure and ion adsorption on variable charge minerals. *Geochim. Cosmochim. Acta* 63 (1999), pp. 3009–3015.

Robertson, F.N.: Arsenic in groundwater under oxidizing conditions, south-west United States. *Environ. Geochem. Health* 11:3/4 (1989), pp. 171–185.

Smedley, P.L. and Kinniburgh, D.G.: A review of the source, behaviour and distribution of arsenic in natural waters. *Appl. Geochem.* 17:5 (2002), pp. 517–568.

Smedley, P.L., Kinniburgh, D.G., Macdonald, D.M.J., Nicolli, H.B., Barros, A.J., Tullio, J.O., Pearce, J.M. and Alonso, M.S.: Arsenic associations in sediments from the loess aquifer of La Pampa, Argentina. *Appl. Geochem.* 20:5 (2005), pp. 989–1016.

Smedley, P.L., Nicolli, H.B., Macdonald, D.M.J., Barros, A.J. and Tullio, J.O.: Hydrogeochemistry of arsenic and other inorganic constituents in groundwaters from La Pampa, Argentina. *Appl. Geochem.* 17:3 (2002), pp. 259–284.

Taylor, S.R., McLennan, S.M. and McCulloch, M.T.: Geochemistry of loess, continental crustal composition and crustal model ages. *Geochim. Cosmochim. Acta* 47 (1983), pp. 1897–1905.

Uriarte, M.G., Paoloni, J.D., Navarro, E., Fiorentino, C.E. and Sequeira, M.: Landscape, surface runoff, and groundwater quality in the district of Puan, Province of Buenos Aires, Argentina. *J. Soil Water Conserv.* 57:3 (2002), pp. 192–195.

Warren, C.: *Hydrogeology and groundwater quality of Los Pereyras, Tucuman, Argentina.* MSc, University College London, UK, 2001.

WHO 2004: *Guidelines for drinking-water quality. Volume 1: Recommendations.* World Health Organization, Geneva, Switzerland, 2004.

Zárate, M.A. and Fasano, J.L.: The Plio-Pleistocene record of the Central Eastern Pampas, Buenos Aries Province, Argentina—the Chapamalal case-study. *Palaeogeogr. Palaeoclimatol. Palaeoecol.* 72 (1989), pp. 27–52.

CHAPTER 4

Arsenic hydrogeochemistry in groundwater from the Burruyacú basin, Tucumán province, Argentina

H.B. Nicolli
Instituto de Geoquímica (INGEOQUI), San Miguel, Prov. de Buenos Aires, Argentina
Consejo Nacional de Investigaciones Científicas y Técnicas (CONICET), Argentina

A. Tineo, C.M. Falcón & J.W. García
Cátedra de Hidrogeología, Facultad de Ciencias Naturales e Instituto Miguel Lillo,
Universidad Nacional de Tucumán, Tucumán, Prov. de Tucumán, Argentina

M.H. Merino
Instituto de Geoquímica (INGEOQUI), San Miguel, Prov. de Buenos Aires, Argentina

M.C. Etchichury
Museo Argentino de Ciencias Naturales Bernardino Rivadavia, Buenos Aires, Argentina

M.S. Alonso & O.R. Tofalo
Departamento de Ciencias Geológicas, Facultad de Ciencias Exactas y Naturales,
Universidad de Buenos Aires, Buenos Aires, Argentina

ABSTRACT: The Burruyacú basin in the Eastern plain, Tucumán province, Argentina, has been filled by Tertiary and Quaternary loess deposits, substantially reworked by fluvial and aeolian processes. Groundwaters, mainly at shallow depth, have significant quality problems due to high concentrations of arsenic (ranging between 15.8 and 1610 μg/l) and associated trace elements (F, V, U, B, Mo, etc.) in sodium bicarbonate waters. These concentrations are significantly lower at large depth. High pH values and bicarbonate concentrations from shallow groundwaters favor the leaching of pyroclastic materials in the loess sediments. Sorption of trace elements onto the surface of Fe- and Al-oxides and oxi-hydroxides control their mobility in aquifers of the basin. Desorption prevails under local geochemical conditions in unconfined groundwaters, with resulting increase of trace-element concentrations. No regional trend is found for trace-element distribution in shallow aquifers because the local geomorphological and topographical characteristics play an important role.

4.1 INTRODUCTION

The Chaco-Pampean plain is the largest and most populated geographical region in Argentina. This vast area, over 1×10^6 km^2, extends from the Paraguay border in the north of the country to the Patagonian plateau in the south, and lies east of the Pampean hills. Its central part comprises the country's most important cities. One of the greatest obstacles for the socioeconomic development of the region is the quality of the water resources available for the rural population. Groundwaters with a high salinity or hardness are the only resources in wide areas and this condition limits their use for human consumption. In addition, high trace-element concentrations (mainly arsenic) often render them toxic; (Nicolli *et al.* 1985, 1989, 1997, 2000, 2001a, b, 2004, 2005a, b, Smedley *et al.* 1998, 2002, 2005, Bundschuh *et al.* 2004, Bhattacharya *et al.* 2006). A series of water-related health problems are linked to high arsenic (As) concentrations in drinking water in various

Chaco-Pampean plain areas. Skin-pigmentation disorders have been reported. Besuschio *et al.* (1980) have described the increased incidence of several types of neoplasms, particularly of the skin, bladder (Bates *et al.* 2004) and digestive tract and, in addition, lung cancer. Bladder cancer mortality associated with As in drinking water in Córdoba province, central Argentina, has been reported (Hopenhayn-Rich *et al.* 1996).

A systematic sampling of groundwater was performed in the Burruyacú basin of Tucumán province, located on the northwestern border of the Chaco-Pampean plain. The purpose of such sampling was to compare the water quality and evaluate the problems related to contaminant trace-element contents and to determine the principal controlling geochemical processes.

4.2   REGIONAL SETTING AND HYDROGEOLOGY

The area covered by the Tucumán eastern plain is close to 10,000 km^2 and it is possible to differentiate two hydrogeological basins, the Burruyacú basin in the northeast and the Salí river basin in the southwest and southeast of the province. The Burruyacú hydrogeological basin is located to the west, in the eastern piedmont area of La Ramada and Del Campo ranges, which exhibit rock outcrops from the metamorphic basement of an Upper Precambrian-Lower Paleozoic age and Cretaceous and Tertiary sedimentary rocks with Quaternary intercalations covering a 2700 km^2 area. The basin spreads north as far as to the Urueña river and south as far as the subsoil extension of La Ramada range, the structural lineament "Dorsal de Tacanas" that defines the southern boundary with the hydrological basin of the Salí river (Fig. 4.1).

The climate is typical for a continental subtropical region, with a large thermal amplitude. The mean annual temperature is 14–18°C. In January (summer), temperatures range between 19 and 34°C and in July (winter), between 8 and 12°C. Precipitation decreases from 900–1000 mm per year at the Subandean hills to 500 mm in the eastern part of the basin; 85% of rainfall takes place between November and April.

On the eastern piedmont of Del Campo range (2000 m a.s.l.) and La Ramada range (1160 m a.s.l.) there are low hills (500–700 m a.s.l.) made up by Tertiary deposits covered by Quaternary sediments (Falcón 2004). Drainage networks have a dendritic or subparallel design and runaway courses are temporary and seep into the piedmont. Below 400 m a.s.l., the Burruyacú plain develops with a smooth eastwards slope where good soils occur, ideal for agricultural uses. The Tajamar river is the only one with an important development in the basin. It feeds a 470 km^2 alluvial fan between 600 and 350 m a.s.l. (Tineo *et al.* 1998).

Cretaceous–Tertiary sediments are found in the piedmont area of the Burruyacú basin, with fanglomerates on top of loams and terraced hills, covered by alluvial sediments and loess. Local water-bearing units with scarce flows develop there. The undulate plain extends east of the piedmont with loess-covered gravel and sands. Here water-bearing units have a medium flow and high salinity. Large alluvial fans and paleochannels develop overlapping that unit. The best aquifers are hosted there and deposits from the depressed plain develop above it. They are made up of clayey-silty sediments and loess deposits that intercalate with fine sands from paleochannels and distal deposits from the Tajamar river alluvial fan. Geophysical studies and drilling data of Burruyacú basin have enabled the determination of three hydrogeological environments:

– *A lower hydraulic system*, consisting of aquifers in fine- to medium-quartzose sands with important clayey-silt intercalations from upper Tertiary (Pliocene?). It is located below a 200 m depth and has a regular distribution throughout the basin. In the depressed plain area it has a natural artesian pressure and negative levels in piedmont areas. Groundwater, which shows thermal anomalies (up to 37.2°C) is used for public water-supply in certain population sites, irrigation and cattle breeding.
– *A middle hydraulic system* in Quaternary aquifers with a good development in the Tajamar river alluvial fan (thickness above 100 m) exhibiting confined levels with high sand- and gravel-permeability with clayey-silt intercalations. Good yields have been found in aquifers

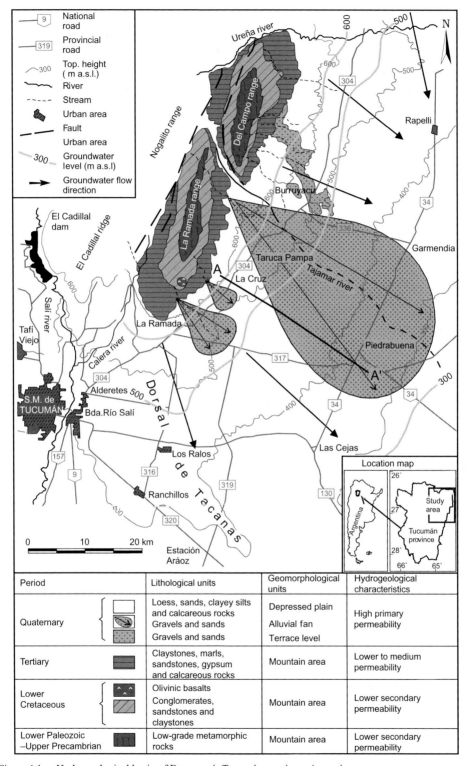

Figure 4.1.   Hydrogeological basin of Burruyacú, Tucumán province, Argentina.

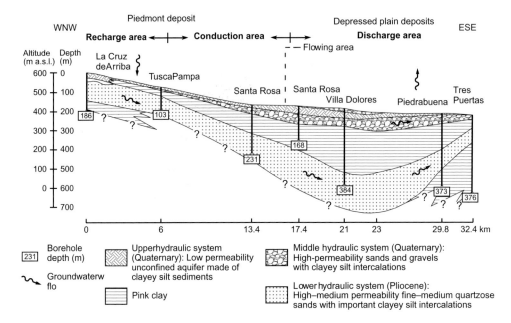

Figure 4.2.   Hydrogeological cross-section A–A′, for location see Figure 4.1.

with negative static water levels with respect to ground surface (−30 to −10 m) which are exploited for public water supply and irrigation.

– *An upper hydraulic system* corresponding to a low-permeability unconfined aquifer (between 10 and 40 m depth) made up of clayey-silt sediments on the edge of the piedmont area. This is used for the water supply of the rural population.

Groundwater samples have been collected from public water-supply boreholes and private boreholes or hand-dug wells. Sixty nine groundwater samples were collected from shallow levels (20 samples), deep levels (21 samples) and artesian aquifers (28 samples), as well as 2 surface-water samples.

4.3   HYDROGEOCHEMISTRY

Waters from the Burruyacú basin show a highly variable chemical composition (Table 4.1). To better assess these variations, shallow groundwater (unconfined levels), deep aquifer, artesian well and surface waters were investigated separately (Table 4.2). Groundwater temperatures vary (19.5–37.2°C; Table 4.1) as a result of significant fluctuations in water levels. The highest temperature values correspond to flowing artesian wells in a geothermal area (Tineo *et al.* 1989).

Chemical data for groundwaters show a large variation in composition. Salinity is highly variable (Table 4.1), with EC values from 413 to 5260 μS/cm and TDS from 451 to 4830 mg/l. The increase in salinity is generated by evaporation and the more saline groundwaters are usually found at shallow depth. EC maximum values for shallow and deep wells (Table 4.2) were 5260 and 2230 μS/cm, respectively; TDS maximum values for shallow and deep wells were 4830 and 1810 mg/l, respectively. Groundwaters were universally oxidizing with DO concentrations of 1.6–8.4 mg/l and redox potentials up to 400 mV (Table 4.1). Nitrate contents were highly variable too, and some high concentrations result from agricultural pollutants (fertilizers) that were restricted to shallow groundwaters. Nitrate maximum content for shallow levels was 64.8 mg/l and for deep levels, 57.8 mg/l. Oxidizing conditions in groundwater, taking into account the concentration range

Table 4.1.   Statistical summary of all analytical data for groundwaters from Burruyacú basin (74 samples).

	Unit	Min.	Max.	Arithmetic mean	Median
Temp.	°C	19.5	37.2	26.9	26.4
pH	–	7.15	8.60	7.74	7.72
Eh	mV	194	400	302	304
DO	mg/l	1.6	8.4	4.4	4.4
EC	µS/cm	413	5260	1590	1510
Ca^{2+}	mg/l	3.85	330	97.5	78.6
Mg^{2+}	mg/l	0.73	73.9	15.5	9.24
Na^+	mg/l	48.4	1040	239	209
K^+	mg/l	1.89	45.2	9.20	6.43
HCO_3^-	mg/l	40.3	1080	316	243
SO_4^{2-}	mg/l	59.7	2240	402	381
Cl^-	mg/l	11.3	547	113	100
SiO_2	mg/l	20.4	160	48.6	38.1
NO_3^-	mg/l	4.54	64.8	17.3	12.3
NO_2^-	mg/l	<5	443	28.7	7.2
TDS	mg/l	451	4830	1260	1130
F	µg/l	83	8740	728	435
As	µg/l	13.8	1610	81.1	30.9
V	µg/l	13.3	1090	72.2	40.3
U	µg/l	0.62	155	15.7	6.6
B	µg/l	244	6740	1160	746
Sb	µg/l	0.03	0.46	0.11	0.08
Mo	µg/l	0.5	90.1	19.3	16.0

For values below the limit of detection, mean values were calculated by substituting a value of half the detection limit.

of DO and the high Eh values (up to 400 mV), resulted in low nitrite concentrations (<5–443 µg/l; median 7.2 µg/l).

The composition of the major ions was controlled by reaction of silicate minerals from loess sediments and carbonate equilibrium. Most groundwaters were hard or very hard. The hardness (as $CaCO_3$) reached up to 1130 mg/l for shallow levels, 808 mg/l for deep levels and 545 mg/l for artesian aquifers. Sodium was the dominant cation (maximum 1040 mg/l; median 209 mg/l) and bicarbonate, the dominant anion (maximum 1080 mg/l; median 243 mg/l). On the other hand, sulfate and chloride were important in a few samples of more saline waters (they reached up to 2240 and 547 mg/l, respectively). Their increase was due to evaporation and agricultural pollution. Groundwaters with the highest sulfate concentrations approach saturation with respect to gypsum. Sodium results from the dissolution of sodium plagioclases and was mainly concentrated by evaporation and cation exchange. Consequently, there were significant differences among shallow groundwaters, deep groundwaters and artesian aquifers (maximum contents 1040, 364 and 392 mg/l, respectively; Table 4.2). Potassium concentrations were often high (maximum value observed 45.2 mg/l) and it may be inferred that such values resulted from pollution, although evaporation also contributes. The increase in potassium contents from deep level groundwaters (median 6.86 mg/l) to shallow levels (median 14.5 mg/l) was thus explained. In addition, reactions of K-bearing minerals and ion-exchange reactions may have been involved. Bicarbonate concentrations and pH values showed a wide range of variation. Since carbonate minerals were important components of sediments, groundwater chemistry was strongly controlled by carbonate reactions. Most of the groundwaters were supersaturated with respect to dolomite, calcite and aragonite (in decreasing order of saturation indices, Table 4.3). The pH values range from

Table 4.2. Statistical summary of analytical data for groundwaters from Burruyacú basin for different hydrogeological units.

	Unit	Shallow wells (20 samples)				Deep wells (21 samples)				Artesian aquifers (28 samples)			
		Min.	Max.	Arit. mean	Median	Min.	Max.	Arit. mean	Median	Min.	Max.	Arit. mean	Median
Temperature	°C	19.5	25.1	21.7	21.3	21.2	30.9	24.8	24.3	26.0	37.2	32.0	32.0
EC	µS/cm	576	5260	1872	1560	586	2230	1440	1450	413	2180	1520	1530
DO	mg/l	1.6	8.4	4.7	4.6	1.7	6.8	5.0	5.2	1.6	6.5	3.5	3.4
pH (in situ)		7.16	8.60	7.66	7.51	7.15	8.20	7.61	7.58	7.56	8.80	7.99	7.88
Redox potential (Eh)	mV	218	400	309	321	237	395	324	327	194	359	276	276
Hardness (CaCO$_3$)	mg/l	14.5	1130	414	341	169	808	343	276	68.9	545	205	193
COD	mg/l	0.38	2.67	1.20	1.15	0.26	1.16	0.60	0.56	0.25	0.85	0.51	0.46
Calcium (Ca)	mg/l	3.85	330	118	81.4	62.9	261	109	91.4	23.2	216	74.7	68.9
Magnesium (Mg)	mg/l	1.41	73.9	29.2	28.9	2.92	54.7	17.5	12.9	0.73	12.2	4.36	3.16
Sodium (Na)	mg/l	48.4	1040	291	212	66.1	364	187	160	90.0	392	241	232
Potassium (K)	mg/l	2.32	45.2	15.7	14.5	4.81	15.2	7.53	6.86	1.89	10.5	5.78	6.06
Carbonate (CO$_3$)	mg/l	24.6	24.6	–	–	–	–	–	–	0.90	0.90	–	–
Bicarbonate (HCO$_3$)	mg/l	306	1080	571	535	153	508	291	278	40.3	258	151	135
Sulfate (SO$_4$)	mg/l	59.7	2240	408	143	78.9	789	369	338	70.7	597	420	453
Chloride (Cl)	mg/l	11.3	721	123	64.1	23.4	204	95.9	84.1	14.7	228	120	117
Nitrate (NO$_3$)	mg/l	4.54	64.8	24.8	19.6	6.75	57.8	19.4	14.4	5.63	17.4	10.3	9.81
Nitrite (NO$_2$)	mg/l	<5	154	25.9	10.4	<5	112	11.5	4.0	<5	443	42.1	5.2
SiO$_2$	mg/l	25.6	160	72.3	72.9	25.4	75.0	45.2	40.8	20.4	50.7	33.9	34.0
TDS	mg/l	593	4830	1650	1430	614	1810	1140	1090	451	1330	1060	1050
CO$_2$	mg/l	8.2	41.1	22.4	21.3	1.95	33.8	10.7	8.36	0.81	12.9	4.00	3.47
F	µg/l	149	8740	1380	683	140	756	338	242	83	1830	557	311
As	µg/l	15.8	1610	160	43.2	13.8	36.6	25.0	22.3	15.7	144	41.1	28.4
V	µg/l	17	1090	148	70.4	13.3	61.2	33.3	30.8	19.2	150	47.4	35.8
U	µg/l	5.9	155	37.0	23.7	3.68	18.2	8.59	6.60	0.622	34.2	5.75	4.43
B	µg/l	244	6740	2130	1410	257	1760	742	708	386	1820	772	689
Sb	µg/l	0.05	0.46	0.16	0.15	0.03	0.13	0.08	0.08	0.03	0.16	0.09	0.08
Mo	µg/l	0.5	90.1	17.2	7.45	4.2	55.2	15.5	12.0	2.2	69.5	23.7	21.2
Al	µg/l	3	108	15.8	8.0	3	145	14.7	8.0	4	40	13.1	13.0
Fe	µg/l	33	921	257	171	52	463	228	205	4.51	240	153	160
Mn	µg/l	0.6	56.5	10.0	2.80	0.4	14.2	3.49	2.3	0.3	19.9	2.85	1.30

For values below the limit of detection, mean values were calculated by substituting a value of half the detection limit.

Table 4.3. Saturation indices from shallow groundwaters (median values).

Phase	SI	log IAP	log KT	Phase	SI	log IAP	log KT
Al (OH)$_3$(a)	−2.09	8.95	8.95	Halite	−6.47	−4.89	1.57
Albite	−0.20	−18.45	−18.24	Hausmannite	−14.60	47.36	61.96
Alunite	−5.46	−6.40	−0.94	Hematite	18.00	14.27	−3.72
Anhydrite	−2.23	−6.58	−4.35	Illite	2.32	−38.45	−40.77
Anorthite	−2.83	−22.65	−19.82	Jarosite-K	−2.92	−11.85	−8.92
Aragonite	0.47	−7.85	−8.31	K-feldspar	1.02	−19.84	−20.86
Ca-Montmor.	2.76	−42.81	−45.56	K-mica	7.87	21.12	13.25
Calcite	0.61	−7.85	−8.46	Kaolinite	3.64	11.40	7.76
Chalcedony	0.34	−3.25	−3.59	Manganite	−5.25	20.09	25.34
Chlorite (14A)	−2.46	67.32	69.78	Melanterite	−9.07	−11.32	−2.26
Chrysotile	−3.67	29.00	32.67	O$_2$ (g)	−32.66	−35.60	−2.94
CO$_2$ (g)	−1.75	−3.17	−1.42	Pyrochroite	−8.04	7.16	15.20
Dolomite	1.09	−15.92	−17.00	Pyrolusite	−8.95	33.02	41.98
Fe(OH)$_3$ (a)	2.25	7.14	4.89	Quartz	0.78	−3.25	−4.04
Fluorite	−1.36	−12.01	−10.64	Rhodochrosite	−1.62	−12.74	−11.12
Gibbsite	0.63	8.95	8.32	Sepiolite	−1.95	13.91	15.86
Goethite	8.00	7.14	−0.87	Sepiolite (d)	−4.75	13.91	18.66
Gypsum	−2.00	−6.58	−4.58	Siderite	−1.72	−12.59	−10.87
H$_2$ (g)	−25.86	−28.99	−3.13	SiO$_2$ (a)	−0.51	−3.25	−2.74
H$_2$O (g)	−1.61	0.00	1.61	Talc	0.66	22.49	21.83

7.16 to 8.80 (Table 4.1) and the highest values were in artesian aquifers. Bicarbonate contents ranged from 40.3 to 1080 mg/l (median 243 mg/l; Table 4.1). The differences between shallow level groundwaters (median 535 mg/l) and deep level groundwaters (median 278 mg/l) were significant. Calcium varied between 3.85 and 445 mg/l and a median at 78.6 mg/l (Table 4.1) with highest contents in the shallow levels. Concentrations of magnesium were generally low (range 0.73–73.9 mg/l; median 9.24 mg/l) and the lowest contents were found in artesian aquifers. SiO$_2$ concentrations were generally very high (20.4–160 mg/l; median 38.1 mg/l; Table 4.1). The wide range reflected different environmental conditions but predominantly high temperature groundwaters (up to 37.2°C) and, in loess sediments of the aquifers, fine grained silicate minerals and a very high proportion of feldspars and volcanic glass which can be easily weathered. Most groundwaters (mainly from shallow levels) were supersaturated with respect to K-mica, kaolinite, Ca-montmorillonite, K-feldspar, albite, quartz and chalcedony. The reaction of silicate minerals also is a key process responsible for the generation of high-pH groundwaters. The reaction of albite to form kaolinite, for example, consumes protons and hence causes a pH rise. This reaction also accounts for the origin of high sodium concentrations in the groundwaters (Smedley *et al.* 2001).

Arsenic concentrations showed a wide range in groundwaters, varying between 13.8 and 1610 µg/l (median 30.9 µg/l) and, of the 74 water samples, all exceeded the WHO (2004) guideline provisional value of 10 µg/l. No definite regional trend had been observed in As distribution since local scale phenomena played an important role. Los Pereyra site is located some 35 km ESE from San Miguel de Tucumán city (Fig. 4.1). It showed the highest As contents for shallow-level groundwaters. From the statistical summary of Table 4.2, significant differences have been found among shallow and deep level groundwaters. Arsenic showed a low correlation coefficient with Al, Fe, and Mn. Low contents of these three trace elements were found in groundwaters from deep levels and in artesian aquifers but they increased significantly in shallow groundwaters. It was found that As closely correlated with HCO$_3$ and pH values. In all cases there was a wide range of trace-element contents (F, V, U, B, etc.) which formed anions and oxyanions in solution. These generally have good positive correlations with As. As a result of high pH and alkalinity of the groundwaters, desorption (a common process in this environment) occurred. Speciation modeling (PHREEQC, version 2: Parkhurst and Appelo 1999) revealed that groundwaters are undersaturated with respect to As minerals (Table 4.3).

Fluoride (F^-) concentration in the shallow groundwaters indicated a wide scale of variation (0.83-8.74 mg/l) and nearly 35% of the samples exceed the WHO guideline value of 1.5 mg/l (see Table 4.3). Concentrations of B ranged between 244 and 6740 μg/l. The median is significantly higher (1410 μg/l) in shallow groundwaters; in deep levels and artesian aquifers the maximum values and the medians were significantly lower, but the maximum values were three times above WHO GV and the medians were higher, in both cases than WHO GV. In shallow levels the exceedance was 85% (Table 4.4). Vanadium concentrations varied between 13.3 and 1090 μg/l. Concentrations showed a strong positive correlation with As and may mobilize in environments with high pH and HCO_3 contents. The dominant anion was vanadate (HVO_4^{2-}, $H_2VO_4^-$) and the mineral sources were similar to arsenic. Molybdenum varied between 0.5 and 90.1 μg/l in groundwaters from shallow levels (median 7.45 μg/l). The rest of the investigated units showed a higher median and a much lower range. In shallow groundwaters, contents exceeded the WHO GV of 70 μg/l in only 5% of the samples (Table 4.4). Uranium concentration in shallow level groundwaters ranged between 5.9 and 155 μg/l (median 23.7 μg/l). The exceedance with respect to WHO GV was 70%. A very good positive correlation was observed between U and As, and between U and HCO_3. For the pH range of the groundwaters, the dominant dissolved species of U are $UO_2(CO_2)_3^{2-}$ and $UO_2(CO_3)_3^{4-}$. Antimony concentrations varied between 0.03 and 0.46 μg/l (median 0.08 μg/l; Table 4.1). The absolute concentrations were low since in none of the groundwater samples exceeded the WHO GV (Table 4.4) but the highest values have been found in shallow levels (up to 0.46 μg/l; Table 4.2).

Table 4.4.   Exceedances of various chemical constituents above WHO (2004) guideline values (health-based) and aesthetic recommendations for groundwaters from Burruyacú basin, shallow levels (20 samples).

Determined	Unit	WHO guideline value (GV)	Number exceeding GV or recommendation	Exceedance (%)
Na[1]	mg/l	200	10	50
Cl[1]	mg/l	250	2	10
SO$_4$[1]	mg/l	250	9	45
NO$_3$	mg/l	50	10	50
NO$_2$	mg/l	3	0	0
As	μg/l	10 (P)	20	100
Al[1]	μg/l	200	0	0
Fe$_T$[1]	μg/l	300	7	35
Mn	μg/l	400	0	0
Ba	μg/l	700	0	0
B	μg/l	500 (P)	17	85
F	μg/l	1500	7	35
Ni	μg/l	20 (P)	0	0
Cu	μg/l	2000	0	0
Zn[1]	μg/l	3000	0	0
Mo	μg/l	70	1	5
Cd	μg/l	4	0	0
Sb	μg/l	20	0	0
Ba	μg/l	700	0	0
Pb	μg/l	10	1	1
U	μg/l	15 (P)	14	70

[1]WHO guideline on aesthetic grounds; P: provisional value.

4.4   LOESS DEPOSITS

The main source of the above mentioned trace elements were the Quaternary loess deposits which have been reworked by aeolian and fluvial processes and fill the Burruyacú basin. Textural analyses have shown that the dominant sediment is a clayey silt and in some places, a silty clay. Among light minerals (grain size larger than 10 μm), the main components were feldspars (dominant pla-gioclases and K-feldspars) and volcanic glass. Feldspars usually ranged between 45 and 70% and volcanic glass shards between 25 and 50%. Decreasing amounts of quartz, muscovite, calcite, and lithic fragments have been found. Opal and chalcedony were present in lesser proportions with an irregular distribution. In the fraction corresponding to the 0.5–0.05 mm size, the most abundant heavy minerals (usually <5%) were pyroxenes and amphiboles (hyperstene, enstatite, hornblende, lamprobolite). Biotite, epidote and Fe- and Mn oxides were less abundant, and garnet, tourmaline, apatite, zircon, chlorite, and rutile were scarce. The X-ray diffraction analysis disclosed the pres-ence of low-crystallinity minerals, with illite prevailing over smectites. Dominant interstratified illite-montmorillonite has been found, as well as scarce pure montmorillonite, kaolinite, and inter-stratified chlorite-montmorillonite. Usually, fine grain-size fractions showed variable amounts of amorphous material (volcanic glass).

The average chemical composition of these sediments usually corresponded to that of a dacite, a rock of common occurrence in the Southern Andes. Table 4.5 gives a statistical summary of

Table 4.5.   Statistical summary of analytical data for loess sediments; Median and mean values have not been quoted where a significant number of samples were below detection limit.

Constituent (%)	Eastern plain (30 samples)				Los Pereyra site (18 samples)			
	Min.	Max.	Arith. mean	Median	Min.	Max.	Arith. mean	Median
SiO$_2$	54.1	75.9	64.1	64.1	60.5	66.8	64.0	64.1
Al$_2$O$_3$	12.1	15.8	14.4	14.6	13.3	15.4	14.5	14.6
Fe$_2$O$_3$	1.33	5.85	4.17	4.23	2.82	5.55	4.44	4.36
FeO	0.24	1.36	0.53	0.46	0.24	0.63	0.41	0.36
MnO	0.02	0.14	0.06	0.04	0.02	0.10	0.05	0.04
MgO	0.87	3.14	2.06	2.07	1.37	2.30	2.02	2.06
CaO	1.49	6.93	2.96	2.47	1.49	6.93	3.07	2.47
Na$_2$O	1.98	3.60	2.60	2.48	2.20	2.71	2.40	2.38
K$_2$O	1.91	4.08	3.19	3.20	1.91	3.51	3.09	3.12
TiO$_2$	0.32	0.71	0.57	0.59	0.43	0.66	0.59	0.59
P$_2$O$_5$	0.05	0.19	0.13	0.13	0.07	0.16	0.13	0.13
LOI (925°C)	1.13	9.91	5.41	5.55	4.14	7.71	5.29	5.07
Trace element (mg/kg)								
F	534	3340	908	791	534	862	712	750
As	6	25	11	11	7	14	10	10
V	51	165	87	86	53	99	84	85
U	3.34	16.0	6.04	4.76	3.62	12.4	4.87	3.88
B	42	128	68	66	42	72	61	62
Co	15	30	20	20	15	30	20	20
Ni	<20	100	–	–	<20	37	–	–
Sb	0.3	1.0	0.8	0.8	0.5	1.0	0.8	0.8
Mo	<2	5	–	–	<2	<2	–	–
Cr	27	81	46	45	29	58	46	45
Cu	13	32	22	24	13	26	21	20
Zn	30	158	84	77	<30	125	78	81
Sr	186	344	271	270	215	344	312	334
Ba	299	582	512	534	417	582	532	538

analytical data for loess sediments from the Tucumán eastern plain and from a well, located in Los Pereyra site, where the highest As content in shallow groundwater had been found. Most samples showed a composition similar to a dacite, some of them to an andesite or a trachyandesite and only one similar to a rhyolite. All samples from Los Pereyra site, taken from the surface down to the water table (-10.20 m), showed a composition similar to a dacite.

The composition of the volcanic glass was typically rhyolitic with SiO_2 ranging between 72.2 and 78.6% (median 73.4%) and LOI (lost on ignition, 925°C) ranging between 3.71 and 8.40% (median 5.22%). The trace-element contents were similar to glass separated from loess from the Pampa plain, Córdoba province (Nicolli et al. 1989).

4.5   DISCUSSION

Arsenic and associated trace-element concentrations in groundwaters from the Burruyacú basin showed a wide range of concentrations, although all investigated groundwater samples were found to be seriously contaminated with As. The As range (13.8–1,610 µg/l) indicated that 100% are above the WHO guideline provisional value (10 µg/l) and its absolute concentrations were high. In the case of fluorine, ranging between 83–8740 µg/l, the exceedance above the WHO GV was lower, particularly in shallow levels (35%). The exceedance for boron (concentrations ranged between 244–6740 µg/l) in the same environments was higher, 85%. Vanadium concentrations were high (13.3–1090 µg/l). Other trace elements associated with As, like Mo varied between 0.5 and 90.1 µg/l and, in shallow levels, exceeded the WHO GV in only 5% of the groundwaters. Uranium varied in concentration between 5.9–155 µg/l in the shallow aquifer, and showed an exceedance of 70% with respect to the WHO provisional GV. The absolute concentrations of antimony were low since no sample exceeded the WHO GV.

No distinct regional trend has been found either in chemical composition of groundwater or in trace-element distribution. This reveals the lack of distinctive geological variations in the aquifers of the Burruyacú basin. Thus, local scale phenomena, such as small slope variations that influence water flow velocities or spatial variations in grain-size and texture of sediments, play an important role in the control of trace-element concentrations. This could be the case of at the Los Pereyra site where all the hand-dug wells exhibit high concentrations of As, F^-, and B in shallow-level groundwaters. Similar phenomena were found in northern La Pampa province (Smedley et al. 2001).

Strong positive correlations ($r > 0.90$) of As with F^- and V, and of F^- with V have been found. Lower values ($r > 0.75$) were typical for the correlation of As with Sb and U. This suggests that those trace elements may have the same source. Lower values ($0.75 > r > 0.60$) in the correlation of HCO_3^- with F^-, As, V, U, B, Sb and Mo were found. On the other hand, the factor controlling their concentrations in groundwater is the dissolution/flow velocity relationship (Nicolli et al. 2001b). In some places there was a high degree of spatial variability in groundwater chemistry and trace-element contents over short distances, indicating restricted groundwater flow with poor mixing and lack of homogenization. Those phenomena, which probably occurred as a result of low permeability in loess sediments, mainly in shallow levels, were similar to those observed in other parts of the Chaco-Pampean plain (Pampa plain, Córdoba province: Nicolli et al. 1989; northern La Pampa province: Smedley et al. 2001, Santiago del Estero province: Bundschuh et al. 2004, Bhattacharya et al. 2006).

It is known that all such trace elements are enriched in volcanic materials (Smedley et al. 2001). The Quaternary loess deposits, reworked by fluvial and aeolian processes, are the source of such trace elements. The average chemical composition of loess is similar to that of a dacite, a common occurrence volcanic rock in the Southern Andes, and As concentration ranges between 6 and 25 mg/kg. The composition of volcanic glass is, on the other hand, similar to that of a rhyolite and the As concentration ranges between <5 and 8 mg/kg, with significant amounts of uranium (4.3–27.7 mg/kg). The high pH values in groundwaters (up to 8.8) favor the volcanic glass dissolution and leaching of volcanic origin material in loess sediments of the aquifers. In the universally oxidizing condition of groundwaters, As is present dominantly as As(V) (arsenate). Hence, most trace elements tend to be mobilized in the form of complex anions or oxyanions: the dominant species are HVO_4^{2-}, $H_2VO_4^-$, $B(OH)_3$ (aq) (neutral species), $UO_2(CO_3)_2^{2-}$, $UO_2(CO_3)_3^{4-}$, and other F complexes with B, Fe(III) and Al. Those groundwaters are mostly bicarbonate dominated with high pH values and controlled by carbonate reactions.

Sorption processes of As and other trace elements onto the surface of Fe- and Al-oxides and oxi-hydroxides (Dzombak and Morel 1990) tend to restrict the mobility of those ionic species and, hence, regulate their distribution in groundwaters. Comparisons among groundwater from shallow and deep levels and artesian aquifers in the Burruyacú basin (Table 4.2) can be performed by calculating the ion mean composition (arithmetic mean) and the median values of trace-element contents. These values were used to obtain the saturation indices by applying the PHREEQC (version 2) program (Parkhurst and Appelo 1999). Results showed 14 supersaturated mineral phases in shallow groundwaters. These are: hematite, goethite, K-mica, kaolinite, Ca-montmorillonite, illite, $Fe(OH)_3$ (a), dolomite, K-feldspar, quartz, talc, gibbsite, calcite, aragonite and chalcedony. Clay minerals have been found (K-mica, kaolinite, Ca-montmorillonite and illite) as well as Fe- and Al-oxides and oxi-hydroxides (hematite, goethite, $Fe(OH)_3$ (a), gibbsite) which, due to their high sorption capacity, are able to adsorb As and other trace elements, restricting their mobility in groundwaters and thus regulating their distribution in those aquifers. Nevertheless, the sorption of As and other anions and oxyanion species is limited in bicarbonate waters with high pH values. In this case, desorption phenomena occur, bringing about an increase in the concentrations of As and other trace elements in groundwaters, particularly in shallow level groundwaters (Nicolli *et al.* 2000, 2001a, b, 2003, 2004, 2005a; Smedley *et al.* 2001).

4.6 CONCLUSIONS

- Groundwaters from the Burruyacú basin, Tucumán province, Argentina, showed high trace-element concentrations, the most remarkable being the high concentrations of As, U, F^-, B, Mo and V. Such phenomena result from desorption processes of these trace elements from oxide-mineral surfaces (mainly Fe- and Al-oxides) in loess aquifers with high bicarbonate concentrations and high pH values.
- A wide range of generally high trace-element concentrations was found. Arsenic stood out, since 100% of the investigated samples exceeded the provisional WHO GV (10 µg/l). The median was 3 times higher than the GV and the maximum value was 160 times the GV. The concentrations were also high for U, F^-, B, Mo and V.
- No particular regional trend has been observed in the trace-element distribution since local scale phenomena play an important role. Small slope variations affect the water flow velocity and spatial variations in grain-size and texture of sediments show certain control on trace element concentrations. A strong positive correlation among As, F^- and V (r > 0.90) has been found, as well as among these trace elements with U, F^- and V, although of lower significance. Such trace elements show the highest concentration in groundwaters with high bicarbonate contents and high pH values, these characteristics being generated by carbonate reaction under closed conditions and advanced silicate dissolution. The groundwaters are universally oxidizing with high concentrations of dissolved oxygen and high redox potential values and As occurs mainly as As(V).
- The mobilization of the trace-element anions and oxyanions takes place as arsenate, $UO_2(CO_3)_2^{2-}$, $UO_2(CO_3)_3^{4-}$, HVO_4^{2-}, $H_2VO_4^-$, $B(OH)_3$ (aq), molybdate and F^- and other complexes of F with B, Fe(III) and Al. Sorption of As and other trace elements onto the surface of Fe- and Al-oxides and oxi-hydroxides (hematite, goethite, $Fe(OH)_3$ (aq), gibbsite, etc.) may restrict such mobilization in aquifers from the Burruyacú basin. Nevertheless, this does not happen in unconfined groundwaters with high bicarbonate concentrations and high pH values. Under these environmental conditions desorption processes occur, releasing the above mentioned anions and oxyanions, and causing the increase of trace-element concentrations in shallow groundwaters.

ACKNOWLEDGEMENTS

The authors wish to acknowledge CONICET (Argentine National Council for Scientific and Technical Research) and ANPCyT (Argentine National Agency for Scientific and Technological Research) that provided additional funding from PIP No. 02374 and PICT No. 7-9525, respectively. A.J. Barros assisted in chemical analysis of waters and sediments. The critical review performed by Dr. J. Bundschuh has been greatly appreciated.

REFERENCES

Bates, M.N., Rey, O.A., Biggs, M.L., Hoppenhayn, C., Moore, L.E., Kalman, D., Steinmaus, C. and Smith, A.H.: Case-control study of bladder cancer and exposure to arsenic in Argentina. *Am. J. Epidemiol.* 159 (2004), pp. 381–389.

Besuschio, S.C., Desanzo, A.C., Pérez, A. and Croci, M.: Epidemiological associations between arsenic and cancer in Argentina. *Biol. Trace Element Res.* 2 (1980), pp. 41–55.

Bhattacharya, P., Claesson, M., Bundschuh, J., Sracek, O., Fagerberg, J., Jacks, G., Martin, R.A., Storniolo A. del R. and Thir, J.M.: Distribution and mobility of arsenic in the Río Dulce alluvial aquifers in Santiago del Estero Province, Argentina. *Sci. Total Environ.* 358 (2006), pp. 97–120.

Bundschuh, J., Farias, B., Martin, R., Storniolo, A., Bhattacharya, P., Cortes, J., Bonorino, G. and Albouy, R.: Groundwater arsenic in the Chaco-Pampean Plain, Argentina: Case-study from Robles County, Santiago del Estero Province. *Appl. Geochem.* 19:2 (2004), pp. 231–243.

Dzombak, D.A. and Morel, F.M.M.: *Surface complexation modeling: hydrous ferric oxide*. J. Wiley and Sons, New York, 1990.

Falcón, C.M.: *Hidrogeología del sector sudoriental de la Sierra de La Ramada, provincias de Tucumán y Santiago del Estero*. PhD Thesis, Facultad de Ciencias Naturales, Universidad Nacional de Tucumán, San Miguel de Tucumán, Argentina, 2004.

Hopenhayn-Rich, C., Biggs, M.L., Fuchs, A., Bergoglio, R., Tello, E., Nicolli, H. and Smith, A.H.: Bladder cancer mortality associated with arsenic in drinking water in Córdoba, Argentina. *Epidemiology* 7 (1996), pp. 117–124.

Nicolli, H.B., O'Connor, T.E., Suriano, J.M., Koukharsky, M.M.L., Gómez Peral, M.A. Bertini, A, Cohen, L.M., Corradi, L.I., Baleani O.A. and Abril, E.G.: Geoquímica del arsénico y de otros oligoelementos en aguas subterráneas de la llanura sudoriental de la provincia de Córdoba. *Miscelánea* 71 (1985), Acad. Nac. Ciencias, Córdoba, Argentina.

Nicolli, H.B., Suriano, J.M., Gómez Peral, M.A., Ferpozzi, L.H. and Baleani, O.A.: Groundwater contamination with arsenic and other trace elements in an area of the Pampa, Province of Córdoba, Argentina. *Environ. Geol. Water Sci.* 14:1 (1989), pp. 3–16.

Nicolli, H.B., Smedley, P.L. and Tullio, J.O.: Aguas subterráneas con altos contenidos de flúor, arsénico y otros oligoelementos en el norte de la provincia de La Pampa: estudio preliminar. *Cong. Int. sobre Aguas y Workshop sobre Química Ambiental y Salud (Resúmenes)*, III-40, Buenos Aires, Argentina, 1997.

Nicolli, H.B., Tineo, A. and García, J.W.: Estudio hidrogeológico y de calidad del agua en la cuenca del río Salí, provincia de Tucumán. *Rev. Asoc. Arg. Geol. Apl. Ing. Amb.* 15 (2000), pp. 82–100. Buenos Aires, Argentina.

Nicolli, H.B., Tineo, A., García, J.W., Falcón, C.M. and Merino, M.H.: Trace-element quality problems in groundwater from Tucumán, Argentina. In R. Cidu (ed): *Water-Rock Interaction* 2. Balkema, Lisse, The Netherlands, 2001a, pp. 993–996.

Nicolli, H.B., Tineo, A., Falcón, C.M. and Merino, M.H.: Movilidad del arsénico y de otros oligoelementos asociados en aguas subterráneas de la cuenca de Burruyacú, provincia de Tucumán, República Argentina. In: A. Medina, J. Carrera, and L. Vives (eds): *Congreso Las Caras del Agua Subterránea I*, Instituto Geológico y Minero de España, Madrid, Spain, 2001b, pp. 27–33.

Nicolli, H.B., Tineo, A., García, J.W., Falcón, C.M., Merino, M.H., Etchichury, M.C., Alonso, M.S. and Tofalo, O.R.: The role of loess in groundwater pollution at Salí River basin, Argentina. In: R.B. Wanty and R.R. Seals II (eds): *Water-Rock Interaction* 2. Balkema, Leiden, The Netherlands, 2004, pp. 1591–1595.

Nicolli, H.B., Tineo, A., Falcón, C.M. and García, J.W.: Distribución del arsénico y otros elementos asociados en aguas subterráneas de la región de Los Pereyra, provincia de Tucumán, Argentina. In: G. Galindo, J.L. Fernández Turiel, M.A. Parada and D. Gimeno Torrente (eds): *Arsénico en aguas: origen, movilidad y tratamiento*. IV Cong. Hidrogeológico Argentino, Río Cuarto, Argentina, 2005a, pp. 83–92.

Nicolli, H.B., Tineo, A., García, J.W. and Falcón, C.M.: Caracterización hidrogeoquímica y presencia de arsénico en las aguas subterráneas de la cuenca del río Salí, provincia de Tucumán, Argentina. In: G. Galindo, J.L. Fernández Turiel, M.A. Parada and D. Gimeno Torrente (eds): *Arsénico en aguas: origen, movilidad y tratamiento*. IV Cong. Hidrogeológico Argentino, Río Cuarto, Argentina, 2005b, pp. 93–102.

Parkhurst, D.L. and Appelo, C.: User's guide to PHREEQC (Version2)—A computer program for speciation, batch-reaction, one-dimensional transport, and inverse geochemical calculations. *U.S. Geol. Survey Water-Resources Investigations Report* 99-4259, Denver: US Geological Survey, 1999.

Smedley P.L., Nicolli, H.B., Barros, A.J. and Tullio, J.O.: Origin and mobility of arsenic in groundwater from the Pampean Plain, Argentina. In: E.B. Arehart and J.R. Hulston (eds): *Water-Rock Interaction*. Balkema, Rotterdam, 1998, pp. 275–278.

Smedley, P.L., Nicolli, H.B., Macdonald, D.M.J., Barros, A.J. and Tullio, J.O.: Hydrogeochemistry of arsenic and other inorganic constituents in groundwaters from La Pampa, Argentina. *Appl. Geochem.* 17:3 (2002), pp. 259–284.

Smedley, P.L., Kinniburgh, D.G., Macdonald, D.J.M., Nicolli, H.B., Barros, A.J., Tullio, J.O. and Alonso, M.S.: Arsenic associations in sediments from the loess aquifer of La Pampa, Argentina. *Appl. Geochem.* 20:5 (2005), pp. 989–1016.

Tineo, A., Iglesias, E., Durán, M., Verma, M., García, J., Falcón, C. and Barragán, M.: Geochemical survey of the llanura Tucumana geothermal area, Argentina. *Geothermal Resources Council Transactions* XIII, 1989, pp. 165–171.

Tineo, A., Falcón, C., García, J., D'Urso, C. and Rodríguez, G.: Hidrogeología. In: *Geología de Tucumán*. 2nd ed., Colegio de Geólogos de Tucumán (Special publ.), S.M. de Tucumán, Argentina, 1998, pp. 259–274.

WHO: Guidelines for Drinking-Water Quality. Vol.1: Recommendations, 3rd ed., World Health Organization, Geneva, Switzerland, 2004.

CHAPTER 5

Mineralogical study of arsenic-enriched aquifer sediments at Santiago del Estero, Northwest Argentina

O. Sracek, M. Novák & P. Sulovský
Institute of Geological Sciences, Faculty of Science, Masaryk University, Brno, Czech Republic

R. Martin
Facultad de Ciencias Exactas y Tecnologias, Universidad Nacional de Santiago del Estero (UNSE), Prov. de Santiago del Estero, Santiago del Estero, Argentina

J. Bundschuh
International Technical Cooperation Program, CIM (GTZ/BA), Frankfurt, Germany
Instituto Costarricense de Electricidad (ICE), San José, Costa Rica

P. Bhattacharya
KTH-International Groundwater Arsenic Research Group, Department of Land and Water Resources Engineering, Royal Institute of Technology (KTH), Stockholm, Sweden

ABSTRACT: A shallow aquifer in an alluvial fan at Santiago del Estero, northwestern Argentina, is enriched in arsenic (As). Sediments from sites with high As concentration were collected by hand auger and studied by X-ray diffraction and by electron microprobe. X-ray diffraction confirmed the presence of quartz and albite. Concentration of total organic carbon (TOC) in soil is low, but concentration of total inorganic carbon (TIC) may be significant. The electron microprobe investigation found abundant glass particles with fluidal structure. The grains show significant weathering features such as voids and dissolution pits. Biotite grains present in sediments are also weathered. Iron oxyhydroxides occur in isolated spots on the surface of silicate minerals, but not as continuous coatings. Heavy minerals were represented by altered ilmenite, monazite, zircon, and garnet with predominant almandine component. The primary source of As could not be determined unequivocally, but volcanic glass and biotite are potential candidates. Ferric oxyhydroxides, which are important adsorbents of As, seem to have formed by precipitation of iron released from minerals like titano-magnetite, and ilmenite. However, the amount of precipitated ferric oxides and hydroxides is low and, furthermore, their As adsorption capacity depends on factors like pH and ionic strength of groundwater and on concentrations of species competing for adsorption sites.

5.1 INTRODUCTION

Natural arsenic (As) enrichment is a serious topic in several countries, including Bangladesh, West Bengal in India, Taiwan, Chile, USA, etc. (Smedley and Kinniburgh 2002). There are several regions with high As concentrations in groundwater in the Pampean region of Argentina (Fig. 5.1). Arsenic was found at Chaco plain at Tucumán (Nicolli *et al.* 1989), at Santiago del Estero (Bundschuh *et al.* 2004, Bhattacharya *et al.* 2006) and at La Pampa (Smedley *et al.* 2005).

A common feature of As-contaminated sites at the Pampean region of Argentina include: (1) Arsenic is found in Na-HCO$_3$ type of groundwater with high pH and EC values and with oxidizing or moderately reducing conditions, (2) aquifer sediments are young, of loessic origin and volcanic ash/volcanic glass material is found in both discrete layers and finally dispersed in sediment, (3) primary source of As is not known so far, but volcanic ash/volcanic glass is suggested

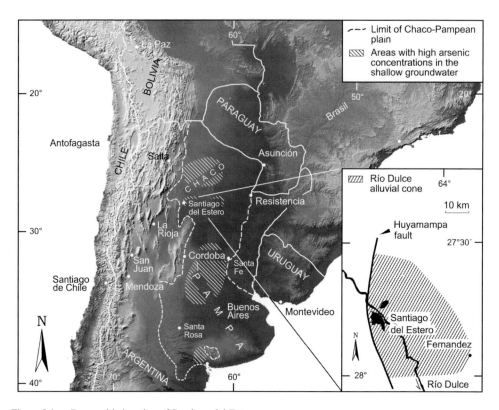

Figure 5.1.   Geographic location of Santiago del Estero.

on the basis of other dissolved species like Mo, V, F, B, etc. present in volcanic material (Smedley *et al.* 2005, Bhattacharya *et al.* 2006).

A conceptual model of As release has been suggested by Bhattacharya *et al.* (2006), that includes As release from volcanic material and its adsorption on ferric oxide and hydroxides in zones with Ca-HCO$_3$ type of groundwater with neutral pH and, in contrast, its high mobility in zones with Na-HCO$_3$ type of groundwater and high pH values. Furthermore, competition for adsorption sites with other dissolved elements like V was suggested as a factor contributing to high As dissolved concentrations (Smedley *et al.* 2005). Factors responsible for the formation of high pH, Na-HCO$_3$ type of groundwater are not completely clear, but it seems that both cation exchange coupled with dissolution of carbonates and dissolution of silicates are under operation (Sracek *et al.* in preparation).

Objectives of this chapter are: (1) to describe the mineralogical transformation of primary and secondary minerals in As-enriched aquifer, and (2) to suggest potential primary source of As and factors resulting in high As concentrations in groundwater.

5.2   MATERIALS AND METHODS

Sampling sites are shown in Figure 5.2. They are located at Nuevo Libano (site 1) and at Cara Pujio (site 2). Both sites are As hot spots with As concentrations above 1.0 mg/l in groundwater. Samples were obtained from a depth of about 2.0 m b.g.s. by hand drilling auger. This was above the water table, but it is probable that the capillary fringe has been reached at site 2. At site 1, the sample corresponded to a silt, at site 2 to a fine sand. Samples were preserved in PTFE bottles and kept under nitrogen atmosphere.

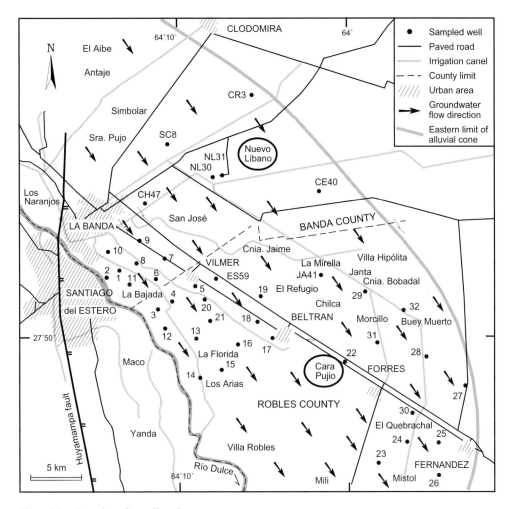

Figure 5.2.   Location of sampling sites.

After transport to Masaryk University, samples were split into three aliquots: the first part was used for X-ray diffraction analyses by STOE Stadi P powder diffractometer.

The second part was used for electron microprobe analyses. The samples were oven-dried, mounted in epoxy resin, and polished with diamond pastes. The resulting polished sections were studied using a CAMECA SX100 microprobe in back-scattered electrons mode (BSE).

The third part was used for determination of total inorganic carbon (TIC) and total organic carbon (TOC). Total carbon was analyzed by LECO infrared detector and TIC was determined as a difference between total carbon before and after treatment of samples with concentrated HCl.

5.3   RESULTS

5.3.1   *X-ray diffraction results and bulk mineral composition*

Representative X-ray diffraction pattern of the sediments is shown in Figure 5.3. Principal minerals were quartz and albite. This is consistent with findings of Bundschuh *et al.* (2004) based on optical

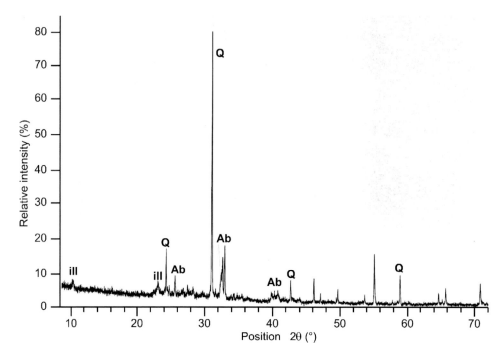

Figure 5.3.   Typical X-ray diffraction pattern of loessic sediments; Q: quartz, Ab: albite.

Table 5.1.   Soil TOC, TIC and calcite content.

Sample/parameter	TOC (wt%)	TIC (wt%)	$CaCO_3$ (wt%)
Nuevo Libano	0.12	0.48	4.0
Cara Pujio	0.22	0.16	1.33

microscopy, which also confirmed albite as the principal plagioclase. Calcite has not been found, but this is not surprising due to the high detection limit of this method (about 5 wt%).

Results of total organic carbon (TOC) and total inorganic carbon (TIC) determinations are given in Table 5.1. It is evident that TOC concentrations are very low, in the range of 0.12 to 0.22 wt%, and soil organic matter cannot be a significant redox driver as it is in Bangladesh and West Bengal (McArthur *et al.* 2001, 2004). On the other hand, when TIC is converted to soil carbonate, assuming that all TIC corresponds to calcite, concentrations of calcite are up to 4.0 wt%. Thus, dissolution of calcite may have an important impact on groundwater chemistry.

5.3.2   *Electron microprobe*

Of the phases that could be considered potential As sources, volcanic glass with rhyolitic composition is the most abundant and at the same time the richest in As. Glass shards often have vesicular nature (see Fig. 5.4a). This increases their specific surface, i.e. the surface open to dissolution and to As release. The As content in glass particles was not possible to determine by EMP, yet the data presented by Smedley and Kinniburgh (2002), indicate that it should be around 6 mg/kg. Considering the abundance of volcanic glass in the majority of samples studied and its relatively low resistance to dissolution under conditions in the aquifer, volcanic glass can be considered a significant As source. The glass also contains abundant apatite inclusions. Another potential source of As are spinel-type minerals of titanomagnetite (ulvöspinel) composition. The grains of

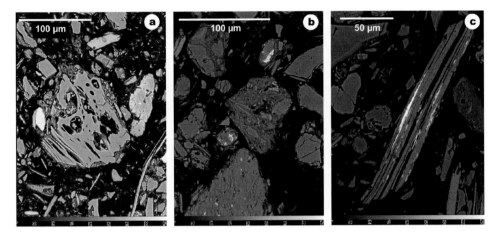

Figure 5.4.   Representative results of electron microprobe analysis in BSE mode: (a) Weathered glass with fluidal structure; (b) iron oxyhydroxides precipitated on primary silicates (bright spots); (c) barite (white) precipitated in cleavage fractures of altered biotite.

this mineral are relatively abundant in the studied samples. They are often partially altered and depleted in iron. Among the aluminosilicate rock-forming minerals, micas were found to be the most As-rich phase (for example, Raimbault and Burnol 1998). However, data about As content in rock-forming aluminosilicates are scarce. Limited alteration of biotite is presumed to have contributed to As release (according to Smedley and Kinniburgh 2002, biotite contains 1.4 mg/kg As). Moreover, the alteration of biotite (Fig. 5.4c), together with titano-magnetite and ilmenite, is a presumed iron source, precipitating as oxyhydroxides in isolated spots on the surface of primary silicates (Fig. 5.4b). Gypsum aggregates with dehydrated margins probably formed from bassanite are also present in sediments. Heavy minerals were represented by altered ilmenite, monazite, zircon, and garnet with predominating almandine component.

5.4   DISCUSSION

Results of mineralogical analyses indicate that the weathering rate of volcanic glass (Fig. 5.4a) and of several minerals like titano-magnetite is relatively fast (Fig. 5.4c). Both types of mineral phases are potential sources of As. Micas were suggested as a potential source of As at Laxmipur in Bangladesh by Dowling *et al.* (2002). Furthermore, Foster *et al.* (2000) studied micaeous sediments from Brahmanbaria in Bangladesh by EXAFS (X-ray absorption fine structure) and As(III) species were found in micas. In contrast, As(V) was found adsorbed on ferric oxide and hydroxides. Also muscovite may be a potential source of As because content of As in muscovite can be up to 30 mg/kg (Raimbault and Burnol 1998).

Another implication of this study is the relatively small amount of secondary ferric oxide and hydroxides precipitating on the surface of primary silicate minerals. These precipitates occur as isolated spots rather than continuous coatings (Fig. 5.4b). In spite of the fact that the number of samples studied by electron microprobe was rather small, this is consistent with the relatively small amount of iron obtained in the oxalate step of sequential extraction, Fe_{ox} from 67 to 141 mg/kg (Bhattacharya *et al.* 2006). This corresponds to 0.63–1.31% of total solid phase Fe determined by Bundschuh *et al.* (2004). Similar results were obtained by Smedley *et al.* (2005), at La Pampa site southeast from Santiago del Estero, where Fe_{ox} was from 600 to 1600 mg/kg and oxalate-extractable Fe was typically less than 5% of the total Fe concentration in samples. These results are very different from Bangladesh, where much higher oxalate-extractable

Fe concentrations were found, for example, up to 2830 mg/kg at Sathkira site (Ahmed *et al.* 2004). This suggests that most Fe at the Pampean region of Argentina is present in primary mafic minerals instead of secondary oxyhydroxides. Thus, the adsorption capacity of solid phases for As in shallow aquifers is limited and competition with other elements forming oxyanions like V, Mo, and P for adsorption sites may be very important.

Gypsum and less hydrated gypsum (probably bassanite) were also found, suggesting that their dissolution may be responsible for relatively high sulfate concentrations at some spots. Strontium seems to behave in a non-conservative way because it co-precipitates with barite. This questions the application of Sr/Ca ratios as an indicator of the origin of Ca (carbonates *vs.* silicates).

Results of TOC and TIC determination support the conceptual model of As mobilization presented by Bhattacharya *et al.* (2006). Values of TOC are low, which indicates that soil carbon cannot be the major redox driver of reductive dissolution of ferric iron minerals as it is in the Bengal delta plain (Ahmed *et al.* 2004, McArthur *et al.* 2004). On the other hand, TIC content converted to calcite indicates calcite content up to 4.0 wt%. This is consistent with the postulated role of cation exchange and calcite dissolution in the evolution of high pH, Na-HCO$_3$ type of groundwater.

5.5   CONCLUSIONS

Shallow aquifer located in an alluvial fan at Santiago del Estero, northwest of Argentina, is contaminated by As. High dissolved As concentrations (more than 1.0 mg/l) were linked to the Na-HCO$_3$ type of groundwater with alkaline pH values. X-ray diffraction confirmed the presence of quartz and albite as principal primary minerals. Calcite had not been found, probably due to its concentration below detection limit. Ferric oxyhydroxides were not detected by X-ray diffraction, probably due to their amorphous or poorly crystalline nature. Samples had low content of total organic carbon (TOC) and relatively high calcite content based on total inorganic carbon (TIC) values. This is consistent with the conceptual model of groundwater chemistry evolution in relatively oxidizing environment, where cation exchange and dissolution of carbonates play important roles.

The electron microprobe investigation found abundant glass particles with fluidal structure and with common apatite inclusions. The glass shows significant weathering features as indicated by the presence of voids and dissolution pits. Biotite grains present in sediments are also partially weathered and strontium-rich barite crystallizes on cleavage fractures of altered biotite. Corroded titano-magnetite grains with trellis structure were also found. Iron oxyhydroxides occur as isolated spots on the surface of silicate minerals but not as continuous coatings. Gypsum aggregates with dehydrated margins probably formed by bassanite were present as well. Heavy minerals were represented by altered ilmenite, monazite, zircon, and garnet with predominating almandine component.

The investigation confirmed significant weathering of volcanic glass and some silicate minerals in sediments. The primary source of As could not be determined unequivocally, but volcanic glass, and biotite are potential candidates. Ferric oxyhydroxides, which are important adsorbents of As, seem to be formed by the precipitation of iron released from minerals like titano-magnetite, and ilmenite. However, the amount of precipitated ferric oxides and hydroxides is very low and, furthermore, their As adsorption capacity depends on factors like pH and ionic strength of groundwater and on concentrations of species competing for adsorption sites. Under high pH conditions observed at As hot spot sites (Bhattacharya *et al.* 2006) As is quite mobile and its concentration in groundwater is high.

ACKNOWLEDGEMENTS

The authors are thankful to the Stategic Environmental Research Foundation (Mistra) for the grant (dnr: 2005-035-137) for the research project "Targeting As-safe aquifers in regions with high

As groundwater and its worldwide implications". Financial support from the Swedish Research Council for Environment, Agricultural Sciences and Spatial Planning (Formas) (dnr: 214-2006-1619) for the International Conference on Natural Arsenic in the Groundwaters of Latin America (As-2006) in Mexico City during June 2006 is gratefully acknowledged. OS was partly supported by the grant MSM0021622412 of the Czech Ministry of Education, Youth, and Sport.

REFERENCES

Ahmed, K.M., Bhattacharya, P., Hasan, M.A., Akhter, S.H., Alam, S.M.M., Bhuyian, M.A.H., Imam, M.B., Khan, O. and Sracek, O.: Arsenic enrichment in groundwater of the alluvial aquifers in Bangladesh: an overview. *Appl. Geochem.* 19 (2004), pp. 181–200.

Bhattacharya, P., Claesson, M., Bundschuh, J., Sracek, O., Fagerberg, J., Jacks, G., Martin, R.A., Storniolo, A.R. and Thir, J.M.: Distribution and mobility of arsenic in the Río Dulce alluvial aquifers in Santiago del Estero Province, Argentina. *Sci. Total Environ.* 358 (2006), pp. 97–120.

Bundschuh, J., Farías, B., Martin, R., Storniolo, A., Bhattacharya, P., Cortes, J., Bonorino, G. and Albouy, R.: Groundwater arsenic in the Chaco-Pampean Plain, Argentina: case study from Robles county, Santiago del Estero Province. *Appl. Geochem.* 19 (2004), pp. 231–243.

Dowling, C.B., Poreda, R.J., Basu, A.R. and Peters, S.L.: Geochemical study of arsenic release mechanisms in the Bengal Basin groundwater. *Water Resour. Res.* 38:9 (2002), pp. 12–18.

Foster, A.L., Breit, G.N., Welch, A.H., Whitney, J.W., Yount, J.C., Islam, M.S., Alam, M.M., Islam, M.K. and Islam, M.N.: In-situ identification of arsenic species in soil and aquifer sediments from Ramrail, Brahmanbaria, Bangladesh. *Eos Trans. Amer. Geophy. Union* 81:48 (2000), Fall Meet. Suppl., Abstract H21D-01.

McArthur, J.M., Ravenscroft, P., Safiullah, S. and Thirwall, M.F.: Arsenic in groundwater: testing pollution mechanism for sedimentary aquifers in Bangladesh. *Water Resour. Res.* 37:1 (2001), pp. 109–117.

McArthur, J.M., Banerjee, D.M., Hudson-Edwards, K., Mishra, R., Purohit, R., Ravenscroft, P., Cronin, A., Howarth, R.J., Chatterjee, A., Talukde, T., Lowry, D., Houghton, S. and Chadha, D.K.: Natural organic matter in sedimentary basins and its relation to arsenic in anoxic ground water: the example of West Bengal and its worldwide implications. *Appl. Geochem.* 19 (2004), pp. 1255–1293.

Nicolli, H.B., Suriano, J.M., Gómez Peral, M.A., Ferpozzi, L.H. and Baleani, O.A.: Groundwater contamination with arsenic and other trace elements in an area of the Pampa, Province of Córdoba, Argentina. *Environ. Geol. Water Sci.* 14 (1989), pp. 3–16.

Raimbault, L. and Burnol, L.: The Richemont rhyolite dyke, Massif Central, France: a subvolcanic equivalent of rare-metal granites. In: P. Černý (ed): Granitic pegmatites. *Can. Mineralogist* 36:2 (1998), pp. 265–282.

Smedley, P.L., Nicolli, H.B., Macdonald, D.M.J., Barros, A.J. and Tullio J.O.: Hydrogeochemistry of arsenic and other inorganic constituents in groundwaters from La Pampa, Argentina. *Appl. Geochem.* 17 (2002), pp. 259–284.

Smedley, P.L. and Kinniburgh, D.G.: A review of the source, distribution and behaviour of arsenic in natural waters. *Appl. Geochem.* 17 (2002), pp. 517–568.

Smedley, P.L., Kinniburgh, D.G., Macdonald, D.M.J., Nicolli, H.B., Barros, A.J., Tullio, J.O., Pearce, J.M. and Alonso, M.S.: Arsenic association in sediments from the loess aquifer of La Pampa, Argentina. *Appl. Geochem.* 20 (2005), pp. 989–1016.

CHAPTER 6

Intermediate to high levels of arsenic and fluoride in deep geothermal aquifers from the northwestern Chaco-Pampean plain, Argentina

M.G. García
CIGeS, Facultad de Ciencias Exactas, Físicas y Naturales (FCEFyN), Universidad Nacional de Córdoba, Córdoba, Prov. de Córdoba, Argentina

C. Moreno, M.C. Galindo & M. del V. Hidalgo
Facultad de Ciencias Naturales and Inst. Miguel Lillo, Universidad Nacional de Tucumán, Tucumán, Prov. de Tucumán, Argentina

D.S. Fernández
Servicio Geológico Minero Argentino, Delegación Tucumán, Tucumán, Prov. de Tucumán, Argentina

O. Sracek
Department of Geological Sciences, Faculty of Science, Masaryk University, Brno, Czech Republic

ABSTRACT: In the southern part of the province of Tucumán (northwestern Chaco-Pampean plain), several wells show intermediate to high concentrations of arsenic (As) (between 26 and 76 µg/l) and high concentrations of fluoride (F^-) (between 0.6 and 6.0 mg/l). These wells penetrate saturated layers as deep as 500 m b.s., and show groundwater temperatures above the annual average in the region. The aquatic geochemistry suggests that the source of both As and F^- is mainly associated with the interaction of meteoric water with the Tertiary volcanic sediments deposited in the deepest part of the sedimentary sequence. Temperature, pH and redox potential are the main factors controlling the mobility of these two elements.

6.1 INTRODUCTION

High levels of naturally occurring arsenic (As) and fluoride (F^-) in groundwaters from the Chaco-Pampean plain have been attributed to the presence of volcanic shards spread within the loess matrix (e.g., Bundschuh *et al.* 2004, Nicolli *et al.* 1989 and 2001, Smedley and Kinniburgh 2002, Smedley *et al.* 1998, 2002 and 2005, Warren *et al.* 2005). The primary source of these elements has not been determined yet but there is a general understanding about the mechanisms that promote their mobilization in the aquifers. Geochemical evidence suggests that, after being released into groundwater, the concentration of As in solution is controlled by pH. Arsenic is preferentially scavenged by adsorption on Fe (hydr)oxide coatings under acidic to neutral pH conditions. The concentration of fluoride depends on fluorite solubility and also on pH-dependent adsorption with adsorption minimum at high pH values (Smith and Martell 1976).

In the province of Tucumán, the loessic layer is restricted to the first 30 m of the Quaternary sequence. As a consequence, most shallow groundwater is frequently contaminated by high levels of As and F^- (Warren *et al.* 2005). Groundwater in deep confined aquifers is generally considered to be suitable for human consumption. However, in the southeastern part of the province, several wells penetrate saturated layers as deep as 500 m b.s., and show intermediate to high concentrations of As and high concentrations of F^-. As these wells show groundwater temperatures above the annual average temperature of the region, they are considered as a resource of thermal waters that could be potentially used for recreational and health purposes. Presently, they are just used for human consumption and irrigation, even though the concentrations of F^- and As in most cases exceed the Argentine drinking water requirements (CAA 1994). The Argentine law establishes the maximum concentrations in drinking water of F⁻ and As as 1.5 and 0.05 mg/l respectively (in 2007 a new law reduced the arsenic limit to 0.01 mg/l). In all cases the value for As is higher than 0.01 mg/l, the maximum concentration for arsenic in drinking water set by the World Health Organization (WHO 1993). This chapter aims to assess the concentrations of As and F⁻ in deep geothermal groundwater from the northwestern Chaco-Pampean plain and to identify some processes that control their distribution in the aquifer.

6.2   STUDY AREA

In the southeastern Tucumán province, NW Argentina, between the Aconquija and Guasayán ranges, there is a structural depression named Central depression of Tucumán (Mon *et al.* 1990), which is filled with up to 3000 m of Tertiary and Quaternary sediments (Mon and Vergara 1987). The sector is affected by a fault system that converges near the city of Termas de Río Hondo (Fig. 6.1), where anomalies in the groundwater temperature have been detected. Using geochemical geothermometers, Vergara *et al.* (1998) have observed a coincidence between the highest temperatures and the fault system, the latter of which was attributed to post-Pleistocene orogeny that folded the Cenozoic sediments and created the conduits for the upflow of endogenous fluids.

Loess and loessic sands blanket a large region of central and northern Argentina and are characterized by a widespread presence of volcanic glass and plagioclase, and minor amounts of quartz and carbonate (Gallet *et al.* 1998). The thickness of loess in the whole area is widely variable, reaching values of up to 30 m in the southeastern part of Tucumán province. Basal Quaternary sediments formed by highly permeable gravel and conglomerates that alternate with layers of silt and clay. The thickness of these deposits ranges from 70 to more than 200 m. Tertiary Andean

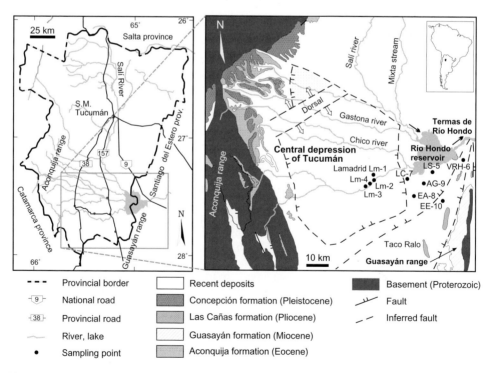

Figure 6.1.   Geology of the Central depression of Tucumán and location of sampling points (modified from Dal Molin 2003).

volcanism has deposited a 1.5 m thick layer of white ash in the southern part of the basin, which was described in the profiles of the deep wells drilled in this area (Battaglia 1982). These ash deposits are made of 80% of amorphous materials; 15% of K-feldspar, quartz, and mica; and 5% of metamorphic fragments; veins of quartz; and fragments of granites and pegmatites (Dal Molin *et al.* 2003).

On the basis of well logging data (DPA files), two aquifers were defined in the geothermal area. The shallow unconfined aquifer consists of loessic sediments that alternate with sand channels, with depth ranging from 0.05 to 30 m. The confined multilayered aquifer consists of saturated levels of sand and gravel that alternate with thick layers of silt and clay.

Flow pattern in the region was defined from the static hydraulic head data (Garcia *et al.* 2001). The piezometric contour map indicates that the recharge area of the confined multilayered aquifer in this part of the basin is located at the foothills of the Aconquija range (towards the west and south). After infiltration, meteoric waters reach the saturated zone and groundwater flow follows the regional topographic slope (W–E). Towards the eastern border of the basin, near the Guasayán range, the fault system allows the up flow of deep groundwater, with temperatures between 25 and 50°C. Baldis *et al.* (1983) proposed that the heat is supplied from a basaltic layer, located 7000 m below the villages of Lamadrid and Taco Ralo, but other authors consider that the temperature is a result of the natural geothermal gradient (R. Mon, *pers. comm.*).

6.3   METHODOLOGY

The present study was carried out along a transect that coincides with the regional flow path in the southern part of the province of Tucumán (Fig. 6.1). Artesian wells with depths ranging between 200 and 480 m b.s. were selected for sampling. At the well head, total depth and static levels were registered as well as the corresponding geographic coordinates.

Field measurements included determination of pH, Eh, electrical conductivity, temperature, and alkalinity. Water samples for the determination of trace elements were filtered using 0.45 μm Millipore cellulose acetate membrane filters and acidified to pH < 2 with ultrapure HNO_3.

Major ions, dissolved silica, and dissolved oxygen were determined following standard recommendations (APHA *et al.* 1989). Trace elements (Fe, Mn, and As) were measured by graphite furnace AAS (Hitachi Z-5000), and F^- by ion chromatography.

The calibration curve for As determinations was made using a sodium arsenite standard (1 g/l As). A 500 mg/l $Ni(NO_3)_2$ matrix was added to the samples before the measurement, following the recommendations by Creed *et al.* (1994). The relative standard deviation (RSD) of the mean As values, was less than 10%.

The program PHREEQC (Parkhurst and Appelo 1999) was used for speciation calculations. The database of the program MINTEQA2 (Allison *et al.* 1991) was used to calculate the speciation of As.

6.4   RESULTS AND DISCUSSION

6.4.1   *General hydrogeochemistry*

The concentrations of major ions, some trace elements and physicochemical characteristics of water samples are shown in Table 6.1. According to the major ionic content, groundwaters are of sodium sulfate/chloride composition. Alkaline and moderately reducing conditions prevail in the samples; the mean pH is 8.0, while the mean Eh value is 189 mV.

Water temperatures vary between 34.2 and 44.5°C along the selected transect, and they tend to increase with increasing well depth. With depth, conditions become slightly more alkaline and reducing. The concentrations of Na^+, Cl^- and SO_4^{2-} tend to increase with depth, while K^+, and Ca^{2+} tend to decrease (Fig. 6.2a). The concentration of HCO_3^- remains almost constant with

Table 6.1.   Major ionic composition and concentrations of F^- and As, well depths and main physicochemical characteristics of samples from the study area.

Sample	Location	Na^+	K^+	Ca^{2+}	Mg^{2+}	HCO_3^-	Cl^-	SO_4^{2-}	SiO_2	F^-	As
		(mg/l)									(μg/l)
Lm-1	Lamadrid (1)	131	7.8	38.5	4.4	167	62.7	168	74.3	0.6	50
Lm-2	Lamadrid (2)	106	9.8	28.6	4.1	145	42.2	166	49.8	nd	nd
Lm-3	Lamadrid camping	121	9.7	31.2	3.8	167	59.6	142	8.9	nd	36
Lm-4	Lamadrid centre	98	9.4	27.6	4.3	153	43.3	136	8.3	1.4	40
LS-5	La Soledad	218	2.1	4.5	0.5	140	132	215	25.2	6.0	74
VRH-6	V° Río Hondo	225	1.6	7.2	bdl	73	94.4	209	19.5	4.4	52
LC-7	Los Cercos	171	12.5	20.8	1.4	171	55.4	137	57.6	2.4	27
EA-8	El Arbolito	285	3.1	4.8	bdl	146	108	245	28.2	3.9	76
AG-9	Árboles Grandes	159	8.2	9.6	bdl	140	66.5	160	52.6	1.6	61
EE-10	El Espinal	110	10.2	9.8	1.3	146	44.3	117	41.3	nd	52

Sample	Location	Depth (m b.s.)	pH	T (°C)	EC (μS/cm)	Eh (mV)	D.O. (mg/l)
Lm-1	Lamadrid (1)	320	7.9	34.2	674	311	1.2
Lm-2	Lamadrid (2)	280	7.9	37.0	nd	nd	nd
Lm-3	Lamadrid camping	321	7.3	38.0	636	nd	1.8
Lm-4	Lamadrid centro	328	7.3	37.0	574	nd	3.9
LS-5	La Soledad	474	8.7	44.5	1890	144	3.8
VRH-6	V° Río Hondo	300	9.1	38.0	1290	133	1.9
LC-7	Los Cercos	203	7.6	35.0	1050	215	1.5
EA-8	El Arbolito	470	8.3	42.2	1740	135	1.3
AG-9	Árboles Grandes	298	7.4	41.0	1290	190	1.3
EE-10	El Espinal	350	8.1	41.5	1050	192	3.9

nd: not determined; bdl: below detection limit.

depth. The variation of major ions concentrations along the flow direction indicates that Na^+, Cl^-, and SO_4^{2-} increase towards the east, while Ca^{2+}, K^+, and HCO_3^- decrease in the same direction (Fig. 6.2b).

6.4.2   *Fluoride and arsenic geochemistry*

All sampled wells show concentrations of As that exceed the 10 μg/l WHO (1993) guideline for As in drinking water, and three of them have concentrations lower than 50 μg/l, which is the old (up to the year 2007) Argentine drinking water requirement (CAA 1994). All samples, except two, show F^- concentrations above the national and WHO requirements for drinking water.

The concentrations of As and F^- vary concordantly with depth and along the flow direction. In general, they tend to increase in depth and along the groundwater flow direction (Figs. 6.2c and d).

The geochemistry of As and F^- in deep aquifers shows certain characteristics that are not completely consistent with those described for the rest of the Chaco-Pampean plain (e.g., Bundschuh *et al.* 2004, Nicolli *et al.* 1989, 2001, Smedley *et al.* 1998, 2002, 2005). Unlike in shallow groundwaters, the concentrations of F^- and As increase with increasing depth (Fig. 6.2c), and with rising temperature (Figs. 6.3a and b).

The major ionic composition found in the studied groundwaters is probably the result of the dissolution of salts that occur in the Tertiary deposits of Las Cañas and Guasayán formations. The first consists of a sequence of siltstones and sandstones that contain volcanic fragments and that alternate with discontinuous layers of green clays. In the upper part of the sequence there is a 1.5 m thick of white tuff. The latter is made of green clays with high contents of nodules and

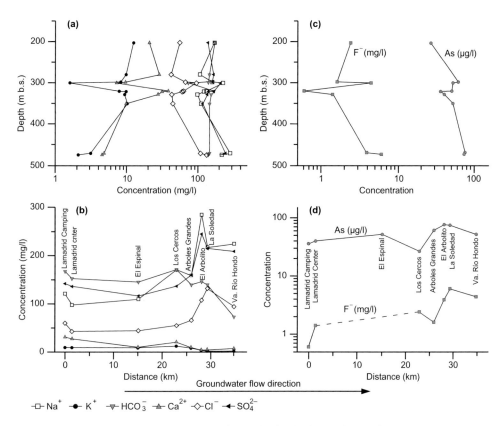

Figure 6.2.   Variation with depth and along the flow path of major ions (a–b), and fluoride and arsenic (c–d).

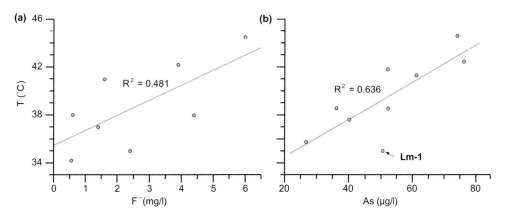

Figure 6.3.   Bivariate plots showing the variation of (a) F⁻, and (b) As with temperature.

banks of gypsum. Below, there is a thin layer of volcanic ash (Battaglia 1982). Concordantly, the concentrations of both As and F⁻ in the groundwater seem to be the result of the weathering of the minerals contained in this sequence. This would explain the positive trends observed between their concentrations and the concentrations of SO_4^{2-}, Cl^- and Na^+ (Fig. 6.4), and the trends observed with temperature (as mineral solubility generally increases with temperature).

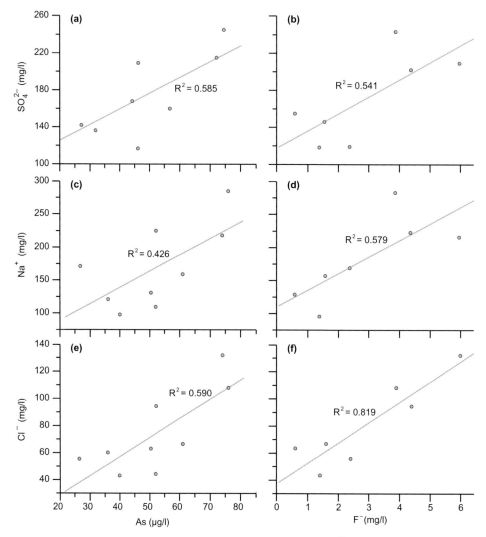

Figure 6.4.   Scatter plots showing the variation of As and F^- with (a–b) SO_4^{2-}, (c–d) Na^+, and (e–f) Cl^-.

In general, geothermal waters are known to contain high concentrations of As and F^-, as well as other trace elements such as B and Li. These waters usually exhibit neutral to alkaline pH and high concentrations of Cl^- (e.g., Wilkie and Hering 1998, Aiuppa *et al.* 2003, Arnórsson 2003, Romero *et al.* 2003). Positive correlation between As and Cl^- in groundwaters flowing through intermediate and silicic volcanic rocks from New Zealand, has been assigned to the occurrence of salts on mineral grain surfaces (Ellis and Mahon 1964) that may have been deposited by sublimation of the volcanic gas-derived volatiles (Arnórsson 2003).

On the other hand, a positive correlation between As and SO_4^{2-} is generally associated with rock-forming and secondary minerals such as As-rich pyrite and arsenopyrite (e.g., Nordstrom 2000, Kolker and Norstrom 2001). The inverse correlation between F^- and Ca^{2+} (Fig. 6.5a), theoretically suggests that fluorite may precipitate from the solution, but supersaturation with respect to fluorite is never reached due to the removal of Ca^{2+} probably by cation exchange and/or the precipitation of calcite (Table 6.2).

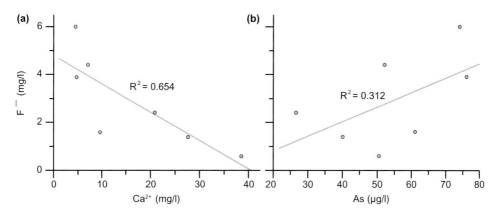

Figure 6.5.   Bivariate plots showing the correlation between (a) As and F^- and (b) Ca^{2+} and F^-.

Table 6.2.   Calcite and Fluorite Saturation Index calculated in samples from the study area, using PHREEQCI 2.8 (Parkhurst and Appelo, 1999).

Sample	Location	Calcite	Fluorite	Sample	Location	Calcite	Fluorite
Lm-1	Lamadrid (1)	0.29	−1.87	VRH-6	V° Río Hondo	0.3	−0.94
Lm-2	Lamadrid (2)	0.16	na	LC-7	Los Cercos	−0.23	−0.92
Lm-3	Lamadrid camping	−0.31	na	EA-8	El Arbolito	−0.22	−1.28
Lm-4	Lamadrid centro	−0.41	−1.28	AG-9	Árboles Grandes	−0.77	−1.67
LS-5	La Soledad	0.2	−0.97	EE-10	El Espinal	−0.01	na

na: not available.

The slightly positive correlation between As and F^- (Fig. 6.5b), may indicate that both elements are released from a common source. Furthermore, the occurrence of F^- and As-enriched groundwater from the study area can be a combination of several factors: (1) the dissolution of salts, precipitated from the sublimation of endogenous vapors on the Tertiary sediment grains; (2) the release of F^- and As-enriched fluids that occur as inclusions within minerals from the volcanic layers; (3) the weathering of primary and/or secondary sulfides in cinerites and volcanic ash; and (4) the mixing of meteoric waters with up flowing endogenous vapors that reach the deep aquifers through the regional structure.

The parameters that mainly control the mobility of As and F^- in the deep aquifers are temperature, pH and redox potential. As described above, temperature enhances the solubility of As- and F-bearing minerals and hence, these elements are released to water. Moreover, the concentrations of F^- increase with increasing pH and decreasing Eh, but this trend is less evident for As (Fig. 6.6).

Speciation carried out with PHREEQC (Parkhurst and Appelo 1999) indicates that the negatively charged oxyanions $H_2AsO_4^-$ and $HAsO_4^{2-}$ account for most of the dissolved As. However, the contribution of As(III), as H_3AsO_3 to the total As concentration is considered to be more important at depth. Arsenate adsorbs strongly to Fe-oxide minerals in neutral or acidic pH conditions, whereas the affinity of arsenite for the surface of Fe-oxides is relatively weak (Dzombak and Morel 1990, Langmuir 1997). A greater contribution of As(III) under alkaline and increasing reductive conditions would explain the increase in As concentration with increasing pH and decreasing Eh.

The concentration of F^- in solution is also associated with the precipitation of calcite. The removal of Ca^{2+} from the solution due to calcite precipitation enhances the dissolution of fluorite, which may explain the negative correlation between Ca^{2+} and F^- (Fig. 6.5) and the increase in F^- concentrations at higher calcite SI (Fig. 6.7).

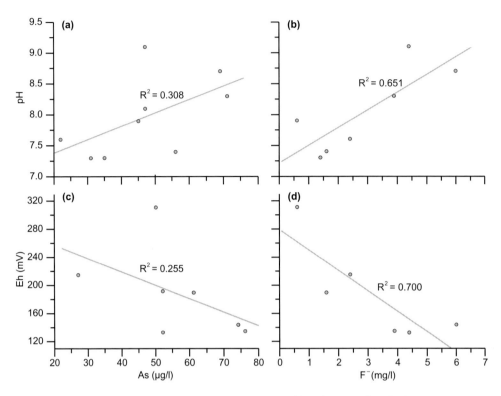

Figure 6.6.   Scatter plots showing the variation of As and F⁻ with (a–b) pH, and (c–d) Eh.

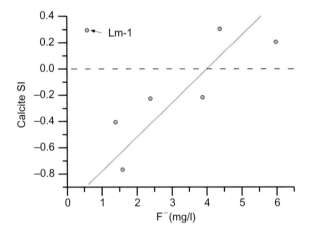

Figure 6.7.   Variation of F⁻ with the calcite SI value in sampled groundwater.

6.5   CONCLUSIONS

Deep groundwaters located between 200 and 500 m b.s. in the southern part of the province of Tucumán, exhibit intermediate to high concentrations of As (between 10 and 76 µg/l) and high concentrations of F⁻ (between 0.6 and 6.0 mg/l), as well as temperatures ranging between 35 and 50°C. These aquifers are considered a resource of thermal waters that could be potentially used for recreational and health purposes. The area comprises a fault system that converges near the city of Termas de Río Hondo and which could be partially responsible for the anomalies in the groundwater temperature (Vergara *et al.* 1998).

Groundwater in the area exhibits high electrical conductivity, with sodium, sulfate and chloride being predominant ions. Moreover, they are mainly alkaline and oxidizing, but Eh tends to decrease with depth. The concentrations of both F^- and As increase with depth, and with increasing temperature. All of these features are typical of a geothermally-influenced groundwater environment (Smedley and Kinniburgh 2002).

Arsenic and fluoride show a positive correlation with Cl^-, SO_4^{2-} and Na^+. Positive correlation between As and Cl^- can be associated with the dissolution of salts precipitated from the sublimation of endogenous vapors on the surface of Tertiary volcanic grains. On the other hand, positive correlation between As and SO_4^{2-} could be assigned to the weathering of sulfides such as arsenian pyrite and arsenopyrite, but the occurrence of these minerals in Tertiary sediments has not yet been confirmed.

The factors that mainly control the mobility of As and F^- in the deep aquifers seem to be temperature, pH and redox potential. Increasing temperature enhances the solubility of As and F-bearing minerals and hence, these elements are released to groundwater. The pH can control the adsorption of As oxyanions (particularly arsenate species) and F^-, because their affinity for Fe (hydr)oxide surfaces is higher under acidic pH. Redox state also has an impact on the adsorption, due to the lower affinity of As(III) species for these surfaces.

The concentration of F^- in solution is also influenced by the precipitation of calcite, because the resulting removal of Ca^{2+} enhances the dissolution of fluorite.

Therefore, the occurrence of F^- and As-enriched groundwater in the study area seems to be a combined result of several factors such as: (1) the dissolution (by meteoric waters) of salts precipitated from the sublimation of endogenous vapors on the Tertiary sediment grains; (2) the release to the solution of F^- and As-enriched fluids that occur as inclusions within minerals from the volcanic layers; (3) the oxidative dissolution of primary and/or secondary sulfides in volcanic materials; and (4) the mixing of meteoric waters with up-flowing endogenous vapors that reach the deep aquifers through the regional structure. Further chemical data (including isotopes and other trace elements) as well as lithological and structural information from the area and other thermal fields located towards the north, are needed for a better understanding of the source and transport of As and F^- in this geothermal system.

REFERENCES

Aiuppa, A., D'Alessandro, W., Federico, C., Palumbo, B. and Valenza, M.: The aquatic geochemistry of arsenic in volcanic groundwaters from southern Italy. *Appl. Geochem.* 18 (2003), pp. 1283–1296.

Allison, J.D., Brown, D.S. and Novo-Gradac, K.J.: MINTEQA2, A Geochemical assessment data base and test cases for environmental systems. US Environmental Protection Agency, Athens, GA, 1991.

American Public Health Association, American Water World Association, Water Pollution Control Federation (APHA): Standard methods for the examination of water and wastewater. 17th ed., Baltimore, Maryland, 1989.

Arnórsson, S.: Arsenic in surface- and up to 90°C waters in a basalt area, N-Iceland: processes controlling its mobility. *Appl. Geochem.* 18 (2003), pp. 1297–1312.

Baldis, B., Demicheli, J., Febrer, J., Fournier, H., García, E., Gasco, J.C., Mamaní, M. and Pomposiello, M.C.: Magnetotelluric results along a 1200 km long deep profile with an important geothermal area and its north-west end in the provinces of Tucumán and Santiago del Estero in Argentina. *Acta Geodact. Geophys et Montanist. Hung.* 18 (1983), pp. 489–499.

Battaglia, A.: Descripción geológica de la Hoja 13f "Río Hondo". Bulletin 186, Servicio Geológico Nacional. Subsecretaría de Minería de la Nación, Buenos Aires, Argentina, 1982.

Bundschuh, J., Farías, B., Martin, R., Storniolo, A., Bhattacharya, P., Cortes, J., Bonorino, G. and Albouy, R.: Groundwater arsenic in the Chaco-Pampean Plain, Argentina: case study from Robles county, Santiago del Estero Province. *Appl. Geochem.* 19:2 (2004), pp. 231–243.

CAA: Art.1 Res. MS y AS No. 494. Ley 18284, Dec. Reglamentario 2126, Anexo I y II, Marzocchi, Buenos Aires, Argentina, 1994.

Creed, J.T., Martin, T.D. and O'Dell, J.W.: Determination of trace elements by stabilized temperature graphite furnace atomic absorption. Environmental Monitoring Systems Laboratory Office of Reasearch and Development, US Environmental Protection Agency, Cincinnati, OH, 1994.

Dal Molin, C., Fernández, D.S., Escosteguy, L. and Villegas, D.: Hoja Geológica No. 2766-IV "Concepción", provincias de Tucumán, Santiago del Estero y Catamarca. Boletín No. 342, Programa Nacional de Cartas Geológicas de la República Argentina, Servicio Geológico Minero Argentino (ed). Buenos Aires, Argentina, 2003.

Dzombak, D.A. and Morel, F.M.M.: *Surface complexation modelling. Hydrous ferric oxide*. John Wiley and Sons, New York, 1990.

Ellis, A.J. and Mahon, W.A.J.: Natural hydrothermal systems and experimental hot water/rock interactions. *Geochim. Cosmochim. Acta* 28 (1964), pp. 1323–1357.

Gallet, S., Jahn, B., Van Vliet Lanoë, B., Dia, A. and Rossello, E.: Loess geochemistry and its implications for particle origin and composition of the upper continental crust. *Earth Planet. Sci. Lett.* 156 (1998), pp. 157–172.

Garcia, M.G., Hidalgo, M. Del V. and Blesa, M.A.: Geochemistry of groundwater in the alluvial plain of Tucumán, Argentina. *Hydrogeol. J.* 9:6 (2001), pp. 597–610.

Kolker, A. and Nordstrom, D.K.: Occurrence and microdistribution of arsenic in pyrite. US Geol. Survey Workshop—Arsenic in the environment, Denver, CO, 2001, available from: www.brr.cr.usgs.gov/Arsenic/.

Langmuir, D.: *Aqueous environmental geochemistry*. Prentice Hall, Upper Saddle River, NJ, 1997.

Mon, R. and Vergara, G.: The geothermal area of the eastern border of the Andes of North Argentina at Tucumán Province. *Bull. Int. Assoc. Eng. Geol.* 35 (1987), pp. 87–92.

Mon, R., Pomposiello, M. and Trinidad Díaz, M.: Estructura de la cuenca de Tucumán de acuerdo a investigaciones gravimétricas. *Proceedings of the XI Congreso Geológico Argentino* 1, San Juan, Argentina, 1990, pp. 251–254.

Nicolli, H.B., Suriano, J.M., Gómez Peral, M.A., Ferpozzi, L.H. and Baleani, O.A.: Groundwater contamination with arsenic and other trace elements in an area of the Pampa, Province of Córdoba, Argentina. *Environ. Geol. Water Sci.* 14:1 (1989), pp. 3–16.

Nicolli, H.B., Tineo, A., García, J.W., Falcón, C.M. and Merino, M.H.: Trace-element quality problems in groundwater from Tucumán, Argentina. In: R. Cidu (ed): *Water-rock interaction 2001* 2. Balkema, Leiden, 2001, pp. 993–996.

Nordstrom, D.K.: Thermodynamic properties of environmental arsenic species: limitations and needs. In: C. Young (ed): *Minor elements 2000: processing and environmental aspects of As, Sb, Se, Te, and Bi*. SME, Littleton, CO, 2000, pp. 325–331.

Parkhurst, D.L. and Appelo, C.A.J.: User's Guide to PHREEQC (Version 2), a computer program for speciation, batch-reaction, one-dimensional transport, and inverse geochemical calculations, Water-Resources Investigation Report 99-4259, US Geological Survey, 1999.

Romero, L., Alonso, H., Campano, P., Fanfani, L., Cidu, R., Dadea, C., Keegan, T., Thornton, I. and Farago, M.: Arsenic enrichment in waters and sediments of the Rio Loa (Second Region, Chile). *Appl. Geochem.* 18 (2003), pp. 1399–1416.

Smedley, P.L. and Kinniburgh, D.G.: A review of the source, behaviour, and distribution of arsenic in natural waters. *Appl. Geochem.* 17 (2002), pp. 517–568.

Smedley, P.L., Nicolli, H.B., Barros, A.J. and Tullio, J.O.: Origin and mobility of arsenic in groundwater from the Pampean Plain, Argentina. In: E.B. Arenhart and J.R. Hulston (eds): *Water-rock interaction*. Balkema, Rotterdam, 1998, pp. 275–278.

Smedley, P.L., Nicolli, H.B., McDonald, D.M.J., Barros, A.J. and Tullio, J.O.: Hydrogeochemistry of arsenic and other inorganic constituents in groundwaters from La Pampa, Argentina. *Appl. Geochem.* 17 (2002), pp. 259–284.

Smedley, P.L., Kinniburgh, D.G., Macdonald, D.M.J., Nicolli, H.B., Barros, A.J., Tullio, J.O., Pearce, J.M. and Alonso, M.S.: Arsenic associations in sediments from the loess aquifer of La Pampa, Argentina. *Appl. Geochem.* 20:5 (2005), pp. 989–1016.

Smith, R.M. and Martell, A.E.: *Critical stability constants*, vol. 4, *Inorganic complexes*. Plenum Press, New York, 1976.

Vergara, G., Hidalgo, M. del V., Balegno de Vergara, M.T. and Masmut, M.P.: Corte geotérmico transversal de la Provincia de Tucumán. In: M. Gianfrancisco, M.E. Puchulu, J. Durango de Cabrera and G. Aceñolaza (eds): *Geología de Tucumán*. Colegio de Graduados en Ciencias Geológicas, Tucumán, Argentina, 1998, pp. 161–178.

Warren, C.J., Burgess, W. and García, M.G.: Hydrochemical associations and depth profiles of arsenic and fluoride in Quaternary loess aquifers of northern Argentina. *Mineral. Mag.* 69:5 (2005), pp. 877–886.

WHO: Guidelines for drinking water quality. Revision of the 1984 guidelines. Final task group meeting, Geneva, 21–25 September 1992, World Health Organization, Geneva, Switzerland, 1993.

Wilkie, J.A. and Hering, J.G.: Rapid oxidation of geothermal arsenic (III) in steam waters of the eastern Sierra Nevada. *Environ. Sci. Technol.* 32 (1998), pp. 657–662.

CHAPTER 7

The origin of arsenic in waters and sediments from Papallacta lake area in Ecuador

L.H. Cumbal
Centro de Investigación Científica, Escuela Politecnica del Ejercito (ESPE), Sangolqui, Ecuador

J. Bundschuh
International Technical Cooperation Program, CIM (GTZ/BA), Frankfurt, Germany
Instituto Costarricense de Electricidad (ICE), San José, Costa Rica

V. Aguirre, E. Murgueitio, I. Tipán & C. Chávez
Scientific Research Center of Escuela Politecnica del Ejercito (ESPE), Sangolqui, Ecuador

ABSTRACT: This chapter focuses on characterizing sources of arsenic (As) in the vicinity of Papallacta lake, Ecuador. Some of these sources are geothermal water discharges. Arsenic concentrations in these geothermal waters range from 1090 to 7853 µg/l and temperatures range from 13.8 to 63.0°C. Arsenic concentrations were determined in water and sediment from the Tambo river, the main tributary of Papallacta lake. Arsenic concentrations are greater than 62 µg/l in water and up to 128 mg/kg in sediments. In comparison, total As concentrations in the shallow lake water during the dry season range from 220 to 369 µg/l. At depth, the As distribution was nearly homogeneous and fluctuated between 289 and 351 µg/l. Analyses of the lake sediments indicate that As concentrations vary between 60 and 613 mg/kg. From this study, it can be concluded that discharges of geothermal waters to the Tambo river are the main natural sources of As in Papallacta lake.

7.1   INTRODUCTION

Arsenic (As) occurrence in natural waters has been discussed extensively during recent years because of its adverse effects on human health. In Ecuador, various sources of drinking water with As concentrations exceeding the WHO-recommended maximum concentration of 10 µg/l have been identified. In many systems, it has been shown that As present in aquifers is of natural rather than anthropogenic origin. Geothermal waters acquire elevated concentrations of As as a result of dissolution of As-containing minerals during long periods of contact with rocks in the subsurface. Arsenic minerals such as realgar (AsS), orpiment (As_2S_3), arsenopyrite (FeAsS), etc. are thermo-dynamically stable under reducing conditions, as are often found in the subsurface (Sengupta 2002). Geothermal waters in contact with these minerals can dissolve them, increasing As concentrations in the aqueous phase. Elevated As concentrations in geothermal waters have been reported by several investigators (Welch *et al.* 2000, Criaud and Fouillac 1989, Romero *et al.* 2000) in different parts of the world, including the United States, Japan, New Zealand, France, Dominica, Bulgaria, Chile, and Argentina. For example, Ball *et al.* (1998) reported As concentrations in excess of 2830 µg/l in the thermal waters from Yellowstone National Park, USA. Arsenic concentrations of 3800 µg/l have been reported in residual thermal waters from the Wairakei geothermal field in New Zealand (Robinson *et al.* 1995). High As concentrations have also been found in thermal waters in the El Tatio system in the Antofagasta region of Chile (100–1000 µg/l) (Caceres 1999, Queirolo 2000).

    In Ecuador, elevated As concentrations in Papallacta lake were found during the course of studies conducted as part of the remediation of a crude oil spill (Fig. 7.1). The spill resulted

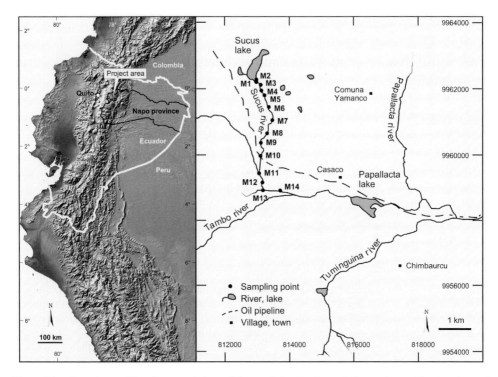

Figure 7.1.   Locations of Papallacta and Sucus lakes and the Tambo and Sucus rivers. The locations of
sampling sites along the Sucus river are also shown.

from the rupture of the Transecuatorian pipeline in 2003. Arsenic concentrations between 39 and
10,560 µg/l were found in geothermal waters and rivers of the lake's watershed (De la Torre
et al. 2004). Arsenic concentrations in Papallacta lake range from 390 to 670 µg/l. The most likely
sources of the high As concentrations are thought to be: (1) mobilization of As from sediments
during their removal from the lake during remediation and (2) elevated As concentrations in the
lake's tributaries (Heredia and Bernal 2003, Ecuavital 2004, Alerta Verde 2004, De la Torre *et
al.* 2004). The company responsible for remediation of the lake carried out leaching tests on the
sediments from the lake. From these results, the company concluded in its final report that the
As contamination in the lake did not come from the sediments but rather from geothermal waters
that discharge into tributaries (Ecuavital 2004). In 2005, our research team, quantified As in water
samples from the lake and found concentrations of about 500 µg/l. The persistence of elevated
concentrations of As in the lake despite the time elapsed since the remediation activities suggests
that the toxic chemical is of natural origin. It is not clear however the location of the geothermal
sources.

   The objectives of this study were: (1) to identify the sources of geothermal waters that discharge
into the Tambo and Sucus rivers, (2) to quantify As concentrations in geothermal springs and in
the Tambo and Sucus rivers, and (3) to determine As concentrations in water and sediments of
Papallacta lake.

7.2   AREA OF STUDY AND ENVIRONMENTAL CONDITIONS

Papallacta lake is located in Quijos county, Napo province, in northeastern Ecuador, at an average
altitude of 3360 m a.s.l. The lake was formed in 1760 by a lava flow known as Antisanilla, which

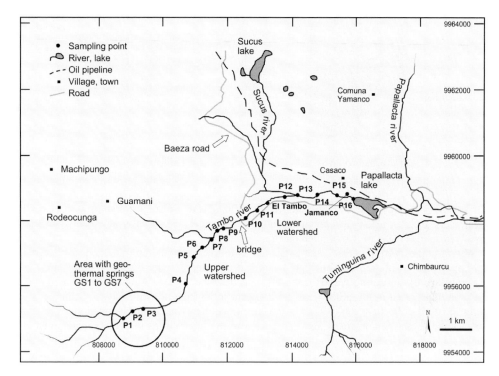

Figure 7.2.   Location of sampling sites in the Tambo river watershed.

dammed a stream (Bourdon *et al.* 2002). The lake covers an area of 310,000 m^2 in the dry season and 440,000 m^2 in the rainy season; the average area is 330,000 m^2. Annual precipitation in the Papallacta region varies between 600 and 1600 mm and contributes significant quantities of water to the lake's watershed. Strong winds towards the northeast provide good aeration of Papallacta lake. The lake receives inflows from the Tambo river; submerged inputs along the northern boundary of the lake; small, cold streams; and uncontrolled residual discharge from the Jamanco hot springs. The Tambo river is fed by several small thermal streams along an approximately 12.8 km stretch (Fig. 7.2). In addition, residual thermal waters from El Tambo hot springs are discharged into the Tambo river. All these sources influence strongly the chemical composition of the Tambo river and Papallacta lake. On the other hand, driven by the steep hydraulic gradient (0.0768 m/m), the Tambo river flows with a substantial destructive force, removing and redistributing sediments along the path of the river. In the lower watershed of the Tambo river its water is used for agriculture development and trout farming.

7.3   SAMPLING AND ANALYTICAL PROCEDURES

7.3.1   *Sampling*

Samples of water and sediments were taken in Papallacta lake, Tambo and Sucus rivers, and geothermal springs and streams (sampling points are shown in Figs. 7.1 and 7.2). In rivers, sampling was carried out every 200 to 600 m or wherever the topographical conditions allowed. Geothermal springs were sampled at their outlets and before their confluence with the river. In the lake, shallow samples were collected at different locations as shown in Figure 7.3a. Water samples for three vertical profiles were collected (Fig. 7.3b). Samples for the vertical profiles were collected every 4 or 8 m down to the bottom at each location using a bathometer (Kemmerer BTL 1.2 L SS).

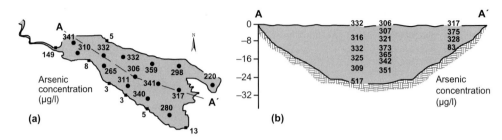

Figure 7.3.   Arsenic distribution (a) on the surface (circles) and (b) at different depths in the Papallacta lake. Squares in (a) show arsenic concentrations in streams flowing into Papallacta lake.

All sampling locations were determined using a handheld global positioning system (GPS). Parameters such as pH, redox potential, dissolved oxygen, electrical conductivity, and total dissolved solids (TDS) were measured on site.

Water samples were processed and preserved as follows: (1) filtered (0.45 μm, for determining concentrations of anions), (2) filtered and acidified samples (for determining total As), and (3) processed by separating As(III) from As(V) as described below. Samples were stored at 4 to 5°C and transported to the ESPE (Escuela Politecnica del Ejercito) for analysis. Sediment sampling involved collecting approximately 0.5 kg of sediments at 5 cm below surface. Samples were stored in plastic containers and transported to the laboratories in coolers maintained at 4 to 5°C.

7.3.2   Separation of arsenic species

Speciation of inorganic arsenic [As(III) and As(V)] in water samples was performed as described by Ficklin (1983) and Clifford *et al.* (1983). A volume of 30 to 50 ml of water was adjusted to pH 4.0 to 5.0, passed through a mini-column packed with an anion exchange resin (A-400) and collected at the column outlet.

7.3.3   Chemical analysis

Arsenic concentrations in acidified and non-acidified water samples, and the effluent from the mini-columns used to separate As(III) from As(V), were determined using a Perkin Elmer hydride generator (HG) coupled to a Perkin Elmer atomic absorption spectrometer, AA100. The results of chemical analyses were validated by comparing As concentrations with standard solutions purchased from Sigma Aldrich.

7.3.4   Extraction of arsenic from sediments

Sediment samples were dried in an oven for 12 hours at a 60°C. Dried sediments were disaggregated using a mortar and pestle and then sieved using a 0.2 mm sieve. For the determination of As, sieved samples (0.25 g) were digested using a solution containing 1 M $Mg(NO_3)_2$, HCl and KI with a volumetric ratio of 1:5:5, respectively (Romero *et al.* 2003). The suspensions were heated in a furnace for 6 h at 450°C and solid residues were dissolved in 10 ml 1 M HNO_3. Then the acidified solution was filled with deionized water up to 25 ml and analyzed by HG-AAS. Since most of the samples became brownish, a procedure to determine As bound to organic matter and sulfide was conducted (Romero *et al.* 2003). Sediment samples were digested using 3 ml of 0.02 M HNO_3 and 5 ml of 30% H_2O_2 at pH 2, at 85°C followed by a second addition of 3 ml aliquot of 30% H_2O_2 at pH 2. Samples were then treated with 5 ml of 3.2 M ammonium acetate (NH_4OAc) in 20% HNO_3.

7.3.5 *Analysis of mineral fraction in sediments*

A portion of each sediment sample was ground further in an agate mortar and used for mineralogical analysis. Samples were sent to Lehigh University for X-ray diffraction analysis in a Rigaku Rotaflex diffractometer.

7.4 RESULTS AND DISCUSSION

7.4.1 *Arsenic at the Sucus river*

Table 7.1 presents the pH, temperature, electrical conductivity, TDS, and total As concentrations at different sampling sites (Fig. 7.1) along the Sucus river. In general, the pH of all sites is near to neutral and the TDS values are similar. The average As concentration of 9.5 µg/l found in this investigation is slightly below the WHO guidance value for drinking water (WHO 2001). These comparatively low As concentrations may be related to the absence of inputs form geothermal sources.

7.4.2 *Arsenic in the Tambo river*

The Tambo river was divided in two zones, as shown in Figure 7.2. The lower-watershed zone corresponded to the region extending from approximately 100 m downstream from the bridge on the Baeza road to the mouth at the river at Papallacta lake. The upper-watershed zone extended from the bridge upstream to the headwaters of the river. Observed values of the altitude, pH, electrical conductivity, TDS, and total As for sites along the river are presented in Table 7.1. The increase in As concentration between sampling points P11 and P12 likely results from the inflow of an As-rich geothermal stream. The As concentration at the mouth of the river is 149 mg/l (P16). Two inputs from residual geothermal waters (with 1544 and 520 µg As/l, respectively) from the Jamanco hot springs are responsible for the elevated As concentration at this sampling point. The headwaters of the Tambo river (P1 to P7) contain geothermal sources that are responsible for the observed elevated concentrations of As (233 to 698 µg As/l).

Arsenic concentrations reported in this study are high compared to the typical values found in rivers crossing geothermal fields, such as 10 to 114 µg As/l in the western United States, New Zealand, and in the central region of Argentina (McLaren and Kim 1995, Nimick *et al.* 1998, Lerda and Prosperi 1996) but lower than those reported in the Loa river in the arid region of Chile, which has an average As concentration of 1400 µg As/l (Romero *et al.* 2003). The characteristics of the hydrothermal system of the Antisana volcano and the composition of the soils in Papallacta lake watershed may determine the concentration of As in the surrounding rivers.

Arsenic is also present in sediments at several sites in the lower watershed of the Tambo river (20–128 mg/kg) (Table 7.2). These As concentrations are smaller than those found in sediments of Loa river, Chile (320 mg/kg) (Romero *et al.* 2003). The relatively low As concentrations in the sediments of the Tambo river could be attributed to their composition. Results of mineralogical analyses show that the sediments are mainly comprised of andesine ($Na_{0.72}Ca_{0.35}Al_{1.52}Si_{2.5}O_8$), sanidine ($KSi_3AlO_8$), enstatite ($Mg_2Si_2O_6$), and quartz ($SiO_2$). Arsenic does not bind significantly to these silicate minerals. However, sediment sequential extraction data demonstrate that As is strongly associated with organic matter, which is plentiful in the river bed, and is transported by the river flow. Wang and Mulligan (2006) report that natural organic matter commonly associated with soils and aquifer sediments can serve as significant As adsorbents.

7.4.3 *Arsenic in geothermal springs*

Arsenic concentrations in geothermal waters are in the range of 1090–7852 µg/l (Table 7.1). High contents of As can be the result of reductive dissolution of and desorption from Fe- and Mn-oxyhydroxides, which can occur at depth in the aquifer as hydrothermal deposits from the

Table 7.1.   Water composition of the Sucus and Tambo rivers and geothermal springs in the upper Tambo watershed (for sampling locations, see Figs. 7.1 and 7.2).

Sample	Altitude (m a.s.l.)	pH	T (°C)	Conductivity (µS/cm)	TDS (mg/l)	t-As (µg/l)
Sucus river						
M1		7.2	15.5	357.0	180	7
M2		7.4	13.7	117.9	56	22
M3		7.2	15.2	121.5	56	22
M4		7.0	13.8	109.8	53	13
M5		6.8	11.7	95.4	45	11
M6		7.1	10.8	98.7	46	7
M7		7.0	10.1	129.2	61	6
M8		7.2	9.3	85.1	39	8
M9		6.9	9.7	121.1	57	25
M10		8.1	9.4	118.6	61	3
M11		7.4	9.5	120.2	56	4
M12		8.0	9.3	117.8	52	6
M13		8.2	9.4	123.6	55	2
Tambo river						
P1	3783	7.1	15.5	341	165	626
P2	3766	7.3	14.6	380	168	538
P3	3802	8.4	14.1	401	206	698
P4	3906	8.2	15.6	588	292	413
P5	3875	8.3	13.2	540	269	233
P6	3906	8.5	12.7	500	249	472
P7	3906	7.1	15.2	517	267	98
P8	3608	7.2	–	242	126	115
P9	3598	7.3	18.0	245	128	108
P10	3580	7.1	14.4	261	132	64
P11	3552	7.4	19.5	882	442	62
P12	3399	7.5	19.0	250	128	115
P13	3397	6.7	21.0	835	440	153
P14	3389	6.8	18.1	897	451	154
P15	3385	6.7	15.0	889	443	148
P16	3381	7.3	17.0	890	448	149
Geothermal springs						
GS1	3834	6.8	45.4	477	238	3152
GS2	3856	6.2	35.3	4270	1835	1090
GS3	3912	7.8	15.6	270	186	3555
GS4	3910	6.3	55.6	4820	3500	4380
GS5	3915	7.4	57.5	285	162	3161
GS6	3910	8.2	13.8	8300	3500	7852
GS7	3917	6.6	63.0	5920	2890	6120
El Tambo	3595	–	–	–	–	2802
Jamanco	3373	6.8	56.0	–	–	3004

Antisana volcano. Volcanic ash deposits, which are widespread in soils of the Ecuadorian Andes, can also contribute to the generation of high concentrations of As in groundwater. Nicolli *et al.* (1989), Smedley *et al.* (2002), and Bundschuh *et al.* (2004) indicate that volcanic rocks, especially ashes, are often implicated in the production of water with elevated As concentrations.

Speciation of As in water collected from the two hot springs GS1 and GS7 indicates the dominance of As(III), which amounts to 74.4 and 61.2%, respectively, of the total As concentration (Table

Table 7.2.   Arsenic in sediments in the lower watershed of the Tambo river.

Sampling point	t-As (mg/kg)	Sampling point	t-As (mg/kg)
P8	20.71	P13	107.24
P9	26.69	P14	128.33
P10	–	P15	104.71
P11	34.62	P16	73.2
P12	–		

Table 7.3.   Arsenic species in some geothermal water sources.

	As(III)		As(V)		Eh
	µg/l	%	µg/l	%	mV
GS1	2344	66.7	808	33.3	−103.8
GS3	1144	32.2	2411	67.8	+9.2
GS6	2632	33.5	5220	66.5	+7.3
GS7	3744	61.2	2376	38.8	−112.2

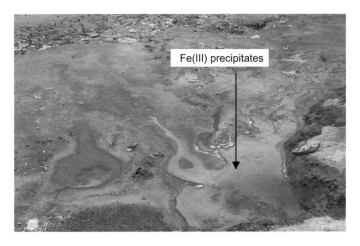

Figure 7.4.   Formation of Fe(III) precipitates on the hot spring surface.

7.2). Total As concentrations for these springs are 3152 and 6120 µg/l, respectively (Tables 7.1 and 7.3). The redox conditions of these springs are highly reducing (−112.2 and −103.8 mV). This is consistent with the predominance of As(III). In addition, iron precipitates found in these geothermal springs are abundant (Fig. 7.4) suggesting that they are rich in Fe. Iron oxidation and precipitation of iron hydroxides occurs as the water comes in contact with the atmosphere. Thus, reductive dissolution of iron and manganese minerals as observed in Bangladesh (e.g., Smedley and Kinniburgh 2002, Ahmed *et al.* 2004) is proposed as the main mechanism of the elevated As concentrations observed in these hot springs. In contrast, As(V) is predominant in the springs GS3 (67.8%) and GS6 (66.5%), whose total As concentrations are 3555 and 7852 µg/l, respectively. This is in agreement with the higher measured redox potential compared to the previous two springs (+9.2 and +7.3 mV). In addition, temperatures of spring GS3 and GS8 are low (15.6 and 13.8°C). Low temperatures compared to the other springs could result from mixing with shallow groundwater or oxygenated recharge water. In general, geothermal waters that reach the Tambo river contain significant concentrations of As(V). This may suggest that the residence time of the oxygenated recharge water within the hydrothermal system is short. Therefore, redox changes are incomplete and the oxic water discharges into the Tambo river. Another possible mechanism is that As might be rapidly oxidized owing to efficient oxygenation of water stimulated by strong winds, which are very frequent in the region (Table 7.1).

7.4.4   Arsenic in Papallacta lake

Seven streams with water flows ranging from 0.3 to 4.5 l/s and the Tambo river (220 to 1508 l/s) discharge at the Papallacta lake. Arsenic concentrations in the streams reaching the lake (Fig. 7.3a, squares) show that the highest value is in the Tambo river (149 µg/l) while the other streams have concentrations in the range of 2 to 13 µg/l.

Arsenic concentrations in surface water samples in Papallacta lake vary from 220 to 359 µg/l (Fig. 7.3a). These samples were collected on April 22, 2006 but samples collected on July 20, 2006 had lower As concentrations (86 to 177 µg/l, data not shown). This temporal variability in As concentration mainly results from a dilution effect. In summer, the volume of the lake is 8,310,000 m^3 while in winter it increases to 13,420,850 m^3. Determination of As concentrations in water samples collected along the northern shore of the lake is underway.

Arsenic concentration along a vertical cross-section of the lake are shown in Figure 7.3b. At depth, the distribution of As is almost homogeneous in the central part of the lake (289–351 µg/l) while, near the margin, a significant decrease in the As concentration in the bottom-most sample is observed (83 µg/l). Water mixing driven by strong winds (with speeds up to 20 m/s) noticeably influences the vertical distribution of As.

The As concentration of 517 µg As/l, observed in the bottom-most sample from the lake (at around 27 m of depth, Fig. 7.3b), is well above the As concentrations of the other samples. The high As concentration in this sample may have resulted from release of As from suspended sediments when it was acidified.

Arsenic concentrations in sediments along the border of Papallacta lake are variable. Samples taken on the eastern and southeastern margins contain more As (540 and 613 mg/kg) than those from the northwestern and southwestern margins (60 and 72 mg/kg). This observed spatial pattern in As concentration may result from the continuous removal and redistribution of sediments caused by Tambo river overflows as well as As leaching from mineral and organic fractions of the sediments. Based on the mineralogical composition of the sediments (74% andesine, 10% of enstatite, and 8.5% of quartz), one would expect As concentrations to be low. However, natural organic matter, present at high concentrations at the surface of these sediments (2.67 to 13.43%), may scavenge As, thus increasing its concentration. Various laboratory studies provide evidence that As is released from sediments after flooding or when anaerobic conditions develop (Smedley and Kinniburgh 2002). Nevertheless, in Papallacta lake, As release without the influence of heavy rains and flooding could be very slow and subject to pH and Eh changes.

7.5   CONCLUSIONS

Water from Papallacta lake is enriched in several chemical elements in addition to As. The high As concentrations (311 µg/l average) are associated with high concentrations of total dissolved solids (436 mg/l), of which chloride is the major component (Cumbal et al. unpubl. data). The lagoon is slightly alkaline (pH = 7.7), possibly due to the concentration of bicarbonate (Cumbal et al. unpubl. data). The vertical distribution of As concentrations in the lake is relatively uniform because of the constant wave action driven by strong winds, which promotes mixing.

Hot geothermal waters that discharge into the Tambo river influence its water chemistry and contribute to the high As concentrations observed in Papallacta lake. Arsenic concentrations in the Sucus river, which lacks significant geothermal input, are smaller.

Sediments along the margin of the lake also accumulate As. Sediment As concentrations along the western border are lower than those elsewhere because of the continuous removal and redistribution of sediments by overflows of the Tambo river. Arsenic concentrations are higher along the eastern border, probably because of precipitation of As-rich solids. Variable composition of As is also present in sediments collected from several sites along the lower watershed of the Tambo river.

ACKNOWLEDGEMENTS

The authors thank the financial support given by National Council of Higher Education of Ecuador (CONESUP) through a project grant. We also express gratitude to volunteer students of the 8th semester of the Geographic and Environment Eng. of ESPE and the Brigada de Fuerzas Especiales "Patria" No. 9 for their help in the field trips.

REFERENCES

Alerta Verde: Papallacta, a punto de cerrarse un caso, no por limpieza, sino por impotencia. *Bol. Acción. Ecol.* 134 (2004), Quito, Ecuador.

Ahmed, K.M., Bhattacharya, P., Hasan, M.A., Akhter, S.H., Alam, S.M.M., Bhuyia, M.A.H., Imam, M.B., Khan, A.A. and Sracek, O.: Arsenic enrichment in groundwater of the alluvial aquifers in Bangladesh, an overview. *Appl. Geochem.* 19:2 (2004), pp. 181–200.

Ball, J.W., Nordstrom, D.K., Jenne, E.A. and Vivit, D.V.: Chemical analyses of hot springs, pools, geysers, and surface waters from Yellowstone National Park, Wyoming, and Vicinity, 1974–1975. USGS Open-File Rep. 98-182, 1998.

Bourdon, E., Eissen, J.P., Monzier, M., Robin, C., Martin, H., Cotton, J. and Hall, M.: Adakite-like lavas from Antisana Volcano (Ecuador): Evidence for slab melt metasomatism beneath Andean northern volcanic zone. *J. Petrology.* 43:2 (2002), pp. 199–217.

Bundschuh, J., Farías, B., Martin, R., Storniolo, A., Bhattacharya, P., Cortes, J., Bonorino, G. and Albouy, R.: Groundwater arsenic in the Chaco-Pampean Plain, Argentina: case study from Robles County, Santiago del Estero Province. *Appl. Geochem.* 19 (2004), pp. 231–243.

Clifford, D., Ceber, L. and Chow, S.: As(III)/As(V) separation by chloride-form ion-exchange resins. *Proceedings of the AWWA WQTC*, Norfolk, VA, 1983.

Criaud, A. and Fouillac, C.: The distribution of arsenic (III) and arsenic (V) in geothermal waters: examples from the Massif Central of France, the Island of Dominica in the Leeward Islands of the Caribbean, the Valles Caldera of New Mexico, USA, and southwest Bulgaria. *Chem. Geol.* 76 (1989), pp. 259–269.

De la Torre, E., Guevara, A., Muñoz, G. and Criollo, E.: Estudio de aguas superficiales y sedimentos de la cuenca de los ríos Sucus, Tambo y Papallacta. Unpublished report, Quito, Ecuador, 2004.

Ecuavital: Informe final de la remediación ambiental de laguna de Papallacta, Quito. Unpublished report, Quito, Ecuador, 2004.

Ficklin, W.H.: Separation of As(III) and As(V) in groundwaters by ion exchange. *Talanta* 30:5 (1983), p. 371.

Heredia, E. and Bernal, C.: Distribución de contaminantes producto del derrame de petroleo del 8 de abril en la laguna de Papallacta. Unpublished report, Quito, Ecuador, 2003.

Lerda, D.E. and Prosperi, C.H.: Water mutagenicity and toxicology in Rio Tercero, Córdoba, Argentina. *Water Res.* 30 (1996), pp. 819–824.

McLaren, S.J. and Kim, N.D.: Evidence for a seasonal fluctuation of arsenic in New Zealand's longest river and the effect of treatment on concentrations in drinking water. *Environ. Pollut.* 90 (1995), pp. 67–73.

Nicolli, H.B., Suriano, J.M., Peral, M.A.G., Ferpozzi, L.H. and Baleani, O.A.: Groundwater contamination with arsenic and other trace-elements in an area of the Pampa, province of Córdoba, Argentina. *Environ. Geol. Water Sci.* 14 (1989), pp. 3–16.

Nimick, D.A., Moore, J.N., Dalby, C.E. and Savka, M.W.: The fate of geothermal arsenic in the Madison and Missouri rivers, Montana and Wyoming. *Water Resour. Res.* 34 (1998), pp. 3051–3067.

Queirolo, F., Stegen, S., Mondaca, J., Cortes, R., Rojas, R., Contreras, C., Munoz, L., Schwuger, M. and Ostapczuk, P.: Total arsenic, lead, cadmium, copper, and zinc in some salt rivers in northern Andes of Antofagasta, Chile. *Sci. Total Environ.* 255 (2000), pp. 85–95.

Robinson, B., Outred, H., Brooks, R. and Kirkman, J.: The distribution and fate of arsenic in the Waikato river system, North Island, New Zealand. *Chem. Spec. Bioavail.* 7 (1995), pp. 89–96.

Romero, L., Alonso, H., Campano, P., Fanfani, L., Cidub, R., Dadea, C., Keegan, T., Thornton, I. and Farago, M.: Arsenic enrichment in waters and sediments of the Rio Loa (Second Region, Chile). *Appl. Geochem.* 18 (2003), pp. 1399–1416.

SenGupta, A.K. (ed): Environmental separation of heavy metals: engineering processes. In: *Arsenic in subsurface water: its chemistry and removal by engineered processes.* Lewis Publishers, Boca Raton, FL, 2002, pp. 265–305.

Smedley, P.L. and Kinniburgh D.G.: A review of the source, behaviour and distribution of arsenic in natural waters. *Appl. Geochem.* 17 (2002), pp. 517–568.

Wang, S. and Mulligan, C.N.: Natural attenuation processes for remediation of arsenic contaminated soils and groundwater. *J. Hazard. Mat.* 138:3 (2006), pp. 459–470.

Welch, A.H., Westjohn, D.B., Helsel, D.R. and Wanty, R.B.: Arsenic in ground water of the United States: occurrence and geochemistry. *Ground Water* 38 (2000), pp. 589–604.

WHO: Arsenic in drinking water. Fact sheet 210, World Health Organization, Geneva, Switzerland, 2001, http://www.who.int/mediacenter/factsheets/fs210/en/print.html.

CHAPTER 8

Arsenic contamination, speciation and environmental consequences in the Bolivian plateau

J. Quintanilla, O. Ramos, M. Ormachea & M.E. García
Instituto de Investigaciones Químicas, Universidad Mayor de San Andrés, La Paz, Bolivia

H. Medina
Instituto Nacional de Salud Ocupacional (INSO-SNS), La Paz, Bolivia

R. Thunvik & P. Bhattacharya
KTH-International Groundwater Arsenic Research Group, Department of Land and Water Resources Engineering, Royal Institute of Technology (KTH), Stockholm, Sweden

J. Bundschuh
International Technical Cooperation Program, CIM (GTZ/BA), Frankfurt, Germany
Instituto Costarricense de Electricidad (ICE), San José, Costa Rica

ABSTRACT: There are no comprehensive studies of geogenic arsenic (As) contamination of water resources and its human impact in Bolivia. A few studies have been conducted in the historic mining areas of the Bolivian plateau, where acid water drainage (locally known as *copagira*) from active and abandoned sites has caused extensive contamination of rivers and soils. This chapter describes As and heavy metal contamination of surface water, groundwater and soils in the Poopó and Uru Uru basins and the Uyuni salt pan in western Bolivia, which are generally attributed to past mining of silver and gold associated with sulfides of Fe, Cu, Cd, Zn, Pb, As and Co. Concentrations of As and Cd at the sites close to mining areas are much higher than in reference areas. Without exception, the rivers that drain mining areas into Poopó lake are chemically contaminated. Some rivers are acid, with a pH of around 3, and polluted with heavy metals and As at concentrations 10 to 100 times above WHO guidelines. Such concentration ranges are also observed in the suspended solid concentrations.

8.1 INTRODUCTION

In many regions of the world the occurrence of geogenic arsenic (As) in water resources limits their use for both drinking and irrigation. People in low-income countries, particularly in south and southeast Asia and Latin America, are most severely affected. Here, the use of groundwater containing high concentrations of As has resulted in severe environmental health problems, and hinders economic development. In Latin America, As occurrence in groundwater has been reported from Argentina, Bolivia, Brazil, Chile, Costa Rica, Mexico, Nicaragua, and Peru; and with severe health effects reported from Argentina (Bundschuh *et al.* 2004, Bhattacharya *et al.* 2006), Chile (Sancha 2003), Mexico (Cebrian *et al.* 1983) and Brazil (Matschullat 2000).

In Bolivia, there are no comprehensive studies of contamination of water resources from either geogenic sources or mining activities, nor their impact on the population. A few studies have been conducted of As occurrences in the Bolivian Altiplano (highlands). Several mining areas in the Bolivian highlands have been exploited for five centuries—from colonial times to the present. Silver and gold deposits are associated with sulfides of Fe, Cu, Zn, Pb, As and Co (SERGEOMIN 1999). Acid water drainage (locally known as *copagira*) from mining areas, both abandoned and active, has caused extensive contamination to the adjoining rivers and soils (Table 8.1).

Table 8.1.   Composition of surface waters affected by acid mine drainage (μg/l).

Surface water body	As	Cd	Cu	Ni	Pb	Sn	Zn
Desaguadero river-Español bridge	150	0.4	20	9	8	3.9	30
San Juan de Sora-Sora river	–	570	–	470	28	50	38000
Poopó river	4.5	370	27	80	2	5.6	195
Antequera river	4.9	480	460	300	9	35	140000
WHO guideline for drinking water	10	3	2000	20	10	5	3000

Source: SERGEOMIN 1999.

The Bolivian highlands (Fig. 8.1) comprise the provinces of La Paz, Oruro and Potosí. La Paz province, located in northwest Bolivia, has an area of 133,985 km^2 and contains the capital city, La Paz, at an altitude of 3640 m a.s.l. It is bounded to the north by Pando province, Oruro province to the south, Cochabamba province to the east, and Peru and Chile to the west. The population of the La Paz province is approximately 1,900,000 (1992 census).

The "Project Pilot Oruro" (PPO 1996a and PPO 1996b) examined the general environmental conditions in the northwest of Oruro province (including Poopó lake). Subsequently the "Alliance of the Lake Titicaca" (ALT 1999) made an evaluation of mining activity in the entire watershed of Lake Titicaca, including Poopó lake, the Desaguadero river and the Coipasa salt pan.

Several mining industries are responsible for major heavy metal and As contamination. Beveridge *et al.* (1985) and Zabaleta (1994) described the effects of mining-related pollution on commercial fish farming in Poopó lake, and Apaza *et al.* (1996) described the impact of the same contamination on ecotoxicity and the food chain.

The rural population of the Bolivian highlands is widely dispersed in small communities and has a very low socio-economic and educational status (Fig. 8.2), and characterized by a general lack of local community participation in solving the environmental problems.

At present, there is an increasing demand for both surface and groundwater in the Bolivian highlands, but the poor quality of water impedes development in the region. As a consequence, most development projects include surface and groundwater resource management components.

The objective of this study is to present a preliminary assessment of the status of As and heavy metal contamination in groundwater, surface water and soils in the provinces of La Paz, and Oruro of the Bolivian highland, its relation to geology, historical mining, and other industrial activities around the major cities, as well as their implications on human health in the region.

8.2   GEOLOGY OF THE BOLIVIAN HIGHLANDS

Geological characteristics influence the quality of surface waters. The Altiplano is surrounded by two cordilleras that have different geological origins. The Eastern Cordillera comprises a Phanerozoic polygenetic fold belt of Paleozoic and Mesozoic schists, skirted by low angle faults (USGS-GEOBOL 1992). These rocks were deposited on Precambrian basement and deformed by at least three orogenic cycles between the Paleozoic and the Cenozoic. The Eastern Cordillera is composed of intensely fractured Ordovician limestone, quartzite, slate and schist, with low angle faulting (YPFB-GEOBOL 1996). The Western Cordillera formed as a consequence of the accumulation of huge amounts of volcanic material in the late Tertiary and early Quaternary. Young volcanic craters, lava cones, domes and extensive flows dominate the landscape along the border between Bolivia and Chile. The bedrock in this region consists mainly of basaltic, dacitic, rhyodacitic and andesitic lavas (YPFB-GEOBOL 1996).

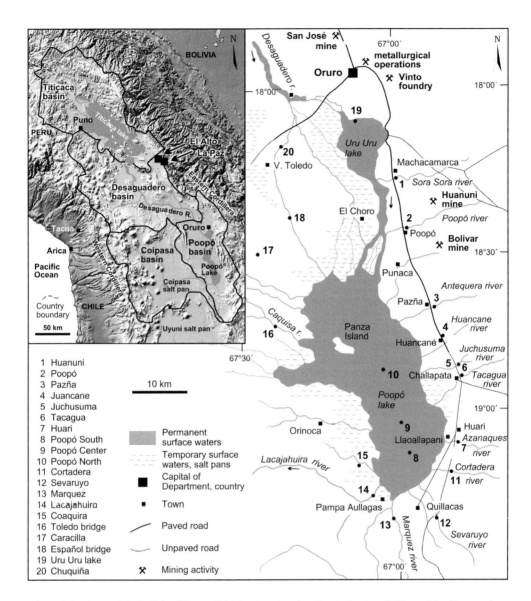

Figure 8.1. Map of the Bolivian Plateau (Altiplano) region showing the basins of Titicaca lake, Desaguadero river, Uru Uru and Poopó lakes and Uyuni salar (salt pan), The study area, located south of Oruro city, comprises the basins of Uru Uru and Poopó lakes.

8.3   ENVIRONMENTAL CHARACTERISTICS OF THE URU URU-POOPÓ LAKE BASIN

The Uru Uru-Poopó basin lies within Oruro province, and includes the counties of Cercado, Poopó, Avaroa, Sebastián Pagador, Ladislao Cabrera, South Carangas and Saucari. Intensive mining in the basin has affected the Desaguadero, Mauri, Márquez and Paso Julian rivers, together with the Poopó basin (Montes de Oca 1997). Lake Poopó is located between latitudes 18°27′ and 19°10′ S and longitudes 66°50′ and 67°24′ W, at an altitude of 3686 m a.s.l. The main contributions of water come from the Desaguadero river in the north and the Márquez river in the south. The lake has a maximum length of 90 km from the mouth of the Desaguadero river to the Márquez river and a maximum width of 53 km.

Lake Poopó is part of the Titicaca-Desaguadero-Poopó-Salares (TDPS) basin, located between latitudes 14°35′ and 22°50′ S and longitudes 66°10′ and 71° W, with an elongated form that extends for approximately 1000 km, with an average width of 200 km (Fig. 8.1). According to Biosca (1998), Poopó lake has an area of 4200 km^2, but according to Montes de Oca (1997) the area is 2650 km^2, while (ALT 1999) report an area of 3084 km^2. These differences arise because the surface area of the lake varies greatly on a seasonal basis, depending on the rains that fall between December and March (Quintanilla 1985, 1994). The basin is very flat, and as a consequence, its area varies significantly due to the irregularity of river influxes, and make this a very unstable environment.

Lake Poopó has a depth of between 0.5 to 2.5 m (average 1.4 m). Lake Uru Uru, located immediately north of Poopó lake, was formed in 1955 following diversion of the Desaguadero river. Lake Uru Uru covers an average area of 260 km^2 and perimeter of 128 km, with maximum dimensions of 32 km long and 11 km wide (UNEP/OEA 1996).

8.4   SOURCES OF ENVIRONMENTAL CONTAMINATION

8.4.1   *Natural sources*

SERGEOMIN (1999) identified both natural and anthropogenic sources of As and heavy metal contamination in the region. The main natural sources, located west of Oruro city, in the Western Cordillera, are the result of intense volcanic activity during Miocene. Extensive pyroclastic and volcanic layers have contributed to the development of a planar landscape. The contamination results from geothermal activities (e.g., geothermal springs, fumaroles) that discharge volatile elements like sulfur, boron and As (SERGEOMIN 1999).

8.4.2   *Mining related sources*

The main sources of anthropogenic contamination are related to mining around San José, Huanuni, Poopó, Avicaya, Itos and Llallagua, and mostly originate from leaching of mine tailings that generates acid drainage waters, and by suspended dust (SERGEOMIN 1999, PPO 1996a). A significant part of the environmental problems can be attributed to the formation of acidic and metal-rich acid effluent at active and inactive mines, which contaminates streams, lakes and groundwater, killing the flora and aquatic fauna along great sections of the rivers and degrading the quality of the water (PPO 1996a).

Figure 8.2.   (a) Typical village well supplying drinking water in the Bolivian highlands (Altiplano); (b) Poopó lake (near Llapallapani) with deposits of salt.

8.4.3 *Industrial sources*

The city of El Alto, in La Paz province (Fig. 8.1), is home to the Calbol, Hormet and Bustos foundries. The Alto Lima II suburb of El Alto has a population of 18,000 people, of whom 43% are less than the 14 years old, and is located next to the metallurgical area of the city. For the past 30 years, Calbol has specialized in manufacturing zinc laminates, producing 400 tonnes per month, using chromium, zinc and lead as raw materials. Although process chemicals are recycled, a part is discharged to a natural channel. The smelting furnace has a 10 m high chimney without any equipment to control the emissions. Since 1960, Hormet has concentrated on refining lead and tin to produce weld from a composite of car batteries and scrap lead and zinc. The process does not generate liquid waste, but the solid wastes are deposited in the vicinity of the industry, which has no equipment to control gas emissions.

The Bustos foundry has operated for 48 years, using wolfram and tin concentrates as raw materials. The raw wolfram concentrate is estimated to contain 4% arsenic. The concentrated wolfram is fused and refined to produce material of high quality. Arsenic is released to the environment through the toasting process, which produces 180 tonnes of fused material a year, with antimony trioxide as a useful byproduct. The Vinto metallurgical plant, operated by the National Company of Foundries since 1971, 7 km from Oruro city (Fig. 8.1), produces tin, antimony, lead and other metals through a smelting process that includes volatilization and electrical furnaces. Vinto has an annual production capacity of 20,000 tonnes of high grade tin, 4300 tonnes of metallic antimony and 1000 tonnes of antimony trioxide. It also produces an unknown amount of As and Pb. A population of around 2500 live in the vicinity of the Vinto plant.

8.5 STATUS OF ARSENIC CONTAMINATION

8.5.1 *Arsenic in soils*

Arsenic concentrations in soils in the Vinto-Oruro region are related to distance from smelters, and range from 39 to 793 mg/kg (Table 8.2). The As concentrations are almost 100 times more than background levels in the basin. The sampling point to the northeast of the plant had the lowest content of As, only 8.5 times greater than the background level, because there is little anthropogenic As influence at this point.

The highest As concentrations in residential dust were recorded in Vinto-1 and Vinto-2, almost double the levels found in the northeast and approximately four times higher than the background (Table 8.3).

8.5.2 *Arsenic concentrations in surface water*

One of the most significant influxes of As to Poopó lake comes from the Desaguadero river. The main distributary channel has As concentrations of between 588 and 1180 μg/l (average 740 μg/l). The As load of the Desaguadero river is estimated to be approximately 0.9–1.8×10^5 kg/year (PPO 1996a). Mining effluents from the San Juan de Sora Sora, Poopó and Pazña river catchments

Table 8.2. Arsenic (mg/kg) in surface soil in Vinto-Oruro region.

Site	n	Average	St. dv.	Range
Vinto-1	8	91.5	25.5	39–132
Vinto-2	3	198.2	121.6	52–350
Vinto-3	4	356.9	260.5	105–793
NE	1	71.1	–	–
Background	2	8.5	–	8.4–8.6

Table 8.3.   Arsenic (mg/kg) in residential dust of Vinto-Oruro region.

Site	n	Average	St. dv.	Range
Vinto-1	6	290.0	153.2	142–613
Vinto-2	4	296.8	57.5	224–385
NE	1	150.6	–	–
Background	2	71.5	–	12–131

Table 8.4.   Seasonal variations of arsenic concentrations in surface waters (for locations see Fig. 8.1).

Sampling sites	Dry period (μg/l)	Rainy period (μg/l)	Intermediate period (μg/l)
Sora-Sora	163	14	–
Poopó	11140	834–725	4666
Pazña	60	212	203
Huancané	<10	20	<10
Juchusuma	5500	<10	<10
Huari	807	<10	<10
Desaguadero	<190	<190	993
Toledo	–	–	2888
Caracilla	–	–	700
Español bridge	–	5455	–
Cortadera	1220	1783	–
Lacajahuira	1850	2456	4040
Poopó lake	<10	31.8–34.7	<10

behave differently to other rivers in the Huanuni sub-basin. Arsenic concentrations were generally less than 170 μg/l, although there are considerable seasonal variations. The dry season (July 2001) As concentration of 118 μg/l decreased to only 14 μg/l in the rainy season, and increased again to 163 μg/l in the dry period of August 2003. The seasonal variability of As concentrations is summarized in Table 8.4.

The As concentration in the Poopó river was 2074 μg/l in the dry period of July 2001, but declined in subsequent wet periods to 834 and 725 μg/l in March and December 2002 respectively, before rising again to 11,140 μg/l in the dry period of August 2003. In the Pazña river, As concentrations in July 2001 were below the detection limit (10 μg/l), then rose to 203 μg/l in October 2001, reduced to 60 μg/l in March 2002, and rose again to 212 μg/l in December 2002.

Rivers draining agricultural areas (Huancané, Juchusuma, Tacagua, Huari and Caquiza) show different behavior. In the Huancané river, As concentrations were below the detection limit, except at the beginning of the wet season (December 2002) when they reached 20 μg/l. However, the Juchusuma river had a concentration of 30 μg/l in the wet period of March 2002, and 5500 μg/l in the following dry season. Similar behavior was observed in the Tacagua and Caquiza rivers, with low concentrations in the wet season and higher concentrations in the dry season. In the Huari river, concentrations were below detection limits in the wet season, but rose to 807 μg/l in the dry period of August 2003.

At Chuquiña, in the Desaguadero river system, the As concentrations were 993 μg/l in intermediate times, but later reduced to around 190 μg/l in the wet season. At Toledo bridge, the concentration at intermediate times was 2888 μg/l , and at Caracilla was 700 μg/l . The highest concentration at Chuquiña (5455 μg/l) was observed in the wet period of December 2002.

South of Poopó lake, the Cortadera river had higher As concentrations during the intermediate periods, October 2001 and August 2003 with values of 1783 and 1220 µg/l, respectively. However during the dry periods of March and December 2002, the concentrations decreased significantly to values of 273 and 60 µg/l respectively. Along the shores of Poopó lake, the highest influent concentration was observed in the intermediate period (4040 µg/l), after which fell to 2456 µg/l in rainy season (December 2002), and 1850 µg/l during the following dry period (August 2003). As concentrations in Poopó lake ranged from 7.9 to 10.1 µg/l during intermediate periods, but increased to between 31.8 and 34.7 µg/l during the rainy period of December 2002.

8.5.3 *Groundwater*

The average As concentration in groundwater is 47 µg/l (n = 22), and ranges from below detection limits to 200 µg/l in Kondo K, and 245 µg/l in Santuario de Quillacas. In the central region and in Pampa Aullagas, As average concentrations are 152 µg/l and 187 µg/l respectively, where 13 of 23 wells exceeded the WHO guideline (10 µg/l).

8.5.4 *Speciation in aquatic environment*

The geochemical model CHIMERE (Coudrain-Ribstein and Jamet 1988) was used to determine the speciation of As and the equilibrium concentrations of the As species. Speciation modeling indicates that five As species dominate: AsO_4^{3-} (32%), $H_2AsO_4^-$ (22%), $HAsO_4^-$ (16.5%), As_2O_5 (17%), $H_3AsO_4^-$ (6%) and others (6.5%) (Martinez 1996).

8.6 SUMMARY AND CONCLUSIONS

Since colonial times, small-scale mining activities in many parts of the Andean highlands, especially the Uru Uru and Poopó basins, have generated wastes due to processing of ores that contain high concentrations of As and heavy metals. Mining of silver and gold, associated with sulfides of As and heavy metals such as Fe, Cu, Zn, Pb and Cd have contaminated the adjoining environment. At most abandoned and operational mines and ore processing plants, drainage of acid mine water continues to contaminate rivers and sediments. In the UUPB project area, As contamination originates from natural sources (leaching of Miocene volcanic rocks) in the west, and from mining activities (leaching of wastes and residual fluids from ore processing) in the east

The ecosystems of lakes Poopó and Uru Uru are subject to strong seasonal climatic variability that induces instability in the biological communities. Without exception, rivers that drain mining areas and flow into Poopó lake are enriched in metaliferous contaminants including As. Some effluents are acidic (pH 3) and contain As and heavy metals with concentrations that are 10 to 100 times above the regulatory standards of Bolivia. In general, the quality of surface waters entering lakes Poopó and Uru Uru vary as follows:

- Discharges entering the Poopó and Uru Uru lakes contain high concentrations of sodium, sulfate and chloride. However, the concentrations are lower at the outlet (Lacajahuira river), probably due to precipitation and sedimentation.
- The discharges contain elevated concentrations of metals including in Cd (0.5 µg/l), Pb (0.3 µg/l) and As (4.6 µg/l), which decrease in the outlet of the lake (Lacajahuira river), which are also probably due to precipitation and sedimentation.
- Only the Huaya Pajchi river, on the Huari slope, and the Huancané river (when they have water) do not contain major elements or heavy metals that exceed permissible limits for any use.

Therefore, only water from the Huancané and Huaya Pajchi rivers can be used for irrigation without restriction, whereas the waters of Poopó and Uru Uru lakes are not suitable for any use without treatment. However, water from other rivers can be used for irrigation with restrictions depending on the crop type and sowing time.

Lake Uru Uru is contaminated with As, but not Fe, even though the Fe concentration in lake water is greater than in rivers. Poopó lake, including its outflow, is contaminated by As. The lake sediments are contaminated by Pb, Cd and Zn mainly in the northeast, where rivers drain mining regions. Zn is also present in both the river and lake water. Arsenic in groundwater of the Poopó basin is attributed to oxidation of sulfide minerals.

In the Desaguadero river, the total As load in the water and sediments is greater than the loads of Zn, Cd and Pb. Local natural mineralization is the main source of Zn, Cd and Pb in the groundwater, but the presence of these metals in surface water sources is dominantly related to mining activity. High As concentrations were found in four wells in the south of the basin, with a steep concentration gradient towards the south. Nevertheless, although contamination from mining activity is found in near-surface waters, deeper aquifers offer potentially safe sources of water.

ACKNOWLEDGEMENTS

The authors gratefully acknowledge the research funding from SIDA-SAREC (Contribution number 7500707606; dnr 2007-001118) for the study.

REFERENCES

ALT: Macrozonificación ambiental del sistema TDPS. Autoridad Binacional Autónoma del Sistema Hídrico del Lago Titicaca, río Desaguadero, lago Poopó y Salar de Coipasa, ALT-OIEA, La Paz-/Puno-Perú, 1999.

Apaza, R., Franjen, M., Osorio, F., Pinto, J. and Marin, R.: Estudio de la contaminación del lago Poopó en relación a metales pesados en la cadena trófica, incluido el hombre. Instituto de Ecología-UMSA-FONAMA, Informe Final TDPS:OEA (Programa Naciones Unidas para el Medio Ambiente), La Paz, Bolivia, 1996.

Beveridge, M.C., Stanford, E. and Coutts, R.: Metal concentrations in the commercially exploited fishes of an endorheic saline lake in the tin-silver province of Bolivia. *Aquacult. Fish. Manage.* 16:1 (1985), pp. 41–53.

Bhattacharya, P., Classon, M., Bundschuh, J., Sracek, O., Fagerberg, J., Jacks, G., Martin, R.A., Storniolo, A. del S. and Thir, J.M.: Distribution and mobility of arsenic in the Río Dulce alluvial aquifers in Santiago del Estero Province, Argentina. *Sci. Total Environ.* 358 (2006), pp. 97–120.

Biosca, A.: *Atlas geográfico de Bolivia y Universal.* Grupo Océano Editorial, Barcelona, Spain, 1998.

Boulange, B., Rodrigo, L. and Vargas, C.: Morphologie, formation et aspectos sedimentologiques du lac Poopó (Bolivie). *Cah. OR-STOM, Ser. Geol.* X:1 (1978), pp. 69–78.

Bundschuh, J., Farías, B., Martin, R., Storniolo, A., Bhattacharya, P., Cortes, J., Bonorino, G. and Albouy, R.: Groundwater arsenic in the Chaco-Pampean Plain, Argentina: case study from Robles county, Santiago del Estero Province. *Appl. Geochem.* 19 (2004), pp. 231–243.

Cebrian, M.E., Albores, A., Aguilar, M. and Blakely, E.: Chronic arsenic poisoning in the north of Mexico. *Hum. Toxicol.* 2:1 (1983), pp. 121–133.

Coudrain-Ribstein, A. and Jamet, P.: Le modele geochimique CHIMERE. Principles et notice d'emploi. Raport CIG/EMP, LHM/RD/88/35, 1988.

Martinez, J.: *Utilización de modelos químicos aplicables a sistemas hídricos de la cuenca endorreica dela Altiplano Boliviano.* Lic. Thesis, Instituto de Investigaciones Químicas (UMSA), La Paz, Bolivia, 1996.

Matschullat, J., Perobelli, B., Deschamps, E., Ribeiro Figueiredo, B., Gabrio, T. and Schwenk, M.: Human and environmental contamination in the Iron Quadrangle, Brazil. *Appl. Geochem.* 15:2 (2000), pp. 181–190.

Montes de Oca, I.: *Geografía y Recursos Naturales de Bolivia.* EDOBOL, 3rd ed., La Paz, Bolivia, 1997.

PPO: Impacto de la minera y el procesamiento de minaerales en cursos de aguas y lagos. Proyecto Piloto Oruro, R-BO-E-9.45-9703-PPO 9612, La Paz, Bolivia, 1996a.

PPO: Impacto de la contaminación minera en industrial sobre aguas subterráneas. Proyecto Piloto Oruro, R-BO-E-9.45-9702-PPO 9616, La Paz, Bolivia, 1996b.

Quintanilla, J.: Estrategia de estudio del sistema fluviolacustre del Altiplano. *Ecología en Bolivia* 7 (1985), pp. 65–74.

Quintanilla, J.: Evaluación Hidroquímica de la cuenca de los lagos Uru Uru y Poopó. IIQ-UMSA, Seminario taller regional sobre el lago Poopó, La Paz, Bolivia, 1994.

Sancha, A.M.: Removing arsenic from drinking water: A brief review of some lessons learned and gaps recognized in Chilean water utilities. In: W.R. Chappell, C.O. Abernathy, R.L. Calderon and D.J. Thomas (eds): *Arsenic exposure and health effects* V. Elsevier Science, Amsterdam, The Netherlands, 2003, pp. 471–481.

SERGEOMIN: Inventariación de recursos naturales renovables (hídricos) y no renovables (minerales e hidrocarburos) del departamento de Oruro. *Boletín del Servicio Nacional de Geología y Minería* 24 (1999).

UNEP/OEA: Diagnostico ambiental del sistema hídrico del lago Titicaca, río Desaguadero, lago Poopó y Salar de Coipasa (TDPS) Bolivia-Perú. Departamento de Desarrollo Regional y Medio Ambiente. Secretaria General de la OEA, Washington, DC, 1996.

USGS and GEOBOL: Geological and mineral resources of the Altiplano and Cordillera Occidental, Bolivia. US Geological Survey and Servicio Geológico de Bolivia, USGS, Boston, MA, 1992.

YPFB, GEOBOL, 1996. Mapa Geológico de Bolivia, p. 27.

Zabaleta, V.L.: Análisis situacional de la pesca en el lago Poopó y su incidencia de los cambios ambientales en las comunidades influenciadas. Tesis para Ing. Agrónomo. Universidad Técnica de Oruro, Facultad de Ciencias Agrícolas y Pecuarias, Oruro, Bolivia, 1994.

CHAPTER 9

Using GIS to define arsenic-anomalous catchment basins considering drainage sinuosity

A.B. Silva
State University of Feira de Santana, Bahia, Brazil

ABSTRACT: Surface drainage systems with lower gradients and sinuous paths are more likely to yield higher arsenic (As) levels as these factors restrict dispersion. In the study area, 431 As-bearing basins situated in an Archaean volcanic-sedimentary sequence, the Itapicuru green-stone belt (Bahia state, northeastern Brazil), were sampled; 121 of these basins were modeled in a geographic information system (GIS) with respect to total lengths of drainage and distances from uppermost to lowermost points. The drainage sinuosity indexes (DSI) were calculated dividing total stream length by the distance between uppermost and lowermost points. The results were reclassified into four ranges for sinuosity (very low, low, medium and high). Empirical values of 1, 1.25, 1.5, and 2 were assigned to each of those ranges. Arsenic values for each basin were weighted by classes of sinuosity and reclassified, resulting in a modified As anomaly map. The basins were automatically extracted from a digital elevation model (DEM). Some basins otherwise not considered first priority were thus emphasized, while others which were originally considered anomalous were de-emphasized. Thus, the resulting As-anomalous map reflects more accurately the factors affecting the dispersion of arsenic and filters out the confounding effects of physical dispersion and accumulation of arsenic in stream sediments which were sampled as part of an exploration program. Basins selected using sinuosity-weighted As anomalies correlate well with those selected by pathfinder associations, and the two approaches are considered complementary. The As-anomalous map showing the anomalous catchment basins and the arsenic sinuosity-averaged anomalous map were correlated using Kappa index of agreement and the prevalence-adjusted bias-adjusted Kappa index.

9.1 INTRODUCTION

The world is infinitely complex and the contents of a spatial database represent a particular view of the world. The measurements and samples contained in a database must represent as complete and accurate view of the world as possible. At the most fundamental level, geographical data can be defined as a collection of facts about sites. The basic unit of spatial data is the tuple <x, y, z>, where x and y define a site, and z contains descriptions about the site, in this case the analytical results of a stream sediment survey. A convenient way to define the difference between spatial and geographical data is to insist that for geographical data, the coordinates (x, y) should be defined by some system of measurement on the Earth, latitude/longitude, for example, or in the UTM (Universal Transverse Mercator) coordinate system. Geographic Information Systems (GIS) are an indispensable technology in the evaluation, control, and management of natural resources. The generation of a geospatial database is an initial and decisive task in a GIS project. The geospatial data stored must have an adequate level of accuracy to ensure that the results obtained after modeling portrays the terrestrial reality as closely as possible (Burrough and McDonnell 1998). The purpose of this work is to define As-anomalous catchment basins considering drainage sinuosity in the Rio Itapicuru greenstone belt (RIGB) situated in the northeastern part of the São Francisco craton located in Bahia state, Brazil (Fig. 9.1).

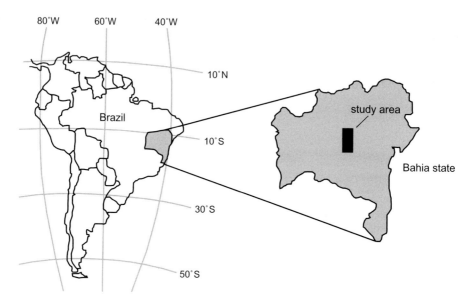

Figure 9.1.   Location of the study area.

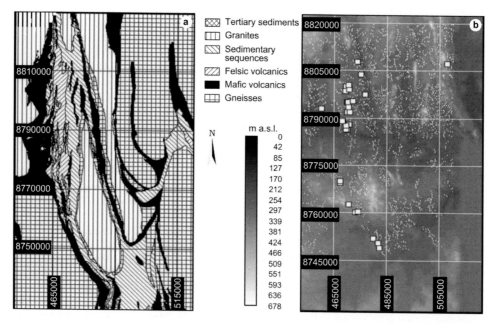

Figure 9.2.   (a) Simplified geological map of the Río Itapicuru greenstone belt (RIGB); (b) Digital elevation model and sample point position (white circles); outliers are shown as squares (coordinates are in m UTM).

The geological framework of the study area corresponds to supracrustal units exposed within an area more than 100 km long and about 60 km wide (Fig. 9.2a). The lithological succession of the low grade supracrustal pile consists, from bottom to top, of a tholeiitic volcanic unit, a calc-alkaline felsic volcanic unit and a sedimentary unit (Kishida and Riccio 1980). Geochemical and petrological data suggest that the RIGB corresponds to a back-arc basin as the geotectonic environment for deposition of the supracrustals (Silva 1987). The supracrustal units are underlain by gneisses and migmatites considered as representative of the Archaean basement. Sequences are also intruded by several granitoids dated around 2.0 to 2.2 Ga and attributed to the Transamazonian cycle (Gaál *et al.* 1987). According to Alves da Silva *et al.* (1995), the three main sequences are described as: (1) the basal mafic sequence comprising massive to schistose tholeiitic basalts sometimes showing pillow structure; (2) the intermediate felsic sequence represented by andesitic lavas with intercalations of pyroclastic lenses and metasediments; (3) the upper sedimentary sequence which is a complex succession of quartz-chlorite-sericite schist, graywacke, conglomerate, quartz-carbonaceous-schist and tuff. Arsenic is found in arsenopyrite hosted in both intermediate and felsic volcanic rocks.

9.2   THE DATA

The use of a catchment basin as the area of influence of a stream sediment source is widespread in the literature and innovative approaches based on geographic information systems (GIS) are giving it new life, especially in areas where no evident control data are available. In stream sediment surveys, the dispersion of elements is controlled by type and locations of sources, Eh-pH conditions, colloidal phenomena, biological and hydrolytic reactions and diffusion, as well as stream flow. Furthermore, some physical characteristics play an important role, such as the different resistances of minerals to alteration and weathering processes. Dispersion processes can reduce the concentration of elements in element associations because of a physicochemical environment and with increasing distance from their source. In drainage reconnaissance surveys the sampling technique is of primary importance. Too many types of rivers and too many lithological units are generally sampled in only one data set. Sampling drainages of different orders can cause severe interference in the concentration of the element because of different controls of dispersion processes and dilution effects. The stream-sediment samples of the study area were collected from small channels draining catchment areas typically in the range of 3 to 8 km^2. Drainages with lower gradients and more sinuous paths are more likely to yield higher As levels as these factors restrict dispersion. Arsenic analysis of the <200 mesh sieve fraction of 2032 stream sediments was carried out by Companhia Vale do Rio Doce. Samples of 1 kg were homogenized, quartered, and 50 g aliquots were analyzed by X-ray fluorescence spectrometry. Topographic sheets mapped by the Brazilian Institute of Geography and Statistics (IBGE) at 1:100.000 scale were converted to a single seamless datum using GIS to provide a base map formatted to DXF files. Using a normalized residual index technique (Berry 1997) the more adequate mathematical algorithm to interpolate the data was chosen; in this case the minimum curvature was used to develop a digital elevation model (DEM). From the DEM (Fig. 9.2b) using GIS procedures, a map of catchment basins was delimitated and each basin was associated with its corresponding samples.

  The reliability of information derived from spatial data is described in order to establish criteria for auditing a spatial database. The database showed that 25 samples had the same coordinates but showed different results. These points were removed from the dataset. Outliers are observations with a unique combination of characteristics identifiable as distinctly different from the other observations. They can be classified into four classes: errors in the procedure, results of an extraordinary event, unexpected observations for which the researcher has no explanation, and observations that fall within the ordinary range of values on each of the variables but are unique in their combination of values across the variables. In all these cases the outliers must be removed from the data set. Outliers were detected using Z-test, the data values were converted to standard scores, which have a mean of 0 and a standard deviation of 1, standard scores higher than 3.5 were considered outliers. As a result of the Z-test, samples with values of 56 mg/kg or greater were removed from the dataset. The Kolmogorov-Smirnov (KS) test is used to decide if a sample comes from a population with a specific distribution. The KS test was carried out showing that the distribution is not normal, but the data behave as a lognormal distribution. Analytical results were associated to sample site locations with maximum errors of 50 m in plan view and 30 m in elevation.

9.3   ARSENIC-ANOMALOUS CATCHMENT BASINS

In the <200 mesh sieve fraction, As values vary from 5 to 304 mg/kg. Using Jenk's optimization, 4 classes of As values are established based on natural breaks (ESRI 1996). This method identifies breakpoints between classes using a statistical formula that minimizes the sum of variance within each of the classes. Table 9.1 shows that As anomalies occur in 167 of the 431 basins sampled and Figure 9.3a presents the As anomalous map showing the anomalous catchment basins. Although the objective of this chapter is not to correlate the geology with the anomalous sites, using GIS it is possible to identify the intersection of catchment basins and areas of dominant lithology.

This type of spatial query involving different layers of information is one of the most important uses of GIS. To model As geochemical anomalies is not an easy task, because the As content will vary as a function of several factors, such as the inherent variability of rock type, the presence of arsenopyrite and scavenging of metal ions by oxides. Physical factors, on the other hand, such as stream gradients and sinuosity of drainages, act to modify dispersion patterns. Drainages with lower

Table 9.1.   Number of catchment basins with samples situated in each class interval, for the four classes of As distribution.

As distribution class interval (mg/kg)	Number of samples in each interval
5–23	1865
24–38	122
39–52	20
>52	25
Total samples	2032
% basins with As >23 mg/kg	9

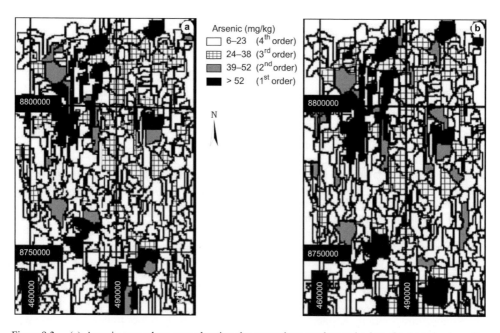

Figure 9.3.   (a) Arsenic anomalous map showing the anomalous catchment basins; (b) Arsenic sinuosity-averaged anomalous map showing the anomalous catchment basins based on DSI (coordinates are in m UTM).

gradients and more sinuous paths are more likely to yield higher As levels, because they restrict down-gradient dispersion. It is therefore desirable to minimize such effects, through modeling, in order to decrease the influence of physical factors that might confuse basin rankings.

To carry this out, basins were modeled with a GIS with respect to total length of the main drainage and its tributaries in the same catchment basin. The drainage segments were extracted from the drainage information layer and associated to these basins in a new information plan. Next, drainage segment lengths in this new layer were summarized by basin, and the total length for each basin annotated. To create an index, coordinates of the most upstream and downstream points of each main drainage segment and tributaries in each basin were determined and the distance between them calculated and summarized by basins. The drainage sinuosity index (DSI) was calculated by dividing total length of drainages within a basin by the total distance of the starting and ending points of the main drainage and its tributaries. Within a given basin, the longest possible combination of consecutive drainage segments is chosen as the main drainage and the other segments are considered as its tributaries. This selection significantly affects the end result, as the distance between uppermost points varies accordingly. Results were classified into four ranges of sinuosity: very low (1.0), low (1.25), medium (1.5), and high (2). Each range was assigned a weighting factor by which As values for each basin were multiplied to generate an As anomalous map showing the anomalous catchment basins (Fig. 9.3b).

9.4 SINUOSITY-AVERAGED ARSENIC ANOMALIES COMPARED TO THE ORIGINAL ARSENIC ANOMALOUS MAP

A contingency matrix was created to compare the two As anomalous maps: the sinuosity-averaged As anomalies and the original As anomalous map. A contingency matrix is a grid on which several related concepts are listed along the x- (horizontal) and y- (vertical) axes. Contingency maps can give you a picture of how to relate different concepts connected with a complex issue. Writing a "yes" when sinuosity-averaged As basin anomaly reinforces or increases original As anomaly basin and writing "no" if you think that more of sinuosity-averaged As basin anomaly reduces original As anomaly basin. Kappa index of agreement (KIA) is a commonly used measure of inter-observer agreement between two observers for dichotomous data. However, because KIA is affected in complex ways by the presence of bias between observers and by the distribution of data across the categories, we also computed the prevalence-adjusted bias-adjusted Kappa (PABAK), the bias index (BI), and the prevalence index (PI), as recommended by Byrt *et al.* (1993). The bias index (BI) is defined as the difference between the proportions of "yes" for the two raters. The prevalence index (PI) is defined as the difference between the probability of "yes" and the probability of "no". A BI close to 0 indicates less bias, while values closer to 1 (absolute value)

Table 9.2.   Measures from a contingency matrix: Sinuosity-averaged As anomalies and the original As anomalous map.

	Agreement among validation		
	1st order	2nd order	3rd order
Po	0.94	0.83	0.92
Pe	0.64	0.37	0.32
KIA	0.83	0.62	0.85
PI	0.64	0.38	0.31
BI	0.05	0.14	0.08
PABAK	0.91	0.71	0.76

Po: observed agreement; Pe: expected agreement; BI: bias index; PI: prevalence index; KIA: Kappa index agreement; PABAK: prevalence-adjusted bias-adjusted Kappa.

indicate greater bias. Similarly, a PI close to 1 (absolute value) indicates high prevalence, while a PI closer to 0 indicates lower prevalence. The BI then measures the degree to which one reviewer tends to identify more or fewer occurrences than the other, while the PI measures the degree to which "yes" agreements or "no" agreements predominate. The PABAK index of agreement between two observers is a measure that adjusts for both bias and prevalence. Although the derivation of the PABAK index is somewhat more complex, in practice it can be calculated as 2Po-1, where Po is the proportion of observed agreement. Consequently, PABAK ranges from −1 to +1 and like KIA, a value of 0 represents no better than a random agreement, while values approaching 1 indicate maximal agreement. Table 9.2 shows the comparison between the sinuosity-averaged As anomalies and the original As anomalous map.

9.5   CONCLUSIONS

It is highly recommended to audit the data to eliminate outliers to decide if a sample comes from a population with a specific distribution and how to filter out the confounding effects of physical dispersion and accumulation of As in stream sediments sampled as part of a geochemical survey. The sinuosity indexes calculated were classified into four ranges of sinuosity and revealed the similarities and the differences between the sinuosity-averaged As anomalies and the original As anomalous map. The resulting modified anomalies map represents more closely the original chemical distribution of As. Trial and error adjustment of the assigned weights were made to assure that conspicuously anomalous samples did not get de-emphasized, as it is a good practice to always follow those up. Likewise, weak anomalies should not be over-emphasized. The real usefulness of the method lies in ranking anomalies of intermediate value. It was possible to correlate the results obtained from the original anomalous map and the sinuosity-averaged anomalous map by Kappa index agreement (KIA). Computing the prevalence-adjusted bias-adjusted Kappa (PABAK), to establish the differences into marginal distributions, shows that there are no significant differences between KIA and PABAK. Using DSI, one 1st order catchment basin was re-defined to 2nd order, eight 3rd order were re-defined to 2nd order, five 2nd order were re-defined to 1st order and nine 4th order were re-defined to 3rd order, as a result a better interpretation of the As-anomalous catchment basins can be achieved.

REFERENCES

Alves da Silva, F.C., Guerrot, C., Chauvet, A. and Faure, M.: Chronological and structural evidences for both archaic and modern-type tectonic styles within the Palaeoproterozoic Rio Itapicuru Greenstone Belt, Brazil. *Proceedings 8th EUG*, Strasbourg, France, Blackwell Scientific Publications, 1995, p. 103.

Berry, J.K.: Justifiable interpolation. *GIS World*, Feb (1997), p. 34.

Burrough, P.A. and McDonnell, R.A.: *Principles of Geographical Information Systems*. Oxford Univ. Press, Oxford, UK, 1998.

Byrt, T., Bishop, J. and Carlin, J.B.: Bias, prevalence and Kappa. *J. Clin. Epidemiol.* 46:5 (1993), pp. 423–429.

ESRI: ArcView, GIS. Redlans, California, 1996.

Gaál, G., Teixeira, J.B.G., Silva, M.G. and Del Rey, J.M.H.: New U-Pb data from granitoids, reflecting Early-Proterozoic evolution in northeast Bahia-Brazil. *Proceedings Symposium on Granites and Associated Mineralizations*, Salvador, Bahia, Brazil, 1987.

Kishida, A. and Riccio, L.: Chemostratigraphy of lava sequences from the Rio Itapicuru greenstone belt, Bahia, Brazil. *Precambrian Res.* 11 (1980), pp. 161–178.

Silva, M.G.: *Geochemie, Petrologie und tektonische Entwicklung eines Proterozoischen Grünsteingürtels: Rio Itapicuru, Bahia, Brazil*. PhD Thesis, Freiburg University, Freiburg, Germany, 1987.

Central America and Mexico

CHAPTER 10

Natural arsenic groundwater contamination of the sedimentary aquifers of southwestern Sébaco valley, Nicaragua

M. Altamirano Espinoza
Centro para la Investigación en Recursos Acuáticos de Nicaragua (CIRA/UNAN), Universidad Nacional Autónoma de Nicaragua, Managua, Nicaragua

J. Bundschuh
International Technical Cooperation Program, CIM (GTZ/BA), Frankfurt, Germany
Instituto Costarricense de Electricidad (ICE), San José, Costa Rica

ABSTRACT: In Nicaragua, arsenic (As) contamination of drinking water resources due to geogenic sources can be identified in the southwestern part of Sébaco valley. The As sources are weathering products of Tertiary volcanic rocks of the Coyol group that form the alluvial aquifer of Sébaco valley. This valley is located at the external eastern margin of the Nicaraguan depression (or Central American graben). The region is characterized by intensive tectonic stress and the presence of different active and non-active fault systems, intensive fracturing and hydrothermal alteration along faults. The volcanic rocks outcrop at the margins and at several locations within the valley overlying the alluvial aquifer. This aquifer is used by several communities for drinking water supply. In 21 of 57 water samples collected from wells of this alluvial aquifer, As concentrations ranged from 10 to 122 µg/l, exceeding the national and the WHO limit of 10 µg/l for As in drinking water. Several hot spots with high groundwater As were identified, e.g., El Zapote village with 122 µg/l. In comparison, the localities of Las Mangas and Tatazcame, were identified as areas with safe groundwater, most suitable for future drinking water supply projects. The high As concentration in El Zapote water is correlated with elevated As concentrations in rocks and soils (14.98 and 57.19 µg/kg, respectively). Arsenic sources are the volcanic hard rocks, the alluvial sediments composing the alluvial aquifer and the unsaturated zone, and the hydrothermally altered volcanic rocks. Chemical mineral weathering and changing redox conditions increase mobility of As and its transfer from solid to dissolved form. Similar Tertiary volcanic rocks cover wide parts of Nicaragua and other Central American countries. It can be expected that many other not yet identified sites with groundwater As problems exist in the region.

10.1 INTRODUCTION

An important environmental concern of Nicaragua is the high (toxic) levels of arsenic (As) concentrations in several groundwater aquifers, which are used often as the only available drinking water source for the rural population. Several affected areas are located in western Nicaragua, where the presence of Tertiary volcanic rocks and their weathering products are the As source. The weathering products form the alluvial aquifers. The mobilization of As from the rocks is favored by the presence of active and non-active tectonic faults related to the Nicaraguan depression. These faults often show hydrothermal alterations suggesting they act as permeable channels for fluid and heat (Fig. 10.1).

Natural occurrences of As in groundwater was documented in Nicaragua for the first time in 1996 (ECO/OPS 1997). The source of As was identified as geogenic due to weathering and dissolution of volcanic rocks and products. The highest recorded concentration of As (about 1320 µg/l) was found in a tube well in El Zapote village (Sébaco valley) (Figs. 10.1–10.3) (Gomez 2009,

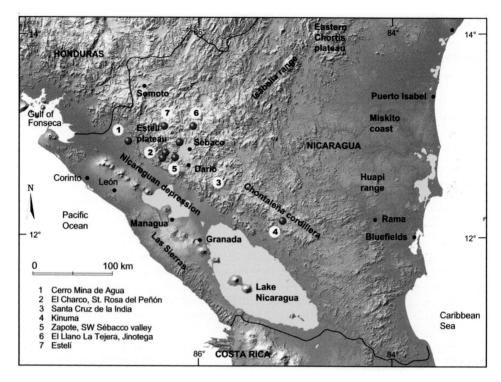

Figure 10.1.   Main areas of Nicaragua with high concentrations of arsenic in groundwater. The study area is located at the southwest Sébaco valley. According to a 1996 study (ECO/OPS 1997), the highest concentration of groundwater arsenic (1320 µg/l) was reported at Zapote village.

Barragne-Bigot 2004). Here, the first symptoms of As-related diseases were detected in 1996 (Gomez 2009) (Fig. 10.3). In order to solve the water supply problem in this area, in March, 2004, UNICEF drilled a well in the community of Las Mangas a neighboring town of El Zapote.

In 2003, UNICEF studied the As concentrations of groundwater from 77 wells in 5 areas (Barragne-Bigot 2004). This study identified 4 areas (Fig. 10.2) with groundwater As concentrations exceeding the limit of safe drinking water (10 µg/l; WHO 2001) as well as the local drinking water standard of 10 µg/l. These areas were: (1) Cerro Mina de Agua (municipality Villanueva), (2) communities El Charco and Santa Rosa del Peñón (municipality Santa Rosa del Peñón), (3) community Cruz de la India (municipality Santa Rosa del Peñón), and (4) community Kinuma (municipality La Libertad) (Fig. 10.1). The waters have a roughly neutral pH (6 to 8) and As is predominantly present in the form of As(V). Total As concentrations range from 10 to 107 µg/l. In these areas a total of 1270 people are exposed to toxic levels of As in drinking water (Barragne-Bigot 2004).

In 2005, UNICEF sampled 54 wells in the community of Llano La Tejera (municipality Jinotega), which has 714 inhabitants and 88% of the houses have their own wells. Arsenic concentrations vary from the detection limit up to 1200 µg As/l (average 100.4 µg/l) making 87% of the wells unsuitable for drinking (Nicaraguan limit 10 µg As/l) (Bundschuh *et al.* 2006, 2009a, Bundschuh and García 2008, Larios *unpubl. data*).

However detailed hydrogeochemical studies of the aquifers in volcanic rocks and derived sediments, and the mobilization mechanisms are still missing. No studies on chemistry of groundwater were performed to understand the mechanisms of As mobilization in these aquifers.

The aim of this study was to investigate the distribution of As in the groundwater of shallow alluvial aquifers of the southwestern basin of Sébaco valley (Figs. 10.1 and 10.2). This study focuses

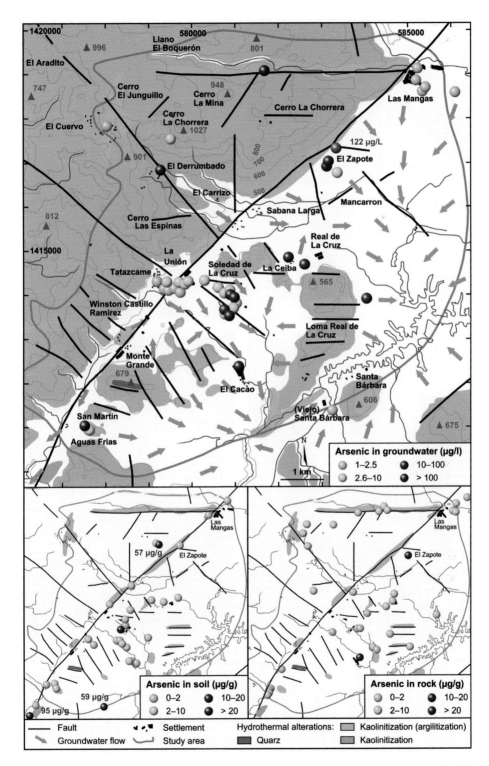

Figure 10.2.   Arsenic concentrations in groundwater, soils and rocks and tectonic setting with related hydrothermal alteration along the faults in the southwestern basin of Sébaco valley, Nicaragua.

Figure 10.3.   Southwestern Sébaco valley, Nicaragua: (a) typical rural houses; (b) water well, which was used
for the water supply of El Zapote village from 1994 to 1996. This well was closed in 1996 by
the authorities after a high As concentration of 1320 μg/l was detected. Sign along the road to
El Zapote village announcing the conclusion of the project to supply several communities of the
area with As-safe drinking water. The project was executed from 2002–2003; (d–f): Arsenic
impacts on the population of El Zapote village, after consuming for 2 years (1994–96) drinking
water with 1320 μg As/l (photo taken in 1996). Woman (72 years old) with diffuse palmar
hyperkeratosis (d, e); Hyperkeratosis on the left hand of a 20 year old woman, who suffers also
from hyperkeratosis of palm and soles, keratosis in the dorsum of the hand (f), and in her feet
and knees.

on the area located SW to El Zapote village, where in 1996 high As concentrations (1230 µg/l) were found in a drinking water tube well. The As concentration in the rocks and soils of the study were also determined in order to better understand the complex hydrogeological and geochemical conditions that are responsible for the mobilization of As in groundwater.

10.2 LOCATION AND GEOLOGICAL SETTING

10.2.1 *The study area*

The study area is located in the southwestern Sébaco valley (Matagalpa department), in the Central Cordillera, at the external eastern margin of the Nicaragua depression (Fig. 10.1), and covers an area of 52 km^2. The area is rural with small agricultural settlements that depend on artificial irrigation. The southern and eastern part of the study area is a plain on the Sébaco valley with elevation about 420–470 m a.s.l. The northern and western parts are hilly regions on the Estelí plateau reaching altitudes of up 1027 m a.s.l. (Cerro La Chorrea). The plain is covered by alluvial sediments, which correspond to the weathering products of the surrounding hills and form the shallow aquifer. This aquifer is an extension of the Sebaco valley aquifer and the main drinking water resource for the 15 communities of the area (Fig. 10.2): El Zapote, Las Mangas, Tatazcame, Roberto Centeno, Sabana Larga, La Union, El Cacao, La Sabaneta, La Montañita, Aguas Frias, El Derrumbado, La Ceiba, Real de la Cruz, Cerro La Mina and San Ramon de la Uva. These villages comprise a total population of 3225 inhabitants (INEC 1995). Most of the towns lack a good water supply. At this time 9 drilled wells are not in use due to the high level of As concentrations. This situation has forced its settlers to use groundwater from dug wells with also high As concentrations.

Hot, rainy season lasting from May to October and very dry season from November to April characterize the climate. The annual mean temperature is 25.1°C (from 1983 to 2000, INETER, Meteorological Station San Isidro, which is a few kilometers outside of the study area), and the average annual precipitation is 873.1 mm in the plain and up to 2000 mm in the hills, for the same time period. Evapotranspiration is very high, especially in the rainy season. The area was originally forested but due to intensive timber harvesting only limited forest areas remain. Vegetation generally consists of low bushes and ground vegetation.

10.2.2 *Geological and hydrogeological characteristics*

The hardrock outcroppings in the area belong to the Tertiary volcanic province, which is predominantly formed by the Coyol group. The rocks of this group are as follows: Coyol Inferior is formed predominantly by dacitic ignimbrites, tobas, agglomerates, and andesites; Coyol Superior is formed predominantly by ignimbrites, basalts, and agglomerates. The rocks show an intense chemical weathering that has produced saprolite, with elevated contents of clays, reaching up to several meters of thickness. This can be observed especially along fault zones. Three fault systems can be distinguished: (1) NW–SE, (2) NE–SW and (3) N–S, which is the youngest. These fault systems have formed a morphology of echelon blocks, with a general dipping towards the Nicaraguan depression, and are also responsible for the parallelogram shape of the actual Sébaco valley. The weathering products of the rocks from the mountainous region form the alluvial sediments (gravels, sands, intercalated with silts and clays) of the plain of the study area. These permeable sediments form the alluvial aquifer that provides the water supply for the region.

10.3 MATERIALS AND METHODS

10.3.1 *Groundwater, rock and soil sampling*

A detailed compilation of geologic, chemical, and hydrogeological studies was carried out in order to study the availability and quality of groundwater in the area and to identify areas suitable for installing wells for water supply.

Groundwater, rock and soil samples were collected during September–October 2004 from the southwestern basin of Sébaco valley. Fifty-seven well sites were selected for sampling. Groundwater was mainly sampled by hand-pumping tube wells penetrating the shallow alluvial aquifers with a maximum depth of 20 m. Sampling wells are shown in Figure 10.2.

The well positions at each of the sampling sites were determined using global positioning system (GPS). The static water level in wells was determined using a light plumb line. The values of pH, redox potential (Eh), dissolved oxygen, temperature and electrical conductivity of groundwater were measured in the field. Water samples collected from each well involved: (1) filtered samples for major anion analysis; (2) filtered, acidified samples for major cation and trace elements analysis in the laboratory.

35 samples of residual soil were taken in the flat part of the study area (Fig. 10.2), from a depth of 30 cm. 25 rocks samples were collected in outcrops of the steep area to study the sediment/rock groundwater interactions.

10.3.2   *Sample preparation and analytical methods*

Water samples: Anions such as Cl^- and SO_4^{2-} were analyzed in a Dionex Dx-100 ion chromatograph with an integrator 4600, and Na^{2+} and K^+ were analyzed using flame photometer CORNING 410. Colorimetric methods were used to analyze nitrite (sulfanilamide method), fluoride (SPANS method), silica (molibdosilicate method), iron (fenantrolina method) using a Perkin Elmer UV/VIS Spectrometer Lambda 35. Calcium and the sum of $Ca^{2+} + Mg^{2+}$ were determined by titration with EDTA. Magnesium was calculated as the difference from Ca^{2+} and the sum $Ca^{2+} + Mg^{2+}$. Total As was determined using a hydride vapor generator (VGA 76) and an atomic absorption spectrophotometer Varían Spectr AA-20. The methodology is described in Rothery *et al.* (1984). The detection limit of the instrument is 2 µg/l in water. Each water sample was analyzed twice.

Rock and soil samples: The samples were ground with a mortar and pestle, and the finer fraction separated with a 30 mesh sieve. The same procedure was used for soils and rocks. 0.250 g of each powdered sample were weighted in a 100 ml volumetric flask. 5 ml of concentrated hydrochloric acid were added and mixed to wet all the sample. Then 0.5 ml of hydrogen peroxide were added each 20 minutes during one hour (three additions). The sample was heated in a water bath at 90°C during 30 minutes with constant stirring. After cooling, the leachate was analyzed for As in the same manner as the water samples. This digestion oxidizes the sediment and converts all the As to As(V). This is a methodology modified by CIRA/UNAN (Nicaraguan Research Center for Aquatic Resources of the National Autonomous University of Nicaragua) that takes into consideration the digestion of sediments described in Moffet (1988), and the instructions for the hydride vapor generator (Varian 1984).

For the water samples and leachate samples from the sediments, the hydride generator converts first the As(V) to As(III) using KI in an acidic environment. Further reduction using $NaBH_4$ generates the arsine hydride AsH_3. This gaseous hydride passes throughout the air-acetylene flame in the atomic absorption spectrophotometer and absorbs the appropriate radiation in proportion to the As concentration (Varian 1984).

10.4   RESULTS

10.4.1   *Geology and hydrogeology*

At the valley, the water table is only a few meters below the ground surface. It can be assumed that the elevation of the groundwater table follows the surface topography and construct a map of the groundwater flow field. Groundwater flow is directed towards the storage lake (Embalse La Virgen) to the SE, which is the topographically lowest elevation of the study area. Several hard rock outcrops are found in the area. These outcrops seem to separate the alluvial aquifer in different partly interconnected sub-basins. The depth to groundwater table of the alluvial aquifer ranges in the plain from 10 to 20 m. However, there is no clear limit between the base of the alluvial aquifer and the highly weathered volcanic rocks below. Hydrothermal alteration is frequently found in the hard rock outcrops along the tectonic faults. An important fraction of groundwater recharge to the aquifer occurs by surface runoff from the hills in the N, W and S of the study area. Major groundwater flow is horizontal from N to SE and W to E (Fig. 10.2). Local groundwater flow patterns based on the measured depths to groundwater table (Table 10.1) and the topographic heights of the sampling points are shown in Figure 10.2. The hydraulic gradient ranges from about 0.007 to 0.020 at the SW Sébaco valley. No information from pumping tests on hydraulic conductivity of the upper-most aquifer is available.

10.4.2 *Groundwater chemistry*

The pH ranged between 6.1 and 7.9 with an average of 7.3 in the 57 groundwater samples collected. Field measured redox potential ranged from +167 to +636 mV with an average value of +271 mV. Electric conductivity (EC) ranged between 128 and 755 μS/cm with an average value of 480 μS/cm.

The 25 samples analyzed for main ions show Ca^{2+} concentrations ranging from 10 to 115 mg/l (average 46 mg/l), and Na^+ from 8 to 96 mg/l (average 42 mg/l) as the main cations, and HCO_3^- (57–435 mg/l, average 253 mg/l) as the dominating anion in groundwaters. The concentration ranges of other main ions are: Mg^{2+}: 2.7–18.7 mg/l (average 10.1 mg/l), K^+: 0.5–18.6 mg/l (average 3.3 mg/l), SO_4^{2-}: 1.1–71.2 mg/l (average 11.9 mg/l), Cl^-: 4.2–25.7 mg/l (average 11.3 mg/l). NO_3^- ranges: 0.03–32.0 mg/l (average 13.2 mg/l). 21 samples have NO_3^- values exceeding 5 mg/l indicating anthropogenic influences. Fluoride concentrations range from 0.02 to 0.64 mg/l (average 0.36 mg/l) and SiO_2 from 31.1 to 98.2 mg/l (average 31.4 mg/l), which are typical values of aquifers composed by materials derived from volcanic rocks (e.g., Bundschuh *et al.* 2004). According to Piper classification diagrams, most water samples (18 samples; 72%) belong to the type "normal alkaline-earth freshwater with high concentrations of alkalis, and dominant hydrogen-carbonate" (area d in Figure 10.4; (Ca, Mg) HCO_3 composition). Fewer samples (5; 20%) belong to "alkaline freshwater with dominant hydrogen-carbonate" (area f in Figure 10.4; $NaHCO_3$ waters), and only 2 samples belong to "highly alkaline-earth freshwater with dominant hydrogen-carbonate" (field a; $CaHCO_3$) (Fig. 10.4). With respect to the main ions, all waters showed concentrations within permissible limits according to the water standard for human consumption (CAPRE 1994).

Total As concentration show considerable spatial variations with an average of 12.4 μg/l (maximum 122 μg/l) for the 57 samples. Forty samples have As concentrations below the national drinking water limit of 10 μg/l. However, 14 of them are in the range 7–10 μg/l and need to be monitored since As concentrations undergo seasonal fluctuations and could exceed the 10 μg/l limit. Seventeen wells exceed the national limit and fall in the 10–31 μg/l concentration range. Only one with a high value of 122 μg/l was found (Fig. 10.2).

10.4.3 *Arsenic in hard rocks and soils*

Geochemical results reveal considerable enrichment of As in the volcanic rocks of the study area (up to 15 μg/g were found at El Zapote village) and for soil samples up to 95 μg/g (Table 10.2). Most of the soil samples (29) show As concentrations below 10 μg/g, but 2 samples have much higher values of 23.4 and 57.2 μg/g, the last belonging to a sample from El Zapote. Two additional soil samples were collected outside the study area, in the entrance of the Mine La India and in Aguas Frias, presenting concentrations greater and similar to El Zapote with values of 59.5 and 95.2 μg/g, respectively.

Table 10.1.　Results of distribution of arsenic in groundwater at the southwestern quadrant of Sébaco valley, Nicaragua.

Nr	Well type	Coordinates (Lambert)	Community	Well owner	As (µg/l)	pH	EC (µS/cm)	O_2 (mg/l)	Eh (mV)	SL (m)	Water type
1	PP	577620E/1410944N	Aguas Frias	Pozo comunal	5.86	7.0	695	3.7	326.5		Ca-HCO$_3$
2	PE	577500E/1411039N	Aguas Frias	La quebrada	13.64	7.2	519	4.5	288.3	29.3	
3	PE	579642E/1414179N	Union/Tatazcame	Escuela	3.98	7.7	523	6.0	258.0	11.9	(Ca)-HCO$_3$
4	PP	579597E/1414162N	Tatazcame	Pozo Comunal 1	<ld	7.3	449	4.5	285.1		Na-HCO$_3$
5	PE	579410E/1414295N	Tatazcame	Anibal Matamoros	3.48	7.5	663	5.8	269.0	18.5	(Ca)-HCO$_3$
6	PP	579191E/1414327N	Tatazcame	Pozo comunal 2	2.33	7.4	755	4.2	281.7	28.5	(Ca)-HCO$_3$
7	PE	579405E/1414138N	Tatazcame		<ld	7.6	571	6.0	235.7	14.9	
8	PE	579718E/1414083N	Tatazcame	Maria Dolores	<ld	7.6	592	3.8	636.0	12.2	
9	PE	579355E/1414153N	Tatazcame	Veronica Macis	<ld	7.5	556	4.5	379.0	13.2	
10	PE	579459E/1414131N	Tatazcame	Bartolome Garcia M	2.22	7.6	571	5.5	336.0	13.0	Na-HCO$_3$
11	PE	579804E/1414309N	Comarca La Union	Juan Angel Espinoza	4.14	7.8	465	5.8	385.5	11.2	
12	PE	579864E/1414298N	Comarca la Union	Adelina Rojas	4.59	7.9	345	5.0	345.6	12.6	(Ca)-HCO$_3$
13	PE	579917E/1414325N	Comarca la Union	Ligia Espinoza	8.11	7.5	580	5.8	304.7		
14	PE	581152E/1412165N	El Cacao	Erminia Matamoro	6.56	7.3	628	5.0	287.2		(Ca)-HCO$_3$
15	PE	581154E/1412228N	El Cacao	Catalina Rivas Macis	7.99	7.1	541	5.8	271.4	18.1	
16	PE	581132E/1412323N	El Cacao	Socorro Matamoro	9.84	7.1	541	5.8	271.4	15.6	
17	PP	581109E/1412420N	El Cacao	Escuela	11.63	7.1	548	4.0	273.8		(Ca)-HCO$_3$
18	PE	581068E/1412393N	El Cacao	Fabio Espinoza M.	11.01	7.2	565	4.0	276.8	16.6	
19	PE	580820E/1413603N	Soledad de la Cruz	Jose Leonardo R.	13.22	7.6	556	5.3	263.2	12.1	
20	PE	580747E/1413620N	Soledad de la Cruz	Ana Julia Matamoro	10.98	7.3	504	2.0	260.5	12.1	(Ca)- HCO$_3$
21	PE	580710E/1413643N	Soledad de la Cruz	Jose Ramon Rivas	3.5	7.9	250	4.9	260.0		
22	PE	580781E/1413651N	Soledad de la Cruz	Maximino Rivas	11.1	7.4	551	5.0	249.2	15.0	(Ca)-HCO$_3$
23	PE	580764E/1413587N	Soledad de la Cruz	Juan Jose Rivas	7.43	7.4	514	4.5	204.4	11.8	
24	PE	580867E/1413671N	Soledad de la Cruz	Perfecto antonio Rivas	8.21	7.3	571	3.5	196.9	11.8	
25	PE	580879E/1413690N	Soledad de la Cruz	Juan Fco Rivas	9.56	7.3	573	4.2	209.6	12.7	
26	PE	580898E/1413821N	Soledad de la Cruz	Finca Guasimo	7.71	7.5	444	4.0	197.2	12.2	
27	PE	580938E/1413833N	Soledad de la Cruz	Valeriano Martinez R.	11.94	7.2	454	4.5	541.7	11.7	
28	PE	580978E/1413800N	Soledad de la Cruz	Pozo comunal	7.1	7.6	427	4.5	317.0	12.1	(Ca)-HCO$_3$

No.	Type	Coordinates	Location	Name	As	pH	EC	SL			Water type
29	PE	580938E/1413797N	Soledad de la Cruz	Fidelina Corea	9.7	7.4	465	3.8	282.8	11.8	
30	PE	580968E/1413887N	Soledad de la Cruz	Eddy Antonio Rivas	10.0	7.6	446	0.8	240.6	27.6	
31	PE	580927E/1413990N	Soledad de la Cruz	Pedro Pablo Rivas	17.0	7.3	553	2.4	235.5	18.6	(Ca)-HCO$_3$
32	PE	580860E/1414037N	Soledad de la Cruz	Iglesia Nazareno	19.0	7.7	511	5.8	221.2	15.7	
33	PE	580882E/1414097N	Soledad de la Cruz	Yelba Garcia	8.9	7.5	477	4.2	204.2	18.5	
34	PE	580757E/1414071N	Soledad de la Cruz	Jose T. Matamoro	9.8	7.6	511	5.3	201.1	14.4	
35	PE	580687E/1414146N	Soledad de la Cruz	Nicolas Martinez	6.4	7.4	484	2.5	208.1	11.6	
36	PE	580620E/1414183N	Soledad de la Cruz	Maria L Rivas	6.8	7.4	500	3.9	201.8	13.4	
37	PE	580284E/1414345N	Soledad de la Cruz	Indalesio Ruiz M	8.1	7.5	501	5.5	188.3	11.6	
38	PE	584004E/1413909N	Real de la Cruz	Pozo ubicado huerta	20.8	7.5	635	2.2		8.7	
39	PE	582573E/1414709N	Real de la Cruz	Javier Rivera	30.7	7.6	683	2.1		15.0	
40	PP	582239E/1414860N	Real de la Cruz	Pozo Comunal	11.6	6.5	215	4.8			(Ca)-HCO$_3$
41	PE	583041E/1416969N	El Zapote	Saturdina Silva	9.3	7.4	373	5.8	229.7	9.1	(Ca)-HCO$_3$
42	PE	583086E/1416951N	El Zapote	Manuel A Osorio	16.0	7.2	412	4.3	166.7	12.5	Na-HCO$_3$
43	PE	583105E/1417020N	El Zapote	Bentura Ruiz Martinez	17.7	7.5	441	4.1		10.3	Na-HCO$_3$
44	PE	583333E/1416791N	El Zapote	Finca HectorArguello	4.9	6.6	186	5.2		7.6	
45	PE	583328E/1417390N	El Zapote	en Una Huerta	122.2	7.3	378	1.2		18.6	Na-HCO$_3$
46	PP	586083E/1418600N	Las Mangas	PP abastecimiento	bd	6.8	619	6.0	261.7	13.1	(Ca)-HCO$_3$
47	PE	585382E/1418744N	Las Mangas	Gilberto Vilchez	2.2	6.8	339	5.9		14.6	
48	PE	585306E/1418705N	Las Mangas	Martin Vilchez	bd	6.9	660	4.5		15.0	
49	PE	585156E/1418895N	Las Mangas	Ermeregildo Vilchez	bd	6.5	157	4.3		6.7	
50	PE	585200E/1419161N	Las Mangas	Emma Espinoza Ruiz	bd	6.6	369	2.5		11.4	
52	PE	581671E/1419117N	Cerro La Mina	PE Cerro la Mina	22.8	6.1	176	6.3	180.0	16.4	(Ca)-HCO$_3$
53	PP	578031E/1417849N	S/Ramon de la Uva	San Ramon de la Uva	bd	7.2	204	8.4	277.0	165.0	(Ca)-HCO$_3$
54		579505E/1417576N	M. Quequisque	Manantial El Quequisq	3.3	7.2	440	6.6	177.3		Ca-HCO$_3$
56	PE		El Derrumbado	El derrumbado	15.6	6.9	557	6.3	238.2	35.6	(Ca)-HCO$_3$
57		583211E/1411392N	Presa la Virgen	Presa la Virgen	bd	6.6	128	5.6	250.0		(Ca)-HCO$_3$

<bd: Below detection limit; EC: Electrical conductivity; SL: Static groundwater level (m below ground surface); PE: Excavated well; PP: Perforated well; M: Spring; P: Dam.

10.5   DISCUSSION AND CONCLUSIONS

10.5.1   *Distribution of arsenic and lateral variability in shallow groundwater*

Arsenic occurs at toxic levels in the groundwaters of the shallow aquifer of SW Sébaco valley in different parts. In 17 of 57 groundwater samples the As concentrations exceeded the Nicaraguan drinking water limit of 10 μg/l. Dug and drilled wells with high As concentrations were identified in the towns of: Real de la Cruz, Aguas Frias, La Ceiba, La Union, Cerro La Mina, El Derrumbado, and El Cacao (Fig. 10.2). The highest As concentration of 122 μg/l was found in a dug well in El Zapote (Fig. 10.2). This is the same locality where in 1996 a high concentration of 1320 μg As/l was found (INAA 1996) in a well drilled into the highly fractured hardrock aquifer. This well supplied drinking water to the population since 1994. As a consequence, the water from this well has produced severe health impacts on the population. The well was closed in June 1996 by the Nicaraguan Institute of Aqueducts and Sewage Systems (INAA) and the Ministry of Health (MINSA).

Lateral variations in the concentrations of As show several high-As zones where concentrations exceed 10 μg/l. A comparison of the data on As distribution in groundwater with the concentrations of major ions in the same wells did not show clear trends. However, only a limited number of wells with data for all the major ions (only 25 wells) were available. A comparison of groundwater As concentrations with EC, pH, and Eh did not show clear relationships. This can be explained by the relative uniform chemical composition of the groundwaters and only relative small variation of pH and Eh values. This behavior is in contrast to other areas of the world with high groundwater As concentrations such as Argentina and Bangladesh (e.g., Bundschuh *et al.* 2004, Bundschuh *et al.* 2009b). In addition, As concentrations in the investigated shallow aquifer are much lower compared to studies from Argentina and Bangladesh. Therefore, pH and Eh may play a less important role as an As mobility control. It should be noted that the highest observed As concentration of 122 μg/l (El Zapote) coincides with a low concentration of dissolved oxygen (1.2 mg/l) and a low redox potential (Eh) of 156 mV compared with the respective average values (Fig. 10.5). Additionally (but statistically not significant due to low number of samples), the 3 samples from Zapote with high concentrations of groundwater As (16.0, 17.7 and 122.0 μg/l) correspond to $NaHCO_3$ waters (field f of Piper diagram), whereas waters with lower As concentration belong to $CaHCO_3$ waters. This observation is in agreement with other studies (Smedley *et al.* 2005, 2009, Bundschuh *et al.* 2004, Bhattacharya *et al.* 2006), which have found that high groundwater As zones belong to aquifers with $NaHCO_3$ waters (e.g., Argentine, Bundschuh *et al.* 2004). $CaHCO_3$ waters are characterized by lower As concentrations. The genesis of $NaHCO_3$ waters is explained by ion exchange, where Ca^{2+} from the water is exchanged for Na^+ from the solid (Bundschuh *et al.* 2004). However, the pH increase, which is expected to result from this ion exchange process is not observed at El Zapote (pH 7.2–7.5).

Two areas in the studied region, Las Mangas and Tatazcame, are characterized by non-detected As in shallow groundwater, and low As concentrations in soil and rock samples. Therefore, both areas are recommended for future drilling of wells.

10.5.2   *Sources and mobility of arsenic in groundwater*

Transport and mobility of As in shallow groundwater of Sébacco valley is governed by factors such as: (1) the availability of leachable As in the sediment/rock/soil, which depends on its total concentration in the solid phase and especially on its chemical speciation which determines its leachability; (2) groundwater recharge; (3) groundwater flow pattern/residence time and (4) mineral equilibria, ion exchange and sorption processes.

In the shallow alluvial aquifer of Sébaco valley As may be leached (1) from the aquifer sediments derived from volcanic rocks, (2) from the volcanic intensely fractured hardrocks (predominantly andesites and dacites) with high secondary hydraulic permeability and forming a deeper aquifer. These rocks are locally hydrothermally altered and are found as outcrops or underlying

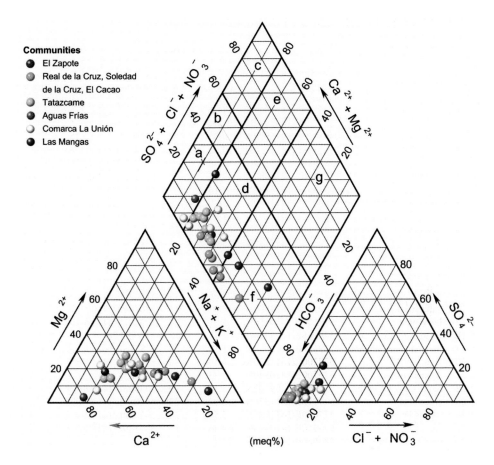

Figure 10.4.   Piper diagram: Chemical classification of groundwaters from southwestern Sébaco valley.

bedrocks of the sedimentary aquifer, and (3) from the unsaturated zone, where it is dissolved by percolating rain and irrigation water. Of all these sources, the primary As source is Tertiary volcanic rocks, implying that the source of the groundwater As is geogenic.

The hardrocks outcroping at several sites of the plain are deeply fractured and locally highly chemically weathered by hydrothermal fluids. Due to the lower permeability of the andesites and basalts forming the lower aquifers, deeper groundwater should have zones with trapped groundwater or at least strongly reduced flow velocity (compared to the shallower zones) resulting in higher contact times between aquifer solid material and fluid favoring mobilization of As into groundwater (Fig. 10.2). Although, there is only one well drilled into the hardrock aquifer (El Zapote 1320 µg As/l), it can be assumed that As concentrations increase with depth.

Intense hydrothermal alteration of Tertiary volcanic rocks, facilitate As release (by dissolution of primary and secondary minerals) from the mineralized structures to groundwater along active and non-active faults and fractures. These faults and fractures act as passageways of groundwater but also for rainwater and surface runoff. Circulating water leaches these hydrothermally altered zones before recharging the shallow sedimentary aquifer. There are no mineralized or hydrothermally altered bodies in contact with the shallow alluvial aquifer due to deep basement (drilled and dug wells do not reach these altered or mineralized bodies, they were only identified in the hard rock outcrops).

The highest As concentration found at El Zapote in the groundwater of the sedimentary aquifer (122.15 µg/l), occurs at the same sites as the highest concentrations of As in rocks and soil found

Table 10.2.   Arsenic in soils and rocks of southwestern Sébaco valley, Nicaragua.

Soils				Rocks		
Nr	Community	Coordinates (Lambert)	As (μg/g)	Nr	Coordinates (Lambert)	As (μg/g)
1	Las Mangas	585604E/1419561N	1.30	1	584826E/1419227N	2.49
2	Las Mangas	584728E/1418942N	0.85	2	581769E/1419006N	4.96
3	Zapote	583377E/1417570N	4.05	3	581641E/1419112N	3.24
4	La Ceiba	581168E/1415750N	3.23	4	581438E/1418922N	6.35
5	La Ceiba	582084E/1415104N	4.66	5	580500E/1418850N	5.20
6	La Ceiba	582630E/1414997N	3.25	6	578122E/1417841N	2.10
7	Real de la Cruz	583300E/1415189N	3.14	7	579439E/1418267N	<dl
8	Street Real de la Cruz	580879E/1413969N	1.49	8	579308E/1418376N	3.09
9	Street Real de la Cruz	580590E/1413273N	1.53	9	578290E/1418281N	2.07
10	Road to Cacao	580510E/1413265N	1.15	10	578268E/1418277N	7.08
11	Road to Cacao	580690E/1413135N	3.43	11	578317E/1417223N	4.48
12	Road to Cacao	580925E/1412950N	2.38	12	582787E/1412351N	6.63
13	El Cacao	581180E/1412340N	0.77	13	581919E/1412974N	6.45
14	El Cacao	581350E/1412230N	0.39	14	580288E/1415019N	5.68
15	Street Real de la Cruz	581085E/1414651N	3.68	15	576678E/1410120N	4.06
16	Sabana Larga-Zapote	580130E/1414900N	2.62	16	576624E/1410022N	8.80
17	La Virgen	579800E/1414550N	3.07	17	577601E/1410942N	1.42
18	Road Aguas Caliente	579145E/1413350N	2.46	18	580601E/1411801N	10.27
19	Road Aguas Caliente	579410E/1413070N	1.18	19	578585E/1415068N	3.08
20	Aguas Frias	577200E/1411276N	1.16	20	581750E/1414289N	6.45
21	Aguas Frias	576700E/1410340N	1.33	21	582025E/1414586N	3.61
22	Santa Cruz La India	575860E/1409300N	95.13	22	583993E/1413908N	6.36
23	Aguas Frias	577620E/1411270N	6.02	23	582087E/1414909N	5.93
24	Aguas Frias	577650E/1411080N	5.54	24	582739E/1416752N	14.98
25	Aguas Frias	579910E/1410320N	59.46	25	584849E/1418883N	6.86
26	near Rio Viejo	580660E/1413770N	23.41			
27	Cacao	581110E/1412305N	2.31			
28	P. Virgen	581930E/1413390N	3.77			
29	P. Virgen	592000E/1412340N	5.11			
30	P. Virgen	502120E/1412620N	4.02			
31	Soledad Cruz	580855E/1414215N	2.82			
32	Zapote	582335E/1417585N	57.19			
33	Orilla de Quebradita	582335E/1417585N	4.59			

<dl: Below detection limit.

in the project area 15 μg/g and 57 μg/g, respectively). However, two soil samples collected outside the study area, in the entrance of the Mine La India and in Agua Fria, also show As concentrations greater and similar to those of El Zapote with values 59 and 95 μg/g, respectively. This indicates that the contamination by As extends beyond the study area. Even more, we can assume that there are many not yet detected sites in Nicaragua and other Central American countries, in which Tertiary volcanic rocks (as well as volcanic rocks of Quaternary or recent age) that cover important areas could be releasing As. Groundwater As could be playing an important role as a natural contaminant of drinking water resources.

Complete geochemical information is not available on the volcanic rocks, the hydrothermally altered products, and the aquifer sediments and on the unsaturated zone. As a consequence, the relative contribution of each one of these sources can not be ascertained. It is not possible to know what is the main source for As in this aquifer. Therefore, further studies are needed to understand the factors that govern the mobilization and release of As into the groundwater and the spatial

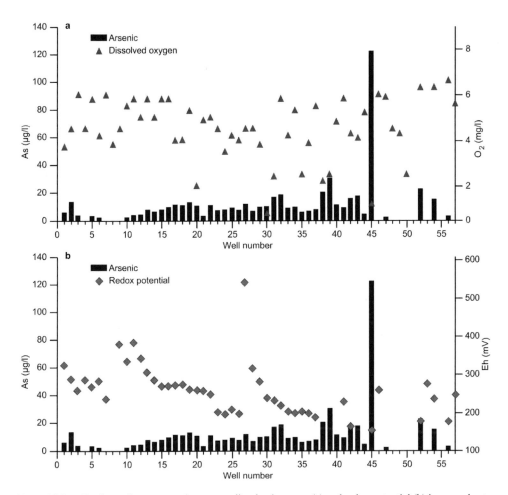

Figure 10.5.   Total arsenic concentration versus dissolved oxygen (a) and redox potential (b) in groundwater of the southwestern Sébaco valley, Nicaragua.

variation of its concentration, and to identify the aquifers that would yield good water for safe drinking in the affected regions.

REFERENCES

Barragne, P.: Contribución al estudio de cinco zonas contaminadas naturalmente por arsénico en Nicaragua. UNICEF, Managua, Nicaragua, 2004.

Bhattacharya, P., Claesson, M., Bundschuh, J., Sracek, O., Fagerberg, J., Jacks, G. Martin, R.A., Storniolo, A. and Thir, J.M.: Distribution and mobility of arsenic in the Rio Dulce alluvial aquifers in Santiago del Estero Province, Argentina. Sci. *Tot. Environ.* 358 (2006), pp. 97–120.

Bundschuh, J., Farias, B., Martin, R., Storniolo, A., Bhattacharya, P., Cortes, J., Bonorino, G. and Albouy, R.: Groundwater arsenic in the Chaco-Pampean Plain, Argentina: case study from Robles county, Santiago del Estero Province. *Appl. Geochem.* 19 (2004), pp. 231–243.

Bundschuh, J., García, M.E. and Bhattacharya, P.: Arsenic in groundwater of Latin America—A challenge of the 21st century. Geological Society of America Annual Meeting, Philadelphia, 22–25 Oct. 2006, *Geological Society of America Abstracts with Programs* 38 (7) 2006, p. 320.

Bundschuh, J. and García, M.E.: Rural Latin America—A forgotten part of the global groundwater arsenic problem? In: P. Bhattacharya, Al. Ramanathan, J. Bundschuh, D. Chandrasekharam, A.K. Keshari and A.B. Mukherjee (eds): Groundwater for sustainable development: problems, perspectives and challenges. Balkema Publisher, Leiden, The Netherlands, 2008.

Bundschuh, J., García, M.E., Birkle, P., Cumbal, L.H., Bhattacharya, P. and Matschullat, J.: Groundwater arsenic in rural Latin America—occurrence, health effects and remediation experiences. In: J. Bundschuh, M.A. Armienta, P. Birkle, P. Bhattacharya, J. Matschullat and A.B. Mukherjee (eds): *Natural arsenic in groundwater of Latin America*. Balkema Publisher, Leiden, The Netherlands, 2009a (This Volume).

Bundschuh, J., Bhattacharya, P., Litter, M.I. and García, M.E.: Arsenic-safe aquifers as innovative socially acceptable source of safe drinking water—What can rural Latin America learn from Bangladesh experiences? In: J. Bundschuh, M.A. Armienta, P. Birkle, P. Bhattacharya, J. Matschullat and A.B. Mukherjee (eds): *Natural arsenic in groundwater of Latin America*. Balkema Publisher, Leiden, The Netherlands, 2009b (This Volume).

ECO/OPS: Epidemilogía ambiental, un proyecto para America Latina y el Caribe. Lima, Peru, 1997.

Gómez, A.: Chronic arseniasis in El Zapote, Nicaragua. In: J. Bundschuh, M.A. Armienta, P. Birkle, P. Bhattacharya, J. Matschullat and A.B. Mukherjee (eds): *Natural arsenic in groundwater of Latin America*. Balkema Publisher, Leiden, The Netherlands, 2009 (This Volume).

INEC: Instituto Nacional de Censos y Estadísticas-Nicaragua. Managua, Nicaragua, 1995.

INAA: Caso del Zapote. Informe Técnico de progresos y soluciones Normación Técnica del INAA, Managua, Nicaragua, 1996.

Moffet, J.: The determination of arsenic in nonsilicate geological ore samples using a vapor generation accessory. Varian Instruments at Work No. AA-78, Varian Techtron Pty. Limited, Mulgrave, Victoria Australia, 1988.

Smedley, P.L., Kinniburgh, D.G., Macdonald, D.M.J., Nicolli, H.B., Barros, A.J. and Tullio, J.O.: Arsenic associations in sediments from the loess aquifer of La Pampa, Argentina. *Appl. Geochem.* 20 (2005), pp. 989–1016.

Smedley, P.L., Nicolli, H.B., Macdonald, D.M.J. and Kinniburgh, D.G.: Arsenic in groundwater and sediments from La Pampa Province, Argentina. In: J. Bundschuh, M.A. Armienta, P. Birkle, P. Bhattacharya, J. Matschullat and A.B. Mukherjee (eds): *Natural arsenic in groundwater of Latin America*. Balkema Publisher, Leiden, The Netherlands, 2009 (This Volume).

Varian: VGA-76. Vapor generation accessory. Operation manual, Varian Techtron, Australia, 1984.

WHO 2001. Arsenic in drinking water: Fact Sheet #210. WHO, Geneve, Switzerland, http://www.who.int/mediacentre/factsheets /fs210/en/print.htm (accessed on 5 November 2006).

CHAPTER 11

Arsenic and water quality of rural community wells in San Juan de Limay, Nicaragua

L. Morales, C. Puigdomènech, A. Puntí, E. Torres, C. Canyellas & J.L. Cortina
Departament d'Enginyeria Química, Universitat Politècnica de Catalunya, Barcelona, Spain

A.M. Sancha
División de Recursos Hídricos y Medio Ambiente, Facultad de Ciencias Físicas y Matemáticas, Universidad de Chile, Santiago de Chile, Chile

ABSTRACT: Many people in the rural areas of Nicaragua lack access to safe drinking water. Tube well drinking water sources from a number of districts are found to contain elevated levels of arsenic (As). Geogenic As reported in groundwaters from wells in several villages of San Juan de Limay at the Estelí department (NW Nicaragua). San Juan de Limay was selected as an area to investigate the presence of As in groundwater used for drinking because studies carried out by the local environmental authorities have indicated occurrences of As. This study was initiated to determine the extent of the problem. Average total As concentration was 14 µg/l with the maximum concentration up to 115 µg/l, with significant spatial variability. In most water samples, As(V) was the dominant species.

11.1 INTRODUCTION

In Nicaragua, natural occurrence of arsenic (As) in groundwater has been reported from different parts of the shallow aquifer of the Coyol formation in the Estelí valley. Early investigations were conducted by government agencies of Nicaragua in 1996, which confirmed elevated As concentrations (>200 µg/l) in groundwater of shallow aquifers (UNICEF 2002). In this area, about 0.5 million people are potentially affected by As through drinking water from groundwater sources, exceeding the WHO guideline of safe drinking water (10 µg/l; WHO 2001) as well as the national drinking water standard of 10 µg/l. The population here lives in rural dispersed settlements where groundwater is used without treatment for drinking. Exposure to As was first reported in 1996 in El Zapote, a rural community of 125 inhabitants, located in the valley of Sébaco, in the northern part of Nicaragua (Gonzalez *et al.* 1998, Aguilar *et al.* 2000, UNICEF 2003). For two years between 1994 and 1996, this population drank water from a public tube well containing 1320 µg/l of inorganic As (Gomez *et al.* 2000). Elevated As (45–66 µg As/l) was also detected in private hand tube wells used before 1994 and after 1996.

A range of clinical and ecotoxicological studies were initiated in 2000 for early diagnosis of diseases associated with As exposure and to determine their characteristics and prevalence in El Zapote (Gomez 2004). El Zapote residents of 1994–1996 went through medical examination. A total of 111 individuals volunteered for the medical examination that included abdominal ultrasonography and others laboratory tests. Participants were divided into two groups according to the average As concentration ingested during a period of eight years: high As ingestion (80–380 µg As/l) and low As ingestion (<80 µg As/l) (UNICEF 2003). Keratosis and hyperpigmentation, typical of chronic arsenicosis, were strongly associated with high As ingestion (Gomez 2004). Respiratory symptoms were related to high As ingestion. The typical cutaneous manifestations shown by these patients, confirmed the diagnosis of these first collective cases of chronic

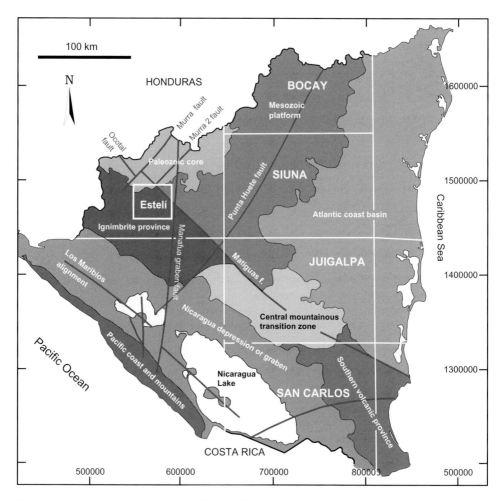

Figure 11.1.   Location of the project area the Estelí department of northwestern Nicaragua.

arsenicosis in Nicaragua and Central America. The relationship between high As ingestion and the respiratory effects is statistically significant.

A more recent study analyzed groundwater in various departments of Nicaragua (UNICEF 2002, 2003, 2004) and detected elevated concentrations of As in groundwater and soils. Wells in San Juan de Limay in the Estelí department of northwestern Nicaragua (Fig. 11.1) were found to contain elevated levels of As and subsequently, most of them were closed for drinking water usage.

In recent years, the groundwater chemistry of the shallow aquifers in Nicaragua (Fig. 11.1) has been studied (Barragne-Bigot 2004) to improve the understanding of As sources and the mechanisms of As mobilization. The aim of the present study was to investigate the presence and the distribution of As in groundwaters of the shallow aquifers of selected parts of San Juan de Limay to determine the needs for remediation of As in drinking water.

11.2   LOCATION AND GEOLOGICAL SETTING OF THE STUDY AREA

The study area is located in the northwestern part of Nicaragua around the city of San Juan de Limay (Fig. 11.1). It covers parts of the alluvial deposits formed by Río Coco river (herein after referred as the Río Coco alluvial cone), where the Estelí valley has an area of about 9787 km^2. The area is rural, but densely populated in comparison with the surrounding countryside due to its fertile soils and irrigation systems with channels distributing water from the Río Coco river. Small agricultural settlements dependent on artificial irrigation are common throughout the Río Coco cone.

11.3   MATERIALS AND METHODS

From July to September 2005, groundwater samples of the shallow aquifer were collected from 40 mostly hand-pumped tube wells of eight different villages of San Juan de Limay in the Estelí department. The depth of most wells was less than 12 m. All sampled wells were geo-referenced using global positioning system (GPS). Temperature, pH, redox potential (Eh), and electrical conductivity (EC) of groundwater were measured directly on site.

Groundwater sampling included the collection of: (1) filtered samples for mayor anion analysis, (2) filtered, acidified samples for major cation, minor and trace element analysis, (3), samples for microbiological analysis, and (4) filtered samples for As(V) and total As (t-As) following the procedure described elsewhere (Vidal *et al.* 1999, Vilano *et al.* 2000).

Most of the analyses were performed at the *Centro para la Investigación en Recursos Acuaticos de Nicaragua* at the *Universidad Nacional Autónoma de Nicaragua*, in Managua. Anions were analyzed by ion chromatography (Cl$^-$, F$^-$, NO$_3^-$ and SO$_4^{2-}$). The major cations, and the minor and trace metals were analyzed by ICP-AES and ICP-MS respectively. As(V) was calculated as a difference between t-As and As(III) (Vidal *et al.* 1999, Ruiz-Chancho 2003). Certified standards and synthetic chemical standards were prepared in the laboratory, and duplicates were analyzed after every 10 samples during the runs. Trace element concentrations in standards were within 90–110% of their true values. In case of wider variations, the standards were recalibrated and the preceding batch of 10 samples reanalyzed.

11.4   RESULTS

The pH of the groundwater ranged from 7.0 to 10.3 (average 7.2), electric conductivity (EC) ranged from 348 to 1147 μS/cm (average 652 μS/cm), and major ion composition indicated Ca^{2+} (average 63 mg/l) and HCO$_3^-$ (average 336 mg/l) as the dominant ions (Table 11.1). Redox conditions in the study are were oxidizing or moderately reducing. This is consistent with the predominantly As(V) species.

The concentration of t-As in the groundwater showed a considerable spatial variability (average 15 μg/l). Some of the wells had high values with a maximum value of 115 μg/l. Speciation of As indicated the dominance of As(V). Regarding other minor and trace elements, Fe concentrations were low (<0.3 mg/l), with the exception of the well with the highest As concentrations with 0.54 mg/l of Fe. All samples were characterized by high SiO$_2$ concentrations (average 53.3 mg/l). Fluorine concentrations were high (average 0.34 mg/l). The microbiological analysis showed that 50% of the analyzed wells had coliforms or *Escherichia coli* levels above the threshold values used by the WHO to consider water as potable.

11.5   DISCUSSIONS AND CONCLUSIONS

The principal source of As seems to be the volcanic ash layer, which is found either as a distinct layer in most of the study area or dispersed in the sediment, and which contains highly soluble volcanic glass. The relation of As with F and Si, most likely due to their common origin in volcanic ash, indicates its importance as a source of As in shallow groundwater.

Table 11.1.   Results of physico-chemical, chemical and bacteriological analysis of groundwaters from selected wells.

Community	Code		TDS	Ca^{2+}	Mg^{2+}	Na^+	K^+	NH_4^+
			(mg/l)					
Jocote Renco	1.2		392.9	80.2	15.5	27.3	1.2	0.03
La Grecia	2.3		459.9	54.5	16.5	84.7	1.6	0.03
La Grecia	2.4		412.4	7.2	2.4	151	1.7	0.02
El Morcillo	3.1		195.4	43.9	6.6	7.01	1.7	0.02
El Morcillo	3.2		383.0	64.1	19.1	34	1.3	0.02
Parcila	4.1		341.1	52.1	15.1	37.9	2.1	0.04
Platanares	5.1		698.5	72.5	7.5	142	5.6	0.06
Platanares	5.2		586.2	40.5	7.5	142	5.4	0.03
San Lorenzo	6.1		356.9	62.1	14.4	33.6	0.9	0.31
Aguafría	7.1		347.8	76.6	14.7	19.4	1.2	0.03
El Palmar	8.3		268.6	10.1	1.5	66.2	13.4	0.69
El Palmar	8.4		335.1	66.1	18.8	16.4	0.7	0.13

Community	Code	t-As (µg/l)	Cl^-	SO_4^{2-}	HCO_3^-	Alk_{tot}	SiO_2	F^-
			(mg/l)					
Jocote Renco	1.2	<2.02	5.3	43.3	317.7	260.4	53.9	0.28
La Grecia	2.3	3.51	4.1	5	492.8	403.9	45.7	0.14
La Grecia	2.4	115.45	4.8	20.9	406.5	333.2	23.9	0.43
El Morcillo	3.1	<2.02	1.4	6.1	189.6	155.4	33.5	0.18
El Morcillo	3.2	10.69	5.4	9.6	386.9	317.2	55.8	0.13
Parcila	4.1	2.46	6.2	9.7	300.2	246.1	67.9	0.25
Platanares	5.1	15.84	7.6	262.8	295.2	241.9	54.5	0.44
Platanares	5.2	8.86	6.5	139	377.8	309.6	58.7	0.35
San Lorenzo	6.1	3.77	4.9	5.1	353.7	289.9	60.5	0.19
Aguafría	7.1	2.92	4.4	3.1	371.6	304.6	41.9	0.27
El Palmar	8.3	5.92	3.4	23.9	185.7	152.3	56.5	0.73
El Palmar	8.4	3.9	8.6	3.2	339.6	278.3	52.7	0.25

			Bacteriological			Physico/chemical	
			Total coliforms	Thermo-tolerant coliforms	E.coli	Turbidity	Fe_{tot}
Community	Code	Well type	(MPN/ 100 ml)	(MPN/ 100 ml)	(MPN/ 100 ml)	(NTU)	(mg/l)
Jocote Renco	1.2	D	23	8	<2	<5	<0.3
La Grecia	2.3	D	50	22	22	<5	<0.3
La Grecia	2.4	D	30	2	<2	37.4	0.54
El Morcillo	3.1	E	800	300	230	<5	<0.3
El Morcillo	3.2	D	230	50	4	<5	<0.3
Parcila	4.1	D	<2	<2	<2	<5	<0.3
Platanares I		D	4	<2	<2	<5	<0.3
Platanares	5.1	D	22	13	13	<5	<0.3
Platanares	5.2	D	<2	<2	<2		
San Lorenzo	6.1	river	13	2	<2	<5	<0.3
Aguafría	7.1	D	<2	<2	<2	<5	<0.3
El Palmar	8.3	E	300	2	2	11.72	<0.3
El Palmar	8.4	E	11000	70	50	8.6	<0.3

D: drilled well; E: excavated well.

Redox conditions in the study area are oxidizing or moderately reducing. This is consistent with the predominance of As(V). Thus, reductive dissolution of ferric minerals observed, for example, in Bangladesh (Bhattacharya *et al.* 2001, Smedley and Kinniburgh 2002, Bhattacharya *et al.* 2002b) can be ruled out as a principal mechanism of As input. In contrast, high pH values seem to promote desorption of As adsorbed onto the amorphous oxides of Al, Mn and Fe.

Wells with high As concentrations generally coincide with wells of high pH values. High pH can be explained by dissolution of carbonates induced by cation exchange.

This is inconsistent with the negative correlation between Ca^{2+} and Na^+ observed in earlier studies (Bundschuh *et al.* 2004) and also with a close relation between Na^+ and HCO_3^-. Another factor contributing to high pH values may be dissolution of silicates in volcanic glass. The coincidence of areas with high concentrations of groundwater As and high pH values was also observed in the plains of La Pampa (Argentina) (Smedley *et al.* 2001).

In areas where the drinking water supply contains unsafe levels of As, the immediate concern is finding a safe source of drinking water. There are two main options: finding a new safe source, and removing As from the contaminated source. For this project the approach focused on the removal option. Although the released As can be immobilized in the soil, it can be easily spread into other regions through the transport of As-contaminated solids and As dissolution occurred by changes in the geochemical environment to a reductive condition (Bhattacharya *et al.* 2002a, b).

ACKNOWLEDGEMENTS

We gratefully acknowledge to M.J. Ruiz-Chancho and R. Rubio, *Departament de Química Analítica. Universitat de Barcelona*, for the help on As-analysis. This work was funded by the Centre per la Cooperació I el Desenvolopament (CCD-UPC) project CCD-2005 and by the Spanish MCYT program (RN2002-1147-C02).

REFERENCES

Aguilar, E., Parra, M., Cantillo, L. and Gómez, A.: Chronic arsenic toxicity in El Zapote-Nicaragua, 1996. *Med. Cután. Iber. Lat. Am.* 28:4 (2000), pp. 168–173.

Barragne-Bigot, P.: *Contribución al estudio de cinco zonas contaminadas naturalmente por arsénico en Nicaragua*. UNICEF, Managua, Nicaragua, 2004.

Bhattacharya, P., Jacks, G., Jana, J., Sracek, O., Gustafsson, J.P. and Chatterjee, D.: Geochemistry of the Holocene alluvial sediments of Bengal Delta Plain from West Bengal, India: Implications on arsenic contamination in groundwater. In: G. Jacks, P. Bhattacharya and A.A. Khan (eds): *Groundwater arsenic contamination in the Bengal Delta Plain of Bangladesh.* Proc. 21–40, KTH-Dhaka University Seminar, University of Dhaka, Bangladesh KTH Special Publication, Stockholm, TRITA-AMI report 3084, 2001.

Bhattacharya, P., Frisbie, S.H., Smith, E., Naidu, R., Jacks, G. and Sarkar, B.: Arsenic in the environment: a global perspective. In: B. Sarkar (ed): *Handbook of heavy metals in the environment.* Marcell Dekker Inc., New York, 2002a, pp. 147–215.

Bhattacharya, P., Jacks, G., Ahmed, K.M., Khan, A.A. and Routh, J.: Arsenic in groundwater of the Bengal Delta Plain aquifers in Bangladesh. *Bull. Env. Cont. Toxicol.* 69 (2002b), pp. 538–545.

Bundschuh, J., Farías, B., Martin, R., Storniolo, A., Bhattacharya, P., Cortes, J., Bonorino, G. and Albouy, R.: Groundwater arsenic in the Chaco-Pampean Plain, Argentina: case study from Robles County, Santiago del Estero Province. In: P. Bhattacharya, A.H. Welch, K.M. Ahmed, G. Jacks and R. Naidu (eds): Arsenic in groundwater of sedimentary aquifers. *Appl. Geochem.* 19:2 (2004), pp. 231–243.

Gómez, A.: Arsenic and cancer in S and SW communities of the valley of Sébaco, Nicaragua 1999. *Proceedings of the XXII Central America Congress of Dermatology*. Panama City, Nov. 21–26, 2000, pp. 143–149.

Gómez, A. and Aguilar, E.: Case of Hydroarsenicosis and cutaneous cancer. El Carrizo, valley of Sebaco-Nicaragua 1952–2000. *Proceedings of Summer Meeting American Academy of Dermatology*, New York, NY, July 31–August 4, 2000, pp. 345–351.

Gomez, A.: Chronic arsenicosis in El Zapote, Nicaragua 1994–2002. In: A.M. Sancha (ed): *Tercer Seminario Internacional sobre Evaluación y Manejo de las Fuentes de Agua de Bebida contaminadas con Arsénico* (proceedings available as CD), Universidad de Chile, November 08–11, 2004, Santiago de Chile, Chile, 2004.

González, M., Provedor, E., Reyes, M., López, N., López, A. and Lara, K.: Arsenic exposition in rural communities of San Isidro, Matagalpa, 1997. Health Studies and Research Center (CIES), Pan American Health Organization/WHO PLAGSALUD-MASICA, 1998.

Martin, A.: Hidrogeología de la provincia de Santiago del Estero. Ediciones del Rectorado Universidad Nacional de Tucumán, Tucumán, Argentina, 1999.

Ruiz-Chancho, M.J.: *Estudio de métodos de extracción de especies de arsénico en muestras ambientales.* Master in Experimental Chemistry, Faculty of Chemistry, University of Barcelona, Barcelona, Spain, 2003.

Smedley, P.L. and Kinniburgh, D.G.: A review of the source, behavior and distribution of arsenic in natural waters. *Appl. Geochem.* 17 (2002), pp. 517–568.

Smedley, P.L., Nicolli, H.B., Macdonald, D.M.J., Barros, A.J. and Tullio, J.O.: Hydrochemistry of arsenic and other inorganic constituents in groundwaters from La Pampa, Argentina. *Appl. Geochem.* 17 (2001), pp. 259–284.

UNICEF: Arsénico y metales pesados en agua de Nicaragua. Ed. O. Moraga, UNICEF, Managua, Nicaragua, 2002.

UNICEF: Monitoreo y atención de intoxicados con arsénico en al comunidad el Zapote, San Isidro Matagalpa (1994–2002). Ed. O. Moraga, UNICEF, Managua, Nicaragua, 2003.

Vidal, M., López-Sánchez, J.F., Sastre, J., Jiménez, G., Dagnac, T., Rubio, R. and Rauret, G.: Prediction of the impact of the Aznalcóllar toxic spill on trace element contamination of agricultural soils. *Sci. Total Environ.* 242 (1999), p. 131.

Vilano, M., Padró, A. and Rubio, R.: Coupled techniques based on liquid chromatography and atomic fluorescence detection for arsenic speciation. *Analytica Chimica Acta* 411:1/2 (2000), pp. 71–79.

WHO: *Guidelines for drinking water quality.* World Health Organization, Geneva, Switzerland, 1993.

WHO: Arsenic in drinking water. Fact sheet 210, World Health Organization, Geneva, Switzerland, http://www.who.int/mediacentre/factsheets/fs210/en/print.htm, 2001.

CHAPTER 12

Volcanic arsenic and boron pollution of Ilopango lake, El Salvador

D.L. López & L. Ransom
Department of Geological Sciences, Ohio University, Athens, OH, USA

J. Monterrosa
Fundacion de Amigos del Lago de Ilopango, San Salvador, El Salvador

T. Soriano, F. Barahona & R. Olmos
*Departamento de Fisica, Facultad de Ciencias y Humanidades, Universidad de El Salvador,
San Salvador, El Salvador*

J. Bundschuh
International Technical Cooperation Program, CIM (GTZ/BA), Frankfurt, Germany
Instituto Costarricense de Electricidad (ICE), San José, Costa Rica

ABSTRACT: More than 200,000 inhabitants live in the drainage basin of Ilopango lake, El Salvador, even though high concentrations of boron (B) and arsenic (As) make this water unsuitable for human consumption. The water and sediments of Ilopango lake have been sampled and analyzed for trace elements and major ions, including As, nutrients (NO_3^-, PO_4^{3-}), SO_4^{2-}, Cl^- and B. For the waters, B concentrations range from 1.5 to 8.7 mg/l and As concentrations from 0.15 to 0.77 mg/l with the higher values to the south and lower values close to Cerros Quemados islands (at the lake center), where the last volcanic activity happened in 1880. Arsenic concentrations in the sediments were elevated at Cerros Quemados (86 mg/kg sediment), which also has the lowest concentration of organic matter. Sulfate and Cl^- concentrations and the pH of Ilopango lake indicate that it is in a quiet phase when compared with other volcanic lakes of the world. However, fluxes of volcanic gases seem to occur to the south of the caldera, where seismic activity is higher, and anomalous concentrations of soil gases, especially radon, are evident. Concentrations of As, Cl^-, B, PO_4^{3-}, and SO_4^{2-} in the water are also higher to the south, but the sediment concentrations of As, B, and Li are lower. Interaction of volcanic-hydrothermal fluids discharged to the south of the caldera, with the lake sediments could be generating the high element concentrations observed in the water.

12.1 INTRODUCTION TO THE ARSENIC PROBLEM IN LATIN AMERICA

Ilopango lake in El Salvador, Central America (Fig. 12.1), is an important fresh water resource for El Salvador. It is the largest lake in the country with 185 km^2 and 240 m maximum depth. The lake is hosted within a volcanic caldera. Its eruption in A.D. 429 (Dull 2001) was responsible for the migration of the Mayan people from El Salvador and the highlands of Guatemala to Tikal (Sheets 1979). The last eruption of this caldera was in 1879–1880 with the emplacement of Cerros Quemados domes in the center of the lake (Goodyear 1880). Ilopango caldera is still considered an important geological hazard for the city of San Salvador, which is located less than 10 km from the lake. Due to its volcanic origin the waters of Ilopango lake are rich in As and B, making them unsuitable for human consumption. However, during the last two decades, due to the social and political problems in the region, as well as the increasingly high population density of the country, the population in the watershed of this lake has increased to more than 200,000. Extensive deforestation, fertilization, and industrialization in the lake basin contribute to the environmental degradation of Ilopango. People living close to the lake's shore rely on Ilopango's water for domestic uses and consumption, and their health is at risk due to the chemical composition of the water.

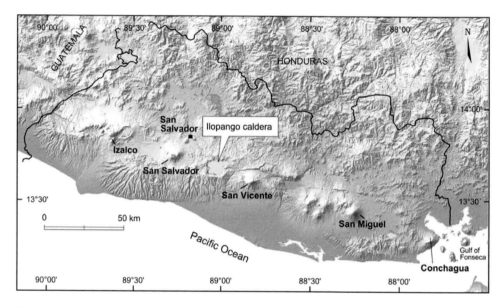

Figure 12.1.   Ilopango caldera is located in central El Salvador. Other important volcanic centers are shown.

In this chapter, we present results of our investigations into the water and sediment chemistry of Ilopango lake, the origin of trace elements present and implications for the use of this water.

12.1.1   *Volcanic setting*

The volcanic front of Central America is the result of the subduction of the Cocos plate under the Caribbean plate, along the Middle American trench (Molnar and Sykes 1969). The Cocos plate is currently being subducted at a rate of 5–7 cm a year (White and Harlow 1993). The chain of volcanoes in El Salvador is surrounded by normal graben faults running north–west to south–east, and is parallel to the subduction zone throughout El Salvador. Ilopango caldera is located at the southern margin of this trough.

Ilopango caldera measures 11 km east to west and 8 km north to south (Fig. 12.2). The lake level is around 400 m a.s.l. (Meyer-Abich 1956). The Chaguite river flows into the lake in the west and the Desaque river drains the lake to the east. Williams and Meyer-Abich (1955) state that the caldera formed via three distinct collapse episodes that produced violent volcanic eruptions. The first collapse occurred during the Plio-Pleistocene formation of the Central American graben. The last caldera collapse generated the Tierra Blanca Joven deposit and occurred approximately 1600 years ago (A.D. 429). This eruption produced about 18 km^3 of rock and a basin of 20 km^3 that now contains the lake (Hart 1981). Petrographic studies indicate that injection of mafic magma into partially crystallized dacite magma triggered the A.D. 429 and the 1879–1880 eruptions (Richer *et al.* 2004).

The volcanic rocks of Ilopango are super-calcic to strong calcic, falling within the calcalkaline series. Williams and Meyer-Abich (1955) identified the majority of the Ilopango caldera exposed rocks as dacite and rhyodacite pumice deposits with some undifferentiated Pliocene volcanic rocks exposed in the southeast portion of the caldera. Analysis of rocks from Ilopango caldera in Richer *et al.* (2004) give the following percentages: SiO_2: 67.74, Al_2O_3: 15.47, Fe_2O_3: 4.02, MnO: 0.12, MgO: 1.41, CaO: 3.83, Na_2O: 4.26, K_2O: 2.08, and TiO_2: 0.42, which defines a dacitic composition. Four tephra deposits have been identified as being associated with eruptions from Ilopango (Hart 1981). The majority of the area is covered with the most recent Tierra Blanca Joven pyroclastic deposits.

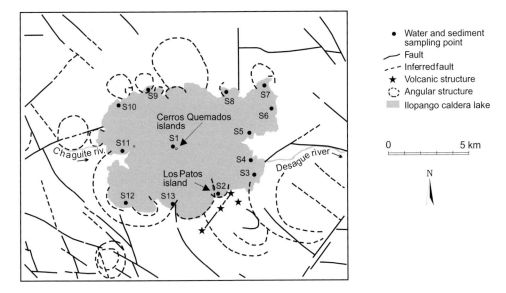

Figure 12.2.   Simplified structural map of Ilopango caldera (after Geographical Institute of El Salvador, 1970). Locations of sediment and water samples collected July 25, 2000.

Petrological studies of the intracaldera stratigraphy indicate that lacustrine deposits were present more than 43,670 years ago, as well as providing evidence of subaqueous eruptive activity and dome growth (Mann *et al.* 2004). Currently Ilopango caldera does not show any subaerial fumarolic activity. However, local divers have reported hot water seeping into the lake water in the south of the lake, close to Cerro Los Patos island (Fig. 12.2). In addition, Ilopango caldera experiences frequent seismicity (López *et al.* 2004) in the area between the center of the lake, and south and southeast margins. Several resurgent domes associated with various eruptive episodes are found outside the caldera boundaries, within the caldera walls, and within the lake (Fig. 12.2). The main faults in the area are also shown in Figure 12.2.

Investigations of diffuse soil degassing of radon, mercury, and carbon dioxide in 2000 (López *et al.* 2004) indicates that the caldera presents low fluxes of carbon dioxide that compare with background levels at San Salvador volcano to the west. Soil radon was higher than 100 pCi/l in several locations. Soil radon and carbon dioxide anomalies were observed to the southwest, northeast, and southeast and northwest of the lake. The strongest anomalies were for the region located between the south and southwest of the lake, close to the region of frequent seismicity. Mercury concentrations were generally low (up to 0.015 µg/l) with anomalies at the east, west and southeast of the lake.

12.1.2   *Precipitation of sulfur at Ilopango lake*

Events of sulfur precipitation on the water surface have been documented at Ilopango lake. Between November and March the lake has an almost isothermal, oxygenated water column. High wind velocities and slightly lower air temperatures during this time of the year produce convective water movement (López and Hernandez 1998). From March to November, the lake shows a slight thermal stratification. Almost every year, fishermen report the precipitation of sulfur on the surface of the lake. This precipitation occurs when waters are oxygenated and undergoing convection. Lopez and Hernandez (1998) observed that elemental sulfur precipitation typically occurs in December or January at the time of maximum wind velocities and low air temperature. When sulfur precipitates appear on the surface of the lake, the oxygen content in all of the water column falls to less than

1 mg/l. Only the fish that can expand their air bladder are able to survive (Fundación Ilopango, *pers. commun.*). Strong convective currents are suspected to perturb the lake sediments, probably releasing sulfide-rich sediment particles into the water column. Sulfide oxidation consumes the oxygen and precipitates elemental sulfur as it rises throughout the water column (Lopez and Hernandez 1998). The complete mechanism of this process is not well understood and should be investigated further. However, the occurrence of this oxidation event does change the oxygen content of the water, and can have implications for the speciation of elements that undergo redox reactions, such as As, and those that could coprecipitate with sulfur.

12.1.3 *Arsenic and boron*

Signorelli (1997) stated that volcanic gases vary considerably in As content, possibly related to the different physicochemical features of magma source and rock properties within a volcanic system. Nriagu (1979, 1989), Lantzy and Mackenzie (1979), and Nriagu and Pacyna (1988) have confirmed that volcanoes are an important natural source of As, especially in the southern hemisphere.

Arsenic is usually present as pentavalent As in well-oxygenated water and as the trivalent species (considered more toxic than the pentavalent species) under reducing conditions, such as those conditions found in deep lake sediments or groundwaters (Irgolic 1982, Lemmo *et al.* 1983, Cui and Liu 1988, Welch *et al.* 1988). In oxidizing environments, As is adsorbed onto Fe(III) oxides and hydroxides with the adsorption process depending on pH and redox state of As (Sracek *et al.* 2004). At low pH, the adsorption affinity is higher for As(V), and at higher pH for As(III). If the environment is very reducing and sulfate is reduced, As can precipitate with sulfide minerals such as orpiment, As-bearing pyrite or marcasite (Sracek *et al.* 2004). Arsenic can also occur in fossil hydrothermal systems as a minor impurity in phosphate minerals or dispersed in limonite (Craw *et al.* 2000).

Arsenic contamination in aquifers hosted in volcanic rocks is a problem in many regions of the world. In Argentina, extensive rural areas have aquifers affected by As contamination (e.g., Bundschuh *et al.* 2004, Bhattacharya *et al.* 2006). Investigations in aquifers of the Chaco-Pampean plain and in Río Dulce alluvial aquifers suggest that As could be released to the water by dissolution of volcanic glass in the volcanic ash present in these regions. The volcanic ash occurs either as one of the stratigraphic layers, or dispersed in the sediments. At the Chaco-Pampean plain, clastic sediments of metamorphic and igneous rocks could also be a source of As (Bundschuh *et al.* 2004).

Boron is not found in elemental form in nature, but does occur as boric acid [$B(OH)_3$] usually in certain volcanic spring waters and as borates in borax and colemanite. Boron partitions into the glassy matrix of rhyolites and andesites during magmatic differentiation (Reyes and Trompetter 1998). Boron concentration in igneous rocks and soil average 10 mg/kg (Lide 2000), ranging from 2 to 100 mg/kg dry weight (Bradford 1966). Volcanic emissions release boric acid vapor and boron trifluoride (BF_3). For this reason, concentrations of boron in surface and groundwater in volcanic regions are elevated (Durocher 1969).

Arsenic and its compounds are poisonous and have been associated with various forms of cancer. Arsenic has been linked to several internal cancers, at exposure concentrations of several hundred micrograms per liter of drinking water (National Research Council 1999). It has also been determined that inorganic As compounds are skin and lung (via inhalation) carcinogens in humans. Some non-carcinogenic effects have also been identified. For example, weeks to months of ongoing ingestion of inorganic arsenic (i-As) at doses of approximately 0.04 mg/l per day or higher can result in gastrointestinal problems, such as diarrhea and cramping, and hematological effects, including anemia and leukopenia (National Research Council 1999). Arsenic toxicity decreases according to its speciation: As(III) > As(V) > organoarsenic (Fergusson 1991). The As and B concentrations found in previous studies of Ilopango lake (0.06 to 0.82 mg/l for As and 0.1 to 10.3 mg/l for B as reported in Ransom (2002)) are cause for concern because they are considerably higher than the USEPA (2001) drinking water standards of 0.01 mg/l As and 0.1 mg/l B. In Ilopango lake, in addition to As and B, other pollutants have been previously reported. High concentration of SO_4^-, NO_3^-, and PO_4^{3-} (Lopez and Hernandez 1998) have been found and attributed to the use of fertilizers on crops surrounding the lake, volcanic input (for sulfate), and leaching of phosphate minerals from the rocks. In this research, the spatial distribution of trace elements in water and sediments, as well as nutrients (NO_3^- and PO_4^{3-}), SO_4^-, B and Cl^-, were investigated to understand and identify the possible sources for these elements.

12.2   METHODOLOGY

Water and sediment samples were collected around the perimeter of the lake and in the center near the Cerros Quemados islands. Water samples at 5 m depth were collected within the oxic zone of the stratified lake in July. At this time of the year the thermocline is around 20–30 m depth (López and Hernandez 1998). Sediment samples (S1–13 on Fig. 12.2) were collected from various depths, between 8 and 37 m. The sites were selected based on accessibility for the divers and lake bottom topography. Water samples were collected at 12 of these sites (at S3 only a sediment sample was available) using a "Science Source" water sampler. Each water sample was filtered with a 0.45 μm filter and nitric acid was added to the cation samples in order to prevent precipitation of minerals. At the same sites shallow sediment samples were collected using a suction-tube sediment sampler. We report in this chapter results for the analysis of Ba, Li, B, Cr, Mn, Fe, Co, Cu, Zn, As, and Pb, and the anions; SO_4^{2-}, Cl^-, NO_3^-, and PO_4^-.

The sediments were digested with nitric and hydrochloric acid (EPA method 3050B) and the water samples were digested according to the EPA Mild Digestion method. Chemical analyses of the sediment and water samples were completed using a 6800 Shimadzu atomic absorption spectrophotometer and a 6500 Shimadzu graphite furnace, except for As which was analyzed using inductively coupled plasma spectroscopy and the hydride vapor generation technique. Flame atomic absorption was used to measure Mn, Fe, and Zn. Electrothermal atomic absorption was used to measure Li, Cr, Co, Cu, Cd, Ba, and Pb. Errors were less than 4% for all the ions, except for Fe and Pb that had standard deviation equal to 0.004 mg/kg on average. Organic matter (%) in sediments was determined using the loss on ignition method, incinerating the dried sample in a muffle furnace at 550°C for 16–20 hours (Dean *et al.* 1974). Ash percent and organic matter content were determined from the difference in weight before and after ignition in the furnace.

Spectrophotometric methods were used to determine boron, nitrate, phosphate, chloride, and sulfate, using a Hach spectrophotometer. Errors were ±0.20 mg/l for boron, ±0.010 mg/l for nitrate, ±0.01 mg/l for phosphate, and ±0.9mg/l for sulfate.

12.3   RESULTS

12.3.1   *Water chemistry*

Table 12.1 displays Ilopango lake water chemistry results. Chloride (range, 36–390 mg/l) and sulfate (range, 3–73 mg/l) results follow the same trend with site S13 having the highest value. Phosphate (range, 0.67–2.28 mg/l) has also the highest value at site S13. Nitrate (range, 0.01–0.44 mg/l) does not vary greatly from site to site with the exception of site S10. Site 10 is located in the northwest, near the most populated area around Ilopango lake. Sulfate *versus* chloride in Figure 12.3a shows a clear increasing linear trend. The test of significance of the correlation coefficient (Swan and Sandilands 1995) gives 100% for the confidence level of this correlation. Chloride and sulfate are also weakly correlated with phosphate concentration at the 93% and 94% confidence level (Fig. 12.3b).

Boron (5–8.7 mg/l), and As (0.15–0.77 mg/l) follow the same spatial trend. Sites S2, S4, S9, S12, and S13 all have high values relative to the other sites. Arsenic *versus* boron concentrations is presented in Figure 12.4a. For this pair the correlation is significant at the 99.999% level. Cobalt (below detection or <2 μg/l), iron (below detection or <105 μg/l), zinc (5–18 μg/l), barium (0.11–0.90 mg/l), and copper (below detection or <54 μg/l) all have peak values at site S4, but the

Table 12.1.   Chemical composition of nutrients (phosphate and nitrate), sulfate, chloride, boron, and trace elements of Ilopango lake waters, at 5 m depth.

Sample location	Cl⁻ mg/l	PO_4^{3-} mg/l	SO_4^{2-} mg/l	NO_3^- mg/l	B mg/l	AS mg/l	Ba mg/l	Co µg/l	Fe µg/l	Zn µg/l	Cu µg/l	Pb µg/l	Ni µg/l	Cr µg/l	Li µg/l
S1	58	1.98	12	0.09	1.5	0.15	0.64	BD	44	11	43	3	4	29.7	90
S2	36	0.94	3	0.01	7.4	0.77	0.82	BD	11	11	39	BD	3	29.7	57
S4	116	0.67	25	0.08	6.8	0.68	0.90	2	105	18	54	2	4	29.8	29
S5	86	1.56	20	0.06	4.8	0.49	0.47	1	BD	5	54	1	34	29.8	6
S6	108	1.28	23	0.06	5.3	0.54	0.42	1	BD	15	39	2	4	29.8	9
S7	140	0.92	25	0.09	1.8	0.29	0.65	BD	28	15	39	2	12	29.8	BD
S8	248	1.24	42	0.06	4.5	0.46	0.69	BD	8	9	54	1	2	29.9	BD
S9	88	0.82	16	0.11	7.1	0.78	0.66	BD	30.0	16	54	2	3	30.0	9
S10	104	0.94	22	0.44	2.6	0.34	0.36	1	11	13	54	2	2	30.1	BD
S11	88	1.08	19	0.10	5.8	0.42	0.11	1	BD	12	BD	3	2	30.1	BD
S12	64	0.94	13	0.08	8.6	0.69	0.45	BD	68	14	39	2	1	30.2	4
S13	390	2.28	73	0.08	8.7	0.75	0.31	1	BD	16	39	1	2	30.2	3

BD = below detection level.

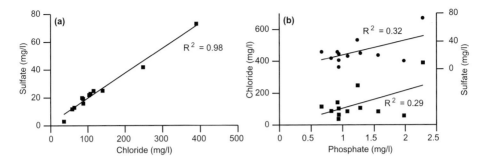

Figure 12.3. Correlation between concentrations of sulfate *vs.* chloride (a); sulfate *vs.* phosphate (•) and chloride *vs.* phosphate (■) (b) for Ilopango lake waters.

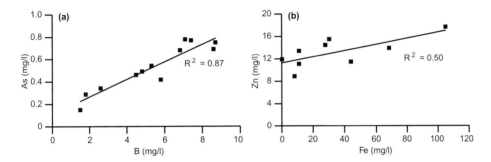

Figure 12.4. Correlation between concentrations of As *vs.* B (a) and Zn *vs.* Fe (b) for Ilopango lake waters.

other sites are randomly variable. However, a positive correlation between zinc and iron is observed in Figure 12.4b at the 99% confidence level. Lead (below detection or <3 μg/l) was very variable and chromium (29.7–30.2 μg/l) did not vary greatly from site to site.

12.3.2 Sediment chemistry

The sediments collected at Ilopango lake varied in color from light gray (S5, S6, S7, S8, S9, S10, S11, S13) to dark brown (S1, S12). All were poorly sorted and had abundant pumice and dacitic rock fragments. Gastropod shells were found in all the samples except S7, with percentages ranging from $<1\%$ (S2, S6, and S12), 2% for S8, 4% for S3 and S13, 20% for S1, 40% for S10 and S11, 50% for S5, 70% for S9, and 90% for S9. The percentage of gastropod shells was determined by visual estimation. Sediment % organic matter ranged from 2.9 to 12.3% (Table 12.2). Organic matter percentage varied around the lake with the highest value at S9 followed by S12.

Table 12.2 displays sediment chemistry results. Boron (1.71–3.55 g/kg), As (5.6–103.4 mg/kg) and Li (12.2–56.7 mg/kg), display similar trends through sites S10 to S13 with S11 being the highest and S13 having the lowest value of the 4 sites. Iron (4.5–23.0 g/kg), Mn (0.13–0.39 g/kg), Cu (0.9–3.4 mg/kg), Pb (1.9–7.7 mg/kg), Co (1.2–3.2 mg/kg), and Cr (0.12–0.99 mg/kg) all have high values for S1, S13 or less commonly S6. In general the trends are quite similar among these six elements, especially from sites S10 to S13.

Strong correlation between As and B is observed at the 99.6% confidence level. Lithium is correlated with As (Fig. 12.5a) at the 97% confidence level. Lithium and As are positively correlated with organic matter at the 100% and 95% confidence level, respectively. The spatial distribution of As and B in the water and sediments of Ilopango lake can be observed in Figures 12.6a to 12.6d. For the waters, As and B concentrations are highest at sites S13 and S9. In comparison for the sediments, the highest concentrations occur at site 11.

Table 12.2.   Chemical composition of trace elements and organic matter of Ilopango lake sediments.

Sample location	% Organic matter	B g/kg	Fe g/kg	Mn g/kg	Zn g/kg	As mg/kg	Cu mg/kg	Pb mg/kg	Ba mg/kg	Co mg/kg	Cr mg/kg	Li mg/kg
S1	3.4	2.73	23.0	0.29	0.54	86.2	2.7	3.3	76.3	2.0	0.3	12.2
S2	2.9	2.01	12.4	0.21	0.46	24.8	1.3	3.4	211.9	2.2	0.2	15.4
S3	3.7	1.71	19.7	0.19	0.49	BD	1.0	2.2	70.3	1.8	0.2	25.3
S4	7.5	2.12	7.3	0.16	0.52	32.1	0.9	3.3	99.2	1.6	0.2	30.0
S5	7.7	2.04	14.2	0.15	0.32	62.9	1.5	2.5	100.4	1.9	0.3	33.4
S6	4.9	3.04	12.8	0.15	0.55	52.4	2.3	4.3	63.7	2.8	0.3	20.9
S7	3.2	3.55	8.7	0.17	0.27	5.6	1.1	1.9	79.9	1.6	0.1	12.7
S8	3.9	2.29	8.5	0.13	0.50	23.4	1.1	2.6	68.6	1.9	0.2	18.6
S9	12.3	2.32	4.5	0.16	0.52	76.6	1.4	3.4	91.3	1.2	0.2	56.7
S10	8.1	2.7	7.0	0.19	0.53	50.8	0.9	3.5	85.8	1.6	0.2	28.5
S11	9.5	3.45	4.6	0.16	0.52	103.4	1.9	3.1	217.4	1.3	0.2	55.0
S12	10.4	2.85	22.2	0.39	0.43	51.9	3.4	7.7	83.2	3.2	1.0	42.6
S13	4.5	2.11	19.7	0.17	0.53	17.2	2.4	6.0	78.2	2.6	0.5	24.1

BD = below detection level.

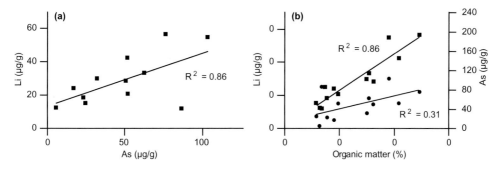

Figure 12.5. Correlation between concentrations of Li *vs.* As (a) and Li (■) and As (•) *vs.* % of organic matter (b) in sediments of Ilopango lake.

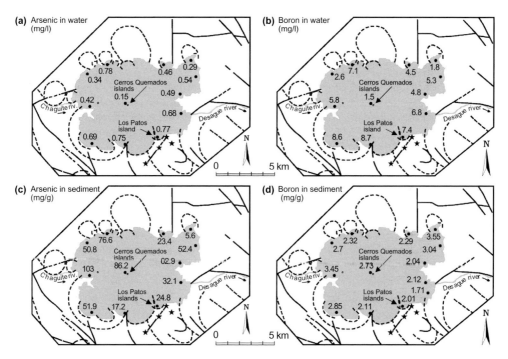

Figure 12.6. Concentration of As and B in water (a and b) and sediments (c and d) of Ilopango lake. Symbols as in Figure 12.2.

12.4 DISCUSSION OF RESULTS

For Ilopango waters, Site S13 is the location of highest concentrations of Cl^-, PO_4^{3-}, SO_4^{2-}, and B. S13 also had relatively high concentrations of zinc. Site S12 had high concentrations in B, As, and Fe. Site S4 also stands out among the other sites with highest concentrations of Zn, Fe, Ba, Co and Cu. Site S4 is located near the outlet of Ilopango lake (Fig. 12.2). These high concentrations could be the result of element transport and deposition at the outlet. The concentrations of the following chemical species exceed water quality standards of the World Health Organization (Howard, 1997): Cl^- (range 36–390 mg/l, standard 250 mg/l), As (range 0.15–0.77 mg/l, standard 0.01 mg/l),

B (range 1.5–8.7 mg/l, standard 0.5 mg/l). Barium (range 0.11–0.82 mg/l) exceeds the Council of European Communities standard (0.1 mg/l; Howard 1997). Phosphate (range 0.67–2.28 mg/l) and As exceed the USEPA standards (Howard 1997), of 0.05 and 0.01 mg/l, respectively. Phosphate is considerably higher than the criterion for aquatic life given by USEPA (0.1 mg/l). These values show that Cl, B, As, Ba, and phosphate have concentrations that should be impacting on the quality of Ilopango lake. The lake has problems of excessive algae growth, especially to the south where the highest concentrations of phosphate were found. The water circulates in the lake with a counterclockwise direction and it seems to slow down at the southwest corner. Arsenic concentrations in water have higher values to the south and lower values close to Cerros Quemados islands (at the lake center), where the last volcanic activity happened in 1880. In comparison, the As concentration in the sediments was high at Cerros Quemados (86 mg/kg sediment), which also has the lowest concentration of organic matter.

The concentrations of Ilopango lake sediments presented in Table 12.2 can be compared with the Canadian sediment quality guidelines (Canadian Council of Ministers of the Environment, 2001). The range of values found in the sediments of this study for As (5.6–103.4 g/kg) and zinc (270–550 g/kg) exceed the PEL (probable effect levels) of 5.9–17.0 g/kg and 123–315 g/kg, respectively. For other elements, such as boron and barium, guidelines have not yet been established. Zinc can be released into the environment through burning of fossil fuels as well as from rocks containing zinc-rich minerals. Arsenic can occur in waters as the result of erosion of natural deposits (USEPA 2001) and is likely coming from the sediment in Ilopango lake and the weathering of rock exposed to air and water within the caldera walls.

Ilopango lake is a volcanic lake, so it was not expected that all of the elements determined in this study would be within their water quality limits. Varekamp et al. (2000) collected data from various volcanic lakes throughout the world, and characterized the different types of volcanic lakes based on Na^+, Mg^{2+}, SO_4^{2-}, and Cl^- concentrations and pH. A linear logarithmic correlation between sulfate and chloride for many volcanic lakes of the world was identified in their work, with values ranging from 10 to 100,000 mg/l for both variables. This large range reflects "dilute" quiescent lakes as well as the crater lakes of active volcanoes, which have very high ion concentrations. For Ilopango lake, a statistically significant linear correlation between sulfate and chloride is observed in Figure 12.3a consistent with its volcanic nature and suggesting a volcanic origin for these chemicals.

Varekamp et al. (2000) also related pH and the sum of chloride and sulfate concentrations to the type of volcanic lake. They found that "CO_2 dominated lakes" have pH ranging from 5 to 9 and $Cl^- + SO_4^{2-}$ ranging from 0.3 to 10 mg/l. Lakes classified as "quiescent lakes" have pH ranging from 1 to 9 and $Cl^- + SO_4^{2-}$ ranging from 10 to 3000 mg/l, while "active crateric lakes" have pH ranging from 1 to 3 and $Cl^- + SO_4^{2-}$ ranging from 3000 to 200,000 mg/l. According to this classification, Ilopango lake with $Cl^- + SO_4^{2-}$ ranging from 39 to 463 mg/l, and pH usually between 8 and 9, is a "quiescent lake". These results are consistent with the low flux of soil CO_2 and mercury found in diffuse degassing studies undertaken within and around the caldera (López et al. 2004). However, the fact that the lake is in a quiescent state does not mean that it cannot have low fluxes of volcanic gases. Soil gases $\delta^{13}C$ values in the caldera (-13 and -20) suggest low magmatic inputs (López et al. 2004). Less negative isotopic values correspond to areas of anomalous carbon dioxide concentration and flux values. In addition, the southern area of the lake where we find the highest values for As, B, Cl^-, and SO_4^{2-}, also showed the highest anomalies for soil carbon dioxide and radon (López et al. 2004).

Chloride and sulfate concentrations found in water samples collected around the perimeter of the lake, and the radon concentration at the soils of the caldera taken close to the lake's shore (López et al. 2004), are plotted in Figure 12.7. The curve for radon has been smoothed using a 3 points moving average. The site numbers for the diffuse soil gas samples as reported in López et al. (2004) are presented in the x-axis. Diffuse soil gas sampling sites are separated by around 200 m. Site numbers for the water and sediment samples are labeled on the graph. Sites are presented in clockwise direction; refer to Figure 12.2 for location of sites. High values of soil radon are observed to the south of the caldera (points 13 to 27 in the diffuse soil degassing study), especially to the southwest (points 20 to 27). Note that even with the sparse data collected, the three variables have similar spatial trends, suggesting a common source.

For the sediments, statistically significant correlations are found for Li *vs.* As (Fig. 12.5a) at the 97% confidence level. A good correlation between Li and As in the sediments is consistent with their common volcanic origin. However, as it can be observed in Tables 12.1 and 12.2, the sites with maximum concentrations in the sediments for As and B, differ from those in the water. To understand this behavior, As and B for the water samples and soil radon were plotted *versus* site number (for diffuse soil gases studies) in Figure 12.8a, and As, Li, and B in the sediment samples versus site number in Figure 12.8b. Note that As and B in the water are high to the south, with B following better the trend of soil radon and As higher at sites S2, S13, and S12. In comparison, the sediment concentrations have an inverse behavior with respect to the water samples for As and B, with lower values to the south. The trends in the sediments followed by As and B are also observed for Li. Li in the water was detectable only in a few points (see Table 12.1).

The spatial distribution of As, B, and Li in the water and sediments of Ilopango lake can be explained in two different ways. First, Figures 12.7 and 12.8 suggest that As and B, and probably Li are leached from the sediments and incorporated into the lake water. Gases such as CO_2 and radon are discharged from the soils of the caldera with high anomalies to the south (López *et al.* 2004). Volcanic gases can contain elements such as As (Signorelli 1997) but they are probably undersaturated with respect to As minerals. Note that Cerros Quemados sample S1 has the highest concentration of As in the sediments, consistent with its younger age that implies less reaction time with water. In the same way, other gases of volcanic-hydrothermal origin such as hydrochloric acid and hydrogen sulfide could be discharging into the soils and lake. These gases or their dissolved species could be transferred throughout the lake sediments, where they could be reacting with the minerals and glass. The reaction products of these leaching reactions should include As, Li, and B, which will be transferred to the lake water, as well as sulfate and chloride.

Phosphate is also high at S13 and has a positive correlation with chloride and sulfate (Fig. 12.3b). Minerals such as apatite, common in volcanic rocks, could be the source for this chemical, in addition to the obvious fertilizer source. Note that the southern part of the lake is less developed and it is covered mainly by forest instead of crops. The second possibility is that the chemical compositions of water and sediments in Ilopango lake could be the result of the water movement and deposition of suspended particles within the lake. The area to the southwest of the lake is characterized by quieter

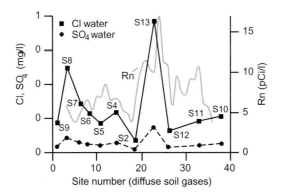

Figure 12.7.   Concentration of chloride, sulfate, and radon soil gas concentration *vs.* distance along the lake's shore. Site numbers are points along the lake's shore that were sampled for diffuse soil degassing. Radon curve has been smoothed with 3 point moving average. Site numbers for the diffuse soil gas samples as reported in López *et al.* (2004) are shown on the x-axis. Diffuse soil gas sampling sites were separated by around 200 m. Sites are presented in clockwise direction; refer to Figure 12.2 for location of water sampling sites.

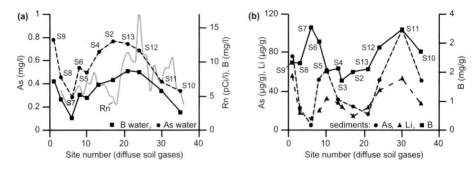

Figure 12.8.   (a) Concentration of As, Li and B in water *vs.* distance along the lake's shore. Site numbers on the x-axis are points along the lake's shore that were sampled for diffuse soil degassing. Diffuse soil gas sampling sites were separated by around 200 m. The radon curve has been smoothed with 3 point moving average. (b) Concentration of As, Li, and B in sediments *vs.* distance along the lake's shore.

water where sediments can settle to the bottom (S12). Statistically significant positive correlations between Li or As, and the organic matter content of the sediments (Figure 12.5b) suggest that this is a possible mechanism. Abundant shells were found in the majority of sediment samples. The live matter that was filling the shells would contribute to the amount of organic matter in the sediment samples. However, high Cl^-, and SO_4^{2-} concentrations in the water column to the south of the lake and the inverse correlation between sediments and water for As, B, Cl, and SO_4^{2-} concentrations cannot be explained only with sediment transport processes. More likely a combination of the two processes is happening at Ilopango lake.

12.5   CONCLUSIONS

Ilopango caldera is currently quiet with respect to its volcanic activity. Water sample locations S12 and S13, which are located within the region of higher radon anomalies, have anomalously high concentrations of Cl^-, PO_4^{3-}, SO_4^{2-}, B, and As in the waters, but low concentrations in the sediments. Contaminants of concern in the lake include Cl^-, PO_4^{3-}, B, As, and Ba. Boron preferentially partitions into the amorphous silicate material during magma fractionation and solidification (Reyes and Trompetter 1998). Leaching and erosion from the volcanic deposits of the lake basin could be the source of contaminants in Ilopango lake. The Tierra Blanca Joven pyroclastic deposit present in the caldera is rich in ash (glass), and it is likely to be the source of B, As, Li, and Ba, as well as other trace elements present in the water and sediments of Ilopango. Chloride, lithium, and boron all increase proportionally with silica content in volcanic rocks (Reyes and Trompetter 1998). During hydrothermal alteration of rhyolites, Cl partitions into solution while Li and B have an affinity for the rock. The distribution coefficients of lithium and boron between water and rock increases with increasing temperature (Reyes and Trompetter 1998).

Even when Ilopango caldera is in quiet state, fluxes of volcanic gases appear to occur at the south of the caldera, where seismic activity is higher, and anomalous concentrations of soil gases, especially radon, are present. Concentrations of As, Cl^-, B, PO_4^{3-}, and SO_4^{2-} in the water are also higher to the south but the sediments concentrations for As, B, and Li seem lower in this region. Leaching of the sediments by the volcanic-hydrothermal gases may enrich the water in these elements. However, these results are preliminary because the distance between sampling points for the soils gases on the margin of the lake is large (around 200 m; Ransom 2002) and relatively few points have been collected for the water and sediments (see Fig. 12.2). In addition, the distance between the bottom sediments and the collected water is between 3 and 32 m (water collected at 5 m depth). Future studies should include collection of water samples close to the bottom and at different depths, and the sediment pore waters. Soil gases, soil and water temperature measurements, water and sediment samples should be collected at distances of less than 100 m to identify the areas most likely to have hydrothermal fluids discharging into the lake. Identification of those areas is important for monitoring of the volcanic activity of the lake as well as for lake water quality management.

The results of this study suggest that the water of Ilopango lake cannot be used for drinking purposes and that the population living on the lake's shore and using lake water are exposed to a high health risk. Fish farms are located to the east of the lake and there are plans to expand these farms to other locations. The fish that are farmed in Ilopango grow rapidly and are assumed to have low accumulation of contaminants due to their short life. A complete assessment of the fish As, B, and Ba concentrations is needed to determine if it they are good for human consumption. New fish farms should not be located to the south of the lake where the concentrations of contaminants appear higher.

If the ash of the last calderic eruption of Ilopango (Tierra Blanca Joven, about 1600 years ago) is one of the sources of As, Li, B, and Ba for the waters of Ilopango lake, a more comprehensive study of this unit is needed. The ash of this eruption covers all El Salvador and part of Central America and could be the source of As contamination for other aquifers.

ACKNOWLEDGEMENTS

We thank the *Fundación de Amigos del Lago de Ilopango* and LaGeo for their logistic support and field assistance.

REFERENCES

Bhattacharya, P., Claesson, M., Bundschuh, J., Sracek, O., Fagerberg, J., Jacks, G., Martin, R.A., Storniolo, A.R. and Thir, J.M.: Distribution and mobility of arsenic in the Rio Dulce alluvial aquifers in Santiago del Estero Province, Argentina. *Sci. Total Environ.* 358 (2006), pp. 97–120.

Bradford, G.R.: Boron. In: H.D. Chapman (ed.): *Diagnostic criteria for plants and soils.* Division of Agricultural Science, University of California, Riverside, 1966, p. 33.

Bundschuh, J., Farías, B., Martin, R., Storniolo, A., Bhattacharya, P., Cortes, J., Bonorino, G. and Albouy, R.: Groundwater arsenic in the Chaco-Pampean Plain, Argentina: case study from Robles county, Santiago del Estero Province. *Appl. Geochem.* 19:2 (2004), pp. 231–243.

Canadian Council of Ministers of the Environment: Canadian sediment quality guidelines for the protection of aquatic life. Canadian Environmental Quality Guidelines. Updated from 1999 edition, Canada, 2001.

Craw, D., Chappell, D. and Reay, A.: Environmental mercury and arsenic sources in fossil hydrothermal systems, Northland, New Zealand. *Environ. Geol.* 39 (2000), pp. 875–887.

Cui, C.G. and Liu, Z.H.: Chemical speciation and distribution of arsenic in water, suspended solids and sediment of Xiangjiang River, China. *Sci. Total Environ.* 77 (1988), pp. 69–82.

Dean, W.E. Jr.: Determination of carbonate and organic matter in calcareous sediments and sedimentary rocks by loss on ignition: comparison with other methods. *J. Sed. Petrol.* 44 (1974), pp. 242–248.

Durocher, N.L.: Preliminary air pollution survey of boron and its compounds. A literature review prepared under Contract No. PH 22-68-25, Public Health Service, National Air Pollution Control Administration, US Department of Health, Education and Welfare, Raleigh, NC, 1969.

Dull, R.A.: *El bosque perdido: A cultural-ecological history of Holocene environmental change in western El Salvador.* PhD thesis, Berkeley, University of California, CA, 2001.

Fergusson, J.E.: *The heavy elements; chemistry, environmental impact and health effects.* Pergamon Press, Oxford, UK, 1991.

Geographical Institute of El Salvador: *Geological map of El Salvador, Central America.* San Salvador, El Salvador, 1970 (1964–1970).

Giggenbach, W.F.: Redox processes governing the chemistry of fumarolic gas discharges from White Island, New Zealand. *Appl. Geochem.* 2 (1988), pp. 143–162.

Goodyear, W.A.: Earthquake and volcanic phenomena, December 1979 and January 1880 in the Republic of El Salvador, Central America. Star and Herald, Panama City, Panama, 1980, p. 56.

Hart, W.J.E.: *The panchimalco tephra, El Salvador, Central America.* MSc Thesis, The State University, Rutgers, NJ, 1981.

Howard, K.W.F.: Impacts of urban development on ground water. In: Environmental geology of urban areas. GEOTEXT3, Geological Association of Canada, St. John's, Canada, 1997.

Irgolic, K.J.: Speciation of arsenic compounds in water supplies. Report No. EPA-600/S1-82-010, US Environmental Protection Agency, Research Triangle Park, NC, 1982.

Lantzy, R.J. and MacKenzie, F.T.: Atmospheric trace metals: global cycles and assessment of man's impact. *Geochim. Cosmochim. Acta* 43 (1979), pp. 511–525.

Lemmo, N.V., Faust, S.O., Belton, T. and Tucker, R.: Assessment of the chemical and biological significance of arsenical compounds in a heavily contaminated watershed. Part 1: The fate and speciation of arsenical compounds in aquatic environments—A literature review. *J. Environ. Sci. Health* A18 (1983), p. 335.

Lide, D.R.: *CRC Handbook of chemistry and physics.* 81 st ed., 2000–2001, CRC Press, Boca Raton, Florida, 2000.

López, D.L. and Hernández, J.: Anthropogenic and volcanic pollution at Ilopango Lake, El Salvador. *18th International Symposium of the North American Lake Management Society* (NAMLS), Banff, Alberta, Canada, 1998, p. 116.

Lopez, D.L., Ransom, L., Perez, N., Hernandez, P. and Monterrosa, J.: Dynamics of diffuse degassing at Ilopango Caldera, El Salvador. In: W.I. Rose, J.J. Bommer, D.L. López, M.J. Carr and J.J. Major (eds): *Natural Hazards in El Salvador.* Geological Society of America Special Paper No. 375, 2004, Boulder, CO, pp. 191–202.

Mann, C.P., Stix, J. and Richer, M.: Subaqueous intracaldera volcanism, Ilopango Caldera, El Salvador, Central America. In: W.I. Rose, J.J. Bommer, D.L. López, M.J. Carr and J.J. Major (eds): *Natural Hazards in El Salvador.* Geological Society of America Special Paper No. 375, 2004, Boulder, CO, pp. 159–174.

Meyer-Abich, H.: Los volcanes activos de Guatemala y El Salvador (America Central). Anales del Servicio Geológico Nacional de El Salvador, San Salvador, El Salvador, 1956.

Molnar, P. and Sykes, L.: Tectonics of the Caribbean and Middle America regions from focal mechanisms and seismicity. *Geol. Soc. Am. Bull.* 80:9 (1969), pp. 16–1684.

National Research Council: *Arsenic in drinking water.* National Research Council, Subcommittee on Arsenic in Drinking Water. National Academy Press, Washington, DC, 1999.

Nriagu, J.O.: Global inventory of natural and anthropogenic emissions of trace metals to the atmosphere. *Nature* 279 (1979), pp. 409–411.

Nriagu, J.O.: A global assessment of natural sources of atmospheric trace metals. *Nature* 338 (1989), pp. 47–49.

Nriagu, J.O. and Pacyna, J.M.: Quantitative assessment of worldwide contamination of air, water and soils by trace metals. *Nature* 333 (1988), pp. 134–139.

Ransom, L.: *Volcanic diffuse soil degassing and lake chemistry of the Ilopango Caldera system, El Salvador, Central America.* MSc Thesis, Athens, Ohio University, OH, 2002.

Reyes, A.G. and Trompetter, W.J.: Lithium, boron, and chloride in volcanics and greywackes in Northland, Auckland and Taupo volcanic zone. Institute of Geological and Nuclear Science, Institute of Geological and Nuclear Science Report 1171-9184, 98/21, 1998.

Richer, M., Mann, C. and Stix, J.: Mafic magma injection triggers eruption at Ilopango Caldera, El Salvador, Central America. In: W.I. Rose, J.J. Bommer, D.L. López, M.J. Carr and J.J. Major (eds): *Natural Hazards in El Salvador.* Geological Society of America Special Paper No. 375, 2004, Boulder, CO, pp. 175–189.

Sheets, P.D.: Volcanic disasters and the archaeological record. In: D.K. Grayson (ed): *Volcanic activity and human ecology.* Academic Press, New York, 1979.

Signorelli, S.: Arsenic in volcanic gases: *Environ. Geol.* 32:4 (1997), pp. 239–244.

Sracek, O., Bhattacharya, P., Jacks, G., Gustafsson, J.P. and Von Brömssen, M.: Behavior of arsenic and geochemical modeling of arsenic enrichment in aqueous environments. *Appl. Geochem.* 19 (2004), pp. 169–180.

Swan, A.R.H. and Sandilands, M.: *Introduction to geological data analysis.* Blackwell Science, Ltd., Cambridge, UK, 1995.

USEPA: National Primary Drinking Water Standards. United States Environmental Protection Agency, EPA 816-F-01-007, 2001.

Varekamp, J.C., Pasternack, G.B. and Rowe Jr., G.L.: Volcanic lake systematics II. Chemical constraints. *J. Volcanol. Geotherm. Res.* 97 (2000), pp. 161–179.

Welch, A.H.: Arsenic in ground water of the western United States. *Ground Water* 26 (1988), pp. 333–347.

White, R. and Harlow, D.: Destructive upper-crustal earthquakes of Central America since 1900. *Bull. Seismol. Soc. Am.* 83:4 (1993), pp. 1114–1115.

Williams, H. and Meyer-Abich, H.: Volcanism in the southern part of El Salvador, with particular reference to the collapse basins of lakes Coatepeque and Ilopango. *University of California Publ. Geol. Sci.* 32 (1955), pp. 1–64.

CHAPTER 13

The abundance of natural arsenic in deep thermal fluids of geothermal and petroleum reservoirs in Mexico

P. Birkle
Instituto de Investigaciones Eléctricas, Gerencia de Geotermia, Cuernavaca, Mor., Mexico

J. Bundschuh
International Technical Cooperation Program, CIM (GTZ/BA), Frankfurt, Germany
Instituto Costarricense de Electricidad (ICE), San José, Costa Rica

ABSTRACT: The lack of chemical similarity between thermal fluids in geothermal and petroleum reservoirs in Mexico indicates a distinct origin for arsenic (As) in both systems. Deep fluids from geothermal reservoirs along the Transmexican volcanic belt are characterized by elevated As concentrations, within an average range of 1 to 100 mg/l at a depth from 600 to 3000 m b.s.l. The lack of correlation between As and salinity reflects the importance of secondary water-rock interaction processes. The predominance of As compared to Fe and Cu concentrations, and the abundance of secondary minerals in temperature-dependent hydrothermal zones, support this hypothesis. Oilfield waters from sedimentary basins in SE-Mexico show maximum As concentrations of 2 mg/l, at a depth from 2900 to 6100 m b.s.l. The linear Cl^-/As correlation for oilfield waters indicates that As input occurs during the mixing between meteoric water and evaporated sea-water, and that there is only minor As derived from interaction with carbonate host rock.

13.1   INTRODUCTION

In general, there is little data available for the metal and non-metal composition of thermal groundwater in Mexican geothermal and petroleum reservoirs. As a basic tool, chemical-physical parameters such as salinity, pH, temperature, and water percentage (the latter in oil reservoirs) are measured primarily to characterize basic reservoir conditions. Infrequently, a sequence of major ions (Na^+, Ca^{2+}, Mg^{2+}, K^+, Cl^-, HCO_3^-, SO_4^{2-}) is determined for the general hydrochemical classification of water types. Sometimes hydrochemical studies include the determination of some minor elements such as Si, in the case of geothermal fluids, and Sr, Rb and Cs, to reconstruct the origin of deep fluids and to analyze scaling problems during reinjection. The analysis of metals and non-metals is mainly limited to Fe and As concentrations. A preliminary comparison of the compositional link between thermal fluids in Mexican deep reservoirs has been presented in Birkle (2005).

This article presents a summary of published data for As concentrations in fluids from deep geothermal reservoirs in Mexico, as well as a published and new analytical data from several petroleum reservoirs. The data for both reservoir types are compared to determine the origin of As in formation water from main geothermal and petroleum reservoirs in Mexico.

13.2   METHODS

Basic chemical data was collected from previously published articles and reports. Mercado *et al.* (1989) and Lippmann *et al.* (1999) present chemical data, while Portugal *et al.* (2000) and Mazor and Mañon (1979) describe both chemical and $\delta^{18}O$ and δD compositions for the Cerro Prieto

Table 13.1. Minimum and maximum arsenic concentrations of formation water and its corresponding extraction depth at the respective geothermal and petroleum reservoir.

	As_{min} (mg/l)	As_{max} (mg/l)	$Depth_{min}$ (m b.s.l.)	$Depth_{max}$ (m b.s.l.)	Reference
Geothermal fields					
Cerro Prieto	0.25	1.5	675	1200	Lippmann *et al.* 1999, Mercado *et al.*1989
Los Azufres	5.1	49.6	628	2188	Birkle 1998, González *et al.* 2000
Los Humeros	0.5	162	1330	3060	González *et al.* 2001, Arellano *et al.* 2003
Petroleum fields					
Cactus-Sitio Grande	<0.003	0.047	3585	4545	Birkle and Portugal 2001, Birkle and Angulo 2005
Luna-Sen	<0.003	0.548	4726	6121	Birkle *et al.* 2002
Jujo-Tecominoacán	<0.003	1.89	5237	6151	Birkle 2004, Birkle *et al.* 2008 (in print)
Pol-Chuc-Abkatún	0.09	2.01	2925	4817	Birkle 2003

fluids. Chemical data from 24 fluids samples from the Los Humeros geothermal field were taken from González *et al.* (2001) and Arellano *et al.* (2001), and extraction depths from Portugal *et al.* (2001) and Arellano *et al.* (2003). Chemical analysis from 17 geothermal wells in Los Azufres were taken from González *et al.* (2000). Chemical composition and interpretations about the origin of formation water from the Activo Luna oil field were taken from Birkle *et al.* (2002). Hydrochemical data from 28 production wells of the Samaria-Sitio Grande reservoir are given in Birkle and Portugal (2001), Birkle and Maruri (2003), Birkle and Angulo (2005) and Birkle *et al.* (2006). The chemical composition of Pol-Chuc fluids from 29 production wells, as well as interpretations of the hydrogeological-isotopic reservoir model, were extracted from Birkle (2003). Hydrochemical data from 19 production wells from the Jujo-Tecominoacán petroleum reservoir taken from Birkle 2004 and Birkle *et al.* 2008 (in print).

The samples for the analysis of As in oil field waters were stored in HDPE bottles, pre-filtered with 0.45 μm Millipore filters, acidified with HNO_3-suprapur, and analyzed by *ACTLABS*, Ontario, Canada, with the ICP-MS technique. Arsenic concentrations in geothermal fluids were generally determined with the atomic adsorption spectrometry (AAS) method. A summary of minimum and maximum As concentrations, as well as the corresponding production depths is shown in Table 13.1.

13.3   LOCATION AND HYDROGEOLOGICAL DESCRIPTION OF THE RESERVOIR

The Cerro Prieto geothermal field in northern Baja California is hosted in deltaic sands and shales of the southern Salton Sea (Fig. 13.1). Fluids are extracted from depths between 800 and 3000 m with an average reservoir thickness of 1900 m and temperatures above 260°C.

Los Azufres is one of several Pleistocene silicic volcanic centers, with active geothermal systems in the E–W trending Transmexican volcanic belt (TMVB). A 2700 m thick interstratification of lava flows and pyroclastic rocks of andesitic to basaltic composition (Dobson and Mahood 1985), provide the main aquifer with fluid flow through fractures and faults that sometimes reach the surface (Birkle *et al.* 2001). The production wells extract vapor and liquid from a depth between 350 and 2500 m. A silicic sequence of rhyodacites, rhyolites and dacites, with a thickness of up to 1000 m (Dobson and Mahood 1985), form a caprock seal to the aquifer, allowing the geothermal system to pressurize. The NaCl-rich fluids reach temperatures as high as 320°C, but temperatures of 240–280°C are normal in the field.

The Los Humeros geothermal field is located in the eastern part of the Transmexican volcanic belt, with a total of 42 wells of which 22 are currently used for electricity generation. Metamorphosed carbonate forms the basement below a low-liquid-saturation reservoir, which is located at

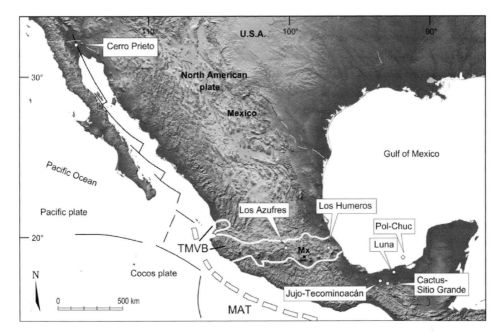

Figure 13.1.   Location of the geothermal and petroleum reservoirs referred to in this study, for which hydrochemical analytical data are available for deep formation water (Mx: Mexico City, TMVB:
Transmexican volcanic belt, MAT: Middle America trench).

a depth between 1950 and 2700 m (850–100 m a.s.l.). Basalt and hornblende andesites of intermediate permeability form the host rock of the lower reservoir, whereas augite andesite hosts
the upper geothermal reservoir (800–1700 m) (Cedillo 1997 and 1999, Arellano *et al.* 2001).
Both aquifers are separated by impermeable vitreous tuff, and overlain by low permeable lithic
tuffs and ignimbrites. At the top, pumice, olivine basalts and andesites form shallow aquifer
systems.

The Luna-Sen and the Cactus-Sitio Grande oil reservoirs are located onshore within the southeastern coastal plain of the Gulf of Mexico. Both oil fields form part of the NW-SE trending
Villahermosa uplift horst structure, which is separated from the Comalcalco basin to the NW, and
from the Macuspana basin to the SE by the Comalcalco and Frontera faults. The Villahermosa
uplift comprises between 5000 m and 10,000 m of Tertiary and Mesozoic sediments in the western and eastern part, respectively (Salvador 1991). Formation water is produced as an undesired
co-product of the petroleum exploitation, mainly from Jurassic to Cretaceous limestone and
dolomite formations. The initial piezometric level of the water-hydrocarbon contact was encountered at a depth between 5900 m and 6100 m at Luna-Sen, and between 3800 m and 4500 m at
Cactus-Sitio Grande. The Jujo-Tecominoacán reservoir is located close to the Cactus-Sitio Grande
reservoir, but produces at greater depth, from 5200 to 6150 m b.s.l. The Pol-Chuc reservoir, together
with the adjacent Abkatún, Batab, Caan, and Taratunich oil fields, is located within the Gulf of
Mexico, about 80 km offshore from the town Ciudad del Carmen. Dolomitized carbonates in
Paleocene breccia and Upper Cretaceous calcareous sandstone are part of the NW–SE trending
anticlines that host the principal Pol reservoir at a production depth between 2900 and 4800 m. The
original water-oil contact was detected at a depth of 3960 m.

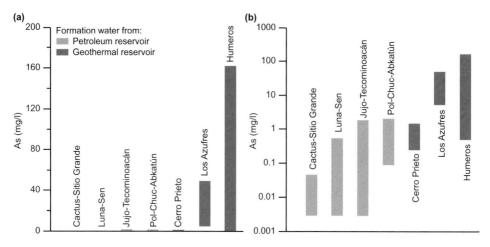

Figure 13.2.   Arsenic concentrations in formation water from petroleum and geothermal reservoirs in Mexico: (a) linear scale, (b) logarithmic scale. The minimum concentration shown of 0.003 mg/l (for Cactus-Sitio Grande, Luna-Sen and Jujo-Tecominoacán fluids) is the detection limit for the analytical method used.

13.4   RESULTS

13.4.1   *Arsenic concentrations in geothermal fluids*

The abundance of hydrothermal minerals in the volcanic host rocks of the Los Azufres geothermal field (state of Michoacán, Central Mexico), as well as elevated As concentrations between 5.1 and 49.6 mg/l in the geothermal brines, indicate the importance of dissolution and exchange processes between hot fluids and As-enriched host rock (Fig. 13.2a, Table 13.1).

Elevated natural As concentrations of up to 3.9 mg/l in surface manifestations, such as hot springs and fumaroles, are probably due to the vertical ascent of convective fluids to the surface (Birkle and Merkel 2000). Deep waters from the Los Humeros geothermal field, located in the eastern part of the Transmexican volcanic belt, show a wide range of As concentrations from 0.5 mg/l (Well H-15) towards 162 mg/l (Well H-12). The latter is the highest As concentration detected to date in deep reservoirs in Mexico. However, As-bearing primary minerals have not been reported from the host rocks. The logarithmic scale in Figure 13.2b illustrates the wide range of As concentrations from Los Humeros fluids compared to other geothermal fluids.

In contrast, the dominance of sandstones in the sedimentary basin of the Cerro Prieto geothermal field (Baja California state, NW-Mexico) explains the relatively low As concentrations from 0.25 to 1.5 mg/l in reservoir fluids, despite bottom-hole temperatures that are extremely elevated (max. 370°C). The sedimentary origin of the reservoir rocks explains the similarity between As concentrations in Cerro Prieto fluids and those of oil reservoir fluids (Fig. 13.2b).

In order to avoid environmental impacts of As-enriched, deep geothermal fluids on the surface environment, it is essential to maintain a closed production cycle between extraction wells, energy generation and reinjection wells (Birkle and Merkel 2000).

13.4.2   *Arsenic concentrations in oil field fluids*

In comparison to geothermal reservoir fluids in Mexico, deep fluids from petroleum reservoirs in SE-Mexico are depleted in As (Fig. 13.2a). The reservoirs of Pol-Chuc, Abkatún, Batab, Caan, and Taratunich (named "Pol-Chuc-Abkatún" in Figs. 13.2a and b), located 80 km off-shore the Gulf coast, have a combined maximum As concentration of 2.01 mg/l, at a depth from 2910 to 4658 m b.s.l. Off-shore fluids are characterized by minimum As concentrations of 0.09 mg/l, whereas on-shore fluids (Cactus-Sitio Grande, Luna-Sen, Jujo-Tecominoacán) have minimum As concentrations below the analytical detection limit (<0.003 mg/l). Formation water from the Cactus, Nispero and Sitio Grande oil reservoirs (state of Tabasco) are located at a depth from 3585 to 4545 m b.s.l. Although these fluids can be hypersaline (up to 257,000 mg/l TDS), As concentrations (<0.003 to 0.047 mg/l) are relatively low. This effect can be attributed to the carbonate host rock type; calcareous sandstone, dolomitized mudstone and brecciated and fractured dolomite units from Early–Late Cretaceous period. These have little ore and only minor hydrothermal mineralization. Relatively low reservoir temperatures of around 130°C, do not favor extensive geochemical reactions.

13.4.3   *Vertical dependence of arsenic*

Different aquifer types from geothermal and petroleum reservoirs in Mexico can be distinguished by plotting the As concentration against depth. In general, geothermal fluids are located in shallower horizons (depth: 675 to 3060 m b.s.l.) than the petroleum reservoirs (depth: 2925 to 6151 m b.s.l.). As shown in Table 13.1 and Figure 13.3, geothermal formation waters have As concentrations of 1 to more than 100 mg/l, whereas petroleum fluids show lower but more variable concentrations of <0.003 to 2 mg/l. Overall, a general decrease in the As concentration with increasing depth can be observed. However, each individual field is characterized by a heterogeneous distribution of As concentrations, also reflecting variability in total salinity, and no depth-related trend can be observed.

The variability in the Los Humeros reservoir can be explained by two fluids types. Low mineralized waters form part of a liquid-dominated, bicarbonate reservoir at a depth from 1330 to 1755 m b.s.l., characterized by a smaller range and lower concentrations of As. In contrast, deeper wells (1985–3060 m b.s.l.) produce a two-phase fluid with maximum As concentrations of 162 mg/l. Main hydrothermal zones are formed by chlorite, epidote, quartz, calcite, and low proportions of leucoxene and pyrite, as well as clays, biotite, zeolite, anhydrite, garnet, diopside and wollastonite (Izquierdo *et al.* 2000). However, no primary As minerals have been recognized.

13.4.4   *Origin of arsenic*

Positive correlation trends between As and other elements facilitate the reconstruction of diagenetic links in the past. Using chloride as a typical non-reactive, conservative constituent, formation waters from petroleum fields show a positive correlation with As concentrations (Fig. 13.4a). The extreme heterogeneity in chloride concentrations is caused by the infiltration of both meteoric water and extremely evaporated seawater during Late Pleistocene–Early Holocene (Birkle and Angulo 2005, Birkle *et al.* 2002). The individual mixing trend between the two end members is shown for each reservoir on Figure 13.4a. The abundance and variable concentration of As in oil field waters is mainly related to the infiltration and mixing of these surface waters. On the other hand, the postulated reduction of the initial seawater volume by up to 30 times during the evaporation process (Birkle and Angulo 2005), does not completely explain the degree of As enrichment in hypersaline fluids from Mexican oil reservoirs (up to 2.01 mg/l As), compared to a typical seawater composition (0.0026 mg/l As). An exclusive seawater origin for As would require a reduction of more than 700 times in the initial seawater volume.

In contrast, the lack of correlation between Cl and As concentrations for formation waters from geothermal reservoirs, indicates that additional secondary processes, such as water-rock interaction, has caused the elevated As concentrations (Fig. 13.4b). An analogous behavior can be observed for most major elements, such as calcium.

Comparing the concentrations of other minor elements in deep reservoir fluids, iron is extremely abundant (maximum concentration: 1029 mg/l Fe) in formation water from on-shore oil fields. Some off-shore waters from Pol-Chuc (closed circles in Fig. 13.5) are also enriched in As (maximum concentration: 2.01 mg/l As). It is likely that the relatively low temperatures (between 120 and

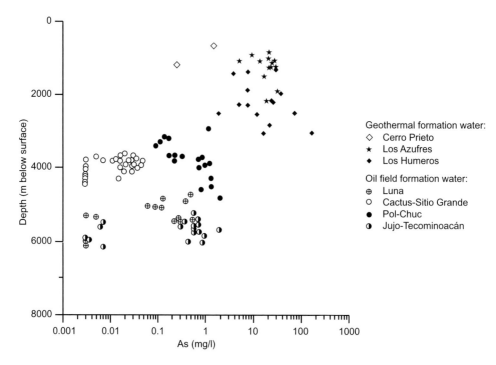

Figure 13.3.   Vertical distribution of As concentrations in formation water from petroleum and geothermal reservoirs.

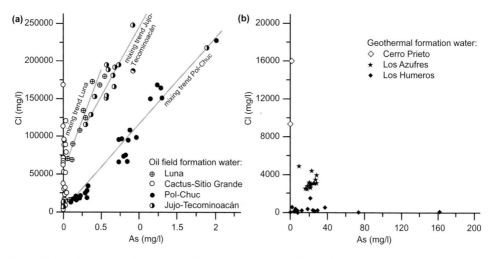

Figure 13.4.   Correlation of arsenic *vs*. chloride concentration of formation waters from petroleum (a) and geothermal (b) reservoirs in Mexico.

170°C) of the sedimentary petroleum basins prevent the hydrothermal formation of secondary Cu-Fe ore minerals, such as a pyrite, arsenopyrite and chalcopyrite. In contrast, copper and iron concentrations in Los Azufres geothermal fluids are very low (maximum concentrations: 0.029 mg/l Cu and 0.8 mg/l Fe), which may be due to the precipitation of secondary Cu-Fe minerals by secondary water-rock interaction.

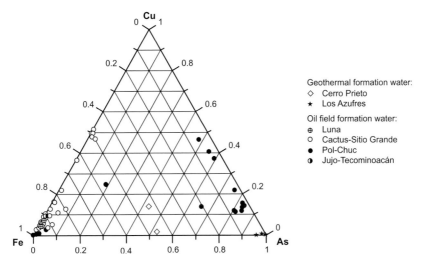

Figure 13.5. Triangular diagram of As, Cu and Fe concentrations in formation water from petroleum and geothermal reservoirs in Mexico. Cu and Fe data is unavailable for Los Humeros fluids.

The elevated temperature conditions in the Los Azufres reservoir (>260°C) probably allow a higher solubility and fluid enrichment of the metaloid As, in comparison to metal ions such as Fe and Cu. Chlorite, epidote, quartz, calcite and sericite are major secondary hydrothermal minerals in the Los Azufres reservoir, as well as minor zeolite, several types of clay minerals, potassium feldspar, albite, prehnite, amphibole, hematite and pyrite (Torres 1996, González *et al.* 2000).

The geothermal fluids from the Cerro Prieto sedimentary basin have a composition intermediate to those of petroleum fluids and Los Azufres geothermal fluids. This demonstrates the effect of high temperature reactivity (as is typical for geothermal reservoirs) on sedimentary host rocks with a low natural content of As (as is typical for petroleum carbonate reservoirs). Also, the infiltration of Colorado river water into the reservoir (Lippmann *et al.* 1991) could lead to dilution of the reservoir with low-As river water. The existence of temperature-dependent hydrothermal zones, from a low temperature diagenetic zone, to a illite-chlorite zone (above 150°C), a calc-aluminum silicate zone (above 230°C) and finally the lowermost biotite zone (above 325°C), reflects post-depositional alteration of the Cerro Prieto reservoir units (Elders *et al.* 1981).

13.5   CONCLUSIONS

The studied geothermal fluids at Los Azufres and especially Los Humeros are strongly affected by water-rock interaction processes under elevated temperatures conditions (>280°C). Although fluid salinity is relatively low (maximum TDS: 15,000 and 2000 mg/l, respectively), the combination of physical and chemical conditions causes an enrichment in As concentrations through the dissolution of volcanic host rock. In contrast, low saline to hypersaline waters in Mexican oil reservoirs (maximum TDS: 257,000 mg/l) are more depleted in As. This concentration of As is instead attributed to As enrichment during evaporation of seawater before infiltration, and to minor interaction processes with sedimentary host rocks under lower temperature conditions (<170°C). Secondary As precipitation may occur, but has yet not been observed as specific primary As minerals in petroleum and geothermal reservoirs.

ACKNOWLEDGEMENTS

We are grateful for the thoughtful review of the manuscript by D. Kirk Nordstrom and Jenny Webster.

REFERENCES

Arellano, V.M., Izquierdo, G., Aragón, A., Barragán, R.M., García, A. and Pizano, A.: Distribución de presión inicial en el campo geotérmico de Los Humeros, Puebla, México. *Ingeniería Hidráulica en México* XVI, 2001, pp. 75–84.

Arellano, V.M., García, A., Barragán, R.M., Izquierdo, G., Aragón, A. and Nieva, D.: An updated conceptual model of the Los Humeros geothermal reservoir (Mexico). *J. Volcanol. Geotherm. Res.* 124:1/2 (2003), pp. 67–88.

Birkle, P.: Herkunft und Umweltauswirkungen der Geothermalwässer von Los Azufres. *Wiss. Mitt.* 6, Institute for Geology, Technical University Freiberg, Germany, 1998.

Birkle, P.: Estudio isotópico y químico para la definición del origen de los acuíferos profundos del Activo Pol-Chuc. Final report, Instituto de Investigaciones Eléctricas, Cuernavaca, Mexico, IIE/11/12093/I02/F, 2003.

Birkle, P.: Caracterización química e isotópica de los acuíferos profundos del campo Jujo-Tecominoacán. Final report, Instituto de Investigaciones Eléctricas, Cuernavaca, Mexico, IIE/11/2473/F, 2004.

Birkle, P.: Compositional link between thermal fluids in Mexican deep reservoirs. *Proc. World Geothermal Congress 2005*, April 24–29, Antalya, Turkey, Session 8: Geochemistry, 2005.

Birkle, P., Martínez, B.G. and Milland, C.P.: Origin and evolution of formation water at the Jujo-Tecominoacán oil reservoir, Gulf of Mexico, Part 1: Chemical evolution and water-rock interaction. *Appl. Geochem.* (2008, in print).

Birkle, P. and Merkel, B.: Environmental impact by spill of geothermal fluids at the geothermal field of Los Azufres, Michoacán, México. *Water Air Soil Pollut.* 124:3/4 (2000), pp. 371–410.

Birkle, P., Merkel, B., Portugal, E. and Torres, I.: The origin of reservoir fluids at the geothermal field of Los Azufres, México—Isotopic and hydrological indications. *Appl. Geochem.* 16 (2001), pp. 1595–1610.

Birkle, P. and Portugal, E.: Caracterización química e isotópica de los acuíferos profundos de los campos petroleros Cactus, Níspero, Río Nuevo y Sitio grande en Chiapas: origen y dinámica de la migración. Final report 11-11840, Instituto de Investigaciones Eléctricas, Cuernavaca, Mexico, 2001.

Birkle, P., Portugal, E., Rosillo, J.J. and Fong, J.L.: Evolution and origin of deep reservoir water at the Activo Luna oil field, Gulf of Mexico, Mexico. *AAPG Bull.* 86:3 (2002), pp. 457–484.

Birkle, P. and Maruri, R.A.: Isotopic indications for the origin of formation water at the Activo Samaria-Sitio-Grande oil field, Mexico. *J. Geochem. Explor.* 78–79 (2003), pp. 453–458.

Birkle, P. and Angulo, M.: Conceptual hydrochemical model of Late Pleistocene aquifers at the Samaria-Sitio-Grande petroleum reservoir, Gulf of Mexico, Mexico. *Appl. Geochem.* 20:6 (2005), pp. 1077–1098.

Birkle, P., Angulo, M. and Lima, S.: Hydrochemical-isotopic tendencies to define hydraulic mobility of formation water at the Samaria-Sitio-Grande oil field, Mexico. *J. Hydrology* 317:3/4 (2006), pp. 202–220.

Cedillo, F.: Geología del subsuelo del campo geotérmico de Los Humeros, Puebla. Internal report HU/RE/03/97, Comisión Federal de Electricidad, Gerencia de Proyectos Geotermoeléctricos, Residencia Los Humeros, Puebla, Mexico, 1997.

Cedillo, F.: Modelo hidrogeológico de los yacimientos geotérmicos de Los Humeros, Puebla, México. *Geotermia* 15 (1999), pp. 159–170.

Dobson, P.F. and Mahood, G.A.: Volcanic stratigraphy of the Los Azufres geothermal area, Mexico. *J. Volcanol. Geotherm. Res.* 25 (1985), pp. 273–287.

Elders, W.A., Hoagland, J.R. and Williams, A.E.: Distribution of hydrothermal mineral zones in the Cerro Prieto geothermal field of Baja California. *Geothermics* 10 (1981), pp. 245–253.

González, E.P., Birkle, P. and Torres-Alvarado, I.: Evolution of the hydrothermal system at the geothermal field of Los Azufres, Mexico, based on fluid inclusion, isotopic and petrologic data. *J. Volcanol. Geotherm. Res.* 104:1–4 (2000), pp. 277–296.

González, E.P., Tello, E.H. and Verma, M.P.: Interacción agua geotérmica-manantiales en el campo geotérmico de Los Humeros, Puebla, México. *Ingeniería Hidráulica en México* XVI(2), 2001, pp. 185–194.

Izquierdo, G., Arellano, V.M., Aragón, A., Portugal, E. and Martínez, I.: Fluid acidity and hydrothermal alteration at the Los Humeros geothermal reservoir, Puebla, Mexico. *Proc. World Geothermal Congress*, Kyushu-Tohoku, Japan, 2000, pp. 1301–1306.

Lippmann, M.J., Truesdell, A.H., Halfman-Dooley, S.E. and Mañon, A.: A review of the hydrogeologic-geochemical model for Cerro Prieto. *Geothermics* 20 (1991), pp. 39–52.

Lippmann, M.J., Truesdell, A.H. and Frye, G.: The Cerro Prieto and Salton Sea geothermal fields—Are they really alike? *Proc. 24th Workshop on Geothermal Reservoir Engineering*, Stanford University, California, SGP-TR-162, 1999.

Mazor, E. and Mañon, A.M.: Geochemical tracing in producing geothermal fluids: A case study at Cerro Prieto. *Geothermics* 8 (1979), pp. 231–240.

Mercado, S., Bermejo, F., Hurtado, R., Terrazas, B. and Hernández, L.: Scale incidence on production pipes of Cerro Prieto geothermal wells. *Geothermics* 18:1/2 (1989), pp. 225–232.

Portugal, E., Izquierdo, G., Barragán, R.M. and Birkle, P.: Isotopy of fluids from Los Humeros system (Mexico). *Proc. 10th Int. Symposium on Water-Rock Interaction* WRI-10, Villasimius, Italy, 2001, pp. 907–910.

Portugal, E., Izquierdo, G., Verma, M.V., Barragán, R.M., Pérez, A. and de León, J.: Isotopic and chemical behaviour of fluids of Cerro Prieto geothermal field, Mexico, *Proc. 21st Annual PNOC-EDC Geothermal Conference*, Manila, Philippines, 2000, pp. 227–232.

Salvador, A.: *The Gulf of Mexico Basin.* In: The geology of North America. Geological Society of America, Boulder, CO, 1991.

Torres, I.A.: *Wasser/Gesteins-Wechselwirkung im geothermischen Feld von Los Azufres, Mexiko: Mineralogische, thermochemische und isotopengeochemische Untersuchungen.* PhD Thesis, Tübinger Geowissenschaftliche Arbeiten, Reihe E, Nr. 2, Germany, 1996.

CHAPTER 14

Development of a geographic information system for Zimapán municipality in Hidalgo, Mexico

E. Ortiz, R. Reséndiz, E. Ramírez & V. Mugica
Universidad Autónoma Metropolitana-Azcapotzalco, Col Reynosa, Tamaulipas, Mexico City, Mexico

M.A. Armienta
Instituto de Geofísica, Universidad Nacional Autónoma de México (UNAM), Mexico City, Mexico

ABSTRACT: A methodology of geographical information system (GIS) was developed to help the identification of sites, and potential environmental arsenic (As) pollution sources, in the municipality of Zimapán Hidalgo in central Mexico. This system incorporates published information at the AGEB level (basic geo-statistic units, according to Spanish abbreviation), for the following variables: geographical location of the study area, land use, geology and hydrogeology, meteorology, water supply sources, human population, industry location, urban services and transport information. The present study gathers information reported on the topic to create GIS-based methodology that allows identification of places bearing higher risks for population exposure to As. The purpose is to establish a methodology for their application in integral studies, and to support decision-making processes. The need for using and analyzing a huge volume of spatial hazards and As exposure data in a fast and reasonably accurate way, makes GIS a powerful tool for effective environmental risk assessment and management.

14.1   INTRODUCTION

Epidemiological studies of several human populations have demonstrated that chronic exposure to high levels of arsenic (As) in drinking water induces cancer in diverse tissues such as skin, liver, lung, bladder and kidney. Arsenic is clastogenic, induces sister chromatid exchanges, causes spindle disturbance, interferes with DNA repair, and is also a co-mutagen but does not produce a mutagenic effect by itself. High concentrations of As in drinking water constitute a public health risk in several countries such as Argentina, Chile, United States, Poland, Hungary, Taiwan, China, Canada, India and Bangladesh (Bhattacharya *et al.* 2006, Bundschuh *et al.* 2005, World Bank 2005).

Natural and anthropogenic As in groundwater and soil has been detected in Zimapán valley, Mexico. About 35% of the analyzed samples from this site exceed the Mexican standard for As concentration in drinking water (25 μg/l) (NOM 2000). Due to the complexity of the geology and hydrogeology of the valley, it is difficult to identify the source of natural As. Interaction of groundwater with As-bearing rocks has been proposed as one of the three main sources of As in Zimapán valley. There is a geological association for As minerals in areas of skarn and chimney/manto in calcareous rocks. The smelter gases in the area of Zimapán carried As, which has contaminated shallow water wells (Armienta *et al.* 1996, 2001). In addition, mining wastes, mainly tailings (*jales*) act as anthropogenic sources, since under oxidizing conditions sulfides are oxidized to sulfates and As is released as arsenate (Rimstidt *et al.* 1994). Identification of the sources and migration mechanisms for As have allowed establishment of low-cost and low-tech remediation procedures (Ongley 2001).

An indirect measurement of As concentration circulating in humans was carried out in a study by Reséndiz and Zuniga (2003) who determined As levels in the hair of the residents of Zimapán.

They found an average concentration of 9.74 mg/kg, whereas a range between 0.3 and 1.75 mg/kg has been reported by the World Health Organization. Likewise, the reports indicated that 56.1% of the sampled population showed some skin symptoms such as hyperkeratosis, hypopigmentation and hyperpigmentation.

Most of the environmental studies use great quantities of information that has to be stored, processed, analyzed and in turn be presented in written or graphical form. The use of GIS addresses and simplifies handling databases. In essence, GIS is a technological platform to pursue spatial analyses capable of providing coherent information storage, simultaneously allowing its upgrade and manipulation. The GIS facilitates maps, the design of cartographic models, and implementation of effective analysis tools, like signaling of corridors from a certain distance to a river or highway, interaction charts between two or more maps, calculation of slopes and display of measures, among other things. All of this begins with the transformation or combination of diverse geographical data.

The goal of this work is to develop a methodology that allows the systematization of information through spatial analysis of the pollution source to drinking water supplies, and their interaction with human population's exposure to As. This information could be useful in the future for mitigation strategies, and for economical, technical and social optimization of resources.

14.2   GEOLOGICAL CHARACTERISTICS OF THE STUDY AREA

The Zimapán valley is located in the western part of the state of Hidalgo about 200 km north of Mexico City. The town of Zimapán is located at 20°44′20″ N and 99°22′58″ W, at an approximate elevation of 1750 m a.s.l. It is surrounded by mountains up to 2720 m a.s.l. (Sierra del Monte, Sierra Daxi). This valley is limited to the north by the municipality of Pacula, to the east by Nicolás Flores, to the southeast by Tecozautla and to the west through the river Moctezuma and the state of Querétaro (Fig. 14.1). The municipality of Zimapán has an area of 800 km^2 constituted by the villages: Benito Juárez, (Detzani), Venustiano Carranza, (San Pedro), Francisco I. Madero (Guadalupe), Alvaro Obregón (Temuthe), Zimapán Centro and other small communities.

The annual rainfall is ~400 mm in Zimapán Centro and increases to the north and west to ~800 mm. The evapotranspiration potential ranges from 365–400 mm. The residents of the municipality of Zimapán depend on groundwater sources which they exploit using chain pump wells (*norias*), and drilled wells. The drilled wells (mostly municipal) are deeper than 100 m. Water for municipal supply is routed to a central mixing tank in which it is chlorinated and then distributed to residents. About 9000 people are served by the municipal water system, 25% of the total municipal population (Resendiz and Zuñiga 2003). Zimapán has been a mining district since the 16th century; currently, ore extraction and processing still are the main economic activities of the region.

The geological formations in the Zimapán valley are Cretaceous age limestones and shales and Tertiary igneous rocks (Simmons and Mapes-Vázquez 1956). Zimapán is located in the fold and thrust belt of central Mexico and lies in the lead-zinc-silver metallogenic province. Upper Jurassic to Upper Cretaceous rocks compose the mountains in most of the district. Tertiary rocks are also present consisting of continental sediments, intermediate volcanics, plutonic and hypabyssal monzonite and abundant dikes of varied composition. Late Tertiary and Quaternary semi-consolidated alluvium covers much of the low lying areas. The oldest rocks belong to Las Trancas formation (Simons and Mapes-Vázquez 1956) of Upper Jurassic–Lower Cretaceous. These rocks are composed of thin-bedded shallow limestone interbedded with laminated shale that conformably lies under the Abra/Tamaulipas. The fractured cretaceous age limestones (mainly Tamaulipas and Soyatal formations) constitute the most productive aquifer exploited by deep wells. Groundwater flow is controlled by a fracture system with a regional NW–SE direction. The El Morro and the volcanic rocks also have secondary porosity and permeability caused by presence of fractures. Shallow and deep wells with low production rates located to the east of the valley tap these younger units. The

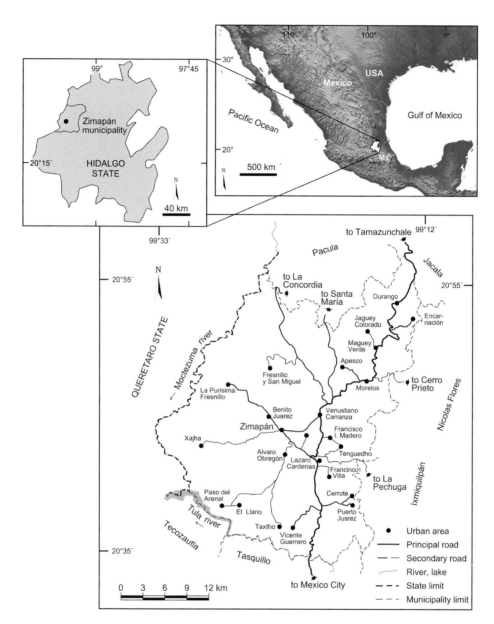

Figure 14.1.   Location of Zimapán in Hidalgo state.

Zimapán fanglomerate acts as an unconfined aquifer in the north-central part of the valley; many dug wells exploit this unit (Armienta *et al.* 1997).

14.3   METHODOLOGY

To detail the relation between the most important factors that define the region regarding pollutant generation, a GIS was developed. The INEGI cartography of Zimapán, Hidalgo was added to the GIS. The minimal unit chosen was at block level. A digital model of land was generated with the topography of the study area. Before analysis, Arcview 3.2, Autocadmap 2000 and Surfer, Dbase, and Excel were selected as the most appropriate formats and programs. It was also necessary to design data processing systems in language C++, and SQL that allowed the use and administration of the information of the databases in different formats.

The methodology features and techniques are:

- The establishment of assessment approaches in the information treatment based on:
    - Cartographic control procedures.
    - Mathematical calculations to standardize units.
- Databases design that feeds GIS:
    - Defining the basic unit of analysis.
    - Adapting the data to the analysis unit.
- Cartographic representation:
    - Determination of the number of levels of information per map.
    - Establishment of statistical ranges for the classifications.
    - Definition of cartographic symbols to represent the variables.
    - Regionalization of the areas that identify the sources and mechanisms of migration of pollutants.

First, it was necessary to gather the information about the locations of mine tailings, geology, hydrology, roads and political divisions (Simmons and Mapes 1956, INEGI 2001, Mendez 2002, Mendez and Armienta 2003). Information about physical and chemical characteristics of water and soils, as well as the content of As was obtained from different scientific studies carried out in the area (Armienta *et al.* 1997, Armienta *et al.* 2001). With the aid of GIS all this information was spatially referred. The Arcview application was selected due to its relative user friendliness and its generalized use by Mexican authorities and research centers.

Software and equipment: The GIS was elaborated in a PC with Pentium IV with a platform Arcinfo 3.5 using as render Arcview 3.2. Also, some processes were carried out through digital incorporation of information starting from a platform Autocadmap 2000.

14.3.1 *Base map*

The information layer known as AGEB contains the limits of the 11 basic geostatistic areas that compose Zimapán, Hidalgo. This layer was used as a fundamental graphical element to establish the reference for all the information layers that make up the system (see Fig. 14.2a). The database was structured adding the following information.

14.3.2 *Population*

The theme population was built with information taken from the year 2000 INEGI's population census (Table 14.1). Population and housing variables were considered for this theme, such as overall population, male and female, age strata, housing types and religion, among others. The items: roads, streets and services constitute the urban trace. This information was supplied by the Municipal President's office.

14.3.3 *Geology and hydrogeology*

All the information contained in this section and in its corresponding register, was technically developed by Armienta *et al.* (1996, 1997), at *Instituto de Geofísica*, UNAM. In this study the fields were digitized and geo-referenced, and the bases were structured for each of the theme charts that support the spatial analysis based on the selection of sites, which potentially contaminate potable water sources.

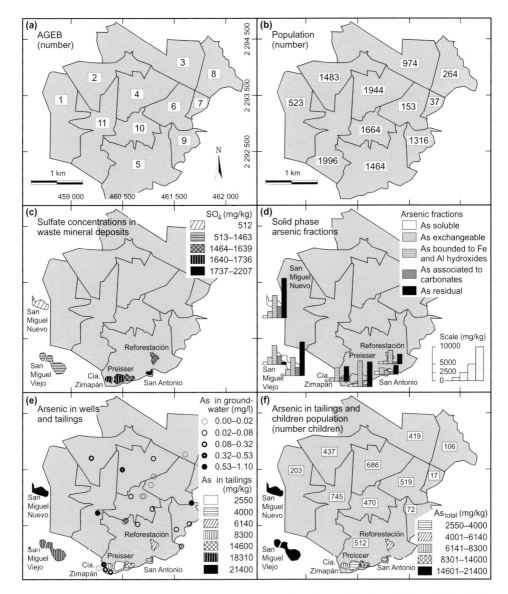

Figure 14.2. (a) Basic geo-statistic units (AGEB's) of the Zimapán municipality, Hidalgo; (b) Population distribution; (c) Sulfate concentrations in waste mineral deposits; (d) Arsenic solid phase distribution in mine tailings; (e) Arsenic concentration in wells and tailings; (f) Concentration of As in mine tailings and children population by AGEB.

14.3.4   *Wells and chain pump wells (norias)*

Specific location of water wells and chain pump wells was performed using as control basis the INEGI's 1:50,000 scale topography charts. Their geographical coordinates were calculated and the data base generated which feeds the GIS for the creation and analysis of the theme charts. This information defines a theme called wells. Chemical, physicochemical and contaminant concentration in the water variables were incorporated into these thematic maps. The data in Table 14.2 is an example of these (Armienta *et al.* 1993).

Table 14.1.   Summary of population database included in the GIS.

ID	Key	Total population	Females	Males	Children
1	017-3	523	251.	272	74
2	018-8	1483	936	547	161
3	023-9	974	470	504	122
4	021-A	1944	893	1051	206
5	019-2	1464	666	798	153
6	022-4	1316	594	722	162
7	024-3	37	15	22	4
8	026-2	264	127	137	38
9	025-8	153	77	76	20
10	009-9	1664	760	904	148
11	020-5	1996	931	1065	234

Table 14.2.   Concentration of As in wells.

ID	Key	As (μg/l)	ID	Key	As (μg/l)
1	Well	bdl	8	Well	57.00
2	Well	52.6	10	Well	bd
3	Well	52.5	13	Well	bd
4	Well	80.0	18	*Noria*	32.13
5	Well	38.50	19	Well	43.70
6	Well	74.30	20	Well	bd
7	Well	77.10	21	*Noria*	23.71

bd: less than 0.50 μg/l.

14.4   RESULTS AND DISCUSSION

Figure 14.2b represents the overall population density, where it can be noted that central and southern part of Zimapán have the largest population density, while the smallest density of inhabitants is located to the north.

The geographic location of the waste mineral deposits (Fig. 14.2c) was done using polygons. This theme contains physicochemical information (pH, electrical conductivity and sulfates) and As concentration (soluble, total, interchangeable, associated to hydroxides, residual and associated to carbonates) in the waste mineral deposits taken from Mendez and Armienta (2003). The waste mineral deposits are located in the southern and eastern Zimapán's zones (Fig. 14.2c). Waste mineral deposits contain the sulfides pyrite and pyrrhotite, and other residual minerals that remain after the processes of metal recoveries. Thus an inadequate handling of these residuals may generate waters of low pH with high SO_4^{2-} concentrations (Blowes *et al.* 1990).

In order to know the sulfate distribution in the waste mineral deposits, plots were made and included in Figure 14.2c. Concentration ranges indicate the development of more intense oxidation processes at San Antonio, and Cía Zimapán deposits.

To visually compare the solid phase As fractions among tailing piles, bar graphics were included in Figure 14.2d. The chart shows that waste mineral deposits "San Miguel Viejo" and "San Miguel Nuevo" have a great proportion of residual arsenic and As-associated with hydroxides, which indicates that As in these deposits located in the west area are mainly in the most stable phases.

Four shallow wells containing As are located near the waste mineral deposits of the Cía Minéra Zimapán. One of them (Fig. 14.2e) had a lower As concentration. This well is located crossing the ravine, showing that it works as a geological barrier.

Arsenic contamination of the wells near Cía Minéra Zimapán and Preisser, along with the sulfate and As concentrations in these wastes, indicates that the well-water has been polluted as a result of As release from these tailings.

The distribution of children population related with As concentrations contained in mine tailings is presented in Figure 14.2f. These results show that an important children population lives close to the aforementioned deposits that contaminate the soil, thus posing a risk to the children's health. Although dwellers are aware of the pollution of wells located near tailings, their location in this area poses a risk, and their use as potable sources must be prohibited by the authorities.

14.5 CONCLUSIONS

A decision-making support system was developed with the aid of GIS by mapping the political division, location of wells and mineral waste deposits as well as the population, As concentrations found at different locations in the municipality, and population exposed to pollution in Zimapán, Hidalgo. This information system allows risk assessment by As exposure of the population and in the future could be an important tool for the study of different scenarios to address decision-making and pollution abatement. To sum up, the development of GIS allows the systematization of information through spatial analysis of the pollution source to drinking water supplies and their interaction with human population's exposure to As. This utilization is proved since the results show that an important children population lives close to the aforementioned deposits that contaminate the soil.

REFERENCES

Armienta, M., Rodríguez, R. and Villaseñor G.: *Estudio de reconocimiento de la contaminación por arsénico en la zona de Zimapán, Hidalgo.* Technical report, Instituto de Geofísica, Universidad Nacional Autónoma México (UNAM), Mexico City, Mexico, 1993.

Armienta, M. and Rodríguez, C.: Arsénico en el Valle de Zimapán, México: Problemática ambiental. *MAPFRE Seguridad* 63 (1996), pp. 33–43.

Armienta, M., Rodríguez, R., Aguayo, A., Ceniceros, N., Villaseñor, G. and Cruz, O.: Arsenic contamination of groundwater at Zimapán, México. *Hydrogeol. J.* 5 (1997), pp. 39–46.

Armienta, M.A., Villaseñor, G., Rodriguez, R., Ongley, L.K. and Mango, H.: The role of arsenic-bearing rocks in groundwater pollution at Zimapán Valley, México. *Environ. Geol.* 40 (2001), pp. 571–581.

Bhattacharya, P., Claesson, M., Bundschuh, J., Sracek, O., Fagerberg, J., Jacks, G., Martin, R.A., Storniolo, A.R. and Thir, J.M.: Distribution and mobility of arsenic in the Río Dulce Alluvial aquifers in Santiago del Estero Province, Argentina. *Sci. Total Environ.* 358 (2006), pp. 97–120.

Blowes, D.W., Jambor, J., Cherry, J.A. and Reardon, E.J.: The formation and potential importance of cemented layers inactive sulfide mine tailings. *Geochim. Cosmochim. Acta* 55 (1990), pp. 965–978.

Bundschuh, J., Bhattacharya, P. and Chandrashekharam, D.: *Natural arsenic in groundwater: occurrence, remediation and management.* Taylor and Francis/Balkema, The Netherlands, 2005.

INEGI: Anuario Estadístico. Hidalgo. Instituto Nacional de Estadística Geografía e Informática, Aguascalientes, Mexico, 2001.

Manahan, S.: *Toxicology Chemistry. A guide to toxic substances in chemistry.* Lewis Publishers, Chelsea, MI, 1989.

Mendez, M. and Armienta, M.: Arsenic phase distribution in Zimapán mine tailings, Mexico. *Geofísica Internacional* 42 (2003), pp. 131–140.

Mendez, M.: Fraccionamiento de arsénico en jales de Zimapán Hidalgo. MSc Thesis, Instituto de Geofísica, Universidad Nacional Autónoma México (UNAM), Mexico City, Mexico, 2002.

NOM: Modificación a la Norma Oficial Mexicana NOM-127-SSA1-1994, Salud ambiental, agua para uso y consumo humano—Límites permisibles de calidad y tratamientos a que debe someterse el agua para su potabilización, México, D.F., Diario Oficial de la Federación, 20 de octubre de 2000, Mexico City, Mexico, 2000.

Ongley, L.K., Armienta, M., Heggeman, K., Lathrop, A., Mango, H., Miller, W. and Pickelner, S.: Arsenic removal from contaminated water by the Soyatal formation, Zimapán mining district, Mexico—a potential low cost low-tech remediate system. *Geochem.-Explor Env. A*. 1 (2001), pp. 23–31.

Reséndiz, R. and Zúñiga J.: *Evaluación de la Exposición al Arsénico en pobladores del municipio de Zimapán, Hidalgo, México.* Lic Thesis, Universidad Tecnológica de México (UNITEC), Mexico City, Mexico, 2003.

Rimstidt, J.D., Chermak, J.A. and Gagen, P.M.: Rates of reaction of galena, sphalerite, chalcopyrite, and arsenopyrite with Fe(III) in acidic solutions. In: C.N. Alpers and D.W. Blowes (eds): *Environmental geochemistry of sulfide oxidation*, vol. 550, American Chemical Society, Washington, DC, 1994, pp. 2–13.

Simons, F. and Mapes-Vazquez, E.: Geology and ore deposits of the Zimapán mining district, State of Hidalgo, Mexico. US Geological Survey Professional Paper 284, 1956.

World Bank: Towards a more effective operational response. Arsenic contamination of groundwater in South and East Asian countries Water and Sanitation Program—South Asia. Designed and printed by: Roots Advertising Services Pvt. Ltd. 55 Lodi Estate, New Delhi, India, 2005, http://siteresources.worldbank.org/INTSAREGTOPWATRES/Resources/ArsenicVolI_WholeReport.pdf.

CHAPTER 15

Determining the origin of arsenic in the Lagunera region aquifer, Mexico using geochemical modeling

C. Gutiérrez-Ojeda
Instituto Mexicano de Tecnología del Agua (IMTA), Jiutepec, Mor., Mexico

ABSTRACT: Elevated arsenic (As) concentrations found in groundwater from the alluvial aquifer in the Lagunera region, northern Mexico, have been attributed to several possible sources. Previous studies have shown that this naturally occurring As likely comes from extinct intrusive hydrothermal activity. Geochemical modeling was used in this study to show that under, natural conditions, the evaporated surface water carried by the Nazas and Aguanaval rivers ($0.00201 <$ As < 0.01766 mg/l) may have contributed to the elevated groundwater As concentrations ($0.003 <$ As < 0.443 mg/l) found in lower parts of the closed basin.

15.1 INTRODUCTION

Since 1962, Mexican agencies have reported elevated arsenic (As) concentrations in extensive areas of the unconfined alluvial aquifer in the Lagunera region (Quiñones *et al.* 1979). The aquifer is the main source of drinking water for more than two million people that inhabit the area. Arsenic in the groundwater has caused adverse health effects in both people and animals (García *et al.* 1991). Arsenic is naturally occurring and the reported concentrations are in the range 0.003 to 0.443 mg/l (IMTA 1990, Gutiérrez-Ojeda 1995). Overexploitation of the aquifer has caused groundwater drawdowns of more than 100 m in less than 50 years at the central part of the aquifer and the migration of groundwater with As concentrations well above the Mexican standard (0.025 mg/l) for human use and consumption (NOM 1996).

The Lagunera region is situated in a broad closed basin located in the central part of northern Mexico (Fig. 15.1), between $102°40'$ and $104°$ W longitude and between $25°15'$ and $26°15'$N latitude. It covers a total area of about 12,000 km^2 in parts of the Mexican states of Coahuila, Durango and Zacatecas. The cities of Torreón, Gómez Palacio and Lerdo represent the main urban areas, in which 70% of the total population lives.

According to the Köppen classification, the Lagunera region has a very dry climate and is semi-warm, with an average temperature in the summer of 25°C and 16°C in winter. The mean annual precipitation in the area is 221 mm and the rainy season is from June to October. Evaporation is high, with a mean annual value of 2406 mm, eleven times the mean annual precipitation.

15.2 WATER RESOURCES

The main source of surface water in the region are the Nazas and Aguanaval rivers. The primary groundwater source is the alluvial aquifer (Fig. 15.1). Water from both sources is used to irrigate almost 90,000 ha. Groundwater is used for domestic, industrial and livestock needs.

The Nazas river catchment covers 63% of the total area, representing the main source of surface water and only carries water during the irrigation cycle (March–August). The river is regulated by two reservoirs: Lázaro Cárdenas and Francisco Zarco (operating since 1946 and 1968), with storage capacities of 2778 and 235 million cubic meters (Mm^3), respectively. Water allocation

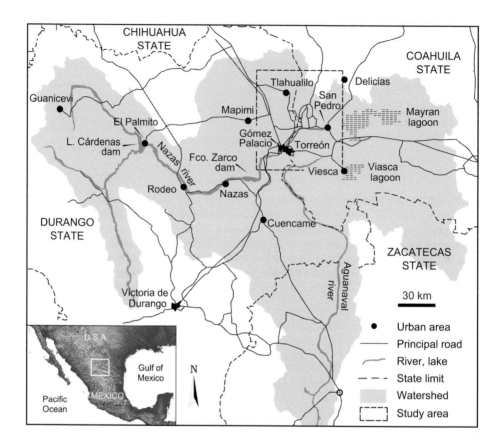

Figure 15.1.   Geographic location of the Lagunera region (IMTA 1990) and Comarca Lagunera watershed.

records indicate that about 600–830 Mm^3 of water from these two reservoirs are used each year in the Lagunera region.

15.3   HYDROGEOLOGY

The aquifer underlies half of the area and contributes 50% of all water used. It is an unconfined aquifer formed by the granular material derived from the Santa Inés formation, and by the alluvial and lacustrine deposits of sedimentary origin that filled the old bolons (closed basins). Gravels, sands, silts and clays from the surrounding mountains are part of these deposits. Its transmissivity ranges from 0.007 m^2/s, in areas close to the rivers, to 0.0005 m^2/s in the flood plains. Its inferred storativity coefficient is about 0.05–0.06. The principal sources of recharge are the Nazas river and infiltration of water from excess irrigation (IMTA 1991, 1992).

Access to groundwater has become a serious problem in recent years because of overexploitation of the aquifer and regulation of the Nazas and Aguanaval rivers. These activities have caused groundwater drawdowns of more than 100 m in less than 50 years, the disappearance of the Mayrán and Viesca lagoons and a decline in groundwater quality. Abstraction is at least two times greater than recharge. The general flow direction is from the borders of the basin to the areas located within the central portion of the region, where an extended cone of depression has formed between the towns of Fco I Madero and Tlahualilo. Depths to the water table range from 40 to more than 110 m (Fig. 15.2a). In some areas water must be pumped from depths greater than 140 m.

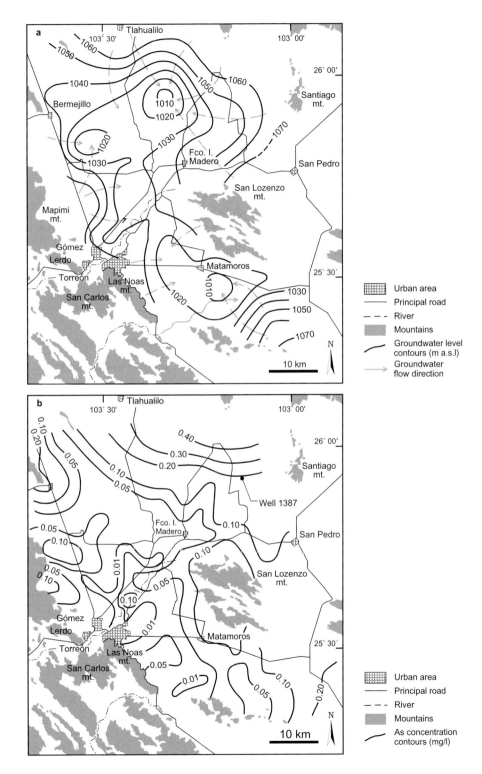

Figure 15.2.   Groundwater table elevations, and flow in 1991; (b) Total As concentrations in 1990.

15.4   ORIGIN OF ARSENIC

The IMTA study performed in 1990 showed that large areas of the Lagunera region aquifer have As concentrations above the Mexican drinking water standard of 0.025 mg/l (Fig. 15.2b). Arsenic concentrations ranged from 0.003 to 0.443 mg/l (mean: 0.074 mg/l; standard deviation: 0.099 mg/l). The highest As concentrations are found in the lagoonal deposits located in the northeastern part of the basin, as well as towards the northwestern and southeastern areas. In general, areas with high As concentrations also have high concentrations of TDS, sulfate, fluoride, chloride, sodium, boron and lithium, above their respective regulatory limits. During the study, water was analyzed from 95 wells distributed over the whole of the main aquifer, of which 41% had concentrations of As above the Mexican drinking water standard.

Samples were collected at each location directly from the water pump. They were stored, without being filtered, in stoppered, polyethylene bottles (previously cleaned with tap water) which were filled to the brim to exclude air. No acids or other preservatives were added. Samples were kept at room temperature (20–25°C) and were analyzed within 4 months of being collected. Field measurements included conductivity, pH and temperature. Samples were analyzed in triplicate and were sent to be analyzed in different laboratories. Arsenic determinations were made with a Varian AA 175 atomic absorption spectrophotometer equipped with a Model 65 vapor generation accessory. No information about the analytical methods used to determine the other parameters was reported.

The presence of As has been related to several potential sources (IMTA 1990): hydrothermal activity, use of arsenical pesticides, mining activities and sedimentary origin. The data analysis showed that As is naturally occurring and that its most probable source is due to extinct, intrusive hydrothermal activity combined with a sedimentary process.

15.5   RESULTS OF GEOCHEMICAL MODELING

Historically, the Nazas and Aguanaval rivers flowed from the upper parts of the basin toward the Mayrán (or Tlahualilo, before 1829) and Viesca lagoons, respectively (Fig. 15.1). The water flowed down to the Lagunera region floodplain and later infiltrated into the aquifer across the permeable layers located along the rivers' courses across the basin. The remaining water reached the lagoons (northern and eastern), located in the lowest parts of the basin.

The Mayrán was a broad lagoon with a radius of approximately 25 km (Instituto de Geología 1937). Floods carried considerable volumes of water, sometimes greater than 2000 Mm^3, to this lagoon.

Given the high mean annual evaporation of 2406 mm, a large fraction of the water from the lagoons was evaporated. The residual, concentrated solution probably infiltrated the aquifer. Isotopic data (IMTA 1990) for groundwater from these areas are inconsistent with high evaporation rates (in contrast to water from the reservoir). However, it is possible that water from the lagoons that had undergone extensive evaporation only infiltrated to shallow depths, considering that the surficial layer of the flood plain is composed mainly of fine sediments (Escolero et al. 1990).

The highest As concentrations are associated with fine-grained, lagoonal deposits located at the northern (Tlahualilo ex-lagoon) and eastern (Mayrán ex-lagoon) margins of the basin (Adams 1993). These areas are where the water table is still shallow.

On the other hand, considering that the geological and hydrochemical conditions have remained unchanged since the formation of the aquifer, it is possible that water reaching the lagoons had a similar composition to that reported for water from the reservoirs.

Areas of volcanic activity in the upper catchments of the rivers and within the Lagunera region are considered to be the most likely sources of As. These areas may provide As to surface waters flowing into the basin (Wallrabe-Adams 1993). Although these areas may have contributed dissolved As, it is likely that sediment-bound As was the predominant form of As transported in suspension by surface water flow. Under the prevailing chemical conditions, the hydrous iron

Table 15.1.   Mean arsenic and chloride concentrations in the aquifer and two reservoirs (from IMTA 1990).

Description	Cl (mg/l)	As (mg/l)
Fco Zarco reservoir	4.80	0.00962
L Cárdenas reservoir	3.80	0.00781
Aquifer	72.57	0.07410
Aquifer/reservoir ratios (%)	126.32	123.81

oxides present in the particulate matter adsorb arsenate extensively. However, the high evaporation rates in the area may have increased significantly As concentrations in solution before infiltration occurred. Indeed the effect of evaporation is evident if we compare solute concentrations determined in water samples from two reservoirs located along the Nazas river (Table 15.1).

The ranges of As and Cl$^-$ concentrations in surface waters from the two reservoirs were 0.00201–0.01766 mg/l and 2.92–15.99 mg/l, respectively (IMTA, 1990, 1992). The ranges in As and Cl$^-$ concentrations in groundwater were 0.00243–0.44300 mg/l and 4.30–709.74 mg/l, respectively.

Thus, it is possible that extensive evaporation in the shallow Tlahualilo and Mayrán ex-lagoons contributed to the high As concentrations found in the northern and eastern regions of the bolson. The possible contribution of surface water evaporation to the observed elevated As concentrations in groundwater was examined as follows.

Calculations were conducted by first choosing water compositions representative of the surface water (the composition of water from reservoirs was chosen) and the groundwater with elevated As concentrations. The evaporation enrichment factor was calculated from the ratio of chloride concentrations of the two waters. A modified version of the aqueous speciation program WateqF was used to calculate the saturation indices of plausible minerals based on the composition of both the evaporated surface water and the groundwater. The geochemical program Netpath (Plummer *et al.* 1991) was then used to determine the net geochemical mass balance reactions between the evaporated surface water (initial water) and the groundwater (final water).

Plummer *et al.* (1991) explain that 'the net geochemical mass balance reactions consists of the masses [per kilogram of water (H_2O)] of plausible minerals and gases that must enter or leave the initial water along the flow path to define the composition of a selected set of chemical constraints observed in the final water'. Netpath examines every possible geochemical mass balance reaction between the initial and final water given a set of chemical constraints and plausible phases (minerals and gases), which should be consistent with the mineralogical composition of the system.

Because of the isotopic signature, as well as the As and Cl$^-$ concentrations, the composition of water from well 1387 was chosen for this analysis. This well is situated in San Pedro county, where high As and Cl$^-$ concentrations have been reported (Fig. 15.2b). The composition of water from sample number 96 (collected from L. Cárdenas reservoir) was considered to be representative for the surface water. A Piper diagram was used to show that well 1387 has sodium-sulfate water while the L. Cárdenas reservoir has calcium-bicarbonate water (IMTA 1990).

The chloride ratio computed from the groundwater (well 1387) and surface water (reservoir sample no. 96) compositions was of K1 = 51.51 (199.87/3.88). Table 15.2 shows the chemical composition of water from well 1387 and reservoir sample no. 96, both determined analytically, and the calculated composition of evaporated surface water (sample no. 96 times K1), called Mod96.

The saturation indices of the plausible minerals, partial pressure of CO_2, ionic strength, I, and the charge imbalance, CI (%), computed from the groundwater (well 1387) and evaporatively concentrated surface water (Mod96) compositions are presented in Table 15.3.

These calculations show that gypsum is subject to dissolution, whereas calcite and fluorite are subject to precipitation. The partial pressure of CO_2 calculated from the composition of Mod96

Table 15.2.   Analytic data of well 1387, sample no. 96 and Mod96 (in mg/l).

Element	Well 1387	Sample no. 96	Mod96
Ca	146.7	27.37	1409.91
Mg	13.21	2.13	109.72
Na	542.26	10.83	557.88
K	7.12	3.84	197.81
Li	0.20	0.02	1.03
HCO_3	168.40	112.20	5779.75
Cl	199.87	3.88	199.87
SO_4	1179.38	10.54	542.95
F	3.22	1.13	58.21
NO_3	43.02	3.38	174.11
CO_3	7.20	–	–
As	0.19250	0.00657	0.34
B	1.43	0.25	12.88
Fe	0.004	0.04	2.06
Pb	0.020	0.20	10.30
pH	8.10	8.10	8.10

Table 15.3.   Geochemical parameters of well 1387 and Mod96 (from WateqF).

Description	Well 1387	Mod96
Calcite	0.666	2.907
Gypsum	−0.634	−0.436
Fluorite	0.295	3.472
pCO_2 (atm)	0.00125	0.0314
Ionic strength I	0.04281	0.11802
CI (%)	−5.58	−8.04

(initial water) is higher than the respective value for the groundwater (well 1387, final water). This suggests that some outgassing of CO_2 may have occurred. Although the extended Debye-Hückel equation is applicable when $I < 0.1$ (Pankow 1991), it was nevertheless used to calculate the activity coefficients for dissolved species in both waters.

Three ion exchange reactions were also considered. These included: the Ca/Na, Ca/K and Ca/Mg exchange reactions. Calcium, which normally dominates the exchange sites of the marine clays in this area, is expected to enter the aqueous solution while excess sodium, potassium and magnesium are expected to be incorporated into the clays. The three ion exchange reactions considered were:

$$\text{Ca/Na EX: } 2Na^+ + \overline{Ca - X_2} \rightarrow Ca^{2+} + \overline{2 \cdot Na - X} \tag{15.1}$$

$$\text{Ca/K EX: } 2K^+ + \overline{Ca - X_2} \rightarrow Ca^{2+} + \overline{2 \cdot K - X} \tag{15.2}$$

$$\text{Ca/Mg EX: } Mg^{2+} + \overline{Ca - X_2} \rightarrow Ca^{2+} + \overline{Mg - X_2} \tag{15.3}$$

Carbon, sulfur, calcium, fluoride, magnesium, potassium and sodium, were selected as chemical constraints. The results of the Netpath calculations indicate that one model satisfied the constraints (Table 15.4).

Table 15.4.   Results from Netpath: First run (in mmol/l).

Initial well: Mod96
Final well: 1387

	Final	Initial
C	2.7290	89.7870
S	12.3060	5.7040
CA	3.6690	35.4990
F	0.1700	3.0920
MG	0.5450	4.5540
K	0.1830	5.1050
NA	23.6410	24.4880

MODEL 1	
CALCITE−	−43.86450
FLUORITE−	−1.46100
GYPSUM+	6.60200
CO_2 GAS	−43.19350
Ca/Na EX	0.42350
Ca/K EX	2.46100
Ca/Mg EX	4.00900

1 model satisfied the constraints.

An additional calculation was performed using Netpath's evaporation option. The composition of surface water from the reservoir (sample No. 96) was used as the initial water. The composition of groundwater from well 1387 was, once again, used as the final water. In contrast to the first set of calculations, fluorite became supersaturated. Calcite and gypsum behaved as described above. The three ion exchange reactions described above were considered again as plausible reactions. The same chemical constraints were considered in this case. Four models were found to satisfy the constraints. Of these, only one model was considered to be possible because it was consistent with the evaporation enrichment factor considered before (based on the chloride ratio). The results were basically the same except that the Ca/Na exchange reaction was no longer required.

15.6   CONCLUSIONS

Evaporation of 51.51 l of surface water from the Nazas river would produce one liter of Mayrán lagoon water requiring the precipitation of calcite (43.86 mmol) and fluorite (1.46 mmol), dissolution of gypsum (6.60 mmol), outgassing of CO_2 (43.19 mmol), and exchange of 1 mmol of calcium (entering the aqueous solution) for 0.42 mmol of sodium, 2.46 mmol of potassium and 4.01 mmol of magnesium (leaving the aqueous solution).

The resulting As concentration (0.34 mg/l), although not exactly the same as observed in well 1389 (0.1925 mg/l), could later be subject to adsorption on iron hydroxides. This could explain the low correlation coefficient between Cl^- and As (r = 0.07). Thus, evaporation can be considered as an important factor in the evolution of high As concentrations in the Lagunera region.

REFERENCES

Escolero, F.O., Mejía, V.R. and Barrera, O.C.: Revisión geológica e interpretación de la geometría del acuífero principal de la Comarca Lagunera, en los estados de Coahuila y Durango, México. Internal report OOA-5/92, Gerencia de Aguas Subterráneas, Comisión Nacional del Agua, Mexico City, Mexico, 1992.

García, V.G., García, R.A., Aguilar, R.M., García, S.J., del Razo, L.M., Otrosky, W.P., Cortinas, N.C. and Cebrián, M.E.: A pilot study on the urinary excretion of prophyrins in human populations chronically exposed to arsenic in Mexico. *Hum. Exp. Toxicol.* 10 (1991): pp. 189–193.

Gutiérrez-Ojeda, C.: *Origin of arsenic in the alluvial aquifer of the Región Lagunera, states of Coahuila and Durango, México*. MSc Thesis, Department of Hydrology and Water Resources, The University of Arizona in Tucson, Tucson, AZ, 1995.

Gutiérrez-Ojeda, C., Ortiz-Flores, G. and Mata-Arellano, I.: Infiltration tests on highly permeable soils of the Comarca Lagunera aquifer, Mexico. 31st Congress on New Approaches to Characterizing Groundwater Flow, 10–14 September, 2001, Munich, Germany, 2001.

IMTA: Estudio hidrogeoquímico e isotópico del acuífero granular de la Comarca Lagunera. Instituto Mexicano de Tecnología del Agua, Comisión Nacional del Agua, Mexico City, Mexico, 1990.

IMTA: Geohidrología de La Laguna, Parte I. Final report, Project SA-9101, Comisión Nacional del Agua, Mexico City, Mexico, 1991.

IMTA: Geohidrología de La Laguna, Parte II. Final report, Project SA-9201, Comisión Nacional del Agua, Mexico City, Mexico, 1992.

IMTA: Sistema de recarga artificial en el acuífero de la Comarca Lagunera, Coahuila, Primera Etapa. Project TH9920, Comisión Nacional del Agua, Mexico City, Mexico, 1999.

Instituto de Geología: Reseña geológica del Estado de Coahuila. Universidad Nacional Autonóma de México, Mexico City, Mexico, 1937.

NOM: Norma Oficial Mexicana NOM-127-SSA1-1994. Official Newspaper, January 18, 1996, Mexico City, Mexico, 1996.

Pankow: *Aquatic chemistry concepts*. Lewis Publishers, Chelsea, MI, 1991.

Plummer, L.N., Prestemon, E.C. and Parkhurst, D.L.: An interactive code (NETPATH) for modeling net geochemical reactions along a flow path. US Geological Survey, Water-Resources Investigations Report 91-4078, 1991.

Quiñones, A., Gosset I.G., Carboney, A., Cortinas de Nava, C. and Ito, F.: Arsénico y salud, Salud Pública de México XXI: 187–197, 1979.

Wallrabe-Adams, H.-J.: Report on a mission to Mexico for the IAEA: Groundwater Resources of the Comarca Lagunera. Technical report, British Geological Survey, London, UK, 1993.

CHAPTER 16

Arsenic mobilization in aquatic sediments of an impacted mining area, north-central Mexico

N.A. Pelallo-Martinez & M.C. Alfaro-De la Torre
Facultad de Ciencias Químicas, Universidad Autónoma de San Luis Potosí (UASLP),
San Luis Potosí, S.L.P., Mexico

R.H. Lara-Castro
Instituto de Metalurgia, Universidad Autónoma de San Luis Potosí (UASLP),
San Luis Potosí, S.L.P., Mexico

J. Castro-Larragoitia
Facultad de Ingeniería/Instituto de Geología, Universidad Autónoma San Luis de Potosí (UASLP),
San Luis Potosí, S.L.P., Mexico

ABSTRACT: In the mining district of Villa de la Paz-Matehuala, polluted soils and aquatic systems been reported to be associated with the dispersion of historical and active tailings impoundments, waste rock dumps and historical refining activities. Arsenic was identified as the most important pollutant at this site. This work focuses on the mechanisms of arsenic (As) mobilization in sediments of aquatic systems (dug wells, springs and channels) polluted with mining and smelting residues. High concentrations of total As were determined in aquatic sediments (393–4914 mg/kg) and in porewater (0.14–123 mg/l). Furthermore, high concentrations of dissolved sulfides were determined in porewater. The results suggest that the As diffusion mechanism responsible for As transport between sediment and water column is regulated by two processes: (1) the oxidation of sulfides from the sediments releasing As into solution, and (2) the accumulation of As by precipitation as a sulfide phase under anoxic conditions.

16.1 INTRODUCTION

Arsenic (As) concentrations in groundwater have been a topic of environmental concern in Mexico for over two decades. There are well studied sites such as the Lagunera region near the city of Torreón, North of Mexico, and Zimapán in the central state of Hidalgo where total concentrations in water have ranged from 8 to 1112 µg/l. Health risk and some effects in humans have been also studied (Cebrian *et al.* 1994, Armienta *et al.* 1997).

In aquatic systems, the flux of substances from sediments to the water column and *vice versa* is an important mechanism controlling the quality of natural waters. The sediment-water interface plays an important role in the cycle of many trace elements in the aquatic systems. At the interface, the organic and inorganic particles as well as the organisms living there contribute to the transport and accumulation of trace elements between the water column and the sediments. Sedimentation of solid particles and molecular diffusion explain the major fluxes of materials across the solid-solution interface (Stumm and Morgan 1996). In comparison with most of the trace elements, As concentrations in natural waters are probably controlled by mechanisms occurring at the solid-solution interface. Metal oxides are the most important minerals binding As in aquatic systems, particularly those of Fe, Al and Mn (Sullivan and Aller 1996, Smedley and Kinniburg 2002, Bhattacharya *et al.* 2002). In the case of As, the processes affecting the Fe redox chemistry are particularly important since they can directly affect the mobility of As in the sediments. One of the principal processes causing high As concentrations in groundwater is the reductive dissolution of hydrous Fe oxides and/or the release of adsorbed or co-precipitated As (Aggett and O'Brien 1985, Bhattacharya *et al.* 1997, Smedley and Kinniburgh 2002).

In reducing sediments with non sulfate limitation, the element distribution and mobilization are controlled by the precipitation of diagenetic metallic sulfides (Brannon 1987, Moore 1988, Huerta-Diaz *et al.* 1998).

This work intends to elucidate the main processes controlling the mobility and distribution of As between the sediments and the water column of aquatic systems (dug wells, springs and channels) polluted by mining and smelting residues. We compare the profiles of total As and total dissolved sulfide concentrations, and pH in porewater, their variation with depth, and the mineralogical characterization of the sediments at three sampling sites.

16.2   SITE STUDIED

The mining district of Villa de la Paz-Matehuala is located in the semiarid Altiplano of central Mexico. At this place, the Pb-Zn-Ag (Cu-Au) skarn ore system has been mined during the last 200 years. Processing and refining of the ores has produced five tailing impoundments installed around Villa de la Paz, and an abandoned slag pile from an old smelter is located in the northeast side of Matehuala. Razo *et al.* (2004) showed that after their disposal, the unsecured residues are mainly dispersed through streams flowing in the W–E direction (Fig. 16.1) and by aeolian transportation. Water, soil and sediments have been impacted by As and heavy metals like Cd and Pb covering an area of 105 km^2. Arsenic concentrations determined in groundwater exceed more than 100–1000 times the Mexican drinking water standard (25 μg/l). Samples of sediment and porewater were taken at the three sampling sites indicated below in the map (Fig. 16.1): S1: dug well located in La Florida club; S2: Cerrito Blanco channel; S3: at 2.5 meter deep in an excavated basin located in Halcones Club; W1: Cerrito Blanco pond, and W2: Cerrito Blanco channel.

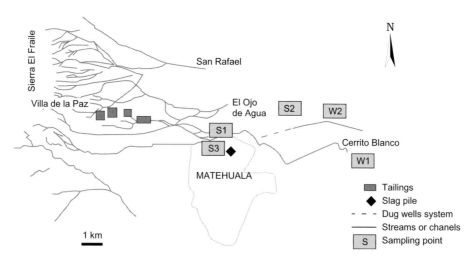

Figure 16.1.   The study area including the location of the sampling sites (S and W), the tailing impoundments and the slag pile in the mining district of Villa de la Paz-Matehuala (modified from Razo 2002).

16.3   METHODS

16.3.1   *Sampling and samples treatment*

Labware and material for sampling porewater, overlying water and sediments were cleaned in diluted HNO_3 and rinsed with ultrapure water prior to use. We used acrylic porewater samplers (peepers) with two columns containing 30 cells of 4 ml (1 cm vertical resolution) covered with a membrane filter of polysulfonate (0.22 μm pore diameter; Pall Co. HT 200). Three peepers were deployed at the sampling sites, left in place for 2 weeks at S1, S2 and S3 sites to determine dissolved As concentration profiles.

The water column was also sampled using an acrylic Kemmerer sampler (Sites W1, W2 and W3). Sediment cores (sites S1, S2 and S3) were taken with a gravity core sampler (Mod 2404-A14 Wildco, Wildlife Supply Co) using plexiglass tubes 7.2 cm in diameter. The cores were removed and sectioned at 0.5–2.0 cm intervals. Core slices were placed in plastic vials and kept at 4°C during their transport to the laboratory. The samples for dissolved sulfides ($\Sigma[H_2S]$) were collected from peepers and treated in the field with 10% w/v zinc acetate and 6% w/v NaOH. Porewater and overlying water were acidified with 0.02 N HNO_3 to reach a pH 2. Dried subsamples of sediment (45–50°C, 7 hours) were completely digested with HF, HNO_3, and $HClO_4$, the residue was dissolved with 5% HNO_3.

16.3.2   *Arsenic, sulfides and pH determination*

Water column and porewater concentration of As and Fe were measured by graphite furnace atomic absorption spectrophotometry (GFAAS; Varian Duo220). Total As was quantified by flame (FAAS) or by hydride generation AAS (Perkin-Elmer, Analyzer 200). As(III) was determined with AAS or by polarography (OSWSV; BAS100 W, 694VA Stand). To ascertain the accuracy of the sediment and water determinations, field blanks and certified reference materials (MESS-2 for sediment; NIST 2604 and TM-DWS for water; NWR-IEC) were used on a regular basis. Total dissolved sulfides ($\Sigma[H_2S]$) were determined by UV-Vis spectrophotometry using the Cline's method (Cline 1969). The pH was measured in field (IQ 150 pH meter).

16.3.3   *Mineralogical characterization*

Mineralogical characterization of sediment samples from the three sampling sites (S1, S2, S3) was performed by X-ray diffraction (RIGAKU DMAX 2200) and by scanning electron microscopy (PHILLIPS XL30) with a BSE detector.

16.4   RESULTS AND DISCUSSION

16.4.1   *Total As concentrations*

Table 16.1 shows the concentrations of total As in porewater, water column and sediments where S1 is a dug well in the La Florida fitness club, S2 is the Cerrito Blanco channel, S3 is a spring in the Halcones club, W1 is the water column at the Cerrito Blanco pond and W2 is the water column at the Cerrito Blanco channel. The concentration of total dissolved As varies among the sites, probably related to the pollution source or sources, but this relationship with the source has not yet been clarified.

16.4.2   *pH profiles*

The pH profiles with depth obtained at S1 and S3 sites (Fig. 16.2) demonstrate different behavior. It seems that carbonates act as a buffer neutralizing any acid production in system S1 (Fig. 16.2a). In contrast, at system S3 (Fig. 16.2b) the dissolution and reduction or oxidation of FeS(s) and

Table 16.1.   Total As concentrations in porewater and sediment in the different sampling sites in Matehuala, S.L.P. (Mexico).

Site	Porewater (μg/l)	Water column (μg/l)	Sediment (mg/kg)
S1	–	–	–
Well 1[1]	–	26.4–21.8	–
Well 2	142–1014	111–201	393–1069
S2	3345–4332	3300–4300	1638–2536
S3	21300–123000	58700	2344–5360
W1[1]	–	1455 ± 490	–
W2[1]	–	3260 ± 640	–
W3[2]	–	Non detected	–

[1] Sediment was not sampled; [2] Reference site.

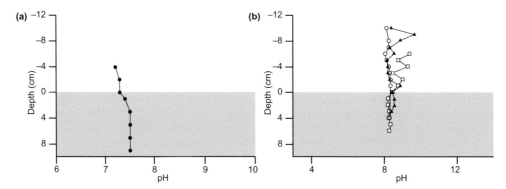

Figure 16.2.   Profiles of pH with depth at sites S1 (a) and S3 (b). The grey zone corresponds to the sediment column. Line at "0" on the depth axis corresponds to the water-sediment interface. On (b) the symbols represent data points obtained from three different peepers installed at the same sampling point.

FeAsS (s) can explain the decreasing pH in reduced sediments. Thus, the pH variation can probably be explained by the following equations:

$$CaCO_3(s) + H^+(aq) = Ca^{2+}(aq) + HCO_3^-(aq) \qquad (16.1)$$

$$FeAsS(s) + 11Fe^{3+}(aq) + 7H_2O(l) = 12Fe^{2+}(aq) + H_3AsO_3(aq) + 11H^+(aq) + SO_4^{2-}(aq) \qquad (16.2)$$

16.4.3   *Concentration profiles of arsenic and sulfide in sediment and porewater*

At site S1, dug well La Florida (Fig. 16.3), concentrations of total dissolved As vary with depth showing two maximums, one of them near the sediment-water interface, and the other in the reduced sediments.

The increase in the concentration of dissolved As (3–5 cm depth; Fig. 16.3) can be related to dissolution and/or desorption processes of Fe mineral phases entering to the system from external sources (e.g., during the tailings dispersion; Razo 2004). Also, As mobilization from the sediments to the water column could be explained by sulfide dissolution or the sulfate/sulfide redox process in the reduced sediments.

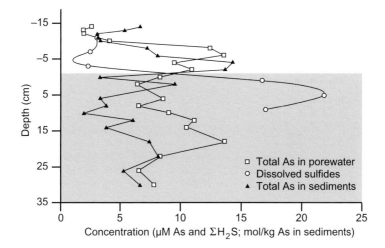

Figure 16.3.   Total As and dissolved sulfide concentration profiles in sediment and porewater at site S1 in La Florida club.

The decreasing concentrations of dissolved As occurring in the zone of sulfide production (5–15 cm) can be explained by precipitation of As sulfides (Fig. 16.3).

High concentrations of dissolved sulfates in porewater (10.45 to 16.22 mM) were determined. In these conditions, As mobilization in the zone of sulfide production can be controlled by diagenetic sulfide production and precipitation (i.e. biogenic pyrite). In reduced sediments, As and other trace elements can be released during the redistribution of iron monosulfides (Moore 1988, Huerta-Díaz 1998, Smedley and Kinniburgh 2002, Bhattacharya *et al.* 2003, O'Day and *et al.* 2004).

The total As concentration in the sediment (solid phase) decreased near the sediment-water interface (2–3 cm). This is probably due to mobilization processes occurring in the sediments and diffusion to the water column. Concentrations increase at the sediment surface (1–2 cm) suggesting an association with another solid phase (i.e. FeOOH(s)).

At site S3 (spring Halcones; Fig. 16.4) high concentrations of dissolved sulfides suggest reduced conditions in the sediments. High concentrations of dissolved As in the water column were determined suggesting its mobilization during sulfide dissolution and the diffusion of As to the water column. The concentration profile of $\Sigma[H_2S]$ suggests that the mobilization of As from sediments to the water column is related to the sulfate/sulfide redox process.

The dissolved As concentration profiles suggest diffusion from the water column to the sediments at site S3 (Fig. 16.4). Mineral phases of amorphous As sulfides were observed in the zone of sulfide production (2–6 cm depth) indicating precipitation of As in the reduced sediments. Concentrations of As(III) were determined in the porewater at the same depth of sulfide production. The concentration of As(III) slowly decreased at the sediment-water interface. Speciation of As in porewater was estimated using the thermodynamic model MINEQL4.5, the concentrations of anions and cations and the pH values. Calculations suggest that the main As dissolved species at sites W1 and W2 are: $HAsO_4^{2-}$ (71.8 ± 3.5%) and $H_2AsO_4^-$ (28.2 ± 3.5%) at site W1; $HAsO_4^{2-}$ (93.0 ± 8.0%) and $Ca_3(AsO_4)_2$ at site W2. Non measurable As dissolved concentrations were detected at the reference site W3. At the site S3 ("Halcones" club), the predominant species are As(III) in porewater (70%) and As(V) in the water column (90%).

16.4.4   *Chemical and mineralogical composition of the polluted sediments*

Calculation of the ionic activity products (IAP) using the porewater concentrations of dissolved As and other constituents can be used to predict the precipitation of solid phases in the sediments.

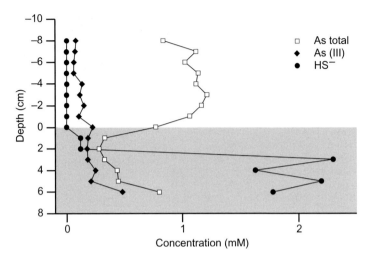

Figure 16.4.   Arsenic (III), total As and total sulfides ($\Sigma[H_2S]$) in the sediment porewater at site S3.

Table 16.2.   Ionic activity products (IAP) estimated using porewater composition at sites S2 ("La Florida" club) and S3 ("Halcones" club).

Solid phase	Log K_s	Log IAP	Reference
am-Fe(OH)$_3$	3.96	13.94	Stumm and Morgan (1996)
Fe(OH)$_2$	12.9	12.96	Stumm and Morgan (1996)
AsS$_3$ (s, orpiment)	–	−80.89 to −71.91*	Webster (1990)
AsS$_3$ (s, am)	−88.91 to −87.51	–	Eary (1992)
FeAsO$_4 \times 2H_2O$	−24.41	12.68 to −9.09*	Yinian Zhu (2001)
Ca$_3$(AsO$_4$)$_2 \times 4H_2O$	−21.257	−11.71	Donahue and Hendry (2003)

Table 16.2 shows some IAP values estimated using the water characteristics determined at site S3 (Halcones club) and compared to the solubility product (K_s) of mineral phases. Estimations for the physicochemical conditions of the sediments (pH, concentrations of anions and cations) have suggested saturated conditions with respect to AsS$_3$ (s, am), FeAsO$_4 \times 2H_2O$, am-Fe(OH)$_3$, Ca$_3$(AsO$_4$)$_2 \times 4H_2O$.

Some sections of the sediments sampled at sites S1 and S3 were characterized by X-ray diffraction and by electronic microscopy. The main mineral phases observed were calcite, quartz and gypsum (Fig. 16.5a). Electronic microscopy has confirmed the presence of amorphous As sulfides (Fig. 16.5b). Also, precipitation As sulfides was suggested by the theoretical calculation of the IAP.

16.5   CONCLUSIONS

Aquatic sediments polluted with residues from the mining district of Villa de la Paz-Matehuala (S.L.P., Mexico) contain high concentrations of As. Also, high concentrations of dissolved sulfides suggest reduced conditions in these sediments. We propose that in the studied sediments, As mobilization is controlled by (1) sulfate/sulfide redox processes and diffusion at the sediment-water interface, (2) dissolution of sulfides, or (3) precipitation and redistribution of sulfides during diagenetic processes. Arsenic precipitation in the zone of sulfide production has suggested the role

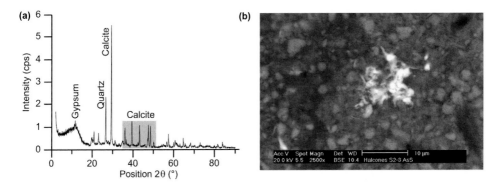

Figure 16.5.   The mineralogical characterizations of the sediments at site S3 ("Halcones" club) show at panel
(a) the main mineral phases observed by X-ray diffraction and at panel (b) a microphotography
of an amorphous arsenic sulfide particle.

of sulfides in the accumulation of dissolved As in the sediments. Mineralogical characterization
confirms the precipitation of As sulfides in the reduced sediments.

REFERENCES

Aggett, J. and O'Brien, G.A.: Detailed model for the mobility of arsenic in lacustrine sediments based on
measurements in Lake Ohakuri. *Environ. Sci. Technol.* 19 (1985), pp. 231–238.
Armienta, M.A., Rodriguez, R., Aguayo, A., Ceniceros, N., Villasenor, G. and Cruz, O.: Arsenic contamination
of groundwater at Zimapan, Mexico. *Hydrogeol. J.* 5 (1997), pp. 39–46.
Bhattacharya, P., Chatterjee, D. and Jacks, G.: Occurrence of As-contaminated groundwater in alluvial aquifers
from the Delta Plains, eastern India: option for safe drinking water supply. *Int. J. Water Res. Dev.* 13 (1997),
pp. 79–92.
Bhattacharya, P., Frisbie, S.H., Smith, E., Naidu, R., Jacks, G. and Sarkar, B.: Arsenic in the environment:
a global perspective. In: B. Sarkar (ed): *Handbook of heavy metals in the environment.* Marcel Dekker,
New York, 2002, pp. 145–215.
Bhattacharya, P., Tandukar, N., Neku, A., Valero, A.A., Mukherjee, A.B. and Jacks, G.: Geogenic arsenic in
groundwaters from Terai Alluvial Plain. *J. Phy.* IV (2002), pp. 173–176.
Brannon, J.M. and Patrick, W.H.: Fixation, transformation and mobilization of arsenic in sediments. *Environ.
Sci. Technol.* 21 (1987), pp. 450–459.
Cebrian, M.E., Albores, M.A., Garcia-Vargas, G., Del Razo, L.M. and Ostrosky-Wegman, P.: Chronic arsenic
poisoning in humans. In: J.O. Nriagu (ed): *Arsenic in the environment, Part II: human health and ecosystem
effects.* John Wiley, New York, 1994, pp. 93–107.
Cline, J.D.: Spectrophotometric determination of hydrogen sulfide in natural waters. *Limnol. Oceanog.*
14 (1969), pp. 454–458.
Donahue, R. and Hendry, M.J.: Geochemistry of arsenic in uranium mine mill tailling, Saskatchewan, Canada
Appl. Geochem. 18 (2003), pp. 1733–1750.
Eary, L.E.: The solubility of amorfous As_2S_3 from 25 to 90°C *Geochim. Cosmochim. Acta* 40 (1992),
pp. 925–934.
Huerta-Diaz, M.A., Tessier, A. and Carignan, R.: Geochemistry of trace metals associated with reduced sulfur
in freshwater sediments. *Appl. Geochem.* 13 (1998), pp. 213–233.
Moore, J.N., Ficklin, W.H. and Johns, C.: Partitioning of arsenic and metals in reducing sulfidic sediments.
Environ. Sci. Technol. 22 (1988), pp. 432–437.
Nriagu, J.O. (ed): *Arsenic in the environment. Ecosystem effects.* John Wiley and Sons, New York, 1994.
O'Day, P.A., Vlassopoulos, D., Root, R. and Rivera, N.: The influence of sulfur and iron on dissolved arsenic
concentrations in the shallow subsurface under changing redox conditions. *PNAS, Proceedings of National
Academic of Sciences of USA* 101, 2004, pp. 13,703–13,708.

Razo, S.I.: *Evaluación de la contaminación por metales y del riesgo en salud en un sitio minero de sulfuros polimetálicos: Caso de Villa de la Paz-Matehuala, S.L.P. (México).* Thesis, Universidad Autónoma de San Luis Potosi, S.L.P., Mexico, 2002.

Razo, I., Carrizales, L., Castro, J., Diaz-Barriga, F. and Monroy, M.: Arsenic and heavy metal pollution of soil, water and sediments in a semi-arid climate mining area in Mexico. *Water Air Soil Pollut.* 152 (2004), pp. 129–152.

Smedley, P.L. and Kinniburgh, D.G.: A review of the source, behaviour and distribution of arsenic in natural waters. *Appl. Geochem.* 17 (2002), pp. 517–568.

Stumm, W. and Morgan, J.J.: *Aquatic chemistry.* 2nd ed., Wiley-Interscience Publication, New York, 1996.

Sullivan, K.A. and Aller, R.C.: Diagenetic cycling of arsenic in Amazon shelf sediments. *Geochim. Cosmochim. Acta* 60 (1996), pp. 1465–1477.

Webster, J.G.: The solubility of As_2S_3 and speciation of As in dilute and sulfide-bearing fluids at 25 and 90°C. *Geochim. Cosmochim. Acta* 54 (1990), pp. 1009–1017.

Zhu, Y. and Merkel, B.J.: The dissolution and solubility of scorodite, $FeAsO_4 : H_2O$ Evaluation and simulation with PHREEQC2. *Wiss. Mitt Inst. für Geologie* 18 (2001), TU Bergakademie Freiberg, Germany, pp. 1–12.

CHAPTER 17

Contamination of drinking water supply with geothermal arsenic in Los Altos de Jalisco, Mexico

R. Hurtado-Jiménez
El Colegio de la Frontera Norte, A.C., Ciudad Juárez, Chih., Mexico

J.L. Gardea-Torresdey
Department of Chemistry, The University of Texas at El Paso, El Paso, TX, USA

ABSTRACT: The main objective of this study was to establish the degree of arsenic (As) contamination in drinking water at Los Altos de Jalisco and estimate the levels of human exposure. Total As concentrations were determined in 129 well water samples. The levels of exposure to As were estimated for babies, children and adults. High concentrations of As (greater than the Mexican guideline value of 25 µg/l) were found in 44 water samples (34%). Nevertheless, only 8% of the water wells meet the WHO guideline value of 10 µg/l. The mean concentration of As ranged from 14.7 to 101.9 µg/l. The highest concentration was found in the city of Mexticacán (262.9 µg/l). The estimated exposure doses ranges were: 1.1–7.6, 0.7–5.1, and 0.4–2.7 µg/kg/day, for babies, children and adults, respectively. Skin diseases, gastrointestinal effects, neurological damages and cardiovascular problems are some of the potential health effects due to the As exposure levels in Los Altos de Jalisco.

17.1 INTRODUCTION

Los Altos de Jalisco is a geographic region located in the northeastern part of the state of Jalisco, Mexico, that comprises 20 counties (Fig. 17.1). The year 2000 census (INEGI 2001) showed that Los Altos de Jalisco had a population of 696,318 inhabitants distributed in an area of 16,410 km^2. Table 17.1 shows both surface area and population of the Los Altos de Jalisco counties. The population of the correspondent county seats (a town or city that is the administrative center of its county) is also shown in Table 17.1.

Geologically, Los Altos de Jalisco is located in the Transmexican volcanic belt (TMVB) (Fig. 17.1), which is a large volcanic province that crosses Mexico between 19° and 21°N latitude, from Veracruz (Gulf of Mexico) to Puerto Vallarta (Pacific Ocean). The TMVB is a Pliocene Quaternary calc-alkaline province characterized by intensive hydrothermal activity (Campos-Enriquez and Garduño-Monroy 1995).

The *Comisión Nacional del Agua* (Mexican Water Commission) has defined 28 geo-hydrological zones for the state of Jalisco. Six of them are located in Los Altos de Jalisco. While all water wells may not have the same stratigraphic column, a large number of them show that the main aquifers occur in fissures and fractures of lava flows, tuffs and related intrusive and extrusive igneous rocks of Tertiary age (INEGI 2000).

It is well known that hot waters have a higher capacity to dissolve minerals. High concentrations of many chemical elements can occur in such waters, depending predominantly on regional geology and rock/water interactions. This is the case of some aquifers located in Los Altos de Jalisco, which are exposed to high temperature deep waters containing relatively high levels of potentially toxic elements such as fluoride (Hurtado *et al.* 2001, Hurtado and Gardea-Torresdey 2004) and arsenic (As). The main problem is that these naturally polluted aquifers are being used to supply potable water to urban areas and agriculture activities. Transportation of chemical elements in hydrothermal

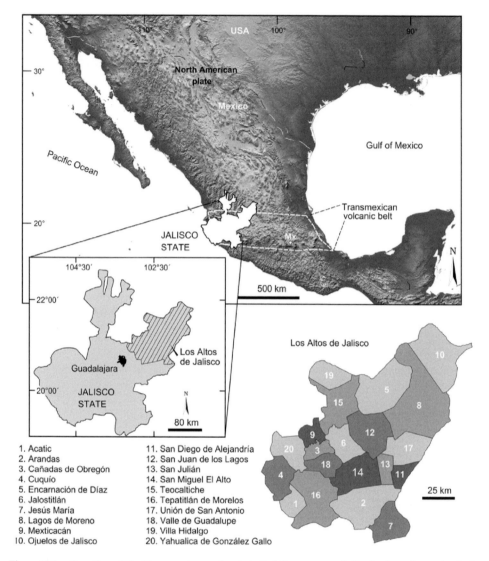

1. Acatic
2. Arandas
3. Cañadas de Obregón
4. Cuquío
5. Encarnación de Díaz
6. Jalostitlán
7. Jesús María
8. Lagos de Moreno
9. Mexticacán
10. Ojuelos de Jalisco

11. San Diego de Alejandría
12. San Juan de los Lagos
13. San Julián
14. San Miguel El Alto
15. Teocaltiche
16. Tepatitlán de Morelos
17. Unión de San Antonio
18. Valle de Guadalupe
19. Villa Hidalgo
20. Yahualica de González Gallo

Figure 17.1.   Location of the Transmexican volcanic belt, Jalisco state and distribution of counties in Los Altos de Jalisco.

solutions to other areas is commonly through faults, fracture systems, or other permeable geologic pathways.

Drinking water in Los Altos de Jalisco has two origins: (1) surface water, and (2) groundwater aquifers. In the case of surface water, the element content is low, and is mainly from rainwater that is stored in surface reservoirs. The element content of rivers, lakes and reservoirs is mainly due to the discharge of spring waters into the main streams. There are two counties in Los Altos de Jalisco where drinking water only comes from surface reservoirs. These counties are Cuquío and Yahualica de González Gallo.

The interest in determining the level of As in the drinking water supplies, and the associated potential health risks in Los Altos de Jalisco, originated from the direct observation of two facts: (1) high temperature in tap water (greater than 30°C) in many houses, and (2) the main source of

Table 17.1.   Population and surface area of counties and county seats in Los Altos de Jalisco.

County	Surface (km^2)	County population	County seat population
Acatic	362.39	19282	11005
Arandas	1238.02	76293	39478
Cañadas de Obregón	471.62	4407	2358
Cuquío	880.96	17554	4101
Encarnación de Díaz	1296.97	46421	20772
Jalostotitlán	481.44	28110	21291
Jesús María	570.00	19842	7852
Lagos de Moreno	2849.36	128118	79592
Mexticacán	204.99	6974	3603
Ojuelos de Jalisco	1317.00	27230	9338
San Diego de Alejandría	432.00	6384	4749
San Juan de los Lagos	874.47	55305	42411
San Julián	268.44	14760	12117
San Miguel El Alto	510.93	27666	21098
Teocaltiche	914.00	37999	21518
Tepatitlán de Morelos	1532.78	119197	74262
Unión de San Antonio	687.79	15664	6317
Valle de Guadalupe	516.12	5958	4178
Villa Hidalgo	510.93	15381	11552
Yahualica de González Gallo	520.75	23773	14225
TOTAL	16440.96	696318	411817

Source: INEGI (2001).

drinking water are aquifers located in the TMVB, which may be contaminated with geothermal waters.

Occurrence of As in drinking water has been reported in several countries, including Argentina, Austria, Bangladesh, Chile, China, Ghana, Greece, Hungary, India, Mexico, Romania, South Africa, Taiwan, Thailand and the United States (WHO 2001).

In Mexico, there are several areas where high levels of As in groundwater have been reported. The most important are: (1) Lagunera region (states of Durango and Coahuila), where the total As concentration in well water ranged from 7 to 740 µg/l (Rosas *et al.* 1999); (2) Guadiana valley (state of Durango), with As concentrations in the range of 5.0 to 167 µg/l (Alarcón Herrera *et al.* 2001); (3) Zimapán valley (state of Hidalgo), located in central Mexico, where concentrations of As in the range of 14 to 1097 µg/l were observed in water pumped from one of the most productive wells (Armienta *et al.* 1997a); and (4) city of Hermosillo (state of Sonora), located in northern Mexico, with As concentrations up to 305 µg/l (Wyatt *et al.* 1998b).

The current national (Mexican) guideline value for As in water delivered to the public water system has recently been set at 25 µg/l (SSA 2000). In 1993, the World Health Organization (WHO) set 10 µg/l as the guideline value for As in drinking water (WHO 1996). In the United States of America, the standard or maximum contaminant level (MCL) for As in drinking water was until recently 50 µg/l, which was set in 1975. A new standard (10 µg/l) was promulgated in 2001 by the US Environmental Protection Agency (USEPA). Water systems had to comply with the new EPA standard by January 2006 (USEPA 2001).

17.2   GEOCHEMISTRY OF ARSENIC

Arsenic is a naturally-occurring chemical element that can be found everywhere. Nevertheless, it mainly occurs combined with other elements such as hydrogen, oxygen and sulfur (inorganic As;

i-As). However if it is combined with carbon, it is known as organic arsenic (As$_{org}$). In general, i-As, which predominates in groundwater, is more dangerous than As$_{org}$ (ATSDR 2000).

Arsenic minerals can be found in a variety of geological environments including igneous, sedimentary and metamorphic rocks (USEPA 2000). Other sources of As include volcanic gases and geothermal waters (Welch *et al.* 1988, 1999).

While the As concentrations in natural waters around the world range from less than 0.5 to 5000 μg/l, typical concentrations in most waters are less than 10 μg/l, and sometimes substantially lower (<1 μg/l) (Smedley and Kinniburgh 2002). High As concentrations occur in a variety of environments including both oxidizing (high pH) and reducing groundwater aquifers (Mandal and Suzuki 2002).

Arsenic contamination of drinking water supplies can result from natural or artificial processes. Generally, natural contamination comes mainly from mineral deposits or geothermal fluids as was previously mentioned. The main sources of artificial or anthropogenic contamination are generally associated with industrial applications, such as wood preservation, petroleum refining, semiconductor manufacturing, production of nonferrous alloys (principally lead alloys used in lead-acid batteries), burning of fossil fuels (especially coal), metal production (such as gold), agricultural applications (pesticides and feed additives) and irrigation with polluted waters (ATSDR 2000, Heinrichs and Udluft 1999, Hudak 2000, Bhattacharya *et al.* 2002, USGS 2004).

The aim of this study was: (1) to determine the levels of As in the public water wells located in the county towns of the Los Altos de Jalisco region; and (2) to estimate the levels of the As exposure via drinking water.

17.3   HEALTH EFFECTS

It is well known that chronic exposure to high levels of As causes a wide variety of serious human health problems including dermal changes (pigmentation, hyperkeratosis and ulceration), gastrointestinal effects (stomach pain, nausea, vomiting and diarrhea), neurological damage, cardiovascular problems (high blood pressure, heart attack and stroke), diverse types of cancer (skin, bladder, lung, kidney and other organs), and respiratory, pulmonary, hematological, hepatic, renal, developmental, reproductive, immunological, genotoxic and mutagenetic effects (NRC 1999, ATSDR 2000, Mandal and Suzuki 2002, WHO 2001).

17.4   MATERIALS AND METHODS

17.4.1   *Well water sampling*

In order to determine the level of As in aquifers located in Los Altos de Jalisco, a total of 129 water samples from public wells were collected. These water wells represent approximately 81% of the total number of wells that are supplying water to 17 county seats in the region. Three water samples were collected in pre-washed polyethylene bottles from each public well under its normal operation conditions (i.e., while delivering water to the public water system). Samples were taken manually from the wellhead sampling port of each well. Before sampling, bottles were rinsed three times using the same well water. In order to get a representative sample (free of any contamination from the pipe line), the sampling port was previously purged for approximately 15 minutes. Water samples were preserved by adding trace metal grade nitric acid (to get a final concentration of 2% nitric acid). The samples were then stored at approximately 5°C. Two series of water samples were taken. The first series was collected during November, 2002 and the second one in October, 2003.

17.4.2   *Physicochemical methods*

Total As concentration in water samples was determined using inductively coupled plasma-optical emission spectrometry (ICP-OES). The instrument used was an Optima 4300 DV ICP-OES manufactured by Perkin Elmer. Four standard solutions were used to obtain calibration curves. The correlation coefficients for the wavelengths used (188.979 and 197.197 nm) were in the range of 0.9970 to 0.9999. Temperature, pH and conductivity were determined at the wellhead using portable instruments (Orion Model 115Aplus for temperature, and conductivity and Orion Model 250 for pH).

17.4.3   *Data reduction*

Statistical values (standard deviation, arithmetical mean, minimum and maximum) were calculated using a standard spreadsheet. The distribution of As in the study area was presented graphically using a geographic information system. The town's average concentrations of As were assigned to polygon coverage of county boundaries. Pie diagrams were used to show the percentage distribution of water wells with different levels of As in each town.

17.4.4   *Arsenic exposure levels*

Daily exposure levels to As in drinking water were estimated for babies (10 kg), children (20 kg), and adults (70 kg) using the same method of calculation described by the authors to estimate the levels of exposure to fluoride in Los Altos de Jalisco (Hurtado-Jiménez and Gardea-Torresdey 2005).

17.5   RESULTS AND DISCUSSION

The results of the As analyses are summarized in Table 17.2 where the correspondent statistical values (mean arithmetic, minima and maxima) are shown. Temperature, pH and electric conductivity (EC) are presented in a similar manner in Table 17.3.

17.5.1   *Arsenic occurrence*

Arsenic concentrations range from 14.7 µg/l to 101.9 µg/l. While all of the 17 towns have a mean concentration of total arsenic (t-As) higher than 10 µg/l (the guideline value set by the WHO), only seven towns exceeded the national guideline value for As of 25 µg/l. These cities are Encarnación de Díaz, Mexticacán, Ojuelos de Jalisco, San Juan de los Lagos, San Miguel El Alto, Teocaltiche and Valle de Guadalupe. With the exception of San Miguel El Alto, the other six cities have a mean groundwater temperature greater than 30°C, indicating origin or contamination from deeper geothermal waters. The cities with the highest concentrations of As are Mexticacán (101.9 µg/l), Teocaltiche (89.5 µg/l) and San Juan de los Lagos (54.5 µg/l).

   Even though the mean groundwater temperature of the cities of Acatic, Lagos de Moreno and Tepatitlán de Morelos is greater than 30°C, total As concentrations for each city are lower than 25 µg/l. Electrical conductivity (EC) in the 112 water wells tested, ranged from 155 to 1888 µS/cm. The highest value was measured in well 1 in San Juan de los Lagos. Other wells with high values of EC were found in Lagos de Moreno (well O with 1084 µS/cm), and Teocaltiche (well 5 with 1074 µS/cm). In general, a positive correlation between As concentration and conductivity was found. The pH of the water wells tested ranged from 5.9 to 8.0. The distribution of the As concentrations in the Los Altos de Jalisco water wells is shown in Figure 17.2.

   Occurrence of As in groundwater depends on three main factors: (1) a source of As (minerals, rocks, sediments and soils); (2) environmental conditions to enhance mobilization of As (temperature, pH, redox potential and chemical species); and (3) physical and chemical phenomena that increase the transport processes (adsorption-desorption) (Smedley and Kinniburgh 2002, Aiuppa *et al.* 2003).

   In natural waters, As can occur in several oxidation states (-3, 0, $+3$ and $+5$) but it is present mainly as arsenite [As(III)] and arsenate [As(V)] oxyanions. While speciation of As is strongly dependent on both redox potential (Eh) and pH, other important factors such as temperature and

Table 17.2.   Total As concentrations (μg/l) determined for the towns of Los Altos de Jalisco where well water samples were collected in November 2002 and October 2003.

Town	Population	Wells	Mean	(Min–Max)
Acatic	11005	4	16.5 ± 6.0	(5.1–30.8)
Arandas	39478	11	14.7 ± 4.6	(3.8–26.1)
Encarnación de Díaz	20772	6	25.2 ± 5.0	(17.9–42.8)
Jalostotitlán	21291	5	23.4 ± 3.7	(18.1–29.8)
Jesús María	7852	4	22.8 ± 4.7	(4.0–52.6)
Lagos de Moreno	79592	18	21.8 ± 6.0	(10.8–43.3)
Mexticacán	3603	4	101.9 ± 85.1	(12.9–262.9)
Ojuelos de Jalisco	9338	3	28.3 ± 3.6	(20.5–33.9)
San Diego de Alejandría	4749	2	20.7 ± 14.6	(4.8–52.2)
San Juan de los Lagos	42411	11	54.5 ± 29.8	(0.5–113.8)
San Julián	12117	3	20.2 ± 16.2	(8.6–62.0)
San Miguel el Alto	21098	6	30.8 ± 21.3	(5.8–74.8)
Teocaltiche	21518	5	89.5 ± 27.9	(54.0–157.7)
Tepatitlán de Morelos	74262	27	22.1 ± 16.4	(4.5–74.0)
Unión de San Antonio	6317	3	20.0 ± 9.8	(6.1–31.8)
Valle de Guadalupe	4178	4	28.1 ± 6.4	(16.5–47.0)
Villa Hidalgo	11552	13	16.3 ± 7.8	(4.0–36.5)

Min: minimum; Max: maximum; Population data from INEGI (2001).

Table 17.3.   Temperature (°C), pH and electric conductivity for the towns of Los Altos de Jalisco where well water samples were collected in November 2002 and October 2003.

Town	Wells	Temperature		pH		EC (μS/cm)	
		Mean	(Min–Max)	Mean	(Min–Max)	Mean	(Min–Max)
Acatic	4	32.4	(31.2–33.7)	6.7	(6.5–7.0)	252	(182–297)
Arandas	11	28.0	(23.3–32.8)	6.5	(6.2–6.7)	279	(230–408)
Encarnación de Díaz	6	30.1	(25.8–36.6)	7.1	(7.0–7.4)	537	(451–661)
Jalostotitlán	5	26.8	(24.7–29.3)	7.0	(6.8–7.0)	459	(417–494)
Jesús María	4	27.3	(24.1–31.3)	6.5	(6.5–6.7)	413	(206–421)
Lagos de Moreno	18	30.2	(22.2–35.2)	7.0	(6.5–7.5)	556	(395–1084)
Mexticacán	4	30.3	(26.6–37.7)	7.2	(7.1–7.3)	563	(489–691)
Ojuelos de Jalisco	3	37.5	(36.4–38.7)	6.9	(6.9–7.0)	548	(548–548)
San Diego de Alejandría	2	27.5	(27.4–27.7)	7.4	(7.3–7.5)	475	(434–517)
San Juan de los Lagos	11	32.2	(26.7–45.2)	7.1	(6.9–7.5)	830	(474–1888)
San Julián	3	26.2	(25.1–27.3)	7.0	(6.8–7.1)	452	(439–467)
San Miguel el Alto	6	25.7	(22.6–28.7)	7.2	(7.0–7.3)	553	(347–922)
Teocaltiche	5	30.6	(26.2–33.9)	6.8	(6.6–7.2)	796	(620–1074)
Tepatitlán de Morelos	27	30.9	(24.0–42.3)	6.7	(5.9–8.0)	310	(177–576)
Unión de San Antonio	3	26.8	(24.6–30.2)	7.2	(7.0–7.4)	429	(427–431)
Valle de Guadalupe	4	35.0	(33.9–36.8)	7.0	(6.8–7.3)	479	(375–548)
Villa Hidalgo	13	28.5	(25.7–32.4)	6.3	(6.0–7.2)	171	(155–204)

the occurrence of metallic sulfide and oxides also affect As speciation. Under reducing conditions, such as those often found in aquifers, arsenite is generally the most common species present (Lemmo *et al.* 1983). The ability of metal oxides to adsorb and desorb arsenite and arsenates, plays an important role in controlling the level of As in natural waters (Calvo Revuelta *et al.* 2003). The oxidation state of As is important from the health perspective, because trivalent arsenic [As(III)] is generally considered to have a higher toxicity.

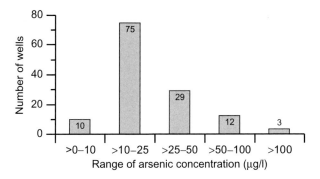

Figure 17.2.   Distribution of As concentrations in water wells located in Los Altos de Jalisco.

The results show that the total As concentrations in groundwater at Los Altos de Jalisco fall at the lower end of the typical range for subsurface natural waters (<0.5 to 5000 µg/l As) (Smedley and Kinniburgh 2002) and for deep geothermal waters (<100 to 50,000 µg/l As) (Ballantyne and Moore 1988). Data obtained in this study did not establish the oxidation state of As in Los Altos de Jalisco groundwaters. Nevertheless, mean pH values (from 6.3 to 7.4), depth of the water wells, and high proportion of volcanic rocks in the aquifers indicate that reducing conditions might exist in Los Altos de Jalisco groundwater supplies (Calvo Revuelta *et al.* 2003, Stüben *et al.* 2003, Saxena *et al.* 2004). Analysis of data from Tables 17.2 and 17.3 suggests that there are two groups of aquifers in Los Altos de Jalisco. One group shows a positive correlation between As and temperature. Another group shows a negative correlation. In general, higher temperature aquifers have shown a positive correlation with As concentration. Correlations between pH and As concentration have also shown both positive and negative patterns.

The occurrence of high levels of As (greater than national and international guideline values) and some other toxic elements such as fluoride and selenium (Hurtado *et al.* 2001, Hurtado and Gardea-Torresdey 2004) in aquifers at Los Altos de Jalisco suggest that several thousand people are consuming contaminated water. This fact might represent a serious human health risk not only to populations living there, but also to any people who are consuming agricultural and livestock products fed with such contaminated waters.

17.5.2   *Human exposure to arsenic*

The estimated exposure doses to As in drinking water are presented in Table 17.4. The results indicate that babies (10 kg) are being exposed to the highest doses.

Ingestion of As at Los Altos de Jalisco comes from three main sources: drinking water, food, and hot beverages (tea and/or coffee). Human exposure to As in drinking water at Los Altos de Jalisco has been estimated from data obtained in this study. While there are many data on the intake of As from hot beverages and food elsewhere, including Bangladesh, Canada, Chile, Croatia, India, Japan, Mexico, Spain and United States (Postruznik *et al.* 1996, Tao and Bolger 1998, Sapunar-Schoof *et al.* 1999, Rosas *et al.* 1999, Queirolo *et al.* 2000, Del Razo *et al.* 2002, Delgado-Andrade *et al.* 2003, Roychowdhury *et al.* 2003, Llobet *et al.* 2003, Das *et al.* 2004), for Los Altos de Jalisco there is no such information. Nevertheless, the data on the daily intake of t-As in food from other similar regions (WHO 1996, Del Razo *et al.* 2002, Roychowdhury *et al.* 2003, Meza *et al.* 2004) have been used to make a preliminary estimation.

It was found, in the above referred studies that the highest levels of As are in seafood, cereals, meat and meat by-products. Approximately 90% of the dietary intake of t-As comes from seafood. Most seafood As (80–99%) is present in organic form, which is one of the less-toxic forms (Tao and Bolger 1998). Estimates of the mean daily intake of t-As in food for adults range from 42 µg in Canada to 286 µg in Spain (Health Canada 2004).

Table 17.4.   Daily exposure doses to arsenic in drinking water in the towns of Los Altos de Jalisco (µg/kg/day).

Town	Babies Mean	(Min–Max)	Children Mean	(Min–Max)	Adults Mean	(Min–Max)
Acatic	1.2	(0.4–2.3)	0.8	(0.3–1.5)	0.4	(0.1–0.8)
Arandas	1.1	(0.3–2.0)	0.7	(0.2–1.3)	0.4	(0.1–0.7)
Encarnación de Díaz	1.9	(1.3–3.2)	1.3	(0.9–2.1)	0.7	(0.5–1.1)
Jalostotitlán	1.8	(1.35–2.2)	1.2	(0.9–1.5)	0.6	(0.5–0.8)
Jesús María	1.7	(0.3–3.9)	1.1	(0.2–2.6)	0.6	(0.1–1.4)
Lagos de Moreno	1.6	(0.8–3.2)	1.1	(0.5–2.2)	0.6	(0.3–1.1)
Mexticacán	7.6	(1.0–19.7)	5.1	(0.6–13.1)	2.7	(0.3–6.9)
Ojuelos de Jalisco	2.1	(1.5–2.5)	1.4	(1.0–1.7)	0.7	(0.5–0.9)
San Diego de Alejandría	1.6	(0.4–3.9)	1.0	(0.2–2.6)	0.5	(0.1–1.4)
San Juan de los Lagos	4.1	(0.0–8.5)	2.7	(0.0–5.7)	1.4	(0.0–3.0)
San Julián	1.5	(0.6–4.6)	1.0	(0.4–3.1)	0.5	(0.2–1.6)
San Miguel El Alto	2.3	(0.4–5.6)	1.5	(0.3–3.7)	0.8	(0.2–2.0)
Teocaltiche	6.7	(4.1–11.8)	4.5	(2.7–7.9)	2.4	(1.4–4.2)
Tepatitlán de Morelos	1.7	(0.3–5.6)	1.1	(0.2–3.7)	0.6	(0.1–2.0)
Unión de San Antonio	1.5	(0.5–2.4)	1.0	(0.3–1.6)	0.5	(0.2–0.8)
Valle de Guadalupe	2.1	(1.2–3.5)	1.4	(0.8–2.4)	0.7	(0.4–1.2)
Villa Hidalgo	1.2	(0.3–2.7)	0.8	(0.2–1.8)	0.4	(0.1–1.0)

The Joint Food and Agriculture Organization/World Health Organization has evaluated As toxicity and established a provisional tolerable weekly intake (PTWI) of 15 µg/kg/week of i-As, which is equivalent to a provisional tolerable daily intake (PTDI) of 2.14 µg/kg/day. PTWI is an estimate of the amount of a contaminant that can be ingested over a lifetime without appreciable risk (FAO/WHO 1989). Using this reference value to calculate exposure doses for the selected cases, the suggested daily As intake for babies (10 kg), children (20 kg) and adults (70 kg) living in Los Altos de Jalisco would be 21.4, 42.8 and 150 µg As/day, respectively (Table 17.4). These values indicate that the As concentration in drinking water for babies, children and adults should be no higher than 28.5, 42.8, and 81.0 µg/l, respectively. Data from Table 17.2 indicate that: (1) there are four towns where the mean concentration of t-As is greater than 28.5 µg/l; (2) in two towns the mean As concentration is higher than 42.8 µg/l; and (3) in one town water has a mean As concentration over 81.0 µg/l. It is important to notice that these numbers have a relative value and should be used only as a reference. Moreover, an intake above the PTDI does not automatically mean that health is at risk.

There are many cases in which the water discharged from one well is directly delivered to a group of houses. In such cases, individual calculations (mainly for those wells exceeding guideline value) should be performed.

In addition to the exposure time and the doses to which a person is exposed, sensitivity to As toxicity varies with each individual and appears to be strongly dependent of their nutritional status and genetic characteristics (Anawar *et al.* 2002). Nevertheless, there are some other factors that play an important role in mitigating the toxic effect of ingested As such as increased awareness, and better education (Chakraborti *et al.* 2002). There are some nutrients that can reduce the As toxicity such as vitamin C and methionine (Chen *et al.* 1997), whereas vitamin A deficiency enhances the toxic effects of As (Roychowdhury *et al.* 2003). Long-term exposure (years) to drinking water at levels as low as 1.0 µg As/kg/day have been associated with skin diseases and skin, bladder, kidney and liver cancer (USEPA 2000).

There are many examples around the world showing the dramatic health effects of ingesting As from groundwater that should motivate health authorities to implement solutions to stop this potential calamity. The most impressive have been reported in India (West Bengal), Bangladesh,

China (Inner Mongolia) and Taiwan (Mandal and Suzuki 2002, Chakraborti *et al.* 2002). Unfortunately, there are no epidemiological studies in Los Altos de Jalisco that would show the impact of groundwater As on the health of people living there.

Most of the scientific studies analyze the toxicological effects of ingesting As alone. Nevertheless, drinking water might contain excesses of several different chemical elements or molecules. This means that synergistic and/or antagonist effects should be considered. In the case of groundwater at Los Altos de Jalisco, there are some cases where both As and F^- are exceeding national guidelines values. While epidemiological studies performed in Xinjiang, China indicated that F^- and As do not have a mutual synergistic action, similar studies in Guizhou, China, suggested that the toxicological effects of fluoride could be enhanced by As (Zheng *et al.* 2002).

Several publications from Mexico (Wyatt *et al.* 1998a, Alarcón *et al.* 2001, Meza *et al.* 2004) and many other countries around the world (Hudak 2000, WHO 2001, Smedley and Kinniburgh 2002, Chakraborti *et al.* 2002, Mandal and Kumar 2002) suggest that guideline values (national or international) for both As and F^- quite frequently exceeded, not only in drinking water supplies, but also in waters that are delivered to the public water system. Additionally, high levels of fluoride were reported in commercial bottled drinking waters in Los Altos de Jalisco (Hurtado and Gardea-Torresdey 2004). Presently, As and F^- are recognized as the most serious inorganic contaminants in drinking water on a worldwide basis (Smedley and Kinniburgh 2002).

In Mexico, there are only few epidemiological studies on the effects ingesting As from drinking water. Several studies from the Comarca Lagunera determined the effect of As on skin cancer and its relation with the human papilloma virus, alteration of the immune system (Salazar *et al.* 2004, Rosales-Carrillo *et al.* 2004, Pineda-Zavaleta *et al.* 2004) and others in the city of Hermosillo, Zimapán valley and Yaqui valley dealt with the excretion of As in urine and the accumulation of it in some parts of the body (Armienta *et al.* 1997b, Wyatt *et al.* 1998a, Meza *et al.* 2004). The As concentrations found in Los Altos de Jalisco were lower than those reported in some other Mexican areas including the Comarca Lagunera, the Zimapán valley and the city of Hermosillo, but there are considerable number of wells with As levels exceeding national and international standards, and consequently this should be a major concern. Skin diseases, gastrointestinal effects, neurological damages, cardiovascular problems, and hematological effects are some of the potential health effects from chronic As ingestion at the exposure doses experienced in Los Altos de Jalisco. While everyone may not be affected, an important fraction of the total population at Los Altos de Jalisco is exposed to a serious health risk.

17.6 CONCLUSION

After reviewing literature, it was found that most developed countries have set the drinking water guideline value for As at 10 μg/l, as suggested by the WHO. It is important to notice that this value is constantly being reviewed, and will be lowered if there is enough information indicating that potential cancer risks remain high at the present guideline (NRC 2001). Regulations are a very important starting point, but there are other factors needed to limit As exposure and toxicity: (1) enforcement of the regulations; (2) improvement of nutrition; and (3) communication of the health risks.

This epidemiologic study generated a database on the level of As in groundwater supplies at Los Altos de Jalisco that can be used as a general indicator of the potential significance of As contamination. While the data are not sufficient to allow detailed analysis of health and environmental risks, they support the need for the following investigations: (1) chemical studies on the oxidation state of As in groundwater to determine its degree of toxicity and to better understand the process of mobilization and transport; (2) geochemical studies to determine transport and fate of As in the environment; (3) determinations of total and i-As in food and beverages to improve estimates of human exposure; (4) epidemiological studies to show the impact of chronic ingestion of As on health, and (5) point of use solutions for those towns with higher levels of As in their drinking water.

REFERENCES

Aiuppa, A., D'Alessandro, W., Federico, C., Palumbo, B. and Valenza, M.: The aquatic geochemistry of arsenic in volcanic groundwaters from southern Italy. *Appl. Geochem.* 18 (2003), pp. 1283–1296.

Alarcón Herrera, M.T., Montenegro, I.F., Romero Navar, P., Domínguez, I.R.M. and Trejo Vázquez, R.: Contenido de arsénico en el agua potable del valle del Guadiana, México. *Ingeniería Hidráulica en México* 15:4 (2001), pp. 63–70.

Anawar, H.M., Akai, J., Mostofa, K.M.G., Safiullah, S. and Tareq, S.M.: Arsenic poisoning in groundwater: Health risk and geochemical sources in Bangladesh. *Enivron. Int.* 27 (2002), pp. 597–604.

Armienta, M.A., Rodriguez, R., Aguayo, N., Ceniceros, N., Villaseñor, G. and Cruz, O.: Arsenic contamination of groundwater at Zimapán, Mexico. *Hydrogeol. J.* 5:2 (1997a), pp. 39–46.

Armienta, M.A., Rodríguez, R. and Cruz. O.: Arsenic content in hair of people exposed to natural arsenic polluted groundwater at Zimapán, Mexico. *Bull. Environ. Contam. Toxicol.* 9 (1997b), pp. 583–589.

ATSDR: *Toxicological profile for arsenic.* US Department of Health and Human Services, Agency for Toxic Substances and Disease Registry, Atlanta, GA, 2000.

Ballantyne, J.M. and Moore, J.N.: Arsenic geochemistry in geothermal systems. *Geochim. Cosmochim. Acta* 2 (1988), pp. 47–483.

Bhattacharya, P., Mukherjee, A.B., Jacks, G. and Nordquist, S.: Metal contamination at a wood preservation site: Characterization and experimental studies on remediation. *Sci. Tot. Environ.* 290:1–3 (2002), pp. 168–180.

Calvo Revuelta, C., Álvarez-Benedi, J., Andrade Benítez, M., Marinero Diez, P. and Bolado Rodríguez, S.: Contaminación por arsénico en aguas subterráneas en la provincia de Valladolid: Variaciones estacionales. In: J. Álvarez Benedeti and P. Marinero (eds): *Estudios de la zona saturada del suelo,* Vol. VI. Universidade da Coruña, Valladolid, Spain, 2003.

Campos-Enriquez, J.O. and Garduño-Monroy, V.H.: Los Azufres silicic center (Mexico): inference of caldera structural elements from gravity, aeromagnetic, and geoelectric data. *J. Volcanol. Geoth. Res.* 67 (1995), pp. 123–152.

Chakraborti, D., Rahman, M.M., Paul, K., Chowdhury, U.K., Sengupta, M.K., Lodh, D., Chanda, C.R., Saha, K.C. and Mukherjee, S.C.: Arsenic calamity in the Indian subcontinent. What lessons have been learned? *Talanta* 8 (2002), pp. 3–22.

Das, H.K., Mitra, A.K., Sengupta, P.K., Hossain, A., Islan, F. and Rabbani, G.H.: Arsenic concentration in rice, vegetables, and fish in Bangladesh: a preliminary study. *Environ. Intern.* 30 (2004), pp. 383–387.

Del Razo, L.M., Garcia-Vargas, G.G., Garcia-Salcedo, J., Sanmiguel, M.F., Rivera, M., Hernandez, M.C. and Cebrian, M.E.: Arsenic levels in cooked food and assessment of adult dietary intake of arsenic in the Region Lagunera, Mexico. *Food Chem. Toxicol.* 40 (2002), pp. 1423–1431.

Delgado-Andrade, C., Navarro, M., López, H. and López, M.C.: Determination of total arsenic levels by hydride generation atomic absorption spectrometry in foods from south-east Spain: estimation of daily dietary intake. *Food Addit. Contam.* 20:10 (2003), pp. 923–932.

FAO/WHO: *Expert committee on food additives: evaluation of certain food additives and contaminants.* 33rd Report, Tech. Rep. Ser. 776, The Food and Agriculture Organization/World Health Organization, Geneva, Switzerland, 1989.

Health Canada: *Arsenic in drinking water.* Document for National Consultation. Federal-Provincial-Territorial Committee on Drinking Water, Water Quality and Health Bureau. Doc. (11/04) Arsenic, Ottawa, Ontario, Canada, 2004, http://www.hc-sc.gc.ca/hecs-sesc/water/pdf/arsenic_drinking_water.pdf (accessed Feb. 2005).

Heinrichs, G. and Udluft, P.: Natural arsenic in Triassic rocks: A source of drinking-water contamination in Bavaria, Germany. *Hydrogeol. J.* 7:5 (1999), pp. 468–476.

Hudak, P.F.: Distribution and sources of arsenic in the southern high plains aquifer, Texas, USA. *Environ. Sci. Health Part A,* 35:6 (2000), pp. 899–913.

Hurtado, R., Gardea-Torresdey, J. and Tiemann, K.J.: Fluoride occurrence in tap water at "Los Altos de Jalisco", in the central Mexico region. In: L.E. Erikson and M.M. Rankin (eds): *Proceedings of the 2000 conference on hazardous waste research: Environmental changes and solutions to resource, development, production, and use.* Kansas State University, Manhattan, KS, 2001, pp. 211–219.

Hurtado, R. and Gardea-Torresdey, J.: Environmental evaluation of fluoride in drinking water at "Los Altos de Jalisco", in the central Mexico region. *J. Toxicol. Environ. Health Part A* 67:20–22 (2004), pp. 1741–1753.

Hurtado-Jiménez, R. and Gardea-Torresdey, J.: Estimación de la exposición a fluoruros en Los Altos de Jalisco, México. *Salud Publica Mex.* 47 (2005), pp. 58–63.

INEGI: Estudio hidrológico del estado de Jalisco: Aguascalientes, Ags. Instituto Nacional de Estadística, Geografía e Informática, Mexico City, Mexico, 2000.

INEGI: XII Censo General de Población y Vivienda 2000: Aguascalientes, Ags. Instituto Nacional de Estadística, Geografía e Informática, Mexico City, Mexico, 2001.

Lemmo, N.V., Faust, S.O., Belton, T. and Trucker, R.: Assessment of the chemical and biological significance of arsenical compounds in a heavily contaminated watershed. Part 1: The fate and speciation of arsenical compounds in aquatic environments—a literature review. *J. Environ. Sci. Health* A18 (1983), pp. 335–387.

Llobet, J.M., Falcó, G., Casas, C., Teixidó, A. and Domingo, J.L.: Concentrations of arsenic, cadmium, mercury, and lead in common foods and estimated daily intake by children, adolescents, adults, and seniors of Catalonia, Spain. *J. Agric. Food Chem.* 51 (2003), pp. 838–842.

Mandal, B.K. and Suzuki, K.T.: Arsenic round the world: a review. *Talanta* 58 (2002), pp. 201–235.

Meza, M.M., Kopplin, M.J., Burgess, J.L. and Gandolfi, A.J.: Arsenic drinking water exposure and urinary excretion among adults in the Yaqui Valley, Sonora, Mexico. *Environ. Res.* 96 (2004), pp. 119–126.

NRC: *Arsenic in drinking water*. National Research Council, National Academy Press, Washington, DC, 1999.

NRC: *Arsenic in drinking water, updated 2001*. National Research Council, National Academy Press, Washington, DC, 2001.

Pineda-Zavaleta, A.P., García-Vargas, G., Borja-Aburto, V.H., Acosta-Saavedra, L.C., Vera Aguilar, E., Gómez-Muñoz, A., Cebrián, M.E. and Calderón-Aranda, E.S.: Nitric oxide and superoxide anion production in monocytes from children exposed to arsenic and lead in region Lagunera, Mexico. *Toxicol. Appl. Pharmacol.* 198 (2004), pp. 283–290.

Queirolo, F., Stegen, S., Restovic, M., Paz, M., Ostapczuk, P., Schwuger, M. and Muñoz, L.: Total arsenic, lead, and cadmium levels in vegetables cultivated at the Andean villages of northern Chile. *Sci. Total Environ.* 255 (2000), pp. 75–84.

Rosales-Carrillo, J.A., Acosta-Saavedra, L.C., Torres, R., Ochoa-Fierro, J., Borja-Aburto, V.H., Lopez-Carrillo, L., Garcia-Vargas, G.G., Gurrola, G.B., Cebrian, M.E. and Calderón-Aranda, E.S.: Arsenic exposure and human papillomavirus response in non-melanoma skin cancer Mexican patients: a pilot study. *Int. Arch. Occup. Environ. Health* 77 (2004), pp. 418–423.

Rosas, I., Belmont, R., Armienta, A. and Baez, A.: Arsenic concentration in water, soil, milk and forage in Comarca Lagunera, Mexico. *Water Air Soil Pollut.* 112 (1999), pp. 133–149.

Roychowdhury, T., Tokunaga, H. and Ando, M.: Survey of arsenic and other heavy metals in food composites and drinking water and estimation of dietary intake by the villagers from an arsenic-affected area of West Bengal, India. *Sci. Total Environ.* 308 (2003), pp. 15–35.

Salazar, A.M., Calderón-Aranda, E., Cebrián, M.E., Sordo, M., Bendesky, A., Gómez-Muñoz, A., Acosta-Saavedra, L. and Ostrosky-Wegman, P.: p53 expresion in circulating lymphocytes of non-melanoma skin cancer patients from an arsenic contaminated region in Mexico. A pilot study. *Mol. Cell Biochem.* 255 (2004), pp. 25–31.

Sapunar-Postruznik, J., Bazulie, D. and Kubala, H.: Estimation of dietary intake of arsenic in the general population of the Republic of Croatia. *Sci. Total Environ.* 191 (1996), pp. 119–123.

Saxena, V.K., Kumar, S. and Singh, V.S.: Occurrence, behaviour and speciation of arsenic in groundwater. *Curr. Sci.* 86:2 (2004), pp. 281–284.

Schoof, R.A., Yost, L.J., Eickhoff, J., Crecelius, E.A., Cragin, D.W., Meacher, D.M. and Menzel, D.B.: A market basket survey of inorganic arsenic in food. *Food Chem. Toxicol.* 37 (1999), pp. 839–846.

Smedley, P.L. and Kinniburgh, D.G.: A review of the source, behavior and distribution of arsenic in natural waters. *Appl. Geochem.* 17 (2002), pp. 517–568.

SSA: Modificación a la Norma Oficial Mexicana NOM-127-SSA1–1994. *Salud ambiental. Agua para uso y consumo humano. Límites permisibles de calidad y tratamientos a que debe someterse el agua para su potabilización.* Secretaría de Salud México, DF, Official Newspaper. Mexico City, Mexico, 2000.

Stüben, D., Berner, Z., Chandrasekharam, D. and Karmakar, J.: Arsenic enrichment in groundwater of West Bengal, India: geochemical evidence for mobilization of As under reducing conditions. *Appl. Geochem.* 18 (2003), pp. 1417–1434.

Tao, S.S.H. and Bolger, P.M.: Dietary arsenic intakes in the United States: FDA total diet study, September 1991–December 1996. *Food Addit. Contam.* 16:11 (1998), pp. 465–472.

Thompson, T.S., Le, M.D., Kasick, A.R. and Macaulay, T.J.: Arsenic in well water supplies in Saskatchewan. *Bull. Environ. Contam. Toxicol.* 63 (1999), pp. 478–483.

USEPA: US Arsenic occurrence in public drinking water supplies. Report EPA-815-R-00-023, US Environmental Protection Agency, Office of Water, Washington, DC, 2000.

USEPA: National primary drinking water regulations. Arsenic and clarifications to compliance and new source contaminants monitoring. Final rule. US Environmental Protection Agency. *Federal Regist.* 66:14 (2001), pp. 6976–7066.

USGS: *Mineral commodity summaries 2004: Arsenic statistic and information.* US Geological Survey, http://minerals. usgs. gov/minerals/pubs/ commodity/arsenic/arsenmcs04. pdf (accessed Sept. 2004).

Welch, A.H., Lico, M. and Hughes, J.: Arsenic in ground water of the western United States. *Ground Water* 26:3 (1988), pp. 333–347.

Welch, A.H., Helsel, D.R., Focazio, M.J. and Watkins, S.A.: Arsenic in ground water supplies of the United States. In: W.R. Chappell, C.O. Abernathy and R.L. Calderon (eds): *Arsenic exposure and health effects*: Elsevier Science, New York, 1999, pp. 9–17.

WHO: *Guidelines for Drinking Water Quality.* 2nd ed., Vol. 1, World Health Organization, Geneva, Switzerland, 1996.

WHO: *Arsenic and arsenic compounds.* World Health Organization (Environmental Health Criteria 224), Geneva, Switzerland, 2001.

Wyatt, C.J., Fimbres, C., Romo, L., Méndez, R.O. and Grijalva, M.: Incidence of heavy metal contamination in water supplies in northern Mexico. *Environ. Res.* 76 (1998a), pp. 114–119.

Wyatt, C.J., Quiroga, V.L., Olivas Acosta, R.T. and Méndez, R.O.: Excretion of arsenic (As) in urine of children, 7–11 years, exposed to elevated levels of As in the city water supply in Hermosillo, Sonora, Mexico. *Environ. Res.* 78 (1998b), pp. 19–24.

Zheng, Y., Wu, J., Ng, J.C., Wang, G. and Lian, W.: The absorption and excretion of fluoride and arsenic in humans. *Toxicol. Lett.* 133 (2002), pp. 77–82.

Other countries and general processes

CHAPTER 18

Geogenic arsenic in an Australian sedimentary aquifer: Risk awareness for aquifers in Latin American countries

B. O'Shea
Department of Geology, Dickinson College, Carlisle, PA, USA

J. Jankowski
School of Biological, Earth and Environmental Sciences, The University of New South Wales, Sydney, NSW, Australia

ABSTRACT: Arsenic concentrations in a coastal sandy aquifer, extensively used for drinking water and irrigation supply, have been reported to deliver more than 45 times the permissible limit in Australian drinking water. The presence of arsenic (As) in the Stuarts Point aquifer was surprising due to its assumed deposition by on-shore sediment supply and thus lack of an obvious As source. Development of an aquifer-specific geomorphic model now suggests that As has been derived from regional erosion of As-rich stibnite deposits. The heterogeneity of the sedimentary aquifer causes groundwater redox conditions to control As mobilization via reductive dissolution of iron oxyhydroxides and precipitation of arsenian pyrite. The valuable lessons learned from an Australian sedimentary aquifer where As occurrence is complex, should be utilized in the assessment and identification of potential As-rich groundwaters of Latin America.

18.1 INTRODUCTION

In the last decade there has been a dramatic increase in the number of studies investigating arsenic (As) in groundwater. Much of this can be attributed to the discovery of elevated As in domestic groundwater supplies in Bangladesh (Nickson *et al.* 2000, Harvey *et al.* 2005). The increased awareness of the ubiquitous nature of As in the environment led other countries to focus on the occurrence of As in their groundwater. Elevated As has since been found in groundwater of Australia (Smith *et al.* 2003 and 2006, O'Shea *et al.* 2006), Taiwan (Chen *et al.* 1994), Vietnam (Berg *et al.* 2003) and many parts of the USA (Schreiber *et al.* 2000, Welch *et al.* 2000, Sidle *et al.* 2001). In recent years, the occurrence of As in aquifers of Latin American countries has become increasingly apparent (Smedley *et al.* 2002, Romero *et al.* 2004, Bundschuh *et al.* 2004, Bhattacharya *et al.* 2006).

The effects of ingesting As-rich groundwater can be devastating both physically and socially. Physical effects include various cancers (Bissen and Frimel 2003), skin lesions, gangrene and other clinical symptoms like joint pain, chronic cough and abdominal pain (Khalequzzaman *et al.* 2005). Unfortunately, many people who depend on As-rich groundwater supplies also live in rural areas where basic necessities are unsatisfied and incidences of poverty can be high, such as the Chaco-Pampean plain of Argentina (Bundschuh *et al.* 2004).

In contrast, the Stuarts Point aquifer of eastern Australia supports small coastal communities where the standard of living is more than satisfactory. Dissolved As has been reported by Smith *et al.* (2003) at concentrations as high as 337 µg/l, more than 45 times the acceptable drinking water limit of 7 µg/l As (NHMRC 1996). As such, groundwater in this aquifer is now being treated to remove dissolved As prior to human consumption.

The purpose of this study is to provide an outline of the investigation conducted at Stuarts Point. Arsenic sources, sinks and controls on mobilization into groundwater have been examined in detail

(O'Shea 2006) and are described herein. By examining the geochemical controls on As occurrence in this aquifer, it is hoped that the outcomes produced in this study can be applied to the assessment of aquifers at risk of As occurrence in Latin American countries.

18.2   ENVIRONMENTAL SETTING

18.2.1   *Aquifer lithology and hydrogeology*

Stuarts Point is located approximately 400 km north of Sydney, on the New South Wales (NSW) mid-north coast (Fig. 18.1). Groundwater for domestic supply and agricultural irrigation is extracted from the Stuarts Point aquifer.

The aquifer is approximately 20–50 m thick and consists of unconsolidated sediment overlying regional bedrock geology. Previous investigations (Eddie 2000, Smith *et al.* 2003) have predominantly described the aquifer as a beach barrier deposit, with little emphasis placed on fluvial contribution. It is now understood (O'Shea 2006) that a combination of fluvial, estuarine and onshore sediment supply have contributed to aquifer formation during the Quaternary period. This combination of depositional conditions accounts for the heterogeneous nature of the aquifer lithology, as seen in the generalized stratigraphy (Fig. 18.1). Four distinct lithologic facies, "beach barrier sand", "fluvial sand", "fluvial sand/estuarine clay" and "bedrock clay" can be identified throughout the aquifer.

Groundwater in the unconsolidated aquifer flows in a south-easterly direction from the topographic high at the base of Mt. Yarrahapinni towards the Macleay Arm estuary located adjacent to the aquifer. Hydraulic conductivity varies through the heterogeneous sand and clay units, ranging from 0.01 to 35.3 m/day (Northey 2001). The aquifer is recharged directly via precipitation and indirectly via leakage from the bedrock aquifer below.

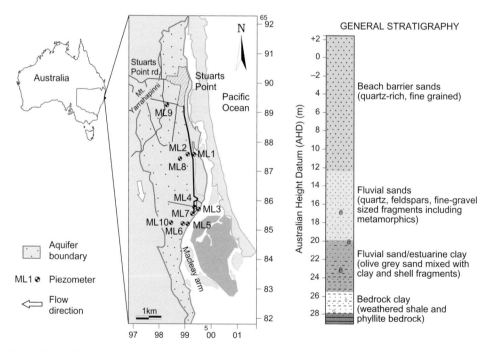

Figure 18.1.   Location of the Stuarts Point aquifer and multi-level piezometers installed for this study. The borelog from ML9 shows the general stratigraphy of the aquifer sediments.

18.2.2   Regional and bedrock geology

Regional and bedrock geology belongs to the New England fold belt (NEFB). Part of the NEFB is drained by the Macleay river, which discharges to the ocean immediately south of the Stuarts Point aquifer. Nearby Mt. Yarrahapinni (Fig. 18.1) is a coastal granitoid with mineralization grading from an outward zoning molybdenum zone, to a silver-lead zone and finally to a silver-As-zone (Gilligan *et al.* 1992). It is possible that these zones may extend beneath the northern portion of the Stuarts Point aquifer.

18.3   METHODOLOGY

18.3.1   Sample collection and preservation

Ten boreholes were drilled by solid flight augers and later converted to multi-level piezometers for groundwater sampling. During drilling (July 2001), 36 sediment samples were collected at 1.5 m intervals via *in situ* split spoon sampling in four multi-levels (ML7–ML10). Sediment samples were field-logged, frozen and kept in the dark during transportation to the laboratory. Groundwater in the ten multi-level piezometers (ranging in depth from 22 to 30 m with a sample point located at 1 m depth intervals) was purged using a peristaltic pump. A total of 227 samples were collected when general parameters (pH, temperature, electrical conductivity) had stabilized to within ±5% (O'Shea 2006). Samples were filtered through 0.45 μm MilliporeTM cellulose acetate membrane filter paper, preserved with concentrated analytical grade nitric acid and kept chilled until analysis in the laboratory.

18.3.2   Sediment and groundwater analysis

X-ray fluorescence (Siemens SRS300) was conducted at the University of New South Wales to determine total elemental composition of the sediments. Scanning electron microscopy (SEM) was performed at the Electron Microscope Unit at the University of New South Wales on a Hitachi S4500 Field Emission SEM (1996), with high resolution (1.5 nanometers), a tilting stage, Robinson back-scatter detector, Oxford cathodoluminescence detector (MonoCL2/ISIS) and a Link ISIS 200 Microanalysis system for chemical determination. The accelerating voltage was set at 20 kV throughout the analysis process. Arsenic was not expected to be detected during SEM analysis given the highest reported As concentration (14 mg/kg) by XRF was well below the As detection limit (approximately 50 mg/kg) for the SEM. A Cameca SX50 electron microprobe engaging four wavelength dispersive spectra's and one energy dispersive spectra, giving it the advantage of a more accurate and sensitive chemical analysis than the standard SEM-EDS analysis, was also employed. Trace elements in groundwater were analyzed by ICP-MS and major ions were analyzed by ICP-AES in the chemical laboratory at the University of New South Wales. Bicarbonate was determined in the field by titration with 0.01 M HCl against bromocresol green indicator (American Public Health Association 1992). Statistical analyses (descriptive statistics and Pearsons correlations) were performed in SPSS version 12.01 statistical software package as described in O'Shea and Jankowski (2006).

18.3.3   Quality assurance/quality control

Duplicate sediment and groundwater samples were collected during the field program. The accuracy of the sediment analyses was assessed by running duplicate XRF analyses. Duplicate results were reported within ±1%. For As, a Japanese Reference Material (JB-1) was used to ensure solid As concentrations were reported within ±1%. Both microprobe instruments were calibrated with Standard Reference Materials for the various elements analyzed. The S4500 uses Guide E1508-98 Standard Guide for Quantitative Analysis by Energy-Dispersive Spectroscopy; while the SX50 employs the data reduction matrix correction procedure based on the methods of Pouchou and

Pichoir (1985). Charge balance error (CBE) values for the groundwater samples were generally in the range of ±2–3%. CBE's above $\pm12\%$ were rejected. The interference between argon and chloride during the analysis of As by ICP-MS can raise reported As concentrations by 1 μg/l for every 100 mg/l of chloride present. Chloride concentrations in groundwater sampled for this study reported an average concentration of 42 mg/l. Interference from the formation of argon chloride should thus be minimal in these samples. Field titrations for bicarbonate were repeated at the well-head and the average of the two results taken.

18.4   ARSENIC OCCURRENCE AND DISTRIBUTION AT STUARTS POINT

18.4.1   *Source of arsenic in the aquifer*

The presence of As in the Stuarts Point aquifer was surprising due to its assumed deposition by on-shore sediment supply and thus lack of an obvious As source. Coastal aquifers along the east coast of Australia have endured similar depositional conditions and eustatic changes in sea level, initially raising concern over the likely presence of As in other coastal aquifers (O'Shea and Jankowski 2002, Smith *et al.* 2006). In the absence of any significant anthropogenic sources the As at Stuarts Point is deemed naturally occurring. Four geogenic sources are proposed:

- Arsenic has been contributed to the aquifer matrix via deposition of regionally eroded geological units containing As mineralization (O'Shea *et al.* 2006);
- Arsenic is derived from remnant seawater trapped in marine clay units deposited during eustatic changes of sea level in the Quaternary (Smith *et al.* 2006);
- The oxidation of arsenian pyrite present in ASS material contributes dissolved As to the groundwater (O'Shea 2006, Smith *et al.* 2006); and/or
- The underlying bedrock contains As, which is being contributed to the aquifer via upwards vertical leakage of groundwater (O'Shea 2006, Smith *et al.* 2003).

The dominant As source supported herein is derivation of As from regional erosion of As-rich stibnite deposits in the upper reaches of the Macleay river (O'Shea *et al.* 2006). A combination of sediment chemistry, statistical analysis, paleontological interpretation and sedimentological analysis allowed the construction of a detailed geomorphic model specific to the Stuarts Point aquifer (Fig. 18.2).

This model confirms fluvial sedimentation occurred in the aquifer and thus prompted the investigation into an upgradient As source. Naturally eroded As-rich stibnite (Sb_2S_3) deposits from the upper catchment have been linked to the As currently present in the Stuarts Point aquifer sediments (O'Shea *et al.* 2006). Minor contributions of As from other sources can be expected in localized areas of the aquifer, particularly as discharge from mineralized bedrock in the northern portion of the aquifer (O'Shea *et al.* 2006).

18.4.2   *Current arsenic sinks*

Mean sediment element concentrations are listed in Table 18.1. Arsenic concentrations in the aquifer matrix range from 1.4 to 14.0 mg/kg and represent average background values. Quartz grains dominate the sand aquifer and are responsible for high concentrations of silica in the matrix. Aluminum and iron concentrations increase in clay facies.

Frequently associated with naturally occurring As are aluminum, silica and potassium (clay minerals); iron and manganese (oxides); and sulfur (commonly delineating a pyrite association). Thus, a correlation between As and one or more of these elements could identify the presence of particular solid phase As sinks. Pearsons correlations (Table 18.2) show statistical associations between As and these elements in each lithologic facies.

There is a strong negative correlation with As and the clay mineral indicators, aluminum ($R^2 = -0.98$) and potassium ($R^2 = -0.98$), and a strong positive correlation with silica ($R^2 = 0.95$).

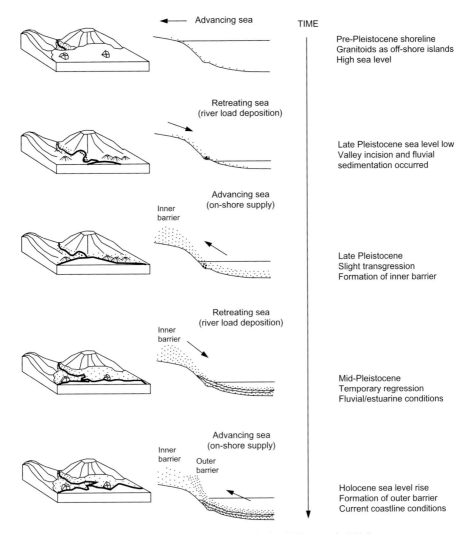

Figure 18.2. Geomorphic model proposed for Stuarts Point by O'Shea *et al.* (2006).

Illite group clay minerals contain Al, K and Si and have been identified as coatings on quartz grains (Fig. 18.3), however no As has been identified associated specifically with illite occurrence in the aquifer matrix. Illite has a moderate surface area available for sorption when compared to other clay minerals such as kaolinite. Positive surface charges for anion adsorption can be generated by protonation on broken Al-OH bonds exposed at the particle surface. Lin and Puls (2000) found illite had moderate As adsorption and subsequently moderate desorption. Therefore, As adsorption to clay minerals may be occurring, but is not considered a dominant sink within the Stuarts Point aquifer matrix.

Arsenic and iron correlate strongly ($R^2 = 0.93$) in the fluvial sands. These sands are orange-brown in color, a characteristic which is frequently indicative of iron oxyhydroxide presence. Smedley and Kinniburgh (2002) note that iron oxides are probably the most important adsorbents in sandy aquifers because of their great abundance and strong binding affinity. SEM identified iron oxyhydroxide coatings on sand grains in the matrix (Fig. 18.4). It is interesting to

Table 18.1.   Selected mean element concentrations for sediments in each lithologic facies at Stuarts Point.

Element % oxide	Barrier sand	Fluvial sand	Fluvial sand/ estuarine clay	Bedrock clay
SiO_2	96.11	93.06	84.89	72.34
Al_2O_3	0.94	1.53	5.24	12.37
Fe_2O_3	0.28	0.54	1.4	4.84
MnO	0.01	0.01	0.02	0.02
MgO	0.21	0.26	0.49	0.87
CaO	0.05	1.59	2.49	0.78
Na_2O	0.37	0.46	0.79	0.92
K_2O	0.25	0.47	1.13	1.78
SO_3	0.04	0.19	0.37	0.59
As[1]	3.99	6.33	6.99	9.33

[1]Reported in mg/kg.

Table 18.2.   Correlation ($p < 0.05$) between sediment arsenic concentrations and major elements for each lithologic facies in the Stuarts Point aquifer.

	Correlation with arsenic			
Element[1]	Barrier sand	Fluvial sand	Fluvial sand/ estuarine clay	Bedrock clay
SiO_2	−0.24	−0.73	−0.57	0.95
Al_2O_3	0.74	0.65	0.73	−0.98
Fe_2O_3	0.70	0.93	0.01	0.68
MnO	−0.12	0.01	0.16	−0.06
MgO	−0.29	0.84	0.14	−0.66
CaO	−0.15	0.53	0.26	−0.09
Na_2O	0.25	0.76	0.22	−0.57
K_2O	0.42	0.70	0.14	−0.98
SO_3	0.08	0.78	0.38	−0.09

[1]Analysed by XRF and reported as % oxide.

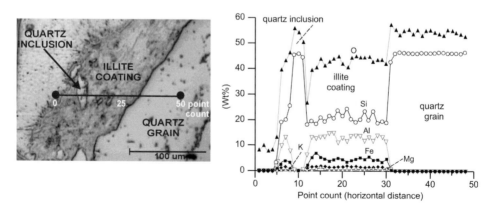

Figure 18.3.   Electron microprobe optical photograph and quantitative analysis (weight %) of an illite coating on a quartz grain found within the fluvial sand/estuarine clay lithologic facies. XRD has also identified illite within the matrix (O'Shea 2006).

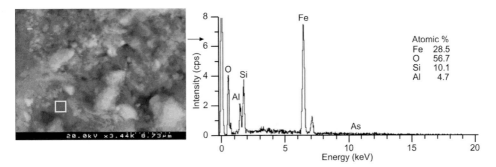

Figure 18.4.   SEM photograph and semi-quantitative analysis (within the spectral box) showing Fe oxy-hydroxide presence within the fluvial sands of the Stuarts Point aquifer.

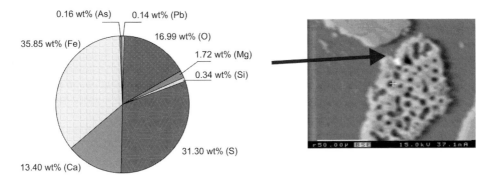

Figure 18.5.   Electron microprobe back scattered electron image and quantitative analysis (weight %) shows a discrete arsenian-pyrite cluster on a shell fragment in the Stuarts Point aquifer.

note that As and iron are not at all correlated ($R^2 = 0.01$) for the fluvial sand/estuarine clays. These sands and clays are characterized by an olive/grey color indicating iron oxide minerals are no longer dominant. The iron present within this unit (Table 18.1) is probably associated with illite (Fig. 18.3) contained within the estuarine clays. There is no correlation ($R^2 < 0.16$) between As and Mn in the aquifer, suggesting arsenic is not associated with Mn oxides at Stuarts Point.

A moderate correlation ($R^2 = 0.78$) exists between As and S in the fluvial sands, suggesting the possibility of an arsenic sulfide relationship. Arsenian pyrite was identified by electron microprobe (Fig. 18.5) as discrete micron sized pyrite clusters on quartz grains and shell fragments. These As-rich pyrites are thought to be precipitating under changes in redox conditions of the groundwater (Section 18.4.3). The numerous statistical correlations observed between As and other elements suggest several geochemical processes are acting on its distribution in the solid phase.

18.4.3   *Controls on arsenic mobilization*

Groundwater is predominantly fresh (generally $< 1000\ \mu$S/cm) with slightly acidic-neutral pH (4.7–7.9). Vertical chemical zonation occurs within the aquifer and is largely controlled by both geochemical processes (atmospheric input, sulfide oxidation, shell dissolution reactions, organic matter flux, seawater intrusion) and lithological controls (clay-water-shell interaction). A full hydrogeochemical analysis can be found in O'Shea (2006). Changes in groundwater Eh occur with depth; water becomes more reducing in deeper parts of the aquifer. This redox change is influenced by

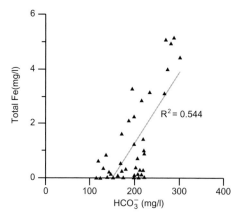

Figure 18.6.   Total Fe versus HCO_3^- in groundwater flowing through the fluvial sands of the aquifer matrix.

organic matter fluxes within the aquifer, which may include input from septic systems, presence of marine clays, and flow of dissolved organic matter through the aquifer from the adjacent wetland.

The successful identification of iron oxyhydroxides and arsenian pyrite phases in the aquifer matrix directs groundwater interpretation towards investigating the validity of these two sinks as As mobilization processes within the aquifer.

Groundwater flowing through the orange-red, iron rich, fluvial sands show a moderate ($R^2 = 0.54$) correlation between dissolved Fe and HCO_3 (Fig. 18.6). A good correlation can be expected (1) if iron oxyhydroxides are dissolving under reducing conditions (-50 to -150 mV in this part of the aquifer):

$$8FeOOH(-As) + CH_3COOH + 14H_2CO_3 \rightarrow 8Fe^{2+} + 16HCO_3^- + As + 12H_2O \quad (18.1)$$

The abundant shell material dissolving in the Stuarts Point aquifer may influence this correlation as excessive amounts of HCO_3 are added during calcite dissolution. Arsenic is expected to be released from dissolving iron oxyhydroxide minerals but re-adsorbed onto fresh surfaces, thus keeping dissolved As concentrations low (mean As for these groundwaters is 10.9 µg/l). This correlation decreases in the underlying olive grey fluvial sand/estuarine clays ($R^2 = 0.44$) where iron oxyhydroxide dissolution approaches completion and As concentrations begin to increase (maximum 61.6 µg/l As) as adsorption sites in the aquifer matrix decrease.

The formation of the identified arsenian pyrite in the aquifer is also a possible control on As mobilization in dissolved phase. Aqueous redox potential is sufficiently reducing (up to -235 mV in some deeper parts of the aquifer) to promote SO_4^{2-} reduction to S^{2-} under natural conditions. Marine estuarine clays may be the source of organic matter and sulfate; with addition of iron and As from the dissolution of iron oxides:

$$4SO_4^{2-} + Fe_2O_3 + 8CH_2O \rightarrow 2Fe(As)S_2 + 8HCO_3^- + 4H_2O \quad (18.2)$$

Once palaeowater salinity has exhausted the supply of sulfate, pyrite precipitation may be limited. This process is bacterially mediated and is the subject of much current research (Oremland and Stolz 2005). Localized seawater intrusion contributes additional sulfate to the fresh water aquifer and increases arsenian pyrite formation.

Several other processes have been found to effect As mobilization in the Stuarts Point aquifer and are currently under further investigation. These include arsenian pyrite oxidation by anthropogenic nitrate and addition of As from the adjacent Macleay Arm estuary under seawater intrusion processes (O'Shea 2006). The overall mobility of As in the aquifer is influenced by both anthropogenic (moderate contribution) and natural processes (dominant contribution) and is largely controlled by redox conditions and aquifer heterogeneity.

18.5   IMPLICATIONS FOR LATIN AMERICAN AQUIFERS

Emerging research suggests multiple geogenic sources may contribute to As occurrence in Latin American aquifers. Bundschuh *et al.* (2004) suggested three sources of As to groundwater in the Chaco-Pampean plain of Argentina:

- Layers of volcanic ash with 90% rhyolitic glass;
- Volcanic glass dispersed throughout the sediments; and
- Clastic sediments of metamorphic and igneous origin.

They concluded that the clastic sediments, which are the source of As at Stuarts Point, contribute little or no As to groundwater in the Chaco-Pampean plain. Also in contrast to the investigations described herein are the redox geochemical conditions observed in Latin American aquifers. Arsenic has been found in oxidizing groundwater environments (Del Razo *et al.* 1990, Rodriguez *et al.* 2004) rather than reducing environments like Stuarts Point and Bangladesh.

What is similar between most As groundwater occurrences are the dominant mobilization processes releasing As to the environment. At Stuarts Point, aquifer heterogeneity causes several of these release mechanisms to occur; oxidation of pyrite (O'Shea 2006), reductive dissolution of iron oxyhydroxides, pH-influenced desorption (Smith *et al.* 2006) and precipitation of arsenian pyrite. While it is important to delineate high As groundwater resources, it is well worth the time to assess the likelihood of an As source being present within the aquifer matrix. Developing a set of 'As source risk factors' according to aquifer depositional history and lithology, may indicate areas at high risk of As contamination prior to fieldwork, which can then be targeted in groundwater surveys. The lessons learned from a small coastal sedimentary aquifer in Australia should be utilized in the assessment and identification of potential As-rich groundwaters of Latin America.

18.6   CONCLUSIONS

At Stuarts Point, mobilization of As occurs when iron oxide surfaces coating sediment grains are dissolved under reducing conditions, releasing their adsorbed As into the groundwater. Increased reducing conditions develop with depth in the aquifer, as do concentrations of As. Anoxic conditions in the aquifer develop due to decreased influx of oxidized rainwater, water-sediment interaction and increased flux of organic matter. Progression of the geochemical conditions into a strongly reducing environment promotes precipitation of As into iron sulfides, with sulfate being supplied by seawater intrusion into the aquifer due to excessive exploitation of the groundwater resource. This removes As from the groundwater and incorporates it into a solid phase sink, preventing further migration of dissolved As.

The low risk of anthropogenic As input in combination with the accepted geomorphic model for the Stuarts Point aquifer indicated the aquifer was not expected to contain elevated concentrations of dissolved As in the matrix. Characterization of the As geochemistry of the Stuarts Point aquifer can be applied as a potential risk assessment tool for fluvial and coastal aquifers of Latin America. A geomorphic approach to aquifer characterization and identification of potential As sources may indicate Latin American aquifers that could be susceptible to elevated As concentrations. In addition, an assessment of hydrogeochemical conditions to determine the probability of As mobilization in the aquifer, can aid in the placement of potentially 'safe' wells for human consumption of groundwater. Heterogeneity of the aquifer chemical conditions and anthropogenic use of the aquifer should also be considered in the determination of low As groundwater environments.

ACKNOWLEDGEMENTS

The authors would like to thank the New South Wales Department of Infrastructure, Planning and Natural Resources for financial assistance with this study. Field and laboratory assistance was provided by Sarah Groves, John Wischusen, Irene Wainwright and Dorothy Yu.

REFERENCES

American Public Health Association: *Standard methods for the Examination of Water and Wastewater*. 18th ed., APHA-AWWA-WET, Washington, DC, 1992.

Berg, M., Tran, H.C., Pham, K.T., Pham, H.V., Schertenleib, R. and Giger, W.: Arsenic pollution of water resources in Vietnam—A plea for early mitigation actions. In: G.R. Gobran and N. Lepp (eds): *7th International Conference on the Biogeochemistry of Trace Elements*, Volume 2., Symposia, June 15–19, 2003 Uppsala, Sweden, 2003, pp. 8–9.

Bhattacharya, P., Claesson, M., Bundschuh, J., Sracek, O., Fagerberg, J., Jacks, G., Martin, R.A., Storniolo, A.R. and Thir, J.M.: Distribution and mobility of arsenic in the Rio Dulce alluvial aquifers in Santiago del Estero Province, Argentina. *Sci. Total Environ.* 358 (2006), pp. 97–120.

Bissen, M. and Frimmel, F.H.: Arsenic—a Review. Part 1: Occurrence, toxicity, speciation, mobility. *Acta Hydrochim. Hydrobiol.* 31:1 (2003), pp. 9–18.

Bundschuh, J., Farías, B., Marin, R., Storniolo, A., Bhattacharya, P., Cortes, J., Bonorino, G. and Albouy, R.: Groundwater arsenic in the Chaco-Pampean Plain, Argentina: case study from Robles county, Santiago del Estero Province. *Appl. Geochem.* 19 (2004), pp. 231–243.

Chen, S.-L., Dzeng, S.R., Yang, M.-H., Chiu, K.-H., Shieh, G.-M. and Chien, M.W.: Arsenic species in groundwaters of the blackfoot disease area. Taiwan. *Environ. Sci. Technol.* 28 (1994), pp. 877–881.

Del Razo, L.M., Arellano, M.A. and Cebrian, M.E.: The oxidation states of arsenic in well-water from a chronic arsenicism area of northern Mexico. *Environ. Pollut.* 64 (1990), pp. 143–153.

Eddie, M.W.: *Soil landscapes of the Macksville and Nambucca 1:100 000 Sheets*. Department of Land and Water Conservation, Sydney, Australia, 2000.

Gilligan, L.B., Brownlow, J.W., Cameron, R.G. and Henley, H.F.: *Dorrigo-Coffs Harbour 1:250000 metallogenic map SH/56–10, SH/56–11: Metallogenic study and mineral deposit data sheets*. New South Wales Geological Survey, Sydney, Australia, 1992.

Harvey, C.F., Swartz, D.H., Badruzzaman, A.B.M., Keon-Blute, N., Yu, W., Ali, M.A., Jay, J., Beckie, R., Niedan, V., Brabander, D., Oates, P.M., Ashfaque, K.N., Islam, S., Hemond, H.F. and Ahmed, M.F.: Groundwater arsenic contamination on the Ganges delta: biogeochemistry, hydrology, human perturbations, and human suffering on a large scale. *C.R. Geoscience* 337 (2005), pp. 285–296.

Khalequzzaman, Md., Faruque, F.S. and Mitra, A.K.: Assessment of arsenic contamination of groundwater and health problems in Bangladesh. *Int. J. Environ. Res. Publ. Health* 2:2 (2005), pp. 204–213.

Lin, Z. and Puls, R.W.: Adsorption, desorption and oxidation of arsenic affected by clay minerals and aging process. *Environ. Geol.* 39:7 (2000), pp. 753–759.

NHMRC: *Australian drinking water guidelines*. National Health and Medical Research Council (NHMRC) and Agriculture and Resources Management Council of Australia and New Zealand (ARMCANZ), National Water Quality Management Strategy, Canberra, Australia, 1996.

Nickson, R.T., McArthur, J.M., Ravenscroft, P., Burgess, W.G. and Ahmed, K.M.: Mechanism of arsenic release to groundwater, Bangladesh and West Bengal. *Appl. Geochem.* 15 (2000), pp. 403–413.

Northey, J.: *A hydrological and hydrogeochemical investigation of Stuarts Point aquifer system, NSW, following the re-introduction of tidal exchange into Yarrahapinni Wetland*. Honours Thesis, University of New South Wales, Australia, 2001.

O'Shea, B.: *Delineating the source, geochemical sinks and aqueous mobilisation processes of naturally occurring arsenic in a coastal sandy aquifer, Stuarts Point, New South Wales, Australia*. PhD Thesis, University of New South Wales, Australia, 2006.

O'Shea, B. and Jankowski, J.: Anthropogenic and geomorphic factors contributing to arsenic distribution in a coastal aquifer, New South Wales, Australia. In: E. Bocanegra, D. Martinez, and H. Massone (eds): *Proceedings of the XXXII IAH and VI ALHSUD Congress, Groundwater and Human Development*, October 21–25, 2002, Mar del Plata, Argentina, 2002, pp. 860–868.

O'Shea, B. and Jankowski, J.: Determining subtle hydrochemical anomalies using multivariate statistics: an example from the Great Artesian Basin, Australia. *Hydrol. Processes* 20:20 (2006), pp. 4317–4333.

O'Shea, B., Jankowski, J. and Sammut, J.: The source of naturally occurring arsenic in a coastal sand aquifer of eastern Australia. *Sci. Total Environ.* 379 (2007), pp. 151–166.

Oremland, R.S. and Stolz, J.S.: Arsenic, microbes and contaminated aquifers. *Trends Microbiol.* 13:2 (2005), pp. 45–49.

Pouchou, J.L. and Pichoir, F.: F (rZ) procedure for improved quantitative microanalysis. In: J.T. Armstrong (ed): *Microbeam analysis*. San Francisco Press Inc., San Francisco, CA, 1985, p. 104.

Rodriguez, R., Ramos, J.A. and Armienta, A.: Groundwater arsenic variations: the role of local geology and rainfall. *Appl. Geochem.* 19 (2004), pp. 245–250.

Romero, F.M., Armienta, M.A. and Carrillo-Chavez, A.: Arsenic sorption by carbonate-rich aquifer material, a control on arsenic mobility at Zimapan, Mexico. *Environ. Contam. Toxicol.* 47 (2004), pp. 1–13.

Schreiber, M., Gotkowitz, M., Simo, J. and Frieberg, P.: Stratigraphic and geochemical controls on naturally occurring arsenic in groundwater, eastern Wisconsin, USA. *Hydrogeol. J.* 8 (2000), pp. 161–176.

Sidle, W.C., Wotten, B. and Murphy, E.: Provenance of geogenic arsenic in the Goose river basin, Maine, USA. *Environ. Geol.* 41 (2001), pp. 62–73.

Smedley, P.L. and Kinniburgh, D.G.: A review of the source, behaviour and distribution of arsenic in natural waters. *Appl. Geochem.* 17 (2002), pp. 517–568.

Smedley, P.L., Nicolli, H.B., Macdonald, D.M.J., Barros, A.J. and Tullio, J.O.: Hydrogeochemistry of arsenic and other inorganic constituents in groundwaters from La Pampa, Argentina. *Appl. Geochem.* 17 (2002), pp. 259–284.

Smith, J.V.S., Jankowski, J. and Sammut, J.: Vertical distribution of As(III) and As(V) in a coastal sandy aquifer: factors controlling the concentration and speciation of arsenic in the Stuarts Point groundwater system, Northern New South Wales, Australia. *Appl. Geochem.* 18 (2003), pp. 1479–1496.

Smith, J.V.S., Jankowski, J. and Sammut, J.: Natural occurrences of inorganic arsenic in the Australian coastal groundwater environment: Implications for water quality in Australian coastal communities. In: R. Naidu, E. Smith, G. Owens, P. Bhattacharya and P. Nadebaum (eds): *Managing arsenic in the environment: from soil to human health*. CSIRO Publishing, Melbourne, Australia, 2006, pp. 129–153.

Welch, A.H., Westjohn, D.B., Helsel, D.R. and Wanty, R.B.: Arsenic in ground water of the United States: occurrence and geochemistry. *Ground Water* 38 (2000), pp. 589–604.

CHAPTER 19

Arsenic in a Triassic sandstone aquifer, Castellón, Spain

M.V. Esteller
Centro Interamericano de Recursos del Agua (CIRA), Universidad Autónoma del Estado de México, Toluca, Mexico

E. Giménez
Área Dptal. Ciencia y Tecnología Agroforestal y Ambiental, Facultad de Ciencias y Artes, Universidad Católica De Ávila (UCAV), Ávila, Spain

I. Morell
Instituto Universitario de Plaguicidas y Aguas, Universitat Jaume I, Castellón, Spain

ABSTRACT: Exploitation from the Triassic sandstone aquifer has begun to provide an alternative water supply for towns along the Spanish Mediterranean coast. These boreholes have revealed an increase over time in arsenic (As) concentrations, exceeding the limit for potable water (10 µg/l). The origin of the As seems to be related to the mineralogy of the aquifer and the hydrodynamics of the fissured sandstone. The rise in As concentrations in recent years is linked to an increase in the water volumes extracted, which has caused mobilization of waters held in the micro-fissures that has had a longer residence time in the aquifer.

19.1   INTRODUCTION

In 1993, the World Health Organization reduced the value of permissible arsenic (As) concentrations in potable water, and established the maximum admissible level at 10 µg/l. This new recommended value took into account the toxicity of As, even at low concentrations. Likewise, the European Community reduced the maximum admissible level of As in potable water. In Spain, Royal Decree RD 140/2003 established that, from 1 January 2004, water destined for human consumption should not contain concentrations of As exceeding this value. The fact that the limit for As in potable water is so low, compared to its abundance in nature has created a widespread problem. In fact, since the new limits were laid down, many public supply waters have been declared unpotable. Such is the case in certain parts of the central Iberian peninsula, as well as in the area of the present study. Within the study area, the Buntsandstein (Triassic) sandstone aquifer is a supply source for various towns on the Mediterranean coast. Exploitation of this aquifer began due to quality problems (farm pollution and salinization caused by marine intrusion), in the surficial, detrital coastal aquifers.

The presence of As in these waters may be related as much to natural processes as to anthropogenic activities, although the majority of environmental problems caused by excess As are due to mobilization of this element under natural conditions (Smedley and Kinniburgh 2002).

Arsenic concentrations in natural waters show a wide variation but the highest concentrations are generally found in groundwater. Water-rock interaction is largely responsible for the presence of As in this environment, and the physico-chemical conditions facilitate its mobilization. Therefore, special attention needs to be paid to the aquifer's mineralogy.

Ravenscroft *et al.* (2005) in a study of the Bengal basin (Bangladesh) offered two explanations for the geological mobilization of As: (1) dissolution of the hydrated iron oxides in a reducing environment, which gives rise to the liberation of adsorbed As and (2) oxidation of pyrite caused by piezometric drawdown resulting from overabstraction.

The importance of these oxides in controlling As concentration in natural waters has been studied over many years (Livesey and Huang 1981, Matisoff *et al.* 1982). In the large majority of aquifers affected by As the most important process seems to be the desorption/dissolution of As from iron and manganese oxides and oxyhydroxides. The factors that control the rate of desorption are mainly regulated by changes in pH and Eh (Smedley and Kinniburg 2002).

Thus, inputs due to local mineralization must be taken into account and, in fact, the geochemical behavior of As closely follows that of sulfur—such that the highest As concentrations are usually found in sulfur minerals (e.g., pyrite, arsenopyrite and arsenic pyrite; Nordstrom 2000, Smedley and Kinniburg 2002). Under oxidizing conditions, pyrite containing As will be weathered, producing Fe-oxides and a large amount of SO_4^{2-} and As.

Smedley and Kinniburgh (2002) reviewed cases of As pollution all over the world, and stated that many waters show elevated concentrations of this trace element as a result of local mineralization. Such is the case of As in waters linked to Triassic sandstone reported in Heinrichs and Udluft (1999), where reference is made to the aquifer system of the Keuper sandstones, which comprises one of the public water supplies in northern Bavaria, and which is affected by high As concentrations varying from 10 to 150 µg/l. The authors relate the presence of As to the geochemical characteristics of specific lithofacies comprising sediments of continental origin. Analyses of core samples from boreholes suggest that As contained in the rock is probably the source of As in the groundwater. Marked As accumulations occur in mudstone layers, where As content reach up to 35 mg/kg, while lower levels are found in most of the sandstones. Finally, they concluded that the As contained in the rocks was deposited in a fluvial-marine transitional environment and that the high As content could be linked to mineralized regions along the fracture zones.

Another reference to sandstone aquifers with As-rich waters is from eastern of Wisconsin (USA), where As concentrations as high as 12 mg/l were recorded in a confined aquifer of Ordovician sandstone (Schreiber *et al.* 2000). The authors proposed that the main source of As is a layer of secondary cementation with an elevated sulfur content. Arsenic appears in pyrites and marcasites as well as in hydrated iron oxides. In this case, it was also confirmed that the isotope type of pyrites and dissolved sulfates were very similar and that there was a close correlation between concentrations of SO_4^{2-}, Fe and As in the water. All this suggests that the oxidation of sulfates is the process that controls As uptake in the groundwater. However, one can not dismiss the possibility that hydrated iron oxides provide a further As source. In addition, the influence of fluctuations in the water table needs to be considered, so that the As contamination may be derived from interaction with water, air and sulfur salts whilst the oxidation of the latter may be influenced by variations in piezometric level.

Within the Iberian cordillera, the Triassic shows a typically germanic facies, and various studies have highlighted localized mineralization. Bustillo *et al.* (1999) describe the lithology of the Lower Triassic in the NW Iberian cordillera and highlight the localized presence of geodes containing a certain amount of iron oxides (hematite) and hydrated iron oxides (principally goethite). The Buntsandstein sandstones also indicate mineralization of Zr, Ba, Hg, As and S^{2-} of hydrothermal origin. Likewise, it seems that barite is associated with quartz, goethite, hematite, calcite and pyrolusite. Other authors mention the presence of dispersed pyrite in the limestone and dolomite of the Middle Triassic. Whatever the case, there is mining activity in this area, such as the mines extracting iron, barite, galena and mercury-barium minerals.

The minerals that can occur in the Buntsandstein aquifer (oxides and sulfo-salts), and that are in contact with the groundwater, have a variable As content. The hematites can contain more than 160 mg/kg (Baur and Onishi 1969), the iron oxides (undifferentiated), more than 20,000 mg/kg (Boyle and Jonasson 1973), the hydrated iron oxides, more than 76,000 mg/kg (Pichler *et al.* 1999), pyrite between 100 and 77,000 mg/kg (Baur and Onishi 1969), galena, between 5 and 10,000 mg/kg (Baur and Onishi 1969), quartz between 0.4 and 1.3 mg/kg (Baur and Onishi 1969), and calcite between 1 and 8 mg/kg (Boyle and Jonasson 1973), as described in Smedley and Kinniburg (2002).

Schreiber *et al.* (2000) considered that references to oxidation of sulfur salts as a source of As in groundwater to be scarce, since sulfur salts in aquifers are usually totally saturated and stable due to the reducing atmosphere and the absence of oxidants. On the other hand, Smedley and Kinniburg (2002) said that oxidation of sulfur salts, in particular pyrite, could be a significant source of As, above all where these minerals are exposed as a consequence of a fall in the water table surface. However, they recognized that when this occurs it could first liberate iron, which would subsequently tend to precipitate as hydrated iron oxide, with a consequent adsorption and co-precipitation of dissolved As. In other words, these authors considered that the oxidation of pyrite or other iron sulfur salts can not be considered to be an efficient mechanism for liberating As into groundwater. Even so, we find references to oxidation of sulfo-salts as a source of As in several studies: (Maccall *et al.* 2002, Slotnick *et al.* 2003, Smith *et al.* 2003, Neal *et al.* 2005, Ravenscroft *et al.* 2005).

As well as the water-rock interactions, the hydrodynamic and hydrogeological conditions need to be considered. It seems that two key factors are implicated in the genesis of As-enriched ground-water on a regional scale (Smedley and Kinniburgh 2002): in the first place, there must be some geochemical chain reaction that liberates As from the solid phase of the aquifer to the water. Sec-ondly, the liberated As must remain in the groundwater and not be transported far away. If the geochemical process responsible for mobilizing the As is suitable, as well as the hydrological regime, significant As concentrations in the water can result.

The objective of the present study was to establish which geochemical process might be respon-sible for the liberation of As into the groundwater and in what way the hydrodynamic regime might determine the concentration of this element.

19.2   METHODS

The methods used focused on two aspects. The first was a review of existing knowledge of the hydrology and hydrogeology of the study area, the characteristics of the boreholes, as well as historical analyses of physico-chemistry of the water. The second was the acquisition of new information from field data and water sampling.

A survey of eight sampling points was undertaken in 2005 in the Buntsandstein Triassic sandstone aquifer. Water samples were taken from springs and water supply wells. In the case of the boreholes, an attempt was made to adopt a pumping time sufficiently long to purge the borehole, whilst direct sampling was done from the springs.

In the city of Nules samples were taken from two wells Mallá 1 and Mallá 2, which are the two water supply boreholes for the city, though they were not in operation at this time. In the Vall d'Uixó municipality the supply borehole (Pipa) was sampled. Water from this well enters supply after treatment by reverse osmosis.

Duplicate samples were taken at each sampling point: 500 ml for analysis of the major ions, and 250 ml (acidified with 2 ml nitric acid) for determining heavy metals, whilst Eh, pH, temperature and electrical conductivity were measured *in situ*.

Chloride, sulfate and nitrate were analyzed using an ALKEMP 501 autoanalyzer. Bicarbonate was determined volumetrically, using a methyl orange indicator. The cations sodium, potassium, magnesium, calcium, strontium and lithium were determined by atomic adsorption. Boron was analyzed using the azomethine-H method, whilst mass ICP was used for SiO_2, Fe, Mn, Ba and As. The fluoride ion was determined using ion chromatography (IC).

19.3   GEOLOGY AND HYDROGEOLOGY

19.3.1   *Geology*

The study area is within the Triassic domain of the Sierra de Espadán (Fig. 19.1). This domain covers the southern part of Castellón province and a small part of Valencia province. It typically contains Triassic deposits, with some outcrops of the Paleozoic basement.

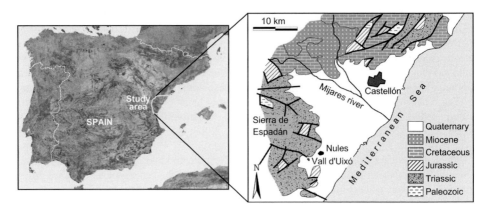

Figure 19.1.   Situation and geology of the study area.

The geological feature that characterizes the entire series is the presence of folding, together with fracturing that runs predominantly NW–SE and coexisting alpine structures having a NE–SW and NNE–SSW orientation.

Within the domain, there is a thick, homogeneous sequence of silicaceous sandstones (quartzoarenites) from the Buntsandstein (Lower Triassic) that has a general cross planar stratification. Locally, and on rare occasions one can observe thin intercalations of very compact red lutites. Another marked feature is the basal conglomerate layer, which may or may not be present; in its absence the series begins with a sandstone band that incorporates some testimonial quartzites.

The principal components of these quartzoarenites are quartz grains, whilst a minor fraction is composed of schist, zircon, feldspar, potassium spar, muscovite, tourmaline and iron oxides. The chloritic matrix comes from the alteration of biotite. Iron is liberated during this process, which is subsequently oxidized and fixed as cement or matrix, and it is this that gives the characteristic red color to the rock series.

Hydrothermal veins can be recognized in this formation. These originate from the mixing of two solutions: a superficial one, rich in sulfates that derives from the leaching of the Keuper evaporites (Upper Triassic), and another, deeper one that would bring metallic elements that are leached out during their upwards migration through the basement.

19.3.2   *Hydrogeology*

These Triassic sandstones confer an interest on this unusual aquifer because of the good quality of its waters and high hydraulic conductivity, provided that the aquifer is exploited where it is not excessively compartmentalized through faulting. In spite of being a detritic sedimentary formation, subsequent transformations have converted it into quartzarenites that lack intergranular porosity because of the complete growth of the silicaceous matrix, which ends up producing interpenetrating, saturated contacts. We are dealing with a relatively narrow band, whose permeability is clearly due to tectonic fissuring, which becomes deeper and more intense in certain sectors. On occasions, the low yield of the boreholes is due to their placement in parts of the aquifer that have been less tectonicized and where the degree of fissuring is less.

A problem that often arises during attempts to abstract from this aquifer is its lack of lateral continuity that results from tectonic causes. The network of fissures that configures this complex mosaic is formed from small wedges. In many cases, this same wedging forms the actual limit of the aquifer. When such circumstances occur and a borehole or well is installed to extract the groundwater, problems such as a continuous drop in piezometric level or scarce recharge are frequently encountered, leading to point over-abstraction.

The tendency towards falling piezometric levels in the medium to long term is a consequence of the lack of order and hierarchization of the flow in the fissure network, and for this reason a double porosity can be defined as a function of the characteristics of the fissuring, the compartmentalization of the aquifer blocks and the scarcity of renewable water resources (Sanchis *et al.* 1984, Garay 2000).

19.4   RESULTS AND DISCUSSION

A basic hydrogeochemical study was undertaken to determine the main chemical features of the groundwater and the reason for the presence of As.

The first stage was a review of the water chemistry based on data collected in previous years and the trend in the piezometric level. Figure 19.2a shows how As concentration has increased over time, reaching values of around 20 μg/l. In the case of the Mallá wells, this increase has coincided with a fall in piezometric level, as is shown in Figure 19.2b. The drop in piezometric level and the lack of recovery would be related to the hierarchization of fissures in the aquifer.

Table 19.1 shows the results corresponding to samples taken during 2005. The samples, which are representative of the aquifer water, indicate a pH from 7.2 to 7.9. Mineralization is moderate (230–670 μS/cm), except in the case of the Pipa well, which gave conductivity values of 974 μS/cm. The iron content is not elevated and is practically the same in all the samples. Manganese concentrations are higher in the wells than in the springs, and the same is true for As.

The oxidation state of As, and therefore its mobility, is basically controlled by redox conditions and pH. In addition, the values of the redox potential highlight the reducing conditions in the well water. Under the conditions found at depth in the aquifer (negative redox potential and a pH of around 7.5) the dominant species will be H_3AsO_3, and the As will be the As(III) species (Fig. 19.3). In contrast, Table 19.1 shows that the springs offer oxidizing conditions.

The difference in temperature between the wells and springs also needs to be highlighted. The springs maintain temperatures of below 20°C, while the wells exceed 23°C, with a maximum recorded from the Pipa borehole, which is 500 m deep. The only exception is at the Vilavella spring, where the temperature was 28.5°C. The high sulfate content recorded in the water from this spring suggests the influence of the large Espadán reverse fault, which would make possible, at a certain depth, the existence of a pinching out of evaporitic deposits (Keuper gypsums) that might explain the elevated sulfate content. This same fault could also facilitate a deep circulation (at around 800 m), which could explain the high temperature observed in water from this source.

The chemical data have been summarized in a Piper diagram (Fig. 19.4). The waters are mostly calcium and calcium-magnesium bicarbonate type, with the exception of samples from the Mallá 1 and Mallá 2 wells, which are magnesium bicarbonate type. The Vilavella spring waters stand out

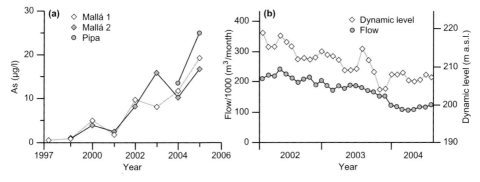

Figure 19.2.   (a) Temporal trend in As concentration in boreholes tapping the sandstones; (b) Changes in dynamic level in the Mallá well and its relationship to pumped flow.

Table 19.1.   Physico-chemical characteristics of water samples taken in 2005.

		EC	pH	T	Eh	Cl^-	SO_4^{2-}	NO_3^-	HCO_3^-	Ca^{2+}	Mg^{2+}	Na^+
	Location	μS/cm	–	°C	mV	mg/l						
S	Vilavella	590	7.26	28.5	543	52.8	18.5	6.9	267	64.3	38.6	20.5
S	Matilde	327	7.46	18.2	185	7.4	10.1	0.4	226	30.2	20.1	6.2
S	S. Josep	305	7.81	16.8	167	6.6	2.9	0.7	207	36.3	16.7	5.6
S	Baseta	234	7.87	17.0	176	6.1	3.1	0.4	100	19.1	11.3	4.0
S	Bosques	244	7.16	19.5	187	6.9	3.3	0.1	198	42.4	15.0	5.6
W	Mallá 1	669	7.60	24.8	−93.7	32.6	4.5	nd	408	43.0	49.8	12.1
W	Mallá 2	643	7.56	23.3	−85.9	25.5	7.2	27.1	372	33.6	49.5	9.6
W	Pipa	974	7.49	30.2	−17.8	24.9	160.1	nd	364	104	54.2	15.3

		K^+	Sr	B	SiO_2	F		Li	As	Fe	Mn	Ba
		mg/l						μg/l				
S	Vilavella	2.5	0.2	nd	13.6	9		9	3.16	206	0.41	388
S	Matilde	1.8	0.13	nd	11.1	6		6	2.08	117	1.73	208
S	S. Josep	2.0	0.07	nd	10.2	9		9	0.68	176	0.38	555
S	Baseta	0.9	0.03	nd	8.3	2		2	0.95	109	0.16	106
S	Bosques	1.0	0.06	nd	8.8	4		4	0.77	145	0.13	216
W	Mallá 1	5.8	0.13	0.01	8.9	8		8	19.50	156	33.8	31.8
W	Mallá 2	1.8	0.15	nd	8.7	10		10	16.80	107	16.3	50.9
W	Pipa	4.5	0.82	0.01	13.8	29		29	25.10	142	20.5	190

S: spring; W: well; EC: electrical conductivity; nd: not detected.

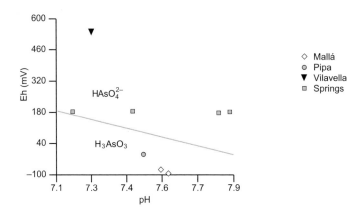

Figure 19.3.   Redox state of arsenic in water samples, according to pH and Eh measurements.

due to their peculiar characteristics (mentioned above). The Pipa water is also noteworthy, with its high sulfate concentrations, and in this respect, it is worth mentioning that the characteristic smell of hydrogen sulfide was detected at this well.

To establish possible relationships between the various parameters analyzed, the correlation matrix between components was calculated (Fig. 19.5). This indicates positive correlations between As and temperature (0.715), As and Mn (0.909), As and Mg (0.914), As and HCO_3^- (0.874). The correlation between depth of the borehole and As content is also positive.

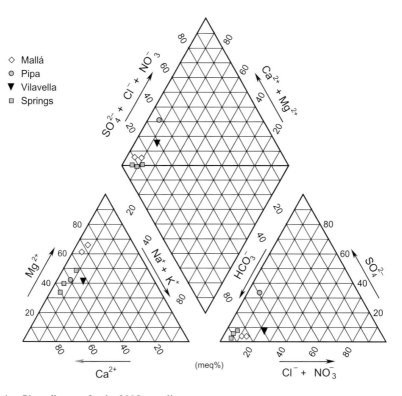

Figure 19.4.   Piper diagram for the 2005 sampling survey.

Correlations between other constituents were of less interest, with the exception of As-SO_4^{2-} (0.642), As-Sr (0.706) and As-Li (0.762). However, with respect to the first, it has to be said that samples with similar sulfate concentrations contained variable As concentrations, as seen in samples from the springs and the Mallá well.

It appears that the liberation of As may be related to the development of a reducing atmosphere and the existence of metallic mineralization, capable of liberating As, within the Lower Triassic sandstones. These factors could explain the presence of As adsorbed onto the mineral fraction and its subsequent liberation under reducing conditions. The lack of correlation between As and Fe, as well as the low iron concentration may be due to precipitation of iron compounds.

Whatever the case, the important point is that the highest concentrations of As are encountered in the wells with the lowest Eh (Mallá 1, Mallá 2 and Pipa). pH does not seem to exercise any significant control in the desorption of As, since it is slightly higher than 7.

If hydrodynamic conditions in the aquifer are taken into account, the possibility exists that As mobility is controlled by water circulation in different fracture systems with differing permeability. In other words, a primary level of well-developed fractures would permit relatively rapid flow, corresponding to the water abstracted during early exploitation of the aquifer, of excellent quality and containing little or no As. In parallel, a second level of fissuring, undoubtedly smaller in size (micro-fissuring) would give rise to a much slower flow. Conditions favorable to As mobilization (i.e., higher temperature, redox) would have developed in this environment, which would facilitate the incorporation of As into the waters held in the micro-fissures. This scenario would explain why, after an initial period, once the water held in the macro-fissures had been exhausted, poorer quality water containing higher As concentrations would be tapped. All these mechanisms may provide the explanation why the As concentrations have increased over recent years in the wells (Fig. 19.2a).

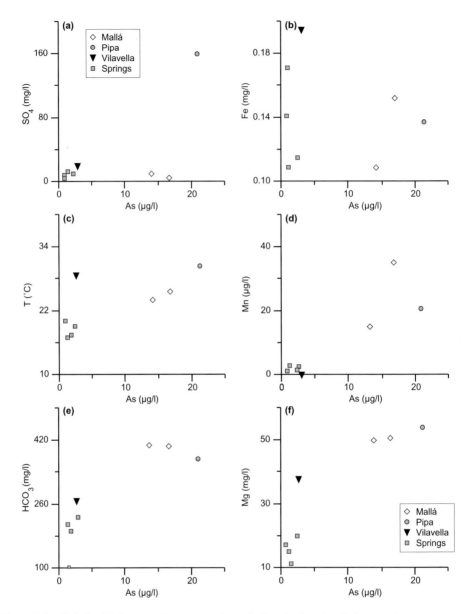

Figure 19.5.   Relationship between As concentration and other physico-chemical parameters.

19.5   CONCLUSIONS

The Triassic sandstone aquifer was chosen as an alternative potable water supply for a number of coastal towns along Spain's Mediterranean coast. These zones have traditionally suffered supply problems related to salinization caused by marine intrusion that affect the most superficial aquifers. With the passage of time, these new supplies have given rise to potability problems because of the presence of As in the water, and this concentration is increasing over time.

The information obtained suggests that the As could be related to the desorption of As from the mineral phases within the aquifer, favored by the reducing atmosphere. The influence of temperature cannot be disregarded in water-rock interactions.

In parallel, the process is accompanied by particular hydrodynamic conditions that have intervened in a definitive way in the evolution of As concentrations in the water, as water abstraction has continued. In the early years, the waters abstracted were those stored in the large fissures—water that is relatively young and scarcely mineralized. Later, the aquifer has released waters that were held in the micro-fissures, which correspond to a slower flowing older water.

The models established in this study are somewhat general, in an attempt to trace the most important features of a clearly complex situation. Nevertheless, they highlight the need to undertake more detailed studies that consider, not only a larger number of sampling points, but also a detailed analysis of the all parameters that may have intervened in the transfer and mobilization of As in the studied aquifer.

REFERENCES

Baur, W.H. and Onishi, B.-M.H.: Arsenic. In: K.H. Wedepohl, (ed): *Handbook of geochemistry*. Springer-Verlag, Berlin, 1969, pp. 33A1–33O5.

Boyle, R.W. and Jonasson, I.R.: The geochemistry of As and its use as an indicator element in geochemical prospecting. *J. Geochem. Explor.* 2 (1973), pp. 251–296.

Bustillo, M.A., García-Guinea, J., Martínez-Frías, J. and Delgado, A.: Unusual sedimentary geodes filled by gold-bearing hematite laths. *Geol. Mag.* 136:6 (1999), pp. 671–679.

Garay, P.: *El dominio triásico Espadán-Calderona. Contribución a su conocimiento geológico-hidrogeológico.* PhD Thesis, Universidad de Valencia, Spain, 2000.

Heinrichs, G. and Udluft, P.: Natural arsenic in Triassic rocks: A source of drinking-water contamination in Bavaria, Germany. *Hydrogeol. J.* 7:5 (1999), pp. 468–476.

Livesey, N.T. and Huang, P.M.: Adsorption of arsenate by soils and its relation to selected chemical properties and anions. *Soil Sci.* 131 (1981), pp. 88–94.

Maccall, P.J., Walter, L.M. and Szramek, K.J.: Arsenic sources and sinks in a surface water/groundwater system: tracking recharge to discharge in glacial drift deposits (Hell, Michigan). *Denver Annual Meeting*, October 27–30, 2002, Paper No. 131–137, The Geological Society of America, Denver, CO, 2002.

Matisoff, G., Khourey, C.J., Hall, J.F., Varnes A.W. and Strain, W.H.: The nature and source of arsenic in northeastern Ohio groundwater. *Ground Water* 20 (1982), pp. 446–456.

Neal, W., Ehman, K.D., Witebsky, S., Nuckolls, H.M. and Harbaugh, D.W.: Natural acidic groundwater-surface water occurrence, Ladera Sandstone, San Mateo County, California. *Geol. Soc. Am. Abstracts with Programs* 37:4 (2005), Cordilleran Section, 101st Annual Meeting, April 29–May 1, 2005, San Jose, California, Paper No. 2–7, 2005.

Nordstrom, D.K.: An overview of arsenic mass poisoning in China. In: C. Young (ed): *Minor elements 2000. Processing and Environmental Aspects of As, Sb, Se, Te and Bi*. Society for Mining, Metallurgy and Exploration Inc. (SME), Littleton CO, UK, 2000, pp. 21–30.

Pichler, T., Veizer, J. and Hall, G.E.M.: Natural input of arsenic into a coral reef ecosystem by hydrothermal fluids and its removal by Fe(III) oxyhydroxides. *Environ. Sci. Technol.* 33 (1999), pp. 1373–1378.

Ravenscroft, P., Burgess, W.G., Ahmed, K.M., Burren, M. and Perrin, J.: Arsenic in groundwater of the Bengal Basin, Bangladesh: Distribution, field relations, and hydrogeological setting. *Hydrogeol. J.* 13:5/6 (2005), pp. 727–751.

Sanchis, E., Delgado, S. and Morell, I.: Consideraciones sobre el aprovechamiento del Buntsandstein medio como acuífero para abastecimiento urbano en el área de la Sierra del Espadán (Valencia). *I Congreso Español de Geología*, April 9–14, 1984, Segovia, Spain, Tomo IV, 1984, pp. 253–364.

Schreiber, M.A., Simo, J.A. and Freiberg, P.G.: Stratigraphic and geochemical controls on naturally occurring arsenic in groundwater, eastern Wisconsin, USA. *Hydrogeol. J.* 8:2 (2000), pp. 161–176.

Slotnick, M.J., Meliker, J. and Nriagu, J.: Natural sources of arsenic in Southeastern Michigan groundwater. *J. Phys. IV France* 107 (2003), p. 1247.

Smedley, P.L. and Kinniburgh, D.G.: A review of the source, behaviour and distribution of arsenic in natural waters. *Appl. Geochem.* 17 (2002), pp. 517–568.

Smith, J.V.S., Jankowski, J. and Sammut, J.: Vertical distribution of As(III) and As(V) in a coastal sandy aquifer: factors controlling the concentration and speciation of arsenic in the Stuarts Point groundwater system, northern New South Wales, Australia. *Appl. Geochem.* 18 (2003), pp. 1479–1496.

CHAPTER 20

Arsenic distribution in the groundwater of the Central Gangetic plains of Uttar Pradesh, India

Parijat Tripathi, AL. Ramanathan, Pankaj Kumar & Anshumali Singh
School of Environmental Sciences, Jawaharlal Nehru University, New Delhi, India

P. Bhattacharya & R. Thunvik
KTH-International Groundwater Arsenic Research Group, Department of Land and Water Resources Engineering, Royal Institute of Technology (KTH), Stockholm, Sweden

J. Bundschuh
International Technical Cooperation Program, CIM (GTZ/BA), Frankfurt, Germany
Instituto Costarricense de Electricidad (ICE), San José, Costa Rica

ABSTRACT: Elevated arsenic (As) concentrations have recently been reported from groundwaters of the Holocene aquifers of the Indo-Gangetic plains in the states of Uttar Pradesh, Bihar and Jharkhand. This chapter reports about the nature of As contamination in groundwater in the Central Gangetic plain around the Ghazipur and Ballia districts, Uttar Pradesh, northern India. Groundwater samples were collected from a 2 km stretch in the flood plains of the Ganga and Ghagra rivers during two seasons for a one-year period. The samples were collected from shallow and deep tube wells and analyzed for hydrogeochemical parameters including the major anions and cations, and dissolved trace elements, including As. Arsenic concentrations in Ballia and Ghazipur were around 200 μg/l, albeit at some places below the detection limit. The As concentrations were very high in locations close to the river basin. They were also very high near an inland lake. Concentration of As were found to be moderate to low in the interior flood plains between these two basins. Groundwater As in these districts showed that intermediate aquifers have more As compared to deep and shallow aquifers. It was observed that As concentrations in groundwater were comparatively low in most places, expect for few locations in Ghazipur, where it exceeded 200 μg/l. The correlation between Fe and As was low, and sulfate concentrations in groundwater were relatively low. The shallow aquifers seem to be particularly at risk, due to the prevailing geochemical conditions in which oxidized and reduced waters mix, where the amount of sulfate, available for microbial reduction, seems to be limited. Further detailed study is needed in these two districts to obtain an insight into the precise geochemical processes occurring and controlling the As concentration in groundwaters.

20.1 INTRODUCTION

Arsenic (As) is a ubiquitous element found in the natural waters and mobilized through a combination of natural processes such as weathering reactions, biological activity and volcanic emissions as well as through a range of anthropogenic activities (Bhattacharya *et al.* 2002a and 2007, Nriagu *et al.* 2007, Smedley and Kinniburgh 2002). However, on a global scale, contaminated drinking water is the principal source of chronic human intoxication (Gabel 2000). The current drinking water quality guideline by WHO is 10 μg/l (WHO 2001), while the drinking water standard for As in India is 50 μg/l (Mukherjee *et al.* 2006).

The Indo-Gangetic plain is the most thickly populated area of India. The principal states located in the northern and central Gangetic plain are Uttar Pradesh (238,000 km^2 area) and Bihar (94,163 km^2 area). The pandemic of As poisoning, due to contaminated groundwater is widely reported from West Bengal, India, and in the adjacent country of Bangladesh, has been thought to be limited to the Ganges delta, where the source of As is geogenic and its mobilization has been attributed mainly to the reduction of iron-oxyhydroxides under anoxic conditions of the aquifers (Acharyya *et al.* 2000, Ahmed *et al.* 2004, Bhattacharya *et al.* 1997, 2002a, b and 2006, Nickson *et al.* 1998 and 2000, Nriagu *et al.* 2007, Zheng *et al.* 2004). Despite the earlier reports of As contamination in groundwater in Union Territory of Chandigarh and its surroundings in the northwestern Gangetic plain (Dutta 1976) as well as the recent findings in the Terai areas of Nepal (Amaya 2002, Bhattacharya *et al.* 2003, Chakraborti *et al.* 2003 and 2004, Tandukar *et al.* 2005 and 2006), there has been a lack of initiative by the local authorities to explore the water quality of the Gangetic plains in northern India. This chapter aims to present a preliminary report on the hydrogeochemical characteristics in two districts of Uttar Pradesh in the central Gangetic plain and their implications on the occurrence of As in groundwater and to understand the mechanism of As mobilization in sedimentary aquifers during development of groundwaters in the region.

20.2   AREAS OF INVESTIGATION

20.2.1   *Location and climate*

Two areas have been investigated in the state of Uttar Pradesh in the districts of Ghazipur and Ballia in northern India. Ghazipur district is situated in eastern part of Uttar Pradesh (Fig. 20.1a) between 25°19′ N to 25°54′ N and 83°04′ E to 83°58′ E. Hence, the climate of Ghazipur is neither very hot nor cold. The coldest months are December–January and the hottest months are May–June. The temperature varies from 5 to 17°C in winter and 30 to 42°C in summer. The south-west monsoon advances and the rainy season begins from around late June and ends in October. Thereafter, winter starts from around the middle of October and continues to late February. Annual rainfall in the district was between 800 and 1200 mm. On average there are 49–55 rainy days with rainfall of 2.5 mm or more per year. Relative humidity peaks are found during July and September with over 70%. During the post-monsoon and winter season, the humidity is high in the morning. In summer, relative humidity decreases to less than 25%.

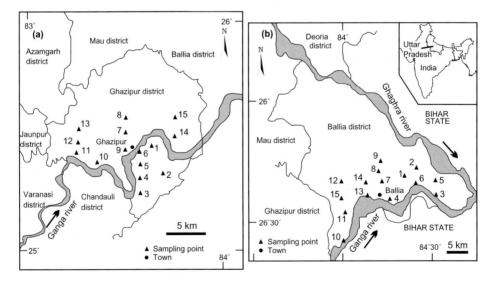

Figure 20.1.   Map showing the study areas around (a) Ghazipur and (b) Ballia district in eastern Uttar Pradesh with the sampling locations. Inset shows the map of India with the location of the Indo-Gangetic plain.

Ballia is situated between 25°33′ N to 26°11′ N and 83°38′ E to 84°39′ E (Fig. 20.1b). Nearly 70% of rainfall is received during three months (July to September). There are wide seasonal temperature variations with mean monthly maximum temperatures as high as 45°C in June and as low as 2°C in January. Most of the district qualifies for aridic (torric) moisture conditions according to the criteria laid in soil taxonomy. Soil moisture regime computations employing the Newhall mathematical model indicate that the area has a 'weak aridic' moisture regime (Van Wambeke 1985). Both districts have extensive agriculture and use surface water from canals and groundwater from the tube wells for irrigation.

20.2.2 Geology and hydrogeology

The districts are situated on the alluvial plains and drained by numerous rivers and streams. Level surface is varied because of the high banks of these rivers and the gentle slope from the central watershed towards Ganga, Ghagra and Chhoti Saryu, the three major rivers draining through the region. The geology of the region is dominated by the Indo-Gangetic alluvial sediments of Quaternary age deposited by the rivers Ganga and Ghagra over the last 1.8 million years. These fluviatile sediments comprise alternating beds of sand, silt and clay and is usually overlain by a thin layer of unstratified loam and lenses of peaty organic matter. Occasionally gravel beds are also found in the sedimentary sequence. Clay occurs in the form of lenticular bodies at various depths. There are small sand dunes in different parts of the districts. These consist of silt and very fine sand, high brownish, yellowish to buff in color (Fig. 20.2). The aquifers in the district belong to huge aquifer system in Indo-Gangetic plain with rich groundwater potential. Shallow unconfined aquifers present in these regions are mostly exploited for drinking water purposes.

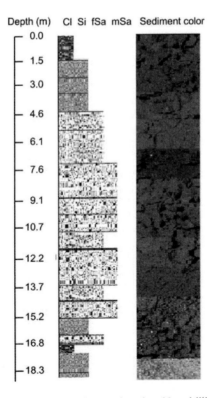

Figure 20.2.   Representative lithology of the study area, based on 20 m drilling.

Groundwater occurs in Holocene sandy sediments and forms extensive unconfined to leaky confined aquifers. Water levels fluctuate with seasonal recharge and discharge from 5–15 m below ground level. Most of the area is flooded during monsoon and water rises to near surface level. The multiple aquifer system of this region has variable hydraulic conductivity and water quality.

20.3   MATERIALS AND METHODS

20.3.1   *Field investigations and sampling*

To understand the general variation in groundwater chemistry in the study area, a well inventory survey was carried out during March 2003, and electrical conductivity (EC) and pH were measured. A global positioning system (GPS) was used for location and elevation reading. This was supported by topographic sheets made available from the Geological Survey of India. These data were used to select the representative wells and hand pumps for groundwater sampling. The selection was to represent different geological formations as well as land-use pattern at varying topography. A total of 62 groundwater samples were collected from the shallow and deep aquifers in the floodplains in 2004 and 2005, during pre- and post-monsoon seasons.

Water samples were collected in clean polyethylene bottles. At the time of sampling, bottles were thoroughly rinsed 2–3 times with groundwater to be sampled. In the case of bore wells and hand pumps, the water samples were collected after purging for 10 minutes. This was done to remove groundwater stored in the well itself and to obtain representative samples. Field measurements included the pH, oxidation-reduction potential (ORP) and electrical conductivity (EC) using a portable Orion Thermo water analyzing kit (Model Beverly, MA, 01915). Samples collected were classified for anions and cation analysis. Each water sample was filtered by 0.45 μm Millipore filter paper and acidified with 7M HNO_3 (Ultrapure Merck) for cation analysis and H_3BO_3 acid was used as preservative for nitrate analysis. The samples were stored at temperatures below 4°C prior to laboratory analysis.

20.3.2   *Laboratory analyses and data handling*

Major cations (Na^+, K^+, Ca^{2+} and Mg^{2+}) were analyzed by Flame Photometry (APHA 1995). Arsenic was analyzed after filtration using the Merck (2007) As field test kit to get an approximation of total As. Chemical parameters, such as nitrate, chloride, bicarbonate, sulfate, dissolved silica and phosphate, were analyzed in the laboratory following the methods of APHA (1995). Trace elements including As and Fe were analyzed at the laboratories of the School of Environmental Sciences, JNU, on Shimadzu AA-6800 atomic absorption spectrophotometer equipped with a graphite furnace (GF) at standard wavelength in absorption mode using different chemicals standards prepared from analytical grade chemicals.

The major ion compositions were plotted on Piper diagrams using Aquachem software. These plots provide better insight into the hydrochemical processes operating in the groundwater flow system that resulted in the observed spatial and temporal variation in the groundwater quality.

20.4   RESULTS AND DISCUSSION

20.4.1   *General groundwater chemistry*

The groundwater was slightly acidic to slightly alkaline in both the investigated regions. In Ghazipur, the pH of pre-monsoon samples ranged from 7.1–7.6 (average 7.33), while a slight increase was observed during post-monsoon, when pH-values varied from 7.3–8.1 (average 7.61). Compared with Ghazipur, groundwater in Ballia was more alkaline, with average pH-values of 7.56 and 7.77

for pre- and post-monsoon samples, respectively. Both values were higher for the respective season as compared to Ghazipur ranging between 7.1–8.2 and 7.2–8.7 during pre-monsoon and post-monsoon seasons, respectively. Precipitation induced dissolution processes might have effected the pH of groundwater. Increase of pH in the post-monsoon suggests that soil is very reactive which eventually enhances dissolution of carbonate minerals in the aquifers (Subramanian and Saxena 1983, Sracek *et al.* 2004). Groundwater EC in Ghazipur ranged from 430–1960 μS/cm and 240–1500 μS/cm in pre and post-monsoon, respectively. Average EC values were higher in the pre-monsoon suggesting the dilution effect. Like pH, EC was higher in Ballia than Ghazipur possibly due to leaching and weathering processes. Very high standard deviation in EC for Ballia suggests local variation in soil type, multiple aquifer systems and agricultural activities. In Ghazipur, the concentration of HCO_3^- ranged from 156–670 mg/l in pre-monsoon and 115–704 mg/l in post-monsoon, while in Ballia it ranges from 252–773 mg/l in March 2003 and 217–963 mg/l in September 2003 with average values of 482 and 502 mg/l in the respective seasons. Bicarbonate concentration was slightly higher in the post-monsoon period, indicating a contribution from chemical weathering. The amount of total dissolved solids (TDS) was high (670–800 mg/l) in deeper wells. Phosphate concentrations were generally higher, possibly due to extensive use of fertilizer. High bicarbonate concentrations were found in the intermediate and deep aquifers, possibly due to the presence of carbonates in the aquifer matrix. Among the cations, Ca^{2+} dominated, followed by Na^+, Mg^{2+} and K^+ ions. The high concentration of Ca^{2+} was due to weathering of calcite minerals, abundant in the aquifer matrix of the flood plain.

The observed variation in seasonal and spatial distribution of the major ions is locally very significant. The high HCO_3^- concentrations in certain locations reflect the possible effects of dissolution of carbonate minerals in the aquifer sediments. Chloride concentrations did not reveal significant variations in Ghazipur with concentration ranges of 4–625 mg/l and 28–674 mg/l for the pre-monsoon and in post-monsoon sampling periods, respectively. Although Cl^- concentration in Ballia groundwaters (16–625 mg/l) were similar to the Ghazipur samples during the pre-monsoon period, a significant increase in the concentrations (58–4466 mg/l) were observed in groundwater during the post-monsoon period. In most natural waters, the concentration of SO_4^{2-} was lower than Cl^-. In Ghazipur, SO_4^{2-} ranged from 10.3–203 mg/l in pre-monsoon and from 4.5–97 mg/l in the post-monsoon samples while in Ballia, it ranged from 22–903 mg/l in pre-monsoon and from 42–3050 mg/l in post-monsoon.

The dominant nutrients in the groundwater were in the order $NO_3^- > H_4SiO_4 > PO_4^{3-}$. The concentration of silica in the groundwater of Ghazipur ranged from 18–44.5 mg/l in pre-monsoon and from 20–51.3 mg/l in post-monsoon. In Ballia, it ranges from 16–82 mg/l in pre-monsoon and from 22–45.3 mg/l in post-monsoon. The concentration of silica in groundwater samples does not show any significant fluctuation over the season. The existence of alkaline environment enhances the solubility of silica. NO_3^- in the groundwater of Ghazipur ranged from 12–106 mg/l in pre-monsoon and from 23–120 mg/l in post-monsoon. In Ballia it ranged from 8–120 mg/l in pre-monsoon and from 15–120 mg/l in post monsoon, indicating the input from agricultural activities.

20.4.2   *Graphical representation of hydrochemical data*

The piper plot of Ghazipur (Fig. 20.3a) shows that almost all the groundwater samples of September 2003 fall in the field of Ca-HCO_3 type of water, indicating sufficient recharge and fresh water. In general, both the pre-monsoon and the post monsoon samples plot in same field few samples plot in the field of mixed Ca-Mg-Cl type of water (Fig. 20.3a). Some samples represented the Ca-Cl types. It is evident from the plot that alkaline earth elements (Ca^{2+} and Mg^{2+}) significantly exceeded the alkaline elements (Na^+ and K^+) and that weak acids (HCO_3^- and CO_3^{2-}) dominated strong acids (Cl^- and SO_4^{2-}). The alkaline earth elements showed higher concentrations than bicarbonate, indicating the exchange of Na^+ for Ca^{2+}.

The major ion composition of the groundwater in Ballia (Fig. 20.3b) shows more variation in water types compared to the Ghazipur. Most of the wells sampled during March 2003 fall in the

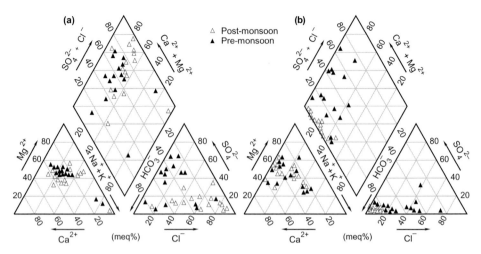

Figure 20.3.   Major ion composition of the groundwater for pre-monsoon and post-monsoon periods from
(a) Ghazipur and (b) Ballia districts of Uttar Pradesh, India.

field of mixed Ca-Mg-Cl type of water and in post-monsoon most of them fall in the region of
Ca-Cl type of water. Major ion composition of some of the samples do not show significant change
and are characterized by young Ca-HCO$_3$ type water or more evolved Na-Cl type of water during
both the sampling seasons. The distribution of the cations in the Ballia groundwaters was similar
to those from Ghazipur. The distribution of anions was strikingly different in Ballia, where HCO$_3^-$
and SO$_4^{2-}$ were the dominant anions in the pre-monsoon samples, but Cl$^-$ is dominant in the
post-monsoon samples and thus indicates a clear shift of water type.

20.4.3   *Distribution of arsenic and iron*

High As concentrations were observed in the groundwaters in both Ghazipur and Ballia districts.
The concentrations ranged from 2.5 to 128 μg/l and from 4 to 210 μg/l in the wells at Ghazipur
and Ballia districts, respectively (Fig. 20.4a, b). In general, 73% of the investigated wells in
Ghazipur and 87% of the wells in Ballia revealed As concentrations above the WHO drinking water
guideline value (10 μg/l). At Ghazipur, the As peaks were noted for wells placed at depths
between 16–19 m, 27–33 m, 42–45 m and 48 m, whereas in Ballia, As concentrations
peaked in wells placed between 15–17 m and 24–28 m. Seasonal variations of the concentrations were observed for all wells, but the changes were more conspicuous in the shallow
wells, especially at depths between 15 and 30 m. A decrease in the As concentrations was
observed in the well waters during the post-monsoon period, however the magnitude of the
changes was generally dependent on the overall hydrogeochemical characteristics of the well
water.

Concentration of Fe in groundwaters closely followed the As distribution in the wells of both
Ghazipur and Ballia districts. The Fe concentrations in the Ghazipur water samples ranged between
10 and 1940 μg/l while those in the wells of Ballia ranged between 5 and 265 μg/l (Fig. 20.5a,
b). Similar to As, the Fe concentration peaks were pronounced for wells at the depths 16–19 m,
27–33 m, 42–45 m and 48 m at Ghazipur, whereas Fe concentrations peaked intermittently in the
wells at depths 15–20 m and 28–40 m. Seasonal variations in Fe concentrations were observed for
all the analyzed wells, and an increased Fe concentration was noted in the well waters, sampled
during the post-monsoon season (Fig. 20.5a, b).

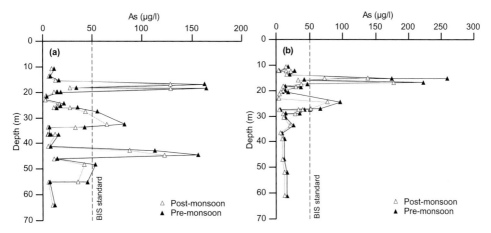

Figure 20.4.   Depth-wise distribution and seasonal variation of As in groundwater samples from (a) Ghazipur, and (b) Ballia.

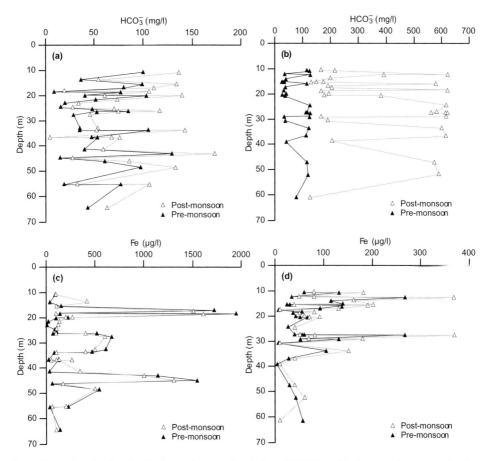

Figure 20.5.   Depth-wise distribution and seasonal variation of HCO_3^- and Fe in groundwater samples from (a), (c) Ghazipur, and (b), (d) Ballia.

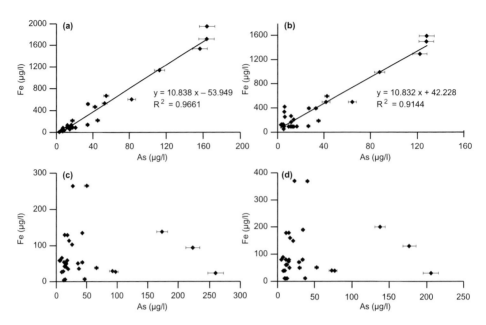

Figure 20.6.   Bivariate plots showing the relationship between As and Fe in the groundwater sampled from Ghazipur, (a) Pre-monsoon; (b) Post-monsoon seasons, and Ballia (c) Pre-monsoon; (d) Post-monsoon. Note strong correlation between As and Fe in the Ghazipur samples.

20.4.4   *Hydrogeochemical variability and salient correlations*

Arsenic concentrations were generally higher in the wells along the Ganga river basin both at shallow and intermediate depths, as compared to the wells in the shallow aquifers of the Ghagra river basin. Strong correlations were observed between the distribution of As and Fe in the groundwater samples from Ghazipur during both pre monsoon ($R^2 = 0.97$; $p \leq 0.01$) and post-monsoon ($R^2 = 0.91$; $p \leq 0.01$) sampling periods (Fig. 20.6a, b). The samples from Ballia district on the contrary did not show any definite correlation between As and Fe (Fig. 20.6c, d).

High concentration of ferrous iron in the groundwater is coupled with the anaerobic condition of most of the shallow and intermediate aquifers. Recent borehole drillings by the local well drillers reveal that high As wells are placed in the aquifers characterized by dark greyish to greyish-black colored, medium to coarser channel sediments. Our ongoing studies indicate that groundwater with low As concentrations are generally abstracted from the aquifers characterized by yellow/orange colored, medium to coarse sand, similar to studies by Jakariya (2007) and von Brömssen *et al.* (2007) in Matlab Upazila, Bangladesh. In such sediments, As could be sorbed in the pentavalent form As (V) on the secondary Fe- and Mn-oxyhydroxides mainly co-precipitated from the colloidal suspensions in the rivers.

Arsenic mobilization in the groundwater seems to be triggered by the increasing demand for irrigation due to needs for rice cultivation as well as for supporting the drinking water supply for the population. Since rice is the main agricultural crop of both Ghazipur and Ballia district, there is a possible risk for the contamination of the rice produced in this region.

20.5   CONCLUSIONS

Groundwater with high concentrations of geogenic As occurs extensively in the Holocene alluvial aquifers of Ghazipur and Ballia districts in the central Gangetic plain of Uttar Pradesh,

India. Hydrochemical data reveal that the aquifers in the two study areas have distinct depositional environments as shown by sediment textures and differential redox status. The pre-monsoon and post-monsoon seasons show significant variations in the hydrogeochemical characteristics. This preliminary study documents elevated As concentrations prevalent in the groundwater from the sedimentary aquifers of the central Ganga plain in Ghazipur and Ballia districts. The As levels in the shallow and intermediate aquifers at depths of 10–30 m as well as at 34–50 m are elevated by a factor of 3 to 25 as compared to the national drinking water standard of India and the WHO drinking water guideline values, respectively. Thus abstraction of groundwater for drinking water supplies and irrigation is a significant health concern in the future. Both rivers, Ganga and Ghagra, flowing from the Himalayan catchment, seem to contribute to As being co-precipitated onto the flood plain sediments in this region. However, our ongoing studies reveal that groundwater with low As concentrations can be targeted from aquifers characterized by yellow/orange coloured, medium to coarse sand and thus be considered safe for drinking purposes.

ACKNOWLEDGEMENTS

The authors like to express their thanks to Jawaharlal Nehru University and the Royal Institute of Technology (KTH) for giving necessary permission and providing facilities for this work. ALR gratefully acknowledges the support received from the Swedish South Asia Studies Network (SASNET) for his travel and support during his visit to Mexico and Sweden during June–July 2006. We thank the Swedish Research Council and the Swedish International Development Agency (VR-Sida) for the research grant (dnr: 346-2006-6005) on the project on targeting safe aquifer in regions with high As in groundwaters of India and the options for sustainable drinking water supply. We deeply appreciate the critical comments of the three anonymous reviewers that helped us to improve the manuscript.

REFERENCES

Acharyya, S.K., Lahiri, S., Raymahashay, B.C. and Bhowmik, A.: Arsenic toxicity of groundwater of the Bengal basin in India and Bangladesh: the role of Quaternary stratigraphy and Holocene sea-level fluctuation. *Environ. Geol.* 39 (2000), pp. 1127–1137.

Ahmed, K.M., Bhattacharya, P., Hasan, M.A., Akhter, S.H., Alam, S.M.M., Bhuyian, M.A.H., Imam, M.B., Khan, A.A. and Sracek, O.: Arsenic contamination in groundwater of alluvial aquifers in Bangladesh: An overview. *Appl. Geochem.* 19:2 (2004), pp. 181–200.

Amaya, A.: *Arsenic in groundwater of alluvial aquifers in Nawalparsasi and Kathmandu districts of Nepal.* MSc Thesis, Dept. of Land and Water Resources Engineering, Kungl Tekniska Högskolan, Stockholm, Sweden, 2002.

APHA: Standard methods for the examination of water and wastewater. 19th ed., American Public Health Association, Washington, DC, 1995.

Bhattacharya, P., Chatterjee, D. and Jacks, G.: Occurrence of arsenic contamination of groundwater in alluvial aquifers from Delta Plain, Eastern India: option for safe drinking supply. *Int. J. Water Res. Dev.* 13 (1997), pp. 79–92.

Bhattacharya, P., Frisbie, S.H., Smith, E., Naidu, R., Jacks, G. and Sarkar, B.: Arsenic in the environment: a global perspective. In: Sarkar, B. (ed): *Handbook of heavy metals in the environment.* Marcell Dekker, New York, 2002a, pp. 145–215.

Bhattacharya, P., Jacks, G., Ahmed, K.M., Khan, A.A. and Routh, J.: Arsenic in groundwater of the Bengal Delta Plain aquifers in Bangladesh. *Bull. Environ. Cont. Toxicol.* 69 (2002b), pp. 538–545.

Bhattacharya, P., Welch, A.H., Ahmed, K.M., Jacks, G. and Naidu, R.: Arsenic in groundwater of sedimentary aquifers. *Appl. Geochem.* 19:2 (2004), pp. 163–167.

Bhattacharya, P., Tandukar, N., Neku, A., Valero, A.A., Mukherjee, A.B. and Jacks, G.: Geogenic arsenic in groundwaters from Terai alluvial plain of Nepal. *Jour. de Physique IV France* 107 (2003), pp. 173–176.

Bhattacharya, P., Ahmed, K.M., Hasan, M.A., Broms, S., Fogelström, J., Jacks, G., Sracek, O., von Brömssen, M. and Routh, J.: Mobility of arsenic in groundwater in part Brahmanbaria district, NE

Bangladesh. In: Groundwater arsenic contamination in India: Extent and severity. In: R. Naidu, E. Smith, G. Owens, P. Bhattacharya and P. Nadebaum (eds): *Managing arsenic in the environment: from soil to human health.* CSIRO Publishing, Melbourne, Australia, 2006, pp. 95–115.

Bhattacharya, P., Welch, A.H., Stollenwerk, K.G., McLaughlin, M.J., Bundschuh, J. and Panaullah, G.: Arsenic in the environment: biology and chemistry. *Sci. Total Environ.* 379 (2007), pp. 109–120.

Chakraborti, D., Mukherjee, S.C., Pati, S., Sengupta, M.K., Rahman, M.M., Chowdhury, U.K., Lodh, D., Chanda, C.R., Chakraborti, A.K. and Basu, G.K.: Arsenic groundwater contamination in middle Ganga plain, Bihar, India: A future danger? *Environ. Health Perspect.* 111 (2003), pp. 1194–1201.

Chakraborti, D., Sengupta, M.K., Rahaman, M.M., Ahamed, S., Chowdhury, U.K., Hossain, M.A., Mukherjee, S.C., Pati, S., Saha, K.C., Dutta, R.N. and Quamruzzaman, Q.: Groundwater arsenic contamination and its health effects in the Ganga-Meghna-Brahmaputra plain. *J. Environ. Monitor.* 6:6 (2004), pp. 74 N–83 N.

Clesceri, L.S., Greenberg, A.E. and Eaton, A.D.: Standard methods for the examination of water and wastewater. 20th ed., American Public Health Association, Washington, DC, 1998.

Gebel, T.: Confounding variables in the environmental toxicology of arsenic. *Toxicology* 144 (2000), pp. 155–162.

Mukherjee, A.B., Bhattacharya, P., Jacks, G., Banerjee, D.M., Ramanathan, A.L., Mahanta, C. Chandrashekharam, D., Chatterjee, D. and Naidu, R.: Groundwater arsenic contamination in India: Extent and severity. In: Groundwater arsenic contamination in India: Extent and severity. In: R. Naidu, E. Smith, G. Owens, P. Bhattacharya and P. Nadebaum (eds): *Managing arsenic in the environment: from soil to human health.* CSIRO Publishing, Melbourne, Australia, 2006, pp. 533–594.

Nickson, R., Mc Arthur, J.M., Ravenscroft, P., Burgess, W.G. and Ahmed, K.M.: Mechanism of arsenic release of groundwater, Bangladesh and West Bengal. *Appl. Geochem.* 16 (2000), pp. 403–413.

Nriagu, J.O., Bhattacharya, P., Mukherjee, A.B., Bundschuh, J., Zevenhoven, R. and Loeppert, R.H.: Arsenic in soil and groundwater: an introduction. In: P. Bhattacharya. A.B. Mukherjee, J. Bundschuh, R. Zevenhoven and R.H. Loeppert (eds): *Arsenic in soil and groundwater environment: biogeochemical interactions, health effects and remediation.* Elsevier, Amsterdam, The Netherlands, 1997, pp. 3–60.

Sracek, O., Bhattacharya, P., Jacks, G., Gustafsson, J.P. and von Brömssen, M.: Behavior of arsenic and geochemical modeling of arsenic contamination. *Appl. Geochem.* 19:2 (2004), pp. 169–180.

Subramanyam, B.V.: Inorganic irritant poisons—Arsenic. In: Modi's textbook of medical juriprudence and toxicology. 22nd ed., Butterworths India Ltd., New Delhi, India, 2000.

Tandukar, N., Bhattacharya, P., Jacks, G. and Valero, A.A.: Naturally occurring arsenic in groundwater of Terai region in Nepal and mitigation options. In: J. Bundschuh, P. Bhattacharya and D. Chandrashekharam (eds): *Natural arsenic in groundwater: occurrence, remediation and management.* Balkema/Taylor and Francis Group, Leiden, The Netherlands, 2005, pp. 41–48.

Tandukar, N., Bhattacharya, P., Neku, A. and Mukherjee, A.B.: Extent and severity of arsenic poisoning in Nepal. In: Groundwater arsenic contamination in India: Extent and severity. In: R. Naidu, E. Smith, G. Owens, P. Bhattacharya and P. Nadebaum (eds): *Managing arsenic in the environment: from soil to human health.* CSIRO Publishing, Melbourne, Australia, 2006, pp. 595–604.

van Wambeke, A.: Calculated moisture regime and temperature regimes of Asia. SMSS Technical Monograph, Vol. 9, Cornell University and US Department of Agriculture, Ithaca, New York, 1985.

von Brömssen, M., Jakariya, M., Bhattacharya, P., Ahmed, K.M., Hasan, M.A., Sracek, O., Jonsson, L., Lundell, L. and Jacks, G.: Targeting low-arsenic aquifers in groundwater of Matlab Upazila, Southeastern Bangladesh. *Sci. Total Environ.* 379:2/3 (2007), pp. 121–132.

WHO: Guidelines for drinking water quality: Recommendation, Vol. 1. 2nd ed., World Health Organization, Geneva, Switzerland, 1993.

WHO: Arsenic in drinking water: Fact Sheet 210. World Health Organization, Geneva, Switzerland, 2001, URL: http://www.who.int/mediacentre/factsheets/fs210/en/print.html (accessed on May 4, 2007).

Younger, P.L.: Low-cost groundwater water quality investigation methods—an example from the Bolivian Altiplano. In: H. Nash and G.J.H. Mc Call (eds): *Groundwater Quality.* 17th Special report, Chapman and Hall, London, UK, 1995, pp. 55–65.

Zheng Y., Stute, M., van Geen, A., Gavrieli, I., Dhar, R., Simpson, H.J. and Ahmed, K.M.: Redox control of arsenic mobilization in Bangladesh groundwater. *Appl. Geochem.* 19:2 (2004), pp. 201–214.

CHAPTER 21

Temporal variations of groundwater arsenic concentrations in southwest Bangladesh

M. Jakariya
NGO Forum for Drinking Water and Sanitation, Lalmatia, Dhaka, Bangladesh

P. Bhattacharya
KTH-International Groundwater Arsenic Research Group, Department of Land and Water Resources Engineering, Royal Institute of Technology, Stockholm, Sweden

M. Manzurul Hassan
Department of Geography and Environment, Jahangirnagar University, Savar, Dhaka, Bangladesh

K. Matin Ahmed, M. Aziz Hasan & Sabiqun Nahar
Department of Geology, Dhaka University, Dhaka, Bangladesh

J. Bundschuh
International Technical Cooperation Program, CIM (GTZ/BA), Frankfurt, Germany
Instituto Costarricense de Electricidad (ICE), San José, Costa Rica

ABSTRACT: The pattern of groundwater arsenic (As) concentrations in Bangladesh is thought to be highly variable with time. To assess the changing pattern of As concentrations over time for developing a sustainable and affordable testing protocol for As, water from 246 sample tube wells of Jhikargachha, southwest Bangladesh, was analyzed using atomic absorption spectrophotometry (AAS) at two different time periods. The local hydrogeology and its relationship with As concentrations were also analyzed. Statistical analysis illustrates relationships between As in groundwater with different well parameters and aquifer properties. A positive correlation was observed between well depth and percentage of As-contaminated wells (R = 0.89). Similarly, the percentage of contaminated wells increased with grain size of the aquifer sediments, with a maximum percentage of contaminated wells in medium sized sand. Coarser sands showed a decrease in the number of contaminated wells. This increasing trend reversed at depths greater than 60 m. Most tube wells showed changes in As concentration with time (either increasing or decreasing). The highest increase was 91 μg/l and lowest decrease was 128 μg/l. However, almost half of the wells showed an increase and half showed a decrease, suggesting that these variations could be produced by experimental errors and/or seasonal variations.

21.1   INTRODUCTION

Arsenic (As) contamination in groundwater has lead to widespread concern in Bangladesh as more and more information has been gathered through a large number of studies (BGS and DPHE 2001, Chakraborti *et al.* 2002, Ahmed *et al.* 2004, Bhattacharya *et al.* 2006). Although there are differences in the numbers reported by different studies, at least one third of the country's domestic hand tube wells yield water at concentrations above the Bangladesh limit of 50 μg/l and more than 60% of the wells exceed the WHO guideline value of 10 μg/l. Arsenic contamination, in fact, has become a global issue of public health as it has been encountered in many countries and regions of the world under varying conditions (Bhattacharya *et al.* 2002a, b, Mandal and Suzuki 2002, Smedley and Kinniburgh 2002, van Geen *et al.* 2002).

Since the detection of As contamination in 1993, various studies have provided information regarding origin, occurrence, distribution and the factors controlling the presence of As in groundwater (BGS and DPHE 2001, Nickson *et al.* 2000, Bhattacharya *et al.* 2002a, b and 2006). Studies have also been undertaken to find sustainable approaches to mitigate the As problem in Bangladesh (Jakariya *et al.* 2007, von Brömssen *et al.* 2007). However, there is no single solution to the problem and mitigation strategies would differ from area to area based on local geology and culture. In the considering remediation approaches, new concerns such as possible change of As concentrations with time, and transfer of As from the water to plants through irrigation, need to be addressed. It has already been reported that uptake by plants irrigated with As-contaminated water might create another pathway of As intake by humans (Huq *et al.* 2001, Chowdhury *et al.* 2003). At the same time there is substantial evidence for As concentrations in well-water increasing with time (BGS and DPHE 2001, Burgess *et al.* 2002a, b, van Geen *et al.* 2003). Therefore, public health impacts of As remain a matter of great concern for those who have already been exposed to high As concentrations (Smith and Rahman 2000).

In December 2001, the Bangladesh Rural Advancement Committee (BRAC) completed a community based As mitigation study in Jhikargachha *upazila* which started in early 1999 (BRAC 2003). One of the main objectives of the project was to test all tube wells of the study area and to develop recommendations about the influence of time series and seasonal fluctuation on As concentrations from tube well water. The current study synthesizes the data generated from BRAC's community-based As mitigation pilot study in Jhikargachha *upazila* with financial support from UNICEF and the Department of Public Health Engineering (DPHE) of the government of Bangladesh. In this chapter, we investigate the occurrence and temporal variation of As concentrations in Jhikargachha.

21.2   MATERIALS AND METHODS

The Jhikargachha *upazila* (subdistrict) of the Jessore district, with an area of 529 km^2, is located in southwestern Bangladesh and lies between 89°00′ to 89°07′ East and 22°55′55″ to 23°12′34″ North (Fig. 21.1a). All samples from tube wells were screened for As using a Merck field kit at two different times (June 1999 and July 2001). At the same time, information on the wells and well-users were collected following a pre-structured questionnaire. Further information on subsurface geology and hydrogeology were collected from the Geological Survey of Bangladesh (GSB) and other organizations. In order to assess changes in As concentration over time, a total of 15,996 tube wells (all of the safe and 10% of the red-marked tube wells) that were tested in phase I (June 1999) were re-tested in phase II (July 2001) with the same field kit. In order to quantify changes in As concentration during that period, a total of 246 tube wells were selected randomly and the collected water samples were analyzed with atomic absorption spectrophotometry (AAS). Acidified (3 drops of 14 M HNO$_3$) bottles (100 ml) were used to collect water samples for laboratory analysis. In order to get reliable concentrations of As in the aquifer, stagnant water stored in the tube wells was purged by pumping for at least five minutes before collecting water in the pre-acidified bottles. We only used the laboratory (AAS) data in our present investigation.

Arsenic concentrations and tube well attributes were analyzed with statistical and spatial methods. Differences in As concentration between June 1999 and July 2001 were calculated. The calculated values were categorized into three different groups: (1) no change in concentration; (2) increases in concentration; and (3) decreases in concentration. Simple frequency distribution tables and deviation and correlation techniques were adopted for this chapter. In addition, GIS analysis was employed for mapping the distribution of As concentrations in the study area.

21.3   RESULTS AND DISCUSSION

21.3.1   *Distribution pattern*

Arsenic concentrations in groundwater from the recent alluvial aquifer in Jhikargacha *upazila* range from less than 10 µg/l to 500 µg/l. A number of hotspots (zones of high As concentrations)

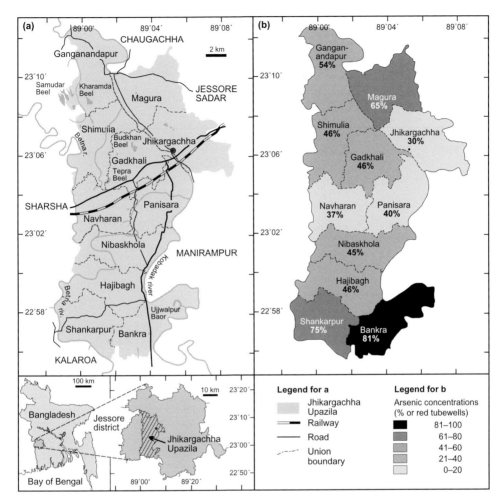

Figure 21.1. (a) Location map of the study area; (b) Distribution of arsenic-contaminated wells in Jhikargachha *upazila*.

were found in the area, but not spatial pattern was apparent. Groundwater from most of the area had As concentrations greater than the drinking water standard for Bangladesh. The southern part of study area had the highest percentage of contaminated wells: 45–81% (Fig. 21.1b). The northern part was moderately affected, with 46–65% of the wells contaminated with As, whereas the central part was less affected with 30–40% of the wells contaminated with As (Fig. 21.1b). Field kit results for 15,996 tube wells in the study site show that 57% of the wells were found to be unsafe for human consumption.

21.3.2 *Arsenic concentrations and depth*

Statistical analyses were performed to investigate the relationship between As concentrations in groundwater with different well and aquifer characteristics. There is a positive correlation between As concentration and well depth (Fig. 21.2a) with an *R*-value of +0.895. Less than 60% of the wells with depth up to 30 m are contaminated with As, whereas more than 60% of the wells with depth ranging between 30 and 76 m are contaminated. Most wells with depth >45 m are contaminated (80%). Figure 21.2a also shows a general increasing trend of percentage of contaminated wells with

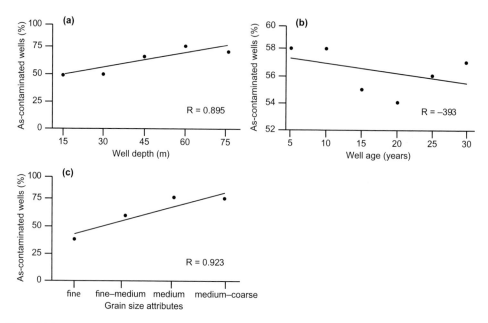

Figure 21.2.   (a) Correlation between percentage of contaminated wells and well depth; (b) Correlation between percentage of contaminated wells and well age; (c) Correlation between percentage of contaminated wells and the grain size attributes of the aquifer sand at the well screen depths.

depth between 30 and 60 m. The percentage of contaminated wells decreases slightly between 60 and 75 m depth. However, for depth larger than 75 m, this decreasing trend cannot be ascertained because of a lack of information below 76 m. Other studies carried out in different parts of Bangladesh reported that the aquifers at greater depth (>100 m) are less contaminated or free of As (BGS and DHPE 2001). The depth distribution is a factor of local geology (Ahmed *et al.* 2004).

21.3.3   *Arsenic concentrations and tube well age*

There appears to be a relationship between As-contaminated wells and tube well age (Fig. 21.2b). However, it is evident from the R-value ($R = -0.393$) that there is no definite correlation between the well age and percentage of contaminated wells. There appears to be a higher percentage of As-contaminated water in wells of 5 to 10 years of age as compared to older wells. However, the percentage falls sharply for wells between 15 and 20 years and increases again for wells >20 years.

Relationships between As contamination and lithology was also investigated (Fig. 21.2c). The percentage of As-contaminated wells increases with increasing grain size, with the medium sand aquifer having the highest percentage of As-contaminated wells (91%). The plot of the percentage of As-contaminated wells and the lithology at the screened depth show a strong positive correlation ($R = +0.923$). These correlations contradict some of the earlier relationships reported in other studies (Burgess and Ahmed 2006).

It has been reported that high As is generally associated with finer sediments (Ahmed *et al.* 2004), but the present study suggests that higher As is related to sediments with a medium grain-size. Disagreement with findings from earlier studies might result from a lack of knowledge of the textural properties of the aquifer sands at the depth of the well screens. In some previous studies (BGS and DPHE 2001) it has also been reported that older wells have higher probability of high concentrations of As in water.

21.3.4   *Hydrogeological characteristics and arsenic distribution*

Groundwater level elevations during the dry and wet seasons show groundwater movement from the central part of the study area towards the peripheral rivers. Restricted groundwater movement occurs in the northern part. The As distribution map in the study area shows that <40% of As-contaminated wells lie in the central part whereas >60% of the contaminated wells lie in the northern and southern part (Fig. 21.1b). Active flushing of groundwater in the central part might have reduced the As concentration but restricted groundwater movement hindered flushing in the northern part. However, the observed pattern of higher As concentrations in groundwater of the southern part of the study area could not be explained adequately without considering the regional pattern of groundwater flow in the study area. For the period of 1973–97, hydrographs of water level data from BWDB (Bangladesh Water Development Board) observation wells within Jhikar-gachha (JES006) depict the pattern of groundwater level fluctuations in the study area (Fig. 21.3). It can be inferred from the data that within the study area there was no long-term variation in water levels, which fluctuated due to seasonal cycles. However, seasonal fluctuations in the water levels have been accentuated in recent years due to the large groundwater withdrawals associated with irrigation.

21.3.5   *Behavior of arsenic concentration changes with time*

21.3.5.1   *Re-testing results of the tube wells*
The majority of the tube wells did not change status (i.e., contaminated to As-safe or *vice versa*) during the study period (Table 21.1). About 93% As-safe and 80% As-contaminated tube wells did not change their status.

A total of 246 water samples were selected randomly from all the functioning tube wells in Jhikargachha for analysis by AAS. Changes in As concentration in tube well water between the two samples showed no clear pattern (Table 21.2). The highest observed increase in concentration was 91 μg/l with a mean increase of 15.7 μg/l. The highest observed decrease in concentration was 128 μg/l with a mean decrease of 18.7 μg/l. The similar and relatively high standard deviations for both the cases indicate no clear temporal variation in As concentrations in the study site. Deviations were almost equally distributed between increases and decreases (47 and 50%), suggesting that these variations are probably related to experimental errors or seasonal effects.

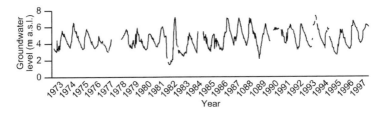

Figure 21.3.   Long term hydrograph of elevation of groundwater level in Jhikargachha.

Table 21.1.   Re-testing results of tube wells.

		Re-testing results	
Type of tube wells	Tube wells	(>50 μg/l)	(≤50 μg/l)
Arsenic safe	10770	708 (7)	10062 (93)
Arsenic contaminated	1092	879 (80)	213 (20)

Figures within parenthesis indicate percentage.

Table 21.2.   Laboratory analysis of water samples.

Category	Frequency	As concentration (μg/l)			
		Min.	Max.	Mean	St. deviation
No change	9 (3)	0	0	0	0
Concentration increased	115 (47)	0.20	91.0	15.7	15.8
Concentration decreased	122 (50)	0.40	128.2	18.7	18.8
Total	246				

Figures within parenthesis indicate percentage.

21.4   CONCLUSIONS AND RECOMMENDATIONS

Supply of drinking water not contaminated with As is the most crucial goal of As mitigation programs. Given the hydrogeological situation of Bangladesh, aquifer lithologies and water chemistry vary from region to region. Water chemistry is particularly important in developing As removal technologies because performance of a number of As removal technologies depends on water quality, especially pH. Various As removal technologies can be used as a short-term strategy for providing safe drinking water. However, to solve the problem permanently, a long-term solution needs to be adopted.

In an area where only a fraction of the tube wells are As-contaminated, people can collect water from those tube wells that are As-free. However, there is a strong need for monitoring these tube wells since they could become As-contaminated over time. Before promoting any safe drinking water remediation strategies, water samples should be analyzed at the community level because many As removal technologies are found to be effective only under ideal conditions. Any water supply options should be acceptable to users, affordable and accessible.

Much time and effort were invested in convincing people to shift from using relatively polluted surface water to groundwater. Now people have a strong dependence on tube wells and it would not be easy to shift drinking water sources for a large population. There is evidence that a large portion of the arsenicosis patients are still drinking As-contaminated water. The situation may be worse in areas where no symptoms of arsenicosis have become visible even after a long duration of As exposure. So, a strong 'behavioral change communication (BCC)' component should be added to water supply programs in order to bring positive changes in the behavior of the people.

Whatever water supply option is chosen, the community needs to perform system operation and maintenance. Without proper operation and maintenance, the water supply options may not remain reliable over the long term. Whatever procedures are used to mitigate the As problem, they should not increase the risk of other water borne contaminants or pathogens. Thus, the risks and benefits of As mitigation options should be assessed.

Both, short and long term solutions may be needed for areas severely affected by As-contaminated groundwater. The provision of piped water might be a better solution in this regard. We do not have a 'golden' solution yet, but there are some steps forward to face the crisis.

REFERENCES

Ahmed, K.M., Bhattacharya, P., Hasan, M.A., Akhter, S.H., Alam, S.M.M., Bhuyian, M.A.H., Imam, M.B., Khan, A.A. and Sracek, O.: Arsenic contamination in groundwater of alluvial aquifers in Bangladesh: An overview. *Appl. Geochem.* 19:2 (2004), pp. 181–200.

BGS and DPHE: Arsenic Contamination of Groundwater in Bangladesh, Vol. 2. Final Report, BGS Technical Report WC/00/19, London, UK, 2001.

Bhattacharya, P., Frisbie, S.H., Smith, E., Naidu, R., Jacks, G. and Sarkar, B.: Arsenic in the environment: A global perspective. In: B. Sarkar (ed): *Handbook of heavy metals in the environment.* Marcell Dekker Inc., New York, 2002a, pp. 145–215.

Bhattacharya, P., Jacks, G., Ahmed, K.M., Khan, A.A. and Routh, J.: Arsenic in the groundwater of the Bengal Delta Plain aquifers in Bangladesh. *Bull. Environ. Cont. Toxicol.* 69 (2002b), pp. 538–545.

Bhattacharya, P., Ahmed, K.M., Hasan, M.A., Broms, S., Fogelström, J., Jacks, G., Sracek, O., von Brömssen, M. and Routh, J.: Mobility of arsenic in groundwater in a part of Brahmanbaria district, NE Bangladesh. In: R. Naidu, E. Smith, G. Owens, P. Bhattacharya and P. Nadebaum (eds): *Managing arsenic in the environment: From soil to human health.* CSIRO Publishing, Melbourne, Australia, 2006, pp. 95–115.

Burgess, W. and Ahmed, K.M.: Arsenic in aquifers of the Bengal Basin: From sediment source to tube wells used for domestic water supply and irrigation. In: R. Naidu, E. Smith, G. Owens, P. Bhattacharya and P. Nadebaum (eds): *Managing arsenic in the environment: From soil to human health.* CSIRO Publishing, Melbourne, Australia, 2006, pp. 31–56.

Burgess, W.G., Burren, M., Perrin, J. and Ahmed, K.M.: Constraints on sustainable development of arsenic-bearing aquifers in southern Bangladesh. Part 1: A conceptual model of arsenic in the aquifer. In: K.M. Hiscock, M.O. Rivett and R.M. Davison (eds): *Sustainable groundwater development.* Geological Society, Special Publications 193, London, 2002a, pp. 145–163.

Burgess, W.G., Ahmed, K.M., Cobbing, J., Cuthbert, M.O., Mather, S.E., McCarthy, E.M. and Chatterjee, D.: Anticipating changes in arsenic concentration at tube wells in alluvial aquifers of the Bengal Basin. In: E. Bocanegra, D. Martinez and H. Massone (eds): *Groundwater and human development.* Proceedings of the 32nd IAH and 6th ALHSUD Congress, October 21–25, 2002, Mar del Plata, Argentina, 2002b, pp. 365–371.

BRAC: The use of alternative safe water options to mitigate the arsenic problem in Bangladesh: community perspectives. Research Monograph Series 24, BRAC, Dhaka, Bangladesh, 2003.

Chakraborti, D., Rahman, M.M., Paul, K., Chowdhury, U.K., Sengupta, M.K., Lodh, D., Chanda, C.R., Saha, K.C. and Mukherjee, S.C.: Arsenic calamity in the Indian subcontinent: What lessons have been learned. *Talanta* 58 (2002), pp. 3–22.

Chowdhury, T.R., Uchino, T., Tokunaga, H. and Ando, H.: Survey of arsenic in food composite for an arsenic-affected area of West Bengal, India. *Food Chem. Toxicol.* 40 (2003), pp. 1611–1621.

Huq, S.M.I., Ara, Q.A.J., Islam, K., Zaher, A. and Naidu, R.: Groundwater Arsenic Contamination in the Bengal Delta Plains of Bangladesh. In: G. Jacks, P. Bhattacharya and A.A. Khan (eds): *Proceedings of the KTH-Dhaka University Seminar*, University of Dhaka, Bangladesh, KTH Special Publication TRITA-AME 3084, Stockholm, Sweden, 2001, pp. 97–108.

Jakariya, M., von Brömssen, M., Jacks, G., Chowdhury, A.M.R., Ahmed, K.M. and Bhattacharya, P.: Searching for sustainable arsenic mitigation strategy in Bangladesh: experience from two upazilas. *Int. J. Environ. Poll.* 31:3/4 (2997), pp. 415–430.

Mandal, B.K. and Suzuki, K.T.: Arsenic around the world: a review. *Talanta* 58 (2002), pp. 201–235.

Nickson, R.T., McArthur, J.M., Ravenscroft, P., Burgess, W.G. and Ahmed, K.M.: Mechanism of arsenic release to groundwater, Bangladesh and West Bengal. *Appl. Geochem.* 15:4 (2000), pp. 403–413.

Smedley, P.L. and Kinniburgh, D.G.: A review of source, behavior and distribution of arsenic in natural waters. *Appl. Geochem.* 17:5 (2002), pp. 517–568.

Smith, A.H. and Rahman, M.: Contamination of drinking-water by arsenic in Bangladesh: a public health emergency. *Bull. World Health Organ.* 78:9 (2002), pp. 1093–1103.

van Geen, A., Ahasan, H, Horneman, A.H., Dhar, R.K., Zheng, Y., Hussain, I., Ahmed, K.M., Gelman, A., Stute, M., Simpson, H.J., Wallace, S., Small, C., Parvez, F., Slakovich, V., Lolacono, N.J., Becker, M., Cheng, Z., Momtaz, H., Shahnewaz, M., Seddique, A.A. and Graziano, J.: Promotion of well-switching to mitigate the current arsenic crisis in Bangladesh. *Bull. World Health Organ.* 80 (2002), pp. 732–737.

van Geen, A., Zheng, Y., Versteeg, R., Stute, M., Horneman, A., Dhar, R., Steckler, M., Gelman, A., Small, C., Ahsan, H., Graziano, J., Hussein, I. and Ahmed, K.M.: Spatial variability of arsenic in 6000 tube wells in a 25 km^2 area of Bangladesh. *Water Resour. Res.* 35:5 (2003), p. 1140.

von Brömssen, M., Jakariya, M., Bhattacharya, P., Ahmed, K.M., Hasan, M.A., Sracek, O., Jonsson, L., Lundell, L. and Jacks, G.: Targeting low-arsenic aquifers in groundwater of Matlab Upazila, Southeastern Bangladesh. *Sci. Total Environ.* 379 (2007), pp. 121–132.

Section III
Analytical methods for arsenic and laboratory studies

CHAPTER 22

Rapid, clean and low-cost assessment of inorganic and total arsenic in food by visible and near-infrared spectroscopy

R. Font & A. De Haro-Bailón
Instituto de Agricultura Sostenible (CSIC), Córdoba, Spain

D. Vélez & R. Montoro
Instituto de Agroquímica y Tecnología de Alimentos (CSIC), Burjassot, Valencia, Spain

M. Del Río-Celestino
Centro IFAPA Alameda del Obispo, Córdoba, Spain

ABSTRACT: This chapter assesses the potential of near-infrared spectroscopy (NIRS) for screening inorganic arsenic (i-As) in rice (*Oriza sativa*) and fiddler crabs (*Uca tangeri*), and i-As and total arsenic (t-As) contents in red crayfish (*Procambarus clarkii*). Samples of these species were freeze-dried and scanned by NIRS. Another portion of the sample was treated with nitric acid and ashing aid suspension ($MgNO_3$ + MgO), evaporated to dryness, ashed at 450°C, then diluted in hydrochloric acid. Inorganic As was separated and determined using the procedure developed by Muñoz *et al.* (1999). Both solutions were analyzed for As using FI-HG-AAS. The i-As and t-As contents of the samples were regressed against different spectral transformations by modified partial least square regression. The derivative transformation equations of the raw optical data resulted in equations showing coefficients of determination and also RPD values in cross-validation that were indicative of equations useful for screening As in such matrices. Spectral information related to chromophores and major cell components were important in modeling the prediction equations for i-As and t-As in the different matrices studied. This pioneering use of NIRS to predict As in foods represents an important saving in time and cost of analysis.

22.1 INTRODUCTION

Concern for food safety currently calls for an exhaustive control of chemical contaminants. The toxicity criteria for metals and metalloids have traditionally been established on the basis of total content. However, for some of them, such as arsenic (As) their toxicity depends on their atomic or molecular form. It is known that total arsenic (t-As) can be found in food in various chemical forms differing in their degree of toxicity and associated pathologies. The most toxic forms are the inorganic ones, As(III) and As(V), and the sum of both forms, which is known as inorganic arsenic (i-As). Arsenic is classified as a human carcinogen (IARC 1987). The i-As contents in some foods are subjected to regulation in a small number of countries, for example: fish and fish products in Australia and New Zealand; seaweed in Australia, New Zealand, and France (ANZFA 1997, Mabeau and Fleurence 1993). However, most existing legislation still bases its limits on the total As content, an ineffective criterion from the viewpoint of food safety. The availability of fast methodologies to quantify t-As and i-As levels in different kinds of foods would contribute to promote legislation to warranty the consumption of healthy food free of this metalloid.

The standard methodologies for analyzing trace metals offer a high level of precision but have some handicaps, such as high cost of analysis, slowness of operation, destruction of the sample, and use of hazardous chemicals. In contrast, near infrared spectroscopy (NIRS) is a valuable technique that offers speed and low cost of analysis (Williams and Norris 1987). In addition, the sample is analyzed without using chemicals. The spectral information can be used for simultaneous prediction of numerous constituents of the sample, once appropriate calibration equations have been prepared from sets of standard samples analyzed by both NIRS and conventional analytical techniques (Malley *et al.* 1996). After calibration, the regression equation allows accurate analysis of many other samples by prediction of results on the basis of the spectra.

NIRS has been applied to analysis of metal content mostly in the environmental field, and to a lesser extent in the agro-food industry. Some environmental applications include: the analysis of heavy metals in lake sediments (Nilsson *et al.* 1996, Malley and Williams 1997), studies concerning the chemical characterization of soils (Krischenko *et al.* 1992, Confalonieri *et al.* 2001, Font *et al.* 2004a), and the determination of heavy metals and As by NIRS in plant tissues (Clark *et al.* 1989, Font *et al.* 2002, Morón and Cozzolino 2002). In the agro-food field the feasibility of this technique for measuring K, Na, Mg, and Ca in white wines (Sauvage *et al.* 2002) was demonstrated. Chemical speciation can be determined with this technique. NIRS has been used for predicting mercurial species in the membrane constituents of living bacterial cells (Feo and Aller 2001). Recently, this technique has been applied for determining i-As in the crayfish *Procambarus clarkii* (Font *et al.* 2004b) and *Oriza sativa* (Font *et al.* 2005). In this chapter we summarize our recently published research concerning the application of NIRS to determine As in different food matrices. Unpublished NRIS results performed over crustacean *Uca tangeri* samples are also presented.

22.2   MATERIALS AND METHODS

22.2.1   *Equipment and software*

Near infrared spectra were recorded on a NIRS spectrometer model 6500 (Foss-NIRSystems, Inc., Silver Spring, MD, USA) in reflectance mode equipped with a transport module. The monochromator 6500 consists of a tungsten bulb and a rapid scanning holographic grating with detectors positioned for transmission or reflectance measurements. To produce a reflectance spectrum, a ceramic standard is placed in the radiant beam, and the diffusely reflected energy is measured at each wavelength. The actual absorbance of the ceramic is very consistent across wavelengths. In this work, each spectrum was recorded once from each sample, and was obtained as an average of 32 scans over the sample, plus 16 scans over the standard ceramic before and after scanning the sample. Reflectance energy readings were referenced to corresponding readings from an internal ceramic disk provided by the instrument manufacturer. The whole time of analysis took about 2 min each sample, approximately.

Mathematical transformations of the spectra as described below, and regressions performed on the spectral and laboratory data were obtained by using the GLOBAL v.1.50 program (WINISI II, Infrasoft International, LLC, Port Matilda, PA, USA).

22.2.2   *Collection and preparation of samples*

Samples of commercial rice were selected at different markets in Valencia (Spain) on the basis of the type of rice (brown or milled, long, medium or short grain). Rice samples were ground and stored at 4°C until analysis. Samples of crayfish were collected from sampling stations situated in Bajo Guadalquivir, Seville (Spain). The entire organism was used for the determination of t-As and i-As. Samples were freeze-dried, ground and stored at 4°C until analysis. Samples of crustacean *Uca tangeri* from the south of Spain were prepared with the aim of analyzing only the edible portion. After this preparation, the samples were freeze-dried, ground and stored at 4°C until analysis. The t-As and i-As were analyzed using hydride generation-atomic absorption spectrometry (HG-AAS) as described below, and NIRS.

22.2.2.1   *Hydride generation-atomic absorption spectrometry (HG-AAS) procedure: Total As and inorganic As reference measurements*

For t-As determination, samples were treated with nitric acid and ashing aid suspension ($MgNO_3$ + MgO), evaporated to dryness and ashed at 450°C with a gradual increase in temperature (Muñoz *et al.* 1999). The white ash was dissolved in hydrochloric acid, pre-reduced with (ascorbic acid + KI) agent, and the As was quantified by flow injection-hydride generation-atomic absorption spectrometry (FI-HG-AAS; model 3300 and model FIAS-400 of Perkin Elmer). The analytical characteristics of the method were: detection limit = 0.026 mg/kg dry weight (wt); precision = 2%. The accuracy was tested with certified references materials of dogfish muscle (DORM-2, CNRC) and rice flour (SRM1568a, NIST).

The determination of i-As was done using the methodology developed previously by Muñoz *et al.* (1999). Deionized water and HCl were added to freeze-dried sample. The mixture was left overnight. After reduction by HBr and hydrazine sulfate, the i-As was extracted into chloroform, and back-extracted into an acid phase. The back-extraction phase was dry-ashed and the i-As was quantified by FI-HG-AAS. The analytical characteristics of the method were: detection limit = 0.013 µg/g dry wt; precision = 3–5%; recovery As(III) 99% and As(V) 96%.

22.2.2.2   *NIRS procedure: recording of spectra and processing of data*

Freeze-dried, ground sub samples used to conduct this work were placed in the NIRS sample holder (3 cm diameter) until it was 3/4 full (weight $\cong$ 3.50 g), and were then scanned. Their NIR spectra were acquired at 2 nm intervals over a wavelength range from 400 to 2500 nm (visible plus near infrared regions).

Samples were recorded as an NIR file, and were checked for spectral outliers [spectra with a standardized distance from the mean (*H*) >3 (Mahalonobis distance)], by using principal component analysis (PCA). The objective of this procedure was to detect and, if necessary, remove possible samples whose spectra differed from the other spectra in the set.

In the second step, laboratory reference values for i-As (all the matrices) and t-As (red swamp crayfish), as obtained from the FI-HG-AAS reference method, were added to the NIR spectra file. Calibration equations were computed in the new file by using the raw optical data (log 1/R, where R is reflectance), or first or second derivatives of the log 1/R data, with several combinations of segment (smoothing) and derivative (gap) sizes. To correlate the spectral information (raw optical data or derived spectra) of the samples and the analyte content determined by the reference method, modified partial least squares (MPLS) was used as regression method, using wavelengths from 400 to 2500 nm every 8 nm. Standard normal variate and De-trending (SNV-DT) transformations (Barnes *et al.* 1989) were used to correct baseline offset due to scattering effects (differences in particle size among samples).

22.2.3   *Cross-validation*

The statistical meaning of the different calibration equations were determined using cross-validation. Cross-validation is an internal validation method seeking to validate the calibration model on independent test data. This method does not waste data for testing as it occurs in external validation. The method is carried out by splitting the calibration set into *M* segments and then calibrating *M* times, each time testing about a (1/*M*) part of the calibration set (Martens and Naes 1989).

The prediction ability of the equations obtained was determined on the basis of their coefficient of determination in the cross-validation (r^2) (Shenk and Westerhaus 1996) (eq. 22.1) and standard deviation (SD) to standard error of cross-validation (SECV) ratio (RPD) (Dunn *et al.* 2002) (eq. 22.2).

$$r^2 = \left(\sum_{i=1}^{N} (\hat{y} - \bar{y})^2 \right) \left(\sum_{i=1}^{N} (y_i - \bar{y})^2 \right)^{-1} \tag{22.1}$$

where: $\hat{y}$ = NIR measured value; $\bar{y}$ = mean "y" value for all samples; y_i = laboratory reference value for the ith sample; N = number of samples.

$$\text{RPD} = SD \left\langle \left[\left(\sum_{i=1}^{N} (y_i - \hat{y}_i)^2 \right) (N - K - 1)^{-1} \right]^{1/2} \right\rangle^{-1} \tag{22.2}$$

where: y_i = laboratory reference value for the ith sample; $\hat{y}$ = NIR measured value; N = number of samples, K = number of wavelengths used in an equation; SD = standard deviation.

The statistics shown in eq. 22.1 and 22.2, give a more realistic estimate of the applicability of NIRS to the analysis than those of the external validation, as cross-validation avoids the bias produced when a low number of samples representing the full range are selected as validation set (Shenk and Westerhaus 1996, Williams and Sobering 1996). The SECV method is based on an iterative algorithm which selects samples from a sample set population, it uses them to develop the calibration equation, and it predicts on the remaining unselected samples. This statistic indicates an estimate of the standard error of prediction (SEP) that may have been found in an external validation (Workman 1992). SEP is calculated as the square root of the mean square of the residuals for $N-1$ degrees of freedom, where the residual equals the actual minus the predicted value.

Cross-validation was computed on the calibration set for determining the optimum number of terms to be used in building the calibration equations and to identify chemical (T values >2.5) or spectral (H value >3.0) outliers. The outlier elimination procedure was set to allow the software to remove outliers twice before completing the final calibration (NIRSystems 1995).

22.3   RESULTS AND DISCUSSION

22.3.1   Arsenic contents in the samples

Samples of rice (n = 40) showed mean content and SD of 0.110 and 0.049 µg/g dry weight (dw), respectively (Table 22.1). The range of i-As found in the samples extended from 0.013 to 0.268 µg/g dw, these values being similar to those contents previously reported in white rice from the United States of America (Lamont 2003). Inorganic As contents were normally distributed with respect to the mean in the whole range. Individuals of *Uca tangeri* showed i-As contents that ranged from 0.20 to 1.65 µg/g dw, with a mean content and SD of 0.81 and 0.40, respectively (Table 22.1). The t-As and i-As contents found in the crayfish samples (n = 62) (Table 22.1) vary from 0.15 to 2.82 µg/g, dw, similar to the concentration range previously reported for this species in the same area (Devesa *et al.* 2002).

22.3.2   Spectral data pre-treatments and equation performances

The transformation of the raw spectra (Log 1/R) to their second derivative (SNV+DT), resulted in substantial correction of the baseline shift caused by differences in particle size and path length variation among samples, which contributed to reduce spectral noise. This fact can be observed in Figures 22.1 and 22.2, which correspond, respectively, to the raw spectra and second derivative spectra of rice samples. Peaks and troughs in second derivative rice spectra correspond to the points of maximum curvature in the raw spectrum. The increase in the complexity of the derivative spectra resulted in a clear separation between peaks overlapping in the raw spectra.

In cross-validation the selected equation showing the highest prediction ability for the different studied species was performed over the visible plus near-infrared segments (Fig. 22.3), except for *U. tangeri* samples, whose calibration was developed over the near-infrared segment alone (Table 22.1). Also, i-As in this species of crayfish was best modeled by using the first derivative transformation of the raw spectra rather than the second derivative, as it occurred with the other two species.

Table 22.1.   Calibration and cross-validation statistics for inorganic and total arsenic for the selected equations at both inorganic and total arsenic considered in this work (μg/g, dry weight).

Species		Calibration							Cross-validation	
		N	Range	Mean	SD	SEC	R^2		RPD	r^2
O. sativa[1]	i-As	40	0.013–0.268	0.110	0.049	0.020	0.80		1.69	0.65
U. tangeri[2]	i-As	32	0.20–1.65	0.81	0.40	0.16	0.82		1.75	0.71
P. clarkii[1]	i-As	62	0.15–2.82	1.26	0.75	0.19	0.93		2.63	0.84
P. clarkii[2]	t-As	60	0.72–5.11	2.22	1.07	0.30	0.91		2.38	0.81

[1] Wavelength range: 400–2500 nm. Mathematical treatment: 2, 5, 5, 2, (SNV+DT); [2] Wavelength range: 1100–2500 nm. Mathematical treatment: 1, 4, 4, 1 (SNV+DT); N: number of samples in the calibration file; range: minimum and maximum reference values in the calibration file; SD: standard deviation of the calibration file; SEC: standard error of calibration; R^2: coefficient of determination in the calibration; RPD: SD to standard error of cross-validation ratio; r^2: coefficient of determination in the cross-validation.

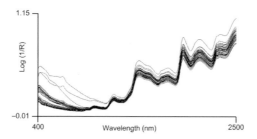

Figure 22.1.   Raw spectra (Log 1/R) of the rice samples (n = 40), in the range from 400 to 2500 nm.

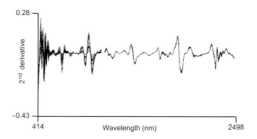

Figure 22.2.   Second derivative spectra (2, 5, 5, 2; SNV+DT) of the raw optical data of rice samples in the range from 400 to 2500 nm.

With respect to rice, the equation showing the highest prediction ability yielded an r^2 of 0.65 (meaning that the 65% of the chemical variability in the data was explained by the model in cross-validation), which was indicative of equations useful for a correct separation of samples with low, medium and high contents (Shenk and Westerhaus 1996) (Fig. 22.3). In accordance with the RPD value (1.67) shown by that equation, and considering the limits for RPD recommended by Dunn *et al.* (2002), this equation was acceptable for i-As prediction in rice.

The equation for predicting i-As in *U. tangeri* showed RPD (1.75) and r^2 (0.71) values, which were slightly higher than those obtained for rice. These values characterize the equation for *U. tangeri* as acceptable for determining i-As, and show good quantitative information (Shenk and Westerhaus 1996) in cross-validation (Fig. 22.3).

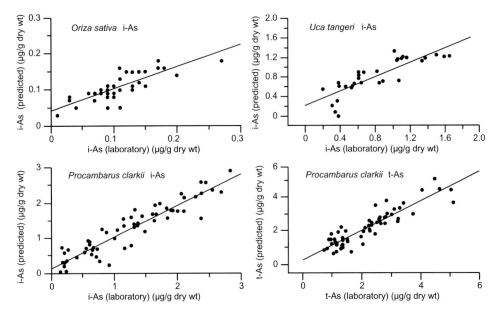

Figure 22.3.   Cross-validation scatter plots of laboratory *vs.* predicted values by NIRS for i-As in *O. sativa* (n = 40), *U. tangeri* (n = 32) and *P. clarkii* (n = 62), and also for total arsenic in *P. clarkii* (n = 60).

The equations modeled for predicting i-As and t-As in *P. clarkii* were by far the ones showing the highest prediction ability of those reported in this work. On the basis of the r^2 statistic, both equations showed good quantitative information in cross-validation (Shenk and Westerhaus 1996) (Fig. 22.3). The high RPDs obtained for i-As (2.63) and for t-As (2.38) characterize those equations as excellent (Dunn *et al.* 2002) for predicting t-As and i-As in such matrix.

22.3.3   *MPLS loading plots*

MPLS regression reduces the spectral information contained in the samples by creating a much smaller number of new orthogonal variables (factors), which are combinations of the original data, and retain the essential information needed to predict the composition. However, metals contained in organic and inorganic matrices do not absorb NIR radiation (Clark *et al.* 1987). Thus, the basis for the success of NIR spectroscopy applications concerning metal determinations remains unclear (Shenk *et al.* 1992). It has been stated that the success of estimation via NIRS of specific mineral elements in some grasses and legumes is usually dependent on the occurrence of those elements in either organic or hydrated molecules (Clark *et al.* 1987). Although it is possible that NIR reflectance spectra of the different matrices shown in this work contain some *primary* information related to the association of i-As with sulfhydryl groups of proteins (Mouneyrac *et al.* 2002) or with other molecules. However, the low concentrations in which these elements are present in the studied matrices do not explain by itself the high correlations obtained.

The role played by the NIR absorbers (organic and inorganic molecules) present in the samples in the modeling of the calibration equations can be interpreted by studying the bands of the MPLS factors (loading plots). These loading plots show the regression coefficients of each wavelength related to the element (i-As) being calibrated, for each factor of the equation. The wavelengths represented in the loading plots that are more important in the development of each factor are those of greater spectral variation and better correlated with the element in the calibration set.

For instance, it was demonstrated (Font *et al.* 2004b) that some chromophores existing in the tissues of the crayfish greatly influenced the first three MPLS loadings of the second derivative transformation (2, 5, 5, 2; SNV+DT) (Fig. 22.4). This fact is in agreement with the correlations existing between i-As content and apparent absorption in our samples that showed high correlations in the visible region of the spectrum. Of the first three factors of the selected equation (2, 5, 5, 2; SNV+DT), the second MPLS loading was the most highly correlated with i-As. The high influence of the band at 712 nm in modeling this second factor should be noted. This behavior could be related to the absorption of radiation by plant matter contained in the digestive tract of the crayfish samples. Absorptions due to C-H combination tones at 2308 and 2348 nm by lipids (Murray 1986) also highly influenced the two first factors of the equation, and together with the previously mentioned wavelengths were the most weighted in the first three MPLS factors.

The same phenomenon is supported by data obtained from calibration for i-As in rice (Font *et al.* 2005). It is concluded from that study that C-H (912 nm) and also O-H (984 nm) groups of starch highly influenced the first three MPLS loadings of the calibration for i-As. In addition, C-H groups of oil and fiber (2308 and 2348 nm) also participated in the modeling of the first term of the equation. In the visible region of the spectrum, chromophores located in the caryopsis, the grain of the rice plant (absorption at 672 nm and shorter wavelengths), also participated actively in constructing the first terms. In spite of the low *r* value shown by the band at 912 nm (Fig. 22.4), this band was selected as important for the first three terms of the equation for i-As because it is displayed due to the high variability in absorbance.

Such phenomenon raises the question to what extent *direct* information provided by bonds of As to organic molecules are important for the development of the NIR equations, or indeed if secondary correlations of As with major cell components determine the success of the method, as it has been previously reported for some minerals in agricultural products (Shenk *et al.* 1992). More work is required to determine the relationship between As and computer-selected calibration wavelengths.

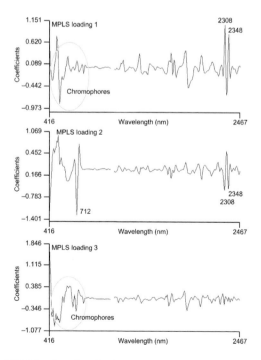

Figure 22.4.   First three MPLS loading plots for the equation (2, 5, 5, 2; SNV+DT) for inorganic arsenic in *P. clarkii* samples.

22.4   CONCLUSIONS

It is shown in this chapter how NIRS is able to predict the i-As and t-As concentration in freeze-dried samples of rice and different animal matrices with sufficient accuracy for screening purposes. Thus, NIRS can be used for identifying samples with low, medium and high i-As and t-As contents. In a second step, after selecting the samples of interest, a more accurate concentration of the element in the selected samples can be obtained by a reference method, such as FI-HG-AAS. NIRS can, therefore, decrease the number of analyses in the laboratory needed for monitoring the As-content in screening programs.

ACKNOWLEDGEMENTS

This research was supported by projects MCyT AGL2001-1789 and CYTED XI-23, for which the authors are deeply indebted.

REFERENCES

Australian New Zealand Food Authority (ANZFA): Food Standards Code, Issue 41, Australia, 1997.

Barnes, R.J., Dhanoa, M.S. and Lister, S.J.: Standard normal variate transformation and de-trending of near-infrared diffuse reflectance spectra. *Appl. Spectrosc.* 43 (1989), pp. 772–777.

Clark, D.H., Mayland, H.F. and Lamb, R.C.: Mineral analysis of forages with near infrared reflectance spectroscopy. *Agron. J.* 79 (1987), pp. 485–490.

Clark, D.H., Cary, E.E. and Mayland, H.F.: Analysis of trace elements in forages by near infrared reflectance spectroscopy. *Agron. J.* 81 (1989), pp. 91–95.

Confalonieri, M., Fornasier, F., Ursino, A., Boccardi, F., Pintus, B. and Odoardi, M.: The potential of near infrared reflectance spectroscopy as a tool for the chemical characterization of agricultural soils. *J. Near Infrared Spectrosc.* 9 (2001), pp. 123–131.

Devesa, V., Súñer, M.A., Lai, V.W.-M., Granchinho, S.C.R., Martínez, J.M., Vélez, D., Cullen, W.R. and Montoro, R.: Determination of arsenical species in freshwater crustacean *Procambarus clarkii. Appl. Organomet. Chem.* 16 (2002), pp. 123–132.

Dunn, B.W., Beecher, H.G., Batten, G.D. and Ciavarella, S.: The potential of near infrared reflectance spectroscopy for soil analysis—a case study from the Riverine Plain of south-eastern Australia. *Aust. J. Exp. Agr.* 42 (2002), pp. 607–614.

Feo, J.C. and Aller, A.J.: Speciation of mercury, methylmercury, ethylmercury and phenylmercury by Fourier transform infrared spectroscopy of whole bacterial cells. *J. Anal. At. Spectrom.* 16 (2001), p. 146.

Font, R., del Río, M. and de Haro, A.: Use of near infrared spectroscopy to evaluate heavy metal content in *Brassica juncea* plants cultivated on the polluted soils of the Guadiamar river area. *Fresenius Environ. Bul.* 11 (2002), pp. 777–781.

Font, R., del Río, M., Simón, M., Aguilar, M. and de Haro, A.: Heavy element analysis of polluted soils by near infrared spectroscopy. *Fresenius Environ. Bull.* 13 (2004a), pp. 1309–1314.

Font, R., Del Río-Celestino, M., Vélez, D., De Haro-Bailón, A. and Montoro, R.: Visible and near-infrared spectroscopy as a technique for screening the inorganic arsenic content in the red crayfish (*Procambarus clarkii* Girard). *Anal. Chem.* 76 (2004b), pp. 3893–3898.

Font, R., Vélez, D., Del Río-Celestino, M., De Haro-Bailón, A. and Montoro. R.: Screening inorganic arsenic in rice by visible and near-infrared spectroscopy. *Microkim. Acta* 151 (2005), pp. 231–239.

IARC: Monographs of evaluation of carcinogenic risks to humans. Supplement 7; International Agency for Research on Cancers, Lyon, France, 1987, pp. 100–106.

Krischenko, V.P., Samokhvalov, S.G., Fomina, L.G. and Novikova, G.A.: Use of infrared spectroscopy for the determination of some properties of soil. In: I. Murray and A. Cowe (eds): *Making light work: advances in near infrared spectroscopy.* VCH, Weinheim, Germany, 1992, pp. 239–249.

Lamont, W.H.: Concentration of inorganic arsenic in samples of white rice from the United Sattes. *J. Food Comp. Anal.* 16 (2003), p. 687.

Mabeau, S. and Fleurence, J.: Seaweed in food products: biochemical and nutritional aspects. *Trends Food Sci. Technol.* 4 (1993), pp. 103–107.

Malley, D.F., Williams, P.C., Hauser, B. and Hall, J.: Prediction of organic carbon, nitrogen and phosphorus in freshwater sediments using near infrared reflectance spectroscopy. In: A.M.C. Davies and P.C. Williams (eds): *Near infrared spectroscopy: the future waves*, Chichester, NIR Publications, 1996, pp. 691–699

Malley, D.F. and Williams, P.C.: Use of near-infrared reflectance spectroscopy in prediction of heavy metal in freshwater sediment by their association with organic matter. *Environ. Sci. Technol.* 31 (1997), pp. 3461–3467.

Martens, H. and Naes, T.: *Multivariate calibration*. John Wiley and Sons, New York, 1989.

Morón, A. and Cozzolino, D.: Determination of macro elements in alfalfa and white clover by near-infrared reflectance spectroscopy. *J. Agric. Sci.* 139 (2002), pp. 413–423.

Mouneyrac, C., Amiard, J.C., Amiard-Triquet, C., Cottier, A., Rainbow, P.S. and Smith, B.D.: Partitioning of accumulated trace metals in the talitrid amphipod crustacean Orchestia gammarellus: a cautionary tale on the use of metallothionein-like proteins as biomarkers. *Aquat. Toxicol.* 57 (2002), pp. 225–242.

Muñoz, O., Vélez, D. and Montoro, R.: Optimization of the solubilization, extraction and determination of inorganic arsenic (AsIII+AsV) in seafood products by acid digestion, solvent extraction and hydride generation atomic absorption spectrometry. *Analyst* 124 (1999), pp. 601–607.

Murray, I.: The NIR spectra of homologous series of organic compounds. In: J. Hollo, K.J. Kaffka and J.L. Gonczy (eds): *Proceedings of the International NIR/NIT Conference*. Akademiai Kiado, Budapest, Hungary, 1986, pp. 13–28.

Nilsson, M.B., Dabakk, E., Korsman, T. and Renberg, I.: Quantifying relationships between near-infrared reflectance spectra of lake sediments and water chemistry. *Environ. Sci. Technol.* 30 (1996), pp. 2586–2592.

NIRSystems: NIRS 2, Routine analysis manual. NIRSystems Infrasoft International, Port Matilda, 1995.

Sauvage, L., Frank, D., Stearne, J. and Millikan, M.B.: Trace metal studies of selected white wines: an alternative approach. *Anal. Chim. Acta* 458 (2002), pp. 223–230.

Shenk, J.S., Workman, Jr. J.J. and Westerhaus, M.O.: Application of NIR spectroscopy to agricultural products. In: D.A. Burns and E.W. Ciurczak (eds): *Handbook of near-infrared analysis*. Dekker Inc., New York, 1992, pp. 383–431.

Shenk, J.S. and Westerhaus, M.O.: Calibration the ISI way. In: A.M.C. Davies and P.C. Williams (eds): *Near infrared spectroscopy: The future waves*. NIR publications, Chichester, 1996, pp. 198–202.

Williams, P.C. and Norris, K.H.: Near-infrared technology in the agricultural and food industries. American Association of Cereal Chemists Inc., St. Paul, MN, 1987.

Williams, P.C. and Sobering, D.C.: How do we do it: a brief summary of the methods we use in developing near infrared calibrations. In: A.M.C. Davies and P.C. Williams (eds): *Near infrared spectroscopy: The future waves*. NIR publications, Chichester, 1996, pp. 185–188.

Workman, J.J. Jr.: Nir spectroscopy calibration basics. In: D.A. Burns and E.W. Ciurczak (eds): *Handbook of near-infrared analysis*. Dekker Inc. New York, 1992, p. 247.

CHAPTER 23

Infield detection of arsenic using a portable digital voltameter, PDV6000

M. Wajrak

Edith Cowan University, Joondalup, WA, Australia

ABSTRACT: There are a growing number of countries in the world where arsenic (As) in ground-water, which is used for drinking and irrigation, has been detected at concentrations above the WHO safe drinking limit of 10 μg/l. These include; Argentina, Bangladesh, Chile, China, Hungary, India, Mexico, Peru, Thailand, and the USA. Of a particular concern is the situation in Bangladesh where it is estimated that there are more than 1 million people drinking As-rich water (above 50 μg/l). It is imperative that people stop drinking from wells where As levels are high. However, as yet, there is no reliable, simple, and field-based method for As detection. This chapter presents some preliminary results from a method being currently developed to detect As in groundwater using portable digital voltameter (PDV6000). The evaluation of this infield method using ICP-MS shows that the voltammetric results are on average within 1.2 μg/l of the laboratory-based spectrometric results.

23.1 INTRODUCTION

23.1.1 *Arsenic chemistry and toxicity*

Arsenic (As) is a brittle, crystalline, grey solid at room temperature. It is a metalloid and thus is a poor conductor of electricity. Arsenic belongs to group V of the periodic table and exists in four valency states, however, most stable states are As(III) (arsenite) and As(V) (arsenate).

Although small amount of As does occur in its free form in nature, most As is found in mineral compounds such as realgar (As_2S_2) and orpiment (As_2S_3). When heated rapidly in air it oxidizes to As-trioxide which has a characteristic garlic-like odor and is very toxic. Arsenic also forms inorganic compounds with most of the other non-metals, such as chlorine and hydrogen, for example $AsCl_3$, and AsH_3. There are also many organic compounds of As such as arsenilic acid, arsenobetaine and arsphenamine, which in 1940s was actually used to treat syphilis.

Inorganic arsenic (i-As) has been recognized as a human poison since ancient times. All inorganic compounds of As are toxic and in particular arsine gas, AsH_3. It is extremely toxic to humans, with headaches, vomiting, and abdominal pains occurring within a few hours of exposure. Inorganic As is found throughout the environment; it is released into the air by volcanoes, the weathering of As-containing minerals and ores, and by commercial or industrial processes (ATSDR 1998, 1990).

General population is exposed to i-As and organic arsenic (As_{org}) through air, water, food and beverages (USEPA 1984). However, for most people, food is the largest source of As exposure (about 25 to 50 μg/day), with lower amounts coming from drinking water and air. Among foods, some of the highest levels are found in fish and shellfish; however, this As exists primarily as organic compounds, which are essentially much less-toxic (ATSDR 1998).

The As-related toxicity is systematic involving a number of organ systems. Acute (short-term) high-level inhalation exposure to As dust or fumes results in gastrointestinal effects (nausea, diarrhea, abdominal pain), which occur within 30 minutes. Central and peripheral nervous system can also be affected leading to heart failure. Chronic (long-term) inhalation exposure to i-As in humans is associated with irritation of the skin and mucous membranes and has been shown to be strongly associated with lung, skin, bladder and liver cancers. Chronic oral exposure results in gastrointestinal effects, anemia, peripheral neuropathy, skin lesions, hyperpigmentation, and liver or kidney damage in humans. EPA has classified i-As as a Group A, human carcinogen (USEPA 1999).

23.1.2　As contamination of groundwater—Bangladesh story

Contamination of groundwater by natural As-rich geological strata in countries such as Bangladesh, Taiwan, India, Mexico, Chile, Argentine and Mongolia has had serious health consequences (Chappell *et al.* 1997). In particular, a decision about 30 years ago by UN and supported by WHO and UNICEF to switch people in Bangladesh from drinking bacteria-laden surface water to groundwater has now created the largest case of mass poisoning in the world (Meharg 2005). It is believed that primary source of As contamination in Bangladesh is the pyritic sedimentary rock laid down over the centuries by the rivers that run down from the Himalayas. The natural sources of arsenic and the processes responsible for the distribution and mobility of arsenic due to weathering and erosion of the primary sulfides (e.g. arsenopyrite) and the effect of oxidation and reduction by micro-organisms in Bangladesh is presented schematically in Figure 23.1.

Although, the latest reports show (New Scientist Magazine 2003) that contamination of groundwater is now more wide spread, where people in 17 countries around the world are currently being exposed to toxic levels of As in drinking water, the situation in Bangladesh is particularly alarming. Latest estimates from World Bank's 'World Development Report 2004' states that ' ... between 25 to 30 million people may be at risk in future' of As poisoning (Meharg 2005).

The BGS estimates that 35 million people are drinking water containing 50 μg/l or more As and WHO in 2000 estimated that 35–77 million of people are potentially exposed to too high level of As in their drinking water, where 10 μg/l is the safe level as set by WHO (Meharg 2005). With over 1 million (as reported in 2003 at the ISEE conference) people already being exposed to As levels of 600 μg/l and 90% of 125 million Bangladeshi population getting their drinking water from groundwater, it is no wonder that this truly is the 'largest case of mass poisoning in the history of humankind' (Hug 2002).

Although, there is considerable research being done around the world to successfully treat contaminated groundwater and remove As or at least reduce its concentration to a safe level,

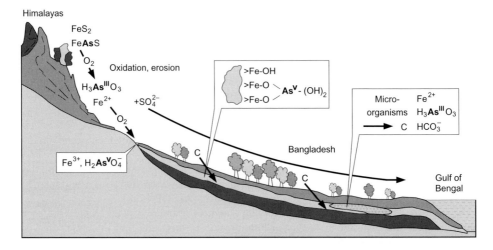

Figure 23.1.　Natural sources and distribution of arsenic in Bangladesh (modified from Hug 2002).

currently, the only way to prevent people's exposure to high levels of As is to test the groundwater and identify wells which have high As content. It is imperative that people stop drinking from wells which contain high concentrations of As in order to reduce their exposure and thus prevent As toxicity. Therefore, what is needed now is a reliable, simple, fast, inexpensive and field-based method for As detection.

The aim of this research project is to investigate the possible use of a portable digital voltammeter as an in-field method for As detection in groundwater and soil.

23.2   ARSENIC DETECTION METHODS

23.2.1   *Brief outline of the methods*

There are currently a number of methods used to detect As in water, ranging from simple methods such as the Guitzheit test to sophisticated atomic spectrometric method. The Guitzheit test is an old method of As detection. It is a relatively cheap and simple test and can be used in-field, however, it is not very reliable or accurate and produces toxic arsine gas:

$$6H_{2(g)} + As_2O_{3(aq)} \rightarrow 2AsH_{3(g)} + 3H_2O_{(l)} \tag{23.1}$$

The concentration of As is determined by observing the color on the strip of paper that has been exposed to arsine gas. White to yellow to reddish-brown spots indicate increasing concentration of As due to formation of AsH_2HgBr. Colorimetric methods of analysis are based on spectrophotometry, where light is shone on a solution and the amount of light absorbed corresponds to the concentration of species in the solution. However, the use of such methods did show that over 45% of the wells previously tested using colorimetric and arsine gas methods were incorrect in their reported concentration of As and many wells had to be re-tested (Meharg 2005).

More accurate and reliable detection methods are atomic spectrometric methods (Ahmed 2003, Heitkemper 2001):

- Inductively coupled plasma mass spectrometer (ICP-MS) has minimum detection limit of less than 2 µg/l.
- Inductively coupled plasma atomic emission spectroscopy (ICP-AES) has minimum detection limit of 8 µg/l.
- Atomic absorption spectroscopy hydride generation (AAS-HG) has minimum detection limit of 0.5 µg/l.
- Atomic absorption spectroscopy graphite furnace (AAS-GF) had minimum detection limit of 0.5 µg/l.

The atomic spectrometric methods of As detection use sophisticated and expensive laboratory based instruments. They require water samples to be collected and then brought to the laboratory where a trained chemist analyzes them. With over 6 million wells spread across Bangladesh, using such methods is not only highly expensive, but also time-consuming.

There is a need for a method which is more accurate and reliable than the colorimetric and Guitzheit test methods, but cheaper, faster and requiring less infrastructure than the AAS or ICP methods.

23.2.2   *Anodic stripping voltammetry*

Stripping analysis is an analytical technique that utilizes a bulk electrolytic step to pre-concentrate the analyte from the sample solution onto the working electrode. This pre-concentration step is followed by an electrochemical measurement of the concentrated analyte. In anodic stripping voltammetry (ASV), at the deposition (pre-concentration) step (see Fig. 23.2a) metals are reduced

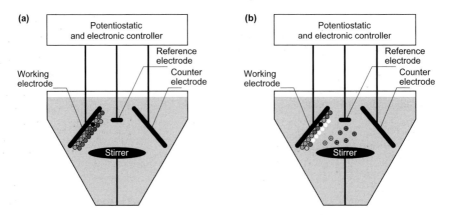

Figure 23.2.   (a) Deposition step (reduction) and (b) stripping step (oxidation).

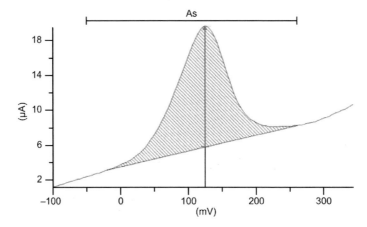

Figure 23.3.   Voltammogram: height or area under the curve corresponds to a particular concentration of arsenic. In this case a 50 µg/l As standard was run for 60 s, which produced a current of 14 µA (height of the voltammogram peak from the base). We can then use this to work out an unknown concentration of arsenic by running a sample at the same deposition time and comparing the peak heights.

at negative (cathodic) potential and concentrated on the mercury film electrode. The amalgamated metals are measured in the second stripping step by applying a positive (anodic) potential scan (see Fig. 23.2b) and measuring the peak currents produced as the system reaches the oxidation potential of the metals (Wang 1985).

Thus, there are three steps in this method:

• Reduction (deposition): concentrates the metal onto the working electrode by reducing the metal ions in solution to the metal:

$$M^{n+}_{(aq)} + e^- \rightarrow M_{(s)} \tag{23.2a}$$

• Oxidation (stripping): ramps the potential at the working electrode in a positive direction at a fixed rate. This oxidizes each metal off the electrode in sequence:

$$M_{(s)} \rightarrow M^{n+}_{(aq)} + e^- \tag{23.2b}$$

Measurement: the electrons released by this process form a current. This is measured and may be plotted as a function of applied potential to give a 'voltammogram' (Fig. 23.3).

23.2.3 The PDV6000 instrument

The instrument that was chosen for the development of an infield voltammetric As detection method was the PDV6000 instrument. The PDV6000 (Fig. 23.4) is a portable, voltammetric analyzer capable of performing either as a standalone unit or in conjunction with a laptop or desktop computer running VAS software (Voltammetric Analysis System, MTI Diagnostics) in the field or at the bench. The apparatus basically consists of the electrochemical cell with three electrodes (working electrode, reference electrode and counter electrode) and a handheld controller. The working electrode is a solid gold electrode.

23.2.4 Method development

The development of the method involved; investigating different working electrodes (carbon with gold film and solid gold electrodes with different diameters), various electrolyte solutions, such as

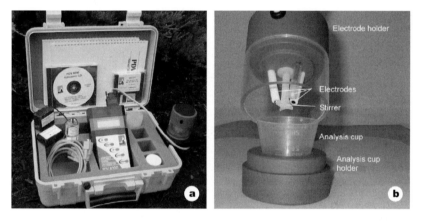

Figure 23.4. Components of the PDV6000 instrument (a) and close up of the electrochemical cell (b).

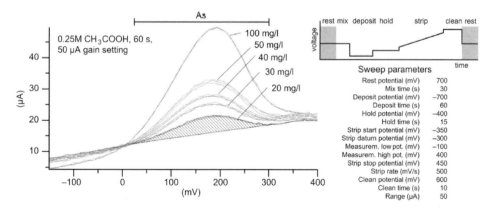

Figure 23.5. Voltammograms showing the linearity of the method between 20 μg/l to 100 μg/l As standard using 0.25 M ethanoic acid as the electrolyte solution and with deposition time of 60 s. On the right are the final experimental parameters used for the detection of As(III).

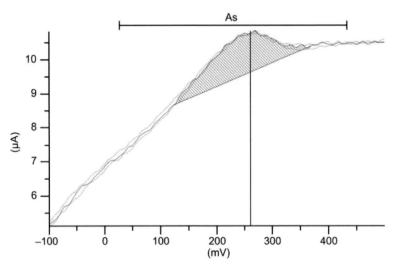

Figure 23.6.   Voltammograms showing the reproducibility of the method. 200 μg/l As standard was run in 0.25 M ethanoic acid electrolyte solution ten times using 60 s deposition time. The first run gave a peak height of 53 μA and the last run gave a peak height of 49 μA. Thus reproducibility is within 7%.

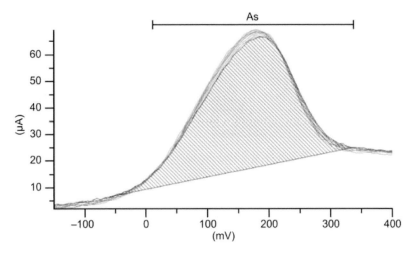

Figure 23.7.   Voltammograms showing the limit of detection. 2.5 μg/l As(III) standard was analyzed using 0.25 M ethanoic acid electrolyte solution at 300 s deposition time, giving a peak height of 1.5 μA.

HCl, NaCl/HCl, HNO$_3$ and CH$_3$COOH, deposit and strip potentials and electrode conditioning techniques. Best results (in terms of linearity—Fig. 23.5, reproducibility—Fig. 23.6 and detection limit—Fig. 23.7) were found using solid gold electrode, 0.25 M CH$_3$COOH electrolyte, conditioning the electrode with 1 mg/l As standard and with the potentials as shown in Figure 23.5.

The detection limit of the instrument was also investigated and it was found that it is possible to detect As down to 2.5 μg/l using 0.25 M ethanoic acid electrolyte solution with 300 s deposition time (Fig. 23.7).

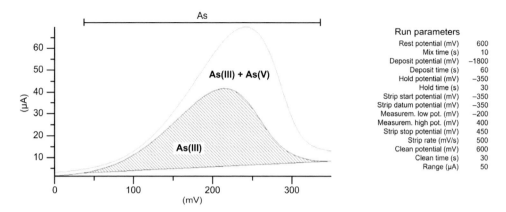

Figure 23.8.   Voltammograms showing As(III) and As(V) being detected in the same solution. On the right are the potential parameters used to detect As(V).

The results shown above are all for As(III) form being detected, however, As in water does exist in two forms. However, spectrometric methods such as ICP and AAS are not able to speciate between the two forms of As in water. Being able to detect the two forms of As is important, because as it was discussed previously in Section 23.1.1, As(III) and As(V) have different toxicity levels in humans. It is possible to detect the two forms As using voltammetry. A standard solution containing As(III) was analyzed and then spiked with As(V) standard. The potential parameters were adjusted until As(V) was detected, as it is not possible to detect As(V) using As(III) parameters (Fig. 23.8).

23.3   VALIDATING VOLTAMMETRIC METHOD FOR ARSENIC-DETECTION

23.3.1   *Comparing ASV results with ICP-MS results*

The ASV method is currently at the stage of being validated using ICP-MS, which is one of the approved methods for As detection in drinking water. For the validation process 'real' groundwater samples were needed for analysis. It has been possible to obtain such groundwater samples in Perth, Western Australia. A decline in the water table due to a long period of low rainfall and the disturbance of sulfidic peat soils, which cover more than 40,000 km^2 of Australia's coastline, by dewatering and excavation in some of Perth's suburbs has resulted in widespread acidification of groundwater and has caused the water to be contaminated by heavy metals, and in particular As (Appleyard *et al.* 2004). A new development in Osborne park, along Cedric street had monitoring wells installed to monitor groundwater As levels every second week for six months. The water samples were found to contain relatively high levels (up to 30 µg/l) of As and therefore those samples were used to compare the ASV method to ICP-MS.

The samples were collected every two weeks over a period from March to May 2005. Duplicate samples were collected by environmental company at the site and one lot was analyzed by ASV method and the other sample was sent to National Association of Testing Authorities (NATA) accredited laboratory for analysis by ICP-MS. Only a handful of results are available, due to the fact that the environmental company suddenly stopped monitoring in May 2005. Nevertheless these results show that on average the ASV method is within 1.2 µg/l of the ICP-MS method (see Table 23.1). This is very encouraging as it demonstrates that ASV method is comparable to the more sophisticated analytical ICP-MS method.

However, further sample analyses are required to complete the validation process. It is hoped that apart from Perth groundwater samples, water samples from Bangladesh or India will also be analyzed as part of the validation process.

Table 23.1.   Comparison of ASV results with ICP-MS data from Osborne park, WA, groundwater.

Sample round (Well ID)	ICP-MS (μg/l)	ASV (μg/l)
2 (MW4)	15	15
3 (MW4)	15	16
4 (MW4)	16	17
5 (MW4)	17	17
2 (MW5)	32	29
3 (MW5)	25	21
4 (MW5)	27	28
5 (MW5)	25	27

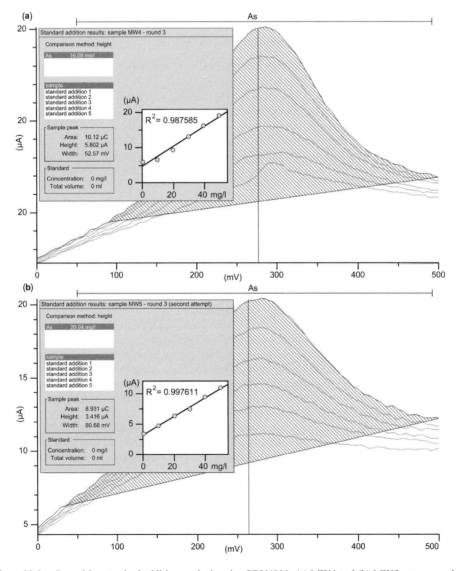

Figure 23.9.   Round 3—standard addition analysis using PDV6000: (a) MW4 and (b) MW5 water sample.

Plotting ASV results against ICP-MS data shows correlation with R^2 value of 0.9. Considering that only eight samples were analyzed that is a good correlation, however, more samples are needed for statistically sound correlation. ASV samples were analyzed using standard addition method, see Figures 23.9a and b for examples of two ASV calculations.

23.4   CONCLUSIONS

The chapter reports on the development of an accurate, reliable and field method for As analysis using ASV. Method development together with some preliminary results using groundwater samples contaminated with As from Perth, WA, indicate that the PDV6000 could be considered as an alternative method for the detection of As in groundwater. The ASV results are in most cases exactly the same as the ICP-MS results and only in one case the results are 4 μg/l in difference. The ASV method is capable of detecting As down to 2.5 μg/l levels and it has been shown to give linear and reproducible results.

Although, more testing still needs to be done to validate the ASV method, the results so far indicate that the portable digital voltameter (PDV6000) could be used for screening contaminated groundwater samples and more importantly that screening can be done in the field and thus provide a cheaper and accurate alternative to the currently used laboratory based techniques. The use of this method could be very important in the Bangladesh situation and also other parts of the world, where groundwater is contaminated with As, such as India, Mexico and Brazil, but it would also be significant for Perth, WA, where the As groundwater contamination is becoming a very wide spread problem.

REFERENCES

Agency for Toxic Substances and Disease Registry (ATSDR): *Toxicological profile for arsenic* (draft). US Public Health Service, US Department of Health and Human Services, Atlanta, GA, 1998.
Agency for Toxic Substances and Disease Registry (ATSDR): *Case studies in environmental medicine. Arsenic toxicity.* US Public Health Service, US Department of Health and Human Services, Atlanta, GA, 1990.
Appleyard, S., Wong, S., Willis-Jones, B., Angeloni, J. and Watkins, R.: Groundwater acidification caused by urban development in Perth, Western Australia: source, distribution, and implications for management. *Aust. J. Soil Res.* 42 (2004), pp. 1–7.
Chappell, W.R., Abernathy, C.O., Calderon, R.L. and Thomas, D.J.: *Arsenic exposure and health effects.* Chapman and Hall, London, 1997.
Heitkemper, D.T., Vela, N.P., Stewart, K.R. and Westphal, C.S.: Determination of total and speciated arsenic in rice by ion chromatorgraphy and inductively coupled plasma mass spectrometry. *J. Anal. Atomic Spectrom.* 16 (2001), pp. 299–306.
Hug, S., Wegelin, M., Gechter, D. and Canonica, L.: Arsenic contamination in groundwater: Disastrous consequences in Bangladesh. *EAWAG News* 49 (2000), p. 18.
Meharg, A. *Venomous Earth.* MacMillan, New York, 2005.
Pearce, F.: Arsenic's fatal legacy grows. *New Scientist Magazine* 9 (2003), p. 4.
USEPA: *Health assessment document for inorganic arsenic.* EPA/540/1-86/020. Environmental Protection Agency, Environmental Criteria and Assessment Office, Office of Health and Environmental Assessment, Office of Research and Development, Washington, DC, 1984.
USEPA: *Integrated risk information system (IRIS) on arsine.* Environmental Protection Agency, National Center for Environmental Assessment, Office of Research and Development, Washington, DC, 1999.
Wang, J.: *Stripping analysis, principles, instrumentation and applications.* VCH Publishers Inc. Weinheim, Germany, 1985.

CHAPTER 24

The use of synchrotron micro-X-ray techniques to determine arsenic speciation in contaminated soils

J.L. López-Zepeda, M. Villalobos, M. Gutiérrez-Ruiz & F. Romero
LAFQA, Instituto de Geografía, Universidad Nacional Autónoma de México (UNAM), Mexico City, Mexico

M. Marcus
Advanced Light Source, Lawrence Berkeley National Laboratory, Berkeley, CA, USA

G. Sposito
Ecosystem Sciences Division, University of California at Berkeley, Berkeley, CA, USA

ABSTRACT: We have found a greatly reduced mobility of oxidized arsenic (As) in soils contaminated by mineral processing and metallurgical residues as compared to As in the original wastes, regardless of their specific origin. Contaminated samples were studied in three different areas of Mexico: Real de Ángeles, Zacatecas, contaminated by mine tailings from mining of Pb, Ag and Zn ores; San Luis Potosí city, in soils near Cu, As and Pb refinement plants; and Monterrey, Nuevo León, in soils surrounding a Pb smelting plant where calcium arsenate wastes were generated. The present work reports the collected data, and procedures followed to determine the molecular-scale speciation of As in heterogeneous soil samples, using micro-X-ray techniques from synchrotron sources, including μ-XRF (X-ray fluorescence) and μ-XAS (X-ray absorption spectroscopy). Suitable soil samples with high total As contents in the solid phase, but low aqueous As levels were processed. Using adequate As standards, the obtained data provided structural information on the correlation and involvement of metal cations such as Fe, Pb, and Zn, in the immobilization processes of As in soils. The latter include formation of adsorbed species and surface precipitates on the surface of mineral oxides in groundwater aquifers.

24.1 INTRODUCTION

Speciation of environmentally relevant trace elements in soils is crucial for a solid scientific understanding of their behavior in natural settings and for prediction of their transport properties, and ultimately their fate in the environment. It involves determination of the associations of these elements with other soil components at the molecular scale. This information allows elucidation of their dominant chemical forms and modes of sequestration (Brown *et al.* 1999). Speciation is particularly essential in the case of As because several stable oxidation states may occur in natural environments. Arsenic shows considerable variability in aqueous mobility and stability depends highly on the physicochemical conditions present. Mining and other anthropogenic activities alter considerably the natural stability of As and other trace elements, potentially increasing the risks of these elements entering biological systems.

Extensive evidence exists on the commanding role of iron oxides in the regulation of As mobility in aqueous environments, such as soil water, through mechanisms that range from adsorption to precipitation (Foster *et al.* 1998, Roussel *et al.* 2000, Savage *et al.* 2000, Craw *et al.* 2002, Tye *et al.* 2002, Néel *et al.* 2003, Paktunc *et al.* 2003, 2004). Recently, the involvement of low-solubility heavy metal arsenates has been proposed as an As natural attenuation mechanism in different soil environments (Gutiérrez-Ruiz *et al.* 2005).

Figure 24.1.   Location of arsenic-contaminated study sites (Mx: Mexico City).

Three different soil sites contaminated with As from mine processing and metallurgical wastes have been studied in the past to characterize and diagnose their contamination level and extent. They are located in the north and north central mining and mineral processing districts of Mexico: (1) Real de Ángeles, Zacatecas (REAL), where the problem is a tailings deposit from Pb, Ag and Zn mining, apparently in the early stages of oxidation. The waste was disposed in an impoundment; (2) an As and Cu plant in San Luis Potosí (SLP), which produces Cu, As_2O_3, and unrefined Pb; it has contaminated soils mainly with As and Pb; and (3) a Pb smelting and refining plant in Monterrey, Nuevo León (MONT), which generates wastes with high contents of calcium arsenates and Pb oxides. The site locations are shown in Figure 24.1.

Despite the fact that the species of the original As-contaminating source is different for each site, all soils show a considerable decrease in the mobility of As (concentration of aqueous As species) relative to that in the original contaminating source. This finding has prompted the current study, with the main hypothesis that formation of highly insoluble heavy metal (e.g., Pb, Zn) arsenates is responsible for this decrease in solubility. X-rays from synchrotron sources are investigated as convenient probes for such determinations.

X-rays have been used for quite some time as non-invasive probes for investigating structural aspects of mineral phases in earth materials. The extremely bright X-ray beams obtained from synchrotron sources allow low detection limits, and, therefore, are well suited for determining species found in trace levels in these materials. Suitable applications of these beams yield very accurate compositional and structural information, such as mineralogy, oxidation state, and nature of local bonding environment, with little to no alteration of the sample (Brown *et al.* 1999, Brown and Sturchio 2002). However, processing of natural samples is problematic and has required pre-concentration procedures due to high chemical heterogeneity, in which the most reactive particles usually occur in the nanometer size range. These particles show a high specificity in sequestration of trace elements and often a multiplicity of removal mechanisms. These difficulties have been successfully resolved in the past year by microscopic X-ray techniques developed at third generation synchrotron facilities (Morin *et al.* 2002, Manceau *et al.* 2002, 2003, Strawn *et al.* 2002), which allow high focusing of bright X-ray beams to obtain spatial resolutions down to the micrometer and even nanometer scales in relatively unaltered natural samples.

Synchrotron-based X-ray techniques have been employed for *in situ* studies of speciation and structural details involving trace element sequestration in model mineral materials, and very recently applied successfully to natural samples (Morin *et al.* 2002, Brown and Sturchio 2002, Manceau *et al.* 2002, 2003, Strawn *et al.* 2002).

24.2   MATERIALS AND METHODS

X-ray fluorescence (XRF) and absorption (XAS) data were collected on different spots of selected soil samples mounted on adequate holders (beryllium planchets). The X-ray beam was

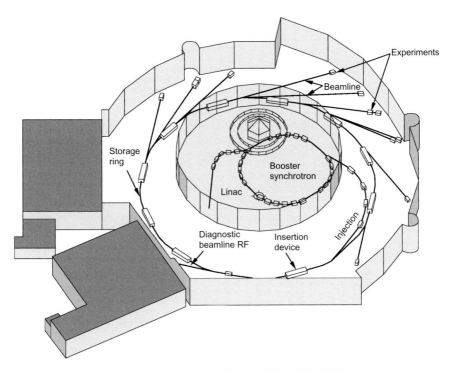

Figure 24.2.   The Advanced Light Source (ALS) synchrotron facility in Berkeley.

micro-focused to yield spot diameters of approximately ≥ 5 μm, highly suitable for the heterogeneity of soil samples. The X-ray source was from the Advanced Light Source (ALS) synchrotron at the Lawrence Berkeley National Laboratory (Fig. 24.2). The number of the beamline used to collect the data was number 10.3.2. Before collecting extended X-ray absorption fine structure (EXAFS) data, the elemental composition of an area of the sample, typically of a few hundred μm^2, was scanned by μ-X-ray fluorescence (μ-XRF), and specific spots rich in As within it were selected for EXAFS measurements on the As K-edge (arising from subtraction of a core electron from the 1s orbital).

Figure 24.3 shows the steps followed for EXAFS data processing. Spectra for a total of 45 standards (Table 24.1) were collected. Fifteen of these were commercial crystalline As minerals [mostly of As(V)], and the rest were prepared in the laboratory, by either homogeneous precipitation of arsenate in the presence of other heavy metals (Pb, Zn, Cu), or similar precipitates in the presence of a relevant metal oxide [goethite, gibbsite, calcite, Cr(III) oxide, or SiO$_2$].

Typically 4–6 spectra were collected in both fluorescence and transmission mode for each sample spot to ensure reliable data at high-energy values, while only 2–3 scans were collected for the As-rich standards. All were processed using the SixPack software (Webb 2005).

Individual energy *vs.* counts spectra were averaged, background subtracted and normalized to k space. Transmission mode data were used to correct for over-absorption phenomena. The whole set of sample spectra (26 in total) was processed through a principal component analysis (PCA), in a range of $k = 2.5$–10.5 1/Å, to determine the minimum set of mathematical orthogonal components that would describe them all. Following this, a target transformation procedure was performed, yielding a list of As standards in an order based on how well they are described by the components resulting from the PCA analysis (Manceau *et al.* 2002). The criterion parameter is called SPOIL, and the lowest SPOIL values are desired.

In a final step, each sample spectrum was linearly fitted with all adequate standards individually based on their SPOIL value, with one and a maximum of two standards, following the fit

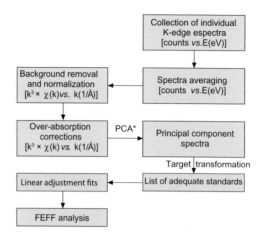

Figure 24.3.　Flowchart of XAS data processing procedure. *PCA = Principal component analysis.

Table 24.1.　List of standards used for XAS data processing.

$AsCuZnCr_2O_3$	$MG\_AsGoepH4^2$
Adamite, $Zn_2 (AsO_4)(OH)^1$	$Cal\_AsPbZn\_05^4$
$MG\_2–50mMAsZnGoepH4^2$	MGAsVGibbsite
Schultenite, $PbHAsO_4{}^1$	$MG\_0–25mMAsZnGoepH4^2$
Chalcophyllite, $Cu_{18}Al_2(AsO_4)_3(SO_4)_3(OH)_{27} \times 33H_2O^1$	MGZnAs(V)precipitate
MGAsV + ZnIIGibbsite	MGAsVGoethite
MGAsZn725	$AsCuGoe^2$
$AsCuCr_2O_3$	$Cal\_05\_Pb\_AsO_4{}^4$
$AsCuZnGib^3$	MGAsZn715
Scorodite, $FeAsO_4 \times 2H_2O^1$	$MG\_AsGoepH7^2$
MGAs7075	$Pb\_Zn\_AsO_4$
CuZnAs(V)precipitate	Allactite, $Mn_7(AsO_4)_2(OH)_8{}^1$
CuAs(V)precipitate	$Goe\_075Pb\_AsO_4{}^2$
$AsZnCr_2O_3$	MGAs7025
$AsCuGib^3$	Orpiment, $As_2S_3{}^1$
$Cal\_05\_Zn\_AsO_4{}^4$	$Na_2HAsO_{4(aq)}$
$AsCuZnGoe^2$	Mimetite, $Pb_5(AsO_4)_3Cl^1$
Sodium arsenate standard, $Na_3AsO_4{}^1$	Mansfieldite, $AlAsO_4 \times 2H_2O^1$
Ojuelaite, $ZnFe_2(AsO_4)_2(OH)_2 \times 4H_2O^1$	PbAsVsalt, $PbHAsO_{4(s)}{}^1$
$Goe\_075Pb\_Zn\_AsO_4{}^2$	AsVZnIISilicaOxide
Olivenite, $Cu_2AsO_4(OH)^1$	Pb_AsO_4
Arsenopyrite, $FeAsS^1$	Yukonite, $Ca_7Fe_{11}(AsO_4)_9O_{10} \times 24.3H_2O^1$
Arseniosiderite, $Ca_2Fe_3(AsO_4)_3O_2 \times 3H_2O^1$	

[1] Commercial crystalline As minerals, [2] Goe: goethite, [3] Gib: gibbsite, [4] Cal: calcite.

criterion that the goodness-of-fit R value, defined as $R = \sum |K^3_{\chi exp} - k^3_{\chi model}|/\sum |k^3_{\chi exp}|$. It had to decrease more than 10% to make the second standard an acceptable addition to the fit (Panfili *et al.* 2005).

24.3　RESULTS

Figure 24.4 shows a typical elemental correlation found between As and Pb contents in many of the soil samples processed, in this case for a SLP sample. This behavior suggests that in general As

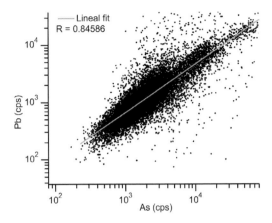

Figure 24.4.   Correlation between arsenic and lead composition for a selected area of a sample of SLP.

species co-contain lead, although it is possible that both elements are co-adsorbed onto the same solid phase, and not necessarily chemically bound to the same compound. Figure 24.5 shows the sequence of spectral data treatment, illustrated for a SLP sample. A matter that complicates greatly the fitting of samples to standards is that, in general, the second shell of near-neighbor atoms around As yields low signals. This behavior translates into small peaks in the Fourier transformed spectra (in R space) in all samples and most standards (Fig. 24.5). This leaves the first shell, of As-O, as major contributor to the total EXAFS spectra, and thus it makes it harder to distinguish among standards and samples.

The results from the principal components analysis (PCA) after processing all 26 samples yielded five relevant components (minimum value of IND of 0.02770). The target transformation procedure assigned SPOIL values to all standard spectra when fit to the abstract components generated by the PCA (Table 24.2). Most values fell in the category of 1.5–3, for good fits from the PCA (Manceau *et al.* 2002). Only seven standards showed SPOIL values in the next category, of 3–4.5, for fair fits. These results are indicative of the large similarities in the spectra of most As standards, as pointed out above.

Nevertheless, a careful linear fitting procedure was performed using spectra of most standards, with a maximum of two standards per fit. Figures 24.6–24.8 show examples of the resulting chi spectral fits from a representative spot at each one of the three sites investigated. The examples show the best fits for each spot, and one illustrative bad fit.

The spot selected for the SLP site was rich in As, Ca, Pb and Zn, but contained some Fe as well. It is no surprise that the best fit (Fig. 24.6) was obtained with a Pb-Zn arsenate precipitate in a combination of homogeneous precipitate (41%) and a precipitate onto calcite (44%). The remaining 15% composition is well within the uncertainty of the PCA (15–20%) and is not statistically different from zero. A standard composed of the same Pb-Zn arsenate precipitate onto goethite yielded a considerably poorer fit of the EXAFS spectrum, and thus was excluded as a possible component. Also, none of the iron-containing arsenate standards (e.g., scorodite, or arsenate adsorbed to goethite) yielded adequate fits to the EXAFS data of this sample. This behavior supports the hypothesis that the extremely lower solubility of Pb, Zn (and Cu) arsenates *vs.* that of Ca and Fe arsenates, yields a high thermodynamic driving force that may result in the favorable formation of the Pb, Zn, or Cu arsenates rather than those of Ca and Fe, given adequate availability of all metallic ions.

The spot selected for the MONT site was rich in calcium, rich to moderate in Pb, moderate in As, and low in Zn. Cu, and Fe. The best single fit (Fig. 24.7) was obtained again with a Pb-Zn arsenate precipitated onto calcite. This time the optimal combination was obtained with an additional homogeneous precipitate of Pb arsenate (20%), and 63% of the Pb-Zn arsenate

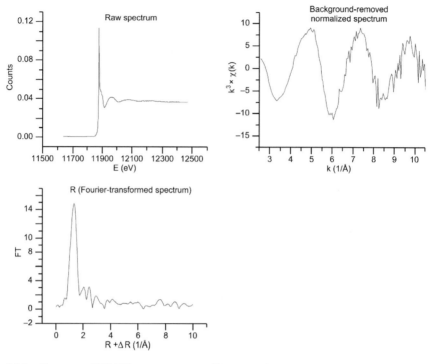

Figure 24.5.   Sequence of EXAFS spectra treatment illustrated with a SLP sample spectrum: Raw spectrum, CHI spectrum, and Fourier transformed, radial structure function, spectrum.

Table 24.2.   SPOIL values from target transformation analysis in increasing order.

Standard	SPOIL	Standard	SPOIL
AsCuZnCr$_2$O$_3$	1.8596	MGAsVGibbsite	2.3950
Adamite	1.9154	MG_0–25mMAsZnGoepH4	2.5113
MG_2–50mMMAsZnGoepH4	1.9373	MGZnAs(V)precipitate	2.6109
Schultenite	1.9402	MGAsVGoethite	2.6434
Chalcophyllite	1.9554	AsCuGoe	2.6906
MGAsV + ZnIIGibbsite	1.9967	Cal_05_Pb_AsO$_4$	2.7013
MGAsZn725	2.0336	MGAsZn715	2.7403
AsCuCr$_2$O$_3$	2.0444	Yukonite	2.7453
AsCuZnGib	2.0668	MG_AsGoepH7	2.7766
Scorodite	2.0857	Pb_Zn_AsO$_4$	2.8036
MGAs7075	2.0994	Allactite	2.8161
CuZnAs(V)precipitate	2.1200	Goe_075Pb_AsO$_4$	2.8256
CuAs(V)precipitate	2.1231	MGAs7025	2.8264
AsZnCr$_2$O$_3$	2.1234	Orpiment	2.8757
AsCuGib	2.2042	Arsenopyrite	2.9338
Cal_05_Zn_AsO$_4$	2.2097	Na$_2$HAsO$_{4(aq)}$	3.0427
AsCuZnGoe	2.2619	Mimetite	3.1905
Sodium arsenate standard	2.2711	Arseniosiderite	3.3173
Ojuelaite	2.2903	Mansfieldite	3.4314
Goe_075Pb_Zn_AsO$_4$	2.3163	PbAsVsalt	3.6079
Olivenite	2.3166	AsVZnIISilicaOxide	3.6870
MG_AsGoepH4	2.3313	Pb_AsO$_4$	4.2037
Cal_AsPbZn_05	2.3728		

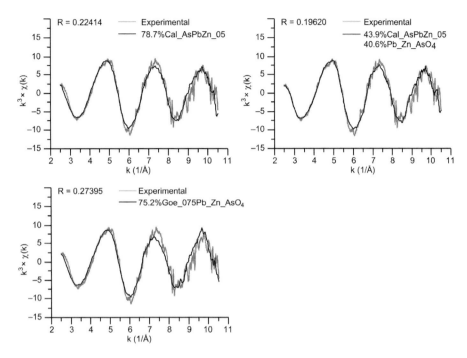

Figure 24.6.   EXAFS spectral fits for a selected sample of the SLP site.

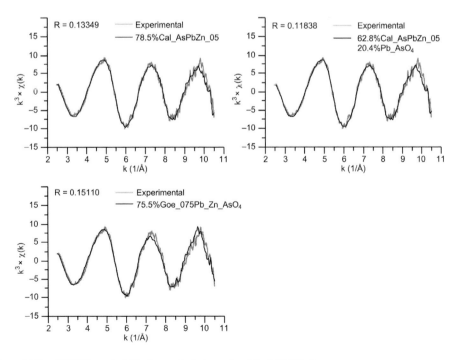

Figure 24.7.   EXAFS spectral fits for a selected sample of the MONT site.

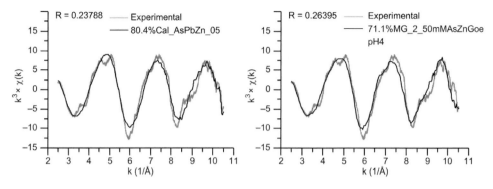

Figure 24.8.   EXAFS spectral fits for a selected sample of the REAL site.

precipitate on calcite. This is consistent with the relatively lower Zn contents for this sample. Despite the fact that calcium arsenate is known to be the source pollutant, not very good fits were obtained with calcium-containing arsenate standards, arseniosiderite and yukonite. As in the SLP sample, a standard composed of the same Pb-Zn arsenate precipitate onto goethite yielded a poorer fit of the EXAFS spectrum, suggesting it is probably not a component. Also, none of the iron-containing arsenate standards (e.g., scorodite, or arsenate adsorbed to goethite) yielded adequate fits to the EXAFS data of this and the other samples.

In the case of the REAL site, less adequate fits were obtained in general, and no double-component fit improved any of the single-component fits (Fig. 24.8). Again, the best fit was obtained with a Pb-Zn arsenate precipitate onto calcite, with an 80% contribution, followed by a Zn arsenate precipitate on goethite, with a 71% contribution. The contaminating source in this site are mine tailings, and therefore, concentrations of unoxidized arsenopyrite may be expected. In this latter mineral, As is found in a low oxidation state, and thus its EXAFS spectrum may be quite different, which is indeed the case (spectra not shown). However, arsenopyrite yielded a considerably poorer fit of the EXAFS data of this sample, and thus was excluded as a probable component.

24.4   CONCLUSION

Soil samples contaminated with As from mine processing wastes contain solid arsenate species associated with lead and zinc, but precipitated onto bulk solids of the soil matrix. Due to its abundance in the sites studied, calcite (calcium carbonate) seems to be the preferred solid matrix onto which the arsenate precipitates are formed. The findings of heavy metal arsenates formation in soils in this study are novel, since most reports identify arsenate bonding with iron phases, either coprecipitated with or adsorbed to iron oxides, or found as iron arsenates (e.g., scorodite, ojuelaite), in natural environments.

Investigations continue in order to identify the exact molecular structure of the formed heavy metal arsenates because the standards that matched the samples were precipitates synthesized in the laboratory and their structures are currently unknown. For this, *ab initio* modeling using FEFF software is necessary (this type of modeling involves both atomic theory and empirical relations between parameters). Additionally, the formation mechanism of these arsenates is being investigated. This is important because we need to distinguish the conditions favoring binding of arsenate minerals to iron oxides instead of other minerals. This distinction can be achieved using wet-chemistry experiments coupled with additional X-ray absorption analysis.

ACKNOWLEDGMENTS

We wish to acknowledge the kind contribution of Professor Donald Sparks and Dr. Markus Grafe, from the Environmental Soil Chemistry Group, Department of Plant and Soil Sciences, University of Delaware, for providing the EXAFS spectra of many of the standards used in this work, as well as for providing two of the lead-arsenate standards that we used to collect X-ray data on at the synchrotron beamline. Also, Markus Grafe was very helpful in discussions of specific aspects of As EXAFS data processing and interpretation. We thank Dr. Rufino Lozano, from the Geology Institute at UNAM, for kindly providing the characterized calcite sample for preparation of standards. John Bargar, Sam Webb, Brandy Toner, and Sirine Fakra, were kind enough to collect the first set of EXAFS data at beamline 10.3.2.

The operations of the Advanced Light Source at Lawrence Berkeley National Laboratory are supported by the Director, Office of Science, Office of Basic Energy Sciences, US Department of Energy under contract number DEAC03-76SF00098.

REFERENCES

Brown, G.E., Foster, A.L. and Ostergren, J.D. 1999. Mineral surfaces and bioavailability of heavy metals: A molecular-scale perspective. *Proceedings of the National Academy of Sciences of the United States of America* 96 (1999), pp. 3388–3395.

Brown, G.E. and Sturchio, N.C.: An overview of synchrotron radiation applications to low temperature geochemistry and environmental science. In: P. Fenter, M. Rivers, N. Sturchio and S. Sutton (eds): *Synchrotron radiation applications in low-temperature geochemistry and environmental science*. Mineralogical Society of America, Washington DC, 2002, pp. 1–115.

Craw, D., Koons, P.O. and Chappell, D.A.: Arsenic distribution during formation and capping of an oxidised sulphidic minesoil, Macraes mine, New Zealand. *J. Geochem. Explor.* 76 (2002), pp. 13–29.

Foster, A.L., Brown, Jr., G.E., Tingle, T.N. and Parks, G.A.: Quantitative arsenic speciation in mine tailings using X-ray absorption spectroscopy. *Am. Mineral.* 83 (1998), pp. 553–568.

Gutiérrez-Ruiz, M., Villalobos, M., Romero, F. and Fernández-Lomelín, P.: Natural attenuation of arsenic in semi-arid soils contaminated by oxidized arsenic wastes. In: P.A. O'Day, D. Vlassopoulos, X. Meng and L.G. Benning (eds): *Advances in arsenic research: integration of experimental and observational studies and implications for mitigation*. American Chemical Society, Washington, DC, 2005, pp. 235–252.

Manceau, A., Marcus, M.A. and Tamura, M.: Quantitative speciation of heavy metals in soils and sediments by synchrotron X-ray techniques. In: P. Fenter, M. Rivers, N. Sturchio and S. Sutton (eds): *Synchrotron radiation applications in low-temperature geochemistry and environmental science*. Mineralogical Society of America, Washington, DC, 2002, pp. 341–428.

Manceau, A., Tamura, N., Celestre, R.S., MacDowell, A.A., Geoffroy, N., Sposito, G. and Padmore, H.A.: Molecular-scale speciation of Zn and Ni in soil ferromanganese nodules from loess soils of the Mississippi Basin. *Environ. Sci. Technol.* 37:1 (2003), pp. 75–80.

Morin, G., Lecoq, D., Juillot, F., Calas, G., Idelfonse, P., Belin, S., Briois, V., Dillmann, P., Chevallier, P., Gauthier, C., Sole, A., Petit, P.-E. and Borensztajn, S.: EXAFS evidence of sorbed arsenic (V) and pharmacosiderite in a soil overlying the Echassieres geochemical anomaly, Allier, France. *Bull. Soc. Geol. Fr.* 3 (2002), pp. 281–291.

Néel, C., Bril, H., Courtin-Nomade, A. and Dutreuil, J.-P.: Factors affecting natural development of soil on 35-year-old sulphide-rich mine tailings. *Geoderma* 111 (2003), pp. 1–20.

Paktunc, D., Foster, A. and Laflamme, G.: Speciation and characterization of arsenic in Ketza River mine tailings using X-ray absorption spectroscopy. *Environ. Sci. Technol.* 37:10 (2003), pp. 2067–2074.

Paktunc, D., Foster, A., Heald, S. and Laflamme, G.: Speciation and characterization of arsenic in gold ores and cyanidation tailings using X-ray absorption spectroscopy. *Geochim. Cosmochim. Acta* 68:5 (2004), pp. 969–983.

Panfili, F., Manceau, A., Sarret, G., Spadini, L., Kirpichtchikova, T., Bert, V., Laboudigue, A., Marcus, M.A., Ahamdach, N. and Libert, M.-F.: The effect of phytostabilization on Zn speciation in a dredged contaminated sediment using scanning electron microscopy, X-ray fluorescence, EXAFS spectroscopy, and principal components analysis. *Geochim. Cosmochim. Acta* 69:9 (2005), pp. 2265–2284.

Roussel, C., Néel, C. and Bril, H.: Minerals controlling arsenic and lead solubility in an abandoned gold mine tailings. *Sci. Total Environ.* 263 (2000), pp. 209–219.

Savage, K.S., Tingle, T.N., O'Day, P.A., Waychunas, G.A. and Bird, D.K.: Arsenic speciation in pyrite and secondary weathering phases, Mother Lode Gold District, Tuolumne County, California. *Appl. Geochem.* 15 (2000), pp. 1219–1244.

Strawn, D., Doner, H., Zavarin, M. and McHugo, S.: Microscale investigation into the geochemistry of arsenic, selenium, and iron in soil developed in pyritic shale materials. *Geoderma* 108 (2002), pp. 237–257.

Tye, A.M., Young, S.D., Crout, N.M.J., Zhang, H., Preston, S., Bailey, H., Davison, W., McGrath, S.P., Paton, G.I. and Kilham, K.: *Environ. Sci. Technol.* 36:5 (2002), pp. 982–988.

Webb, S.M.: SIXPACK: A graphical user interface for XAS analysis using IFEFFIT. *Phys. Scr.* T115 (2005), pp. 1011–1014.

CHAPTER 25

Arsenic speciation study using X-ray fluorescence and cathodic stripping voltammetry

L.A. Valcárcel, A. Montero, J.R. Estévez & I. Pupo
Centro de Aplicaciones Tecnológicas y Desarrollo Nuclear (CEADEN), Havana, Cuba

ABSTRACT: Two methods for the determination of total arsenic (t-As) concentration and its inorganic species by 'energy dispersive X-ray fluorescence' (EDXRF) and 'cathodic stripping voltammetry' (CSV) were developed. The effect of pH on As(III) recovery after precipitation was studied using ammonium pyrrolidine dithiocarbamate (APDC) and EDXRF measurements. Quantification of As was done using the thin layer approach. A reduction of As(V) to As(III) with sodium thiosulfate was needed in order to determine the total As concentration. The effect of the amount of reducing agent on the recovery was also studied. As(V) concentration was calculated by difference between total As and As(III) concentration. In addition, a polarographic method, using the cathodic stripping mode was implemented. As(III) deposition on the electrode was enhanced by addition of Se(IV). Factors affecting As determination (selenium concentration, deposition potential, deposition time) were studied. Detection limits for both methods are lower than the maximum level established by international agencies for total As concentration ($10\,\mu g/l$) in drinking water, although APDC-EDXRF shows the lowest. The polarographic method is well suited for the analysis of a high sample number.

25.1 INTRODUCTION

Arsenic (As) is an element that is present in soils, rocks, the atmosphere, water and organisms. Its mobility in the environment occurs through a combination of biological processes, dissolution of rocks and volcanic emissions. Most environmental problems related to As are due to its transport under natural conditions. Nevertheless, human activity had contributed greatly through mining, fuel burning, the use of arsenical pesticides, herbicides, and the use of this element as feed additive for animal food (WHO 2001).

Arsenic in drinking water is of greater concern. Underground water flowing throughout As-rich rock may become contaminated with high concentrations of toxic As, which can make its way into private and public water supply wells. Some adverse health effects have been attributed to chronic As exposure, primarily from drinking water (NRC 1999, WHO 2001). Following the accumulation of evidence for the chronic toxicological effects of As in drinking water, the recommended and regulatory limits for As concentration in drinking water has being reduced. The WHO recommended guideline value for As in drinking water is $10\,\mu g/l$. The USEPA limit was reduced from 50 to $10\,\mu g/l$ in January 2001. The European Community maximum admissible concentration for As in drinking water has been also reduced to $10\,\mu g/l$.

It is well known that the determination of total arsenic (t-As) concentration is insufficient to evaluate the risk associated with the presence of this element. Arsenic toxicity, bioavailability, transport and distribution in living organism and the environment is dependent on its chemical species present in the sample (Jain and Kali 2000, Morita and Edmonds 1992). Thus, additional information about the different species and their concentration is needed to determine the risk of toxicity.

Several methods had been employed for the determination of t-As concentration and its chemical species. The most used speciation techniques often involve a combination of chromatographic separation with spectrometric detection (Barra *et al.* 2000, Vilanó *et al.* 2000, Brisbin *et al.* 2002, Gong *et al.* 2002, Montes-Bayon *et al.* 2003). Nevertheless, these coupled techniques are expensive and require trained personnel.

Separation techniques based on selective preconcentration procedures followed by detection of the analytes are valid and inexpensive alternatives for speciation studies. Chemical compounds derived from dithiocarbamic acid are known to chelate a large number of metals (Jansen 1958, Watanabe *et al.* 1972, Belzile *et al.* 1997, Takahashi *et al.* 2000, Sato and Ueda 2000). Ammonium pyrrolidine dithiocarbamate (APDC) is one of the dithiocarbamate compounds widely used as chelating agent for preconcentration and separation of traces metals from aqueous solution. (Wai and Mock 1985). A relevant feature of dithiocarbamates is that the stability of metal-dithiocarbamate complexes depends on the oxidation state of the metal. Thus, the selectivity of APDC for chelating As(III) allows its separation from As(V). Reduction of As(V) to As(III) allows determination of t-As concentration. Pentavalent As can be calculated by the difference between t-As and As(III).

Electrochemical techniques, especially stripping voltammetry, have also been used for As determination (Barra *et al.* 2000, Ferreira and Barros 2002, Cavicchioli *et al.* 2004). As(III) can be reduced to elemental As in acidic solution, deposited onto an electrode and then stripped off using anodic or cathodic stripping voltammetry (ASV and CSV). There are often problems associated with the use of solid electrodes, such as "memory" effects, limited sensitivity, and poor precision, which makes this approach inconvenient for routine analysis. To avoid these problems, cathodic stripping voltammetry at a hanging mercury drop electrode (HMDE) has been used to determine As. This method uses the reaction between As and Cu or Se to form an intermetallic compound that can be pre-concentrated on the HMDE and then stripped cathodically (Holak 1980, Henze *et al.* 1997, Barra *et al.* 2000, Ferreira and Barros 2002). As(V) is electroinactive and must be reduced to As(III) before its determination by CSV.

The aim of this work was to develop a method to determine t-As concentration and inorganic As(III) and As(V) concentration in water, based on the selective precipitation of As(III) with APDC followed by its measurement by energy dispersed X-ray fluorescence (EDXRF). A cathodic stripping method was also developed.

Several compounds had been used for the reduction of As, with variable degree of effectiveness: KI, $Na_2S_2O_3$, $NaHSO_3$, Na_2SO_3, $SO_2(g)$, L-cystein, ascorbic acid, hydrazine, and thiourea (Wai and Mok 1985, Sproal *et al.* 2002, Machado *et al.* 2004). Sodium thiosulfate was often used, combined with other compound or alone, with good results.

25.2   EXPERIMENTAL

25.2.1   *Apparatus and reagents*

The EDXRF spectrometer included a Si(Li) detector ($r\xi = 180$ eV for Mn-K_α), multichannel analyzer (Camberra S30) coupled to a computer. Samples were irradiated with a [109]Cd radio-isotope annular source (9 MBq). The fitting of the spectrum was performed with the AXIL program (IAEA 1995). SAX software was used for the quantification of As in the filters using the thin layer approach (Torrez *et al.* 1998).

Voltammograms were obtained with a PA-4 polarographic analyzer interfaced with a system composed by a saturated calomel reference electrode (SCE), a platinum rod auxiliary electrode, and a hanging mercury dropping electrode (HMDE) as working electrode. The voltammograms were recorded by a X-Y recorder.

As(III) standard was purchased from Fluka (1000 mg/l for AA analysis). The 1000 mg/l As(V) stock solution was prepared by dissolving 0.4223 g $Na_2HAsO_4 \times 7H_2O$ in 1 l water. Se(IV) was also purchased from Fluka (1000 mg/l for AA analysis). Working solutions were prepared from these stock solutions daily by appropriate dilution.

Ammonium pyrrolidine dithiocarbamate (APDC) from SIGMA was used to precipitate As from solutions. Aqueous APDC solutions were prepared fresh daily. Filtrations were done using 0.45 μm nitrate cellulose filter membranes (Whatman).

25.2.2 *Determination of As(III)*

A synthetic solution (1 l) of 50 μg/l As(III) was adjusted to pH 4 with buffer acetate. After the addition of 5 ml of cadmium solution (100 μg/ml), As(III) was coprecipitated using 25 ml of 0.25% (w/V) APDC. The turbid solution was gently swirled and filtered, 30 minutes later, through a nitrate cellulose membrane. The filtrate was let dry for 10 minutes and measured by XRFED. A clean filter was used as the blank.

25.2.3 *Determination of total arsenic*

A reduction step is necessary to reduce all the As(V) to the trivalent state, allowing the determination of total As concentration. One liter of As(V) working solution (50 μg/l) was acidified to pH = 2 and 0.6 g of sodium thiosulfate were added. After 15 minutes, the reaction was completed. Then, the same procedure described above for As(III) precipitation was followed. Again, a clean filter was used as the blank.

25.2.4 *Cathodic stripping procedure*

10 ml of samples and standards, adjusted to pH = 2 with 20% HCl, were placed in the cell and 20 μl of 20 mg/l Se(IV) solution were added. The solution was purged for 5 minutes with nitrogen. The parameters settings are shown below.

Deposition potential: −0.550 V
Final potential: −0.900 V
Current range: 0.05–0.1 mA
Deposition time: 40 s
Amplitude: 5 mV
Scan rate: 5 mV/s

25.3 RESULTS AND DISCUSSION

25.3.1 *Selective precipitation procedure*

The working conditions used for As(III) precipitation were those reported by Montero *et al.* (2000) for the determination of Fe, Co, Ni, Cu, Zn and Pb in rainwater samples, reporting recoveries higher than 94%. Similar recovery values, about 95%, were obtained for As(III) in the concentration range of 5–100 μg/l. The relationship between the initial concentration of As(III) added and the concentration measured with the experiment is illustrated in Figure 25.1. The graph shows a linear relationship.

The influence of the pH (range from 2 to 6) was studied to select the best conditions for quantitative precipitation. The results are illustrated in Figure 25.2a. Recovery values were constant until pH = 5, but at pH = 6 recovery decreases drastically. Probably, this is due to the sudden change of redox potential of As at pH values between 5 and 6. Thus, appreciable oxidation of As(III) to As(V) occurs, decreasing As-precipitation (Van Elteren *et al.* 2002). A pH value of 4 was selected for other experiments.

The experiments of reduction of As(V) solutions with thiosulfate, followed by precipitation with APDC, and concentration measurement by XRFED showed recoveries between 93 and 95%.

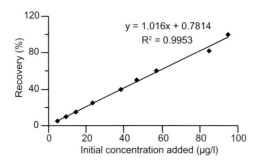

Figure 25.1.   Recovery of As(III) using the cathodic stripping procedure.

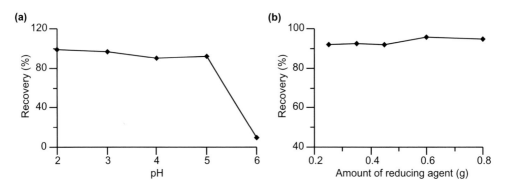

Figure 25.2.   (a) Effect of the pH on recovery of As(III); (b) Effect of the amount of reducing agent on As-recovery.

This result was not better when a mix of $Na_2S_2O_3/KI$, at different proportions, were used. This result agrees with those reported by Wai and Mok (1985).

At the pH of the solution in the reduction step (pH $= 2$), significant amount of colloidal sulfur is formed because of a thiosulfate disproportion reaction (Harris 1992). The presence of a high amount of sulfur in the filters could interfere with the As thin layer approach, making impossible the quantification of the analyte. In order to evaluate the influence of sulfur precipitation on the recovery values of As, different quantities of reducing agent (0.25–0.8 g) were assayed for the reduction of As(V). Different concentrations of thiosulfate were used. Figure 25.2b shows that the amount of precipitate in the filter did not affect the recovery values, which remained almost constant. Thus, using up to 0.8 g of sodium thiosulfate, the sulfur in the filtrates does not interfere with the formation and determination of the As thin layer. A concentration of 0.6 g/l of reducing agent was selected.

The same procedure was applied to the determination of t-As, As(III) and As(V) concentrations in synthetic solutions, that were prepared by appropriated dilution of As solutions used for As analyses by atomic absorption spectroscopy. A known volume of the sample was analyzed following the procedure for As(III) determination. Another portion of the same volume was treated with sodium thiosulfate before precipitation with APDC, allowing the determination of total As concentration. The result of the analysis (n $=$ 3, 95% confidence) for total As concentration and the inorganic species is shown in Table 25.1.

Satisfactory recovery values for both inorganic species (>95%) were obtained. The detection limit, calculated based on Curie criteria (Curie 1968) and a measuring time of 56,500 seconds, was 0.47 μg/l.

Table 25.1.   Speciation of As(III) and As(V) from a spiked water sample.

	As(III) (μg/l)	As(V) (μg/l)	As$_{total}$ (μg/l)
Spike	50.0 ± 0.5	50.0 ± 0.6	100.0 ± 0.8
Experimental value	47.5 ± 4.2	48.2 ± 7.7	95.7 ± 6.5

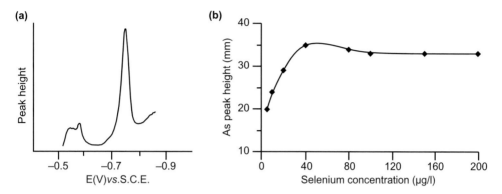

Figure 25.3.   (a) Electrochemical behavior of As(III) in presence of Se(IV); (b) Effect of Se(IV) concentration on the arsenic peak height.

25.3.2   *Cathodic stripping analysis*

The As determination by CSV was carried out in two steps: the concentration of As(III) on the HMDE as elemental As, followed by the cathodic stripping to obtain a peak due to the formation of the arsine. The low solubility of As on Hg is a drawback for its determination using this method. The addition of Se(IV) increases the As deposition on the mercury electrode. The formation of the arsine takes place at more negative potential. The behavior of the As in solution, adjusted to pH $= 2$ with HCl, containing 50 μg/l As(III) and 40 μg/l Se(IV) is shown in Figure 25.3a. The peak at -0.57 V is due to the formation of hydrogen selenide. At the potential of -0.76 V, the As, deposited on the working electrode as an intermetallic compound with selenium, is reduced and arsine is formed.

The deposition potential is critical to ensure the highest sensitivity for the determination of As. At potentials greater than -0.57 V, the formation of hydrogen selenide proceeds immediately and the deposition of As on the electrode decreases. In order to obtain the desired effect, selenium must be present on mercury. The variation of the As peak height at different deposition potentials (range from -0.4 to -0.6 V) was studied. The As peak height increases until -0.55 V. More negative potential lowered the peak height. Thus, a deposition potential of -0.55 V was chosen.

Figure 25.3b shows the effect of Se(IV) concentration on the cathodic stripping current of As(III). Arsenic peak height increases and has a maximum value at about 40 μg/l Se(IV). Further increments in the concentration of the selenium show a mild decrease in peak high and finally a steady high after a selenium concentration of around 100 μg/l.

The influence of deposition time on peak height was also studied. The peak height of As increases with the increase in deposition time (40, 120, 240, 360 s). This effect is useful for determination of very low concentration of As. For higher concentration of the analyte, deposition time must be shorter in order to decrease the time of the analysis.

The calibration curve, obtained from solutions of As(III) (concentrations from 20–200 μg/l) adjusted to pH $= 2$ and analyzed using a deposition potential of -0.55 V, deposition time of 40 s and 40 μg/l Se(IV), showed the following equation: Peak height (mm) $= 0.568$ [As(III) conc.] -0.122, with $R^2 = 0.9991$. As As(III) is the only electro-active specie under the experimental conditions, reduction of As(V) is needed for the determination of total As concentration. The procedure of As(V) reduction using sodium thiosulfate was applied in this method for the reduction of 10 ml of solution.

The cathodic stripping analysis of the same synthetic solution used in the determination of As by the selective precipitation procedure gave the following results: As(III) 48.3 ± 3.9 µg/l, As(V) 48.1 ± 7.4 µg/l and As$_{total}$ 96.4 ± 6.3 µg/l. Results are similar to those obtained with the selective precipitation method. Detection limit, under the conditions of the analysis, was 5 µg/l. Both methods were applied to a sample of drinking water. Arsenic was not detected in that sample.

25.4   CONCLUSIONS

A highly sensitive methodology for As speciation analysis in water samples using APDC precipitation and EDXRF measurement was implemented, allowing the concentration of the element from high sample volumes. Cathodic stripping method is fast, consumes low volume of sample and could be used for the determination of concentrations of As that are not very low. This method is well suited for the analysis of a high sample number.

The comparison of the methods shows good agreement between them. The precision values (expressed as relative standard deviation) are similar, lower than 9%. The achieved detection limits for both methods are lower than the guideline level for t-As concentration (10 µg/l) in drinking water. Although the precipitation method (APDC-EDXRF) is time consuming, significantly better detection limit was achieved.

REFERENCES

Barra, C.M., Santelli, R.E., Abrão, J.J. and de la Guardia, M.: Especiação de arsênio—uma revisão. *Quimica Nova* 23:1 (2000), pp. 58–70.
Belzile, N., Chen, H., Huang, J. and Chen, Y.: Determination of trace metals in lake waters by X-ray fluorescence after a precipitation preconcentration. *Can. J. Anal. Sci. Spec.* 42:2 (1997), pp. 49–56.
Brisbin, J.A., B'Hymer, C. and Caruso, J.A.: A gradient anion exchange chromatographic method for the speciation of arsenic in lobster tissue extracts. *Talanta* 58 (2002), pp. 133–145.
Cavicchioli, A., La-Scalea, M.A. and Gutz, I.G.R.: Analysis and speciation of traces of arsenic in environmental, food and industrial samples by voltammetry: a review. *Electroanalysis* 16:9 (2004), pp. 697–711.
Curie, L.: Limits for qualitative detection and quantitative determination. *Anal. Chem.* 40:3 (1968), pp. 585–593.
Ferreira, M.A. and Barros, A.: Determination of As(III) and As(V) in natural waters by cathodic stripping voltammetry at a hanging mercury drop electrode. *Anal. Chim. Acta* 459:1 (2002), pp. 151–159.
Gong, Z., Lu, X., Ma, M., Watt, C. and Chris, X.: Arsenic speciation analysis. *Talanta* 58 (2002), pp. 77–96.
Harris, D.C.: *Análisis químico cuantitativo.* Grupo Editorial Iberoamérica, S.A. de C.V., Col. Cuauhtémoc, Mexico, D.F., 1992.
Henze, G., Wagner, W. and Sander, S.: Speciation of arsenic (V) and arsenic (III) by cathodic stripping voltammetry in fresh water samples. *Fresenius J. Anal. Chem.* 358 (1997), pp. 741–744.
Holak, W.: Determination of arsenic by cathodic stripping voltammetry with a hanging mercury drop electrode. *Anal. Chem.* 52 (1980), pp. 2189–2192.
IAEA: QXAS/AXIL User's manual (Version 3.2). International Atomic Energy Agency, Vienna, Austria, 1995.
Jain, C.K. and Ali, I.: Arsenic: occurrence, toxicity and speciation techniques. *Water Res.* 34 (2000), pp. 4304–4312.
Jansen, M.J.: The stability constants, solubilities and solubility products of complexes of cooper with dialkyldithiocarbamic acids. *J. Inorg. Nucl. Chem.* 8 (1958), pp. 340–345.

Machado, L., Cisero do Nascimento, P., Bohrer, D., Scharf, M. and da Silva, M.: Especiação analítica de compostos de arsênio empregando métodos voltamétricos e polarográficos: uma revisão comparativa de suas principais vantagens e aplicações. *Quimica Nova* 27:2 (2004), pp. 261–269.

Montero, A., Estévez, J.R. and Padilla, R.: Heavy metal analysis of rainwaters by nuclear related techniques: application of APDC precipitation and energy dispersive X-ray fluorescence. *J. Radioanal. Nucl. Chem.* 245:3 (2000), pp. 485–489.

Montes-Bayon, M., DeNicola, K. and Caruso, J.A.: Review: Liquid chromatography-inductively coupled plasma mass spectrometry. *J. Chromatogr A.* 1000 (2003), pp. 457–476.

Morita, M. and Edmons, J.S.: Determination of arsenic species in environmental and biological samples. *Pure and Appl. Chem.* 64:4 (1992), pp. 575–590.

National Research Council: *Arsenic in drinking water*. National Academy Press, Washington, DC, 1999.

Sato, H. and Ueda, J.: Electrothermal atomic absorption spectrometric determination of cadmium after coprecipitation with nickel diethyldithiocarbamate. *Anal. Sci.* 16 (2000), pp. 299–301.

Sproal, R., Turoczi, N. and Stagnitt, F.: Chemical and physical speciation of arsenic in a small pond receiving gold mine effluent. *Ecotoxicol. Environ. Saf.* 53 (2002), pp. 370–375.

Takahashi, A., Igarashi, S. and Ueki, Y.: X-ray fluorescence analysis of trace metal ion following a preconcentration of metal-diethyldithicarbamate complexes by homogeneous liquid-liquid extraction. Fresenius. *J. Anal. Chem.* 368 (2000), pp. 607–610.

Torres, E.L., Fuentes, M.V. and Greaves, E.D.: SAX, software for the analysis of X-Ray fluorescence spectra. *X-Ray Spectrom.* 27 (1998), pp. 161–165.

Van Elteren, J.T., Stibilj, V. and Slejkovec, Z.: Speciation of inorganic arsenic in some bottled Slovene mineral waters using HPLC-HGAFS and selective coprecipitation combined with FI-HGAFS. *Water Res.* 36 (2002), pp. 2967–2974.

Vilanó, M., Padró, A. and Rubio, R.: Coupled techniques based on liquid chromatography and atomic fluorescence detection for arsenic speciation. *Analitica Chimica Acta* 4:1/2 (2000), pp. 71–79.

Wai, C.M. and Mok, W.M.: *Arsenic speciation and water pollution associated with mining in the Coeur D'Alene mining district, Idaho*. Research technical completion report, Idaho Water Resource Research Institute, Idaho University, ID, 1985.

Watanabe, H., Berman, S. and Russell, D.S.: Determination of trace metals in water using X-ray fluorescence spectrometry. *Talanta* 19 (1972), pp. 1363–1375.

WHO: *Arsenic and arsenic compounds*. EHC (Environmental Health Criteria) 224, World Health Organization, WHO, Geneva, Switzerland, 2001.

CHAPTER 26

Dissolution kinetics of arsenopyrite and its implication on arsenic speciation in the environment

J. Cama, M.P. Asta & P. Acero
Institute of Earth Sciences "Jaume Almera" (CSIC), Barcelona, Catalonia, Spain

G. De Giudicci
Department of Earth Science, University of Cagliari, Cagliari, Italy

ABSTRACT: Arsenopyrite oxidative dissolution contributes arsenic (As) to waters. The effects that environmental factors (e.g., pH, iron and sulfate content, dissolved oxygen, and surface reactivity) exert on arsenopyrite decomposition were studied by means of non-stirred flow-through experiments at pH 1 and 3 and at variable dissolved oxygen contents. Steady state dissolution rates were calculated based on the release of As and iron into solution. In the investigated pH range, variation in proton concentration had no or little influence on arsenopyrite dissolution. On the contrary, the decrease in dissolved oxygen concentration from 8.7 to 1.8 mg/l strongly diminished the arsenopyrite dissolution rate at pH 3. Similarly, a content of 0.01 M Fe(II) caused a decrease in the arsenopyrite dissolution rate. In the output solutions As was present as arsenite and arsenate and the respective amount depended on the oxygen availability during arsenopyrite oxidation. In order to assess the variation of dissolving arsenopyrite surfaces, atomic force microscope (AFM) experiments were performed both *in situ* and *ex situ*.

26.1 INTRODUCTION

The most common natural occurrences of arsenic (As) are associated with hydrothermal mineral deposits, which are further exposed by mining and processing. In particular, As-rich discharges from gold mine tailings or from groundwater interaction with As-rich rocks are well known. Very frequently, these processes produce an elevation in background of dissolved As concentrations (Smedley and Kinniburgh 2002).

Arsenopyrite (FeAsS) is the dominant As mineral in most As-bearing natural occurrences and is therefore the main mineral responsible for elevated As concentrations at surface sites (Smedley *et al.* 1996). This mineral is present in sulfide ores associated with sediment-hosted Au deposits. It tends to be the earliest-formed mineral derived from hydrothermal solutions and formed at medium to high temperatures. Furthermore, it can occur in metamorphic deposits and pegmatites, as well as authigenic arsenopyrite in sediments (Rittle *et al.* 1995).

Arsenopyrite under oxidizing conditions produces arsenite (AsO_3^{3-}), arsenate (AsO_4^{3-}) and sulfate (SO_4^{2-}) (Nesbitt *et al.* 1995, 1998, Richardson and Vaughan 1989), thus contributing to the acidification of water as well as to the release of soluble As species. In such waters some of the solutes released, such as Fe, As or S, will be adsorbed on solid phases near the site of dissolution or will find their way into the surrounding environment. Dissolved As occurs dominantly as inorganic As(III) (arsenite) and As(V) (arsenate) ions, and if they are consumed in large amounts, they may poison the nervous and digestive systems and cause a variety of skin diseases. The acute health risk is particularly high if arsenite is dominant, as this form of dissolved As is 60 times more toxic than arsenate (Bottomley 1984). Therefore, arsenopyrite presents a serious health hazard if large amounts of the mineral are exposed to oxidizing waters that are subsequently used for human consumption. Oxygen availability and pH are two environmental variables that affect both reactivity and apparent solubility of arsenopyrite. According to previous works on sulfide dissolution, this process accelerates when dissolved oxygen (DO) is present in high concentrations (Domènech et al. 2002 and references therein). On the other hand, pH variation also seems to influence the sulfide dissolution mechanism. A rate law that accounts for the effect of pH and DO on sulfide dissolution has been proposed (Domènech et al. 2002, McKibben and Barnes 1986):

$$\text{Rate} = k \cdot S_{\text{min}} \cdot [O_2(aq)]^n \cdot (a_{H^+})^m \tag{26.1}$$

where Rate is in mol/m^2/s, k is the dissolution rate constant, S_{min} is the reactive surface area (m^2), [O$_2$(aq)] and a_{H^+} are the activities of dissolved oxygen and protons in solution (M), and n and m are empirical factors.

The estimation of the arsenopyrite dissolution rate in the laboratory under conditions that match those in the field (e.g., acidic pH and oxic conditions) is a first step towards the elaboration of a rate law. The use of flow-through experiments allows us to study the effects that the different factors involved in the reaction exert on the dissolution rate.

In the present study we sought to obtain the steady state dissolution rates of arsenopyrite at pH 1 and 3, with dissolved oxygen ranging from 8.7 to 1.8 mg/l and at 25°C with non-stirred flow-through experiments. The effect of electrolyte (H$_2$SO$_4$ or HCl) and the Fe^{2+} content in solution has also been assessed. Moreover, in situ atomic force microscope (AFM) experiments were conducted to characterize changes in dissolving arsenopyrite surfaces. The obtained results are the basis for future experiments designed to fully characterize the complex mechanism ruling the dissolution of arsenopyrite, and to eventually propose a dissolution rate law as expressed in eq. (26.1).

26.2   EXPERIMENTAL METHODOLOGY

26.2.1   *Sample characterization*

The samples of arsenopyrite used in this study are from the Cerdanya region (Catalan Pyrenees, Spain). Mineral fragments were ground in an agate mortar and sieved to obtain a powder fraction between 10 and 100 µm to be used in the flow-through experiments. Based on electron microprobe analysis (EMP) the arsenopyrite atomic composition is Fe 33.5 ± 0.1%, As 32.1 ± 0.3% and 34.4 ± 0.3%. The BET-determined initial surface area is 0.5 ± 0.05 m^2/g using 5-point N$_2$ adsorption isotherms.

26.2.2   *Flow-through experiments*

Experiments were carried out using non-stirred flow-through Lexan reactors with a reaction chamber of 35 ml in volume, as shown in Figure 26.1. All the experiments were conducted at a dissolved oxygen concentration of 8.6 mg/l corresponding to 21% (atmospheric conditions) of oxygen, except one that was carried out at 1.8 mg/l (4.5% O$_2$). One experiment was conducted in the presence of 0.01 M Fe(II) in solution. To ensure the desired concentrations, the experiments were carried out in a glove box purged with the corresponding O$_2$/N$_2$ gas mixture. The pH of feed solutions was 1 and 3. Input solutions were supplied to the reactor by a peristaltic pump at a flow rate of 0.035–0.040 ml/min. The flow rate and feed solution concentration were held constant for each experiment until steady state conditions were achieved. Steady state was considered to be attained when differences in the metal concentration in the output solutions in consecutive leachate samples were within ±10% for at least 200 hours.

26.2.3   *In situ AFM experimental setup*

The AFM investigation was carried out using a molecular imaging microscope. Arsenopyrite surfaces were imaged using triangular, 200 mm Si$_3$N$_4$ levers in contact mode. A full scan lasted

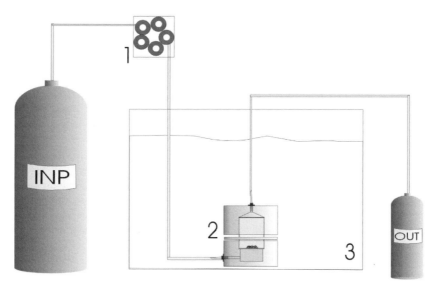

Figure 26.1.   Schematic illustration of the flow-through set up: (1) peristaltic pump; (2) non stirred flow-
through reactor immersed in the water bath; (3) water-bath held at constant temperature
$(25 \pm 0.1°C)$.

approximately 3 minutes. Arsenopyrite fragments were glued with epoxy (Buelher) to the cell to
remain immovable when scanning. The images obtained were processed with Digital Instrument
(DI) software to obtain height mode, topography mode, cross-section of topography and vertical
measurements.

26.2.4   Solutions

Input solutions at pH 1 and 3 were prepared with Millipore MQ water (18.2 MΩ/cm) and analytical-
grade HCl and H_2SO_4 reagents. $FeSO_4 \cdot 7H_2O$ and H_2SO_4 (95–97%) were used to prepare a
0.01 M Fe^{2+} solution. Total concentrations of As, S, and Fe in input and output solutions were
determined by inductively coupled plasma atomic emission spectroscopy (ICP-AES). Analyses of
As speciation were carried out by means of liquid chromatography coupled to hydride generation
atomic fluorescence (LC-HG-AFS) (Vilanó *et al.* 2000). Input and output solution pH was measured
with a Crison combined glass electrode at room temperature $(22 \pm 2°C)$.

26.2.5   Calculation of the reaction rates

In the flow-through experiments, once steady state was attained, the dissolution rate, R (mol/m^2/s)
was calculated from the release of As according to the expression:

$$R = \frac{q(c_i - c_i^o)}{v_i A} \tag{26.2}$$

where q is the flow rate (m^3/s) through the system, v_i is the stoichiometric coefficient of i in the
mineral, A is the surface area (m^2) and c_i^o and c_i are the concentrations (mol/m^3) of component i
in the input and output solution, respectively. The error associated with the calculated dissolution
rates in the flow-through experiments is approximately 20%, estimated using the Gaussian error
propagation method described in Barrante (1998).

26.3   RESULTS

26.3.1   *Flow-through experiments*

The experimental conditions in all the experiments are shown in Table 26.1. The variation of output Fe, As and S concentrations in one of the flow-through experiments as a function of time is shown in Figure 26.2. Duration of experiments was between 1300 and 2015 h. High Fe, As, and S concentrations were observed at the onset of most of the experiments. Afterwards, Fe, As, and S concentrations decrease until steady state is approached after 500 to 1200 h from the beginning of the experiment (Fig. 26.2).

The stoichiometric ratio between two elements (e.g., Fe and S) is defined as the ratio between the release of one element (Fe) and the release of the other one (S) at steady state. Figures 26.3a and 3b show the variation of the ratio Fe/As and As/S as a function of time. For most experiments the Fe/As ratio was 1.11 ± 0.05 and the As/S ratio was 1.76 ± 0.10 (Table 26.1). In the experiments with very high content of sulfur and iron in solution (0.01 M), the sulfur and iron concentration in the input solutions was very high and similar to the output concentration, and therefore the stoichiometric ratio was not calculated.

The average Fe/As ratio was very close to the ratio in the solid (0.96 ± 0.02, from the microprobe analysis of the arsenopyrite). However, the As/S or the Fe/S ratios at steady state were much

Table 26.1.   Experimental conditions and steady state values used to calculate the dissolution rates.

	Flow rate (ml/min)	pH	DO (mg/l)	Mass (g)	Electrolyte	Fe	As	S	Fe/As	As/S	Rate As (mol/m²/s)
						µM					
ASP1[1]	0.04	1.01	8.7	0.501	H₂SO₄	32.76	29.38	99877	1.11	–	7.9×10^{-11}
ASP2[1]	0.04	2.74	8.7	0.503	H₂SO₄	38.39	34.66	962	1.10	–	1.0×10^{-10}
ASP3	0.04	1.19	8.7	0.504	HCl	37.75	34.50	18.37	1.09	1.88	8.8×10^{-11}
ASP4	0.04	3.10	8.7	0.508	HCl	30.96	29.67	18.37	1.04	1.62	7.6×10^{-11}
ASP5	0.03	1.26	1.8	0.503	HCl	9.16	7.69	4.34	1.19	1.79	1.8×10^{-11}
ASP6[1]	0.04	1.16	8.7	0.494	FeSO₄	10072	21.95	124371	–	–	5.7×10^{-11}

[1] Experimental data were not used to calculate solution saturation conditions due to high total sulfur concentration.

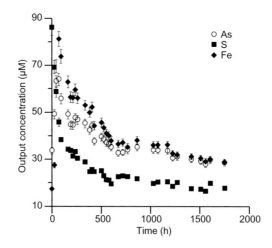

Figure 26.2.   Variation in the output concentrations of Fe, As and S as a function of time. Output steady state concentrations are reached after 1200 h approximately.

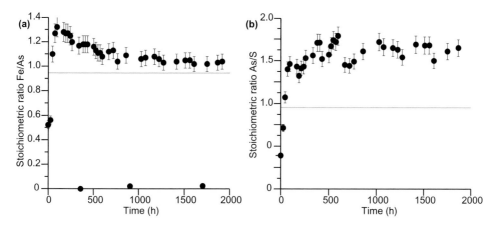

Figure 26.3.   Variation in the stoichiometric ratios of Fe/As (a) and As/S (b) as a function of time. The grey solid line denotes the respective ideal stoichiometric ratio in the arsenopyrite (Fe/As/S = 0.96).

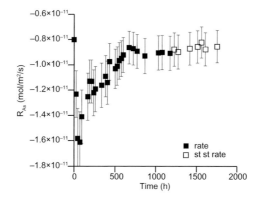

Figure 26.4.   Variation in the normalized dissolution rate based on As release (R_{As}) as a function of time. The open symbols correspond to steady state dissolution rate values.

higher than the mineral stoichiometric ratios of 0.95 and 0.99, respectively. Therefore a deficit in dissolved sulfur was observed. Speciation analyses of total As (t-As) in the output solutions show the existence of As(III) and As(V) in solution. The amount of As(III) increased with decreasing O_2 in the glove box during arsenopyrite oxidation.

The steady state dissolution rates were normalized using the initial BET surface area. Due to the deficit in aqueous sulfur, dissolution rates were only calculated based on As steady state concentrations using eq. 26.2 (Table 26.1). Figure 26.4 shows that steady state dissolution rate values were obtained after about 1000 hours.

26.3.2   *Arsenopyrite surface studied by in situ AFM*

Figures 26.5a and b show surface regions of $3.1 \times 3.1\ \mu m$ of a mineral fragment during arsenopyrite dissolution at pH 1 (HCl). Cleaved surfaces appeared to be rugged. During dissolution surface roughness did not vary significantly. The cross-section analyses along two selected flatter regions show the vertical distances in the topography (Figs. 26.5c and d).

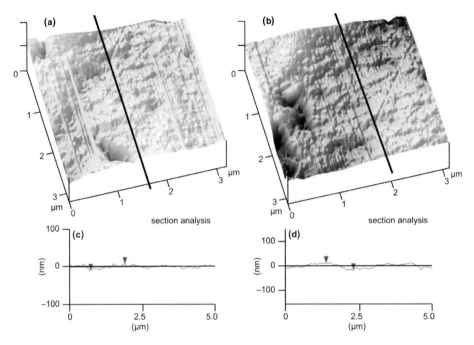

Figure 26.5.   Two regions of 3.1 × 3.1 μm of 5 × 5 μm AFM images of the reacted arsenopyrite surfaces after
3.45 h (a, c) and 5 h (b, d) are shown: Topography mode diagram (a, b) and topography along
the cross-section (black solid line in the upper images) (c, d). Vertical distances are less than
100 nm. Triangles show the highest and lowest surface points in the flat region of the images.
The vertical distances between the highest and lowest points are 18 nm (c) and 30 nm (d).

26.4   DISCUSSION

26.4.1   Flow-through experiments

The high concentrations observed at the onset of the experiments are likely caused by the high
reactivity of ultra-fine particles attached to the arsenopyrite grains (Fig. 26.6a). SEM images of
the arsenopyrite powder retrieved after the experiments show that the amount of microparticles
attached onto the grains has diminished markedly (Fig. 26.6b). Thus, after the microparticles are
dissolved the steady state concentrations of As, Fe, and S are achieved.

The lack of dissolved sulfur in highly acidic solutions has been observed in previous studies
related to sulfhide dissolution (Buckley $et\ al.$ 1989, Lochman and Pedlik 1995, Janzen $et\ al.$ 2000,
De Giudici and Zuddas 2001, Weisener $et\ al.$ 2003, 2004, Cama and Acero 2005, Cama $et\ al.$
2005). Based on the As, Fe, and S output concentrations, the saturation states of the output solutions were calculated using the PHREEQC code (Parkhurst and Appelo 1999) at pH 1 and 3 (HCl)
at 25°C. Calculations show that dissolution took place under highly undersaturated conditions with
respect to any S-bearing phase in the Minteq.dat database. Since supersaturation with respect to S
or S-bearing phases at pH 1 and 3 is not obtained, sulfur deficit may be attributed to the loss of
H_2S (aq) via gasification (e.g., $H_2S(g)$ production) instead of precipitation of S or S-bearing phases.

Dissolution rates at pH 1 and 3 are the same within error (approx. $8.6 \pm 0.9 \times 10^{-11}$ mol/m²/s).
Hence it appears that in this pH range the variation in proton concentration does not have a significant effect on arsenopyrite dissolution. At pH 1 and 3 the effect of the electrolyte (SO_4^{2-} or
Cl^-) on arsenopyrite dissolution is negligible (Table 26.1). However, at pH 1 and 0.01 M Fe^{2+}
the arsenopyrite dissolution rate decreased by approximately 40% (Table 26.1). Therefore, large
amounts of Fe(II) in solution may cause a decrease in the dissolution rate.

Figure 26.6.   SEM photographs of the samples used in the experiments. (a) Freshly ground arsenopyrite grains; (b) The amount of microparticles attached onto the arsenopyrite grains after the experiment at pH 1 with H_2SO_4 has decreased.

On the other hand, at pH 1 the decrease in dissolved oxygen (1.8 mg/l) decreases the arsenopyrite dissolution to 1.8×10^{-11} mol/m^2/s. The DO effect on the dissolution of several sulfides has been reported by Williamson and Rimstdit (1994), Lengke and Temple (2001), Domènech *et al.* (2002) and Tallant and McKibben (2005).

26.4.2   *In situ AFM experiments*

Arsenopyrite dissolution at pH 1 for 5 h did not cause marked changes in surface topography as is illustrated by the similarity between the topographic profiles (Figs. 26.5c, d). Nonetheless, more and longer *in situ* AFM experiments are to be conducted to examine possible changes in surface topography during dissolution, taking into account that at this pH range the arsenopyrite dissolution rate is significant.

26.5   CONCLUSIONS

The dissolution rates of FeAsS were determined by means of non-stirred flow-through experiments at pH values of 1 and 3, and two different dissolved oxygen concentrations (1.8 and 8.7 mg/l). *In situ* AFM experiments show that cleaved surfaces of arsenopyrite fragments are rugged and that topographic variations are minor after 300 minutes.

The results showed that at the pH values studied (1) the arsenopyrite oxidative dissolution rate is pH-independent, (2) the electrolyte used (HCl or H_2SO_4) does not influence the dissolution rate, (3) the presence of 0.01 M Fe^{2+} seems to diminish the dissolution rate by approximately 40%, and (4) the decrease in dissolved oxygen is the only environmental variable that affects the oxidation rate (arsenopyrite dissolution rate increases when DO increases). Accordingly, with an atmospheric oxygen content, As leaching is higher than in a reducing atmosphere.

The respective amounts of As(III) and As(V) in solution appear to depend on the O_2 content during arsenopyrite dissolution, and on oxygen availability during atmosphere-solution interaction. Therefore, the toxicity of an As-rich solution will depend on the oxygen content interacting with either the arsenopyrite surface or with As leachates.

ACKNOWLEDGEMENTS

This work was partially funded by the Research and Development contract with the Spanish Government (REN 2003-09590-C04-02). We are indebted to Àngels Canals who supplied the arsenopyrite.

The analytical assistance of Javier Pérez, Rafael Bartolí, Mercè Cabanes, María José Chancho and Xavier Llovet is gratefully acknowledged. We thank Josep M. Soler and an anonymous reviewer for helpful suggestions to improve the original manuscript.

REFERENCES

Barrante, J.R.: *Applied mathematics for physical chemistry*. Prentice Hall, New Jersey, 1974.

Bottomley, D.J.: Origins of some arseniferous groundwater in Nova Scotia and New Brunswick. *Can. J. Hydrol.* 69 (1984), pp. 223–257.

Buckley, A.N., Wouterlood, H.J. and Woods, R.: The surface composition of natural sphalerites under oxidative leaching conditions. *Hydrometallurgy* 22 (1989), pp. 39–56.

Cama, J. and Acero, P.: Dissolution of minor sulphides present in a pyritic sludge at pH 3 and 25°C. *Geol. Ac.* 3 (2005), pp. 15–26.

Cama, J. Acero, P., Ayora, C. and Lobo, A.: Galena surface reactivity at acidic pH and 25°C based on flow-through and in-situ AFM experiments. *Chem. Geol.* 214 (2005), pp. 309–330.

De Giudici, G. and Zuddas, P.: In situ investigation of galena dissolution in oxygen saturated solution: Evolution of surface features and kinetic rate. *Geochim. Cosmochim. Acta* 65 (2001), pp. 1381–1389.

Domènech, C., De Pablo, J. and Ayora, C.: Oxidative dissolution of pyritic sludge from Aznalcóllar mine (SW Spain). *Chem. Geol.* 190 (2002), pp. 339–353.

Janzen, M.P., Nicholson, R.V. and Scharer, J.M.: Pyrrhotite reaction kinetics: reaction rates for oxidation by oxygen, ferric iron, and for non-oxidative dissolution. *Geochim. Cosmochim. Acta* 64 (2000), pp. 1511–1522.

Lengke, M.F. and Temple, R.G.: Kinetics rates of amorphous As_2S_3 oxidation at 25 to 40°C and initial pH 7.3 to 9.4. *Geochim. Cosmochim. Acta* 65 (2001), pp. 2241–2255.

Lochmann, J. and Pedlik, M.: Kinetic anomalies of dissolution of sphalerite in ferric sulphate solution. *Hydrometallurgy* 37 (1995), pp. 89–96.

McKibben, M.A. and Barnes, H.L.: Oxydation of pyrite in low temperature acidic solutions: rate laws and surface textures. *Geochim. Cosmochim. Acta* 50 (1986), pp. 1509–1520.

Nesbitt, H.W. and Muir, I.J.: Oxidation states and speciation of secondary products on pyrite and arsenopyrite reacted with mine waste waters and air. *Miner. Petrol.* 62 (1998), pp. 123–144.

Nesbitt, H.W., Muir, I.J. and Pratt, A.R.: Oxidation of arsenopyrite by air and air-saturated, distilled water, and implications for mechanism of oxidation. *Geochim. Cosmochim. Acta* 59 (1995), pp. 1773–1786.

Parkhurst, D.L. and Appelo, C.A.J.: *User's Guide to PHREEQC (Version 2), a computer program for speciation, batch reaction, one-dimensional transport, and inverse geochemical calculations*. US. Geological Survey Water-Resources, Water Resources Research Investigations Report 99–4259, 1999.

Richardson, S. and Vaughan, D.J.: Arsenopyrite: a spectroscopic investigation of altered surfaces. *Mineral. Mag.* 53 (1989), pp. 223–229.

Rittle, K.A., Drever, J.I. and Colberg, P.J.S.: Precipitation of arsenic during bacterial sulfate reduction. *Geomicrobio. J.* 13 (1995), pp. 1–11.

Smedley, P.L., Edmunds, W.M. and Pelig-Ba, K.B.: Mobility of arsenic in groundwater in the Obuasi gold-mining area of Ghana: Some implications for human health. *Spec. Publ. Geol. Soc. London* 113 (1996), pp. 163–181.

Smedley, P.L. and Kinniburgh, D.G.: A review of the source, behaviour and distribution of arsenic in natural waters. *Appl. Geochem.* 17 (2002), pp. 517–568.

Tallant, B.A. and McKibben, M.A.: Arsenic mineral kinetics: arsenopyrite oxidation. *Goldschmidt Conference Abstracts* 69:10 (2005), p. 820.

Vilanó, M., Padró, A., and Rubio, R.: Coupled techniques based on liquid chromatography and atomic fluorescence detection for arsenic speciation. *Anal. Chim. Acta* 411 (2000), pp. 71–79.

Weisener, C.G., Smart, R.S. and Gerson, A.R.: Kinetics and mechanisms of the leaching of low Fe sphalerite. *Geochim. Cosmochim. Acta* 67 (2003), pp. 823–830.

Weisener, C.G., Smart, R.S. and Gerson, A.R.: A comparison of the kinetics and mechanism of acid leaching of sphalerite containing low and high concentrations of iron. *Int. J. of Min. Proc.* 74:1–4 (2004), pp. 239–249.

Williamson, M.A. and Rimstidt, J.D.: The kinetics and electrochemical rate-determining step of aqueous pyrite oxidation. *Geochim. Cosmochim. Acta* 58 (1994), pp. 5443–5454.

Section IV
Arsenic in soil, plants and food chain issues

Arsenic in soils

CHAPTER 27

Geogenic enrichment of arsenic in histosols

T.R. Rüde
Institute of Hydrogeology, RWTH Aachen University, Aachen, Germany

H. Königskötter
Institute of Geography, Ludwig-Maximilians-University, Munich, Germany

ABSTRACT: There is an increasing awareness of natural arsenic (As) enrichment in aquifers
and overlying soils worldwide. We investigated natural As enrichment in histosols with As up to
4000 mg/kg. Our results demonstrate that As is not enriched by shrinking and decomposition of
the organic matter in the soils set off in the past by draining the soils to cut peat and for agricultural
use. Arsenic is enriched in iron, and to a lesser amount manganese, oxides. The original source of
the As is probably a deeper aquifer with well known geogenic enrichment of As. There is a risk
that the pedogenic sink of As will be dissolved during changes of land use especially an artificial
rise of the water table to restore the original bogs and wetlands.

27.1   INTRODUCTION

In contrast to anthropogenic contaminated sites, geogenic enrichment of arsenic (As) in aquifers
and soils occurs over large areas, and cause an often unforeseen threat to humans. The case of
the Bengal basin is the most important, but the number of reports of groundwaters with high
concentrations of As is increasing (e.g., Smedley and Kinniburgh 2002, Bundschuh *et al.* 2005).

Geogenic enrichment of As in aquifers is well known from parts of southern Germany. In the
Franconian Upper Triassic sandstones, As occurs at concentrations of more than 100 μg/l in
groundwater, and up to 1400 μg/g in the sandstone (Heinrichs 1996). The importance of these
aquifers for regional drinking water supply has led to intense investigations, and research into
treatment technologies to deal with this challenge (e.g., Driehaus 2005, Jekel 1994). However,
high As concentrations are also found in the Tertiary Badenian sands of the Southern German
Molasse trough (Bayer and Henken-Mellies 1998). Arsenic is present at concentrations of up to
1900 μg/g, adsorbed to iron oxyhydroxides coatings on the sand. Arsenic is effectively immobile
in the oxic environment of this aquifer, with concentrations of only 0.3–1.8 μg/l (1st quartile to 3rd
quartile), but with a few exceptions of anoxic areas (Wagner *et al.* 2003). Unlike the Franconian
sandstones, the geogenic As is not regarded as a serious risk to groundwater quality in the Southern
German Molasse trough.

Discoveries of high As in soils north of the Bavarian city of Munich at the end of the year 2002
inspired investigations into the distribution of As, its transfer to grass and crops, its source, and into
the processes leading to its accumulation (Bavarian Geological Survey 2003). The soils, especially
histosols, have been cultivated for about a hundred years. It has been suggested that As may be
enriched in organic matter due to adsorption on humic substances (Thanabalasingam and Pickering
1986), which might occur during drainage, and accompanied shrinking and decomposition of these
horizons, enriching As in the residual matter (Bavarian Geological Survey 2003). Alternatively,
As can be enriched on pedogenic iron and/or manganese oxides (e.g., Inskeep *et al.* 2002). This
chapter reports studies of three soil profiles to test both hypotheses.

27.2   SITE DESCRIPTION

27.2.1   *Location and geology*

This study is located in southeastern Germany, north of the city of Munich. The area is part of the South German Molasse trough, and the Tertiary sediments are overlain by Quaternary tills, periglacial gravel plains and postglacial fluvial sediments (Freudenberger and Schwerd 1996). After the retreat of the Quaternary glaciers, bogs formed at the northern edge of a 1500 km² gravel plain around Munich. Discharge of groundwater gave rise to the development of eutric histosols and gleysols during the Holocene (Fetzer *et al.* 1986). These soils cover an area of approximately 380 km². For more than a hundred years the soils have been partially drained to cut peat, and for agriculture and pasture, the latter especially on histosols. The cultivation of the soils leads to shrinking and mineralization of the organic matter.

27.2.2   *Arsenic in the soils of the area*

The Bavarian Geological Survey (2003) investigated the distribution of As in gleysols and histosols along the northern edge of the gravel plain of Munich. At 45 locations, up to eight core samples were taken to represent areas of about 180 m² each. Samples were taken separately from upper and lower soil horizons. This survey measured median As concentrations, in the upper horizons of gleysols, of 17.9 µg/g beneath agricultural land, and 35.1 µg/g beneath pasture. The lower horizons had much lower median As concentrations of 7.5 µg/g and 5.8 µg/g, respectively. The highest concentration was 170 µg/g, but even the median values are well above typical values of soils of 5–7 µg/g (e.g., Woolson 1983, Matschullat 2000).

The histosols have higher median As concentrations of 47.4 and 31.1 µg/g in the upper and lower horizons, with a maximum of 1600 µg/g. With respect to German soil protection legislation, these values are of concern for plants and grass grown on the soils, however, there was only one report of excessive As concentrations in shoots or grass.

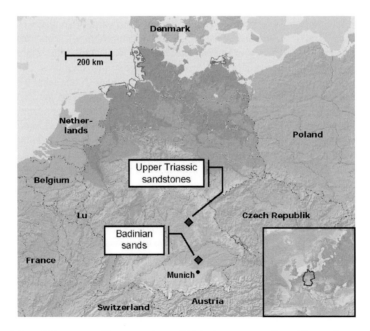

Figure 27.1.   Map of Germany with two regions of high As concentrations. The study area lies north of Munich.

27.2.3   *The investigated soil profiles*

We chose three profiles for detailed investigation of the distribution of As with depth and its correlation with soil properties from sites of known high As concentrations. The selection was done in collaboration with the Bavarian Geological Survey.

Profile 1 was taken from an agricultural area and is a calcic gleysol with 6 horizons. The two lower most are calcareous gleyic with an anoxic environment. Peat forms the two middle horizons, with less than 30% of organic matter, and the upper part is a humic topsoil with a ferric layer. Details are given in Table 27.1. Profile 2 is a calcic gleysol with the water table located 47 cm below surface (31.03.2004). The calcareous gleyic horizon above the water table is well aerated.

Table 27.1. Description of the soil profiles.

Profiles	Depth [cm]	Type[1]	Abbr.[2]	Description
No 1				
	18	Ah	1-Ah	dark brown (7.5YR3/3), numerous roots
	26	Ah-Go	1-AGo	light brown, yellowish (10YR3/6)
	43	nHv	1-Hv	black, densely
	55	nHw	1-Hw	black, few roots, wood pieces
	80	Gcr1	1-Gr1	blue-grey (2.5Y4/2), no roots, calcareous, remnants of organic matter
	100	Gcr2	1-Gr2	blue-grey, calcareous (10Y5/1)
No 2				
	15	Ah	2-Ah	brownish-black (10YR2/2), roots, quartz gravels
	34	nHv	2-Hv	black to dark brown (5R2/2), densely packed, few roots
	47	Gor	2-Go	yellowish-grey (5Y5/2), angular gravel, calcareous, remnants of peat
	100	Gcr	2-Gr	yellowish light grey (5Y7/2), rounded gravel, calcareous
No 3				
	13	nHvm	3-Hm	brown-black (5YR2/2), numerous roots
	28	nHcv	3-Hc	black (10R2/2), densely packed, calcareous
	48	eHw	3-Hw	dark brown-black (5YR3/2), calcareous, densely packed
	78	Gcr	3-Gr	yellowish-light grey (10YR6/2), gravels, calcareous

[1]Soil type according to the German soil taxonomy; [2]Abbreviations used in this chapter.

Table 27.2.   Chemical data of the soils. Concentrations are calculated for dry soils (105°C).

Horizon	pH	EC (μS/cm)	H$_2$O (m/m-%)	C$_{carb}$ (%)	C$_{org}$ (%)	Fe$_{tot}$ (%)	Fe$_o$/Fe$_d$ (−)	Mn$_{tot}$ (μg/g)	As$_{tot}$ (μg/g)
1-Ah	7.34	228	47.5	2.91	13.2	9.13	0.19	1589	1303
1-AGo	7.33	179	54.4	1.21	11.3	30.9	0.18	2160	4063
1-Hv	7.18	101	74.0	0.05	26.7	2.31	0.54	197	291
1-Hw	7.07	72	62.4	0.02	15.8	2.15	0.24	126	146
1-Gr1	7.29	122	28.1	3.20	1.63	1.43	0.47	117	18.1
1-Gr2	7.55	132	20.8	4.63	0.70	1.41	0.58	211	9.14
2-Ah	6.82	149	47.2	0.86	16.9	6.26	0.29	1495	349
2-Hv	6.91	102	68.0	0.38	25.3	8.34	0.21	1379	312
2-Go	7.63	94	14.6	5.02	1.70	2.60	0.60	1505	18.5
2-Gr	7.82	84	9.90	6.59	0.40	1.45	0.85	255	10.2
3-Hm	7.18	281	69.8	0.65	33.8	3.32	0.07	460	227
3-Hc	7.15	264	73.1	0.71	34.1	4.65	0.38	505	247
3-Hw	7.32	131	71.6	7.24	14.6	0.37	0.39	121	15.4
3-Gr	7.52	196	15.8	8.58	2.12	0.38	0.42	151	5.03

Note: Fe$_o$/Fe$_d$ is the ratio of less crystallized iron oxides (Fe$_o$) to the total amount of iron oxides (Fe$_d$).

This horizon is overlain by peat (less than 30% of organic matter) and a humic topsoil. The third profile is taken from pasture, like profile 2, and is a histosol on top of calcareous gravel. The water table is 78 cm below the surface (20.04.2004).

All three profiles were described in the field and samples of each horizon were taken for laboratory studies. The pH of non-dried samples was determined in a 0.01 mol/l CaCl$_2$ background electrolyte. For all other laboratory tests the samples were air dried and the grain size smaller than 2 mm was taken. The following parameters were determined: water content; carbonate; organic carbon content; total concentrations of As, Fe and Mn; solubility of As in water after shaking for 24 h; pH; and the electrical conductivity of the aqueous eluates.

A sequential extraction scheme, similar to that of Zeien and Brümmer (1989), was used to dissolve different soil fractions of As, Fe and Mn. Soil samples of 3 g were extracted by 30 ml each of 0.5 M NH$_4$F (anion exchange), 1 M NH$_4$-acetate (carbonates), 0.1 M NH$_2$OH · HCl + 1.0 M NH$_4$-acetate (Mn oxides), 0.5 M EDTA (organically bounded), 0.2 M NH$_4$-oxalic buffer in the dark (Fe-oxides of low crystallinity) and hot 0.2 M NH$_4$-oxalic buffer + 0.1 M ascorbic acid (Fe-oxides of high crystallinity). Finally, the residues were dissolved in aqua regia.

Phosphate was used to determine the exchangeable As adsorbed to the surfaces of soil particles. Soil samples of 5 g were shaken for 24 h with 15 ml of 0.25 M NaH$_2$PO$_4$ × H$_2$O, centrifuged and the solution decanted. The exchange procedure was repeated for one hour and both solutions were combined for analyses of As and phosphate. Analytical data are summarized in the Table 27.2 and discussed below.

27.3   RESULTS AND DISCUSSION

27.3.1   *Data from soil profiles*

The humic layers of the profiles have a very high field capacity, which is demonstrated by water contents of 62–74% and which is much higher than that of the mineral horizons with values of 10–28%. Leachates from the laboratory batch tests had low values of electrical conductivity of 84–280 μS/cm. These results indicate that the materials are fairly inert, releasing only small amounts of ions into the soil waters or that the water fluxes were high enough to wash out most of the products of water-rock interactions, leaving only the insoluble compounds and the ions from ongoing processes. The latter seems more plausible because the deeper horizons consist mostly of carbonates with up to 71% $CaCO_3$ (2-Gr). Nevertheless, the EC of these horizons is well below typical values of carbonaceous porous aquifers and more similar to aquifers with high fluxes.

Due to carbonate, the pH of all samples is neutral to slightly alkaline, except for the two uppermost horizons of profile 2. Both are at least partially influenced by a former henhouse at the location where the quartz pebbles in sample 2-Ah were observed (see Table 27.2).

The organic carbon (C_{org}) content was calculated from analyses of total C by combustion and substracting the C_{carb} (HCl-CO_2-method). The plausibility of the data was tested by wet washing and annealing loss. Although these methods give a wide spread of results, the general distribution of the C_{org} in the soils is confirmed. Herein the values of the difference-method are used as the most reliable ones. C_{org} is around 13% in top soils and up to 34% in peat. The lower carbonaceous layers have very low values of 0.4–2%.

27.3.2   *Quality of the analytical data of As, Fe and Mn*

The concentrations of Fe, Mn, and As shown in Table 27.2 were calculated by adding the values of the seven steps of the sequential extraction scheme. Analyses of Fe and Mn were done by flame AAS and flow-injection hydride-generation AAS, was used for As. Samples were also analyzed by a Canadian laboratory using ICP-MS after dilution of the complete (<2 mm) milled sample in aqua regia to give total As concentrations (e.g., Göd 1994, Rüde 1995).

Figure 27.2 shows the relative difference of our results to the mean of both analyses. The differences are around 5–6% and less except three samples of which two have differences of 13% and one outlier is about 28%. There is no trend with concentrations and no trend in positive or negative false values. Besides laboratory uncertainties, the differences may also be the result of sample heterogeneity. The overall distribution of As in the profiles is not influenced by these differences between both labs.

Fe shows similar results in comparison to As except that most of the data of our lab are lower than the Canadian ones. With respect to Mn the differences are much higher with most of the samples having a negative difference to the mean of 10–30%. There is no obvious reason for this and we use the results of our lab herein for consistency.

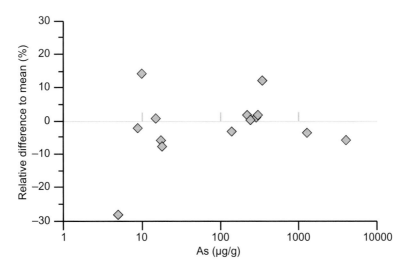

Figure 27.2.   Comparison of As concentrations of the lab in Munich and another one by the relative difference of our values to the mean of both labs.

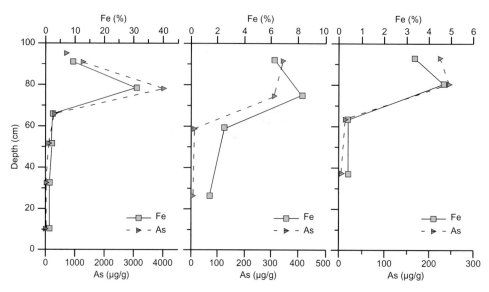

Figure 27.3.   Profiles of As and Fe in the soils. From left to right: profiles 1, 2 and 3. Symbols mark the center of each horizon.

27.3.3 *Correlation of As with Fe and Mn and organic carbon*

Iron concentrations are the highest in the oxic topsoils, typically 6–9% with the highest values in ferric layers such as 1-AGo (30.9%). Iron concentrations decrease with depth and are below 1% in the redoximorphic lowest part of the profiles (Fig. 27.3). The distribution of manganese with depth is similar to iron but the changes from the peat layers to the redoximorphic horizons are less sharp. Sequential extractions demonstrate that pedogenic iron oxides are the dominant iron minerals, especially in the humic horizons where they contribute up to 80% of iron.

The fifth step of the extraction scheme, which uses oxalic acid buffer in the dark (Schwertmann 1964), defines the content of poorly-crystalline iron oxides (Fe_o). The sum of the fifth and the sixth steps equal the total amount of iron oxides (Fe_d). Although the hot ascorbic acid-oxalic acid buffer we used is not equal to the dithionite-citrate buffer conventionally used for Fe_d, Fischer and Fechter (1982) have shown that both solutions are in good agreement. The Fe_o/Fe_d ratios (Table 27.2) are low in the upper horizons and indicate the dominance of poorly crystallized iron oxides in those layers. In the horizons that are influenced by groundwater and the capillary fringe the values go up to 0.85 (2-Gr). There exist high amounts of freshly precipitated, poorly crystallized Fe-oxyhydroxides that can easily be dissolved.

Figure 27.3 shows that total Fe and As have similar distributions with depth, and are well correlated ($r^2 = 0.661$). Arsenic concentrations in the profiles not only follow the distribution of total-Fe but are also well correlated to iron oxides ($r^2 = 0.683$, Fig. 27.4). Two samples (1-Ah and 1-AGo) are not shown in Figure 27.4 because their high concentrations would distort the least squares calculations. At low As concentrations, the correlation is weak, and may be influenced by chemical processes or the result of different timing and concentrations in the waters which transported the ions through the soils. There is no such distinct correlation of As with C_{org} (Fig. 27.4) which seems to cluster into two groups.

27.3.4 *Mobility of arsenic in the soils*

Water-soluble As contents are shown in Figure 27.5 as loads and as percentages of the total concentrations. The highest loads, up to 11 µg/g, occur in the horizons with highest total

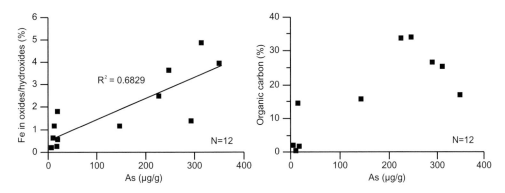

Figure 27.4.   Scatter plots of As and the Fe released from iron hydroxides and oxides (left). The right
diagram shows As and the total organic carbon content.

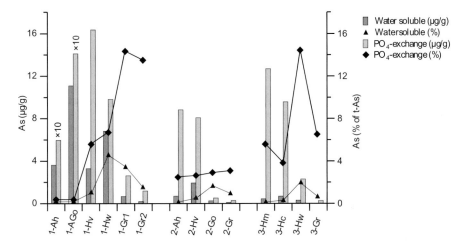

Figure 27.5.   Loads and percentages of soluble and exchangeable arsenic. The columns of PO4-exchangeable
arsenic of the horizons 1-Ah and 1-AGo have to be multiplied by ten to give correct values.

concentrations, and are lowest in the gleyic horizons. The highest phosphate-exchangeable As
contents was 140 µg/g in the ferric horizon 1-Ago, and 59.6 µg/g in 1-Ah. Both are extremely
high compared to the other samples and related to the very high total concentrations in both samples. Due to the high concentration of phosphate in the experiment these values can be regarded as
a maximum of exchangeable As. Phosphate exchanges only 0.3–2 µg/g As in the gleyic horizons.
It seems the loads of exchangeable As decrease with depth in the humic horizons but the numbers
of samples is too small to draw a firm conclusion.

Despite the high loads in the upper horizons, the percentage of As which is water soluble or
exchangeable is small. Only 0.2–0.6% is water soluble in the upper two horizons of all three
profiles, and exchangeable As concentrations are around 0.3–5.6%. The fraction of water-soluble
and exchangeable As has a maximum near to the contact of humic and redoxomorphic horizons
with up to 4.6% soluble and 14.3% exchangeable (Fig. 27.5).

27.4   CONCLUSIONS

Concentrations of As up to 4000 µg/g in gleysols and histosols north of the city of Munich
were unexpected, and caused public concern. Although the geogenic origin of As is assured, the

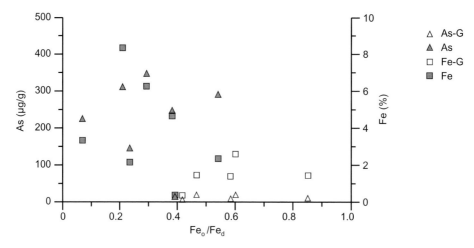

Figure 27.6.   Total concentrations of As and Fe related to the ratio of less crystallized iron oxides (Fe$_o$) to the total amount of iron oxides (Fe$_d$). The "G" marks the 5 samples from the gleyic horizons of the profiles.

previous assumptions about the causes of high As, due to adsorption on humic substances and later enrichment due to the cultivation and mineralization of the organic matter, have been rejected. Analyses from three profiles show that As is associated with iron oxides and hydroxides. Their distribution with depth is generally, but not always, similar to that of the C$_{org}$, which masks the relation of As to the Fe minerals. The original source of As is probably the Badenian sandstone aquifer. This aquifer and the Quaternary sediments covered by the investigated soils are connected by hydraulic windows (e.g., Bayer and Henken-Mellies 1998) which allow the transport of As into the upper aquifer and enrichment of As during pedogenesis.

The mobile fraction of As is very small, less than 1% of the total concentrations in most samples, especially the uppermost horizons with the highest total concentrations. A few have values up to 4.6%. This is consistent with reports of low concentrations in many groundwater samples in the area, and the few spots of increased As concentrations in grass and crops (Bavarian Geological Survey 2003).

The concentrations of As in the lower gleyic horizons are low, with a maximum of 18 µg/g, but water-soluble As is highest in these horizons, and especially at the transition to the upper organic layers. A high percentage (up to 14.3%) of exchangeable As indicates that As is adsorbed onto accessible surfaces. Figure 27.6 compares the concentrations of As and Fe in the upper horizons and the gleyic ones to the fraction of less crystallized iron oxides, i.e., high Fe$_o$/Fe$_d$ ratios. The high ratios indicate ongoing processes of oxide dissolution and precipitation related to changes in the water table, and hence shifting of oxic and suboxic environments.

There is a risk that the pedogenic sink of As will dissolve during changes of land use, especially artificial raising the water table to restore bogs and wetlands. The low concentrations of As in the gleyic horizons of the profiles and the high amount of less crystallized Fe hydroxides indicate the potential for such remobilization.

ACKNOWLEDGEMENT

We acknowledge the support of the Bavarian Geological Survey (now a division of the Bavarian Environmental Agency), especially Dr. W. Martin for his cooperation in the soil sampling, and the former president of the Survey, Prof. Dr. H. Schmid, for interest in the work and fruitful discussions about the data and interpretations. We thank the staff of the hydrochemical laboratory of the Department of Earth and Environmental Sciences of the Ludwig-Maximlians-University for the many samples analyzed.

REFERENCES

Bavarian Geological Survey (ed): Bericht über Untersuchungen auf Arsengehalte in Böden des Dachauer, Freisinger und Erdinger Mooses. *Report 415.3.2-53-1793*, Bavarian Geological Survey, Munich, Germany, 2003.

Bayer, M. and Henken-Mellies, W.-U.: Untersuchungen arsenführender Sedimente in der Oberen Süßwasser-molasse Bayerns. In: U. Kleeberger, H. Frisch and G. Heinrichs (eds): *Arsen im Grund- und Trinkwasser Bayerns*. Sven von Loga, Köln, Germany, 1998, pp. 43–66.

Bundschuh, J., Bhattacharya, P. and Chandrasekharam, D. (eds): *Natural arsenic in groundwater.* Balkema, Leiden, The Netherlands, 2005.

Driehaus, W.: Technologies for arsenic removal from potable water. In: J. Bundschuh, P. Bhattacharya and D. Chandrasekharam (eds): *Natural arsenic in groundwater*. Balkema, Leiden, The Netherlands, 2005, pp. 189–203.

Fetzer, K.D., Grottenthaler, W., Hofmann, B., Jerz, H., Rückert, G., Schmidt, F. and Wittmann, O.: *Stan-dortkundliche Bodenkarte von Bayern 1:50,000 München, Augsburg und Umgebung.* Bavarian Geological Survey, Munich, Germany, 1986.

Fischer, W.R. and Fechter, H.: Analytische Bestimmung und Fraktionierung von Cu, Zn, Pb, Cd, Ni und Co in Böden und Unterwasserböden. *Zeitschrift für Pflanzenernährung und Bodenkunde* 145 (1982), pp. 151–160.

Freudenberger, W. and Schwerd, K.: *Erläuterungen zur Geologischen Karte von Bayern 1:500,000.* Bavarian Geological Survey, Munich, Germany, 1996.

Göd, R.: Geogene Arsengehalte außergewöhnlichen Ausmaßes in Böden, nördliche Saualpe—ein Beitrag zur Diskussion um Grenzwerte von Spurenelementen in Böden. *Berg- und Hüttenmännische Monatshefte* 139 (1994), pp. 442–449.

Heinrichs, G.: Geogene Arsenkonzentrationen in Keupergrundwässern Frankens/Bayern. *Hydrogeologie und Umwelt* 12 (1996), pp. 1–193.

Inskeep, W.P., McDermott, T.R. and Fendorf, S.: Arsenic (V)/(III) cycling in soils and natural waters: chemical and microbiological processes. In: W.T. Frankenberger Jr. (ed): *Environmental chemistry of arsenic*. Marcel Dekker, New York, 2002, pp. 183–215.

Jekel, M.R.: Removal of arsenic in drinking water treatment. In J.O. Nriagu (ed): *Arsenic in the environment, Part I: Cycling and characterization*. John Wiley, New York, 1994, pp. 119–132.

Matschullat, J.: Arsenic in the geosphere—a review. *Sci. Total Environ.* 249 (2000), pp. 297–312.

Rüde, T.R.: Beiträge zur Geochemie des Arsens. *Karlsruher Geochemische Hefte* 10 (1996), pp. 1–206.

Schwertmann, U.: Differenzierung der Eisenoxide des Bodens durch Extraktion mit Ammoniumoxalatlösung. *Zeitschrift für Pflanzenernährung und Bodenkunde* 105 (1964), pp. 194–202.

Smedley, P.L. and Kinniburgh, D.G.: A review of the source, behaviour and distribution of arsenic in natural waters. *Appl. Geochem.* 17 (2002), pp. 517–568.

Thanabalasingam, P. and Pickering, W.F.: Arsenic sorption by humic acids. *Environ. Pollut.* B12 (1986), pp. 233–246.

Wagner, B., Töpfner, C., Lischeid, G., Scholz, M., Klinger, R. and Klaas, P.: Hydrogeochemische Hintergrundwerte der Grundwässer Bayerns. *GLA Fachberichte* 21 (2003), pp. 1–250.

Woolson, E.A.: Emissions, cycling and effects of arsenic in soil ecosystems. In: B.A. Fowler (ed): *Biological and environmental effects of arsenic*. Elsevier, Amsterdam, 1983, pp. 51–139.

Zeien, H. and Brümmer, G.W.: Chemische Extraktion zur Bestimmung von Schwermetallbindungsformen in Böden. *Mitteilungen der Deutschen Bodenkundlichen Gesellschaft* 59 (1989), pp. 505–510.

CHAPTER 28

Sorption and desorption behavior of arsenic in the soil

S. Tokunaga & M.G.M. Alam
National Institute of Advanced Industrial Science and Technology, Tsukuba, Ibaraki, Japan

ABSTRACT: To better understand the difference in the behavior of As(III) and As(V) ions in soil, comparative studies were conducted, on a laboratory scale, on the sorption and desorption of As(III) and As(V) ions in Kuroboku soil. The percentage sorption was measured on 1.60 mg/l As(III) and 14.9 mg/l As(V) solutions over the pH range 2 to 11. The sorption of As(III) ion was less pH-dependent and a relatively high percentage sorption was attained in the pH range 8 to 10. The sorption of As(V) ion was highly pH-dependent, attaining the highest percentage sorption in the pH range 2 to 5. The sorption rate for As(V) ion at pH 4 was much higher than that of As(III) ion at pH 8. Desorption behavior was measured using model soils contaminated by either As(III) or As(V) ion. Rate of As desorption was determined by controlling pH at 4, 7, and 9. The As concentrations desorbed from the As(III)-contaminated soil were much higher than those from the As(V)-contaminated soil. The kinetic data for sorption and desorption were analyzed using eight different kinetic models, namely, zero-order, first-order, second-order, third-order, parabolic diffusion, two-constant rate, Elovich-type, and differential rate models, which were evaluated by least-square regression.

28.1   INTRODUCTION

Although the industrial use of arsenicals has rapidly declined in recent years due to their strong toxicity and carcinogenicity, an increasing number of arsenic (As) contaminated sites are still being reported from many countries (Rubin 1999, Sierra *et al.* 2000, Moore *et al.* 2000). A considerable number of people in various areas of the world have been afflicted with chronic poisoning by As derived from geological materials (Wang *et al.* 1997, Rahman *et al.* 1999, Karim 2000). Arsenic bears +III and +V valences in soil environments. In general, As(III) species are predominant under anaerobic conditions, while As(V) species predominate under aerobic conditions. Both As(III) and As(V) ions are subject to sorption by soil constituents such as hydrous iron(III) oxides and aluminum oxides (Jacobs *et al.* 1970). The structures of As(III) and As(V) complexes on the surface of goethite (α-FeOOH) have been elucidated by Fourier transform infrared spectroscopy (Sun and Doner 1996) and extended X-ray absorption fine structure spectroscopy (Fendorf *et al.* 1997, Manning *et al.* 1998). The difference of valence significantly affects the mobility of As in soils and its toxicity. In order to better understand the problems of soil contamination by As and to develop an effective remediation technology, differences in the sorption and desorption behavior of As(III) and As(V) ions in Kuroboku soil were studied on a laboratory scale. The pH-dependence of As sorption was analyzed over the pH range 2 to 11. The rates of As sorption were measured at optimum pH. A kinetic study was conducted on the sorption of As(III) and As(V) ions by the soil at optimum pH. Similarly, another kinetic study was conducted on the desorption of As from an As(III)- or As(V)-contaminated soil by controlling pH at 4, 7, and 9.

28.2   EXPERIMENTS

28.2.1   *Kuroboku soil*

Kuroboku soil (andosol) was collected from horizon A of a forest land in Ibaraki, Japan. Kuroboku soil is one of the typical soils in Japan which originates from volcanic ash. The soil was sieved through a 2 mm opening sieve, and the fine fraction was air-dried for one week and stored in a plastic air-tight container. The chemical and physical characteristics of the soil are shown in Table 28.1. The pH of the soil was measured by equilibrating it in deionized water at a soil-to-water ratio of 1:5. The chemical composition was determined by X-ray fluorescence spectroscopy (SXF-1200, Shimadzu Corp.) using geological reference materials (Imai *et al.* 1996) for calibration. The content of organic carbon was determined by the dry combustion method (Nelson and Sommers 1996) using a TOC analyzer (TOC-5000, Shimadzu Corp.). The cation exchange capacity (CEC) was measured with ammonium acetate solution. The surface area was measured by the BET method (Flow Sorb II 2300, Micromeritics). The particle size distribution was measured using standard sieves.

28.2.2   *Sorption of As(III) and As(V) by Kuroboku soil*

A series of 25 ml of 1.60 mg/l As(III) or 14.9 mg/l As(V) solutions were prepared in 40 ml polycarbonate centrifuge tubes. The pH was adjusted to values ranging from 2 to 11 either with 0.5 M HCl or 0.5 M NaOH solution. Approximately 0.25 g of Kuroboku soil were added and the suspension was shaken for 16 h in a 20°C thermostat. The suspension was centrifuged at 9800 G for 20 min using a refrigerated centrifuge, and the supernatant was further filtered through a 0.45 μm membrane filter. The filtrate was analyzed for pH and residual As concentration. Arsenic was determined by graphite furnace atomic absorption spectrometry (Z-5710, Hitachi Ltd.). To prepare As(III) and As(V) solutions, $NaAsO_2$ and $Na_2HAsO_4 \cdot 7H_2O$, respectively, were dissolved in deionized water. Nitrogen was bubbled in the As solutions to remove dissolved oxygen.

Separately from the above experiment, 500 ml of 1.50 mg/l As(III) (pH 8) or 15.0 mg/l As(V) (pH 4) solution was prepared in a 1 l beaker and approximately 5 g of the soil were added. Nitrogen was bubbled in the suspension. The suspension was continuously agitated with a magnetic stirrer in a 20°C thermostat and the pH of the suspension was kept constant by adding either 0.5 M HCl or 0.5 M NaOH solution. At arbitrary time intervals, an aliquot was taken with a plastic syringe and filtered through a membrane filter. The filtrate was analyzed for As(III) and As(V) by ICP-MS (7500 Series, Agilent Technologies) connected with HPLC (1100 Series, Agilent Technologies). At the time of sampling, oxidation reduction potential (ORP) of the suspension was measured.

28.2.3   *Desorption of As from contaminated Kuroboku soil*

Approximately 1.5 kg of Kuroboku soil was intermittently shaken for 3 weeks with 3 l of As(III) or As(V) solution. Meanwhile the pH of the suspension was kept at approximately 8 or 4 for As(III)

Table 28.1.   The chemical and physical characteristics of Kuroboku soil.

Soil type	Kuroboku soil	CEC (meq/100 g)	5.55
Order	andosol	Surface area (m^2/g)	59.6
pH-water	5.24	Particle size distribution (%)	
Water (%)	20.2	2000–850 μm	44.3
Chemical composition (%)		850–500 μm	15.8
Fe_2O_3	14.1	500–250 μm	17.8
Al_2O_3	25.6	250–125 μm	12.6
SiO_2	40.4	125–63 μm	9.1
Organic carbon	5.1	<63 μm	0.4

or As(V), respectively, by adding a 0.5 M HCl solution. Thereafter, the resultant suspension was filtered with a No. 2 filter paper and the resultant soil was rinsed with deionized water, which was repeated three times. The recovered soil was dried under vacuum at room temperature and stored in an air-tight container. The resultant As-contaminated soils were used as the model samples for the desorption study. Their As contents were determined by ICP after digesting the soils with HNO_3 and H_2O_2 (US Environmental Protection Agency 1986) and found to be 1622 mg/kg As(III) or 3190 mg/kg As(V) on dry-weight basis.

A 500 ml of water was placed in a beaker, of which pH was adjusted to 4, 7, or 9, and approximately 5 g of the model contaminated soil were added. The suspension was continuously agitated with a magnetic stirrer in a 20°C thermostat. At arbitrary time intervals, an aliquot was taken and the above procedure was followed. Also for this experiment, nitrogen was continuously bubbled in the suspension to keep oxygen-free conditions and the pH was kept constant.

28.3 RESULTS AND DISCUSSION

28.3.1 *Sorption of As(III) and As(V) by Kuroboku soil*

The soil sample used in this study, Kuroboku soil, can be characterized by weakly acidic pH, relatively rich iron and aluminum contents, and sandy particle size distribution. The pH-dependence of the sorption of As(III) and As(V) ions by Kuroboku soil is shown in Figure 28.1. The degree of sorption is expressed in percentage of the amount of As sorbed by the soil to the initial amount of As in the solution. The As(III) sorption was incomplete despite the low initial As(III) concentration, 1.60 mg/l. The percentage sorption of As(III) ion was highest in the pH range 8 to 10 where approximately 75% of As(III) ion was sorbed. The pH-dependence of As(III) sorption was relatively small, 60 to 70% sorption being attained in the neutral to weakly acidic pH range.

On the other hand, the sorption of As(V) ion was highly pH-dependent, where the initial As(V) concentration was 14.9 mg/l which was approximately ten times higher than that of As(III). Almost 100% sorption was attained in the pH range 2 to 5, while, in the neutral to alkaline pH range, the percentage sorption significantly decreased. Thus it can be said that the optimum sorption pH range for As(V) ion is 2 to 5, where As(V) ion is subject to sorption by Kuroboku soil. Such pH-dependence of the sorption of As(III) and As(V) ions by Kuroboku soil was similar to those of other soils (Tokunaga 2006).

A kinetic study was conducted on the sorption of As(III) and As(V) ions by controlling pH at 8 and 4, respectively, where the highest percentage sorption was obtained, and the results are

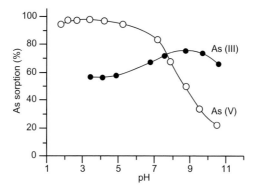

Figure 28.1.   pH-dependence of sorption of As(III) and As(V) ions by Kuroboku soil. Initial concentration: 1.60 mg/l As(III); 14.9 mg/l As(V).

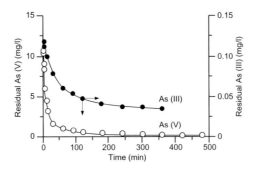

Figure 28.2.   Rate of sorption of As(III) and As(V) ions by Kuroboku soil. Initial concentration: 1.50 mg/l As(III); 15.0 mg/l As(V).

Table 28.2.   Analysis of the kinetic data of the sorption and desorption of As in Kuroboku soil.

Reaction	pH	Eh[1] (mV)	Kinetic model	Constants	r^2
Sorption of As(III) by soil	8	388–334	Elovich-type	$\alpha = 0.251$, $\beta = 5.16$	0.977
Sorption of As(V) by soil	4	509–631	Second-order	$b = 0.00902$	0.995
Desorption from As(III)-contaminated soil	4	536–608	Elovich-type	$\alpha = 0.951$, $\beta = 2.23$	0.997
Desorption from As(III)-contaminated soil	7	385–383	Elovich-type	$\alpha = 0.437$, $\beta = 2.84$	0.977
Desorption from As(III)-contaminated soil	9	341–293	Two-constant	$a = -0.732$, $b = 0.270$	0.970
Desorption from As(V)-contaminated soil	4	540–624	None		
Desorption from As(V)-contaminated soil	7	429–421	Two-constant	$a = -4.96$, $b = 0.246$	0.996
Desorption from As(V)-contaminated soil	9	370–282	Two-constant	$a = -5.74$, $b = 1.07$	0.991

[1] Initial Eh–Final Eh.

shown in Figure 28.2. Only As(III) or As(V) species was detected in the As(III) or As(V) solution, respectively, indicating that no oxidation-reduction of As took place while As ion was contacted with Kuroboku soil.

Figure 28.2 also shows that the sorption of As(III) ion was incomplete, the residual As(III) concentration being at 0.03 mg/l or higher even after 360 min. The As(V) sorption rate was much higher than that of As(III), the residual As(V) concentrations reaching almost zero in 300 min. These kinetic data were analyzed using eight different models, namely, zero-order, first-order, second-order, third-order, parabolic diffusion, two-constant rate, Elovich-type, and differential rate models (Chien and Clayton 1980, Onken and Matheson 1982), which were evaluated by least-square regression. The results are summarized in Table 28.2 in which change in ORP during sorption process is also given. The kinetic data for As(III) at pH 8 and those for As(V) at pH 4 were found to be best described by the following Elovich-type (eq. 28.1) and second-order kinetic models (eq. 28.2), respectively, in terms of coefficient of correlation:

$$C_0 - C_t = (1/\beta) \ln(\alpha\beta) + (1/\beta) \ln t \qquad (28.1)$$

$$1/C_t = 1/C_0 + bt \qquad (28.2)$$

where C_0 = initial concentration of As in solution at time zero (mg/l), C_t = concentration of As in solution at any time t (mg/l), α and β: constants (mg/l and min $\times$ l/mg, respectively), t: reaction time (min), and b: constant (l/mg/min).

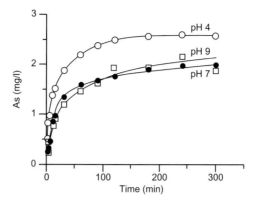

Figure 28.3.   Rate of As desorption from As(III)-contaminated soil at pH 4, 7, and 9.

28.3.2   *Desorption of As from contaminated Kuroboku soil*

In most countries, As-containing soils are regulated by As concentration in the leachate obtained from the leaching test, and the guideline value is usually set at as low as 0.01 mg/l (Ministry of Environment 2000). In order to elucidate the leaching risk of soils contaminated by As(III) or As(V) ion, rates of As desorption were measured on the model contaminated Kuroboku soil on which either As(III) or As(V) ion was artificially loaded. The desorption rates were determined by maintaining pH at 4, 7, or 9 to elucidate the effects of pH change in the soil environment on the As desorption characteristics.

The rates of As desorption from the As(III)-contaminated soil are shown in Figure 28.3, where As(III) concentrations in the solution are plotted for the runs at pH 4 and 7, while, concentrations of total As [As(III) + As(V)] are plotted for pH 9. There was no significant difference in the As desorption rates among the pH values, indicating small pH-dependent desorption characteristics. The concentrations of desorbed As in the solution were significantly higher than those from the As(V)-contaminated soil which will be discussed later, indicating that As(III)-contaminated soil is more mobile and, therefore, has higher leaching risk than As(V)-contaminated soil. Consideration of the above-mentioned guideline value for As in soils indicates that such As(III)-contaminated soils can hardly satisfy the environmental regulation. From a remediation point of view, Figure 28.3 shows that washing with aqueous solutions is effective in removing As from As(III)-contaminated soil. In the resultant solutions from the runs at pH 4 and 7, only the As(III) species was detected. On the other hand, As desorption at pH 9 was complicated, not only As(III) species but also As(V) species were detected. The amounts of As(V) was 14 to 16% of As(III) in the first 5 min, but no As(V) was detected during the time period of 10 to 60 min. Arsenic (V) was again detected from after 90 min and its concentration reached 49.0% of As(III) concentration at 240 min. These facts indicate the occurrence of oxidation of As(III) to As(V) by dissolved soil constituents and subsequent sorption of As(V) on to soil. Similarly to the kinetics of sorption, the kinetic data of desorption from the As(III)-contaminated soil were analyzed using the above eight kinetic models and the results are given in Table 28.2 together with ORP change. The As desorption rates at pH 4 and 7 were best described by the following Elovich-type model (eq. 28.3), while those of pH 9 by two-constant kinetic model (eq. 28.4).

$$C_t = (1/\beta) \ln(\alpha\beta) + (1/\beta) \ln t \tag{28.3}$$

$$\ln C_t = a + b \ln t \tag{28.4}$$

where a and b = constants (mg/l and mg/l/min, respectively). The denotation for other symbols are the same as those of (28.1) and (28.2).

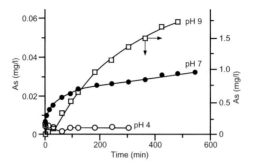

Figure 28.4.   Rate of As desorption from As(V)-contaminated soil at pH 4, 7, and 9.

The results of the kinetic study on the As(V)-contaminated soil are shown in Figure 28.4. Only the As(V) species was detected at every pH and thus As(V) concentrations are plotted in Figure 28.4. The rate of As desorption from the As(V)-contaminated soil was significantly affected by pH. At pH 4, the desorption of As from the soil was insignificant. The highest As concentration, as low as 0.006 mg/l, was attained after 5 min, and then the concentrations were lower than 0.005 mg/l, indicating an occurrence of resorption of As in the solution. Thus, As(V) in the soil was very stable against desorption at pH 4 and only trace amounts of As desorbed. But when placed under anaerobic condition, As(V) in soils is subject or reduction to As(III), raising the leaching risk. At pH 7, the concentrations of desorbed As increased with the passage of time, attaining 0.03 mg/l or higher after 420 min. At pH 9, the desorption of As was much more significant than those at pH 4 and 7, the As concentrations exceeding 1 mg/l after 240 min. Such significant As desorption at pH 9 supports the effectiveness of the alkaline washing method for remediation of As(V)-contaminated soils (Legiec 1997). The desorption of As from the As(V)-contaminated soil increased with increasing pH. These kinetic data were analyzed by the eight models and the results are shown in Table 28.2. None of the eight models best described the As desorption at pH 4. The rates of As desorption at both pH 7 and 9 were best described by two-constant rate model (28.4).

28.4   CONCLUSIONS

Sorption of As(III) ion by Kuroboku soil was less pH-dependent and incomplete even at 1.60 mg/l of an initial As(III) concentration. The percentage As(III) sorption was highest in the pH range 8 to 10. On the other hand, sorption of As(V) ion was highly pH-dependent, attaining almost 100% sorption in the pH range 2 to 5 at 14.9 mg/l of an initial As(V) concentration. The percentage As(V) sorption significantly decreased in the neutral to alkaline pH range. These facts show that As(V) ion is much more subject to sorption by Kuroboku soil than As(III) ion particularly in the weak acidic pH range. The kinetic data for As(III) sorption at pH 8 and those for As(V) at pH 4 were found to be best described by Elovich-type and second-order kinetic models, respectively.

Desorption behavior was determined at pH 4, 7, and 9 using model soils contaminated by either As(III) or As(V) ion. From the As(III)-contaminated soil, only As(III) species was detected at pH 4 and 7, while at pH 9, significant amount of As(V) was detected. From the As(V)-contaminated soil, only As(V) species was detected at every pH. For the As(III)-contaminated soil, there was no significant difference in the As desorption rates among the pH values. On the other hand, the desorption rate increased with increasing pH value for the As(V)-contaminated soil. The concentrations of As desorbed from the As(III)-contaminated soil were much higher than those from the As(V)-contaminated soil. These kinetic data for As desorption were best described by Elovich-type and two-constant kinetic models.

REFERENCES

Chien, S.H. and Clayton, W.R.: Application of Elovich equation to the kinetics of phosphate release and sorption in soils. *Soil Sci. Soc. Am. J.* 44 (1980), pp. 265–268.

Fendorf, S., Eick, M.J., Grossl, P. and Sparks, D.L.: Arsenate and chromate retention mechanisms on goethite. 1. Surface structure. *Environ. Sci. Technol.* 31:2 (1997), pp. 315–320.

Imai, N., Terashima, S., Itoh, S. and Ando, A.: 1996 compilation of analytical data on nine GSJ geochemical reference samples, "sedimentary rock series". *Geostandards Newsletter* 20:2 (1996), pp. 165–216.

Jacobs, L.W., Syers, J.K. and Keeney, D.R.: Arsenic sorption by soils. *Soil Sci. Soc. Am. Proc.* 34 (1970), pp. 750–754.

Karim, M.M.: Arsenic in groundwater and health problems in Bangladesh. *Water Res.* 34:1 (2000), pp. 304–310.

Legiec, I.A., Griffin, L.P., Walling, P.D. Jr., Breske, T.C., Angelo, M.S., Isaacson, R.S. and Lanza, M.B.: DuPont soil washing technology program and treatment of arsenic-contaminated soils. *Environ. Progress* 16:1 (1997), pp. 29–34.

Manning, B.A., Fendorf, S.E. and Goldberg, S.: Surface structures and stability of arsenic(III) on goethite: Spectroscopic evidence for inner-sphere complexes. *Environ. Sci. Technol.* 32:16 (1998), pp. 2383–2388.

Ministry of Environment, Japan: *Survey and countermeasure guidelines for soil and ground water contamination* (English version). Geo-Environmental Protection Center, Tokyo, 2000.

Moore, T.J., Rightmire, C.M. and Vempati, R.K.: Ferrous iron treatment of soils contaminated with arsenic-containing wood-preserving solution. *Soil Sediment Contam.* 9:4 (2000), pp. 375–405.

Nelson, D.W. and Sommers, L.E.: Total carbon, organic carbon, and organic matter. In: D.L. Sparks (ed): *Methods of soil analysis,* part 3: *chemical methods.* Soil Science Society of America, Inc. and American Society of Agronomy, Madison, Wisconsin, 1996, pp. 961–1010.

Onken, A.B. and Matheson, R.L.: Dissolution rate of EDTA-extractable phosphate from soils. *Soil Sci. Soc. Am. J.* 46 (1982), pp. 276–279.

Rahman, M., Tondel, M., Ahmad, S.A., Chowdhury, I.A., Faruquee, M.H. and Axelson, O.: Hypertension of arsenic exposure in Bangladesh. *Hypertension* 33:1 (1999), pp. 74–78.

Rubin, E.S.: Toxic releases from power plants. *Environ. Sci. Technol.* 33:18 (1999), pp. 3062–3067.

Sierra, J., Montserrat, G., Martí, E., Garau, M.A. and Cruañas, R.: Contamination levels in the soils affected by the flood from Aznalcóllar, Spain. *Soil Sediment Contam.* 9:4 (2000), pp. 311–329.

Sun, X. and Doner, H.E.: An investigation of arsenate and arsenite bonding structures on goethite by FTIR. *Soil Sci.* 16:12 (1996), pp. 865–872.

Tokunaga, S. and Hakuta, T.: Acid washing and stabilization of an artificial arsenic-contaminated soil. *Chemosphere* 46 (2002), pp. 31–38.

Tokunaga, S.: Removal of arsenic from water by adsorption process. In: R. Naidu, E. Smith, G., Owens, P. Bhattacharya and P. Nadebaum (eds): *Managing arsenic in the environment: from soil to human health.* Melbourne, Australia, 2006, pp. 393–397.

US Environmental Protection Agency: *Test methods for evaluating solid waste, physical/chemical methods,* SW-846/3050B Washington D.C., 1986.

Wang, G.Q., Huang, Y.Z., Xiao, B.Y., Qian, X.C., Yao, H., Hu, Y., Gu, Y.L., Zhang, C. and Liu, K.T.: Toxicity from water containing arsenic and fluoride in Xianjiang. *Fluoride* 30:2 (1997), pp. 81–84.

CHAPTER 29

Effect of wastewater irrigation on arsenic concentration in soils and selected crops in the state of Hidalgo, Mexico

C.A. Lucho-Constantino
Depto. de Biotecnología y Bioingeniería, Centro de Investigación y de Estudios Avanzados del Instituto Politécnico Nacional (CINVESTAV-IPN), Mexico City, Mexico
Depto. de Ingeniería en Biotecnología, Universidad Politécnica de Pachuca, Zempoala, Hgo., Mexico

H.M. Poggi-Varaldo
Depto. de Biotecnología y Bioingeniería, Centro de Investigación y de Estudios Avanzados del Instituto Politécnico Nacional (CINVESTAV-IPN), Mexico City, Mexico

L.M. Del Razo & M.E. Cebrian
Sección Externa de Toxicología, Centro de Investigación y de Estudios Avanzados del Instituto Politécnico Nacional (CINVESTAV-IPN), Mexico City, Mexico

I. Sastre-Conde
Conselleria d'Agricultura i Pesca, Gov. of Islas Baleares, Palma de Mallorca, Islas Baleares, Spain

R.I. Beltrán-Hernández & F.R. Prieto-García
Centro de Investigaciones Químicas (CIQ), Universidad Autónoma del Estado de Hidalgo (UAEH), Pachuca, Hidalgo, Mexico

ABSTRACT: The accumulation and distribution of arsenic (As), selected heavy metals and several major elements in 31 agricultural soils of the Irrigation District 03 (DR03) in the state of Hidalgo, Mexico, irrigated with raw wastewater for up to 90 years, were evaluated. Also, two control soils (rain-fed) were analyzed. Samples of topsoils (0–30 cm depth) were extracted using a modified Tessier method. Arsenic and heavy metal concentrations were determined in selected crops harvested in soils of DR03. Total concentrations of Cd and Pb fell in the following ranges: 0.06–5.6 mg Cd/kg, 4.0–660 mg Pb/kg. Concentrations of As were below the detection limit of the method (0.03 mg/kg) in all soils. Mercury was detected in the following soil samples: 0.77 mg/kg in soil S3 (Ixmiquilpan1), 0.03 mg/kg in soil S4 (Ixmiquilpan2), 5.59 mg/kg in soil S11 (Rosario) and 3.15 mg/kg in soil S18 (San Juan Tepa). Total concentrations of Cd were generally below the maximum permissible levels set by the regulations of the European Union. However, total concentrations of Cd and Pb in several soils were in the middle to the high end of European Union maximum ranges. Percentages of Pb in the most mobile fractions (soluble and exchangeable) were in the range of 3 to 28%. These percentages translate into concentrations of soluble plus exchangeable Pb of 2 mg Pb/kg or higher in several soils, significantly higher than the Swiss tolerance limit of 1.0 mg Pb/kg for mobile fractions of Pb in soils. Concentrations of As in alfalfa (*Medicago sativa* L.) ranged between 0.003 and 0.09 mg/kg, whereas As concentrations in 'nopal' (young stem-segments of the prickly pear *Opuntia robusta*) grown in wastewater-irrigated soils were as high as 0.69 mg/kg. It seems that long term irrigation with wastewater of agricultural soils in Irrigation DR03 of Mexico does not pose an immediate risk from the standpoint of As accumulation in soils. However, high Pb concentrations in alfalfa and nopal are of concern. Although further monitoring and research are required to gain a more complete assessment, it appears that Pb accumulation would justify the implementation and enforcement of crop restriction regulations in some soils.

29.1   INTRODUCTION

The term 'trace elements' refers to elements present at low concentrations (in the order of mg/kg) in agroecosystems. Some elements, such as As, Hg, Cd, Pb, Cr and Ni, have toxic effects on living organisms and are often considered contaminants. Soil contamination with heavy metals and other toxic elements resulting from point sources often occurs in a limited area and is easy to identify (He *et al.* 2005). Over the last century, industrial and agricultural activities have contributed to increases in trace element concentrations in soils. Increased loading of metals in soils through irrigation varies markedly from location to location and could become more significant than contamination from point sources.

In countries with extended arid and semi-arid regions, wastewater is a valuable resource and its use in agriculture is practiced widely. For example, the amount of wastewater currently used in Israel for agricultural purposes amounts to 1200 million m^3/year, which corresponds to 162 m^3 per capita per year (Haruvy 1997). In Mexico, approximately 3132 million m^3/year of wastewater are used for agricultural irrigation (31.3 m^3/capita/year) (CNA 1998, Peasey *et al.* 2000).

In Mexico City and its metropolitan region, raw wastewater discharged at a rate of 60 m^3/s has been used to irrigate agricultural land in the Mezquital valley for nearly 100 years. The Mezquital valley is the largest area in the world irrigated with raw wastewater from a major urban area (Tortajada and Castelán 2003, British Geological Survey 1995, INEGI 1999). Wastewater is used to irrigate the 83,000 ha of the Irrigation Districts 03 and 100 (DR03 and DR100, respectively) (Siebe and Cifuentes 1995, Jiménez and Landa 1998).

The DR03 has a surface area ca. 45,000 ha with 500,000 inhabitants and 27,500 farmers. Its annual agricultural production in 1998 consisted of 3,104,434 metric tonnes (t) of alfalfa, 526,650 t of corn, 97,153 t of forage oat, 24,030 t of bean, 15,410 t of marrow, 14,688 t of green chili, 17,941 t of tomatoes (green plus red varieties), among others (INEGI 1999). Crops grown in this area are mainly consumed by the populations of Mexico City and Pachuca (capital of Hidalgo state) since they are the principal market outlets for the production. The annual agricultural production form this area is valued at 100 million US$.

In the Mezquital valley, wastewater is used as a source of irrigation water as well as a source of organic matter (soil conditioner and humus replenishment) and plant nutrients (such as nitrogen, phosphorus, and potassium), allowing the farmers to minimize costs associated with application of mineral fertilizers. Evaporation (approximately 2100 mm/year) greatly exceeds precipitation (approximately 450 mm/year) in the valley (British Survey 1995) and therefore, irrigation with wastewater has become an indispensable practice for sustaining the agricultural production since 1896 (Hernández Silva *et al.* 1994, British Geological Survey 1995). At the DR03, irrigation is primarily applied by flooding. Application rates range from 80 to 140 cm/year for most crops (British Geological Survey 1995) and up to 200 cm/year for alfalfa (*Medicago sativa* L.) (Siebe and Cifuentes 1995). Most of the irrigated water is raw wastewater consisting of a mixture of municipal sewage with a substantial contribution from industrial effluents. This is of concern because the wastewater contains several potentially toxic trace organic and inorganic substances, including heavy metals, boron (B), salts, etc., that may negatively affect the long-term productivity of the soil as well as animal and human health. (Bhari 1999).

According to a recent review by Hossain (2006), information on As content in foods, crops and beverages in Mexico is scanty or non-existent. Therefore, the purpose of this work was to evaluate the accumulation and distribution of As, B, selected heavy metals and several major elements in 31 agricultural soils of the DR03 in the state of Hidalgo, Mexico, irrigated with raw wastewater for up to 90 years. Also, two control soils (rain-fed) were analyzed. In addition, As and heavy metal concentrations were determined in selected crops harvested from soils of DR03.

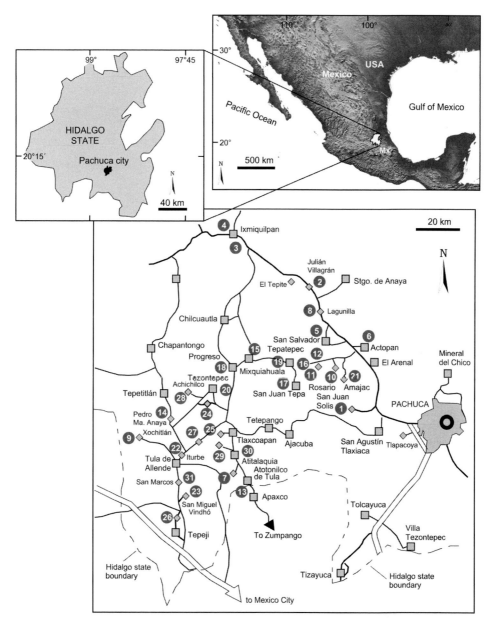

Figure 29.1.   Map showing locations S1–S31 at which soil and plant samples were collected in the DR03, Hidalgo.

29.2   METHODS

29.2.1   *Soil and plants analyses*

Soil samples were collected from the Ap horizon (topsoil to 30 cm depth). Samples were air dried and sieved (mesh 10 ASTM, <2 mm) before analysis. The metals in the soil were sequentially extracted following a slightly modified version of the method described elsewhere (Tessier *et al.* 1979, Lucho-Constantino *et al.* 2005). Plants grown on soils at selected sites (Fig. 29.1), were collected and washed in 3% HCl, to eliminate air-borne pollutants in the laboratory after sampling, and then dried in an oven at 65°C for 24 h.

Two different types of plants were assayed for trace-element uptake: alfalfa (*Medicago sativa* L.) and nopal (*Opuntia robusta* L.). Samples were ground in a stainless steel grinder. Ten milliliters of concentrated nitric acid were added to 0.25 g of a homogenized sample in a Teflon vessel and digested for 10 min in a microwave oven (Mars X). Soil and plant extracts were assayed for As and Hg using a Zeeman flame atomic absorption spectrophotometer equipped with a hydride generation unit for As and a cold vapor unit for Hg (Varian SpectrAA 800). Quality control assessment for total metals included the analysis of a standard reference of soil (SRM 2709). Recoveries ranged from 96.5 to 104% and coefficient of variation was between 3% and 10% for triplicate analyses.

29.3   RESULTS AND DISCUSSION

29.3.1   *Total concentration of arsenic and selected trace elements in soils of DR03, Hidalgo*

Arsenic concentrations in all soils assayed were below the method detection limit (0.03 mg/kg) (Table 29.1). It follows that concentrations of this metalloid were also below the maximum allowable concentrations (MAC) proposed in 1977 (50 mg/kg, Kabatas-Pendias 1995) (Table 29.1), although it is important to recognize that other countries such as Russia and the United Kingdom have other As MAC's of 2 and 10 mg/kg, respectively (Kabata-Pendias and Pendias 2001).

The low concentrations of As in soils observed in this study are consistent with the fact that the As concentration in the irrigation wastewater is very low (<0.0004 mg/l, Lucho *et al.* 2005a). It also suggests that other mechanisms of As pollution to soils, such as atmospheric deposition, fertilization, etc. (Adriano 1992), are not very significant. In addition, chemical conditions in the irrigation wastewater may be conducive to mobilization of As, which could explain why As concentrations were below those reported for uncontaminated soils. For example, our results are lower than ranges of As concentrations reported for European and Asiatic non-polluted soils (0.5 to 25 mg/kg) (Kabatas-Pendias 1992, Hossain 2005), and much lower than ranges reported for As-polluted soils (10 to 2500 mg/kg).

Concentrations of Pb and Cd found in soils were above detection in most samples. Cadmium concentrations were below maximum allowable concentrations for agricultural soils specified in regulations of the European Union countries and the USA. In some cases (S5 and S6 with 1.23 and 1.89 mg/kg, respectively), however, total Cd concentrations were in the middle of the maximum range allowed by the European Community (1 to 3 mg/kg). All soils except one had Pb concentrations below 600 mg/kg. However, a significant percentage of Pb was found in the easily soluble and exchangeable fractions of several soils. It was found that mobile Pb concentrations in several soils of zone 1 were greater than that allowed by Swiss standards for mobile heavy metal contents (Lucho-Constantino *et al.* 2005b).

29.3.2   *Total concentration of As and selected trace elements in crops of DR03, Hidalgo*

Arsenic concentrations in tissues of *Medicago sativa*, (Table 29.2) were below 0.003 mg/kg for plants from soils S3 and S10. However, alfalfa from soils S8 and S31 had higher concentrations of As: 0.06–0.09 mg/kg in stems and 0.02–0.04 mg/kg in leaves. These values fall in the normal range for As concentrations in plant tissues, although they are higher than those observed for green plants in UK (0.003–0.007 mg/kg). A recent compilation reports that As concentration in plants grown

Table 29.1.   Total concentrations of arsenic, mercury, lead and cadmium in soils of DR03, Hidalgo, Mexico.

Soil	Irrigation time (years)	t-As	t-Hg	t-Pb	t-Cd
		(mg/kg dry soil)			
S1 San Juan Sólis (rainfed)	0	<0.03	<0.01	29.87/2.03[1]	0.87/0.08[1]
S2 Julian Villagrán	6	<0.03	<0.01	3.99/0.18[1]	0.51/0.06[1]
S3 Ixmiquilpan	25	<0.03	0.77	16.48/0.66[1]	0.58/0.04[1]
S4 Ixmiquilpan	25	<0.03	0.03	19.46/0.86[1]	0.92/0.08[1]
S5 San Salvador	25	<0.03	<0.01	19.10/1.15[1]	0.96/0.05[1]
S6 Actopan	25	<0.03	<0.01	46.86/0.87[1]	1.23/0.02[1]
S7 Cardonal	25	<0.03	<0.01	44.08/0.00[1]	2.84/0.19[1]
S8 Lagunilla	32	<0.03	<0.01	14.08/0.35[1]	1.89/0.14[1]
S9 Xochitlán	37	<0.03	<0.01	35.80/0.42[1]	0.96/0.01[1]
S10 Rosario1	39	<0.03	5.59	47.48/0.53[1]	1.06/0.03[1]
S11 Rosario2	39	<0.03	<0.01	43.05/0.21[1]	2.22/0.34[1]
S12 Rosario3	39	<0.03	<0.01	24.03/0.68[1]	<0.06
S13 Atotonilco de Tula	40	<0.03	<0.01	37.06/0.00[1]	2.24/0.18[1]
S14 Pedro Ma. Anaya	40	<0.03	<0.01	32.25/0.00[1]	1.28/0.10[1]
S15 Progreso	42	<0.03	<0.01	27.07/0.18[1]	<0.06
S16 Represo	42	<0.03	<0.01	41.53/0.09[1]	2.10/0.33[1]
S17 San Juan Tepa	69	<0.03	3.15	41.39/0.24[1]	<0.06
S18 Mixquiahuala	69	<0.03	<0.01	659.83/72.65[1]	0.97/0.01[1]
S19 Tepatepec	69	<0.03	<0.01	22.55/3.22[1]	2.67/0.32[1]
S20 Tezontepec	76	<0.03	<0.01	60.54/5.69[1]	4.59/0.53[1]
S21 Amajac	76	<0.03	<0.01	20.17/2.12[1]	>0.06
S22 Iturbe	76	<0.03	<0.01	52.60/4.73[1]	1.65/0.20[1]
S23 San Miguel Vindhó	76	<0.03	<0.01	47.25/3.69[1]	1.21/0.10[1]
S24 Huitel	76	<0.03	<0.01	63.21/7.02[1]	1.81/0.22[1]
S25 Doxey	76	<0.03	<0.01	52.25/4.21[1]	5.59/0.60[1]
S26 Presa Requena	76	<0.03	<0.01	42.03/3.28[1]	1.25/0.15[1]
S27 Teocalco	76	<0.03	<0.01	48.90/4.78[1]	2.07/0.23[1]
S28 Achichilco	89	<0.03	<0.01	49.75/3.69[1]	1.91/0.12[1]
S29 Dendhó	89	<0.03	<0.01	34.75/3.47[1]	1.30/0.11[1]
S30 Atitalaquia	89	<0.03	<0.01	44.52/2.89[1]	1.80/0.15[1]
S31 San Marcos	89	<0.03	<0.01	24.83/1.74[1]	2.12/0.20[1]

Maximum allowable concentrations of trace elements in agricultural soils (mg/kg dry soil).
(adapted from Kabata-Pendias and Pendias 2001)

Counter/year					
USA[2]/1993		nr[4]		8	20
EC[3]/1986		nr[4]		1–1.5	1–3
Austria/1977		50		5	5

[1] Standard deviation; [2] USA: United States of America; [3] EC: European Community; [4] Data not reported.

in non-polluted soils ranges from 0.009 to 1.50 mg/kg dry weight (Kabata-Pendias and Pendias 2001). Alam *et al.* (2003) found that crops cultivated in Bangladesh and irrigated with groundwater contaminated with As (0.24 mg/l) had total As concentrations in the range 0.306–0.489 mg/kg, significantly higher than corresponding values in crops grown in non-polluted soils.

The concentration of As in *Opuntia robusta* grown in a control, rain-fed soil, was <0.003 mg/kg in pulp and peels, similar to As concentrations found in *Opuntia robusta* from soils S3 and S9, which had been irrigated with wastewater for 25 and 37 years, respectively (Table 29.2). Nopal plants from soils S6, S11 and S12 had As concentrations between 0.09–0.54 mg/kg in the pulp. Nopal grown in soils with irrigated with wastewater for 39 years showed As concentrations as high

Table 29.2.   Concentrations of trace elements in alfalfa (*Medicago sativa* L.) and nopal (*Opuntia robusta* L.) grown in agricultural soils of DR03, Hidalgo, Mexico.

Soil	Irrigation (years)	Cd	Pb	As	Hg
		(mg/kg dry weight)			
Medicago sativa L.					
S3 Ixmiquilpán	25 Stem	$<10^{-3}$	$10.27/0.42^1$	$<10^{-3}$	$1.08/4 \times 10^{-2^1}$
	Leaves	$<10^{-3}$	$9.98/0.69^1$	$<10^{-3}$	$0.76/2 \times 10^{-2^1}$
S8 Lagunilla	32 Stem	$<10^{-3}$	$10.57/0.86^1$	$6 \times 10^{-2}/2 \times 10^{-3^1}$	$0.45/1 \times 10^{-2^1}$
	Leaves	$<10^{-3}$	$3.68/0.11^1$	$4 \times 10^{-2}/3 \times 10^{-3^1}$	$0.43/1 \times 10^{-2^1}$
S10 Rosario1	39 Stem	$<10^{-3}$	$22.12/1.01^1$	$<10^{-3}$	$<8 \times 10^{-3}$
	Leaves	$<10^{-3}$	$46.46/2.88^1$	$<10^{-3}$	$<8 \times 10^{-3}$
S31 San Marcos	89 Stem	$<10^{-3}$	$29.17/0.12^1$	$9 \times 10^{-2}/1 \times 10^{-2^1}$	$1.22/9 \times <10^{-2^1}$
	Leaves	$<10^{-3}$	$8.93/0.23^1$	$2 \times 10^{-2}/1.8 \times 10^{-3^1}$	$1.92/3 \times 10^{-2^1}$
Average	46 Stem	$<10^{-3}$	18.03	4×10^{-2}	0.86
	Leaves	$<10^{-3}$	17.26	1.6×10^{-2}	0.78
Opuntia robusta					
S Cero Progreso	0 Pulp	$<10^{-3}$	$<4 \times 10^{-2}$	$<10^{-3}$	$<8 \times 10^{-3}$
	Peel	$<10^{-3}$	$<4 \times 10^{-2}$	$<10^{-3}$	$<8 \times 10^{-3}$
S3 Ixmiquilpán	25 Pulp	$<10^{-3}$	$6.57/1.01^1$	$<10^{-3}$	$<8 \times 10^{-3}$
	Peel	$<10^{-3}$	$7.17/0.49^1$	$<10^{-3}$	$0.14/1 \times 10^{-2^1}$
S6 Actopán	25 Pulp	10.94 $6 \times 10^{-2^1}$	$11.11/0.17^1$	$9 \times 10^{-2}/1 \times 10^{-2}$	$9 \times 10^{-2}/3 \times 10^{-3^1}$
	Peel	5.02 $9 \times 10^{-2^1}$	$22.43/0.38^1$	$<3 \times 10^{-3}$	$1.02/1 \times 10^{-2^1}$
S9 Xochitlán	37 Pulp	$<10^{-3}$	$18.39/1.15^1$	$<3 \times 10^{-3}$	$<8 \times 10^{-3}$
	Peel	$<10^{-3}$	$27.48/1.11^1$	$<3 \times 10^{-3}$	$0.14/1 \times 10^{-2^1}$
S11 Rosario2	39 Pulp	$<10^{-3}$	$14.09/0.51^1$	$0.54/3 \times 10^{-2^1}$	$<8 \times 10^{-3}$
	Peel	$<10^{-3}$	$14.90/0.41^1$	$0.69/4 \times 10^{-2^1}$	$<8 \times 10^{-3}$
S12 Rosario3	39 Pulp	$<10^{-3}$	$15.37/1.54^1$	$0.54/3 \times 10^{-2^1}$	$<8 \times 10^{-3}$
	Peel	$<10^{-3}$	$17.54/2.76^1$	$0.69/3 \times 10^{-2^1}$	$<8 \times 10^{-3}$
Average	27 Pulp	1.00	13.1	0.20	$1,1 \times 10^{-2}$
	Peel	2.18	17.91	1.3×10^{-2}	0.19
Typical concentration of trace elements in plants (mg/kg)					
Bohn *et al.* (1985)		0.02–0.08	0.1–10	nr[3]	nr[3]
Mengel and Kirkby (1982)[2]		0.1–7.6	0.5–5.3	nr[3]	nr[3]
Kabata-Pendias and Pendias (2001)		–	–	–	0.001–0.10

[1] Standard deviation; [2] Adapted from Assadian *et al.* (1998); [3] Data not reported.

as 0.69 mg/kg in the peel, considerably higher than concentrations reported for crops irrigated with groundwater contaminated with 0.24 mg/l As in Bangladesh (Alam *et al.* 2003). These values yield As in these nopal plants concentration factors higher than 20.

Lead concentrations in several plants harvested from the same sites as those used in this study were found to be at higher than the MAC allowed by Dutch and German standards for cereals, forage, and fresh vegetables (Lucho-Constantino *et al.* 2005b).

Results from DR03 suggest that long-term irrigation with wastewater does not pose an immediate risk from the standpoint of As accumulation in soils. However, As concentrations found in some tissues of nopal plants grown in soils in DR03 are similar to those found in crops in Bangladesh that were irrigated with groundwater with 0.25 mg/l As. High lead concentrations in alfalfa and nopal also could be of concern. Further monitoring and research are required to have a more complete assessment, although that the observed lead accumulation might justify the implementation and enforcement of crop restriction regulations in some soils.

ACKNOWLEDGEMENTS

The authors wish to thank CONACYT for funding the Research Project (SIZA-CONACYT No. 200008006030) and granting a graduate scholarship (165550) to one of the authors (CALC).

REFERENCES

Adriano, D.C.: Biochemistry of trace metals. Lewis Publisher, Boca Raton, FL, 1992.

Alam, M.G.M.. Snow, E.T. and Tanaka, A.: Arsenic and heavy metal contamination of vegetables grown in Samta village, Bangladesh. *Sci. Total Environ.* 308 (2003), pp. 83–96.

Bahri, A.: Agricultural reuse of wastewater and global water management. *Water Sci. Technol.* 40 (1999), pp. 339–346.

British Geological Survey and Comisión Nacional del Agua: Impact of wastewater reuse on groundwater in the Mezquital Valley, Hidalgo State, Mexico. Phase I report, 1995.

Comisión Nacional del Agua and Secretaria de Recursos Hidráulicos: Características del Distrito de riego 03 Tula y 100 Alfajayucan. 00024-RVM/5100/C65/00024. 9–11, Mexico D.F., Mexico, 1990.

Haruvy, N.: Agricultural reuse of wastewater: nation-wide cost-benefit analysis. *Agric. Ecosyst. Environ.* 66 (1997), pp. 113–119.

He, L.Z., Yang, E.X. and Stoffella, J.P.: Trace elements in agroecosystems and impacts on the environment. *J. Trace Elem. Med. Biol.* 19 (2005), pp. 125–140.

Hernández, S.G., Flores, D.L., Maples, V.M., Solorio, M.J.G. and Alcalá, M.J.R.: Riesgo de acumulación de Cd, Pb, Cr, y Co en tres series de suelos del DR03, Estado de Hidalgo, México. *Revista Mexicana de Ciencias Geológicas* 11 (1994), pp. 53–61, Editorial Instituto de Geología, UNAM, Mexico, D.F., Mexico.

Hossain, M.F.: Arsenic contamination in Bangladesh—An overview. *Agric. Ecosyst. Environ.* 113 (2006), pp. 1–16.

Instituto Nacional de Estadística Geografía e Informática (INEGI): Estadísticas económicas de volumen de producción agrícola y superficie cosechada por cultivos seleccionados según año agrícola y entidad federativa: 1–3, Mexico D.F., Mexico, 1999.

Jiménez, C.B. and Landa, V.H.: Physical-chemical and bacteriological characterization of wastewater from Mexico City. *Water Sci. Technol.* 37 (1998), pp. 1–8.

Kabata-Pendias, A.: Agricultural problems related to excessive trace metals contents of soil. In: W. Salomons, U. Förstner and P. Mader (eds): *Concerning heavy metals: Problems and solutions*. Springer Verlag, Berlin, 1995, pp. 19–31.

Kabata-Pendias, A. and Pendias, H. 2001. *Trace elements in soils and plants*. 3rd ed., CRC Press, Boca Raton, FL, USA.

Lucho-Constantino, C.A., Prieto-García, F., Del Razo L.M., Rodríguez, V.R. and Poggi-Varado, H.M.: Chemical fractionation of boron and heavy metals in soils irrigated with wastewater in central Mexico. *Agric. Ecosyst. Environ.* 108 (2005a), pp. 57–71.

Lucho-Constantino, C.A., Álvarez, S.M., Beltrán-Hernandez, R.I., Prieto-García, F. and Poggi-Varado, H.M.: A multivariate analysis of the accumulation and fractionation of selected cations and heavy metals in agricultural soils in Hidalgo State, Mexico irrigated with raw wastewater. *Environ. Internat.* 3 (2005b), pp. 313–323.

Peasey, A., Blumenthal, U., Mara, D. and Ruiz, P.G.: A review of policy and standards for wastewater reuse in agriculture: a Latin American perspective. WELL Report 68 Part II, London School of Hygiene and Tropical Medicine and Loughborough University, UK, 2000.

Siebe, C. and Cifuentes, E.: Environmental impact of wastewater irrigation in central Mexico: an overview. *Int. J. Environ. Health Res.* 5 (1995), pp. 161–173.

Tessier, A., Campbell, P.G.C. and Bisson, M.: Sequential extraction procedure for the speciation of particulate traces metals. *Anal. Chem.* 51 (1979), pp. 844–851.

Tortajada, C. and Castelán, E.: Water management for a megacity: Mexico City metropolitan area. *Ambio* 32 (2003), pp. 124–129.

CHAPTER 30

Arsenic determination in soils from a mining zone in the eastern Pyrenees, Catalonia (Spain)

M.J. Ruiz-Chancho, J.F. López-Sánchez & R. Rubio
Departament de Química Analítica, Universitat de Barcelona, Barcelona, Spain

ABSTRACT: Seven soil samples were collected in a contaminated mining zone in the eastern Pyrenees. After the samples had been air-dried and sieved to 2 mm and 90 μm, the total arsenic content was determined in aqua regia extracts by ICP-AES and HG-AFS. Arsenic species were extracted by a mixture of phosphoric and ascorbic acids under focused microwaves, and measured by the coupled HPLC-HG-AFS technique. Arsenic levels ranged from 51 to 38,000 mg/kg. Only inorganic arsenic [As(III) and As(V)] was detected and quantified.

30.1 INTRODUCTION

Arsenic concentrations in soils may be elevated either because of anthropogenic activity or because of a high natural abundance. The normal content of arsenic (As) in soils ranges from 1 to 40 mg/kg, with levels significantly higher in mining zones (Burguera *et al.* 1997).

Mining activity in the north of Spain was highly significant in the eighteenth and nineteenth centuries. The main industry was the extraction of metals like copper, lead and tin, but gold and silver were also mined. This activity led to high levels of metals in several locations. The presence of As in some of these sites makes their in-depth study important.

Arsenic in soils and sediments is mainly present as the inorganic forms (arsenate and arsenite), although the methylated compounds MA and DMA may also be present in lower amounts (Yehl *et al.* 2001). The identification and quantification of the As species is necessary to investigate the behavior of this element in As contaminated soils. Arsenic extraction and speciation in contaminated soils is a topic of current interest, the major difficulty being the extraction of As species without altering their chemical identity. For this reason, soft extractants and extraction procedures must be applied in order to avoid any modification of As species throughout the entire analytical procedure. Phosphoric acid was revealed as a good extractant due to chemical similarities of phosphor and arsenic.

Liquid chromatography coupled with hydride generation and detection by atomic absorption spectroscopy, atomic emission spectrometry and atomic fluorescence spectrometry have been shown as suitable techniques for measuring As species. The use of a derivatization step to reach good sensitivity is commonly used with these couplings (Garcia-Manyes *et al.* 2002).

Sample pre-treatments have a great importance in speciation analysis as a possible source of analytical changes when the pre-treatment procedure is aggressive. Some authors have studied the effect of temperature during drying and storing as a possible negative effect. A common procedure used as pre-treatment is the grinding of samples. No data for As species transformation during the soil grinding process is reported although an oxidation is observed for other metals such as iron.

30.2 SAMPLING AND SAMPLE PRE-TREATMENT

The Vall de Ribes is located in the eastern Pyrenees, Catalonia (Spain). The region was an active mining district at the beginning of nineteenth century where As and Sb veins with subordinate

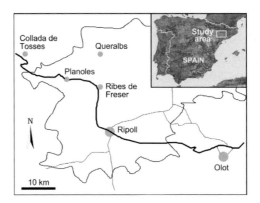

Figure 30.1.   Map of the Ripollés district in the eastern Pyrenees, Spain.

amounts of Cu, Pb and Ag were mined (Ayora *et al.* 1981, 1986). Although the presence of Au and Ag was also discovered at that time, the irregularity of the veins as well as the geographic location of the mines made them uneconomical and in a short time led to their closure.

Seven (S1 to S7) soil samples were collected in October 2004 after selecting three differentiated sampling locations. Figure 30.1 shows a map of the Ripollés district with the selected sampling points. Surface (0–15 cm depth) soil samples were collected around mine tailings. Samples S1 to S4 were taken near Ribes de Freser, a village in the Ripollés district, where small abandoned Sb and Zn mines are located. Soil samples S5 and S6 were taken around the mine spoil of an old As and Sb mine next Planoles, at the northwest of Ribes de Freser. Finally, sample S7 was collected in the tailing of an abandoned As mine near Queralbs, a village at the north of Ribes de Freser.

In a second sampling, in April 2005, two soil samples (B1 and B2) of a non-contaminated area were taken in order to obtain natural background As contents. The Collada de Tosses is located at around 18 km from Ribes de Freser and it is situated in a zone with the same geological characteristics.

30.2.1   *Pre-treatment*

Soil samples S1 to S7 were separately air-dried at room temperature and divided into two portions; one of these portions was sieved to 2 mm and the other one to 90 μm mesh. Sub-samples were stored in plastic containers at room temperature until analysis. Soil samples B1 and B2 were also air-dried as the same way and sieved to 2 mm mesh.

Portions of samples S1, S5 and S7 (2 mm) and samples representing the background level (B1 and B2) were ground with a tungsten carbide disc mill. B1 sample corresponds to a soil with similar characteristic of S1 to S6 and B2 to a soil with similar characteristics of soil sample S7.

30.3   SOIL SAMPLE CHARACTERIZATION

In order to evaluate the characteristics of the previously collected samples, the major components were determined by X-ray fluorescence. Table 30.1 shows the results obtained both for the soils from the contaminated zone and the soils from the non-contaminated site. Similar compositions and a relatively high percentage of SiO_2, Fe_2O_3 and Al_2O_3 can be observed in the soils. Major components were determined in the 90 μm fraction as well as in the ground 2 mm portion, yielding similar results except for Fe_2O_3 that gave higher values in the small particle size fraction.

The results from the study of the composition of B1 and B2 were in agreement with those corresponding to the contaminated samples, except for the Fe_2O_3 percentage. The lower values

Table 30.1.   Major components of soil samples obtained by X-ray fluorescence (n = 2).

	Fe$_2$O$_3$ (%)		CaO (%)		K$_2$O (%)		SiO$_2$ (%)		Al$_2$O$_3$ (%)	
	2 mm	90 μm	2 mm	90 μm	2 mm	90 μm	2 mm	90 μm	2 mm	90 μm
S1	5.69	5.25	0.32	0.35	3.12	3.64	73.32	70.78	13.51	16.25
S2	–	4.86	–	0.51	–	3.66	–	71.67	–	16.35
S3	–	5.90	–	0.61	–	4.72	–	66.83	–	18.49
S4	–	5.86	–	0.81	–	3.23	–	68.29	–	17.30
S5	8.07	8.39	0.34	0.46	3.95	3.99	61.05	60.11	19.74	21.02
S6	–	9.61	–	0.55	–	5.69	–	53.67	–	21.92
S7	12.88	18.34	1.00	1.95	4.00	4.00	55.74	42.19	16.28	18.24
B1	7.56	–	0.33	–	3.57	–	63.31	–	18.75	–
B2	6.09	–	0.31	–	3.44	–	65.02	–	17.79	–

Table 30.2.   pH values, nitrogen, organic carbon and sulfur percentages.

	pH	Nitrogen (%)	Organic carbon (%)	Total sulfur (%)
S1	4.60	0.34	4.75	0.03
S2	5.67	0.47	5.98	0.04
S3	4.80	0.73	10.9	0.08
S4	5.85	0.60	6.15	0.05
S5	5.70	0.36	3.60	0.04
S6	5.50	0.40	3.86	0.25
S7	7.79	0.17	1.25	0.29

of iron in B1 and B2 are attributable to the absence of pyrite and arsenopyrite in these samples. Among the contaminated samples, S5, S6 and S7 showed the highest percentage of Fe$_2$O$_3$.

The pH values in soil following the Spanish official analytical methods (water soil suspension) were determined in all samples, and the results are shown in Table 30.2. A pH range between 4.6 and 7.8 can be observed for these soils. Table 30.2 also shows the results obtained for elemental analysis. An elemental analyzer, equipped with a flash combustion furnace and a thermal conductivity detector, was used. Low values of organic carbon were obtained, the highest being the value reported for sample S3, which was located inside the forest. Low percentages of nitrogen and sulfur were obtained for all the samples analyzed.

30.4   PSEUDO-TOTAL ARSENIC CONTENT

The pseudo-total As content was determined after aqua regia extraction (ISO/CD 11466, 1995) followed by both ICP-AES (inductive coupled plasma atomic emission spectrometry) and HG-AFS (hydride generation-atomic fluorescence spectroscopy) measurement. The As content in the highest concentrated extracts (S1, S6 and S7) was measured by ICP-AES using external curve. For the quantification of the lower concentration samples, the more sensitive HG-AFS technique was used. In this case, a pre-reduction step must be applied to assure the total reduction to As(III) prior to measurement. The quantification was performed with an external calibration curve.

Table 30.3 shows the results obtained for the soil samples both in 2 mm and 90 μm fractions. The results showed high concentration of As for the samples collected near Queralbs and Planoles (S5, S6 and S7), indicating a high degree of As contamination in this area. Also observable are significant differences in As concentration among the most concentrated samples. For this reason,

Table 30.3.   Total arsenic content in aqua regia extracts (mean ± sd, n = 3).

	2 mm (mg/kg)	90 μm (mg/kg)	2 mm grinded (mg/kg)
S1[1]	53.1 ± 2.9	59.5 ± 0.7	51.3 ± 2.8
S2[1]	62.9 ± 2.1	66.3 ± 0.4	–
S3[1]	51.9 ± 4.5	50.7 ± 0.6	–
S4[1]	66.8 ± 1.0	80.1 ± 1.1	–
S5[2]	1886 ± 49	2539 ± 49	1888 ± 41
S6[2]	6114 ± 670	9072 ± 144	–
S7[2]	21174 ± 750	38214 ± 323	18477 ± 807
B1[1]	15.2 ± 0.7	–	14.3 ± 2.7
B2[1]	6.5 ± 0.5	–	7.7 ± 0.3

[1] Measurement with HG-AFS; [2] Measurement with ICP-AES.

and in order to check the efficiency of aqua regia extraction with respect to the particle size, some of the samples were ground with a tungsten carbide disc mill and digested in the same way. The results obtained are presented in Table 30.3. The values agreed with those obtained before demonstrating the efficiency of aqua regia extraction, irrespective of the sample particle size. The higher As concentration in 90 μm fraction samples S5, S6 and S7 with respect to the 2 mm fraction could be attributed to the association of As with small size particle compounds as iron oxides.

30.5   ARSENIC SPECIATION

For the As species determination, 0.1 g of both 90 μm and 2 mm ground soil subsamples were extracted with 15 ml of a mixture of 1 M phosphoric acid and 0.5 M ascorbic acid previously purged with an argon stream for 15 min under focused microwaves. The extract must be purged with an argon stream for five minutes in order to decrease the kinetics of As(III) oxidation. This extraction procedure was optimized in previous studies by our research group (Ruiz-Chancho et al. 2005) after observing an instability of As(III) species in the soil extracts (Garcia-Manyes et al. 2002).

The As species measurement in the extracts was carried out by the coupled technique LC-HG-AFS (liquid chromatography-hydride generation-atomic fluorescence spectroscopy) within 24 hours after the extraction.

The As species in the 90 μm fraction extracts are shown in Table 30.4, where only the presence of inorganic arsenic (i-As) can be observed in all samples. The table also shows the percentage of As(V) with respect to the total As content, and percentages higher than 95% were obtained in all samples. We also studied the proportion between As(III) and As(V) in the extracts, and in all cases except for the samples of Planoles (S5 and S6) we obtained the same proportion, which did not depend on the total As concentration.

Figure 30.2 shows an example of three chromatograms of soil extracts in which the presence of i-As and the major presence of As(V) is confirmed in all cases.

When the speciation analysis was performed on the extracts of 2 mm ground soil samples, in all cases only i-As was found, however a decrease in As(III) concentration was observed with respect to the extracts from 90 μm. This fact indicated the transformation of the original species in the soil due to the grinding procedure in spite of the short time of grinding (less than 10 seconds). Figure 30.3 shows, as an example, a comparison between the species in phosphoric extracts before and after the sample grinding. For both samples the same behavior is observed, with a significant decrease of As(III) concentration.

Table 30.4.   Arsenic species content in phosphoric/ascorbic soil extracts (mean ± sd, n = 3).

	As(III) (mg/kg)	As(V) (mg/kg)	As(V)/As(III)	As(V)(%)
S1	1.87 ± 0.62	56.0 ± 4.3	30.0	96.6
S2	3.29 ± 0.90	69.1 ± 5.6	21.0	95.5
S3	2.09 ± 0.19	55.5 ± 8.9	26.6	96.4
S4	3.85 ± 1.45	93.5 ± 5.2	24.2	96.0
S5	27.7 ± 9.8	2808 ± 85	101.3	99.0
S6	53.7 ± 6.0	9300 ± 214	173.3	99.4
S7	1201 ± 321	32852 ± 6033	27.3	96.5

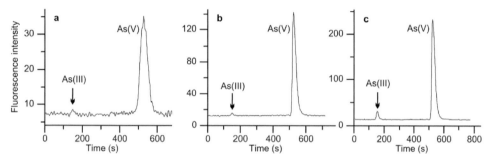

Figure 30.2.   Example chromatograms of arsenic species in soil extracts (a) S1 soil sample, (b) S5 soil sample and (c) S7 soil sample.

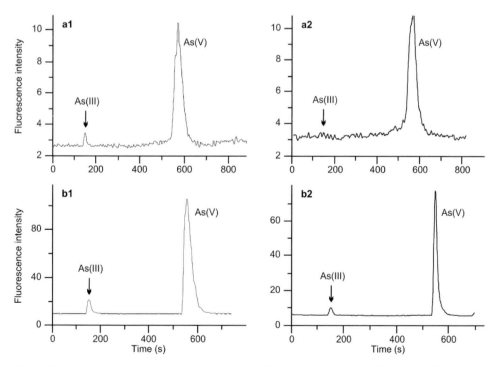

Figure 30.3.   Example chromatograms of samples (a1) S2 before grinding, (a2) S2 after grinding, (b1) S7 before grinding, (b2) S7 after grinding.

30.6   CONCLUSIONS

The characterization of seven soil samples which belonged to a contaminated area in the eastern Pyrenees, Catalonia (Spain) was performed and high percentages of Fe_2O_3 were obtained, mainly due to the presence of pyrite and arsenopyrite.

Because of the pseudo total As determination in the soil samples, it can be concluded that a high degree of contamination is present in several locations near Ribes de Freser.

When As species measurement in the extracts was carried out by LC-HG-AFS only i-As was found in all samples, arsenate being the main species present.

During sample pre-treatment a transformation of As species was observed in phosphoric acid extracts after soil samples grinding with a tungsten carbide disc mill.

REFERENCES

Ayora, C. and Phillips, R.: Natural occurrences in the systems PbS-Bi_2S_3-SbS_3 and PbS-Sb_2S_3 from Vall de Ribes, Eastern Pyrenees, Spain. *Bul. Minéral.* 104 (1981), pp. 556–564.

Ayora, C. and Casas, J.M.: Strata-bound As-Au mineralization in pre-Capadocian rocks from the Vall de Ribes, Eastern Pyreenes, Spain. *Miner. Deposita* 21 (1986), pp. 278–287.

Burguera, M. and Burguera, J.L.: Analytical methodology for speciation of arsenic in environmental and biological samples. *Talanta* 44 (1997), pp. 1581–1604.

Garcia-Manyes, S., Jiménez, G., Padró, A., Rubio, R. and Rauret, G.: Arsenic speciation in contaminated soils. *Talanta* 58 (2001), pp. 97–109.

Ruiz-Chancho, M.J., Sabé, R., López-Sánchez, J.F., Rubio, R. and Thomas, P.: New approaches to the extraction of arsenic species from soils. *Microchim. Acta* 151 (2005), pp. 241–248.

Yehl, M.P., Grleyuk, H., Tyson, J.F. and Uden, P.C.: Microwave-assisted extraction of monomethyl arsonic acid from soil sediment standard reference materials. *The Analyst* 126 (2001), pp. 1511–1518.

Arsenic in plants and food

CHAPTER 31

Bioavailability of arsenic species in food: Practical aspects for human health risk assessments

J.M. Laparra, D. Vélez & R. Montoro
Instituto de Agroquímica y Tecnología de Alimentos (CSIC), Burjassot, Valencia, Spain

R. Barberá & R. Farré
Facultad de Farmàcia, Universitat de València, Burjassot, Valencia, Spain

ABSTRACT: This work on the bioavailability of arsenic (As) in food shows that bioavailability of As species from cooked food should be considered in As health risk assessments. Seaweed, rice and fish, raw and cooked, were the foods studied, because of their major dietary contribution to As intake. The research has evaluated (1) the effect of *in vitro* gastrointestinal digestion, from which one can ascertain the maximum quantity of soluble As from a particular food, and (2) uptake and transport of As species present in the bioaccessible fraction by a Caco-2 cell line, a model of the intestinal epithelium. Results obtained for total arsenic, inorganic arsenic, arsenosugars and other organoarsenical species generally indicate high bioaccessibility and low uptake and cell transport.

31.1 INTRODUCTION

Food and drinking water are the main sources of arsenic (As) for humans. Arsenic appears in these matrices in different organic and inorganic chemical forms of varying toxicity. Of the chemical forms so far detected in foods, the inorganic species As(III) and As(V) are the most toxic. The sum of these two species, known as inorganic arsenic (i-As), has been classified as a human carcinogen by the International Agency for Research on Cancer (IARC 1987), and the World Health Organization (WHO) has established a Provisional Tolerable Weekly Intake of 15 µg/week/kg body weight (WHO 1989).

In recent decades, the study of As present in foods has been approached from various viewpoints: characterization of As species, the effect of cooking on the transformation of As species, estimation of As intake, studies of human toxicity, and evaluation of the toxicity of As species in experiments with animals and cell cultures. However, for a better understanding of the implications of food consumption for the assessment of As-related health risks, the bioavailability for the human being of the As species present in food must be considered.

Bioavailability can be defined as the As fraction that is solubilized and finally absorbed from the gastrointestinal tract into the systemic circulation of humans and animals (Caussy 2003). Bioavailability is the result of the integral sum of a four-part process: (1) ingestion, (2) bioaccessibility, (3) absorption and (4) first-pass effect, a process that takes place in the liver immediately after absorption (Caussy 2003). Bioaccessibility is defined as the fraction of As that dissolves in the stomach and is available for absorption during transit through the small intestine (Ruby *et al.* 1996). It depends, among other things, on the ability of digestive enzymes to release the element in the intestinal gut, as well as on As solubility and behavior in the gastrointestinal tract, which in turn is a function of the chemical form released from the food (Hocquellet and L'Hotelier 1997, Eckmekcioglu 2002).

To estimate mineral and trace element bioavailability, *in vivo* and *in vitro* studies are used. Evaluation in humans would be the ideal tool, but human studies are time-consuming, costly to perform and impractical for large-scale applications (Jovaní *et al.* 2001). Moreover, with toxic metals such as As, ethical limitations restrict their use in humans, while animal studies cannot always be extrapolated to human beings.

In vitro bioavailability methods offer a good alternative to *in vivo* and are generally based on simulation of gastrointestinal digestion followed by determination of how much of the element is soluble or dialyses through a membrane of a certain pore size (Wienk *et al.* 1999). Solubility or dialysability can be used to establish trends in the bioavailability values of a particular element. In fact, these methods only estimate the fraction of the element available for absorption (bioaccessibility), which is the first step of the *in vivo* process of mineral absorption.

After As intake, the mucosa of the epithelium is the first physiological barrier that the exogenous toxic agent has to cross in order to be absorbed. This fact might reduce the potential harmful effects for the organism. Despite the part played by the intestinal epithelium, most studies on As have been documented with other cell lines of target organs for As (Kitchin 2001, Thomas *et al.* 2001, Bode and Dong 2002, Carter *et al.* 2003) and there is a lack of studies with enterocytes. *In vitro* methods have been improved by incorporation of a human colon carcinoma cell line (Caco-2) presenting many of the functional and morphological properties of mature enterocytes (Pinto *et al.* 1983). The system is able to mimic and estimate uptake and/or transport of mineral elements (Ekmekcioglu 2002). The low cost, ease of use and widespread acceptance of the Caco-2 cell line make this model system an attractive alternative to animal studies and a tool available for use in conjunction with human trials (Glahn *et al.* 1998).

31.2   BIOAVAILABILITY OF ARSENIC IN FOOD

Previous research concentrated on studying water as the medium for conveying As to humans. In fact, oral toxicity values for As were derived from epidemiological studies with drinking water. Studies investigating the absorption of soluble As ingested by humans suggest that nearly 100% of water-soluble i-As is absorbed from the gastrointestinal tract (Ruby *et al.* 1996).

The first As bioavailability studies for solid samples were carried out with samples of soils, applying *in vitro* digestion methods simulating human gastrointestinal digestion to study i-As bioaccessibility (Ruby *et al.* 1996 and 1999, Hamel *et al.* 1999, Rodríguez *et al.* 1999, Oomen *et al.* 2002). Studies with foods are still scarce; they have only begun recently and have mostly been carried out by our research group (Laparra *et al.* 2003, 2004 and 2005, Almela *et al.* 2005). Foods are complex samples containing a wide variety of As species. In samples of vegetables grown on land, i-As and monomethylarsonic (MMA) and dimethylarsinic acid (DMA) have been detected. In animal and vegetable samples from the sea, the species present are mostly organic, such as arsenobetaine (AB), arsenocholine (AC), trimethylarsine oxide (TMAO), ion tetramethylarsonium (TMA^+), DMA, MMA and arsenosugars. These species, of varying toxicity, must be considered in order to obtain As-specific data for risk assessments, in which the bioavailability of each species has to be examined. In fact, part of the i-As absorbed is biotransformed by methylation of As into monomethylarsenic and dimethylarsenic forms, which are then excreted in urine (Tice *et al.* 1997), whereas other species, such as arsenobetaine, are eliminated in urine without any alteration. Although part of these transformations may take place after absorption by the intestinal mucosa, gastrointestinal digestion and absorption and transport through the epithelium could also contribute to modify the species present in the food ingested, both quantitatively and qualitatively.

Among the foods that contribute As to the diet, seaweed, rice and fish are currently the foods that are most important in dietary intake of As. Our laboratory has carried out bioavailability studies of As and As species in these foods. Edible seaweed customarily forms part of the diet in many Asian countries, and its consumption as a dietary supplement has increased in Europe. Arsenic contents in edible seaweed are high. Arsenosugars are the most abundant species in most varieties of seaweed, but *Hizikia fusiforme* accumulates very high i-As contents (Almela *et al.* 2002).

Rice, a cereal containing a relatively high amount of As as compared with other agricultural products, largely in the form of i-As (42–100%) (Schoof *et al.* 1998 and 1999), is a food staple for millions of people. In rice-based subsistence diets, the contribution of i-As from rice can represent over 50% of the tolerable daily intake recommended by the WHO (Williams *et al.* 2005).

Fish are the food products that contribute most As to the diet in western countries (as much as 90%; Urieta *et al.* 1996). They contain a preponderance of organic species, especially arsenobetaine, and i-As contents that rarely exceed 0.2 μg/g wet weight.

31.2.1 *Seaweed*

Raw and cooked seaweed bought in Spanish retail establishments were analyzed. After characterizing their total arsenic (t-As), inorganic arsenic (i-As) and arsenosugar contents, a study was made of bioavailability (Laparra *et al.* 2003 and 2004, Almela *et al.* 2005), defined as the percentage of bioaccessible t-As and i-As with respect to the total t-As or i-As contents in the seaweed.

Enteromorpha sp., *Porphyra* sp. and *H. fusiforme* were the varieties of seaweed selected for the study, which showed that bioaccessibility depends on the type of seaweed analyzed. The bioaccessibility of t-As was slightly higher ($p < 0.05$) in *Porphyra* sp. seaweed (67%) than in *H. fusiforme* (62%), and in both cases the bioaccessibility value was double than that of *Enteromorpha* sp. (32%). We did not observe the same tendency for i-As in raw seaweed, where bioaccessibility was higher in *H. fusiforme* (75%) and *Enteromorpha* sp. (77%) than in *Porphyra* sp. (49%).

After the seaweed had been cooked, there were changes in bioaccessibility. For t-As, a significant increase in bioaccessibility after cooking was only observed in *Porphyra* sp. (80% *vs.* 67%). For i-As, a significant ($p < 0.05$) increase was observed after cooking, both in *Porphyra* sp. (73%) and in *H. fusiforme* (88%), with a greater increase in *Porphyra* sp., which doubled the bioaccessibility of the raw product. The differences between these two seaweeds might be due to the different thermal treatments used, baking in the case of *Porphyra* sp. and boiling for *H. fusiforme*. The effect that each of these treatments has on proteins, to whose sulfydryl groups i-As bonds predominantly, might have affected the efficiency with which the digestive enzymes employed released As in each of the samples. It is noteworthy that in both of these seaweeds over 70% of the i-As present in the cooked product was bioaccessible.

The study of the influence of the gastric and intestinal stages of *in vitro* digestion carried out in *H. fusiforme*, showed that the gastric stage limits the maximum content of solubilized t-As and i-As, in both raw and cooked sample. This was also shown by Hamel *et al.* (1999) in a study on bioavailability of As from soils, in which they indicated that the stomach is the region of the gastrointestinal tract that is considered to have the greatest influence on As bioavailability. The low pH in the gastric stage seems to be a prerequisite for solubilizing As (Oomen *et al.* 2002). In fact, in studies carried out with contaminated soils, a linear correlation ($r = 0.82$) was obtained between the As solubilized in the gastric stage after *in vitro* gastrointestinal digestion and excretion of As in the urine of immature swine exposed to soils treated with As (Rodriguez *et al.* 1999). The solubility of As in the acid environment of the stomach might, therefore, be predictive for the oral bioavailability of this element in animal models (Ruby *et al.* 1999). There are no similar studies for humans.

The As(III) and As(V) contents found in each of the stages of the *in vitro* digestion of the various batches of raw and cooked *H. fusiforme* were evaluated. No general pattern was observed in the solubilization of either As(III) or As(V). In cooked *H. fusiforme*, the As(III) and As(V) solubilized contents (gastrointestinal stage) represented 20–70% and 21–68% of the solubilized i-As, respectively.

The levels of i-As in the raw *H. fusiforme* analyzed, considerably exceeded the limits established by the regulations for this food, existing in France, the USA, Australia, and New Zealand (ANZFA 1997, Mabeau *et al.* 2002). For i-As it is usual to establish toxicological considerations on the basis of the content in the raw product, although this may not be the correct option. On this basis, assuming the mean i-As content in the raw *H. fusiforme* analyzed (mean value: 83.2 μg/g), consumption of 3 g (minimum average daily consumption of brown seaweed by the Japanese) (Sakurai *et al.* 1997) of this seaweed would provide 250 μg of i-As, 67% more than the tolerable daily intake (TDI; 150 μg i-As/day for an adult with a body weight of 70 kg, WHO 1989). However, a more realistic approximation can be obtained including both the cooking of the product and the bioaccessibility of i-As. Assuming the mean bioaccessible i-As content found in the cooked *H. fusiforme* (35.5 μg/g), after consumption of 3 g of seaweed, 107 μg of i-As could remain available for absorption, which is 71% of the TDI. The difference between the percentage of TDI represented by ingestion of this seaweed in the two approximations, indicates the need to take cooking and bioaccessibility into account when evaluating the food safety of *H. fusiforme* with respect to i-As.

As indicated earlier, when studying bioaccessibility from the viewpoint of speciation, one should not rule out the possibility that the *in vitro* digestion method employed may bring about some transformation of the As species present in the initial product. For example, arsenosugars, the major organoarsenical species generally present in seaweed, are excreted in urine as DMA, a metabolite that can produce an increase in toxicity. The references in the literature do not indicate at what stage of the metabolism the production of this metabolite occurs. In order to evaluate whether degradation of arsenosugars takes place during gastrointestinal digestion, the bioaccessibility of these species was evaluated in four raw and cooked edible seaweeds: kelp powder, sample, a commercially available algae product from Eastern Canada, *H. fusiforme*, *Undaria pinnatifida* and *Porphyra* sp. Bioaccessibility of arsenosugars was greater than 80% in the raw and cooked samples, indicating that they are species that easily become available for subsequent absorption by the gastrointestinal epithelium, exceeding the i-As bioaccessibility reported in edible seaweeds.

In the bioaccessible fraction we did not detect species other than those present in the sample, which seems to indicate that during the *in vitro* digestion applied there was no degradation of the arsenosugars released. In the literature, with the use of simulated gastric juice (pepsin and HCl), degradation of glycerol, phosphate, sulfate and sulfonate ribose to another dimethylarsinoylribose has been shown in purified arsenosugars (Gamble *et al.* 2002) and *Laminaria* sp. (Van Hulle *et al.* 2004). Degradation was slow for arsenosugar standards (1.5%/hour) and much greater for the arsenosugars present in boiled *Laminaria* (changes in relative arsenosugar concentration between 32% and 86% after 4 h of incubation). The studies not evaluate intestinal digestion stage. As gastrointestinal digestion does not seem to contribute to the generation of DMA either in our study or in the two works mentioned, it remains to be seen in what stage of metabolism this urinary metabolite is generated.

The results reported indicate that the bioaccessibility of i-As and arsenosugars in cooked edible seaweed is over 70%, and that cooking does not reduce it. In the estimation of retention and transport of As species in Caco-2 cells from the bioaccessible fraction of cooked *H. fusiforme*, transport was only detected for As(V) and glycerol ribose, in percentages of 10% and 24% respectively. This transport could be considered as an indicator of relative bioavailability, allowing approximate evaluation of the risk for *H. fusiforme*, including the effects of cooking, gastrointestinal digestion and Caco-2 total uptake. Assuming a consumption of *H. fusiforme* of 3 g/day, the contribution of i-As from this seaweed would be less than 1% of the tolerable daily intake, substantially altering the risk, which until now has been evaluated on the basis of the content in the raw product.

31.2.2   Rice

Bioavailability was studied in rice (white and brown rice) cooked with water contaminated with As(V) in the laboratory, at concentrations usual in As-endemic areas in Asia (Laparra *et al.* 2005). The As contents in cooked rice ranged between 0.88 and 4.2 μg/g dry weight (dw) of which a large proportion (74–95%) was i-As. Over 90% of the As in the cooked rice was bioaccessible, with i-As contents of 0.84–3.1 μg/g, dw, in the bioaccessible fraction. Speciation analysis of i-As in the bioaccessible fraction showed that As(V) was the main chemical form (0.6–2.9 μg/g, dw), as would be expected, since it was the species added to the cooking water and the major species in drinking water. A variable relationship between As(V) and As(III) was found (2.8–19.2).

In Caco-2 cells, As retention, transport and total uptake (retention + transport) varied between 0.6–6%, 3.3–11% and 4–18%, respectively. Additions to Caco-2 cells of bioaccessible rice fractions with similar As contents gave different total uptake percentages, indicating that other soluble components in rice may cause differences in the extent of absorption.

If we consider total uptake as a measure of relative As bioavailability and take the lowest (4%) and highest (18%) As uptake values obtained, consumption of 5.7 and 1.2 kg of cooked rice/day, respectively, would be required to reach the tolerable daily intake established by the WHO for i-As (150 μg/day for a person weighing 70 kg). In Asian As-endemic areas where the population depends heavily on rice for caloric intake, an average adult male consumption of 1.5 kg cooked rice/day has been reported (Bae *et al.* 2002). Consequently, the population might reach the tolerable daily intake with a single food.

31.2.3 *Fish*

The bioavailability of t-As and As species were evaluated in raw and cooked fish (sole and halibut) and a DORM-2 reference sample (dogfish muscle). Cooking did not affect the bioaccessibility of t-As in sole (98% *vs.* 102%), but in halibut an increase in t-As bioaccessibility (79% *vs.* 100%) was observed. The high t-As bioaccessibility in the raw and cooked samples showed the feasibility of using digestive enzymes in the simulated gastrointestinal process to release As from these foods.

In all the samples, AB was the species with greatest solubility (68–100%). The solubility of other species, such as dimethylarsinic acid and trimethylarsine oxide, was lower, and the TMA^+ present in the sample was not detected in the bioaccessible fraction. It is noteworthy that in DORM-2 and sole the bioaccessible MMA content was much higher than that quantified in the sample using methanol/water extraction, the solvent mixture used more extensively for the species extraction in biological samples. This might indicate that the *in vitro* digestion process releases a greater quantity of MMA from the sample than methanol/water extraction.

After exposure of the Caco-2 cell culture to the DORM-2 bioaccessible fraction, of the As species present (AB, MMA, DMA and TMA^+), transport was only detected for the major species, AB, for which it was of the order of 12%. It should be noted that transport of this species from the DORM-2 soluble fraction was more efficient than the transport for standards of this species (12% *vs.* 2%). This indicates that transport differs between standard and food solutions, showing the importance of evaluating bioavailability of As in foods with a view to improving risk evaluation.

31.3 CONCLUSIONS

Any evaluation of the risk associated with As ingestion should consider not only the content in the product but also As species bioavailability for humans. After human gastrointestinal digestion the soluble As species do not necessarily coincide with the species present in the raw product, as quantitative and/or qualitative modifications of species may be brought about both by cooking of the product (Devesa *et al.* 2001) and by the digestive process (Gamble *et al.* 2002), aspects which can condition its bioavailability. It has been indicated that many metal toxicity studies show a lack of correlation between the risk for humans or animals and exposure to an external dose, mainly because bioavailability is not included in the risk estimation (Caussy 2003).

There is a need to study the bioavailability of As species in foods with a view to obtaining a more realistic estimate of the possible risk resulting from dietary intake of this element. Given the restrictions on the use of *in vivo* methods with toxic metals, *in vitro* methods, despite their limitations, can be used to make relative estimations of bioavailability, permitting comparison of foods and evaluation of dietary factors that may affect bioavailability. The release of As from a food depends on its nature, so that it is not advisable to establish a general percentage of As bioavailability in risk evaluation; if a general percentage is established, caution should be shown when making extrapolations from one food to another (Caussy 2003). Nowadays, the inclusion of Caco-2 cells in *in vitro* studies provides a valid model of the intestinal epithelium and therefore a better approximation to the *in vivo* situation.

In mineral elements of nutritional interest such as Ca, Fe and Zn, the study of their bioavailability in foods has yielded valuable information about absorption in the intestinal epithelium and the factors affecting this process. The bioavailability data obtained have had an impact on food formulations. It is to be hoped that for toxic elements such as As bioavailability studies will provide valuable information for estimating risks and regulating maximum permissible concentrations.

ACKNOWLEDGEMENTS

This research was supported by projects MCyT AGL2001-1789 and CYTED XI.23, for which the authors are deeply indebted. J.M. Laparra received a Personnel Training Grant from this project to carry out this study.

REFERENCES

Almela, C., Algora, S., Benito, V., Clemente, M.J., Devesa, V., Súñer, M.A., Vélez, D. and Montoro, R.: Heavy metals, total arsenic and inorganic arsenic contents of algae food products. *J. Agric. Food Chem.* 50 (2002), pp. 918–923.

Almela, C., Laparra, J.M., Vélez, D., Montoro, R., Barberá, R. and Farré, R.: Arsenosugars in raw and cooked edible seaweeds: characterization and bioaccessibility. *J. Agric. Food Chem.* 53 (2005), pp. 7344–7351.

ANZFA: Food standards code. Australian New Zealand Food Authority, Issue 41, Canberra, Australia, 1997.

Bae, M., Watanabe, C., Inaoka, T., Sekiyama, M., Sudo, N., Bokul, M.H. and Ohtsuka, R.: Arsenic in cooked rice in Bangladesh. *Lancet* 360 (2002), pp. 1839–1840.

Bode, A.M. and Dong, Z.: The paradox of arsenic: molecular mechanisms of cell transformation and chemotherapeutic effects. *Crit. Rev. Oncol. Hematol.* 42 (2002), pp. 5–24.

Carter, D.E., Vasken Aposhian, H. and Gandolfi, J.: The metabolism of inorganic arsenic oxides, gallium arsenide, and arsine: a toxicochemical review. *Toxicol. Appl. Pharmacol.* 193 (2003), pp. 309–334.

Caussy, D.: Case studies of the impact of understanding bioavailability: arsenic. *Ecotoxicol. Environ. Saf.* 56 (2003), pp. 164–173.

Caussy, D., Gochfeld, M., Gurzau, E., Neagu, C. and Ruedel, H.: Lessons from case studies of metals, investigating exposure, bioavailability and risk. *Ecotoxicol. Environ. Saf.* 56 (2003), pp. 45–51.

Devesa, V., Martínez, A., Súñer, M.A., Vélez, D., Almela, C. and Montoro, R.: Effect of cooking temperatures on chemical changes in species of organic arsenic in seafood. *J. Agric. Food Chem.* 49 (2001), pp. 2272–2276.

Ekmekcioglu, C.: A physiological approach for preparing and conducting intestinal bioavailability studies using experimental systems. *Food Chemistry* 76 (2002), pp. 225–230.

Fleurence, J.: Seaweed proteins: biochemical, nutritional aspects and potential uses. *Trends Food Sci. Technol.* 10 (1999), pp. 25–28.

Gamble, B.M., Gallagher, P.A., Shoemaker, J.A., Weis, X., Schwegel, C.A. and Creed, J.T.: An investigation of the chemical stability of arsenosugars in simulated gastric juice and acidic environments using IC-ICP-MS and IC-ESI-MS/MS. *Analyst* 127 (2002), pp. 781–785.

Glahn, R.P., Lai, C., Hsu, J., Thompson, J.F., Guo, M. and Van Campen, R.J.: Decreased citrate improves iron availability from infant formula: application of an in vitro digestion/Caco-2 cell culture model. *Nutrition* 128 (1998), pp. 257–264.

Hamel, S.C., Ellickson, K.J. and Lioy, P.J.: The estimation of the bioaccessibility of heavy metals in soils using artificial biofluids by two novel methods: mass-balance and soil recapture. *Sci. Total Environ.* 243 (1999), pp. 273–283.

Hocquellet, P. and L'Hotelier, M.D.: Bioavailability and speciation of mineral micronutrients: the enzymolysis approach. *J. AOAC Int.* 80 (1997), pp. 920–927.

IARC (International Agency for Cancer Research): IARC Monographs on the evaluation of carcinogenic risks to humans. Overall evaluations of carcinogenicity: an updating of IARC monographs. Volumes 1 to 42, Suppl. 7, International Agency for Cancer Research, Lyon, France, 1987.

Jovaní, M., Barberá, R., Farré, R. and Martín de Aguilera, E.: Calcium, iron and zinc uptake from digests of infant formulas by Caco-2 cells. *J. Agric. Food Chem.* 49 (2001), pp. 3480–3485.

Kitchin, K.T.: Recent advances in arsenic carcinogenesis: modes of action, animal model systems, and methylated arsenic metabolites. *Toxicol. Appl. Pharmacol.* 172 (2001), pp. 249–261.

Laparra, J.M., Vélez, D., Montoro, R., Barberá, R. and Farré, R.: Estimation of arsenic bioaccessibility in edible seaweed by an in vitro digestion method. *J. Agric. Food Chem.* 51 (2003), pp. 6080–6085.

Laparra, J.M., Vélez, D., Montoro, R., Barberá, R. and Farré, R.: Bioaccessibility of inorganic As species in raw and cooked Hizikia fusiforme seaweed. *Appl. Organomet. Chem.* 18 (2004), pp. 662–669.

Laparra, J.M., Vélez, D., Barberá, R., Farré, R. and Montoro, R.: Bioavailability of inorganic arsenic in cooked rice: Practical aspects for human health risk assessments. *J. Agric. Food Chem.* 51 (2005), pp. 6080–6085.

Mabeau, S. and Fleurence, J.: Seaweed in food products: biochemical and nutritional aspects. *Trends Food Sci. Technol.* 4 (1993), pp. 103–107.

Oomen, A.G., Hack, A., Minekus, M., Zeijdner, E., Cornelis, C., Schoeters, G., Verstraete, W., Van de Wiele, T., Wragg, J., Rompelberg, C.J.M., Sips, A. and Van Wijnen, J.H.: Comparison of five in vitro digestion models to study the bioaccessibility of soil contaminants. *Environ. Sci. Technol.* 36 (2002), pp. 3326–3334.

Pinto, M., Robine-Leon, S., Appay, M.D., Kedinger, M., Triadou, N., Dussaulx, E., Lacroix, B., Simon-Assmann, P., Haffen, K., Fogh, J. and Zweibaum, A.: Enterocyte-like differentiation and polarization of the human colon carcinoma cell line caco-2 in culture. *Biology of the Cell* 47 (1983), pp. 323–330.

Ruby, M.V., Davis, A., Schoff, R., Eberle, S. and Sellstone, C.M.: Estimation of lead and arsenic bioavailability using a physiologically based extraction test. *Environ. Sci. Technol.* 30 (1996), pp. 422–430.

Ruby, M.V., Schoff, R., Brattin, W., Goldade, M., Post, G., Harnois, M., Mosby, D.E., Casteel, S.W., Berti, W., Carpenter, M., Edwards, D., Cragin, D. and Chappell, W.: Advances in evaluating the oral bioavailability of inorganics in soils for use in human health risk assessment. *Environ. Sci. Technol.* 33 (1999), pp. 3697–3705.

Rodríguez, R., Basta, N.T., Casteel, S.W. and Pace, L.W.: An in vitro gastrointestinal method to estimate bioavailable arsenic in contaminated soils and solid media. *Environ. Sci. Technol.* 33 (1999), pp. 642–649.

Sakurai, T., Kaise, T., Ochi, T., Saitoh, T. and Matsubara, C.: Study of in vitro cytotoxicity of a water soluble organic arsenic compound, arsenosugar, in seaweed. *Toxicology* 122 (1997), pp. 205–212.

Schoof, R.A., Yost, L.J., Crecelius, E.A., Irgolic, K., Goessler, W., Guo, H.R. and Greene, H.: Dietary arsenic intake in Taiwanese district with elevated arsenic in drinking water. *HERA* 4 (1998), pp. 117–135.

Schoof, R.A., Yost, L.J., Eickhoff, J., Crecelius, E.A., Cragin, D.W., Meacher, D.M. and Menzel, D.B.: A market basket survey of inorganic arsenic in food. *Food Chem. Toxicol.* 37 (1999), pp. 839–846.

Thomas, D.J., Styblo, M. and Lin, S.: The cellular metabolism and systemic toxicity of arsenic. *Toxicol. Appl. Pharmacol.* 176 (2001), pp. 127–144.

Tice, R.R., Yager, J.W., Andrews, P. and Crecelius, E.: A chemical hypothesis for arsenic methylation in mammals. *Chem. Biol. Interact.* 88 (1997), pp. 89–114.

Urieta, I., Jalon, M. and Eguileor, I.: Food surveillance in the Basque country (Spain). 2: Estimation of the dietary intake of organochlorine pesticides, heavy metals, arsenic, aflatoxin M(1), iron and zinc through the total diet study, 1990/91. *Food Addit. Contam.* 13 (1996), pp. 29–52.

Van Hulle, M., Zhang, C., Zhang, X. and Cornelis, R.: Arsenic speciation in Chinese seaweeds using HPLC-ICP-MS and HPLC-ES-MS. *Analyst* 17 (2002), pp. 634–640.

WHO: Evaluation of certain food additives and contaminants. 33rd report of the Joint FAO/WHO Expert Committee on Food Additives, WHO Technical Report Series 759, World Health Organization, Geneva, Switzerland, Geneva, Switzerland, 1989.

Wienk, K.J.H., Marx, J.J.M. and Beynen, A.C.: The concept of iron bioavailability and its assessment. *Eur. J. Clin. Nutr.* 38 (1999), pp. 51–75.

Williams, P.N., Price, A.H., Raab, A., Hossain, S.S., Feldmann, J. and Meharg, A.A.: Variation in arsenic speciation and concentration in paddy rice related to dietary exposure. *Environ. Sci. Technol.* 39 (2005), pp. 5531–5540.

CHAPTER 32

Determination of arsenic content in seafood products in the school meals distribution program, *Junta Nacional de Auxilio Escolar y Becas,* region VII, Chile

S. Vilches & G. Andrade
Escuela de Ingeniería en Alimentos, Universidad del Bío-Bío, Chillán, Chile

O. Muñoz
Instituto de Ciencia y Tecnología de los Alimentos, Universidad Austral de Chile, Valdivia, Chile

J.M. Bastías
Departamento de Ingeniería en Alimentos, Universidad del Bío-Bío, Chillán, Chile

ABSTRACT: The objective of this study was to determine the total arsenic (t-As) and inorganic arsenic (i-As) content in seafood products in the *Junta Nacional de Auxilo Escolar y Becas* (JUNAEB) school meals program (PAE), in region VII, Chile. The t-As and i-As determination was carried out on 35 samples (24 hake fillets, 5 salmon pulps, 5 canned jack mackerel and 1 dehydrated bag of shellfish) obtained from concessionary companies that supply the PAE in region VII. The t-As concentration in seafood products was between 0.38 and 1.40 μg/g wet weight (ww). Six samples exceeded the maximum limit established by Chile's Food Sanitary regulations. On the other hand, for seafood, the i-As concentration obtained was from 0.01 to 1.13 μg/g (ww), no samples exceeding the maximum limit established by Chilean legislation, the dehydrated shellfish being the sample with the greater concentration.

32.1   INTRODUCTION

Arsenic is widely distributed in the earth's crust, soil, water, vegetable tissue and animal tissue (Figueroa 2001). The effects of chronic As exposure on health are multiple; gastrointestinal, kidney malfunctions, hematological and neurological alterations (Saha and Diksihit 1999). In addition, it can cause chronic queratodermis plantar handspam, melanodermis and Bowen's disease (Biagini *et al.* 1995). The International Agency for Research on Cancer (IARC) has assigned arsenic, inhaled or ingested, to Group I human carcinogens, because there is sufficient evidence from epidemiological studies to support a causal association between As exposure and cancer (Tsuda *et al.* 1992).

Studies conducted by the United Kingdom by the Ministry of Agriculture, Fisheries and Food (MAFF) showed that the highest t-As contents are found in fish products 4.3 μg/g wet weight (ww) the intake of total As is 120 μg/day (Ysart *et al.* 1999). In the USA, the Food and Drug Administration (FDA) has also determined that the main source of arsenic in the diet are fish products, with an intake of 38.6 μg/day of t-As (Gunderson 1995). A study made in the Basque country (Spain) found in fishing products an average As content 3.2 μg/g (ww), which together with high fish consumption resulted in an average t-As ingestion rate of 286 μg/day, being one of the most elevated ingestions in the world for this element (Urieta *et al.* 1996).

Worldwide, only a few countries regulate the t-As content in products of marine origin. with a limit of 1 μg/g (ww). Hong Kong has the most liberal legislation at 6 μg/g (ww) for fish products, and 10 μg/g (ww) for other seafood products (British Food Manufacturing Industries Research Association 1993).

In Chile the food As content is regulated by the Food Sanitary Regulation (RSA) (Gonzalez 2004), which establishes the limits for individual foods, such as mollusks, crustaceans and gastropods must contain less than 2 μg/g (ww) (as inorganic arsenic). Whereas for fresh, chilled, frozen and canned fish this standard is 1 μg/g (ww) (as total arsenic).

Chile has a widespread school meals distribution program (PAE), administered by the Ministry of Education. The program's goal is: "to facilitate the incorporation and permanence of the scholastic population in the educational system, therefore allowing equality of opportunities in education" (www.junaeb.cl 2004). Currently the PAE gives 1,400,000 daily rations of breakfasts and lunch nationally. In region VII 109,000 rations are distributed daily (www.junaeb.cl 2004). Food consumption containing toxic polluting agents is of special concern for children, since they are at a vulnerable stage of development. This present study's objective is to determine the total and i-As content in sea products given by the PAE.

32.2   MATERIALS AND METHODS

32.2.1   *Sample collection campaign*

The samples of different seafood products (fish and shellfish), were collected from the warehouses of the four companies that provide the services to the educational establishments of the PAE region VII (JUNAEB 2002). A total of 35 samples were analyzed, with the samples 1 to 24 being hake fillets; 5 samples of frozen salmon pulp (samples 25 a 29), 5 samples of jack mackerel (samples 30 a 34) and one assorted frozen shellfish (sample 35). The frozen canned samples were defrosted at room temperature, the canned jack mackerel was drained and the dehydrated samples of shellfish were reconstituted according to manufacturer specifications. The sample obtained was homogenized in a food processor. All the samples were analyzed in duplicate.

32.2.2   *Instrumentation*

Arsenic determination was performed with an atomic absorption spectrometer (AAS) model Spectra A55 (Varian) and a flow injection (FI) hydride generation system (VGA77, Varian). Other equipment used included a domestic food processor (Moulinex), and muffle model FB 1410M-26 (Thermolyne).

32.2.3   *Reagents*

Deionized water, 18.2 MΩ/cm, was used for the preparation of reagents and standards. All chemicals were of pro analysis quality or better. Commercial standard solutions (1000 mg/l) of As(V) were used (Merck).

32.2.4   *Arsenic determination*

The dried samples (0.25 ± 0.01 g) were treated with an ashing aid suspension (1 ml of 20% w/v Mg $(NO_3)_2 \cdot 6H_2O$ + 2% w/v MgO) and nitric acid (5 ml of 50% v/v), evaporated to dryness, and mineralized at 450°C with a gradual increase in temperature. The ash was dissolved in hydrochloric acid (6 mol/l) and pre-reduced (5% w/v ascorbic acid + 5% w/v KI). After 30 min, it was diluted to volume with water and filtered through Whatman No. 1 filter paper into a 25 ml or 10 ml volumetric flask (Muñoz *et al.* 2000).

The analysis conditions used for As determination by FI-HG-AAS were the following: loop sample, 0.5 ml; reducing agent, 0.2% (w/v) $NaBH_4$ in 0.05% (w/v) NaOH, 5 ml/min flow rate; HCl solution 10% (v/v), 10 ml/min flow rate; carrier gas argon, 100 ml/min flow rate; wavelength 193.7 nm; spectral band-pass 0.7 nm; electrodeless discharge lamp system 2; lamp current setting 400 mA; cell temperature 900°C (Muñoz *et al.* 2000). Calibration standard solutions of As(III) were prepared from a reduced standard solution of As(V), using as reducing solution a mixture containing 5% (w/v) KI and 5% (w/v) ascorbic acid. Triplicate analyses were performed for each sample.

32.2.5   *Determination of inorganic arsenic*

The method used has been described in a previous paper (Muñoz *et al.* 1999a). The dried sample $(0.50 \pm 0.01$ g) was weighed into a 50 ml screw-top centrifuge tube, 4.1 ml of water were added, and the sample was agitated until it was completely moistened. Then 18.4 ml of concentrated HCl were added, and the sample was agitated again for 1 hr, and then left to stand for 12–15 hours (overnight). The reducing agent (1 ml of 1.5% w/v hydrazine sulfate solution and 2 ml of HBr) was added and the sample was agitated for 30 s. Then 10 ml of $CHCl_3$ were added and the sample was agitated for 3 min. The phases were separated by centrifuging at 2000 rpm for 5 min. The chloroform phase was separated by aspiration and poured into another tube. The extraction process was repeated two more times. The chloroform phases were combined and centrifuged again. The remnants of the acid phase were completely eliminated by aspiration (acid phase remnants in the chloroform phase cause substantial overestimates of inorganic arsenic). Possible remnants of particulate organic material in the chloroform phase were eliminated by passing it through Whatman GD/X syringe filters with a 25 mm PTFE membrane.

The i-As in the chloroform phase was back-extracted by agitating for 3 min with 10 ml of 1 mol/l HCl. The phases were separated by centrifuging at 2000 rpm, and the aqueous phase was then aspirated and poured into a beaker. This stage was repeated once again and the back-extraction phases obtained were combined. When the back-extraction phase generated emulsions that could not be broken by centrifuging at over 2000 rpm, the emulsion was transferred to the beaker. Ashing aid suspension and HNO_3 were added and the result was heated gently in the sand bath for not more than 30 s. The emulsion was then broken and the chloroform phase formed was removed by aspiration. The determination of i-As in the back-extraction phase was performed by means of the following procedure: 2.5 ml of ashing aid suspension and 10 ml of concentrated HNO_3 were added to the combined back-extraction phases. The result was evaporated and treated in the same way as for total arsenic.

To the validation of the analytic method the parameters used were exactitude, precision and detection limit, those are the ones considered as primary limits for a methodology validation. (Compañó and Rios 2002). The statistical analyses of the results were made through the analysis of simple variance to find statistically significant difference between the analyzed groups and Turkey's test with a level of confidence of 95%. To identify which groups differ significantly, an experimental design data grouped in blocks of randomized incomplete data was used.

32.3   RESULTS AND DISCUSSION

32.3.1   *Methodology validation*

The analytical characteristics such as detection limit, precision and recovery were evaluated in samples of seafood. The detection limit (wet weight) was established as the total arsenic, and i-As concentration in the seafood products that provides an absorbance reading statistically different from that of the blank. It was calculated by dividing three times the standard deviation of the absorbance readings of nine reagent blanks by the slope of the calibration curve of As(III) and taking into account the sample mass and dilution employed. The precision of the method, expressed as the mean of the precision of the method for the three samples analyzed, was 7%. For each sample the precision was calculated from the relative standard deviation of six sub-samples. The recovery percentage was evaluated by spiking three subsamples of each sample with As(V). In relation to the percentage recovery for spiked samples, this was 105.6% for t-As and 89.31% for i-As, the latter possibly giving a lower value than that of the extraction methodology. In any case results are within acceptable limits (80–120%). The coefficient of variation was 7.00 and 7.04% for t-As and i-As, respectively, being within acceptable standards since it should be inferior of 10% (Directiva CEE 2001).

32.3.2   *Total arsenic content*

The t-As concentrations present a minimum value of 0.4 μg/g (ww) (sample 28, frozen hake fillet), and the maximum value is of 2.6 μg/g (ww) (sample 35, assorted dehydrated shellfish). According to these results, there are five frozen samples of hake (numbers 3, 4, 5, 13 and 14) and one of frozen salmon pulp (sample 21) that exceed the limit of 1 μg/g established by the RSA, see Figure 32.1.

The limit established by the RSA of 1 μg/g (ww) for t-As corresponds only to fresh, frozen and canned fish, however, for mollusks and crustaceans it is 2 μg/g (ww) expressed as i-As (Gonzalez 2004). This leads to confusion in the interpretation of the RSA, so it has been suggested that national legislation should be unified on the basis of inorganic arsenic, in an attempt to reduce the harm that can be done by this type of food, as i-As is considered more toxic than t-As. Although studies show that there are also trivalent organoarsenical compounds (monomethylarsonic acid (III)) that have been re-evaluated as having a higher toxicity than the inorganic As(III) and As(V) (Yamauchi and Fowler 1994, Hughes 2002).

32.3.3   *Inorganic arsenic content*

The minimum concentration found for i-As corresponds to sample 19 (frozen hake fillet) with 0.012 μg/g (ww) (Fig. 32.2) and the maximum concentration of 1.1 μg/g (ww) corresponds to sample 35 (assortment of dehydrated shellfish). Although the RSA establishes i-As concentrations only for crustaceans and mollusks (2 μg/g ww), the shellfish sample would be within the Chilean norm (Gonzalez 2004).

32.3.4   *Comparison between the organic and inorganic arsenic content*

Figure 32.2 shows that the i-As content is very low compared to the t-As content (Figure 32.1). The average inorganic As content is representing only 5.8% of the sample's total arsenic. Literature on the subject shows i-As values in marine products to be between 1.5 and 22% of the t-As (Urieta *et al.* 1996), and close to 11% (Muñoz *et al.* 2000). Therefore, the results of this study have a percentage

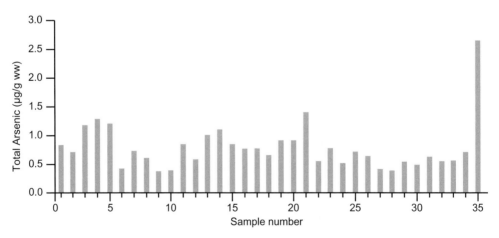

Figure 32.1.   Total Arsenic concentration in sea food products. Limit by RSA (1 μg/g ww). The samples correspond to: Hake fillet sample 1 to 24, frozen salmon samples 25 to 29, jack mackerel samples 30 to 34 and one assorted frozen shellfish sample 35.

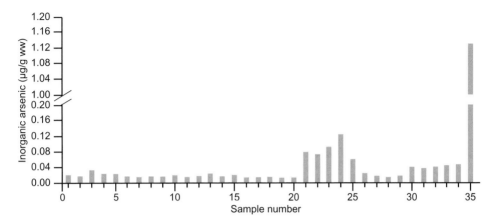

Figure 32.2.   Inorganic arsenic concentration (μg/g ww) in sea products. Established for the RSA are 2 μg/g ww for seafood and crustaceous as determined product. The samples correspond to: Hake fillet samples 1 to 24, frozen salmon samples 25 to 29, jack mackerel samples 30 to 34 and one assorted frozen shellfish sample 35.

Table 32.1.   Average of t-As and i-As concentration (μg/g wet weigh w/w) in different types of food product.

Products	Total (t-As) μg/g	Inorganic (i-As) μg/g	i-As as share of t-As %
Frozen hake (FH)	0.76	0.02	2.3
Frozen salmon (FS)	0.80	0.08	10.2
Canned jack mackerel (CJM)	0.60	0.04	6.8
Assorted frozen shellfish sample (ADS)*	2.64	1.04	39.3

* The ADS sample, was not included in the statistical analysis because it had only one sample, nevertheless it is presented in the table like a reference value for this kind of product.

of i-As similar to what was found in literature (Brooke and Evans 1981, Larsen *et al.* 1993, Branch *et al.* 1994, López *et al.* 1994, Šlejkovec *et al.* 1996, Muñoz *et al.* 1999a, b). However, the fact that i-As is found in low concentrations (0.01–1.13 μg/g ww) does not mean that the food will not cause health problems in humans.

Statistical analysis of the t-As content using the ANOVA test shows that there are no significant differences between the "FH" (frozen hake), "FS" (frozen salmon), and "CJM" (canned jack mackerel) groups. In relation to the statistical analysis (analysis of variance) for inorganic arsenic, according to the type of product it is possible to observe in Table 32.1, that there is a significant difference among the i-As concentrations in the three sea food products.

32.4   CONCLUSIONS

In relation to t-As content, six samples of fish exceeded the limits established by the RSA (five of frozen hake and one of frozen salmon). According to the i-As content no sample exceeded the established limits, nevertheless, the RSA only establishes limits of i-As for shellfish, which leaves a flaw concerning the i-As limits in fish. In case of the total and i-As contents, the statistical analysis showed that the type of species influences significantly in the concentrations obtained.

ACKNOWLEDGMENTS

The authors are grateful to the *Dirección de Investigación* of the Universidad del Bío-Bío (DIUBB research grant No. 03-4217-3/R); and Project CYTED XI.23.

REFERENCES

Biagini, R.E., Salvador, M.A., de Qüerio, R.S., Torres Soruco, C.A., Biagini, M.M. and Diez Berrantes, A.: Hidroarsenisismo crónico: Comentario de casos diagnosticados en el período 1972–1993. *Arch. Argent. Dermatol.* 45 (1996), pp. 47–52.

British Food Manufacturing Industries Research Association: Food legislation surveys 6: Metallic contaminants in food—a survey of international prescribed limits. 3rd edition, Leatherhead Food International, Leatherhead, Surrey, British Food Manufacturing Industries Research Association, UK, 1993.

Brooke, P.J. and Evans, W.H.: Determination of total inorganic arsenic in fish, shellfish and fish products. *Analyst* 106 (1981), pp. 514–520.

Compañó, B.R. and Ríos, C.A.: Garantía de la calidad en los laboratorios analíticos. SINTESIS Publisher, Madrid, Spain, 2002, pp. 217–243.

European Community: Directiva 2001/22/CEE de la Comisión 8, marzo, 2001. Official newspaper of the European Community, Brussels, Belgium, 2001.

FAO/WHO: Evaluation of certain food additives and contaminants; Technical report series 837; World Health Organization, Geneva, Switzerland, 1993.

Figueroa, L.: *Arica inserta en una región arsenical: El arsénico en el ambiente que la afecta y 45 siglos de arsenicismo crónico.* Ediciones Universidad de Tarapacá, Santiago de Chile, Chile, 2001.

González C.: Nuevo reglamento sanitario de los alimentos. *Ediciones Publiley*, Santiago, Chile, pp. 63–69.

Gunderson, E.L.: FDA total diet study, July 1986–April 1991, dietary intakes of pesticides, selected elements and other chemicals. *Journal of the AOAC International* 78:6 (1995), pp. 1353–1363.

Hughes, M.F.: Arsenic toxicity and potential mechanisms of action. *Toxicol. Lett.* 133 (2002), pp. 1–16.

Junta de Auxilio Escolar y Becas (JUNAEB): Bases administrativas y técnicas, propuesta pública N° 1, Santiago de Chile, Chile, 2002.

Junta de Auxilio Escolar y Becas (JUNAEB): www.junaeb.cl (accessed July 2007), Santiago de Chile, Chile, 2004.

Larsen, E.H., Pritzl, G. and Hansen, S.H.: Arsenic speciation in seafood samples with emphasis on minor constituents—an investigation using high performance liquid chromatography with detection by inductively coupled plasma mass spectrometry. *J. Anal. At. Spectrom.* 8 (1993), pp. 1075–1084.

López, J.C., Reija, C., Montoro, R., Cervera, M.L. and De la Guardia, M.: Determination of inorganic arsenic in seafood products by microwave assisted distillation and atomic absorption spectrometry. *J. Anal. At. Spectrom.* 9 (1994), pp. 651–656.

Muñoz, O., Devesa, V., Súñer, M.A., Vélez, D., Montoro, R., Urieta, I., Macho, M.L. and Jalón, M.: Total and inorganic arsenic in fresh and processed fish products. *J. Agri. Food Chem.* 48 (2000), pp. 4369–4376.

Muñoz, O., Vélez, D. and Montoro, R.: Optimization of the solubilization, extraction and determination of inorganic arsenic [As(III)+As(V)] in seafood products by acid digestion, solvent extraction and hydride generation atomic absorption spectrometry. *Analyst* 124 (1999a), pp. 601–607.

Muñoz, O., Vélez, D., Cervera, M.L. and Montoro, R.: Rapid and quantitative release, separation and determination of inorganic arsenic [As(III)+As(V)] in seafood products by microwave-assisted distillation and hydride generation atomic absorption spectrometry. *J. Anal. At. Spectrom.* 14 (1999b), pp. 1607–1613.

Reilly, C.: *Metal contamination of food.* Applied Science Publisher Ltd., London, UK, 1980.

Robson, M.: Methodologies for assessing exposures to metals: Human host factors. *Ecotoxicol. Environ. Safety* 56 (2003), pp. 104–109.

Saha, J. and Diksihit, A.: A review of arsenic poisoning and its effects in the human healt. *Cri. Rev. Environ. Sci. Technol.* 29 (1999), pp. 281–313.

Šlejkovec, Z., Byrne, A.R., Smodiš, B. and Rossbach, M.: Preliminary studies on arsenic species in some environmental samples. Fresenius *J. Anal. Chem.* 354 (1996), pp. 592–595.

Suñer, M.A., Devesa, V., Rivas, I., Velez, D. and Montoro, R.: Speciation of cationic arsenic species in seafood by coupling liquid chromatography with hydride generation atomic fluorescence detection. *Jaas.* 15 (2000), p. 1501.

Tsuda, T., Babazono, A., Ogawa, T., Hamada, H., Mino, Y., Aoyama, H., Kurumatani, N., Nagira, T., Hotta, N., Harada, M. and Inomata, S.: Inorganic arsenic: A dangerous enigma for mankind. *Appl. Organomet. Chem.* 6 (1992), pp. 309–322.

Urieta, I., Jalón, M. and Eguileor, I.: Food surveillance in the Basque country (Spain) II. Estimation of the dietary intake of organochlorine pesticides, heavy metals, arsenic, aflatoxin M1, iron and zinc throug the total diet study. *Food Addit. Contam.* 13 (1996), pp. 29–52.

Yamauchi, H. and Fowler, B.A.: Toxicity and metabolism of inorganic and methylated arsenicals. In: J.O. Nriagu (ed): *Arsenic in the environment. Part II: Human and ecosystem effects.* Wiley and Sons Inc., New York, 1994, pp. 35–53.

Ysart, G., Miller, P., Crews, H., Robb, P., Baxter, M., De L'Argy, C., Lofthouse, S., Sargent, C. and Harrison, N.: Dietary exposure estimates of 30 elements from the UK total diet study. *Food Addit. Contam.* 16:9 (1999), pp. 391–403.

CHAPTER 33

Arsenic contamination from geological sources in environmental compartments in a pre-Andean area of Northern Chile

O. Díaz & R. Pastene
Facultad de Química y Biología, Universidad de Santiago de Chile, Santiago de Chile, Chile

N. Núñez
Programa Agrícola CODELCO Chile, Calama, Chile

E. Recabarren G.
Facultad de Ingeniería, Universidad de Santiago de Chile, Santiago de Chile, Chile

D. Vélez & R. Montoro
Instituto de Agroquímica y Tecnología de Alimentos (CSIC), Burjassot, Valencia, Spain

ABSTRACT: The aim of this work was to study the distribution of total (t-As) and inorganic arsenic (i-As) in the water of Loa river, and in soil and living organisms (freshwater alga *Gracilaria* sp., trout *Orcorhynchus mykiss*, and vegetables, carrot, asparagus, lettuce and corn) in the Second Region, Chile, an As-endemic area. The river Loa water has high levels of As (up to 0.92 mg/l). *Gracilaria* sp. has the highest levels of total arsenic (t-As); 98.03 µg/g (dry basis), with the i-As contents representing more than 46% of the t-As. Trout shows lower As concentrations than algae, and organic arsenic (As_{org}) represents most of t-As. The irrigation water does not fulfill the national regulations. The soil has high t-As concentrations (32.6–68.2 µg/g), whereas the As bioavailable from the soil has far lower values (up to 0.33 µg/g). This might explain why the t-As concentrations are distinctly lower in vegetables (0.08–0.45 µg/g).

33.1 INTRODUCTION

Arsenic (As), the king of poisons, has probably influenced human history more than any other element or toxic compound. It is present in the environment in inorganic forms, As(III) and As(V), which are toxic to humans, terrestrial and aquatic species, and also in organic forms, some of which are considered to be innocuous [arsenobetaine (AB), arsenocholine (AC) and trimethylarsine oxide (TMAO)], while others are toxic [dimethylarsinic acid (DMA), monomethylarsonic acid (MMA) and tetramethylarsonium ion (TMA^+)].

Because of geological factors, the Second Region in Chile, located in the pre-Andean area in the north of the country, has an environment with high concentrations of As. The volcanic bedrock in this area has a high content of As associated with pyrite minerals, and subsequent processes of disintegration and lixiviation distribute the As in the atmosphere, soil and water (Vather *et al.* 1995). The course of the river Loa, located in the Second Region in Chile, has special ecological conditions, due particularly to high As concentrations in the water, salinity and drainage in the soil, high evaporation and transpiration, and a desert climate that reduces the growth of plants (Pastenes *et al.* 1992). In the shallow waters of the region's rivers, mostly used for irrigation, As concentrations even higher than 2 mg/l have been measured. This situation also affects the soil, which increases the As concentration (Pastenes *et al.* 1992). In farming soil samples collected in the villages of Chiu-Chiu and Lasana (Second Region, Chile), the As concentration ranged between 50 and 70 mg/kg (Muñoz *et al.* 2002).

Numerous authors have studied the ability of some aquatic organisms, which are grouped into different trophic levels, to assimilate i-As dissolved in seawater. There are fewer papers dealing with studies carried out on freshwater organisms. For this reason, only limited information is available about As levels in freshwater environments. In seawater, phytoplankton and algae assimilate As and reduce it to As(III), which is methylated into MMA and DMA or into more complex species such as arsenosugars (Cullen and Reimer 1989, Eisler 1994). Recently, Murray *et al.* (2003) showed that the freshwater alga *Chlorella vulgaris* metabolizes As in a similar manner to marine macroalgae. *Chlorella vulgaris* can survive in environments with levels exceeding 10,000 mg As/l, and its ability to bioaccumulate As increases with the As(V) concentration in the water (Cullen and Reimer 1989). Nevertheless, in freshwater food-chain models the t-As concentrations in organisms decrease by an order of magnitude for each step (Suhendrayatna and Maeda 2001).

On the other hand, vegetables and cereals can become a path by which As may enter the food chain, because they can reflect the levels of As that exist in the environment in which they are cultivated (soil, irrigation water and atmosphere). Greenhouse experiments have revealed that an increase in As in cultivated soils leads to an increase in As levels in edible vegetables. However, many complex factors affect this situation, including bioavailability, uptake and phytotoxicity of As (Carbonell-Barrachina *et al.* 1999).

The aim of this study is to report the distribution of As in the following environmental compartments: river Loa irrigation water, algae, freshwater trout, soils and vegetables. We consider all these compartments as a system.

33.2   MATERIALS AND METHODS

33.2.1   *Study area*

The research area was restricted to the sector of the river Loa between the village of Lasana and the town of Calama and the agricultural region between the villages of Chiu-Chiu and Lasana, in the pre-Andean area of the Second Region in Chile (Fig. 33.1).

33.2.2   *Biological materials*

Algae (*Gracilaria* sp.), freshwater trout (*Orcorhynchus mykiss*) and vegetables (carrots, asparagus, lettuce and corn) were collected. The algae were obtained from five study areas on the riverbank, which where they were the most abundant resource (Fig. 33.1). This alga species was selected because it is the first and most important link in the whole aquatic system, and it is one of the main freshwater species that exist in the river Loa. The samples were collected in the month of November in the years 2000 and 2001. Once the algae had been collected, they were packed into polyethylene bags, properly identified and stored in a cool box with ice (Encina *et al.* 1995).

The trout samples consisted of five specimens collected by a capturing net in the Poma sector of the river Loa, in November 2001 (Fig. 33.1). Between December and March of each year there is a notable increase in precipitation in the river Loa area in the Andes highlands. These circumstances were especially intense in 2001, notably affecting the availability of trout, leading to a low number of specimens. Once the fish had been collected, the intestine was removed and the head and tail were separated. The head, tail and trunk were packed separately into polyethylene bags, properly identified, and preserved in a cool box with ice.

The vegetables were collected during November 2000 in agricultural smallholdings corresponding to the communities of Chiu-Chiu and Lasana, located in the interior of the Second Region in Chile (Fig. 33.1). The samples were selected on the basis of production and consumption criteria, and were obtained in the field with a random collection procedure. They were collected by hand and afterwards carefully packed into polyethylene bags (Muñoz *et al.* 2002).

All biological materials were washed with distilled water to remove remaining soil, frozen at $-20°C$, and then freeze dried. The lyophilized samples were ground in a domestic apparatus and the

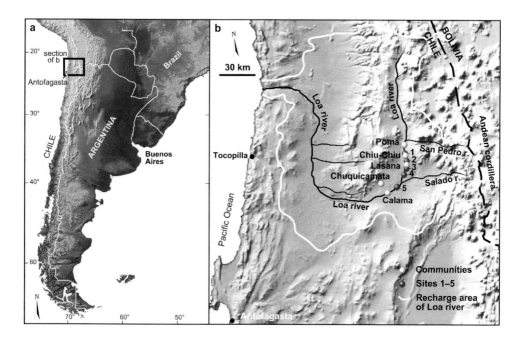

Figure 33.1.   Study areas in the sector of the Loa river (Second Region, Chile).

resulting powder was vacuum packed and kept at refrigeration temperature (+4°C) until analysis (Encina *et al.* 1995).

33.2.3   *Water samples*

Surface water samples (500 ml) were collected at five points along the river Loa, which correspond with the same places where the algae and vegetables were obtained. To preserve the As contents, 0.5 ml of HCl (0.01 mol/l) were added. In the laboratory, the samples were stored at +4°C until analysis (Díaz *et al.* 2004).

33.2.4   *Soil samples*

Surface soil samples (0–20 cm) were obtained during November 2001 from the area adjacent to vegetable roots and afterwards packed into polyethylene bags. The samples were then dried in the air, sieved to obtain a particle size less than 2 mm and stored at ambient temperature until the time of the As analysis (Díaz *et al.* 2004).

33.2.5   *Instruments and reagents*

The determination of t-As and i-As in solid samples was performed with an atomic absorption spectroscope (AAS, model 3300, Perkin Elmer), equipped with an autosampler (model AS-90, Perkin Elmer) and a flow injection system (model FIAS-400, Perkin Elmer) in order to provide hydride generation in continuous flow mode.

   A GBC model 903 AAS was used to determine t-As in surface water samples. Other equipment used included a PL 5125 sand bath (Raypa, Scharlau), a K 1253 muffle furnace equipped with a Eurotherm Controls 902 control program (Heraeus), a KS 125 basic mechanical shaker (IKA Labortechnik), and an Eppendorf 5810 centrifuge (Merck).

Deionized water (18 MΩ) was used for preparation of the reagents and standards. All chemicals were pro analysis quality. Commercial standard solutions (1000 mg/l) of As(V) were used. Calibration standard solutions of As(III) were prepared from a reduced standard solution of As(V).

All glassware was treated with 10% v/v HNO_3 for 24 h and then rinsed three times with deionized water before use. The following certified reference materials were employed: *Fucus* sp. (algae; International Atomic Energy Agency, Vienna, Austria), Dorm-2 (dogfish muscle, Institute for Reference Materials and Measurements, Brussels, Belgium), and Montana Soil (NIST 2711, National Institute of Standards and Technology, Gaithersburg, USA).

33.2.6   Determination of total arsenic

Samples of soil and biological materials were analyzed by flow injection-hydride generation-atomic absorption spectrometry (FI-HG-AAS) after a dry-ashing step. Waters were directly analyzed by FI-HG-AAS. For the dry-ashing step, the sample (0.25 g) was treated with an ashing aid suspension [20% m/v of $Mg(NO_3)_2$ + 2% m/v of MgO] and nitric acid (5 ml of 50% v/v) evaporated to dryness, and mineralized at +450°C with a gradual increase in temperature. The ash was dissolved in 6 mol/l HCl and prereduced (5% m/v of ascorbic acid + 5% m/v of KI).

The analytical conditions used for As-determination by FI-HG-AAS were the following: loop sample, 0.5 ml; reducing agent, 0.2% m/v of $NaBH_4$ in 0.05% m/v of NaOH, 5 ml/min flow rate; HCl solution 10% v/v, 10 ml/min flow rate; carrier gas argon, 100 ml/min flow rate; wavelength, 193.7 nm; spectral band-pass, 0.7 nm; electrodeless discharge lamp system 2; lamp current setting, 400 mA; cell temperature, +900°C.

33.2.7   Determination of inorganic arsenic

Water (4.1 ml) and concentrated HCl (18.4 ml) were added to 0.50 g of sample (Muñoz *et al.* 2000). The mixture was left overnight. The reducing agent was then added (1 ml of 1.5% m/v of hydrazine sulfate solution and 2 ml of HBr) and the sample was agitated for 30 s. Chloroform (10 ml) was then added, and after 3 min of shaking and 5 min of centrifuging (2000 rpm) the chloroform phase was separated. The extraction process was repeated twice, and the chloroform phase was combined and filtered. The i-As in the chloroform phase was back-extracted by shaking for 10 min with 10 ml of 1 mol/l HCl. The phases were separated by centrifuging at 2000 rpm and the aqueous phase was then aspirated and poured into a beaker. This stage was repeated once again, and the back-extraction phases obtained were combined.

The i-As in the back-extraction phase was determined by means of the followed procedure: 2.5 ml of ashing aid suspension and 10 ml of concentrated HNO_3 were added to the combined back-extraction phases, dry ashed, and quantified by FI-HG-AAS in the conditions described previously for the determination of t-As.

33.2.8   Determination of bioavailable arsenic in soil

The soil samples were extracted with 0.5 mol/l $NaHCO_3$ at pH 8.5 (NaOH 7% m/v) (Clemente *et al.* 2006) in a 1/10 relationship (soil/$NaHCO_3$). The solution was agitated for 20 hours and then the aqueous phase was transferred to centrifugal tubes, centrifuged at 2500 rpm (10 min) and filtered through Whatman no. 1 paper. 1.0 ml of filtrate, 2.5 ml of ashing aid suspension [20% m/v of MgO + 2% m/v of $Mg(NO_3)_2$], and 10 ml of concentrated HNO_3 were added. The mixture was evaporated to total dryness. Once the sample was dry it was redissolved with 5 ml HCl (50% v/v) and 5 ml of reducing solution (5% m/v of ascorbic acid + 5% m/v of KI) and filtered through Whatman no. 1. The As was quantified by FI-HG-AAS in the conditions described previously.

Quality assurance-quality control: The suitability of the analytical methods employed for total and inorganic As determination has been checked previously by evaluating their analytical characteristics (limit of detection, precision and accuracy) (Muñoz *et al.* 2000). The precision, expressed as the relative standard deviation for three independent analyses, was less than 10% for the determination of t-As or i-As in environmental and biological samples. Consequently, in the tables of the analysis results only the average values are shown.

33.3 RESULTS

The concentration of As measured in the water of the river Loa varied according to the place of sample collection. The highest part of the river was the place with the highest levels (0.28 mg/l and 0.92 mg/l in the years 2000 and 2001, respectively) (Figs. 33.2a and b). The concentration of As in the river Loa downstream from where it is joined by the river El Salado (collection site number 5) increased notably, especially in the sample collected in November 2001 (0.40 mg As/l) (Fig. 33.2b). This result can be explained by the proximity of one of the tributaries of the river Loa, the river San Pedro, with water that contains As-levels as high as 0.23 mg/l (Alonso 1992). After the junction with the river El Salado, the As-concentration in the river Loa also increases sharply (from 0.27 mg/l to 0.80 mg/l) (Alonso 1992), similar to our results. Pastenes *et al.* (1992) observed significant differences between the sampling stations in the Loa-El Salado sector. They found the highest values in the river El Salado (1.01 mg As/l) and in the river Loa after it receives the El Salado tributary (0.78 mg As/l), similar to our results.

Similar studies performed in other countries have also shown high concentrations of As with variations in the surface water of rivers affected by natural pollution, due to the effects of typical volcanic systems (De Sastre *et al.* 1992, Quintanilla 1992).

In *Gracilaria* sp., the highest concentrations of t-As [98.03 µg/g dry weight (dw)] and i-As (100.55 µg/g dw) were found in the samples collected at site 2 in November 2000 (Fig. 33.2a). This was also the site where the highest concentration of t-As in water (0.28 mg/l) was found. In November 2001, the differences between the t-As and i-As concentrations at the various sites were not as marked as in 2000, and they did not have such a clear relationship with the As concentrations detected in water. At least 46% of the t-As in this alga is inorganic (Table 33.1). This high proportion of i-As has also been detected in the seaweed *Hizikia fusiforme*, which has a very high concentration of i-As (83–88 mg/kg, dw), representing 60–72% of the t-As present in samples (Almela *et al.* 2002).

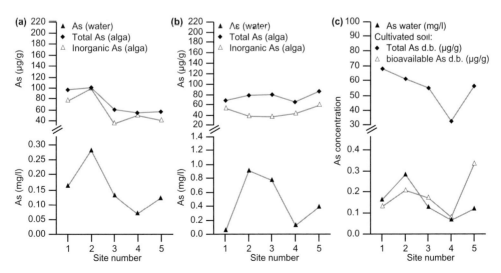

Figure 33.2. Arsenic concentration (total and inorganic, dry base) in water and alga (*Gracilaria* sp.) (a) November 2000; (b) November 2001; (c) Arsenic concentration in irrigation water and cultivated soil (dry basis).

Table 33.1.   Bioaccumulation factor for total arsenic (t-As) and inorganic arsenic (i-As) and i-As/t-As ratio in *Gracilaria* sp. algae.

| Station | November 2000 | | | November 2001 | | |
	i-As/t-As (%)	$BF_{t\text{-}As}$[1]	$BF_{i\text{-}As}$	i-As/t-As (%)	$BF_{t\text{-}As}$	$BF_{i\text{-}As}$
1	81	599	485	78	1281	1004
2	103	350	359	49	86	42
3	59	462	272	46	104	48
4	93	772	719	68	540	365
5	74	466	346	69	216	149

[1]BF = Bioaccumulation factor.

A bioaccumulation factor has been calculated as the relation between the concentration for a given As species in living organisms (algae, vegetables and trout) and the environment in which they grow (soil and/or water) (Pizarro *et al.* 2003). *Gracilaria* sp. has a high ability to bioaccumulate As, as shown by the high values of the bioaccumulation factor in both sampling periods (Table 33.1).

Algae, especially macroalgae, bioaccumulate high concentrations of several toxic elements and are affected by temporary changes in concentrations and integrate temporal changes in concentrations that occur in the environment, due to their high tolerance to metals in their tissues. This might provide a record of long- and medium-term accumulation and encourage the use of algae as indicators of water pollution (Díaz *et al.* 1989, Encina *et al.* 1995, Almela *et al.* 2002). Variations in metal concentrations have been demonstrated and could be associated with both the origin of the product, which reflects the pollution found in its natural habitat, and the growth phase and time of year when samples are collected (Hou and Yan 1998). Algae in continental environments have been shown to be capable of accumulating various trace metals and As, with a bioaccumulation factor that can vary between 240 and 2800 (Suhendrayatna *et al.* 1999). The bioaccumulation factors found in the present study lie within that range. Some metals and metalloids such as As, once they have bioaccumulated in alga tissues, can be transferred to other trophic levels, including humans, who may use them directly as food and indirectly as additives in the food and cosmetic industries (Encina *et al.* 1995, Almela *et al.* 2002).

Trout showed lower concentrations of t-As and i-As than algae, and also a lower bioaccumulation factor (Table 33.2). The highest t-As and i-As concentrations measured in trout were found in the trunk and tail, and lower concentrations were found in the head. The only previous study of freshwater fish in this area is the work reported by Pizarro *et al.* (2003) in trout collected from the river Loa. They detected extremely high concentrations of i-As [As(III): 30.0 ± 1.1 µg/g; As(V): 16.0 ± 0.9 µg/g; dw], which they attributed to the effect of serious As contamination of the river.

Not more than 25% of the t-As measured in trout was i-As (Table 33.2). In the literature, a high percentage (30%) has also been reported for frogs in the river Danube (Schaeffer *et al.* 2006). However, the percentages reported in marine fish are lower, 0.15–11% (Muñoz *et al.* 2000). Arsenobetaine tends to be the predominant species of As in marine fish, but recent studies show that this species may be lacking in freshwater fish, and in some cases arsenosugars may be the predominant species (Schaeffer *et al.* 2006). The evident differences between marine and freshwater fish should be investigated further.

In the trout analyzed, the bioaccumulation factor in trunk and tail represented between 200% and 300% of the values obtained in the head. This might be explained by the large proportion of muscular mass that is located in the fish's trunk, together with the richness in protein of this mass and the chemical affinity of proteins to join several elements such as As (S-S groups). In fact, interaction between As and proteins has been considered to be one of the main toxic mechanisms.

These results show that the algae collect more As, particularly the highly toxic inorganic form, than the trout. Consequently, the chemical constituents of aquatic organisms may provide useful

Table 33.2.   Total arsenic (t-As) and inorganic arsenic (i-As) concentration (dw) in trout (*Orcorhynchus mykiss*), bioaccumulation factor and i-As/t-As ratio.

	t-As (μg/g)	BF_{t-As}[1]	t-As (μg/g)	BF_{i-As}	i-As/t-As (%)
Head	5.69	6.69	1.42	1.67	25
Trunk	16.0	18.8	2.40	2.82	15
Tail	10.1	11.8	1.71	2.01	17

[1]BF = Bioaccumulation factor.

Table 33.3.   Total arsenic (t-As) and inorganic arsenic (i-As) concentrations (ww) in vegetables, bioaccumulation factor and i-As/t-As ratio.

Vegetable	t-As[1] (μg/g)	BF_{t-As}[2]	i-As[1] (μg/g)	BF_{i-As}	i-As/t-As (%)
Carrot	0.14	1.08	0.13	1.00	93
Asparagus	0.08	0.38	0.06	0.29	75
Lettuce	0.45	2.65	0.39	2.29	87
Corn	0.15	1.88	0.11	1.38	73

[1]Muñoz *et al.* 2002; [2]BF = Bioaccumulation factor; ww = wet weight.

environmental and toxicological information about biological species that enter the human food chain (Schaeffer *et al.* 2006).

Fig. 33.2c shows As concentrations in irrigation water and cultivated soils. With the exception of one sample of water (0.07 mg As/l), all the remaining water did not comply with the National Regulation which establishes a maximum limit of 0.10 mg As/l for water used for sprinkling (Instituto Nacional de Normalización 1979). This situation has already been reported in a study carried out by Pastenes *et al.* (1994). In the present research, samples of irrigation water collected from the river Loa had an average concentration of 0.17 mg As/l, while irrigation water collected in the agricultural sector of Colina (Metropolitan Region, Chile), not affected by As contamination, had an average level of 0.0005 mg As/l.

In the samples of agricultural surface soil collected in 2001 (Fig. 33.2c) the t-As concentrations varied between 32.6 μg/g and 68.2 μg/g, with a concentration profile very similar to that found in water. There are previous studies of As contents in soils in the area (17–31 mg/kg) (Pastenes *et al.* 1994, Pizarro *et al.* 2003). In general, soil with As concentrations of 20 mg/kg is not recommended for cultivation of vegetables for human consumption (Helgesen and Larsen 1998).

In spite of the high levels of t-As in the soil obtained in our study, the concentrations measured in vegetables were distinctly lower (Table 33.3). For a valid comparison between As in soils and As in vegetables, the comparison should be made in terms of bioavailable As, the total fraction available for living organisms (Ginocchio and Narváez 2002). This bioavailable fraction would correspond to the soluble chemical form present in the soil solution (i.e. ions and organic complexes of less than molecular weight), and to the adsorption of labile forms to places of specific desorption present in organic matter and in clay particles of soil (Sauvé *et al.* 2000). In the soils in the present study the bioavailable As varied between 0.08 and 0.33 μg/g, which is less than 1% of the t-As quantified in the soil. There was only a small difference between the values of t-As in water and those measured in the As bioavailable from soil, indicating that in soils of this kind the soluble fraction in water is practically the same as that extracted with $NaHCO_3$. These results are a further indication of the suitability of using As bioavailability values and not t-As contents in order to evaluate the consequences that this contaminant may have on crops grown in the area.

A previous study carried out by Figueroa and Gonzalez (1987) gave a quantitative description of the As concentration in a soil-irrigation water system in the Lluta, Azapa and Camarones valleys (First Region, Chile). In the Azapa valley the water used for irrigation contained only 0.009 mg As/l, whereas surface soil (0–30 cm) contained 0.16 mg/l of soluble As. In the Camarones valley, As in the water was high (0.92 mg/l), and so was soluble As in the soil (1.29 mg/l). The results obtained in our study for the bioavailable As concentration in soil samples are much lower than those obtained by Alonso (1992), who conducted a study to measure the concentration of As in the water-soil-alfalfa system in the Second Region in Chile. A concentration of 8.34 µg of As/g for extractable As was measured in soils collected from the village of Lasana, the place also selected for our study. The notable difference in the results may be due to the fact that the methodologies used in the two cases were different. In the Alonso study the As was analyzed by the silver diethyldithio-carbamate method, while in our study the bioavailable As in soils was extracted with $NaHCO_3$.

In soil highly polluted with copper in the central area of Chile, 0.002% of the total copper is in ionic form in the soil solution, and 0.04% is in labile forms potentially usable by plants (Ginocchio and Narváez 2002). The percentages of bioavailable As fluctuated between 0.10 and 0.58% of the t-As in the soil. Consequently, the bioavailable fraction of many toxic elements would not be constant in soils with different physico-chemical characteristics, because the solubility of these elements or the buffering capacity of the soil are governed by a number of variables such as pH, content of organic matter, and content and type of clays (Sauvé *et al.* 2000).

One of the highest values of t-As in surface soil (61.2 µg/g) and bioavailable As (0.21 µg/g) was obtained at site 2, where the As concentration in water was highest (0.28 mg/l), indicating the influence of the sampling location and particularly of the water on the concentrations of As measured in the soil.

Total and i-As concentrations were measured in edible portions of vegetables, carrot (horticultural root species), lettuce (horticultural leaf species), asparagus, (horticultural stem species) and corn (horticultural fruit species) (Table 33.3). The highest concentrations of total and i-As were detected in lettuce leaves.

All of the vegetables analyzed had t-As contents below the maximum limits permitted by Chilean legislation, which establishes two maximum limits for As. For cereals, legumes and leguminous plants, the maximum concentration permitted is 0.5 mg/kg of (ww). Other vegetables must be compared with the value established by Chilean legislation for the food group "other solid products" (1 mg/kg of ww) (Diario Oficial de la República de Chile 1997). Pastenes *et al.* (1994) measured the concentration of t-As in some horticultural species, such as lettuce (*Lactuca sativa*) and carrot (*Daucus carota*), cultivated in an agricultural sector known as Salar del Carmen (Second Region, Chile), where As is eliminated by a water treatment plant. The concentration of t-As in lettuce leaves varied between 5.01 µg/g and 31.13 µg/g, while the average concentration in carrot was 3.50 µg/g. These values are higher than those obtained in our study, probably owing to the strong influence of the sampling place. In a more recent study, Queirolo *et al.* (2000) determined that the t-As concentration in different horticultural species varied according to the place where they were grown. Potatoes grown in Socaire (Second Region, Chile) contained 0.86 µg As/g wet weight (ww), while in potatoes grown in Talabre (Second Region, Chile) the level of t-As was 0.24 µg/g (ww).

Over 70% of the t-As in the vegetables corresponded to inorganic species. There have been a few reports of As speciation in vegetables grown in natural or As-contaminated soils. Arsenic speciation in carrots, growing in soils contaminated by a wood preservation plant, showed that they contained As(III) and As(V) and traces of trimethylarsine oxide (TMAO) (Helgesen and Larsen 1998). In a more recent study carried out by Pizarro *et al.* (2003), which consisted in studying the distribution of As species in environmental samples collected in a small area located in the agricultural smallholdings of the community of Chiu-Chiu (Second Region, Chile), the same area as in our study, in which carrots were growing, the main As content was made up of toxic species, As(III) and As(V). These species represented 45% and 31% of the t-As, respectively. In other vegetables grown in Chiu-Chiu and Lasana and analyzed by Muñoz *et al.* (2002), i-As represented between 28 and 114% of t-As. In the present study, the highest concentrations of total and i-As were found in edible roots and leaves, whereas the lowest levels appeared in fruits. Agricultural practices, water quality, crops and soils play an important part in the uptake of i-As and t-As by plants.

The bioaccumulation factors for the vegetables in our study are low when they are calculated with the t-As content in the soil (Table 33.3). However, they are not low when compared with the bioavailable As concentration in soil samples. In all the species of vegetables there was a great ability to accumulate i-As. Lettuce show the highest bioaccumulation factor value for t-As and i-As. The results obtained confirm that absorption and subsequent accumulation of As in vegetables is influenced by numerous factors, such as the species of vegetable, the bioavailability of As from the soil, and the water used for irrigation. However, further research on the mechanisms of As uptake by different plants in a wide range of As-polluted and non-polluted areas would be valuable for assessing human exposure to As as a result of consumption of vegetables from domestic gardens and agricultural fields (ATSDR 2005).

33.4 CONCLUSION

Contamination of the environment by As in Chile's Second Region affects not only drinking water but also the food chain. The cultivation of vegetables and cereals in contaminated soils and the use of water with high As contents for irrigation produces crops that accumulate appreciable quantities of i-As, which must be taken into account when evaluating the toxicological risk for the communities that consume them. The aquatic systems also reflect the contamination of the environment. There is a need to carry out more studies in As-endemic areas in Chile to evaluate the potential risk of As in the environment to agriculture, livestock and fisheries.

ACKNOWLEDGEMENTS

We are grateful for support and the helpful suggestions made by the following institutions: University of Santiago of Chile (DICYT), Latin American Cooperation Program (CYTED), Agricultural Programme (CODELCO, Chile), Institute of Agrochemistry and Food Technology (IATA-CSIC, Spain) and Dr Pilar Bernal (CEBAS-CSIC). We would also like to thank Prof. J. Bundschuh and Dr A.B. Mukherjee for editing the manuscript.

REFERENCES

Almela, C., Algora, S., Benito, V., Clemente, M., Devesa, V., Suñer, M., Vélez, D. and Montoro, R.: Heavy metal, total Arsenic and inorganic arsenic contents of algae food products. *J. Agr. Food Chem.* 50 (2002), pp. 918–923.

Alonso, H.: Arsenic enrichment in superficial waters. II Region Northern Chile. *Proceedings International Seminar. Arsenic in the Environment and its Incidence on Health.* Universidad de Chile, Santiago de Chile, Chile, 1992, pp. 21–27.

ATSDR (Agency for Toxic Substances and Disease Registry): Toxicological profile for arsenic. US Dept. of Health and Human Services. Atlanta, GA, 2005.

Carbonell-Barrachina, A.A., Burló-Carbonell, F., Valero, D., Lopez, E., Martinez-Romero, D. and Martinez-Sanchez, F.: Arsenic toxicity and accumulation in turnip as affected by arsenic chemical speciation. *J. Agr. Food Chem.* 47 (1999), pp. 2288–2294.

Clemente, R., Almela, C. and Bernal, M.P.: A remediation strategy based on active phytoremediation followed by natural attenuation in a soil contaminated by pyrite waste. *Environ. Pollut.* 143 (2006), pp. 397–406.

Cullen, W.R. and Reimer, K.J.: Arsenic speciation in the environment. *Chem. Rev.* 89 (1989), pp. 713–764.

De Sastre, M.S.R., Varillas, A. and Kirschbaum, P.: Arsenic content in water in the northwest area of Argentina. *Proceedings International Seminar Arsenic in the Environment and its Incidence on Health.* Universidad de Chile, Santiago de Chile, Chile, 1992, pp. 91–99.

Diario Oficial de la República de Chile, Martes 13 de mayo de 1997, Decreto 977, Reglamento Sanitario de los Alimentos, Título IV, De los contaminantes, Párrafo I, De los metales pesados, Artículo 160, Arsénico. Santiago de Chile, Chile, 1997.

Díaz, O., Recabarren, E., Ward, J. and Villalobos, J.: Metales pesados: aspectos ecológicos y tecnológico-alimentarios. *Contribuciones Científicas y Tecnológicas, Área Ambiente* 84 (1989), pp. 5–10.

Díaz, O., Muñoz, O., Recabarren, E., Montes, S., Martínez, S., Soto, H., Yánez, M., Núñez, N., Vélez, D. and Montoro, R.: Bioremediation of soil and irrigation water contaminated with arsenic in the Second Region of Chile: Preliminary results. In: I. Gaballah, B. Mishra, R. Solozabal and M. Tanaka (eds): *Proceedings Global Symposium on Recycling, Waste Treatment and Clean Technology (REWAS)*. Spain, 2004, pp. 2465–2474.

Eisler, R.: A review of arsenic hazards to plants and animals with emphasis on fishery and wildlife. In: J.O. Nriagu (ed): Arsenic in the environment: Part II: Human health and ecosystem effects. John Wiley and Sons, New York, 1994, pp. 185–259.

Encina, F., Chuecas, L. and Díaz, O.: Metodología analítica base para la determinación de metales pesados en microalgas. In: K. Alveal, M. Ferrario, E.C. Oliveira and E. Sar (eds): Manual de métodos ficológicos. Universidad de Concepción, Concepción, Chile, 1995, pp. 763–777.

Figueroa, L. and González, M.: Traslocación de arsénico desde el sistema suelo-agua-forraje, una visión cuantitativa. *V Simposio sobre Contaminación Ambiental Orientado a los Alimentos. Tomo I. Resúmenes*. Santiago de Chile, Chile, 1987, pp. 59–62.

Ginocchio, R. and Narváez, J.: Importancia de la forma química y de la matriz del sustrato en la toxicidad por cobre en *Noticastrum sericeum* (Less.) Less. ex Phil. *Revista Chilena de Historia Natural* 75 (2002), pp. 603–612.

Helgesen, H. and Larsen, E.H.: Bioavailability and speciation of arsenic in carrots grown in contaminated soil. *Analyst* 123 (1998), pp. 791–796.

Hou, X. and Yan, X.: Study of the concentration and seasonal variation of inorganic elements in 35 species of marine algae. *Sci. Total Environ.* 47 (1998), pp. 141–156.

Instituto Nacional de Normalización: Requisitos de calidad de agua para diferentes usos. Norma Chilena Oficial (NCH 1333–1978). Santiago de Chile, Chile, 1979.

Muñoz, O., Díaz, O., Leyton, I., Núñez, N., Devesa, V., Súñer, M.A., Vélez, D. and Montoro, R.: Vegetables collected in the cultivated Andean area of Northern Chile: Total and inorganic arsenic contents in raw vegetables. *J. Agr. Food Chem.* 50 (2002), pp. 642–647.

Murray, L.A., Raab, A., Marr, I.L. and Feldmann, J.: Biotransformation of arsenate to arsenosugars by *Chlorella vulgaris. Appl. Organomet. Chem.* 17 (2003), pp. 669–674.

Pastenes, J., Salgado, M., Jofré, V., Romero, A. and Portilla, L.: Arsenic incidence in the water-soils-plants system in Loa river course, Antofagasta, Chile. *Proceedings International Seminar. Arsenic in the Environment and its Incidence on Health. Proceedings*. Universidad de Chile, Santiago de Chile, Chile, 1992, pp. 21–27.

Pastenes, J., Salgado, H., Illanes, A., López, J. and Olmos, E.: Efecto de la acumulación de arsénico en el sector agrícola de la ciudad de Antofagasta, Chile. *Segunda Jornada sobre Arsenicismo Laboral y Ambiental*. Antofagasta, Chile, 1994.

Pizarro, I., Gómez, M.M., Camara, C. and Palacios, M.A.: Distribution of arsenic species in environmental samples collected in Northern Chile. *Inter. J. Environ. Anal. Chem.* 83 (2003), pp. 879–890.

Queirolo, F., Stegen, S., Restovic, M., Paz, M., Ostapczuk, P., Schwuger, M.J. and Muñoz, L.: Total arsenic, lead, and cadmium levels in vegetables cultivated at the Andean villages of Northern Chile. *Sci. Total Environ.* 255 (2000), pp. 75–84.

Quintanilla, J.: Evaluation of arsenic in bodies of superficial water of the South Lipez of Bolivia (South-West). *Proceedings International Seminar Arsenic in the Environment and its Incidence on Health*. Universidad de Chile, Santiago de Chile, Chile, 1992, pp. 109–121.

Sauvé, S., Hendershot, W.H. and Allen, H.E.: Solid-solution partitioning of metals in contaminated soils: Dependence on pH and total metal burden. *Environ. Sci. Technol.* 34 (2000), pp. 1125–1130.

Schaeffer, R., Francesconi, K. A, Kienzl, N., Soeroes, C., Fodor, P., Varadi, L., Raml, R., Goessler, W. and Kuehnelt, D.: Arsenic speciation in freshwater organisms from the river Danube in Hungary. *Talanta* 69:4 (2006), pp. 856–865.

Suhendrayatna, O., Kuroiwa, T. and Maeda, S.: Arsenical compounds in fresh water green microalga Chlorella vulgaris after exposure to arsenite. *Appl. Organomet. Chem.* 13 (1999), pp. 127–133.

Suhendrayatna, O.K. and Maeda, S.: Biotransformation of arsenite in freshwater food-chain models. *Appl. Organomet. Chem.* 15 (2001), pp. 277–284.

Vather, M., Concha, G., Nermellet, B., Nilsson, R., Dulout, F. and Natarajan, A.T.: A unique metabolism of inorganic arsenic in native Andean women. *Eur. J. Pharmacol. Environ. Toxicol. Pharmacol. Sect.* 293 (1995), pp. 454–462.

CHAPTER 34

Total arsenic content in vegetables cultivated in different zones in Chile

A.M. Sancha & N. Marchetti

*División de Recursos Hídricos y Medio Ambiente, Facultad de Ciencias Físicas y Matemáticas,
Universidad de Chile, Santiago de Chile, Chile*

ABSTRACT: The objective of this study was to gain preliminary data concerning total arsenic (t-As) levels in vegetables cultivated in the northern, central and southern zones of Chile. Samples of vegetables were gathered from markets and produce stands from throughout the country. Analyses were made of the raw edible parts of these samples. The highest t-As contents were found in the northern zone. Vegetables cultivated by the indigenous peoples of the Altiplano were those with the highest contents, caused by the presence of arsenic (As) in the soil and irrigation water in which the vegetables were grown. Other cultivated areas of the northern zone also demonstrated high t-As levels, but lower than for the Altiplano. In contrast, in the central and southern zones of the country, t-As contents generally fall within the normal ranges reported in the international literature. The results of the analyses of t-As in vegetables demonstrate that arsenic is present in higher concentrations in the tissues of leaves, roots, bulbs and tubers; and in lower concentrations in fruits, pulses and grains. In the case of fruits, moreover, the results show that As is concentrated more in the skin than in the flesh of the fruit.

34.1 INTRODUCTION

Studies in recent years have demonstrated the deleterious health impacts that exposure to arsenic (As) can cause in human populations (Smith *et al.* 1992, Ferreccio *et al.* 2000). In the north of Chile, between $20°5'$ and $26°5'$, there are high As concentrations in the environment caused by Quaternary volcanic and geothermal activity in the Andes (Enríquez 1978, Romero *et al.* 2003). There is extensive mining activity throughout the area. In addition to great aridity, the soils of this northern zone are characterized by the presence of arsenic, iron, molybdenum, lead and zinc (González *et al.* 1997, De Gregori *et al.* 2004). Northern soils are neutral or slightly alkaline, sandy-loams with low organic matter and iron contents (Luzio and Alcayata 1986). The soils in the zone are not suitable for the development of extensive agriculture. However, there are some small areas cultivated by native farmers for the consumption of some 4000 indigenous peoples, mainly Atacameños and Quechuans. Very little of this production reaches larger towns like Antofagasta or Calama.

Unlike the north, the soils of the central zone of Chile offer favorable environmental conditions, a Mediterranean climate, and ideal conditions for the cultivation of fruits and vegetables. In the southern zone, in contrast, the production of animal fodder and the cultivation of wheat, potatoes and sugar beets predominates, due to both geological and climatological factors (ODEPA 2002). For these reasons, the Chilean population is supplied with fruits and vegetables grown principally in the central zone and also from certain parts of the northern zone (Arica, Azapa, Coquimbo, La Serena), while meat, dairy products, potatoes and wheat come from the southern zone.

Vegetables cultivated in environments with elevated arsenic contents may contain harmful levels of arsenic (Elliot and Shields 1988, Sims and Kline 1991, Ma and Rao 1997, Velez and Montoro 2001). The bioavailability of As in soil can be affected by pH, redox potential, texture, clay content and type; organic mater content; iron-, manganese- and aluminum-oxides (Rieuwertz *et al.* 1998, Jiang and Singh 1994, Silviera *et al.* 2003).

34.2   MATERIALS AND METHODS

The principal point of reference for this study was a Canadian arsenic in vegetable study (Dabeka *et al.* 1993). The sampling was undertaken from 1994 to1995 within the framework of a research program designed to gather data concerning the presence of As in the Chilean environment in order to establish standards for quality and emissions.

Samples of vegetables were taken from markets and produce stands throughout the country. For the indigenous populations of the north, along with vegetables, soil and irrigation water samples were also taken. Vegetable samples were washed with distilled water to remove any traces of soil. The samples were frozen at −20°C and then freeze-dried. The lyophilized samples were stored at 4°C prior to being analyzed. The parts of the vegetables analyzed were the edible parts. None were cooked. The number of samples analyzed, although not statistically representative, do allow of estimations of t-As contamination. Vegetables and soils samples were analyzed for t-As after a nitric/sulfuric/percloric acid digestion. Total As (t-As) was analyzed using hydride generation-atomic absorption spectrometry (Perkin Elmer 2100-MHS 20). Analytical quality control for water samples was provided by accurate analysis of SLRS-2 Riverine Water Reference Material for Trace Metals, Marine Analytical Chemistry Standard Program, National Research Council Canada (0.77 ± 0.09 µg/l). All the results were based on peak area measurements.

34.3   RESULTS AND DISCUSSION

Table 34.1 shows the results obtained in the sampling of vegetables cultivated by indigenous peoples in the Altiplano as well as As in irrigation water and soils. It is interesting to note that the level of As content for each crop corresponds proportionately to the level of As contents in the irrigation water and soil in which the crop grew. These results suggest tendencies only, and do not present a clear pattern of behavior.

The results are in agreement with other studies (Pastenes *et al.* 1983, Queirolo *et al.* 2000a, Queirolo *et al.* 2000b, Flyn *et al.* 2002, Diaz *et al.* 2004). High t-As concentrations were found in all soils tested in the northern zone, especially at sites located in the Loa valley.

Table 34.1 and Figure 34.1 show the results obtained from samples taken from markets and produce stands from throughout the country. If these contents are compared with those reported

Table 34.1.   Arsenic levels (ng/g fresh base) in vegetables.

		Foliage	Bulbs and roots	Grains and pulses	Fruits
North Chile	N (samples)	34	37	31	23
	Maximum	207	550	93	61
	Minimum	<2	<2	<2	<2
	Mean	135	124	23	11
Central Chile	N (samples)	34	51	5	17
	Maximum	260	93	48	47
	Minimum	<2	<2	<2	<2
	Mean	88	14	17	11
South Chile	N (samples)	11	14	1	3
	Maximum	125	10	16	<2
	Minimum	<2	<2	16	<2
	Mean	48	<2	16	<2
Reference Canada[1]	Mean	<0.1–10	<0.1–20	<0.1–14	<0.1–14

Foliage: beet, lettuce, spinach, parsley, cabbage. Bulbs and roots: garlic, swiss chard, carrot, onion, potatoes. Grain and pulses: broad bean, corn, kidney bean; [1]Dabeka *et al.* (1993).

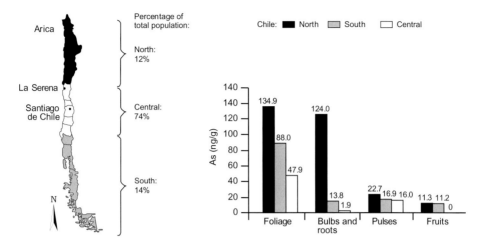

Figure 34.1.   Arsenic levels (ng/g fresh base) in vegetables cultivated in Chile.

Table 34.2.   Arsenic in water, soil and vegetables grown in northern Chile.

Vegetable	Irrigation water (mg/l)	Soil (mg/kg)	Edible portion (ng/g fresh wt.)
Foliage			
Cabbage	0.17	220	54
	0.22	108	33
	0.62	448	715
Chard	<0.002	64	218
	0.17	220	282
	0.62	448	718
Bulbs and roots			
Beetroot	0.22	108	156
	0.62	448	520
Radish	0.17	220	207
	0.62	448	938
Garlic	0.01	86	18
	0.22	108	50
Onion	<0.002	64	36
	0.17	220	106
Potato	0.17	220	40
	0.22	108	44
Pulses			
Broad beam	<0.002	64	30
	0.17	220	152
	0.22	108	44

for Canada (Dabeka *et al.* 1993), it can be observed that in the northern zone the t-As values found in vegetables both from markets and produce stands and from the Altiplano (see Table 34.2) greatly exceed those reported in Canada. In the central and southern zones, median concentrations for bulbs and roots (garlic, beets, carrots, onions, potatoes), and for fruits, are within reported Canadian ranges. However, the Chilean values are slightly higher than those of Canada for greens (swiss chard, lettuce, parsley, spinach) and legumes and grains (broad beans, peas, corn).

Table 34.3.   Distribution of arsenic in fruits (ng/g fresh wt.).

| Fruits (n of samples) | Chile | | Canada |
	Flesh	Skin	
Papaya (1)	1.4	5.6	
Lemon (3)	11.0–20.0	25.0	
Peach (3)	1.0–5.8	2.0–19.0	<0.5–7.6
Pepino (3)	5.6–12.0	5.3–33.0	
Apple (4)	<2–15.0	19.0	0.3–8.5
Grape (3)	<2–24.0		<3.3–5.9
Orange (4)	5.6–11.0		0.9–14.0
Kiwi (1)	6.7		
Melon (4)	3.4–56.0	12.0	
Peal (2)	<2–4.9	10.0	0.7–6.2
Mango (1)	27.0		
Guavapple (1)	33.0		

Studies of t-As in vegetables obtained from As-endemic areas have increased in recent years (Muñoz *et al.* 2002, Roychowdhury *et al.* 2002, Alam *et al.* 2003, Schoof *et al.* 1999, Díaz *et al.* 2004). Nonetheless, comparing such results is not easy because, as some of these studies warn, As concentrations in vegetables will depend, among other factors, on the bioavailability of As in the soil. Bioavailability is difficult to predict because it depends on factors such as pH, organic matter content, and the content and types of clay in the soil (Sims and Kline 1991, Ma and Rao 1997, Rieuwertz *et al.* 1998, Silviera *et al.* 2003). Another factor which could affect some of these findings is the irrigation technique utilized; good drainage could diminish bioavailability (Ayers and Wescot 1994).

Table 34.3 reports t-As contents in the pulp and skin of some fruits cultivated in Chile. Comparing these results with those reported by Canadian researchers (Dabeka *et al.* 1993), it is noted that the As levels we encountered fall within the ranges reported by them. In the majority of this limited number of crops t-As contents are higher in the skin than in the fleshy part of the fruits. The finding could indicate varying As distributions according to the type of tissue involved or external contamination from dust.

34.4   CONCLUSIONS

The results of this study demonstrate that, in general, arsenic is not distributed homogeneously in all the different vegetables grown in the same type of soil nor irrigated with the same type of water, nor within the different tissue structures of the same vegetable. From these results it could be inferred that As appears to be concentrated in greater amounts in leaves, bulbs, tubers and roots and less in grains, pulses and fruits. This in turn seems to suggest that subject to more extensive future investigations, certain measures could be undertaken to diminish the risks associated with the use of As-contaminated soil and water by selecting for cultivation only those crops whose edible tissues present the lowest levels of contamination.

ACKNOWLEDGEMENTS

The authors gratefully acknowledge the financial support provided by FONDEF/CONICYT for this study. We also thank the staff at the *Instituto de Salud Pública de Chile* (Public Health Institute of Chile) who performed the analysis, and Drs. C. Ferreccio and P. Frenz for their valuable suggestions.

REFERENCES

Alam, M.G.M., Snow, E.T. and Tanaka, A.: Arsenic and heavy metal contamination of vegetables grown in Samta Village, Bangladesh. *Sci. Total Environ.* 308 (2003), pp. 83–96.

Ayers, R.S. and Wescot, D.W.: Water quality for agriculture. FAO irrigation and drainage paper 29 Rev. 1; Rome, Italy, 1994, reprinted 1989.

Dabeka, R., McKenzie, A., Lacroix, G., Cleroux, Ch., Bowe, S., Graham, R. and Conacher, H.: Survey of arsenic in total diet food composites and estimation of the dietary intake of arsenic by Canadian adults and children. *Journal of AOAC International* 76:1 (1993), pp. 14–25.

De Gregori, I., Fuentes, E., Olivares, D. and Pinochet, H.: Extractable copper, arsenic and antimony by EDTA solution from agricultural Chilean soils and its transfer to alfalfa plants (*Medicago sativa* L.) *J. Environ. Monit.* 6 (2004), pp. 1–11.

De Gregori, I., Fuentes, E., Rojas, M., Pinochet, H. and Potin-Gautier, M.: Monitoring of copper, arsenic and antimony levels in agricultural soils impacted and non-impacted by mining activities, from three region in Chile. *J. Environ. Monit.* 5 (2003), pp. 287–295.

Diaz, O.P., Leyton, I., Muñoz, O., Núñez, N., Devesa, V., Súñer, M.A., Vélez, D. and Montoro, R.: Contribution of water, bread, and vegetables (raw and cooked) to dietary intake of inorganic arsenic in a rural village of northern Chile. *J. Agric. Food Chem.* 52 (2004), pp. 1773–1779.

Enriquez, H.: Relación entre el contenido de arsénico en agua y el volcanismo cuaternario en Chile, Bolivia y Perú. Documentos Técnicos en Hidrología, UNESCO, Montevideo, Uruguay, 1978.

Ferreccio, C., Gonzalez, C., Milosavjlevic, V., Marshall, G., Sancha, A.M. and Smith A.: Lung cancer and arsenic concentrations in drinking water in Chile. *Epidemiology* 11:6 (2000), pp. 673–679.

Flyn, H., Mc Mahon, V., Chong, G., Demergasso, C., Corbisier, P., Meharg, A. and Paton, G.: Assessment of bioavailable arsenic and copper in soils and sediments from the Antofagasta region of northern Chile. *Sci. Total Environ.* 286 (2002), pp. 51–59.

Gonzalez, M., Ite, R. and Galvez, X.: Heavy metal profiles in Chilean agricultural soils: a transect from the III Region of Atacama to the XI Region of Aysen. In: I.K. Iskandar, S.E. Hardy, A.C. Chang and G.M. Pierzynski (eds). *Fourth International Conference on the Biogeochemistry of Trace Elements, Extended abstracts*, Berkeley California U.S.A., 1997, pp. 43–44.

Jiang, Q.Q. and Singh, B.R.: Effect of different forms and sources of arsenic on crop yield and arsenic concentration. *Water Air Soil Pollut.* 74 (1994), pp. 321–343.

Luzio, W. and Alcayata, S.: Clasificación taxonómica de los suelos de regiones deserticas y áridas del norte de Chile. *Sociedad Chilena de las Ciencias del Suelo Boletin* 5 (1986), pp. 141–145.

Ma, L.Q. and Rao, G.: Chemical fractionation of cadmium, copper, nickel and zinc in contaminated soils. *J. Environ. Qual.* 26 (1997), pp. 259–264.

Muñoz, O., Diaz, O., Leyton, I., Nunez, N., Devesa, V., Súñer, M.A., Vélez, D. and Montoro, R.: Vegetables collected in the cultivated Andean area of northern Chile: Total and inorganic arsenic contents in raw vegetables. *J. Agric. Food Chem.* 50 (2002), pp. 642–627.

ODEPA: Agricultura Chilena. Rubros segun tipo de productor y localización geográfica. Working document 8, http//www.odepa.gob.cl, Santiago de Chile, Chile, 2002.

Pastenes, J., Acevedo, E., Valladares, I. and Irachet, R.: Relaciones del contenido de arsénico en el sistema agua-suelo-planta en el valle de Chiu Chiu. *Bol. Soc. Chil. Quim.* 28 (1983), pp. 480–482.

Queirolo, F., Stegen, S., Mondaca, J., Cortes, R., Rojas, R., Contreras, C., Muñoz, L., Schwuger, M.J. and Ostapczuk, P.: Total arsenic, lead, cadmium, copper, and zinc in some salt rivers in northern Andes of Antofagasta, Chile. *Sci. Total Environ.* 255 (2000a), pp. 85–95.

Queirolo, F., Stegen, S., Restovic, M., Paz, M., Ostapczuk, P., Schwuger, M.J. and Muñoz, L.: Total arsenic, lead, and cadmium levels in vegetables cultivated at the Andean villages of northern Chile. *Sci. Total Environ.* 255 (2000b), pp. 75–84.

Rieuwerts, J.S., Thonton, I., Farago, M.E. and Ashmore, M.R.: Factors influencing metals bioavailability in soils: preliminary investigations for the development of a critical loads approach for metals. *Chem. Speciation Bioavailability* 10:2 (1998), pp. 61–75.

Romero, L., Alonso, H., Campano, P., Fanfani, L., Cidu, R., Dabea, C., Keegan, T., Thornton, I. and Farago, M.: Arsenic enrichment in waters and sediments of the Rio Loa (second region, Chile). *Appl. Geochem.* 18 (2003), pp. 1399–1416.

Roychowdhury, T., Uchino, T., Tokunaga, H. and Ando, M.: Survey of Arsenic in food composites from an arsenic-affected area of West Bengal, India. *Food Chem. Toxicol.* 40 (2002), pp. 1611–1621.

Schoof, R.A., Yost, L., Eickhoff, J., Crecelius, E.A., Cragin, D.W., Meacher, D.M. and Menzel, D.B.: A market basket survey of inorganic arsenic in food. *Food Chem. Toxicol.* 37 (1999), pp. 839–846.

Silviera, M.L.A., Alleoni, L.R.F. and Guilherme, L.R.G.: Biosolids and heavy metals in soils. *Scientia Agricola* 60:4 (2003), pp. 793–806.

Sims, J.T. and Kline, J.S.: Chemical fractionation and plant uptake of heavy metals in soils amended with co-composted sewage sludge *J. Environ. Qual.* 20 (1991), pp. 387–395.

Smith, A.H., Hopenhayn-Rich, C., Bates, M.N. *et al.*: Cancer risks from arsenic in drinking water. *Environ. Health Perspect.* 97 (1992), pp. 259–267.

Velez, D. and Montoro, R.: Inorganic arsenic in foods: current overview and future challenges. *Recent Res. Devel. Agricultural and Food Chem.* 5 (2001), pp. 55–71.

CHAPTER 35

Assimilation of arsenic into edible plants grown in soil irrigated with contaminated groundwater

I.M.M. Rahman, M. Nazim Uddin & M.T. Hasan
Applied Research Laboratory, Department of Chemistry, University of Chittagong, Chittagong, Bangladesh

M.M. Hossain
Institute of Forestry and Environmental Sciences, University of Chittagong, Chittagong, Bangladesh

ABSTRACT: The arsenic (As) calamity of Bangladesh is the largest known mass poisoning in the history of mankind. Edible vegetables, medicinal and aromatic plants grown in As-contaminated soil may uptake and accumulate significant amount of As in their tissue which has been studied in this chapter. The plants studied during the present investigation were *Lablab niger* (bean), *Lycopersicon esculentum* (tomato), *Solanum melongena* (brinjal), *Cucurbita maxima* (sweet gourd), *Amaranthus gangeticus* (red amaranth), *Carica papaya* (green papaya), *Capsicum* sp. (chilli), *Lagenaria siceraria* (bottle gourd), *Momordica charantia* (bitter gourd), *Mentha viridis* (mint), *Vigna sesquipedalis* (string bean), *Abelmoschus esculentus* (okra), *Trichosanthes dioica* (palwal), *Basella alba* (Indian spinach). Mean As concentration in the selected plants in the area studied was 0.113 μg/g fresh weight. The minimum was found in *T. dioica* (0.026 μg/g) and the maximum in *M. viridis* (0.566 μg/g) followed by *V. sesquipedalis*, *Capsicum* sp. (0.400, 0.200 μg/g, respectively) while As contents in *A. esculentus*, *B. alba* and *C. papaya* were below detectable limit. The average dietary intake of As from the plants in the study area was estimated to be 14.69 μg/day. Correlation with the groundwater As status and statistical significance of variations has also been determined.

35.1   INTRODUCTION

Groundwater in Bangladesh contains arsenic (As) above acceptable limit, an exposure which brought millions of people under the threat of lethal diseases (Milton *et al.* 2004, 2003, Khan *et al.* 2003, Mitra *et al.* 2002, Kadono *et al.* 2002, Rahman *et al.* 2001, Smith *et al.* 2000). In addition to direct consumption of As-contaminated groundwater for drinking, it is also used for irrigation and cooking. Numerous greenhouse studies revealed that an increase in As concentration in cultivated soils leads to an increase in the levels of As in edible vegetables (Buat-Menard *et al.* 1987, Burlo *et al.* 1999, Carbonell-Barachina *et al.* 1999, Helgensen and Larsen 1998, Larsen *et al.* 1992). Significant As uptake by rice and a range of vegetable crops that were irrigated by contaminated water has been reported (Meharg 2004, Meharg and Rahman 2003, Alam and Rahman 2003, Alam *et al.* 2003). In Bangladesh, vegetables production in home gardens is a traditional practice. According to Hassan and Ahmad (1984), a Bangladeshi person, regardless of gender, consumes an average of 130 grams vegetables per day (leafy and non-leafy) and in the total diet, the proportion varied from 12 to 21%. Home garden soils are usually irrigated with water from shallow tube wells, which often contains elevated arsenic concentrations. Therefore, As transfer into the food chain is a sheer possibility and may pose a long-term effect to the public health in Bangladesh. The present study focuses the extent of As in groundwater as well as in some edible plants grown in the homestead gardens of Feni district (an administrative block) of Bangladesh, irrigated mostly by As-contaminated groundwater.

35.2   MATERIAL AND METHODS

35.2.1   *Study area*

Water and vegetable samples were collected from Feni district of Bangladesh. Feni is in the southwestern part of Bangladesh with an area of 928.34 km^2 and consists of six *upazilas* (small administrative units) namely Sonagazi, Feni Sadar, Daganbhuiyan, Parshuram, Chhagalnaiya and Phulgazi (Fig. 35.1). Total population is 1,196,219 (male 42.92%, female 57.08%). Occupation of 48.21% of the inhabitants is agricultural production and in terms of land control, only 1% of the total population can be considered rich (Anon 2003).

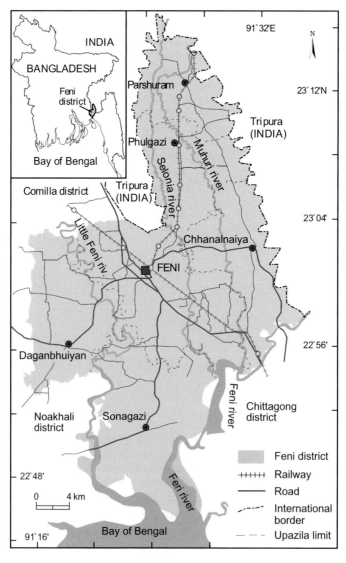

Figure 35.1.   Map of the study area in Feni district (Bangladesh).

35.2.2 *Sampling, preservation, pretreatment and determination of arsenic*

35.2.2.1 *Water*

Groundwater samples, 80 from Feni Sadar and 50 from each of Sonagazi, Daganbhuiyan, Parshuram, Chhagalnaiya and Phulgazi, were randomly extracted from shallow tube wells. Samples were preserved in pre-washed polyethylene bottles, adding 0.01% HNO_3 and kept at 4°C before analysis. Total arsenic (t-As) in the water samples was measured by Ag-DDTC-hexamethylenetetramine-chloroform method, with a detection limit of 0.020 µg/l of As (APHA 1971, Sandhu and Nelson 1979).

35.2.2.2 *Plants*

Fresh random-samples of 14 edible plant species (vegetables) were collected from home gardens of the study area with replications for each *upazila* (Table 35.1). For analysis, the edible parts of each plant were severed by hand using vinyl gloves, carefully packed into polyethylene bags and weighed *in situ*. The samples were washed three times with distilled water and finally rinsed with de-ionized water to eliminate the pollutants, dried in an oven at 65°C for 24 h, reweighed to determine water content, and then, for metal analysis, grinded using a ceramic-coated grinder.

Plant parts (10–25 g) were taken into a 100 ml Microkjeldhal flask with a glass bead and 15 ml concentrated nitric acid. The flask was then placed on the digester and gently heated. The solution was removed and cooled after the initial brisk reaction. Concentrated sulfuric acid (4 ml) was then added carefully to the solution followed by the addition of 2 ml of 70% perchloric acid. Heating of the solution was continued till the formation of dense SO_3 fumes, repeating nitric acid addition, if necessary. The solution was then refluxed at 110–120°C. The residue was dissolved in distilled water and was filtered into 100 ml volumetric flask and made up to the mark. The digested sample solutions were injected by an automatic sampler and analyzed by using air acetylene flame with combination as well as single element hollow cathode lamps into an atomic absorption spectrophotometer (Model-Shimadzu, AA-6401F). The detection limit was 0.002 mg/l of As.

35.2.2.3 *Statistical analysis*

SPSS for Windows (version 11) was used for all statistical analyses. Statistical significance was considered valid only at 5% level.

Table 35.1.   English, scientific and family name of the sampled vegetables.

Family	Scientific name	English name
Amaranthaceae	*Amaranthus gangeticus*	Red amaranth
Basellacease	*Basella alba*	Indian spinach
Caricaceae	*Carica papaya*	Green papaya
Cucurbitaceae	*Cucurbita maxima*	Sweet gourd
Cucurbitaceae	*Lagenaria siceraria*	Bottle gourd
Cucurbitaceae	*Momordica charantia*	Bitter gourd
Cucurbitaceae	*Trichosanthes dioica*	Palwal
Labiateae	*Mentha viridis*	Mint
Leguminoseae	*Lablab niger*	Hyacinth bean
Leguminoseae	*Vigna sesquipedalis*	String bean
Malvaceae	*Abelmoschus esculentus*	Okra
Solanaceae	*Capsicum* sp.	Chilli
Solanaceae	*Lycopersicon esculentum*	Tomato
Solanaceae	*Solanum melongena*	Brinjal

35.3   RESULT AND DISCUSSION

All groundwater samples have been analyzed for the concentration of As and As content as low as 0.002 mg/l and as high as 0.305 mg/l have been observed. *Upazila* based results (Table 35.2) substantiate that 39.71% of the samples have As content above Bangladesh Guideline Standard (BGS) of 0.05 mg/l and it is 55.88% when WHO recommended guideline value of 0.01 mg/l is considered i.e. more than half of the screened tube wells have As content above this guideline value.

A groundwater monitoring report for the two adjacent localities—Mirsharai and Sitakundu *upazillas* of Chittagong district showed As content above BGS in approximately 94% and 83% of the studied samples, respectively (Rahman 2003). Due to the absence of community water supply systems or better alternatives, the inhabitants of the area use tube well water not only for drinking but also for other household purposes and irrigation. As natural water recovery process is insufficient to cope with the withdrawal rate, excessive pumping of groundwater for miscellaneous purposes results in lowering of water table, as well as appears to trigger frequent As mobilization into the groundwater (Mandal *et al.* 1996, Mallik and Rajagopal 1996, Acharyya *et al.* 2000, Bhattacharya *et al.* 1997, Nickson *et al.* 2000, McArthur *et al.* 2001, Dowling *et al.* 2002, Anawar *et al.* 2003). Our randomized study indicates most of the tube wells (84.12%) of Feni district are

Table 35.2.   Arsenic status at different *upazilas* of Feni district.

Upazila name	Arsenic concentration range (mg/l)			
	BDL[1] %	<0.010[2] %	0.010–<0.05[3] %	0.05–>0.05[4] %
Sonagazi	31.67	11.67	15	41.67
Feni Sadar	28.75	12.5	17.5	41.25
Daganbhuiyan	22	16	14	48
Parshuram	36	18	14	32
Chagalnaiya	26	16	20	38
Phulgazi	38	10	16	36

[1]BDL: Below detectable limit; [2]Safe; [3]Above WHO standard; [4]Above Bangladesh standard.

Table 35.3.   Tube well depth-dependent distribution of arsenic contamination at Feni district.

Depth (m)	Arsenic concentration range (mg/l)			
	BDL[1] %	<0.010[2] %	0.010–<0.05[3] %	0.05–>0.05[4] %
<15	1.76	1.47	0.88	5.88
15 to 20	4.41	2.06	2.94	8.53
>20 to 25	7.35	3.82	4.12	9.41
>25 to 30	3.82	1.18	2.35	7.06
>30 to 35	5	2.65	3.82	7.35
>35 to 40	2.94	2.06	1.76	1.47
>40 to 45	0.29	0	0	0
>45 to 50	0.29	0	0.29	0
>50 to 60	0.59	0.59	0	0
>60 to 200	1.18	0	0	0
>200	2.65	0	0	0

[1]BDL: Below detectable limit; [2]Safe; [3]Above WHO standard; [4]Above Bangladesh standard.

in the depth between 15 m and 40 m and for that particular depth zone, 48.81% of the samples are found to contain As above WHO safety limit (Table 35.3). It also posts a notice of caution because a number of studies on groundwater As-depth relation in different parts of Bangladesh showed that maximum As concentration occur at depths between 20 and 50 m (BGS and DPHE 2001, NRECA 1997, Broms and Fogelstrom 2001).

Mean As concentrations in vegetables were found to be higher than of those grown on untreated or uncontaminated soils. To estimate the risks from As-laden diets, Feni is selected as a model for Bangladesh because in this area As contamination of groundwater is yet to show its visual existence through arsenicosis suffered patients though contamination exists. Also, home gardening is a common practice in this locality and our field observations show that the inhabitants find it is convenient to irrigate their home garden with tube well water. Arsenic concentrations in fourteen different vegetable species from Feni were studied, the maximum was observed in the mint (0.566 μg/g fresh weight) and the minimum was in Palwal (0.026 μg/g FW) while it was below detectable limit in okra, Indian spinach and green papaya. As reported in literature, total As contents in food products of vegetable origin ranged <0.004–0.303 μg/g FW (Urieta *et al.* 1996, Dabeka *et al.* 1993, Schoof *et al.* 1999, Ysart *et al.* 1999) which is lower than the range of values found in the present study. Average As concentration in plants of Feni was 0.113 μg/g FW and it was higher than that of United Kingdom, 0.003, μg/g FW (MAFF 1997) and Croatia 0.0004 μg/g FW (Sapunar-Postruznik *et al.* 1996). The highest As contaminations were observed in plants from Phulgazi (0.184 μg/g FW) and the lowest in Chhagalnaiya (0.048 μg/g FW). In Sonagazi and Daganbhuiyan, highest mean As content was observed in bean while in Feni Sadar, Parshuram, Chhagalnaiya and Phulgazi highest As content was respectively in tomato, mint, brinjal and string bean.

Duncan's multiple range test (DMRT) showed that green papaya, okra, palwal, sweet gourd, red amaranth had the lowest As concentrations among the plants analyzed. Though the order of mean As concentration among them was green papaya < okra < palwal < sweet gourd < red amaranth but the differences were not statistically significant (P > 0.05). Arsenic concentration in string bean was significantly higher than all other plants (P < 0.05). The other plants can be arranged according to their mean As concentration as tomato < brinjal < bottle gourd < bean < bitter gourd < chilli, however the differences were not significant. Their As content was slightly higher than the previous five plants and marginally lower than string bean. Distribution of As in different vegetable species without any momentous pattern has also been observed previously for Samta village of Bangladesh (Alam *et al.* 2003). The t-values and the associated significance values indicate that there was significant variation in As concentrations among the samples of bean at 5% level but not at 1% level.

In light of legislation and health considerations, the vegetable products of Feni are safe to consume because the average As concentrations in the vegetables are much lower than the country limit (1 mg/kg). It is also safe considering legislations from other countries: 1 mg/kg in Guyana, Jamaica, Trinidad and Tobago, Kenya, Zambia, Malaysia, Singapore and the United Kingdom; 1.5 mg/kg in Papua New Guinea. Only one sample of mint (0.566 mg/kg) analyzed in the present study exceeded the maximum allowed limit of 0.5 mg/kg as set by Bulgaria, Czech Republic, Slovak Republic and Hungary (Anon 1993).

During our survey, we have tried to amass the information regarding the financial condition and food habit of inhabitants of Feni to measure the dietary consumption pattern of vegetables. Our study shows that most of them are poor and rice and vegetables are their main dish. They take fish once or twice in a month while meat is a dish of festival only. Excluding the contribution of rice, pulses, meats, fishes and spices to dietary exposures, the average dietary intake of t-As from vegetable by the inhabitants of Feni was estimated to be 14.69 μg/day. Daily dietary intake of As as estimated are higher than that of Belgium: 12 μg/day (Buchet *et al.* 1983) and Croatia: 11.7 μg/day (Sapunar-Postruznik *et al.* 1996) but lower than the Netherlands: 15 μg/day (De Vos *et al.* 1984), Canada: 59.2 μg/day (Dabeka *et al.* 1993), Sweden: 60 μg/day (Jorhem *et al.* 1998), Japan: 160–280 μg/day (Tsuda *et al.* 1995) and Spain: 291 μg/day (Urieta *et al.* 1996). Present study only includes the determination of total As content without speciation in the vegetables.

As found in the literature, inorganic As species content in diets, so far, as follows: 40% (USEPA 1988), 65% (Dabeka *et al.* 1993), 95–96% (Chowdhury *et al.* 2001) and 100% (Tao and Bolger 1998). Based on those reports, we can assume that at least 50% of the t-As in the samples studied is inorganic. Then the daily dietary intake of inorganic arsenic (i-As) from vegetables in area investigated is 7.34 µg. From the toxicological point of view, inorganic As compounds are most toxic and according to WHO (1992), a daily intake of 2 µg of i-As/kg body weight should not be exceeded to minimize the risk to humans. However, nutritional status of diets is also an important factor in such cases. People eating nutritious foods can tolerate As up to certain range in spite of high dietary As consumption (Harrington *et al.* 1978, USEPA 1988, Das *et al.* 1995). As surveyed, most of the locals of Feni are poor and can hardly avail nutritious food. From the legislation point of view, consumption of the vegetables from Feni are proved safe but there may still be a definite health-risk for the inhabitants, if the present rate of dietary consumption pattern exists in combination with drinking As-contaminated water over a long period of time.

35.4   CONCLUSION

Effects of As-rich irrigation-water to the vegetables grown in the homestead gardens were investigated in the present study. Elevated levels of As in the analyzed water and some plant samples were observed and the phenomenon of soil-crop-food transfer was confirmed. Dietary As intake was calculated and the average dietary intake pattern of the inhabitants in the studied area showed that they are potentially at risk of As-related hazards. Over exploitation of groundwater initiated the release As into the groundwater and excessive use of groundwater for crop irrigation could instigate a new dimension in existing risk from groundwater As in Bangladesh.

REFERENCES

Acharyya, S.K., Lahiri, S., Raymahashay, B.C. and Bhowmik, A.: Arsenic toxicity of groundwater of the Bengal basin in India and Bangladesh: the role of Quaternary stratigraphy and Holocene sea-level fluctuation. *Environ. Geol.* 39 (2000), pp. 1127–1137.
Alam, M.G.M., Snow, E.T. and Tanaka, A.: Arsenic and heavy metal contamination of vegetables grown in Samta Village Bangladesh. *Sci. Total Environ.* 308 (2003), pp. 83–96.
Alam, M.Z. and Rahman, M.M.: Accumulation of arsenic in rice plant from arsenic contaminated irrigation water and effect on nutrient content. *Water Sci. Technol.* 42 (2003), pp. 131–135.
Anawar, H.M., Akai, J., Komaki, K., Terao, H., Yoshioka, T., Ishizuka, T., Safiullah, S. and Kato, K.: Geochemical occurrence of arsenic in groundwater of Bangladesh: sources and mobilization processes. *J. Geochem. Explor.* 77 (2003), pp. 109–131.
Anon: British Food Manufacturing Industries Research Association: *Metallic contaminants in foods—A survey of international prescribed limits*. 3rd ed., Food Legislation Surveys 6, Leatherhead Food R.A.: Leatherhead, UK, 1993.
Anon: *Banglapedia: National encyclopedia of Bangladesh*, Volume 4. Asiatic Soc. Bangladesh, Dhaka, 2003, p. 87, http://banglapedia.search.com.bd/HT/F_0055.htm and http://banglapedia.search.com.bd/Maps/MF_0055.GIF.
APHA: *Standard methods for the examination of water and waste water*. 13th ed., American Public Health Association (APHA), Washington, DC, 1971.
BGS and DPHE: Arsenic Contamination of Groundwater in Bangladesh, Vol. 2. Final Report, BGS Technical Report WC/00/19, London, UK, 2001.
Bhattacharya, P., Chatterjee, D. and Jacks, G.: Occurrence of arsenic-contaminated groundwater in alluvial aquifers from delta plains, eastern India: options for safe drinking water supply. *J. Water Resour. Dev.* 13 (1997), pp. 79–92.
Broms, S. and Fogelstrom, J.: *Field investigations of arsenic-rich groundwater in the Bengal Delta Plain, Bangladesh*. MSc Thesis, Series 2001:18, Department of Land and Water Resources Engineering, KTH, Stockholm, Sweden, 2001.
Buat-Menard, P., Peterson, P.T., Havas, M., Steinnes, E. and Turner, D.: Group report: Arsenic. In: T.C. Hutchison and K.M. Meema (eds): *Lead, mercury, cadmium and arsenic in environment*. John Wiley and Sons Ltd, India, 1987, pp. 43–47.

Buchet, J.P., Lauwerys, R., Vanderwoorde, A. and Pycke, J.M.: Oral daily intake of cadmium, lead, manganese, copper, chromium, chromium, mercury, calcium, zinc and arsenic in Belgium: a duplicate meal study. *Food Chem. Toxicol.* 21 (1983), pp. 19–24.

Burlo, F., Guijarro, I., Carbonell-Barrachina, A.A., Valero, D., Martinez-Romero, D. and Martinez-Sanchez, F.: Arsenic species: effects on and accentuation by tomato plants. *J. Agric. Food Chem.* 47 (1999), pp. 1247–1253.

Carbonell-Barachina, A.A., Burlo, F., Valero, D., Lopez, E., Martinez-Romero, D. and Martinez-Sanchez, F.: Arsenic toxicity and accumulation in turnip as affected by arsenic chemical speciation. *J. Agric. Food Chem.* 47 (1999), pp. 2288–2294.

Chowdhury, U.K., Rahman, M.M., Mandal, B.K., Paul, K., Lodh, D., Biswas, B.K., Basu, G.K., Chanda, C.R., Saha, K.C., Mukherjee, S.C., Roy, S., Das, R., Kaies, I., Barua, A.K., Palit, S.K., Quamruzzaman, Q. and Chakraborti, D.: Groundwater arsenic contamination and human suffering in West Bengal, India and Bangladesh. *Environ Sci.* 8 (2001), pp. 393–415.

Dabeka, R.W., McKenzie, A.D., Lacroix, G.M.A., Cleroux, C., Bowe, S., Graham, R.A., Conacher, H.B.S. and Verdier, P.: Survey of arsenic in total diet food composites and estimation of the dietary intake of arsenic by Canadian adults and children. *J. AOAC Int.* 76 (1993), pp. 14–25.

Das, D., Chatterjee, C., Mandal, B.K., Samanta, G. and Chakraborti, D.: Arsenic in groundwater in six Districts of West Bengal, India: the biggest arsenic calamity in the world, Part 2: Arsenic concentration in drinking water, hair, nails, urine, skin-scale and liver tissue (biopsy) of the affected people. *Analyst* 120 (1995), pp. 917–924.

De Vos, R.H., Van Dokkum, W., Olthof, P.D.A., Quiruns, J.K., Muys, T. and Vander Poll, J.M.: Pesticides and other chemical residues in Dutch total diet samples (June 1976–July 1978). *Food Chem. Toxicol.* 22:1 (1984), pp. 11–21.

Dowling, C.B., Poreda, R.J., Basu, A.R. and Peters, S.L.: Geochemical study of arsenic release mechanisms in the Bengal Basin groundwater. *Water Resour. Res.* 38 (2002), pp. 1173–1190.

Harrington, J.M., Middaugh, J.P., Morse, D.L. and Housworth, J.: A survey of a population exposed to high concentrations of arsenic in well water, in Fairbanks, Alaska. *Am. J. Epidemiol.* 108 (1978), pp. 377–385.

Hassan, N. and Ahmad, K.: Intra-familial distribution of food in rural Bangladesh. *Food and Nutrition Bulletin* 6 (4), The United Nations University Press, Tokyo, Japan, 1984, http://www.unu.edu/unupress/food/8F064e/8F064E05.htm

Helgensen, H. and Larsen, E.H.: Bioavailability and speciation of arsenic in carrots grown in contaminated soil. *Analyst* 123 (1998), pp. 791–796.

Jorhem, L., Becker, W. and Slorach, S.: Intake of 17 elements by Swedish women, determined by a 24-h duplicate portion study. *J. Food Comp. Anal.* 11 (1998), pp. 32–46.

Kadono, T., Inaoka, T., Murayama, N., Ushijima, K., Nagano, M., Nakamura, S., Watanabe, C., Tamaki, K. and Ohtsuka, R.: Skin manifestations of arsenicosis in two villages in Bangladesh. *Int. J. Dermatol.* 41:12 (2002), pp. 841–846.

Khan, M.M., Sakauchi, F., Sonoda, T., Washio, M. and Mori, M.: Magnitude of arsenic toxicity in tube well drinking water in Bangladesh and its adverse effects on human health including cancer: evidence from a review of the literature. *Asian Pac. J. Cancer Prev.* 4:1 (2003), pp. 7–14.

Larsen, E.H., Moseholm, L. and Nielsen, M.M.: Atmospheric deposition of trace elements around point sources and human health risk assessment: II. Uptake of arsenic and chromium by vegetables grown near a wood preservation factory. *Sci. Total Environ.* 126 (1992), pp. 263–275.

Mallik, S. and Rajagopal, N.R.: Groundwater development in the arsenic-affected alluvial belt of West Bengal—Some questions. *Curr. Sci.* 70 (1996), pp. 956–958.

Mandal, B.K., Roy Chowdhury, T., Samanta, G., Basu, G.K., Chowdhury, P.P., Chanda, C.R., Lodh, D., Karan, N.K., Dhar, R.K., Tamili, D.K., Das, D., Saha, K.C. and Chakraborti, D.: Arsenic in groundwater in seven districts of West Bengal, India: the biggest arsenic calamity in the world. *Curr. Sci.* 70 (1996), pp. 976–986.

McArthur, J.M., Ravenscroft, P., Safiullah, S. and Thirlwall, M.F.: Arsenic in groundwater: testing pollution mechanism for sedimentary aquifers in Bangladesh. *Water Resour. Res.* 37 (2001), pp. 109–117.

Meharg, A.A.: Arsenic in rice-understanding a new disaster for South-East Asia. *Trends Plant Sci.* 9:9 (2004), pp. 415–417.

Meharg, A.A. and Rahman, M.M.: Arsenic contamination of Bangladesh paddy field soils: Implications for rice contribution to arsenic consumption. *Environ. Sci. Technol.* 37:2 (2003), pp. 229–234.

Milton, A.H., Hasan, Z., Rahman, A. and Rahman, M.: Non-cancer effects of chronic arsenicosis in Bangladesh: preliminary results. *J. Environ. Sci. Health A: Tox. Hazard Subst. Environ. Eng.* 38:1 (2003), pp. 301–305.

Milton, A.H., Hasan, Z., Shahidullah, S.M., Sharmin, S., Jakariya, M.D., Rahman, M., Dear, K. and Smith, W:
Association between nutritional status and arsenicosis due to chronic arsenic exposure in Bangladesh. *Int. J. Environ. Health Res.* 14:2 (2004), pp. 99–108.

MAFF: Total diet study: aluminium, arsenic, cadmium, chromium, copper, lead, mercury, nickel, selenium, tin and zinc. Food surveillance information sheet, No. 191, HMSO, Ministry of Agriculture, Fisheries and Food London, UK, 1997.

Mitra, A.K., Bose, B.K., Kabir, H., Das, B.K. and Hussain, M.: Arsenic-related health problems among hospital patients in southern Bangladesh. *J. Health Popul. Nutr.* 20:3 (2002), pp. 198–204.

Nickson, R.T., McArthur, J.M., Ravenscroft, P., Burgess, W.G. and Ahmed, K.M.: Mechanism of arsenic release to groundwater, Bangladesh and West Bengal. *Appl. Geochem.* 15 (2000), pp. 403–413.

NRECA: Report of study of the impact of the Bangladesh Rural Electrification Program on groundwater quality. Prepared for Bangladesh Rural Electrification Board by NRECA International with personnel provided by The Johnson Company Inc. (USA) and ICDDRB (Dhaka) for USAID, 1997.

Rahman, I.M.M., Majid, M.A., Nazimuddin, M. and Huda, A.S.M.S.: Status of arsenic in groundwater of some selected areas of Chittagong District. *Chitt. Univ. J. Sci.* 27 (2003), pp. 7–12.

Rahman, M.M., Chowdhury, U.K., Mukherjee, S.C., Mondal, B.K., Paul, K., Lodh, D., Biswas, B.K., Chanda, C.R., Basu, G.K., Saha, K.C., Roy, S., Das, R., Palit, S.K., Quamruzzaman, Q. and Chakraborti, D.: Chronic arsenic toxicity in Bangladesh and West Bengal, India—a review and commentary. *J. Toxicol. Clin. Toxicol.* 39:7 (2001), pp. 683–700.

Sandhu, S.S. and Nelson, P.: Concentration and separation of arsenic from polluted water by ion-exchange. *Environ. Sci. Technol.* 13:4 (1979), pp. 476–478.

Sapunar-Postruznik, J., Bazulic, D. and Kubala, H.: Estimation of dietary intake of arsenic in the general population of the Republic of Croatia. *Sci. Total Environ.* 191 (1996), pp. 119–123.

Schoof, R.A., Yost, L.J., Eickhoff, J., Crecelius, E.A., Cragin, D.W., Meacher, D.M. and Menzel, D.B.: A market basket survey of inorganic arsenic in food. *Food Chem. Toxicol.* 37 (1999), pp. 839–846.

Smith, A.H., Lingas, E.O. and Rahman, M.: Contamination of drinking-water by arsenic in Bangladesh: a public health emergency. *Bull. World Health Organ.* 78:9 (2000), pp. 1093–1103.

Tao, S.H. and Bolger, P.M.: Dietary intakes of arsenic in the United States. Paper presented at the 3rd International Conference on Arsenic Exposure and Health Effects, 12–15 July, San Diego, CA (as cited by Alam *et al.*, 2003. *Sci. Total Environ.* 308: 83–96), 1998.

Tsuda, T., Inoue, T., Kojima, M. and Aoki, S.: Market basket and duplicate portion estimation of dietary intakes of cadmium, mercury, arsenic, copper, manganese, and zinc by Japanese adults. *J. AOAC Int.* 78 (1995), pp. 1363–1368.

Urieta, I., Jalon, M. and Eguileor, I.: Food surveillance in the Basque country (Spain). II. Estimation of the dietary intake of organochlorine pesticides, heavy metals, arsenic, aflatoxin M1, ironand zinc through the Total Diet Study, 1990/91. *Food Addit. Contam.* 13 (1996), pp. 29–52.

USEPA: Special report on ingested inorganic arsenic. Skin cancer, nutritional essentiality (EPA/625/3-87/013). Environmental Protection Agency, Washington, DC, 1988.

WHO: Inorganic arsenic compounds other than arsine: health and safety guide. Health and safety guide no. 70, World Health Organization, Geneva, Switzerland, 1992, http://www.inchem.org/documents/hsg/hsg/hsg070.htm.

Ysart, G., Miller, P., Crews, H., Robb, P., Baxter, M., De L'Argy, C., Lofthouse, S., Sargent, C. and Harrison, N.: Dietary exposure estimates of 30 elements from the UK Total Diet Study. *Food Addit. Contam.* 16 (1999), pp. 391–403.

CHAPTER 36

Investigation of arsenic accumulation by vegetables and ferns from As-contaminated areas in Minas Gerais, Brazil

H.E.L. Palmieri & M.A.B.C. Menezes
Centro de Desenvolvimento da Tecnologia Nuclear, Comissão Nacional de Energia Nuclear, Belo Horizonte, Minas Gerais, Brazil

O.R. Vasconcelos
Departamento de Engenharia Sanitária e Ambiental, Universidade Federal de Minas Gerais (UFMG), Belo Horizonte, Minas Gerais, Brazil

E. Deschamps
Fundação Estadual do Meio Ambiente (FEAM), Belo Horizonte, Minas Gerais, Brazil

H.A. Nalini, Jr.
Departamento de Geologia, Universidade Federal de Ouro Preto, Ouro Preto, Minas Gerais, Brazil

ABSTRACT: Soil and sediments around gold ore deposits and mining areas in the Iron Quadrangle presented positive arsenic (As) anomalies (median concentrations >100 mg/kg) and wide range of concentrations (<20 to 2000 mg/kg) even in densely populated areas. This study aims at investigating the presence of As in vegetables and in the ferns *Pteris vittata* and *Pityrogramma calomelanos* as well as the availability of As to ferns in these areas. All vegetables from private gardens investigated showed significant As uptake in both edible and non-edible parts. Enrichments differed among species showing elevated values. The ferns actually extract As from the soil and translocate it into their fronds showing higher As concentrations in the leaves (91–2295 μg/g) than in the rhizoids (25–139 μg/g). Difference in the uptake of As among the sites was related to different mineral contents in the soils, especially Fe, Mn and P. Rubidium and zinc accumulation in leaves and rhizoids was also observed in some samples of *Pteris vittata*.

36.1 INTRODUCTION

The Iron Quadrangle, located in the Brazilian state of Minas Gerais, is one of the richest and best-known mineral deposit structures worldwide. A great number of active and ancient gold mines can be found in this region. The active mines include Morro Velho, Raposos, Cuiabá and São Bento, while among the disused ones is the famous Passagem de Mariana in the town of Ouro Preto-Mariana. The gold ore from these mines is rich in arsenic (As) with the As/Au ratios ranging from 300 to 3000 among the several deposits. In the Iron Quadrangle region, the highest As concentrations in water and sediment occur near the gold mining areas where the river sediments had been contaminated by waste materials discharge as early as the colonial times (Borba 2000). Although present mining operations may no longer contribute significantly to the contamination of soils and sediments, there are many potential risks for As intoxication induced by the dispersion of old tailings, human exposure to polluted soils, and the consumption of contaminated surface and groundwater.

This work was conducted in the districts of Santa Barbara and Ouro Preto-Mariana, where As enrichments have been detected recently in environmental compartments (soil, sediments, water, drinking water and dusts) as well as in human urine samples (Matschullat 2000, Borba 2000,

Deschamps and Matschullat 2007). Vegetables from private gardens were sampled in the city of Santa Barbara in Minas Gerais. These local food products are the basis of human nutrition in this region and of great relevance to human health.

Arsenic is a toxic element to humans and numerous studies have been conducted in order to assess the amount of As and its chemical species present in food and biological samples. According to Koch *et al.* (2000), the predominant As species in terrestrial plants are inorganic forms, which are from a toxicological point of view, the most toxic As species. Determination of As in vegetables is particularly important because this metalloid is not biodegradable and can accumulate in human vital organs, producing progressive toxicity.

The bioavailability of As to plants depends on several physical and chemical factors in the soil. The available As content in the soil is a better indicator of phytotoxicity than the total As concentration. The quantity of soluble or potentially soluble As in soil varies widely with pH, Eh and the presence of other soil components such as Fe, Al and clay minerals and organic matter (O'Neill 1995).

As most trace elements, the uptake rate of As from soil to plants varies widely between plant species. In As-contaminated soil the uptake of As by the plant tissue can occur, particularly in vegetables and edible crops (Larsen *et al.* 1992). Large amounts of inorganic arsenic (i-As) have been found in carrots (*Daucus carota* L.) and other vegetables grown in As-contaminated soil (Pyles and Woolson 1982, Helgesen and Larsen 1998).

Ma *et al.* (2001) and Francesconi *et al.* (2002) have demonstrated that the fern species, *Pteris vittata* and *Pityrogramma* calomelanos are As-hyperaccumulating plants and recommend them for the phytoremediation of As-contaminated soils. These species were suggested for phytoremediation due to their high bioaccumulation factors, short life cycle, high propagation rates, wide distribution, large shoot mass and their ability to tolerate high As concentrations in soils (Visoottiviseth *et al.* 2002). Phytoremediation, an emerging plant-based technology for the removal of toxic elements from the soil and water has been receiving renewed attention. Ferns were collected together with their respective soils in order to evaluate the bioaccumulation factor (BF) at each site. BF is defined as the concentration ratio of As in the plant to the soil medium. Other parameters such as Fe, Mn, P were also determined in ferns and in soils in order to evaluate the availability of As to ferns.

36.2   EXPERIMENTAL

Some vegetables, like yam, bean, sweet potato, lettuce, kale and mustard from private vegetable gardens were sampled in the city of Santa Barbara. The samples were weighed, washed with distilled water to eliminate airborne pollutants and dried in an oven at 65°C for 24 h. Nitric acid was used to digest the samples on a laboratory hot plate. Arsenic was determined by graphite furnace atomic absorption spectrometric (GF-AAS) and inductively coupled plasma mass spectrometry (ICP-MS).

Fern samples were collected in February and March, 2003, at sites along the Carmo river (between Mariana and Monsenhor Horta) and along Agua Suja stream, in Antonio Pereira district (Table 36.1). The samples (*Pteris vittata*) were divided into leaves and rhizoids. The leaves and rhizoids were washed thoroughly with tap water, rinsed with deionized water and sliced in small pieces. After freeze dried, the samples were ground before analysis. Soil samples were air-dried, sieved and the fine particles (<0.062 mm), consisting of silt and clay, were used for analysis. The moisture of soil samples was determined by weight loss in a separate sub-sample dried at 110°C for 2 h.

Arsenic and other element concentrations in the soil and fern tissue were determined using neutron activation analysis (NAA), specifically the k_0-standardization method. The irradiation was performed in the reactor TRIGA MARK I IPR-R1 at CDTN, at 100 kW, under a thermal flux 6.6×10^{11} neutrons/cm^2/s. The gamma spectroscopy was performed in a HPGe detector at 15% of efficiency. The gamma spectra were obtained and evaluated by HyperLab PC software and the concentration was calculated using the KAYZERO/SOLCOI software.

Table 36.1. Site (S), UTM coordinates, As, Fe, Mn, P, Rb and Zn concentrations in soil and in the leaves (L) and rhizoids (R) of *Pteris vittata* (*PV*) and *Pityrogramma calomelanos* (*PC*) and the bioaccumulation factor (BF) at six study sites.

S	UTM	Sample	As	BF	Fe	Mn	P	Rb	Zn
			µg/g	–	%	µg/g			
1	652 101	PV1 (L)	128	2.6	0.046	54	536	67	109
	7744961	PV1 (R)	77		0.27	75	131	12	59
		Soil 1	49		22.2	12014	1071	78	209
2	651 237	PV2 (L)	137	7.6	0.011	37	327	43	54
	7745313	Soil 2	18		26.4	2938	410	65	46
3	676 762	PV3 (L)	91	7.0	0.017	29	746	56	63
	7749122	PV3 (R)	25		0.018	6	242	9	36
		Soil 3	13		5.3	344	558	25	70
4	658 496	PV4 (L)	329	27.4	0.19	58	592	51	52
	7755208	PV4 (R)	139		0.46	116	190	13	74
		Soil 4	12		29.4	1143	325	106	30
5	658 395	PV5 (L)	180	1.8	0.54	293	765	75	52
	7755080	PV5 (R)	76		0.78	451	564	25	43
		Soil 5	101		45.7	6753	728	10	38
6	663 945	PV6 (L)	2295	2.4	0.020	75	444	107	83
	7746302	Soil 6	966		32.6	3754	515	24	46
		PC (L)	1744	1.8	0.032	47	688	141	55

36.3   RESULTS AND DISCUSSION

The concentrations of As, Fe, Mn, P, Rb and Zn in the leaves and rhizoids of *Pteris vittata* and *Pityrogramma calomelanos* and in the soil at six sites are shown in Table 36.1. The bioaccumulation factor (BF) was calculated to characterize the absorption of As by the ferns from the soil they grow on. The soil sample concentrations are expressed in dry weight and the results of metal concentrations in leave samples are fresh weight corrected.

The bioaccumulation factor obtained for the samples varied among the different sites (Table 36.1). This difference in the uptake of As at the sites may be related to different mineral content in the soils, especially Fe and Mn. The highest values of the BF were obtained at sites 2, 3 and 4, where Fe and Mn concentrations were lower compared to the other sites.

According to Kabata-Pendias and Pendias (1984), iron content in soil has a strong negative influence on As availability and this has been explained by the observation that iron strongly adsorbs As. It also was demonstrated by Deschamps *et al.* (2003), that soils enriched in Mn and Fe minerals from the Iron Quadrangle, have a significant uptake of both the trivalent and pentavalent As species contributing to the reduction of the As concentration in waters and consequently of its bioavailability. Site 4, which presented the lowest P concentration, showed the highest BF. According to Woolson *et al.* (1971), phosphate in soil may compete with arsenate in its uptake by plants due to the chemical similarities of the two anions. Thus, a low content of phosphate in soils may result in a high uptake of arsenate.

Besides As, Rb and Zn accumulation were also observed in leaves and rhizoids in some *Pteris vittata* samples (Table 36.1). According to Campbell *et al.* (2005), Rb, a rarely studied alkali metal, may be an essential ultra-trace element for humans and other organisms.

The As results in vegetable samples, commonly cooked in Brazil, are shown in Table 36.2. The results show that in the As-enriched soil all vegetables present As concentrations higher than the background value for vegetables (1.0 mg/kg As) in both their edible and non-edible parts.

Table 36.2.   Arsenic concentrations (dry weight) in vegetable samples from Santa Barbara district, Minas
Gerais, Brazil.

Sample	As (mg/kg)	Sampling site (sampling date)
Yam		Carrapato St., 947-Barra Feliz (07/03/2004)
Leaves	2.78 ± 0.71	
Stalk	2.43 ± 0.47	
Tubercle	4.81 ± 0.94	
Roots	207 ± 17	
Soil	247 ± 17	
Yam		Carrapato St., 917-Barra Feliz (07/03/2004)
Leaves	2.57 ± 0.21	
Stalk	4.55 ± 0.81	
Tubercle	1.46 ± 0.21	
Roots	60 ± 4	
Soil	130 ± 25	
Yam (young plant)		Carrapato St., 947-Barra Feliz (29/04/2004)
Leaves	2.2 ± 0.09	
Stalk	<0.2	
Tubercle	<0.2	
Roots	105 ± 7	
Soil	75 ± 14	
Sweet potato (tubercle)	1.41 ± 0.13	Carrapato St., 947-Barra Feliz (08/06/2004)
Bean (grain)	8.3 ± 0.7	Carrapato St., 439-Barra Feliz (09/06/2004)
Mustard (leaves)	1.59 ± 0.32	Carrapato St., 917-Barra Feliz (09/06/2004)
Kale (leaves)	6.29 ± 0.44	Carrapato St., 985-Barra Feliz (08/06/2004)
Kale (leaves)	4.50 ± 0.06	Carrapato St., 947-Barra Feliz (08/06/2004)
Kale (leaves)	1.86 ± 0.18	Carrapato St., 363-Barra Feliz (09/06/2004)
Kale (leaves)	1.81 ± 0.09	Carrapato St., 917-Barra Feliz (09/06/2004)
Lettuce (leaves)	1.24 ± 0.05	Carrapato St., 917-Barra Feliz (28/04/2005)
Lettuce		Mr. Nilo's Garden (22/06/2005)
Leaves	<0.2	
Roots	34 ± 3	
Soil	39 ± 4	
Lettuce		Mr. Moço's garden-house 15 (22/06/2005)
Leaves	1.09 ± 0.09	
Roots	54 ± 5	
Soil	59 ± 6	

36.4   CONCLUSIONS

All vegetables from private gardens investigated presented significant As uptake in both edible
and non-edible parts. Enrichments differed among species showing elevated values. The results
obtained for the *Pteris vittata* and *Pityrogramma calomelanos* confirm literature data. These
ferns actually extract As from the soil and translocate it into their fronds showing higher As
concentrations in the leaves (91–2295 µg/g) than in the rhizoids (25–139 µg/g). Difference in the
uptake of As among the sites could be related to different mineral contents in the soils, especially
Fe, Mn and P. Rb and Zn accumulation in leaves and rhizoids was also observed in some samples
of *Pteris vittata*.

ACKNOWLEDGMENTS

The authors acknowledge the support from the National Fund for the Environment—FNMA/MMA.

REFERENCES

Borba, R.P., Figueiredo, B.R., Rawlins, B.G. and Matschullat, J.: Arsenic in water and sediment in the Iron Quadrangle, Minas Gerais state, Brazil. *Rev. Bras. Geoc.* 20 (2000), pp. 554–557.

Campbell, L.M., Fisk, A.T., Wang, X., Köck, G. and Muir, D.C.G.: Evidence for biomagnification of rubidium in freshwater and marine food webs. *Can. J. Fish. Aquat. Sci.* 62 (2005), pp. 1161–1167.

Deschamps, E., Ciminelli, V.S.T., Weidler, P.G. and Ramos, A.Y.: Arsenic sorption onto soils enriched with manganese and iron minerals. *Clays Clay Min.* 51 (2003), pp. 197–204.

Deschamps, E. and Matschullat, J.: *Arsênio antropogênico e natural: um estudo em regiões do Quadrilátero Ferrífero*. Gráfica e Editora Sigma Ltda., Belo Horizonte: Fundação Estadual do Meio Ambiente, 2007, pp. 175–240.

Francesconi, K., Visoottiviseth, P., Ksridokchan, W. and Goessler, W.: Arsenic species in a hyperaccumulating fern, *Pityrogramma calomelanos*: a potencial phytoremediator of arsenic-contaminated soils. *Sci. Total Environ.* 284 (2002), pp. 27–35.

Helgesen, H. and Larsen, E.H.: Bioavailability and speciation of arsenic in carrots grown in contaminated soil. *Analyst* 123 (1998), pp. 791–796.

Kabata-Pendias, A. and Pendias, H.: *Trace elements in soils and plants*. CRC Press, Boca Raton, FL, 1984.

Koch, I., Wang, L., Olloson, C.A., Cullen, W.R. and Reimer, K.J.: The predominance of inorganic arsenic species in plants from Yellowknife, Northwest Territories, Canada. *Environ. Sci. Technol.* 34 (2000), pp. 22–26.

Larsen, E.H., Moseholm, L. and Nielsen, M.M.: Atmospheric deposition of trace elements around point sources and human health risk assessment: II. Uptake of arsenic and chromium by vegetables grown near a wood preservation factory. *Sci. Total Environ.* 126 (1992), pp. 263–275.

Ma, L.Q., Komart, K.M., Tu, C., Zhang, W., Cai, Y. and Kennelly, E.D.: A fern that hyperaccumulates arsenic. *Nature* 409 (2001), p. 579.

Matschullat, J., Borba, R.P., Deschamps, E., Figueiredo, B.R., Gabrio, T. and Schwenk, M.: Human and environmental contamination in the Iron Quadrangle, Brazil. *Appl. Geochem.* 15 (2000), pp. 181–190.

O'Neill, P.: Arsenic. In: B.J. Alloway (ed): *Heavy metals in soils*. Blackie Academic and Professional, London, UK, 1995, pp. 105–121.

Pyles, R.A. and Woolson, E.A.: Quantification and characterization of arsenic compounds in vegetables grown in arsenic acid treated soil. *J. Agric. Food Chem.* 30 (1982), pp. 866–870.

Visoottiviseth, P., Francesconi, K. and Ksridokchan, W.: The potential of Thai indigenous plant species for the pytoremediation of arsenic contaminated land. *Environ. Poll.* 118 (2002), pp. 453–461.

Woolson, E.A., Axley, J.H. and Kearney, P.C.: Chemistry and phytotoxicity of arsenic in soils. I. Contaminated Field Soils. *Soil Sci. Soc. Am. Proc.* 35 (1971), p. 101.

CHAPTER 37

Arsenic in plant samples from a contaminated mining area in the eastern Pyrenees, Catalonia (Spain)

M.J. Ruiz-Chancho, J.F. López-Sánchez & R. Rubio
Departament de Química Analítica, Universitat de Barcelona, Barcelona, Spain

ABSTRACT: Plant samples were collected in a contaminated area in the eastern Pyrenees, Catalonia (Spain). About 40 samples of 20 plant species were collected. Total arsenic (As) content was measured by ICP-MS and HG-AFS after acidic digestion in a microwave oven. As concentrations in plant samples ranged from 0.1 to 5240 mg/kg. These were higher than the concentrations found in the uncontaminated area, which ranged from 0.06 to 0.58 mg/kg. Three extractant solutions were tested for speciation, which was performed using the coupled technique LC-UV-HG-AFS.

37.1   INTRODUCTION

Mining activities generate a large amount of waste rocks and tailings which are deposited at the surface. The degraded soils, and the waste rocks and tailings are often very unstable and become sources of pollution. Populations of a variety of plants can colonize these polluted environments, responding by exclusion, indication or accumulation (Baker 1981).

Traditionally, only the total As content in environmental samples has been reported. However, the toxicity of As compounds depends on the chemical nature of the element. Therefore, it is essential to discern the various As compounds in environmental samples to understand and monitor the fate of As in the environment (Quaghebeur *et al.* 2005).

As concentrations in terrestrial organisms are generally much lower than in the marine environment. Uncontaminated terrestrial plants usually contain between 0.2 and 0.4 mg As/kg (Cullen *et al.* 1989).

For the total As determination in plant samples, a range of digestion techniques based on strong acids or oxidizing agents are applied. Milder extractants must be used for species extraction in order to preserve the original species. The most common extractants used are water, mixtures of methanol/water and dilute acidic solutions. The extraction procedures are usually sonication or mechanical agitation, sometimes with microwave-assisted heating or accelerated solvent extraction systems (Bohari *et al.* 2002).

Modern sensitive and selective analytical techniques such as high performance liquid chromatography coupled with specific detectors such as AFS and ICP-MS have facilitated the separation and selective detection of As species in environmental samples (Francesconi *et al.* 2004).

In this study, we measured the total As content of a group of samples collected in a contaminated mining area. We tested three extractants for As speciation, and analyzed the species by the coupled technique LC-UV-HG-AFS.

37.2   SAMPLING AND SAMPLE PRE-TREATMENT

Arsenic and Sb veins with subordinate amounts of Cu, Pb and Ag can be found in the Vall de Ribes in the eastern Pyrenees, Catalonia (Spain). The Vall de Ribes was an active mining district at the beginning of the nineteenth century, but the irregularity of the veins and the geographical situation of the mines made them uneconomical and they were soon closed.

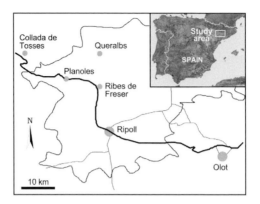

Figure 37.1.   Map of the Ripollés district in the eastern Pyrenees, Spain.

Figure 37.1 shows a map of the Ripollés district, where 48 plant samples were collected at eight selected sites on two separate occasions. The samples included representatives of the plant community at these sites. Entire vascular plants, mosses and tree leaves were collected around mine tailings. For the first sampling, sample points P1 to P4 were located near Ribes de Freser, a village in the Ripollés district, where small abandoned Sb and Zn mines are located. P5 and P6, where the samples were collected around the mine spoil, were located above an As and Sb mine near Planoles, to the northwest of Ribes de Freser. Finally, P7 was located near Queralbs, a village to the North of Ribes de Freser where an abandoned As mine can be found.

During the second sampling, in April 2005, seven plant samples representative of those collected on the first occasion were collected at two sites (P1B and P2B) in a non-contaminated area, Collada de Tosses, in order to measure the natural As contents in similar samples. The two sampling areas have similar geological characteristics.

Pre-treatment: Plant samples were carefully washed in deionized water and dried in an oven at 40°C. They were then pulverized with a tungsten carbide disc mill.

37.3   TOTAL ARSENIC DETERMINATION

Acidic digestion method comparison: Plant samples were digested in nitric acid and hydrogen peroxide in a microwave oven. To check whether part of the As was retained in the samples by silicon, HF was also used in some extraction procedures. A Mileston MLS-1200 microwave with closed vessels and also a Prolabo A301 with opened reflux vessels were used for the acidic digestion. The acidic extracts were analyzed for total As by ICP-MS (inductive coupled plasma mass spectrometry) and HG-AFS (hydride generation-atomic fluorescence spectroscopy).

Table 37.1 shows the results obtained for three selected samples from different species when each method was used. Similar results were obtained for the total content given by either of the extraction procedures (HF + HNO_3 + H_2O_2 or HNO_3 + H_2O_2).

Total As content: Plant samples were digested in nitric acid and hydrogen peroxide in open reflux vessels under focused microwaves prior to total As determination. Levels of total As in several samples from the contaminated area were higher than those obtained from the control area (Table 37.2). The higher levels were found at P6.

37.4   ARSENIC SPECIATION

Three extractant solutions and two extraction methods were tested for As speciation. The extractants most frequently used for As speciation in terrestrial plants are mixtures of methanol and water. Some authors also have tested phosphoric acid, with good results and high extraction

Table 37.1.   Total As content expressed as arsenic (mean $\pm$ sd, n $=$ 3).

	HF-HNO$_3$-H$_2$O$_2$[1] mg/kg	HNO$_3$-H$_2$O$_2$[2] mg/kg
P1-S3	10.70 $\pm$ 0.70	10.90 $\pm$ 0.70
P2-S1	1.07 $\pm$ 0.36	1.35 $\pm$ 0.07
P4-S3	6.05 $\pm$ 1.53	7.79 $\pm$ 0.46

[1]ICP-MS; [2]HG-AFS.

Table 37.2.   Total arsenic content for plant samples growing in contaminated and control areas. Results expressed as mg/kg (dry weight) of arsenic (mean $\pm$ sd, n $=$ 3).

	Species	Common name	As (mg/kg)
Contaminated area			
P1-S3	*Hydnum cupressiforme* Hedw.	Moss	10.95 $\pm$ 0.75
P1-S7	*Buxus sempervirens* L.	Boxtree	0.28 $\pm$ 0.03
P2-S5	*Dryopteris filix-max* (L.) Schott.	Fern	1.65 $\pm$ 0.30
P2-S6	*Stellaria halostea*	Stitchwort	4.66 $\pm$ 0.55
P3-S3	*Quercus pubescens* Mill.	Downy oak	0.84 $\pm$ 0.02
P3-S5	*Hydnum cupressiforme* Hedw.	Moss	1.67 $\pm$ 0.08
P4-S3	*Dryopteris filix-max* (L.) Schott.	Fern	5.79 $\pm$ 0.46
P4-S6	*Rubus ulmifolius* (L.) Schott.	Blackberry	0.14 $\pm$ 0.02
P5-S3	*Dryopteris filix-max* (L.) Schott.	Fern	14.6 $\pm$ 1.9
P5-S4	*Quercus pubescens* Mill.	Downy oak	4.03 $\pm$ 0.31
P6-S1	*Juncus inflexus* L.	Hard rush	107 $\pm$ 11
P6-S6	*Brachythecium* cf. *reflexum* (F.Weber and D.Mohr) Schim.	Moss	5240 $\pm$ 1460
Control area			
P1B-1	*Sarothamnus scoparius* (L.) Wimm. ex Koch	Broom	0.35 $\pm$ 0.01
P1B-2	*Hydnum cupressiforme* Hedw.	Moss	0.51 $\pm$ 0.01
P2B-1	*Sarothamnus scoparius* (L.) Wimm. ex Koch	Broom	0.06 $\pm$ 0.04
P2B-2	*Hylocomium splendens* (Hedw.) Schimp.	Moss	0.58 $\pm$ 0.08

yields (Bohari *et al.* 2002). Phosphoric acid (0.3 mol/l) and mixtures of methanol and water (1:1) and (1:9) were tested as extractants for the present study. We used also end-over-end shaking and ultrasonic extraction. For end-over-end shaking, 0.3 g of sample P1-S3 were placed in a 25 ml recipient and shaken for 16 h at room temperature. The same portion was weighed and extracted for 2 h when the samples were extracted ultrasonically. After the extraction, speciation was carried out by the coupled technique LC-UV-HG-AFS (liquid chromatography-photo-oxidation-hydride generation-atomic fluorescence spectroscopy). A derivatization step by UV irradiation was applied after chromatographic separation to oxidize organic species which do not form the volatile hydride. Phosphoric acid was found to be the most efficient extractant (Table 37.3). There was no significant difference between the two extraction methods. Extraction yields were higher with end-over-end shaking only when using mixtures of methanol and water.

Two chromatograms of sample P1-S3, obtained after ultrasonic extraction, show As(III) and a small peak corresponding to an unknown species with the same retention time as As(III) (Fig. 37.2).

The unidentified peak only appeared when the photo-oxidation step was applied (Fig. 37.3). We conclude that this peak at the same retention time as As(III), corresponds to an organic As species that does not form a volatile hydride.

Table 37.3.   Extraction efficiencies expressed as percentage (n = 2).

	End-over-end	Ultrasonication
Phosphoric acid 0.3 mol/l	48.8	50.7
MeOH/H$_2$O (1:1)	7.0	4.1
MeOH/H$_2$O (1:9)	12.2	8.0

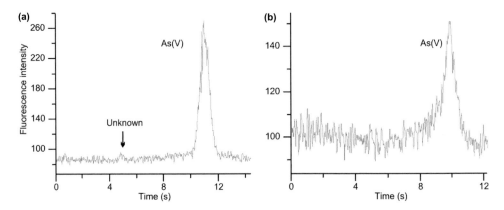

Figure 37.2.   Example chromatograms of As species in plant extracts of sample P1-S3 after ultrasonic extraction with (a) phosphoric acid 0.3 mol/l and (b) MeOH/H$_2$O (1:9).

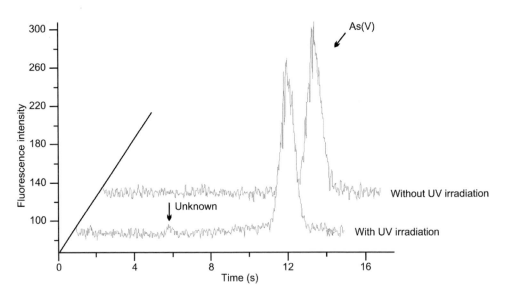

Figure 37.3.   Comparison of sample P1-S3 chromatograms with and without photo-oxidation step.

37.5   CONCLUSIONS

Total As content was determined in about forty terrestrial plants collected in a contaminated min- ing area. Similar results regarding the total As content were obtained by either of the digestion procedures (HF + HNO_3 + H_2O_2 or HNO_3 + H_2O_2). Levels of total As in several samples from the contaminated area were higher than those from the control area. Phosphoric acid was found to be the most efficient extractant for speciation. All samples were found to contain only As(V), except when phosphoric acid was used. In this case, an unknown species with the same retention time as As(III) was detected.

REFERENCES

Baker, A.J.M.: Accumulators and excluders-strategies in the response of plants to heavy metals. *J. Plant Nutrit.* 3 (1981), pp. 643–654.

Bohari, Y., Lobos, G., Pinochet, H., Pannier, F., Astruc, A. and Potin-Gaultier, M.: Speciation of arsenic in plants by HPLC-HG-AFS: extraction optimization on CRM materials and application to cultivated samples. *J. Environ. Monit.* 4 (2002), pp. 596–602.

Cullen, W.R. and Reimer, K.J.: Arsenic speciation in the environment. *Chem. Rev.* 89 (1989), pp. 713–764.

Francesconi, K.A. and Kuehnelt, D.: Determination of arsenic species: A critical review of methods and applications, 2000–2003. *The Analyst* 129 (2004), pp. 373–395.

Quaghebeur, M. and Rengel, Z.: Arsenic speciation governs arsenic uptake and transport in terrestrial plants. *Microchim. Acta* 151 (2005), pp. 141–152.

CHAPTER 38

Soil-to-leaf transfer factor for arsenic in peach (*Prunus persica* L.)

D.L. Orihuela, J.C. Hernández & R.J. López-Bellido
Escuela Politécnica Superior, Universidad de Huelva, Huelva, Spain

S. Pérez-Mohedano & L. Marijuán
Huntsman-Tioxide, Madrid, Spain

N.R. Furet
Instituto Superior de Ciencias y Tecnologías Superiores, La Habana, Cuba

ABSTRACT: Transfer factors (TF) are conceptually defined in the scientific literature as the relation between the concentration of a certain element either in a plant, or in its organs and the concentration of that element in the related soil. From a mathematical point of view, this definition would be a deceivingly simple linear model. The mathematical models that express the experimental data of the TF are, in general, more complex, but always have the remarkable advantage of quantitatively and temporally helping to understand part of the process of soil-to-plant translocation of the nutrient elements. The aim of this work is to study the TF of arsenic (As) in peaches (*Prunus persica* L.) when its solubility in soil changes with pH correction. We conclude that the As transfer models in calcareous soil toward the leaves of a peach tree, expressed by the TF, are in general linear models, almost horizontal, showing the fact that the As concentrations in leaves are independent of its concentration in soil, and that the pH-alterations in the soil scarcely modify the TF values.

38.1 INTRODUCTION

Plants are involved in sophisticated strategies for the absorption of relatively scarce micro-nutrients from soil. These essential micro-nutrients tend to be highly reactive, and some are potentially toxic; their uptake, transport and accumulation therefore needs to be well coordinated and well regulated (Kochian *et al.* 2002).

Deficiencies due to inadequate coordination or to an inability to absorb certain micro-nutrients from the soil and the related symptoms are widely addressed in the literature. Moreover, measures to correct Fe-deficiency (Fe-chlorosis) and Zn-deficiency, for example, are routinely implemented by farmers. Certain ions, such as Cu^{2+}, Zn^{2+} and Mn^{2+}, are essential to plant metabolism; however, they pose a metabolic dilemma, in that the slightest excess of these elements may prompt serious toxicity problems.

In this respect, a number of key questions remain to be answered regarding the uptake, transport (Clement *et al.* 2002) and metabolism of these metals. The related knowledge is crucial for soil amendment (Fatiga *et al.* 2004), plant breeding and plant protection, and the mechanisms deployed by plants to prevent the build-up of toxic levels of some metals. In order to improve the feeding process and/or avoid toxic damage, it is essential to determine how plants ensure that tissues receive an adequate supply of each metal without reaching toxic levels.

Scientific concern regarding these questions is increased by the fact that the soil-plant route is the most common means of entry of trace elements, such as As, into the food chain. The present chapter neither addresses the intracellular processes by which correct distribution of these elements is ensured, nor the known mechanisms used by the plant to prevent excessive uptake of certain metals.

It rather seeks to determine the ratio between As concentrations in soil *versus* leaves. Soil-plant transfer is a highly-complex process governed by several mechanisms, both natural and affected by man (Kabata-Pendias 2004). The uptake and transfer of a given element from the roots to the aerial portion of the plant is determined by a number of factors as summarized below (Carini 2001):

- The chemico-physical characteristics of the soil (moisture, pH, cation exchange capacity, redox potential, fertilizer application, soil texture and structure, organic matter).
- Root interception and uptake (root surface, root growth rate, genotype, etc.).
- Ion transport across root membranes (membrane type, ion type).
- Ion translocation (unidirectional transport in the xylem, bidirectional in the phloem).
- Remobilization of ion from storage organs (transport from leaves or woody parts to sink areas such as fruits and buds).
- Root exudation (modification of pH or exudates for mycorrhiza).
- Mycorrhiza (plant-mycorrhiza associations or total rhizosphere activity).

Phytoavailability of trace elements is governed (Gommers 2005, Kabata-Pendias 2004) by the following factors (in order of importance): pH and redox potential, texture, organic matter, soil mineral composition and water regime.

Iron chlorosis (inability to absorb Fe) is associated with a high pH, in limestone soils and sensitive crops (vines, citrus, peaches, etc.). Remediation of Fe-deficiency using chelates is a widespread farming practice, with the only constraint generally being of an economic nature. Remediation of soil pH, the primary cause of chlorosis, using more affordable products offers a possible solution to this problem.

Lowering the pH of basic soils generally implies all-round benefits for plants and for the microbial environment, but it may also have undesirable side-effects. One of these is the increased solubility of certain metals which were insoluble in a basic medium. Increased solubility, whilst meeting the aim of correcting the deficiency, may at the same time give rise to toxic levels of previously-insoluble elements.

Some tree species are unable to adapt to a sudden increase in soil solubility, an obstacle that can only be overcome by the use of certain ecotypes or by breeding aimed at developing resistant stock (Kahle 1993).

By contrast, other species display no symptoms or in the case of crops, fall-off in production when grown in soils with high concentrations of trace metals; this considerably enhances their agricultural value. These species share a restricted ability to absorb trace metals through their root systems—together with low levels of translocation and storage in leaves and fruits, although the development of tolerance may require a more precise physiological and genetic mechanism (Dickinson 2000). Because individual evaluation of all the processes involved is a highly complex issue, they tend to be systematized and grouped into more simple concepts. One such concept is the transfer factor (TF).

The TF relates the concentration of an element in the soil to its concentration either in the whole plant or part of the plant (e.g., leaf, fruit), and may be expressed as a ratio of element per dry weight soil to element per dry weight plant (IAEA 1994), or element per soil surface area (at a depth of 20 cm for crops and 10 cm for grass) to element per plant dry weight (Frissel 1997). With fruit, it is usually expressed in terms of soil dry weight to fresh weight. The simplest definition of TF in the literature is "the ratio of the concentration of a given element in a plant, or plant organ, to the concentration of that element in the soil".

$$C_{plant} = TF \cdot C_{soil} \tag{38.1}$$

TF has also been defined as " ... *the expected amount of an element entering a plant from a soil in equilibrium conditions*" (Chojnacka *et al.* 2005). Here however, equilibrium conditions prove difficult to define. This basic concept of translocation or uptake has been applied, amongst other things, to trace metals (Chamberlain 1983, Gast *et al.* 2003), pesticides (Trapp *et al.* 1990), and radionuclides (Ehlken and Kirchner 2002).

The earlier definition rests on two unlikely assumptions: that the relationship between the two concentrations is linear, and that it is constant. Numerous published studies have shown that in many cases the relationship is neither linear nor constant, and that given considerable variability of concentrations within a single plant and a single soil, this exact linear relationship is in fact no more than a chance encounter in both space and time. This variability is to be expected, given the numerous factors (e.g., climatic, biological, genetic) and parameters (e.g., pH, soil CEC, moisture, inter-ion competition) governing soil-plant relations.

The rather simplistic formulation of a linear model to account for plant uptake of a given element in a given soil-water system must therefore be viewed with considerable caution. Published studies assume linear models mainly for elements present in very low concentrations in the soil, and widely-ranging values are reported (Blanco *et al.* 2002). In a linear model, a horizontal line would indicate that the soil-plant transfer of a given element is not governed by the soil concentration of that element. Additionally, a wide scattering of TF values with respect to the mean would indicate that the element was present in the form of either highly-insoluble salts (not readily absorbed by the plant) or, conversely, highly soluble (readily absorbed) salts.

More elaborated linear transfer models formulate the concentration ratio as the equation for a straight line intersecting the Y axis at a point such that:

$$C_{plant} = a \cdot C_{soil} + b \tag{38.2}$$

where it is assumed that the element enters the plant only through the roots. If b is zero, then this equation corresponds to equation (38.1) and the discussion regarding the angular coefficient of the line (a = TF) would be the same as before.

More complex models reported in the literature express the uptake ratio by means of a hyperbolic equation of the type:

$$C_{plant} = m \cdot C_{soil}^{n} \tag{38.3}$$

If n = 1 then the curve becomes a straight line with an angular coefficient equal to m. When n = 0, the straight line is horizontal; plant concentration is constant and equal to m. If n < 1, plant concentration values fall as soil concentration values rise.

Other TF models, such as that used by Ambe *et al.* (1999), take plant transpiration and soil moisture into account:

$$TF = ST_c/(\theta + K_d) \tag{38.4}$$

where S is the coefficient of selective uptake of the element; T_c the transpiration coefficient (cc/g), i.e., water required for the production of one gram of plant; θ is soil water content (cc/g) and K_d the coefficient of distribution (cc/g).

More complex models have been used in various fields of research, ranging from theoretical conceptual models such as the Mitscherlich model (Tudoreanu and Phillips 2004), a number of dynamic and static models reviewed by Kabai *et al.* (2004), and the dynamic model proposed by Kabai himself; the latter is considerably more complex, since it incorporates a large number of variables, including atmospheric deposition, plant surface area, interstitial water, soil surface area, and concentration of the element in the root zone.

TF or transfer models are widely used, especially in research on the transfer of radionuclides and trace metals, and phytorestoration. Application of these models to edible crops, i.e., to leaves, fruit, etc., may help us to chart the movement of heavy metals at low concentrations through soil-plant systems. However, quantitative and/or temporal application of these models to nutritive macroelements may not be feasible.

The International Union of Radioecologists is currently working on the development of transfer models for cereals to be used as units applicable, via a conversion factor, to other crops (Frissel *et al.* 2002, Nisbet and Woodman 2000).

38.2   MATERIAL AND METHODS

38.2.1   *Experimental plot*

A peach crop was selected, grown at an experimental plot in Cartaya (Huelva, Spain). The area of the experiment plot was 36×36 m. The planting distance used was 6×3 m. Alternate trees were sampled in each row, giving a square sampling grid of 6×6 m (36 trees). The soil is a Pliocene clay soil classified as Palexeralf Aquic.

38.2.2   *Field sampling technique and analysis*

The first soil samples were collected in September of the first study year (2004); samples were designated As1s; further samples were collected in September of the second year (As3s). All samples were bagged and labeled, identifying the tree from which soil was collected and the date prior to being sent to the laboratory for analysis. The first leaf samples (As1f) were collected in September of the first study year, the second in May of the second year (As2f). In September of the second year the third leaf samples (As3f) were collected.

The agent used to lower soil pH was iron sulfate monohydrated (FSM), obtained from ilmenite by the titanium industry. The mean pH value for the first year's soil samples (i.e., prior to pH remediation) was 8.1, while the mean for the second year's samples (i.e., after pH remediation was 6.2, significance 0.000 (95%).

FSM was applied at a rate equivalent to 600 kg/ha, close to the drip irrigation outlet for each tree. A total of 36 trees were sampled. Leaf samples were taken in May and September of each of the two years.

Chemical analysis was based on standard methods, using inductively-coupled plasma-mass spectrometry (ICP-MS; Hewlett-Packard 4500 Series), spectrophotometry (Perkin-Elmer Lambda-2), and a pH-meter with fitted calomel electrode (WTW). Leaf samples were crushed and washed in 18.2 MW ultrapure water (Milli-Q RO); ICP-MS measurements were made using this solution. Acid digestion was performed using 10 ml aqua regia per 1 g leaf sample. pH-measurements were made every 10 min. Data obtained were processed using the SPSS 12.0 statistical software package. Analytical results and statistical evaluations are shown in Tables 38.1 and 38.2.

The iron sulfate monohydrated (FSM), containing other trace elements and sulfuric acid, was supplied in crystalline form. This product, used to prevent Fe-chlorosis in all kinds of crops, acts as a soil-acidifying agent and prevents trace-element deficiency. Technical specifications (as supplied by Huntsman-Tioxide S.L.) were as follows: water-soluble Fe 20.5%; water-soluble SO_3 40.0%; water-soluble Ti 1.5%; water-soluble Mn 1.0% and water-soluble Zn 0.5%.

38.3   RESULTS AND DISCUSSION

38.3.1   *Soil pH values*

pH1 values (8.1 ± 0.29) and pH3's (6.2 ± 1.4) were compared (Table 38.1). It was observed that treatment with FSM decreased pH significantly ($t = 9130$, Sig $= 0.000$).

The decrease of soil pH can be explained by the sulfuric acid contained in the FSM. P-soil and pH are the two main influential factors in As absorption in plants. Nevertheless As-P interactions are difficult to interpret by means of their chemical analogies (Tu and Ma 2003). Difficulties increase as remarkable quantities of FSM are added to soil. There are many papers in the scientific literature about melioration of polluted soils with As. There are multiple recovery techniques for these soils like clay application or Fe-hydroxides. All of them absorb and immobilize As. Fe-hydroxides exhibit a high absorption capacity and reduce As mobility.

Some studies (Carbonel-Barrachina *et al.* 1999) show that Fe-hydroxides play an important role in As adsorption-desorption processes, decreasing in plant absorption. In fact, Fe-sulfate is known

Table 38.1.   Descriptive statistics of the experimental values for pH, As, and TF.

Parameter	N	Minimum	Maximum	Mean	Std. Dev.
pH1s	36	7.3	8.8	8.1	0.2880
pH3s	36	3.3	7.8	6.2	1.3994
As1f (mg/kg)	36	0.68	5.2	2.2	1.0086
As2f (mg/kg)	36	0.28	2.6	0.8	0.5821
As3f (mg/kg)	36	0.38	1.4	0.7	0.2290
As1s (mg/kg)	36	7.7	12.3	10.7	1.1644
As3s (mg/kg)	36	5.9	12.0	9	1.6629
TF1	36	0.08	0.5	0.2	0.0098
TF3	36	0.03	0.2	0.008	0.0030

pH1s: pH-soil September first year; pH3s: pH-soil September second year; As1f: As-leaf September first year; As2f: As-leaf May second year; As3f: As-leaf September second year; As1s: As-soil September first year; As3s: As-soil September second year; TF1: Transfer factor As-soil-to-leaf September first year; TF3: Transfer factor As-soil-to-leaf September second year.

as a highly As absorbent compound in the scientific literature. Other minerals, such as goethite, birnessite and amorphous hydroxides are also strong As adsorbents.

The adsorption process depends on electric charges at the oxides' surface; adsorption quantity depends on pH. For most oxide minerals, surface charge changes from positive (low pH) to negative (high pH) in the pH range of 7 to 10. This justifies a pH decrease and consequently the reduction of As levels. A decrease in pH is also favored by reducing soil conditions (that would be helped by humidity from drip irrigation).

As mobility in calcareous soils could be inferred in the presence of Na resulting in the formation of Na-arsenate, highly toxic for plants. Therefore, pH is a very important parameter to estimate As mobility.

38.3.2   *Soil arsenic concentrations*

Average values for total As in earth-soils vary between 0.5 and 2.5 mg/kg. In this case these values were exceeded. Initial value was As1s = 10.7 ± 1.2 mg/kg, and final one As3s = 9.0 ± 1.7 mg/kg. This decrease is probably due to FSM application since Fe-compounds can lead to adsorption processes as formerly described with As. The high As values are probably caused by the loamy and sedimentary soils. Similar values have been described in some podzols in Canada or chernozem in Bulgaria.

Some authors report that a decrease in soil pH increases As levels in soil. This does not agree with the results obtained in this experiment. Decreasing As levels can be related to two factors: (1) pH decline in soil decreases total As because part of the As is transferred to the soil solution; (2) due to FSM application with resulting partial immobilization of total As by Fe-hydroxide.

38.3.3   *Leaf arsenic concentrations*

At the beginning of the experiment, the As leaf to soil concentration ratio fitted a linear regression whose equation is expressed as follows: $C_{leaf} = 0.1709C_{soil} + 0.4227$. At the end of the experiment, the line equation was $C_{leaf} = -0.0052C_{soil} + 0.7335$. The initial straight line showed a slight curve, indicating that an increase of As in soil was related to a weak increase of As in the leaves. After FSM application, the regression line was virtually horizontal, indicating that As leaf levels were independent of As values in soil.

The decrease of As levels from the first (2.2 ± 1.0 mg/kg) to second (0.69 ± 0.23 mg/kg) sampling was backed by a general improvement of the trees. Better leaf development prompted lower As concentrations in leaves (dissolution effect) without modifying the absorption ratio. This

Table 38.2.   TF-model values.

DV	IV	Model	Rsq.	f.d.	F	Sig.	b0	b1	b2	b3
As1f	As1s	Linear	0.039	34	1.38	0.249	0.4227	0.1709		
As1f	As1s	Logarithmic	0.034	34	1.21	0.28	−1.6108	1.6332		
As1f	As1s	Cuadratic	0.069	34	1.22	0.308	11.4596	−1.9948	0.104	
As1f	As1s	Cubic	0.073	34	1.31	0.285	5.1834	−0.1028		0.0071
As3f	As3s	Linear	0.001	34	0.05	0.826	0.7335	−0.0052		
As3f	As3s	Logarithmic	0	34	0.01	0.912	0.7358	−0.0226		
As3f	As3s	Cuadratic	0.04	34	0.69	0.511	−0.5300	0.2906	−0.0167	
As3f	As3s	Cubic	0.039	34	0.67	0.519	−0.1104	0.1437		−0.0006

DV: dependent variable; IV: independent variable; Rsq.: residual square; f.d.: degree of freedom; F: Fisher; Sig.: Significance; b0, b1, b2 and b3: model coefficients.

agrees with studies by Baroni *et al.* (2004). The ability of some plants to absorb As (expressed by the biological accumulation coefficient) was independent from As content in soil.

38.3.4   *Soil-to-leaf TF*

There is an apparent lack of the soil-to-leaf TF values in *Prunus persica* L. in the literature. The most recent studies discuss about rice and give values from 0.36 to 3.34 (Roychowdhury *et al.* 2005), which exceed ours.

First year soil to old leaves TF1 value (0.21 ± 0.01) was much higher (26.5 times) than the same value in the following year, TF3 (0.008 ± 0.003)—see Table 38.2. The FSM application did not only lead to lower As concentrations in soil but also to a general improvement of the plantation that led to a TF decrease as a result of a dilution effect (As transferred to leaf stratum to be distributed in higher leaf volume. It would be a "tree-improvement effect".

On the other hand, it is well known that old leaves generally contain higher concentrations of elements (trace metals and others) than young ones because they have less cellular liquid. This leaf-dilution effect is well documented in the literature. In this case pH-decrease processes resulted in second year soil/old leaf concentration relationships almost the same as TF value for soil/new leaf one.

Models formulation, especially the linear ones (Blanco *et al.* 2002) for TF elements in low concentration in vegetables is quite common. In our case, four models were tested: linear, algorithmic, quadratic and cubic.

TF from the ratio $TF = C_{leaf}/C_{soil}$ was initially fitted to a linear model of equation $TF = -0.0061 \, C_{soil} + 0.2768$ while at the end of experiment equation was $TF = -0.0094 \, C_{soil} + 0.1639$ including a descending slope. After soil pH modification (second sampling), TF descended as As levels increased. Modification of pH slightly altered TF models.

Initially, As was quite unstable in terms of solubility. A wide range of scattering for values around its average (initial mean $TF = 0.2106 \pm 0.0986$; 46% scatter) is shown. In the range of decreasing pH values, they show a higher scattering range (final mean 0.0786 ± 0.03136; 39% scatter), and solubility processes are less ordered. Thus, As solubility varied from low to very high in a narrow range of pH variation, including high values of TF scattering.

38.4   CONCLUSIONS

We did not find TF-values of As for peaches in calcareous soils in the scientific literature, to compare them with our results. The best approximation is the revision of Carini (2001). According to the formerly presented discussion we conclude that:

- Application of FSM to calcareous soils entailed a significant pH decrease, which implies a decrease of total As levels in soil.

- Decrease of soil pH improved vegetative development of trees with clear chlorosis symptoms which had a better leaf development, decreasing As levels in leaves.
- Decrease of As in leaves is caused by a double effect of As decrease in soil and general improvement of the leaf stratum.

REFERENCES

Ambe, S., Shinonaga, T., Ozaki, T., Enomoto, S., Yasuda, H. and Uchida, S.: Ion competition effects on the selective absorption of radionuclides by komatsuna (*Brassica rapa* var. perviridis). *Environ. Exp. Bot.* 41 (1999), pp. 185–194.

Baroni, F., Boscagli, A., Di Lella, L.A., Protano, G. and Riccobono, F.: Arsenic in soil and vegetation of contaminated areas in southern Tuscany (Italy). *J. Chem. Explor.* 81 (2004), pp. 1–14.

Blanco, P., Vera, F. and Lozano, J.C.: About the assumption of linearity in soil-to-plant transfer factor from uranium and thorium isotopes. *Sci. Total Environ.* 284 (2002), pp. 167–175.

Carbonel-Barrachina, A.A., Aarabi, M.A., DeLaune, R.D., Gambrell, R.P. and Patrick, W.H.: The influence of arsenic chemical form and concentration on *Spartina patens* and *Spartina alterniflora* growth and tissue arsenic concentration. *Plant Soil* 198 (1999), pp. 33–43.

Carini, F.: Radionuclide transfer from soil to fruit. *J. Environ. Radioact.* 52 (2001), pp. 237–279.

Chamberlain, A.C.: Fallout of lead and uptake by crops. *Atmos. Environ.* 17 (1983), pp. 693–706.

Chojnacka, K., Chojnacki, A., Górecka, H. and Górecki, H.: Bioavailability of heavy metals from polluted soils to plants. *Sci. Total Environ.* 337 (2005), pp. 175–182.

Clement, S., Palmgren, M.G. and Krämer, U.: A long way ahead: understanding and engineering plant metal accumulation. *Trends in Plant Sci.* 7 (2002), pp. 309–315.

Dickinson, N.M.: Strategies for sustainable woodland on contaminated soils. *Chemosphere* 41 (2000), pp. 259–263.

Ehlken, S. and Kirchner, G.: Environmental processes affecting plant root uptake of radioactive trace elements and variability of transfer factor data: a review. *J. Environ. Radioact.* 58 (2002), pp. 97–112.

Fatiga, A.O., Ma, L.Q., Cao, X. and Rathinasabapathi, B.: Effects of heavy metals on growth and arsenic accumulation in the arsenic hyperaccumulator *Pteris vittata* L. *Environ. Pollut.* 132 (2004), pp. 289–296.

Frissel, M.J.: Protocol for experimental determination of soil to plant transfer factors (concentration ratios) to be used in radiological assessment models. *UIR Newsletter* 25 (1997), pp. 5–8.

Frissel, M.J., Deb, D.L., Fathony, M., Lin, Y.M., Mollah, A.S., Ngo, N.T., Othman, I., Robison, W.L., Skarlou-Alexiou, V., Topcuoglu, S., Twining, J.R., Uchida, S. and Wasserman, M.A.: Generic values for soil-to-plant transfer factor of radiocesium. *J. Environ. Radioact.* 58 (2002), pp. 113–118.

Gast, C.H., Jansen, E., Bierling, J. and Haanstra, L.: Heavy metals in mushrooms and their relationship with soil characteristics. *Chemosphere* 17 (2003), pp. 789–799.

Gommers, A., Gäfvert, T., Smoldres, E., Merckx, R. and Vandehove, H.: Radiocaesium soil-to-wood transfer in commercial willow short rotation coppice in contaminated farm land. *J. Environ. Radioact.* 78:3 (2005), pp. 267–287.

IAEA: Handbook of parameter values for the prediction of radionuclide transfer in temperate environments. Technical Report Series 364, International Atomic Energy Agency, Vienna, Austria, 1994.

Kabai, E., Zagyvai, P., Láng-Lázi, M. and Oncsik, M.B.: Radionuclide migration modelling through the soil-plant system as adapted for Hungarian environment. *Sci. Total Environ.* 330 (2004), pp. 199–216.

Kabata-Pendias, A.: Soil-plant transfer of trace elements—an environmetal issue. *Geoderma* 122 (2004), pp. 143–149.

Kahle, H.: Response of roots of trees to heavy metals. *Environ. Exp. Bot.* 33 (1993), pp. 99–119.

Kochian, L.V., Pence, N.S., Letham, D.D.L., Pineros, M.A., Magalhaes, J.V., Hoekenga, O.A. and Garvin, D.F.: Mechanisms of metal resistance in plants: aluminium and heavy metals. *Plant Soil* 247 (2002), pp. 109–119.

Nisbet, A.F. and Woodman, R.F.: Soil-to-plant transfer factor for radiocesium and radiostroncium in agricultural systems. *Health Phys.* 78:3 (2000), pp. 279–288.

Riddell-Black, D.A.: Review of the potential for the use of trees in the rehabilitation of contaminated land. WRc. Report CO 3467, Water Research Centre, Medmenham, UK, 1993.

Roychowdhury, T., Tokunaga, H., Uchino, T. and Ando, M.: Effect of arsenic contaminated irrigation water on agricultural land soil and plants in West Bengal, India. *Chemosphere* 58 (2005), pp. 799–810.

Trapp, S., Matthies, M., Scheunert, I. and Topp, E.M.: Modelling the bioconcentration of organic chemicals in plants. *Environ. Sci. Technol.* 24 (1990), pp. 1246–1252.

Tu, S. and Ma, L.Q.: Interactive effects pf pH, arsenic and phosphorus on uptake of As and P growth of the arsenic hyperaccumulator *Pteris vittata* L. under hydroponic conditions. *Environ. Exp. Bot.* 50 (2003), pp. 243–251.

Tudoreanu, L. and Phillips, C.J.C.: Modelling cadmium uptake and accumulation in plants. *Adv. Agron.* 84 (2004), pp. 121–157.

CHAPTER 39

Arsenic uptake and distribution in broccoli, cauliflower and radish plants grown on contaminated soil

M. Del Río-Celestino, M.M. Villatoro-Pulido & M.I. De Haro-Bravo
Centro IFAPA Alameda del Obispo, Córdoba, Spain

R. Font & A. De Haro-Bailón
Instituto de Agricultura Sostenible (CSIC), Córdoba, Spain

ABSTRACT: Arsenic uptake by cauliflower (*Brassica oleracea* L. var. *botrytis* L.), broccoli (*Brassica oleracea* L. var. *italica* Plenck) and radish (*Raphanus sativus* L.) plants, grown in contaminated soil, were studied. The main objectives of this work were to study the distribution of the accumulated As among root, shoots and edible parts and to establish whether As concentrations in edible parts of these vegetables are potentially dangerous to human health. Cauliflower and broccoli plants accumulated As mainly in the root system, in contrast, radish plants accumulated As mainly in shoots. As concentrations in the edible parts of the three plant species were always below the maximum limit set for As in vegetables and ranged from 0.2 to 1.2 mg/kg (dry weight) for plants grown in contaminated soils. The results presented demonstrate that there is not a risk associated with consumption of cauliflower, broccoli and radish grown in the conditions of this work.

39.1   INTRODUCTION

Arsenic (As) is one of the most important global environmental toxicants and is present in the environment from both natural and anthropogenic sources. Natural pathways of As include weathering, biological activity and volcanic activity. The primary anthropogenic input derives from combustion of municipal waste, fossil fuels in coal- and oil-fired power plants, release from metal smelters, and direct use of As-containing herbicides by industry and agriculture. There are a number of ways by which the human population can become exposed to As. The most important one is probably through ingestion of As in drinking water or food (National Research Council 1999, Lee *et al.* 2000, USEPA 2001).

Chronic As poisoning can cause serious health effects including cancers, melanosis (hyperpigmentation or dark spots and hypopigmentation or white spots), hyperkeratosis (hardened skin), restrictive lung disease, peripheral vascular disease (blackfoot disease), gangrene, diabetes mellitus, hypertension, and ischemic heart disease (Chen *et al.* 1996, Morales *et al.* 2001, Rahman 2002, Srivastava *et al.* 2001).

The toxicity of As to biological systems has made it a useful constituent of insecticides, herbicides, fungicides, desiccants, and wood preservatives (Johnson and Hitbold 1969, Marin 1995). However, indiscriminate use of these As substances has led to elevated concentrations of plant-available As in many soils, which may reduce soil productivity (Liebig 1966, Marin 1995) and be toxic to plants (Deuel and Swoboba 1972, Marin 1995).

Typical uncontaminated agricultural soils contain 1–20 mg of As/kg of soil (Wauchope 1983), but contaminated soils associated with mineralized zones may contain levels as high as 2600 mg of As/kg of soil (Meharg *et al.* 1994). Soluble As concentrations vary from 0.007 mg of As/kg in uncontaminated upland soils to 48 mg of As/kg in highly contaminated mining areas (Wei and Chen 2006).

Generally, in unpolluted environments, ordinary crops do not accumulate enough As to be toxic to man. However in As-contaminated soil, the uptake of As by the plant tissue is significantly elevated, particularly in vegetables and edible crops (Larsen *et al.* 1992). There is, therefore, concern regarding accumulation of As in agricultural crops and vegetables grown in the As-affected areas.

Some of the most likely plant species for phytoremediation are members of the Brassicaceae family. *Thlaspi caerulescens*, for instance, is a known Zn hyperaccumulator and some ecotypes can tolerate as much as 40,000 mg/kg Zn dry weight in the shoots (Chaney 1983). Some members of the mustard family from the genus *Brassica*, though, may show promise for use in remediation of metal-contaminated sites. *Brassica juncea* (Indian mustard), for example, has been shown to accumulate moderate levels of Se (Banuelos and Meek 1990, Banuelos *et al.* 1993), Pb, Cr, Cd, Ni, Zn and Cu (Nanda Kumar *et al.* 1995). Other *Brassica* spp., namely *B. napus* and *B. rapa*, have shown a similar tendency to accumulate moderate levels of heavy metals (Nanda Kumar *et al.* 1995, Ebbs and Kochian 1997).

In Spain, an indiscriminate application of inorganic arsenicals as pesticides led to a buildup of As residues in many agricultural soils and reduction of their productivity; these polluted soils are now frequently used for vegetable growing, including turnips, tomatoes, beans, radishes, cauliflower, broccoli, etc. Besides, at the present moment, there is a great concern about As pollution in Spain due to the environmental accident in a pyrite mine located in the city of Aznal-cóllar, Sevilla (southern Spain). In this accident, approximately 4.5×10^6 m^3 of pyretic sludges were spilled into the Agrio and Guadiamar rivers and the surrounding agricultural areas. Immedi-ately after the spill, the Autonomous Council of Andalucía began soil reclamation activities in order to reduce the impact caused by leaching of the toxic heavy metals in the affected area to a minimum. After physically removing the sediments, the soils remained polluted by trace metals such as Pb, Cu, Zn, Cd, Tl, Sb and As (Simón *et al.* 1999). Some of the pollutants, including As, reached Doñana national park, the largest wetland area in Europe and affected soils, plants, and even animals.

Arsenic from these soils may accumulate in any of the agricultural species being grown on them and enter the human food chain through their edible parts. We designed a greenhouse experiment that allowed the study of As accumulation and distribution in the parts of the plant. The main objective of this study was to determine whether As could accumulate in the edible part of radish, cauliflower and broccoli plants in concentrations potentially dangerous for human health. The distribution of the accumulated As among root, shoots and edible parts is also reported here.

39.2   MATERIAL AND METHODS

39.2.1   *Plant material and greenhouse experiments*

The plant species studied during the present investigation were the following: two varieties of radish namely Round-red with white tip "ICEBERG" and Middle East Giant of *Raphanus sativus* L., two varieties of broccoli namely Chevalier and Romanesco of *Brassica oleracea* L. var. *italica* Plenck and two varieties of cauliflower namely Thasca and Premia of *Brassica oleracea* L. var. *botrytis* L.

For studying the effect of soil pH on the uptake and accumulation of As in plants, two soils (namely S1 and S2) were selected. S1 and S2 contaminated soils were obtained from two sites located in the experimental area "El Vicario" (37°26′21″ N, 6°13′00″ W) within the Green Corridor, close to the pyrite mine of Aznalcóllar (Fig. 39.1).

The soils were classified as Typic Haploxeralf. One week before planting, soil from two sites was mixed with commercial potting mixture (1:1 vol.). Soil samples (2 mm) were analyzed for pH in saturated paste (Hesse 1971).

The commercial potting mixture was used as control. The soil was a sandy loam soil (sand 50%, silt 33% and clay 17%) and chemical characteristics were: pH $= 6.0$, $C_{org} = 35\%$, $N_T = 0.3\%$ and organic matter $= 60\%$.

For heavy metal (Pb and Cd) analysis, soil samples (<60 μm) were digested with concentrated nitric and hydrochloric acid to dryness, and re-dissolved in 4% nitric acid. Arsenic concentration

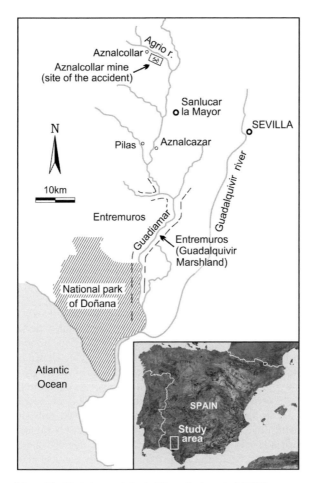

Figure 39.1.   Area of the spill-affected zone (adapted from Santos *et al.* 2002).

in the soil was determined using the same method described for plant analyses. For the determination of plant-available heavy metal fractions, diethylenetriaminepentaacetate (DTPA) was used as extracting agent (Lindsay and Norwell 1969).

Pots were placed in the greenhouse under natural light, temperature of 27/18°C (day/night) and a relative humidity of 50/70% (day/night). Seeds were germinated in Petri dishes for 48 h and when the plants had reached adequate height (8–12 cm) one plant of each species was transferred to plastic pots containing 3 kg of S1 and S2 soils.

In order to study the As accumulation, a complete random design was used for 40, 100 and 140 days of exposure of As in varieties of radish, broccoli and cauliflower, respectively. Controls with unpolluted soil were also included. All treatments were replicated 10 times for S1 and S2 soils.

39.2.2   *Sample preparation and chemical analysis*

The plants were separated into shoots, roots and edible parts washed with tap water, rinsed several times with distilled water and weighed to assess their biomass. To prepare the samples for As analysis, the lyophilized sample (0.25 ± 0.01 g) was weighed, and 1 ml of ashing aid suspension and 5 ml of 50% (v/v) HNO_3 were added, and the mixture was evaporated on a sand bath until total dryness. The ash from the mineralized samples was dissolved in 5 ml of 50% (v/v) HCl and 5 ml of reducing solution (KI-ascorbic acid). After 30 minutes the resulting solution was diluted to volume with 50% (v/v) HCl and filtered through Whatman No. 1 filter paper into a 25 ml calibrated flask. The As concentration was determined by FIA-HG-AAS (Muñoz *et al.* 2000).

The accuracy and precision of the analytical method was assessed by carrying out analyses of the BCR (Community Bureau of Reference) reference sample CMR 279 (sea lettuce) (Griepink and Muntau 1988). The values obtained for the reference sample by FIA-HG-AAS, were concordant with the certified values (data not shown). For each plant species, we calculated the shoot/root metal concentration quotient (M_S/M_R) as a measure to know the As uptake strategy of all species of plants.

39.2.3 *Statistical analyses*

Duncan's multiple range test, applied to data of dry weight and As contents of plant species, was used to detect differences between plant species grown in contaminated and control soils. A significance level of $p < 0.05$ was used throughout the study. The program SPSS Version 10.0 software (SPSS Inc., 1989–1999) was used to perform the statistical analyses.

39.3   RESULT AND DISCUSSION

39.3.1 *Soil analysis*

Table 39.1 shows values of pH and concentrations of heavy metals (Pb and Cd) and As of S1 and S2 soils. The lowest mean values of pH in the studied soils were found in the S2 soil (4.8), while the pH of the S1 soil was neutral (6.8). Mean concentrations of As in S1 and S2 contaminated soils were higher than the upper limits of the ranges of normal soils which are shown in Table 39.1 (Bowen 1979). Mean concentration values of Pb and Cd in contaminated soils were within the ranges of normal soils. Available data in the literature show that values of As concentrations in contaminated soils (<20 mg/kg) can be considered toxic for plant growth (Ross 1994, Singh and Steinnes 1994). The levels of heavy metals and As in the S1 and S2 contaminated soils showed no significant differences, but in contrast they showed significant differences for pH values (Table 39.1).

Plant available metal was very low (Table 39.1) due to the previous chemical treatments (soil amendments with calcium carbonate and ferric oxides) used by regional authorities to fix metals in the polluted soils of Aznalcóllar (Junta de Andalucía 2001).

39.3.2 *Biomass production of cauliflower, broccoli and radish plants*

In the present study, dry weight of shoots decreased significantly for all varieties of cauliflower, broccoli and radish when grown in the presence of As in S1 and S2 polluted soils (Table 39.2). With a single exception, root biomass was significantly lower for the varieties of cauliflower and broccoli in the presence of polluted soils compared to those of control plants (Table 39.2). For the two varieties

Table 39.1.   Soil characteristics, pH, and total and extractable in DTPA concentrations (mean ± SD, n = 6) of heavy metals and arsenic (mg/kg dry matter) of the two S1 and S2 sampling sites. Normal ranges in soils are also shown (Bowen 1979, Wauchope 1983).

	S1			S2			Normal soil[1]
	pH	Total	Extractable	pH	Total	Extractable	
	6.8 ± 0.0			4.3 ± 0.0			
Pb (mg/kg)		80.6 ± 2.2	14.8 ± 0.1		83.2 ± 0.1	14.1 ± 0.2	2–300
Cd (mg/kg)		0.7 ± 0.0	0.1 ± 0.0		0.4 ± 0.0	0.08 ± 0.0	0.01–2
As (mg/kg)		80.9 ± 0.0	4.3 ± 0.1		74.8 ± 0.1	3.8 ± 0.1	0.1–20

[1]Bowen 1979.

Table 39.2.   Shoot, root and dry biomass (g) of plants grown in contaminated soils (mean ± SD of ten samples per species).

Variety	Part	Dry weight (g)		
		Control	S1	S2
Broccoli Chevalier	shoot	23.5 ± 0.3a[1]	3.8 ± 0.2b	5.5 ± 1.0b
Broccoli Chevalier	root	1.6 ± 0.1a	0.8 ± 0.1b	0.6 ± 0.3b
Broccoli Chevalier	edible part	7.3 ± 0.1a	1.4 ± 0.3b	1.6 ± 0.6b
Broccoli Romanesco	shoot	40.1 ± 0.4a	4.0 ± 1.1b	5.8 ± 1.7b
Broccoli Romanesco	root	3.5 ± 0.1a	0.4 ± 0.1b	0.4 ± 0.07b
Broccoli Romanesco	edible part	2.0 ± 0.0a	0.5 ± 0.1b	0.9 ± 0.2ab
Cauliflower Thasca	shoot	22.5 ± 0.3a	8.1 ± 0.5b	9.9 ± 1.4b
Cauliflower Thasca	root	2.3 ± 0.0a	0.9 ± 0.1b	0.7 ± 0.2b
Cauliflower Thasca	edible part	13.1 ± 0.1a	2.2 ± 0.8b	2.9 ± 0.4b
Cauliflower Premia	shoot	38.3 ± 0.2a	11.4 ± 1.9b	23.9 ± 3.1b
Cauliflower Premia	root	2.2 ± 0.1a	1.8 ± 0.2a	2.5 ± 0.5a
Cauliflower Premia	edible part	6.9 ± 0.2a	–	2.3 ± 0.5a
Radish "ICEBERG"	shoot	3.4 ± 0.5a	2.5 ± 0.1b	2.6 ± 0.2b
Radish "ICEBERG"	edible part	2.3 ± 0.9b	3.6 ± 0.4ab	4.7 ± 0.4a
Radish Middle East Giant	shoot	2.2 ± 0.5a	0.9 ± 0.1b	1.5 ± 0.3ab
Radish Middle East Giant	edible part	2.8 ± 0.9b	3.1 ± 0.2a	5.1 ± 0.7a

[1] Mean values with different letters (a, b) within the same row are significantly different from each other using Duncan Multiple Range Test ($p < 0.05$).

of radish, roots (edible part) was significantly higher than roots of control plants. The reason for the observed positive growth response is unclear but may be linked with P nutrition. Phosphate and arsenate are taken into plant roots by a common carrier; however, both high- and low-affinity phosphate/arsenate plasma membrane carriers have a much higher affinity for phosphate than arsenate (Meharg and Macnair 1990). Arsenate/phosphate uptake can be suppressed in plant roots if the plants are P sufficient, resulting possibly in an increased plant growth (Carbonell-Barrachina *et al.* 1998).

Dry weight of edible parts decreased significantly for varieties of cauliflower and broccoli when grown in both S1 and S2 polluted soils compared to those of control plants (Table 39.2).

39.3.3   *Accumulation of arsenic by cauliflower, broccoli and radish plants*

The As concentrations in plants grown in S1 and S2 contaminated soils are shown in Table 39.3. Due to the low bioavailability of metals in soils (Table 39.1) the As accumulation in the tissues of the plants studied was low (Table 39.3). Significant differences were found for As in roots and/or shoots for all accessions (except for the variety Broccoli Chevalier) at contaminated and uncontaminated sites (Table 39.3).

The As concentration in varieties of cauliflower, broccoli and radish plants was not significantly affected by the different pH of the S1 and S2 soils. Genotypical differences in tolerance to heavy metals and As are well known in some species and ecotypes of natural vegetation (Marschner 1997). Thus, in both S1 and S2 contaminated soils, As content in roots of Broccoli Romanesco was higher than Broccoli Chevalier; Cauliflower Thasca showed higher As concentrations in roots than Cauliflower Premia and also it was found that Radish Middle East Giant accumulated higher As content in shoots than Radish Iceberg (Table 39.3).

The As concentrations found in shoots of plant species (Table 39.3) in this work are in agreement with those reported by Soriano and Fereres (2003) in shoots of rapeseed, which found mean concentrations of 1.5 mg As/kg (dw) for plants grown in field experiments on the contaminated soils of Aznalcóllar. On the other hand, the As concentrations of this work were lower than those

Table 39.3.   Arsenic concentrations (mg/kg dry matter) in shoots, roots and edible parts (mean and SD of ten samples per species) of the studied plant species.

Variety	Part	As concentration (mg/kg)		
		Control	S1	S2
Broccoli Chevalier	shoot	$0.6 \pm 0.0a^1$	$0.7 \pm 0.1b$	$0.5 \pm 0.0b$
Broccoli Chevalier	root	$2.0 \pm 0.0a$	$2.6 \pm 1.3a$	$7.3 \pm 0.1a$
Broccoli Chevalier	edible part	$0.4 \pm 0.0a$	$0.2 \pm 0.0b$	$0.3 \pm 0.0b$
Broccoli Romanesco	shoot	$1.6 \pm 0.0a$	$1.8 \pm 0.4b$	$0.3 \pm 0.0b$
Broccoli Romanesco	root	$1.8 \pm 0.1a$	$19.7 \pm 4.0a$	$12.1 \pm 3.1a$
Broccoli Romanesco	edible part	$0.5 \pm 0.1a$	$0.4 \pm 0.2b$	$0.5 \pm 0.3b$
Cauliflower Thasca	shoot	$1.4 \pm 0.0a$	$1.0 \pm 0.1b$	$0.7 \pm 0.1b$
Cauliflower Thasca	root	$1.2 \pm 0.1a$	$9.1 \pm 2.3a$	$7.3 \pm 1.8a$
Cauliflower Thasca	edible part	$0.2 \pm 0.0b$	$0.4 \pm 0.0b$	$0.4 \pm 0.0b$
Cauliflower Premia	shoot	$0.6 \pm 0.0b$	$1.9 \pm 0.3b$	$1.0 \pm 0.2b$
Cauliflower Premia	root	$2.6 \pm 0.1a$	$5.0 \pm 0.6a$	$4.9 \pm 0.6a$
Cauliflower Premia	edible part	$0.4 \pm 0.0b$	–	$0.4 \pm 0.1b$
Radish "ICEBERG"	shoot	$0.6 \pm 0.2a$	$1.5 \pm 0.6a$	$1.7 \pm 1.0a$
Radish "ICEBERG"	edible part	$0.4 \pm 0.1a$	$0.8 \pm 0.3b$	$1.0 \pm 0.2b$
Radish Middle East Giant	shoot	$0.7 \pm 0.0a$	$2.3 \pm 1.0a$	$3.9 \pm 0.8a$
Radish Middle East Giant	edible part	$0.2 \pm 0.2a$	$0.8 \pm 0.1b$	$1.2 \pm 0.2b$

[1] For each plant variety, mean values with different letters (a, b) within the same column are significantly different using Duncan Multiple Range Test ($p < 0.05$).

reported by Del Río et al. (2002), which found concentrations of 7 and 3.3 mg As/kg (dw), respectively, in shoots of *Raphanus raphanistrum* L. and *Sinapis alba* L. collected from Aznalcóllar.

When a toxic metal or metalloid has been absorbed by plants, the most extended mechanism involved in plant tolerance is limiting the upward transport, resulting in accumulation primarily in the root system (Meharg and Macnair 1990). From the data on Table 39.3, it seems that the strategy developed by the two varieties of cauliflower and broccoli plants to tolerate As was avoidance, limiting As transport to shoots. The differences between root and shoot concentrations for the majority of the plants indicated an important restriction of the internal transport of metals from the roots towards shoots and edible parts. Such metal immobilization in root cells, as emphasized by the M_S/M_R (shoot/root metal concentration quotient) quotient $\ll 1$ (range 0.08–0.47), is related to an exclusion strategy (Baker 1981).

In contrast, when radish plants were grown in polluted soils, a little portion of As remained in the roots and a major portion of the element was translocated to shoots (Table 39.3). The behavior of the two varieties of radish with respect to As was characterized in the majority of the cases by $MS/MR \gg 1$ (range 0.74–10.61).

Tlustos et al. (1998) found that the distribution of As among plants was affected by the amount of the bioavailable fraction of As in soils. Plants grown on lower As soil accumulated more As in leaves than roots whereas those grown on higher As soil had more As in roots than leaves. Carbonell-Barrachina et al. (1999) also reported that radish plants grown on soils with higher As concentrations had higher As content in the roots than the shoots.

The statutory limit set for As content in fruits, crops and vegetables is 1.0 mg/kg (fresh weight) (Mitchell and Barr 1995). Since the average water content of the edible part of the cauliflower, broccoli and radish in our experiment was 80, 80 and 95% respectively, the statutory limit for each of these 3 species on a dry weight (dw) basis is calculated to be 5, 5 and 20 mg/kg, respectively. In our study, As concentrations in the edible parts of three species were always below this maximum limit and ranged from 0.2 to 1.2 mg/kg (dw) for plants grown in contaminated soils.

Since the bioavailability of As in the soil is low therefore if plants grow on higher level of contamination could result in higher As in the plant tissues. Further research is needed to be carried out with higher soil As concentrations, to determine the plant ability to accumulate this element in their edible parts.

39.4 CONCLUSIONS

Cauliflower and broccoli plants accumulated As mainly in the root system, but in contrast radish plants accumulated As mainly in shoots. As concentrations in the edible parts of the three plant species were always below the maximum limit set for As in vegetables and fruits. Our results suggest that there should not be a risk associated with the consumption of cauliflower, broccoli and radish grown on As-contaminated soil in conditions as described in this work. Whether higher As concentrations in plant tissues could be reached in a different set of As content in the growth medium needs further studies.

ACKNOWLEDGEMENTS

This research was supported by the *Consejería de Agricultura y Pesca* (*Junta de Andalucía*), Project C03-070, for which the authors are deeply indebted. M. del Río is financed by the *Ramón y Cajal* program (*Ministerio de Ciencia y Tecnología*). M.I. de Haro is financed by *Consejería de Agricultura y Pesca* (*Junta de Andalucia*). The authors are very grateful to Gloria Fernández (IAS-CSIC, Córdoba) for technical assistance in performing the analyses of plants. The authors also wish to thank Dr. Moreno Rojas, (University of Córdoba) for their contribution in soil chemical analysis.

REFERENCES

Baker, A.J.M.: Accumulators and excluders—strategies in the response of plants to heavy metals. *J. Plant Nutrition* 3 (1981), p. 643.
Banuelos, G.S. and Meek, D.W.: Accumulation of selenium in plants grown on selenium-treated soil. *J. Environ. Qual.* 19 (1990), pp. 772–777.
Banuelos, G.S., Cardon, G., Mackey, B., Ben-Asher, J., Wu, L., Beuselinck, P., Akohoue, S. and Zambrzuski, S.: Boron and selenium removal in boron laden soils by four sprinkler irrigated plant species. *J. Environ. Qual.* 22 (1993), pp. 786–792.
Bowen, H.J.M.: *Environmental chemistry of the elements.* Academic Press, London, UK, 1979.
Carbonell-Barrachina, A.A., Burlo, F., López, E. and Martínez-Sánchez, F.: Arsenic toxicity and accumulation in radish as affected by arsenic chemical speciation. *Environ. Sci. Health* B34 (1999), pp. 661–679.
Chaney, R.L.: Plant uptake of inorganic waste. In: J.F. Parr, P.B. Marsh and J.M. Kla (eds): *Land treatment of hazardous wastes.* Noyes Data Corp., NJ, 1983, pp. 50–76.
Chen, C.J., Chiou, H.Y., Chiang, M.H., Lin, T.M. and Tai, T.Y.: Dose-response relationship between ischemic heart disease mortality and long-term arsenic exposure. *Arterioscler Tromb. Vasc. Biol.* 16 (1996), pp. 504–510.
Del Río, M., Font, R., Almela, C., Velez, D., Montoro, R. and De Haro, A.: Heavy metals and arsenic uptake by wild vegetation in the Guadiamar river area after the toxic spill of the Aznalcóllar mine. *J. Biotechnol.* 98 (2002), pp. 125–137.
Deuel, L.E. and Swoboda, A.R.: Arsenic solubility in a reduced environment. *Soil Sci. Soc. Am. Proc.* 36 (1972), pp. 276–278.
Ebbs, S.D. and Kochian, L.V.: Toxicity of zinc and copper to *Brassica* species: Implications for phytoremediation. *J. Environ. Qual.* 26 (1997), pp. 776–781.
Griepink, B. and Muntau, H.: The certification of the contents (mass fractions) of As, B, Cd, Cu, Hg, Mn, Mo, Ni, Pb, Sb, Se and Zn in rye grass, CRM 281. Report no. EUR 11839 EN, Luxembourg, 1987.
Griepink, B. and Muntau, H.: The certification of the contents (mass fractions) of As, Cd, Cu, Pb, Se and Zn in a sea lettuce (*Ulva lactuca*), CRM 279. Report no. EUR 11185 EN, Luxembourg, 1988.
Hesse, P.R.: *A textbook of soil chemical analysis.* John Murray Publishers, London, 1971.
Johnson, L.R. and Hitbold, A.E.: Arsenic content of soil and crops following use of methanearsonate herbicides. *Soil Sci. Soc. Am. Proc.* 33 (1969), pp. 279–282.

Junta de Andalucía: Corredor Verde del Guadiamar. In: Edition of Consejería de Medio Ambiente de la Junta de Andalucía, Ed. Corredor Verde del Guadiamar, Junta de Andalucía, Sevilla, Spain, 2001, pp. 1–70.

Larsen, E.H., Moseholm, L. and Nielsen, M.M.: Atmospheric deposition of trace elements around point sources and human risk assessment. II. Uptake of arsenic and chromium by vegetables grown near a wood preservation factory. *Sci. Total Environ.* 126 (1992): 263–275.

Lindsay, W.L. and Norwell, W.A.: Development of a DTPA micronutrients soil test. *Agron. Abstr.* 6 (1969), p. 84.

Lee, B.G., Griscom, S.B., Lee, J.S., Choi, H.J., Koh, C.H., Luoma, S.N. and Fisher, N.S.: Influences of dietary uptakes and reactive sulfides on metal bioavailability from aquatic sediments. *Science* 287 (2000), pp. 282–284.

Liebig, G.F.: Arsenic. In: H.D. Chapman (ed): *Diagnostic criteria for plants and soils.* University of California, Division of Agric. Sci., Riverside, CA, 1966, pp. 13–22.

Lindsay, W.L. and Norvell, W.A.: Development of DTPA micronutrient soil test. *Agron. Abstr.* (1969), p. 84.

Marin, A.R.: *Effect of soil redox potential and pH on nutrient uptake by rice with special reference to arsenic forms and uptake.* PhD thesis, Louisiana State University, Baton Rouge, LA, 1995.

Marschner, H.: Functions of mineral nutrients: Micronutrients. In: *Mineral nutrition of higher plants.* 2nd ed., Academic Press: Harcourt Brace and Company Publishers, London, 1997, pp. 313–404.

Máthé-Gaspár, G. and Antón, A.: Heavy metal uptake by two radish varieties. *Acta Biol. Szeged* 46 (2002), pp. 113–114.

Meharg, A.A. and Macnair, M.R.: An altered phosphate uptake system in arsenate tolerant *Holcus lanatus. New Phytol.* 116 (1990), pp. 29–35.

Meharg, A.A., Naylor, J., and Macnair, M.R.: Phosphorus nutrition of arsenate-tolerant and nontolerant phenotypes of velvetgrass. *J. Environ. Qual.* 23 (1994), pp. 234–238.

Mitchell, P. and Barr, D.: The nature and significance of public exposure to arsenic: A review of its relevance to southwest England. *Environ. Geochem. Health* 17 (1995), pp. 57–82.

Morales, K.H., Ryan, L., Kuo, T.L., Wu, M.M. and Chen, C.J.: Risk of internal cancers from arsenic in drinking water. *Environ. Health Perspect.* 108 (2000), pp. 655–661.

Muñoz, O., Devesa, V., Suñer, M.A., Vélez, D., Montoro, R., Urieta, I., Macho, M.L. and Jalón, M.: Total and inorganic arsenic in fresh and processed fish products. *J. Agric. Food Chem.* 48 (2000): 4369–4376.

Nanda Kumar, P.B.A., Dushenkov, V., Motto, H. and Raskin, I.: Phytoextraction: the use of plants to remove heavy metals from soils. *Environ. Sci. Technol.* 29 (1995), pp. 1232–1238.

National Research Council: *Arsenic in drinking water.* National Academy Press, Washington, DC, 1999.

Rahman, M.: Arsenic and contamination of drinking water in Bangladesh: a public health perspective. *J. Health Popul. Nut.* 20 (2002), pp. 193–197.

Ross, S.M.: Sources and forms of potentially toxic metals in soil-plant systems. In: S.M. Ross (ed): *Toxic metals in soil-plants system.* John Wiley and Sons Ltd., Chichester, 1994, pp. 3–25.

Santos, A., Alonso, E., Callejón, M. and Jiménez, J.C.: Heavy metal content and speciation in groundwater of the Guadiamar river basin. *Chemosphere* 48 (2002), pp. 279–285.

Simón, M., Ortiz, I., García, I., Fernández, E., Fernández, J., Dorronsoro, C. and Aguilar, J.: Pollution of soils by the toxic spill of a pyrite mine (Aznalcóllar, Spain). *Sci. Total Environ.* 242 (1999), pp. 105–115.

Singh, B.R. and Steinnes, E.: Soil and water contamination by heavy metals. In: R. Lal and B.A. Stewart (eds): *Soil processes and water quality.* Lewis Publ, CRC Press, Boca Raton, FL, 1994, pp. 233–270.

Soriano, M.A. and Fereres, E.: Use of crops for in situ phytoremediation of polluted soils following a toxic flood from a mine spill. *Plant and Soil* 256 (2003), pp. 253–264.

SPSS for Windows, version 10.0. SPSS Inc, Headquarters, Chicago, 1989–1999.

Srivastava, M., Ahmad, N., Gupta, S. and Mukhtar, H.: Involvement of Bcl-2 and Bax in photodynamic therapy-mediated apoptosis. Antisense Bcl-2 oligonucleotide sensitizes Rif 1 cells to photodynamic therapy apoptosis. *J. Biol. Chem.* 276 (2001), pp. 15,481–15,488.

Tlustos, P., Balik, J., Szakova, J. and Pavlikova, D.: The accumulation of arsenic in radish biomass when different forms of As were applied in the soil (Czech). *Rostlinna Vyroba* 44 (1998), pp. 7–13.

USEPA 2001. Proposed Revision to Arsenic Drinking Water Standard, 815-F-00-012. United States Environmental Protection Agency, Office of Water. 2001a, http://www.epa.gov/safewater/ars/proposalfs.html (accessed June 2006).

Wauchope, R.D.: Uptake, translocation and phytotoxicity of arsenic in plants. In: W.H. Lederer and R.J. Fensterheim (eds): *Arsenic: Industrial, biomedical, environmental perspectives.* Van Nostrand Reinhold, NY, 1983, pp. 348–375.

Wei, C.Y. and Chen, T.B.: Arsenic accumulation by two brake ferns growing on an arsenic mine and their potential in phytoremediation. *Chemosphere* 63 (2006), pp. 1048–1053.

CHAPTER 40

Arsenic mobility in the rhizosphere of the tolerant plant *Viguiera dentata*

R. Briones-Gallardo, G. Vázquez-Rodríguez & M.G. Monroy-Fernández
Facultad de Ingeniería-Instituto de Metalurgia, Universidad Autónoma de San Luis Potosí (UASLP), San Luis Potosí, S.L.P., Mexico

ABSTRACT: In the rhizospheric soil, the exposure of plants to arsenic (As) is determined by the chemical equilibrium between soluble, mineral and biological phases. In this work, the physico-chemical and biological accessibility of As in rhizospheric soil of *Viguiera dentata*, an As-tolerant plant, collected from a mining site in Mexico, was studied using two different tests of sequential extraction. The results have shown that 1% and 14.8% of total As concentration (t-As) in soil are available by cationic and anionic exchange, respectively. The anionic exchanges correspond to 83.4% of As, which is probably adsorbed on secondary phases and mobilized after a reduction step. In addition, 3.8% of As is in a bioaccessible fraction, as deduced by its extraction with organic acids. These results indicate that As tolerance in *Viguiera dentata* in this site may be related to the low availability of As in the soil medium studied.

40.1 INTRODUCTION

In Mexico, soils near mines require remediation to remove contaminants introduced by mining activities. In the mining district of Villa de la Paz-Matehuala in San Luis Potosí, Mexico, distribution contour maps of total arsenic (t-As) in soil according to a high density sampling of 5 cm of depth demonstrated a t-As concentration in soil in values ranging from 29 to 28,000 μg/g (Razo 2004). The physiological predominance of *Viguiera dentata*, an Astaraceae family plant, was observed in the area near the highest As concentration. The t-As concentration on dry total biomass (DB) of this plant was 94.8 μg As/g_{DB}, distributed as 82, 12 and 6% in leave, stem and root, respectively. At this site, a t-As concentration in rhizospheric soils (RS) of 3867 μg As/g_{RS} was found. The observed accumulation of As in the plants could be associated to the t-As concentration in soil and to their intrinsic tolerance mechanisms (Lasat 2002). The accumulation mechanisms are normally related to the symbiotic association, intracellular biosynthesis of phytochelatins or physicochemical interaction between available As and functional organic groups expressed in the biological surface of root (i.e., bioavailability fraction). However, few studies have examined the tolerance of plant *versus* the available fractions related to the several physicochemical processes in the rhizospheric soil (Li 2001, Wan 2003).

The As in the rhizospheric soil may be present as sulfide mineral (typically primary phases) and as species adsorbed on iron and manganese oxides (OFM like secondary phases in this mining site). The As mobilization a result of equilibrium among all phases. The As mobilization in primary phases occurs by chemical or biological oxidation. For secondary phases As mobilization associated with the anionic exchange, OFM reduction processes or local acidification (i.e., microbial activities and biological exudation) (Schilling 1998, Klein 1988). In this case, the As is mainly associated to carbonate minerals or secondary phases like OFM. The destruction of these secondary phases can occur in an environment with low redox potential, mobilizing the adsorbed As. Furthermore, organic complexations may modify the As distribution in rhizosphere by increasing organic acid concentration during sugar biodegradation by microbial activity in rhizospheric soil. All these processes govern the mobility and distribution of As in the soil rhizosphere.

The aim of this work was to determine the distribution and accessibility of As in the rhizosphere as a function of the geochemical phases in soil and two different size fractions. The physicochemical mobilization in the rhizospheric soil was analyzed by two different tests of sequential extraction (SE). Different sequential extraction procedures for As and metallic ions have been proposed and successively modified, to describe the As fractions as cation exchangeable (CE), associated to carbonates (CA), iron and manganese oxides (OFM), organic matter and sulfides phases (MOS) or residual (RES) (Tessier 1979, Morera 2001). Assuming that the As on OFM surface may be desorbed by anion exchange (AE), a modified extraction by phosphate soluble salt has been used (Keon 2001).

Besides the physicochemically accessible fractions, in the rhizospheric soil there is a chemical equilibrium between the plant exudates and As of primary or secondary mineral phases. These exudates are mainly characterized by low-molecular-weight organic acids (LMWOA), such as some carboxylic acids, which are secreted or produced by the biological activity of the plants (Jones 1995, Neumann 1999). The exudates composition and contents of carboxylic acids determine the quantity of bioaccesible As in the rhizospheric soil associated to its equilibrium with organic complexes. The carboxylic acids composition in the rhizophere *Viguiera dentata* was not analyzed, so the bioaccessible fraction of As was estimated using a solution that simulates the composition of the organic acids in the wheat rhizosphere (Cieslinski 1998).

40.2   MATERIALS AND METHODS

40.2.1   *Soil sampling*

Two rhizospheric soil samples (RS), with t-As concentration of 14 μg As/g_{RS} and 3867 μg As/g_{RS} were collected from *Viguiera dentata* rhizosphere. The first sample was used as a reference of soil non impacted soil by mining activity (NIS) and the second sample was used to represent a mining impacted soil (MIS). The RS samples were oven dried at 40°C for 72 h and were passed through two sieves. The first size partition was between 2 mm and 600 μm (B), and the second partition corresponds to particles with a diameter lower than 600 μm (C). The As concentration in dry total biomass of plants (DB) for NIS was below detection limit (<0.05 μg As/g_{DB}) while for MIS it was 94.8 μg As/g_{DB}.

40.2.2   *Physicochemical accessibility of arsenic in RS by sequential extraction protocols*

Two different tests of sequential extraction were used to determine the physicochemical accessibility of As. The first protocol (SE1) is a 5-step protocol developed by Tessier (1979). The second (SE2) includes one additional step, corresponding to anionic exchangeable fraction (AE), a modification proposed by Keon (2001) for As fractions. The ES1 procedure defines the geochemical distribution as CE, CA, OFM, MOS and RES. Triplicated analyses were performed for the two size partitions B and C.

In the ES1 protocol, 1 g of soil was used and treated with different chemicals in each step. Solutions extracted from each step were recovered by centrifugation at 800 g for 10 minutes (Beeckman, Allegra 21) and filtration. The solid residue was used in the next step. In step 1 (CE), the soil was reacted with 16 ml of 0.5 M $MgCl_2$ under continuous agitation at 300 g for 20 minutes (Heidolph REAX20). The solution was filtered (2.5 μm Whatman filter) and preserved with two drops of concentrated nitric acid. In step 2 (CA) the residues were reacted with 16 ml of 1 M CH_3COONa at pH 5, for 5 h. In Step 3 (OFM), the soil residues were extracted with 16 ml of 0.04 M $NH_4OH \times HCl$ and 25% v/v acetic acid in open system at 96°C for 6 h. In step 4 (OMS), the residual soil was oxidized at 85°C for 2 h with 3 ml of 3% H_2O_2 at pH 2 and adjusted with 37% HNO_3. After the solution had cooled, 3 ml of 30% H_2O_2 at pH 2 were added and the reaction was maintained for 3 h. Extracted As in this step was stabilized with a solution consisting of 5 ml

Table 40.1.   Composition of low molecular weight organic acids (LMWOA) characterized in rhizosphere of wheat (Cieslinski 1998).

Organic acid compounds	Concentration (10^{-3} mM)
Succinic acid	1940
Oxalic acid	43
Fumaric acid	12
Malic acid	39.8
Citric acid	0.6
Acetic acid	2898

of 3.2 M of CH_3COONH_4 in 20% v/v HNO_3 for a further 20 minutes under continuous agitation. In step 5 (RES), the last residue was digested with a mixture of HNO_3/HCl.

The second protocol (ES2) is a modification of ES1. The variation on step 3 consisted in letting the residual soil from step 2 react twice with 20 ml of 1 M NaH_2PO_4 for 24 h. The combined extract is defined as the anionic exchangeable fraction.

40.2.3   *Bioaccessibility of arsenic by organic acids extraction (OAE)*

Determinations of As-bioaccesible fractions in all RS samples were carried out using a LMWOA solution, which was previously reported (Cieslinski 1998; Table 40.1). For the two rhizospheric soils and for both size partitions (B and C), 1 g of dry RS was placed in contact with 16 ml of LMWOA solution for 5 h with continuous agitation. LMWOA solutions after extraction were then centrifuged and filtered. Finally, all solutions were digested with 37% HNO_3 on a hotplate at 50°C for 5 h. During the first 4 h the solutions were heated under reflux and in last hour, the solution was heated to complete dryness. The digested residues were dissolved with a 0.02% HNO_3 solution for chemical analysis.

40.2.4   *Digestion for total arsenic concentration analysis (TOT)*

Total As concentrations in RS were determined by digestion of 1 g of soil with 15 ml of HNO_3-HCl (3:2 v/v) in a microwave oven (CEM MARSX 3100) at 30 psi maximum pressure and 150°C, for 10 minutes. After digestion, the extracted solution was filtered (2.5 μm Whatman). All analyses were performed on triplicated samples.

40.2.5   *Arsenic chemical analyses and quality control*

Arsenic analyses were performed with a Varian Spectra AA 220 spectrometer when As concentrations in solution were higher than 3 mg/l and for samples with lower concentration with a flow injection analysis system (FIAS) coupled to Analyst 200 Perkin Elmer 2380 hydride generator. The reliability of As in the total digestion procedure was made using two standard reference materials (SRM) from the National Institute of Standard and Technology. For the low and high As concentrations SRM 2709 (San Joaquín soil) and SRM 2710 (Montana soil) were used, respectively. A quality control was performed for 10 batch samples. The As recovery percent values were 98.8% and 100.6% for SRM 2709 and SRM 2710, respectively.

40.3   RESULTS AND DISCUSSION

In general, the t-As concentrations obtained by microwave total digestion, for NIS were 6.65 and 8.75 μg/g for B and C size partition, respectively. For MIS, the t-As concentrations were

3966.7 µg/g$_{RS}$ and 4166.7 µg/g$_{RS}$ for B and C size partition, respectively (Table 40.2). In both cases, the As is more concentrated in the finest particles fraction (Figs. 40.1 and 40.2). In the MIS case, this observation can be attributed to the typically higher concentration of sulfide mineral particles in the smallest size fractions (Fig. 40.2).

Arsenic distribution mobilization for NIS is represented in Figure 40.1. Arsenic in NIS samples was distributed in a major proportion in the residual fraction, 42.5% and 34.3% for B and C size partition, respectively. The As extracted by ES1 in NIS and MIS for fractions B and C are presented in Figures 40.1 and 40.2, respectively. In NIS it was observed that the As linked to OFM can be mobilized under reducing conditions for both fractions. For NIS in fraction B the quantified As was 18% (1.2 µg As/g$_{RS}$) while for fraction C it was 12% (1.05 µg As/g$_{RS}$) from the total concentration. For the NIS sample, it was observed that the As associated to OFM in fraction B is 1.3 times higher than in fraction C. This can be attributed to limitations in diffusion processes in small pore diameters.

Arsenic mobilization associated to the alteration of carbonates phase in NIS is the second geochemical phase most available and was higher for B compared with C size partition. In the case of ES1 protocol a loss of As was observed systematically in B and C size partition. In this case,

Table 40.2.   Total arsenic concentration (TOT) and total arsenic bioaccessible by LMWOA solution (OAE) in NIS and MIS samples for B and C partition (in µg/g$_{RS}$).

	NISB	NISC	MISB	MISC
TOT	6.65 ± 2.47	8.75 ± 0.49	3866.7 ± 907.4	4166.7 ± 873.7
OAE	0.76 ± 0.14	0.69 ± 0.04	147.3 ± 10	116.7 ± 25.7

In all cases the values correspond to the mean of three replicates with ± standard deviations. NIS-non impacted soil samples. MIS-mining impacted soil samples.

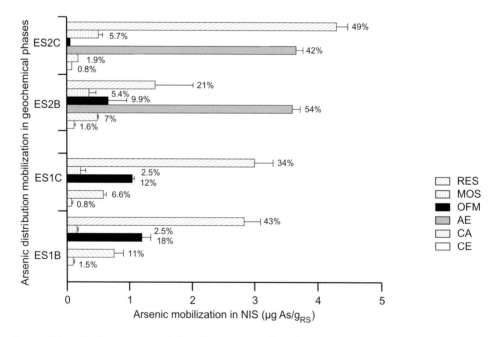

Figure 40.1.   Distribution of arsenic in different geochemical phases from NIS samples with ES1 and ES2 protocols. CE: cation exchange; CA: associated to carbonates; AE: desorbed by anion exchange; OFM: associated to iron and manganese oxides; MOS: associated to organic matter and sulfides phases; and RES: residual fraction.

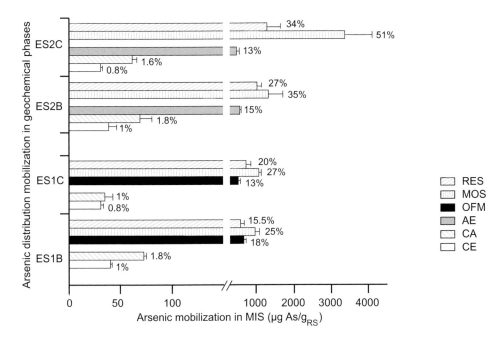

Figure 40.2.   Distribution of arsenic in different geochemical phases from MIS samples with ES1 and ES2
protocols. CE: cation exchange; CA: associated to carbonates; AE: desorbed by anion exchange;
OFM: associated to iron and manganese oxides; MOS: associated to organic matter and sulfides
phases; and RES: residual fraction.

a loss As is observed (24 to 42.8%) referred to the concentration obtained by total digestion. This
result suggests that the As losses may be related to the endothermic steps (MOS fractions), which
induce the As volatilization before the residual digestion in ES1. In the case of MIS sample with
ES1 there were also observed differences between total and fractional digestions. These differences
were 38.6% to 42.1% for B and C partition, respectively.

In the NIS sample, the As distribution was around 5% for MOS fraction independently of size
partition while for OFM fraction it was proportional to size partition (Fig. 40.1). It was also observed
an inverse correlation between As in residual fraction and size partition. These results highlight
the importance in size partition of soil in the evaluation of physicochemical accessibility.

The ES2 results are also showed in Figure 40.2. Arsenic losses observed in the ES1 were lower
for ES2. In the NIS sample, a good correlation between total digestions and different geochemical
fraction was obtained. In the case of MIS, the As losses were 18% of that observed for ES1 for the
B size partition while no losses were observed for C size partition (Fig. 40.2).

In ES2 procedure, for the NIS samples (B and C size partition), As distribution is mainly
associated to any positive mineral charge (in primary or secondary phases), probably an arsenate
anion. However, these positive surfaces do not correspond exclusively to iron and manganese
oxides. It can be also explained because in ES1 protocol the destruction of OFM corresponds to
only 9.9% of total reduction reaction and in ES2 procedure the anionic exchanges with phosphate
were 54% for B and 41.8% for C size partition, probably due to aluminosilicate minerals present
in soil matrices (Fig. 40.1).

In MIS the As is mainly found as sulfide minerals and can be mobilized by chemical or biological
oxidation reactions. For MIS it contained 974.7 $\mu g/g_{RS}$ (25.2%) in fraction B and 1074 $\mu g/g_{RS}$
(25.8%) of As in fraction C. Moreover, in ES2 protocol, the MOS fraction was around 35% of t-As
concentration for B size partition and around 64% for C size partition.

However, in this soil a contribution of 15% of t-As concentration may be accessible by anionic exchange. The mobilized As corresponds to 83.4% of As adsorbed on iron and manganese oxides and probably a negligible fraction associated to aluminosilicate minerals. This effect was also observed in NIS. However, the As adsorbed on OFM is limited by steric effects, which are characterized by a low diffusivity of arsenates into the internal sites.

Furthermore, 2.5% of available As is associated to cation exchange and partial carbonates phase destruction by local acidification in rhizosphere by microbial activity. The decrease of pH and ionic exchange, which are typical processes in rhizospheric soils, could explain that the As concentration in contact with *Viguiera dentata* is directly associated to the mobilization of As by means of the local acidification interactions. The available As in MIS samples, 17% of t-As concentration, corresponds to the maximum quantity in contact with biological surface. The rest of As in these soils was present as non available forms (MOS and RES fractions).

The obtained results with organic acid extraction showed that the most bioaccessable fraction is present in the NIS sample, being 11.5% and 8% for B and C size partition. For MIS sample, the obtained values were 3.8% and 2.8% of TOT for B and C size partition, respectively. However, the higher t-As concentration in MIS sample resulted in higher bioaccessible As. The exposure of plants to As was estimated as the bioaccessible fraction in rhizosphere soil in MIS. This value was $118 \pm 1.6\ \mu g/g_{RS}$ in both size partitions (Table 40.2). In addition, the bioaccessible fraction is mainly related to cationic exchange (1%) and carbonate phases (1.8%). The observed difference (1%) may be attributed to contribution of anionic exchange fraction and probably associated to the formation of a secondary soluble complex of As in solution between As anion and a cationic calcium organic carboxylate complex. This last hypothesis requires further studies of the secondary soluble complex formation between anionic As and soluble organic acids with different cations in solution.

40.4   CONCLUSION

The results have shown that the t-As concentrations determined by total digestion in NIS were of $6.65\ \mu g/g_{RS}$ and $8.75\ \mu g/g_{RS}$ for the B and C size partition. In this soil the available As concentration is associated to cationic exchange (1.5%), anionic exchange (54%), acid dissolution (11.3%) and reduction condition (18%). The As mobilization by anionic exchange is associated to the As fraction adsorbed in OFM and very probably to aluminosilicate phases (36%). Finally 42.6% of total As were found in the residual fraction, which is the most stable As fraction in this soil because it corresponds to the As in silicate minerals.

On the contrary, in the MIS case, independently to size partition, the most stable As fraction is associated to sulfide phases (25%) and silicate minerals phases (around $16.8\% \pm 1$). The available contribution of As in B size partition was distributed in 1% in CE, 1.95% in CA, 14.8% in AE and 17.7% in OFM. In the C size partition the As distributions were 2% for CE and CA fractions, 9.4% for AE and 12.3% for OFM.

The As mobilized in NIS and MIS by the organic acids was $6.65\ \mu g/g_{RS}$ (11.5%) in fraction B and $8.75\ \mu g/g_{RS}$ (7.9%) in fraction C. These results suggest that the reference soil has a very low As concentration available for *Viguiera dentata*. On the other hand, for MIS in presence of organic acids it was observed that As mobilized $147.3\ \mu g/g_{RS}$ (3.81%) for fraction B and $116.7\ \mu g/g_{RS}$ (2.8%) for fraction C. These values are higher than those obtained by acid solubilization of CA. It can be attributed to the fact that the pH decrease attributed to the presence of organic acids completely solubilizes the CA phases and provides organic links for the stabilization of soluble complex of As adsorbed on secondary phases.

The determination of As fractions in rhizospheric soils indicates that As tolerance of *Viguiera dentata* in MIS site may be related to the low availability of As. The As accumulation may be a result of a progressive exposure to low available As concentration in rhizospheric soil (around $118\ \mu g/g_{RS}$). The mining impacted rhizospheric soil may be seen as a continuous As source that controls the accumulation mechanisms in *Viguiera dentata*, avoiding interfacial toxic concentration that limited their growth in the studied site.

ACKNOWLEDGEMENTS

This work was funded by Grant SEMARNAT-CONACYT 2002-C01-0362 and PROMEP. We are also grateful to the Mexican National Council for Science and Technology (CONACyT) for the fellowship of G. Vázquez-Rodríguez. We thank Dr. Brent E. Handy for improving the English and Dr. Irmene Ortiz Lopez for her helpful comments.

REFERENCES

Cieslinski, G. and Van Ress, C.J.: Low-molecular-weight organics acids in rhizosphere soils of durum wheat and their effect on cadmium bioaccumulation. *Plant Soil* 203 (1998), pp. 109–117.

Jones, D.L.: Influx and exflux of organic acids across the soil-root interface of *Zea mays* L. and its implications in rhizosphere C flow. *Plant Soil* 173 (1995), pp. 103–109.

Keon, N.E.: Validation of arsenic sequential extraction method for evaluating mobility in sediments. *Environ. Sci. Technol.* 35 (2001), pp. 2778–2784.

Klein, D.A.: Rhizosphere microorganisms effects on soluble amino acids, sugars and organic acids in the root zone of *Agropyron cristatum, A. Smithii* and *Bouteloua gracilis. Plant Soil* 110 (1988): 19–25.

Lasat, M.M.: Phytoextraction of toxic metals: a review of biological mechanisms. *J. Environ. Qual.* 31 (2002), pp. 109–120.

Li, X.D. and Thornton, I.: Chemical partitioning of trace and major elements in soils contaminated by mining and smelting activities. *Appl. Geochem.* 16 (2001), pp. 1693–1706.

Morera, M.T.: Isotherms and sequential extraction procedures for evaluating sorption and distribution of heavy metals in soils. *Environ. Poll.* 113 (2001), pp. 135–144.

Neumann, G. and Romheld, V.: Root excretion of carboxylic acids protons in phosphorus deficient plants. *Plant Soil* 211 (1999), pp. 121–130.

Razo, I., Téllez, J., Monroy, M., Carrizales, L., Díaz-Barriga, F. and Castro, J.: Arsenic and heavy metal pollution of soil, water and sediments in a semi-arid climate mining area in Mexico. *Water, Air Soil Poll.* 52 (2004), pp. 129–152.

Schilling, G.: Phosphorus availability, root exudates and microbial activity in the rhizosphere. *Z. Pflanzenernähr. Bodenk.* 161 (1998), pp. 465–478.

Tessier, A.: Sequential extraction procedure for the speciation of particulate trace metals. *Anal. Chem.* 51:7 (1979), pp. 844–851.

Wan, W,: Relationship between the extractable metals from soils and metals taken up by maize roots and shoots. *Chemosphere* 53 (2003), pp. 523–530.

Section V
Toxicology and metabolism

CHAPTER 41

Survey of arsenic in drinking water and assessment of the intake of arsenic from water in Argentine Puna

S.S. Farías, G. Bianco de Salas & R.E. Servant
Gerencia Química, Comisión Nacional de Energía Atómica, San Martín, Prov. de Buenos Aires, Argentina

G. Bovi Mitre, J. Escalante, R.I. Ponce & M.E. Ávila Carrera
Grupo InQA-Investigación Química Aplicada, Facultad de Ingeniería, Universidad Nacional de Jujuy, S.S. de Jujuy, Prov. de Jujuy, Argentina

ABSTRACT: This chapter describes occurrence and risks related to the presence of arsenic (As) from geogenic sources in superficial and groundwaters in a 30,000 km^2 area of two regions of Argentine Altiplano (Puna) and Sub Andine valleys counties. In these regions, with prevailing rural population, levels of As and related elements in drinking water are much higher than limits established by Argentine and international regulations. Forty-six water samples from Puna counties and sixteen water samples from Sub Andine valleys counties were analyzed. Levels that exceeded 2 mg/l were detected in 5% of both superficial and groundwater samples; levels in 61% of the samples exceeded the WHO limit of 10 μg/l. Results of this study that examined water consumed by 355,000 individuals were used for a preliminary risk assessment. Eight percent of the study population in Puna and Sub Andine valleys counties are exposed to levels of As higher than those recommended for drinking water. The intake of As from drinking water ranged from 13 to 374 μg As/day for children and from 20 to 561 μg As/day for adults.

41.1 INTRODUCTION

Inorganic arsenic (i-As) is a multi-site carcinogen that may be present in drinking water. Epidemiological studies performed in United States, European Union, Taiwan, Bangladesh, Mexico, Argentina and Chile have linked exposure to only 10 μg As/l of "natural contaminated" drinking water to an increased risk of skin, bladder, liver, kidney and lung cancers (Borgoño 1977, Besuschio *et al.* 1980, Hopenhayn-Rich *et al.* 1996, 1998, Smith *et al.* 1998, Bates *et al.* 2004).

In Argentina, As studies were performed mainly in the Chaco-Pampean plain (Biagini *et al.* 1967, Farías *et al.* 2003, Bundschuh *et al.* 2004, Bhattacharya *et al.* 2006). Only a few reports refer to As levels in drinking water in La Puna (Concha *et al.* 2006) even though it has been recognized that the prevalence of arid climate, oxidizing conditions, and high salinity of waters in this region favor anomalous concentrations of anion forming elements like As, boron, fluoride, molybdenum, selenium, uranium and vanadium (Criaud and Fouillac, 1989, Thompson *et al.* 1996, Smedley and Kinninburgh 2002). Most reports have focused on skin cancer (Argüello, 1942, 1943, Bergoglio 1963, Astolfi 1971, Astolfi *et al.* 1981, Biagini 1972, 1975, 1977, Besuschio *et al.* 1980) and skin diseases (Tello 1975, 1979, Astolfi *et al.* 1981). Only a few studies suggested an increased risk of lung and bladder cancers (Bergoglio 1964, Biagini *et al.* 1978, Tello 1979, 1988, Besuschio *et al.* 1980, Astolfi *et al.* 1981, Hopenhayn-Rich *et al.* 1996, 1998). Epidemiological studies carried out in affected communities (Besuschio *et al.* 1980, Hopenhayn Rich *et al.* 1996, Smith *et al.* 1998, Steinmaus *et al.* 2000) provided evidence of risks associated with As present in drinking water.

The Argentine Puna represents the southern end of the Peruvian and Bolivian Altiplano. It is a strip of land situated between latitudes 21°45′ S and 26°15′ S, covering approximately 200 km in an E–W direction, and 550 km in a N–S direction (100,000 km^2). Its average altitude is 3800 m a.s.l.

Towards the west, the main chain of the Central Andes high volcanoes separates the Argentine Puna from the Chilean Puna de Atacama, 1000 m lower.

From a geographical point of view, the region may be defined as a high level desert. Its main features are aridity, high mountain chains that correspond to fault blocks, high volcanic peaks (up to 6000 m), lack of external drainage, and basins occupied by evaporites. The climate varies from dry and cold in summer to very cold in winter. Rainfall ranges from 50 to 300 mm per year.

Late Precambrian rocks occur in southern and western areas, and marine Paleozoic sediments (predominantly Ordovician) reach thicknesses of up to 3000 m. Mesozoic marine and continental sediments (2000 m thick) occur and Cenozoic continental rocks that reach 6000 m thickness. Two types of Cenozoic volcanic rocks characterize the region. The older unit is made up of lava flows and ignimbrites (most of them of dacitic composition) and Pliocene age dacitic tuffs. The younger unit occurs as large stratovolcanoes made up of lava, ignimbrites, and Pleistocene age tuffs of andesitic composition (Turner and Méndez 1979). During the Quaternary, large deposits of piedmont fanglomerates occur with transitional lacustrine and evaporitic facies.

Volcaniclastic sediments reworked by fluvial and subordinated aeolic processes developed permeable non-consolidated deposits that favor water infiltration and groundwater flow towards the depressed areas. Groundwater collection is performed by means of shallow wells from unconfined aquifers, but in some places fine sediment intercalations occur confining the aquifer. Sedimentological and geochemical features of the formations, weathering and solubilization processes and subsequent concentration, are responsible for the high As and other trace-element concentrations in the waters (Welch *et al.* 1988).

Generally, the Puna region is related to the presence of volcanic activity occurring during the Quaternary, when large volcano chimneys were formed; their products were scattered around a huge area in the region. This activity, nowadays restricted but not completely extinguished, is also the source of thermal manifestations, and it is responsible for a constant supply of compounds arising from dissolution of the minerals present in volcaniclastic sediments (Thompson *et al.* 1985, Nimick *et al.* 1998). Other factors regulating the distribution of As and other toxic trace-elements are the pH values of groundwater, and in aquifer sediments, sorption onto Fe- and Al-oxide and hydroxide surfaces may regulate the distribution of trace elements. However, in groundwater with high pH values and high HCO_3 concentrations desorption occurs, increasing trace-element concentrations significantly (Nicolli *et al.* 2004).

In the present study, surface and groundwaters were considered the only environmental factors related to the transport of contaminants (including As) for a preliminary risk assessment of As intake from contaminated water.

The aims of this work were (1) the development and validation of analytical methods for evaluation of As levels in superficial and groundwaters used for irrigation and for human and animal consumption in rural communities of La Puna and in Sub Andine valleys where the major part of population and economic activities are concentrated; (2) the application of collected data for evaluation of the daily intake (DI) and for comparison with the WHO reference value and the provisional tolerable weekly intake (PTWI) value, and (3) the evaluation of toxicological risk associated with use of water in La Puna.

41.2   EXPERIMENTAL

41.2.1   *Sampling sites*

The 30,000 km^2 large study area included the southern and western regions of the Jujuy province and the northwestern region of the province of Salta, including five scarcely populated La Punas counties: Tumbaya, Susques, and Cochinoca in Jujuy province, and La Poma and Los Andes in Salta province, and three densely populated counties in the valleys of Sub Andine mountains, including Manuel Belgrano, Palpalá, and San Antonio, all of them in Jujuy province (Table 41.1, Fig. 41.1). Sampling was performed in June 2003, June 2004, and July 2004.

Table 41.1.   Sampling areas and number of inhabitants.

Region	County	Number of inhabitants[3]	Total number of inhabitants
Sub Andine valleys (SAV)	San Antonio[1]	3700	
	M. Belgrano[1]	238000	326700
	El Carmen[1]	85000	
La Puna (LP)	Tumbaya[1]	4600	
	Susques[1]	3700	
	Cochinoca[1]	12100	28000
	La Poma[2]	2000	
	Los Andes[2]	5600	
SAV + LP			354700

[1] Counties in the province of Jujuy; [2] Counties in the province of Salta; [3] According to 2001 census data.

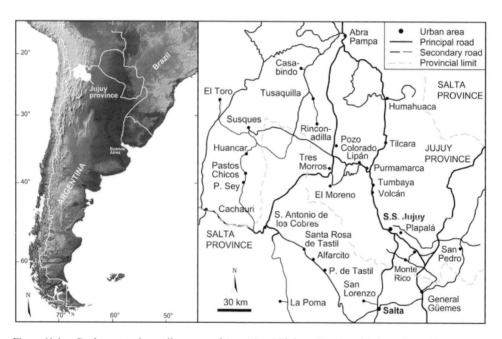

Figure 41.1.   Study area and sampling areas of Argentine Altiplano (Puna) and Sub Andine valleys.

Superficial waters and groundwaters used for watering and for animal and human consumption were collected in cities, small towns, and villages in the two studied zones. It should be noted that the power for pumps that provide groundwater from wells all around La Puna is supplied by solar panels. In Sub Andine valleys, samples were collected either in large cities like San Salvador de Jujuy, Perico and Palpalá, or in small towns and villages. In both cases, special attention was paid to schools where students (who represent the highest risk group because of their physical features and age) daily consume 'naturally contaminated' water.

Although scarce reports on As in water in San Antonio de los Cobres in the province of Salta have been published (Concha *et al.* 2006), no information is available from the area of La Puna located in the Jujuy province. This work which also includes risk assessment contributes to the knowledge of As contamination in the study area. The sampling area was very large and some counties were scarcely populated. The goal was to sample all water sources in every settlement, no matter how small it was.

41.2.2 Sampling

Sampling procedures were developed, tested, and standardized prior to the actual sampling campaign, following the quality assurance sampling protocols developed in our laboratory. These procedures included records of sample code, date, time, sampling area, coordinates, sampling site name, type of the groundwater supply (i.e., windmills, boreholes, artesian well, deep and shallow wells and wells equipped with electrical pumps), sample volume, tools for sampling, depth, and climate conditions. *In situ* analyses, e.g., pH and temperature determination, were also included.

Satellite images were used to select the sampling sites (a total of 62), taking into account the morphology of La Puna. Water samples were filtered through 0.45 μm Millipore® filters, placed in clean 100 ml Nalgene® narrow-mouth bottles, and 0.2 ml concentrated HNO_3 or HCl were added. After arrival to the laboratory, samples were stored in a refrigerator (4°C). Samples were thermostabilized before analysis. The sampling campaign was carried out during May 2003, June, and July 2004.

41.2.3 Analytical determinations

41.2.3.1 Equipment
A Perkin-Elmer (Norwalk, CT, USA) inductively coupled plasma-atomic emission spectrometer (ICP-OES) Optima 3100 XL (axial view) was used for the trace level As determinations and an ICP-OES Perkin-Elmer (Norwalk, CT, USA) ICP 400 coupled with a hydride generation (HG) cell was used to determine As at ultra trace levels.

41.2.3.2 Reagents
Unless otherwise specified, all reagents were of an analytical reagent grade. An ultrapure Ar from Indura (Buenos Aires, Argentina) was used for the ICP-OES analyses. Welding argon from L'Air Liquide (Buenos Aires, Argentina) was used for As analyses by HG-ICP-OES. Deionized distilled water (DDW) was produced by feeding Nanopure® Barnstead purification system (Dubuque, USA) with distilled water. A commercially available 1000 mg As/l standard solution (CertiPUR® Merck, Darmstadt, Germany) was used. Diluted working solutions were prepared daily by serial dilutions of stock solutions.

Nitric acid 70% (Merck, Darmstadt, Germany) and hydrochloric acid 65% (Merck, Darmstadt, Germany) were used after additional purification by sub-boiling distillation in quartz still. A 3% (w/v) sodium tetrahydroborate solution was prepared by dissolving $NaBH_4$ powder (Riedel de Häen, Germany) in DDW. The solution was stabilized by 1% (w/v) NaOH (Merck) and filtered through an SandS Weißband (Schleicher and Schull, Germany) filter paper to eliminate turbidity. The solution was stored in a polyethylene flask at 4°C. A diluted working solution of $NaBH_4$ (0.5% w/v) was prepared daily before use. A 20% (w/v) KI (Merck) solution in DDW stored in polyethylene flask at 4°C was used to pre-reduce As (V) to the hydride generating As (III); the working solution was prepared daily.

41.2.3.3 Quantification of As
Quantitative determination of As was carried out using methodologies developed, optimized, and validated in our laboratories ('in house validation') in accordance with international guidelines (EURACHEM Guide: The Fitness for Purpose of Analytical Methods, 1998). Calibration curves were generated for standards prepared by serial dilution of the 1000 mg As/l standard stock solution. One sample of the reference material and blanks (reagent blanks) were included in each analytical batch (every 10 samples). All samples were analyzed by ICP-OES. All samples with As levels near or below the ICP-OES quantification limit were also analyzed by HG-ICP-OES. Two calibration curves were obtained, one for the trace levels of As (ICP-OES analysis: 20 to 1000 μg As/l) and the other for the ultra trace levels of As (HG-ICP-OES analysis: 5–20 μg As/l).

Table 41.2. Basic characteristics of the analytical methods.

	ICP-OES	HG-ICP-OES
Detection limit[1] (μg As/l)	3	1
Precision %[2] (20 μg As/l)	5	5
Precision %[2] (50 μg As/l)	3	–

[1]3σ, n = 10, [2]n = 10 (three times).

Table 41.3. The certified and measured arsenic concentrations for two standard reference materials, NIST SRM 1640 'Trace elements in natural water' and NIST SRM 1643d 'Trace elements in water'.

		Measured value (μg/l)	
Reference material	Certified value (μg/l)	ICP-OES	HG-ICP-OES
NIST 1640	26.67 ± 0.41	–	27.7 ± 1.6
NIST 1643d	56.02 ± 0.73	58.6 ± 3.2	–

Limit of detection is usually calculated as 3σ for repeated analyses of a blank. However, in this study, the limit of detection was determined as 3σ for 10 independent replicates of a 3 μg As/l standard solution for HG-ICP-OES and of a 10 μg As/l standard solution for ICP-OES, in accordance with the As levels required by national legislation* (50 μg/l for drinking water, *Ley* 18284 *Código Alimentario Argentino* 1994* and *Ley* 24051 *Ley de Residuos Peligrosos* 1993*) and with the WHO recommended value (10 μg As/l, WHO 1996) (Table 41.2).

Precision of the method was determined as repeatability (% standard deviation, %SD); three independent analyses of ten replicates of a 20 μg As/l standard solution for HG-ICP-OES and of 20 μg As/l and 50 μg As/l standard solution for ICP-OES were performed. No significant statistical differences between the three series of results were found by F-test for each of the above methods. Average %SD were low, providing good repeatability for both methods (Table 41.2).

Accuracy of measurements was tested using two standard reference materials: NIST SRM 1643d 'Trace elements in water' for ICP-OES determinations and NIST SRM 1640 'Trace elements in natural water' for HG-ICP-OES determinations. No significant statistical differences were found by t-test between the certified and experimental values for both NIST SRM 1643d and NIST SRM 1640 (4.4% and 3.7% bias, respectively). Ten replicated instrumental measurements were performed for each of the three SRM aliquots. The procedure was carried out during three different days and the results and the corresponding uncertainties were averaged. Results for both SRMs are summarized in Table 41.3.

41.3 RESULTS AND DISCUSSION

41.3.1 *As contents in samples from the Puna and Sub Andine valleys*

Average results obtained for As levels in superficial and groundwaters of the studied regions are listed in Table 41.4. Results obtained in every sampling site and sampling campaign characteristics are described in Tables 41.5 and 41.6.

In La Puna, groundwaters and surface waters are generally used for human consumption, animal consumption, or irrigation. The use of water supplies is indiscriminate and is determined by the amount of resources seasonably available and by the prevailing climate. Springs may be

---

*NOTE: The limit for As in drinking water in Argentina was modified in June 2007. The new maximum allowed concentration is 10 μg As/l. The text was not modified as the mentioned legislation was in force when the work was performed and written.

Table 41.4.   Average arsenic levels in waters in the study areas in the Salta and Jujuy provinces.

Zone	Number of samples	As (µg/l) Mean	Median	Range
Sub Andine valleys	16	6.5	6.0	6.0–10
La Puna	46	578	35	10–10000
	43[1] (46)	187[1,2]	30[1]	10–1000[1]

[1]Results obtained after exclusion of data from Puesto Sey (i.e., two samples with almost 10,000 µg As/l) and a sample from Huancar greenhouse in El Toro (437 ± 22 µg As/l).
[2] The average As concentration used for DI calculation in La Puna.

only used for watering because previous studies on water quality have demonstrated chemical or bacteriological contamination, e.g., the case of Huancar greenhouse in El Toro where levels of As in waters reached 437 ± 22 µg As/l. People were notified about high As concentrations in water supplies and the consequences of using water from these supplies. Borderline cases, including the source of El Toro's greenhouse (437 ± 22 µg As/l) or drinking waters of Puesto Sey's school (9560 ± 500 As µg/l) were not considered for evaluation of As intake in this study.

   Collas (native Indians) live in small isolated settlements located in the mountains. Because they are mainly shepherds, they drink water from a variety of groundwater and surface water sources they find during their daily walks.

   The highest As levels were detected in water samples from La Puna, particularly in samples of superficial waters which are easily contaminated by contact with As rich soils and sediments. In La Puna, where 28,000 people reside, Susques, Pastos Chicos, Castro Tolay, Puesto Sey (near Susques), Tres Morros and San Antonio de los Cobres were the villages with higher As levels in drinking water. In Toro (near Pastos Chicos), irrigation water was also highly contaminated. Similarly, extremely high levels of As up to almost 10,000 µg/l, have been determined at Puesto Sey school.

   As concentrations were also surprisingly high in rivers, springs, and wells which are used as drinking water sources in Pastos Chicos and Susques, with As concentrations ranging from 400 to 800 µg/l. In San Antonio de los Cobres, As levels in treated drinking waters reached 200–250 µg/l and As concentration in rivers and streams used as drinking water supplies by *Collas* ranged from 1000 to 2000 µg/l.

   Local factors such as changes in geomorphological features or in the sediment composition of each sampled area influence As distribution. This is more remarkable in basin edge facies where seasonal variations in As concentrations should be highlighted. Different As concentrations in groundwaters, 10–30 µg/l and 170–270 µg/l were found in the neighboring communities El Moreno and Tres Morros, respectively. Most of the Puna water samples exceeded Argentine legislation limits (50 µg/l)* and the WHO recommended values (10 µg/l) (WHO 1996) for As in drinking water. As discussed by Smedley and Kinninburgh (2002), the commonly high As levels in groundwaters may be related to both the geochemical environment and the previous and present hydrogeology. No information is available about geochemistry of aquifers and sources of As or about As levels in water sources in Puna region.

   Fortunately, surface and groundwaters from the area of Sub Andine valleys where 327,000 people reside, and almost all cattle raising and agricultural activities take place, show As levels far below the limits stated by Argentine regulations*: 50 µg/l for drinking water; 500 µg/l for animal drinking water; 100 µg/l for irrigation (*Ley* 18284 *Código Alimentario Argentino* 1994* and *Ley* 24051 *Ley de Residuos Peligrosos* 1993. Thus, population of this area is not affected by the presence of As in drinking water (Table 41.4).

---

* See the footnote in the preceeding page.

Table 41.5.   Results from sampling Campaign in La Puna (where not otherwise stated, one sample was taken).

Sampling site	Water source/supply	(As) µg/l mean ± SD	Sampling date
Train station, S. Antonio de los Cobres	Treated GW (ep)	248 ± 30	May-03
Perez family, S. Antonio de los Cobres	Treated GW (ep)	200 ± 30	May-03
Elementary school, S.A. de los Cobres	Treated GW (ep)	179 ± 25	May-03
San Antonio de los Cobres river	SW	2030 ± 100	May-03
Quebrada del Toro stream, S.R. Tastil	SW	10 ± 1	May-03
El Moreno reservoir	GW (windmill)	30 ± 2	May-03
El Moreno windmill	GW (windmill)	30 ± 2	May-03
Drinking water school N° 251, El Moreno	GW (windmill)	30 ± 2	May-03
El Moreno spring	GW	34 ± 2	May-03
El Moreno river	SW	14 ± 1	May-03
Greenhouse watering water, El Moreno	GW (windmill)	30 ± 2	June-04
Padilla family, Tres Morros	GW (ep-solar panels)	167 ± 10	May-03
Ponce family, Tres Morros	GW (ep-solar panels)	243 ± 20	June-04
Water source for animals, Tres Morros	GW (ep-solar panels)	75 ± 7	June-04
Spring Tres Morros	GW	36 ± 2	May-03
Spring Cuesta de Lipán	GW	6.0 ± 0.6	May-03
Spring Puesto Sey 1	GW	710 ± 21	June-04
Spring Puesto Sey 2	GW	720 ± 22	June-04
Reservoir Puesto Sey 1	GW	9770 ± 290	June-04
Reservoir Puesto Sey 2	GW	9150 ± 275	June-04
Drinking water Puesto Sey 1	GW	200 ± 6	June-04
Drinking water Puesto Sey 2	GW	200 ± 6	June-04
Water source, Susques (46 samples)	GW (ep)	22 ± 2	June-04
Susques river	SW	335 ± 30	June-04
Rosario river, El Toro	SW	952 ± 50	June-04
Toro river	SW	28 ± 2	June-04
Water source, El Toro	GW (ep)	24 ± 2	June-04
Pastos Chicos river	SW	669 ± 30	June-04
Water source "old", Huancar	GW	12 ± 2	June-04
Water source "Rastrojos", Huancar	GW	16 ± 2	June-04
Water source greenhouse, Huancar	GW	372 ± 20	June-04
Reservoir, Huancar	GW (ep)	18 ± 2	June-04
Drinking water school N° 365	GW (ep)	14 ± 2	June-04
Reservoir Pozo Colorado	GW	6.0 ± 0.6	July-04
Water source Pozo Colorado	GW	26 ± 1	July-04
Reservoir Rinconadilla	GW	17 ± 1	July-04
Water source Rinconadilla	GW	157 ± 5	July-04
Water source El Codito, Rinconadilla	GW	61 ± 2	July-04
Water source Lagunilla, Rinconadilla	GW	94 ± 3	July-04
Spring Cocha Los Rosales, Rinconadilla	GW	21 ± 1	July-04
Water source S.R. Hídricos, S.J. Cerrito	GW (ep)	58 ± 2	July-04
Water source San Francisco de Alfarcito	GW	30 ± 2	July-04
Reservoir San Francisco de Alfarcito	GW (ep)	22 ± 1	July-04
Water source Purmamarca	Treated GW (ep)	65 ± 2	July-04
Water source Tumbaya	Treated GW (ep)	10.0 ± 1.0	July-04
Water source Volcán	Treated GW (ep)	8.1 ± 1.0	July-04

GW: groundwater, SW: surface water, ep: electrical pumps.

In the Puna study area, the health of population, as well as their economical resources, cattle raising and agriculture, may be severely threatened by the environment. People might suffer of adverse effects of As in drinking water and possibly effects associated with As that contaminates food and beverages. Here, the risk related to the presence of As and other toxic anion-forming

Table 41.6.   Results from sampling campaign in Sub Andine valleys (where not otherwise stated, one sample was taken).

Sampling site	Water source/supply	(As) µg/l mean ± SD	Sampling date
Tap water hotel Augustus, S.S. Jujuy	Treated SW	10 ± 1	May-03
Mineral water "Palau", S.S. Jujuy	Treated GW (ep)	10 ± 1	May-03
Perico river	SW	6.0 ± 0.6	June-04
Chico Juan Galán river	SW	6.0 ± 0.6	June-04
Dique Los Alisos	SW	5.0 ± 0.6	June-04
del Movado river (Los Alisos river)	SW	6.0 ± 0.6	June-04
Dique Las Maderas	SW	6.0 ± 0.6	June-04
Vertiente (GA5)	SW	5.0 ± 0.5	June-04
Dique La Ciénaga (16 samples)	SW	6.0 ± 0.6	June-04
Arroyo Burrumayo	SW	6.0 ± 0.6	June-04
Drinking water 1, Palpalá	Treated GW	6.0 ± 0.6	July-04
Drinking water 2, Palpalá	Treated GW	8.0 ± 0.8	July-04
Drinking water 1, S.S. Jujuy	Treated GW	6.0 ± 0.6	July-04
Drinking water 2, S.S. Jujuy	Treated GW	6.0 ± 0.6	July-04
Drinking water Alto Comedero	GW (ep)	5.0 ± 0.5	July-04
Xibi-Xibi river, S.S. Jujuy	SW	5.0 ± 0.5	July-04

GW: groundwater, SW: surface water, ep: electrical pumps.

Table 41.7.   The daily As intake in the study area expressed as µg As/day and comparison with the WHO reference value (WHO 1996), considering water as the only source of exposure.

Region/season	Water intake (l/day)	As in water (µg/l) Range	As in water (µg/l) Mean	As intake (µg/day) Range	As intake (µg/day) Mean	As reference value (µg/day)
Adults						
S.A. valleys (summer/winter)	3	6.0–10	6.5	18–30	19.5	–
La Puna (summer)	3	10–1000	187	30–3000	561	146
La Puna (winter)	1.5	10–1000	187	15–1500	280.5	–
Children						
S.A. valleys (summer/winter)	2	6.0–10	6.5	12–20	13	–
La Puna (summer)	2	10–1000	187	20–2000	374	54
La Puna (winter)	1	10–1000	187	10–1000	197	–

elements in drinking water has never been assessed. Thus, results of this work may raise awareness of the authorities of these risks and initiate future studies of As in diets and food.

41.3.2   *Evaluation of the provisional tolerable weekly intake (PTWI) of As*

The population in La Puna is continuously exposed to high concentrations of As in drinking water. In comparison, the residents of the Sub Andine valleys who served as a reference population in this study are exposed to low As concentrations. However, no information about the presence of As in water or about health risks related to the intake of As from water sources has been available. The distinctive features of this region include changes in climatic conditions with remarkable differences between day and night, summer and winter. Days are excessively hot in summer and relatively warm in winter, but nights are always very cold. That is why the water consumption depends on the season. It should be pointed out that in winter La Puna inhabitants drink groundwater almost exclusively as other sources remain frozen until spring. In the Sub Andine valleys the climate is usually moderate and warm in summer time. In this study, data on water intake were obtained from personal interviews with local inhabitants who revealed that the predominant source of drinking water was local water and that adults drink about 1.5–3.0 l of water daily, depending on the season.

To determine the water intake of children, school authorities were interviewed. Results of the inquiry revealed that in the case of children the average daily water consumption in winter and summer was about 1.0 and 2.0 l, respectively. Local water was the only source of drinking water. Thus, in La Puna the average water consumption in summer was estimated to be about 3 l/day for adults and 2 l/day for children; in winter, the average water intake may decrease by 50% (1.5 and 1 l/day, respectively). Because people in the Sub Andine valleys are accustomed to drinking *mate* (a kind of green tea), which is prepared with boiling water, all day long, the estimated water consumption may vary between 2 and 3 l/day for children and adults, respectively. Papers published by others (Standing Committee on the Scientific Evaluation of Dietary Reference Intake 2004, Watanabe *et al.* 2004) indicate that the adult water intake ranges between 2 and 6 l/day, depending on physical activity. The water intake in a warm climate or during a high-level physical activity may vary between 2 and 7 l/day (Ruby *et al.* 2002).

Maximum and minimum concentrations of As found in drinking waters (assuming that total As is represented by inorganic As) and information on water consumption were used to calculate the daily As intake (DI) and to obtain information about related health risks for exposed population and reference groups. The extreme values found in Puesto Sey (two drinking water samples with almost 10,000 µg As/l) were not included in the calculation, as they are not representative of the whole sampling area. An average body weight of 68 kg for an adult and a 25 kg body weight for a child were used in these calculations. The 'provisional tolerable weekly intake' (PTWI) was selected to evaluate As intake and to compare this As intake with FAO/WHO 'reference intake value'. (PTWI is the parameter defined by FAO-WHO for inorganic As (i-As) intake, i-As is the main species present in water). Given the above body weights, the PTWI reference intake stated by the FAO/WHO (1996) (15 µg inorganic As/week/kg body weight) is equivalent to a reference daily intake of 146 µg i-As/day for adults and 54 µg i-As day for children. These reference values were compared with i-As intakes from drinking water by the residents of the study area (Table 41.7). In the Sub Andine valleys, the daily i-As intake from drinking water was 14–20% and 22–37% of the FAO/WHO reference intake value for adults and children, respectively. Thus, As did not represent a problem for people living in the Sub Andine valleys or for agriculture and cattle farmers. However, the results were different for La Puna. Calculated from the average As concentration (187 µg As/l), the i-As intake for adult population during winter was two-fold greater than the FAO/WHO recommended value and it was four-fold greater during summer. If the extreme values are included in the calculation the average daily intakes would exceed 10 and 20 times the reference value in winter and summer, respectively. Risk for children is even greater, considering i-As average value of 187 µg As/l. Here, the As intake would be 3.5- and 7-fold higher than the FAO/WHO recommended value for winter and summer, respectively. If higher i-As concentrations were considered, the reference value would be 18.5 and 37 times higher for winter and summer, respectively. These values are estimated strictly for As intake from drinking water. If food is taken into account, the daily intake of As would surely be much higher because preparation of cooked dishes and soups can concentrate As.

41.4   CONCLUSIONS

The daily As intake for the majority of the population in La Puna is higher than the limit indicated by FAO/WHO. A complete survey of La Puna would have to be undertaken in order to evaluate the magnitude of As contamination. In order to improve people's health, risks associated with this element should be evaluated and authorities should develop policies to mitigate this natural contamination. It must be stressed that some drinking water supplies in these regions do not comply with maximum levels stated by Argentine legislation*. Actions must be taken to ensure that proper remediation procedures are carried out in a very near future. Other contaminants originating either from mining activities or from natural sources should be examined in La Puna and toxicological risk should be assessed, especially for children.

ACKNOWLEDGEMENTS

This research was partially supported by Acción CYTED 105PI0272; the authors are deeply grateful.

REFERENCES

Argüello, A.R.: Consideraciones sobre el Asismo crónico regional endémico y la enfermedad de Bowen. *Rev. Arg. Dermat.* 7 (1942), pp. 313–320.

Argüello, A.R.: Cáncer y Asismo regional endémico. *Rev. Arg. Dermat.* 9 (1943), pp. 152–153.

Astolfi, E.A.N.: Estudio de Arsénico en agua de consumo. *Prensa Méd. Arg.* 14 (1971), pp. 1342–1343.

Astolfi, E.A.N., Maccagno, A., García Fernández, J.C., Vaccaro, R. and Stimola, R.: Relation between As in drinking water and skin cancer. *Biolog. Trace Element Res.* 3 (1981), pp. 133–143.

Bates, M.N., Rey, O., Biggs, M.L., Hopenhayn, C., Moore, L., Kalman, D., Steinmaus, C. and Smith, A.H.:Case-control study of bladder cancer and exposure to As in Argentina. *Amer. J. Epidemiol.* 159 (2004), pp. 381–389.

Bhattacharya, P., Claesson, M., Bundschuh, J., Sracek, O., Fagerberg, J., Jacks, G., Martin, R.A., Storniolo, A. and Thir, J.M.: Distribution and mobility of arsenic in the Rio Dulce alluvial aquifers in Santiago del Estero Province, Argentina. *Sci. Total Environ.* 358 (2006), pp. 97–120.

Bundschuh, J., Farías, B., Martin, R., Storniolo, A., Bhattacharya, P., Cortes, J., Bonorino, G. and Albouy, R.: Groundwater arsenic in the Chaco-Pampean Plain, Argentina: case study from Robles county, Santiago del Estero Province. *Appl. Geochem.* 19 (2004), pp. 231–243.

Bergoglio, R.M. Cancer mortality in zones of arsenical waters of the Province of Córdoba. Argentine Republic contribution to regional pathology of cancer. *Prensa Med. Argent.* 51 (1964), p. 994.

Besuschio, S.C., Perez Desanzo, A.C. and Crocci, M.: Epidemiological association between As and cancer in Argentina. *Biolog. Trace Element Res.* 3 (1980), pp. 41–55.

Biagini, E.R. and Vázquez, C.A.: Agua de consumo con alto contenido Asal en la Ciudad de Córdoba. *Rev. Derm. Arg.* 51 (1967), p. 43.

Biagini E.R.: HidroAsismo crónico y muerte por cánceres malignos. *Sem. Méd.* 25 (1972), pp. 812–816.

Biagini E.R.: HidroAsismo crónico en la República Argentina. *Méd. Cut.* I.L.A. 6 (1975), pp. 423–432.

Biagini E.R.: HidroAsismo crónico y leucoplasia. *Arch. Arg. Dermat.* 22 (1977), pp. 53–58.

Biagini E.R., Rivero, M., Salvador, M. and Córdoba, S.: HidroAsismo crónico y cáncer de pulmón. *Arch. Arg. Dermat.* 28 (1978), pp. 151–157.

Borgoño, M.J.: As in the drinking water of the city of Antofagasta. Epidemiological and clinical study before and after the installation of a treatment plant. *Environ. Health Perspect.* 19 (1977), pp. 103–105.

Concha, G., Nermell, B. and Vahter, M.: Spatial and temporal variations in As exposure via drinking-water in northern Argentina. *J. Health Pop. Nutr.* 24 (2006), pp. 317–326.

Criaud, A. and Fouillac, C.: The distribution of As (III) and As (V) in geothermal waters: Examples from the Massif Central of France, the Island of Dominica in the Leeward Islands of the Caribbean, the Valles Caldera of New Mexico, USA and southwest Bulgaria. *Chem. Geol.* 76 (1989), pp. 259–269.

EURACHEM Guide: The fitness for purpose of analytical methods. A laboratory guide to method validation and related topics. LGC, Teddington, Middlesex, 1998.

Farías, S.S., Casa, V.A., Vazquez, C., Ferpozzi, L., Pucci, G.N. and Cohen. I.M.: Natural contamination with As and other trace elements in ground waters of Argentine Pampean Plain. *Sci. Total Environ.* 309 (2003), pp. 387–399.

Hopenhayn-Rich, C., Biggs, M.L., Fuchs, A., Bergoglio, R., Tello, E., Nicolli, H.B. and Smith, A.H.: Bladder cancer mortality associated with As in drinking water in Córdoba, Argentina. *Epidemiology* 7 (1996), pp. 117–124.

Hopenhayn-Rich, C., Biggs, M.L. and Smith, A.H.: Lung and kidney cancer mortality associated with As in drinking water in Córdoba, Argentina. *Int. J. Epidemiol.* 27 (1998), pp. 561–569.

Ley 18284: *Código Alimentario Argentino*. Sancionada y promulgada el 18/7/69. B.O. 28/7/69, Modificación 1988 y Modificación 1994. Buenos Aires, Argentina, 1969.

Ley 24051: *Ley de Residuos Peligrosos*. Sancionada el 17/12/91, promulgada de hecho el 8/1/92, B.O. 17/1/92. Decreto 831/93. Reglamentación Ley 24051, B.O. 3/5/93. Sec. 1, 14–19. Buenos Aires, Argentina, 1993.

Nicolli, H.B., Tineo, A., García, J.W., Falcón, C.M., Merino, M.H., Etchichury, M.C., Alonso, M.S. and Tofalo, O.R.: The role of loess in groundwater pollution at Salí River basin, Argentina. In: R.B. Wanty and R.R. Seals II (eds): *Water-Rock Interaction* 2. Balkema, Leiden, The Netherlands, 2004, pp. 1591–1595.

Nimick, D.A., Moore, J.N., Dalby, C.E. and Savka M.W.: The fate of geothermal As in the Madison and Misouri Rivers, Montana and Wyoming. *Water Resour. Res.* 34 (1998), pp. 3051–3067.

Ruby, B.C., Shriver, T.C., Zderic, T.W., Sharkey, B.J., Burks, C. and Tysk, S.: Total energy expenditure during arduous wildfire suppression. *Med. Sci. Sports Exerc.* 34 (2002), pp. 1048–1054.

Smedley, P.L. and Kinninburgh, D.G.: A review of the source, behaviour and distribution of As in natural waters. *Appl. Geochem.* 17 (2002), pp. 517–568.

Smith, A.H., Goycolea, M., Hayne, R. and Biggs, M.: Marked increase of bladder and lung cancer mortality in a region of northern Chile due to As in water. *Am. J. Epidemiol.* 147 (1998), pp. 660–669.

Standing Committee on the Scientific Evaluation of Dietary Reference Intake: Dietary reference intake for water, potassium, sodium, chloride and sulfate. Panel on dietary references intakes for electrolytes and water, Chapter 4: Water. The National Academic Press, Washington, DC, 2004, pp. 73–185.

Steinmaus, C., Moore, L., Hopenhayn-Rich, C., Biggs, M.L. and Smith, A.H.: As in drinking water and bladder cancer. *Cancer Invest.* 18 (2000), p. 82.

Tello, E.E.: El HidroArsenisismo Crónico Regional Endémico (HACRE) y los carcinomas cutáneos. *Arch. Arg. Dermatol.* 25 (1975), pp. 199–204.

Tello, E.E.: Los Cánceres de los órganos internos coincidentes con epiteliomas cutáneos en el hidroAsismo crónico regional endémico argentino (HACREA), *Rev. Méd. Córdoba* 67 (1979), pp. 28–35.

Tello, E.E.: Los carcinomas de los órganos internos y su relación con las aguas Asales de consumo en la República Argentina. *Med. Cut. I.L.A.* 16 (1988), pp. 497–501.

Thompson, J.M. and Demonge, J.M.: Chemical analysis of hot springs, pools and geysers of Yellowstone National Park, Wyoming and vicinity, 1980–1993. US Geol. Surv. Open File Rep. 96–98, 1996.

Thompson, J.M. and Keith, T.E.C.: Water chemistry and mineralogy of Morgan and Growler hot spring, Lassen KGRA, California. *Transactions of the Geothermal Research Council* 9 (1985), pp. 357–362.

Turner, J.C.M. and Méndez, V.: Puna. In: J.C.M. Turner (ed): *Geología regional Argentina*. Academia Nacional de Ciencias, Córdoba, Argentina, 1979, pp. 13–56.

Watanabe, C., Kawata, A., Sudo, N., Sekiyama, M., Inaoka, T., Bae, M. and Ohtsuka, R.: Water intake in an Asian population living in As-contaminated area. *Toxicol. Appl. Pharmacol.* 198 (2004), pp. 272–282.

Welch, A.H., Lico, M.S. and Hugues, J.L.: As in groundwater of the Western United States. *Ground Water* 26 (1988), pp. 333–347.

WHO: International Programme on Chemical Safety. *Guidelines for drinking water quality*. Vol. 2, *Health criteria and other supporting information*. 2nd ed., World Health Organization, Geneva, Switzerland, 1996.

CHAPTER 42

Chronic arseniasis in El Zapote, Nicaragua

A. Gómez
UNICEF-Nicaragua, Managua, Nicaragua

ABSTRACT: The first cases of arseniasis in Central America were reported in 1996 in El Zapote, rural community of 125 inhabitants, located in the valley of Sébaco, north of Nicaragua. During the years 1994 to 1996, the population ingested arsenic-contaminated water from a public tube well containing 1320 μg/l of inorganic arsenic (i-As). Contamination was also detected in private hand-pumped wells used before 1994 and after 1996 (45–66 μg/l As). This chapter presents the results of a clinical and laboratory examination of 111 individuals that lived in this community and consumed As-contaminated water. Both the skin and the respiratory system were significantly affected among the inhabitants with higher As exposure. Clinical signs of hydroarsenicism such as hyperkeratosis and hyperpigmentation were observed in some individuals confirming the diagnosis of a collective case of chronic arseniasis in Nicaragua and Central America.

42.1 INTRODUCTION

42.1.1 *Exposure history*

The first cases of arseniasis Central America were reported in May 1996 in El Zapote, a rural community of 125 inhabitants located in the municipality of San Isidro, in the Sébaco valley, in northern Nicaragua. Between 1994 and 1996, some members of this small community ingested water of a tube well contaminated with 1320 μg/l of arsenic (As). The well was installed in 1994 with standard water quality testing procedures at that time, which did not include the determination of As (Aguilar *et al.* 2000). Although the contaminated well was closed in the middle of 1996, exposure is known to have continued because later studies measured concentrations of 45–66 μg/l of As in private hand-wells of El Zapote, that were used by the whole population before 1994 and after 1996 (Gomez 2000).

In 1997, a multidisciplinary team of the Health Studies and Research Center (CIES) rejected agrochemicals as the cause of As contamination, and suggested a relation between the contamination and geologic faults located in W and SW side of the valley of Sébaco, where alluvial deposits come into contact with volcanic rocks of the Coyol group (Gonzalez *et al.* 1998). Arsenic-contaminated water was also identified in other communities of the Sébaco valley and its margins, including a spring in El Carrizo that contained 110 μg/l As. According to Gomez and Aguilar (2000) a man who lived in this last community between 1952 and 1959 and had ingested water from the contaminated spring, showed signs of chronic As toxicity including keratosis and hyper-pigmentation with multiple cutaneous cancer, which was detected in 1983, and caused his death in 2000.

Among the population of El Zapote As-related cancer has not been registered. Nevertheless the high levels of As in the water that the population of El Zapote ingested for two years and the continuity of the contamination, although in smaller scale, placed this population in high risk of developing symptoms of chronic As poisoning.

42.1.2 *Study design*

The present cross-sectional study was made between July and October 2002, by a team of medical specialists in co-ordination with local authorities of the Nicaraguan Ministry of Health. Clinical and laboratory investigations of the inhabitants of El Zapote, and those who had lived there between 1994 and 1996, were carried out to diagnose symptoms associated with arseniasis, and to be able to prevent worse consequences in the health of this population.

42.2 METHODOLOGY

42.2.1 *Study design*

In July 2002, a census of the inhabitants was carried out of El Zapote, accompanied by interviews to identify those families who lived in the community between 1994 and 1996 when the contaminated well was in use. Some of these families were living in nearby towns, where they were visited and added to the census. The field team was composed of a pediatrician, an internal medical doctor, a radiologist and a dermatologist, all with more than ten years of professional experience. The dermatologist had six years of experience in the diagnosis of arseniasis. An epidemiologist, a laboratory technician and an auxiliary nurse, who were local health workers, collaborated with the field team, which also included the support of a health promoter. The participants were questioned about their socio-demographic characteristics, smoking habits, water consumption and medical symptoms. Vital statistics, weight and size were measured; and a complete physical examination was performed by the pediatrician and the internal medical doctor on each of the participants. The physicians did not know the level of As exposure of the patients. Later, the dermatologist made a complete skin inspection in search of cutaneous symptoms of arseniasis. The files of the patients who had previously been hospitalized in regional and national hospitals were reviewed, including the two cases that were referred during this study.

42.2.2 *Criteria for skin lesions and other tests*

Cutaneous As lesions were identified as: keratosis, characterized by a diffuse bilateral thickening of palms and/or plants, with or without nodules; and hyperpigmentation, consisting of mottled bilateral dark brown spots on the trunk, with affectation or not on the extremities. This classification of arsenical skin lesions was based on studies by Guha Mazumder *et al.* (1998c, 2000a) and Centeno *et al.* (2006). Also, an abdominal ultrasonography was performed on each participant using portable ultrasound equipment (Mysono, model 201, mode B, real time) with 5 transducers of 5 MHz for adults, and of 7.5 MHz for children. These studies were performed by the same investigator, with the patient in an oblique-supine position, and without knowledge of the patient's level of As exposure. The evaluation of internal organs (liver, biliary vesicle, spleen, pancreas, kidneys, bladder and portal vein) was according to criteria of normality already established in literature (Rumack *et al.* 1999). In children, the evaluation used the criteria of the Radiology Department of the "La Mascota" reference children's hospital.

Glucose levels during fast were determined for each participant using the Accuchek Advantage II test. A blood sample was taken from the fingertip for hematocrit determination in a local laboratory, and a urinary cytology test was also performed, where the urine samples transferred in thermos to a laboratory in the city of Managua, according to established procedures.

According to the clinical findings or ultrasound results, the following laboratory determinations were made: a general urine analysis, the levels of serum creatinine, transaminases, hepatitis B surface antigen, malaria and platelet count. All the blood samples were centrifuged at local level and transported in a cooling device the same day to the laboratory of reference in the city of Managua.

For all the respiratory symptomatic patients, three bacteriologic examination of the sputum were performed for differential diagnosis with tuberculosis. The analysis was made in the health

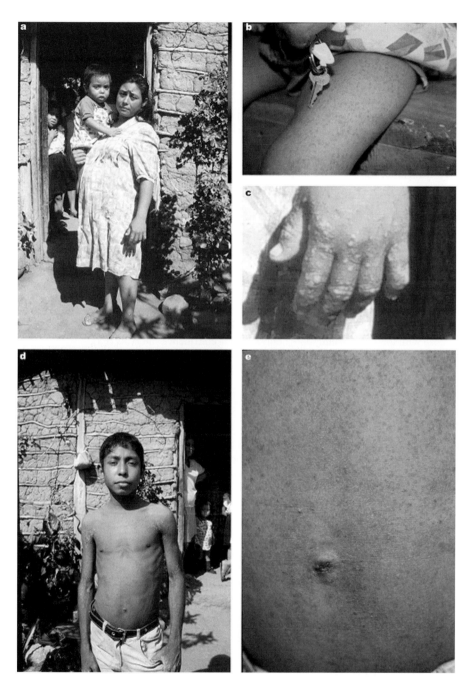

Figure 42.1.   Photos (year 1996) from inhabitants of El Zapote village after 2 years of exposure to drinking
water with 1320 µg/l As: (a) Pregnant 20-year old, female patient with her 2-year old son.
Facial edema was present in both. The child presented hypermelanosis in trunk and extremities
while the mother had hyperkeratosis on palms and soles and keratosis in the dorsum of the hand,
feet and knees; (b) shows severe spotted hypermelanosis on the thigh of the 2 year old child,
whereas (c) shows the hyperkeratosis on the left hand of his mother; (d) and (e) show a thirteen
year old male patient with hypermelanosis in the trunk, especially in the abdomen. He also had
Norwegian scabies with hyperkeratosis in the arms.

center located in the town of San Isidro, with quality control in the national center of diagnosis and reference of the Ministry of Health, according to established procedures. In this last center a Mycobacterium culture (Lowestein-Jensen) was performed to eleven sputum samples. The results were reported nine weeks later.

Chest X-rays were performed on patients who showed clinical indications of respiratory affects. These plates were taken at the hospital of La Trinidad, in the neighbor municipality of San Isidro. Their interpretation was made by the radiologist and the pediatrician members of the field team, having confirmed the results with other specialists with experience.

42.2.3 *Data analysis*

The data obtained were collected in five formats as described above. The participants were divided in two groups according to the level of exposure. The "high-As exposure" group included those people who ingested water of the well containing 1320 µg/l As for a period of six months to two years, which was the time in which this water source was in use. The second, "low-As ingestion" group included the persons who did not ingest water of the well mentioned before or did so for less than six months, but they still took water from excavated wells containing 45–66 µg/l As. These last wells also were the water source of the first group, before and after the highly contaminated well functioned.

The prevalence per 100 for each symptom was determined according to the level of As ingestion. The odds ratio of clinical manifestations was calculated by comparing those with high-As ingestion with those with low-As consumption, applying 95% confidence intervals (CI).

42.3 RESULTS

42.3.1 *Sample size*

In July 2002, El Zapote had 17 houses and 95 inhabitants. A further thirty-five people lived in this community between 1994 and 1996, but later moved to nearby communities. Out of 116 people, only 111 agreed to be examined, of whom 47% were under 15 years old (Table 42.1).

42.3.2 *As exposure determination*

Considering the period of time during which they lived in El Zapote and the water sources that they consumed from, it was estimated that mean As consumption between 1994 and 2002 for the high-As ingestion group was 251 µg/l As. The average calculated for the same period for the low-As ingestion group was 88 µg/l As. The age and gender characteristics of both groups are presented in Table 42.1.

Table 42.1. Distribution of the participants by age, sex and arsenic level consumption.

Arsenic exposure	<15 years		≥15 years		Total
	Males	Females	Males	Females	
High As ingestion[1]	16	13	15	25	68
Low As ingestion[2]	15	10	10	7	43
Total	31	23	25	32	111

[1] Average of consumption for the last 8 years (1994–2002): 251 µg/l As (range: 45–1320).
[2] Average of consumption for the last 8 years (1994–2002): 88 µg/l As (range: 45–1320).

42.3.3 Clinical manifestations

Within the "high As ingestion" group, the earliest As skin lesions reported were those of a seven year old boy. According to his mother, he initiated the hyperpigmentation 12 days after birth. She consumed water from the most contaminated well during pregnancy. On the other hand a 92 year old patient, the oldest one of the study, did not show arsenical hyperpigmentations or keratosis. Although the most contaminated well had been closed six years ago, the nodules on the hands and feet not only persisted, but continued growing. None of those examined had skin cancer. The patients of the high-As ingestion group that had respiratory manifestations commonly called it "incurable influenza". Nine of these patients were hospitalized on repeated occasions with diagnoses of pneumonia. Higher prevalences of weakness, headache, burning of the eyes, nausea and edema, and especially paresthesia and the cutaneous and respiratory manifestations, were statistically significant (Table 42.2).

No participant under 15 years old was a smoker and among the older ones, the relationship between smoking and the respiratory manifestations was not statistically significant (Table 42.3).

42.3.4 Ultrasonography study

Three patients in the "low As ingestion" group were diagnosed with hepatomegaly. One was recovering from an acute poisoning with an organo-phosphate chemical, although a second had arsenical keratosis. Splenomegaly was detected in four patients, all with "high As ingestion" and accompanied by respiratory and digestive As manifestations. Two of them had to be transferred to the reference hospitals. Both patients also presented thickening of the portal vein, along with

Table 42.2. Prevalence of clinical manifestations according to As exposure.

Symptoms and signs	High As ingestion n = 68		Low As ingestion n = 43		Total		Odds ratios
	# of cases	%	# of cases	%	# of cases	%	95% C.I.
General							
Weakness	31	45.6	12	27.9	43	38.7	2.16 (0.95–4.91)
Headache	46	67.7	19	44.2	65	58.6	2.64 (1.20–5.81)
Burning of the eyes	38	55.9	13	30.2	51	27.9	2.92 (1.30–6.56)
Nausea	21	30.9	5	11.6	26	23.4	3.39 (1.17–9.85)
Anorexia	24	35.3	18	41.9	42	37.8	0.76 (0.35–1.66)
Diahorrea	11	16.2	6	14.0	17	15.3	1.19 (0.41–3.50)
Abdominal pain	47	69.1	21	48.8	68	61.3	2.34 (1.07–5.16)
Hyporeflex	10	14.7	7	16.3	17	15.3	0.89 (0.31–2.54)
Paresthesis	23	33.8	3	7.0	26	23.4	6.81 (1.90–24.42)
Pain on inferior M	3	4.4	6	14.0	9	8.1	0.28 (0.07–1.21)
Oedema on inferior M	11	16.2	0	0	11	9.9	–
Peripheral vascular disease	2	2.9	0	0	2	1.8	–
Respiratory							
Cough	47	69.1	20	46.5	67	60.4	2.57 (1.17–5.67)
Hemoptisis	7	10.3	0	0	7	6.3	–
Chest sound	19	27.9	2	4.7	21	18.9	7.95 (1.75–36.16)
Skin							
Cutaneous sting	38	55.9	13	30.2	51	46.0	4.66 (1.48–14.69)
Hyperpigmentation	45	66.2	0	0	45	40.5	–
Keratosis	45	66.2	1	2.3	56	41.4	82.17 (10.62–635.66)

See text for explanation of exposure classes.

Table 42.3.   Association between respiratory manifestations and cigarette smoking in participants ≥15 yr old.

Symptoms and signs	Smokers (n = 14)		Non-smokers (n = 43)		Odds ratios
	# of cases	%	# of cases	%	95% C.I.
Cough	7	50.0	22	51.2	0.95 (0.29–3.19)
Hemoptisis	0	0.0	2	4.7	–
Chest sound	3	21.4	8	18.6	1.19 (0.27–5.29)

Table 42.4.   Chest X-ray results by according to age and arsenic exposure.

Chest X-ray diagnosis	High arsenic ingestion		Low arsenic ingestion	
	<15 years	≥15 years	<15 years	≥15 years
# Tests performed	10	8	1	2
Normal	0	4	0	2
Inflammatory infiltration	8	0	1	0
Air trapped	5	0	1	0
Calcifications	3	0	0	0
Diffuse fibrosis	7	4	1	0

other two patients of the "low As ingestion" group, one of whom had recently been poisoned by agrochemicals. No significant disorders were found in other abdominal organs.

42.3.5   Laboratory tests

Pancytopenia was confirmed in both patients referred to Managua. The results of the sputum bacteriologic examination performed to 20 patients and the culture for Mycobacterium in 11 of them were all negatives. Nevertheless out of 21 to whom chest X-rays were taken, 15 presented the alterations described in Table 42.4. No significant results were found in other laboratory tests performed.

42.3.6   Patients referred to hospitals

Two patients whose health had deteriorated were referred to hospitals for specialized attention and further studies. The first one was a seven year old boy with 12.9 cm splenomegaly, portal hypertension, antecedents of digestive bleeding, diarrhea and abdominal pain for last three years. He also presented weakness, headache, chest sounds and a dry cough which had been persistent for six years. The chest X-ray showed the existence of inflammatory infiltration and diffuse pulmonary fibrosis. He had been hospitalized on a number of occasions by severe anemia, hematemesis and pneumonia. This patient lived with his family in the Zapote since birth until his first year of life, having ingested water with 1320 μg/l As. In addition, his mother drank from this same water source during his period of gestation.

The second patient transferred was an 18 year old adolescent that lived with his family in El Zapote, from his birth until 1998. He ingested water with 1320 μg/l As for two years. He presented 6 cm splenomegaly accompanied by portal hypertension, abdominal pain, anorexia, in addition to hyporeflex and paresthesia. An electrocardiogram showed incomplete blockage of the right side. The patient and his relatives showed keratosis, moderate to severe hyperpigmentation and chronic cough accompanied by chest sounds (rhonchi and crepitations). This patient refused to take chest X-ray, however, three of his brothers presented inflammatory infiltration and diffuse pulmonary fibrosis. In addition, his mother had splenomegaly.

42.4   DISCUSSION AND CONCLUSIONS

1. The high prevalence of cutaneous injuries among people in the high-As ingestion group was to be expected (Table 42.2) since it has been reported in diverse studies (Das *et al.* 1995, Guha Mazumder *et al.* 1997 and 1998c, Mukherjee and Bhattacharya 2001, Chakraborti *et al.* 2001, Bejarano and Nordberg 2003). Nevertheless it is notable that, in El Zapote, the majority of symptoms were initiated after six months of ingesting water containing 1320 μg As/l. Most authors have indicated a latency period of at least two to three years (Tseng *et al.* 1968, Rosenberg 1974). However, in West Bengal (India), Guha Mazumder *et al.* (1997) reported a latency period of less than a year in some cases.

   The premature cutaneous injuries in the boy with hyperpigmentation twelve days after birth supports the evidence of the transplacentar exposure to As (Goyer 1996, Concha *et al.* 1998) and contrasts with the old woman of 92 years without arsenicals skin lesions. This difference suggests the influence of a genetic factor, as it has been mentioned in other studies (Alain *et al.* 1993, BCAS 1997).

   The relatively small amounts of As that these patients continued to consume (<66 μg As/l) after abandoning the highly contaminated well/or switching to wells with low As, could have contributed to the persistence, growth and appearance of new arsenical skin manifestations.

2. The presence of coughs in 69.1% in the high-As ingestion group (Table 42.2) was a little greater than reported by Guha Mazumder *et al.* (1997) in a study of 156 cases (prevalence of 57%) in West Bengal, one of the first reports of nonmalignant respiratory effects in India. Previously Borgono *et al.* (1977) in Chile, detected the presence of cough in 39% of 144 individuals with "abnormal pigmentation" who had ingested water containing 800 μg/l As.

   The hemoptisis reported (10.3%) in the high As ingestion group (Table 42.2) was similar to the 8% reported by Guha Mazumder *et al.* (1997).

   The prevalence of chest sounds was significantly higher in the high-As ingestion group (27.9%) compared to those with low As ingestion (4.65%) (Table 42.2) (OR 7.95, CI 1.75–35.16). A similar result was observed by Guha Mazumder *et al.* (2000b) in India among 6864 males and females studied, the prevalence of chest sounds in the lungs rose with increasing As concentration in drinking water ($\geq$500 μg/l As), most pronounced in individuals with skin lesions (females: prevalence OR 9.6, CI 4.0–22.9; males: OR 6.9, CI 3.1–15.0).

   In addition to the cough, the hemoptisis and the chest sounds, the radiological diagnosis of diffuse fibrosis in 12 of the 21 chest X-rays was particularly worrisome because it was pre-dominant in patients younger than 15 years old (Table 42.4). In 1974 Rosenberg had discovered interstitial fibrosis in two of four lungs that were examined during autopsies on children of Antofagasta, Chile who presented skin hallmark of arseniasis (Rosenberg 1974). Moreover, Figueroa *et al.* (1992) studied hundreds of years old mummies of an area in Chile that contained water sources contaminated by As. They found that the greater As contents were deposited in kidney, liver, nail and pulmonary tissue, in even greater amounts than the ones found in skin and hair, supporting the idea that lungs are also organs with a high accumulation of As.

   The absence of cases of pulmonary tuberculosis registered in the study area and the negative results of all the sputum tests performed in the symptomatic respiratory discards this differential diagnosis. In addition, according to the literature, although pulmonary tuberculosis can cause the appearance of nodules with or without fibrotic scars, which is usually well delimited and located in the hilar area or upper lobes of the lungs (ATS 2000).

   The statistically significant association between high ingestion of As and respiratory mani-festations observed in this study is similar to that in India (e.g., Guha Mazumder *et al.* 1997 and 2000b, Guha Mazumder 1998a) and supports the strong presumption that the As ingestion is playing a determining role in the appearance of these respiratory processes. Spirometric testing and other studies have to be performed to confirm the current findings. On the other hand, the damage that As could have caused in the pulmonary tissue and the immunosuppressive effect of this metal (Ostrosky *et al.* 1991, Gonsebatt *et al.* 1994, NCR 1999, Soto Peña *et al.* 2006) could be causing a greater susceptibility to the respiratory infections.

   Greenberg *et al.* (2002) indicated that pulmonary fibrosis is a condition that can be considered preneoplastic. Hubbard *et al.* (2000) found that the idiopathic pulmonary fibrosis increases seven times the incidence of lung cancer. This situation resembles the pneumoconiosis such as asbestosis or silicosis (Hubbard *et al.* 2000, Samet 2000). In the case of As, diverse studies have demonstrated the association of this element with the lung cancer (Hopenhayn-Rich *et al.* 1998, Smith *et al.* 1998) but future research is needed to confirm pulmonary fibrosis As-induced.

3. The low prevalence of hepatomegaly was not expected (3/111). It was very different from the 76.9% reported in studies made in West Bengal (Guha Mazumder *et al.* 1997 and 1998b). In the study made in El Zapote in 1996, in spite of not having made a systematic examination of the population in search of internal anomalies, five of the people who had ingested water of the well containing 1320 μg/l As presented hepatomegaly (Aguilar *et al.* 2000). In India 24 cases were followed-up during 2–10 years revealing 86% of hepatomegaly (Guha Mazumder 1998a).

4. It is to expect that neoplastic processes in skin or other internal organs have not yet been detected. Periods of latency of more than 20 years are described in the literature (Smith *et al.* 1992). This also happened in El Carrizo a neighbor community of El Zapote, where a patient was reported to have consumed water containing 110 μg/l As between 1952 and 1959, but it was not until 24 years later that he was diagnosed with the first of multiple malignant cutaneous tumors (Gómez and Aguilar 2000).

5. The seven year old boy and the 18 year old adolescent showed portal hypertension and the splenomegaly. According to NRC (1999), noncirrhotic portal hypertension, although uncommon, is a gastro-intestinal manifestation strongly associated with chronic As poisoning. This same finding was reported by Nevens *et al.* (1990) in eight patients in Belgium who had ingested Fowler solution to treat psoriasis, and by Guha Mazumder *et al.* (1998b) in West Bengal based on 41 out of 45 biopsies of patients suffering from chronic As poisoning.

ACKNOWLEDGEMENTS

Funds for this work came from UNICEF-Nicaragua and from the Italian Assistance Mission "Humanitarian support project for Child, Adolescent and Women protection in response to the Central American dry season".

REFERENCES

Aguilar, E., Parra, M., Cantillo, L. and Gómez, A.: Chronic arsenic toxicity in El Zapote-Nicaragua, 1996. *Med. Cután. Iber. Lat. Am.* 28:4 (2000), pp. 168–173.

Alain, G., Tousignant, J. and Rozenfarb, E.: Chronic arsenic toxicity. *Int. J. Dermatol.* 32:12 (1993), pp. 899–901.

ATS American Thoracic Society: Diagnostic standards and classification of tuberculosis in adults and children. Official Statement of the American Thoracic Society, The Centers for Disease Control and Prevention and endorsed by the Council of the Infectious Disease Society of America. *Am. J. Respir. Crit. Care Med.* 161:4 (2000), pp. 1376–1395.

BCAS: Arsenic Special Issue. BCAS Newsletter 8:1 (1997), Bangladesh Centre for Advanced Studies, http://bicn.com/acic/resources/infobank/asi_bcas.htm.

Bejarno-Sifuentes, G. and Nordberg, E.: *Mobilisation of arsenic in the Rio Dulce Alluvial Cone, Santiago del Estero Province, Argentina.* MSc Thesis, Royal Institute of Technology, Stockholm, Sweden, 2003. http://www.lwr.kth.se/Publikationer/PDF_Files/LWR_EX_03_6.PDF

Borgono, J.M., Vicent, P., Venturino, H. and Infante, A.: Arsenic in the drinking water of the city of Antofagasta: Epidemiological and clinical study before and after the installation of the treatment plant. *Environ. Health Perspect.* 19 (1977), pp. 103–105.

Centeno, J.A., Tchounwou, P.B., Paltota, A.K., Mullick, F.G., Murakata, L., Meza, E., Todorov, T., Longfellow, D. and Yedjou, C.G.: Environmental pathology and health effects of arsenic poisoning. In: R. Naidu, E. Smith, G. Owens, P. Bhattacharya and P. Nadebaum (eds): *Managing arsenic in the environment: From soil to human health.* CSIRO Publishing, Melbourne, Australia, 2006, pp. 311–327.

Chakraborti, D., Basu, G.K., Biswas, B.K., Chowdhury, U.K., Rahman, M.M., Paul, K., Chowdhury, T.R., Chandra, C.R., Lodh, D. and Ray, S.L.: Characterization of arsenic bearing sediments in Gangetic delta of West Bengal—India. In: W.R. Chappell, C.O. Abernathy and R.L. Calderon (eds): *Arsenic exposure and health effects IV.* Elsevier Science, New York, 2001, pp. 27–52.

Concha, G., Vogler, G., Lezcano, D., Nermell, B. and Vahter, M.: Exposure to inorganic arsenic metabolites during early human development. *Toxicol. Sci.* 44:2 (1998), pp. 185–190.

Das, D., Chatterjee, A., Mandal, B.K., Samanta, G., Chakraborti, D. and Chanda, B.: Arsenic in ground water in six districts of West Bengal, India: the biggest arsenic calamity in the world. Part 2: Arsenic concentration in drinking water, hair, nails, urine, skin-scale and liver tissue (biopsy) of the affected people. *Analyst* 120:3 (1995), pp. 917–924.

Figueroa, L., Razmilic, B. and González, M.: Corporal distribution of arsenic in mummied bodies owned to an arsenical habitat. In: A.M. Sancha (ed): *Proceedings International Seminar Arsenic in the Environment and its Incidences on Health.* Universidad de Chile, Santiago de Chile, Chile, 1992, pp. 77–82.

Gómez, A. and Aguilar, E.: Case of Hydroarsenicosis and cutaneous cancer. El Carrizo, valley of Sebaco-Nicaragua 1952–2000. *Proceedings of Summer Meeting American Academy of Dermatology.* New York, July 31–August 4, 2000.

Gómez, A.: Arsenic and Cancer in S and SW communities of the valley of Sébaco, Nicaragua 1999. *Proceedings of the XXII Central America Congress of Dermatology.* Panama City, Nov. 21–26, 2000.

Gonsebatt, M.E., Vega, L., Montero, R., García, G., Del Razo, L.M., Albores, A., Cebrián, M.E. and Ostrosky-Wegman, P.: Lymphocyte replication ability in individuals exposed to arsenic via drinking water. *Mutation Res.* 313 (1994), pp. 293–299.

González, M., Provedor, E., Reyes, M., López, N., López, A. and Lara, K.: Arsenic exposition in rural communities of San Isidro, Matagalpa, 1997. *Health Studies and Research Center (CIES), Pan American Health Organization/*WHO PLAGSALUD-MASICA, 1998.

Goyer, R.A.: Toxic effects of metals. In: C.D. Klaassen (ed): Casarett and Doull's Toxicology, the basic science of poisons. 5th ed., McGraw-Hill, New York, 1996, pp. 696–698.

Greenberg, A.K., Herman, Y. and Rom, W.: Preneoplastic lesions of the lung. *Respir. Res.* 3:1 (2002), p. 20.

Guha Mazumder, D.N., Das Gupta, J., Santra, A., Pal, A., Ghose, A., Sarkar, S., Chattopadhaya, N. and Chakraborti, D.: Non cancer effects of chronic arsenicosis with special reference to liver damage. In: W.R. Chappell, C.O. Abernathy and R.L. Calderon (eds): *Arsenic exposure and health effects II.* Elsevier Science, New York, 1997, pp. 112–123.

Guha Mazumder, D.N.: Chronic arsenic toxicity: Dose related clinical effect, its natural history and therapy. In: W. Chappell, C. Abernathy and R. Calderon (eds): *Proceedings of 3rd International Conference on Arsenic. Exposure and Healths Effects*, San Diego, CA July 12–15 1998. Elsevier Science Ltd., New York, 1998a, pp. 335–347.

Guha Mazumder, D.N., Das Gupta, J., Santra, A., Pal, A., Ghose, A. and Sarkar, S.: Chronic arsenic toxicity in Bengal—the worst calamity in the world. *J. Indian Med. Assoc.* 96:1 (1998b), pp. 4–7 and 18.

Guha Mazumder, D.N., Haque, R., Ghosh, N., De, B., Santra, A., Chakraborty, D. and Smith, A.: Arsenic levels in drinking water and the prevalence of skin lesions in West Bengal, India. *Int. J. Epidemiol.* 27:5 (1998c), pp. 871–877.

Guha Mazumder, D.N.: Diagnosis and treatment of chronic arsenic poisoning. In: *United Nations synthesis report on arsenic in drinking water.* New York, 2000a.

Guha Mazumder, D.N., Haque, R., Ghosh, N., De, B.K., Santra, A., Chakraborti, D. and Smith, A.H.: Arsenic in drinking water an the prevalence of respiratory effects in West Bengal, India. *Int. J. Epidemiol.* 29 (2000b), pp. 1047–1052.

Hopenhayn-Rich, C., Biggs, M.L. and Smith, A.H.: Lung and kidney cancer mortality associated with arsenic in drinking water in Córdoba, Argentina. *Int. J. Epidemiol.* 27 (1998), pp. 561–569.

Hubbard, R., Venn, A., Lewis, S. and Britton, J.: Lung cancer and cryptogenic fibrosing alveolitis. A population based cohort study. *Am. J. Respir. Crit. Care Med.* 161:1 (2000), pp. 5–8.

Mukherjee, A.B. and Bhattacharya, P.: Arsenic in groundwater in the Bengal Delta Plain: slow poisoning in Bangladesh. *Environ. Rev.* 9 (2001), pp. 189–220.

National Research Council: Health effects of arsenic. In: *Arsenic in drinking water*. National Academy Press. Washington DC, 1999, pp. 83–149.

Nevens, F., Fevery, J., Van Steenbergen, W., Sciot, R., Desmet, V. and De Groote, J.: Arsenic and non-cirrhotic portal hypertension. A report of eight cases. *J. Hepatol*. 11:1 (1990), pp. 80–85.

Ostrosky-Wegman, P., Gonsebatt, M.E., Montero, R., Vega, L., Barba, H., Espinosa, J., Palau, A., Cortinas, C., García-Vargas, G., del Razo, L.M. and Cebrián, M.: Lymphocyte proliferation kinetics and genotoxic findings in a pilot study on individuals chronically exposed to arsenic. *Mutation Res*. 250 (1991), pp. 477–482.

Rosenberg, H.G.: Systemic arterial disease and chronic arsenicism in infants. *Arch Pathol*. 97:6 (1974), pp. 360–365.

Rumack, C.M., Wilson, S.R. and Charboneau, J.W.: *Diagnostic Ultrasound*. 3th ed., Mosby, Inc., St. Louis, Missouri, 2005.

Samet, J.M.: Does idiopathic pulmonary fibrosis increase lung cancer risk? *Am. J. Respir. Crit. Care Med*. 161:1 (1999), pp. 1–2.

Smith, A.H., Goycolea, M., Haque, R. and Biggs, M.L.: Marked increase in bladder and lung cancer mortality in a region of Northern Chile due to arsenic in drinking water. *Am. J. Epidemiol*. 147:7 (1998), pp. 660–669.

Smith, A.H., Hopenhayn-Rich, C., Bates, M.N., Goeden, H.M., Hertz-Picciotto, I., Duggan, H.M., Wood, R., Kosnett, M.J. and Smith, N.T.: Cancer risks from arsenic in drinking water. *Environ. Health Perspect*. 97 (1992), pp. 259–267.

Soto-Peña, G.A., Luna, A.L., Acosta-Saavedra, L., Conde, P., Lopez-Carrillo, L., Cebrian, M.E., Bastida, M., Calderon-Aranda, E.S. and Vega, L.: Assessment of lymphocyte subpopulations and cytokine secretion in children exposed to arsenic. *FASEB J*. 6 (2006), pp. 779–781.

Tseng, W.P., Chu, H.M., How, S.W., Fong, J.M., Ling, C.S. and Yeh, S.: Prevalence of skin cancer in an endemic area of cronic arsenism in Taiwan. *J. Natl. Cancer Inst*. 40 (1968), pp. 453–463.

CHAPTER 43

Transfer of arsenic from contaminated dairy cattle drinking water to milk (Córdoba, Argentina)

A. Pérez-Carrera, C. Moscuzza & A. Fernández-Cirelli
Centro de Estudios Transdisciplinarios del Agua, Facultad de Ciencias Veterinarias,
Universidad de Buenos Aires, Buenos Aires, Argentina

ABSTRACT: The Chaco-Pampean plain of central Argentina represents one of the largest regions with high levels of arsenic (As) in groundwaters. The aim of the present study was to determine As content in livestock drinking water and milk from dairy farms located in the southeast of the Córdoba province. Arsenic concentrations in all samples collected from the shallow unconfined aquifer were above 0.15 mg/l, the level that has been suggested to cause chronic intoxication in cattle. In 55% of collected samples, the concentrations of As exceeded the upper limit recommended for livestock drinking water (0.5 mg/l). Arsenic concentrations in milk ranged from 0.5 to 7.8 μg/l. If drinking water is considered the only source of the exposure to As the biotransfer factor (BTF) for milk would range from 1.5×10^{-5} to 4.3×10^{-4}. BTF provides a simple tool for estimation of As levels in the milk using results of As analysis in livestock drinking water.

43.1   INTRODUCTION

The Chaco-Pampean plain of central Argentina covering approximately 1×10^6 km^2 is one of the largest known regions with high levels of arsenic (As) in groundwaters. High concentrations of As in groundwaters have been documented in several densely populated provinces, particularly in Córdoba and Santa Fe (Smedley and Kinniburgh 2002) where agriculture is one of the main activities.

Most problems caused by environmental As are associated with mobilization of As as a result of natural processes. However, anthropogenic sources, including industrial production or use of As as a wood preservative and as a feed additive for livestock, particularly poultry, significantly contribute to these problems. The greatest threat to human health is from As in drinking water. However, As intake from food may also significantly contribute to human exposures to As (Mandal and Suzuki 2002).

The source of the high As concentrations in groundwater from the Pampean plain are the Quaternary deposits of loess (mainly silt) reworked by aeolian and fluvial processes with a high proportion of volcanic glass shards of rhyolitic composition, while the average chemical composition of sediments is similar to a dacite (Nicolli *et al.* 1989, Smedley *et al.* 1998 and 2002). In some places, discrete volcanic ash layers can be found. The prevalent form of As in groundwater is As(V). Metal oxides in sediments (especially Al, Fe and Mn oxides and hydroxides) are thought to be the main secondary source of dissolved As produced by desorption at high pH, while volcanic class is the principal primary source.

It is well recognized that consumption of As, even at low levels, leads to carcinogenesis (Bates *et al.* 1992, Hopenhayn-Rich *et al.* 1996). An endemic disease linked to drinking water with high As levels was first described in Argentina early in the last century. This disease known as HACRE (*Hidroarsenicismo Crónico Regional Endémico* or Chronic Endemic Regional Hydroarsenism) has been associated with a specific form of skin cancer (Astolfi *et al.* 1981). One of the areas most affected by HACRE is the southeast of Córdoba, which is also one of the most important milk production zones of Argentina. Milk in its original form and as an ingredient of various dairy products is a basic component of human diet. In Argentina, consumption of dairy products has reached 192 equivalent milk l/inhabitant/year. As a product of mammary gland, milk can carry various xenobiotics, which constitute a technological risk factor for dairy products and more importantly, a health risk for consumers (Licata *et al.* 2004).

To this date, only few studies have examined effects of high As levels in livestock drinking water on livestock health and on As content in the milk (Stevens 1991, Rosas *et al.* 1999). The aim of the present study was to determine As concentrations in livestock drinking water and milk from dairy farms located in the southeast of the Córdoba province. The main goal was to establish a representative range for the biotransfer factor (BTF) that characterizes transfer of As from drinking water to the milk produced by these farms.

43.2   MATERIAL AND METHODS

43.2.1   *Sampling*

The study area was situated in a rural southeast region of the Córdoba province, Argentina (between 62°33′ and 62°57′, W longitude and 32°12′ and 32°50′, S latitude) and included four locations: Bell Ville, Morrison, Cintra and San Antonio de Litín (Fig. 43.1).

Eighteen dairy farms were randomly selected from 103 farms located in the study area. Samples of drinking water and milk were collected from each of the selected farms in the winter (August) of 2004. Most of the selected dairy farms (11) used water from the unconfined aquifer from 3 to 8 m deep wells. The remaining 7 farms used deep wells (80–150 m).

Water: To determine physicochemical parameters, 500 ml samples were collected in polyethylene bottles rinsed with deionized water. For As analysis, 100 ml samples were collected in bottles that were rinsed with 10% nitric acid and deionized water prior to collection. Samples were placed into an ice box (but not frozen) for transportation to the laboratory.

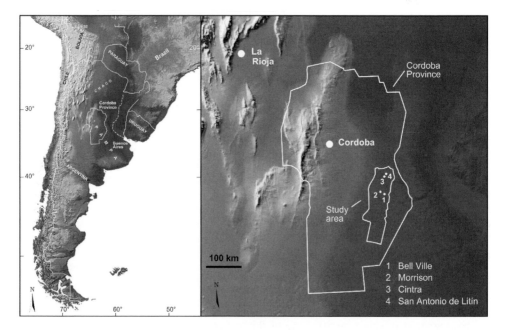

Figure 43.1.   Maps of the Córdoba province (Argentina) and of the study area (modified from Bundschuh *et al.* 2009).

Milk: Two milk samples were obtained from each of the selected dairy farms. Both samples (500 ml each) were collected from the bulk milk storage tanks and placed into polyethylene bottles that were rinsed with 10% nitric acid in deionized water prior to collection. Samples were placed into an ice box (but not frozen) for transportation to the laboratory.

43.2.2 *Analysis*

43.2.2.1 *Water*
Physicochemical analyses: pH and electrical conductivity were determined instrumentally *in situ*. Total dissolved solids (TDS) and hardness were analyzed by conventional techniques (APHA 1993, Brown *et al.* 1970, Rodier 1981). All measurements were performed in duplicates with relative errors below 1.0%. Total As contents in water samples were determined by inductively coupled plasma optical emission spectrometry (ICP-OES), using Perkin Elmer Optima 2000 spectrometer and following the APHA (1993) recommendations. All samples were analyzed in duplicates with relative error below 1.0%. Fluoride (F) concentrations were determined, using a Thermo ORION 96-09 ion selective electrode (EPA Method 340) and sodium fluoride GR (Merck 6449) as a reference material. Detection limit of the method was 0.02 mg/l.

43.2.2.2 *Milk*
Milk samples (approximately 100 g each) were placed into 400 ml heat resistant glass beakers and dry ashed as previously described (Cervera *et al.* 1994, Pérez-Carrera and Fernández-Cirelli 2005). White ash was moistened with a reagent grade water, dissolved in 10 ml 6 M HCl, filtered through Whatman No. 1 paper into a 25 ml volumetric flask, and diluted to original volume with 6 M HCl. Duplicate blanks were prepared by treating the ashing solution with the same digestion procedure. Because As concentrations in cow's milk are generally low, a highly sensitive hydride generation-atomic absorption spectrometry with detection limit of 0.1 ng/g (Perkin Elmer 1979) was used for analysis of As in collected milk samples. All samples were analyzed in duplicates with relative error below 1.0%. The NIST-8435 reference material was used for quality control.

43.2.3 *Data analysis*

The concentrations of As in drinking water and milk from farms using water from the shallow unconfined aquifer and farms using water from deep wells were compared by Analysis of Variance (ANOVA). Differences between these values were considered statistical significant when $P < 0.05$. Normality of distribution and homogeneity of variances were tested prior to ANOVA, using Kolmogorov-Smirov and Levene tests, respectively (Sokal and Rohlf 1995, Zar 1999). BTF values for the transfer of As from water to the milk were calculated according to Stevens (1991), assuming steady state conditions and water intake of 75 l/day:

BTF (day/l) = concentration of As in milk (mg/l)/daily animal intake of As (mg/day).

43.3 RESULTS AND DISCUSSION

Groundwater is used as a livestock drinking water in all dairy farms included in this study. Sixty one percent of wells on these farms collect water from the unconfined aquifer from the depth of 3 to 8 m. The remaining wells (deep wells) range in depth from 80 to 150 m. Results of the groundwater analyses are summarized in Table 43.1. Minimum, maximum, median, and average values for As and F concentrations and the physicochemical characteristics are shown. The results of the physicochemical analyses of the water samples collected in this study are similar to those previously reported for water from dairy farms in the same area (Pérez-Carrera and Fernández-Cirelli 2004). Samples from the shallow unconfined aquifer were alkaline (pH 8.4); 37% of these samples were

Table 43.1.   Physicochemical characteristics and concentrations of As and F in groundwater samples from dairy farms included in this study.

	pH (−)	Conductivity (mS/cm)	TDS (mg/l)	Hardness (mg/l CaCO₃)	Arsenic (ug/l)	Fluoride (mg/l)
Shallow wells						
Minimum	7.5	1.15	924	10	188	1.3
Maximum	9.2	9.6	6720	543	2592	10
Average	8.4	3.1	2281	98	1415	5.4
Median	7.8	1.5	1064	115	1550	5.3
SD	0.5	2.7	1773	156	957	3.3
Deep wells						
Minimum	7.6	0.8	532	63	36	0.4
Maximum	8.3	3.3	2324	268	101	0.6
Average	7.9	1.7	1180	122	64	0.5
Median	7.8	1.6	1127	118	63.8	0.5
SD	0.2	0.8	550	69	23	0.1

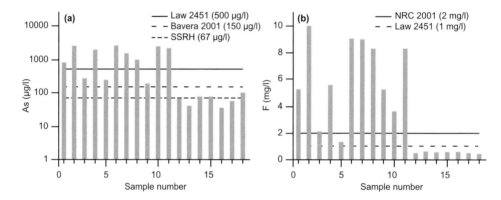

Figure 43.2.   Concentrations of As (a) and F (b) in groundwater samples.

moderately saline and 63% slightly saline. Waters from deep wells were slightly alkaline (pH 7.9); 17% non-saline, 77% slightly saline, and only 5% moderately saline. Fifty-eight per cent of all samples were slightly hard waters.

The concentrations of As and F in samples from the shallow aquifer were in accordance with the values previously reported in this study area (Pérez-Carrera and Fernández-Cirelli 2004). Arsenic concentrations in 73% of the shallow groundwater samples were higher than the upper limit recommended by local regulations for livestock drinking water (500 µg/l, Law 24051 1993). In addition, As concentrations in all these samples exceeded the levels shown to result in chronic intoxication in cattle: 67 µg/l (SSRH 2004) or 150 µg/l (Bavera *et al.* 2001) (Fig. 43.2a).

In contrast, As concentrations in all water samples from deep wells were below the upper limit recommended for livestock drinking water (Bavera *et al.* 2001). ANOVA showed a statistically significant difference between the average As concentrations in samples of phreatic water and water from deep wells ($F_{(1,16)} = 13.65$; $p = 0.001$). A similar difference was found for concentrations of F (Fig. 43.2b). In the shallow groundwater, F concentrations ranged from 1.3 to 10 mg/l. In groundwater from deep wells, F concentrations were lower, ranging from 0.4 to 0.6 mg/l. The concentrations of F in all shallow groundwater samples collected in this study exceeded the upper limit (1 mg/l) suggested by Argentine regulations.

Information about As content in cow's milk is scarce (Cervera *et al.* 1994: 0.14–0.77 ng/g; Rosas *et al.* 1999: <0.9–27.4 ng/g; Licata *et al.* 2004: 37.9 ng/g). In a previous study carried out by our laboratory, As concentrations were determined in samples of milk from four dairy farms where As concentrations in drinking water ranged from 0.23 to 2.54 mg/l. Two of these farms were medium size (100–120 dairy cows) and two were small (10–20 dairy cows). One medium size dairy farm with a low As concentration in water (0.04 mg/l) was also included for comparison. In all cases, milk samples were collected from six or seven year old Holstein dairy cows born and bred on the farms. Concentrations of As in milk ranged from 2.8 to 10.5 ng/g for dairy farms using phreatic groundwater. In contrast, As concentration in the milk from farms using deep wells averaged 0.5 ng/g (Pérez-Carrera and Fernández-Cirelli 2005).

In 1991, Stevens carried out a pioneering study in which a steady state bovine milk BTF value was estimated using a pharmacokinetic approach based on previous long term feeding studies (Vreman *et al.* 1986, Marshall *et al.* 1963). One of those studies (Vreman *et al.* 1986) reported increased levels of As in milk. Analytical methods used in other studies were not sensitive enough to detect As in milk. A maximum BTF value of 3.0×10^{-5} day/l was calculated. In contrast, Rosas *et al.* (1999) determined BTF value for As in milk from As concentrations in water, soil, milk and forage collected on major dairy farms in the Comarca Lagunera (Mexico), a region naturally rich in As. The concentrations of As found in milk from these farms ranged from less than 0.9 to 27.4 ng/g; BTF values ranged from 3.2×10^{-5} to 6.7×10^{-4} day/l.

BTF values determined in our previous study (5.2×10^{-5} to 1.8×10^{-4} day/l) are in a good agreement with values reported by Rosas *et al.* (1999). Because natural forage or alfalfa grown without irrigation is used to feed livestock in this area, drinking water is considered the main source of As for cattle. In order to validate the previously determined BTF values and to provide a reliable tool for estimation of As content in milk, the area of interest was extended in the present study to include a representative number of dairy farms. Results of the analysis of As in drinking water and in the milk samples collected from bulk storage tanks on these farms, as well as the corresponding BTF values are shown in Table 43.2.

Concentrations of As in milk samples ranged from 0.5 to 7.8 µg/l for dairy farms using phreatic groundwater and from 0.5 to 2.5 µg/l for farms using deep well water. The average concentration of

Table 43.2.   Estimated BTF values for As in the milk collected from dairy farms included in this study.

Dairy farm	Depth of water source (m)	As in water (µg/l)	As in milk (µg/l)	BTF (day/l)
1	3–8	829.0	5.5	8.878×10^{-5}
2	3–8	2540.0	7.8	4.068×10^{-5}
3	3–8	266.0	3.9	1.952×10^{-4}
4	3–8	1900.0	6.9	4.872×10^{-5}
5	3–8	231.0	0.5	2.597×10^{-5}
6	3–8	2592.2	4.9	2.541×10^{-5}
7	3–8	1549.3	5.3	4.553×10^{-5}
8	3–8	958.4	3.1	4.285×10^{-5}
9	3–8	188.4	1.7	1.203×10^{-4}
10	3–8	2420.0	4.2	2.288×10^{-5}
11	3–8	2095.8	2.4	1.508×10^{-5}
12	80–150	63.8	0.9	1.860×10^{-4}
13	80–150	40.0	0.5	1.667×10^{-4}
14	80–150	76.5	2.5	4.270×10^{-4}
15	80–150	73.9	0.5	9.743×10^{-5}
16	80–150	35.5	0.6	2.103×10^{-4}
17	80–150	55.9	0.6	1.336×10^{-4}
18	80–150	100.9	0.7	9.620×10^{-5}

As in milk from dairy farms that use shallow groundwater was significantly higher than the average concentration of As in milk collected in farms with deep wells ($F_{(1.16)} = 14.45$; $p = 0.001$).

The formula used in this study for calculation of BTF was also used in our previous study. In this study, BTF values ranged from 1.5×10^{-5} to 4.3×10^{-4} day/l (Table 43.2). These values are in the same order as BTF values reported in our previous study carried out in the same area. Determination of As concentration in water is easier than analysis of As in milk. Thus, BTF values validated in the present study will facilitate assessment of As content in milk using values of As concentration in drinking water consumed by cattle in the milk producing farms.

REFERENCES

APHA: Standard methods for the examination of water and wastes. American Public Health Association. Washington, DC, 1993.

Astolfi, E.A., Maccagno, A., García Fernández, J.C., Vaccaro, R. and Stimola, R.: Relation between arsenic in drinking water and skin cancer. *Biol. Trace Elem. Res.* 3 (1981), pp. 133–143.

Bates, M., Smith, A. and Hopenhayn-Rich, C.: Arsenic ingestion and internal cancers: a review. *Am. J. Epidemiol.* 135 (1992), pp. 462–476.

Bavera, G., Rodriguez, E., Beguet, H., Bocco, O. and Sanchez, J.: *Water and watering places.* Ed. Hemisferio Sur, Buenos Aires, Argentina, 2001.

Brown, E., Skougstad, M. and Fishman, M.: Methods for collection and analysis of water samples for dissolved mineral and gases. *Techniques of Water Resources Investigations* 5 (A1), US Geological Survey, Washington DC, 1970.

Bundschuh, J., García, M.E., Birkle, P., Cumbal, L., Bhattacharya, P. and Matschullat, J: Groundwater arsenic in rural Latin America—Occurrence, Health effects and remediation experiences. In: J. Bundschuh, M.A. Armienta, P. Birkle, P. Bhattacharya, J. Matschullat and A.B. Mukherjee (eds): *Natural arsenic in groundwater of Latin America.* Taylor and Francis/Balkema. Leiden, The Netherlands, 2009 (this volume).

Cervera, M.L., Lopez, J.C. and Montoro, R.: Arsenic content of Spanish cows' milk determined by dry ashing hydride generation atomic absorption spectrometry. *J. Dairy Res.* 61 (1994), pp. 83–89.

Hopenhayn-Rich, C., Biggs, M.L., Fuchs, A., Bergoglio, R., Tello, E., Nicolli, H. and Smith, A.: Bladder cancer mortality associated with arsenic in drinking water in Argentina. *Epidemiology* 7 (1996), pp. 117–123.

Law 24051: Regime for hazardous wastes. Argentine. Government report, Buenos Aires, Argentina, 1993.

Licata, P., Trombetta, D., Cristiani, M., Giofre, F., Martino, D., Calo, M. and Naccari, F.: Levels of "toxic" and "essential" metals in samples of bovine milk from various dairy farms in Calabria, Italy. *Environ. Internat.* 30 (2004), pp. 1–6.

Mandal, B. & Suzuki, K.: Arsenic round the world: a review. *Talanta* 58 (2004), pp. 201–235.

Marshall, S.P., Hayward, F.W. and Meager, W.R.: Effects of feeding arsenic and lead upon their secretion in milk. *J. Dairy Sci.* 46 (1963), pp. 580–581.

National Research Council: *Nutrients requirements of dairy cattle.* 7th ed., Nat., Acad, Press, Washington, DC, 2001.

Nicolli, H., Suriano, J., Gomez-Peral, M., Ferpozzi, L. and Baleani, O.: Groundwater contamination with arsenic and other trace elements in an area of the Pampa, Province of Córdoba, Argentina. *Environ. Geol. Water Sci.* 14 (1989), pp. 3–16.

Pérez-Carrera, A. and Fernández-Cirelli, A.: Niveles de arsénico y flúor en agua de bebida animal en establecimientos de producción lechera (Bell Ville, Pcia. de Córdoba). *Rev. Invest. Veterin. (INVET)* 6:1 (2004), pp. 51–59.

Pérez-Carrera, A. and Fernández-Cirelli, A.: Arsenic concentration in water and bovine milk in Córdoba, Argentina: Preliminary results. *J. Dairy Res.* 72 (2005), pp. 122–124.

Perkin-Elmer: *Analytical methods using the MHS-10 Mercury/Hydride System.* Perkin Elmer Corp., Norwalk, CT, 1979.

Rodier, J.: *Water analysis.* Omega Ed., Barcelona, Spain, 1981.

Rosas, I., Belmont, R., Armienta, A. and Baez, A.: Arsenic concentrations in water, soil, milk and forage in Comarca Lagunera, Mexico. *Water Air Soil Pollut.* 112 (1999), pp. 133–149.

Smedley, P.L., Nicolli, H.B., Barros, A.J. and Tullio, J.O.: Origin and mobility of arsenic in groundwater from the Pampean Plain, Argentina. In: G.B. Arehart, and, J.R. Hulston (eds): *Water-rock interaction.* Balkema, Rotterdam, 1998, pp. 275–278.

Smedley, P.L., Nicolli, H.B., Macdonald, D.M., Barros, A.J. and Tullio, J.O.: Hydrogeochemistry of arsenic and other inorganic constituents in groundwaters from La Pampa, Argentina. *Appl. Geochem.* 17 (2002), pp. 259–284.

Smedley, P.L. and Kinniburgh, D.: A review of the source, behaviour and distribution of arsenic in natural waters. *Appl. Geochem.* 17 (2002), pp. 517–568.

Stevens, J.: Disposition of toxic metals in the agricultural food chain. 1. Steady-state bovine milk biotransfer factors. *Environ. Sci. Technol.* 25 (1991), pp. 1289–1294.

Sokal. R. and Rohlf, F.: *Biometry*. Freeman, New York. 1995. SSRH: Development of guide levels of arsenic content for livestock. Subsecretary of Hidric Resources, Buenos Aires, Argentina, 2004.

Vreman, K., van der Veen, N.J., van der Molen, E.J. and de Ruig, W.G.: Transfer of cadmium, lead, mercury and arsenic from feed into milk and various tissues of dairy cows: Chemical and pathological data. *Netherl. J. Agric. Sci.* 34 (1986), pp. 129–144.

Za, J.H.: *Biostatistical analysis*. 4th ed., Prentice-Hall, New York, 1999.

CHAPTER 44

Molecular mechanisms of arsenic-induced carcinogenesis

K.K. Singh, M. Vujcic & M. Shroff
Department of Cancer Genetics, Roswell Park Cancer Institute, Buffalo, NY, USA

ABSTRACT: Epidemiological studies suggest that inorganic arsenic (i-As) exposure leads to carcinogenesis. Arsenic-induced cancers include lung, skin, liver, bladder and kidney cancers. Arsenic is widely distributed in food, water, air and soil. Arsenic-contaminated groundwater has been found in many places around the world, including the United States, but the largest reported population exposed to arsenic (As) is in Bangladesh. The mechanism(s) by which As causes human cancers is poorly understood. This review describes the current status of molecular mechanism involved in As-induced carcinogenesis. These mechanisms include As-induced cell proliferation, altered gene expression, increased oxidative stress, inhibition of DNA repair, and induction of chromosomal instability.

44.1 INTRODUCTION

44.1.1 *Global problem of arsenic exposure*

Arsenic (As), a metalloid and a natural component of the earth's crust, is classified as a human carcinogen by the International Agency for Research on Cancer and the US Environmental Protection Agency (IARC 2004, USEPA 2000). It is present ubiquitously in the environment due to natural and anthropogenic reasons (Bettley *et al.* 1975, Waalkes *et al.* 2000). Arsenic occurs in both organic and inorganic form, inorganic being the more common and toxic form. Arsenic contamination in drinking water is a major public health concern throughout the world. Contaminated groundwater has been found in Taiwan, Argentina, Vietnam, Mexico, Peru, USA and Canada, just to mention a few, but the largest reported population exposed to As is in Bangladesh and India (Pearce 2003). This situation has been described as "Humanity's Biggest Mass Poisoning" (Bhattacharjee 2007). It is estimated that of the 125 million people in Bangladesh, approximately 50 million people are at the risk of drinking contaminated water (Mandal *et al.* 1996, Pearce 2003). At the same time, nearly 1 million are affected in the neighboring state of West Bengal in India (Bagla and Kaiser 1996, Das *et al.* 1994). In the United States, nearly 20 million people are exposed to drinking water with As concentrations greater than 10 μg/l, the EPA standard (USEPA 2001, Frazer 2005). Together, over 2 million people are exposed in Mexico, Chile and Argentina (Smith *et al.* 1998).

44.1.2 *Arsenic metabolism in humans*

Metabolism of As plays an important role in its toxic effects in humans (Thomas *et al.* 2001, Vahter 2002). The metabolic pathway for inorganic As in humans involves two types of enzymatic reactions: (1) the reduction of pentavalent arsenicals to trivalency and (2) the oxidative methylation of trivalent arsenicals to yield pentavalent methylated species. Therefore, both pentavalent and trivalent methylarsonic acid (MMA) and dimethylarsinic acid (DMA) species are intermediates or products of this pathway. An S-adenosylmethionine-dependent As (+3 oxidation state) methyltransferase (AS3MT) that is expressed in human hepatocytes has been shown to catalyze both types of the enzymatic reactions in this pathway (Thomas *et al.* 2001, Vahter 2002).

44.2   MECHANISM OF ARSENIC INDUCED CARCINOGENESIS

Epidemiological studies have demonstrated increased risk of lung, skin, liver, urinary bladder, and kidney cancers associated with exposure to As (Leonard and Lauwerys 1980, Chappell *et al.* 1997). Arsenic exposure also leads to cardiovascular and peripheral vascular disease, and neurological and neurobehavioral disorders (Tchounwou *et al.* 2003). Compelling evidence suggests that inorganic As and trivalent organic species are more acutely toxic than the pentavalent organic As species, MMA and DMA (Petrick *et al.* 2000, Styblo *et al.* 2000). Arsenite is considered toxic due to its strong affinity for and ability to bind to key sulfhydryl groups (SH) in proteins, which leads to inhibition of cellular functions, including respiration and metabolism (Squibb and Fowler 1983). A good example of As toxicity is inhibition of pyruvate dehydrogenase (PDH) complex. Arsenite binds to -SH groups of lipoic acid of enzyme complex 1 which are important for its enzymatic activity (Aposhian 1989). Arsenate is a phosphate analog and, therefore, competes with phosphate in transport and during ATP synthesis resulting in uncoupling of oxidative phosphorylation (Gresser 1981). Given the fact that arsenate is reduced to arsenite in the detoxification process, it is likely that exposure to arsenate (similar to arsenite) may result in protein damage in cells.

The mechanisms of As-induced carcinogenesis are not properly understood. To date, the described mechanisms include (1) increased cell proliferation, (2) altered gene expression, (3) increased oxidative stress, (4) inhibition of DNA repair, and (5) induction of chromosomal abnormalities (Fig. 44.1). These mechanisms are proposed to explain the As-induced carcinogenesis in humans.

44.2.1   *Arsenic effects on cell proliferation*

Evidence, both *in vivo* and *in vitro* suggests increased cell proliferation as a mode of action for As-induced carcinogenesis (Cohen and Ellwein 1991). In humans, exposure to high levels of As leads to keratosis of the palms and soles of the feet, which further develop into invasive squamous cell carcinomas (Schaumburg-Lever *et al.* 1986). Keratosis is characterized by abnormal development in the squamous epithelium which includes decrease in differentiation and an increase in the number of cell divisions. It was observed that oral administration of DMA without prior exposure to other genotoxic carcinogens in rodents, resulted in increased cell proliferation in the bladder, liver, and kidney; whereas oral administration of DMA after exposure to one or more genotoxic carcinogen led to increased incidences of tumors of the lung, bladder, liver, kidney and thyroid (Murai *et al.* 1993, Yamamoto *et al.* 1995, Wanibuchi *et al.* 1996). Increased cell proliferation has been reported following addition of certain As-compounds to the culture medium. Mechanism through which As exposure induces increased cell proliferation is not clear, but cytotoxicity followed by regenerative proliferation seems to be the most probable pathway (Murai *et al.* 1993, Wanibuchi *et al.* 1996).

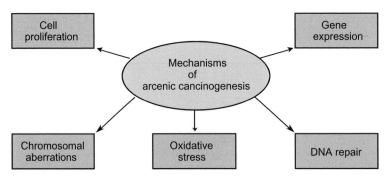

Figure 44.1.   Various cellular processes known to be involved in arsenic induced carcinogenesis.

44.2.2 *Arsenic effects on gene expression*

Changes in gene expression due to abnormal cytosine methylation of DNA are a common feature of human tumors (Counts and Goodman 1995, Jones 1996, Issa *et al.* 1997). Consistently, As-induced changes in methylation of the p53 promoter have been reported (Mass and Wang 1997). This study suggests that DNA methylation may play a role in mechanistic action of As carcinogenesis. However, such abnormalities have been observed in spontaneous as well as mutagen induced tumors (Issa *et al.* 1996). Also, promoter gene hypermethylation has not been confirmed as an exclusive mode of action for As carcinogenesis (Issa *et al.* 1996). Unfortunately, to date no study has examined DNA methylation changes in As-induced tumors.

44.2.3 *Arsenic effects on DNA repair*

Arsenite seems to act as a co-mutagen at low concentrations. It has been shown to increase the mutagenic effect of UV in *E. coli* (Rossman 1981). In Chinese hamster cells, arsenite has been found to enhance the mutagenesis of UV, methylmethanesulfonate (MMS), and methylnitrosourea (MNU) (Lee *et al.* 1985, Li and Rossman 1989a, Yang *et al.* 1992). Arsenic has also been shown to inhibit X-ray and UV-induced DNA damage repair (Snyder and Lachmann 1989). Evidence suggests that As enhances the effects of X-ray and UV-induced chromosomal damage in peripheral human lymphocytes and fibroblasts (Jha *et al.* 1992), alters the mutational spectrum of UV-irradiated Chinese hamster ovary cells (Yang *et al.* 1992), and also increases the chromosomal aberrations induced by diepoxybutane (Wiencke and Yager 1992). Arsenite inhibits the removal of pyrimidine dimers in human SF34 cells post UV treatment (Okui and Fujiwara 1986). This may be due to effect on DNA ligation in the excision repair process (Li and Rossman 1989b, Lee-Chen *et al.* 1994). However, DNA ligases have not be shown to be inhibited at concentrations much higher than that found to inhibit DNA repair in cells which suggests that arsenite does not mediate its effects via inhibition of DNA repair enzymes.

44.2.4 *Arsenic induced chromosomal abnormalities*

Arsenic has been shown to induce chromosomal abnormalities such as changes in chromosomal aberrations, aneuploidy, and sister chromatid exchanges (IARC 1987, Rasmussen and Menzel 1997). Arsenite causes induction of micronuclei formation in both rodents and humans (Warner *et al.* 1994). Micronuclei formation was observed in exfoliated bladder cells, buccal cells, sputum cells and lymphocytes in humans (Warner *et al.* 1994). Arsenic has also been shown to cause sister chromatid exchanges in human fibroblasts and CHO cells.

Arsenic compounds have demonstrated clastogenicity in various cell types such as human fibroblasts (Oya-Ohta *et al.* 1996) and have shown to cause cell transformation and cytogenetic damage (Barrett *et al.* 1989). Arsenite also induced gene amplification at the dihydrofolate reductase (*dhfr*) locus in human and rodent cells; however it did not produce amplification in SV40 sequences of SV40-transformed human keratinocytes (Rossman *et al.* 1992). This indicates that arsenite does not activate signaling pathways distinctive of DNA-damaging agents but may act on the checkpoint pathways which involve p53, resulting in gene amplification (Livingstone *et al.* 1992).

44.2.5 *Arsenic induced oxidative stress*

Exposure to As has been shown to generate reactive oxygen (ROS) and nitrogen species (RNS), resulting in increased oxidative stress and alteration in signaling pathways involved in tumor promotion and progression. Oxidative stress plays a significant role in As-induced carcinogenicity (Rossman 2003, Huang *et al.* 2004, Shi *et al.* 2004, Tchounwou *et al.* 2004). The peroxyradicals and other active oxygen species produced during the metabolism have been shown to cause DNA strand breaks both *in vitro* and in rodents (Yamanaka and Okada 2001). Recently, Hei *et al.* (1998) and Liu *et al.* (2001) demonstrated that mutagenicity of As in human-hamster hybrid (A_L) cells was mediated by hydroxyl radicals that are produced by superoxide-driven process involving hydrogen peroxide. Liu *et al.* (2005) also showed that the mitochondrial damage plays an important role in induction of mutations by As. Mitochondria are metabolic centers of the cell and major producers of ROS, mainly superoxides and hydrogen peroxides. The electron transport system converts approximately 2–4% of total oxygen taken up by the mitochondria to ROS (Chance *et al.* 1979), therefore, mitochondria are more prone to oxidative damage.

As described before, As-induced oxidative stress has been shown to cause DNA damage through the production of ROS. Consistent with these observations, *in vitro* studies have shown that antioxidants such as superoxide dismutase, catalase, dimethyl sulfoxide (DMSO), glutathione, N-acetylcysteine (NAC), and vitamin E can protect against genotoxic effects of As, including DNA mutations, production of high levels of superoxide, and As-induced chromosomal damage (Hei *et al.* 1998, Nordenson *et al.* 1991). Mutagenicity of human lymphocytes and A_L cells (Hei *et al.* 1998) due to arsenite was effectively blocked by addition of superoxide dismutase in culture medium. Induction of antioxidant proteins like metallothionein and heme oxygenase-1 which protect cells against oxidative stress has been observed in response to arsenite (Del Razo *et al.* 2001).

Genotoxicity can also occur via oxidative actions other than ROS formation. For example, depletion of glutathione seems to increase the toxicity of arsenite in cell cultures (Oya-Ohta *et al.* 1996). Glutathione is a key component of the As detoxification in cells where glutathione is required for conversion of arsenate to arsenite and in the methylation of arsenite to DMA (Scott *et al.* 1993, Ghosh *et al.* 1999). In addition, inorganic arsenite as well as organic As, inhibits glutathione reductase (Styblo *et al.* 1997), which would also result in oxidative stress. These studies suggest the involvement of oxidative damage in As-induced genotoxicity.

44.3   CONCLUSIONS AND PERSPECTIVES

Arsenic is a serious human carcinogen affecting millions of people worldwide. However, the molecular mechanism of As-induced carcinogenicity is poorly understood. To date, the described mechanisms which are known to contribute to carcinogenesis in humans include (1) increased cell proliferation, (2) altered gene expression, (3) increased oxidative stress, (4) inhibition of DNA repair, and (5) induction of chromosomal instability (Fig. 44.1). It is likely that coming years will reveal other mechanism that may also be involved. One of the approaches to further elucidate other mechanisms is to take advantage of newly developed, genome-wide screens of simpler eukaryotes, such as yeast, that can point the search in the right direction by identifying human homologs of yeast genes. Indeed this approach has yielded a number of interesting genes inactivation of which leads to As toxicity (Vujcic *et al.* 2007). Some of these identified genes are involved in mitochondria-to-nucleus cross talk suggesting that mitochondrial pathways must play an important role in As-induced carcinogenesis in humans.

ACKNOWLEDGMENT

We thank Ellen Sanders-Noonan for editing the manuscript. Research in our laboratory is supported by grants from National Institute of Health (RO1-CA121904, RO1-CA 113655 and RO1-CA 116430).

REFERENCES

Aposhian, H.V.: Biochemical toxicology of arsenic. *Rev. Biochem. Toxicol.* 10 (1989), pp. 265–299.
Bagla, P. and Kaiser, J.: India's spreading health crisis draws global arsenic experts. *Science* 274 (1996), pp. 174–175.

Barrett, J.C., Lamb, P.W., Wang, T.C. and Lee, T.C.: Mechanisms of arsenic-induced cell transformation. *Biol. Trace Elem. Res.* 21 (1989), pp. 421–429.

Bettley, F.R. and O'Shea, J.A.: The absorption of arsenic and its relation to carcinoma. *Br. J. Dermatol.* 92:5 (1975), pp. 563–568.

Bhattacharjee, Y.: A sluggish response to humanity's biggest mass poisoning. *Science* 315 (2007), pp. 1659–1661.

Chance, B., Sies, H. and Boveris, A.: Hydroperoxide metabolism in mammalian organs. *Physiol. Rev.* 59:3 (1979), pp. 527–605.

Chappell, W.R., Beck, B.D., Brown, K.G., Chaney, R., Cothern, R., Cothern, C.R., Irgolic, K.J., North, D.W., Thornton, I. and Tsongas, T.A.: Inorganic arsenic: a need and an opportunity to improve risk assessment. *Environ. Health Perspect.* 105:10 (1997), pp. 1060–1067.

Chen, C.J. and Wang, C.J.: Ecological correlation between arsenic level in well water and age-adjusted mortality from malignant neoplasms. *Cancer Res.* 50:17 (1990), pp. 5470–5474.

Cohen, S.M. and Ellwein, L.B.: Genetic errors, cell proliferation, and carcinogenesis. *Cancer Res.* 51:24 (1991), pp. 6493–6505.

Counts, J.L. and Goodman, J.I.: Hypomethylation of DNA: a nongenotoxic mechanism involved in tumor promotion. *Toxicol. Lett.* 82/83 (1995), pp. 663–672.

Del Razo, L.M., Quintanilla-Vega, B., Brambila-Colombres, E., Calderón-Aranda, E.S., Manno, M. and Albores, A.: Stress proteins induced by arsenic. *Toxicol. Appl. Pharmacol.* 177 (2001), pp. 132–148.

Frazer, L.: Metal attraction: an ironclad solution to arsenic contamination? *Environ. Health Perspect.* 113:6 (2005), pp. A398–401.

Gresser, M.J.: ADP-arsenate. Formation by submitochondrial particles under phosphorylating conditions. *J. Biol. Chem.* 256:12 (1981), pp. 5981–5983.

Haugen, A.C., Kelley, R., Collins, J.B., Tucker, C.J., Deng, C., Afshari, C.A., Brown, J.M., Ideker, T. and Van Houten, B.: Integrating phenotypic and expression profiles to map arsenic-response networks. *Genome Biol.* 5:12 (2004), p. R95.

Hei, T.K., Liu, S.X. and Waldren, C.: Mutagenicity of arsenic in mammalian cells: role of reactive oxygen species. *Proc. Natl. Acad. Sci. USA.* 95:14 (1998), pp. 8103–8107.

Huang, C., Ke, Q., Costa, M. and Shi, X.: Molecular mechanisms of arsenic carcinogenesis. *Mol. Cell Biochem.* 255:1/2 (2004), pp. 57–66.

IARC: *Monographs on the evaluation of acrcinogenic risk to humans: some metals and metallic compounds.* IARC Monographs, Supplement 7, International Agency for Research on Cancer, Lyon, France, 1987.

IARC: Some drinking-water disinfectants and contaminants, including arsenic related nitrosamines. IARC Monographs 84, International Agency for Research on Cancer, Lyon, France, 2004.

Issa, J.P., Baylin, S.B. and Herman, J.G.: DNA methylation changes in hematologic malignancies: biologic and clinical implications. *Leukemia* 11 (Suppl. 1) (1997), pp. S7–11.

Issa, J.P., Vertino, P.M., Boehm, C.D., Newsham, I.F. and Baylin, S.B.: Switch from monoallelic to biallelic human IGF2 promoter methylation during aging and carcinogenesis. *Proc. Natl. Acad. Sci. USA* 93:21 (1996), pp. 11,757–11,762.

Jha, A.N., Noditi, M., Nilsson, R. and Natarajan, A.T.: Genotoxic effects of sodium arsenite on human cells. *Mutat. Res.* 284:2 (1992), pp. 215–221.

Jones, P.A.: DNA methylation errors and cancer. *Cancer Res.* 56:11 (1996), pp. 2463–2467.

Landolph, J.R.: Molecular mechanisms of transformation of C3H/10T1/2 C1 8 mouse embryo cells and diploid human fibroblasts by carcinogenic metal compounds. *Environ. Health Perspect.* 102 (Suppl. 3) (1994), pp. 119–125.

Lee-Chen, S.F., Yu, C.T., Wu, D.R. and Jan, K.Y.: Differential effects of luminol, nickel, and arsenite on the rejoining of ultraviolet light and alkylation-induced DNA breaks. *Environ. Mol. Mutagen.* 23:2 (1994), pp. 116–120.

Lee, T.C., Huang, R.Y. and Jan, K.Y.: Sodium arsenite enhances the cytotoxicity, clastogenicity, and 6-thioguanine-resistant mutagenicity of ultraviolet light in Chinese hamster ovary cells. *Mutat. Res.* 148:1/2 (1985), pp. 83–89.

Li, J.H. and Rossman, T.G.: Mechanism of comutagenesis of sodium arsenite with n-methyl-n-nitrosourea. *Biol. Trace Elem. Res.* 21 (1989a), pp. 373–381.

Li, J.H. and Rossman, T.G.: Inhibition of DNA ligase activity by arsenite: a possible mechanism of its comutagenesis. *Mol. Toxicol.* 2:1 (1989b), pp. 1–9.

Liu, S.X., Athar, M., Lippai, I., Waldren, C. and Hei, T.K.: Induction of oxyradicals by arsenic: implication for mechanism of genotoxicity. *Proc. Natl. Acad. Sci. USA* 98:4 (2001), pp. 1643–1648.

Liu, S.X., Davidson, M.M., Tang, X., Walker, W.F., Athar, M., Ivanov, V. and Hei, T.K.: Mitochondrial damage mediates genotoxicity of arsenic in mammalian cells. *Cancer Res.* 65:8 (2005), pp. 3236–3242.

Livingstone, L.R., White, A., Sprouse, J., Livanos, E., Jacks, T. and Tlsty, T.D.: Altered cell cycle arrest and gene amplification potential accompany loss of wild-type p53. *Cell* 70:6 (1992), pp. 923–935.

Mandal, B.K., Chowdhury, T.R., Samanta, G., Basu, G.K., Chowdhury, P.P., Chanda, C.R., Lodh, D., Karan, N.K., Dhar, R.K., Tamili, D., Das, D., Saha, K.C. and Chakraborti, C.: Arsenic in groundwater in seven districts of West Bengal, India—The biggest arsenic calamity in the world. *Current Science* 70:11 (1996), pp. 976–986.

Martzen, M.R., McCraith, S.M., Spinelli, S.L., Torres, F.M., Fields, S., Grayhack, E.J. and Phizicky, E.M.: A biochemical genomics approach for identifying genes by the activity of their products. *Science* 286 (5442) (1999), pp. 1153–1155.

Mass, M.J. and Wang, L.: Arsenic alters cytosine methylation patterns of the promoter of the tumor suppressor gene p53 in human lung cells: a model for a mechanism of carcinogenesis. *Mutat. Res.* 386:3 (1997), pp. 263–277.

Murai, T., Iwata, H., Otoshi, T., Endo, G., Horiguchi, S. and Fukushima, S.: Renal lesions induced in F344/DuCrj rats by 4-weeks oral administration of dimethylarsinic acid. *Toxicol. Lett.* 66:1 (1993), pp. 53–61.

Nordenson, I. and Beckman, L.: Is the genotoxic effect of arsenic mediated by oxygen free radicals? *Hum. Hered.* 41:1 (1991), pp. 71–73.

Okui, T. and Fujiwara, Y.: Inhibition of human excision DNA repair by inorganic arsenic and the co-mutagenic effect in V79 Chinese hamster cells. *Mutat. Res.* 172:1 (1986), pp. 69–76.

Oya-Ohta, Y., Kaise, T. and Ochi, T.: Induction of chromosomal aberrations in cultured human fibroblasts by inorganic and organic arsenic compounds and the different roles of glutathione in such induction. *Mutat. Res.* 357:1/2 (1996), pp. 123–129.

Pearce, F.: Arsenic's fatal legacy grows worldwide. *New Scientist* 2407 (2003), pp. 1–3.

Petrick, J.S., Ayala-Fierro, F., Cullen, W.R., Carter, D.E. and Vasken Aposhian, H.: Monomethylarsonous acid (MMA (III)) is more toxic than arsenite in Chang human hepatocytes. *Toxicol. Appl. Pharmacol.* 163:2 (2000), pp. 203–207.

Rasmussen, R.E. and Menzel, D.B.: Variation in arsenic-induced sister chromatid exchange in human lymphocytes and lymphoblastoid cell lines. *Mutat. Res.* 386:3 (1997), pp. 299–306.

Rosen, B.P.: Biochemistry of arsenic detoxification. *FEBS Lett.* 529:1 (2002), pp. 86–92.

Rossman, T.G.: Enhancement of UV-mutagenesis by low concentrations of arsenite in E. coli. *Mutat. Res.* 91:3 (1981), pp. 207–211.

Rossman, T.G.: Mechanism of arsenic carcinogenesis: an integrated approach. *Mutat. Res.* 533:1/2 (2003), pp. 37–65.

Schaumburg-Lever, G., Alroy, J., Ucci, A. and Lever, W.F.: Cell surface carbohydrates in proliferative epidermal lesions. II: Masking of peanut agglutinin (PNA) binding sites in solar keratoses, Bowen's disease, and squamous cell carcinoma by neuraminic acid. *J. Cutan. Pathol.* 13:2 (1986), pp. 163–171.

Scott, N., Hatlelid, K.M., MacKenzie, N.E. and Carter, D.E.: Reactions of arsenic(III) and arsenic(V) species with glutathione. *Chem. Res. Toxicol.* 6:1 (1993), pp. 102–106.

Shi, H., Shi, X. and Liu, K.J.: Oxidative mechanism of arsenic toxicity and carcinogenesis. *Mol. Cell Biochem.* 255:1/2 (2004), pp. 67–78.

Smith, A.H., Goycolea, M., Haque, R. and Biggs, M.L.: Marked increase in bladder and lung cancer mortality in a region of Northern Chile due to arsenic in drinking water. *Am. J. Epidemiol.* 147:7 (1998), pp. 660–669.

Snyder, R.D. and Lachmann, P.J.: Thiol involvement in the inhibition of DNA repair by metals in mammalian cells. *Mol. Toxicol.* 2:2 (1989), pp. 117–128.

Squibb, K.S. and Fowler, B.A.: *Biochemical mechanisms of arsenical toxicity. Biological and environmental effects of arsenic.* Elsevier Press, Amsterdam, 1983.

Styblo, M., Del Razo, L.M., Vega, L., Germolec, D.R., LeCluyse, E.L., Hamilton, G.A., Reed, W., Wang, C., Cullen, W.R. and Thomas, D.J.: Comparative toxicity of trivalent and pentavalent inorganic and methylated arsenicals in rat and human cells. *Arch. Toxicol.* 74:6 (2000), pp. 289–299.

Styblo, M., Serves, S.V., Cullen, W.R. and Thomas, D.J.: Comparative inhibition of yeast glutathione reductase by arsenicals and arsenothiols. *Chem. Res. Toxicol.* 10:1 (1997), pp. 27–33.

Tchounwou, P.B., Centeno, J.A. and Patlolla, A.K.: Arsenic toxicity, mutagenesis, and carcinogenesis—a health risk assessment and management approach. *Mol. Cell Biochem.* 255:1/2 (2004), pp. 47–55.

Tchounwou, P.B., Patlolla, A.K. and Centeno, J.A.: Carcinogenic and systemic health effects associated with arsenic exposure—A critical review. *Toxicol. Pathol.* 31:6 (2003), pp. 575–588.

Thomas, D.J., Styblo, M. and Lin, S.: The cellular metabolism and systemic toxicity of arsenic. *Toxicol. Appl. Pharmacol.* 176:2 (2001), pp. 127–144.

USEPA: *Arsenic in drinking water.* US Environmental Protection Agency, 2001, http://www.epa.gov/safewater/arsenic/index. html (accessed December 2006).

Vahter, M.: Mechanisms of arsenic biotransformation. *Toxicology* 181/182 (2002), pp. 211–217.

Vujcic, M., Shroff, M. and Singh, K.: Genetic determinants of mitochondrial response to arsenic in yeast *Saccharomyces cerevisiae. Can. Res.*67:20 (2007), pp. 9740–9749.

Waalkes, M.P., Fox, D.A., States, J.C., Patierno, S.R. and McCabe, M.J., Jr.: Metals and disorders of cell accumulation: modulation of apoptosis and cell proliferation. *Toxicol. Sci.* 56:2 (2000), pp. 255–261.

Wanibuchi, H., Yamamoto, S., Chen, H., Yoshida, K., Endo, G., Hori, T. and Fukushima, S.: Promoting effects of dimethylarsinic acid on N-butyl-N-(4-hydroxybutyl)nitrosamine-induced urinary bladder carcinogenesis in rats. *Carcinogenesis* 17:11 (1996), pp. 2435–2439.

Warner, M.L., Moore, L.E., Smith, M.T., Kalman, D.A., Fanning, E. and Smith, A.H.: Increased micronuclei in exfoliated bladder cells of individuals who chronically ingest arsenic-contaminated water in Nevada. *Cancer Epidemiol. Biomarkers Prev.* 3:7 (1994), pp. 583–590.

Wiencke, J.K. and Yager, J.W.: Specificity of arsenite in potentiating cytogenetic damage induced by the DNA crosslinking agent diepoxybutane. *Environ. Mol. Mutagen.* 19:3 (1992), pp. 195–200.

Yamamoto, S., Konishi, Y., Matsuda, T., Murai, T., Shibata, M.A., Matsui-Yuasa, I., Otani, S., Kuroda, K., Endo, G. and Fukushima, S.: Cancer induction by an organic arsenic compound, dimethylarsinic acid (cacodylic acid), in F344/DuCrj rats after pretreatment with five carcinogens. *Cancer Res.* 55:6 (1995), pp. 1271–1276.

Yamanaka, K. and Okada, S.: Induction of lung-specific DNA damage by metabolically methylated arsenics via the production of free radicals. *Environ. Health Perspect.* 102 (Suppl. 3) (1994), pp. 37–40.

Yang, J.L., Chen, M.F., Wu, C.W. and Lee, T.C.: Posttreatment with sodium arsenite alters the mutational spectrum induced by ultraviolet light irradiation in Chinese hamster ovary cells. *Environ. Mol. Mutagen.* 20:3 (1992), pp. 156–164.

CHAPTER 45

Early signs of immunodepression induced by arsenic in children

L. Vega, G. Soto, A. Luna, L. Acosta, P. Conde, M. Cebrián & E. Calderón
Sección Externa de Toxicología, Centro de Investigación y de Estudios Avanzados del Instituto Politécnico Nacional (CINVESTAV-IPN), Mexico City, Mexico

L. López
Instituto Nacional de Salud Pública, Cuernavaca, Mor., Mexico

M. Bastida
Jurisdicción Sanitaria, Secretaria de Salubridad y Asistencia, Zimapán, Hgo., Mexico

ABSTRACT: Immunotoxic effects of arsenic (As) have not yet been investigated in children. In this study, peripheral blood mononuclear cells (PBMC) from ninety 6 to 10 year old children exposed to As in drinking water were examined. We found significant or marginally significant negative correlations between the concentration of As in urine and the proliferative response of PBMC to phytohemagglutinin (PHA) stimulation ($p = 0.005$), IL-2 secretion by PHA-stimulated cells ($p = 0.003$), percentage of CD4 cells ($p = 0.092$), and CD4/CD8 ratio ($p = 0.056$). In contrast, a significant positive correlation ($p < 0.001$) was found between the urinary As level and secretion of granulocyte-macrophage colony-stimulating factor (GM-CSF) by mononucleated cells. We did not find significant associations between the urinary As and relative amounts of CD8, B, or NK cells or secretion of IL-4, IL-10, or IFN-γ by PHA-stimulated PBMC. These data indicate that chronic As exposure could result in an immunodepressed status and thus, in an increased susceptibility of children to opportunistic infections and to development of cancer. In addition, increased GM-CSF secretion may be associated with chronic inflammation.

45.1 INTRODUCTION

Arsenic is a well-known human carcinogen with multiple cellular effects (Thomas *et al.* 2001). Human populations chronically exposed to arsenic (As) via drinking water are in an increased risk of developing bladder, kidney, liver and particularly, skin cancer (Kitchin 2001). It has been postulated that the carcinogenic effect of As in humans is associated with its potential genotoxicity (Ostrosky-Wegman *et al.* 1991, Vega *et al.* 1995, Gonsebatt *et al.* 1997) and with the ability of As to act as a co-carcinogen with UV light exposure (Rossman *et al.* 2005). An alternative mechanism of action suggests that As participates in tumor development by inducing damage to immune cells and by suppressing the ability of these cells to attack transformed cells and to prevent chronic and opportunistic infections (Andres 2005). Although none of these effects has been properly documented, previous reports have shown that As exposure increases incidence of autoimmune diseases such as diabetes mellitus (Tseng 2004) and other diseases associated with immunodepression. For example, As induces skin cancer similar to that diagnosed in individuals that are immunodepressed as a result of organ transplantation or HIV infection (Andres 2005, Lim *et al.* 2005). Integrity of the immune system is necessary to guarantee adequate immuno-surveillance and response to infectious agents. Therefore, the aim of the present study was to determine the toxic effects of As exposure on the immune system in children.

It has been demonstrated that a single dose of gallium arsenide (GaAs; 200 mg/kg) inhibits T cell proliferation, macrophage activity, and IgM and IgG production in B6C3F1 female mice (Sikorsky *et al.* 1989). Immunization of GaAs-treated mice with sheep red blood cells produced a 50% decrease in CD4+ splenic cells after 24 h (Burns and Munson 1993). In addition, exposure to sodium arsenite in drinking water (200 mg As/l) for 4 weeks was shown to decrease contact hypersensitivity response to 2,4-dinitrofluorobenzene (DNFB) in female Balb/c mice (Patterson *et al.* 2004).

Compared with relatively abundant information on immunotoxic effects of As exposure in various animal models, very little is known about immunotoxic effects of As in human populations. It has been reported that, depending on the dose, As can either inhibit or induce proliferative responses in animal and human cells. Very low (nM) concentrations of As have been shown to induce lymphocyte proliferation (Vega *et al.* 1999). In contrast, high (μM) concentrations of As inhibit lymphocyte proliferation (Meng and Meng 2000, Gonsebatt *et al.* 1992 and 1994, Vega *et al.* 2004). Inhibition of lymphocyte proliferation in response to phytohemagglutinin (PHA) stimulus has been reported in adult human populations exposed to 412 μg/l of As in drinking water (Gonsebatt *et al.* 1994). In addition, *in vitro* exposure to 1 μM of sodium arsenite has been shown to inhibit IL-2 secretion by human lymphocytes (Vega *et al.* 1999, Galicia *et al.* 2003).

In 1997, Armienta *et al.* reported high levels of As (4.6 to 10 mg/kg) in hair of residents in Zimapán region in the Mexican state of Hidalgo who drink water with As levels ranging from 0.014 to 1.09 mg/l. The present study examined immunotoxic effects of As exposure in children from families residing in this region.

45.2   METHODS

45.2.1   *Study design*

The study population comprised of 90 children, 6 to 10 year old, including 38 girls and 52 boys. Eligible children were recruited from 7 elementary schools which represent 9% of all schools in the study area. The study design (as described by Soto-Peña *et al.* 2006) and the informed consent letter used in this study were approved by the Ethics Committee of the National Institute of Public Health (INSP). The inclusion criteria for study subjects were as follows: at least one year of residence in the study area, overall healthy status (each child was examined by a physician), no drug or antibiotic use for at least two weeks before the study, no history of immune diseases or related treatments, and an informed consent signed by parents. Parents were asked to provide information about the child socio-economic background, medical history, exposure to other xenobiotics, lifetime exposure to As, and other relevant information. Data provided in questionnaires were used to determine the socio-demographic status of each study subject as previously described (Bronfman *et al.* 1988). Confounding factors examined in this study included age, sex, nutritional status, exposure to lead, and exposure to pesticides.

45.2.2   *Sample collection*

Venous blood was collected by venipuncture into EDTA-containing vacutainers. First morning urine samples were collected and stored in plastic bottles at 4°C until analyzed for As content. The urine samples were protected from light during the storage.

45.2.3   *Determination of arsenic concentration in urine*

Concentrations of inorganic As and As metabolites in urine were determined by hydride generation-atomic absorption spectrometry (HG-AAS), using procedures described by Del Razo *et al.* (2001). The following metabolic ratios were calculated: i:M, dividing the amount of inorganic arsenic (i-As) by the amount of monomethyl arsenic acid (MMA); M:D, dividing the amount of MMA by the amount of dimethyl arsinic acid (DMA), and i:D, dividing the amount of i-As by the amount of DMA (Meza 2005).

45.2.4   *Isolation and culture of peripheral blood mononuclear cells*

Plasma was separated from blood by centrifugation. Peripheral blood mononuclear cells (PBMC) were isolated by centrifugation in Ficoll-Hypaque density gradient and washed with phosphate buffered saline (PBS). Cells were cultured as described elsewhere (Vega *et al.* 2004). Isolation and culture of adhesive cells were carried out as follows: PBMC (2×10^6 cell/well) were incubated in culture plates at 37°C in a 5% CO_2 atmosphere for 2 h. Plates were then washed with cold RPMI medium and fresh medium containing lipopolysaccharides from *Escherichia coli* (LPS; 30 ng/ml) and interferon-γ (IFN-γ, 200 U/ml) was added. After a 30 h incubation, media were collected and frozen at −20°C.

45.2.5   *Evaluation of mononuclear cell subpopulations*

Following procedures described by Lima and Vega (2005), PBMC were washed by PBS supplemented with fetal calf serum and sodium azide. Cells were incubated with labeled primary antibodies for 2 h (1 µg/ml, anti-CD3-cy-chrome, CD4-RPE, and CD8-FITC for triple staining, and CD57-FITC or CD19-FITC in washing solution). Cells were washed and fixed with 1% *p*-formaldehyde in washing solution. Fixed cells were analyzed using a FACS can flow cytometer (Becton Dickinson). Results are presented as percentage of positively stained cells in a sample of 10,000 cells. The Th/Tc ratio was calculated dividing the amount CD4+ positive cells by the amount of CD8+ positive cells.

45.2.6   *Cell proliferation assay*

Cell proliferation was examined as described elsewhere (Lima and Vega 2005). Briefly, 2×10^5 cells were incubated in 96-well plates for 72 h under the same conditions as described above and stimulated with PHA (1µg/ml). Culture medium was collected and frozen at −20°C. Cells were then incubated for 18 to 20 h in a fresh medium containing [^{3}H] thymidine (6.7 Ci/mmol, Dupont, Boston, MA, USA): 0.5 µCi/well. The amount of [^{3}H] thymidine incorporated in the cultured cells was measured by liquid scintillation, using a Betaplate counter (Wallac, Turku, Finland). The stimulation index (SI) for PHA-treated cells was calculated dividing the radioactivity (cpm) detected in PHA-stimulated cells by the radioactivity detected in untreated cells.

45.2.7   *Cytokine analysis*

Analysis of IL-2, IL-4, IL-10, and IFN-γ in culture media from PHA-stimulated lymphocytes and analysis of GM-CSF in media from LPS/IFN-γ-activated macrophages was performed by ELISA as previously described (Vega *et al.* 2001). Briefly, ELISA plates were incubated with corresponding capture antibodies at 4°C overnight. Plates were then washed with PBS/Tween (PBS with 0.05% Tween 20) and blocked for 2 h with PBS-containing 3% bovine serum albumin. Samples of culture media were added and incubated for 2 h. After washing, plates were treated with biotinylated anti-cytokine detection antibodies, washed once, and incubated with an avidin-HRP conjugate. Finally, ABTS solution (2-2azino-bis-[3-ethylbenzothiazolin-6-sulphonic acid] in anhydrous citric acid with 3% H_2O_2) was added, and optical density (OD) was measured, using a microplate reader with a 405 nm filter. Cytokine concentrations in culture media (pg/ml) were calculated using calibration curves for known concentrations of recombinant human cytokines.

45.2.8   *Statistical analysis*

For statistical analysis, the study population was divided into two exposure groups based on the Mexican reference value for As in urine of 50 µg/l (NOM-127-SSAI-1994): low exposure group (urinary As $< 50\,\mu g/l$) and high exposure group (urinary As $> 50\,\mu g/l$). General characteristics of individuals in these two groups were evaluated using X^2 and t test statistics. The t test was also used to evaluate gender related differences in mean values for urinary metabolites of As. Contribution of confounding factors was assessed for the following variables: age, sex, time of residence, socio-economic status, nutritional status, exposure to other xenobiotics such as pesticides or lead, history of asthma, allergies, or parasitic infections. Variables influencing by more than 10% the crude associations of interest were considered in further analysis. Multiple lineal regression models were used to evaluate effects of the distribution of As metabolites in urine on the parameters examined. Logarithmic transformation was used to normalize distribution of SI values, IFN-γ and GM-CSF concentrations, Th/Tc ratio, B and NK cell numbers; square transformation was used to normalize distribution of IL-4 and IL-10 concentrations.

45.3   RESULTS

45.3.1   *Study population*

No significant differences were found for general characteristics of the study subjects in the low and high exposure groups (Table 45.1). No association was found between the urinary As levels and exposure to other xenobiotics ($p = 0.346$), socio-economic status ($p = 0.678$) or age ($p = 0.345$). We found a marginally significant ($p = 0.055$) increase in the incidence of asthma, allergies and parasitic infections among children in the high exposure group as compared to the low exposure group (Table 45.1).

45.3.2   *Total arsenic and arsenic metabolites in urine*

Total urinary As was calculated as sum of all As species found in urine. The main urinary metabolite of As was DMA (74%). MMA and i-As represented in average 15.5 and 10.5% of total urinary As, respectively. For most children (86%), the total urinary As level exceeded $50\,\mu g/l$ (Fig. 45.1a). The average concentration of total urinary As was higher for girls (252.1 $\mu g/l$) as compared to boys (144.5 $\mu g/l$) ($p = 0.047$). Despite this difference, no statistically significant gender related differences were found in relative amounts (percentages) of individual As metabolites found in urine. In contrast, we found a statistically significant effect of the exposure level on the ratios of

Table 45.1.   General characteristics of the study subjects in the low and high exposure groups.

	Arsenic urine level	
	Low mean (SD)	High mean (SD)
Girls (%)	11.8	88.2
Boys (%)	15.6	84.4
Age (years)	7.5 (1.2)	8.0 (1.5)
Weight (kg)	25.4 (6.8)	27.9 (6.7)
Body mass index	27.5 (5.7)	27.4 (6.4)
Haemoglobin (g/dl)	14.2 (1.3)	14.2 (1.4)
Pb ($\mu g/dl$)	9.9 (11.6)	8.2 (4.4)
Asthma, allergies, parasites	1/11	21/79**
Total urinary arsenic ($\mu g/l$)	29.3 (14.8)	194.9 (135.3)*

SD, standard deviation; Results of the statistical evaluation of differences between the low and high exposure groups using t test: *$p < 0.05$, **$0.1 > p > 0.05$.

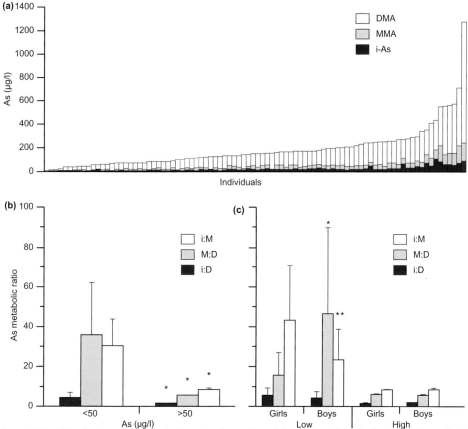

Figure 45.1.   Distribution of As metabolites in urine collected from all subjects included in this study (a), ratios of urinary metabolites of arsenic in high and low exposure groups (b), and differences in the metabolite ratios in urine of girls and boys in the low and high exposure groups (c). Results of the statistical evaluation of differences between the low and high exposure groups or between girls and boys using Student's t test: *p < 0.05, **0.1 > p > 0.05.

urinary metabolites of As. Specifically, the i:D and M:D ratios were greater in the low exposure group as compared to high exposure group (Fig. 45.1b). There were no significant difference in the ratios of urinary metabolites between boys and girls (Fig. 45.1c).

45.3.3   *PBMC subpopulations*

Proportions of total peripheral T cells (T, CD3+), T helper lymphocytes (Th, CD4+), T cytotoxic lymphocytes (Tc, CD8+), B lymphocytes (B, CD19+), and natural killer cells (NK, CD57+) were analyzed in freshly collected blood, using flow cytometry (Fig. 45.2). Associations between the relative amounts of T cell subpopulations and urinary As content (an indicator of the As exposure level) were evaluated, using a multivariate analysis. In general, the relative amounts of total T cells, Tc, B or NK cells were not affected by As exposure. Negative correlations between the total urinary As and CD4 subpopulation or Th/Tc lymphocyte ratio were only marginally significant (Fig. 45.2d).

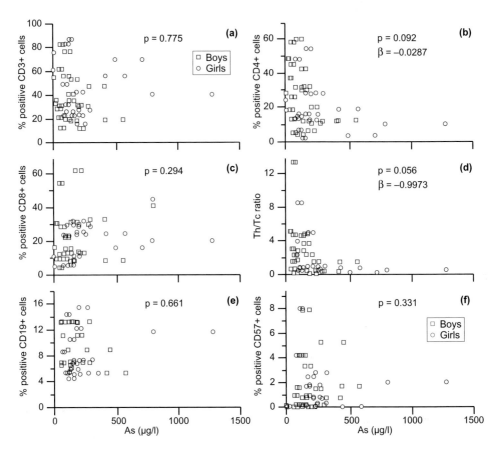

Figure 45.2.　Associations between the total urinary As and relative amounts (%) of subpopulations of mononucleated cells isolated from peripheral blood of the study subjects, including total T (a), Th (b), Tc (c), B (e), and NK (f) cells; (d) association between the urinary As and the ratio of Th/Tc cells. Multivariate analyses were used after adjustment for sex, age, and incidence of asthma, allergies, and parasitic infections.

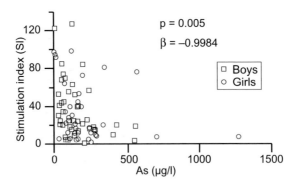

Figure 45.3.　Association between the total urinary arsenic and stimulation index (SI) for PHA-stimulated lymphocytes. Multivariate analyses were used after adjustment for sex, age, and incidence of asthma, allergies, and parasitic infections.

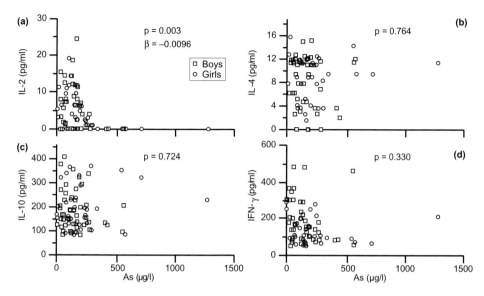

Figure 45.4. Associations between the total urinary arsenic and secretion of IL-2 (a), IL-4 (b), IL-10 (c), and IFN-γ (d) by PHA-stimulated lymphocytes. Multivariate analyses were used after adjustment for sex, age, and incidence of asthma, allergies, and parasitic infections.

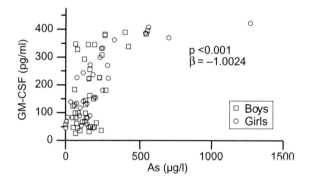

Figure 45.5. Associations between the total urinary arsenic and GM-CSF secretion by LPS/IFN-γ-stimulated monocytes. Multivariate analyses were used after adjustment for sex, age, and incidence of asthma, allergies, and parasitic infections.

45.3.4 *Responses to in vitro stimulation*

In vitro proliferative responses of PBMC to the PHA stimulus were evaluated with respect to the total urinary As levels (Fig. 45.3). A statistically significant negative correlation was found between the total urinary As and the SI value ($\beta = -0.9984$, p = 0.005). Association between the urinary As and secretion of cytokines that are relevant to the process of activation by PHA was also examined (Fig. 45.4). The results show a statistically significant correlation between the total urinary As levels and IL-2 secretion ($\beta = -0.0096$, p = 0.003). However, there was no significant association between the PBMC proliferative response to PHA stimulus and IL-2 secretion (p = 0.368, $R^2 = 0.0133$), indicating that these two processes were independent. No statistically significant correlations were found between the urinary levels of As and secretion of IL-4, IL-10 or IFN-γ by PHA-activated lymphocytes. Notably, PHA-stimulated PBMC from boys secreted more IFN-γ (167.8 pg/ml) than PHA-stimulated PBMC from girls (118.1 pg/ml) (p =

0.035). In contrast, *in vitro* secretion of GM-CSF by LPS/IFN-γ-stimulated macrophages was in a direct proportion with the concentration of total As urine ($\beta = 1.0024$, p < 0.001) (Fig. 45.5).

45.4   DISCUSSION

Results of the present study indicate that high exposures to As in drinking water may be associated with an increased incidence of asthma, allergies, and parasitic infections among children in the Zimapán region of Mexico. However, the number of children involved in this study was relatively small. Thus, studies carried out in larger populations will be needed to confirm this association. In general, our findings are in agreement with previous reports that show As exposure to be immunotoxic in animal models or *in vitro* in human cells (Patterson *et al.* 2004, Vega *et al.* 2001 and 2004). Systemic immunodepression may be caused by several mechanisms and is associated with several endpoints. One of the most common characteristics of the immunodepressed status is an alteration in proportions of lymphocyte subpopulations. Burns and Munson (1993) reported that exposure of mice to GaAs caused a decrease in the number of Th and B cells, but not in the number of Tc cells. A similar trend was found in this study, although the negative correlation between the Th cell population and the concentration of total As in urine was only marginally significant.

A marginally significant negative correlation was also found between the total As concentration in urine and the ratio of Th/Tc cells. This is a potentially important result because this ratio is used for diagnosis of immunodepression in patients with various pathologies, including cancer. It has been reported that As exposure is associated with an increased incidence of Bowen's disease. Notably, Hayashi and associates (1997) reported that Th/Tc ratio is decreased in adult patients with this disease. Moreover, a similar decrease in Th/Tc ratio has been reported for other pathologies, including Kaposi's sarcoma (McPartlin *et al.* 1999), melanoma (Hernberg *et al.* 1998), skin carcinoma (Deng *et al.* 1998), and HIV infection (Schofer and Roder 1995). Thus, the Th/Tc ratio is an important indicator of immunodepression. This ratio has been shown to range from 3 to 5 among healthy individuals and to decrease below 2 for immuno-compromised patients (Deng *et al.* 1998, Hernberg *et al.* 1998, McPartlin *et al.* 1999, Schofer and Roder 1995). In the present study, all individuals with urinary As levels above 200 µg/l exhibited the Th/Tc ratios lower than 2. These results suggest that a low Th/Tc ratio may serve as an indicator of immunodepression associated with chronic exposures to As and that immunodepression may be in part responsible for typical clinical manifestations of chronic As toxicity, including skin lesions. Based on the results of our study, the Th/Tc ratio may be particularly useful as an early marker of adverse affects in children exposed to moderate or high levels of As in drinking water, resulting in urinary As concentrations at or above 150 µg/l.

Another important parameter used to estimate the capacity of the immune system to respond to various challenges is the proliferative response of T cells to a mitogenic stimulus. Our results show that As exposure is associated with inhibition of T lymphocyte proliferation in response to *in vitro* activation with PHA. These findings are in agreement with results of previous studies in adult human populations and of *in vitro* studies (Gonsebatt *et al.* 1992 and 1994, Vega *et al.* 1999 and 2004). Secretion of cytokines is a key event associated with proliferation of activated T cells. Previous studies in our laboratory showed that in the presence of sodium arsenite, mitogen-stimulated T cells secrete smaller amounts of IL-2 than control cells (Vega *et al.* 1999). This could be explained in part by inhibition of proliferation in the arsenite-treated T cell culture. In this study, we found that IL-2 secretion by PHA-stimulated cells negatively correlates with the total As concentration (t-As) in urine (i.e., with the level of As exposure). Previous work has shown that proliferation of PHA-stimulated lymphocytes from adults chronically exposed to As in drinking water is characterized by slow cell cycle kinetics (Gonsebatt *et al.* 1992). This is probably due to reduction in IL-2 secretion and arrest in the G_0/G_1 phase (Vega *et al.* 1999). Consistent with these findings are results of a parallel study that found a negative correlation between the IL-2 mRNA levels in T cells and concentration of t-As in urine of children chronically exposed to As in drinking water (Luna *et al.* 2003). Taken together, these data suggest that As exposure not only reduces the number of T cells entering cell cycle after stimulation with PHA, but also slows down the rate of proliferation for those cells which enter the cell cycle. As a consequence, the production and secretion of IL-2 is decreased, further disrupting the proliferation and activation of PBMC as previously demonstrated in human lymphocytes treated *in vitro* with sodium arsenite (Vega *et al.* 1999).

Other cells, beside T lymphocytes, that are important for adequate immune response are monocytes (peripheral undifferentiated macrophages). These cells secrete several cytokines that are responsible for maturation, recruitment, and activation of vicinal cells. One of these cytokines, GM-CSF, stimulates leukocyte expansion and monocyte differentiation. It has been shown that GM-CSF, in association with other growth factors, plays an important role in cancer induced by As exposure (Germolec *et al.* 1996, 1997 and 1998). In this study, we found a statistically significant positive correlation between *in vitro* GM-CSF secretion by LPS/IFN-γ-activated macrophages and the t-As concentration in urine of the study subjects. This result is consistent with previous reports (Germolec *et al.* 1998, Vega *et al.* 2001) that described stimulation of transcription and secretion of growth factors, including GM-CSF, in human keratinocytes treated with As. GM-CSF that is produced by keratinocytes participates in inflammatory processes in human skin (Braunstein *et al.* 1994). By mediating inflammatory cell influx and increasing numbers of dark cells, GM-CSF may facilitate tumor promotion in mouse skin (Vasunia *et al.* 1994). In addition, increased expression of GM-SCF that is rapidly induced in a paracrine fashion by cytokines, including IL-1, IL-2, TGF-α, and TNF-α, has been associated with various skin diseases (Ansel *et al.* 1990). Other studies have shown increased GM-CSF mRNA levels in skin lesions (Lontz *et al.* 1995) and lesion-free psoriatic skin (Uyemora *et al.* 1993), as well as in skin from patients with atopic dermatitis (Pastore *et al.* 1997). A role for GM-CSF in dermal carcinogenesis has also been postulated based on evidence that shows dose- and time-dependent increases in GM-CSF in skin of mice after topical treatment with tumor-promoting agents (Germolec *et al.* 1998, Spalding *et al.* 1993). Thus, elevated levels of GM-CSF may serve as an early and sensitive biomarker for skin diseases associated with As exposure.

45.5   CONCLUSIONS

Chronic exposures of children to As in drinking water may result in alterations of immune parameters that are consistent with immunodepression described for AIDS or skin cancer patients. Some of these parameters (e.g., Th/Tc ratio, lymphocyte proliferation, IL-2 secretion by T cells or GM-CSF secretion by macrophages) could be used as early markers of adverse effects in human populations exposed to As. These parameters should be validated in future studies, focusing mainly on children chronically exposed to As and on the relationship between As exposure and incidence of opportunistic or chronic infections and other diseases that are associated with compromised immune status.

REFERENCES

Andres, A.: Cancer incidence after immunosuppressive treatment following kidney transplantation. *Crit. Rev. Oncol. Hematol.* 280 (2005), pp. 33, 885–33, 894.

Ansel, J., Perry, P., Brow, J., Damm, D., Phan, T., Hart, C., Luger, T. and Hefeneider, S.: Cytokine modulates keratinocytes. *J. Invest. Dermatol.* 94 (1990), pp. 101–107.

Armienta, M.A., Rodríguez, R. and Cruz, O.: Arsenic content in hair of people exposed to natural arsenic polluted groundwater at Zimapan, Mexico. *Bull. Environ. Contam. Toxicol.* 59 (1997), pp. 583–589.

Braunstein, S., Kaplan, G., Gottliet, B., Schwartz, M., Walsh, G., Abalos, R., Fajardo, T.T., Guido, L.S. and Krueger, J.G.: GM-CSF activates regenerative epidermal growth and stimulates keratinocytes proliferation in human skin in vivo. *J. Invest. Dermatol.* 103 (1994), pp. 601–604.

Bronfman, M., Guiscafré, H., Castro, V., Castro, R. and Gutiérrez, G.: La medición de la desigualdad: una estrategia metodológica, análisis de las características socioeconómicas de la muestra. *Arch. Invest. Med.* 19 (1988), pp. 351–360.

Burns, L.A. and Munson, A.E.: Gallium arsenide selectively inhibits T cell proliferation and alters expression of CD25 (IL-2R/p55). *J. Pharmacol. Exp. Ther.* 265 (1993), pp. 178–186.

Del Razo, L.M., Styblo, M., Cullen, W.R. and Thomas, D.J.: Determination of trivalent methylated arsenicals in biological matrices. *Toxicol. Appl. Pharmacol.* 174 (2001), pp. 282–293.

Deng, J.S., Falo, L.D. Jr., Kim, B. and Abell, E.: Cytotoxic T cells in basal cell carcinomas of skin. *Am. J. Dermatopathol.* 20:2 (1998), pp. 143–146.

Galicia, G., Leyva, R., Tenorio, E.P., Ostrosky-Wegman, P. and Saavedra, R.: Sodium arsenite retards proliferation of PHA-activated T cells by delaying the production and secretion of IL-2. *Int. Immunopharmacol.* 3:5 (2003), pp. 671–682.

Germolec, D., Spalding, J., Boorman, G., James, L., Wilmer, J., Yoshida, T., Simeonova, P.P., Bruccoleri, A., Kayama, F., Gaido, K., Tennant, R., Burleson, F., Dong, W., Lang, R.W. and Luster, M.I.: Arsenic can mediate skin neoplasia by chronic stimulation of keratinocytes-derived growth factors. *Mutat. Res.* 386 (1997), pp. 209–218.

Germolec, D., Spalding, J., Yu, H.S., Chen, G.S., Simeonova, P.P., Humble, M.C., Bruccoleri, A., Boorman, G.A., Foley, J.F., Yoshida, T. and Luster, M.I.: Arsenic enhancement of neoplasia by chronic stimulation of growth factors. *Am. J. Pathol.* 156:6 (1998), pp. 1775–1785.

Germolec, D., Yoshida, T., Gaido, K., Wilmer, J., Simeonova, P., Kayama, F., Burleson, F., Dong, W., Lange, R.W. and Luster, M.I.: Arsenic induces overexpression of growth factors in human keratinocytes. *Toxicol. Appl. Pharmacol.* 141 (1996), pp. 308–318.

Gonsebatt, M.E., Vega, L., Herrera, L.A., Montero, R., Rojas, E., Cebrián, M.E. and Ostrosky-Wegman, P.: Inorganic arsenic effects on human lymphocyte stimulation and proliferation. *Mutat. Res.* 283 (1992), pp. 91–95.

Gonsebatt, M.E., Vega, L., Montero, R., García-Vargas, G., Del Razo, L.M., Albores, A., Cebrián, M.E. and Ostrosky-Wegman, P.: Lymphocyte replicating ability in individuals exposed to arsenic via drinking water. *Mutat. Res.* 313 (1994), pp. 293–299.

Gonsebatt, M.E., Vega, L., Salazar, A.M., Montero, R., Guzmán, P., Blas, J., Del Razo, L.M., García-Vargas, G., Albores, M., Cebrián, M.E., Kelsh, M. and Ostrosky-Wegman, P.: Cytogenetic effects in human exposure to arsenic. *Mutat. Res.* 386 (1997), pp. 219–228.

Hayashi, T., Hinoda, Y., Takahashi, T., Adachi, M., Miura, S., Izumi, T., Kojima, H., Yano, S. and Imai, K.: Idiopathic CD4+ T-lymphocytopenia with Bowen's disease. *Intern. Med.* 36:11 (1997), pp. 822–824.

Hernberg, M., Turunen, J.P., Von Boguslawsky, K., Muhonen, T., and Pyrhonen, S.: Prognostic value of biomarkers in malignant melanoma. *Melan. Res.* 8:3 (1998), pp. 283–291.

Kitchin, K.T.: Recent advances in arsenic carcinogenesis: modes of action, animal model systems, and methylated arsenic metabolites. *Toxicol. Appl. Pharmacol.* 172 (2001), pp. 249–261.

Lim, S.T. and Levine, A.M.: Non-AIDS-defining cancers and HIV infection. *Curr. Infect. Dis. Rep.* 7:3 (2005), pp. 227–234.

Lima, A. and Vega, L.: Methyl-parathion and organophosphorous pesticide metabolites modify the activation status and interleukin-2 secretion of human peripheral blood mononuclear cells. *Toxicol. Lett.* 158 (2005), pp. 30–38.

Lontz, W., Sirsho, A., Liu, W., Lindberg, M., Rollman, O. and Torma, H.: Increased mRNA expression of manganese superoxide dismutase in psoriasis skin lesions and in cultured human keratinocytes exposed to IL-1β and TNF-α. *Free Rad. Biol. Med.* 18 (1995), pp. 349–355.

Luna, A.L., Acosta-Saavedra, L.C., Conde, P., Vera, E., Cruz, M.B., Bastida, M., Gómez-Muñoz, A., López-Carrillo, L., Cebrián, M.E. and Calderón-Aranda, E.S.: Functional activity of Th1 and macrophages from children environmentally exposed to arsenic. *Toxicol. Sci.* 72:S1 (2003), p. 377.

McPartlin, D.W., Ghufoor, K., Patel, S.K. and Jayraj, S.: A rare case of cutaneous kaposiform haemangioendothelioma. *J. Clin. Pract.* 53:7 (1999), pp. 562–563.

Meng, Z.Q. and Meng, N.Y.: Effects of arsenic on blast transformation and DNA synthesis of human blood lymphocytes. *Chemosphere* 41:1/2 (2000), pp. 115–119.

Meza, M.M., Yu, L., Rodriguez, Y.Y., Guild, M., Thompson, D., Gandolfi, A.J. and Klimecki, T.: Developmentally restricted genetic determinants of human arsenic metabolism: association between urinary methylated arsenic and CYT19 polymorphisms in children. *Environ. Health Perpect.* 113 (2005), pp. 775–781.

Ostrosky-Wegman, P., Gonsebatt, M.E., Montero, R., Vega, L., Barba, H., Espinosa, J., Palao, A., Cortinas, C., García-Vargas, G., Del Razo, L.M. and Cebrián, M.E.: Lymphocyte proliferation kinetics and genotoxic findings in a pilot study on individuals chronically exposed to arsenic in Mexico. *Mutat. Res.* 250 (1991), pp. 477–482.

Pastore, S., Fanales-Belasio, E., Albanesi, C., Chinni, L., Giannetti, A. and Girolomoni, G.: Granulocyte macrophage colony-stimulating factor is overproduced by keratinocytes in atopic dermatitis. *J. Clin. Invest.* 99 (1997), pp. 3009–3017.

Patterson, R., Vega, L., Bortner, C. and Germolec, D.: Arsenic-induced alterations in contact hypersensitivity response in balb/c mice. *Toxicol. Appl. Pharmacol.* 198 (2004), pp. 434–443.

Rossman, T.G., Uddin, A.N. and Burns, F.J.: Evidence that arsenic acts as a co-carcinogen in skin cancer. *Toxicol. Appl. Pharmacol.* 198:3 (2004), pp. 394–404.

Schofer, H. and Roder, C.: Kaposi sarcoma in caucasian women. Clinical, chemical laboratory and endocrinologic studies in 8 women with HIV-associated or classical kaposi sarcoma. *Hautarzt* 46:9 (1995), pp. 632–637.

Sikorsky, E.E., McCay, J.A., White, K.I., Bradley, S.G. and Munson, A.E.: Immunotoxicity of the semiconductor gallium arsenide in female B6C3F1 mice. *Fundam. Appl. Toxicol.* 13 (1989), pp. 843–858.

Soto-Peña, G., Luna, A.L., Saavedra-Acosta, L.,Conde, P., Lopez-Carrillo, L., Cebrian, M.E., Bastida, M., Calderon-Aranda, E.S. and Vega, L.: Assessment of lymphocyte subpopulations and cytokine secretion in children exposed to arsenic. *FASEB J.* 20 (2006), pp. 779–781.

Spalding, J.W., Momm, J., Elwell, M.R. and Tennant, R.W.: Chemically induced skin carcinogenesis in a transgenic mouse line (TG. AC) carrying a v-Ha-ras gene. *Carcinogenesis* 14 (1993), pp. 1335–1341.

Thomas, D.J., Styblo, M. and Lin, S.: The cellular metabolism and systemic toxicity of arsenic. *Toxicol. Appl. Pharmacol.* 176:2 (2001), pp. 127–144.

Tseng, C.H.: The potential biological mechanisms of arsenic-induced diabetes mellitus. *Toxicol. Appl. Pharmacol.* 197:2 (2004), pp. 67–83.

Uyemora, K., Yamamura, M., Fivenson, D.F., Modlin, R.L. and Nickoloff, B.J.: The cytokine network in lesional and lesion-free psoriatic skin is characterized by a T-helper 1 cell-mediated response. *J. Invest. Dermatol.* 1001 (1993), pp. 701–705.

Vasunia, B., Miller, L., Puga, A. and Baxter, S.: Granulocyte-macrophage colony-stimulating factor (GM-CSF) is expressed in mouse skin in response to tumor-promoting agents and modulates dermal inflammation and epidermal dark cell numbers. *Carcinogenesis* 15 (1994), pp. 653–660.

Vega, L., Gonsebatt, M.E. and Ostrosky-Wegman, P.: Aneugenic effect of sodium arsenite on human lymphocytes in vitro: an individual susceptibility effect detected. *Mutat. Res.* 334 (1995), pp. 365–373.

Vega, L., Montes de Oca, P., Saavedra, R. and Ostrosky-Wegman, P.: Helper T cell subpopulations from women are more susceptible to the toxic effect of sodium arsenite *in vitro*. *Toxicology* 199 (2004), pp. 121–128.

Vega, L., Ostrosky-Wegman, P., Fortoul, T.I., Díaz, C., Madrid, V. and Saavedra, R.: Sodium arsenite reduces proliferation of human activated T-cells by inhibition of the secretion of interleukin-2. *Immunopharmacol. Immunotoxicol.* 21 (1999), pp. 203–220.

Vega, L., Styblo, M., Patterson, R , Cullen, W,, Wang, C. and Germolec, D.: Differential effects of trivalent and pentavalent arsenicals on cell proliferation and cytokine secretion in normal human epidermal keratinocytes. *Toxicol. Appl. Pharmacol.* 172 (2001), pp. 225–232.

CHAPTER 46

Evaluation of human arsenic contamination in the district of Santa Bárbara, Minas Gerais, Brazil

N.O.C. Silva, C.A. Rocha & T.V. Alves
Laboratório de Contaminantes Metálicos, Divisão de Vigilância Sanitária, Instituto Octávio Magalhães, Fundação Ezequiel Dias (FUNED), Belo Horizonte, Brazil

E. Deschamps & S.M. Oberdá
Fundação Estadual do Meio Ambiente (FEAM), Belo Horizonte, Minas Gerais, Brazil

J. Matschullat
Interdisciplinary Environmental Research Centre, TU Bergakademie Freiberg, Freiberg, Germany

ABSTRACT: To investigate the arsenic (As) exposure of the local population and to delineate the related As sources at Brazilian mining sites, a study was performed between 2002 and 2005. Five sub-districts were selected: Barra Feliz, Brumal, Sumidouro, Santana do Morro, and "Rua do Carrapato", between Brumal and Barra Feliz. Five sampling campaigns were performed in these communities to collect human urine from the inhabitants, with a focus on children since they are more vulnerable towards As pollution. All samples were analyzed by atomic absorption spectrometry with hydride generation, and the results corrected for the creatinine content, determined by UV/VIS spectrometry. Based on the obtained results, it became obvious that the lowest concentrations occurred at "Rua do Carrapato", with maximum concentrations of 25 µg As/g creatinine. All other areas showed higher concentrations up to 60 µg As/g creatinine at values above the upper threshold permitted by Brazilian legislation.

46.1 INTRODUCTION

Most gold deposits are associated with arsenopyrite and sometimes to a lesser extend with pyrite. Soil and sediments around gold ore deposits and mining sites present positive arsenic (As) anomalies, related to geological structures and the additional contamination dissipation due to centuries of mining and smelting activities. The As enrichments have consequences for all environmental compartments including human health (Deschamps 2002, Matschullat *et al.* 2000).

To verify the influence of elevated As levels on residents of the area research, financed by the National Environmental Fund, was set up to perform several sampling campaigns: urine, water, soil, dust, vegetables and green vegetables from some places of the area (Matschullat *et al.* 2000). In the beginning, urine samples were collected from children and adolescents only in campaigns performed at municipal schools. These campaigns were authorized by the Education and Health Authorities, and engaged support of teachers and school principals. The last sampling campaigns were extended to all age groups, and the sampling performed directly at people's homes with the general consent of the population. During collection, the volunteers were requested to answer an epidemiological questionnaire on lifestyle, nutrition and health, followed by the signature of a consent term authorizing the use of the samples for research purposes. Over four years, a total of 522 urine samples from the local population plus 28 control samples from people not exposed to unusual As concentrations, were processed.

46.2   METHODOLOGY

The samples were kept refrigerated in insulated coolers and send to the laboratory. For As analysis, a slightly modified methodology after Guo *et al.* (1997) was used. This method determines the toxicologically relevant forms of As in urine [As (III), As (V), monomethylarsonate, MMA, and dimethylarsinate, DMA]. In this method, samples are diluted in 0.03 mol/l HCl p.a., and treated with 5% L-cystein p.a. After the reaction period, a 500 µl aliquot was injected automatically in the flow injection system (FIAS400, Perkin Elmer), reacting with a mixed solution of 0.5% sodium tetrahydroborate p.a. and 0.5% sodium hydroxide p.a., to form As-hydride that is transferred by argon to the atomic absorption spectrometer (AAnalyst 300) equipped with an automatic sampler (AS-90) and electrodeless discharge lamps (380 mA, λ: 193.7 nm) for As detection (Perkin Elmer 1996). All reagents were prepared using Milli-Q purified water. A calibration curve was repeatedly generated with blanks and As concentrations between 1 and 20 µg/l. The results obtained were corrected by the creatinine levels, determined by UV/VIS spectrometry in the visible range, according to the Brazilian legislation (Brazil 1994, Couto *et al.* 2007).

To test the accuracy of the As determinations, certified reference materials (NIST SRM 2670 and Lyphochek) were analyzed with each sample sequence and inter-laboratory comparisons were run with 18 samples yielding unknown As content. The Figures 46.1a and b represent the results form the analysis of the certified reference samples (NIST and Lyphochek) that stay within the expected limits (Mattos *et al.* 2007).

To further improve quality control, 18 sample aliquots were determined in parallel in a laboratory of the pubic network in Brazil. The correlation of the results was calculated with 0.99015, as presented in Figure 46.2 (Mattos *et al.* 2007).

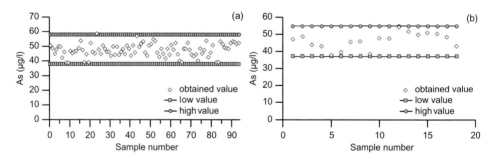

Figure 46.1.   Arsenic concentrations in the NIST SRM 2670 (a) and in the Lyphocheck CRM (b).

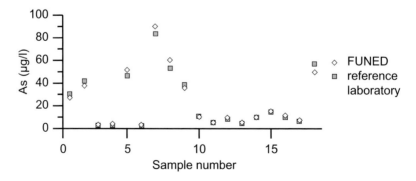

Figure 46.2.   Results from the parallel inter-laboratory analysis of 18 aliquot samples.

46.3 RESULTS

The Brazilian legislation defines a reference value (RV) of 10 μg As/g creatinine as a safe value to which no excessive As exposure occurred, and a maximum permissible value of 50 μg As/g creatinine (MPV). Results between the RV and MPV indicate an elevated corporal As exposition without a real contamination of the investigated person, while concentrations above the MPV indicate a contamination. The Table 46.1 shows the regional results.

These results indicate that the general averages in four of the five studied areas, are above the RV according to Brazilian legislation, with maximum concentrations varying from 25 to 60 μg As/g creatinine (Table 46.1). The lowest values were encountered in the "Rua do Carrapato" area, with an average value of 8.7 μg As/g creatinine and a maximum value of 25 μg As/g creatinine, and in Sumidouro with an average of 14 μg As/g creatinine and a maximum value of 41 μg As/g creatinine. In the latter region of Sumidouro, individual persons showed values above the RV but not a single sample showed values in excess of MPV (Table 46.1). The areas of Barra Feliz, Brumal and Santana do Morro were represented by average values above the RV, with some individuals showing values above the MPV, suggesting an important source of As exposition, and the potential for human contamination (Table 46.1). All control samples showed normal As levels with an average of 3.5 μg As/g creatinine, and not a single person showing values above the RV. The As values obtained from people in the investigated areas are shown in Figure 46.3. These figures represent the compiled data from all four years of analysis—including the control population.

Table 46.1. Analytical results for arsenic in human urine from the years 2002 to 2005.

	Barra Feliz	Brumal	Sumidouro	Rua do Carrapato	Santana do Morro	Control samples
	(μg As/g creatinine)					
General average	14.2	13.3	14.2	8.7	12.9	3.5
Maximum value	51.8	58.9	40.6	25.0	60.1	8.1
Children						
average	15.1	13.4	18.1	10.6	9.7	–
>RV[1] (%)	59.43	54.1	57.1	54.5	54.5	–
>MPV[2] (%)	0.7	0.0	0.0	0.0	0.0	–
samples	143	159	14	22	11	–
Adolescent						
average	11.1	19.9	10.3	9.4	22.0	–
>RV[1] (%)	46.9	43.3	37.7	66.7	61.5	–
>MPV[2] (%)	0.0	4.5	0.0	0.0	7.7	–
samples	32	67	14	3	13	–
Adult						
average	–	–	–	4.9	12.2	3.5
>RV[1] (%)	–	–	–	11.1	50.0	0.0
>MPV[2] (%)	–	–	–	0.0	0.0	0.0
samples	–	–	–	9	24	28
Elderly						
average	–	–	–	4.2	5.5	–
>RV[1] (%)	–	–	–	0.0	11.1	–
>MPV[2] (%)	–	–	–	0.0	0.0	–
samples	–	–	–	2	9	–

[1] RV: Reference value; [2] MPV: Maximum permissible value (biological threshold limit value).

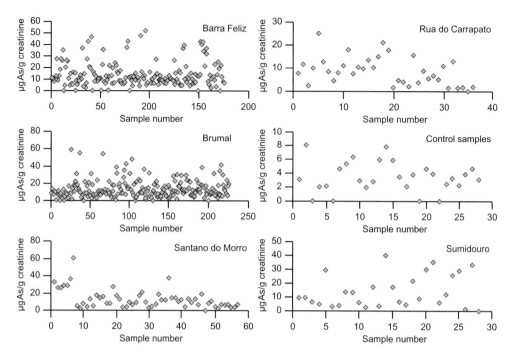

Figure 46.3.   Arsenic distribution in human urine from the investigated areas.

46.4   CONCLUSIONS

The applied methodology for As determination in human urine proved adequate with good precision and accuracy of the results, as shown by the fine replication of certified reference values and the inter-laboratory comparison. All five localities yielded individuals with As values above the reference value, but only at three of the sites were As concentrations encountered that surpass the maximum permissible As value.

The results show that the population of the studied areas has been exposed to elevated As levels, and that a human contamination on different levels exists, depending on the region and the age group of the people. It could be shown that children and teenagers were most affected by the As exposure, most likely because of the habits of this age group, such as walking barefoot, playing on the ground, etc.—apart from the fact that the areas are economically poor and without a proper infrastructure, like paved roads and safe water supply.

Through the collaboration with State Environmental Agency (FEAM), it became possible to construct a water treatment station that supports the community of Santana do Morro. Untreated water from local springs that represents a potential pollution source with elevated As values (and elevated levels of *Escherichia coli*), is now being treated properly.

In a general way, the execution of this project was important because it demonstrated the existing As contamination in the studied communities and helped to solve the problem. The local population took part in meetings and lectures, organized by the project group to communicate the As-related risks.

ACKNOWLEDGMENTS

The authors acknowledge the support from the National Fund for the Environment—FNMA/ MMA.

REFERENCES

Brasil, Secretaria de Segurança e Saúde no Trabalho 1994. Portaria n° 24, de 29 de dezembro de 1994. Aprovação da Norma Regulamentadora n° 7 que estabelece os parâmetros mínimos e diretrizes gerais a serem observados na execução do Programa de Controle Médico de Saúde Ocupacional—PCMSO, Brasília, Brazil, 1994.
Couto, N., Mattos, S. and Matschullat, J.: Amostragem e procedimentos analticos. In: E. Deschamps and J. Matschullat (eds): *Arsenio antropogenico e natural: Um estudo em regioes do Quadrilatero Ferrifero*. Chapter 5.1. Fundacao Estadual de Meio Ambiente, Belo Horizonte, Brazil, 2007, pp. 243–251.
Deschamps, E., Ciminelli, V.S.T., Frank L.F.T., Matschullat, J., Raue, B. and Schmidt, H.: Soil and sediment geochemistry of the Iron Quadrangle, Brazil: the case of arsenic. *J. Soils Sedim.* 2 (2002), pp. 216–222.
Guo, T., Baasner, J. and Tsalec, D.L.: Fast automated determination of toxicologically relevant arsenic in urine by flow injection-hydride generation atomic absorption spectrometry. *Anal. Chim. Acta.* 349 (1997), pp. 313–318.
Matschullat, J., Borba, R.P., Deschamps, E., Figueiredo, B.R., Gabrio, T. and Schwenk, M.: Human and environmental contamination in the Iron Quadrangle, Brazil. *Appl. Geochem.* 15 (2000), pp. 181–190.
Mattos, S., Couto, N. and Matschullat, J.: Resultados e interpretacao. In: E. Deschamps and J. Matschullat (eds): *Arsenio antropogenico e antural. Um estudo na regiao do Quadrilatero Ferrifero*. Chapter 5.2. Fundacao Esta dual de Meio Ambiente, Belo Horizonte, Brazil, 2007, pp. 252–269.
Perkin Elmer: Flow injection mercury/hydride analyses. Recommended analytical conditions and general information. Release 4.0, Germany, 1996.

CHAPTER 47

Effects of fluoride and arsenic on the central nervous system

D.O. Rocha-Amador, L. Carrizales & J. Calderón
Facultad de Medicina, Universidad Autónoma de San Luis Potosí (UASLP),
San Luis Potosí, S.L.P., Mexico

R. Morales & M.E. Navarro
Facultad de Psicología, Universidad Autónoma de San Luis Potosí (UASLP),
San Luis Potosí, S.L.P., Mexico

ABSTRACT: Experimental and epidemiological studies, support the evidence that fluoride (F) and arsenic (As) are neurotoxic. A cross-sectional study to evaluate the effect of the simultaneous exposure to F and As on neurological functions in children was designed. One-hundred thirty two children from 6 to 10 years were included in the study. Children were residents from four communities with different As and F concentrations in drinking water: (1) F 0.7 ± 0.3 mg/l; As $6.7 \pm 1.2 \mu$g/l; (2) F 1.1 ± 0.01 mg/l; As $4.5 \pm 1.5 \mu$g/l; (3) F 5.3 ± 0.18 mg/l; As $169.5 \pm 0.2 \mu$g/l and (4) F 9.4 ± 1.1 mg/l; As $200 \pm 84 \mu$g/l. The Weschler Intelligence Scale revised version for Mexican children was used to evaluate verbal, performance and full IQ. As biomarkers of exposure urinary F and As were used. As confounding factors lead in blood, socioeconomic status and nutritional evaluation were evaluated. After adjusting by confounding factors, negative associations were obtained between F and As in urine and Full, Verbal and Performance IQ. These results suggest that chronic exposure to both elements affects the higher brain functions in these children.

47.1   INTRODUCTION

Fluoride (F) and arsenic (As) exists naturally in water sources. Many countries have reported concentrations of either F and As often exceeding the World Health Organization (WHO) guideline values of 1.5 mg/l and 10 μg/l, respectively (WHO 2004). Several studies in As endemic areas have demonstrated the relationship between the exposure to this element and skin, liver, kidney and bladder cancer, as well as to peripheral neuropathies and vascular disorders (Yoshida *et al.* 2004), whereas living in F-endemic areas the association has been established with dental fluorosis, skeletal fluorosis, endocrine and reproductive effects (CDC 1991). In addition, there are studies that suggest that both elements affect the Central Nervous System (CNS).

In children exposed to F or As there are data that support the hypothesis that both elements diminish the scores of the intelligence quotient (IQ), being a diminution of 10 points between exposed children to concentrations of F in water greater to 3 mg/l compared with non exposed populations, affecting mainly verbal and execution abilities (Li *et al.* 1995, Zhao *et al.* 1996, Lu *et al.* 2000, Calderón *et al.* 2001b, Xiang *et al.* 2003). There has also been reported that the processes associated with memory, visuospatial organization and attention can be modified by the exposure to As (Calderón *et al.* 2001a, b, Tsai *et al.* 2003, Wasserman *et al.* 2004). The adverse effects observed in humans can be supported with experimental data which indicate that both elements cross the blood brain barrier and accumulate in the brain (Ghafgazi *et al.* 1980, Geeraerts *et al.* 1986, Mullenix *et al.* 1995, Foulkes and Abbotsforb 1996, Shivarajashankara *et al.* 2002a, b, ATSDR 2003a, b), affecting locomotive activity and presenting cognitive deficit, learning and in the behavior (Mullenix *et al.* 1995, Rodriguez *et al.* 2001, 2003).

453

The integrity of the CNS in the humans can be evaluated through different electrophysiological and image techniques. In the case of alterations associated with the behavior or learning, the neuropsychological evaluation is a tool that can give information integrated about the CNS to evaluate slight changes in the cognitive functions (attention, memory, language and visuospatial abilities). This tool can be used to evaluate the risks of sites contaminated with toxic substances (ATSDR 1995). In zones where the concentration of F in water exeeds the allowed limit, there are often also concentrations of As in excess of safe levels for human health. In Mexico, it has been considered that approximately 6 million of people resident in the center-north zone of the country, are supplied by water contaminated with both elements (Del Razo *et al.* 1993, Wyatt *et al.* 1998). There are no reports about the effects on the CNS associated with the simultaneous exposure to F and As in children and, therefore, a cross-sectional study was designed to evaluate the possible associations between the simultaneous exposure to F and As and the IQ in children of 6 to 10 years of age resident in zones with different concentrations of As and F in drinking water.

47.2   METHODS

47.2.1   *Study population*

A cross-sectional study was designed. Children from four rural communities in Mexico with different level of F and As in drinking water were selected. Children were from three communities from San Luis Potosí (SLP) and one from Durango (Dgo) states: (1) Soledad de Graciano Sánchez, SLP (F 0.7 ± 0.3 mg/l; As 6.7 ± 1.2 µg/l); (2) Moctezuma, SLP (F 1.1 ± 0.01 mg/l; As 4.5 ± 1.5 µg/l); (3) Salitral de Carrera, Villa de Ramos, SLP (F 5.3 ± 0.2 mg/l; As 169.5 ± 16 µg/l) and (4) 5 de Febrero, Dgo (F 9.4 ± 1.1 mg/l; As 200 ± 84 µg/l). All children attending 1st to 3rd grade at elementary schools, located in the selected areas were screened for study eligibility by a questionnaire. Children who met the inclusion criteria were selected (n = 132).

47.2.2   *Biological and environmental monitoring and analytical methodology*

Urine samples were collected in plastic bottles. F in urine was analyzed according to the method number 8308 from the National Institute of Occupational Safety and Health. F levels were quantified using a sensitive specific ion electrode. As quality control, a reference standard from the National Institute of Standards and Technology was assessed (NIST 2671a). The accuracy was $98 \pm 6\%$. For determination of As in urine, samples were digested according to Cox (1980) and As was quantified by hydride generation-atomic absorption spectrophotometry (HG-AAS) using a Perkin-Elmer model Analyst 100. As quality control we use a freeze-dried urine standard (NIST SRM-2670). The accuracy was $98 \pm 4\%$.

The concentrations of F and As in urine were standardized to urinary creatinine concentration. F in tap water was analyzed following the same procedure described for F in urine. Instead of EDTA, water samples were diluted with TISAB 1:1. As quality control for F in water, NIST SRM-3138 was analyzed; the accuracy was $98 \pm 4\%$. As in tap water was analyzed according the method AOAC, the NIST 1640 for As in water was employed as quality control. The accuracy was $99 \pm 6\%$. Blood samples were obtained by venous puncture using tubes containing EDTA as anticoagulant. Lead in blood was analyzed according to Subramanian (1987) with a Perkin-Elmer 3110 AAS using a graphite furnace. Our laboratory participated in the blood lead proficiency testing program conducted by CDC. The precision was $99 \pm 9\%$.

47.2.3   *Nutritional and socioeconomic status assessment*

Height by age, weight by height as indexes of chronic and acute undernutrition, were calculated using reference tables from the United States National Center for Health Statistics, and as recommended by World Health Organization (Dibley *et al.* 1987). Transferrin saturation was the ratio of serum iron to total iron binding capacity. It was quantified with a SERA-PAK kit from Bayer as control SERA-CHECK serum was analyzed. The Bronffam index of socioeconomic status (SES) was constructed from five socioeconomic variables: crowding within the home, housing conditions, potable water availability, drainage and father's education (Bronffman *et al.* 1988).

47.2.4 *Neuropsychological evaluation*

All tests were administered at school by a trained neuropsychologist who was not informed about each participant's urinary levels of F and As. A standardized version of the Wechsler Intelligence Scale for Children Revised Version for Mexico (WISC-RM) was administered (Gómez *et al.* 1983). Ten different subtests were given to each child, five predominantly verbal (information, similarities, arithmetic, vocabulary and comprehension) and five predominantly performance oriented (picture completion, picture arrangement, block design, object assembly and coding). Raw scores were age-adjusted and summed to yield conventional measures of Full, Verbal and Performance IQ estimates.

47.2.5 *Statistical analysis*

The statistical treatment of the results was carried out by ANOVA tests. Gender differences between groups were evaluated by overall χ^2 test. Univariate, bivariate and multivariate analysis were conducted with SPSS statistical software package version 10.0. Significance level was fixed at 0.05.

47.3 RESULTS

In Table 47.1 the mean concentrations of F and As in drinking water of the evaluated populations are given. Two groups were observed, one of low exposure to F and As that include Soledad de Graciano Sanchez and Moctezuma and another one of high exposure that includes Salitral de Carrera and 5 de Febrero. The concentrations of both elements in these two last communities are superior to the limits established for F and As in water in drinking water in the Mexican official norm (NOM-127-SSA1-1994).

In Table 47.2, the socioeconomic and nutritional characteristics and lead levels in blood of the studied populations are shown. There is no difference between age, sex, mother's education and anthropometric characteristics between the four populations. The level of transferrin saturation was higher in the children of 5 de Febrero compared to the concentrations of the other three communities and the SES was slightly greater in Soledad. The concentration of lead in blood was lower in the children of 5 de Febrero compared to the concentrations of the other three communities; in general 10% of the children exceeded the limit established by the CDC of 10 μg/dl of lead in blood, and 63% had concentrations of lead in blood greater to 5 μg/dl.

Table 47.1. Levels of F and As in drinking water in four rural areas of Mexico.

Parameter	Soledad (n = 34)	Moctezuma (n = 18)	Salitral (n = 20)	5 de Febrero (n = 60)
F in drinking water (mg/l)[1]	0.7 (0.3–1.4)	0.9 (0.9–1.3)	5.3[2] (5.0–5.6)	9.4[2] (8.1–15.7)
As in drinking water (μg/l)[1]	6.9 (4.2–8.9)	4.3 (2.9–7.1)	170[2] (148–186)	194[2] (141–794)

[1] Values are geometric mean and range is shown in parenthesis; [2] Significantly different from Soledad, p < 0.001, by ANOVA test.

Table 47.2. Characteristics of Mexican children living in areas with different levels of F and As in drinking water.

Parameter	Soledad (n = 34)	Moctezuma (n = 18)	Salitral (n = 20)	5 de Febrero (n = 60)
Age (yrs)[1]	8.2 (5.5–10.6)	8.4 (6.8–10.7)	7.7 (5.9–9.4)	8.3 (5.7–10.9)
Socioeconomic status[1]	7.5 (6.0–9.0)	6.1 (3.0–9.0)	6.3 (5.0–7.0)	5.9[4] (2.0–10.0)
Mother education (year)[1]	6.4 (0.0–9.0)	5.6 (0.0–10.0)	4.7 (0.0–16.0)	5.6 (0.0–15.0)
Lead in blood (µg/dl)	7.0[5] (4.2–12.6)	7.3[5] (3.1–11.9)	6.8[5] (2.2–10.5)	4.8 (0.2–15.7)
Gender				
Boys (%)	53	56	50	48
Girls (%)	47	44	50	52
Transferrin saturation (% <20)[2]	26.9[5] (8.1–79.4)	23.9[5] (12.8–34.7)	20.5[5] (6.2–46.7)	33.0 (11.7–72.4)
Weigh for height index (% <2 SD)[3]	0.4 (−1.3–3.0)	−0.3 (−1.9–2.9)	0.2 (−1.7–1.9)	0.5 (−1.3–3.9)
Height for age index (% <2 SD)[3]	0.3 (−1.9–2.0)	0.06 (−1.8–2.6)	−0.3 (−1.6–2.9)	−0.01 (−2.1–2.0)

[1] Values are geometric mean and range is shown in parenthesis; [2] CDC action level; [3] National Health Survey (1999) reference value; [4] Significantly different from Soledad, p < 0.001; [5] Significantly different from 5 de Febrero, p < 0.05.

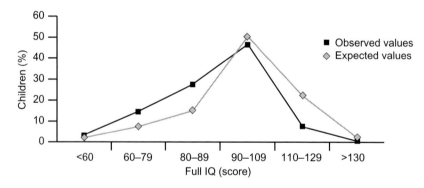

Figure 47.1. Full IQ distribution in Mexican children living in rural areas with different levels of F and As in drinking water.

Figure 47.1 shows the distribution of full IQ obtained in total children included in the study. The distribution of full IQ of the zones in general show a change in the distribution, a greater proportion of children with IQ smaller to 90 points (scores low to the normal) and we also observed that in both zones the percentage of children with scores greater than 110 (scores superior to the normal) was smaller than 25% (expected values).

To test the association between F and As in urine and IQ scores (Full, Verbal and Performance) models of multiple regression linear (MRL) adjusted to mother's education and lead in blood were calculated. In order to evaluate the simultaneous effect of F and As in urine on the scores of IQ, we obtained models of MRL between F in urine and scores of IQ adjusted by concentrations of As in urine, mother's education and lead in blood. 24% of the diminution in IQ was explained by the concentration of F in urine adjusted by As in urine and the other confounding variables (B = −0.52 p < 0.0001). This value was compared with the model of only F in urine (B = −0.47 p < 0.0001). The results obtained were summarized with the standardized parameters B of the models (Fig. 47.2).

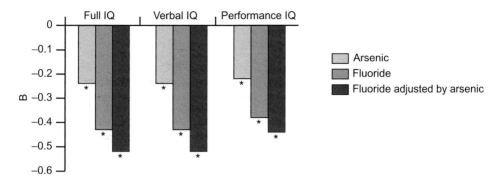

Figure 47.2.   Models of MRL between F and As in urine and scores of IQs adjusted by the mother's education and lead in blood. Models of MRL were summarized with the standardized parameters B, $^*p < 0.01$.

47.4   CONCLUSIONS

We found that exposure to F and As in urine was associated with reduce Full, Verbal and Performance IQ (F Full B $= -0.43^*$, Verbal B $= -0.43^*$ and Performance B $= -0.38^*_{p^* < 0.0001}$; As Full B $= -0.24^*$, Verbal B $= -0.24^*$ and Performance B $= -0.22^*_{p^* < 0.01}$). Nevertheless the effect of the F was harnessed by the presence of As (Full B $= -0.52^*$, Verbal B $= -0.52^*$ and Performance B $= -0.44^*_{p^* < 0.0001}$). In the last 30 years, the CDC has diminished the levels of lead in blood, because it has been shown that each increase of 10 µg/dl of lead in blood diminishes the IQ by 4 to 7 points. According to the results obtained in the present study and other reported studies regarding neurological effects of F exposure (Zhao *et al.* 1996, Lu *et al.* 2000), this xenobiotic is more toxic than lead and presents a synergistic effect in presence of As. These results make it necessary to create strategies to permit the decrease or total elimination of the exposure to these elements as a final alternative for the reduction of their deleterious effects on SNC.

ACKNOWLEDGEMENTS

This project was supported by *Consejo Nacional de Ciencia y Tecnología* CONACYT grant number J-37584-M and by the *Universidad Autónoma de San Luis Potosí* grant Number CO1-FRC-12.12.

REFERENCES

ATSDR: Adult environmental neurobehavioral test battery. Agency for Toxic Substances and Disease Registry, US Department of Health and Human Services, Atlanta, GA, 1995.

ATSDR: Toxicological profile for arsenic. Agency for Toxic Substances and Disease Registry, US Department of Health and Human Services, Atlanta, GA, 2003a.

ATSDR: Toxicological profile for fluoride. Agency for Toxic Substances and Disease Registry, US Department of Health and Human Services, Atlanta, GA, 2003b.

Bronffman, M.G., Castro, H. and Gutiérrez, G.: Medición de la desigualdad: una estrategia metodológica. Análisis de las características socioeconómicas de la muestra. *Arch. Invest. Med.* 19 (1988), pp. 351–360.

Calderón, J., Navarro, M.E., Jiménez-Capdeville, M.E., Santos-Diaz, M.A., Golden, A., Rodríguez-Leyva, I., Borja-Aburto, V. and Díaz-Barriga, F.: Exposure to arsenic and lead and neuropsychological development in Mexican children. *Environ. Res.* 85 (2001a), pp. 69–76.

Calderón, J., Machado, B., Navarro, M.E., Carrizales, L. and Díaz-Barriga, F.: Influence of fluoride on reaction time and organization visuospatial in children. *Epidemiol.* 2 (2001b), p. 153.

CDC: Review of fluoride "benefits and risks". Department of Health and Human Services, 1991, pp. 1–87.

Cox, D.H.: Arsine evolution-electrothermal atomic absorption method for the determination of nanogram levels of total arsenic in urine and water. *J. Anal. Toxicol.* 4 (1980), pp. 207–211.

Del Razo, L.M., Corona, J.C., García-Vargas, G., Albores, A. and Cebrian, M.E.: Fluoride levels in well-water from a chronic arsenicism area of northerm Mexico. *Environ. Pollut.* 80 (1993), pp. 91–94.

Dibley, M.J., Goldsby, J.B., Staehling, N.W. and Trowbridge, F.L.: Development of normalized curves for the international growth reference: historical and technical considerations. *Am. J. Clin. Nutr.* 46 (1987), pp. 736–748.

Foulkes, G.R. and Abbotsforb, B.C.: The fluoride connection. *Fluoride* 29 (1996), pp. 230–238.

Ghafgazi, T., Ridlintong, J.W. and Fowler, B.A.: The effects of acute and subacute sodium arsenite administration on carbohydrate metabolism. *Toxicol. Appl. Pharmacol.* 55 (1980), pp. 126–130.

Geeraerts, F., Gijs, G., Finne, R. and Crokaert, R.: Kinetics of fluoride penetration in liver and brain. *Fluoride* 19 (1986), pp. 108–112.

Gómez, A., Palacios, P. and Padilla, E.: WISC-R Mexicano. Manual de aplicación adaptado y estandarizado en México. Editorial Manual Moderno, Mexico Ciy, Mexico, 1983.

Li, X.S., Zhi, J.L. and Gao, R.O.: Effects of fluoride exposure on intelligence in children. *Fluoride* 28 (1995), pp. 189–192.

Lu, Y., Wu, L.N., Wang, X., Lu, W. and Liu, S.S.: Effects of high-fluoride water on intelligence in children. *Fluoride* 33 (2000), pp. 74–78.

Mexican Official Norm, NOM-127-SSA1-1994: Salud ambiental, agua parauso consumo humano. Límites permisibles de calidad y tratamientos a que se debe someterse el agua para su potabilización. Official Newspaper, first section, Mexico City, Mexico, 2000, pp. 48–54.

Mullenix, P.J., Denbesten, P.K., Schunior, A. and Kernan, W.J.: Neurotoxicity of sodium fluoride in rats. *Neurotoxicol. and Teratol.* 17 (1995), pp. 169–177.

National Institute for Occupational Safety and Health (NIOSH): Fluoride in urine. US Departament of Health and Human Services, Manual of Analytical Methods, 3rd ed., 11, 1984, pp. 8308-1–8308-3.

Rodriguez, V.M., Carrizales, L., Jímenez-Capdville, M.E., Dufour, L. and Giordano, M.: The effects of sodium arsenite exposure on behavioral parameters in the rat. *Brain Res. Bull.* 55 (2001), pp. 301–308.

Rodriguez, V.M., Jímenez-Capdville, M.E. and Giordano, M.: The effects of arsenic exposure on the nervous system. *Toxicol. Lett.* 145 (2003), pp. 1–18.

Shivarajashankara, Y.M., Shivashankara, A.R., Gopalakrishna, B.P., Muddanna, R. and Hanumanth R.S.: Histological changes in the brain of young fluoride-intoxicated rats. *Fluoride* 35 (2002a), pp. 12–21.

Shivarajashankara, Y.M., Shivashankara, A.R., Gopalakrishna, B.P. and Hanumanth, R.S.: Brain lipid peroxidation and antioxidants systems of young rats in chronic fluoride intoxication. *Fluoride* 35 (2002b), pp. 197–203.

Subramanian, K.S.: Determination of lead in blood: comparison of two GFASS methods, *Atomic Spectrosc.* 8 (1987), pp. 7–14.

Tsai, S.Y., Chou, H.Y., The, H.W., Chen, C.M. and Chen, C.J.: The effects of chronic arsenic exposure from drinking water on the neurobehavioral development in adolescence. *Neurotoxicol.* 24 (2003), pp. 747–753.

Wasserman, G.A., Liu, X., Parvez, F., Ahsan, H., Factor-Litvak, P., Van Geen, A., Slavkovich, V., Lalacono, N.J., Cheng, Z., Hussain, I., Momotaj, H. and Graziano, J.H.: Water arsenic exposure and children's intellectual function in Araihazar, Bangladesh. *Environ Health Persp.* 112 (2004), pp. 1329–1333.

WHO: Guidelines for drinking water quality: recommendations, 3rd ed., Vol. 1, World Health Organization, Geneva, Switzerland, 2004.

Wyatt, C.J., Lopez-Quiroga, V., Olivas Acosta, R.T. and Mendez, R.O.: Excretion of arsenic (As) in urine of children 7–11 years, exposed to elevated levels of As in the city waters supply in Hermosillo Sonora México. *Environ. Res.* 78 (1998), pp. 19–24.

Xiang, Q., Liang, Y., Chen, L., Wang, C., Chen, B. and Zhou, M.: Effects of fluoride in drinking water on children's intelligence. *Fluoride* 36 (2003), pp. 84–94.

Yoshida, T., Hiroshi, Y. and Fan, Sun, G.: Chronic health effects in people exposed to arsenic via drinking water: dose-response relationship in review. *Toxicol. Appl. Pharmacol.* 198 (2004), pp. 243–252.

Zhao, L.B., Liang, G.H., Zhang, D.N. and Wu, X.R.: Effects of a high fluoride water supply on children intelligent. *Fluoride* 29 (1996), pp. 190–192.

CHAPTER 48

Neurotoxicity of arsenic

M.E. Gonsebatt, J. Limón-Pacheco, E. Uribe-Querol & G. Gutiérrez-Ospina
*Instituto de Investigaciones Biomédicas, Universidad Nacional Autónoma de México (UNAM),
Mexico City, Mexico*

V.M. Rodríguez
*Environmental and Community Medicine, The University of Medicine and Dentistry of New Jersey
and Rutgers, Piscataway, NJ, USA*

M. Giordano
Instituto de Neurobiología, Universidad Nacional Autónoma de Mexico (UNAM), Querétaro, Mexico

L.M. Del Razo & L.C. Sánchez-Peña
*Sección Externa de Toxicología, Centro de Investigación y de Estudios Avanzados del
Instituto Politécnico Nacional (CINVESTAV-IPN), Mexico City, Mexico*

ABSTRACT: Neurotoxicity caused by drinking well water contaminated with arsenic (As) or accidental exposure to As compounds has been described in adults and infants in previous studies. Biomethylation of As generates toxic metabolites that could help explain the adverse effects observed following As exposure in human and in animal models. Here dose related accumulation of methylated As metabolites in mouse brain is studied which shows that brain metabolizes arsenite to monomethyl arsonic acid (MMA) and dimethyl arsinic acid (DMA), DMA being the main metabolite in this tissue. Glutathione reductase activity was inhibited at the highest dose tested in liver and brain.

48.1 INTRODUCTION

48.1.1 *Inorganic arsenic as a human neurotoxicant*

Inorganic arsenic (i-As) exposure via drinking water has been associated with cancer but also with peripheral neuropathy and diverse effects in the circulatory and nervous system (Rodriguez *et al.* 2003). During subacute or chronic exposure, i-As can occasionally result in subclinical or overt peripheral neuropathy (Yip *et al.* 2002). Diverse neurotoxic effects have been reported in children and adults receiving or exposed to different doses of As. Cerebellar symptoms and mental retardation associated with brain atrophy have been described in adults and infants drinking well water contaminated with diphenylarsinic compounds (Ishii *et al.* 2004). Children exposed to low levels of As during their whole-life showed hearing impairment (Bencko and Symon 1977) and lower scores in verbal intelligence quotient (IQ) (Calderon *et al.* 2001). Water i-As exposure was associated with reduced intellectual function, in a dose-dependent manner (Wasserman *et al.* 2004); children with water i-As levels >50 μg/l achieved significantly lower Performance and Full-Scale scores than did children with water i-As levels <5.5 μg/l. Adolescents exposed to high levels of i-As presented alterations in memory and attention processes (Tsai *et al.* 2003) or mental deterioration (Brower *et al.* 1992), while in adults acutely exposed to high amounts of i-As, impairments in learning, memory, and concentration and severe encephalopathy have been described (Bolla-Wilson and Bleecker 1987, Franzblau and Lilis 1989, Morton and Caron 1989, Berbel-Garcia *et al.* 2004).

48.1.2   *Neurotoxicity of inorganic arsenic in experimental models*

Rodents have been a animal model frequently used to investigate the mechanisms of As neurotoxicity. Behavioral alterations described in this model include deficits in operant learning (Nagaraja and Desiraju 1994), alterations in locomotor activity and increases in errors in a delayed alternation task tested in the T maze (Rodriguez *et al.* 2001, 2002, Chattopadhyay *et al.* 2002a, b, Pryor *et al.* 1983, Itoh *et al.* 1990). Some of these alterations could be related to changes in the cholinergic, monoaminergic, GABAergic and glutamatergic systems in brain regions such as striatum, midbrain, hypothalamus, and hippocampus of animals exposed to arsenicals (Valkonen *et al.* 1983, Nagaraja and Desiraju, 1993, 1994, Tripathi *et al.* 1997, Mejia *et al.* 1997, Delgado *et al.* 2000, Kannan *et al.* 2001, Rodriguez *et al.* 2001). Increased oxidative DNA damage in cerebral and cerebellar cortices has been observed in i-As treated mice (Piao *et al.* 2005). *In vitro* studies have demonstrated that arsenite causes apoptosis in cortical and cerebellar neurons, while dimethyl arsinic acid (DMA) induces apoptosis in cerebellar neurons only. In both types of cell cultures, apoptosis was induced by caspase activation (Namgung and Xia 2000, 2001). An *in vivo* study, reported the formation of hydroxyl radicals in the rat striatum, induced by direct infusion of sodium arsenite via a microdialysis probe (García-Chavez *et al.* 2003).

48.1.3   *Inorganic arsenic metabolism*

Once incorporated into the organism, i-As is biotransformed through metabolic conversions that significantly modify the toxicity of this metalloid (Thomas *et al.* 2004). In humans, As is converted to methyl arsonic acid (MMA) and dimethyl arsinic acid (DMA) through the reduction of the pentavalent species in reactions that require glutathione (GSH) e.g., reduced monomeric glutathione or thioredoxin and by an oxidative methylation that requires S-adenosylmethionine (SAM) as the methyl donor (Thomas *et al.* 2004). In rodents such as rats and hamsters DMA is further methylated to trimethyl arsine oxide (TMAO) (Yoshida *et al.* 1998, Devesa *et al.* 2004). The trivalent methylated metabolites have been shown to be more potent enzyme inhibitors and more cytotoxic than the inorganic forms (Styblo *et al.* 2002). In mice, tissue levels of the metabolites are both tissue specific and dose-dependant (Kenyon *et al.* 2005). However, there are no *in vivo* studies relating the adverse effects of arsenical exposure to the presence of methylated As species in brain, which could be to some extent responsible for the neurotoxic alterations observed. GSH is extensively involved in metabolism of i-As specifically in the reduction of pentavalent arsenicals to their trivalent forms (Thomas *et al.* 2001, Waters *et al.* 2004); being also the main antioxidant in brain. Arsenite and its methylated metabolites have been shown to be potent inhibitors of glutathione reductase (GR) *in vitro* (Styblo *et al.* 1995, 1997; Chouchane and Snow, 2001), thus exposure to arsenicals could compromise the anti-oxidant mechanisms by consuming GSH and inhibiting the enzyme responsible for its recycling. To investigate if these effects could be observed *in vivo*, and to document the formation of methylated metabolites in mouse brain we treated mice with three doses of sodium arsenite and measured the presence of i-As and its methylated metabolites and the activity and levels of GR.

48.2   MATERIALS AND METHODS

48.2.1   *Exposure protocol*

Four groups of CD-1 male mice received 0, 2.5, 5 or 10 mg of sodium arsenite per kg of body weight per day by intragastric route for a total of 9 days. Sodium arsenite solutions were prepared daily with deionized water. The control group received deionized water as treatment. Animals had free access to water and food before and during arsenite exposure. Body weight was recorded daily during exposure.

Day after the last treatment, mice were sacrificed by cervical dislocation, followed by decapitation. Brain and liver were extracted and washed in ice-cold isotonic saline solution to remove debris and blood.

48.2.2   *Determination of methylated arsenic species, enzyme activities and immunoblots*

Liver and brain homogenates were used to measure the levels of As species by hydride generation-atomic absorption spectroscopy after column chromatographic separation of i-As and its metabolites (MMA, DMA and TMAO) as described by Hughes *et al.* (2003). To measure GR activity, homogenates were thawed at room temperature, 20 μl of supernatant of homogenate centrifuged at $700 \times g$ for 10 min were added to quartz cuvettes containing a fresh solution of 0.44 mM GSSG, 0.30 M EDTA, in 0.1 M phosphate buffer −pH 7.0, and 0.036 M NADPH was added just before the enzymatic determination as the starting reagent. Protein content was determined using the Bradford assay (BioRad). For immunoblotting PAGE separated aliquots of homogenates containing equal concentration of proteins were visualized by Western blotting with specific antibodies against GR (Rodríguez *et al.* 2005). The analysis of immunoblots was performed using a densitometer and analyzed with the Quantity One (BioRad) software.

48.2.3   *Culture of brain slices*

After cervical dislocation, adult mice were plunged into a 70% alcohol solution, decapitated, and the brain was rapidly removed and placed in a vibratome chamber filled with sterile artificial cerebro-spinal fluid (ACSF; 126 mM NaCl; 3 mM KCl; 1 mM $MgCl_2$; 26 mM $NaHCO_3$; 2 mM $CaCl_2$; 10 mM glucose; ascorbic acid 200 μl and thiourea 200 μl; oxygenated at least 1 hour before used). Eight 400 μm longitudinal slices were obtained from each brain and stabilized for 1 hour in ACSF before culturing. Two slices were then placed on humidified membranes (30 mm diameter; 0.4 μm pores; Millicell-CM, Millipore) in 6 well cluster plates containing 1.1 ml of culture media (50% DMEM+Hepes, 25% horse serum, 25% Hank's solution and penicillin and streptomycin (all from GIBCO), 6.5 mg/ml glucose final concentration, (Sigma), pH 7.2). Sodium arsenite was dissolved in the culture medium to obtain a non-cytotoxic (Styblo *et al.* 2002) final concentration of 0.1 μM. Cultures were placed in an incubator at 36°C with a 5% CO_2-enriched atmosphere for 48 and 72 hours. Non-treated slices were used as controls. At the end of the experiment, the tissue and culture medium were assessed for methylated forms of As as described above.

48.3   RESULTS

48.3.1   *Mice brain arsenic methylation*

Total arsenic (t-As) concentration, which is the sum of the amounts of i-As, MMA, DMA and TMAO, in brain and liver homogenates is presented in Table 48.1. Arsenic species were determined in tissue homogenates from five animals per dose. Total As increased in a linear mode with dose in both tissues. There was a clear formation of methylated metabolites in brain: DMA increased linearly with the dose. Mice exposed to 10 mg/kg/day showed the highest concentration of DMA. Significant accumulation of MMA was also observed (Table 48.1).

In liver, i-As increased with dose, which was different from brain; also TMAO was detected only in the liver of the animals treated with the highest dose. The formation of methylated metabolites in mouse brain was confirmed by culturing brain slices in the presence of 0.1 μM of sodium arsenite during 72 h. The formation of methylated metabolites was consistent with our *in vivo* observations.

Table 48.1.   Inorganic As and its methylated metabolites in mouse brain and liver (ng/g of tissue).

Oral Dose mg/kg/day	0	2.5	5	10
In brain				
i-As	113.4	133.2	116.0	67.3^2
MMA	3.9	6.7	13.0	8.0
DMA	12.8	30.3	47.6	185.8
TMAO	<3.0	<3.0	<3.0	<3.0
t-As	130.1	170.2^2	176.6^2	261.1^1
In liver				
i-As	103.1	169.0^1	181.0	277.2^1
MMA	2.3	9.6	15.5	9.3
DMA	2.5	97.2	80.5	43.5^2
TMAO	<3.0	<3.0	<3.0	26.0
t-As	107.89	275.7^1	277.0	356.1^1

Abbreviations: i-As (inorganic arsenic); MMA: monomethyl arsonic acid; DMA: dimethyl arsinic acid; TMAO: trimethyl arsine oxide; Total As (t-As): the sum of i-As + MMA + DMA + TMAO. One-way ANOVA followed by a Tukey-Kramer multiple comparison test: [1]$p < 0.001$, [2]$p < 0.01$, vs. controls (brain) and [1]$p < 0.001$, [2]$p < 0.01$ vs. controls (liver).

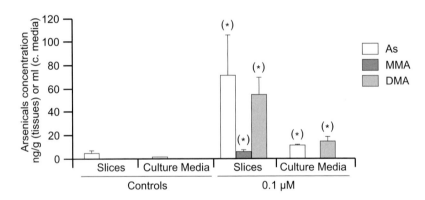

Figure 48.1.   Formation of methylated metabolites in cultured mouse brain slices. Bars represent means ± standard deviation. *Significantly different from control group according to Dunn's post hoc test; $p < 0.05$.

48.3.2   Glutathione reductase activity and immunoblots

Inhibition of the activity of GR was only evident in liver and brain of the animals treated with the highest concentration of sodium arsenite. This inhibition was associated with a significant reduction in the amount of the enzyme observed in Western blots (Figs. 48.2 and 48.3).

48.4   CONCLUSIONS

The results confirm the entrance of As into the brain and demonstrate the formation and, dose-related accumulation of methylated metabolites, mainly DMA, in mouse brain. A similar selective accumulation of DMA has also been observed in mouse lung and bladder; which adequately reflected the species present in urine but not in liver, blood or kidney (Kenyon et al. 2005). The

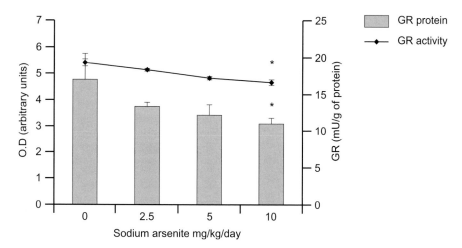

Figure 48.2.   Effect of sodium arsenite on GR in mouse brain. GR protein (bars) was determined by Western blotting and GR activity (line) was determined as described in Section 48.2. Bars and lines represent means ± standard deviation from 5 animals. *Significantly different from control group according to Dunn's post hoc test; p < 0.05. O.D; (optical density).

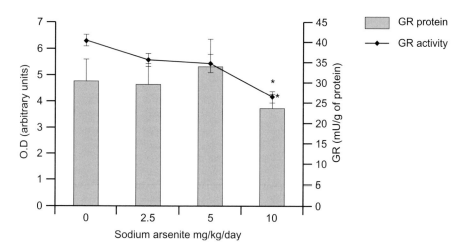

Figure 48.3.   Effect of sodium arsenite on GR in mouse liver. GR protein (bars) was determined by Western blotting and GR activity (line) was determined as described in Section 84.2. Bars and lines represent means ± standard deviation from 5 animals. *Significantly different from control animals according to Dunn's post hoc test; p < 0.05; O.D; (optical density).

methylated As forms in brain tissue may inhibit important antioxidant enzymes such as GR, and glutathione peroxidase among others, therefore GSH levels in brain could be affected by i-As exposure. It is important to remember that GSH has specific functions in the nervous system; it is mainly involved in antioxidant defense and cell detoxification. GSH declines with age, and in neurodegenerative diseases such as Parkinson's, Huntington's, Alzheimer's and cerebellar degeneration (Shulz *et al.* 2000). Also, the formation of MMA and DMA in brain slices shows that brain is capable of As methylation. This process is considered in most mammal species the mayor route for As biotransformation and elimination (Tice *et al.* 1997). Animal models of methionine, choline and folate deficiency reported alterated methylation and elimination of i-As (Marafate and Vather

1986, Vather and Marafate 1987, Tice *et al.* 1997, Spielgestein *et al.* 2003). Similarly, a deficiency in the methylene-tetrahydrofolate reductase (MTHFR) enzyme was correlated to the appearance of mental deterioration, paraparesis, areflexia and bilateral Babinsky signs in a clinical case of a patient exposed to copper acetate arsenite (Brower *et al.* 1992). Thus, the biotransformation of As in brain tissue could alter GSH and SAM homeostasis, with a consequent decrement in the antioxidant system which in turn increases As neurotoxicity.

REFERENCES

Bencko, V. and Symon, K.: Test of environmental exposure to arsenic and hearing changes in exposed children. *Environ. Health Perspect.* 19 (1977), pp. 95–101.

Berbel-García, A., González-Aguirre, J.M., Botia-Paniagua, E., Ortis-Castro, E., López-Zuazo, I., Rodríguez-García, J.L. and Gil-Madre, J.: Acute polyneuropathy and encephalopathy caused by arsenic poisoning. *Rev. Neurol.* 38 (2004), pp. 928–930.

Bolla-Wilson, K. and Bleecker, M.L.: Neuropsychological impairment following inorganic arsenic exposure. *J. Occup. Med.* 29 (1987), pp. 500–503.

Brouwer, O.F., Onkenhout, W., Edelbroek, P.M., de Kom, J.F., de Wolff, F.A. and Peters, A.C.: Increased neurotoxicity of arsenic in methylenetetrahydrofolate reductase deficiency. *Clin. Neurol. Neurosurg.* 94 (1992), pp. 307–10.

Calderon, J., Navarro, M.E., Jimenez-Capdeville, M.E., Santos-Diaz, M.A., Golden, A., Rodriguez-Leyva, I., Borja-Aburto, V. and Diaz-Barriga, F.: Exposure to arsenic and lead and neuropsychological development in Mexican children. *Environ. Res.* 85 (2001), pp. 69–76.

Chattopadhyay, S., Bhaumik, S., Nag Chaudhury, A. and Das Gupta, S.: Arsenic induced changes in growth development and apoptosis in neonatal and adult brain cells in vivo and in tissue culture. *Toxicol. Lett.* 128 (2002a), pp. 73–84.

Chattopadhyay, S., Bhaumik, S., Purkayastha, M., Basu, S., Nag Chaudhuri, A. and Das Gupta, S.: Apoptosis and necrosis in developing brain cells due to arsenic toxicity and protection with antioxidants. *Toxicol. Lett.* 136 (2002b), p. 65.

Chouchane, S. and Snow, E.T.: In vitro effect of arsenicals compounds on glutathione-related enzymes. *Chem. Res. Toxicol.* 14 (2001), pp. 517–522.

Devesa, V., Del Razo, L.M., Adair, B., Drobná, Z., Waters, S.B., Hughes, M.H. Styblo, M. and Thomas, D.J.: Comprehensive analysis of arsenic metabolites by pH-specific hydride generation atomic absorption spectrometry. *J. Anal. Atomic. Spect.* 19 (2004), pp. 1460–1467.

Delgado, J.M., Dufour, L., Grimaldo, J.I., Carrizales, L., Rodríguez, V.M. and Jiménez-Capdeville, M.E.: Effects of arsenite on central monoamines and plasmatic levels of adenocorticotropic hormone (ACTH) in mice. *Toxicol. Lett.* 117 (2000): 61–67.

Franzblau, A. and Lilis, R.: Acute arsenic intoxication from environmental arsenic exposure. *Arch. Environ. Health* 44 (1989), pp. 385–390.

García-Chavez, E., Santamaría, A., Díaz-Barriga, F., Mandeville, P., Juarez, B.I. and Jimenez-Capdeville, M.E.: Arsenite-induced formation of hydroxyl radical in the striatum of awake rats. *Brain Res.* 976 (2003), pp. 82–89.

Hughes, M.F., Kenyon, E.M., Edwards, B.C., Mitchell, C.T., Del Razo, L.M. and Thomas, D.J.: Accumulation and metabolism of arsenic in mice after repeated oral administration of arsenate. *Toxicol. Appl. Pharmacol.* 191 (2003), pp. 202–210.

Ishii, K., Tamaoka, A., Otsuka, F., Iwasaki, N., Shin, K., Matsui, A., Endo, G., Kumagai, Y., Ishii, T., Shoji, S., Ogata, T., Ishizaki, M., Doi, M. and Shimojo, N.: Diphenyl arsinic acid poisoning from chemical weapons in Kamitsu, Japan. *Ann. Neurol.* 56 (2004), pp. 741–745.

Itoh, T., Zhang, Y.F., Murai, S., Saito, H., Nagahama, H., Miyate, H., Saito, Y. and Abe, E.: The effect of arsenic trioxide on brain monoamine metabolism and locomotor activity of mice. *Toxicol. Lett.* 54 (1990), pp. 345–53.

Kannan, G.M., Tripathi, N., Dube, S.N., Gupta, M. and Flora, S.J.: Toxic effects of arsenic (III) on some hematopoietic and central nervous system variables in rats and guinea pigs. *J. Toxicol. Clin. Toxicol.* 39 (2001), pp. 675–682.

Kenyon, E.M., Del Razo, L.M. and Hughes M.F.: Tissue distribution and urinary excretion of inorganic arsenic and its methylated metabolites in mice following acute oral administration of arsenate. *Toxicol. Sci.* 85 (2005), pp. 468–475.

Marafante, E. and Vahter, M.: The effect of dietary and chemically induced methylation deficiency on the metabolism of arsenate in the rabbit. *Acta Pharmacol. Toxicol.* (Copenh.) 59 Suppl. 7 (1986), pp. 35–38.

Mejia, J.J., Diaz-Barriga, F., Calderon, J., Rios, C. and Jimenez-Capdeville, M.E.: Effects of lead-arsenic combined exposure on central monoaminergic systems. *Neurotoxicol. Teratol.* 19 (1997), pp. 489–497.

Morton, W.E. and Caron, G.A.: Encephalopathy: an uncommon manifestation of workplace arsenic poisoning? *Am. J. Ind. Med.* 15 (1989), pp. 1–5.

Nagaraja, T.N. and Desiraju, T.: Regional alterations in the levels of brain biogenic amines, glutamate, GABA, and GAD activity due to chronic consumption of inorganic arsenic in developing and adult rats. *Bull. Environ. Contam. Toxicol.* 50 (1993), pp. 100–107.

Nagaraja, T.N. and Desiraju, T.: Effects on operant learning and brain acetylcholine esterase activity in rats following chronic inorganic arsenic intake. *Hum. Exp. Toxicol.* 13 (1994), pp. 353–356.

Namgung, U. and Xia, Z.: Arsenite-induced apoptosis in cortical neurons is mediated by c-Jun N-terminal protein kinase 3 and p38 mitogen-activated protein kinase. *J. Neurosci.* 20 (2000), pp. 6442–6551.

Namgung, U. and Xia, Z.: Arsenic induces apoptosis in rat cerebellar neurons via activation of JNK3 and p38 MAP kinases. *Toxicol. Appl. Pharmacol.* 174 (2001), pp. 130–138.

Piao, F., Ma, N., Hiraku, Y., Murata, M., Oikawa, S., Cheng, F., Zhong, L., Yamauchi, T., Kawanishi, S. and Yokoyama, K.: Oxidative DNA damage in relation to neurotoxicity in the brain of mice exposed to arsenic at environmentally relevant levels. *J. Occup. Health* 47 (2005), pp. 445–449.

Pryor, G.T., Uyeno, E.T., Tilson, H.A. and Mitchell, C.L.: Assessment of chemicals using a battery of neurobehavioral tests: a comparative study. *Neurobehav. Toxicol. Teratol.* 5 (1983), pp. 91–117.

Rodriguez, V.M., Carrizales, L., Jimenez-Capdeville, M.E., Dufour, L. and Giordano, M.: The effects of sodium arsenite exposure on behavioral parameters in the rat. *Brain Res. Bull.* 55 (2001), pp. 301–308.

Rodriguez, V.M., Carrizales, L., Mendoza, M.S., Fajardo, O.R. and Giordano, M.: Effects of sodium arsenite exposure on development and behavior in the rat. *Neurotoxicol. Teratol.* 24 (2002), pp. 743–750.

Rodríguez, V.M., Jiménez-Capdeville, M.E. and Giordano M.: The effects of arsenic exposure on the nervous system. *Toxicol. Lett.* 145 (2003), pp. 1–18.

Rodríguez, V.M., Del Razo, L.M., Limon-Pacheco, J.H., Giordano, M., Sanchez-Pena, L.C., Uribe-Querol, E., Gutierrez-Ospina, G. and Gonsebatt, M.E.: Glutathione reductase inhibition and methylated arsenic distribution in CD1 mice brain and liver. *Toxicol. Sci.* 84 (2005), pp. 157–166.

Schulz, J.B., Lindenau, J., Seyfried, J. and Dichgans, J.: Glutathione, oxidative stress and neurodegeneration. *Eur. J. Biochem.* 267 (2000), pp. 4904–4911.

Spiegelstein, O., Lu, X., Le, X.C., Troen, A., Selhub, J., Melnik, S., James, J.S. and Finnell, R.H.: Effects of dietary folate intake and folate binding protein-1 (Folbp1) on urinary speciation of sodium arsenate in mice. *Toxicol. Lett.* 145 (2003), pp. 167–174.

Styblo, M. and Thomas, D.J.: In vitro inhibition of glutathione reductase by arsenotriglutathione. *Biochem. Pharmacol.* 49 (1995), pp. 971–974.

Styblo, M., Serves, S.V., Cullen, W.R. and Thomas, D.J.: Comparative inhibition of yeast glutathione reductase by arsenicals and arsenothiols. *Chem. Res. Toxicol.* 10 (1997), pp. 27–33.

Styblo, M., Drobna, Z., Jaspers, I., Lin, S. and Thomas, D.J.: The role of biomethylation in toxicity and carcinogenicity of Arsenic: a research update. *Environ. Health Persp.* 100 (2002), pp. 767–771.

Thomas, D.J., Waters, S.B. and Styblo, M.: Elucidating the pathway for arsenic methylation. *Toxicol. Appl. Pharmacol.* 198 (2004), pp. 319–326.

Thomas, D.J., Styblo, M. and Lin, S.: The cellular metabolism and systemic toxicity of arsenic, *Toxicol. Appl. Pharmacol.* 176 (2001), pp. 127–144.

Tice, R.R., Yager, J.W., Andrews, P. and Crecelius, E.: Effect of hepatic methyl donor status on urinary excretion and DNA damage in B6C3F1 mice treated with sodium arsenite. *Mutation Res.* 386 (1997), pp. 315–334.

Tripathi, N., Kannan, G.M., Pant, B.P., Jaiswal, D.K., Malhotra, P.R. and Flora, S.J.: Arsenic-induced changes in certain neurotransmitter levels and their recoveries following chelation in rat whole brain. *Toxicol. Lett.* 92 (1997), pp. 201–208.

Tsai, S.Y., Chou, H.Y., The, H.W., Chen, C.M. and Chen, C.J.: The effects of chronic arsenic exposure from drinking water on the neurobehavioral development in adolescence. *Neurotoxicology* 24 (2003), pp. 747–753.

Valkonen, S., Savolainen, H. and Jarvisalo, J.: Arsenic distribution and neurochemical effects in peroral sodium arsenite exposure of rats. *Bull. Environ. Contam. Toxicol.* 30 (1983), pp. 303–308.

Vather, M. and Maranfate E.: Effects of low dietary intake o methionine, choline or proteins on the biotransformation of arsenite in the rabbit. *Toxicol. Lett.* 37 (1987), pp. 41–46.

Wasserman, G.A., Liu, X., Parvez, F. Ahsan, H., Factor-Litvak, P. van Geen, A., Slavkovich, V., LoIacono, N.J., Cheng, Z., Hussain, I., Momotaj, H. and Graziano, J.H.: Water arsenic exposure and children's intellectual function in Araihazar, Bangladesh. *Envrion. Health Persp.* 112 (2004), pp. 1329–1333.

Waters, S.B., Devesa, V., Del Razo, L.M., Styblo, M. and Thomas, D.J.: Endogenous reductants support the catalytic function of recombinant rat cyt19, an arsenic methyltransferase *Chem. Res. Toxicol.* 17 (2004), pp. 404–409.

Yip, S.-F., Yeung Y.-M. and Tsui E.-Y.-K.: Severe neurotoxicity following arsenic therapy for acute promyelocytic leukemia: potentiation by thiamine deficiency. *Blood* 99 (2002), pp. 3481–3482.

Yoshida, K., Inoue, Y., Kuroda, K., Chen, H., Wanibuchi, H., Fukushima, S. and Endo, G.: Urinary excretion of arsenic metabolites after long-term oral administration of various arsenic compounds to rats. *J. Toxicol. Environ. Health A.* 54 (1998), pp. 179–192.

CHAPTER 49

Mouse liver cytokeratin 18 (CK18) modulation by sodium arsenite

P. Ramírez
Facultad de Estudios Superiores Cuautitlán, Universidad Nacional Autónoma de México (UNAM), Mexico City, Mexico

L.M. Del Razo
Sección Externa de Toxicología, Centro de Investigación y de Estudios Avanzados del Instituto Politécnico Nacional (CINVESTAV-IPN), Mexico City, Mexico

M.E. Gonsebatt
Instituto de Investigaciones Biomédicas, Universidad Nacional Autónoma de México (UNAM), Mexico City, Mexico

ABSTRACT: Cytokeratins (CK) are important intermediate filaments that participate in the cytoskeleton. Besides providing mechanical cell stability to epithelial cells, they also exhibit protective roles. Abnormal organization and function of liver CK are effects related to liver pathologies often seen in chronic arsenic poisoning such as non-alcoholic steatohepatitis, cirhosis and hepatocellular carcinoma. Sodium arsenite induces changes in CK18 expression in mouse hepatocytes as well as in human liver cell lines. This induction seems to be an oxidative stress response since the addition of antioxidants diminished the levels of mRNA CK18 in arsenite treated cells.

49.1 INTRODUCTION

49.1.1 *Arsenic effects in liver*

The liver is the organ that maintains the metabolic homeostasis of the body. It is also the first organ to encounter the ingested toxicants. Mayor functions of the liver can be altered by chronic exposure to toxicants such as alcohol or metals which could lead to diabetes and hypertension (Tseng 2005, Izquierdo-Vega *et al.* 2006) or more severely to cirrhosis and cancer (Patrick 2003). The liver is considered a major site of arsenic (As) methylation (Vahter 2002) where As is methylated through metabolic conversions that consume antioxidants such as glutathione or thioredoxin. An increased number of humans suffer the effects of chronic As exposure resulting from environmental release of As in water, crops or food (Centeno *et al.* 2002). The effects of As on the liver include inhibition of important enzymes (Gonsebatt *et al.* this volume) and in the case of chronic exposure the development of fibrosis, hepatocellular carcinoma, angiosarcoma and hepatoportal sclerosis (Chen and Wang 1990).

   CK constitute important intermediate filament (IF) proteins that are typically expressed in epithelial cells and serve as cell-type specific markers. In simple type epithelia such as liver the two major IF proteins are cytokeratin polypeptides 8 and 18 (CK8/18) (Moll *et al.* 1982, 1993, Calnek and Quaroni 1993). The accumulation of CK in response to liver toxicants suggests that these proteins may behave as stress proteins, similar to heat stress proteins (Liao *et al.* 1995) and could be part of the early changes associated with liver injury. Moreover, CK 8 and 18 are the main constituents of Mallory or Mallory-Denk bodies which are common characteristic morphological features of hepatocellular neoplasms, drug-induced liver diseases, idiopathic copper toxicosis and other liver diseases (Zatloukal *et al.* 2007).

49.1.2   *Arsenite induces oxidative stress and CK18 expression in a human hepatic cell line*

It has been shown that As induce oxidative stress through the production of superoxide and hydrogen peroxide radicals (Hughes and Kitchin 2006). Accumulation of reactive oxidative species (ROS) can alter protein conformation causing the loss of their biological function. They also act as signaling molecules to activate several stress-associated pathways. The addition of antioxidants prevented the formation of free radicals protecting cultured cells from the oxidative damage of As treatments (Hughes and Kitchin 2006). Sodium arsenite disrupted cytoskeleton proteins such as microtubule polimerization during cell division, generating cells with abnormal chromosome number (Ramirez *et al.* 1997). Also, sodium arsenite induced the accumulation of CK18 filaments in WRL 68 human hepatic cell line. A disruption of the filament organization was also observed in some of the cells treated with micromolar concentrations of sodium arsenite (Ramírez *et al.* 2000). To investigate if the expression of CK18 was an early event in As liver toxicity, we treated mice with different concentrations of sodium arsenite and observed an increase in the levels of CK18 mRNA and protein expression in liver homogenates. Also, to investigate the participation of oxidative stress in the induction of the IF we added a known antioxidant N-acetyl-L-cysteine (NAC) to liver organotypic cultures. The addition of the antioxidant diminished the accumulation of the mRNA and protein.

49.2   MATERIALS AND METHODS

49.2.1   *Exposure protocol*

Inbred male BALB/c mice (5–6 weeks old, weighing 22–25 g) were habituated during one week to the vivarium conditions. Subsequently, they received 0, 2.5 and 5 mg of sodium arsenite per kg of body weight per day by intragastric route for a total of 2 days. Sodium arsenite solutions were prepared daily and dissolved in deionized water. The control group received deionized water as treatment. Animals had free access to water and food before and during arsenite exposure. Body weight was recorded daily during exposure.

One day after the last treatment, mice were sacrificed by cervical dislocation, followed by decapitation. Livers were extracted and washed in ice-cold isotonic saline solution to remove debris and blood.

49.2.2   *Organotypic cultures*

After cervical dislocation, the liver was quickly removed and placed into ice-cold phosphate buffer saline pH 7.2. Liver slices were precisely cut to obtain slices of approximately 5000 μm of diameter and 250–500 μm of thickness weighted and placed in culture plaques (2 slices per well) containing DMEM culture media supplemented with SFB 8% (Invitrogen, Carlsbad, CA) 100 μg/10 ml ampicillin and 1% streptomicin. Organotypic cultures were stabilized for 2 h in an incubator at 37°C with a 5% CO_2-enriched atmosphere for 2 h to establish culture conditions. The liver slices were exposed to 0.01, 1 and 10 μM sodium arsenite during 3 h. Also, after preincubation with arsenite for 1 h, slices were concomitant treated with the corresponding arsenite concentration and 2.5 mM of NAC for 2 h. At the end of the experiments, tissues were washed in ice-cold PBS pH 7.4 containing protease inhibitor cocktail and homogenized in the same buffer. The viability of liver slices was ascertained at the end of the 3 h incubation period determining intracellular K^+ levels using a method described by Azri *et al.* (1990) with some modifications. Briefly, slices were washed in ice-cold PBS and homogenized, an aliquot of tissue homogenate was added to 0.02 ml of concentrated perchloric acid. The mixture was gently shaken and centrifuged at 12000 rpm during 10 min at 4°C. The supernatant fractions were analyzed by flame photometer for K^+ levels and reported per gram of tissue.

49.2.3 *Immunoblots and reverse transcriptase-PCR*

Protein content was determined using the Bradford assay (BioRad). For immunoblotting aliquots of homogenates containing equal concentration of proteins were separated by PAGE and proteins were visualized by Western blotting with specific antibodies against CK18 (Santa Cruz Biotechnology, Santa Cruz, CA).

The CK18 mRNA levels were quantified by RT-PCR. RNA samples were transcribed into first strand cDNA using RT-MLV retrotranscriptase (Invitrogen). The cDNA was amplified by PCR with primers designed based in previous reports (Zhong *et al.* 2004): Sense primer was amplified by PCR with the following primers: 5′-GACGCTGAGACCACACT and 5′-TCCATCTGTGCCTTGTAT. Actin was used as protein load control during immunoblot analyses. Image analysis was performed using a Kodak Gel Logic 100 Imaging System.

49.3 RESULTS AND DISCUSSION

49.3.1 *Expression and synthesis of CK18*

The quantification of mRNA and protein by RT-PCR and immunoblotting respectively, indicated that mice given orally 2.5 and 5 mg/kg of sodium arsenite for 2 days had significantly higher levels of both CK18 mRNA in liver than untreated animals (Fig. 49.1) When organotypic liver cultures were treated with micromolar concentrations of sodium arsenite during 3 h, an induction in both mRNA and protein was also observed (Figs. 49.1 and 49.2), indicating that CK18 synthesis is an early response to As toxicity.

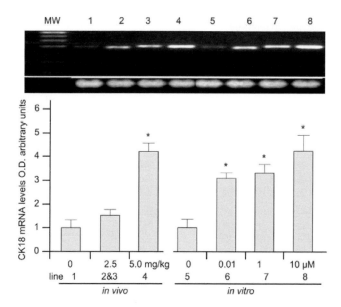

Figure 49.1.   Sodium arsenite induce mRNA CK18 levels in mouse liver *in vivo* and *in vitro*. Bars above the left line show the dose related increased expression of CK18 mRNA. Bars above the right line show the effect on CK mRNA levels in organotypic cultures. Densitometric analysis was performed using actin mRNA as loading control. CK18 mRNA levels are expressed as a fraction of controls. Bars represent mean ± SD of triplicate cultures. A (*) indicates significance (p < 0.05) compared to control cultures according to Dunn's post hoc test.

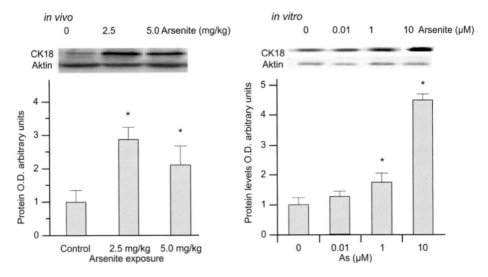

Figure 49.2.   Sodium arsenite induce protein synthesis in mouse liver *in vivo* and organotypic culture. Densitometric analysis was performed using actin as loading control. Protein levels are expressed as a fraction of controls. Bars represent mean ± SD of triplicate cultures. A (*) indicates significance ($p < 0.05$) compared to control cultures according to Dunn's post hoc test.

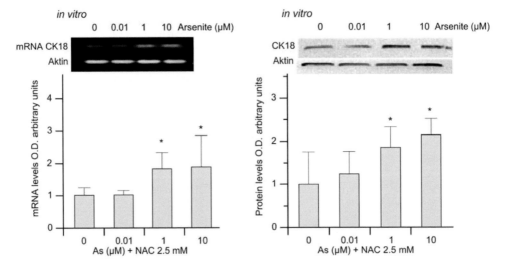

Figure 49.3.   The addition of NAC to organotypic cultures diminish the levels of CK 18 mRNA induced by sodium arsenite. Densitometric analysis was performed using actin as loading control. CK18 mRNA levels and protein levels are expressed as a fraction of controls. Bars represent mean ± SD of triplicate cultures. A (*) indicates significance ($p < 0.05$) compared to control cultures according to Dunn's post hoc test.

49.3.2   NAC diminished CK18 induction

The presence of NAC an antioxidant agent diminished the levels of mRNA in cultured liver slices, suggesting that the induction of CK18 transcription and protein synthesis is probably due to a ROS sensitive pathway, thus acting as a stress-induced protein (Figure 49.3).

The arsenicals compounds have high affinity for thiol groups such as those present in NAC. It is therefore possible that some arsenite may be trapped by NAC so that its delivery to liver cells may be impaired. However, the cysteine analog is readily taken up by cells and can directly scavenge ROS (Kelly 1998, De Flora *et al.* 2001). Moreover, a number of studies have demonstrated that NAC is able to reduce or counteract oxidative damage. Electron spin resonance and confocal microscope studies have showed that As(III) stimulated ROS generation and Hsp70 expression in human pulmonary epithelial and MDA231 cells, and these effects were inhibited by NAC (Han *et al.* 2005, Kim *et al.* 2005). In addition, NAC was able to inhibit the cytotoxicity of the i-As metabolites, monomethylarsonous acid [MMA(III)], dimethylarsinic acid [DMA(V)], dimethylarsinous acid [DMA(III)], and trimethylarsine oxide (TMAO) in rat bladder cells (Wei *et al.* 2005).

CK participates in the network of IF performing a variety of important cellular functions including cell division, motility, maintenance of cellular mechanical integrity, stress responses, and vesicle transport and playing an essential "guardian" role in the liver that is unmasked after exposure to environmental stresses (Omary *et al.* 2002) such as As exposure. Our observations demonstrate that the overexpression of CK18 is an early response to the oxidative stress induce by As in liver. Accumulation of CK in cytoplasmic inclusions is observed in the liver from individuals chronically exposed to As (Centeno *et al.* 2002).

CK synthesis in liver cells is tightly correlated with differentiation programs and with several cellular processes such as apoptosis and cell proliferation and seems to be a substrate of a variety of protein kinases involved in mitosis, apoptosis, and stress. Thus during chronic As exposure the altered CK expression could modify differentiation patterns in this tissue compromising the cellular physiology by impairing the protective role of CK and inducing hepatic susceptibility to further toxic injury (Ku *et al.* 2003, Omary *et al.* 2002).

REFERENCES

Azri, S., Gandolfi, A.J. and Brendel, K.: Carbon tetrachloride toxicity in precision-cut tissue slices. *In Vitro Toxicol.* 3 (1990), pp. 127–138.

Calnek, D. and Quaroni, A.: Differential localization by in situ hybridization of distinct keratin mRNA species during intestinal epithelial cell development and differentiation. *Differentiation* 53 (1993), pp. 95–104.

Centeno, J.A., Mullick, F.G., Martinez, L., Page, N.P., Gibb, H., Longfellow, D., Thompson, P. and Ladich, E.R.: Pathology related to chronic arsenic exposure. *Environ. Health Perspect.* 110 (Supl 5) (2002), pp. 883–886.

Chen, C.J. and Wang, C.J.: Ecological correlation between arsenic level in well water and age-adjusted mortality from malignant neoplasms. *Cancer Res.* 50 (1990), pp. 5470–5474.

De Flora, S., Izzotti, A., D'Agostini, F. and Balansky, R.M.: Mechanisms of N-acetylcysteine in the prevention of DNA damage and cancer, with special reference to smoking-related end-points. *Carcinogenesis* 22 (2001), pp. 999–1013.

Han, S.G., Castranova, V. and Vallyathan, V.: Heat shock protein 70 as an indicator of early lung injury caused by exposure to arsenic. *Mol. Cell Biochem.* 277 (2005), pp. 153–164.

Hughes, M.F. and Kitchin, K.T.: Arsenic, oxidative stress and carcinogenesis. In: K.K. Singh (ed): *Oxidative stress, disease and cancer.* Imperial College Press, London, UK, 2006, pp. 825–850.

Izquierdo-Vega, J.A., Soto, C.A., Sanchez Pena, L.C., De Vizcaya-Ruiz, A. and Del Razo, L.M.: Diabetogenic effects and pancreatic oxidative damage in rats subchronically exposed to arsenite. *Toxicol. Lett.* 160 (2006), pp. 135–142.

Kelly, G.S.: Clinical applications of N-acetylcysteine. *Altern. Med. Rev.* 3 (1998), pp. 114–127.

Kim, Y.H., Park, E.J., Han, S.T., Park, J.W. and Kwon, T.K.: Arsenic trioxide induces Hsp70 expression via reactive oxygen species and JNK pathway in MDA231 cells. *Life Sci.* 77 (2005), pp. 2783–2793.

Ku, N.O., Darling, J.M., Krams, S.M., Esquivel, C.O., Keeffe, E.B., Sibley, R.K., Lee, Y.M., Wright, T.L. and Omary, M.B.: Keratin 8 and 18 mutations are risk factors for developing liver disease of multiple etiologies. *Proc. Natl. Acad. Sci. U.S.A.* 100 (2003), pp. 6063–6068.

Liao, J., Lowthert, L.A., Ku, N.O., Fernandez, R. and Omary, M.B.: Dynamics of human keratin 18 phosphorylation: polarized distribution of phosphorylated keratins in simple epithelial tissues. *J. Cell Biol.* 131 (1995), pp. 1291–1301.

Moll, R.: Cytokeratins as markers of differentiation. Expression profiles in epithelia and epithelial tumors. *Veröff. Pathol.* 142 (1993), pp. 1–197.

Moll, R., Franke, W.W., Schiller, D.L., Geiger, B. and Krepler, R.: The catalog of human cytokeratins: patterns of expression in normal epithelia, tumors and cultured cells. *Cell* 31 (1982), pp. 11–24.

National Academy of Sciences: Arsenic in drinking water. National Academies Press, Washington, DC, 2001, pp. 75–202.

Omary, M.B., Ku, N.O. and Toivola, D.M.: Keratins: guardians of the liver. *Hepatology* 35 (2002), pp. 251–257.

Patrick, L.: Toxic metals and antioxidants: Part II: The role of antioxidants in arsenic and cadmium toxicity. *Altern. Med. Rev.* 8 (2003), pp. 106–128.

Ramirez, P., Eastmond, D.A., Laclette, J.P. and Ostrosky-Wegman, P.: Disruption of microtubule assembly and spindle formation as a mechanism for the induction of aneuploid cells by sodium arsenite. *Mut. Res.* 386 (1997), pp. 291–298.

Ramírez, P., Del Razo, L.M., Gutierrez-Ruíz, M.C. and Gonsebatt, M.E.: Arsenite induces DNA-protein crosslinks and cytokeratin expression in the WRL-68 human hepatic cell line. *Carcinogenesis* 21 (2000), pp. 701–706.

Rhodes, K. and Oshima, R.G.: A regulatory element of the human keratin 18 gene with AP-1-dependent promoter activity. *J. Biol. Chem.* 273 (1998), pp. 26,534–26,542.

Tseng, C.H.: Blackfoot disease and arsenic: a never-ending history. *J. Environ. Sci. Health C Environ. Carcinogen Ecotoxicol. Rev.* 23 (2005), pp. 55–74.

Vahter, M.: Mechanisms of arsenic biotransformation. *Toxicology* 181/182 (2002), pp. 211–217.

Wei, M., Arnold, L., Cano, M. and Cohen, S.M.: Effects of co-administration of antioxidants and arsenicals on the rat urinary bladder epithelium. *Toxicol. Sci.* 83 (2005), pp. 237–245.

Zatloukal, K., French, S.W., Stumptner, C., Strnad, P., Harada, M., Toivola, D.M., Moniqu, C. and Omary, M.B.: From Mallory-Denk bodies: What, how and why? *Exp. Cell Res.* 313 (2007), pp. 2033–2049.

Zhong, B., Zhou, Q., Toivola, D.M., Tao, G.Z., Resurrección, E.Z. and Omary, M.B.: Organ-specific stress induces a mouse pancreatic keratin overexpression in association with NF-Kappa B activation. *J. Cell Sci.* 117 (2004), pp. 1709–1719.

CHAPTER 50

Effects of selenium deficiency on diabetogenic action of arsenite in rats

J.A. Izquierdo-Vega, L.C. Sánchez-Peña & L.M. Del Razo
Sección Externa de Toxicología, Centro de Investigación y de Estudios Avanzados del Instituto Politécnico Nacional (CINVESTAV-IPN), Mexico City, Mexico

C. Soto
Depto. de Sistemas Biológicos, Universidad Autónoma Metropolitana, Xochimilco (UAM-X), Mexico City, Mexico

ABSTRACT: Selenium (Se), an essential element for animals and humans, plays an important role in protection against oxidative damage. In addition, because of its insulin-like properties, Se may affect glucose homeostasis. We have previously shown that a subchronic exposure to arsenite induces insulin resistance in rats. Our aim for this study was to determine whether Se deficiency can further exacerbate the insulin resistance caused by subchronic arsenite exposure. Male Wistar rats were fed a Se deficient diet (0.02 mg Se/kg) and dosed with sodium arsenite (1.7 mg/kg/12 h) by gavage for 90 consecutive days. Data collected at the end of the treatment suggest that consumption of the Se-deficient diet resulted in oxidative damage in the pancreas, but had no effect on insulin resistance induced by arsenite exposure. Thus, in spite of the induction of oxidative damage to the pancreas, Se deficiency does not increase hyperinsulinaemia associated with subchronic exposure to arsenite.

50.1 INTRODUCTION

Arsenic (As) is a ubiquitous element found in several forms in food and environmental media such as soil, air, and water. The predominant form of As in drinking water is inorganic As (i-As), which is both highly toxic and readily bioavailable. i-As is a recognized human carcinogen (NRC 2001). Chronic ingestion of drinking water contaminated with i-As is, therefore, considered a major risk for human health. It has been estimated that 200 million people worldwide are at risk of adverse health effects associated with consumption of drinking water with high concentrations of As (NRC 2001). In humans, chronic ingestion of i-As (>500 µg/day As) has been associated with cardiovascular, neurological, hepatic, and renal disorders, diabetes mellitus, as well as cancer of the skin, bladder, lung, liver, and prostate (ATSDR 2000). The association between i-As exposure and the risk of non-insulin-dependent diabetes mellitus is a relatively new finding (Tseng 2004). A potential relationship between mortality from diabetes and chronic exposure to As in drinking water has been described by Tsai and associates (1999). Associations between As exposure and diabetes mellitus have also been reported in studies of copper smelters (Rahman and Axelson 1995) and art glass workers (Rahman *et al.* 1995). We have previously shown that a subchronic exposure to arsenite produces insulin resistance in rats (Izquierdo-Vega *et al.* 2006).

Generation of reactive oxygen species (ROS) is the major mechanism by which i-As exerts its toxicity (Del Razo *et al.* 2001). Cytotoxic effects associated with accumulation of ROS include oxidative damage to membrane phospholipids, leading to changes in permeability and loss of membrane integrity. ROS are also responsible for conformational changes in proteins and loss of their biological functions.

On the other hand, selenium (Se), an essential trace element for animals and humans, plays an important role in protection against oxidative damage. Se is found in several enzymes, including glutathione peroxidase (GPx), thioredoxin reductase (TR), and deiodinases. Diet is the main source of Se for the human body. Both the chemical form of Se and the amount of Se in the diet are important for maintenance of Se status and functions. Selenate [Se(VI)], but not selenite [Se(IV)], has been shown to relieve insulin resistance in type II diabetic animals and has, therefore, been considered the insulinomimetic form of Se (Muller *et al.* 2003). Se deficiency is one of the factors linked to diabetes, malnutrition, low body weight, muscular dystrophy, rheumatoid arthritis, and diseases associated with oxidative stress (Beckett and Arthur 2005).

Dietary Se has been shown to influence As excretion in animal models. Gregus *et al.* (1998) noted that Se facilitates excretion of i-As metabolites in rats. Toxicity of high doses of inorganic Se (i-Se) is reduced by exposure to i-As (Levander *et al.* 1977). On the other hand, genotoxicity associated with i-As exposure is reduced by co-exposure to Se (Biswas *et al.* 1999). Metabolic and toxicological interactions between Se and As have been reported in many experimental studies (Kenyon *et al.* 1999). Addition of i-Se to primary cultures of rat hepatocytes increases the cellular retention of i-As and decreases formation of i-As metabolites, monomethyl and dimethyl arsenicals (MA and DMA, respectively) (Styblo and Thomas 2001). In addition, Se compounds have been shown to modulate activity of AS3MT, the enzyme that catalyzes i-As methylation reaction (Walton *et al.* 2003).

Co-exposure to As and Se may occur due to the presence of i-As in the environment and Se in daily diet. Therefore, it is possible that diabetogenic effects associated with chronic exposure to i-As are affected by Se intake from food. Assessment of the relationship between Se intake, regulation of glucose metabolism, and pancreatic oxidative damage in rats exposed subchronically to i-As may help to characterize the role of Se in endocrine effects of chronic i-As exposures. The aim of this study was to compare glucose tolerance and oxidative damage in the pancreas of Se-deficient (Def-Se) and Se sufficient (Suf-Se) rats exposed to i-As.

50.2   METHODS

50.2.1   *Procedures*

The experimental design for this study is outlined in Table 50.1. Male Wistar rats weighing 120–130 g were divided into the following four groups (10 animals per group): A), *Suf-Se-Control* rats fed a Se sufficient diet (0.2 mg Se/kg) and gavaged with deionized water (DIW, 2.5 ml/kg) every 12 h for 90 consecutive days; B) *Suf-Se-i-As* rats fed the Se-sufficient diet and gavaged with sodium arsenite (1.7 mg arsenite/kg) every 12 h for 90 days; C) *Def-Se-Control* rats fed a Se deficient diet (0.02 mg Se/kg) and gavaged with DIW (2.5 ml/kg) every 12 h for 90 days, and D) *Def-Se-i-As* rats fed the Se deficient diet and gavaged with sodium arsenite (1.7 mg/kg) every 12 h for 90 days. All rats were kept in animal facilities with a 12/12 h light/dark cycle, temperature of

Table 50.1.   The treatment groups (N = 10) and experimental design.

Study day	Treatment groups			
1–22	Weaning time			
	Suf-Se diet (0.2 mg Se/kg)		Def-Se diet (0.02 mg Se/kg)	
23–57	Rats were fed with corresponding Se diet			
	A	B	C	D
	2.5 ml DIW/kg/12 h	1.7 mg i-As /kg/12 h	2.5 ml DIW/kg/12 h	1.7 mg i-As/kg/12 h
58–147	Rats were fed with corresponding Se diet and dosed with DIW or i-As every 12 h. (7:00 A.M. and 7:00 P.M.)			
148	Sacrifice; blood and tissue collection			

$22 \pm 1°C$, humidity of $50 \pm 5\%$, and free access to food and DIW. The pelleted diets were prepared by Bio-Serv Co. (Frenchtown, NJ) from a basal diet consisting of 14.6% protein, 5.1% fat, 2.2% fibber, 5.1% ashes, 67.2% carbohydrate, and 5.8% moisture. Basic inspection of animals was performed on a daily basis. Food and water consumption and body weights were monitored daily. After 90 days of the treatment, all rats were euthanized by cardiac puncture under sodium pentobarbital anesthesia (60 mg/kg, ip). Rats were fasted 6 h before sacrifice. Blood was collected for analyses of glucose, insulin and glucagon. Pancreas was removed, rinsed in a saline solution, and homogenized in a phosphate buffer (pH 7.4). The homogenates were used for lipid peroxidation and thioredoxin reductase (TrxR) analyses.

50.2.2 Glucose and insulin analyses in blood

The glucose hexokinase kit (Randox®, San Francisco, CA) and a Vitalab Eclipse automated clinical analyzer (Merck) recording at 340 nm were used to determine glucose levels in serum isolated from fasting blood. Insulin levels in serum were determined by Enzyme Linked Immunosorbent Assay (ELISA), using the Ultra-Sensitive rat insulin kit obtained from ALPCO Diagnostics. (Windham, NH).

Insulin Resistance: Fasting insulin, the fasting insulin-to-glucose ratio (Legro *et al.* 1998), and a homeostasis model assessment (HOMA-IR) were used to evaluate insulin resistance. HOMA-IR was calculated using the formula described by Thirunavukkarasu *et al.* (2004):

$$\text{Fasting serum insulin (mU/l)}/(22.5\ e^{-\ln\text{ Fasting plasma glucose (mmol/l)}})$$

Low HOMA-IR values indicate high insulin sensitivity, whereas high HOMA-IR values indicate low insulin sensitivity or insulin resistance (Bonora *et al.* 2002).

50.2.3 Glucagon analysis

Serum glucagon levels were measured using commercially available radioimmunoassay kits (ICN Pharmaceuticals, Diagnostic Division; CostaMesa, CA).

50.2.4 Pancreatic TrxR activity and lipid peroxidation

TrxR activity was measured as the rate of the NADPH-dependent reduction of DTNB to 2-nitro-5-thiobenzoic acid (Smith and Levander, 2002). The following assay mixture (200 µl) was prepared freshly on ice: 1 M of sodium phosphate buffer (pH 7), 0.2 mg/ml NADPH, 0.2 mg/ml bovine serum albumin, 5 mM DTNB, 0.01M EDTA, 1% ethanol; 5 µl of pancreatic homogenate. The assay was carried out at 37°C and absorbance was monitored at 412 nm.

Pancreatic malondialdehyde (MDA), a product of lipid peroxidation, was determined according to the method of thiobarbituric acid reactive substances (TBARS) (Buege and Aust 1978). Tissue homogenate (0.5 ml) was added to a test tube containing 1 ml of 0.5% thiobarbituric acid, 5 µl deferoxamine mesylate, and 5 µl of 3.75% butylated hydroxytoluene. The reaction mixture was heated at 100°C for 20 min. After cooling, the mixture was centrifuged at 4000 rpm for 15 min and absorbance of the supernatants was measured at 532 nm by a spectrophotometer. Results were expressed as the concentration of TBARS: nmol TBARS/mg tissue.

50.2.5 Statistical analysis

Results are expressed as means $\pm$ standard error (SEM). Differences between the experimental groups were evaluated by paired or unpaired two-tailed Student's *t*-test. Differences with p values <0.05 are considered significant. All analyses were performed using the Stata 8.0 statistical software.

50.3   RESULTS

50.3.1   *Insulin resistance*

Results of the fasting serum glucose and insulin analyses are shown in Table 50.2. The glucose and insulin levels in serum from the arsenite-treated rats fed Se sufficient diet (Suf-Se-i-As) were 1.8 and 2.6 fold higher, respectively, as compared to the Suf-Se-Control group. Se deficient control (Def-Se-Control) rats had serum glucose level somewhat lower than Suf-Se-Control rats. However, this difference was not statistically significant. There was no significant difference between glucose levels found in Def-Se-i-As and Def-Se-Control rats. However, insulin levels in serum of Suf-Se-i-As and Def-Se-i-As rats were 2.6- and 5.7-fold higher than insulin levels in serum of the respective controls. Treatment with i-As resulted in a 2.2- and 2.8-fold decrease in the fasting glucose/insulin ratio in rats fed Se sufficient or Se deficient diet, respectively. HOMA-IR values for rats exposed to i-As were significantly higher than for control rats: 3.9-fold for Se sufficient and 7.6-fold for Se deficient groups. The higher HOMA-IR values indicate low insulin sensitivity or insulin resistance due to i-As treatment. However, Se deficiency did not exacerbate insulin resistance in rats treated with i-As as indicated by similar HOMA-IR values for Suf-Se-i-As and Def-Se-i-As rats: 13.17 ± 4.79 and 14.10 ± 5.04, respectively.

50.3.2   *Glucagon in serum*

In this study, we also evaluated the role of glucagon in Se deficient rats subchronically exposed to i-As. Glucagon concentration in serum of rats in the Def-Se-i-As group was significantly higher as compared to the Suf-Se-Control or Def-Se-Control groups (Fig. 50.1). Nevertheless, the i-As-induced increase in serum concentration of glucagon did not reflect the extent of the increase in serum glucose levels.

Table 50.2.   Fasting serum glucose and insulin levels, and indicators of insulin resistance in rats from the four treatment groups (mean $\pm$ SE).

Parameter	Suf-Se-Control	Suf-Se-i-As	Def-Se-Control	Def-Se-i-As
Glucose (mmol/l)	5.42 ± 0.25	$9.99 \pm 0.41^*$	4.75 ± 1.93	6.47 ± 2.08
Insulin (ng/ml)	0.75 ± 0.20	$1.93 \pm 0.44^*$	0.295 ± 0.11	$1.7 \pm 1.0^{**}$
Glucose: Insulin ratio	10.67 ± 2.07	$4.82 \pm 0.78^*$	17.72 ± 4.01	$6.19 \pm 2.01^{**}$
HOMA-IR[1]	3.34 ± 0.59	$13.17 \pm 2.39^*$	1.84 ± 1.35	$14.10 \pm 5.04^{**}$

[1]Homeostasis model assessment of insulin resistance (HOMA-IR). n = 10; $^*p \leq 0.05$, *vs.* Suf-Se-control; $^{**}p \leq 0.05$, *vs.* Def-Se-control.

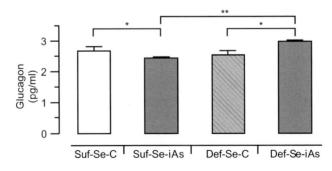

Figure 50.1.   Effects of subchronic exposure to i-As on serum glucagon levels in Suf-Se and Def-Se rats. Each bar represents mean $\pm$ SEM for n $=$ 10; $^*p = 0.001$; $^{**}p \leq 0.001$.

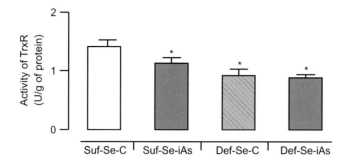

Figure 50.2.   Pancreatic TrxR activity in control and i-As-treated rats fed Se sufficient or Se deficient diet. Each bar represents mean ± SEM for n = 10. *p < 0.045 *vs.* Suf-Se Control.

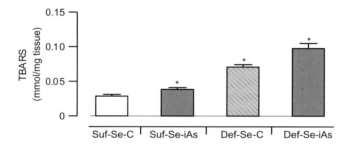

Figure 50.3.   TBARS concentration as a marker of lipid peroxidation in pancreas of control and i-As-treated rats fed Se sufficient or Se deficient diet. Each bar represents mean ± SEM for n = 10; *p < 0.01 *vs.* Suf-Se Control.

50.3.3   *Stress and oxidative damage in pancreas*

As expected, consumption of a Se deficient diet resulted in a decrease in the pancreatic TrxR activity in control rats; an additional decrease was associated with exposure to arsenite (Fig. 50.2). Inhibition of TrxR activity by arsenicals has previously been reported in both *in vitro* (Lin *et al.* 2001) and *in vivo* studies (Nikaido *et al.* 2003).

In addition to the reduced TrxR activity, increased TBARS levels were found in the pancreas of rats from Se deficient as compared to rats from Se sufficient group (Fig. 50.3). Notably, i-As exposure resulted in a statistically significant increase in pancreatic TBARS levels, particularly in Se deficient rats.

50.4   DISCUSSION AND CONCLUSIONS

The main function of the pancreatic beta cells is the storage and secretion of insulin, the key hormone involved in the regulation of glucose homeostasis. The results of this study show that subchronic exposure to i-As caused hyperglycemia in rats with normal Se intake. However, i-As did not induce hyperglycemia in Se deficient rats. Regardless of Se intake, the exposure to i-As resulted in insulin resistance. Table 50.2 shows that Se deficient rats treated with i-As had serum insulin levels 5.7-fold higher than control rats. In contrast, the difference in glucose levels was not statistically significant: 4.75 ± 1.93 *vs.* 6.47 ± 2.08 mmol/l for Def-Se-Control and Def-Se-i-As rats, respectively (p = 0.082). The difference in fasting insulin level and the lack of the difference in fasting blood glucose concentration are consistent with symptoms associated with insulin resistance.

In addition, in this study we evaluated the role of glucagon which major role is to stimulate increase in the blood glucose concentration. Serum glucagon concentration was decreased in rats subchronically exposed to i-As and fed Se-Suf diet (Fig. 50.1). This decrease suggests that hyperglycemia observed in this treatment group was independent of the glucagon action. In contrast, in Se-Def rats, treatment with i-As increased glucagon concentration, although no significant changes in the glucose level were observed (Table 50.2). Notably, serum glucagon levels were higher in Se-Def rats treated with i-As as compared to i-As-treated Se-Suf rats (Fig. 50.1). Previous work has shown that stimulation of lipolysis associated with Se deficiency is a result of elevated glucagon levels (Beckett and Arthur 2005). Increased serum glucagon concentration can also be related to homocysteine (HCys) metabolism. Glucagon regulates metabolism of sulfur aminoacids, including HCys (Jacobs *et al.* 2001). High plasma concentrations of HCys have been reported among residents of arsenic-endemic areas (Gamble *et al.* 2005). In contrast, Se deficiency has been shown to result in a decrease in HCys levels (Uthus *et al.* 2002). We have not measured serum HCys concentrations in this study. Thus, although it is unclear whether HCys alters glucose metabolism, potential effects of HCys on beta-cell function merit further investigation.

Type 2 diabetes mellitus (DM) is characterized by insulin resistance, resulting in a decreased uptake of glucose from blood to target tissues. Consequently, blood glucose levels increase and more insulin is released producing hyperinsulinaemia, which is an early indicator of type 2 DM. In addition, many studies have shown that the increased secretion of insulin is the primary defect in type 2 DM and that insulin resistance develops as a result of chronic hyperinsulinaemia (DeFronzo 1997). The insulin resistance causes progressive damage to pancreatic beta cells.

It has been recognized that i-Se has insulin-mimetic properties *in vitro* and *in vivo*. Insulin-stimulated glucose metabolism is impaired in adipocytes isolated from Se-deficient rats (Souness *et al.* 1983). The insulin-like effects of i-Se in cultured rat adipocytes include stimulation of glucose transport and phosphodiesterase activity (Ezaki 1990). It has been suggested that stimulation of tyrosine kinases involved in the insulin signaling cascade may underlie the insulin-like effects of i-Se (Stapleton *et al.* 1997). However, the insulinomimetic properties of i-Se depends on its chemical form. Supranutritional intake of Se(VI) has been shown to enhance insulin sensitivity in diabetic mice by inhibiting protein tyrosine phosphatases (Mueller *et al.* 2003).

In this study, Se deficiency caused stress and oxidative damage in the pancreas that was further exacerbated by exposure to i-As. Systems that regulate the intracellular redox status by scavenging ROS include the thioredoxin (Trx) system. Trx and its endogenous regulator, TrxR, represent important targets for diseases associated with oxidative stress and are also involved in various cellular functions, including cell signaling (Nordberg and Arner 2001). Trx and TrxR represent candidate molecular targets for agents that induce oxidative stress. Trx plays an essential role in cell function by limiting oxidative stress directly through its antioxidant action and indirectly through the protein-protein interactions with key signaling molecules such as Trx-interacting protein (TrxIP). It has been shown that hyperglycemia and diabetes induce TrxIP and decrease Trx activity (Schulze *et al.* 2004). Arsenicals, particularly methylated trivalent arsenicals formed in the course of cellular metabolism of i-As are potent inhibitors of TrxR activity. In the present study, reduction in the pancreatic TrxR activity was observed in i-As-treated rats regardless of Se intake (Fig. 50.2). The decrease in TrxR activity could be interpreted as an indicator of oxidative stress. Oxidants are able to modulate signaling pathways to regulate insulin secretion in pancreatic beta cells (Maechler *et al.* 1999).

Induction of lipid peroxidation in the pancreas of i-As-treated rats is documented in Figure 50.3. The TBARS concentrations in i-As-treated groups represented ~135% and ~200% of TBARS concentrations found in the corresponding Se-Suf and Def-Se control groups. The increase in lipid peroxidation in i-As-treated rats may be in part due to decreased activity of TrxR. Our data suggest that Se deficiency exacerbated the oxidative damage induced by i-As in the pancreas and modified the ratio of HOMA-IR (i-As treated rats)/HOMA-IR (control rats): 3.9 for Se-Suf and 7.6 in Se Def rats.

In summary, assessment of Se status of individuals exposed to i-As should be considered in future studies. Only then the health effects of dietary Se on i-As-exposed populations can be accurately determined.

ACKNOWLEDGMENTS

The authors thank Angel Barrera for taking care of animals and for his assistance during dissections. This study was partially supported by grant 38471-M from Conacyt-Mexico and by CYTED, XI.23. JAIV was recipient of scholarship from Conacyt-Mexico.

REFERENCES

ATSDR: Toxicological profile for arsenic. Agency for Toxic Substances and Disease Registry, Atlanta, GA, 2000.

Beckett, G.J. and Arthur, J.R.: Selenium and endocrine systems. *J. Endocrinology* 184 (2005), pp. 455–65.

Buege, J.A. and Aust, S.D.: Microsomal lipid peroxidation. *Meth. Enzymol.* 52 (1978), pp. 302–310.

Biswas, S., Talukder, G. and Sharma, A.: Prevention of cytotoxic effects of arsenic by short-term dietary supplementation with selenium in mice in vivo. *Mutat. Res.* 441 (1999), pp. 155–160.

Bonora, E., Targher, G., Alberiche, M., Formentini, G., Calcaterra, F., Lombardi, S., Marini, F., Poli, M., Zenari, L., Raffaelli, A., Perbellini, S., Zenere, M.B., Saggiani, F., Bonadonna, R.C. and Muggeo, M.: Predictors of insulin sensitivity in Type 2 diabetes mellitus. *Diabet. Med.* 19 (2002), pp. 535–542.

DeFronzo, R.A.: Pathogenesis of type 2 diabetes: metabolic and molecular implications for identifying diabetes genes. *Diabetes Rev.* 5 (1997), pp. 177–269.

Del Razo, L.M., Quintanilla-Vega, B., Brambila-Colombres, E., Calderon-Aranda, E.S., Manno, M. and Albores, A.: Stress proteins induced by arsenic. *Toxicol. Appl. Pharmacol.* 177 (2001), pp. 132–148.

Ezaki, O.: The insulin-like effects of selenate in rat adipocytes. *J. Biol. Chem.* 265 (1990), pp. 1124–1128.

Gamble, M.V., Ahsan, H., Liu, X., Factor-Litvak, P., Ilievski, V., Slavkovich, V., Parvez, F. and Graziano J.H.: Folate and cobalamin deficiencies and hyperhomocysteinemia in Bangladesh. *Am. J. Clin. Nutr.* 81 (2005), pp. 1372–1377.

Gregus, Z., Perjesi, P. and Gyurasics, A.: Enhancement of selenium excretion in bile by sulfobromophthalein: elucidation of the mechanism. *Biochem. Pharmacol.* 56 (1998), pp. 1391–1402.

Izquierdo-Vega, J.A., Soto Peredo, C.A., Sanchez-Peña, L.C., De Vizcaya-Ruiz, A. and Del Razo, L.M.: Diabetogenic effects and pancreatic oxidative damage in rats subchronically exposed to arsenite. *Toxicol. Lett.* 160 (2006), pp. 135–142.

Jacobs, R.L., Stead, L.M., Brosnan, M.E. and Brosnan. J.T.: Hyperglucagonemia in rats results in decreased plasma homocysteine and increased flux through the transsulfuration pathway in liver. *Biol. Chem.* 276 (2001), pp. 43, 740–743, 747.

Kenyon, E.M, Hughes, M.F., Del Razo, L.M., Edwards, B.C., Mitchell, C.T. and Levander, O.A.: Influence of dietary selenium on the disposition of arsenate and arsenite in the female B6C3F1 mouse. *Environ. Nutr. Interact.* 3 (1999), pp. 95–113.

Legro, R., finegood, D. and Dunaif, A.: Fasting glucose to insulin ratio is a useful measure of insulin sensitivity in women with polycystic ovary syndrome. *J. Clin. Endocrinol. Metab.* 83 (1998), pp. 2694–2698.

Levander, O.A.: Metabolic interrelationships between arsenic and selenium. *Environ. Health Perspect.* 19 (1977), pp. 159–164.

Lin S., Del Razo, L., Styblo, M., Wang, C., Cullen, W. and Thomas, D.: Arsenicals inhibit thioredoxin reductase in cultured rat hepatocytes. *Chem. Res. Toxicol.* 14 (2001), pp. 305–311.

Maechler, P., Jornot, L. and Wollheim, C.B.: Hydrogen peroxide alters mitochondrial activation and insulin secretion in pancreatic beta cells. *J. Biol. Chem.* 274 (1999), pp. 27, 905–27, 913.

Mueller, A.S., Pallauf, J. and Rafael, J.: The chemical form of selenium affects insulinomimetic properties of the trace element: investigations in type II diabetic dbdb mice. *J. Nutr. Biochem.* 14 (2003), pp. 637–647.

Nikaido, M., Pi, J., Kumagai, Y., Yamauchi, H., Taguchi, K., Horiguchi, S., Sun, Y., Sun, G. and Shimojo, N.: Decreased enzyme activity of hepatic thioredoxin reductase and glutathione reductase in rabbits by prolonged exposure to inorganic arsenate. *Environ. Toxicol.* 18 (2003), pp. 306–311.

Nordberg, J. and Arner, E.S.: Reactive oxygen species, antioxidants, and the mammalian thioredoxin system. *Free Radic. Biol. Med.* 31 (2001), pp. 1287–1312.

NCR (National Research Council): Arsenic in Drinking Water, National Academy Press, Washington, DC, 2001.

Rahman, M. and Axelson, O.: Diabetes mellitus and arsenic exposure: A second look at case-control data from a Swedish copper smelter. *Occup. Environ. Med.* 52 (1995), pp. 773–774.

Rahman, M., Wingren, G. and Axelson, O.: Diabetes mellitus among Swedish art glassworkers: an effect of arsenic exposure? *Scand. J. Work Environ. Health* 22 (1995), pp. 146–149.

Stapleton, S.R., Garlock, G.L., Foellmi-Adams, L. and Kletzien, R.F.: Selenium: potent stimulator of tyrosyl phosphorylation and activator of MAP kinase. *Biochim. Biophys. Acta* 1355 (1997), pp. 259–269.

Styblo, M. and Thomas, D.J.: Selenium modifies the metabolism and toxicity of arsenic in primary rat hepatocytes. *Toxicol. Appl. Pharmacol.* 172 (2001), pp. 52–61.

Smith, A.D. and Levander, O.A.: High-throughput 96-well microplate assays for determining specific activities of glutathione peroxidase and thioredoxin reductase. *Methods Enzymol.* 347 (2002), pp. 113–121.

Souness, J.E., Stouffer, J.E. and Chagoya de Sanchez, V.: The effect of selenium deficiency on rat fat-cell glucose oxidation. *Biochem. J.* 214 (1983), pp. 471–417.

Schulze, P.C., Yoshioka, J., Takahashi, T., He, Z., King, G.L. and Lee, R.T.: Hyperglycemia promotes oxidative stress through inhibition of thioredoxin function by thioredoxin-interacting protein. *J. Biol. Chem.* 279 (2004), pp. 30, 369–30, 374.

Thirunavukkarasu, V., Anitha Nandhini, A.T. and Anuradha, C.V.: Lipoic acid attenuates hypertension and improves insulin sensitivity, kallikrein activity and nitrite levels in high fructose-fed rats. *J. Comp. Physiol. B.* 174 (2004), pp. 587–592.

Tsai, S.-M., Wang, T.-N. and Ko, Y.-C.: Mortality for certain diseases in areas with high levels of arsenic in drinking water. *Arch. Environ. Health* 54 (1999), pp. 186–200.

Tseng, C.H.: The potential biological mechanisms of arsenic-induced diabetes mellitus. *Toxicol. Appl. Pharmacol.* 197 (2004), pp. 67–83.

Uthus, E.O., Yokoi, K. and Davis, C.D.: Selenium deficiency in fisher-344 rats decreases plasma and tissue homocysteine concentrations and alters plasma homocysteine and cysteine redox status. *J. Nutr.* 132 (2002), pp. 1122–1128.

Walton, F.S., Waters, S.B., Jolley, S.L., LeCluyse, E.L., Thomas, D.J. and Styblo, M.: Selenium compounds modulate the activity of recombinant rat AsIII-methyltransferase and the methylation of arsenite by rat and human hepatocytes. *Chem. Res. Toxicol.* 16 (2003), pp. 261–265.

CHAPTER 51

Histological characteristics of sural nerves in rats exposed to arsenite

E. García-Chávez & L.M. Del Razo
Sección Externa de Toxicología, Centro de Investigación y de Estudios Avanzados del Instituto Politécnico Nacional (CINVESTAV-IPN), Mexico City, Mexico

B. Segura
Facultad de Estudios Superiores Iztacala (FES), Universidad Nacional Autónoma de México (UNAM), Mexico City, Mexico

H. Merchant
Instituto de Investigaciones Biomédicas (IIB), Universidad Nacional Autónoma de México (UNAM), Mexico City, Mexico

I. Jiménez
Depto. de Fisiología, Biofísica y Neurociencias, Centro de Investigación y de Estudios Avanzados del Instituto Politécnico Nacional (CINVESTAV-IPN), Mexico City, Mexico

ABSTRACT: The aim of this study was to analyze histological alterations of the rat sural nerve and lipid peroxidation (LPO) induced by arsenite exposure and to evaluate possible protective effects of α-tocopherol (α-TOC) against the oxidative damage. Two groups of male Wistar rats received sodium arsenite by gavage (10 mg/kg bw/day) for 30 days. One of the groups also received α-TOC by gavage (125 mg/kg bw/day) starting 10 days after the first dose of arsenite. LPO and distribution of arsenic (As) were evaluated at the end of the exposure period. Untreated rats or rats receiving only α-TOC were used as controls. LPO positively correlated with As concentration in sural nerves ($R^2 = 0.9972$, $p < 0.001$). Compared to controls, the axons of sural nerves from arsenite-exposed rats showed a significant reduction in the transversal area ($\sim45\%$) and myelin sheath thickness ($\sim56\%$). Supplementation with α-TOC significantly decreased the toxic effects induced by arsenite exposure in sural nerves.

51.1 INTRODUCTION

Exposure to inorganic arsenic (i-As) has long been known to affect the central and peripheral nervous systems (NRC 2001, Rodríguez *et al.* 2003). Encephalopathy, impairment of superior neurological functions such as learning, recent memory or concentration capacity are some of the effects of i-As on the central nervous system (NRC 2001). Evidence of peripheral neuropathy associated with repeated exposure to i-As has also been reported (Gerr *et al.* 2000, Mukherjee *et al.* 2003, Berdel-Garcia *et al.* 2004). Initial neuropathic symptoms typically consist of sensory dysesthesias in symmetric stocking-glove distribution that can be in some cases accompanied by growing weakness and flaccid paralysis.

It is generally understood that metabolic conversions of i-As significantly modify the toxic and cancer promoting effects of this metalloid (Thomas *et al.* 2004). Biomethylation is the major pathway for the metabolism of i-As. In humans, i-As is converted to methylarsenic (MA) and dimethylarsenic (DMA) that contain As in trivalent or pentavalent oxidation state (Thomas *et al.* 2004). In rats and hamsters, DMA is further methylated to yield trimethylarsine oxide (TMAO) (Devesa *et al.* 2004). Accumulation of i-As and its metabolites in the peripheral nerve represent a possible mechanism for its neurotoxicity. These arsenicals are potentially cytotoxic and can facilitate formation of reactive oxygen species (ROS). ROS generation is the major mechanism for i-As toxicity in a variety of tissues (Del Razo *et al.* 2001a). ROS have been detected in brains of rats exposed to i-As (García-Chavez *et al.* 2003). Although, the effects of i-As exposure on the peripheral nervous system are not fully understood, it is possible that the presence of methylated As species is to some extent responsible for the previously reported neurotoxic alterations. Because oxidative stress could be responsible for injury associated with i-As exposure, supplementation with antioxidants might mitigate i-As-induced toxicity.

Alpha tocopherol (α-TOC) is the most abundant and active form of vitamin E (Niki and Noguchi 2004) and the most important lipophilic radical-scavenging antioxidant. It reacts with peroxyl radicals 10,000-fold faster than do polyunsaturated lipids. Therefore, α-TOC is useful as a therapeutic agent in treatments of various disorders associated with oxidative damage. It could potentially be used to ameliorate lipid peroxidation (LPO) induced by i-As exposure. In spite of the extraordinary interest in toxicological properties of i-As, limited information is available about the capacity of this metalloid to interact with and to alter the structure and function of the peripheral nerves. This study was carried out to improve our understanding of the toxic effects of i-As exposure on peripheral nervous system. Here, we examined the distribution of i-As and its metabolites, oxidative damage, and histological characteristics of the peripheral sensory sural nerve in rats exposed to i-As. We also evaluated protective effects of α-TOC against toxic effects of i-As on the sural nerve.

51.2   METHODS

51.2.1   *Procedures*

Male Wistar rats weighing 200–210 g were housed in animal facilities with a 12/12-h light/dark cycle, temperature of $22 \pm 1°C$, and humidity of $50 \pm 5\%$. All animals were provided food (LabDiet®, 5053) containing less than 1 mg/kg As and deionized water *ad libitum*. For experiments, rats were divided into four groups (n = 8) as follows: (1) *Control rats* received a dose of deionized water (2.0 ml/kg bw) by gavage daily for 30 days; (2) *α-TOC rats* received a dose of α-TOC (125 mg/kg bw; i.e., 230 IU/kg bw) by gavage daily for 20 days; (3) *Sodium arsenite-treated rats* received a dose of sodium arsenite (10 mg/kg of bw) by gavage daily for 30 days; (4) *Sodium arsenite + α-TOC rats* received a dose of sodium arsenite (10 mg/kg bw) by gavage daily for 30 days and a dose of α-TOC (125 mg/kg bw) by gavage daily for 20 days, starting 10 days after administration of the first dose of arsenite. When treatment was completed rats were sacrificed by cervical dislocation, followed by decapitation. The sural nerves from both hind limbs were carefully dissected. One nerve segment was used for histological studies; the other segment was washed and immediately homogenized in an ice-cold phosphate buffer (pH 7.4) for LPO and As species analyses.

51.2.2   *Arsenic analysis in sural nerves*

Arsenical species (i-As, MA, DMA, TMAO) were analyzed by hydride generation-atomic absorption spectrometry (HG-AAS), using cryotrapping for separation of arsines and a Perkin Elmer 3100 spectrometer for detection (Del Razo *et al.* 2001b). Prior to analysis, samples of sural nerve were homogenized and homogenates were digested in 3 ml of 2 M ultrapure phosphoric acid at 90°C for 4 hours. Sample aliquots spiked with As standards were used for quality control during the speciation analysis. The accuracy of the method was 99.6% with the variation coefficient <10%. Relative amounts of As species were calculated as percentages of the sum of all As species detected. The sum of As species (i-As + MA + DMA + TMAO) for each sample is referred to as total As (t-As) and is expressed in $\mu g/g$ of tissue.

51.2.3   *Lipid peroxidation*

LPO was evaluated according to the method of Buege and Aust (1978) that measures thiobarbituric acid reactive substances (TBARS). Here, the tissue homogenate (0.5 ml) was added to a test tube containing 1 ml of 0.5% thiobarbituric acid, 5 µl deferoxamine mesylate, and 5 µl of 3.75% butylated hydroxytoluene. The reaction mixture was heated at 100°C for 20 min. After cooling, the mixture was centrifuged at 4000 rpm for 15 min and absorbance of the supernate was measured at 532 nm by a spectrophotometer. Results were expressed as nmol of TBARS/g tissue.

51.2.4   Histological analysis

Sural nerves segments (2–3-mm long) were fixed in a Karnousky solution (1% buffered formalin, 25% glutaraldehyde, 0.2 M sodium cacodylate, pH 7.5), embedded in EPON 812 resin, and stained with 5% toluidine blue in 4% sodium borate. To determine the histological characteristic of axons in the sural nerves, a selected nerve section was photographed and amplified and the nerve diameter, axon diameter, and myelin sheath thickness were measured, using the SigmaScan Pro 4 program. The resulting values were adjusted by the factor used for the amplification of the nerve section.

51.2.5   Statistical analysis

The data are expressed as mean ± SD. Statistical analyses were carried out using a one-way analysis of variance (ANOVA) followed by Bonferroni test for multiple comparisons. Differences were considered statistically significant when $p < 0.05$. All analyses were performed using the STATA 8 statistical program (Stata Corp., College Station, TX).

51.3   RESULTS

51.3.1   Arsenic species and LPO in sensory sural nerves

Relative amounts of i-As and its metabolites in nerve tissues are shown in Figure 51.1. DMA was the major metabolite present in nerves from both control (≈99%) and exposed (≈96%) rats. TMAO was not detected.

Table 51.1 shows average values (± SD) for t-As and TBARS concentrations in the sural nerves from control and exposed animals. Nerves from rats treated with α-TOC and control (untreated) rats contained similar amounts of t-As and TBARS. Exposure to arsenite resulted in a significant increase in t-As concentration in the sural nerves. Similar levels of t-As were found in the nerves from rats treated with both arsenite and α-TOC and rats treated only with arsenite. However, in spite of the similar t-As contents, a significantly lower TBARS concentration was found in the nerves from the arsenite treated rats that received α TOC as compared to the arsenite-only-treated rats (Table 51.1).

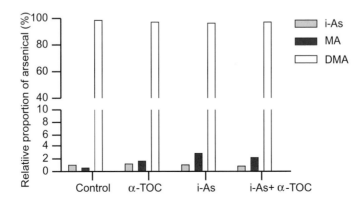

Figure 51.1.   Relative amounts of arsenic species detected in rat sural nerves.

Table 51.1.   Concentrations of t-As and lipid peroxidation in peripheral sensory sural nerves.

Groups	Control	α-TOC	i-As	i-As + α-TOC
t-As (μg/g)	1.3 ± 0.6	1.3 ± 0.3	194.6 ± 64.5^{1}	147.1 ± 53.0
TBARS (nmol/g)	136.2 ± 11.0	92.8 ± 17.6	747.9 ± 420.8^{1}	138.5 ± 50.7

[1] $p < 0.001$ versus control; $n = 8$.

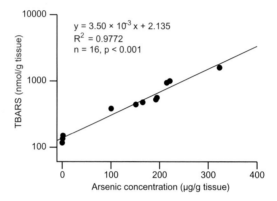

$$y = 3.50 \times 10^{-3} x + 2.135$$
$$R^2 = 0.9772$$
$$n = 16, p < 0.001$$

Figure 51.2.   Linear correlation between t-As and TBARS concentrations in the sural nerve of control and arsenite-treated rats.

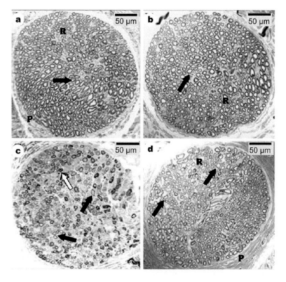

Figure 51.3.   Transversal histological sections of sural nerves from (a) control, (b) α-TOC, (c) i-As, and (d) i-As + α-TOC groups of rats. In panels A, B and C, arrows indicate axons with thicker myelin sheaths, R, Schwann cells plus unmyelinated axons and P, the perineurom; (c) shows collapsed axons and balls of myelin (white arrow); thinly myelinated axons with a considerable deviation from the characteristic circular appearance (black arrows).

As shown in Figure 51.2, a positive linear correlation ($y = 3.5e^{-3}x + 2.13$; $R^2 = 0.9772$; $p < 0.001$) was found between the t-As concentration and TBARS concentration in the sural nerve of control and i-As-treated animals. This correlation suggests that i-As and its metabolites are directly responsible for oxidative damage in peripheral nerve tissues.

51.3.2 Morphometric analysis of sural nerves

Cross-sectional photographs of the sural nerves from control and treated rats are shown in Figure 51.3. The sub-chronic exposure to arsenite caused morphological derangement of the sensory sural nerve; a considerable deviation from circular appearance, thin myelin sheath thickness, and a partial axonal degeneration represented by collapsed fibers was found in axons (Figure 51.3c). In comparison, nerves from rats treated with both arsenite and α-TOC were also characterized by thin myelin sheaths, but maintained circular appearance (Fig. 51.3d). There were no apparent differences in morphology of nerves from control (Fig. 51.3a) and α-TOC-treated rats (Fig. 51.3b).

The sural nerves from rats exposed to i-As displayed a significant reduction in their mean cross-sectional area as compared to nerves from control rats ($34,076 \pm 950\ \mu m^2$ *vs.* $64,025 \pm 927\ \mu m^2$, $p = 0.007$) (Fig. 51.4a). The sural nerve area of rats treated with both arsenite and α-TOC was not significantly different from that of rats treated only with α-TOC ($48,050 \pm 10,943\ \mu m^2$ and $59,058 \pm 1,844\ \mu m^2$, respectively; $p = 0.087$). In both cases, the sural nerve areas were larger than the sural nerve area of rats treated only with arsenite. However, this difference was not statistically significant.

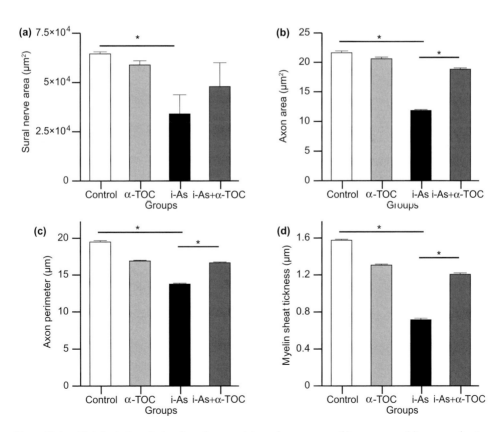

Figure 51.4.   Histological analysis of sural nerve. (a) sural nerve area; (b) axon area; (c) axon perimeter; (d) myelin sheath thickness (mean $\pm$ SD; $^*p < 0.01$).

Compared to controls, a significant reduction in the axonal area was observed for nerves from arsenite-treated rats: $11.8 \pm 3.4 \ \mu m^2$ *vs.* $21.6 \pm 5.7 \ \mu m^2$ ($p < 0.001$) (Fig. 51.4b). Notably, similar axon areas were found for nerves from rats exposed to both arsenite and α-TOC and to α-TOC alone: 18.8 ± 3.8 and $20.6 \pm 4.8 \ \mu m^2$, respectively. For rats in both these treatment groups, the axonal areas were significantly greater than for arsenite-treated rats ($p < 0.001$).

Compared to control rats, the axons of sural nerves from rat exposed to arsenite were characterized by a noticeable reduction in perimeter as (19.4 ± 3.7 *vs.* $13.7 \pm 2.4 \ \mu m$; $p < 0.001$) (Fig. 51.4c); the average perimeter of axons from rats treated with both arsenite and α-TOC was significantly greater: $16.6 \pm 1.8 \ \mu m$ ($p < 0.001$).

As shown above, sural nerve axons from rats exposed to arsenite had significantly ($p < 0.001$) thinner myelin sheaths than axons from control rats (0.7 ± 0.3 *vs.* $1.6 \pm 0.2 \ \mu m$; $p < 0.001$) and than axons from rats treated with α-TOC or with both arsenite and α-TOC ($1.3 \pm 0.04 \ \mu m$ and $1.2 \pm 0.2 \ \mu m$, respectively) (Fig. 51.4d). These data indicate that the supplementation of arsenite-treated rats with α-TOC prevented the loss of axonal myelin sheath.

51.4   DISCUSSION AND CONCLUSIONS

Total As concentration in sural nerves of rat exposed to i-As was 150 fold higher than in nerves of control rats. Speciation of As in nerve tissue showed that administered arsenite was converted mainly to DMA (Fig. 51.1). Although the methodology employed for analysis of methylated metabolites cannot differentiate between the trivalent and pentavalent forms, it is plausible to assume that trivalent methylated arsenicals present in nervous tissues might inhibit important antioxidants enzymes such as glutathione reductase (GR) or glutathione peroxidase (GPx).

The data obtained in the present study suggest that arsenite exposure is associated with an increase in TBARS concentration in peripheral sensory nerves, thus, linking i-As and its metabolites to LPO. Interestingly, the concentration of t-As (represented mainly by DMA) positively correlated with TBARS concentration in peripheral nerves (Fig. 51.2). These results suggest that oxidative damage induced by i-As exposure is due to either an increased production of ROS or a weakened antioxidant defense.

Because As compounds exert their toxicity by generating ROS, we examined whether α-TOC can protect sural nerves of rats exposed to arsenite. Our results show that oxidative damage to sural nerves from rats exposed to arsenite was reduced by α-TOC treatment (Table 51.1), even though the t-As concentration in the nerves did not change (Table 51.1). These results suggest that treatment with α-TOC averted oxidative damage, probably through its capacity to quickly and efficiently scavenge lipid peroxyl radicals before these attack membrane lipids. This is consistent with the previous observation that lipid peroxyl radicals react more efficiently (by four orders of magnitude) with α-TOC than with membrane lipids (Halliwell and Gutteridge 2002).

In order to provide more detailed information about morphological changes produced by i-As in the peripheral nervous system, we evaluated the general histological characteristics of the sensory sural nerve. Our data indicate that i-As exposure results in a significant reduction in the thickness of axon myelin sheath (by 56.3%) and of the axonal area and perimeter (by 45% and 29% respectively; see Figs. 51.3 and 51.4). The observed morphological alterations are similar to those described for cases of demyelization neuropathy (Gabriel *et al.* 2000). This observation may suggest that demyelization processes occur in association with oxidative damage in axons exposed to i-As and its metabolites. This may be due to generation of ROS by i-As or its metabolites, resulting in cell injury and myelin breakdown as a result of LPO. This explanation is supported by the observation that α-TOC supplementation exerts a protective effect against the morphological alterations induced in sural nerves by arsenite exposure.

In summary, sensory sural nerves of i-As-exposed animals are characterized by (1) accumulation of DMA as the major metabolite of i-As; (2) occurrence of oxidative damage; (3) decreases in the axonal area and perimeter; and (4) reduction in the thickness of axonal myelin sheath which is probably related to the oxidative damage. This work shows for the first time that α-TOC supplementation prevents the morphological changes and LPO induced by arsenite in the peripheral sural nerve. An electrophysiological analysis of the effects of arsenite exposure on generation and propagation of electrical impulses in the sural nerve is currently on the way in this laboratory.

ACKNOWLEDGMENTS

The authors are thankful to Luz del Carmen Sanchez-Peña for her technical assistance in the determination of As species and lipid peroxidation and to José Carlos Guadarrama and Angel Barrera Hernández for their technical assistance in the maintenance of animals and dissection of tissues. EGC was a recipient of a scholarship from Conacyt-Mexico. This study was partially supported by CYTED, XI.23.

REFERENCES

Berbel-García, A., González-Aguirre, J.M., Botia-Paniagua, E., Ortis-Castro, E., López-Zuazo, I., Rodríguez-García, J.L. and Gil-Madre, J.: Acute polyneuropathy and encephalopathy caused by arsenic poisoning. *Rev. Neurol.* 38 (2004), pp. 928–930.
Buege, J.A. and Aust, S.D.: Microsomal lipid peroxidation. *Methods Enzymol.* 52 (1978), pp. 302–310.
Del Razo, L.M., Quintanilla-Vega, B., Brambila-Colombres, E., Calderón-Aranda, E.S., Manno, M. and Albores, A.: Stress proteins induced by arsenic. *Toxicol. Appl. Pharmacol.* 177 (2001a), pp. 132–148.
Del Razo, L.M., Styblo, M., Cullen, W.R. and Thomas, D.J.: Determination of trivalent methylathed arsenicals in biological matrices. *Toxicol. Appl. Pharmacol.* 174 (2001b), pp. 282–293.
Devesa, V., Del Razo, L.M., Adair, B., Drobná, Z., Waters, S.B., Hughes, M.H., Styblo, M. and Thomas, D.J.: Comprehensive analysis of arsenic metabolites by pH-specific hydride generation atomic absorption spectrometry. *J. Anal. Atomic Spect.* 19 (2004), pp. 1460–1467.
Gabriel, C.M., Howard, R., Kinsella, N., Lucas, S., McColl, I., Saldanha, G., Hall, S.M. and Hughes, R.A.: Prospective study of usefulness of sural nerve biopsy. *J. Neurol. Neurosurg. Psychiatry* 69 (2000), pp. 442–446.
García-Chávez, E., Santamaría, A., Díaz-Barriga, F., Mandeville, P., Juarez, B.I. and Jiménez-Capdeville, M.E.: Arsenite-induced production of hydroxyl radical in the striatum of awake rats. *Brain Res.* 976 (2003), pp. 82–89.
Gerr, F., Letz, R. and Green, R.C.: Relationships between quantitative measures and neurologist's clinical rating of tremor and standing steadiness in two epidemiological studies. *Neurotoxicology* 21 (2000), pp. 753–760.
Halliwell, B. and Gutteridge, J.M.C.: *Free Rad. Biol. Med.* Oxford University Press Inc., New York, 2002, pp. 105–245.
Mukherjee, S.C., Rahman, M.M., Chowdhury, U.K., Sengupta, M.K., Lodh, D., Chanda, C.R., Saha, K.C. and Chakraborti, D.: Neuropathy in arsenic toxicity from groundwater arsenic contamination in West Bengal, India. *J. Environ. Sci. Health A. Tox. Hazard Subst. Environ. Eng.* 38 (2003), pp. 165–183.
National Research Council (NRC): *Arsenic in Drinking Water*. Nat. Academy Press, Washington, DC, 2001.
Niki, E. and Noguchi, N.: Dynamics of antioxidant action of vitamin E. *Acc. Chem. Res.* 37 (2004), pp. 45–51.
Rodríguez, V.M., Jiménez-Capdeville, M.E. and Giordano, M.: The effects of arsenic exposure on the nervous system. *Tox. Lett.* 145 (2003), pp. 1–18.
Thomas, D.J., Waters, S.B. and Styblo, M.: Elucidating the pathway for arsenic methylation. *Toxicol. Appl. Pharmacol.* 198 (2004), pp. 319–326.

CHAPTER 52

Arsenic-induced p53-DNA binding activity in epithelial cells

M. Sandoval, M. Morales, A. Ortega & E. López-Bayghen
Depto. de Genética y Biología Molecular, Centro de Investigación y de Estudios Avanzados del Instituto Politécnico Nacional (CINVESTAV-IPN), Mexico City, Mexico

P. Ostrosky-Wegman
Depto. de Genética y Toxicología Ambiental, Instituto de Investigaciones Biomédicas, Universidad Nacional Autónoma de México (UNAM), Mexico City, Mexico

ABSTRACT: Exposure to inorganic arsenic (i-As) is associated with an increased incidence of skin pathologies as hyperkeratosis and cancer. Arsenic effects include genotoxicity, cell-proliferation changes, alterations in DNA repair and methylation patterns. Proliferation and differentiation coordinated expression of structural and regulatory proteins is probably lost in As-associated skin pathologies. We evaluated whether the changes in epithelial cells, including human keratinocytes after As exposure, are mediated through modifications in the activity of the tumor suppressor p53, a potent mediator of responses against genotoxic damage, and a pivotal proliferation-controller. We tested if i-As modulates the abilities of p53 as a transcription factor in epithelial cells treated with increasing concentrations of sodium arsenite for up to 24 hours. DNA-binding assays were performed using a p53-consensus sequence. A time and dose-dependent response in p53-DNA binding occurs under As exposure, favoring the notion that p53-regulated genes are important to understand modifications in the proliferation-differentiation balance severely impaired during skin carcinogenesis.

52.1 INTRODUCTION

Inorganic arsenic (i-As), a ubiquitous element, has represented a human heath concern for centuries. The contamination of water supplies with this element has resulted in a very high incidence of skin lesions and cancers as demonstrated in a large amount of epidemiologic studies (Vega *et al.* 2001). Benign skin lesions related to chronic As exposure act as early warning signals, including hyperpigmentation and hyperkeratosis (Chen *et al.* 1985). The molecular mechanisms underlying As carcinogenicity have begun to be explored, noticing exposure-associated effects over genotoxicity, cell proliferation changes and altered DNA repair, but the exact mechanism in cancer development is still poorly understood (Tondel *et al.* 1999).

In response to genotoxic stress and DNA damage, the levels of the tumor suppressor protein p53 are increased (Colman *et al.* 2000). This protein mediates genotoxic stress responses and has been regarded as a key element in maintaining genomic stability (Appella and Anderson 2001). There is controversy of the pathways involved in p53 activation, although Ser15 p53 phosphorylation through the DNA-dependent protein kinase (DNA-PK) is critically involved it its interaction with the transcriptional machinery (Woo *et al.* 1998). Phosphorylated p53 accumulates in the nucleus and transactivates a number of key genes involved in DNA repair processes and cell proliferation.

Our current knowledge of p53 response after As-exposure is variable. Depending on cell type, previous p53 stage and timing, responses are different (Salazar *et al.* 1997, Hamadeh *et al.* 1999 and 2002, Hernandez-Zavala *et al.* 2005). Particularly, the role of p53 as gene activator has been barely explored. In this contribution, we characterized the biological DNA binding activity of this transcription factor when epithelial cells are exposed to sodium arsenite in varying times and doses. We found marked differences in p53-DNA binding depending on the dose and exposure periods.

52.2   MATERIALS AND METHODS

52.2.1   *Primary keratinocyte cultures*

Primary cultures from neonatal human foreskin keratinocytes were obtained as described before (Pirisi *et al.* 1987) and grown in KSFM medium (Life Science Technologies), with the appropriate antibiotic mix at 37°C in a 5% CO_2 atmosphere. Cultures medium was replaced every 2 days.

52.2.2   *Cell lines culture and cell proliferation assay*

HeLa cells (HPV18 positive adenocarcinoma derived cell line) and C33-A (HPV negative cervical carcinoma derived cell line; Yee *et al.* 1985) were routinely cultured in Dulbecco's modified Eagle's medium (DMEM, Invitrogen Gaithersburg, MD) supplemented with 10% fetal calf serum, with the appropriate antibiotic mix at 37°C in a 5% CO_2 atmosphere. Culture media was replaced every two days.

Cell viability was measured by the MTT (3-(4,5-dimethylthiazolyl-2)-2, 5-diphenyl-tetrazolium bromide) reduction assay, performed as first described by Mosmann (1983). After incubation of the cells with the MTT reagent (1 mg/ml) for approximately 4 h, an isopropanol: HCl solution was added to lyse the cells and solubilize the colored crystals. The samples were analyzed with an ELISA plate reader (wavelength of 630 nm) Opsys MR (Dynex Technologies).

52.2.3   *Arsenic treatment*

An aqueous sterile stock solution of sodium arsenite (10 mg/ml) was prepared and appropriate volumes were diluted to obtain the desired final concentrations. Cultures were approximately 80–85% in confluence when treated with various concentrations of freshly prepared sodium arsenite (0.1, 0.5, 1.0, 1.5, 5 µM) for 30 min or 24 h.

52.2.3.1   *Electrophoretic Mobility Shift Assays (EMSA)*
Nuclear extracts were prepared as described previously (Lopez-Bayghen *et al.* 1996). All buffers contained complete protease inhibitor cocktail (Boehringer-Mannheim) to prevent proteolysis. Protein concentration was measured by the Bio-Rad protein assay system. Nuclear extracts (15–20 µg) from epithelial cells were incubated on ice with 1 µg poly (dI-dC) as non-specific competitor (Pharmacia Biotech) and 1 ng of ^{32}P-labeled double oligonucleotide corresponding to the p53 response elements identified in p53-regulated genes as p21 (Gohler *et al.* 2002), for 10 min (5′-CTAGTACAGAACATGTCTAAGCATGCTGGGGACT-3′) or the AP-1/SV40 probe (5′-CTAGRATAAATATGACTAAGCTGTG-3′). The reaction mixtures were electrophoresed in a low ionic strength 0.5 × TBE buffer in 7% polyacrylamide gels. The gel was dried and exposed to a X-ray film.

For competitive studies, the reaction mixtures were preincubated with different amounts of indicated unlabeled competitor oligonucleotides, including the oligonucleotide SP1 (5′-CTAGATTCGATCGGGGCGGGGCGA-3′), as a heterologous competitor, before adding labeled DNA.

52.3   RESULTS

Changes in p53 levels are associated with As exposure in several cell lines. As a transcription factor, one of the first activities performed by p53 after genotoxic or stress damage is to activate the transcription of several downstream promoter targets such as p21 (Kim 1997, Liu *et al.* 2003, Saramaki *et al.* 2006). In order to do so, its ability to recognize a specific DNA sequence in DNA is dramatically increased both through phosphorylation and its nuclear accumulation. Using the p53 recognition sequence from p21 promoter as a probe, we explored p53-DNA binding activity

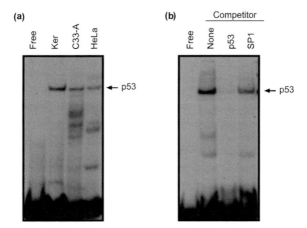

Figure 52.1. DNA binding comparison in epithelial cells. Complexes obtained using nuclear extracts from normal primary human keratinocytes, C33-A or HeLa carcinoma cells and p53 ^{32}P-labeled probe (a). Complex denoted as p53 (arrow) is the only constant complex obtained when different cell batches were tested and also is the specific complex not affected by heterologous competitors as SP-1 (b). Free labeled: probe alone.

after As exposure. In Figure 52.1, a comparison of the protein-DNA complexes, obtained when a p53-labeled probe is tested, with nuclear extracts, obtained from three different epithelial cells: primary human keratinocytes, the cervical derived C33-A and HeLa cell lines, is shown. As can be noticed, the same upper complex is coincident in the three epithelial sources tested, albeit at different abundance. Competition experiments shown in Fig. 52.1b, confirmed the specificity of the p53 complex. Furthermore, we have recently corroborated the identity of this complex in human keratinocytes by immunoEMSA (Sandoval *et al.* 2007).

It was decided to characterize the p53 complex in C33-A cells, since a mutated version of p53 is expressed at higher levels than in HeLa cells in this system. The point mutations present in the p53 version of C33-A cells result in an amino acid change of Arg -* Cys in codon 273 and it is not involved in the protein-DNA interactions (Scheffner *et al.* 1991). Responses to As are varied and occur at different regulatory levels. As a first approach we exposed C33-A cells to As for different time periods, using a fixed concentration of 1.5 μM. This concentration was used since it has been reported to be equivalent to that present in drinking water in areas where people are endemically exposed (Tseng *et al.* 1968, Germolec *et al.* 1998). As depicted in Figure 52.2, As-treatment causes a rapid increase in p53-DNA binding augmentation that not only is not sustained, but is even decreased as compared to control levels after 12 h and continues to lower after 24 h of exposure (60% of original non-treated control, Fig. 52.2b), raising the possibility of a cytotoxic effect. In order to rule out this possibility, we decided to expose C33-A cells to increasing As concentrations for 24 h and test cell viability by a cell proliferation/MTT assay. The results are presented in Fig. 52.2c. No significant effect of As in terms of cell viability was found. It is interesting that the As effects over p53 are different in the short compared to the long time.

To characterize these differences, we exposed C33-A cells to increasing As concentrations for 30 min and for 24 h (Figs. 52.3a and b, respectively). As depicted in the upper panels of Figure 52.3, long term exposure results in significant decrease in p53-DNA binding activity that has already reached its maximal level at a rather low As concentration (0.1 μM). In contrast, 30 min of As treatment results in a sustained, dose-dependent increase in p53-DNA binding activity, that is not saturated even at 1.5 μM. A detailed comparison of this effect is presented in Fig. 52.3c. After a while the constant presence of As prevents p53 to bind DNA even at low As concentrations. Other transcription factors are also activated by As exposure; one of them is AP-1 (c-jun/ c-fos). This factor has been widely reported as one of those activated by As in different cell types

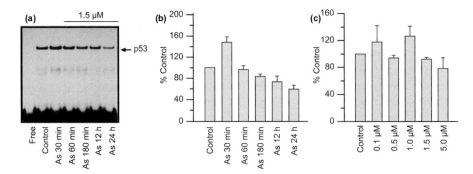

Figure 52.2.   Sodium arsenite exposure changes p53 DNA binding. C33-A cells (90% confluent) were treated with/without sodium arsenite for indicated times and the indicated final concentrations. After treatment completion, cells were harvested and processed for nuclear extracts. Equal amounts of nuclear protein (20 μg) were used in DNA binding and resolved in non-denaturing gels. A typical autoradiography is shown (a) together with plotted densitometryc data (b) reflecting the complex intensity. (c) C33-A cells viability was assessed by the MTT method, cells were treated with indicated sodium arsenite doses for 24 h, differences are not significant when compared with control in all cases (Student's t-test $p < 0.005$). Control: non-treated cells; Free: probe without nuclear protein added.

(Burleson *et al.* 1996). As a control we show the complexes associated with AP-1 transcription factor in C33-A cells, using the very same nuclear extracts tested with p53 and the very well characterized AP-1/SV40 probe (Fig. 52.3d). Clearly, the As effect in this case is quite different after 24 h exposure. As induced a strong AP-1 binding, providing evidence that a particular series of events is governing the lowering in p53 binding after a long and continuous exposure to As. These results strongly suggest that a biphasic effect in terms of signal transduction pathways is present upon As exposure, leading in the long term to a decrease in p53-DNA binding activity that might reflect either a degradation of p53 or an export from the nucleus.

52.4   DISCUSSION

Although As is a well known human carcinogen, and that the number of individuals chronically exposed to this element is elevated, the molecular mechanisms by which As induce cellular transformation is far from being deciphered. Mutations that inactivate or alter the function of p53 are present in different human cancers. Mutations in the p53 gene have been similarly detected in a high percentage of colon, breast, lung, brain, and esophageal human cancers (Nigro *et al.* 1989). When cells are under several kinds of stressors, controlling what happens next mostly relies in p53 activation. It is tempting to speculate that in the short term, p53 is activated in response to As-dependent DNA damage, most likely through Ser15 phosphorylation via DNA-PK resulting in the documented dose-dependent augmentation of the p53-DNA complex (Fig. 52.3a, c). In contrast, on the long term, the decrease of the p53-DNA complex (Fig. 52.3b) might be the result of a MDM-2 induced p53 nuclear export and degradation. This interpretation is favored by the fact that the *mdm-2* gene is transcriptionally activated by p53 (Moll and Petrenko 2003). This model implies that As-dependent p53 activation drives the transcription of its target genes including p21, cyclin D1, PTEN and *mdm-2* among others (Stambolic *et al.* 2001, Rocha *et al.* 2003).

The majority of the transcription factors studied so far, harbor as a primary property a DNA binding ability that may be changed by several parameters for example, the factor abundance, changes in affinity, competition for binding, sequestration outside the nuclear compartment, etc. Therefore, as a closing approach it is important to characterize if the overall DNA binding is affected under stress conditions like As exposure.

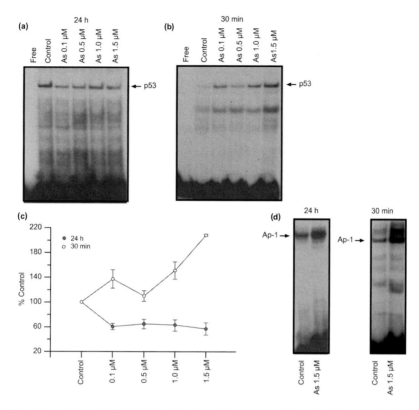

Figure 52.3. Comparison of effects induced by increasing concentrations of sodium arsenite in p53 DNA binding between short and long exposure times. Cells were treated with increasing sodium arsenite doses for 24 h (a) or for 30 min (b). Exposure time in autoradiography was different between panel (a) and (b) in order to obtain an image in which differences with control extracts were better appreciated; (c) densitometryc analysis of p53 complexes for both times are plotted (data from four independent quantifications); (d) same nuclear extracts obtained with cells treated with the highest concentration were used to test AP-1 binding by using an AP-1/SV40 ^{32}P-labeled probe.

52.5 CONCLUSION

The results presented herein, call for an important role for p53 in the signaling events triggered by the continuous exposure to As. Moreover, p53 provides a molecular platform to ascertain the pathological effects of As in epithelial cells. Work currently in progress in our laboratories is aimed at this objective. In epithelial cells, responses to As damage seem to be different depending on cell previous stage.

REFERENCES

Appella, E. and Anderson, C.W.: Post-translational modifications and activation of p53 by genotoxic stresses. *Eur. J. Biochem.* 268 (2001), pp. 2764–2772.
Burleson, F.G. Simeonova, P.P., Germolec, D.R. and Luster, M.I.: Dermatotoxic chemical stimulate of c-jun and c-fos transcription and AP-1 DNA binding in human keratinocytes. *Res. Commun. Mol. Pathol. Pharmacol.* 93 (1996), pp. 131–148.

Chen, C.J., Chuang, Y.C., Lin, T.M. and Wu, H.Y.: Malignant neoplasms among residents of a blackfoot disease-endemic area in Taiwan: high-arsenic artesian well water and cancers. *Cancer Res.* 45 (1985), pp. 5895–5899.

Colman, M.S., Afshari, C.A. and Barrett, J.C.: Regulation of p53 stability and activity in response to genotoxic stress. *Mutat. Res.* 462 (2000), pp. 179–188.

Germolec, D.R., Spalding, J., Yu, H.S., Chen, G.S., Simeonova, P.P., Humble, M.C., Bruccoleri, A., Boorman, G.A., Foley, J.F., Yoshida, T. and Luster, M.I.: Arsenic enhancement of skin neoplasia by chronic stimulation of growth factors. *Am. J. Pathol.* 153 (1998), pp. 1775–1785.

Gohler, T., Reimann, M., Cherny, D., Walter, K., Warnecke, G., Kim, E. and Deppert, W.: Specific interaction of p53 with target binding sites is determined by DNA conformation and is regulated by the C-terminal domain. *J. Biol. Chem.* 277 (2002), pp. 41,192–41,203.

Hamadeh, H.K., Trouba, K.J., Amin, R.P., Afshari, C.A. and Germolec, D.: Coordination of altered DNA repair and damage pathways in arsenite-exposed keratinocytes. *Toxicol. Sci.* 69 (2002), pp. 306–316.

Hamadeh, H.K., Vargas, M., Lee, E. and Menzel, D.B.: Arsenic disrupts cellular levels of p53 and mdm–2: a potential mechanism of carcinogenesis. *Biochem. Biophys. Res. Commun.* 263 (1999), pp. 446–449.

Hernandez-Zavala, A., Cordova, E., Del Razo, L.M., Cebrian, M.E. and Garrido, E.: Effects of arsenite on cell cycle progression in a human bladder cancer cell line. *Toxicology* 207 (2005), pp. 49–57.

Kim, T.K.: In vitro transcriptional activation of p21 promoter by p53. *Biochem. Biophys. Res. Commun.* 234 (1997), pp. 300–302.

Liu, G., Xia, T. and Chen, X.: The activation domains, the proline-rich domain, and the C-terminal basic domain in p53 are necessary for acetylation of histones on the proximal p21 promoter and interaction with p300/CREB-binding protein. *J. Biol. Chem.* 278 (2003), pp. 17,457–17,565.

Lopez-Bayghen, E., Vega, A., Cadena, A., Granados, S.E., Jave, L.F., Gariglio, P. and Alvarez-Salas, L.M.: Transcriptional analysis of the 5′-noncoding region of the human involucrin gene. *J. Biol. Chem.* 271 (1996), pp. 512–520.

Moll, U.M. and Petrenko, O.: The MDM–2-p53 interaction. *Mol. Cancer Res.* 1 (2003), pp. 1001–1008.

Mosmann, T.: Rapid colorimetric assay for cellular growth and survival: application to proliferation and cytotoxicity assays. *J. Immunol. Methods* 65 (1983), pp. 55–63.

Nigro, J.M., Baker, S.J., Preisinger, A.C., Jessup, J.M., Hostetter, R., Cleary, K., Bigner, S.H., Davidson, N., Baylin, S., Devilee, P., Glover, T., Collins, F.S., Weslon, A., Modali, R., Harris, C.C. and Vogelstein, B.: Mutations in the p53 gene occur in diverse human tumour types. *Nature* 342 (1989), pp. 705–708.

Pirisi, L., Yasumoto, S., Feller, M., Doniger, J. and DiPaolo, J.A.: Transformation of human fibroblasts and keratinocytes with human papillomavirus type 16 DNA. *J. Virol.* 61 (1987), pp. 1061–1066.

Rocha, S., Martin, A.M., Meek, D.W. and Perkins, N.D.: p53 represses cyclin D1 transcription through down regulation of Bcl-3 and inducing increased association of the p52 NF-kappaB subunit with histone deacetylase 1. *Mol. Cell Biol.* 23 (2003), pp. 4713–4727.

Salazar, A.M., Ostrosky-Wegman, P., Menendez, D., Miranda, E., Garcia-Carranca, A. and Rojas, E.: Induction of p53 protein expression by sodium arsenite. *Mutat. Res.* 381 (1997), pp. 259–265.

Sandoval, M., Morales, M., Tapia, R., Alarcon, L.C., Sordo, M., Ostrosky-Wegman, P., Ortega, A. and Lopez-Bayghen, E.: p53 Response to arsenic exposure in epithelial cells: protein kinase B/Akt involvement. *Toxicol Sci.* 99 (2007), pp. 126–140.

Saramaki, A., Banwell, C.M., Campbell, M.J. and Carlberg, C.: Regulation of the human p21 (waf1/cip1) gene promoter via multiple binding sites for p53 and the vitamin D3 receptor. *Nucleic Acids Res.* 34 (2006), pp. 543–554.

Scheffner, M., Munger, K., Byrne, J.C. and Howley, P.M.: The state of the p53 and retinoblastoma genes in human cervical carcinoma cell lines. *Proc. Natl. Acad. Sci. USA* 88 (1991), pp. 5523–5527.

Stambolic, V., MacPherson, D., Sas, D., Lin, Y., Snow, B., Jang, Y., Benchimol, S. and Mak, T.W.: Regulation of PTEN transcription by p53. *Mol. Cell* 8 (2001), pp. 317–325.

Tondel, M., Rahman, M., Magnuson, A., Chowdhury, I.A., Faruquee, M.H. and Ahmad, S.A.: The relationship of arsenic levels in drinking water and the prevalence rate of skin lesions in Bangladesh. *Environ. Health Perspect.* 107 (1999), pp. 727–729.

Tseng, W.P., Chu, H.M., How, S.W., Fong, J.M., Lin, C.S. and Yeh, S.: Prevalence of skin cancer in an endemic area of chronic arsenicism in Taiwan. *J. Natl. Cancer Inst.* 40 (968), pp. 453–463.

Vega, L., Styblo, M., Patterson, R., Cullen, W., Wang, C. and Germolec, D.: Differential effects of trivalent and pentavalent arsenicals on cell proliferation and cytokine secretion in normal human epidermal keratinocytes. *Toxicol. Appl. Pharmacol.* 172 (2001), pp. 225–232.

Woo, R.A., McLure, K.G., Lees-Miller, S.P., Rancourt, D.E. and Lee, P.W.: DNA-dependent protein kinase acts upstream of p53 in response to DNA damage. *Nature* 394 (1998), pp. 700–704.

Yee, C., Krishnan-Hewlett, I., Baker, C.C., Schlegel, R. and Howley, P.M.: Presence and expression of human papillomavirus sequences in human cervical carcinoma cell lines. *Am. J. Pathol.* 119 (1985), pp. 361–366.

CHAPTER 53

Microbial volatilization of arsenic

S. Čerňanský
Department of Ecosozology and Physiotactics, Faculty of Natural Sciences, University in Bratislava, Bratislava, Slovakia

M. Urík & J. Ševc
Institute of Geology, Faculty of Natural Sciences, Comenius University in Bratislava, Bratislava, Slovakia

ABSTRACT: Biomethylation plays an important role in the biogeochemical cycle of arsenic (As) in the environment. Filamentous fungi biomethylate inorganic As, forming both volatile and nonvolatile compounds. The quantification of production of volatile arsenicals *in vitro* is discussed in this article. Heat resistant fungi *Neosartorya fischeri, Talaromyces wortmannii, T. flavus, Eupenicillium cinnamopurpureum* were cultivated in 40 ml Sabouraud (SAB) medium enriched by 0.05, 0.25, 1.0 or 15.0 mg of inorganic arsenic acid (H_3AsO_4). After 30 day cultivation under laboratory conditions, total As was determined in mycelium and SAB medium using HG-AAS analytical method. Filamentous fungi volatilized 0.015–1.7 mg of As from the cultivation system, depending on the applied As concentration and fungal species.

53.1 INTRODUCTION

Transformation of arsenic (As) in the environment plays an important role in its toxicity and bioavailability. Speciation of As is influenced not only by abiotic factors (Hiller 2003, Jurkovič *et al.* 2005), but also by metabolic activity of organisms (Sadiq 1997).

Biomethylation and bioreduction are mechanisms that lead to As volatilization (Cullen *et al.* 1984, Michalke *et al.* 2000, Turpeinen *et al.* 2002) and represent the important mechanisms that influence the mobility of As in the environment (Gadd 2000). Biomethylation is a natural intracellular process that involves enzymatic conversion of inorganic As to its organic forms, (Thompson-Eagle and Frankenberger 1992). The process of biomethylation has been established to remove As from laboratory microcosms inoculated with volatilizing microbes (Visoottiviseth and Panviroj 2001). Initial step in As metabolization is the enzymatic reduction of pentavalent arsenate to trivalent arsenite (Mukhopadhyay and Rosen 2002). Once arsenite is in the cytosol it can be enzymatically methylated (Mukhopadhyay and Rosen 2002).

Biomethylation is catalyzed by S-adenosylmethionine dependent methyltransferases in the presence of glutathione and transforms inorganic As into its methyl derivates (Stýblo *et al.* 2002) such as trimethylarsine, dimethylarsinic acid, dimethylarsine, monomethylarsonic acid and dimethylarsinous acid. Methylated As species dimethylarsine and trimethylarsine are volatile, and can be indicated by characteristic garlic smell (Gosio gas).

Application of filamentous fungi for volatilization of toxic metalloids in the contaminated land and water environment has received increasing attention in recent years. This has been proved to be effective especially in the areas, where concentrations are elevated naturally or as a result of mining activities (Thompson-Eagle and Frankenberger 1992). The advantage of usage of heat-resistant fungi in soil remediation is in possible combination of thermal desorption of polluted soils and biological remediation. This may lead to lower costs for these two remediation methods and to higher efficiency of As removal.

This chapter presents the comparison of four different common fungal heat-resistant strains, isolated from a As contaminated site, to biovolatilize As under laboratory conditions. The experiments were designed to quantify amount of volatilized arsenic and thereby to evaluate the possible application of these strains in bioremediation of sites contaminated with arsenic.

53.2   EXPERIMENTAL

53.2.1   Isolation of fungal strains

Strains of *Neosartorya fischeri, Talaromyces wortmannii, T. flavus* and *Eupenicillium cinnamop-urpureum* were isolated from soil samples after heating at 70°C in Sabouraud agar (SAB-HiMedia Laboratories Ltd., Mumbai, India) with Rose Bengal (Jesenská *et al.* 1993, Piecková *et al.* 1994) and identified according to identification keys. Fungi were maintained on Sabouraud agar. Soils were collected from a locality highly contaminated with As (Pezinok, Slovakia). These four fungal strains represent the only indigenous heat-resistant strains isolated from soils and sediments from locality Pezinok in western Slovakia.

53.2.2   Indirect quantification of volatilized arsenic by fungi

To study the production of volatile As compounds, 100 ml Erlenmeyer flasks containing 40 ml of Sabouraud (SAB) medium were inoculated with 5 ml spore suspension. This suspension was prepared from cultures of *Neosartorya fischeri, Talaromyces wortmannii, T. flavus* or *Eupenicillium cinnamopurpureum* grown at room temperature in tubes containing Sabouraud agar. SAB medium was autoclaved for 20 minutes at 121°C before inoculation. SAB medium was then enriched with 0.05, 0.25, 1.0 or 15.0 mg pentavalent As as a solution of arsenic acid (Merck, Germany, H_3AsO_4 in 0.5 mol/l HNO_3).

After 30 day cultivation there were compact mycelia of selected fungi separated from SAB medium mechanically and than were washed with 50 ml of distilled water. This water containing As, which was previously mechanically sorbed on mycelia was than added to SAB medium. Dry fungal matter was than determined by drying mycelia at 40°C to constant weight.

Mycelium and the SAB medium were analyzed for total As concentration by HG-AAS as described below (see Section 53.2.4). There were three replicate runs for each experiment and non-inoculated controls (SAB and desired concentration of As) to determine non spontaneous evolution of volatile As; and fungal controls cultivated without presence of As.

53.2.3   Direct determination of volatilized arsenic in headspace of Neosartorya fischeri strain

Volatile As products from mycelial headspace of *Neosartorya fischeri* strain were captured on the 22nd and the 29th day of cultivation using selective sorbent sample tubes (Anasorb CSC with sorption material made from coconut shell charcoal, SKC Inc. USA, selective for methylated species of As). For this procedure Pocket Pump 210-1002 (SKC Inc., USA) was used with defined flow rate (100 ml/min) and sorption time (1 h). We used three replicate runs for each selected day. The controls were sorbent tubes, which sorbed a laboratory air for 1 h; and unused sorbent tubes.

53.2.4   Sample preparation and analytical methods

Decomposition of samples: Each sample of mycelium (0.1–0.5 g) and sorption material (0.17 g) was digested with 5 ml of concentrated HNO_3 in stainless-steel coated PTFE pressure bomb at 160°C in an electric oven for 6 h. The cool digest was transferred into a 50 ml volumetric flask, filled up to the mark and mixed. A 25 ml portion of solution was transferred into a PTFE beaker, 5 ml of concentrated H_2SO_4 were added, covered and digested on the sand bath (3 h). The solution was evaporated to approximately 5 ml, transferred to 15 ml volumetric flask and made up to the volume.

Pre-reduction with KI: 5 ml of sample, standard and SAB solution were transferred to a 50 ml volumetric flask, 20 ml of re-distilled water, 5 ml of concentrated HCl and 2 ml of 20% m/v KI solution were added. After 15 m in, 2 ml of 10% m/v ascorbic acid solution were added and left for 60 min (Čelková *et al.* 1996).

The reliability of this procedure was tested by soil certified reference material SO-4 (CANMET, Canada), with high content of organic matter. Synthetic As standard was prepared by a dilution of stock solutions of As 1000 g/l (Merck, Germany, H_3AsO_4 in 0.5 mol/l HNO_3). Total As concentration in mycelium and SAB medium was analyzed by using a Perkin-Elmer Atomic Absorption Spectrometer model 1100 (USA) with a laboratory-made hydride generator.

53.3 RESULTS AND DISCUSSION

Mechanisms of biomethylation and bioreduction as part of the biogeochemical As cycle are significant because of affecting physical, chemical and biological properties of As compounds, especially As toxicity and mobility (Oremland and Stolz 2003, Slaninka *et al.* 2006). Fungi play an important role in arsenic biogeochemical cycle as a step that involves all of these transformations, including immobilization (sorption) and volatilization of arsenic. The ability of filamentous fungi to volatilize As is shown in Tables 53.1–53.3. The metabolic transformation of arsenic into its volatile derivatives by microorganisms was observed also by other authors (Pearce *et al.* 1998, Michalke *et al.* 2000, Visoottiviseth and Panviroj 2001). Visoottiviseth and Panviroj (2001) used the same indirect calculation of volatilized arsenic, as the authors of this chapter did. Results of the amount of volatilized As (Tables 53.1–53.3) were calculated as a difference between the content of total As before cultivation and sum of total As in mycelium and SAB medium. Amounts of volatilized As varied between 0.025 and 0.028 mg of As for cultivation systems originally enriched with 0.25 mg of inorganic As for species *T. wortmannii, T. flavus* and *E. cinnamopurpureum*. The amount of volatilized As by these species was higher (0.088 to 0.090 mg of As) for systems enriched with 1 mg of inorganic As. These results are comparable to these published by Visoottiviseth and Panviroj (2001), who isolated As-hypertolerant fungal *Penicillium* sp. strain. *N. fischeri* volatilized 0.180, 0.321 and 1.613 mg As on average from systems originally enriched by 0.25, 1.0 and 15 mg of inorganic As, respectively. This fungal strain with highest ability to volatilize arsenic (the highest biovolatilization factor) was then used for direct determination of production of volatile arsenicals on 22nd and 29th day of its cultivation in a cultivation system enriched by 0.05 mg of inorganic As. The total amount of volatilized As after cultivation (30 days) was 0.0381 mg As, representing 76.2% of initial As content in the system. Direct determination of volatilized As was carried out by capturing volatile arsenicals on sorption material. The amount of As captured on sorption material was 33.8 ± 4.7 ng (22nd day of cultivation) and 55.0 ± 3.3 ng (29th day of cultivation). No As was found in either type of controls without sorption material.

There was no detectable As in fungal control samples, which were cultivated in a As-free medium. We determined no loss of As (data not shown) in cultivation systems without microorganisms. These systems included only 45 ml SAB medium enriched only with the desired amount of As.

Table 53.1. As volatilization by fungi after 30 day cultivation period (initial As content 0.25 mg) ($n = 3$).

Fungal species	As content in SAB mg	Mycelial dry weight g	As accumulated in mycelium mg	Calculated amount of volatilized As mg
Eupenicillium cinnamopurpureum	0.199 ± 0.012	0.300 ± 0.072	0.023 ± 0.004	0.028 ± 0.013
Talaromyces wortmannii	0.194 ± 0.010	0.435 ± 0.024	0.029 ± 0.003	0.027 ± 0.012
Talaromyces flavus	0.199 ± 0.011	0.510 ± 0.051	0.025 ± 0.003	0.025 ± 0.010
Neosartorya fischeri	0.067 ± 0.011	0.500 ± 0.010	0.003 ± 0.000	0.180 ± 0.011

Table 53.2.   As volatilization by fungi after 30 day cultivation period (initial As content 1.00 mg) ($n = 3$).

Fungal species	As content in SAB mg	Mycelial dry weight g	As accumulated in mycelium mg	Calculated amount of volatilized As mg
Eupenicillium cinnamopurpureum	0.814 ± 0.048	0.317 ± 0.029	0.099 ± 0.017	0.088 ± 0.036
Talaromyces wortmannii	0.796 ± 0.018	0.495 ± 0.007	0.114 ± 0.009	0.090 ± 0.026
Talaromyces flavus	0.801 ± 0.020	0.561 ± 0.029	0.111 ± 0.010	0.088 ± 0.017
Neosartorya fischeri	0.619 ± 0.010	0.455 ± 0.049	0.060 ± 0.022	0.321 ± 0.015

Table 53.3.   As volatilization by fungi after 30 day cultivation period (initial As content 15.00 mg) ($n = 2$).

Fungal species	As content in SAB mg	Mycelial dry weight g	As accumulated in mycelium mg	Calculated amount of volatilized As mg
Neosartorya fischeri	12.390 ± 0.320	0.233 ± 0.014	1.002 ± 0.430	1.613 ± 0.105

Another aspect of our experiment was bioaccumulation, which is represented by the sum of As remaining in mycelium after cultivation period. Fungal biomass is a relatively cheap sorbent that can be easily obtained by industrial fermentation and seems to be successful in sorption of trace metals (Šimonovičová *et al.* 2002).

Volatilization of metalloids, such as selenium, was already applied at highly contaminated sites (Thompson-Eagle and Frankenberger 1992) by optimization of environmental parameter affecting fungal growth (such as aeration, addition of nutrients or regulation of pH). Pearce *et al.* (1998) also suggested that volatilization of As and Se is comparable. Our previous research has shown that As volatilization occurs naturally (unpublished data). These suggestions proved volatilization of arsenic as a promising method for bioremediation of sites contaminated with As.

53.4   CONCLUSIONS

This experiment showed that microscopic filamentous fungi volatilize As at high levels under laboratory conditions. The calculated results of volatilized As compare the ability of fungal species to produce volatile arsenicals. This loss of As was measured indirectly by determination of sum of total As in mycelium and SAB. Our results suggest possible application of fungi in bioremediation of sites contaminated with arsenic. While there are many works dealing with As-contamination of water using fungal biomass as a sorbent, there is lack of literature that deals with remediation of As-contaminated substrates by metabolic transformation of arsenic into volatile derivatives. Our research has shown that fungi should play an important role in remediation of contaminated soils through their metabolic pathway of arsenic.

ACKNOWLEDGEMENTS

This work was financially supported by VEGA 1/3462/06 and VEGA 1/4361/07. We thank Miroslav Vančo for treatment of samples for analytical measurements. We thank the comments from the anonymous reviewers that helped in considerable improvement of the earlier drafts of the manuscript.

REFERENCES

Čelková, A., Kubová, J. and Streško, V.: Determination of arsenic in geological samples by HG AAS. *Fresenius J. Anal. Chem.* 355 (1996), pp. 150–153.

Čerňanský, S., Ševc, J. and Urík, M.: Microbial processes of arsenic methylation under laboratory and natural conditions. In: Fečko, P. (ed): *Proceedings 8th Conference on Environment and Mineral Processing Part I.* Ostrava, VŠB-TU, 2004, pp. 227–232.

Cullen, W.R., McBride, B.C., Pickett, A.W. and Reglinski, J.: The wood preservative chromated copper arsenate is a substrate for trimethylarsine biosynthesis. *Appl. Environ. Microbiol.* 47 (2000), pp. 443–444.

Gadd, G.M.: Bioremedial potential of microbial mechanisms of metal mobilization and immobilization. *Curr. Opin. Biotechnol.* 11 (2000), pp. 271–279.

Hiller, E.: Adsorption of arsenates on soils: kinetic and equilibrium studies. *J. Hydrol. Hydromech.* 51 (2003), pp. 288–297.

Jesenská, Z., Piecková, E. and Bernát, D.: Heat resistance of fungi from soil. *Int. J. Food Microbiol.* 19 (1993), pp. 187–192.

Jurkovič, Ľ., Slaninka, I. and Kordík, J.: Geochemické štúdium arzénu a jeho mobilita sekundárne ovplyvnenom povodí potoka Kyjov. *Geochémia 2005*: 33–36 (2005), Bratislava: ŠGÚDŠ.

Michalke, K., Wickenheiser, E.B., Mehring, M., Hirner, A.V. and Hensel, R.: Production of volatile derivates of metal(loid)s by microflora involved in anaerobic digestion of sewage sludge. *Appl. Environ. Microbiol.* 66 (2000), pp. 2791–2796.

Mukhopadhyay, R. and Rosen, B.P.: Arsenate reductases in prokaryotes and eukaryotes. *Environ. Health Persp.* 110 (2002), pp. 745–748.

Oremland, R.S. and Stolz, J.F.: The ecology of arsenic. *Science* 300 (2003), pp. 939–944.

Pearce, R.B., Callow, M.E. and Macaskie, L.E.: Fungal volatilization of arsenic and antimony and the sudden infant death syndrome. *FEMS Microbiol. Lett.* 158 (1998), pp. 261–265.

Piecková, E., Bernát, D. and Jesneská, Z.: Heat resistant fungi isolated from soil. *Int. J. Food Microbiol.* 22 (1994), pp. 297–299.

Sadiq, M.: Arsenic chemistry in soils: An overview of thermodynamic predictions and field observations. *Water Air Soil Pollut.* 93 (1997), pp. 117–136.

Šimonovičová, A., Ševc, J. and Iró, S.: Trichoderma viride Pers. ex Gray as biosorbent of heavy metals (Pb, Hg and Cd). *Ecology (Bratislava)* 21 (2002), pp. 298–306.

Slaninka, I., Jurkovič Ľ, and Kordík, J.: Ekologická záťaž vodného ekosystému arzénom v oblasi odkaliska Poša (Východné Slovensko). *Vodní hospodářství* 11 (2006), pp. 275–277.

Stýblo, M., Drobná, Z., Jasper, I., Lin, S. and Thomas, D.J.: The role of biomethylation in toxicity and carcinogenicity of arsenic: A research update. *Environ. Health Persp.* 110 (2002), pp. 767–771.

Thompson-Eagle, E.T. and Frankenberger, W.T.: Bioremediation of soils contaminated with selenium. In: R. Lal and B.A. Stewar (eds): Advances in soil science. Springer-Verlag, New York, 1992, pp. 261–309.

Turpeinen, R., Pantsar-Kallio, M. and Kairesalo, T.: Role of microbes in controlling the speciation of arsenic and production of arsenic in contaminated soils. *Sci. Total Environ.* 285 (2002), pp. 133–145.

Visoottiviseth, P. and Panviroj, N.: Selection of fungi capable of removing toxic arsenic compounds from Liquid Medium. *Science Asia* 27 (2001), pp. 83–92.

Section VI
Treatment and remediation of arsenic-rich groundwater

Natural geological materials—available locally and regionally

CHAPTER 54

Feasibility of arsenic removal from contaminated water using indigenous limestone

M.A. Armienta

Instituto de Geofísica, Universidad Nacional Autónoma de México (UNAM), Mexico City, Mexico

S. Micete & E. Flores-Valverde

Posgrado en Ciencias e Ingeniería Ambientales, Universidad Nacional Autónoma de México (UNAM), Mexico City, Mexico

ABSTRACT: Zimapán, a small town in central Mexico, is supplied with water from wells that indicate elevated concentrations of arsenic (As). The mobilization of As in the public supply wells are mainly triggered by geochemical processes mainly due to oxidation of arsenopyrite and dissolution of scorodite. Due to the lack of surface water bodies, motivated a study of alternatives to provide safe water to the population. Batch and column tests were used to evaluate the feasibility of using indigenous limestone to produce clean water from a deep well currently supplying drinking water to the town. Results showed a 90% decrease in As concentration when 1 liter of water was treated with 10 g rock particles <0.5 mm. The same rocks may be used five times to clean new batches of contaminated water. Packing columns with rocks 0.84 mm to 1.00 mm size allowed an adequate water flow, and produced a concentration of 0.025 mg/l As in the outflow. Waste rocks could be mixed with tailing piles near the town to raise their pH and control acid mine drainage.

54.1 INTRODUCTION

Arsenic (As) concentrations exceeding the Mexican drinking water standard have been measured in deep wells used for potable supply at Zimapán, Mexico. Arsenic contamination in these wells is produced by natural processes, mainly oxidation of arsenopyrite and dissolution of scorodite (Armienta *et al.* 1997a, 2001, Rodríguez *et al.* 2004). The Zimapán valley is located to the west of the Sierra Madre Oriental ravine. The oldest geologic unit is the Las Trancas formation of Jurassic age, which is composed of calcareous shale. This formation is overlied by Cretaceous limestone formations and by Cenozoic volcanic and continental deposits. The Late Cretaceous Soyatal formation consists of thinly bedded limestone with marl and interbedded calcareous shales. It outcrops on a large area northwest of Zimapán town. The lack of productive non-contaminated wells or surface water bodies in the area, give few alternatives to As pollution. Potable water supply has relied on mixing water pumped from wells containing various As concentrations, producing variable As contents. Currently, good quality water is pumped to Zimapán from a well located 25 km far and 400 m lower altitude, and mixed with water from a well (Z5) located in the town of Zimapán containing 0.5 mg/l As on average, to supply potable water. Variations in the proportion from each source result in variable As concentrations. In July 2005, the As concentration was 0.2908 mg/l. Health effects, mainly hyper- and hypo-pigmentation and thickening of the outer layer of palms and soles, have been related to the consumption of As-polluted water at Zimapán (Armienta *et al.* 1997b, Resendiz and Zúñiga 2003, Valenzuela *et al.* 2005).

Common methods to treat As-polluted waters include: alum and iron coagulation, lime softening, sorption on activated charcoal, alumina, iron coated sand and ion exchange resins, reverse osmosis, and electrodialysis. Due to their low cost, chemical methods like coagulation or sorption, may be an option to As pollution in low-income communities. Iron oxides and zeolites, are the geological materials most used to remove As through sorption at other polluted sites. However, the geology of Zimapán with abundant limestone outcrops, prompted a study of their As removal potential. Besides, previous studies have shown the capacity of limestones from this area to remove As (Ongley *et al.* 2001, Romero *et al.* 2004). The aim of this work was to determine the feasibility of the Soyatal limestone to produce clean water from the deep well Z5.

54.2   METHODS

Batch and column experiments were performed to determine the capability of Soyatal rocks, to remove As from the polluted water discharging from Z5 well. Batch tests were carried out varying time, rock/water ratio, and rock particle size in Erlenmeyer flasks. There was one control for each experiment. Column tests were performed using a hose packed with 0.84–1.00 mm rock particles.

54.2.1   *Sampling and analysis*

Rock samples were collected from an outcropping of Soyatal rocks located NE of Zimapán town. Samples were chosen based on their similar response to HCl addition and the absence of visible alteration products. In the laboratory, the rocks were dried, crushed, milled and sieved through US standard sieves to collect different particle sizes. Arsenic was determined by hydride generation atomic absorption spectrophotometry at the *Instituto de Geofísica*, UNAM, and by FIA-HG-AAS at *Universidad Autónoma Metropolitana*. Analytical determinations were made in duplicate with each method; accuracy was tested through the analysis of As in the same samples at both laboratories. Chemical and mineralogical characterization of rock samples were performed by XRF (SRS3000) and XRD (SIEMENS D5000) at the Chemistry Faculty, UNAM.

54.3   RESULTS

54.3.1   *Arsenic in water*

Concentrations of the deep well Z5 (Zimapán 5), located in Zimapán town, and of the municipal potable water supply P, resulting from mixing of various wells, for 12 years are shown in Figure 54.1. Arsenic content in the water from both sources has been always above the Mexican drinking water standard.

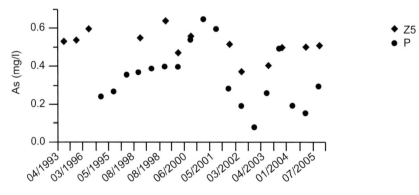

Figure 54.1.   Arsenic concentrations measured in Zimapán 5 (Z5) deep well, and municipal potable water supply (P) since 1993. Several wells containing different As concentrations have been mixed to supply potable water, resulting in variable As contents.

54.3.2   *Limestone characterization*

Chemical and mineralogical analyses of Soyatal rocks used for the experiments were performed to determine a possible cause of As retention. The bulk chemical analysis of the rock revealed a composition as follows: 41.39% CaO, 1.53% Fe_2O_3, 18.32% SiO_2, 2.98% Al_2O_3, 0.157% TiO_2, 0.038% MnO, 0.599% MgO, 0.117% Na_2O, 0.652% K_2O, and 34.1% LOI. Mineralogical analyses evidenced only the presence of calcite and quartz as main minerals; additionally, small peaks corresponding to strontium aluminum oxide were also identified in the diffractograms.

54.3.3   *Arsenic sorption batch experiments*

Experiments were carried out using Erlenmeyer flasks as batch reactors with continuous agitation. An oxidation step before the treatment with rocks was not included, since As is mostly (>90%) As(V). In the first experiment 10 g, and 100 g of the largest particles (>2 mm), were added to 1 l water from Zimapán 5 well containing 0.456 mg/l As, and agitated for 90 minutes. The water was filtered through 0.45 μm microfilter and analyzed. Arsenic concentrations after treatment with 10 g, and 100 g of rock were 0.419 mg/l and 0.365 mg/l, respectively, indicating a low capacity for As removal. Although formation of fines was observed after agitation, the amount was less than 10% of the total mass of coarse rocks. Variable amounts of rock particles less than 0.5 mm were added to 100 ml water and agitated for 1, 3, and 5 hours in the second experiment. Results shown in Table 54.1 indicate that these tests render a high removal efficiency. The amount of As removed from the solution increased with the amount of rock.

The possibility of using the same rock to treat the polluted water several times was tested in the third experiment. Ten grams of particles <0.5 mm were mixed with 20 ml water and agitated for 30 min, 1 h, and 2 h. After those times, water was poured and filtered, and new As-polluted water was added to the rock. This procedure was repeated 5 times. Arsenic concentrations in the filtrate are shown in Figure 54.2. Similar As removal efficiencies were obtained at 30 min, 1 h, and 2 h in this experiment. A slight decreasing removal capacity was observed from the third to the fifth cycle. However, the rock still retained more than 90% of As.

Larger rock particles (0.84–1 mm) were tested for As removal efficiency in the third batch experiment. Ten grams of rock were added with 100 ml water (0.3598 mg/l As), and agitated for 30 min, and 1 h. Fresh water was added to the same rock five times. Results showed a decrease in As removal efficiency compared to the <0.5 mm experiments (Figure 54.2). Approximately 85% of As was removed in the first cycle. However, a decrease in the As retention efficiency was observed from the second cycle for both treatment times.

The largest particles (1–1.7 mm) were used in the fourth experiment. Again, 100 ml of fresh waters (0.3598 mg/l As) were added to 10 g rock and agitated for 30 minutes. The procedure was repeated with fresh water five times. Results shown in Figure 54.3 indicate a much lower removal capacity compared to other particle sizes.

54.3.4   *Arsenic sorption column experiments*

Rock particles <5 mm were used to pack the first column based on batch experiments' results. A PVC column 7.6 cm diameter, 1.1 m high, packed with sorted sized rock particles from the

Table 54.1.   Arsenic concentrations after treatment of water (100 ml, 0.456 mg/l As). Rock particles <0.5 mm.

Rock	1 g		5 g		10 g	
Time (h)	As (mg/l)	% Removal	As (mg/l)	% Removal	As (mg/l)	% Removal
1	0.045	90.1	0.026	94.3	0.016	96.5
3	0.052	88.6	0.025	94.6	0.015	96.7
5	0.048	89.5	0.018	96.0	0.014	96.9

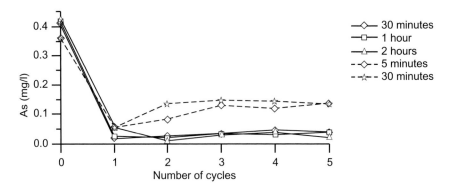

Figure 54.2.   Arsenic concentration at different treatment times in batch experiments: Using 10 g rock particles (<0.5 mm) with 20 ml water (solid lines), and using 10 g rock particles (0.84–1 mm)
with 100 ml water (dashed lines). Fresh As-polluted water was added to the same rock sample
(cycles) up to 5 times.

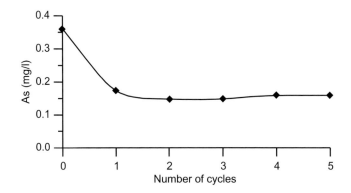

Figure 54.3.   Arsenic concentration at different treatment times in batch experiments using 10 g rock particles
(1.0–1.7 mm) with 100 ml water (0.3598 mg/l As) and agitating 30 minutes. Fresh As-polluted
water was added to the same rock sample (cycles) up to 5 times.

Table 54.2.   Column characteristics.

	Length (m)	Volume (l)	Packed bed length (m)	Packed bed volume (cm^3)	Rock mass (g)	Flow (l/min)
Column 2	4.20	0.532	1.48	187	160	0.060
Column 3	1.85	0.234	0.56	70.9	60	0.064

bottom to 15 cm, and covered by 30 cm packing with particles <0.5 mm, was used in the first
experiment. This test was performed at Zimapán town, next to the polluted well outflow. Water was
poured from the top, but it did not flow through to the bottom. The clay-size of the rocks formed
an impermeable barrier preventing water flow.

Larger rock particles (0.84–1 mm) were used in the second and third experiments. Experimental
parameters are shown in Table 54.2.

Conventional polyethylene hose (1.27 cm diameter) coiled around a column was used to perform
the subsequent column experiments. This material was chosen on the basis of its accessibility in
small towns. Results shown in Figures 4a and b indicate a lower removal capacity for column
2 compared to column 3. However, similar outflow As concentrations were obtained from both

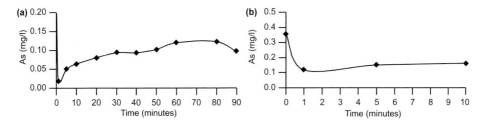

Figure 54.4.   Arsenic concentration in packed column 2 (a) and 3 (b) outflow.

columns. Concentrations were also similar to those achieved in batch experiments with the same rock particles' size.

54.4   WASTE DISPOSAL OPTIONS

Adequate disposal of waste rocks should be considered when proposing a treatment system. Specific landfills may be used to dispose the Soyatal waste rocks. Besides, at Zimapán several tailing piles are located on the outskirts of town. Chemical and mineralogical studies of these wastes indicated high As levels (up to 8.25%). However, not all the As mobilizes to the environment. Sequential extraction showed that most of the As is associated with the most stable residual and hydrous ferric oxides' fractions (Méndez and Armienta 2003). Arsenic and other heavy metals are also retained in secondary minerals formed within tailings. Presence of calcite increases the pH, aids to control acid mine drainage, and promotes hydrous ferric oxides and secondary minerals' formation (Méndez and Armienta 2003, Romero *et al.* 2006). Waste rocks used to treat the water, may thus be mixed with the tailings aiding to increase the pH and immobilize heavy metals.

54.5   CONCLUSIONS

Batch and column tests showed that rocks from the Soyatal limestone outcropping at Zimapán, may be used to produce clean water from the As contaminated deep well Z5. Smaller rock particles removed As more efficiently than larger ones. A 90% decrease in the As concentration was obtained in batch experiments treating 1 liter of water with 10 g rocks of a size <0.5 mm. Experiments were run for several hours, but removal took place within the first minutes. Use of rock particles 0.84–1.00 mm decreased As content to 0.057 mg/l or less, within 5 minutes. The same batch of rocks may be used at least five times to treat the water without losing efficiency. Use of particles <0.5 mm hindered the water flow in column experiments. Packing with rocks 0.84 mm to 1.00 mm size, allowed an adequate water flow, and produced an As concentration of 0.025 mg/l in the outflow.

   Use of the Soyatal limestone rocks at Zimapán has several advantages including their abundance in the area, simple harvesting and milling. Limestones may be used for domestic water-treatment as proposed by Ongley *et al.* (2001). The same rocks may be used five times to clean new batches of contaminated waters. Packed columns may be used on site to treat the water flowing from the polluted well. Waste rocks may be disposed of in tailings located at Zimapán town, and help to increase their pH values and control acid mine drainage.

   Chemical composition and mineralogy of Soyatal rock samples used for the experiments indicated that calcite is the only main mineral able to retain As. Results confirm the retention capacity of calcite determined from pH_{zpc} and batch experiments at various pH values, performed on samples from this formation by Romero *et al.* (2004). Limestones with similar mineralogy to Soyatal rocks may be also used to remove As at other sites, but experiments must be developed to assess their actual capacity.

ACKNOWLEDGEMENTS

The authors acknowledge O. Cruz, N. Ceniceros and A. Aguayo from the *Instituto de Geofísica* UNAM, and M.R. Valladares from the *Universidad Autónoma Metropolitana*, for their support in analytical determinations. *Consejo Nacional de Ciencia y Tecnología* (CONACYT) and *Secretaría de Medio Ambiente y Recursos Naturales* (SEMARNAT) are acknowledged for funding (Project C01-0017-2002).

REFERENCES

Armienta, M.A., Rodriguez, R., Aguayo, A., Ceniceros, N., Villaseñor, G. and Cruz, O.: Arsenic contamination of groundwater at Zimapán, México. *Hydrogeol. J.* 5 (1997a), pp. 39–46.

Armienta, M.A., Rodriguez, R. and Cruz, O.: Arsenic content in hair of people exposed to natural arsenic polluted groundwater at Zimapán, México. *Bull. Environ. Contam. Toxicol.* 59 (1997b), pp. 583–589.

Armienta, M.A., Villaseñor, G., Rodríguez, R., Ongley, L.K. and Mango, H.: The role of arsenic-bearing rocks in groundwater pollution at Zimapán Valley, México. *Environ. Geol.* 40 (2001), pp. 571–581.

Méndez, M. and Armienta, M.A.: Arsenic phase distribution in Zimapán mine tailings, Mexico. *Geofísica Int.* 42 (2003), pp. 131–140.

Ongley, L.K., Armienta, M.A., Heggeman, K., Lathrop, A., Mango, H., Miller, W. and Pickelner, S.: Arsenic removal from contaminated water by the Soyatal Formation, Zimapán mining district, Mexico—a potential low-cost low-tech remediation system. *Geochem. Explor. Environ. Anal.* 1 (2001), pp. 23–31.

Resendiz, M.R.I. and Zúñiga, L.J.C.: *Evaluación de la exposición al arsénico en pobladores del municipio de Zimapán, Hidalgo.* Engineering Chemistry Bachelor's Thesis, Universidad Tecnológica de México, Mexico City, Mexico, 2003.

Rodríguez, R., Ramos, J.A. and Armienta, A.: Groundwater arsenic variations: the role of local geology and rainfall. *Appl. Geochem.* 19 (2004), pp. 245–250.

Romero, F.M., Armienta, M.A. and Carrillo-Chavez, A.: Arsenic sorption by carbonate-rich aquifer material, a control on arsenic mobility at Zimapán, Mexico. *Arch. Environ. Contam. Toxicol.* 47 (2004), pp. 1–13.

Romero, F.M., Armienta, M.A., Villaseñor, G. and González, J.L.: Mineralogical constraints on the mobility of arsenic in tailings from Zimapán, Hidalgo, Mexico. *Int. J. Environ. Pollut.* 26 (2006), pp. 23–40.

Valenzuela, O.L., Borja-Aburto, V.H., Garcia-Vargas, G.G., Cruz-Gonzalez, M.B., Garcia-Montalvo, E.A., Calderon-Aranda, E.S. and Del Razo, L.M.: Urinary trivalent methylated arsenic species in a population chronically exposed to inorganic arsenic. *Environ. Health Perspect.* 113 (2005), pp. 250–254.

CHAPTER 55

Characterization of Fe-treated clays as effective As sorbents

B. Doušová, A. Martaus, D. Koloušek, L. Fuitová & V. Machovič
Institute of Chemical Technology in Prague, Prague, Czech Republic

T. Grygar
Institute of Inorganic Chemistry, Řež, Czech Republic

ABSTRACT: Two methods using Fe(II) and Fe(III) salts were applied to the pre-treatment of natural clay minerals to improve their sorption efficiency for As(V) and As(III) species. Different types of clays, natural kaolin from the Merkur quarry, Czech Republic, calcined at 550°C, and a raw bentonite from a mineral deposit in Hajek, Czech Republic, were used. In the first process the initial material was exposed to a concentrated solution of a Fe(II) salt (0.6 M $FeSO_4 \times 7H_2O$) for 24 hours. In the second process the sorbents were treated overnight with partly hydrolyzed Fe(III)-bearing solutions [0.025 M $Fe(NO_3)_3 \times 9H_2O$; 0.05 M $NaOH$]. The results showed, that the method with Fe(II) was more appropriate for the kaolin treatment [sorption efficiency of kaolin/bentonite for As(V) $\approx 98/28\%$; for As(III) $\approx 30/33\%$], whereas for the bentonite treatment the procedure with Fe(III) was superior [$>99\%$ efficiency for both the As(III) and As(V) sorption]. In general, arsenates [As(V) species] showed a higher sorption affinity to all sorbents prepared than arsenites [As(III) species]. The properties of pre-treated sorbents in aqueous systems strongly depended on the mechanism of Fe-loading process, pH, and reaction time.

55.1 INTRODUCTION

Arsenic has been known as a toxic element for centuries. The most abundant mineral sources of As are arsenian pyrite and arsenopyrite (Cullen and Reimer 1989). Weathering of these minerals in oxidizing environments solubilizes arsenic (As) as As(III) and ultimately As(V). Naturally occurring As in groundwaters is of great environmental concern for example in India, Bangladesh and Nepal, but also in Mexico and Argentina due to alarming health problems of inhabitants relying on strongly contaminated groundwater as a source of drinking and/or irrigation water (Wagner *et al.* 2005, Acharyya and Shah 2005, Tandukar *et al.* 2005, Bhattacharya *et al.* 2005, Cole *et al.* 2005, Rodriguez *et al.* 2005). Arsenic is unique among the heavy metalloids and oxyanion-forming elements due to its sensitivity to redox potential and to mobilization at the pH values typically found in groundwaters (6.5–8.5). Under natural conditions, groundwaters show the greatest range and the highest concentrations of As among the natural waters (Smedley and Kinniburgh 2002). Various techniques have been developed for As removal from aqueous systems. Precipitation, adsorption, membrane processes, ion exchange, ion flotation and biological processing are frequently employed. Adsorption methods currently reach a significant position among the above mentioned due to the great variability of applicable sorbents, both natural and synthetic. The research is focused on production of easily available, inexpensive, but also ecological and effective sorbents for As removal. A number of papers have studied the strong adsorption affinity of As-oxyanions to hydrated iron oxides forming very stable inner-sphere bidentate As(V)-Fe(III) complexes (Lin and Puls 2003, Bruce *et al.* 1998, Sherman and Randall 2003, Doušová *et al.* 2005, Antelo *et al.* 2005, Deliyanni *et al.* 2003). In fact, iron oxides and oxyhydroxides represent the most widespread and effective sorbents for inorganic As forms in natural environments (i.e. in soils, lake and river bottom sediments). On the other hand, their syntheses from iron salts require very specific reaction conditions, therefore they could be assigned to expensive and hardly available sorbents for wide use.

Zhang et al. (2004) tested natural iron ores rich in hematite as sorbents for the removal of As(V) from drinking water. They investigated water systems with As concentrations lower than 1 mg/l. The efficiency of aluminum based sorbents (e.g., red mud, activated alumina, α-Al_2O_3) for treatment of As-contaminated waters has also been studied (Altundogan et al. 2002, Halter and Pfeifer 2001, Singh and Pant 2004). Doušová et al. (2003) compared different types of sorbents for removal of As(V) from model aqueous systems.

The adsorption of As species on clay surfaces is in general very important for economical reasons. Most of the considered natural clays are low-cost and easily available. Clays are not selective sorbents for anionic contaminants due to the low isoelectric point. Elizade-González et al. (2001) tested some types of natural zeolites for As sorption. Several papers have been focused on the metal pre-treatment of carrier materials (clays, sands etc.) to improve their sorption affinity to As oxyanions (Xu 2002, Min and Hering 1998, Izumi et al. 2005, Gupta et al. 2005).

The main objectives of this work were to investigate two simple pre-treating methods [with Fe(II) and Fe(III) salts] and to apply them to different types of raw clays (bentonite and calcined kaolin). The results allowed comparison of both the mentioned methods and to suggest the optimal preparation technique for easily available, low-cost removal of As(III) and/or As(V) from aqueous systems applying highly effective sorbents.

55.2   EXPERIMENTAL

55.2.1   Sorbents

Two sorbents used in this work came from the Czech Republic and represent natural materials. A washed natural metakaolin from the Merkur quarry, West Bohemia, calcined at 500°C for 3 hours (Fig. 55.1a) belongs to the kaolinite group. The primary structural unit of this group is a layer composed of one octahedral sheet condensed with one tetrahedral sheet. In the dioctahedral minerals (e.g., kaolinite and halloysite) the octahedral sites are occupied by aluminum. Kaolinite and halloysite are single-layer nonexpanding structures (1:1) (Moore and Reynolds, 1997).

A raw bentonite from a mineral deposit in Hajek, northwest Bohemia (Fig. 55.1b) contains mostly smectite. The basic structural unit is a layer consisting of two inward-pointing tetrahedral sheets with a central alumina octahedral sheet. The layers are continuous in the a and b directions, but the bonds between layers are weak and have excellent cleavage, allowing water and other molecules to enter between the layers causing expansion in the c direction (i.e. expanding structure, 2:1) (Grim 1968). The semiquantitative analysis of sorbents used is shown in Table 55.1.

55.2.2   Model solutions

A 2×10^{-3} mol/l As(V) solution was prepared from $NH_4H_2AsO_4$ or KH_2AsO_4 of analytical-grade quality and distilled water. The As(III) solution with analogous concentration was prepared from $NaAsO_2$ of analytical-grade quality and distilled water.

55.2.3   Treating processes

In the first procedure both sorbents (20 g) were mixed with 1 l of 0.6 M $FeSO_4 \times 7H_2O$ solution in sealed polyethylene bottles at laboratory temperature (20°C) and agitated for 24 hours. Then the solid phases of samples were filtered off, washed with distilled water, dried at 75°C and homogenized (Bonnin 2000).

In the second procedure the sorbents were treated overnight with a solution of partly hydrolyzed $Fe(NO_3)_3$ (0.025 M $Fe(NO_3)_3 \times 9H_2O$ and 0.05 M NaOH), (30 g in 0.5 l solution). The resulting reddish brown solid was filtered off, washed with distilled water, dried and homogenized.

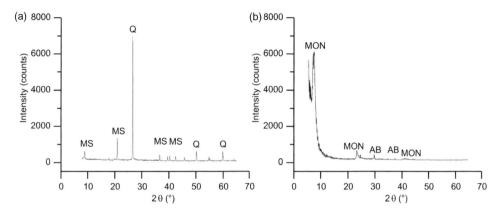

Figure 55.1.   X-ray diffractograms of raw sorbents: (a) metakaoline, (b) bentonite. MS: muscovite; Q: quartz; AB: albite; MON: montmorillonite.

Table 55.1.   XRF analysis of used sorbents (% wt.).

Sample	Al_2O_3	SiO_2	Na_2O	CaO	K_2O	MgO	Fe_2O_3	TiO_2	SO_3	P_2O_5
Metakaolin	37.75	56.37	0.12	0.15	1.94	0.38	2.68	0.27	0.21	0.06
Bentonite	17.68	55.36	0.14	3.16	0.99	3.28	12.76	5.05	0.11	0.97

55.2.4   *As(III)/As(V) sorption*

The model solution at initial pH value [≈5.5 for As(V) and ≈8.8–9 for As(III)] and the defined amount of sorbent were shaken in sealed polyethylene bottles at laboratory temperature (20°C) for 24 (72) hours. The product was filtered off; the filtrate was analyzed for residual As content, while the solid part was examined by IR spectroscopy and voltammetry of microparticles.

55.2.5   *Analytical methods*

Powder X-ray diffraction patterns were recorded using a Seifert XRD 3000P instrument with CoKα radiation ($\lambda = 0.179026$ nm, graphite monochromator, goniometer with Bragg-Brentano geometry) in 2θ range 12–75°, and step size 0.05° 2θ. The concentration of Fe in the aqueous solutions was determined by AAS using SpectrAA-880, unit VGA 77 (Varian) for measuring in flame and SpectrAA-300 (Varian) for hydride process. The concentration of As as AsO_2^- or AsO_4^{3-} in the aqueous solutions was determined by photometry using UV/spectrophotometer UNICAM 5625. The amount of As(V) was determined by the molybdenum blue method at 825 nm. The IR spectra were measured on a Nicolet 740 Fourier transform infrared spectrometer equipped with TGS detector. The KBr pellet technique was used at the resolution of 2 cm and 32 accumulations of the spectra. Voltammetry of microparticles was performed with a conventional paraffin impregnated graphite electrode in a 1:1 acetate buffer with a total acetate concentration of 0.2 M. The peak potentials E_P are given with respect to a saturated calomel reference electrode (SCE) (Grygar *et al.* 2002).

55.3   RESULTS AND DISCUSSION

55.3.1   *Fe (II) treating process*

In the case of metakaolin the efficiency of As(V) sorption increased from about 15% (untreated material) to about 30% and more than 95% [Fe(II) treated sorbent], strongly depending on the reaction time (Fig. 55.2a). As(III) was adsorbed only partially, the sorption efficiency did not

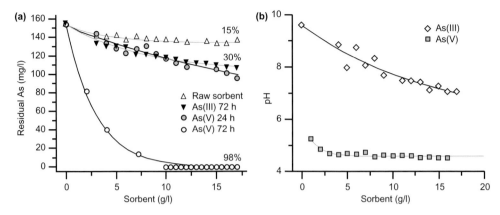

Figure 55.2.   Sorption on Fe(II) treated metakaolin: (a) residual As content (its final fraction is given at curves), (b) pH value.

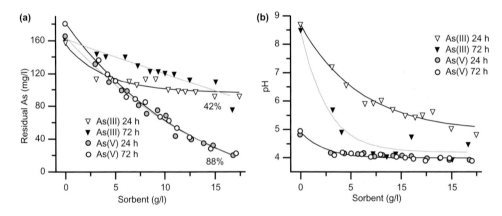

Figure 55.3.   Sorption on Fe(II) treated bentonite: (a) residual As content, (b) pH value.

exceed 30% at the longer reaction time of 72 hours (Fig. 55.2a). pH values related to the mass of sorbent for As(III) and As(V) sorption series are shown in Figure 55.2b. The curves display a slight influence of treated metakaolin on the acidity/basicity of initial solution. During As(III) sorption the pH decreased by about 2 pH units at maximum, due to the strong basicity of initial arsenite solution.

The results of As(III)/As(V) sorption on Fe(II) treated bentonite are shown in Figure 55.3. The sorption efficiency for As(III) achieved about 40%, while almost 90% of As(V) was removed. The reaction time sufficient to attain equilibrium was 24 hours for both, As(III) and As(V) sorption (Fig. 55.3a). The development of pH in Figure 55.3b indicates the higher acidification of initial arsenite solution compared to the above mentioned metakaolin (comparison of Figs. 55.2b and 55.3b).

55.3.2 *Fe(III) treating process*

The results represented in Figure 55.4a show a poor efficiency of As(III)/As(V) sorption onto Fe(III) treated metakaolin (about 30% at 72 hours). In our opinion, the problem might be related to the steric effect of hydrolyzing particles in the Fe(III) treating solution. The FeOOH aggregates originating from a ferric salt solution (unlike the small single particles that formed in ferrous salt

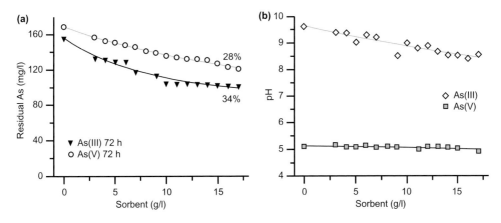

Figure 55.4.   Sorption on Fe(III) treated metakaolin: (a) residual As content, (b) pH value.

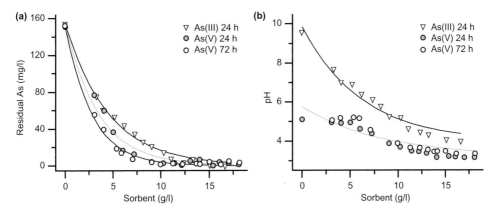

Figure 55.5.   Sorption on Fe(III) treated bentonite: (a) residual As content, (b) pH value.

solution) were probably too large for bonding to active surface sites of metakaolin, therefore, the treatment process was ineffective. The influence of treated metakaolin on the acidity of the sorption system was negligible for both, As(III) and As(V) solutions (Fig. 55.4b).

On the other hand, the combination Fe(III) treatment—bentonite proved to be the most effective among the investigated systems. Both, As(III) and As(V) species were removed almost quantitatively from the aqueous solutions, the sorption efficiency exceeded 99% (Fig. 55.5a). It is obvious, that the bentonite surface was a much more convenient carrier for FeOOH binding than metakaolin. This result corresponds well with different kinds and structures of investigated clays (Douillard and Salles 2004). The pH developments (Fig. 55.5b) are comparable with that for the Fe(II) treated bentonite (Fig. 55.3b). I.e. bentonite increased the acidity of initial As solutions, particularly As(III).

55.3.3   *Efficiency of As removal*

The best results among the investigated pre-treating methods and used materials in connection with As(III)/As(V) sorption efficiency are shown in Figure 55.6. According to the obtained data, metakaolin was more sensitive to the kind of pre-treatment, therefore only the Fe(II) treatment was useful. Both investigated methods were applicable to bentonite, while the Fe(III) treatment was

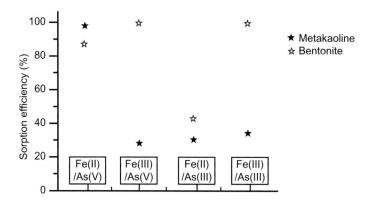

Figure 55.6.   Comparison of the efficiency of Fe treating methods related to the sorbent used.

Table 55.2.   Voltammetric peak potentials of Fe^{3+} species in V *vs.* SCE.

Sample	Fe(II)	Fe(II)-As(III)	Fe(II)-As(V)	Fe(III)	Fe(III)-As(III)	Fe(III)-As(V)
Metakaoline	+0.04	−0.11	x	−	−	−
Bentonite	+0.03	−0.12	x	+0.05	−0.12	−0.12

x: no voltammetric peak.

much more effective for sorption of As(III) and As(V). In general, arsenates were better adsorbed in all systems than arsenites.

55.3.4   *Characterization of solid phases*

The Fe treated sorbents were investigated by means of voltammetry of microparticles to show the changes of Fe-bearing species during the treatment process. Both sorbents indicated a significant accumulation of Fe^{3+} in ion-exchangeable sites during the interaction with Fe(II) or Fe(III) as follows from the presence of voltammetric peaks at $\sim +0.05$ V *vs.* SCE. However, after the adsorption of As(III)/As(V) on both treated sorbents, weak or no corresponding peaks were found (Table 55.2). This fact can be explained by the As(III/V)-Fe(III) sorption mechanism, when the majority of highly reactive ion-exchangeable Fe^{3+} are transformed to more stable inner-sphere Fe(III)-As(III/V) complexes or ferrihydrite, as follows from newly appearing voltammetric peaks at $\sim$0.1 V *vs.* SCE.

The most relevant infrared spectra of metakaolin are shown in Figure 55.7. A and B spectra represent Fe(II) and Fe(III) treated metakaolin with typical sulfate bands at about 1000/cm and the bands at about 500/cm assigned to Fe oxides. The subtracted spectra C, D and E indicate As sorption. The C spectrum concerning As(V) sorption indicates arsenate bands at 826 and 475/cm. As(III) sorption on Fe(II) and Fe(III) treated metakaoline is characterized by D and E subtracted spectra. The dominant bands at about 830, 700 and 600/cm can be assigned to arsenite bonds, therefore, As(III) must have kept its initial valence after the sorption onto metakaolin.

Figure 55.8 illustrates the bentonite spectra; similarly, A and B spectra represent Fe(II) and Fe(III) treated bentonite with characteristic sulfate bands at about 1000/cm and Fe oxides bands at about 500/cm. The subtractions C and D concerning As(V) and As(III) sorption on the Fe(III) treated bentonite indicate the typical arsenate bands at about 488, 800, 840 and 885/cm. The bands at about 870 and 730/cm in spectrum E which represent As(III)-Fe(II) sorption can be assigned to Fe(III)-As(III) bonds that indicate that As again remains trivalent.

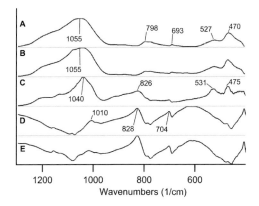

Figure 55.7.   Infrared spectra of Fe treated metakaoline A: Fe(III) treated metakaoline; B: Fe(II) treated metakaoline; C: subtraction after As sorption Fe(II)-As(V); D: Fe(II)-As(III); E: Fe(III)-As(III).

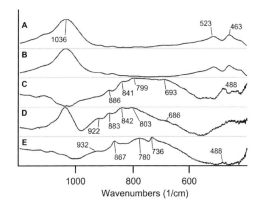

Figure 55.8.   Infrared spectra of Fe treated bentonite. A: Fe(II) treated bentonite; B: Fe(III) treated bentonite; C: subtraction after As sorption Fe(III)-As(V); D: Fe(III)-As(III); E: Fe(II)-As(III).

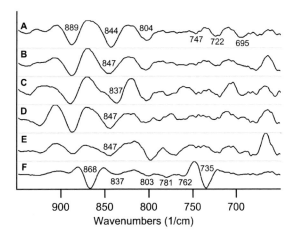

Figure 55.9.   Second derivatives of bentonite subtracted spectra: A: Fe(III)-As(V); B: Fe(III)-As(III); C: Fe(III)-As(III); D: Fe(II)-As(V); E: Fe(II)-As(V); F: Fe(II)-As(III).

The second derivatives of subtracted spectra (Fig. 55.9) illustrate probable differences in the sorption mechanisms onto bentonite owing to the distinct treatment processes. In the case of Fe(III) treated bentonite (spectra A, B and C), As was entirely bonded in arsenate form regardless of the initial As valence. D, E and F spectra assigned to Fe(II) treated bentonite demonstrate the difference between the As(III) and As(V) adsorption processes. D and E spectra concerning As(V) adsorption are similar to what has been mentioned above with the typical arsenate bands at about 800, 840 and 890/cm. Spectrum F which represents Fe(II)-As(III) sorption shows dominant bands at 868 and 735/cm, which can prove the presence of Fe(III)-As(III) bonds on the sorbent surface.

55.4 CONCLUSIONS

The pre-treatment of clays with Fe ions proved to be a very simple method that opens new possibilities of effective and cheap decontamination of As-polluted aqueous systems. Both investigated clay materials allowed Fe pre-treatment and subsequent As sorption. Metakaolin was more sensitive to the type of pre-treatment, therefore only the Fe(II) treatment was useful for As(V) sorption. Both treatment methods were applicable to bentonite, while Fe(III) treatment was much more effective. Arsenates were better adsorbed in all systems than arsenites. The appropriate contact time for reaching equilibrium was 24 hours for the sorption onto bentonite and 72 hours for the sorption onto metakaoline.

The mechanism of Fe-loading and subsequent sorption processes depended on the structure of the raw clay material, kind of Fe treatment, initial solution properties [pH, As(III)/As(V) concentration] and reaction time. It is evident, that As(V) was entirely bonded as arsenate tetrahedrons to available Fe^{3+} particles (ion-exchangable or ferrihydrite) forming stable inner-sphere surface complexes (Sherman and Randall 2003). The adsorption of As(III) oxyanions developed in a different manner; in the case of all metakaolin sorbents arsenite species did not change its initial state of oxidation during the sorption process and were bonded in the trivalent form. The Fe(II) treated bentonite gave the same results, while the Fe(III) treated bentonite bound As only in a pentavalent form, which implies the oxidation of arsenites to arsenates during the sorption process. The final oxidation state of adsorbed arsenites is probably connected with the properties of available Fe^{3+} particles on the sorbent surface and the sorption kinetics; small single Fe^{3+} particles appeared to be more suitable for arsenites binding and keeping the trivalent state than FeOOH aggregates. The unoccupied active Fe^{3+} particles on the sorbent surface could be transformed to more stable and/or crystalline oxide forms during the sorption process.

ACKNOWLEDGMENTS

This work was part of research program MSM 6046137302 (CR) and was supported by the Grant Agency of CR (GA 103/03/0506).

REFERENCES

Acharyya, S.K. and Shah, B.A.: Genesis of arsenic contamination of groundwater in alluvial Gangetic aquifer in India. In: J. Bundschuh, P. Bhattacharya and D. Chandrasekharam (eds): *Natural arsenic in groundwater: occurence, remediation and management.* Balkema, Leiden, The Netherlands, 2005, pp. 17–23.
Altundogan, H.S., Altundogan, S., Tümen, F. and Bildik, M.: Arsenic adsorption from aqueous solutions by activated red mud. *Waste Manage.* 22 (2002), pp. 357–363.
Antelo, J., Avena, M., Fiol, S., López, R. and Arce, F.: Effects of pH and ionic strength on the adsorption of phosphate and arsenate at the goethite-water interface. *J. Colloid Interface Sci.* 285 (2005), pp. 476–486.

Bhattacharya, P., Claesson, M., Fagerberg, J., Bundschuh, J., Storniolo, A.R., Martin, R.A., Thir, J.M. and Sracek, O.: Natural arsenic in the groundwater of the alluvial aquifers of Santiago del Estero Province, Argentina. In: J. Bundschuh, P. Bhattacharya and D. Chandrasekharam (eds): *Natural arsenic in groundwater: occurence, remediation and management.* Balkema, Leiden, The Netherlands, 2005, pp. 57–75.

Bonnin, D.: Method of removing arsenic species from an aqueous medium using modified zeolite minerals. U.S. Patent No. 6,042,731, 2000.

Bruce, A.M., Scott, E.F. and Sabine, G.: Surface structures and stability of arsenic (III) on goethite: spectroscopic evidence for inner-sphere complexes. *Environ. Sci. Technol.* 32 (1998), pp. 2383–2388.

Cole, J.M., Ryan, M.C., Smith, S. and Bethune, D.: Arsenic source and fate at a village drinking water supply in Mexico and its relationship to sewage contamination. In: J. Bundschuh, P. Bhattacharya and D. Chandrasekharam (eds): *Natural arsenic in groundwater: occurence, remediation and management.* Balkema, Leiden, The Netherlands, 2005, pp. 67–75.

Cullen, W.R. and Reimer, K.J.: Arsenic speciation in the environment. *Chem. Rev.* 89 (1989), pp. 713–764.

Deliyanni, E.A., Bakoyannakis, D.N. and Matis, K.A.: Sorption of As(V) ions by akaganéite-type nanocrystals. *Chemosphere* 50 (2003), pp. 155–163.

Douillard, J.M. and Salles, F.: Phenomenology of water adsorption at clay surfaces. In F. Wypych and K.G. Satyanarayana (eds): *Clay surfaces: fundamentals and applications:* Elsevier, Amsterdam, 2004, pp. 118–153.

Doušová, B., Machovič, V., Koloušek, D., Kovanda, F. and Dorničák, V.: Sorption of As(V) species from aqueous systems. *Water, Air and Soil Poll.* 149 (2003), pp. 251–267.

Doušová, B., Koloušek, D., Kovanda, F., Machovič, V. and Novotná, M.: Removal of As(V) species from extremely contaminated mining water. *Applied Clay Science* 28(1–4) (2005), pp. 31–43.

Elizade-González, M.P., Mattusch, J. and Wennrich, R.: Application of natural zeolites for preconcentration of arsenic species in water samples. *J. Environ. Monit.* 3 (2001), pp. 22–26.

Grim, R.E.: Bentonite, kaolin and selected minerals. In: H. Beutelspacher and H.W. van der Marel (eds): *Atlas of electron microscopy of clay minerals and their admixtures.* Elsevier, Amsterdam , 1968.

Grygar, T., Bezdička, P., Hradil. D., Doménech-Carbó, A., Marken, F. Pikna, L. and Cepriá, G.: Voltametric analysis of iron oxide pigments. *Analyst* 127 (2002), pp. 1100–1107.

Gupta, V.K., Saini, V.K. and Jain, N.: Adsorption of As(III) from aqueous solution by iron oxide-coated sand. *J. Colloid Interface Sci.* 288 (2005), pp. 55–60.

Halter, W.E. and Pfeifer, H.R.: Arsenic (V) adsorption onto -Al_2O_3 between 25 and 70°C. *Appl. Geochem.* 16 (2001), pp. 793–802.

Izumi, Y., Masih, D., Aika, K. and Seida, Y.: Characterization of intercalated Iron(III) nanoparticles and oxidative adsorption of arsenite on them monitored by X-ray absorption fine structure combined with fluorescence spectrometry. *J. Phys. Chem. B* 109 (2005), pp. 3227–3232.

Lin, Z. and Puls, R.W.: Potential indicators for the assessment of arsenic natural attenuation in the subsurface. *Adv. Environ. Res.* 7 (2003), pp. 825–834.

Min, J.H. and Hering, J.G.: Arsenate sorption by Fe(III)-doped alginate gels. *Water Res.* 32:5 (1998), pp. 1544–1552.

Moore, D.M. and Reynolds, R.C.: *X-Ray diffraction and the identification and analysis of clay minerals.* 2nd ed., Oxford University Press, New York, 1997.

Rodriguez, R., Armienta, M.A. and Mejia Gómez, J.A.: Arsenic contamination of the Salamanca aquifer system in Mexico: a risk analysis. In: J. Bundschuh, P. Bhattacharya and D. Chandrasekharam (eds): *Natural arsenic in groundwater: occurence, remediation and management.* Balkema, Leiden, The Netherlands, 2005, pp. 77–83.

Sherman, D.M. and Randall, S.R.: Surface complexation of arsenic(V) to iron(III) (hydr)oxides: Structural mechanism from ab initio molecular geometries and EXAFS spectroscopy. *Geochim. Cosmochim. Acta* 67:22 (2003), pp. 4223–4230.

Singh, T.S. and Pant, K.K.: Equilibrium, kinetics and thermodynamic studies for adsorption of As(III) on activated alumina. *Sep. Purif. Technol.* 36 (2004), pp. 139–147.

Smedley, P.L. and Kinniburgh, D.G.: A review of the source, behaviour and distribution of arsenic in natural waters. *Appl. Geochem.* 17 (2002), pp. 517–568.

Tandukar, N., Bhattacharya P., Jacks, G. and Valero, A.A.: Naturally occurring arsenic in groundwater of Terai region in Nepal and mitigation options. In: J. Bundschuh, P. Bhattacharya and D. Chandrasekharam (eds): *Natural arsenic in groundwater: occurence, remediation and management.* Balkema, Leiden, The Netherlands, 2005, pp. 41–48.

Wagner, F., Berner, Z.A. and Stüben, D.: Arsenic in groundwater in the Bengal Delta Plain: geochemical evidences for small scale redox zonation in the aquifer. In: J. Bundschuh, P. Bhattacharya and D. Chandrasekharam (eds): *Natural arsenic in groundwater: occurence, remediation and management.* Balkema, Leiden, The Netherlands, 2005, pp. 3–15.

Xu, Y., Nakajima, T. and Ohki, A.: Adsorption and removal of arsenic (V) from drinking water by aluminum-loaded Shirasu zeolite. *J. Hazard. Mater.* B92 (2002), pp. 275–287.

Zhang, W., Singh, P., Paling, E. and Delides, S.: Arsenic removal from contaminated water by natural iron ores. *Minerals Engineering* 17 (2004), pp. 517–524.

CHAPTER 56

Natural red earth: An effective sorbent for arsenic removal from Sri Lanka

M. Vithanage
Department of Geology and Geography, University of Copenhagen, Denmark
International Water Management Institute (IWMI), Battaramulla, Sri Lanka

K. Mahatantila & R. Chandrajith
Department of Geology, University of Peradeniya, Peradeniya, Sri Lanka

R. Weerasooriya
Institute of Fundamental Studies, Kandy, Sri Lanka

ABSTRACT: This study compared the adsorption of arsenite and arsenate onto natural red earth (NRE). The retention of arsenic (As) species, both As(III) and As(V) (0.385 μmol/l), was examined as a function of pH in NRE. Desorption of As species from NRE was examined at the same concentrations. Over 95% of both As(III) and As(V) were adsorbed on NRE. No pH dependency was observed in NRE. Grain size distribution data indicated that NRE has a high hydraulic conductivity. Therefore NRE has potential to remove As from aqueous systems by modifying available filter systems.

56.1 INTRODUCTION

Arsenic (As) is a carcinogenic substance with a complex chemistry in the environment due to its various oxidation states. Arsenic is found in both surface and subsurface waters, and its occurrence can be due to both natural phenomena and anthropogenic activities. However, in the case of groundwater As contamination it is likely to be the result of natural processes. Arsenic concentrations at elevated levels are encountered in groundwater from aquifers in many countries including Bangladesh, India, China, Thailand (Bhattacharya and Jacks 2000, Bhattacharya *et al.* 2002), Vietnam (Agusa *et al.* 2005), Argentina (Smedley *et al.* 2002, Bundschuh *et al.* 2004, Bhattacharya *et al.* 2006), Pakistan (Nickson *et al.* 2005), Japan (Kondo *et al.* 1997). However, the magnitude of the problem is most severe in Bangladesh and West Bengal, India, where As concentrations in groundwater greatly exceed the WHO guideline (10 μg/l) for safe drinking water.

Due to the adverse impact of As on human health, there is a pressing need to develop cost-effective As removal technologies. The methods which have been applied include coagulation-filtration, adsorption on activated alumina, ion-exchange, adsorption onto ferric oxides and hydroxides, and adsorption and filtration by manganese greensand (Jakariya 2000, Thirunavukkarasu *et al.* 2003). Arsenic removal by adsorption has been studied in depth by various researchers and most results have showed that both arsenite and arsenate removal is pH-dependent, and arsenite is efficiently removed at pH > 7, while arsenate is efficiently removed below pH 7 (Goldberg 2002, Goldberg and Johnston 2001, Mohan and Pittman 2007). Bench-scale studies were conducted with various iron oxides to remove As from drinking water (Pierce and Moore 1982, Wilkie and Hering 1996, Joshi and Chaudhuri 1996, Raven *et al.* 1998, Driehaus *et al.* 1998). These technologies have varying degrees of effectiveness in terms of As removal, and depend mostly on manufactured reagents, which will probably result in an increase in the cost of As removal. However, little research (Genç-Fuhrman *et al.* 2004, Zhang *et al.* 2004) has been conducted on the As adsorption capacity of natural substances.

The red earth investigated is a naturally occurring material that outcrops along the northwestern coast of Sri Lanka. The main deposits are found at Aruwakkalu ($8°14'50''N$; $79°45'45''E$) in the Northwestern province, and in pockets along the northwestern coastal belt. The NRE bed overlies Miocene limestone, and has a thickness of 12 m. NRE is an aeolian, iron coated sand (Dahanayake and Jayawardhana 1979). NRE occurs as rounded and well sorted quartz sand in a red clayey matrix, with accessory ilmenite and magnetite. The brick red color indicates oxidizing conditions and the formation of hematite. The NRE contains little Fe^{2+} (0–1%), however, the Fe^{3+} content is typically $>2.0\%$, sometimes reaching up to 6% (Dahanayake and Jayawardhana 1979). XRD analysis confirmed that crystalline silica is the dominant phase in NRE. The XRF results showed that Al (as Al_2O_3) and Fe (as Fe_2O_3) are also present in significant proportions, probably as an amorphous coating around silica grains, to the extent that the surface properties of silica are completely masked (Vithanage et al. 2006).

Iron oxide minerals are known to adsorb As well. It was decided to explore the As adsorption capacity of NRE because it is a natural material that contains both Fe and Al oxides, and has been found to effective remove both As(V) and As(III) (Vithanage et al. 2006). Mineralogical characteristics of NRE reported in detail by Vithanage et al. 2006, Vithanage et al. 2007a. Experiments were carried out to determine the grain size distribution and hence to estimate the hydraulic conductivity; and to analyze the desorption behavior of the NRE to assess how effective NRE might be in a household filter.

56.2   MATERIALS AND METHODS

For sieve analysis, 750 g of NRE was passed through 5 mesh sizes: 1.5, 0.25, 0.18, 0.125 and 0.063 mm. Sieving was carried out with a mechanical sieve shaker (TYLER). The Krumbein and Monk equation and the Hazen approximation (Pfannkuch and Paulson 2005) were used to estimate NRE permeability.

Adsorption and desorption studies were carried out using the <0.063 mm fraction of NRE with 0.01 M $NaNO_3$ ionic strength. The suspension (5 g/l) was purged with argon gas to minimize the CO_2 contamination, and was allowed to equilibrate for 2 hr. The NRE suspensions were spiked with equal concentrations (0.385 μmol/l) of arsenite and arsenate, as single sorbate systems. 20 ml aliquots were taken into centrifuge tubes while changing the pH from pH 4.0 to pH 9.0. The system was purged continuously with argon, then allowed to equilibrate for 24 h (EYELA B603 shaker). During Ar purging, pH was measured, recorded and the suspension was centrifuged twice (2000 rpm for 15 min) before extracting the supernatant for adsorption measurements. The residue was washed three times with 0.01 M $NaNO_3$, and then 20 ml of water containing 0.01 M $NaNO_3$ were added, which was more or less equal to the previous pH of the system. Again the tubes were allowed to equilibrate in the same shaker for 24 h. The suspension was centrifuged, and filtered after recording the final pH, and the supernatant was taken for As desorption measurements by atomic absorption spectrophotometer (GBC 933 AA, Australia).

All the experimental results were modeled using Diffuse Layer surface complexation model. The parameters used for the model was similar in Vithanage et al. (2006).

56.3   RESULTS AND DISCUSSION

56.3.1   *Adsorption and desorption studies*

Both arsenite and arsenate were adsorbed onto NRE surfaces, but arsenate adsorption was higher ($>98\%$) than that of arsenite (98–95%) in the pH range 4 to 7. With increasing pH, adsorption showed a decrease (pH 7 to 10) up to 95% for arsenate and 85% for arsenite. Since NRE contains both Fe and Al, both arsenate and arsenite are strongly adsorbed on to Fe under acidic conditions and Al is adsorbing the As species in the basic conditions. As the pH increases, competition among ions decreases the As adsorption capacity of NRE. However, NRE is strongly adsorbed by both arsenate and arsenite under environmentally significant pH conditions (Figs. 56.1a and b).

Desorption experiments showed that arsenate desorption is less than that of arsenite desorption. Arsenate desorption was about 3.5% at pH 4, but increased to 10% with increasing pH. Arsenite desorption was highest (32%) at acidic pH's but decreased to 12% with increasing pH (Fig. 56.1b). However, arsenate bonding with NRE is stronger than that of arsenite. This proves the strong bonding behavior of arsenate with Fe sites of NRE (Vithanage *et al.* 2006, Vithanage *et al.* 2007b). Since the arsenite-NRE bonding is less strong, desorption is higher than desorption of arsenate.

56.3.2 *Data modeling*

Modeling results showed that Fe sites of the NRE adsorbed 90% of arsenite below pH 7, while arsenate adsorption on Fe sites was high and constant throughout the pH range (Fig. 56.2). However, arsenite adsorption on Al sites of NRE was low (0–25%) below pH 7 and increased up to 45% and remained same from pH 7 to 10. According to the modeling results, arsenate adsorption was either very low (<10%) on to Al sites of NRE, or all the arsenate was adsorbed on to Fe sites of NRE.

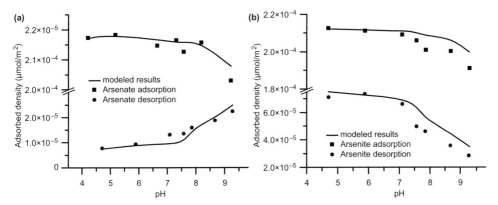

Figure 56.1.   Arsenate (a) and arsenite (b) adsorption and desorption on 5 g/l of NRE. 0.01 M NaNO$_3$. Initial As(V) = 0.385 μmol/l and As(III) = 0.385 μmol/l, respectively. Lines represent modeled results and symbols represent experimental values.

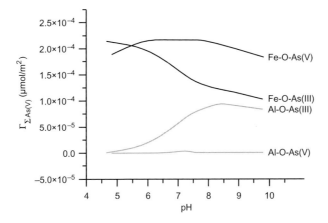

Figure 56.2.   Adsorption of arsenite and arsenate on Fe and Al sites of NRE.

56.3.3   *Grain size distribution and permeability*

The grain size distribution shows that the NRE is a poorly sorted sand. The inclusive (graphic) standard deviation (Pfannkuch and Paulson 2005) is 1.153 phi units, and the NRE has a normal kurtosis, which indicates that sorting is equal through out the curve. The hydraulic conductivity (K) was estimated from two equations. According to the Krumbein and Monk equation, the K is 9.81×10^{-1} cm/s, and according to the Hazen approximation, K is 3.39×10^{-2} cm/s (the coefficient 'C' was taken as 80, for a poorly sorted medium sand). Generally, poorly sorted sands have low permeability due to filling of pore spaces from fine particles. However, these estimates indicate high permeability, but should be confirmed by hydraulic testing.

56.4   CONCLUSIONS

NRE strongly adsorbs both arsenite and arsenate, and is therefore a promising alternative material for use in groundwater treatment plants. However, arsenite desorption is greater than that of arsenate due to weaker bonding between arsenite and the NRE surface. Before NRE can be employed either in household filter systems or large scale waste water treatment plants, more work is needed to determine breakthrough characteristics, the effects of competition from ions such as P, Si and organic acids. If these tests are successful, the NRE should be tested in controlled and carefully monitored field trials, and the practicality of regeneration, and ways of safe disposal of the residue should be studied.

ACKNOWLEDGEMENT

MV likes to thank Devika Wijesekera and Prosun Bhattacharya for their various support given. Reviewers' comments helped to enhance the manuscript quality. Special thanks go to Peter Ravenscroft for his support given.

REFERENCES

Agusa, T., Kunito, T., Fujihara, J., Kubota, R., Minh, T.B., Trang, P.T.K.T., Iwata, H., Subramanian, A., Viet, P.H. and Tanabe, S.: Contamination by arsenic and other trace elements in tube well water and its risk assessment to humans in Hanoi, Vietnam. *Environ. Pollut.* 139 (2006), pp. 95–106.

Bhattacharya, P., Claesson, M., Bundschuh, J., Sracek, O., Fagerberg, J., Jacks, G., Martin, R.A., Storniolo, A. and Thir, J.M.: Distribution and mobility of arsenic in the Rio Dulce alluvial aquifers in Santiago del Estero Province, Argentina. *Sci. Total Environ.* 358 (2006), pp. 97–120.

Bhattacharya, P., Frisbie, S.H., Smith, E., Naidu, R., Jacks, G. and Sarkar, B.: Arsenic in the environment: A global perspective. In: B. Sarkar (ed): *Handbook of heavy metals in the environment.* New York: Marcell Dekker; 2002. pp. 145–215.

Bhattacharya, P. and Jacks, G.: Arsenic contamination in groundwater of the sedimentary aquifers in the Bengal Delta Plains: A review. In: P. Bhattacharya and A.H. Welch (eds): *Arsenic in groundwater of sedimentary aquifers.* Pre-Congress Workshop Abstract Volume, 31st International Geological Congress, Rio de Janeiro, Brazil, 2000, pp. 19–21, http://amov.ce.kth.se/people/Prosun/Rio-abstract.pdf.

Bundschuh, J., Farias, B., Martin, R., Storniolo, A., Bhattacharya, P., Cortes, J., Bonorino, G. and Albouy, R.: Groundwater arsenic in the Chaco-Pampean Plain, Argentina: Case study from Robles county, Santiago del Estero Province. *Appl. Geochem.* 19 (2004), pp. 231–243.

Dahanayake, K. and Jayawardana, S.K.: Study of red and brown earth deposits of north-west Sri Lanka. *J. Geol. Soc. India* 20 (1979), pp. 433–440.

Driehaus, W., Jekel, M. and Hildebrand, T.U.: Granular ferric hydroxide—a new adsorbent for the removal of arsenic from natural water. *J. Water Supply Res. and Technol. Aqua* 47 (1998), pp. 30–35.

Genç-Fuhrman, H., Tjell, J.C. and McConchie, D.: Adsorption of arsenic from water using activated neutralized red mud. *Environ. Sci. Technol.* 38 (2004), pp. 2428–2434.

Goldberg, S. and Johnston, C.T.: Mechanisms of arsenic adsorption on amorphous oxides evaluated using macroscopic measurements, vibrational spectroscopy and surface complexation modelling. *J. Colloid Interface Sci.* 234 (2001), pp. 204–216.

Goldberg, S.: Competitive adsorption of arsenate and arsenite on oxides and clay minerals. *Soil Sci. Soc. Am. J.* 66 (2002), pp. 413–421.

Jakariya, M.: The use of alternative safe water options to mitigate the arsenic problem in Bangladesh: a community perspective. MSc Thesis, Department of Geography, University of Cambridge, UK, 2000, http://bicn.com/acic/resources/infobank/jakariya/sec07.htm.

Joshi, A. and Chaudhuri, M.: Removal of arsenic from ground water by iron oxide coated sand. *J. Environ. Eng.* 122:8 (1996), pp. 769–771.

Kondo, H., Ishiguro, Y., Ohno, K., Nagase, M., Toba, M. and Takagi, M.: Naturally occurring arsenic in the ground water in the southern region of Fukuoka Prefecture, Japan. *Water Res.* 33:8 (1999), pp. 1967–1972.

Mohan, D. and Pittman, C.U.: Arsenic removal from water/wastewater using adsorbents—A critical review. *J. Hazard. Mater.* 142:1/2 (2007), pp. 1–53.

Nickson, R.T., McArthur, J.M., Shrestha, B., Kyaw-Myint, T.O. and Lowry, D.: Arsenic and other drinking water quality issues, Muzaffargarh District, Pakistan. *Appl. Geochem.* 20 (2005), pp. 55–68.

Pfannkuch, H.O. and Paulson, R.: Grain size distribution and hydraulic properties. http://www.cs.pdx.edu/~ian/geology2.5.html.

Pierce, L.M. and Moore, B.C.: Adsorption of arsenite and arsenate on amorphous iron hydroxide. *Water Res.* 16 (1982), pp. 1247–1253.

Raven, K.P., Jain, A. and Loeppert, R.H.: Arsenite and arsenate adsorption on ferrihydrite: kinetics, equilibrium, and adsorption envelopes. *Environ. Sci. Technol.* 32 (1998), pp. 344–349.

Smedley, P.L., Nicolli, H.B., Macdonald, D.M.J., Barros, A.J. and Tullio, J.O.: Hydrochemistry of arsenic and other inorganic constituents in groundwater from La Pampa, Argentina. *Appl. Geochem.* 17 (2002), pp. 259–284.

Thirunavukkarasu, O.S. and Viraraghavan, T.: Arsenic in drinking water: health effects and removal technologies. In: T. Murphy and J. Guo (eds): *Aquatic arsenic toxicity and treatment*. Backhuys, Leiden, The Netherlands, 2003, pp. 129–138.

Vithanage, M., Chandrajith, R., Bandara, A. and Weerasooriya, R.: Mechanistic modelling of arsenic retention on natural red earth in simulated environmental systems. *J. Colloid Interface Sci.* 294 (2006), pp. 265–272.

Vithanage, M., Chandrajith, R. and Weerasooriya, R.: Role of natural red earth in arsenic removal in drinking water; comparison with synthetic gibbsite and goethite. In: P. Bhattacharya, A.B. Mukherjee, J. Bundschuh, R. Zevenhoven and R.H. Loeppert (eds): *Arsenic in soil and groundwater environments: biogeochemical interactions*. In: *Trace metals and other contaminants in the environment* 9. Elsevier, Amsterdam, 2007a, pp. 587–602.

Vithanage, M., Seneviratne, W., Chandrajith, R. and Weerasooriya, R.: Arsenic binding mechanisms on natural red earth; A potential substrate for pollution control. *Sci. Total Environ.* 379 (2007b), pp. 244–248.

Wilkie, J.A. and Hering, J.G.: Adsorption of arsenic onto hydrous ferric oxide: effects of adsorbate/adsorbent ratios and co-occuring solutes. *Colloids Surf. A: Physicochem. Eng. Asp.* 107 (1996), pp. 97–110.

Zhang, W., Singh, P., Paling, E. and Delides, S.: Arsenic removal from contaminated water by natural iron ores. *Miner. Eng.* 17:4 (2004), pp. 517–524.

CHAPTER 57

Adsorption of As(V) onto goethite: Experimental statistical optimization

M. Alvarez-Silva, A. Uribe-Salas, F. Nava-Alonso & R. Pérez-Garibay
Unidad Saltillo, Centro de Investigación y de Estudios Avanzados del Instituto Politécnico Nacional (CINVESTAV-IPN), Ramos Arizpe, Coah., Mexico

ABSTRACT: The removal of As(V) from arsenic contaminated water with a chemical composition similar to that of the Comarca Lagunera (Mexico) was experimentally studied. The aim was to optimize the variables that affect As adsorption onto goethite, produced by hydrolysis of Fe(III). Both ferric sulfate and ferric chloride were used to obtain Fe(III) solutions. The study was performed making use of a statistical experimental design known as the method of steepest ascent. Water temperature and Fe(III)/As(V) molar ratio were the independent variables, while the concentration of As remaining in the treated water was the response variable. The results obtained showed that Fe(III)/As(V) molar ratio is the variable that most affects As removal, while temperature slightly enhances the removal. In the range of temperature studied (10 to $30°C$), a Fe(III)/As(V) molar ratio around 20 proved to be effective for removing As to concentrations below the limit recommended by the Word Health Organization (e.g., 0.01 mg/l).

57.1 INTRODUCTION

Arsenic occupies the 20th place in order of element abundance in the earth crust. Since its synthesis in 1250 by Albertus Magnus, arsenic (As) has been the center of controversy due to its particular physical and chemical properties and to its toxic characteristics. In several regions around the world, As concentration in natural aquifers is above the limit recommended for potable drinking waters, according to worldwide regulations (e.g., WHO), thus representing a potential problem to public health (Mandal and Suzuki 2002). Long-term consumption of water containing As has been associated with diseases such as hyperkeratosis, blackfoot disease, cardiovascular risk, abortion and skin cancer (Smith 1997, Rahman *et al.* 1999, Ahmad *et al.* 2001).

In the Comarca Lagunera region, located in the central-north of Mexico comprises the southwestern part of the state of Coahuila and northeastern part of the state of Durango, aquifers with As concentrations above 0.60 mg/l have been located (Del Razo *et al.* 1990). In this region, rural settlements exist where endemic intoxication is commonly found due to the presence of As in drinking water, and where other associated illnesses have been detected (Cebrián *et al.* 1983).

Arsenic toxicity varies considerably depending on the chemical nature of the species: inorganic arsenite As(III) and arsenate As(V) are highly toxic, organic species such as methylarsonic and methylarsonous acid [MMA(V) and MMA(III)] are less harmful, and As species present in some crustacean shellfish (for example, arsenocoline and arsenobetaine), are practically harmless (Le 2002). Thus, establishing a direct relationship between total arsenic (t-As) present in the water and toxicity is not exact. Nevertheless, the more abundant species in natural aquifers are the highly toxic inorganic species, which represent more than 95% of t-As (Ferguson and Gavis 1972, Le 2002).

It is quite alarming to be aware of the existence of rural settlements where the only water supply is contaminated with As, and that, in the best scenario, some palliative measures have been undertaken to distribute potable water transported in tank cars from "safe" artesian wells, which occasionally present As concentrations above 0.025 mg/l (i.e., above the maximum limit recommended by the Mexican regulation for potable water, NOM-127-SSA1-1994). According to the above, the implementation of a method simple enough to be operative at the domestic level to remove As from drinking water in rural areas has become an imperative necessity.

57.2   BACKGROUND

Arsenic species present in natural waters may be removed by a variety of physical-chemical methods: reverse osmosis, electrocoagulation, adsorption onto oxides and hydroxides, among others. In general, it has been observed that removal of As(V) species is easier and more efficient than removal of As(III) species, and due to this, it is recommended that a previous stage of oxidation of As(III) to As(V) is considered (Hsia *et al.* 1994). Nevertheless, in aquifers of Comarca Lagunera, the predominant species is As(V), according to reports by Del Razo *et al.* (1990).

Removal of As by adsorption/co-precipitation using ferric salts is an attractive alternative due to the reasonable compromise that exists between cost and benefit (Clifford and Ghurye 2002). According to the US Environmental Protection Agency, this technique has been classified within a group of seven that showed fairly good performance and proved efficiency in a series of comprehensive studies performed (USEPA 2001). Among its advantages are the fact of being a flexible technique, which permits its application on-site, and its simplicity, which allows its application with very basic infrastructure and minimum capital cost (Meng *et al.* 2001). According to the above, this technique represents an outstanding alternative for treating As-contaminated water in rural areas.

The removal of As from water by adsorption/co-precipitation in ferric oxi-hydroxides consists of adding a ferric salt (typically sulfate or chloride) to the water to be treated, which dissolves, hydrolyzes and precipitates as insoluble ferric oxi-hydroxide that adsorbs/co-precipitates the As (Pierce and Moore 1982, Clifford and Ghurye 2002). There is sufficient evidence that the responsible mechanism for As removal is chemisorption of As species onto the metal sites of the oxi-hydroxide (Hsia *et al.* 1994, Sun and Doner 1996 and 1998, Fendorf *et al.* 1997). Temperature shows an effect on As adsorption: in general, chemisorption is favored by increasing temperature, since the required activation energy for the reaction becomes available. Relatively high concentrations of silicate and, particularly, phosphate, may dramatically reduce As adsorption, since they compete with As species for adsorption sites on the oxi-hydroxide surface (Meng *et al.* 2000 and 2001). Furthermore, it is known that cations such as calcium and magnesium have a beneficial effect on As adsorption since they neutralize the superficial negative-charge that results from As-anion adsorption; other common ions present in natural waters do not have any significant effect on As adsorption and removal (Meng *et al.* 2000).

This work has been aimed to the optimization of the temperature and ferric salt addition to achieve the best removal of inorganic As(V) species present in water of chemical composition similar to that of typical aquifers of Comarca Lagunera contaminated with As. This was performed by using a statistical experimental design known as the method of steepest ascent (Montgomery 2001). The aim of the experimental design was that the concentration of As remaining in the water, which was the response variable of the statistical method, must fulfill the current international regulations for drinking water. The study is intended to define: (1) the amount of ferric salt necessary to remove the As from water to allowable concentrations; (2) to find out the effect of type of ferric salt used (e.g., sulfate and chloride); and (3) to determine the effect of temperature on As adsorption, in order to generate the necessary knowledge for the development and eventual application of a domestic method of reasonable cost, that may be of use in rural communities whose geographical scattering and economical circumstances do not justify the building of water treatment facilities.

57.3   EXPERIMENTAL PART

57.3.1   *Statistical experimental design*

The method of steepest ascent permits the variation of two or more variables at a time, to predict the magnitude of the response or dependent variable, in the whole range of values tested for

Table 57.1. First block of experiments of the 2^2 factorial design.

Original variables		Coded variables	
Temperature °C	Fe/As molar ratio	Factor A: Temperature	Factor B: Fe/As molar ratio
10	10	−1	−1
30	10	+1	−1
10	20	−1	+1
30	20	+1	+1
20	15	0	0
20	15	0	0
20	15	0	0

the independent variables. The two independent variables considered in this work were ferric salt addition (e.g., Fe/As molar ratio) and temperature, while the response variable was the concentration of As remaining in the water. The study was performed for two ferric salts: ferric sulfate and ferric chloride. According to the statistical methodology (Montgomery 2001), the study was divided in two parts: the first consists of a 2^2 factorial design with three central points, whose results were adjusted to a polynomial first-order equation of the form:

$$y = b_0 + b_1X_1 + b_2X_2 \tag{57.1}$$

where y = the concentration of As remaining in the water; X_1 = the temperature; X_2 = the Fe/As molar ratio; and b_0, b_1 and b_2 = regression coefficients.

The high and low levels of ferric salt addition were selected based on results obtained in preliminary tests; in the case of temperature, these two levels corresponds to the average temperature in the Comarca Lagunera during winter and summer seasons (Table 57.1). The experimental design has the advantage of determining whether the flat model is adequate and supplies an estimation of the experimental error involved.

In the second part of the experimental design, the 2^2 factorial design was complemented with four axial points plus two additional central points. The results obtained in this experimental region were adjusted to a second-order polynomial ot the form:

$$y = b_0 + b_1X_1 + b_2X_2 + b_{11}X_1^2 + b_{22}X_2^2 + b_{12}X_1X_2 \tag{57.2}$$

The software STATISTICA'99 was used to generate the experimental design and to analyze the results in order to find out the effect of the independent variables on the response variable.

57.3.2 *Materials, reagents and methodology*

An arsenate stock solution containing 1000 mg/l As(V) (13.35 mM) was prepared from As_2O_5 regent grade. Two ferric stock solutions containing 200 mg/l Fe(III) (3.58 mM) were prepared with $FeCl_3 \times 6H_2O$ and $Fe_2(SO_4)_3 \times 4H_2O$, reagent grade. Potable water from Ramos Arizpe, state of Coahuila (Mexico), was used as background electrolyte; this water has a chemical composition (e.g., species concentration) very similar to that of typical groundwater from the Comarca Lagunera. Physical and chemical characteristics of water samples from Ramos Arizpe and Ejido de Salamanca, this latter of the municipality of Lerdo, state of Durango, are summarized in Table 57.2. The initial As concentration in all the experiments performed was 0.25 mg/l (3.34×10^{-3} mM), which is an As concentration commonly found in contaminated aquifers of Comarca Lagunera.

The experimental methodology consisted of adding the desired amount of Fe(III) (from the stock solution) to the water to be treated, thus promoting goethite precipitation. The resulting suspension

Table 57.2.   Physical and chemical properties of water from Ramos Arizpe (Coahuila) and Salamanca, (municipality of Lerdo, Durango[1]). Values in mg/l, except when other unit is specified.

Sample	pH	Ca^{2+}	Mg^{2+}	Cl^-	SO_4^{2-}	Fe_{Total}	SiO_2	PO_4^{3-}
Ramos Arizpe	7.6	362.8	123.9	315	1101.5	0.109	26.72	<0.35
Ejido de Salamanca	7.0	271.3	94.6	27.2	1005.9	<0.025	–	–

[1]Data provided by the National Water Commission, Mexico.

Table 57.3.   Arsenic concentration in the treated water, obtained in the first block of experiments, when using ferric sulfate.

pH	Random experimental order	Factor A: Temperature	Factor B: Fe/As molar ratio	Response: Remaining arsenic, mg/l
6.9	2	−1	−1	0.064
7.2	1	+1	−1	0.080
6.6	4	−1	+1	<0.010
6.6	3	+1	+1	0.012
6.9	6	0	0	0.013
6.9	7	0	0	0.013
6.8	5	0	0	0.016

was magnetically stirred for 15 minutes (according to preliminary tests, As adsorption equilibrium is reached in less than 15 minutes). Once this period of time had elapsed, the solutions were filtered through a 2.5 μm pore size filter paper. The filtered water samples were acidified and analyzed for residual As using inductively coupled plasma atomic emission spectrometry (Thermo ICP-AES Mod. Iris Intrepid II XSP).

Goethite was characterized by measuring its zeta potential at the working pH (around −10 mV) and particle size distribution (mean diameter around 12 μm and coefficient of variation around 37%).

57.4   RESULTS

57.4.1   Experimental design using ferric sulfate

Table 57.3 reports the As concentration in the treated water obtained in the first block of experiments.

Due to limits in sensitivity of the ICP spectrometer, which cannot measure As concentrations below 0.01 mg/l, for the purpose of adjustment to the polynomial model, As concentrations in the treated water below such detection limit were considered equal to 0.01 mg/l. Adjusting the experimental results to the first-order polynomial model, give rise to the following equation:

$$y = 1.07 \times 10^{-1} + 5.00 \times 10^{-4}X_1 - 6.10 \times 10^{-3}X_2 \tag{57.3}$$

where y = the concentration of residual As in mg/l; X_1 = the temperature in Celsius degrees; and X_2 = the Fe/As molar ratio. The estimated experimental error obtained by calculating the standard deviation of the three central measurements (eq. 57.4), was 0.0017 mg/l:

$$s = \left[\frac{\sum_{i=1}^{n} (\bar{y} - y_i)^2}{n - 1} \right]^{1/2} \tag{57.4}$$

Table 57.4. Arsenic concentration in the treated water, obtained in the second block of experiments, when using ferric sulfate.

		Original variables		Coded variables		Response:
pH	Random experimental order	Temperature °C	Fe/As molar ratio	Factor A: Temperature	Factor B: Fe/As molar ratio	Remaining arsenic, mg/l
7.0	13	6	15	$-\sqrt{2}$	0	0.022
6.7	8	34	15	$+\sqrt{2}$	0	0.025
7.0	11	20	8	0	$-\sqrt{2}$	0.043
6.4	9	20	22	0	$+\sqrt{2}$	0.025
6.6	10	20	15	0	0	<0.010
6.9	12	20	15	0	0	0.011

where s = standard deviation; $\bar{y}$ = average of response variable; y_i = response variable; and n = number of measurements.

The results showed that the second-order effects and the interaction between temperature and Fe/As molar ratio were considerably large, indicating that the first-order polynomial model was not completely adequate to describe the system, and that a second-order model should be taken into account.

To estimate the six regression coefficients of the second-order model, the experimental design was complemented with four axial points and two central points whose temperature and molar ratio coordinates are reported in Table 57.4. The selected design is orthogonal, meaning that an arbitrary constant may be added to any block and the computed regression coefficients should not vary (Montgomery 2001). In this way, an incommensurable external variable affecting the responses of any of the two blocks may not affect the estimated model. Arsenic concentrations in the treated water obtained in the second block of experiments are presented in Table 57.4. Equation (57.5) is obtained by adjusting the experimental data to the second order equation.

$$y = 2.08 \times 10^{-1} - 2.33 \times 10^{-3}X_1 + 9.20 \times 10^{-5}X_1^2 - 1.97 \times 10^{-2}X_2$$
$$+ 5.80 \times 10^{-4}X_2^2 + 7.00 \times 10^{-5}X_1X_2 \tag{57.5}$$

The standard deviation of the replicated measurements was 0.0004 mg/l, and the standard deviation of the central point of both experimental blocks was 0.0014 mg/l.

Figure 57.1 presents the contour map of the water treatment using ferric sulfate as a goethite precursor, showing the As removal as a function of temperature and Fe/As molar ratio. From the results presented in the figure, it may be established that the Fe/As molar ratio is the most important variable of the process, and that As removal increases as the molar ratio increases up to values around 18; larger molar ratios appear to decrease the removal. Apparently, this may be attributed to the complexity of the process, basically due to the interaction of several phenomena: goethite precipitation, arsenate chemisorption and non-specific adsorption of other ionic species present in the water.

For Fe/As molar ratios below 10 and above 25, temperature appears to have no significant effect on As removal. However, in the region of highest removal, temperature shows certain beneficial effects on As removal up to about 20°C. Further increments appear to have a detrimental effect. Nevertheless, it became evident that As removal from contaminated water to achieve concentrations permitted by international regulations is possible in the whole range of temperatures tested, provided that Fe/As molar ratio is around 18.

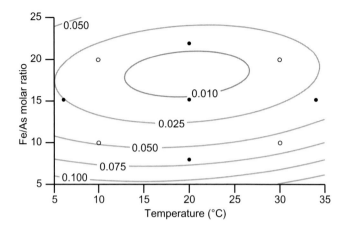

Figure 57.1.   Contour map showing predictions of the 2nd order equation obtained for the case of ferric sulfate as goethite precursor. Empty circles stand for the results of first block of experiments while black circles stand for those of the second block. Iso-concentration lines of arsenic (mg/l) remaining in the treated water are shown.

Table 57.5.   Experimental results obtained in the two experimental blocks, expressed as concentration of arsenic that remained in the treated water, when ferric chloride was used as goethite precursor.

| | | | Experimental variables | | Coded variables | | Response: |
pH natural	Random order	Experimental block	Temp. °C	Fe/As molar ratio	Factor A: Temp.	Factor B: Fe/As	Remaining arsenic, mg/l
7.7	1	1	10	10	-1	-1	0.051
7.8	7	1	30	10	$+1$	-1	0.041
7.2	4	1	10	20	-1	$+1$	0.014
7.5	3	1	30	20	$+1$	$+1$	<0.010
7.6	5	1	20	15	0	0	<0.010
7.5	2	1	20	15	0	0	<0.010
7.5	6	1	20	15	0	0	<0.010
7.2	11	2	6	15	$-\sqrt{2}$	0	0.023
7.6	9	2	34	15	$+\sqrt{2}$	0	0.028
7.6	12	2	25	8	0	$-\sqrt{2}$	0.068
7.3	10	2	25	22	0	$+\sqrt{2}$	<0.010
7.5	8	2	25	15	0	0	0.011
7.4	13	2	25	15	0	0	0.011

57.4.2   Experimental design for ferric chloride as goethite precursor

The methodology described above was also used to obtain the response surface of As removal when using ferric chloride as goethite precursor. The results obtained for the first and second block of experiments are shown in Table 57.5.

Equation (57.6) was obtained by analyzing the results in Table 57.5 according to the analytical methodology of the experimental design:

$$y = 2.27 \times 10^{-1} - 3.22 \times 10^{-3}X_1 + 6.70 \times 10^{-5}X_1^2 - 2.07 \times 10^{-2}X_2$$
$$+ 5.44 \times 10^{-4}X_2^2 + 3.00 \times 10^{-5}X_1X_2 \qquad (57.6)$$

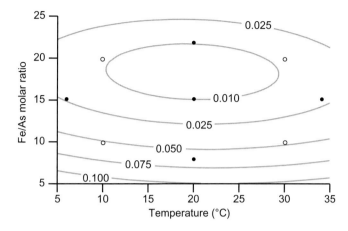

Figure 57.2.   Contour map of the 2nd order equation obtained when ferric chloride was used as goethite precursor. Empty circles stand for the results of first block of experiments while black circles stand for those of the second block. Iso-concentration lines of arsenic remaining in the treated water (mg/l) are shown.

Table 57.6.   Adsorption/co-precipitation of arsenic in water from a contaminated well located in El Mayrán, San Pedro de las Colonias (Coahuila, Mexico). Goethite precursors: ferric sulfate and ferric chloride.

pH	T °C	Fe/As molar ratio	Goethite precursor	Remaining arsenic, mg/l
7.8	22	20	Ferric chloride	<0.010
7.9	23	20	Ferric sulfate	0.016

Figure 57.2 presents the contour map of the second order equation obtained for ferric chloride as goethite precursor. The response surface is similar to that obtained when using ferric sulfate, thus suggesting the goethite produced during Fe(III) hydrolysis has similar surface properties and particle size distribution, and therefore, similar As adsorption properties.

The best combination of temperature and Fe/As molar ratio identified for As removal, namely, 20°C and a molar ratio of 20, was used to test As removal from water of a well of the community El Mayrán, located in the municipality of San Pedro de las Colonias (Coahuila, Mexico). The experiments were performed using both ferric sulfate and ferric chloride as goethite precursors. The total As concentration in this well is 0.21 mg/l, 3% of which is As(III). The results obtained are presented in Table 57.6.

From the results of Table 57.6, it may be concluded that it is possible to remove As from contaminated water to concentrations below those established by both the Mexican and international environmental regulations. The removal process consists of adsorption/co-precipitation of As species onto goethite obtained by hydrolysis of Fe(III) from either ferric sulfate or ferric chloride at 20°C and using a Fe/As molar ratio of 20.

57.5   CONCLUSIONS

It was determined that Fe/As molar ratio is the variable that most affects As removal by adsorption/co-precipitation onto ferric oxi-hydroxides. The removal efficiency is slightly affected by temperature, the optimum temperature value being 20°C. Both higher and lower temperatures appear to slightly increase the concentration of As remaining in the treated water.

It was demonstrated that the developed method works fairly well in the removal of As from contaminated water by using ferric sulfate or ferric chloride as goethite precursors. When a Fe/As molar ratio of around 18 is used at a temperature around 20°C, the concentration of As remaining in the water is below that established by both Mexican and international environmental regulations (e.g., <0.01 mg/l).

ACKNOWLEDGMENTS

The authors thank Mr. Antonio González-Anaya (MSc) for his technical support in the laboratory and Mrs. Socorro García-Guillermo (TLQ) for her excellent assistance with the chemical assays.

REFERENCES

Ahmad, S.A., Salimullan Sayed, M.H., Barva, S., Khan, M.H., Faruquee, M.H., Jalil, A., Hadi, S.A. and Talukder, H.K.: Arsenic in drinking water and pregnancy outcomes. *Environ. Health Persp.* 109 (2001), pp. 629–631.

Cebrián, M.E., Albores, A., Aguilar, M. and Blakely, E.: Chronic arsenic poisoning in north of Mexico. *Human Toxicol.* 2 (1983), pp. 121–133.

Clifford, D.A. and Ghurye, G.L.: Metal-oxide adsorption, ion exchange, and coagulation-microfiltration for arsenic removal from water. In: W.T. Frankerberger (ed): *Environmental chemistry of arsenic.* 217–245. Marcel Dekker, New York, 2002.

Del Razo, L.M., Arellano, M.A. and Cebrián, M.E.: The oxidation status of arsenic in well-water from a chronic arsenicism area of northern Mexico. *Environ. Pollut.* 64 (1990), pp. 143–153.

Fendorf, S., Eick, M.J., Grossl, P. and Sparks, D.L.: Arsenate and chromate retention mechanisms on goethite. 1: Surface structure. *Environ. Sci. Technol.* 31 (1997), pp. 315–320.

Ferguson, J.F. and Gavis, J.: A review of the arsenic cycle in naturals waters. *Water Res.* 6 (1972), pp. 1259–1274.

Hsia, T.H., Lo, S.L., Lin, C.F. and Lee, D.Y.: Characterization of arsenate adsorption on hydrous iron oxide using chemical and physical methods. *Colloids Surf. A: Physicochem. Eng. Asp.* 85 (1994), pp. 1–7.

Le, X.C.: Arsenic speciation in the environmental and humans. In: W.T. Frankerberger (ed): *Environmental chemistry of arsenic.* Marcel Dekker, New York, 2002, pp. 95–116.

Mandal, B.K. and Suzuki, K.T.: Arsenic round the world: a review. *Talanta* 58 (2002), pp. 201–235.

Meng, X., Bang, S. and Korfiatis, G.P.: Effects of silicate, sulfate, and carbonate on arsenic removal by ferric chloride. *Water Res.* 34 (2000), pp. 1255–1261.

Meng, X., Korfiatis, G.P., Christodoulatos, C. and Bang, S.: Treatment of arsenic in Bangladesh well water using a household co-precipitation and filtration system. *Water Res.* 35 (2001), pp. 2805–2810.

Modificación a la Norma Oficial Mexicana NOM-127-SSA1-1994. Salud Ambiental. Agua para uso y consumo humano. Límites permisibles de calidad y tratamientos a los que debe someterse el agua para su potabilización. Diario Oficial de la Federación, 22 de noviembre de 2000, Mexico City, Mexico, 2000.

Montgomery, D.C.: *Design and analysis of experiments.* John Wiley and Sons, New York, 2001.

Pierce, M.L. and Moore, C.B.: Adsoption of arsenite and arsenate on amorfous iron hydroxide. *Water Res.* 16 (1982), pp. 1247–1253.

Rahman, M., Tondel, M., Ahmad, S.A., Chowdhury, I.A., Faruquee, M.A. and Axelon, O.: Hypertension and arsenic exposure in Bangladesh. *Hypertension* 33 (1999), pp. 74–78.

Smith, A.H.: Report and action: plan for arsenic in drinking water focusing on health, Bangladesh. World Health Organization, Geneva, Switzerland, 1997.

Sun, X. and Doner, H.E.: An investigation of arsenate and arsenite bonding structures on goethite by FTIR. *Soil Sci.* 161 (1996), pp. 865–872.

Sun, X. and Doner, H.E.: Adsorption and oxidation of arsenite on goethite. *Soil Sci.* 163 (1998), pp. 278–287.

USEPA: National primary drinking water regulations: arsenic and clarifications to compliance and new source contaminant monitoring. US Environmental Protection Agency, 2001.

Chemical methods

CHAPTER 58

Subsurface treatment of arsenic in groundwater—experiments at laboratory scale

H.M. Holländer, P.-W. Boochs, M. Billib & T. Krüger
Institute of Water Resources Management, Hydrology and Agricultural Hydraulic Engineering,
University of Hanover, Hanover, Germany
Now at: Chair of Hydrology and Water Resources Management, Brandenburg University of Technology,
Cottbus, Germany

J. Stummeyer & B. Harazim
Federal Institute of Geosciences and Natural Resources of Germany (BGR), Hanover, Germany

ABSTRACT: High arsenic (As) concentrations have been found in groundwater (up to 9000 µg/l) at a military site in Northern Germany. Most of the As is organically bound. Laboratory investigations were carried out to identify the possibility to build an *in situ* treatment plant to immobilize the As in the subsurface. Investigations at batch scale, using different iron compounds, showed that ferrous iron (Fe^{2+}) and zerovalent iron (Fe^0 particles) have the best capability to immobilize As. The inorganic As compounds were significantly reduced and organic As at least by 50%. Experiments at soil column scale using iron chloride proved the immobilization for an artificial aquifer at laboratory scale. It is demonstrated that the subsurface treatment of As in groundwater can be successfully conducted. Based on the laboratory experiments, a pilot plant is planned to carry out experiments on field scale size.

58.1   INTRODUCTION

58.1.1   *General*

Chemical weapons containing arsenic (As) compounds have been produced and tested at a military site of 6500 ha size in Northern Germany since 1917. The rest of the ammunition was destroyed after World War I but due to accidents and technically low standards, large contaminations were produced during research, production, storage and destruction of the ammunition. New chemical weapons were produced during World War II and have also been destroyed after the war. This resulted in high As concentration in the groundwater. At several specific contaminated sites an As concentration of more than 9 mg/l can be found.

The groundwater is currently treated using a cost intensive pump-and-treat plant. In order to investigate the possibility of a more cost effective system to reduce the As concentration in the groundwater, a preliminary study to evaluate the possibility of an *in situ* treatment was carried out.

58.1.2   *Arsenic immobilization and remobilization at batch scale*

Arsenic concentration in the aqueous phase can be reduced by some metal salts. In the process the As compounds are adsorbed by metal complexes like iron hydroxide [$Fe(OH)_3$]. Water contaminated with inorganic As compounds was analyzed in detail during batch-tests (Scott *et al.* 1995, Hering *et al.* 1996, Wilkie and Hering 1996, Rott and Meyer 2000, Bissen and Frimmel 2003). Different metal salts, based on aluminum and iron, were used as coagulants for removing inorganic arsenic (i-As). The results showed that iron and aluminum were both able to immobilize i-As by an adsorption process. However, iron was generally a better adsorption partner for the As compared to aluminum.

Iron compounds used for As adsorption are elementary iron (Fe^0) as well as ferrous iron [Fe(II), Fe^{2+}] and ferric iron [Fe(III), Fe^{3+}].

The usage of iron chloride ($FeCl_3$) for the adsorption process results in a clear reduction of As concentration (Scott *et al.* 1995, Wilkie and Hering 1996, Bissen and Frimmel 2003). Using this technique Meng *et al.* (2001) reduced the As concentrations from about 300 µg/l to less than 50 µg/l.

Generally it is easier to immobilize arsenate (AsO_4^{3-}) by adsorption than arsenite (AsO_3^{3-}) (Hering *et al.* 1996, Wilkie and Hering 1996, Meng *et al.* 2001). There is no available information on the adsorption of organic As compounds. However, the presence of sulfate, phosphate, silicate and ammonium reduces the efficiency of the adsorption process (Hering *et al.* 1996, Wilkie and Hering 1996, Meng *et al.* 2001, Su and Puls 2003).

The As compounds can be remobilized under special circumstances from the metal-As complex. Ramaswami *et al.* (2001) did not detect any remobilization of the As which was adsorbed by zerovalent iron during a leaching-test for 3 hours. Jezierski *et al.* (1999) reported that more As was remobilized from aluminum than from iron compounds. Meng *et al.* (2001) determined that As which was in iron sludge yielded low leaching rates. This refers to the fact that As can be remobilized under a reducing environment. Arsenic can also be remobilized from iron compounds at pH > 12. The iron hydroxide complex is reduced in this environment and As compounds are released to the mobile phase.

58.1.3   *Arsenic immobilization at field scale*

Rott *et al.* (1996) and Rott and Meyer (2000) published results of a pilot plant that allows the *in situ* immobilization of As compounds inside the aquifer. A method to immobilize As compounds was developed based on the well known technique to precipitate iron and manganese in a subterranean environment (Rott *et al.* 1978, Henning and Rott 2005). The technique is based on two wells which can infiltrate air-enriched water and also pump water. During a cycle, one well infiltrates and the other is pumping, and in the next cycle, it is reversed. Addition of oxygen resulted in the precipitation of iron hydroxide and adsorption of inorganic As compounds.

Henning (2004) described several phases to establish a well working situation to immobilize As compounds by iron hydroxide. The first phase of this process was to establish an effective reactive body around the two wells which were a distance of 150 m from each other. During this phase, the oxygen increased the redox potential of the groundwater. This resulted in oxidation of the dissolved iron [Fe(II)] to iron hydroxide [$Fe(OH)_3$]. Iron hydroxides are dissolved particles which precipitate on the aquifer matrix. The iron hydroxide complexes allow dissolved iron to adsorb so that dissolved iron from areas outside of the oxidized zone around the wells can also absorb during a pumping phase. This results in reactive zones around both wells.

Problems can be present if manganese is also available in a dissolved form inside the aquifer. Since oxidized manganese will be reduced by iron, the manganese will also precipitate. This happens after the iron is already immobile. This results in higher oxygen consumption.

The second phase allows the adsorption of As compounds by the iron hydroxide if the pumping situation allows the contact of the immobile iron hydroxide with dissolved As.

Therefore, the system described by Henning (2004) introduces two cycles: an infiltration period for oxygen enriched water and a pumping period during which As is moved into the reaction zone. Rott *et al.* (1996) and Rott and Meyer (2000) showed that the pumping cycles were always longer than the infiltration cycles so that more water is pumped for drinking water supply than was infiltrated.

The deposition of iron hydroxide inside the aquifer can lead to clogging. This was not observed during the experiment. It is assumed that the particles accumulate in dead-end-pores or even in pores with low flux rates (Rott and Meyer 2000). Henning (2004) showed that the voluminous iron hydroxide complexes are dewatered to more compact forms like hematite.

During the experiments of Rott *et al.* (1996) and Rott and Meyer (2000) As concentrations below 10 µg/l were obtained (initial concentration of 38 and 25 µg/l, respectively) which is the maximum permissible concentration in drinking water (TVO 2001, USEPA 2001, WHO 2004). After stopping the aeration, a remobilization test was conducted in which groundwater was continuously surveyed over a period of one month. No remobilization was observed during that period. The As concentration in the pumped water never reached concentrations above 10 µg/l during this month. The reactive body was still able to immobilize As although oxidation of iron had stopped. The adsorption capacity was still high and declines only by As sorption and the aging process of the iron hydroxide complex.

58.2 MATERIALS AND METHODS

58.2.1 *Sampling and on-site sample treatment*

A mixture of the As species was expected at the investigated military site. To test a potential subsurface groundwater treatment process, the actual concentration of each As compound has to be known. The As(III)/As(V) equilibrium depends strongly on the chemical conditions of the water (e.g., pH, oxygen concentration, redox potential).

A reliable method for the *in situ* determination of all As species immediately after sampling does not exist. The samples have to be treated on-site as fast as possible, to ensure a stabilization by separation of the different species. Selective analysis of the As compounds can be performed later in the laboratory, using selective instrumental analytical methods. Water samples were split and treated on-site following a stabilization procedure (Stummeyer *et al.* 1996, Bednar *et al.* 2004).

58.2.2 *Separation and stabilization of As(V) and As(III) at sampling site*

The pH of a sample was adjusted with a buffer solution to a value of ≥ 8. Figure 58.1 shows the distribution of the As species depending on pH. Arsenate is an anion and arsenite is uncharged at pH 8. The alkaline water samples were then passed through a small pre-concentration column filled with an anion exchange resin (Dowex 1 × 8). Under the adjusted conditions, arsenate [As(V)] is retained on the resin in anionic form ($HAsO_4^{2-}$), due to its chemistry in alkaline solution (H_3AsO_4: $pK_2 = 6.9$; $pK_3 = 11.5$). Under these conditions, arsenite [As(III)] passes through the pre-concentration column and remains in the alkaline solution in the uncharged form of arsenous acid (H_3AsO_3: $pK_1 = 9.2$).

After separation, no further stabilization steps are required. The concentration of As(V) and As(III) was determined by measuring As concentration in both the resin-phase after elution with HCl and the alkaline solution.

The sample pre-treatment procedure was developed for the analysis of natural waters with normally low As concentrations. In case of the sampled military site, high amounts of organic

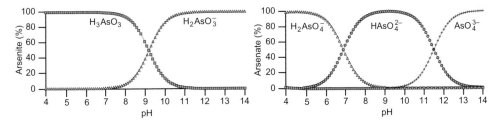

Figure 58.1. Arsenite and arsenate speciation as a function of pH.

As compounds were expected, but unfortunately no information about organic species were available. Depending on the chemical behavior of the As(org) compounds, its distribution between resin-phase and aqueous solution is unknown. A differentiation of organic and inorganic As compounds in both phases can be performed by using different analytical procedures (Hung *et al.* 2004).

58.2.3   *Analytical procedures*

58.2.3.1   *ICP-optical emission spectrometry (ICP-OES)*
The determination of total arsenic (t-As) was performed with inductively coupled plasma-optical emission spectrometry (ICP-OES). Depending on the high temperature of the plasma, all forms of As are atomized and ionized, so that the instrument response does not vary with As species of a sample.

A Spectro (Model Ciros) ICP-spectrometer was used. The instrument parameters were optimized for multi-element analysis of water samples. Total As was determined at a wavelength of 189.42 nm simultaneous with other trace elements. The optical path of the instrument was flushed with Ar, to allow measurements at UV spectral range. Under multi-element conditions (no special optimization for As determination) limit of detection was approximately 20 μg/l As.

58.2.3.2   *Flow-injection hydride-generation AAS (FI-HG-AAS)*
Hydride generation atomic absorption spectrometry (HG-AAS) coupled with flow-injection analysis (FI) can only be used for the determination of inorganic (reducible) As compounds. Samples are pre-reduced using acidic KI solution. After pre-reduction, the i-As is completely reduced to As(III) while persistent organic compounds remain unchanged. The hydride generation is performed with tetrahydroborate and hydrochloric acid:

$$As(OH)_3 + 3BH_4{}^- + 3H^+ \longrightarrow AsH_3 + 3BH_3 + 3H_2O \tag{58.1}$$

A Perkin Elmer Analyst 100 spectrometer coupled with a Perkin Elmer flow-injection device PE FIAS 100 was used for the determination of i-As. The system was optimized to achieve a sensitivity of 0.170 absorbance units for a 10 μg/l As(III) solution. Under these conditions, the limit of detection was approximately 0.5 μg/l As. Arsenic hydrides were separated from solution with a PE membrane gas-liquid separator and then transferred using Ar to an electrically heated quartz cell.

58.2.4   *Batch investigations*

All batch investigations were carried out by adding iron compounds corresponding to 50 mg of iron. This amount of iron was added to a sample of 200 ml groundwater. Then the batch reactor was aerated so that iron hydroxide settled. For pH values less than pH 6, 100–500 μl of 2.5 M sodium hydroxide solution were added.

The first batch investigations were undertaken with groundwater from the observation well ObsW 69/01-02 from a depth of 7 m. In these batch investigations several kinds of iron compounds were tested to determine the best chemical compound of iron to immobilize the As. A second row of batch tests was carried out to determine the effects of $FeCl_2$ at different sites of observation and pumping wells.

58.2.5   *Experiments at a soil column*

Laboratory experiments at a soil column were carried out (Fig. 58.2). The soil column had a length L of 1 m and a diameter D of 58 mm using sand fractionated from 1 mm to 200 μm with a uniformity coefficient U of 2.6 in the soil column. This resulted in a hydraulic conductivity k_f

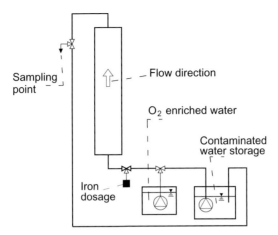

Figure 58.2.   Layout of the laboratory set up.

of 1.3×10^{-3} m/s and porosity n of 0.33. The hydraulic gradient was assumed to be constant throughout the experiment and determined to be about 0.1. This resulted in an initial flux Q of 1 l/h.

The test procedure was conducted in cycles by infiltration of groundwater with the natural As concentration (duration 1 h) before infiltrating Fe(II) enriched water with a concentration of 9.9 g Fe(II)/l (duration 5 min). After another no-flow phase of 5 minutes, oxygen enriched water was infiltrated into the soil column (duration 15 min). The next cycle started after another no-flow phase of 5 minutes. This means that every cycle had a duration of 1.5 hours. During the laboratory experiments a total of 10 g iron chloride [3.7 g Fe(II)] were used. Water samples were taken after every third cycle to determine the As concentration at the outflow side. The samples were tested by the methods described in Section 58.2.2.

58.3   RESULTS

58.3.1   *Arsenic composition*

Arsenic concentrations up to 9.05 mg/l were found at the main contamination site (observation well ObsW 69/01-02). A second major contamination site was found next to observation well ObsW 2/89 (7.83 mg/l). At larger distances from these observation wells the As concentrations were smaller but As was found at greater depths.

The As concentrations are shown in Table 58.1. The results clarify that the largest amount of the As is part of an organic compound (ca. 50–90%). Due to the absence of oxygen in the groundwater and due to the high concentration of organic compounds, arsenate was found only in a small amount (<10%). The rest of the As was arsenite (10–45%).

58.3.2   *Batch investigations*

The first batch investigation showed that Fe(II) allows the best immobilization of As (Table 58.2). Due to admixture of Fe(II) compounds and subsequent aeration with atmospheric oxygen the As concentration in the mobile phase was reduced by 80% (experiment 2 and 3). Using Fe(III) as reagent the efficiency of the immobilization is reduced 10 to 40% (experiment 1, 4 and 5). Zerovalent iron allows nearly no reduction up to a bisection of the As concentration in aqueous phase (experiment 6 and 7). The results are dependent mainly on the size of the particles. The control experiment

Table 58.1.   Arsenic composition at test site.

Sample	As(III) [mg/l]	As(V) [mg/l]	org. As [mg/l]	t-As cal. [mg/l]	t-As ICP [mg/l]
ObsW 69/01-02	2.44	0.28	6.08	8.80	9.05
ObsW 69/02-02	0.24	0.04	0.36	0.55	0.56
Well FB 21A/95	0.10	0.01	0.66	0.77	0.77
ObsW 26/91	0.21	0.02	2.14	2.37	2.40
ObsW 5/89	0.003	0.002	0.076	0.081	0.08
ObsW 70/01-02	0.71	0.13	0.27	1.11	1.18
ObsW 70/02-02	0.10	0.03	0.02	0.15	0.03
ObsW 24/91	0.13	0.04	1.49	1.66	1.61
ObsW 2/89	2.37	0.15	5.24	7.76	7.83
Well FB 1A/94	0.16	0.02	0.18	0.36	0.26

Table 58.2.   Arsenic immobilization by different iron compounds.

Experiment No.	Iron compound	Amount	Aeration time [h]	As [mg/l]	Immobili-zation [%]
1	FeCl$_3$-solu. (10000 mg/kg/0.1n HCl)	5 ml	22	5.4	38.6
2	FeSO$_4$ × 7H$_2$O	250 mg	22	1.9	78.4
3	FeCl$_2$ × 4H$_2$O	180 mg	4.5	1.8	79.5
4	FeCl$_3$ × 6H$_2$O	240 mg	20	8.0	9.1
5	NH$_4$Fe(SO$_4$)$_2$ × 12H$_2$O	430 mg	144	5.2	41.9
6	Fe (Ferrum reductum <150 µm)[1]	50 mg	144	8.3	5.7
7	Nano particles Fe^0 + Ni$_3$[1]	100 mg	20	3.8	56.8
8	Only atmospheric oxygen		27	8.4	1.2

[1] without oxygen.

Table 58.3.   Arsenic immobilization with iron chloride at different sites.

Sample	pH	2.5 M NaOH [µg]	Aeration time [h]	i-As [mg/l]	org. As [mg/l]	Immobili-zation inorg. As [%]	Immobili-zation org. As [%]
ObsW 69/01-02	6.8	–	4	0.032	2.183	99	65
ObsW 69/02-02	5.3	500	over night	0.011	0.105	92	76
Well FB 21A/95	5.9	200	over night	(<0.005)	0.159	94	67
ObsW 70/01-02	5.8	100	over night	0.08	0.134	~100	50
ObsW 70/02-02	5.8	200	4	(<0.005)	0.040	na	na
ObsW 24/91	6.7	100	over night	0.008	0.353	~100	76
ObsW 2/89	6.1	–	8	0.042	1.825	98	65
Well FB 1A/94	4.6	200	8	0.004	0.045	95	75

shows that the exclusive use of atmospheric oxygen does not result in a significant reduction of the As concentration in the aqueous phase.

Comparing the results of the Fe(II) experiment (Table 58.2) with the results of the arsenic composition at the test site (Table 58.1) it is shown that more As is immobilized than inorganic bound As is available in the water. This confirms the fact that organic As compounds have been also immobilized. The second row of the batch tests showed that nearly all groundwater samples needed an increasing pH value using sodium hydroxide to allow precipitation of iron hydroxide compounds after adding iron chloride (Table 58.3). The i-As compounds were immobilized by 92–100% by

Table 58.4.   Arsenic concentrations in the outflow of the soil column.

Sample	Time [h]	t-As [mg/l]	% of initial concentration
0	0	9.05	100.0
1	1.5	5.51	60.9
2	4.5	1.36	15.0
3	9	0.93	10.3
4	24	0.71	7.8

iron hydroxide (66.7 mg Fe(II) per 200 ml sample) in the batch tests. Organic As compounds were reduced using this method mainly between 65 and 76%. Since organic As compounds are the main fraction of the t-As (Table 58.1), in nearly all samples significantly high As concentrations are still measurable.

58.3.3   *Soil column*

The Fe(II) infiltration during iron enrichment was too high and resulted in a reduction of the flux through the soil column to nearly zero after 24 hours. This happened due to the precipitation of iron hydroxide on the aquifer matrix. The As concentration of the water during the first 24 h can be seen in Table 58.4.

Sample No. 1 shows a significant reduction of the As concentration. This means that iron hydroxide precipitated very rapidly and As adsorption started very early. After 24 hours the As concentration was reduced to 7.8% of the initial concentration due to the high iron concentration. The largest reduction was taking place during the first 3 cycles (4.5 h). During the following 19.5 h the concentration was only reduced by another 48%.

58.4   CONCLUSIONS AND SUGGESTIONS FOR A TECHNICAL PLANT

The groundwater from the military site contained inorganic and organic As compounds. During the investigations two major contamination sites, independent of each other, were found with concentrations of 9.05 mg/l and 7.83 mg/l, respectively. The techniques used for detection of the As compounds do not allow special predictions about the form and the toxicity of the organic compounds in the samples. However, the high fraction of organic compounds found in water suggests that the organic As compounds at the site are still released unchanged into the groundwater.

The laboratory experiments showed that an *in situ* remediation of groundwater which contains As is generally possible. Several iron compounds to remove As from contaminated groundwater have been analyzed in batch-tests. Particularly, Fe(II) compounds show very good results in immobilization of As compounds with site specific groundwater from a military site. Since the technique will be tested in the pilot plant at the contaminated site, environmental aspects had to be taken into account. High sulfate contents in groundwater have to be avoided so iron chloride was selected for further investigations.

Additional batch experiments using iron chloride showed that inorganic As compounds can be immobilized by more than 90% of the initial concentration. The organic As compounds were reduced by at least 50%. An optimization of the Fe(II) in relation to the As composition and concentration was not undertaken.

Soil column experiments showed that adsorption of As compounds by iron hydroxide is working well. The immobilization of the As content was measured up to 92%. This means that inorganic

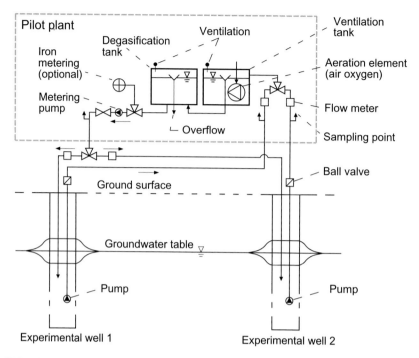

Figure 58.3.   Layout of the planned pilot plant.

and organic As compounds must have been immobilized. The adsorption process starts very early in the soil column. Problems have been found with clogging processes of the artificial aquifer.

Therefore it can be assumed that a technical remediation plant directly at the site is able to effectively immobilize the As concentration in groundwater using a subsurface treatment. However, the laboratory experiments showed also that not all As can be immobilized—especially not all organic bound As compounds. Since most of the As is organic, there is the need to find additional arrangements to secure a transfer from organic to inorganic As compounds so that most of the As can be effectively immobilized in the subsurface.

The pilot plant shall be built next to the observation well ObsW 69/01-02 since the As concentrations there are very high. The batch experiments allowed a reduction of the total As concentration from 9.05 mg/l to 2.215 mg/l. This means more than 75% was immobilized and the possibility of reducing the costs during treatment seems to be very good.

The pilot plant is designed using two wells which pump and infiltrate water alternately (Fig. 58.3). The aquifer in between the two wells is the important reaction area where the groundwater shall be As-free. Since more groundwater shall be taken from the wells than shall be infiltrated, groundwater from outside the main reaction zone area flows inside the zone and is also freed from As compounds. Rott *et al.* (1996), Rott and Meyer (1996) and Henning (2004) described only two phases to reach an immobilization of As. Since batch experiments showed that by aeration only nearly no As is immobilized the planned pilot plant needs three phases. First results of the laboratory tests have also shown this.

Since the concentration of dissolved iron (Fe^{2+}) (<50 µg/l) in the groundwater was very low, an additional phase had to be added to infiltrate water with iron chloride. The other two phases were the infiltration of oxygen-enriched water and a pumping phase. In between these phases short no-flow phases shall be included so that the mixing of the different injected water is more efficient.

REFERENCES

Bednar, A.J., Garbarino, J.R., Burkhardt, M.R., Ranville, J.F. and Wildeman, T.R.: Field and laboratory arsenic speciation methods and their application to natural-water analysis. *Water Res.* 38 (2004), pp. 355–364.

Bissen, M. and Frimmel, F.H.: Arsenic—a review; Part II: Oxidation of arsenic and its removal in water treatment. *Acta Hydroch. Hydrob.* 31:2 (2003), pp. 97–107.

Henning, A.-K.: Biologische Mechanismen bei der unterirdischen Aufbereitung von Grundwasser am Beispiel des Mangans. *Oldenbourg Industrieverlag GmbH, Stutt. Ber. Siedlungswasserwirt.* 176 (2004).

Henning, A.-K. and Rott, U.: Untersuchungen zur Manganoxidation bei der In-situ-Aufbereitung von reduzierten Grundwässern. *Grundwasser* 4 (2003), pp. 238–247.

Hering, J.G., Chen, P.-Y., Wilkie, J.A., Elimelech, M. and Liang, S.: Arsenic removal by ferric chloride. *Am. Water Works Assoc., e-Journal* 88:4 (1996), pp. 155–167.

Hung, D.Q., Nekrassova, O. and Compton, R.G.: Analytical methods for inorganic arsenic in water: a review. *Talanta* 64 (2004), pp. 269–277.

Jezierski, H., Karschunke, K. and Plöthner, D.: Arsenkontamination von Trinkwasser in Bangladesh. *Wasser und Abfall* 10 (1999), pp. 19–23.

Meng, X., Korfiatis, G.P. and Christodoulatos, C.: Treatment of arsenic in Bangladesh well water using a household co-precipitation and filtration system. *Water Res.* 35:12 (2001), pp. 2805–2810.

Ramaswami, A., Tawachsupa, S. and Isleyen, M.: Batch-mixed iron treatment of high arsenic waters. *Water Res.* 35:18 (2001), pp. 4474–4479.

Rott, U., Boochs, P.-W. and Barovic, C.: Unterirdische Grundwasseraufbereitung durch Einleitung von sauerstoffhaltigem Wasser in den Boden. *Vom Wasser* 82 (1978), pp. 201–208.

Rott, U., Meyerhoff, R. and Bauer, T.: In situ-Aufbereitung von Grundwasser mit erhöhten Eisen-, Mangan-, und Arsengehalten. *Wasser und Abfall* 7 (1996), pp. 358–363.

Rott, U. and Meyer, C.: Die unterirdische Trinkwasseraufbereitung—ein Verfahren zur rückstandsfreien Entfernung von Arsen. *Wasser und Abfall* 10 (2000), pp. 36–43.

Scott, K.N., Green, J.F., Do, H.D. and McLean, S.J.: Arsenic removal by coagulation. *Am. Water Works Assoc., e-Journal* 87:4 (1995), pp. 114–126.

Stummeyer, J., Harazim, B. and Wippermann, T.: Speciation of arsenic in water samples by high-performance liquid chromatography-hydride generation-atomic absorption spectrometry at trace levels using a post-column reaction system. *Fresenius J. Anal. Chem.* 354 (1996), pp. 344–351.

Su, C. and Puls, R.W.: In Situ Remediation of Arsenic in Simulated Groundwater Using Zerovalent Iron: Laboratory Column Tests on Combined Effects of Phosphate and Silicate. *Environ. Sci. Technol.*, 37:11 (2003), pp. 2582–2587.

TVO: Verordnung über die Qualität von Wasser für den menschlichen Gebrauch, TrinkwV 2001—Trinkwasserverordnung vom 21. Mai 2001; BGBl. I Nr. 24 vom 28.5.2001 S.959, Bundesanzeiger Verlag, Köln, 2001.

USEPA: National primary drinking water regulations. Arsenic and clarifications to compliance and new source contaminants monitoring. Final Rule, Fed. Reg., 66:14: 6976 (Jan. 22, 2001), US Environmental Protection Agency, Washington, DC, 2001.

Wilkie, J.A. and Hering, J.G.: Adsorption of arsenic onto hydrous ferric oxide: effects of adsorbate/adsorbent ratios and co-occurring solutes. *Colloids Surf. A: Physiochem. Eng. Asp.* 107 (1996), pp. 97–110.

WHO: Guidelines for drinking-water quality. 3rd edition, World Health Organization, Geneva, Switzerland, 2004.

CHAPTER 59

Two-step *in situ* decontamination of mine water enriched with arsenic and iron

B. Doušová, T. Brůha, A. Martaus, D. Koloušek, R. Pažout & V. Machovič
Institute of Chemical Technology in Prague, Prague, Czech Republic

ABSTRACT: The two-step decontamination method allows the removal of arsenic (As) from acid mining water with a high content of As (50 mg/l) and Fe (5000 mg/l). The first treatment step for the raw mining water comprises a partial precipitation with a defined amount of alkaline agent (NaOH, Na_2CO_3 or $Ca(OH)_2$) at pH $\sim$5.0. During this first precipitation more than 90% of As is adsorbed as As (V) onto the iron oxyhydroxide surface forming inner-sphere complexes. The toxic "As" mass from the first step is then separated by decantation and/or filtration. In the second step the liquid residue is subjected to a second precipitation by means of lime $Ca(OH)_2$ at pH $\sim$8.5. While As was removed in the first precipitation, the other components including residual iron, manganese, zinc and sulfates are precipitated quantitatively during the second step.

59.1 INTRODUCTION

Arsenic (As) is of increasing environmental attention due to its risk to plant, animal, and human health. The most abundant mineral sources of As are arsenian pyrite and arsenopyrite (Cullen and Reimer 1989). Weathering of these minerals in oxidizing environments dissolves arsenic (As) as As(III) and rarely As(V). Anthropogenic As particularly proceeds from industrial wastes including manufacture of insecticides, pesticides and fertilizers as well as mining and smelting industries (Lin and Puls 2000). Stock-piles still exist and can greatly increase the concentration of As in groundwaters. Mining activities are responsible for As poisoning in Thailand, while natural occurring As is of great concern in groundwaters from some regions of Bangladesh, India and Nepal, and also Mexico and Argentina, where hazardous concentrations of As have appeared as a result of strong water-rock interactions and the physical and geochemical conditions for As mobilization in aquifers (Smedley and Kinniburgh 2002).

Arsenic mobility and transport in the environment are strongly influenced by associations of the As species with solid phases in soil and sediments (Keon *et al.* 2001). Arsenic is unique among the heavy metalloids and oxyanion-forming elements due to its sensitivity to redox conditions and to mobilization at the pH value close to 7.0 (typical for groundwaters). In natural waters it is mostly found in inorganic form as trivalent arsenites [As(III)] which dominate under reducing conditions or pentavalent arsenates [As(V)] in oxidizing environments, while organic As forms are rarely quantitatively important.

It is well known that As oxyanions, especially arsenates have a strong adsorption affinity for hydrated iron oxides forming inner-sphere bidentate, binuclear As(V)-Fe(III) complexes (Randall *et al.* 2001). This fact can influence the enrichment with soluble As as both As(III) and As(V) in natural sources (e.g., geothermal springs) (Cullen and Reimer 1989). On the other hand, it can be helpful with respect to natural detoxification of As-contaminated waters.

The most common Fe(III) hydroxides, oxides and oxyhydroxides include ferrihydrite ($Fe_5HO_8 \times 4H_2O$, often written as $Fe(OH)_3$, which transforms to hematite (α-Fe_2O_3) and/or goethite (α-FeOOH) depending on solution composition, temperature and pH (Refait *et al.* 2001).

Ferrous and ferric hydroxides, called green rusts represent very important phases for trace metal mobility due to their large reactive surface area and reducing abilities (Lin and Puls 2003).

Green rusts are layered structures containing anions such as CO_3^{2-}, SO_4^{2-} or Cl^- in the interlayers (Hansen 2001). They are usually found in a mildly reducing environment and have the general chemical formula $[Fe_{6-x}^{2+}Fe_x^{3+}(OH)_{12}]^{x+}[(A)_{x/n} \cdot yH_2O]^{x-}$, where A is an *n*-valent anion, e.g., CO_3^2, SO_4^{2-} and Cl^-. They are formed by the positive charged layers of green-blue Fe(II)-Fe(III) hydroxides, whose charge is balanced by the inclusion of anions between the layers. The stability and forms of Fe species significantly depend on geochemical and thermodynamic conditions (Furukawa *et al.* 2002), which is very important for their reactivity and sorption properties (Sherman and Randall 2003). It is evident, that As-oxyanions [both As(III) and As(V)] indicate a high affinity to all above mentioned Fe phases, while ferrihydrite and green rust are preferred.

In this work, the method of effective As removal from strongly contaminated mining water enriched with As (50 mg/l) and iron (more than 5000 mg/l) is evaluated. The first step of the elimination from raw mining water comprises a partial precipitation with a small amount of alkaline agent to pH ~5.0, by which As is removed almost quantitatively. The final treatment of mining water develops in the second step by precipitation applied to the residue of the first treatment by means of lime [Ca(OH)$_2$] at a pH value of about 8.5 (Doušová 2005a). The study of As(V)-Fe-SO_4^{2-} variations in relation to the pH value allows estimations about the optimum conditions of the process, i.e., to produce a minimum mass of toxic precipitate while keeping ecological limits of treated water.

59.2   EXPERIMENTAL

59.2.1   *Mining water*

For these experiments the strongly contaminated mining water from a former ore mining activity close to the historical mining town Kutna Hora in central Bohemia was used (Kopřiva *et al.* 2005). The average chemical composition of raw water is presented in Table 59.1.

59.2.2   *Alkaline solutions*

Alkaline solutions were prepared from NaOH, Na$_2$CO$_3$ and CaO of p.a. quality and CO$_2$-free distilled water in the concentrations: 1.0 M NaOH, 0.5 M Na$_2$CO$_3$ and 1.0 M Ca(OH)$_2$, respectively.

59.2.3   *Two-step precipitation*

The first treatment step applied to the raw mining water was a partial precipitation of As and metals by addition of a defined amount of the alkaline agents at 25°C. A 500 ml sample was stirred (750 rpm) and aerated in a batch reactor, while the Na$_2$CO$_3$or NaOH solution was added dropwise

Table 59.1.   Average composition of the raw mining water from Kank locality (pH = 3.5–4.1).

Parameter	Concentration (g/l)
Fe	5,752
Zn	1.589
Cu	2.7×10^{-5}
As	0.054
Mn	0.166
Cd	2.27×10^{-4}
SO_4^{2-}	17.7
Insoluble comp.	0.295

to a pH of about 5.5. The suspension was stirred and aerated for 30 minutes to achieve the maximum oxidation state and then was allowed to mature under aerobic conditions for the next 30 minutes. Subsequently, the sample was filtered off; the solid phase was dried, weighed and examined with XRD diffraction and Raman spectroscopy. The filtrate was analyzed for As and Fe contents using AAS and UV-VIS spectroscopy before being used for the second step of decontamination.

In the second step a defined amount of $Ca(OH)_2$ was added to portions of the liquid phase from the first step (within the 1st and the 3rd minute of reaction). The reaction mixture was stirred (750 rpm) and aerated at 25°C for 20 minutes while the pH value was continuously monitored. Subsequently, the suspension matured under aerobic conditions for about 30 minutes. Then it was filtered off; the solid phase was treated analogously to the first step. In the liquid residue the concentrations of As, Fe and SO_4^{2-} were determined.

59.2.4 *Analytical methods*

Raman spectra were collected using the LabRam HR system (Jobin Yvon). The 532 nm line of laser was used for excitation. An objective ($\times 100$) was used to focus the laser beam on the sample placed on an X-Y motorized sample stage. The scattered light was analyzed by a spectrograph with a holographic grating (600 gm/mm), a slit width of 150 µm and an opened confocal hole (1000 µm). Adjustment of the system was regularly checked using a silicon sample and by measurement in the zero-order position of the grating. Small power levels up to 4 mW at the entrance optics were used in order to avoid laser induced damage of the sample. The time of acquisition of a particular spectral window was optimized for individual sample measurements from 200 to 400 s. Three accumulations were co-added to obtain a spectrum.

The DRIFR spectra were measured on a Nicolet 740 Fourier transform infrared spectrometer equipped with the TGS detector. Approximately 50 mg symplex in a near form were measured with a diffuse reluctance accessory (Spectra Tech). Each sample spectrum was an average of 512 scans. All spectra were recorded at the resolution of 2/cm. Powder X-ray diffraction patterns were recorded using a Seifert XRD 3000P-instrument with CoKα radiation ($\lambda = 0.179026$ nm, graphite monochromator, goniometer with Bragg-Brentano geometry) in 2θ range 12–75°, step size 0.05°2θ. The semi-quantitative analysis of the solid phase samples was done on an ARL 9400 by the X-ray fluorescence method using a UniQuant 4 program. The concentration of As and Fe in the aqueous solutions was determined by AAS using SpectrAA-880, unit VGA 77 (Varian) for measuring in flame and SpectrA A-300 (Varian) for hydride process. The amount of SO_4^{2-} ions was evaluated by gravimetric analysis as $BaSO_4$.

59.3   RESULTS AND DISCUSSION

59.3.1   *First precipitation step*

As was mentioned above, the mining water was subjected to a precipitation with a defined amount of alkali agent (NaOH, Na_2CO_3, or $NaOH/Na_2CO_3$). The reaction in the first step follows equations (59.1) and/or (59.2):

$$4Fe^{2+} + 8OH^- + O_2 \rightarrow 4FeOOH + 2H_2O \tag{59.1}$$

$$4Fe^{2+} + 4CO_3^{2-} + O_2 + 2H_2O \rightarrow 4FeOOH + 4CO_2 \tag{59.2}$$

The pH value of the filtrates from the first step varied from 3.5 to 6.0 depending on the amount of alkali addition. The residual As and Fe concentrations in filtrates are shown in Table 59.2 and Figure 59.1.

The results presented in Figure 59.1 confirmed the high sorption affinity of As-oxyanions to the active surface of hydrated Fe-oxide hydroxide phases. More than 95% of the As was adsorbed in a

Table 59.2.   Concentration of As and Fe in solution after the first step (mg/l).

Alkali added (g/l)	NaOH		NaOH/Na$_2$CO$_3$		Na$_2$CO$_3$		Average		Treatment efficiency (%)	
	As	Fe	As	Fe	As	Fe	As	Fe	As	Fe
0	54.00	5752	54.00	5752	54.00	5752	54.00	5752	0.0	0.0
1	7.01	3800	3.07	3650	3.03	3555	4.37	3668	91.8	36.2
2	4.32	3550	3.03	3570	1.05	3542	2.80	3554	94.6	38.2
3	3.20	3360	2.93	3462	0.73	3382	2.31	3401	95.8	40.9
4	1.97	3110	1.76	3120	<bd	3210	1.20	3147	97.7	45.3

<bd: Below detection limit.

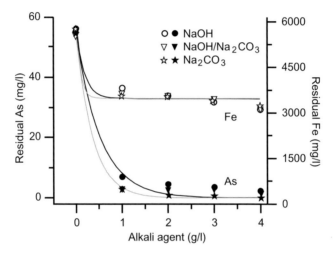

Figure 59.1.   Development of As and Fe removal during the first treatment step.

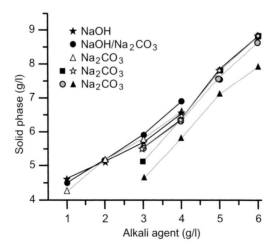

Figure 59.2.   Solid mass after the 1st step.

relatively small portion of precipitated iron (about 40%). This corresponds well with results of Lin and Puls (2003), who investigated the adsorption, desorption, reduction and oxidation of As by goethite, lepidocrocite and green rust and compared the attenuation capacity of iron hydroxides, clays and feldspars for As. They also discussed the transformation of green rust under oxidizing and reducing conditions. Their results confirmed that As was more strongly bound by iron hydroxides than by clays and feldspars, in an order: iron hydroxides > clays > feldspars.

The precipitates from the first step were dried, weighed and analyzed by XRD and Raman spectroscopy. The solid phases mass varied from 4.4 to 8.6 g/l depending strongly on the added alkali but negligibly on its kind (Fig. 59.2). X-ray diffractograms illustrated dominant amorphous phases of Fe oxhydroxides in the precipitates. Raman spectra (Fig. 59.3) showed the typical ferrihydrite bands about 300, 550 and 720/cm. The bands assigned to green rust were found at 350 and 400/cm (Fe^{2+}-OH), (Refait *et al.* 2001), which corresponded to a reduced or semireduced reaction environment (Doušová 2005b). The sulfate S-O bonds appeared at 1000/cm.

59.3.2   *Second precipitation step*

In the second step the final treatment of mining water from residual Fe, other metals and sulfates was performed corresponding to the method described above. The residual As concentration in the

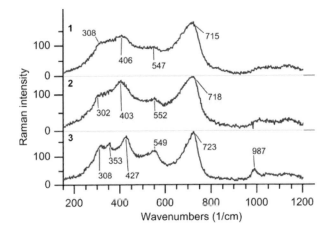

Figure 59.3.   Raman spectra of selected solid phases after the first treating step.

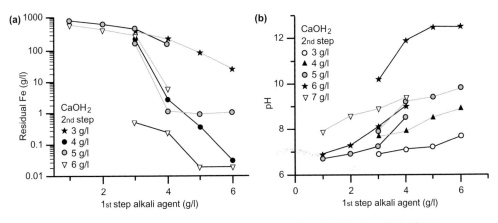

Figure 59.4.   Final treatment of mining water (a) Fe in liquid residue; (b) pH value of liquid residue.

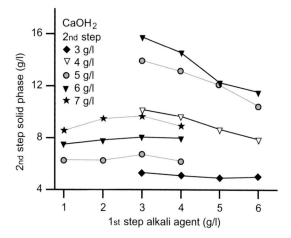

Figure 59.5.   Residual solid mass after the 2nd step.

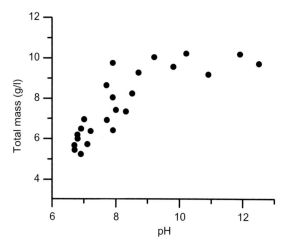

Figure 59.6.   Total solid mass.

remaining liquid volumes of samples was always below the detection limit, the Fe content varied from 890 to 0.5 mg/l, and pH ranged between 6.7 and 12.1. It is evident, that As was essentially removed in the first step, and only negligible amounts of As remained after the second precipitation. The residual Fe content and pH strongly depended on the amount of added $Ca(OH)_2$ (Fig. 59.4).

The solid phases from the second precipitation were treated similarly as those from the first step; their mass varied from 6.3 to 14.0 g/l depending on the amount of $Ca(OH)_2$ added in the second step (Fig. 59.5). Contrary to the first treatment step, the amount of the precipitates of step 2 slightly decreased with increasing amount of the alkali added in the first step (Figs. 59.2 and 59.5).

Therefore, the total mass of precipitate depended simply on the total amount of alkali added during the whole process, which was related to the residual pH. At pH of about 8.5–9.0 all particles were precipitated almost quantitatively; above this pH range the solid phase mass had not changed significantly (Fig. 59.6).

X-ray diffraction of solid phases after the second treatment step confirmed the predominance of gypsum ($CaSO_4$), typical peaks of the basanite modification ($CaSO_4 \times 0.5\ H_2O$) have also been observed (Fig. 59.7). Raman spectra showed the typical bands of goethite (to be compared with the

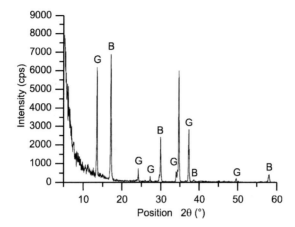

Figure 59.7.   X-ray diffraction of the selected 2nd solid phases; G-gypsum, B-basanite.

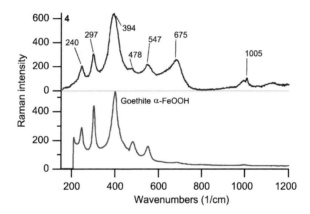

Figure 59.8   Raman spectra of the 2nd step solid phase.

standard spectrum in Fig. 59.8). The strong band at 675/cm can be assigned to ferrihydrite while the others interfered with goethite bands. The S-O bonds at 1000/cm belonged to sulfates.

59.4   CONCLUSIONS

The two-step treatment process allowed *in situ* removal of As from strongly contaminated mining water. In the first step the raw water was treated with a partial precipitation with a defined amount of alkaline agent (NaOH and/or Na_2CO_3) to pH $\sim$5.0. During the first precipitation more than 90% of As originally present in the water is immediately adsorbed as As(V) species onto iron oxyhydroxide surfaces, forming the inner-sphere complexes. About 30–40% of precipitated iron enables the quantitative removal of As from mining water. The As-bearing precipitate from the first step is separated by decantation and/or filtration.

The final treatment of mining water developed in the second step. $Ca(OH)_2$ was added to the liquid phase from the first step to precipitate further dissolved species. While As was substantially removed during the first precipitation, the other components including residual iron, manganese, zinc and sulfates were quantitatively precipitated during the second step. The mass of the total precipitate depended strongly on the amount of alkaline agent used in the whole process, which also affected the residual pH value. Above the pH range of 8.5–9.0 the total quantity of solid mass remained almost constant.

The study of As(V)-Fe-SO$_4^{2-}$ changes as a function of pH allows estimation of the optimum conditions of the process, i.e., to produce minimum amounts of As-bearing toxic precipitate, while the ecological limits of treated water had been maintained. The above described method promises a significant ecological and economical improvement due to a decrease of the quantity of toxic waste.

ACKNOWLEDGMENTS

This work was part of research program MSM 6046137302 (CR).

REFERENCES

Cullen, W.R. and Reimer, K.J.: Arsenic speciation in the environment. *Chem. Rev.* 89 (1989), pp. 713–764.

Doušová, B., Koloušek, D. and Kovanda, F.: Two-step decontamination of mining water with a high content of arsenic and iron ions. *CZ Patent* No. 295 150, Prague, Czech Republic, 2005a.

Doušová, B., Koloušek, D., Kovanda, F., Machovič, V. and Novotná, M.: Removal of As(V) species from extremely contaminated mining water. *Appl. Clay Sci.* 28:1–4 (2005b), pp. 31–43.

Furukawa, Y., Kim, J.W., Watkins, J. *et al.*: Formation of ferrihydrite and associated iron corrosion product in permeable reactive barriers of zero-valent iron. *Environ. Sci. Technol.* 36:24 (2002), pp. 5469–5475.

Hansen, H. and Ch. B.: Evironmental chemistry of iron(II)-iron(III) LDHs. In: V. Rives (ed): *Layered double hydroxides: present and future.* Nova Sci. Publisher, Inc., New York, 2001, pp. 413–434.

Keon, N.E., Swartz, C.H., Brabander, D.J., Harvey, C. and Hemond, H.F.: Validation of an arsenic sequential extraction method for evaluating mobility in sediments. *Environ. Sci. Technol.* 35 (2001), pp. 2778–2784.

Kopřiva, A., Zeman, J. and Šráček, O.: High arsenic concentrations in mining waters at Kaňk, Czech Republic. In: J. Bundschuh, P. Bhattacharya and D. Chandrasekharam (eds): *Natural arsenic in groundwater: occurence, remediation and management.* Balkema, Leiden, The Netherlands, 2005, pp. 49–55.

Lin, Z. and Puls, R.W.: Adsorption, desorption and oxidation of arsenic affected by clay minerals and aging process. *Environ. Geol.* 39:7 (2000), pp. 753–759.

Lin, Z. and Puls, R.W.: Potential indicators for the assessment of arsenic natural attenuation in the subsurface. *Adv. Environ. Res.* 7:4 (2003), pp. 825–834.

Randall, S.R., Sherman, D.M. and Ragnarsdottir, K.V.: Sorption of As(V) on green rust (Fe$_4$(II)Fe$_2$(III)(OH)$_{12}$SO$_4$ · 3H$_2$O) and lepidocrocite (γ-FeOOH): Surface complexes from EXAFS spectroscopy. *Geochim. Cosmochim. Acta* 65:7 (2001), pp. 1015–1023.

Refait, P., Abdelmoula, M., Trolard, F., Génin, J.-M.R., Ehrhardt, J.J. and Bourrié, G.: Mössbauer and XAS study of a green rust mineral; the partial substitution of Fe^{2+} by Mg^{2+}. *Am. Mineral.* 86 (2001), pp. 731–739.

Sherman, D.M. and Randall, S.R.: Surface complexation of arsenic(V) to iron(III) (hydr)oxides: Structural mechanism from ab initio molecular geometries and EXAFS spectroscopy. *Geochim. Cosmochim. Acta* 67 (2003) pp. 4223–4230.

Smedley, P.L. and Kinniburgh, D.G.: A review of the source, behaviour and distribution of arsenic in natural waters. *Appl. Geochem.* 17 (2002), pp. 517–568.

CHAPTER 60

Arsenic removal from groundwater using ferric chloride and direct filtration

R.G. Fernández
Centro de Ingeniería Sanitaria (CIS), Facultad de Ciencias Exactas, Ingeniería y Agrimensura, Universidad Nacional de Rosario, Rosario, Prov. de Santa Fe, Argentina

B. Petrusevski, J. Schippers & S. Sharma
UNESCO-IHE, International Institute for Infrastructural, Hydraulic and Environmental Engineering, Delft, The Netherlands

ABSTRACT: Arsenic (As) is a carcinogenic metalloid that is currently regulated in drinking water. One of the available technologies for As removal from groundwater is adsorption onto coagulated flocs and in this field, ferric chloride is the most commonly used coagulant. This research was conducted to explore a suitable conventional treatment technology in order to reduce the filtrate As concentration to less than 10 μg/l. Bench scale jar test experiments and pilot-scale investigations were carried out. Direct filtration with iron doses of 2 mg/l at pH value about 7.0, could reduce As(V) levels from 50 μg/l to 4 μg/l or less without any risk of iron or turbidity increasing in the filtered water. Direct filtration using ferric chloride as coagulant, could be an appropriate technology to reduce As levels below 10 μg/l for the given groundwater as is recommended by the World Health Organization (WHO).

60.1 INTRODUCTION

Arsenic (As) is the 20th most abundant element present in the earth's crust. It is a common pollutant in groundwater and industrial wastes. It occurs in mineral form and mainly as impurities in other materials. Moreover, it may be present in water, air and biota.

Due to its oxidation state As(III) or As(V), it will represent more or less risk for human health. Most of the As naturally occurs in the As(V) form, which has relatively low toxicity. The health effects have a large diversity, ranging from skin disorders such as skin pigmentation, skin lesions and keratosis to different forms of cancer such as respiratory and bladder cancers. Arsenic may be found in water originating from As-rich rocks. Severe health effects have been observed throughout the globe in populations drinking As-rich water over long periods (WHO 2001).

One of the available technologies for As removal from groundwater is adsorption onto coagulated flocs (Edwards 1994, Hering 1996, 1997, Chen *et al.* 1999). Alum and ferric salts are examples of coagulants in use. However, there are some uncertainties in terms of the best coagulation process to be used when As is present. Research work on available technologies for As removal reported that ferric chloride seems to be the most appropriate coagulant for As removal (Gupta and Cheng 1978, Cheng *et al.* 1994, Hering *et al.* 1997). The selection of a technology for As removal must consider some limiting factors. In the present research ferric chloride was used as coagulant.

The goal of this research is to propose a suitable conventional treatment technology for As removal from given groundwater in order to reduce the filtrate As concentration to less than 10 μg/l. This study, based on bench scale and pilot scale experiments, outlines some preliminary design parameters for full-scale treatment plant. The proposed technology should be able to achieve a removal efficiency that can fulfill the current regulations, and anticipate the new restriction planned for the Maximum Contamination Level (MCL) of As.

The specific objectives of this research are:

- To optimize conditions for coagulation-flocculation process using ferric chloride for the given groundwater.
- To examine, at pilot scale, the performance of direct filtration to minimize residual As concentration.

60.2   LITERATURE REVIEW

60.2.1   *Arsenic removal technologies*

Various treatment methods have been adopted to remove As from drinking water. These methods include co-precipitation, sorption techniques and membrane filtration techniques. Most of them can be implemented at domestic, small, medium and large sized facilities. The technologies under review perform most effectively when treating As in the form of As(V). Moreover, As(III) must be converted through pre-oxidation to As(V) before these methods can be effectively used. The Table 60.1 shows the efficiency of different methods. It is clearly appreciated that the pentavalent form of As is easier to remove than the trivalent one.

60.2.2   *Arsenic removal by coagulation*

60.2.2.1   *Speciation/oxidation*

Inorganic As speciation is an important factor in the efficiency of the process for removal of As from water. As(V) is much more effectively removed than As(III) because the former exists in natural waters as mono or bivalent anion and the latter exists, predominantly, in a nonionic form. For this reason, previous oxidation of As(III) to As(V) may be an important stage of the treatment process in order to achieve more efficient removal (Sancha 2000).

The coagulation technology is based on the principle that As(V) absorbs onto coagulated flocs which are then removed from solution by filtration. Coagulation is an effective treatment process for removal of As(V). The type of coagulant and dosage used affects the efficiency of the process. Aluminum and iron salts can be used as coagulants to achieve efficient As(V) removal (Hering *et al.* 1997).

Although the potential to obtain high efficiency As removal by coagulation is unquestioned, an understanding of the mechanisms and optimization strategies is currently lacking. Addition of iron or aluminum coagulants to water can facilitate the conversion of soluble As(V) and As(III) into insoluble reaction products. These products might form through precipitation, co-precipitation and adsorption mechanisms.

Regarding the coagulant dosages it is illustrative to present the dosages required to obtain a given effluent As concentration. Edwards (1994) reported that for an initial As(V) concentration of 20 μg/l, a coagulant dose of 7 mg/l as $FeCl_3$ (14 mg/l as alum) would meet a 10 μg/l standard for most samples. However, a wide range of dosages-removal values could be referred since several

Table 60.1.   Efficiency of arsenic removal using different technologies.

As form	Coagulation		Ion exchange	Adsorption (alumina, carbon)	Osmosis reverse	Electrodialysis
	$Al_2(SO_4)_3$	$FeCl_3$				
As(III)	0–20%	40–70%	20–40%	40–70%	70–80%	70–80%
As(V)	70–80%	80–100%	80–100%	80–100%	80–100%	80–100%

Source: Chwirka *et al.* 1999.

factors such as turbidity, pH, initial concentrations and another ionic compounds present in the water to be treated affect the As removal efficiency.

60.2.2.2   *Flocs separation systems*
Coagulation-flocculation can only be used successfully when the flocs which are formed can be separated out reliably and when most of the coagulant is also removed. In general, coagulation-filtration method of As removal could mainly be of the following three types (Fields *et al.* 2000, Chwirka *et al.* 2000, Madiec *et al.* 2000, Jekel and Seith 2000, Sancha 2000): (1) coagulation, sedimentation, rapid sand filtration; (2) direct filtration (coagulation-rapid sand filtration); (3) coagulation followed by microfiltration.

60.2.2.3   *Factors affecting arsenic removal by coagulation*
The efficiency of As removal by coagulation filtration processes depends on water quality and process conditions applied. The water quality includes pH, temperature, initial As concentration and speciation and the presence of other competing ions. The process conditions include type and dose of coagulant, flocculation conditions and method of floc separation applied.

60.2.2.4   *Other ions*
Adsorption of As(V) and As(III) oxyanions by ferric chloride may be adversely affected by anions such as phosphate, silicate, sulfate, carbonate and natural organic matter. The effects of competing ions in As removal can be summarized as follows (WHO 2001):

- Effects of co-occurring solutes will be more pronounced when adsorption density is approaching saturation.
- As(III) forms weaker bonds than As(V) with metal oxides, and is thus more likely to be displaced by competing anions.
- Surface complexation chemistry is complex, particularly when multiple anions are present.

60.3   MATERIALS AND METHODS

The research was carried out in two phases. Firstly bench scale tests were performed to represent different coagulation conditions by means of a jar-test. Subsequently based on the results obtained from the jar-test experiments, a pilot plant representing a direct filtration process was also evaluated.

60.3.1   *Bench studies*

60.3.1.1   *Experimental set up*
A series of 20 jar tests was performed for As removal evaluation from November 2001 to February 2002. The main objective of the series was to optimize conditions for maximum As removal efficiency. The experiments represented the present conditions at second filtration step in the

Table 60.2.   Adjusted parameters to represent case study water.

Parameter	Unit	RSF (1 + 2)	Model water	By adding
Temperature	°C	27	20/30	Heater
pH	mg/l	6.7–7.1	7.50	HCl
Calcium	mg/l	40	40	$CaCl_2 \times 2H_2O$
Magnesium	mg/l	13	13	$MgCl_2 \times 6H_2O$
Bicarbonate	mg/l	225	225	$NaHCO_3$
Arsenic	µg/l	50	50	As(V)

treatment line, so, the model water to be analyzed should have the same characteristics as the samples obtained at the outlet of filters 1 and 2 (RSF 1 + 2). To fulfill this aim model water was prepared by adjusting some parameters as shown in the Table 60.2.

60.3.1.2   *Jar test apparatus*
The jar-test experiments were conducted using a modified jar test apparatus containing six jars of 1.8 liters. It consists of six paddle stirrers that keep a constant uniform agitation in all beakers. The device is provided with a double wall to let warm water pass through to allow experiments at different temperatures. The glass jars contain baffles to avoid vortices and a vacuum system that allows samples to be taken of the supernatant of each beaker at the same time. The different dosages of coagulant were added to all jar tests in parallel.

60.3.1.3   *Standards and reagents*
All chemicals were reagent-grade and were used without purification. All solutions were prepared using de-mineralized water. All glass wares were acid-washed. As(V) standard was prepared from As_2O_5 and As(III) standard was prepared from As_2O_3, both dried at 105°C and kept in dessicator. Primary stock solutions for As(V) and As(III) (1000 mg/l) were prepared, from which a secondary stock (10 mg/l) was prepared to make the model water. For coagulation experiments, ferric chloride stock solution (0.1 M) was prepared from $FeCl_3 \times 6H_2O$ and stored in a dark bottle, the pH of this solution was maintained below 1.70. Background electrolyte solutions were prepared from the salts $CaCl_2 \times 2H_2O$, $MgCl_2 \times 6 H_2O$, $NaHCO_3$ and stock solutions (0.1 M) were used for addition to de-mineralized water to make model water. HCl was used to adjust the pH of the model water. A vacuum device with 0.45 μm pore size membrane filter was used to obtain filtered samples from the jar test experiments.

60.3.1.4   *Experimental equipment*
The experimental equipment comprised jar test set up, an electronic balance with a sensitivity of 0.00001 g, a vacuum flask assembly, a pH meter (WTW pH 340), and a turbidimeter (Lange LTP4).

60.3.1.5   *Process conditions*
Mixing conditions: For all jar tests performed in this research rapid mixing conditions (300 rpm, $G = 750$ l/s, t = 1 min) were applied to ensure the uniform mixing of the coagulant into the beakers.
   Flocculation conditions: Three flocculation conditions were used in this research work. For jar tests a G value lower than 10 l/sec was applied for those jar tests where effects of different equilibrium pH was evaluated. Another G value (50 l/s) was used for evaluation of effects of As speciation on removal efficiency. And last, a tapered flocculation conditions was applied as follows:

- $G = 70$ l/s, t = 5 min + $G = 50$ l/s, t = 5 min + $G = 20$ l/s, t = 10 min.
- *pH conditions*: A total of ten pH conditions were tested in this study. The equilibrium pH value (pH measured after settling) ranged between 6 and 8.
- *Arsenic speciation*: Different jar tests were performed using As(V) spiked model water and As(III) spiked model water. The pH for these jar tests was adjusted to 7.0 and the initial concentration for both As species was around 50 μg/l.
- *Temperature conditions*: All jar test were performed at 20°C, except one of them performed at 32°C in order to compare As(V) removal efficiency at different temperature conditions.
- *Initial As(V) concentration conditions*: An experiment to observe the influence of initial As concentration on As removal efficiency was carried out. The coagulant dose used in that case was 6.9 mg/l Fe^{3+} (20 mg/l $FeCl_3$) and pH was maintained at 7.0.
- *Settling conditions for As(III)*: In order to observe an improvement in the As(III) removal efficiency an experiment with a longer settling time was performed. The settling time was increased from 30 min to 2 hours.

- *Fe^{2+} present in model water*: A preliminary experiment was carried out in order to observe As removal efficiency in waters containing ferrous iron at high concentration. The reagent used to prepare this model water was ferrous sulfate ($FeSO_4 \times 7H_2O$) and the initial concentrations were fixed at 2, 5, 10 and 20 mg/l Fe^{2+}.

60.3.1.6 *Experimental procedure*

The general experimental procedure involved adding ferric chloride coagulant to As-containing model water, applying a certain G value by means of stirring, settling and taking samples (filtered and unfiltered ones). The experiments differed in the G value applied, the initial and final pH, As oxidation state, the initial As concentration and the temperature conditions. The doses of ferric chloride were in the range from 2 mg/l to 40 mg/l (0.7 mg/l to 13.8 mg/l Fe^{3+}) for a given jar test condition.

The coagulants were added to each jar, mixed at 300 rpm (G = 750 l/s) for 1 min, flocculated (for standard conditions) at 50 rpm (G = 50 l/s) for 20 min, and 30 min quiescent settling. After the settling period, approximately 300 ml of solution were collected from 3 cm below the liquid surface; 100 ml of this solution were then filtered through a 0.45 μm membrane filter using a vacuum filter flask assembly.

The pH was adjusted in the model water prepared specifically for each jar test. It was measured in the container of model water, after addition to the beakers (initial pH) and after settling (final or equilibrium pH). The different jar tests were characterized and named by the equilibrium pH.

60.3.1.7 *Parameters to be analyzed*

For both filtered and unfiltered samples from each beaker and model water, total iron, total arsenic and turbidity were determined.

60.3.2 *Filtration column studies*

60.3.2.1 *Experimental set up*

From the jar tests experiments it was clear that high As removal efficiencies can be achieved using ferric chloride as coagulant for the given groundwater under lab conditions, when the flocs are separated by means of filtration through 0.45 μm pore size membrane. Thus, a direct filtration process to achieve the flocs separation was carried out to collect operational data on this filtration process. The main goal of the filtration column studies was to analyze the feasibility of applying coagulation-direct filtration process under similar and extreme conditions relevant for the case study treatment plant.

The model water used for filtration experiments was tap water spiked with different chemicals to represent filtrate characteristics of RSF (1 + 2) in the existing GWTP. The coagulant was dosed in line to the stream before it passed to the filter (Fig. 60.1).

60.3.2.2 *Column set up*

The pilot plant consisted of one filter column, 100 mm in diameter, with associated flow measuring and regulating devices and chemical dosing and mixing equipment (in-line static mixer). The filter column was fitted with sampling points and manometer points along its length for measurement. The filter was operated with a constant supernatant depth of 0.80 m. A constant filtration rate (6 m/h) was maintained with help of a flow regulating hand-operated needle valve and a rotameter installed on the filtrate line. The filter was operated for a run length of approximately 8 hours before backwashing.

The model raw water was prepared in a 100 l plastic tank with continuous mixing at 200 rpm. The addition of HCl, $NaHCO_3$ and As were performed by means of peristaltic pump WT 503S and for the addition of ferric chloride a peristaltic pump WT 101 U/R was used. Fresh stock dilutions of 4 l of chemicals and fresh stock dilution of 1l of ferric chloride were prepared for each run.

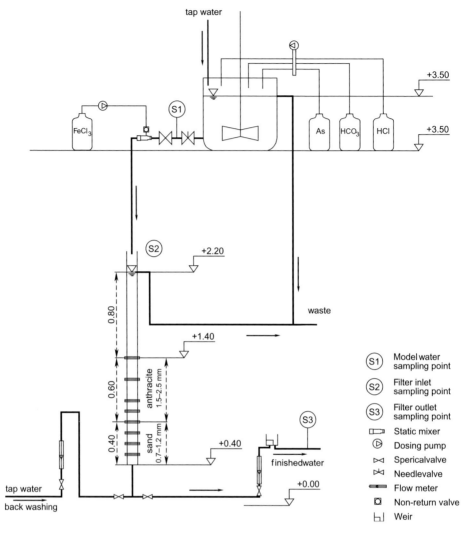

Figure 60.1. Scheme of pilot plant used for filter column experiments.

60.3.2.2.1 Process conditions
The pilot plant was operated under the following conditions:

pH inlet:	6.8 to 7.1
Water temperature:	20°C
Backwash frequency:	8–10 hours
Break-through by:	head losses
Filter bed depth:	1 m (0.6 m anthracite + 0.4 m sand)
Coagulant dose:	2 mg/l Fe^{3+}
Filter control:	constant raw water level, constant filtration rate
Backwash:	Media expansion: 15%, VB: approx. 60 m/h, t: 6 min.

60.3.2.2.2 Experimental procedure
The general experimental procedure consisted of running the filter with certain conditions of pH and coagulant dose, taking samples from the sampling points (see scheme pilot plant) every hour,

analyzing turbidity, arsenic and iron concentration. At the beginning of the run (first hour, ripening period) the samples were taken every 20 min. After the daily run was finished, a backwashing procedure with tap water was applied. It consisted of 6 minutes of stabilized velocity of approximately 60 m/h which guaranteed an expansion of the filter from 15 to 20%.

60.3.2.2.3 Analytical methods
Arsenic samples were analyzed by means of an atomic absorption spectrometer graphite furnace (AAS-GF), Perkin-Elmer 1100 B: range of measurement: 0 to 60 μg/l (of total As), that is As(V) + As(III), detection limit (for total As): 4 μg/l, sensitivity: ±3 μg/l. Iron, calcium and manganese were measured using an AAS-Flame Perkin Elmer 3110, with detection limit 0.03 mg/l for Fe (Standards Methods 1995).

60.4 RESULTS AND DISCUSSION

60.4.1 *Bench studies*

60.4.1.1 *Effect of pH*
Over the initial As concentration range examined, removal efficiencies were independent of initial pH (Fig. 60.2).

60.4.1.2 *Effect of initial arsenic concentration*
Over the initial As concentration range examined, removal efficiencies were independent of initial As(V) concentration (Fig. 60.3). This result is consistent with previous works (Cheng *et al.* 1994,

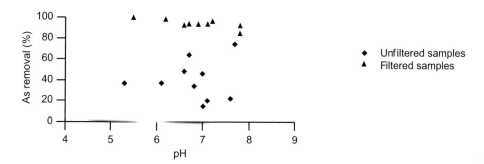

Figure 60.2. As(V) removal efficiency; initial As(V) concentration: 40–50 μg/l; pH equilibrium: 5.5/7.8; temperature: 20°C; dose of Fe^{3+}: 3.4 mg/l (10 mg/l $FeCl_3$); G < 10 1/s.

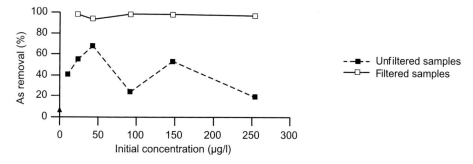

Figure 60.3. As(V) removal efficiency for different initial concentrations; initial As(V) concentration: 10–250 μg/l; pH equilibrium: 7.0; temperature: 20°C; dose of Fe^{3+}: 6.9 mg/l (20 mg/l $FeCl_3$); G < 10 1/s.

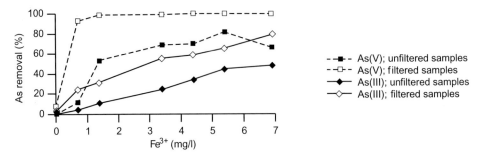

Figure 60.4.   As(V/III) removal efficiency for different dose of Fe^{3+}; initial As(V/III) concentration: 50 µg/l; pH equilibrium: 7.0; temperature: 20°C; G = 50 1/s.

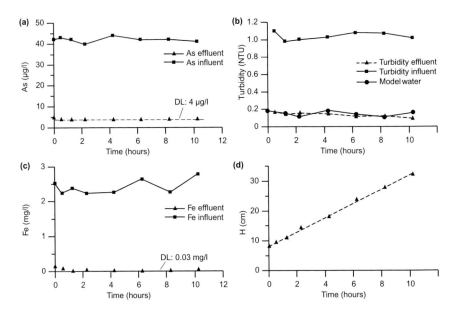

Figure 60.5.   Changes of principal parameters during filtration experiments: (a) As(V) concentration corresponding to As moval efficiency; (b) Turbidity; (c) Iron concentration; (d) Head loss development. Inlet As(V) concentration: 42 µg/l; pH model water: 6.6/7.1; temperature: 20°C; dose of Fe^{3+}: 2.4 mg/l; filtration rate: 6 m/h).

Hering et al. 1997). This lack of dependence is consistent with the expected adsorption behavior of As when the surface of freshly precipitated amorphous metal hydroxide formed upon coagulant addition is not saturated.

60.4.1.3   Effect of coagulant dose and arsenic speciation
Figure 60.4 shows that As(V) removal by ferric chloride was improved by increasing coagulant dose. At a coagulant dose of 0.7 mg/l Fe^{3+}, As(V) removal was 93% for filtered samples. Complete removal of As(V) was achieved for doses above 4.0 mg/l Fe^{3+}, which is consistent with previous research work (Hering et al. 1996). A removal better than 97% could be reached for filtered samples with a coagulant dose >2.0 mg/l Fe^{3+}. This value is relevant for the present study since a removal efficiency of 90% is required. For lower values of coagulant doses (<1.0 mg/l Fe^{3+}), the removal efficiency increased rapidly (from 0 to 80%). With respect to unfiltered samples, it is clear that an improvement in removal efficiency can be achieved by increasing coagulant dose. A removal

efficiency higher than 60% could be achieved for As(V) with a settling time of 30 minutes and for coagulant doses higher than 2.0 mg/l Fe^{3+}.

60.4.1.4   *Filtration column studies*

A total of 6 runs were performed in the filtration column studies. From those runs, two of the most representative ones were analyzed. Average values of the changes of principal parameters during the experiments are presented in Figures 60.5a–d. The results reveal efficient removal of As from the model water and Fe concentrations in the column effluent below detection limit.

60.5   CONCLUSIONS AND RECOMMENDATIONS

An increase in pH from 5.5 to 8.0 had no significant effect on As(V) removal efficiency. However, optimal pH was close to 7.0. During coagulation-flocculation experiments, As(V) removal was independent of initial As(V) concentration. This is helpful when the WTP deals with wells with different raw water quality. Under similar conditions of pH, temperature, coagulant dose and flocculation, As(V) is easier to be removed than As(III) for the given model water. At the lowest coagulant dose applied for filtered samples (1.4 mg/l Fe^{3+}), As(V) removal efficiency was higher than 95% at pH value 7.0. At the highest coagulant dose applied for unfiltered samples (13.8 mg/l Fe^{3+}), As(V) removal efficiency was higher than 90% at pH value 6.8, with a settling time of 30 minutes. However iron concentration in supernatant was above 1 mg/l. Marginal improvement was observed in As(V) removal when water temperature was increased from 20°C to 32°C, for unfiltered samples.

There was no detectable improvement in As(V) removal efficiencies for different flocculation conditions applied (tapered flocculation and constant G value). It was not possible to achieve complete As(V) removal by coagulation-sedimentation process without exceeding iron concentration of 0.1 mg/l in supernatant. Complete removal of As(V) was achieved for iron doses higher than 2 mg/l in samples filtered through 0.45/μm membrane. That suggests direct filtration or microfiltration process as a feasible technique to achieve residual As(V) concentrations <5 μg/l. Pilot tests indicate that under these conditions (pH inflow 7.0, filtration rate 6 m/h, initial As(V) 50 μg/l, iron doses 2 mg/l, length of filter run 10 hours) As removal efficiency was higher than 90%. Iron concentration in the filtrate was consistently <0.10 mg/l. Direct filtration could be an appropriate technique to achieve efficient As removal, specifically when the formed flocs are difficult to be removed by settling.

Filter bed design (depth, composition), filter rate and backwashing conditions require further optimization. Direct filtration experiments should be conducted at higher initial As (for example 200 μg/l) in order to asses the capability limit of the direct filtration process for As removal.

REFERENCES

Chen, H., Frey, M., Clifford, D., Mc Neil, L. and Edwards, M.: Arsenic treatment considerations. *J. AWWA* 91:3 (1999), pp. 74–85.
Cheng, R., Liang, S., Wang, H. and Beuhler, M.: Enhanced coagulation for arsenic rem. *J. AWWA* 86:9 (1994), pp. 79–90.
Chwirka, J., Thomsom, B. and Stomp, J.: Removing arsenic from groundwater. *J. AWWA* 92:3 (2000), pp. 79–88.
Edwards, M.: Chemistry of arsenic removal during coagulation and Fe-Mn oxidation. *J. AWWA* 86:9 (1994), pp. 64–78.
Fields, K., Chen, A. and Wang, L.: Arsenic removal from drinking water by coagulation-filtration and lime softening plants. EPA/600/R-00/063, 2000.
Gupta, S. and Chen, K.: Arsenic removal by adsorption. *J. Water Pollut. Control Federation* 50:3 (1978), pp. 493–506.

Hering, J., Chen, P., Wilkie, J., Elimelech, M. and Liang, S.: Arsenic removal by ferric chloride. *J. AWWA* 88:4 (1996), pp. 155–167.

Hering, J., Chen, P., Wilkie, J. and Elimelech, M.: Arsenic removal from drinking water during coagulation. *J. Environ. Engin.* 123:8 (1997), pp. 800–807.

Jekel, M. and Seith, R.: Comparison of conventional and new techniques for the removal of arsenic in a full scale water treatment plant. *J. Water Supply* 18:1 (2000), pp. 628–631.

Madiec, H., Cepero, E. and Mozziconacci, D.: Treatment of arsenic by filter coagulation: a South American advanced technology. *J. Water Supply* 18:1 (2000), pp. 613–618.

Sancha, A.M.: Removal of arsenic from drinking water supplies: Chile experience. *J. Water Supply* 18:1 (2000), pp. 621–625.

Standard Methods for Examination of Water and Wastewater: 19th Edition, American Public Health Association/American Water Works Association/Water Environment Federation, Washington, DC, 1995.

WHO: Arsenic in drinking water. World Health Organisation, Geneve, Switzerland, http://www.who.int/waters-anitationhealth/Arsenic/arsenic.html, 2001.

CHAPTER 61

The use of iron-coated LECA for arsenic removal from aqueous solutions under batch and flow conditions

I. Cano-Aguilera, A.F. Aguilera-Alvarado, G. de la Rosa, R. Fuentes-Ramírez,
G. Cruz-Jiménez, M. Gutiérrez-Valtierra & M.L. Ramírez-Ramírez
Facultad de Química, Universidad de Guanajuato, Guanajuato, Gto., Mexico

N. Haque
*Environmental Science and Engineering PhD Program, The University of Texas at El Paso,
El Paso, TX, USA*

ABSTRACT: This study investigated the removal of arsenic (As) from aqueous solutions by iron-coated-light-expanded-clay-aggregates (Fe-LECA) under batch and flow conditions. Batch experiments were performed to investigate the development of As sorption onto Fe-LECA. More than 80% of As was sorbed to Fe-LECA within 1 hour of contact. Column experiments demonstrated that a suitable bed depth/flow rate ratio design as well as a highly hydraulic retention time greatly influenced the column efficiency for As removal in flow conditions. The maximum As sorption capacity was observed to be about 3.3 mg/g which was reached when the iron content on LECA was 0.93 mg/g, and the column was operated at a flow rate of 10 ml/min.

61.1 INTRODUCTION

Drinking water contamination with arsenic (As) is a worldwide health concern. Arsenic poisoning with contaminated water has been reported in different parts of the world. It has been shown that, for example, in Bangladesh (Dhar *et al.* 1997, Bhattacharya and Mukherjee 2002), West Bengal (Mandal *et al.* 1996), Argentina (Borgono *et al.* 1977), Taiwan (Chen *et al.* 1994), and Mexico (Del Razo *et al.* 1990) millions of people are at potential risk of developing various As-related diseases due to chronic As poisoning. This element is released into the groundwater by both, natural and anthropogenic processes, including weathering of As-containing minerals, mining, and the application of organo-arsenic-bearing pesticides.

Due to its high toxicity and carcinogenic effects, in 1993 the World Health Organization (WHO) recommended a maximum contaminant level (MCL) for As in drinking water of 10 µg/l. The European Commission and the United States Environmental Protection Agency (USEPA) adopted this value as their MCL in 1998 and 2001, respectively (Drinking Water Directive 1998, USEPA 2001). Furthermore, by January 2006, for all the water treatment systems in USA it was mandatory to follow this new standard. Therefore, continuous research directed to either improve the established methods or to develop low-cost techniques and materials to remove As from industrial effluents or drinking water is a very important issue.

Although many different methods such as precipitation, co-precipitation, ion-exchange, ultrafiltration, and reverse osmosis have been used for As removal, the sorption from solution has received more attention due to its high efficiency. Many types of sorbents have been used including activated carbon (Budinova *et al.* 2006) and activated alumina (Vagliasindi *et al.* 1996), gibbsite (Manning *et al.* 1996), aluminum-loaded materials (Xu *et al.* 1998), and natural solids (Elizalde-González 2001), among others. However, sorption or co-precipitation with iron compounds and iron coated materials (Jain *et al.* 1999, Manning *et al.* 1998, Grossl *et al.* 1997, Johnston and Heijnen 2001) are more efficient and cheaper than other methods.

In this study, batch experiments were conducted to determine the development for As sorption onto iron-coated light expanded clay aggregates (Fe-LECA), and column experiments were conducted to determine conditions for maximum As sorption to Fe-LECA in continuos system as well.

61.2   MATERIAL AND METHODS

61.2.1   *Reagents and apparatus*

The chemicals used for analysis and preparation of standards and samples were obtained from Merck and Sigma. Glass materials and plastic bottles were washed, exposed overnight to 5% HNO_3 and rinsed with enough distilled water. Standards for calibration as well as As solutions for batch and column experiments were prepared from the standard solution (1000 mg/l As from H_3AsO_4 in 0.5 M of HNO_3). Iron solutions for coating were prepared from $FeCl_3 \cdot 6H_2O$ dissolved in deionized water. Deionized water was obtained from a Millipore Purification System.

61.2.2   *Sorbent preparation*

Light expanded clay aggregates (LECA) are gardening material produced by Hassel Forsgarden, Sweden. The chemical properties of LECA are listed in Table 61.1.

The LECA were ground to particles with an average diameter of 0.50 mm. These particles (0.5 porosity) were washed with deionized water and air-dried at 40°C. In order to perform the coating with iron, 50 g of dry LECA were placed in a beaker and stirred for 24 h with 500 ml of 0.1 M $FeCl_3 \cdot 6H_2O$ (this step was repeated several times in order to get enough modified sorbent for all experiments). At this point, HCl 0.01 M was added to maintain the acidic conditions. Subsequently, iron-coated LECA (0.93 mg of Fe/g of LECA) was separated from the solution and washed with deionized water (until iron could no longer be detected), and dried overnight by a hot air flow oven (Heraeus UT 6060) at 60°C. This material was utilized for all batch and column experiments as iron-coated-light-expanded-clay aggregates (Fe-LECA) unless otherwise stated.

61.2.3   *Development for arsenic sorption to Fe-LECA in batch experiments*

A batch experiment was carried out to determine the development for As sorption onto Fe-LECA. All the batch experiments were carried out in a 800 ml Erlenmeyer flasks containing 500 ml of a 1 mg/l As(V) (0.01 mmol) and 10 mg of Fe-LECA. The flasks were stirred at 160 rpm for 24 h. The pH of the solution was adjusted to 6 by using HCl solution. Solutions without As and without Fe-LECA were also tested as controls. After the reaction, samples were collected at different time intervals (0, 0.5, 1, 2, 3, 4, 5, 6, 8, 12, 24 h), and centrifuged at 3000 rpm for 5 min. The final pH of the supernatants was recorded by a pH meter (Corning, Pinnacle 540) and As sorption onto Fe-LECA was estimated as a function of time.

61.2.4   *Arsenic sorption to Fe-LECA in column experiments*

To investigate the As sorption capacity of Fe-LECA under flow conditions, experiments were carried out in columns (50 cm height and 10 cm diameter) packed with Fe-LECA (300 g) and connected

Table 61.1.   Chemical composition of LECA.

Compound	(%)
SiO_2	70
Al_2O_3	20
CaO + MgO	1.3
FeO	8.7

to a peristaltic pump (Cole Palmer, 7553–70). The columns have withdrawal ports reaching the center of the bed to prevent sampling from areas close to the wall. The As-feed solution (1 mg/l) was pumped in upflow direction across the packed bed at different flow rates (10, 20, and 40 ml/min). At the beginning of experiments, samples were collected at different heights (15, 30, and 50 cm at these sampling ports) every half-hour, later on samples were taken with less frequency. The pore volume of the column was 500 ml and samples of same volume were collected at each port.

61.2.5 *Analytical methods for As and Fe determination*

The determination of total arsenic (t-As) in the collected samples was performed by hydride-generation atomic absorption spectrometry (HG-AAS, Perkin-Elmer MHS 15, Perkin-Elmer Analyst 100). The method is based on the conversion of soluble As to the respective volatile arsines, which are subsequently determined by atomic flame absorption. The reduction of As was achieved by KI addition (10% w/v) and 30 min storage in darkness. The pretreatment of the samples also involved the addition of 10 ml 1.5% HCl per ml sample. Subsequent inline hydride genera-tion with argon and sodium borohydride reductants (3% $NaBH_4$ in 1% NaOH solution) and AAS detection allowed determination of t-As species down to 10 $\mu g/l$ at a wavelength of 193.7 nm. Calibrations were performed for the removal range of analysis (0.1 to 5 mg/l). Liquid samples were also analyzed for the determination of iron content by means of a flame atomic absorption spectrometer (FAAS, Perkin-Elmer Analyst 100) at a wavelength of 248.3 nm (the solids samples were previously digested with concentrated HNO_3).

61.3   RESULTS AND DISCUSSION

61.3.1   *Development for arsenic sorption onto Fe-LECA in batch experiments*

The results of the development of As sorption onto Fe-LECA are shown in Figure 61.1. The exper-iment was conducted in order to determine the time required for the system to reach equilibrium. This figure shows that more than 80% of As is sorbed within 1 h. Vaishya and Prasad (1991) demonstrated that when metal sorption occurs within 5 min, only the surface of the adsorbent is highly active. Equilibrium is then slowly approximated within the next 5 h. No iron leaching was observed throughout the experiment and equilibrium may have been attained fast due to the occu-pation of a large proportion of sorption sites possibly due to the iron-arsenic interaction (Volesky and Holan 1995).

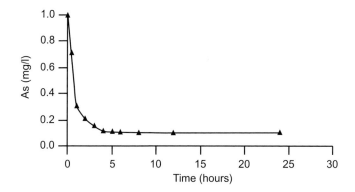

Figure 61.1.   Development for As(V) sorption onto Fe-LECA. A solution of arsenic 1 mg/l, pH 6 was in contact with 10 mg of Fe-LECA in continuous agitation. Samples were collected at different time intervals for the residual As content.

61.3.2   *As sorption onto Fe-LECA in column experiments*

The batch experiments showed that Fe-LECA has the ability to adsorb As species from aqueous solutions. However, a batch system would not be practical for removing As from contaminated waters. Therefore, column experiments were performed to investigate the conditions for maximum As sorption onto Fe-LECA in continuos system.

Figure 61.2 displays the breakthrough curves for As sorption onto Fe-LECA at different bed depths (ports at 15, 30 and 50 cm height) and 10 ml/min upflow rate of a 1 mg/l As-bearing solution at pH 6 in samples of 500 ml (column pore volume). As can be expected, the breakthrough curves are different for different bed depths, and the last breakthrough curve occurred after a throughput of 400 pore volumes in a bed depth of 50 cm.

The effect of flow rate on As sorption onto Fe-LECA was investigated by varying the flow rates from 10 to 40 ml/min in samples collected at a 50 cm bed depth (bottom), 1 mg/l initial As(V) concentration, and pH = 6. Figure 61.3 shows the breakthrough curves for As sorption to Fe-LECA at different flow rates.

Figure 61.3 shows a breakthrough curve occurring first (dots) when flow rate was high due possibly to the insufficiency of time for inter- and intra-particle diffusion of solute (Leupin and Hug 2005). The next two breakthrough curves (squares and triangles) support furthermore this concept because the lower feed flow rates allow more diffusion of solute into the Fe-LECA-packed

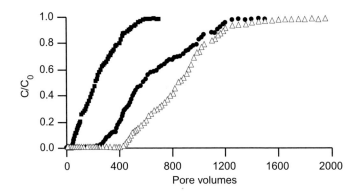

Figure 61.2.   Breakthrough curves from As(V) sorption onto Fe-LECA in packed column. Feed concentration: 1 mg/l As(V), pH = 6, flow velocity: 10 ml/min. Samples were collected at equal time intervals from different ports, (•) − 15, (■) − 30, and (△) − 50 cm.

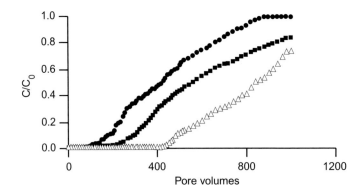

Figure 61.3.   Breakthrough curves from As(V) sorption onto Fe-LECA in packed column. Feed concentration: 1 mg/l As, pH = 6, flow rates: (△) − 10, (■) − 20, and (•) − 40 ml/min. Samples were collected at same time intervals from 50 cm bed depth.

column. At the ultimate flow condition, the maximum As sorption capacity was observed to be approximately 3.3 mg/g of Fe-LECA. These results denote that the Fe-LECA material has higher As sorption capacity than those previously reported (Kanel *et al.* 2006, Nguyen *et al.* 2006).

61.3.3 *Conclusions*

Fe-LECA proved to have a high capacity to adsorb As under batch and flow conditions. More than 80% of As was sorbed to Fe-LECA within 1 hour of contact. Column experiments demonstrated that a suitable bed depth/flow rate ratio design as well as a high hydraulic retention time greatly influenced the column efficiency for As removal in flow conditions. The maximum As sorption capacity was observed to be about 3.3 mg/g which was reached when the iron content on LECA was 0.93 mg/g, and the column was operated at a flow rate of 10 ml/min.

The use of Fe-LECA for the removal of As from contaminated groundwater for drinking purposes, represents advantages because the As content in such contaminated water is usually at low concentration ranges.

ACKNOWLEDGMENTS

The authors are gratefully acknowledge The STINT Foundation from Sweden, CONACYT (grant J110.427/2005), and The University of Guanajuato from Mexico, for their support.

REFERENCES

Bhattacharya, P. and Mukherjee, A.B.: Management of arsenic contaminated groundwater in the Bengal Delta Plain. In: M. Chatterji, S. Arlosoroff and G. Guha (eds): *Conflict management of water resources.* Ashgate Publishing, Hampshire, UK, 2002, pp. 308–348.

Borgono, J.M., Vincent, P., Venturino, H. and Infante, A.: Arsenic in the drinking water of the city of Antofagasta: epidemiological and clinical study before and after the installation of a treatment plant. *Environ. Health Perspect.* 19 (1977), pp. 103–105.

Budinova, T., Petrov, N., Razvigorova, M., Parra, J. and Galiatsatou, P.: Removal of arsenic(III) from aqueous solution by activated carbons prepared from solvent extracted olive pulp and olive stones. *Ind. Eng. Chem. Res.* 45:6 (2006), pp. 1896–1901.

Chen, S.L., Dzeng, S.R., Yang, M.H., Chiu, K.H., Shieh, G.M. and Wai, C.M.: Arsenic species in ground waters of the blackfoot disease area, Taiwan. *Environ. Science Technol.* 28:5 (1994), pp. 877–881.

Del Razo, L.M., Arellano, M.A. and Cebrian, M.E.: The oxidation states of arsenic in well water from a chronic arsenicism area of northern Mexico. *Environ. Pollut.* 64 (1990), pp. 143–153.

Dhar, R.K., Biswas, B.Kr., Samanta, G., Mandal, B.Kr., Chakraborti, D., Roy, S., Jafar, A., Islam, A., Ara, G., Kabir, S., Khan, A.W., Ahmed, S.A. and Hadi, S.A.: Groundwater arsenic calamity in Bangladesh. *Curr. Sci.* 73:1 (1997), pp. 48–58.

European Commission: Drinking Water Directive 98/83/EEC: EC Directive on drinking water quality intended for human consumption. European Commission, Brussels, Belgium, 1998.

Elizalde-González, M.P., Mattusch, J., Einicke, W.D. and Wennrich, R.: Sorption on natural solids for arsenic removal. *Chem. Eng. J.* 81:1 (2001), pp. 187–195.

Grossl, P.R., Eick, M., Sparks, D.L., Goldberg, S. and Ainsworth, C.C.: Arsenate and chromate retention mechanisms on goethite. 2. Kinetic evaluation using a pressure-jump relaxation technique. *Environ. Sci. Technol.* 31 (1997), pp. 321–326.

Jain, A., Raven, K.P. and Loeppert, R.H.: Arsenite and arsenate sorption on ferrihydrite: surface charge reduction and net OH-release stoichiometry. *Environ. Science Technol.* 33:8 (1999), pp. 1179–1184.

Johnston, R. and Heijnen, H.: Safe water technology for arsenic removal. In M.F. Ahmed, M.A. Ali and Z. Adeel (ed): *Technologies for arsenic removal from drinking water.* Bangladesh University of Engineering and Technology, Dhaka. Bangladesh, 2001.

Kanel, S.R., Heechul, Ch., Ju-Yong, K., Saravanamuthu, V. and Wang G.S.: Removal of arsenic(III) from groundwater using low-cost industrial by-products-blast furnace slag. *Water Qual. Res. J. Canada* 41:2 (2006), pp. 130–139.

Mandal, B.K., Chowdhury, T.R., Samanta, G., Basu, G.K., Chowdhury, P.P., Chanda, C.R., Lodh, D., Karan, N.K., Dhar, R.K., Tamili, D.K., Das, D., Saha, K.C. and Chakraborti, D.: Arsenic in groundwater in seven districts of West Bengal, India—the biggest arsenic calamity in the world. *Curr. Sci.* 70:11 (1996), pp. 976–986.

Manning, B.A., Fendorf, S.E. and Goldberg, S.: Surface structures and stability of arsenic(III) on goethite: spectroscopic evidence for inner-sphere complexes. *Environ. Science Technol.* 32:16 (1998), pp. 2383–2388.

Manning, B.A. and Goldberg, S.: Modeling competitive sorption of arsenate with phosphate and molybdate on oxide minerals. *Soil Sci. Soc. America J.* 60:1 (1996), pp. 121–131.

Nguyen, T.V., Saravanamuthu, V., Huu Hao, N., Damoda, P. and Viraraghavan, T.: Iron-coated sponge as effective media to remove arsenic from drinking water. *Water Qual. Res. J. Canada* 41:2 (2006), pp. 164–170.

USEP: National primary drinking water regulations—Arsenic and clarifications to compliance and new source contaminants monitoring. US Environmental Protection Agency, Federal Register 66:14, pp. 6976–7066.

Vagliasindi, G.A.F., Henley, M., Schultz, N. and Benjamin, M.M.: Sorption of arsenic by ion exchange resins, activated alumina and iron-oxide coated sands. *Proceedings of the Water Quality Technology Conference,* Denver, CO, 1996, pp. 1829–1853.

Vaishya, R.C. and Prasad, S.C.: Sorption of Cu(II) on sawdust. *Indian J. Environ. Prot.* 11:4 (1991), pp. 284–289.

Volesky, B. and Holan, Z.R.: Biosorption of heavy metals. *Biotechnol. Progr.* 11:3 (1995), pp. 235–250.

Xu, Y., Ohki, A. and Maeda, S.: Sorption of arsenic(V) by use of aluminium-loaded Shirasu-zeolites. *Chem. Lett.* 10 (1998), pp. 1015–1016.

CHAPTER 62

Polymer-supported Fe(III) oxide particles: An arsenic-selective sorbent

L.H. Cumbal
Centro de Investigación Científica, Escuela Politecnica del Ejercito (ESPE), Sangolqui, Ecuador

A.K. SenGupta
Department of Civil and Environmental Engineering at Lehigh University, Bethlehem, PA, USA

ABSTRACT: Nanoscale hydrated Fe(III) oxide (HFO) particles show excellent properties conducive to selective removal of target compounds from contaminated water bodies. However, these nanoparticles cannot be used in plug flow configurations due to excessive pressure drops and poor durability. Harnessing these HFOs within polymeric beads offers new opportunities that are amenable to rapid implementation in the area of environmental separation and control. In this investigation commercially available cation and anion exchangers were used as host materials for dispersing HFO particles within the polymer phase. The major finding of this study reveals that an anion exchanger as support for dispersed HFO particles offered considerably higher arsenate removal capacity compared to a cation exchanger. Hybrid anion exchanger-macroporous (HAIX-M) beads were amenable to efficient regeneration thus assuring their reuse for several cycles. In addition, rubbing tests demonstrated that HAIX-M particles did not lose mechanical resistance and there was no fines formation.

62.1 INTRODUCTION

It is well recognized that a number of nanoscale inorganic particles (NIPs) and their aggregates offer favorable properties in regard to selective separation and/or chemical transformation of target contaminants from contaminated water bodies. For example: (1) hydrated Fe(III) oxides or HFO particles can selectively sorb dissolved heavy metals such as zinc, copper or metalloids like arsenic oxyacids or oxyanions (Pierce and Moore 1982, Laxen 1983, Slavek and Pickering 1986, Music and Ristic 1992, Slavek and Pickering 1986, Tochiyama *et al.* 1995, Manning *et al.* 1998, Roberts *et al.* 2004); (2) Mn(IV) oxides are fairly strong solid phase oxidizing agents employed in water treatment for oxidation of As(III) to As(V) and for organic compound transformation (Scott and Morgan 1995, Lequart *et al.* 1998, Nesbitt *et al.* 1998, Tournassat *et al.* 2002, Zhang and Huang 2003); (3) magnetite (Fe_3O_4) crystals are capable of imparting magnetic activity on nonmagnetic materials like chitosan and polymeric resins, which in turn can be used as forensic monitors in rivers or lakes (Koseko 1994, Surinenaite *et al.* 1997, Liu *et al.* 2000, Leun and SenGupta 2000); (4) elemental Zn^0 or Fe^0 are excellent reducing agents for both organic and inorganic contaminants present in natural waters (Matheson and Tratnyek 1994, Wang and Zhang 1997, Cheng *et al.* 1997, Arnold *et al.* 1999, Kim and Carraway 2003). Figure 62.1 shows the properties of several NIPs. The methodology of preparation of these NIPs and their aggregates is environmentally safe, operationally simple, and inexpensive. Extremely high surface area to volume ratio of these tiny particles ($8.6 \times 10^7 - 1.0 \times 10^8$ m^2/m^3) offers favorable kinetics for selective sorption and oxidation-reduction reactions (Cumbal *et al.* 2003). However, these particles cannot be used in fixed-bed columns, in groundwater reactive barriers or in any plug-flow type configuration due to excessive pressure drops. Also, these NIPs are not durable and lack mechanical strength.

Harnessing these NIPs within polymeric beads offers new opportunities that are amenable to rapid implementation in the area environmental separation and control. While the NIPs retain their

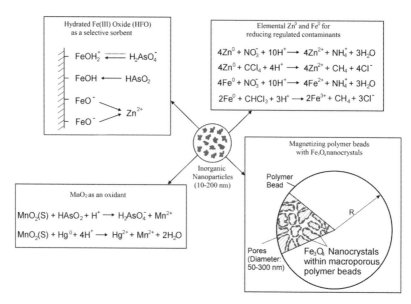

Figure 62.1.   Properties of inorganic nanoparticles.

intrinsic sorption/desorption, redox, acid-base or magnetic characteristics, the robust polymeric support offers excellent mechanical strength, durability, and favorable hydraulic properties (Cumbal *et al.* 2003). In this investigation, we have successfully dispersed nanoscale hydrated Fe(III) oxides (HFO) within different polymeric ion exchanger hosts.

The general objective of the study is to provide experimental evidence that the nature of functional groups in the polymer host material and resulting Donnan co-ion exclusion effects greatly influence the sorption behaviors of the hybrid material. By taking advantage of this property, we have for the first time synthesized a hybrid sorbent that can selectively remove As from contaminated waters.

62.2   EXPERIMENTAL TECHNIQUES: METHODS AND PROCEDURES

62.2.1   *Preparation of hybrid cation exchanger*

Table 62.1 includes the salient properties of the commercially available cation and anion exchangers used for the preparation of the hybrid sorbent materials. The preparation of hybrid cation exchangers or HCIX consisted of the following three steps (SenGupta *et al.* 2000, DeMarco *et al.* 2003, Greenleaf *et al.* 2003): first, loading of Fe(III) onto the sulfonic acid sites of the cation exchanger by passing 4% $FeCl_3$ solution at an approximate pH of 2.0 through a fixed-bed; second, desorption of Fe(III) and simultaneous precipitation of Fe(III) hydroxides within the gel and pore phase of the exchanger through passage of a solution containing both NaCl and NaOH, each at 5% (w/v) concentration; and third, rinsing and washing with 50/50 ethanol-water solution followed by a mild thermal treatment (50–60°C) for 12 hours.

Figure 62.2 illustrates the major steps of the process. Hybrid anion exchangers were subsequently made following a different procedure (SenGupta and Cumbal 2005) not included in this chapter.

62.2.2   *Rubbing test*

Mechanical strength and durability of sorbents were evaluated through rubbing experiments. For these tests, samples of hybrid anion exchanger-macroporous (HAIX-M) beads and granular ferric hydroxide (GFH) granules were exposed to a smooth rubbing by applying a force of approximately

Table 62.1.   Polymeric ion exchangers as host materials.

	Purolite A-400/A-500P	Purolite C-100/C-145
Structure (repeating unit)		
Resin type	Anionic	Cationic
Functional group	Quaternary ammonium	Sulfonic acid
Capacity (meq/g resin)*	1.3 (A-400)	1.5 (C-100)
	0.8 (A-500P)	2.0 (C-145)
Manufacturer	Purolite Inc.	Purolite Inc.
	Philadelphia	Philadelphia

*Provided by manufacturer.

Step 1. Loading with $FeCl_3$ solution at pH < 2

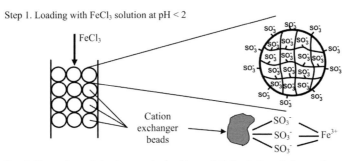

Step 2. Desorption and simultaneous hydroxide precipitation in the gel phase and pore surface

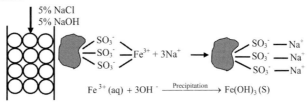

$$Fe^{3+} (aq) + 3OH^- \xrightarrow{\text{Precipitation}} Fe(OH)_3 (S)$$

Step 3. Alcohol wash and mild thermal treatment

$$Fe(OH)_3 (S) \xleftarrow[\text{12 hours}]{50\text{-}60^{\circ}C} FeOOH + \text{Amorphous HFO Particles}$$
crystalline

Figure 62.2.   Illustration of a three-step procedure to disperse amorphous HFO nanoparticles inside the spherical cation exchanger beads.

one pound with the hand fingers. Then samples were immersed in water for particle size distribution using a Coulter LS100Q particle counter with binocular optical system. The particle counter both counted and sized individual hybrid beads and GFH particles as they flew through the sensing zone. It operates by light scattering principles to statistically determine ranges of particle size distribution.

62.2.3 *Materials and methods*

Fixed-bed column runs were conducted using glass columns (11 mm in diameter and 250 mm length), constant-flow stainless steel pumps and Eldex fraction collectors to investigate the sorption behaviors of the hybrid sorbents. The ratio of column diameter to sorbent bead diameter was approximately 10:1; earlier work on chromate and phosphate removal with similar setup and under identical conditions showed no premature leakage due to wall effects (SenGupta and Lim 1988, Zhao and SenGupta 1998). The superficial liquid velocity (SLV) and the empty bed contact time (EBCT) were recorded for each column run. For the tests, fixed-bed columns were fed with aqueous solutions containing As as a trace species (50 and 100 µg/l) and competing anions of 120 mg/l SO_4^{2-}, 90 mg/l Cl^-, and 100 mg/l HCO_3^- at a pH around 7. Concentrations of anions and pH were chosen to similar values found in natural waters. For analysis, effluent samples of different volumes were collected at the column exit.

Analyses of samples for As(V) were conducted using a Perkin Elmer atomic absorption spectrometer (AAS) with graphite furnace accessories (Model SIMAA 6000), electrodeless discharge lamp (EDL). The stock solutions for water feed and calibration were prepared using analytical grade KH_2AsO_4, Na_2SO_4, NaCl, and $NaHCO_3$. Analysis was performed at a wavelength of 193.7 nm and each sample injection included palladium matrix modifier. Furnace conditions included a 110°C injection temperature with a 40 s hold time, a pretreatment temperature between 1200 to 1500°C with a 30 s hold time, and a 2300°C atomization temperature for 5 s. For quality control, As concentrations of stock solutions were intermittently checked against an As standard purchased from Fisher Scientific Inc. Sulfate and chloride were analyzed using the ion chromatograph, Dionex DX-120 IC, fitted with a conventional column for anions and a conductivity detector. Sulfate and chloride response peaks were printed using a Hewlett Packard Agilent 3395 integrator. Bicarbonate or inorganic carbon was determined using a Shimadzu carbon analyzer (Model 5050A).

For XRD tests, samples of freshly prepared and used hybrid anion exchangers-macroporous were carefully sliced and placed in aluminium frames with the cross-section facing up. Vacuum grease was employed to keep the sliced beads in a fixed position for the test. Samples were then placed inside the chamber of an X-ray Rigaku Rotaflex diffractometer and after being exposed to X-ray beams, the output data were recorded in a computer.

62.3 RESULTS AND DISCUSSION

62.3.1 *Column runs with hybrid cation exchangers*

Two separate column runs were conducted with similar influent solutions and under identical hydrodynamic conditions. In both cases, As(V) or arsenate was a trace species compared to competing anions. Gel-type hybrid cation exchanger (HCIX-G) and macroporous-type cation exchanger (HCIX-M) were used for the fixed-bed column runs. The general characteristics of hybrid materials are summarized in Table 62.2. Figure 62.3a shows As(V) effluent history using HCIX-G and Figure 62.3b provides the breakthrough of As with HCIX-M. The performance of the two sorbents is quite apparent. Note that despite greater HFO content, HCIX-G does not remove As and breaks through almost immediately after the start of the column run. On the contrary, HCIX-M removes more As and reaches 22 µg/l after treating 5000 bed volumes of water.

Table 62.2.   Ion exchanger supported HFO particles.

Designation	Type of ion exchanger	Pore structure	Fe loading (mg Fe/g)
HCIX-G	cation	gel	70
HCIX-M	cation	macroporous	80
HAIX-G	anion	gel	70
HAIX-M	anion	macroporous	100

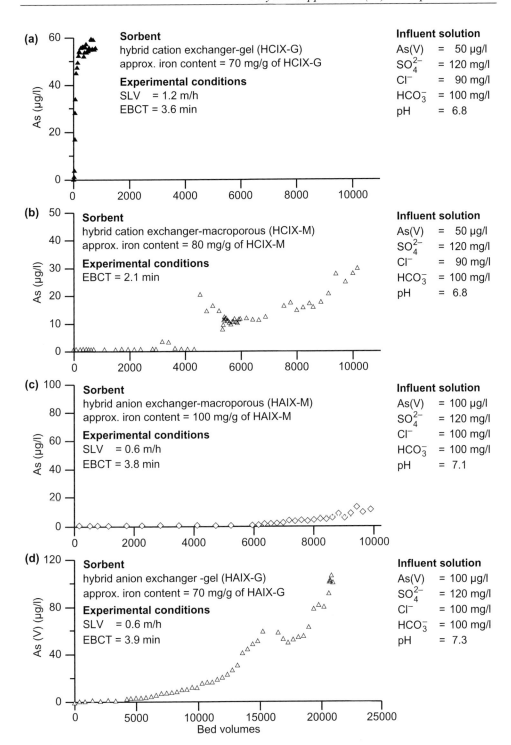

Figure 62.3.   Effluent history of arsenic during a column test using (a) HCIX-G, (b) HCIX-M, (c) HAIX-M and (d) HAIX-G.

62.3.2   *Column runs with hybrid anion exchangers*

In this portion of the study, two separate column runs containing hybrid anion exchangers, HAIX-M and HAIX-G, were conducted; all conditions except As(V) in the influent composition remained identical as above. Figure 62.3c provides the As effluent history using HAIX-M. It is observed that As concentration in the effluent is approximately 10 µg/l after treating ten thousand bed volumes. Figure 62.3d shows the breakthrough for As with HIAX-G, 10 µg/l is reached nearly at eight thousand bed volumes. Comparing the removal efficiency of four hybrid materials, it can be readily inferred that: (1) HAIX-M performs better than HAIX-G; (2) both hybrid anion sorbents exhibit higher As removals compared to hybrid cation polymers and (3) HAIX-M outperforms both HCIX-G and HCIX-M.

Total iron content as hydrated Fe(III) oxide or HFO is nearly the same for all four hybrid materials and varied between 7–10% by mass as Fe. However, the enhanced sorption capacity for As(V) shown by HAIX-M compared to HAIX-G is due to the presence of HFO particles on both pore surface and gel phase that are available for As sorption. And, the difference in As selectivity and removal capacity between HAIX and HCIX can be attributed to the Donnan co-ion exclusion effect resulting from the surface charges (Helfferich 1963). Positively charged functional groups of HAIX allow the passage of arsenates into the gel phase where HFO particles sorb selectively As(V) ions through the formation of inner-sphere complexes or Lewis acid-base interaction while the permeation of arsenate anions onto HFO will be hindered by sulfonic acid groups of HCIX. Earlier studies with ligand polymeric exchangers (parent material: anion exchangers) reported similar sorption behaviors towards positively charged ions (Zhao 1997). Also, in nanofiltration applications, positively charged membranes, repelled more arsenates compared to arsenites at pH of 6.5 (Seidel *et al.* 2001).

62.3.3   *Efficiency of regeneration and reusability*

Following several trials, a mixture of brine and sodium hydroxide was found to be an efficient regenerant for HAIX sorbed with arsenate. Figure 62.4 shows the concentration profile of As during desorption with 2% NaOH and 3% NaCl. Note that 95% As recovery is achieved in 15 bed volumes. A similar regeneration efficiency was obtained for the HIAX column used for As(III) removal (data not shown). The observation that As desorption was completed in 15 bed volumes, demonstrates that sorption sites of HFO nanoparticles are easily accessible through the network of pores, i.e. pore blockage and consequent increase in tortuosity of dissolved solutes did not result from the dispersion of the submicron HFO particles within the porous beads. At high alkaline pH,

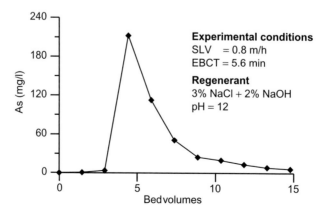

Figure 62.4.   Dissolved arsenic concentration profile during desorption of HIAX using 3% (w/v) NaCl + 2% (w/v) NaOH as regenerant.

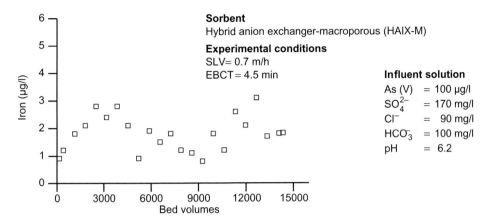

Figure 62.5.   Dissolved iron leakage during a lengthy column run in the absence of oxygen in the feed.

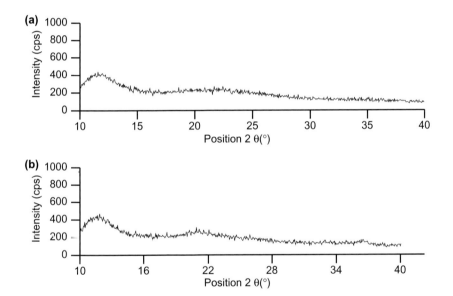

Figure 62.6.   X-ray diffractograms: (a) for sliced freshly prepared HIAX beads and (b) for sliced HIAX beads after a long usage.

HFO sorption sites are all deprotonated and negatively charged, so are all arsenite and arsenate species; thus regeneration is very efficient.

It is recognized that any significant loss of HFO nanoparticles from the hybrid ion exchanger during lengthy column runs would adversely affect the reusability of the material. Figure 62.5 shows total iron concentration at the exit of the fixed-bed column for nearly 15,000 bed volumes. Although there is no specific trend, total iron concentration in the treated water never exceeded 5 µg/l thus confirming that HFO particles are permanently fixed within the matrix of the sorbent and do not leach out. Similar results were reported in previous investigations (SenGupta *et al.* 2000, DeMarco *et al.* 2003, Greenleaf *et al.* 2003).

With aging and changes in environmental conditions, amorphous HFO nanoparticles may get transformed into crystalline forms; namely, goethite, hematite, or magnetite. However, X-ray diffractograms (XRDs) for sliced HAIX beads after prolonged usage did not show any noticeable increase in the crystalline forms of iron oxide particles. Figure 62.6a and b show a XRD for

Table 62.3.   Particle size distribution of HAIX and GFH for a rubbing test.

Sorbent	<10%	<25%	<50%	<75%	<90%
	Size, μm				
HAIX (before rubbing)	583	663	763	853	910
HAIX (after rubbing)	579	676	774	859	913
% change	0.7	−1.9	−1.4	−0.7	−0.3
GFH (before rubbing)	305	436	610	792	896
GFH (after rubbing)	70	312	585	799	887
% change	77.0	28.4	4.1	−0.9	1.0

freshly prepared HIAX-M beads and a XRD for used HIAX beads. Since there is no change in the crystallography of Fe(III) oxide particles, their surface area remains the same, thus sorption capacity of HIAX is not decreased. It has been reported that goethite, hematite, or magnetite mineral phases have less adsorption capacity compared to hydrous iron oxides (Deliyanni *et al.* 2003).

62.3.4   *Mechanical resistance and durability of hybrid sorbents*

The mechanical resistance and durability for HAIX sorbents was assessed through a rubbing test as described before. For comparison a sample of granular ferric hydroxide (GFH) was also exposed to an identical rubbing test. Samples of HAIX beads and GFH granules before and after the rubbing experiments were counted and sized in the particle counter for particle size distributions. The results of such a test are also included in Table 62.3. Note that HAIX particle distribution is fairly similar for all measurements, thus confirming that hybrid sorbents are mechanically strong and durable. For GFH, the particle size distribution changes between the measurements as more fine particles are generated during the rubbing (change of particle size distribution from 1–77%).

62.4   CONCLUSIONS

- HFO particles dispersed inside the gel phase of cation exchangers cannot be used for sorption of As(V) anions because the Donnan co-ion exclusion effect predominates and practically excludes arsenate uptake which explains why the hybrid cation exchanger gel-type (HCIX-G) does not exhibits sorption capacity for As(V) compounds.
- The macroporous hybrid anion exchanger, HAIX-M, performs better compared to the gel hybrid anion exchanger, HAIX-G, in the removal of As(V) because hydrated Fe(III) oxides deposited on both pore and gel phase are available for As sorption.
- The macroporous hybrid anion sorbent, HAIX-M, outperforms both hybrid cation exchangers. The HAIX-M has mainly quaternary ammonium as functional groups while sulfonic acids are the functional groups that predominate in the hybrid cation exchanger. Positively charged fixed groups of HAIX-M allow the passage of arsenates into the gel phase where HFO particles selectively sorb As(V) anions through the formation of inner-sphere complexes or Lewis acid-base interaction while the permeation of arsenate anions onto HFO will be hindered by sulfonic acid groups in hybrid cation. Thus, the difference in selectivity towards As(V) and its removal capacity is attributable to the Donnan co-ion exclusion effect resulting from the surface charges.
- There is no generation of fines during the rubbing tests for HAIX-M particles whereas for GFH the size distribution changes with rubbing; therefore, hybrid beads are mechanically strong and durable.

ACKNOWLEDGEMENTS

We gratefully acknowledge the partial financial support received from the United States Environmental Protection Agency (USEPA) through a STAR grant (No. R82816301). Thanks are also to the reviewers whose constructive comments helped to improve the quality of the manuscript.

REFERENCES

Arnold, W.A., Ball, W.P. and Roberts, A.L.: Polychlorinated ethane reaction with zerovalent zinc: pathways and rate control. *J. Contam. Hydrol.* 40:2 (1999), pp. 183–200.

Cheng, F., Muftikian, Q., Fernando, S. and Korte, N.: Reduction of nitrate to ammonia by zerovalent iron. *Chemosphere* 35:11 (1997), pp. 2689–2695.

Cumbal, L., Greenleaf, J., Leun, D. and Sengupta, A.K.: Polymer supported inorganic nanoparticles: characterization and environmental applications. *React. Funct. Polym.* 54 (2003), pp. 167–180.

Deliyanni, E.A., Bakoyannakis, D.N., Zouboulis, A.I. and Matis, K.A.: Sorption of As(V) ions by akagane'ite-type nanocrystals. *Chemosphere* 50 (2003), pp. 155–163.

DeMarco, M.J., SenGupta, A.K. and Greenleaf, J.E.: Arsenic removal using a polymeric/inorganic hybrid sorbent. *Water Res.* 37 (2003), pp. 164–176.

Greenleaf, J.E., Cumbal, L., Staina, I. and Sengupta, A.K.: Abiotic As(III) oxidation by hydrated Fe(III) oxide (HFO) microparticles in a plug flow columnar configuration. *Trans IChemE* 81:3 Part B (2003), pp. 87–98.

Helfferich, F.: *Ion exchange*. Dover Publications Inc., New York, 1963.

Kim, Y.H. and Carraway, E.R.: Dechlorination of chlorinated phenols by zero valent zinc *Environ. Technol.* 24:12 (2003), pp. 1455–1463.

Koseko, S., Hisamatsu, M. and Yamada, T.: A study on the conversion of glutamine to glutamic acid by an immobilized glutaminase on a support having magnetic sensitivity. I: Preparation of magnetic support for glutaminase immobilization. *Nippon Shokuhin Kogyo Gakkaishi* 41:1 (1994), pp. 31–36.

Laxen, D.P.H.: Adsorption of lead, cadmium, copper, and nickel onto hydrous iron oxides under realistic conditions. *Proceedings 4th Heavy Met. Environ. Int. Conf.* 2, 1983, pp. 1082–1085.

Lequart, C., Kurek, B., Debeire, P. and Monties, B.: MnO_2 and oxalate: an abiotic route for the oxidation of aromatic components in wheat straw. *J. Agri. Food. Chem.* 46:9 (1998), pp. 3868–3874.

Leun, D. and SenGupta, A.K.: Preparation and characterization of magnetically active polymeric particles (MAPPs) for complex environmental separations. *Environ. Sci. Technol.* 34 (2000), pp. 3276–3282.

Liu, C., Honda, H., Ohshima, A., Shinkai, M. and Kobayashi, T.: Development of chitosan-magnetite aggregates containing *Nitrosomonas Europaea* cells for nitrification enhancement. *J. Biosci. Bioeng.* 89:5 (2000), pp. 420–425.

Manning, B.A., Fendorf, S.E. and Goldberg, S.: Surface structures and stability of arsenic (III) on goethite: spectroscopic evidence for inner-sphere complexes. *Environ. Sci. Technol.* 32:16 (1998), pp. 2383–2388.

Matheson, L.J. and Tratnyek, P.G.: Reductive dehalogenation of chlorinated methanes by iron oxides. *Environ. Sci. Technol.* 28 (1994), pp. 2045–2053.

Music, S. and Ristic, M.: Adsorption of zinc (II) on hydrous iron oxides. *J. Radioanalyt. Nucl.* 162:2 (1992), pp. 351–362.

Nesbitt, H.W., Canning, G.W. and Bancrot, G.M.: XPS study of reductive dissolution of $7A°$-birnessite by H_3AsO_3, with constraints on reaction mechanism. *Geochim. Cosmochim. Acta* 62:12 (1998), pp. 2097–2110.

Pierce, M.L. and Moore, C.B.: Adsorption of arsenite and arsenate on amorphous iron oxide. *Water Res.* 16 (1982), pp. 1247–1253.

Roberts, L.C., Hug, S.J., Ruettimann, T., Billah, M., Khan, A.W. and Rahman, M.T.: Arsenic removal with iron (II) and iron (III) in waters with high silicate and phosphate concentrations. *Environ. Sci. Technol.* 38:1 (2004), pp. 307–315.

Scott, M.J. and Morgan, J.J.: Reactions at oxide surface. 1: Oxidation of As(III) by synthetic birnessite. *Environ. Sci. Technol.* 29 (1995), pp. 1898–1905.

Seidel, A., Waypa, J.J. and Elimenech, M.: Role of charge (Donnan) exclusion in the removal of arsenic from water by a negatively charged porous nanofiltration membrane. *Environ. Eng. Sci.* 18:2 (2001), pp. 1105–1113.

SenGupta, A.K. and Lim, L.: Modeling chromate ion-exchange processes. *AIChE J.* 34:12 (1988), pp. 2019–2020.

SenGupta, A.K., DeMarco, M. and Greenleaf, J.: A new polymeric/inorganic hybrid sorbent for selective arsenic removal. In J.A. Greig (ed): *Proceedings of IEX 2000: Ion Exchange at the Millenium*, Churchill College, Cambridge University, July 16–21, 2000, Imperial College Press, London, UK, 2000, pp. 142–149.

SenGupta, A.K.: *Environmental separation of heavy metals engineered processes*. 1st ed., Lewis Publishers, Boca Raton, FL, 2001.

SenGupta, A.K. and Cumbal, L.: Method of manufacture and use of hybrid anion exchanger for selective removal of contaminating ligands from fluids. United States Patent, 069825, 2005.

Slavek, J. and Pickering, W.F.: Extraction of metal ions sorbed on hydrous oxides of iron (III). *Water Air Soil Pollut.* 28:1/2 (1986), pp. 151–62.

Surinenaite, B., Simaityte, V., Bendikiene, V. and Juodka, B.: Investigation of the new type lipase from *Pseudomonas Mendocina* 3121-1. Adsorption onto magnetic derivatives of chitin and chitosan. *Biologia* 1 (1997), pp. 65–70.

Tochiyama, O., Endo, S. and Inoue, E.: Sorption of neptunium (V) on various iron oxides and hydrous iron oxides. *Radiochim. Acta* 68:2 (1995), pp. 105–111.

Tournassat, C., Charlet, L., Bosbach, D. and Manceau, A.: As(III) oxidation by birnessite and precipitation of manganese (II) arsenate. *Environ. Sci. Technol.* 36 (2002), pp. 493–500.

Wang, C.G. and Zhang, W.: Synthesizing nanoscale iron particles for rapid and complete dechlorination of TCE and PCBs. *Environ. Sci. Technol.* 31:7 (1997), pp. 2154–2156.

Zhang, H. and Huang, C.-H.: Oxidative transformation of triclosan and chlorophene by manganese oxides. *Environ. Sci. Technol.* 37:11 (2003), pp. 2421–2430.

Zhao, D.: *Polymeric ligand exchange: a new approach toward enhanced separation of environmental contaminants*. PhD Thesis, Lehigh University, Bethlehem, PA, 1997.

Zhao, D. and SenGupta, A.K.: Ultimate removal of phosphate from wastewater using a new class of polymeric ion exchanger. *Water Res.* 23:5 (1998), pp. 1613–1625.

CHAPTER 63

Application of coagulation-filtration processes to remove arsenic from low-turbidity waters

A.M. Sancha & C. Fuentealba
*División de Recursos Hídricos y Medio Ambiente, Facultad de Ciencias Físicas y Matemáticas,
Universidad de Chile, Santiago de Chile, Chile*

ABSTRACT: Coagulation followed by direct filtration (CF) without flocculation was investigated for the removal of arsenic (As) from groundwaters in the northern and central zones of Chile. Arsenic is removed from water by sorption on ferric hydroxides and the filtering of these through a bed of sand-carbon or sand-anthracite. The variables measured included the untreated water's pH, iron dose, flux, and backwash interval. The pH and ferric dose were found to be the most important variables to ensure effective As removal. The backwash sludge passed the toxicity characteristic leaching procedure (TCLP) test as non-hazardous waste. The excellent results obtained in the northern zone pilot project led to the construction of a full-scale system (32 l/s) at Taltal. Advantages of the technology described herein include: achieving WHO-recommended drinking water standards (0.01 mg As/l, WHO (1993)) at low cost with locally available materials without the need for neither extensive pretreatment areas nor expensive construction outlays, and operated and monitored daily by personnel requiring only intermediate levels of training.

63.1 INTRODUCTION

Arsenic (As) occurs naturally in water in many parts of the world (Ferguson and Gavis 1972). Due to concerns about the long-term health effects of arsenic (Smith *et al.* 1992, NRC 1999, Chen *et al.* 1994, Cantor *et al.* 1996, Ferreccio *et al.* 2000) the World Health Organization set the maximum contamination level (MCL) for As at 10 µg/l (WHO 1993).

In natural water resources, arsenic may be present in inorganic and organic forms or species (Cullen and Reimer 1989, Edwards *et al.* 1998). Organic arsenic is, in general, of little interest (Irgolic 1982). Inorganic arsenic may be present in oxidized or reduced forms. The coexistence of both forms, observed in some groundwaters, may be attributed to a combination of factors, amongst which the actions of microorganisms in oxide-reduction reactions and their kinetics are most important. Organic forms may be present either as the result of production *in situ*, mediated by bacteria based on inorganic forms (biomethylation), or as the result of contamination derived from human activities (Anderson and Bruland 1991).

From a public health perspective arsenite [As(III)] is more toxic than arsenate [As(V)], and of the two, arsenite also is more difficult to remove from drinking-water supplies. Therefore public health authorities and water managers may need to evaluate the scope of the As problem in regard not only to As concentrations, but also to As speciation.

When a source of drinking water contaminated with arsenic is underground, As removal can be a difficult and costly task. These waters, with typically low turbidity, do not require special treatment for purification and thus, in most cases, are only disinfected prior to distribution. In the case of groundwater with As concentrations greater than 10 µg/l, supplying water for human consumption becomes considerably more complicated because water treatment processes are required that involve significant investment expenditures in treatment infrastructure, equipment and specialized personnel (Frey *et al.* 1998, Frost *et al.* 2002).

The current literature cites many technologies to remove As from water. The choice of the most suitable technology will depend on the composition of the water in question, the amounts of water to be treated, concentrations and forms of arsenic, the presence of other ionic constituents, and the degree of sophistication the treatment may involve and its costs. Furthermore, careful consideration must be given to the characteristics of the residues generated by the technology chosen, its handling and ultimate disposal.

The composition of the water matrix to be treated is one of the key elements to be considered in the choice of technology used to remove any contaminant. In the case of As removal, the principal requirements with respect to the quality of the matrix are known for each available technology. Another important factor to be considered in the selection of technologies is local experience and "know-how" in the use of the various technologies and their associated costs (Sancha 2003).

The United States Environmental Protection Agency (USEPA) has identified several of the best available technologies (BAT) for removing arsenic from drinking water: ion exchange, activated alumina, reverse osmosis, modified coagulation-filtration, modified lime softening, electrodialysis reversal, and oxidation-filtration (USEPA 2000). In the case of groundwater, the USEPA recommends reverse osmosis, ion exchange and lime softening. Many of these treatments are very costly, and their effectiveness may be diminished by the presence in the water treated of ions competing for adsorption/exchange sites. Other technologies can only be used with a specific quality of water (Gupta and Chen 1978, Joshi *et al.* 1996, Mc Neill and Edwards 1997, Subramanian *et al.*1997, Waypa *et al.* 1997, Driehaus *et al.* 1998, Meng *et al.* 2000, Wang *et al.* 2002, Holm 2002, Clifford *et al.* 2003, Kim *et al.* 2003). In any case, removal will be costly unless innovative new technologies are found.

In Chile, the Superintendency of Sanitary Services (SISS) recommends different types of As removal processes depending on As concentration in the water (SISS 1999). For water with low As concentrations SISS recommends modified coagulation-filtration; for high As concentrations, it recommends oxidation-coagulation-flocculation-filtration processes.

Since the 1970s in Chile, coagulation followed by flocculation and filtration has been demonstrated to be an effective technology for removing arsenic from surface waters with high As concentrations (Sancha and Ruiz 1984, Cheng *et al.* 1994, Hering *et al.* 1996, Sancha 1999, Karcher *et al.* 1999, Sancha 2002). For groundwater characterized by very low levels of turbidity, dissolved iron and arsenic, investigations have been carried out concerning the use of these same processes, but eliminating flocculation and decantation, thereby reducing the size of the area needed as well as the cost of construction, operation and process maintenance (Barahona and Gonzalez 1987, Ruiz *et al.* 1992, Fuentealba 2003). Other groups of researchers have studied the same modified process (Han *et al.* 2003, Ghurye *et al.* 2004, Chwirka *et al.* 2004).

Arsenic removal by a simplified coagulation-filtration process (CF) requires the addition of a small dose of a coagulant which generates the formation *in situ* of metallic hydroxides upon which the arsenic is adsorbed. These "arsenic-flocs" can be separated from water by filtering them through a bed of sand-anthracite or sand-unprocessed carbon, without the need for flocculation and decantation processes to separate out the flocs. Success in As removal lies upon the quality of the untreated water (its pH and the oxidation state of arsenic), as well as in the efficiency of the filtration process which means that the residual presence of turbidity can be considered to be a good indirect indicator (surrogate indicator) of residual arsenic in the water. Any improvement in the filtration process will translate into better As removal. The handling and ultimate disposal of residues generated in the treatment process will continue to be an important challenge.

63.2   MATERIALS AND METHODS

Arsenic removal using ferric hydroxide coagulation followed by direct filtration without flocculation was investigated for groundwaters from the northern and central zones of Chile, whose physicochemical properties are shown in Table 63.1. The experiments were made to the scale of a

Table 63.1.   Typical physicochemical water properties for the sources tested.

Water constituents and parameters	Chile-northern zone	Chile-central zone
pH	7.0–8.0	8.0–9.0
Arsenic (mg/l)	0.06–0.8	0.04–0.05
Total dissolved solids (mg/l)	730–790	250–300
Alkalinity (as mg/l CaCO$_3$)	50–60	50–60
Manganese (mg/l)	0.2	<0.038
Hardness (as mg/l CaCO$_3$)	350–400	20–40
Turbidity (NTU)	<2.0	<2.0
Silica (as mg/l SiO$_2$)	20–30	15–20
Sulfate (mg/l SO$_4^{2-}$)	–	45
Phosphorus, total (mg/l P)	–	<0.092

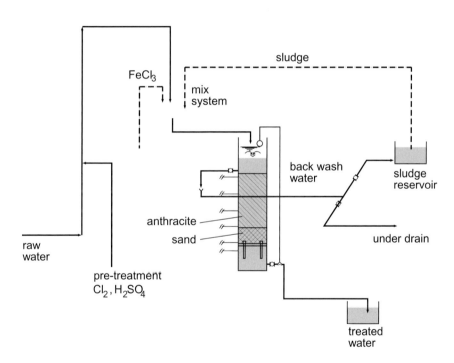

Figure 63.1   Pilot plant, simplified coagulation.

pilot plant (Fig. 63.1), consisting of a mixing tank and a pilot filter whose operation conditions are shown in Table 63.2. The optimal dose of coagulant to be used, pretreatments of oxidation and pH adjustments had been previously investigated at the level of jar tests. Arsenic determination was done by graphite furnace atomic absorption spectrophotometry (GBC model 932A); turbidity by the nephelometric method (Hach Co., Loveland, Colorado) and pH by the electrometric method (Orion PerpHect pH/ISE/mV/T meter from Thermo Electron).

The experiments on the northern zone water were conducted in the 1990s (Ruiz *et al.* 1992) and those on the central zone water in 2003 (Fuentealba 2003). In the north, due to limited availability of water in a desert zone, backwash water from filters was recirculated. In the central zone this recirculation was not necessary.

Table 63.2.   Arsenic removal conditions during simplified iron coagulation process (pilot scale studies).

	Northern zone water	Central zone water
Influent		
pH	7.7	9.01
As(mg/l As)	0.070	0.055
Pre-treatment		
Oxidation	Yes	No
Adjust pH	No	Yes
Coagulation		
Coagulant dose (mg/l FeCl$_3$)	4	10
	8	5
Filtration		
Flow rate (m^3/m^2/d)	150	264–312
Backwash interval (h)	70	8
Backwash flow rate (l/s m^2)	10	15
Time of backwash (min)	10	16
Filter material	sand-natural carbon	sand-anthracite
Filter operation	continuous	discontinuous
Effluent water		
As residual (mg/l As)	0.03 (4 mg/l FeCl$_3$)	0.004 (10 mg/l FeCl$_3$)
	0.005 (8 mg/l FeCl$_3$)	0.011 (5 mg/l FeCl$_3$)

63.3   DISCUSSION OF RESULTS

The results obtained in the CF pilot-scale studies shown in Figures 63.2 and 63.3 demonstrate that iron coagulation-filtration is a feasible technology for achieving the arsenic MCL of 10 µg/l for both groundwaters tested. The coagulation-direct filtration process demonstrates the ability to consistently achieve low levels of As residual. The stability of As removal improves if there is no need to recirculate backwash filter water.

In the case of the northern zone water, the results obtained in the 1990s in the pilot plant lead to the construction of a treatment plant of 32 l/s capacity which has been in operation since 1998 utilizing processes of preoxidation-coagulation-filtration (Fig. 63.4). This treatment results in water with As <0.007 mg/l at an FeCl$_3$ dose of 8 mg/l. The northern zone waters plant located in Taltal consists of four filters 2.5 m high by 12 m in diameter whose filtering bed is composed by 0.4 m of sand and 1.0 m of an unprocessed carbon instead of anthracite. Since its installation in 1998, operating conditions have varied in order to optimize the efficiency of As removal (Fig. 63.5).

The central zone water pilot plant has consistently achieved residual arsenic <0.01 mg/l. The treatment processes are also being optimized and evaluated economically in accordance with Chilean regulations. Treated water consumers will bear the cost of such treatment when the water treatment plant is in operation.

The results of the complete study demonstrate the effectiveness of the simplified coagulation-filtration (CF) process to remove arsenic from low-turbidity waters, such as groundwaters. Success of these processes depends mainly on variables such as raw water pH, iron dose, mixing time, detention time, filtrate flux and backwash interval. Some of these factors in turn depend on the design of the treatment system, which should assure the *in situ* formation of microflocs.

A sand and carbon (natural or anthracite) filtering medium was used in these studies instead microfiltration membranes as have been used in other studies (Ghurye *et al.* 2004). Using a natural carbon filtering bed helps to lower the costs of As removal.

The prior conditioning of raw water pH is not always necessary, given that high-arsenic groundwater tends to have high pH ($\geq$8) as do the waters considered in this study. Oxidation with Cl$_2$ is

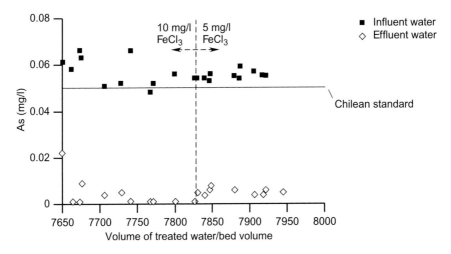

Figure 63.2   Results obtained in central zone pilot plant.

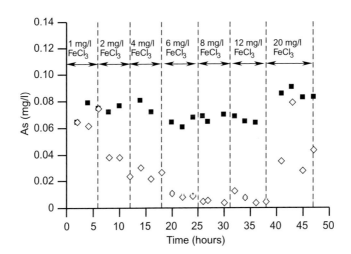

Figure 63.3   Results obtained in northern zone pilot plant.

a pretreatment which will always be recommended in order to assure the presence of As(V) and to protect the filtering medium from biological growth. Speciation analysis revealed that arsenic in Chilean groundwaters was present predominantly as As(V). Analyses confirmed that As concentrations in filtered and unfiltered samples were similar, findings that would suggest that As in these groundwaters is present only in soluble form, and not associated with colloids or fine particulate matter. Pre-oxidation is used as a pretreatment for greater assurance in the efficiency of the removal process.

The process of simplified coagulation utilizes reagents in smaller doses than those normally used in water treatment. Such treatment generates sludge whose As concentrations will be in direct proportion to the volume of water treated, the volume of water used in washing, and the efficiency of removal achieved. The toxicity characteristics of the leaching procedure (USEPA 1992) would indicate that the sludge produced by the CF process does not imply handling

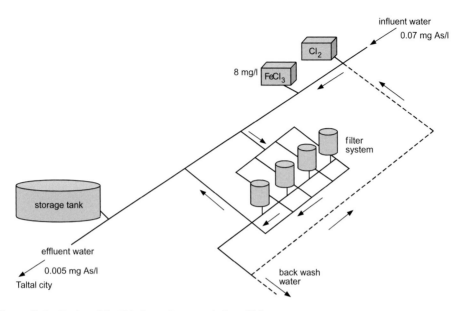

Figure 63.4   Design of the Taltal arsenic removal plant, Chile.

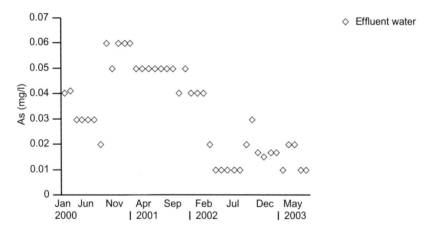

Figure 63.5   Results obtained in northern water plant treating raw water with 0.070 mg/l As, 2000–2003.

as hazardous substances. Similar results have been reported by other investigators (Han *et al.* 2003).

An additional advantage of this method of As removal is that it requires operators whose training is at an intermediate level, not above. Operations can be controlled on a daily basis by means of field kits to measure As levels, and by on-line equipment to measure residual turbidity.

ACKNOWLEDGMENTS

This research was funded by University of Chile, ESSAN of Antofagasta and HIDROSAN and AGUAS ANDINA of Santiago.

REFERENCES

Anderson, L. and Bruland, K.: Biogeochemistry of arsenic in natural waters: the importance of methylated species. *Envir. Sci. and Techn.* 25:3 (1991), pp. 420–427.

Barahona, J. and Gonzalez, Z.: Estudio a nivel de planta piloto sobre abatimiento de arsénico del agua de Taltal. *Proceedings VII Congreso Chileno de Ingenieria Sanitaria y Ambiental*, Chile, 1987, pp. 880–896.

Cantor, K.: Arsenic in drinking water—how much is too much? *Epidemiology* 7 (1996), pp. 113–115.

Clifford, D.A., Ghurye, G.L. and Tripp, A.R.: As removal using ion exchange with spent brine recycling. *J. AWWA* 95:6 (2003), pp. 119–130.

Cullen, W.R. and Reimer, K.J.: Arsenic speciation in the environment. *Chem. Rev.* 89 (1989), pp. 713–764.

Chen, S.-L., Dzeng, S.R., Yang, M.-H., Chiu, K.-H., Shieh, G.-M. and Wai, C.M.: Arsenic species in groundwaters of the blackfoot disease area, Taiwan. *Envir. Sc. and Techn.* 28:5 (1994), pp. 877–881.

Cheng, R.C., Liang, S., Wang, H.-C. and Beuhler, M.D.: Enhanced coagulation for arsenic removal. *J. AWWA* 86:9 (1994), pp. 79–90.

Chwirka, J.D., Colvin, C., Gomez, J.D. and Mueller, P.A.: Arsenic removal from drinking water using the coagulation/microfiltration process. *J. AWWA* 96:3 (2004), pp. 106–114.

Driehaus, W., Jekel, M. and Hildebrandt, U.: Granular ferric hydroxide: a new adsorbent for the removal of arsenic from natural water. *J. Water SRT-Aqua* 47:1 (1998), pp. 30–35.

Edwards, M., Patel, S., McNeill, L., Chen, H.-W, Frey, M., Eaton, A.D., Antweller, R.C. and Taylor, H.E.: Considerations in As analysis and speciation. *J. AWWA* 90:3 (1998), pp. 103–113.

Ferguson, J.F. and Gavis, J.: A review of the arsenic cycle in natural waters. *Water Res.* 6:11 (1972), pp. 1259–1274.

Ferreccio, C., Gonzalez, C., Milosavjlevic, V., Marshall, G., Sancha, A.M. and Smith, A.H.: Lung cancer and arsenic concentrations in drinking water in Chile. *Epidemiology* 11:6 (2000), pp. 673–679.

Frey, M.M., Owen, D.M., Chowdhury, Z.K., Raucher, R.S. and Marc, A.E.: Cost to utilities of a lower MCL for arsenic. *J. AWWA* 90:3 (1998), pp. 89–102.

Frost, F.J., Tollestrup, K., Craun, G.F., Raucher, R., Stomp, J. and Chwirka, J.: Evaluation of costs and benefits of a lower arsenic MCL. *J. AWWA* 94:3 (2002), pp. 71–80.

Fuentealba, C.: *Pilot plant to remove arsenic from a groundwater source of potable water*. Undergraduate Thesis, Civil Engineering Department, Universidad de Chile, Santiago de Chile, Chile, 2003.

Ghurye, G., Clifford, D. and Tripp, A.: Iron coagulation and direct microfiltration to remove arsenic from groundwater. *J. AWWA* 96:4 (2004), pp. 143–152.

Gupta, S. and Chen, K.: Arsenic removal by adsorption. *JWPCF* 30:3 (1978), pp. 492–506.

Han, B., Zimbron, J., Runnells, T.R., Shen, Z. and Wickramasinghe, S.R.: New arsenic standard spurs search for cost-effective removal techniques. *J. AWWA* 95:10 (2003), pp. 109–118.

Hering, J.G., Chen, P.-Y., Wilkie, J.A., Elimelech, M. and Liang, S.: Arsenic removal by ferric chloride. *J. AWWA* 88:4 (1996), pp. 155–167.

Holm, T.R.: Effects of CO_3/bicarbonate, Si, and PO_4 on arsenic sorption to HFO. *J. AWWA* 94:4 (2002), pp. 174–180.

Irgolic, K.: Speciation of arsenic in water supplies. EPA600/1-82-010, US-EPA, Washington, 1982.

Joshi, A. and Chaudhuri, M.: Removal of arsenic from groundwater by iron oxide-coated sand. ASCE *J. Envir. Engrg.* 122:8 (1996), pp. 769–771.

Karcher, S., Caceres, L., Jekel, M. and Contreras, R.: Arsenic removal from water supplies in northern Chile using ferric chloride coagulation. *J. Inst. Water Envir. Mngmt.* 13:3 (1999), pp. 164–169.

Kim, J., Benjamin, M.M., Kwan, P. and Chang, Y.: A novel ion exchange process as As removal. *J. AWWA* 95:3 (2003), pp. 77–85.

McNeill, L.S. and Edwards, M.: Arsenic removal during precipitative softening. *J. Envir. Engrg.* 123:5 (1997), pp. 453–460.

Meng, X., Bang, S. and Korfiatis, G.P.: Effects of silicate, sulfate, and carbonate on arsenic removal by ferric chloride. *Water Res.* 34:4 (2000), pp. 1255–1261.

NRC (National Research Council): Arsenic in drinking water. Subcommittee on Arsenic in Drinking Water, Committee on Toxicology, Washington, D.C., 1999.

Ruiz, G., Perez, O. and Sancha, A.M.: Direct filtration for the treatment of groundwater with arsenic in Chile. *Proceedings XXIII Congreso Interamericano de Ingeniería Sanitaria y Ambiental*, La Habana, Cuba, 1992, pp. 729–739.

Sancha, A.M.: Removal of arsenic from drinking water supplies: Chile experience. *Water Supply* 18:1 (2002), pp. 621–625.

Sancha, A.M.: Removing arsenic from drinking water: A brief review of some lessons learned and gaps arisen in Chilean water utilities, In: W. Chappel, C. Abernathy and R. Calderon (eds): *Proceedings SEGH, 5th International Conference on Arsenic Exposure and Health Effects*. Elsevier Science Ltd., 2003, pp. 469–473.

Sancha, A.M. and Ruiz, G.: Estudio del proceso de remoción de arsénico de fuentes de agua potable empleando sales de aluminio. *XIX Congreso Interamericano de Ingeniería Sanitaria y Ambiental*: Chile, 1984, pp. 379–410.

Sancha, A.M.: Full-scale application of coagulation processes for arsenic removal in Chile: A successful case study. In: W. Chappel, C. Abernathy and R. Calderon (eds): *Proceedings SEGH, International. Conference on Arsenic Exposure and Health Effects*. Elsevier Science, 1999, pp. 373–378.

SISS: Resolución 1745. Instructivo sobre calidad de las fuentes de agua potable. Santiago de Chile, Chile, 1999.

Smith, A.H., Hopenhayn-Rich, C., Bates, M.N., Goeden, H.M., Hertz-Picciotto, I., Duggan, H.M., Wood, R., Kosnett, M.J. and Smith, M.T.: Cancer risks from arsenic in drinking water. *Envir. Health Perspectives* 97 (1992), pp. 259–267.

Subramanian, K.: Manganese greensand for removal of arsenic in drinking water. *Water Qual. Res. J. Canada* 32:3 (1997), pp. 551–553.

USEPA: Toxicity characteristics leaching procedure (TCLP), EPA Method 1311, 40 CFR Part 261, USEPA, Washington, DC, 1992.

USEPA: Technologies and costs for removal of arsenic from drinking water. USEPA, Washington, DC, http://www.epa.gov/safewater/ars/treatments-and-costs.pdf, EPA815/R-00/028, 2000.

Wang, L., Chen, A.S.C., Sorg, T.J. and Fields, K.A.: Field evaluation of As removal by IX and AA. *J. AWWA* 94:4 (2002), pp. 161–173.

Waypa, J.J., Elimelech, M. and Hering, J.G.: Arsenic removal by RO and NF membranes. *J. AWWA* 89:10 (1997), pp. 102–114.

World Health Organization (WHO). Guidelines for drinking water quality, vol 1: Recommendations. 2nd ed., World Health Organization, Geneve, Switzerland, 1993.

CHAPTER 64

Arsenic removal from groundwater by coagulation with polyaluminum chloride and double filtration

R.G. Fernández, A.M. Ingallinella & L.M. Stecca
Centro de Ingeniería Sanitaria (CIS), Facultad de Ciencias Exactas, Ingeniería y Agrimensura, Universidad Nacional de Rosario, Rosario, Prov. de Santa Fe, Argentina

ABSTRACT: The system called ARCIS-UNR developed at Rosario National University, Argentina comprises two stages of filtration: oxidation with chlorine, addition of coagulants on line, roughing up-flow filtration and rapid filtration. The coagulant used is polyaluminum chloride which removed arsenic (As) and fluorine (F^-) simultaneously if the As concentration in raw water is in the range of 100–300 μg/l and fluorine concentration in the range of 1.5–2.5 mg/l. Several laboratory tests and pilot plant tests were made and on the basis of the results obtained, two water treatment plants were projected to supply 11,000 inhabitants and 1000 inhabitants, respectively. The concentrations of As and F^- of the treated water in both plants fulfilled the drinking water standards, and the sludges generated are not classified as hazardous.

64.1 INTRODUCTION

The province of Santa Fe and a wide area of the Argentine territory have high natural concentrations of arsenic (As) in their groundwater, and in many cases it is common to find As associated with fluorine (F^-). In 1998 a research project was initiated to develop an appropriate technology to remove As and F^- by coagulation-adsorption processes with the following aims:

- To find a coagulant that allowed efficient removal of As and F^-.
- To evaluate the up-flow roughing filtration process followed by rapid filtration as a suitable method for the separation of the flocs formed using the coagulant selected.
- To determine the efficiency in the removal of As and F^- in real scale plants.

This chapter shows the results obtained in tests carried out in a laboratory and in a pilot plant and the results obtained in two treatment plants located in the province of Santa Fe, Argentina.

The removal of As through coagulation is based on the principle that As (V) is adsorbed onto the flocs formed, which are later separated by filtration. The kind of coagulant and the dose used affect the efficiency of the process. It has been thoroughly proved (Azbar and Turkman 2000, Scott et al. 1971, Sancha et al. 1992, Cheng et al. 1994, Edwards 1994, Hering et al. 1997) that:

- Ferric chloride is more efficient than aluminum sulfate for As removal.
- It is advisable to introduce a previous oxidizing process, since As (V) is more easily adsorbed than As (III).
- The optimum pH for coagulation-adsorption, when ferric chloride is used, is around 6.9–7.0.
- For fluorine removal, aluminum salts must be used.

Aluminum salts are very commonly used in water treatment. Recently, polymerized forms of aluminum have been used more frequently for water treatment. In this sense, polyaluminum chloride (PACl) is the one most generally applied although polyaluminum sulfates have also been tested. This coagulant is a partially hydrolyzed aluminum chloride which incorporates small amounts of sulfate. The results obtained by using this coagulant are similar to results obtained using aluminum sulfate together with a polyelectrolyte. However, neither the nature of the product nor the reason for the improvement in its performance as a coagulant has been completely clarified (Bratby 1980). Its empirical formula is $Al_n(OH)_mCl_{(3n-m)}$, with $0 < m < 3n$, its approximate density is 1.33 to 1.35 kg/l and it is commercialized as a liquid. This product is increasingly used in the water industry in Germany, Japan, France and at present also in the USA, and it is identified by the initials PACI. The product used in the tests on which this work provides information (PAC 23 of Cloretil) is authorized for use in making water potable by the *Instituto Nacional de Alimentos* of Argentina.

64.2   METHODS

64.2.1   *Laboratory test*

Tests were carried out on samples taken at the outlet of the water supply tank of the town of Villa Cañás which contains in average 2.0 mg/l of fluorine and 0.200 mg/l of As. The water had been previously disinfected with sodium hypochlorite, which would ensure total oxidization of the As (III) to As (V). Jar tests were carried out as follows: 1 minute rapid mixing, 10 minutes slow mixing, 15 minutes sedimentation. To carry out the initial research and to study the variables coagulant dose, pH and initial As concentration jointly, the balanced factorial technique for test design was used. This allows researchers to study the factors considered important simultaneously (Romero-Villafranca and Zunica 1993). A design in which all the factors are studied on two levels, it is called Design 2^k. The statistical analysis of data was carried out with the ANOVA technique. The levels for each one of the factors adopted in each treatment are described in Table 64.1. The response variables were the final As concentration, the fluorine concentration, and the aluminum concentration. Final turbidity and pH were also measured.

The initial pH levels and PACl dose were chosen on the basis of results obtained in preliminary tests. The initial fluorine concentration was that of natural water equal to 1.9 mg/l. Once the tests

Table 64.1.   Conditions of the test in the first stage.

	Factors	Levels
A	Coagulant dose	100 mg/l and 50 mg/l
B	pH	8.2 and 6.9
C	Initial As concentration	570 µg/l and 130 µg/l

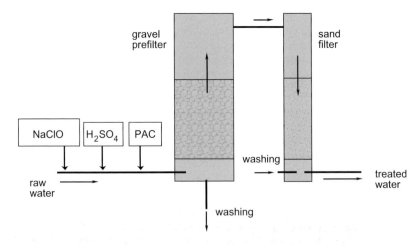

Figure 64.1.   Pilot plant scheme.

in the first phase were carried out a second stage was started to study the influence of pH on the removal of As and F^-.

64.2.2   *Pilot plant tests*

Figure 64.1 shows a plan of the pilot plant. An up-flow roughing filtration stage was chosen because it is a process with a high efficiency in the removal of light flocs as has been shown in previous tests (Ingallinella *et al.* 1998). As a second stage a conventional rapid filter was adopted.

The pilot plant was placed on the base of the potable water tank and was fed with water from the tank outlet to which sodium hypochlorite had been previously added. Runs were carried out with different granular media and filtration rates. Although in the jar tests the efficiencies required were achieved with 120 mg/l of PACl, in the pilot plant that value was adjusted in 100 mg/l in order to achieve a lower consumption of the product. The pH was adjusted with sulfuric acid to an initial pH of 7.0. The following analyses were carried out on samples of raw water, outlet of roughing filter and outlet of rapid filter: pH, turbidity, As (silver diethyldithiocarbamate 3500 As C), fluorine (4500 F^- D SPADNS) and aluminum (Eriochrome Cyanine R 3500 Al D) with techniques following Standard Methods (APHA 1995).

64.3   RESULTS AND DISCUSSION

64.3.1   *Laboratory tests*

The results obtained in the first stage are presented in Table 64.2. After analyzing the results and applying ANOVA, it may be concluded that:

- The three variables studied have significant effects on the final As concentration.
- The removal of As increases as pH diminishes and decreases as it increases.
- Removal is greater with a larger dose of PACl.
- For an initial As concentration of 130 µg/l and for a PACl dose of 100 mg/l and pH = 6.9 the lowest final concentration (20 µg/l) was obtained. The same is applicable to fluorine and the most significant variable for fluorine is pH.
- In all cases the concentrations of residual aluminum in the filtered samples were below 40 µg/l.

Bearing in mind the important influence of pH in the removal of As and F^- and that it is not economical to use doses above 100 mg/l, a series of tests were planned for that dose and varying the initial pH in the samples. Four jar tests were carried out (A, B, C and D) under similar conditions to

Table 64.2.   Tests carried out to determine significant factors.

Test	Factors A	B	C	Arsenic (µg/l) W/filt.	Filtered	Fluorine (mg/l)	Aluminum (mg/l) W/filt.	Filtered	Turb. (NTU)	Final pH
1	100	8.2	130	50	40	1.6	<0.040	<0.040	1.39	8.1
2	100	6.9	130	20	10	1.0	0.900	<0.040	8.33	7.3
3	100	8.2	570	170	160	1.6	0.090	<0.040	1.94	8.1
4	100	6.9	570	70	60	1.4	0.220	<0.040	7.84	7.4
5	50	8.2	130	80	100	1.5	0.240	<0.040	1.33	8.5
6	50	6.9	130	40	30	1.2	0.370	<0.040	8.5	7.4
7	50	8.2	570	250	270	1.5	0.060	<0.040	1.55	8.3
8	50	6.9	570	170	160	1.2	0.410	<0.040	7.76	7.4

Factor A: Dose of PAC, in mg/l; Factor B: pH; Factor C: Initial concentration of arsenic, in µg/l; Turb: turbidity; W/filter: Measurement carried out in unfiltered sample; Filtered: Measurement carried out in filtered sample.

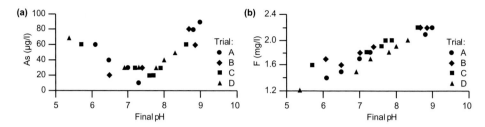

Figure 64.2.   Removal of arsenic (a) and fluorine (b) using PAC: influence of pH.

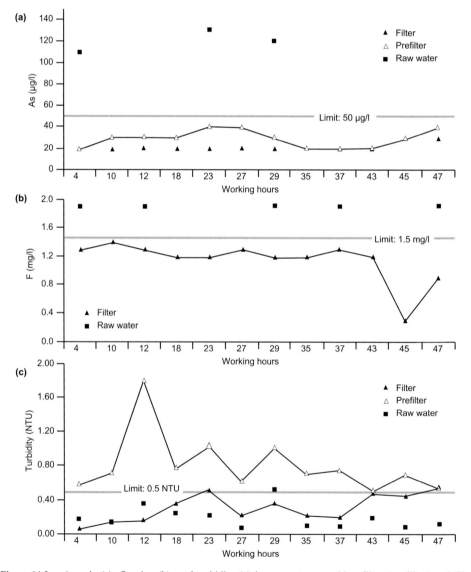

Figure 64.3.   Arsenic (a), fluorine (b), and turbidity (c) in raw water, roughing filter (prefilter) and filter outlets of pilot plant.

have repetitions. The results are shown and discussed below. In Figure 64.2 the results of residual As and F^- in relation to pH are shown for the four tests carried out.

64.3.1.1   *Conclusions of laboratory tests*
- For the conditions under which the tests were carried out, values of As $<50\,\mu g/l$ can be obtained in a pH range of 6.5 to 8.0. In this pH range positively charged insoluble polymeric species are formed on which As is adsorbed. (Van Benschoten and Edzwald 1990).
- The lowest As concentrations were obtained for a pH 7.3.
- Fluorine removal increases as pH diminishes and final concentrations below 1.5 mg/l may be obtained for a pH = 6.0 with PAC doses of 100 mg/l.
- In case an As concentration of 10 $\mu g/l$ in filtered water has to be achieved then pH of water must be controlled carefully.

64.3.2   *Pilot plant tests*

Figure 64.3 are shows the average results of turbidity, As concentration and fluorine concentration of the runs carried out with the conditions indicated above.
   From Figure 64.3 the following conclusions can be derived:

- The highest As removal is produced in the roughing filter.
- Although As removal and turbidity generated during the treatment by rapid filter seems to be unrelated, a turbidity removal is a good indicator of As removal.
- Concentrations of As, F^- and Al as well as turbidity values were below the limits set by rules in force.

64.4   REAL SCALE PLANTS

Bearing in mind the results obtained in the laboratory tests and those in the pilot plant, the treatment plant of the town of Villa Cañás was planned (start up in April 2002) (Ingallinella *et al.* 2002). Subsequently, a similar plant was built in the town of López (start up in January 2003). Table 64.3 shows the characteristics of both plants and Table 64.4 shows the average values obtained during their operation.
   The operation of washing the filters is carried out once a day, for the roughing as well as for the rapid filters. The operation of washing in the gravel roughing filters is carried out through gravity discharge, and the rapid filters are washed countercurrent using the pressure of the existing distributing tank. The wash water from roughing filters and rapid filter is collected in a sedimentation tank and the supernatant is recycled to the plant inlet. The sludges are dewatered in evaporation ponds in Villa Cañás and by filter press in López. The dewatered sludges are disposed in sanitary land fills because they are not hazardous wastes.

Table 64.3.   Characteristics of the Villa Cañás and López plants.

Town	Population supplied (inhab.)	Flow (m^3/day)	Treatment process	Kind of construction
López	1200	150	Addition of PACl + roughing filtration + rapid filtration	Modular plant built of steel
Villa Cañás	11000	1200	Sulfur acidic + PACl + roughing filtration + rapid filtration	Two modules built of concrete; partially automatized

Table 64.4.   Quality of raw water and of filtered water at the treatment plants in Villa Cañás and López.

Parameter	Raw water	Filtered water
Treatment plants in Villa Cañás		
Turbidity (NTU)	0.15	0.20
Arsenic (μg/l)	200	20
Fluorine (mg/l)	2.3	1.5
pH	8.2	7.2
Aluminum (mg/l)	–	0.10
Treatment plants in López		
Turbidity (NTU)	0.15	0.20
Arsenic (μg/l)	70	20
pH	8.2	7.2
Aluminum (mg/l)	–	0.1

64.5   CONCLUSIONS

The process of up flow roughing filtration followed by rapid filtration is suitable for the separation of the flocs formed using polyaluminum chloride. By means of this process it is possible to achieve efficiency in the removal of As (87%) and of F^- (35%), under real operating conditions, without finding residual aluminum concentrations in the water treated above the rule in force (0.20 mg/l).

REFERENCES

APHA: *Standard methods for the examination of water and waste water*. 19th ed., American Public Health Association (APHA), Washington, DC, 1995.

Azbar, N. and Turkman, A.: Defluoridation in drinking waters. IWA *Water Science and Technology* 42:1/2 (2000), pp. 403–407.

Cheng, R., Liang, S., Wang, H. and Beuhler, M.: Enhanced coagulation for arsenic removal. *J. AWWA* 86:9 (1994), pp. 79–90.

Edwards, M.: Chemistry of arsenic removal during coagulation and Fe-Mn oxidation. *J. AWWA* 86:9 (1994), pp. 64–78.

Hering, J., Chen, P., Wilkie, J. and Elimelech, M.: Arsenic removal from drinking water during coagulation. *J. Environ. Eng.* 123:8 (1997), pp. 800–807.

Ingallinella, A., Stecca, L. and Wegelin, M.: Up-flow roughing filtration: rehabilitation of a water treatment plant in Tarata, Bolivia. IWA *Water Sci. Technol.* 37:9 (1998), pp. 105–112.

Ingallinella, A.M., Fernández, R.G. and Stecca, L.M.: Proceso Arcis-UNR para la remoción de arsénico y flúor en aguas subterráneas: una experiencia de aplicación. *Rev. Ingen. Sanit. Ambiental* 66 and 67 (2002), pp. 53–58, AIDIS, Buenos Aires, Argentina.

Romero Villafranca, R. and Zúnica Ramajo, L.R.: *Estadística*. Universidad Politécnica de Valencia, Valencia, Spain, 1993.

Sancha, A., Vega, F. and Fuentes, S.: Efficiency in removing arsenic from water supplies for large towns. Salar del Carmen planta, Antofagasta, Chile. *Proceedings International Seminar Arsenic in the Environment and its Incidence on Health*. CD, Universidad de Chile, Santiago de Chile, Chile, 1992.

Scott, K., Green, J., Do, H. and S. McLean: Arsenic removal by coagulation. *J. AWWA* 87:4 (1995), pp. 114–126.

Van Benschoten, J. and Edzwald, J.: Chemical aspects of coagulation using aluminium salts. *Water Res.* 24:12 (1990), pp. 1519–1526.

CHAPTER 65

A simple electrocoagulation set up for arsenite removal from water

P.D. Nemade & S. Chaudhari
Centre for Environmental Science and Engineering, Indian Institute of Technology IIT Bombay, Powai, Mumbai, India

K.C. Khilar
Department of Chemical Engineering, Indian Institute of Technology IIT Bombay, Powai, Mumbai, India

ABSTRACT: Electrocoagulation (EC) is an emerging water treatment technology that has been applied successfully to treat various wastewaters. The available arsenic (As) removal methods have quite low efficiency for As(III) removal and thereby chemical oxidation of As(III) followed by adsorption is suggested. Electrocoagulation offers possibility of anodic oxidation and *in situ* generation of adsorbents (such as hydrous ferric oxides). In EC process with iron electrodes, As(III) is oxidized to As(V) and is being adsorbed onto iron hydroxides (hydrous ferric oxide), generated in the process. Therefore, electrocoagulation seems to be a better choice for As removal from water. A simple setup of electrocoagulation along with cloth filter was used for arsenite removal. The As-laden water was passed through a container (volume 2 l and 16 l) in which electrocoagulation (current: 0.2–0.3 A) was done and then the water passes through a double layer cotton cloth at a flow rate of 44 to 90 ml/min. Experiments were carried out with an initial As concentration of 1 and 2 mg/l with varying current flow. Results show that As levels below 50 μg/l could be achieved, which is the drinking water standard in India and Bangladesh.

65.1 INTRODUCTION

Arsenic (As) is a ubiquitous element found in the environment (Ng *et al.* 2003). Its toxicity is well known and extremely detrimental to human beings all over the world. High concentrations of arsenic (As) in water has caused symptoms of chronic As poisoning in local populations of many countries like India, Bangladesh, Taiwan, Mongolia, China, Japan, Poland, Hungary, Belgium, Chile, Argentina and Mexico. Manifestation of higher doses of inorganic As compounds in the human body leads to the disease called arsenicosis. Arsenic is a carcinogen and its ingestion may deleteriously affect the gastrointestinal tract, cardiovascular system, central nervous system and diseases like skin lesions, hyperkeratosis, hyperpigmentation (Farrell *et al.* 2001, Jain and Ali 2000). Due to adverse effects on humans, the USEPA has lowered the maximum contaminant level in drinking water from 50 μg/l to 10 μg/l (USEPA 2001). Leaching from geological formations and anthropogenic activities such as uncontrolled industrial discharge from metallurgical, smelting and mining industries, and organo-arsenical pesticides is a source of As in the groundwater (Krishna *et al.* 2001). In groundwater and surface water environments, As is present as As(V) (arsenate) in oxidizing environments, while As(III) (arsenite) is the predominant in reducing environments (Ferguson and Gavis 1972). Arsenic exists in groundwater predominantly as inorganic arsenite, As(III) (H_3AsO_3, $H_2AsO_3^-$, $HAsO_3^{2-}$), and arsenate, As(V) (H_3AsO_4, $H_2AsO_4^-$, $HAsO_4^{2-}$) (Ferguson and Gavis 1972). Since As(III) is more mobile and more toxic than As(V), it is advantageous to convert As(III) to As(V). Due to poor As(III) removal from water by many conventional processes, oxidation of As(III) to As(V) is recommended followed by subsequent removal of As(V).

Adsorption on low-cost media is an attractive means for trace metal removal from water. Adsorption-based processes are reliable and efficient for removal of complex inorganic and organic contaminants as compared to many conventional treatment methods (Benjamin *et al.* 1996). Amorphous iron oxide or ferrihydrite is a common coating of subsoil particles and has a high capacity for different anions including As (Pierce and Moore 1982). Iron containing salts have also been used to coat quartz sand for the removal of As from groundwater (Joshi and Chaudhuri 1996). Electrocoagulation has been reported to offer various advantages over conventional coagulation in conjunction with other processes (Mills 2000), such as dissolved air flotation (Pouet and Grasmick 1995, Chen *et al.* 2000) and has been successfully applied to a wide range of pollutants in an even wider range of reactor designs (Vik *et al.* 1984). Electrocoagulation is widely used to treat different types of wastewater such as domestic grey water reuse (Lin *et al.* 2005), laundry wastewater (Ge *et al.* 2004), domestic wastewater (Vlyssides *et al.* 2002), separation of pollutants from restaurant wastewater (Chen *et al.* 2000), urban wastewater (Pouet and Grasmick 1995), defluoridation of water (Mameri *et al.* 2001), and nitrate removal (Koparal and Ogutveren 2002). Hence, it is expected that the electrocoagulation would be an ideal choice for removal of As from water (Kumar *et al.* 2004).

In the present study, As is removed by electrocoagulation in an innovative process using iron as an electrode which is cheaply available and non toxic to humans. As(III) was chosen for this study due to its predominant occurrence in the Bengal delta basin in India.

65.1.1   *Theoretical consideration*

Electrocoagulation offers the possibility of anodic oxidation and *in situ* generation of adsorbents (such as hydrous ferric oxides (HFO)). The EC process operates on the principle that the cations produced electrolytically from iron anodes enhance the coagulation of contaminants from an aqueous medium. Electrophoretic motion tends to concentrate negatively charged particles in the region of the anode and positively charged ions in the region of the cathode. The consumable, or sacrificial metal anodes are used to continuously produce polyvalent metal cations in the vicinity of the anode. These cations neutralize the negative charge of the particles carried toward the anodes by electrophoretic motion, thereby facilitating coagulation. In the flowing EC techniques, the production of polyvalent cations from the oxidation of the sacrificial anodes (Fe and Al) and the electrolysis gases (H_2 and O_2) work in combination to flocculate the coagulant materials (Parga *et al.* 2005). Thus the removal mechanisms in EC may involve oxidation, reduction, decomposition, deposition, coagulation, absorption, adsorption, precipitation and flotation (Parga *et al.* 2005).

According to Kumar *et al.* (2004), As(III) might be oxidized to As(V) during electrocoagulation and get adsorbed onto the metal hydroxides generated. Therefore, it is expected that electrocoagulation would be a better choice for As removal from water. Previous researchers have mostly considered current density as an important design variable for EC process. Whereas, as per Faraday's law dissolution of electrode is related to the total charge passed. Thereby the amount of adsorbent produced in the electrochemical reactor would be proportional to the charge density (total charge passed through the solution) and serve as design parameter for EC process.

65.2   MATERIALS AND METHODS

65.2.1   *Reagent preparation*

The experiments were performed at the Centre for Environmental Science and Engineering laboratory, IIT Bombay, at ambient temperatures ranging from 26 to 28°C. The chemicals were of analytical reagent grade and were used without any further purification. All glassware was cleaned with water and 1 N H_2SO_4 and then rinsed with distilled water. Stock solution of arsenite (1000 mg/l) was prepared by dissolving appropriate quantity of As-trioxide, As_2O_3, (S.D. Fine Chem. Ltd, India) in distilled water containing 1% (w/w) NaOH and the solution was then diluted to 1 liter with distilled water before use. After calibration working solutions were prepared with tap water by dissolving appropriate amount of As from stock solutions; they were tested for the pH,

Table 65.1.   Experimental conditions employed in continuous mode electrocoagulation.

Expt. run no.	Arsenic type	As conc. (μg/l)	Current (mA)	Current density (mA/cm^2)	pH maintained in EC reactor	Flow rate (ml/min)	Reactor volume (ml)
1	As(III)	1000	200	1.66	7 ± 0.2	44	2000
2	As(III)	1000	240	2.00	7 ± 0.2	44	2000
3	As(III)	2000	200	1.66	7 ± 0.2	44	2000
4	As(III)	2000	230	1.91	8 ± 0.2	88	16000
5	As(III)	1000	300	2.5	7.6 ± 0.2	90	16000

alkalinity, and the presence of As, Fe, and phosphate. It was found that the pH of the water varied from 7.2 to 7.5, bicarbonate alkalinity was approximately 45–50 mg/l as $CaCO_3$, the dissolved iron, phosphate and As concentrations were not detectable in tap water.

65.2.2   *Experimental plan*

Preliminary experiments were conducted with iron electrodes. Two (mild steel) iron electrodes were placed 0.5 cm apart in a 2 l beaker and As-containing water was added in this beaker. The total submerged surface area of each electrode was 36 cm^2. Before each experiment, the electrodes were abraded with sand paper to remove scale and then cleaned with successive rinses of water and 1 N H_2SO_4.

A direct current by stabilized power supply (0–15 V, 2 A) was applied to the terminal electrodes in which electrical current was controlled by a variable transformer. Experiments were conducted at pH 7.0. The pH of the solution was adjusted by adding either dilute HCl or NaOH. The summary of experimental conditions employed is presented in Table 65.1.

65.2.3   *Method of analysis*

The residual As in water sample was determined using molybdenum blue complex method (Johnson and Pilson 1972; detection limit is 1 μg/l). The method was used to estimate As(III) and As(V) concentrations in treated water samples to assess the efficiency of the oxidation step and the subsequent removal of As. Spectrophotometric measurements were made at a wavelength of 865 nm using absorbance cells of 5 cm optical path length for As determination. Calibration curve for total As was prepared using solutions containing As(III) dissolved in distilled water. Total iron analysis was performed by 1,10 phenanthroline method (detection limit is 10 μg/l) as described in standard methods (APHA 1998).

65.3   RESULTS AND DISCUSSION

In any electrocoagulation process, electrode material has significant effect on the treatment efficiency. Most common electrode materials used in EC are aluminum and iron because of their low cost, availability and proven effectiveness (Chen *et al.* 2000, Mollah *et al.* 2001, Larue *et al.* 2003). The electrode material for drinking water treatment should also be non-toxic to human health. Hence iron was chosen as electrode material as it is non-toxic and readily available (Kumar *et al.* 2004).

A study (Fig. 65.1) was conducted to evaluate the As removal by the electrocoagulation using 2 l glass beakers and double cotton cloth to assess the feasibility of this process in domestic filters in As-affected areas. Water samples with arsenic concentrations up to 2000 μg/l were treated by EC cloth filter with flow rate 44 ml/min for 3 hours. The result shows that treated water of As concentration was below 50 μg/l. This finding is significant for As removal in Bengal delta plain (BDP) areas.

Hydraulic retention time (HRT) was calculated by

$$\text{HRT} = \text{volume of reactor/flow rate} = 2000\,\text{ml}/44\,\text{ml} = 45\,\text{minutes}. \qquad (65.1)$$

In electrocoagulation, iron dissolves from the anode and hydrogen gas is developed at cathode. Faraday's law can be used to describe the current density (A/cm^2) and the amount of iron goes into the solution (g Fe/cm^2) (Vik *et al.* 1984),

$$w = \frac{itM}{ZF} \qquad (65.2)$$

where, w = metal dissolving (g Fe/Fe/cm^2); i = current density (A/cm^2); t = time (s); M = molecular weight of Fe (55.84); Z = number of electrons involved in the oxidation/reduction reaction ($Z = 3$); F = Faraday's constant, 96500 coulomb.

Figure 65.1 shows total As removal with time. Initially, around 90% of As was removed after 45 minutes. As time progressed, the residual As concentration was below 50 μg/l after 2 hours. At steady state condition after 2 hours, total As concentration removed was below 50 μg/l achieving adsorbing capacity (qe) of 96 μg As/mg iron at equilibrium As concentration of 45 μg/l. This curve remains horizontal at the end of experiment. Pinisakul *et al.* (2002) found that 1.0–1.1 g Fe is

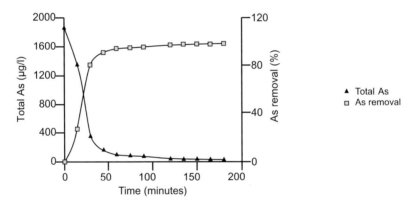

Figure 65.1.   Effect of time on arsenic removal (experimental run 3); Initial As = 2000 μg/l, double layer cloth; Current = 0.2 A, Volume = 2 l in beaker, pH = 7, flow rate 44 ml/min.

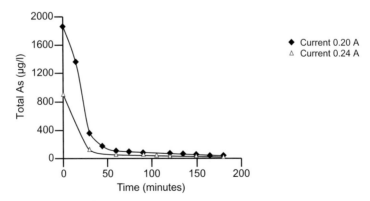

Figure 65.2.   Effect of current density on arsenic removal (experimental run 2 and 3); (Δ) initial As = 1000 μg/l and current = 0.24 A; (♦) initial As = 2000 μg/l and current 0.2 A, double layer cloth; Volume 2 l, pH = 7, flow rate 44 ml/min.

required in the electrochemical process to remove 1000 mg of Arsenic. Arsenic removal capacity of zero-valent iron was found to be approximately 7.5 mg As/gFe (Lien and Wilkin 2004) which is much lower than the As sorption capacity obtained in this study.

Figure 65.2 shows that up to 85–90% of the initial concentration decreased within 30 min of the process and the residual As concentration in water were 358 μg/l and 122 μg/l at pH 7, for currents of 0.2 A and 0.24 A, respectively. After 2 hours, residual concentrations were found to be 48 μg/l and 44 μg/l, respectively, which is below the Indian drinking water standard of 50 μg/l. Figure 65.3 shows that after 55 minutes the As removal was found to be <50 μg/l and achieved 99% As removal from the water. Similarly Figure 65.4 shows that the residual As concentration of 43 μg/l after 5 hours is below 50 μg/l standard for As in drinking water. Also residual iron concentrations varied from 0.033 to 0.048 mg/l which were well below the 0.3 mg/l standard for iron in drinking water.

At the beginning of the process, As removal was rapid and later it decreased gradually over the entire process. Arsenic ions are more abundant at the beginning of the EC process, and the generated iron hydroxides due to corrosion of the anode at that time will form complexes with As and therefore rapid removal of As was observed (Kumar *et al.* 2004). As the experiment proceeded the aqueous phase As concentration went on reducing and simultaneously hydrous ferric oxides concentration increased, so the curves were nearly horizontal at the end of experiment.

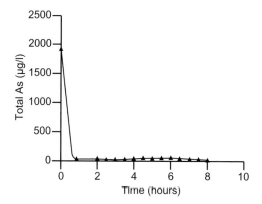

Figure 65.3. Effect of time on arsenic removal (experimental run 4); Initial total As = 2000 μg/l and current = 0.23 A, double layer cloth; Volume 16 l, pH = 7, flow rate 88 ml/min.

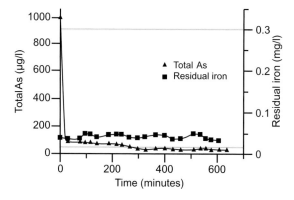

Figure 65.4. Effect of time on arsenic removal (experimental run 5); Current = 0.30 A, double layer cloth; Volume 16 l, pH = 7.6, flow rate 90 ml/min.

The pH increase in electrocoagulation is attributed to the formation of hydrogen gas at the cathode (Vik *et al.* 1984, Chen *et al.* 2000). Also a slight increase in pH may be expected because of sorption reactions of As(V) and As(III), which release OH^- groups from sorbents as a result of ligand exchange (Arienzo *et al.* 2002).

Kumar *et al.* (2004), reported that there is no effect of pH on both As(III) and As(V) removal up to 97–99%. Therefore, the filter runs at pH 7.0.

65.3.1   *Practical considerations*

How much iron is needed for the removal of As(III) is a question of practical relevance. Lien and Wilkin (2005) reported the arsenite removal capacity of zero valent iron was 7.5 mg As/gFe. Meng *et al.* (2001) reported that removal of 90% As in phosphate and silicate-rich water (1.9 mg/l P, 18 mg/l Si) required a Fe/As ratio of at least 40, after the As(III) has been oxidized to As(V) with hypochlorite. Similar to these findings, Leupin and Hug (2005) found that 15–18 mg Fe(III) are needed to remove 90% of 500 µg/l As(V) (with 3 mg/l P and 20 mg/l Si). As(III) removal under the same conditions required over 80 mg/l Fe(III) (Roberts *et al.* 2004).

Similarly electrocoagulation process requires 40 mg of iron needed to remove 99% of 2 mg/l of As(III) and As(V), which is much less than in any other processes.

65.4   CONCLUSIONS

Electrocoagulation process was able to remove 99% of As from As-contaminated water and met the drinking water standard of 50 µg/l using Fe-electrode. As(III) was more efficiently removed in electrocoagulation than with conventional methods. The amount of electricity required per liter of water is 0.36 Watt/h and total operating cost of EC is 10 cent US$/per 1000 liters of As-contaminated water. Therefore electrocoagulation process is very cheap and simple. It can be used as household unit in developing countries like India, Bangladesh and other parts of world, due to its robustness and convenience in preparation, EC filter could become a viable and cost-effective alternative as a filter medium for As removal from contaminated drinking water.

REFERENCES

APHA-AWWA: Standard methods for the examination of water and wastewater. 20th ed. Am. Publ. Hlth. Assoc./Am. Wat. Works Assoc., Washington, DC, 1998.

Arienzo, M., Paola, A., Chiarenzelli, J., Maria, R.B. and Martino, A.D.: Retention of arsenic on hydrous ferric oxides generated by electrochemical peroxidation. *Chemosphere* 48 (2002), pp. 1009–1018.

Benjamin, M.M., Slatten, R.S., Bailey, R.P. and Bennet, T.: Sorption and filtration of metals using iron-oxide-coated sand. *J. Environ. Eng.* 125 (1996), pp. 782–784.

Chen, X., Chen, G. and Po, L.Y.: Separation of pollutants from restaurant wastewater by electrocoagulation. *Sep. Purif. Technol.* 19 (2000), pp. 65–76.

Farrell, W.J., Peggy, O.D. and Colklin, M.: Electrochemical and spectroscopic study of arsenate removal from water using zero-valent ion media. *Environ. Sci. Technol.* 35 (2001), pp. 2026–2032.

Ferguson, J.F. and Gavis, J.: Review of the arsenic cycle in natural waters. *Water Res.* 6 (1972), pp. 1259–1274.

Ge, J., Qu, J., Lei, P. and Liu, H.: New bipolar electrocoagulation-electroflotation process for the treatment of laundry wastewater. *Sep. Purif. Technol.* 36 (2004), pp. 33–39.

Jain, C.K. and Ali, I.: Arsenic: occurrence, toxicity and speciation techniques. *Water Res.* 34 (2000), pp. 4304–4312.

Johnson, D.L. and Pilson, M.: Spectrophotometric determination of arsenite, arsenate and phosphate in natural waters. *Anal. Chim. Acta* 58 (1972), pp. 289–299.

Joshi, A. and Chaudhuri, M.: Removal of arsenic from ground water by iron-oxide coated sand. *J. Environ. Eng.* 122:8 (1996), pp. 769–776.

Koparal, A.S. and Ogutveren, Ü.B.: Removal of nitrate from water by electroreduction and electrocoagulation. *J. Hazard. Mater.* B89 (2002), pp. 83–94.

Krishna, M.V.B., Chandrasekaran, K., Karunasagar, D. and Arunachalam, J.: A combined treatment approach using Fenton's reagent and zero-valent iron for the removal of arsenic from drinking water. *J. Hazard. Mater.* B 84 (2001), pp. 229–240.

Kumar, P.R., Chaudhari, S., Khilar, K.C. and Mahajan, S.P.: Removal of arsenic from water by electrocoagulation. *Chemosphere* 55 (2004), pp. 1245–1252.

Larue, O.E., Vorobiev, E., Vub, C. and Durand, B.: Electrocoagulation and coagulation by iron of latex particles in aqueous suspensions. *Sep. Purif. Technol.* 31 (2003), pp. 177–192.

Leupin, O.X. and Hug, S.J.: Oxidation and removal of arsenic (III) from aerated groundwater by filtration through sand and zerovalent iron. *Water Res.* 39 (2005), pp. 1729–1740.

Lien, H.-L. and Wilkin, R.T.: High-level arsenite removal from groundwater by zero-valent iron. *Chemosphere* 59:3 (2005), pp. 377–386.

Lin, C.-J., Lo, S.-L., Kuo C.-Y. and Wu C.-H.: Pilot-Scale electrocoagulation with bipolar aluminum electrodes for on-site domestic grey water reuse. *J. Environ. Eng.* 131:3 (2005), pp. 491–495.

Mameri, N., Lounici, H., Belhonice, D., Grib, H., Piron, D.L. and Yahiat, Y.: Defluoridation of Sahara water by small plant electrocoagulation using bipolar aluminum electrodes. *Sep. Purif. Technol.* 24 (2001), pp. 13–119.

Meng, X.G., Korfiatis, G.P., Christodoulatos, C. and Bang, S.B.: Treatment of arsenic in Bangladesh well water using a household co-precipitation and filtration system. *Water Res.* 35 (2001), pp. 2805–2710.

Mills, D.: A new process for electrocoagulation. *J. Am. Water Works Assoc.* 92 (2001), pp. 35–43.

Mollah, M.Y.A., Morkovsky, P., Gomes, J.A.G., Kesmez, M., Parga, J. and Cocke, D.L.: Fundamentals, present and future perspectives of electrocoagulation. *J. Hazard. Mater.* B114 (2005), pp. 199–210.

Pierce, M.L. and Moore, C.B.: Adsorption of arsenite and arsenate on amorphous iron hydroxide. *Water Res.* 16:7 (1982), pp. 1247–1253.

Ng, J.C., Wang, J. and Shraim, A.: A global health problem caused by arsenic from natural sources. *Chemosphere* 52 (2003), pp. 1353–1359.

Parga, J.R., Cocke, D.L., Valenzuela, J.L., Gomes, J.A., Kesmez, M., Irwin, G., Moreno, H. and Weir, M.: Arsenic removal via electrocoagulation from heavy metal contaminated groundwater in La Comarca Lagunera Mexico. *J. Hazard. Mater.* 124:1–3 (2005), pp. 247–254.

Pinisakul, A., Polprasert, C., Parkplan, P. and Satayavivad, J.: Arsenic removal efficiency and mechanisms by electro-chemical precipitation process. *Water Sci. Technol.* 46:9 (2002), pp. 247–254.

Pouet, M.T. and Grasmick, A.: Urban wastewater treatment by electrocoagulation and flotation. *Water Sci. Technol.* 31 (1995), pp. 275–283.

Roberts, L.C., Hug, S.J., Ruettimann, T., Billah, Md. M., Khan, A.W. and Rahman, M.T.: Arsenic removal with iron (II) and iron (III) in waters with high silicate and phosphate concentrations. *Environ. Sci. Technol.* 38 (2004), pp. 307–315.

Vik, E.A., Carlson, D.A., Eikum, A.S. and Gjessing, E.T.: Electrocoagulation of potable water. *Water Res.* 18 (1984), pp. 1355–1360.

Vlyssides, A.G., Karlis, P.K., Rori, N. and Zorpas, A.A.: Electrochemical treatment in relation to pH of domestic wastewater using TI/Pt electrodes. *J. Hazard. Mater.* B95 (2002), pp. 215–226.

Other technologies

CHAPTER 66

Arsenic in the environment and its remediation by a novel filtration method

T.R. Roth & K.J. Reddy
Department of Renewable Resources and School of Energy Resources, University of Wyoming, Laramie, WY, USA

ABSTRACT: The emergence of arsenic (As) in groundwater in many parts of the world, which adversely affects the health of millions of people, has raised awareness for effective As removal systems. A flow-through filtration system was tested to remove As from groundwater. The introduction of a simple method using ARTI-64[tm] technology is discussed. The results using ARTI-64[tm] particles with a flow-through apparatus suggest that the removal method is efficient and effective in removing As under natural conditions and high As concentrations. This method shows high removal rates in the presence of arsenate and arsenite, common competing ions and requires no pH treatments. A summary of existing removal techniques is also presented.

66.1 INTRODUCTION

As we progress into the millennium, forthcoming generations will be faced with a multitude of challenges that will arise due to the projected exponential growth of the world's population, particularly in developing nations. Today the world population is at 6.5 billion people with the estimation that the number will rise to over 9 billion by the year 2050 (US Census 2006). Rapid global population growth is creating immense pressure on natural resources such as groundwater for clean drinking water supplies.

Earth has abundant water however, the vast majority of this water is too saline (oceans), or inaccessible (ice caps, glaciers) to utilize. The available supply of fresh water is less than 1% of the total amount of water on earth. Of that 0.61% is groundwater, while only 0.0091% is surface water (USGS 1984). Pollutants vary from water body to water body and vary in their source. Urbanization, agricultural use, and industry wastes contribute to contamination but other non-anthropogenic sources also are factors polluting waters, such as naturally occurring copper, selenium or fluoride in certain parts of the world. Recently, arsenic (As), another naturally occurring pollutant has caused serious health concerns in several regions throughout the world, most predominately in Bangladesh and India (Bagla and Kaiser 1996, Smith *et al.* 2000). Elevated levels of As have also been shown to exist in Chile, Taiwan, and western USA. Over 200 million people are known to be consuming water that exceeds the World Health Organization's (WHO) recommended limit of 10 micrograms per liter (μg/l) for As, with 30 million exposed in Bangladesh alone. Levels within Bangladesh, where the prevalence of As is most severe, vary between <1 μg/l to 2500 μg/l (Nordstrom 2002).

The purpose of this article is to discuss health effects from As consumption, the chemistry of As in natural waters, existing As removal techniques, and an introduction to a novel As filtration method. Additionally, a discussion on the effects of pH, oxidation state, and competing anions in removal of As using a new filtration method is presented.

66.2 HEALTH EFFECTS OF ARSENIC

The US Environmental Protection Agency (USEPA) has set a 10 μg/l maximum contaminant limit (MCL) for As in human drinking water, which became effective January 2006, down from

50 μg/l MCL. The carcinogenic nature of As leads to various forms of cancer developing in the human body after prolonged ingestion of high concentrations of As in drinking water (Bates *et al.* 1992). Exposure to As can occur via inhalation, primarily in industrial settings, or through ingestion. Because drinking water is one of the primary routes of exposure, standards set in 1942 established an MCL of 50 μg/l in drinking water. In a 1984 health assessment, the (USEPA) classified As as a class A human carcinogen, based primarily on epidemiologic evidence, and produced quantitative risk estimates for both ingestion and inhalation routes of exposure (USEPA 1984). Recent studies have shown considerable focus on the association between As and skin cancer, and there is also substantial evidence that exposure to As in drinking water increases the risk of occurrence for several internal cancers (Morales *et al.* 2000, Anawar 2002).

The incidents of cancer from As exposure vary with toxicity and duration, along with innate differences in susceptibility in people. According to the National Research Council (NRC) the sensitivity to As's toxic effects, including carcinogenic effects, varies with each individual and appears to be influenced by such factors as nutrition and genetics (NRC 1999). Numerous epidemiological case studies that have been completed across the globe in As-affected areas, including India and Bangladesh, report that As exposure is associated with diffuse melanosis, spotted melanosis, leucomelanosis, diffuse keratosis, hyperkeratosis, gangrene, skin, kidney, lung, liver, and bladder cancer (Anawar 2002).

66.3   CHEMISTRY OF ARSENIC IN NATURAL WATERS

To understand the case of groundwater As poisoning where it is most acute, on the Bangladesh delta, it is important to understand the chemical and geological processes involved before addressing the problem. A host of possible sources are cited throughout the literature of how As entered the groundwater, but few are plausible. The two central ideas regarding As contamination stem from the simple oxidation of As-rich pyrite which releases the As into the groundwater. Drawdown of the groundwater table allows atmospheric O_2 into the aquifer and so allows the oxidation to take place (Das *et al.* 1995). However, Nickson *et al.* (2000), suggest that the reason for this As release can be attributed not by simple abstraction of water but rather as a result of microbial degradation. The reductive dissolution of As-rich iron oxyhydroxide is manifested by the microbial degradation of sedimentary organic matter and is the reduction process that occurs after microbial oxidation of the organic matter has consumed dissolved O_2 and NO_3^-.

The presence of As in groundwater, however, is not cause for disagreement. The predominant forms of As in water are as inorganic arsenate [As (V)] and arsenite [As (III)]. Under oxidized and aerobic conditions, the arsenate species predominates and exists as oxyanions of arsenic acid ($H_2AsO_4^-$ and $HAsO_4^{2-}$). The arsenite species exists as undissociated arsenious acid ($H_3AsO_3^0$) below pH 9.2 and predominates under moderate reducing conditions in sediments, groundwater, and soils. Arsenate is less toxic to humans than arsenite and is more readily removed than arsenite.

66.4   EXISTING REMOVAL METHODS

Various chemical and physical processes are available to remove As from water. In addition, plants are also used extensively to remove As (phytoremediation) from contaminated soils (Ma *et al.* 2001).

66.4.1   *Precipitation processes*

Coagulation/filtration, enhanced coagulation, coagulation assisted microfiltration, iron/manganese oxidation, and lime softening are the most frequent forms of precipitation removal techniques. Of these methods, coagulation/filtration, iron/manganese oxidation, and lime-softening are used most commonly in water treatment facilities and are examined below.

66.4.2 *Coagulation/filtration*

Coagulation/filtration is the most common form of water treatment and has been used in water systems throughout the world, mostly to remove solids from drinking water. The process involves an alteration to the chemical or physical properties of dissolved colloids or suspended matter within the solution so that an agglomeration occurs and the resulting particles settle out and are removed by filtration.

The disadvantages to this method are, much like many other existing techniques, dependant on the As oxidation state, pH, and presence of competing ions. Arsenite removal has been shown to be less efficient than arsenate removal under comparable conditions (Shen 1973, Gulledge and O'Conner 1973, Sorg and Logsdon 1978, Edwards 1994, Hering *et al.* 1996). Thus a pre-oxidation step is necessary if arsenite is the predominant form present. The effect of pH is also principal in the effectiveness of removal. According to studies, the optimal pH range for As removal by alum and ferric coagulants is between 5 and 8 (Sorg and Logsdon 1978). Sulfates present in water have been shown to significantly reduce arsenite removal and only slightly affect arsenate removal. The presence of calcium actually facilitates removal of arsenate when the pH is above 7 (Hering *et al.* 1996).

66.4.3 *Iron/manganese oxidation*

This coagulation process involving oxidation, typically used to treat groundwater, allows for the formation of hydroxides that remove As by precipitation or adsorption reactions. Initially implemented to remove iron and manganese in groundwater, this method has been shown to be fairly effective in both removing iron and manganese while also efficiently removing As (Edwards 1994). Recent research has focused primarily on greensand filtration with results showing substantial As removal (Subramanian *et al.* 1997). The ability to remove both arsenite and arsenate with this method is by-passed due to the oxidative nature of the manganese. The manganese surface converts arsenite to arsenate and arsenate is then adsorbed onto the surface (Scott and Morgan 1995). However, due to the mass balance between manganese hydroxides and arsenic, reduced manganese, Mn(II), is released from the surface into solution where it will be readily readsorbed onto the remaining manganese hydroxide surface (Moore *et al.* 1990, Scott and Morgan 1995, Subramanian *et al.* 1997).

Disadvantages of the oxidation filtration technologies are primarily dependant on the influent water quality and the presence of competing ions. The concentration of reduced iron [Fe(II)] has shown a strong correlation with As removal, with increased As removal rates when the Fe/As ratio increases (Subramanian *et al.* 1997). Competing ions, in this case sulfates, in the oxidation method only slightly affects As removal (McNeill and Edwards 1995). However, it was also reported that divalent ions, such as calcium, can also compete with As for adsorption sites and thus reduce the removal efficiency.

66.4.3.1 *Lime softening, split lime treatment, lime-soda softening*

Typically used as a water softener in municipal water systems for hard water caused by magnesium and calcium compounds, lime softening, split lime treatment, and lime-soda softening are all effective in reducing As concentrations in water. The introduction of lime into solution raises the pH of the water and subsequently converts bicarbonates into carbonates. As a result, calcium is precipitated as calcium carbonate. Lime-soda softening is required if the solution does not have adequate bicarbonates. Precipitation of calcium occurs in the pH range 9–9.5, whereas magnesium precipitation, in the form of magnesium hydroxide is accomplished at a level higher than pH 10.5.

Arsenic valence, pH, and competing ions are the most common problems with this technique. Studies have shown that arsenate is generally more effectively removed than arsenite (Sorg and Logsdon 1978). Lime softening removal of As is dependent on pH and form of As in water, either arsenite or arsenate; pre-oxidation may be required to increase efficiency if arsenite is most prevalent. The most effective pH range for arsenite removal is approximately 11 and for arsenate approximately 10.5 (Logsdon *et al.* 1974, Sorg and Logsdon 1978). However, levels as high as this requires a reduction in pH for potable use. The presence of phosphates has been shown to affect the removal efficiency and effectively lower the ability to adsorb As. The presence of phosphate significantly decreases arsenate removal at pH less than 12 (McNeill and Edwards 1997).

66.4.4 *Adsorption process*

The most widely used adsorption technique for As removal is the activated alumina (AA) process. AA is considered an adsorption process; however it is more readily defined as an exchange of ions (AWWA 1990). AA requires a physical/chemical process that sorbs ions in the feed water onto the oxidized AA surface. Feed water is continuously passed through the bed to remove contaminants such as F, As, Si, and natural organic matter (NOM). The reaction of contaminated ions exchanging onto the alumina by surface hydroxides is the driving mechanism behind this removal process. Over time the adsorption sites on the AA will become filled and then must be regenerated.

Factors such as pH, As oxidation state and the presence of competing ions all affect the As removal efficiency of AA (USEPA 2000). The zero point charge (ZPC) of AA is at pH 8.2. At a pH below 8.2 the AA will have a net positive charge and have an electrostatic attraction to anions in water, including all forms of As. However, lowering the pH below 8.2 significantly increases As removal and is generally considered optimum at acidic levels, pH of 5.5–6 (Rosenblum and Clifford 1984). Post treatment may be needed to raise the pH to desirable levels. More recently, Mohan and Pittman (2007) provided an excellent and through review of existing As removal methods using different sorbents. These authors also discussed various aspects such as removal efficiency, cost economics, desorption, regeneration, and disposal of spent sorbents.

66.4.5 *Ion exchange process*

Among this process the foremost method is ion exchange (IE) which, much like AA, must have feed water passed through a bed of ion exchange resin beads until the resin is exhausted. Similarly, it is a chemical/physical process by which an ion on the solid phase, typically a synthetic resin that has a specific affinity to the particular contaminant of concern, is exchanged for an ion in the feed water.

Careful consideration regarding pH, resin type, affinity of the resin for the contaminant, disposal requirements, and competing ions must be given to achieve effective and efficient removal of As from feed water. Resins used in this process have a high affinity for As in the arsenate form. However, previous studies have shown that in the presence of high total dissolved solids (TDS) and sulfate levels, competition exists between arsenate and can reduce removal efficiency (AWWA 1990). Competition from background ions for IE sites can greatly affect removal efficiency of IE systems (USEPA 2000). Recent studies have also suggested that when Fe (III) is present, As may form complexes with iron, which are not removed by the IE resins, and therefore As is not removed (Clifford *et al.* 1998).

66.4.6 *Membrane processes*

This removal method involves the use of a membrane, which when water is forced through, allows some constituents to pass while blocking the passage of others. Various membranes exist, classified by pore size, which are separated into four categories: microfiltration (MF), ultrafiltration (UF), nanofiltration (NF), and reverse osmosis (RO). Membranes are also classified according to the pressure ranges, low or high. Each pressure rate removes constituents differently. For example low pressure processes (MF and UF) remove constituents through physical sieving while high pressure processes (NF and RO) remove constituents through chemical diffusion, but typically require more energy output (Aptel and Buckley 1996). The various membrane processes can remove As through filtration, electric repulsion, and adsorption onto As-bearing compounds.

The initial As concentration affects removal rates, along with the oxidation state of the As present, pH, size, shape, and chemical characteristics, particularly the charge and hydrophobicity of both the membrane material and the feed water constituents (USEPA 2000). Effects of pH are also prevalent when dealing with the existing techniques and depending on the conditions, many techniques requires pre and post-treatment for pH (AWWARF 1998).

66.5   A NOVEL ARSENIC FILTRATION METHOD

66.5.1   *Introduction*

Recent studies have shown that ARTI-64[tm] (copper based compound that is widely available) particles can effectively remove As from natural waters across a wide range of water chemistries (Reddy and Viswatej 2005). The ARTI-64[tm] is based on an adsorption process which traps As onto its surface. This adsorption process relies on pH, but unlike many other sorbents, it has a relatively high ZPC, pH of 9.5, (Parks 1964) which allows for natural waters, both surface and groundwater, to have a pH below the ZPC and thus brings a strong positive charge to the material. The uniqueness of ARTI-64[tm] lies in the fact that during numerous tests conducted under a myriad of conditions, results suggest that water pH, competing ions (phosphates, silicates, and sulfates) and As oxidation states have no effect in removing As from natural waters (Roth 2006).

Batch studies completed so far with ARTI-64[tm] provide an understanding of how water pH, oxidation state of As, and concentrations of phosphate, silica, and sulfate do not impact the removal of As from water. However, batch experiments are difficult to apply for field conditions. Thus, the objective of this research was to design a continuous flow column experiment for As removal.

66.6   MATERIALS AND METHODS

The preparation and initial chemical and physical properties of ARTI-64[tm] particles are published elsewhere. Figure 66.1 shows the design of the experimental continuous flow column to remove As from water. We collected a 20 liter natural groundwater sample from Laramie, Wyoming and spiked it with approximately 175 µg/l of both arsenate and arsenite using laboratory reagent grade sodium As salts (Table 66.1).

As shown in Figure 66.1. 10 grams of ARTI-64[tm] were placed in the column and 300 grams of pre-washed sand were placed on top of ARTI-64[tm] particles. The water sample was pumped through the column by the use of a peristaltic pump at a rate of 63 ml/minute. Samples were taken

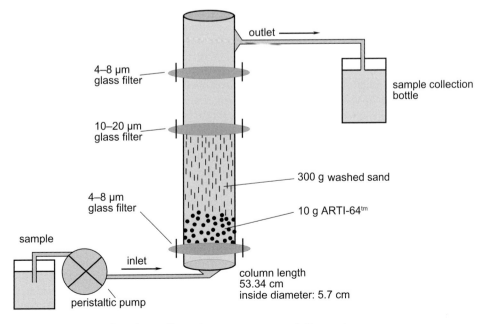

Figure 66.1.   Design of continuous flow column to remove arsenic from water.

Table 66.1.   Initial characteristics of the groundwater sample.

Constituent	Concentration
Na^+	16.8 mg/l
Mg^{2+}	16.8 mg/l
Si	7.2 mg/l
K^+	2.1 mg/l
Ca^{2+}	73.5 mg/l
Cr	7.2 µg/l
Mn	0.3 µg/l
Fe	0.3 mg/l
pH and Eh	7.43 and 65 mV
Cu	45.3 µg/l
As[1]	375.7 µg/l
Se	1.1 µg/l
Pb	1.2 µg/l
F^-	0.1 mg/l
Cl^-	22.1 mg/l
NO_3^-	58.9 mg/l
PO_4^{3-}	<0.1 mg/l
SO_4^{2-}	54.8 mg/l

[1] Spiked with approximately 175 µg/l of both arsenate and arsenite.

directly from the outlet at ten minute intervals for two hours. After two hours samples were then taken at half hour intervals. A final sample was taken from the sample collection bottle. Prior to taking the final sample, the collection bottle was mixed thoroughly. Samples were then analyzed for As, pH, major cations, anions, and trace metals. The pH was measured with the use of a HI 8314 membrane pH meter. Arsenic, major cations and trace metals were analyzed with the inductively coupled plasma-mass spectrometer (ICP-MS). Anions were analyzed by ion chromatography (IC).

66.7   RESULTS

The effect of ARTI-64[tm] treatment on groundwater pH and As concentrations is shown in Figure 66.2a. These results suggest that groundwater pH remained almost the same. Since there is no significant change to the pH, post treatment for pH adjustments is not required. After 120 minutes when 7.5 liters was pumped through the column, total As concentration from samples exceeded 10 µg/l (11.4 µg/l) and 50 µg/l (51.7 µg/l) after 180 minutes when 12 liters were pumped through the apparatus. From this we found that a ratio of 5 grams of ARTI-64[tm] particles to every 4 liters of untreated water produced water with <10 µg/l of As. However, after completion of the study a sample was taken from the total treated water and analyzed for As. The As concentration was 51.5 µg/l (Fig. 66.2a).

The treatment effects on major minerals (cations) are represented in Figure 66.2b. We tested Na, Ca, Mg, and K to show the possible removal of such cations when in contact with ARTI-64[tm] particles. There are no significant effects on the major minerals in water after the treatment. While most other existing techniques also have this same quality, it is important to demonstrate and observe the results.

Figure 66.2c represents the treatment effects on trace metals such as Cr, Mn, Se, and Pb. From the results, only Cr showed any variations. This can be attributed to the extremely low levels within the sample which are close to the ICP-MS detection limit. Since ARTI-64[tm] is a Cu based compound, dissolved Cu concentrations were also monitored in treated water. Dissolved Cu concentrations ranged between 0.23 and 0.12 mg/l and stabilized at 0.11 mg/l. Figure 66.2d shows the treatment

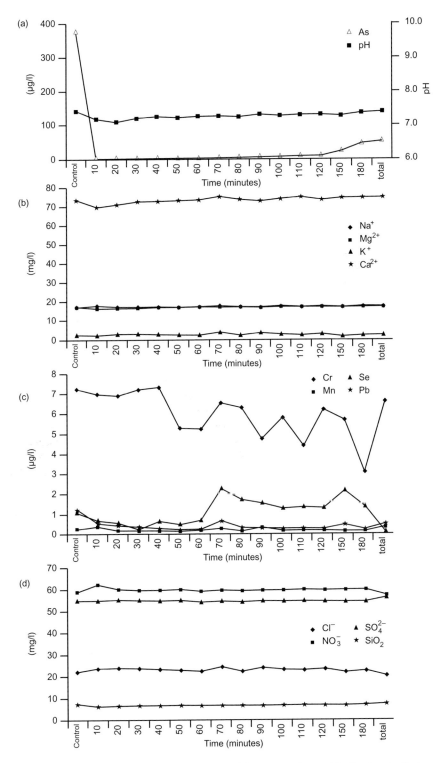

Figure 66.2.   Effect of time on pH and arsenic removal from groundwater (a); Treatment effect on major minerals (cations) (b); on trace metals (c) and on major anions (d).

effect on major anions including NO_3^-, SO_4^{2-}, SiO_2, and Cl^-. Again, there were no changes between the initial concentrations and the final measurements taken. This is encouraging based on the fact that no drastic alterations occur when using ARTI-64[tm] as a removal method to any of the major anions, cations, or minerals.

66.8   DISCUSSION

A fair amount of literature exists on removal of As species using aluminum, iron, manganese, titanium, and ferric phosphate sorbents (Kartinen and Martin 1995, Pierce and Moore 1982, Bajpai and Chaudhuri 1999, Bang *et al.* 2005, Lenoble *et al.* 2005). However, most of the As removal methods require sample acidification and/or a preoxidation step and are influenced by the common competing ions (e.g., phosphate, silica) of water (Jain and Loeppert 2000, Meng *et al.* 2000). Studies have also attributed the difficulty in effective removal of arsenite to $H_3AsO_3^0$ neutral species in natural water (Dambies 2004).

Results of this study suggest that ARTI-64[tm] particles effectively removed As species from groundwater under natural chemical conditions without sample acidification or preoxidation. A possible explanation is that the ZPC of ARTI-64[tm] particles in water occurs at pH 9.5 (Parks 1964) and below this pH the net electrical charge on the ARTI-64[tm] surface will be positive (+). Since the pH of groundwater tested in this study is below 9.5, the ARTI-64[tm] particle surface becomes positive (+) and complexes $HAsO_4^{2-}$. If $H_3AsO_3^0$ exists in water, it is possible that this species may decomplex to $H_2AsO_3^-$ in the presence of ARTI-64[tm] particles, which in turn complexes to the surface of ARTI-64[tm] particles. Effective adsorption of both As oxidation states by ARTI-64[tm] particles strongly suggests inner-sphere complexation and high selectivity of the adsorption process (Stumm and Morgan 1996).

Because no considerable pH alterations are required, this technique provides a reasonable alternative for As removal. The post treatments of many other techniques can be quite difficult because any additive to raise pH may result in alterations to other elements within the water and reduce the water quality. The results from Figure 66.2a suggest that As is indeed removed by the ARTI-64[tm] particles. The removal of As by this technique is noteworthy, in that we spiked this sample with equal amounts of both oxidation states, arsenate and arsenite, suggesting that the ARTI-64[tm] particles are not oxidation-state specific. The treatment of ARTI-64[tm] particles, as shown in Figures 66.2a–d suggest that this As removal technology does not affect pH, nor major anions, trace metals, or minerals of natural waters. These findings suggest that we may have a potential technique to remove As effectively from most natural water sources.

Over time the adsorption sites on ARTI-64[tm] particles become saturated with As. When this occurs, a simple regeneration of the material is required. This process involves rinsing the ARTI-64[tm] particles with sodium hydroxide (NaOH), which raises the pH above its ZPC and in turn releases the As back into solution. Studies have shown that regenerated ARTI-64[tm] was as effective as initial ARTI-64[tm] in removal of As species (Viswatej 2005).

66.9   CONCLUSIONS

Arsenic poisoning due to the consumption of water is of grave concern throughout the world and needs immediate attention from the scientific community to establish a viable technique to alleviate the health risks that are associated with As exposure. This article has attempted to give a brief description along with the limitations of each process, while introducing a novel method that has shown encouraging results from laboratory studies. The results of this study suggest that we may have a viable As removal method that works under natural conditions across a wide range of water chemistries. The traditional problems of As removal (e.g., pH adjustments, oxidation state of As, and competing common anions) do not seem to interfere with the effectiveness of

ARTI-64^{tm}. The continuous flow model presented in this study is the logical progressive step towards field applications. However, further research is needed to determine the adsorption capacity of ARTI-64^{tm} particles and cost economics of this process versus other existing processes.

ACKNOWLEDGEMENTS

We would like to thank Dr. Steve Boese, Dr. Rich Olson, and Viswatej Attili for their contributions to this project. This project and the resulting article would not have been possible without their help and guidance.

REFERENCES

Anawar, H.M., Akai J., Mostofa, K.M.G., Safiullah, S. and Tareq, S.M.: Arsenic poisoning in groundwater: health risk and geochemical sources in Bangladesh. *Environ. Int.* 27 (2002), pp. 597–604.

Aptel, P. and Buckley, C.A.: Categories of membrane operations.In: *Water treatment membrane processes*. American Water Works Research Foundation, Lyonnaise des Eaux, Water Research Commission of South Africa, McGraw Hill, New York, 1996.

AWWA: *Water quality and treatment. A handbook of community water systems*. McGraw-Hill, New York, 1990.

AWWARF: Arsenic treatability options and evaluation of residuals management issues. In: G.L. Amy, M. Edwards, M. Benjamin, K. Carlson, J. Chwirka, P. Brandhuber, L. McNeill and F. Vagliasindi: *Draft Report*, 1998, http://www.awwarf.org/research/topicsandprojects/execSum/153.aspx, accessed June 2006.

Bagla, P. and Kaiser, J.: Epidemiology: India's spreading health crisis draws global arsenic experts. *Science* 274 (1996), pp. 174–175.

Bajpai, S. and Chaudhuri, M.: Removal of arsenic from ground water by manganese dioxide-coated sand. *J. Envir. Eng.* 125 (1999), pp. 782–784.

Bang, S., Patel, M., Lippincott, L. and Meng, X.: Removal of arsenic from groundwater by granular titanium dioxide adsorbent. *Chemosphere* 60 (2005), pp. 389–397.

Bates, M.N., Hopenhayn, C. and Smith, A.H.: Arsenic ingestion and internal cancers: a review. *Am. J. Epidemiol.* 135 (1992), pp. 462–476.

Clifford, D., Ghurye, G. and Tripp, A.: Arsenic removal by ion exchange with and without brine reuse. *AWWA Inorganic Contaminants Workshop*, San Antonio, TX, February 23–24, 1998.

Dambies, L.: Prospective and existing technologies for the removal of arsenic in water. *Sep. Sci. Technol.* 39 (2004), pp. 599–623.

Das, D., Samanta, G., Mandal, B.K., Chowdhury, T.R., Chanda, C.R., Chowdhury, P.P., Basu, G.K. and Chakraborti, D.: Arsenic in groundwater in six districts of West Bengal, India: the biggest arsenic calamity in the world. *Analyst* 120 (1995), pp. 643–650.

Edwards, M.A.: Chemistry of arsenic removal during coagulation and Fe-Mn oxidation. *J. AWWA* 86 (1994), pp. 64–77.

Gulledge, J.H. and'Connor, J.T.O.: Removal of arsenic (V) from water by adsorption on aluminum and ferric hydroxides. *J. AWWA* 8 (1973), pp. 548–552.

Hering, J.G., Chen, P., Wilkie, J.A., Elimelech, M. and Liang, S.: Arsenic removal by ferric chloride. *J. AWWA* 88 (1996), pp. 155–167.

Jain, A. and Loeppert, R.H.: Effect of competing anions on the adsorption of arsenate and arsenite by ferrihydrite. *J. Environ. Qual.* 29 (2000), pp. 1422–1430.

Kartinen, E.O. and Martin, C.J.: An overview of arsenic removal processes, *J. Desalination* 103 (1995), pp. 79–88.

Lenoble, V., Laclautre, C., Deluchat, V., Serpaud, B. and Bollinger, J.C.: Arsenic removal by adsorption on iron (III) phosphate. *J. Hazard. Mater.* 123 (2005), pp. 262–268.

Logsdon, G.S., Sorg, T.J. and Symons, J.M.: Removal of heavy metals by conventional treatment. Proc. 16th Water Quality Conference—Trace Metals in Water Supplies: Occurrence, Significance, and Control. *University Bulletin* 71 (1974), University of Illinois, pp. 111–133.

Ma, L.Q., Komar, K.M., Tu, C., Zhang, W., Cai, Y. and Kennelley, E.D.: A fern that hyperaccumulates arsenic. *Nature* 409 (2001), p.579.

McNeill, L.S. and Edwards, M.A.: Soluble arsenic removal at water treatment plants. *J. AWWA* 87 (1995), pp. 105–113.

McNeill, L.S. and Edwards, M.: Predicting arsenic removal during metal hydroxide precipitation. *J. AWWA* 89 (1997), pp. 75–86.

Meng, X., Bang, S. and Korfiatis, G.P.: Effects of silicate, sulfate, and carbonate on arsenic removal by ferric chloride. *Water Res.* 34 (2000), pp. 1255–1261.

Mohan, D. and Pittman, C.U.: Arsenic removal from water/wastewater using adsorbents—A critical review. *J. Hazard. Mater.* 142 (2007), pp. 1–53.

Moore, J., Walker, J. and Hayes, T.: Reaction scheme for the oxidation of As (III) to As (V) by birnessite. *Clays Clay Miner.* 38 (1990), pp. 549–555.

Morales, K.H., Ryan, L., Kuo, T., Wu, M. and Chen, C.: Risks of internal cancers from arsenic in drinking water. *Environ. Health Perspect.* 108 (2000), pp. 655–661.

Nickson, R., McArthur, J.M., Ravenscroft, P., Burgess, W.G. and Ahmed, K.M.: Mechanism of arsenic release to groundwater, Bangladesh and West Bengal. *Appl. Geochem.* 15 (2000), pp. 403–413.

Nordstrom, D.K.: Worldwide occurrences of arsenic in ground water. *Science* 296 (2002), pp. 2143–2145.

National Research Council: *Arsenic in drinking water*. National Academy Press, Washington, DC, 1999.

Parks, G.A.: The isoelectric points of solid oxides, solid hydroxides, and aqueous hydroxo complex systems. Stanford University Press, Stanford, CA, 1964.

Pierce, M.L. and Moore, C.B.: Adsorption of arsenite and arsenate on amorphous iron hydroxide. *Water Res.* 16 (1982), pp. 1247–1253.

Reddy, K.J. and Viswatej, A.: A novel method to remove arsenate and arsenite from water. *Proceedings of 8th International Conference on Biogeochemistry of Trace Elements, Symposium on Arsenic in the Environment: Biology and Chemistry*, April 3–7, 2005, Adelaide, Australia, 2005.

Rosenblum, E.R. and Clifford, D.A.: The equilibrium arsenic capacity of activated alumina. *PB* 84/10 527, NTIS, Springfield, 1984.

Roth, T.R.: *Groundwater and arsenic: a regional assessment of private wells and a novel point-of-use removal system*. MSc Thesis, Department of Renewable Resources, University of Wyoming, Laramie, WY, 2006.

Scott, M.J. and Morgan, J.: Reactions at oxide surfaces. 1: Oxidation of As (III) by birnessite. *Environ. Sci. Technol.* 29 (1995), pp. 1898–1905.

Shen, Y.S.: Study of arsenic removal from drinking water. *J. AWWA* 8 (1973), pp. 543–548.

Smith, A.H., Lingas, E.O. and Rahman, M.: Contamination of drinking-water by arsenic in Bangladesh: a public health emergency. *Bull. of the World Health Organiz.* 78 (2000), pp. 1093–1103.

Sorg, T.J. and Logsdon, G.S.: Treatment technology to meet the interim primary drinking water regulations for inorganics: Part 2. *J. AWWA* 7 (1978), pp. 379–392.

Stumm, W. and Morgan, J.I.: *Aquatic chemistry*. John Wiley and Sons, New York, 1996.

Subramanian, K.S., Viraraghavan, T., Phommavong, T. and Tanjore, S.: Manganese greensand for removal of arsenic in drinking water. *Water Quality Res. J. Can.* 32 (1997), pp. 551–561.

US Census Bureau: World population information 1950–2050. US Department of Commerce, Washington, DC., http://www.census.gov/ipc/www/world.html (accessed March 8, 2006).

US Geological Survey: The hydrologic cycle. Pamphlet, 1984, National Center, USGS, Reston, VA, ttp://www.usgs.gov/, accessed June 2006.

USEPA: Health assessment document for inorganic arsenic. EPA 600/8-83/021F, US Environmental Protection Agency, Cincinnati, OH, 1984.

USEPA: Technologies and costs for removal of arsenic from drinking water. EPA 815-R-00–028, US Environmental Protection Agency, Office of Water, Cincinnati, OH, 2000.

Viswatej, A.: *A method to remove arsenite and arsenate from water*. MSc Thesis, Department of Renewable Resources, University of Wyoming, Laramie, WY, 2005.

CHAPTER 67

Arsenic removal by solar oxidation in groundwater of Los Pereyra, Tucumán province, Argentina

J. d'Hiriart & M. del V. Hidalgo
Facultad de Ciencias Naturales e Instituto Miguel Lillo, Universidad Nacional de Tucumán,
San Miguel de Tucumán, Prov. de Tucumán, Argentina

M.G. García
CIGeS, Facultad de Ciencias Exactas, Físicas y Naturales (FCEFyN), Universidad
Nacional de Córdoba, Córdoba, Prov. de Córdoba, Argentina

M.I. Litter & M.A. Blesa
Gerencia Química, Comisión Nacional de Energía Atómica, and Escuela de Posgrado, Universidad
de Gral. San Martín, San Martín, Prov. de Buenos Aires, Argentina

ABSTRACT: Shallow groundwater from Los Pereyra, Tucumán, Argentina has arsenic concentrations exceeding the Argentine and international standards for drinking water. Validation tests of the modified SORAS method were performed. Due to the low amounts of iron in natural well waters, iron needs to be added externally. The main objective of this work was to find a low-cost and viable source of iron provision. Packing wire was found to be the optimum iron source. Tests carried out in synthetic waters similar to the natural ones showed good removal efficiency, ranging from 60–90%.

67.1 INTRODUCTION

Shallow groundwater of the east of the Tucumán province (Chacopampean region, northern Argentina) is naturally polluted with arsenic (As). The region is located inside the physiographic unity of the Salí river hydrogeological basin, a structural depression filled with up to 3000 m of Quaternary and Tertiary sediments (Mon and Vergara 1987). The area is mostly occupied by dispersed rural villages, and one of them, Los Pereyra, was chosen as the study site.

Los Pereyra, a small settlement with around 2500 inhabitants, is located on the N° 327 Provincial road, 49 km Southeast of San Miguel de Tucumán City (26°56′51″ S and 64°53′09″ W, 383 m a.s.l.). The village is located in the Northeastern part of the basin, occupied by the Salí river flood plain deposits. The constructions are mainly in masonry, although precarious houses (*ranchos*) still subsist in many sectors. The main use of the land is for farming, wheat and soybean being the most abundant cultivated species. The climate of the region is subtropical with a dry season between May and September. The annual mean precipitation is 800 mm/year and the mean temperature is 19°C. There are no important surface water streams in the region. People do not have access to a water network supply or sewage and natural gas provision. The lack of a drinking water network distribution and the absence of surface water bodies makes it necessary to extract water from deep (more than 100 m) and shallow wells (up to 20 m). In general, water quality is poor, because of the bacteriological content and high levels of nitrates, boron, fluoride, and trace elements like manganese, fluorine, as well as As (García *et al.* 2001, 2003, Nicolli *et al.* 2001, Warren *et al.* 2005). In particular, As occurs in concentrations widely surpassing the 10 μg/l limit established by the Argentine standard requirement for drinking water, CAA (Código Alimentario Argentino 2007), and by WHO. The poor water quality and the poverty and malnutrition conditions cause the incidence of water-borne diseases, including the Chronic Regional Endemic Hydroarsenicism (HACRE).

Various methods have been used to remove As from drinking water, including anion exchange, precipitation, ion flotation, and adsorption (Edwards 1994, Hering *et al.* 1997, Harper *et al.* 1992). However, most of them are not applicable to poor, rural isolated regions, where the As problem is dramatic. Generally, in these rural areas, sunlight is highly available and methods based on the use of solar energy can be employed. SORAS (Wegelin *et al.* 2000) is a simple method of sunlight irradiation of the water that is contained in a colorless transparent polyethylene terephthalate (PET) bottle. The photochemical oxidation of As(III) to As(V) at pH between 6.5 and 8 is promoted under solar irradiation in the presence of citrate, added to the bottle in the form of a few drops of lemon juice. Oxidation is enhanced by the action of reactive oxygen species formed in Fe/citrate containing systems, and the process is completed by As(V) adsorption onto the precipitated iron (hydr)oxides. Equations (67.1–67.6) show the probable mechanisms taking place during the SORAS process:

$$[Fe(III) - Cit^{2+}] + h\nu \rightarrow Fe^{2+} + Cit^{\bullet} \tag{67.1}$$

$$Cit^{\bullet} \rightarrow HO -^{\bullet} CR_2 + CO_2 \tag{67.2}$$

$$HO -^{\bullet} CR_2 + O_2 \rightarrow 3\text{-}OGA + HO_2^{\bullet} \tag{67.3}$$

$$HO_2^{\bullet}(O_2^{\bullet-}) + Fe^{2+} + H^+(2H^+) \rightarrow Fe^{3+} + H_2O_2 \tag{67.4}$$

$$HO_2^{\bullet}(O_2^{\bullet-}) + Fe^{3+} \rightarrow Fe^{2+} + O_2 \tag{67.5}$$

3-OGA is oxoglutaric acid, one of the decomposition products of citric acid. In addition, Fenton reaction can take place:

$$Fe^{2+} + H_2O_2 \rightarrow Fe^{3+} + OH^- + HO^{\bullet} \tag{67.6}$$

Reactive oxygen species (ROS: H_2O_2, $O_2^{\bullet-}$, $HO_2^{\bullet}$, $HO^{\bullet}$) reoxidize Fe^{2+} and oxidize As(III) to As(V). Then, precipitation of iron hydr(oxides), adsorption of As(V) and flocculation take place.

The SORAS method provides an economical technology that can reduce in principle As levels to values below the allowable limits. However, the efficiency of the method is very dependent on the water matrix, and tests must be done to adapt the technology to each region (Cornejo *et al.* 2004).

In previous works (García *et al.* 2003, 2004a, b), the efficiency of the SORAS method was tested in well waters of Los Pereyra. Tests in synthetic waters of controlled ionic composition (CIC), similar to that of natural well waters were performed, looking for a way to improve the method. The optimal conditions, which guaranteed a rather efficient As removal (>90%), were 3 mg/l Fe(III) and 750 μl/l lemon juice, under a solar exposure longer than 3 hours. In real well samples, however, the removal efficiency was much lower (around 30%). It was found that the main factor of failure of the method was the low iron concentration of the real waters, not enough to ensure the removal of As with the precipitated (hydr)oxides. The efficiency was also affected in unpredictable ways by changes in the chemical matrix, or by changes in the operative conditions that could affect the nature of the generated oxides, the oxidation of As(III) and the incorporation of As(V) into the solid in formation. For example, alkalinity (bicarbonate contents) was important for adequate precipitation. Addition of small amounts of citric acid (lemon juice) was beneficial, but larger concentrations were detrimental, probably because of interference in the formation of the solid. The effect of solar irradiation was variable, depending on other experimental factors; generally, the procedure was better under sunlight, because light accelerates oxidation of As(III) to As(V) and affects the nature and sorptive properties of the solid. These results provide evidence that two main factors influence As removal, both linked to the precipitation of iron (hydr)oxides: the chemical water matrix and the source of iron. Therefore, work in this direction was done in the present studies.

Results related to the search of a low-cost source of iron provision and dosification are presented in this work. Different typical materials of the region, like rocks and sediments, as well as packing iron wire and iron nails were tested and compared with synthetic goethite and standard $FeCl_3$ solution.

67.2   EXPERIMENTAL

Cretaceous red sandstone, and Cretaceous and Tertiary red pelites extracted from outcrops in Lules and Potrero de las Tablas counties (about 60 km towards the west of Los Pereyra) were tested as possible sources of iron. The presence of iron-bearing minerals in some of these samples was determined by the X-ray diffractograms (XRD). Illite, kaolinite and quartz were identified in the clay fraction of the sandstone sample, whereas hematite and quartz were identified in the total sample. Other tested iron sources were synthetic goethite (α-FeOOH), non-galvanized packing iron wire, iron nails and Fe(III) in solution ($FeCl_3$, 1 g/l Merck standard).

Tests were performed in CIC waters and in real well waters. CIC waters to test the efficiency of different materials as an iron source were prepared with bidistilled water containing a prefixed amount of As(III), adding $MgSO_4$ (1.3×10^{-4} M), $CaCl_2$ (2.3×10^{-4} M), NH_4Cl (4.0×10^{-5} M), $FeCl_3$ (9.0×10^{-6} M), and NaOH to adjust pH to the average of the well waters.

Natural waters were collected from wells of Los Pereyra. The sample points were geo-referenced using a Magellan Pioneer GPS instrument, and water-table levels were measured with a piezometer.

Initial and final values of the following parameters were recorded in natural and synthetic waters: pH (Metrohm 704 pHmeter, with a temperature probe), conductivity (Metrohm E 583 conductimeter), turbidity (Orbeco-Hellige 966 turbidimeter), temperature and dissolved oxygen (DO, Winkler method), alkalinity, iron and As in solution. Determination of the main ionic composition of shallow and deep well waters was performed according to the standard methods recommended by APHA (APHA 1989).

Water was bidistilled with a Millipore Simplicity system. Sodium arsenite and ferric chloride standards (1 g/l, Merck) were used. All other reagents were at least of analytical grade.

For SORAS experiments, PET transparent bottles ("Villa de los Arroyos" mineral gas water, 1.5 l capacity) were used. Typical experiments were performed as follows: 1.0 or 1.5 l of water was placed in the bottle. Fe(III) was added as $FeCl_3$ or as an amount of the material to be tested as Fe source. Namely, around 5 g of pelites and sandstones, 1 g of synthetic goethite, 1–1.5 g of nails and 4–8 g of packing wire, were used. In the corresponding cases, 750 μl/l lemon juice (calculated considering 0.4 mol/l citric acid in average) were added. The bottle was placed horizontally on the ground for 4 hours, and exposed to direct sunlight. Bottles were not stirred during the tests. Control tests in the dark were done wrapping the PET bottles with aluminum foil; experiments in the absence of lemon juice were also performed in some cases. After exposure, samples were taken, filtered through a 0.45 μm cellulose acetate membrane, acidified with concentrated HNO_3 and analyzed for As and Fe. Initial values of pH, conductivity, turbidity, bicarbonate content, temperature and DO were taken after the addition of iron and lemon juice (in the corresponding cases). It is important to mention that freshly precipitated iron oxides can have sizes smaller than 0.45 μm and thus part of the iron measured in the filtered sample can be not in solution but in colloidal form.

A Hitachi Z-5000 atomic absorption spectrometer was used to measure Fe and As concentrations. Total As content was determined by AAS with graphite furnace. A 500 mg/l Ni $(NO_3)_2$ matrix was added to the samples before the measurement, following the reported methodology (Creed *et al.* 1994). For the determination of total Fe, AAS with flame detection was used.

Average direct solar radiation was measured at 312 nm with a Series 9811 Cole-Parmer radiometer. Values in one day of November 2004, taken in the city of San Miguel de Tucumán between 10:00 am and 16:30 pm, ranged 0.1–0.65 mW/cm^2, with an average of 0.19 mW/cm^2. XRD patterns were obtained with a Philips PW 3710 BASED diffractometer.

67.3   RESULTS AND DISCUSSION

67.3.1   *Composition of well waters*

In Table 67.1, average, maximum and minimum values of the major ionic composition and quality variables of the sampled deep and shallow waters of Los Pereyra are presented. The chemical composition of shallow and deep groundwater differs appreciably in most variables, including As, and the concentrations are much higher in the shallow aquifer.

The main components of shallow groundwater are sodium and bicarbonate, but other major ions such as sulfate, chloride and calcium can also be important. Waters are mainly oxic. Electrical conductivity is high, with values ranging from 429 to more than 3000 μS/cm. Salinity and total solids are also very variable and pH varies between 6.8 and 8.6, with an average value of 7.8. Water quality is poor because of an important bacteriological content and high levels of nitrate, boron, fluoride, as well as As (Navntoft *et al.* 2007). In some wells, the concentration of nitrate exceeds the 45 mg/l limit established by CAA, with an average of 100 mg/l for the area. Arsenic concentrations surpass the values established by CAA and WHO, fluctuating between 70 and 1000 μg/l. Concentrations higher than 400 μg/l are restricted to the first 20 m of depth.

67.3.2   *Results of SORAS tests in CIC water: Influence of the initial As concentration*

SORAS tests at the best conditions found before (750 μl lemon and 3 mg/l Fe, García *et al.* 2003, 2004a, b) were performed to assess the influence of the initial As concentration, a factor not evaluated before. CIC water was prepared simulating the average of the shallow waters of Los Pereyra. Results are shown in Table 67.2. In the optimized conditions, As removal was more efficient at a lower initial As concentration. A kinetic study should be performed to assess the effect of initial As concentration on the reaction rate. Work is underway.

67.3.3   *Results of SORAS tests with different materials as iron source and provision in synthetic waters*

SORAS tests in CIC water containing different initial amounts of As were designed in order to investigate the removal efficiency of different materials as a source of iron. The content of iron was not determined in the materials because only exploratory tests were carried out to define the suitability of each one. The amount of Fe used in each case was chosen only in terms of

Table 67.1.   Maximum, minimum and average values of quality variables of deep and shallow waters of Los Pereyra, Tucumán, Argentina.

	Shallow aquifer			Deep aquifer		
	Maximum	Minimum	Average	Maximum	Minimum	Average
pH	8.63	6.76	7.78	7.78	6.64	7.12
DO (mg/l)	8.1	0.2	4.6	12.1	0.2	4.6
TDS (mg/l)	5990	375	2190	1400	562	717
HCO_3^- (mg/l)	1350	205	707	578	92	208
Cl^- (mg/l)	1860	6	343	347	102	185
NO_3^- (mg/l)	1020	1.4	100	30	0	7.6
SO_4^{2-} (mg/l)	2830	12	550	255	90	160
Na^+ (mg/l)	1770	120	680	490	155	205
K^+ (mg/l)	140	3.7	32	30	5	9.4
Ca^{2+} (mg/l)	370	2	65	62	18.3	49
Mg^{2+} (mg/l)	115	1	24	16	2.3	8.5
As (μg/l)	1022	20	279	70	0.6	13.8

Table 67.2.   Results of SORAS tests in CIC waters: Influence of the initial As concentration. 1.5 l CIC water, 750 μl/l lemon juice and 3 mg/l standard Fe(III). Sunlight irradiation time: 4:05 h.

	As μg/l	pH	Conductivity μS/cm	Turbidity NTU	T °C	Fe mg/l	% As removed
Initial	504	9.41	225	0.68	26.5	3.54	
Final	293	8.73	280	0.70	50	3.05	42
Initial	642	8.01	225	0.79	26.6	3.34	
Final	488	6.79	300	1.75	50.5	2.44	24
Initial	936	8.15	230	0.74	26.3	2.95	
Final	847	6.88	310	0.77	50	3.37	10

estimated costs and trying to avoid sludge formation in the final system. The optimal conditions of lemon juice and irradiation time were used. Table 67.3 shows the results of some selected tests.

In general, the removal efficiency was better with sunlight and in the presence of lemon, as found in previous works (García *et al.* 2004a, 2004b). In the case of the addition of standard FeCl$_3$, the low efficiency can be attributed to the low amount of iron present in solution (0.3 mg/l), used to simulate the concentration naturally found in the waters. This concentration is well below the optimal concentration needed in the modified SORAS technology for well waters of Los Pereyra (García *et al.* 2004a, b). A final yellow color was observed in this case. Removal in the presence of pelites was reasonably good; however, the treated water changed its color after the test and it was not possible to recover the translucent and colorless condition by traditional filtration. As can be observed, sandstone was not a good material. Although synthetic goethite gave excellent removal results, the precipitation of suspended material took some days. In addition, the high turbidity and color degraded the quality of the final water for human consumption; final Fe in solution largely exceeded the limits for water consumption (0.3 mg/l, CAA). In contrast, although As removal with packing wire was not as high as in other cases, the final water did not present turbidity or high concentrations of Fe in solution. However, in the filtration treatment previous to As and Fe determination, the presence of particulate material was observed in the membranes. In these tests, an increase of 2–5% of the final mass of the wire was verified, due to the formation of iron oxides on the surface, as indicated by XRD analysis of the obtained particles (not shown, see later).

Additional tests with iron nails were also performed, but the percentage of removal was less than 5%, indicating the unsuitability of the material as iron source for the SORAS technology.

The behavior of the different materials is very difficult to explain in this simple work and a thorough study would require a long analysis and a complete description of each material, which is beyond the scope of this chapter.

67.3.4   *Results of SORAS tests with different materials as iron source and provision in real well waters*

Although sandstones were not suitable for As removal in CIC waters, tests in real conditions with 750 μl/l lemon juice were carried out to see if some improving effect could be found. Figure 67.1 shows results of tests in well waters of Los Pereyra in field conditions, comparing synthetic goethite and sandstone.

Results indicate again that sandstone is not a suitable material, and that goethite under sunlight and with citric acid in low amounts gives a very good removal yield; however, an orange coloration was observed, together with very fine suspended material; good filtration and discoloration to improve the water quality were not possible. In contrast with tests in CIC water, final Fe in solution (average < 0.2 mg/l) did not exceed the allowable limits. The better removal in waters of the Jerez

Table 67.3.   SORAS tests with different materials as Fe source in CIC water. 750 µl/l lemon juice were used when indicated. Sunlight irradiation time: 4 h.

Description		pH	Conductivity µS/cm	Turbidity NTU	HCO_3^- meq/l	T °C	DO mg/l	Fe mg/l	As µg/l	% As removed
1.0 l CIC +	Initial	8.74	75/80	0.64	2.49	23.5	7.41	0.375	969	
0.3 mg/l st.	Final	3.41	330/309	0.82	0	36	2.5	0.556	909	6
Fe + lemon,										
no hν										
1.0 l CIC +	Initial	8.74	75/80	0.64	2.49	23.5	7.41	0.375	969	
0.3 mg/l st.	Final	4.35	147	1.38	0.19	41	1.0	0.629	932	4
Fe + lemon + hν										
1.5 l CIC +	Initial	7.5	245	236	1.04	29	3.32	3.53	319	
5.2 g pelites[1] + hν	Final	6.5	250	423	1.18	51.5	nd	0.13	173	45.8
1.5 l CIC +	Initial	6.5	210	253	1.04	29	3.6	3.54	248	
5.5 g pelites[1] +	Final	6.5	210	419	1.23	52	nd	3.53	24	90.3
lemon + hν										
1.5 l CIC +	Initial	8.5	93	26.1	0.57	28.5	8.4	0.91	352	
5.0 g pelites[2] + hν	Final	9.74	130.5	81.5	0.62	57	nd	1.76	353	0
1.5 l CIC +	Initial	9.0	43	10.4	0.24	25.5	6.28	0.24	275	
5.0 g pelites[2] +	Final	5.1	75	41.6	0.43	55.0	nd	0.19	80	71.0
lemon + hν										
1.0 l CIC +	Initial	8.74	75/80	0.64	2.49	23.5	7.41	0.375	969	
5 g red	Final	4.65	114	60.7	0.14	36.5	3.0	0.295	913	6
sandstone +										
lemon, no hν										
1.0 l CIC +	Initial	8.74	75/80	0.64	2.49	23.5	7.41	0.375	969	
5 g red	Final	4.35	135	35.3	0.097	42	1.9	0.368	898	7
sandstone +										
lemon + hν										
1.0 l CIC +	Initial	8.74	75/80	0.64	2.49	23.5	7.41	0.375	969	
1 g goethite +	Final	3.30	360	50.1	0	36.5	2.8	1.219	300	69
lemon, no hν										
1.0 l CIC +	Initial	8.74	75/80	0.64	2.49	23.5	7.41	0.375	969	
1 g goethite +	Final	3.69	240	59.1	0	44	0.6	5.260	90	91
lemon + hν										
1.5 l + 1 g	Initial	nd[3]	39	64.6	0.19	28.5	3.28	0.13	347	
goethite + hν	Final	3.83	66	5.81	0.38	59	nd	0.15	64	81.6
1.5 l + 1 g	Initial	9	120	123.6	0.89	28.5	3.37	0.49	16	
goethite +	Final	8.99	210	41.8	0.62	59.5	nd	0.04	31	–
lemon + hν										
1.5 l CIC +	Initial	8.5	280	1.7	1.14	29	3.42	0.08	283	
2.4 g packing	Final	7	280	2.11	1.37	47.5	nd	0.10	196	31
wire + hν										
1.5 l CIC +	Initial	8.5	270	1.62	1.18	29	3.23	0.09	285	
2.4 g packing	Final	6.5	320	1.93	1.33	48	nd	0.18	198	30.5
wire + lemon + hν										

[1] Potrero Las Tablas; [2] Lules; [3] nd: not determined.

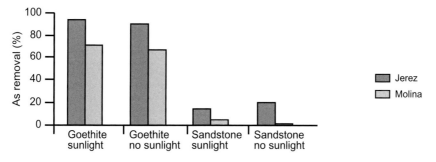

Figure 67.1. Results of SORAS tests in Los Pereyra well waters (Jerez and Molina families) in field conditions. Sunlight irradiation time: 4 hours.

Table 67.4. Results of SORAS tests with 1.0 l CIC water and 4 g packing wire in the absence of lemon juice. Sunlight irradiation time: 4 h.

Description		pH	Conductivity µS/cm	Turbidity NTU	T °C	DO mg/l	Fe mg/l	As µg/l	% As removal
CIC water +	Initial	6.73	16	1.26	15.8	3.8	<0.05	537	
packing wire + hν	Final	6.11	31	2.33	26.9	2.78	1.75	402	25
CIC water +	Initial	6.87	15	1.63	15.8	3.8	<0.05	537	
packing wire + hν	Final	6.68	28	1.98	26.8	2.69	1.70	346	36
CIC water +	Initial	6.57	17	2.53	15.7	3.8	<0.05	537	
packing wire + hν	Final	6.28	29	2.28	26.9	2.87	1.65	273	49
CIC water +	Initial	6.50	21	0.99	11.8	8.30	<0.05	531	
packing wire, no hν	Final	6.03	24	2.06	13.8	6.40	3.4	333	37
CIC water +	Initial	6.74	19	0.62	11.7	8.30	<0.05	531	
packing wire, no hν	Final	6.41	19	2.41	13.7	6.96	1.68	450	15

family than in waters of the Molina family can be attributed to a lower amount of As in the first one (96 *vs.* 709 µg/l), in accordance with results of Section 67.3.2.

67.3.5 *Results of SORAS tests with packing wire in CIC waters in different initial conditions*

Taking into account that packing wire was the best typical material of the region to be used in the modified SORAS method in Los Pereyra, yielding waters of rather good quality, the influence of amount of wire and initial pH on the removal was studied with this material in CIC waters. Tests were performed in the absence of lemon juice because results of Table 67.3 indicated no effect in the removal efficiency. Table 67.4 shows simultaneous experiments with 4 g packing wire. After 4 h sunlight exposure, all samples were colorless.

Although special care was taken in order to reproduce the initial conditions, the results could not be replicated. Control tests carried out without solar exposure did not replicate the original results either. In the samples exposed to solar irradiation, a temperature increment of 11°C was observed. Efficiency in As removal was random (15–49%), and there was no remarkable difference between experiments with and without sunlight irradiation. No important changes in the turbidity of the water were observed.

Results with 6 and 8 g of packing wire were done in similar conditions of sunlight irradiance (not shown). Again, all samples were colorless after sunlight exposure, but membranes used to filter samples before As and Fe determination presented a yellow tint. A better removal (ranging 72–87%) and reproducibility were obtained. The temperature increase in both sets of tests was around 14°C.

Figure 67.2 shows the observed trend in maximum removal under sunlight with different amounts of wire (4, 6 and 8 g) at the typical pH of the waters of the study zone (6–7.2). The maximum efficiency corresponds to a wire mass of 6 g. Consumption of dissolved oxygen and proton release were also measured, both parameters increasing with the wire mass.

In Tables 67.5 and 67.6, results of simultaneous tests performed at pH around 5, 8 and 9, with 4 and 6 g of packing wire, are presented. Experiments were performed in duplicate, with good reproducibility. Only one of them is presented in the tables. In tests at pH 9, a yellow color was observed at the end of the solar exposure.

In Figure 67.3, results of As removal with packing wire at various pH are shown. As can be observed, the highest As removal occurred at pH 5 with 6 g of wire. However, efficiencies at the average pH (7–8) of the waters are also reasonable, ranging 60–80%.

XRD patterns of the suspended particulate material found at the end of the removal tests with packing wire indicated goethite and hematite as main peaks. The better efficiency of the wire compared with the rest of materials can be due to the fact that it is composed of almost pure iron, and that in contrast to goethite, which forms a powdered suspension deposited in the bottom of the bottle, it provides a neat surface to be exposed to sunlight. As seen by XRD, this surface is altered and transformed to iron hydr(oxides) during the treatment, a process beneficial for inclusion of As(V) in the growing solid (equations (67.7)–(67.10)). In fact, at the end of the experiments, a change in the aspect of the wire surface was observed, presenting higher roughness and apparent higher porosity. The increase of the final mass of the wire as well as the oxygen consumption and the proton release during the reaction are also evidences of these changes, although these two last variations can be attributed also to other chemical reactions taking place during this complicated modified SORAS process.

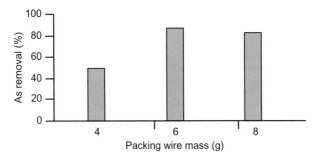

Figure 67.2.   Maximum percentages of As removal under sunlight for 4, 6 and 8 g packing wire. [As]$_0$ range $\approx$ 550–700 μg/l. Sunlight irradiation time: 4 h.

Table 67.5.   Results of SORAS tests at pH 5, 8 and 9 with 4 g of packing wire in CIC water, in the absence of lemon. Sunlight irradiation time: 4 h.

Description		pH	Conductivity μS/cm (25°C)	Turbidity NTU	T °C	DO mg/l	Fe mg/l	As μg/l	% As removal
CIC water +	Initial	5.29	39	0.9	16.4	7.3	0.572	733	
4.34 g packing wire	Final	5.11	45	1.41	28.5	4.64	2.51	159	78
CIC water +	Initial	8.02	47	0.68	16.4	7.3	0.572	733	
4.47 g packing wire	Final	7.59	67	1.14	29.5	4.18	2.49	89	88
CIC water +	Initial	8.93	193	1.14	16.4	7.3	0.572	733	
4.59 g packing wire	Final	7.25	416	3.03	29.3	5.29	1.38	46	94

Table 67.6. Results of SORAS tests at pH 5, 8 and 9 with 6 g of packing wire in CIC water, in the absence of lemon. Sunlight irradiation time: 4 h.

Description		pH	Conductivity μS/cm (25°C)	Turbidity NTU	T °C	DO mg/l	Fe mg/l	As μg/l	% As removal
CIC water +	Initial	5.25	4	0.73	16.5	7.79	0.05	551	
6.46 g packing wire	Final	5.95	48	1.94	28.3	4.92	6.39	76	86
CIC water +	Initial	7.88	33	0.85	16.6	7.79	0.05	551	
6.44 g packing wire	Final	6.54	55	1.30	28.3	4.45	1.42	235	57
CIC water +	Initial	8.68	48	0.92	16.6	7.79	0.05	551	
6.56 g packing wire	Final	7.83	73	2.31	28.5	5.23	1.45	256	54
CIC water, no	Initial	6.95	12	0.52	16.6	7.79	0.05	551	
iron	Final	6.20	21	0.55	28.3	5.48	0.37	534	3

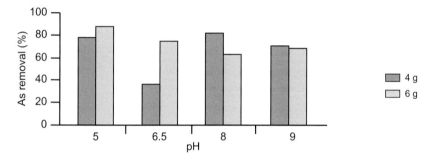

Figure 67.3. Results of simultaneous tests for As removal at different pH, with 4 and 6 g packing wire.

$$Fe(0) + \tfrac{1}{2}O_2 + H_2O \rightarrow Fe(II) + 2HO^- \tag{67.7}$$

$$Fe(II) + \tfrac{1}{4}O_2 + 1.5H_2O \rightarrow Fe(III) + HO^- + ROS, Fe(IV) \tag{67.8}$$

$$Fe(III) + 3H_2O \rightarrow Fe(OH)_3 + 3H^+ \tag{67.9}$$

$$As(III) + intermediates \rightarrow As(IV) \rightarrow As(V) \tag{67.10}$$

Tests with wire under more prolonged irradiation and in real well waters are underway.

67.4 CONCLUSIONS

Different sources of iron and typical low-cost materials were tested for use in the modified SORAS method in Los Pereyra well waters: sandstone, pelites, nails and packing wire. The materials were compared with Fe(III) in solution and synthetic goethite. From all tested materials, packing wire was the best source of iron because a rather good As removal was achieved together with a good water quality (no color or turbidity observed in the final water). Packing wire is a commonly used material in agricultural activities in the region; it is cheap and can be easily handled by the population. No lemon juice must be added and the method can work at the pH of the studied well waters. This work indicates that the SORAS method, conveniently modified according to the composition of waters used can be a suitable procedure to be applied in rural isolated localities where no water network is available.

ACKNOWLEDGEMENTS

Work performed as part of *Comisión Nacional de Energía Atómica* P5-PID-36-4 Program, OEA/AE/141Organization of the American States Project. Part of 26/G211 CIUNT Project, Tucumán National University. To Eng. E.A. Quaia for As determinations. To CYTED for financing 406RT0282 IBEROARSEN Network.

REFERENCES

American Public Health Association, American Water World Association, Water Pollution Control Federation (APHA): *Standard methods for the examination of water and wastewater.* 17th Edition. Baltimore, Maryland, 1989.

Código Alimentario Argentino, modification of articles 982 and 983, May 22, 2007, Buenos Aires, Argentina, 2007, http://www.anmat.gov.ar/normativa/normativa/Alimentos/Resolucion_Conj_68-2007_96-2007.pdf

Cornejo, L., Mansilla, H.D., Arenas, M.J., Flores, M., Flores, V., Figueroa, L. and Yáñez, J.: Removal of arsenic from waters of the Camarones river, Arica, Chile, using the modified SORAS technology. In: M.I. Litter and A. Jiménez González (eds): *Advances in low-cost technologies for disinfection, decontamination and arsenic removal in waters from rural communities of Latin America (HP and SORAS methods).* AOS Project AE/141, Digital Grafic, La Plata, Argentina, 2004, pp. 85–97.

Creed, J.T., Martin, T.D. and O'Dell, J.W.: Determination of trace elements by stabilized temperature graphite furnace atomic absorption. Environmental Monitoring Systems Laboratory Office of Reasearch and Development, USEPA, Cincinnati, Ohio, 1994.

Edwards, M.: Chemistry of arsenic removal during coagulation and Fe-Mn oxidation. *J. Am. Water Works Assoc.* 86 (9): 64–78.

García, M.G., Hidalgo, M. del V. and Blesa, M.A.: Geochemistry of groundwater in the alluvial plain of Tucumán, Argentina. *Hydrogeology J.* 9:6 (2001), pp. 597–610.

García, M.G., Hidalgo, M. del V., Litter, M.I. and Blesa, M.A.: Arsenic removal by RAOS in Los Pereyra, Province of Tucumán, Argentina. In: M.I. Litter and H.D. Mansilla (eds): *Solar light assisted arsenic removal in rural communities of Latin America.* AOS Project AE/141, Digital Grafic, La Plata, Argentina, 2003, pp. 9–33.

García, M.G., Lin, H.J., Custo, G., d'Hiriart, J., Hidalgo, M. del V., Litter, M.I. and Blesa, M.A.: Advances in solar oxidation removal of arsenic in waters of Tucumán, Argentina. In: M.I. Litter and A. Jiménez González (eds): *Advances in low-cost technologies for disinfection, decontamination and arsenic removal in waters from rural communities of Latin America (HP and SORAS methods).* AOS Project AE/141, Digital Grafic, La Plata, Argentina, 2004a, pp. 43–63.

García, M.G., d'Hiriart, J., Giullitti, J., Lin, H., Custo, G., Hidalgo, M. del V., Litter, M.I. and Blesa, M.A.: Solar light induced removal of arsenic from contaminated groundwater: the interplay of solar energy and chemical variables. *Solar Energy* 77:5 (2004b), pp. 601–613.

Harper, T.R. and Kingham, N.W.: Removal of arsenic from wastewater using chemical precipitation methods. *Water Environ. Res.* 64 (1992), pp. 200–203.

Hering, J.G., Chen, P.-Y. Wilkie, J.A. and Elimelech, M.: Arsenic removal from drinking water during coagulation. *J. Environ. Eng.* 8 (1997), pp. 800–807.

Mon, R. and Vergara, G.: The geothermal area of the eastern border of the Andes of north Argentina at Tucumán Province. *Bull. Int. Assoc. Eng. Geol.* 35 (1987), pp. 87–92.

Navntoft, C., Araujo, P., Litter, M., Apella, M.C., Fernández, D., Puchulu, M.E., Hidalgo, M. del V. and Blesa, M.A.: Field tests of the solar water detoxification *Solwater* reactor in Los Pereyra, Tucumán, Argentina. *J. Solar Energy Eng.* 129 (2007), pp. 127–134.

Nicolli, H.B., Tineo, A., García, J.W., Falcón, C.M. and Merino, M.H.: Trace-element quality problems in groundwater from Tucumán, Argentina. In: R. Cidu (ed): *Water-Rock Interaction.* Vol 2. Balkema/Swets and Zeitlinger, Leiden, The Netherlands, 2001, pp. 993–996.

Warren, C.J., Burgess, W. and García, M.G.: Hydrochemical associations and depth profiles of arsenic and fluoride in Quaternary loess aquifers of northern Argentina. *Mineral. Mag.* 69:5 (2005), pp. 877–886.

CHAPTER 68

Removal of arsenic from groundwater using environmentally reactive iron nanoparticles

S.R. Kanel
Department of Civil Engineering, Auburn University, Auburn University, Auburn, CA, USA

H. Choi
Department of Environmental Science and Engineering, Gwangju Institute of Science and Technology (GIST), Gwangju, Korea

ABSTRACT: Recognized as one of the highly toxic elements, arsenic (As) is abundant in our environment both by natural and anthropogenic sources. It is carcinogenic and causes different diseases such as skin diseases, hyperpigmentation, melanoma, circulatory disorders and neurological damage. However, there is no definitive cure for the diseases; hence As remediation is the only one option to save the lives of millions of people. Among different As removal methods, the adsorption/precipitation method has received more attention due to its high efficiency and cost-effectiveness, and is considered as the most suitable technology for As removal. Therefore, it is very important to find new, effective and sustainable adsorbents to remove As from groundwater. Hence, this chapter discusses the opportunity of novel adsorbents such as nanoscale zerovalent iron (NZVI) and surface-modified NZVI (or surface stabilized iron nanoparticles (S-INP)) and their application for *in situ* groundwater remediation.

68.1 INTRODUCTION

Recognized as one of the highly toxic elements (Ferguson and Gavis 1972), As is abundant in our environment both by natural and anthropogenic sources (Mandal and Suzuki 2002). Naturally, it is introduced to aquatic systems by weathering of As-containing minerals and enters the atmosphere through volcanic emissions, wind erosion, sea spray, forest fires and biological formation of volatile arsenicals (Cullen and Reimer 1989). In case of anthropogenic sources, mining activities, ore smelting, sulfuric acid manufacturing process and pesticides are the most predominant (Bhumbla and Keefer 1994). Since it is extremely toxic to humans, long-term exposure causes several health disorders such as cancer, hyperpigmentation, circulatory disorders and neurological damage at aqueous concentrations as low as 0.1 mg/l (Tseng *et al.* 1968, Smith 1992, Pichler *et al.* 1999, Balaji *et al.* 2000, Chunming and Robert 2001). Hence, World Health Organization (WHO) has set the maximum guideline concentration of As in drinking water (WHO 1993) at 0.01 mg/l.

Arsenic exists in groundwater predominantly as inorganic arsenite, As(III), ($H_3AsO_3, H_2AsO_3^-$, $HAsO_3^{2-}$) and arsenate, As(V) ($H_3AsO_4, H_2AsO_4^-, HAsO_4^{2-}$) (Ferguson and Gavis 1972, Manning *et al.* 2002). Greater attention is required for the removal of As(III) from groundwater due to its higher toxicity (Cullen and Reimer 1989) and mobility (Manning and Martens 1997, Dixit and Hering 2003), which mainly arises from its neutral state ($HAsO_3^0$) in groundwater as compared to the charged As(V) species ($H_2AsO_4^-, HAsO_4^{2-}$), which predominate near pH 6–9 (Gulens *et al.* 1973, Clifford and Zang 1994). This also correlates with the less efficient removal of As(III) by conventional water treatment processes (Chiu and Hering 2000). Attention has recently focused on zerovalent iron (ZVI) that has become one of the most common adsorbents for rapid removal of As(III) and As(V) in the subsurface environment (Lackovic *et al.* 2000,

Bang *et al.* 2005, Su and Puls 2001, 2001a, Farrell *et al.* 2001, Dixit and Hering 2003, Olivier *et al.* 2005). However, due to its large size, lower surface area and lack of mobility, use of ZVI was limited only for shallow groundwater treatment. To overcome these problems and to take best advantage of good redox properties and sorption capacity of iron, nanoscale zerovalent iron (NZVI) has been introduced. Due to high surface area and reactivity (Schrick *et al.* 2002), it has already shown great potential for groundwater treatment such as TCE, PCE (Schrick *et al.* 2002, Zhang 2003, Nurmi *et al.* 2005), nitrate (Choe *et al.* 2000), Cr(VI) (Schrick *et al.* 2000, Manning *et al.* 2007), natural organic matter (NOM) (Gaisuddin *et al.* 2006) and PCBs (Lowry and Johnson 2004). Recently, Kanel *et al.* (2005, 2006) reported removal of As [As(III) and As(V)] by NZVI and demonstrated its application for As remediation of groundwaters from developing countries. A number of excellent reviews have summarized many different aspects of As. Here in we focus on recent development of As remediation based on nanotechnology.

68.2   REMEDIATION STRATEGY OF ARSENIC FROM GROUNDWATER

Many different methods including precipitation-coagulation, co-precipitation, ion exchange, electro-coagulation, oxidation and adsorption are being used for As remediation (EPA 1988, Nriagu 1994, Jiang 2001, Manning *et al.* 2002, Bissen and Frimmel 2003a, b). Table 68.1 summarizes the current approach for groundwater remediation and Table 68.2 summarizes the advantages and disadvantages of As remediation techniques.

Among different techniques used, the adsorption method has received more attention due to its high efficiency and cost-effectiveness, and is also considered as the most suitable technology in developing countries (Jiang 2001). Adsorption is the accumulation of solute on the surface

Table 68.1.   Arsenic removal technologies.

Conventional technologies	Innovative-non conventional technologies
Oxidation	Nanoscale adsorbents
Coagulation-precipitation	Iron and iron oxide coated adsorbents
Membrane separation	
Adsorption (blast furnace slag)	
Adsorption (zerovalent iron)	
As-waste stabilization	

Table 68.2.   Comparison of main As removal technologies.

Technology	Advantage	Disadvantage
Oxidation		
a) Air oxidation	Simple low cost	Slow, low efficiency
b) Chemical oxidation	Relatively simple and rapid, oxidizes other impurities also	Kills microorganisms, costly
Coagulation-precipitation		
a) Alum coagulation	Relatively low cost	Produces toxic sludge, low efficiency in As(III)
b) Iron coagulation	Simple, chemical commonly available	Produces toxic sludges, low efficiency in As(III)
c) Lime softening	High efficiency	pH dependent, produces sludges

of adsorbent solids (at which substance is collected). Absorption is the penetration of collected substance into the solid. Adsorbtion and absorption occur simultaneously and the phenomenon is called sorption, and the unit operation is usually referred as adsorption. Physical adsorption occurs due to van der Waals forces and is reversible in nature, whereas chemical adsorption occurs due to chemical reaction and is irreversible in nature. The adsorbents used for As removal include activated alumina, activated carbon, iron oxides and clay.

There are different groundwater remediation techniques that use adsorption and precipitation mechanisms to remove As from groundwater in *in situ* condition, which are described as follows.

68.2.1 *Pump and treat*

This is a conventional method in which groundwater is pumped to the surface, then treated and refilled in groundwater. This expensive method requires surface structures, and can not be used in densely populated residential urban areas, where groundwater is polluted.

68.2.2 *Permeable Reactive Barriers (PRB)*

PRBs constructed of elemental iron have emerged as an effective passive remediation method for groundwater contaminated with a variety of contaminants such as As, chromium, and chlorinated hydrocarbons (CHCs) (Gillahm and O'Hannesin 1994, Agrawal and Tratnyek 1996, Nam and Tratnyek 2000, Hozalski *et al.* 2001, Alowitz and Scherer 2002). A PRB is an *in situ* engineered zone of reactive material placed across the path of contaminated groundwater. The major advantages of PRBs over other conventional groundwater remediation approaches are the lack of structures on the surface, the low operation and maintenance cost and the high remediation efficiency, particularly compared with pump-and-treat systems. Often, iron PRBs are installed within existing contaminant plumes and therefore elevated concentrations of contaminants are observed down gradient of PRBs for some time after the system has been installed, depending on the extent of initial contamination, groundwater flow rates, desorption rates and type of the aquifer material (VanStone *et al.* 2005). PRB can be used for *in situ* treatment of contaminants for shallow groundwater treatment up to 30 m depth. Micron adsorbent can be used for this purpose. The disadvantage of this technique is that it requires excavation and is not useful to treat groundwater deeper than 30 m.

68.2.3 *Colloidal Reactive Barrier (CRB)*

Efficient injection and dispersion of colloidal or particulate amendments represent a formidable technical hurdle when considering an *in situ* groundwater remediation technology. NZVI can play a main role as CRB for *in situ* groundwater remediation.

68.2.4 *Mobile Reactive Barrier (MRB)*

Reactive barrier itself can move and remove groundwater pollutants by surface modified iron nanoparticles INP (S-INP). S-INP can be injected in a subsurface environment; some of it can act as CRB and some like MRB. The iron nanoparticles can theoretically reach areas of contamination that are inaccessible to many conventional methods (e.g., beneath buildings and airport runways) (Elliott and Zhang 2001). In this novel concept the reactive barrier itself can move, hence it is termed as MRB. The prerequisite of this technology is the development of mobile adsorbents which can act as a reactive barrier. In this aspect, S-INP can play an important role as MRB since S-INP is a highly mobile material in subsurface environments.

68.3   FUNDAMENTALS OF IRON NANOPARTICLE (INP)

68.3.1   *Synthesis*

68.3.1.1   *Chemical precipitation method*
The NZVI material can be synthesized by drop wise addition of 1.6 M $NaBH_4$ aqueous solution to a Ne gas-purged 1 M $FeCl_3 \cdot 6H_2O$ aqueous solution at ~23°C with magnetic stirring as described by Wang and Zhang (Wang and Zhang 1997a). Ferric iron (Fe^{3+}) can be reduced according to the following reaction (Glavee *et al.* 1995):

$$Fe(H_2O)_6^{3+} + 3BH_4^- + 3H_2O = Fe^0 \downarrow + 3B(OH)_3 + 10.5H_2 \qquad (68.1)$$

After stirring the solution for 20 min, it is centrifuged at 6000 g for 2 min and the supernatant solution is replaced by acetone (Kanel *et al.* 2005). Acetone-washing prevents the immediate rusting of NZVI during purification and leads to a fine black powder product after freeze-drying. A picture showing synthesis of NZVI is presented in Figure 68.1.

68.3.1.2   *Electro-chemical method*
This method is a novel technology combining electrochemical and ultrasonic methods to produce NZVI by electroplating iron particles and removing the nanoscale iron particles into the solution instantaneously. This technique has been recently introduced by Chen *et al.* (2004). Hence the iron particle was plated on the cathode by adding ferric chloride in solution to reduce the ferric ion to iron particle according to the following equation:

$$Cathode : Fe^{+3} + 3e^- + stabilizer = Fe^0 \text{ nanoparticle}$$

Therefore, according to Faraday's theory, iron atoms were gradually formed on the cathode. As the size of an iron atom is smaller than nanoscale, the iron particle formed on the cathode would lie in the nanoscale range if these particles were quickly removed from cathode into solution before clustering. Therefore, ultrasonic vibrators with 20 kHz were used simultaneously during the reaction to provide physical energy to remove the iron particle from the cathode. Schematic diagram for experimental set up is shown in Figure 68.1.

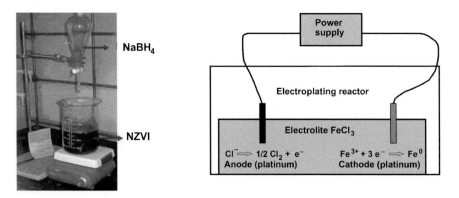

Figure 68.1.   Synthesis of nanoscale zerovalent iron by chemical precipitation method (left) and schematic diagram (right) to produce nanoscale iron by electrochemical method (Shiao-Shing *et al.* 2004).

68.4   INP FOR THE ENVIRONMENT

68.4.1   *Zerovalent iron and iron oxide*

The adsorption of As by iron oxides is a very important natural process, as it is to a large extent responsible for preventing widespread As toxicity problems in nature. Arsenic is very strongly adsorbed by iron oxides, especially under oxidizing and slightly acidic conditions. This is the basis for As removal during water treatment. Once this is known for a wide range of conditions then many properties can be calculated, including the percentage of As adsorbed as a function of pH, the likely efficiency of As removal plants and the extent of retardation of As during movement through soils and aquifers.

While there have been many studies of As adsorption by iron oxides, many of the details that are important for understanding the groundwater As problem remain unclear, including the best mathematical formulation of the isotherm and adsorption changes with pH and redox conditions. The most popular basis at present for calculating the amount of As adsorbed by iron oxides is the diffuse double-layer model of Dzombak and Morel (1990) and its associated database. However, this model does not account for the competitive interactions found in nature very well and the database is based on a limited amount of experimental data. The CD-MUSIC model of Hiemstra and van Riemsdijk (1999) is more promising but it is more complex and at present is not coded into the popular geochemical modeling packages. Huang *et al.* (1984) have also investigated carbon adsorption and metal ion doping as a means of enhancing As extraction with various metals e.g., barium, copper, ferrous and ferric ions. Ferrous ion was found to be highly effective among the other metals. They found that point of zero charge (PZC) of treated carbons was raised to higher pH levels, thereby, providing a positive surface charge on the carbon for exposure to the As-bearing solution species (negatively charged). Treated carbon may adsorb 200 mg/gC compared to 20 mg/gC for untreated carbon.

Fe(III) -bearing materials such as goethite (Matis et al 1997, Fendorf *et al.* 1997), hematite (Singh *et al.* 1996), iron-oxide-coated sand (Jain *et al.* 1999), ferrihydrite (Chanda *et al.* 1988), and Fe(III)- loaded resins (Chanda *et al.* 1988, Matsunaga *et al.* 1996, Rau *et al.* 2000) are the effective non-conventional adsorbents used to remove As. Fe(III) bearing materials have great future potential for As removal since most of the Fe(III) oxides present high As adsorption capacities, Fe(III)- loaded chelating resins are not economically suitable for their use in a full-scale process. Arsenic adsorption on an economic, non-conventional ion exchange material (Forager Sponge, Dynaphore Inc.) based on an open-celled cellulose sponge incorporating a chelating polymer with selective affinity for dissolved heavy metals in both cationic and anionic states is reported. Forager Sponge and other adsorbent sponges have been successfully used in the treatment of heavy metal solutions (Baghai and Bowen 1976). Munoz *et al.* 2002 conducted a basic study of the adsorption of inorganic As species from aqueous solutions using an open celled cellulose sponge (Forager Sponge) with anion-exchange and chelating properties.

Pierce and Moore (1982) reported As removal by amorphous ferrous hydroxide. Additionally, if oxides and hydroxides form strong bonds with As, it is difficult to desorb from the adsorbent (Ford 2002, Lafferty and Loeppert 2005). This result is very favorable for *in situ* groundwater remediation. Singh *et al.* (1989) studied the removal of As(III) by natural hematite (particle size below 200 μm) and found a Langmuir adsorption isotherm, but the maximum capacity was only about 2.6 μmol/g or about 0.2 mg/g As(III). This study demonstrates that even As(III) is removable to some extent. In most areas of developing countries, the addition of mineral oxides to small batches of water may be a feasible process to remove As. Limited information is available on the treatment of industrial effluents rich in As. There has been also great interest in the *in situ* remediation of certain organic and inorganic contaminants in groundwater using zerovalent iron (Fe^0) as a PRB medium (Chunming and Robert 2001). The Fe^0 has been used to effectively destroy numerous chlorinated hydrocarbon compounds via reductive dehalogenation (Johnson *et al.* 1996, Roberts *et al.* 1996, Lackovic *et al.* 1999); Fe^0 could effectively remove inorganic contaminants from aqueous solution. Surface precipitation or adsorption appeared to be the predominant removal mechanism for both As(V) and As(III) by Fe^0 (Chunming and Robert 2001). Joshi and Chaudhari (1996) used iron-based adsorption media and found As removal capacities both in laboratory and pilot scale tests. Chunming and Robert (2001) reported that As removal was largely affected by reaction time, pH, and redox potential. Pierce and Moore (1982) showed that more than 50% arsenite was removed at $Fe(OH)_3$ concentration of 4.45 mg/l and 92% arsenate was removed at 1.33 m/l initial concentration of As at pH 4.

Recently, zerovalent iron (ZVI) has become one of the most common adsorbents for the rapid removal of As(III) and As(V) in the subsurface environment (Lackovic *et al.* 2000, Su and Puls 2001, 2001a, Farrell *et al.* 2001, Olivier *et al.* 2005, Bang *et al.* 2005). The kinetics of As removal by ZVI is within hours to days (Fig. 68.3). However, due to its large size, lower surface area and lack of mobility, its use was limited only to shallow groundwater treatment. To overcome these problems and to take advantage of favorable redox properties and sorption capacity of iron, nanosize, nanoscale zerovalent iron (NZVI) has been introduced. Due to high surface area and reactivity (Schrick *et al.* 2002), it has already shown great potential for treatment of groundwater contaminants (Table 68.3, Fig. 68.2). Recently, removal of As(III) and As(V) by NZVI has been demonstrated for real groundwater remediation in developing countries (Kanel *et al.* 2005, 2006).

68.4.2   *Zerovalent iron for As [As(III) and As(V)] removal*

Many methods are currently in use for removing As from drinking water supplies including anion exchange, reverse osmosis, lime softening, microbial transformation, chemical precipitation, and adsorption (Nriagu 1994, Bissen and Frimmel 2003a, 2003b). Attention has recently focused on zerovalent iron (ZVI) for rapid As(III) and As(V) removal in the subsurface environment (Lackovic *et al.* 2000, Chiu and Hering 2000, Su and Puls 2001, 2001a, Farrell *et al.* 2001, Manning 2002).

Table 68.3.   Common environmental contaminants that can be transformed by nanoscale iron particles (Zhang 2003).

Chlorinated methanes	Trihalomethanes
Carbon tetrachloride (CCl_4)	Bromoform ($CHBr_3$)
Chloroform ($CHCl_3$)	Dibromochloromethane ($CHBr_2Cl$)
Dichloromethane (CH_2Cl_2)	Dichlorobromomethane ($CHBrCl_2$)
Chloromethane (CH_3Cl)	Chlorinated ethenes
Chlorinated benzenes	Tetrachloroethene (C_2Cl_4)
Hexachlorobenzene (C_6Cl_6)	Trichloroethene (C_2HCl_3)
Pentachlorobenzene (C_6HCl_5)	*cis*-Dichloroethene ($C_2H_2Cl_2$)
Tetrachlorobenzenes ($C_6H_2Cl_4$)	*trans*-Dichloroethene ($C_2H_2Cl_2$)
Trichlorobenzenes ($C_6H_3Cl_3$)	1,1-Dichloroethene ($C_2H_2Cl_2$)
Dichlorobenzenes ($C_6H_4Cl_2$)	Vinyl chloride (C_2H_3Cl)
Chlorobenzene (C_6H_5Cl)	Other polychlorinated hydrocarbons
Pesticides	Pesticides PCBs
DDT ($C_{14}H_9Cl_5$)	Dioxins
Lindane ($C_6H_6Cl_6$)	Pentachlorophenol (C_6HCl_5O)
Organic dyes	Other organic contaminants
Orange II ($C_{16}H_{11}N_2NaO_4S$)	N-nitrosodimethylamine (NDMA) ($C_4H_{10}N_2O$)
Chrysoidine ($C_{12}H_{13}ClN_4$)	TNT ($C_7H_5N_3O_6$)
Tropaeolin O ($C_{12}H_9N_2NaO_5S$)	Inorganic anions
Acid Orange	Dichromate ($Cr_2O_2^{7-}$)
Acid Red	Arsenic (AsO_4^{3-})
Heavy metal ions	Perchlorate (ClO_4^-)
Mercury (Hg^{2+})	Nitrate (NO_3^-)
Nickel (Ni^{2+})	
Silver (Ag^+)	
Cadmium (Cd^{2+})	

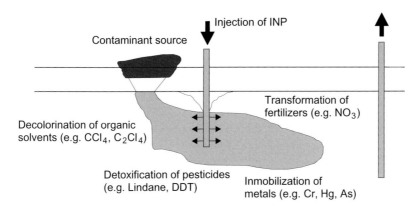

Figure 68.2.   Nanoscale iron particles for *in situ* remediation. Recent research has suggested that as a remediation technique, nanoscale iron particles have several advantages: (1) effective for the transformation of a large variety of environmental contaminants, (2) inexpensive, and (3) nontoxic (Zhang 2003).

The As(III) removal mechanism is mainly due to spontaneous adsorption and co-precipitation of As(III) with Fe(II) and Fe(III) oxides/hydroxides which form *in situ* during ZVI oxidation (corrosion) (Charlet and Manceau 1993, Su and Puls 2001, 2001a, Farrell *et al.* 2001, Manning *et al.* 2002, Olivier *et al.* 2005). The oxidation of ZVI by water and oxygen produces ferrous iron (Ponder *et al.* 2000):

$$Fe^0 + 2H_2O \rightarrow Fe^{2+} + H_2 + 2OH^- \tag{68.2}$$

$$2Fe^0 + O_2 + 2H_2O \rightarrow 2Fe^{2+} + 4OH^- \tag{68.3}$$

Fe(II) further reacts to give magnetite (Fe_3O_4), ferrous hydroxide [$Fe(OH)_2$], and ferric hydroxide [$Fe(OH)_3$] depending upon redox conditions and pH:

$$6Fe^{2+} + O_2 + 6H_2O \rightarrow 2Fe_3O_4(s) + 12H^+ \tag{68.4}$$

$$Fe^{2+} + 2OH^- \rightarrow Fe(OH)_2(s) \tag{68.5}$$

$$6Fe(OH)_2(s) + O_2 \rightarrow 2Fe_3O_4(s) + 6H_2O \tag{68.6}$$

$$4Fe_3O_4(s) + O_2(aq) + 18H_2O \longleftrightarrow 12Fe(OH)_3(s) \tag{68.7}$$

Heterogeneous reactions at the corroding ZVI surface are complex and result in a variety of potential adsorption surfaces for As(III) and As(V). Despite this complexity, studies using X-ray absorption spectroscopy showed the products after reaction of As(III) and As(V) with ZVI were inner-sphere As(III) and As(V) surface complexes on Fe(III) oxides/hydroxide corrosion products (Farrell *et al.* 2001, Manning *et al.* 2002). Figure 68.3 shows the As removal by ZVI in the scale of 10–50 hours (Su and Puls 2001).

Recently, the versatility of nanometer-scale zerovalent iron (NZVI) material has been demonstrated for potential use in environmental engineering. Due to the extremely small particle size, large surface area, and high *in situ* reactivity, these materials have great potential in a wide array of environmental applications such as soil, sediment and groundwater remediation (Lien and Zhang 1999, Ponders *et al.* 2000). In addition, due to small size and capacity to remain in suspension, NZVI can be transported effectively by groundwater (Zhang 2003) and can be injected as sub-colloidal metal particles into contaminated soils, sediments, and aquifers (Cantrell and Kaplan 1997, Lien and Zhang 1999).

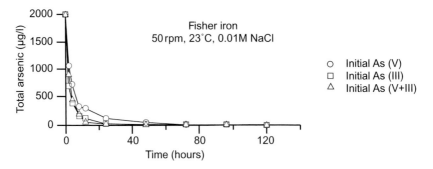

Figure 68.3.   Kinetics of As removal by Fisher iron (Su and Puls 2001).

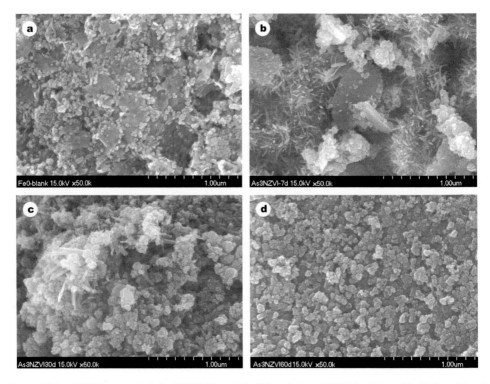

Figure 68.4.   SEM image of pristine NZVI (a), and As(III) sorbed on NZVI for 7 days (b), 30 days (c) and 60 days (d), respectively. Reaction conditions: 100 mg/l As(III) adsorbed on 50 g/l NZVI in 0.01 M NaCl at pH 7, 25°C (Kanel *et al.* 2005).

Hence, recently the application of laboratory-synthesized NZVI for the remediation of As(III) and As(V) in groundwater has been investigated (Kanel *et al.* 2005, 2006). The main objectives were to (1) characterize NZVI and its reaction products using spectroscopic techniques, and to (2) determine the rate and extent of As adsorption by NZVI.

68.4.2.1   *SEM images and EDX spectra*

Solid samples collected from pristine NZVI and 100 mg/l As(III)-treated NZVI after 7, 30 and 60 days and imaged by SEM are shown in Figure 68.4. It was confirmed that synthetic NZVI particles were in the size range of 10–100 nm with a pore size of ∼20 nm. It shows different surface texture and different pore sizes with respect to time of adsorption and precipitation of As onto NZVI. Aggregation of particles increases with reaction time due to Fe(III) oxide/hydroxide precipitation. SEM pictures clearly show a growth of a fine urchin-like crystallite, which may lead to an apparent amorphous phase. The very thin crystallites (about 100 nm long by 20 nm wide) are expected to be energetically unstable and should in turn disappear to be replaced by more stable phases according to the Gay-Lussac-Oswald ripening rule (Kanel *et al.* 2005).

68.4.2.2   *X-ray diffractograms*
The XRD analysis of NZVI, commercial ZVI (Kanto Chemical Co.) and As(III)-treated NZVI samples (1, 7, 30 and 60 days) is shown in Figure 68.5. The zero valence state and crystalline structure of NZVI were confirmed by X-ray diffraction analysis by comparing with Kanto Chemical Co. ZVI material (Fig. 68.5). X-ray diffractograms demonstrate that the NZVI corrosion products are a mixture of amorphous Fe(III) oxide/hydroxide, magnetite (Fe$_3$O$_4$) and/or maghemite (-Fe$_2$O$_3$), and lepidocrocite (-FeOOH). These Fe(II)/(III) and Fe(III) corrosion products indicate that Fe(II) formation is an intermediate step in the NZVI corrosion process.

In the 24 h reaction product an amorphous domain is seen amongst magnetite, lepidocrocite peaks, and a predominant ZVI (Fe^0) peak. Amorphous products were replaced by magnetite and lepidocrocite over a 2 month period (Fig. 68.5). After 60 d the As(III)-NZVI corrosion product had predominantly magnetite and lepidocrocite crystalline composition. Similar results were reported by Manning *et al.* (2002) for corrosion products from ZVI powder and Richmond *et al.* (2004) in As removal via ferrihydrite crystallization control. SEM images (Fig. 68.4) collected after 60 d also show rounded particle morphology indicative of magnetite/maghemite, Fe$_3$ + 2O$_3$ (Davis and Kent 1990, Zhang *et al.* 1992, Charlet *et al.* 1998).

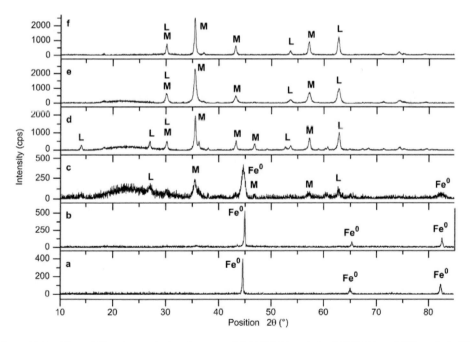

Figure 68.5.   X-ray diffraction analysis of commercial NZVI (a), pristine NZVI (b) and 100 mg/l As(III) sorbed on 50 g/l NZVI in 0.01 M NaCl at pH 7, 25°C for 1 day (c), 7 days (d), 30 days (e) and 60 days (f), respectively. Measure peaks are due to magnetite-maghemite (Fe$_3$O$_4$/γ-Fe$_2$O$_3$) and lepidocrocite (γ-FeOOH). Peaks are referred to magnetite/maghemite (M) (Fe$_3$O$_4$/γ-Fe$_2$O$_3$/), lepidocrocite (γ-FeOOH) (l) and NZVI (Fe^0), respectively (Kanel *et al.* 2005).

68.4.2.3   *Kinetics of As(III) adsorption*
The influence of NZVI concentrations (0.5, 2.5, 5, 7.5, 10 g/l) on the rate of adsorption of As(III) was investigated using 1 mg/l of As(III) at pH 7 (Fig. 68.6). For all treatments except the 0.5 g/l treatment more than 80% As_T was adsorbed within 7 min and ~99.9% within 60 min. An optimum concentration of NZVI (1 g/l) was used in the remaining experiments unless otherwise specified. The As(III) adsorption kinetics data were examined using a pseudo-first-order reaction kinetics expression:

$$Rate = -\frac{d\left(As_{T,t}\right)}{dt} = k_{obs}\,[NZVI]$$

(68.8)

where $(As)_{T,t}$ is the concentration of As (g/l) at time t (min), [NZVI] is the concentration of NZVI (g/l) and k_{obs} is the pseudo-first-order rate constant of As (1/min). For 20 ml of 1 mg/l initial As concentration and 0.5 to 10.0 g/l NZVI, the initial rates were 0.07 to 1.3/min (Table 68.4). An initial faster rate of As(III) disappearance from aqueous solution (~80% As_T) takes place within 7 min followed by a slower uptake reaction. This sequence of reactivity was consistent with the results of pseudo-first order rate of As(V) adsorption by ZVI (Farrell *et al.* 2001, Melitas *et al.* 2002) and chromium (VI) and lead (II) adsorption by nanoscale zerovalent iron (Ponder *et al.* 2000, Ponder *et al.* 2001).

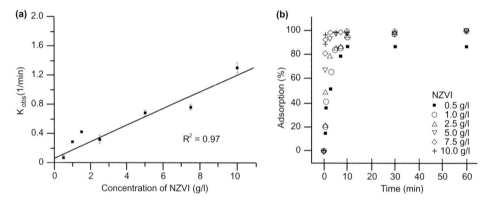

Figure 68.6.   (a) Kinetics of As(III) adsorption. The reaction is pseudo first order with respect to the total NZVI concentrations. Reaction conditions: 1 mg/l As(III) adsorbed on 1.0 g/l NZVI concentration in 0.01 M NaCl at pH 7; 25°C; (b) As(III) adsorption with respect to time, initial As(III): 1 mg/l in 0.01 M NaCl, pH 7 at 25°C (Kanel *et al.* 2005).

Table 68.4.   Pseudo-first-order rate constants (k_{obs}) and their surface area normalized rate constants (k_{sa}) for As-removal by NZVI, As(III): 1 mg/l in 0.01 m M NaCl at pH 7, 25°C (Kanel *et al.* 2005).

NZVI (g/l)	k_{obs} (1/min)	k_{sa} (l/m²/min)	R^2
0.5	0.07	0.0057	0.85
1.0	0.28	0.0115	0.97
2.5	0.32	0.0052	0.95
5.0	0.68	0.0056	0.95
7.5	0.76	0.0042	0.77
10.0	1.30	0.0053	0.77

Su and Puls (2001, 2001a) reported adsorption of As(III) by micron size ZVI and found for 24 g/l ZVI the surface normalized rate constant (k_{sa}) for As(III) was 68.2, 0.59, 0.44 and 1.56 ml/m^2/h for Fisher, Peerless, Master Builders and Aldrich, respectively. In our experiment, the k_{sa} for As(III) was 165, 689, 787, 167.2, 186.9 and 319.7 ml/m^2/h at 0.5, 1.0, 2.5, 5.0, 7.5, 10.0 g/l NZVI, respectively. Hence, the k_{sa} for NZVI was much higher than that of micron size ZVI.

Nanoscale zerovalent iron (NZVI) was synthesized and tested for the removal of As(III), which is a highly toxic, mobile and predominant As species in anoxic groundwater. We used SEM-EDX, AFM, and XRD to characterize particle size, surface morphology, and corrosion layers formed on pristine NZVI and As(III)-treated NZVI. AFM results showed particle size ranged from 1–120 nm. XRD and SEM results revealed that NZVI gradually converted to magnetite/maghemite corrosion products mixed with lepidocrocite over 60 d. Arsenic (III) adsorption kinetics were rapid and occurred on a scale of minutes following a pseudo-first-order rate expression with observed reaction rate constants (k_{obs}) of 0.07 to 1.3/min (at varied NZVI concentration). These values are about 1000× higher than k_{obs} literature values for As(III) adsorption on micron size ZVI. Batch experiments were performed to determine the feasibility of NZVI as an adsorbent for As(III) treatment in groundwater as affected by initial As(III) concentration and pH (3 to 12). The maximum As(III) adsorption capacity in batch experiments calculated by Freundlich adsorption isotherm was 3.5 mg As(III) per g of NZVI. Laser light scattering (electrophoretic mobility measurement) confirmed NZVI-As(III) inner-sphere surface complexation. Competing anions (HCO_3^-, $H_4SiO_4^0$, and $H_2PO_4^{2-}$) showed potential interferences with the As(III) adsorption reaction. Our results suggest that NZVI is a suitable candidate for both *in situ* and *ex situ* groundwater treatment due to its high reactivity.

68.4.3 *As(V) removal by NZVI*

The removal of As(V) was studied by synthetic nanoscale zerovalent iron (NZVI). Batch experiments were performed to investigate As(V) removal by NZVI. As(V) adsorption kinetics were rapid and occurred on a scale of minutes following a pseudo-first-order rate expression with observed reaction rate constants (k_{obs}) of 0.02–0.71/min at varied NZVI concentrations (Kanel *et al.* 2006). Table 68.5 summarizes a comparison of As (V) removal by micron- and nanosized ZVI.

68.4.4 *Overall comparison of micron size ZVI and NZVI for groundwater remediation*

The micron size ZVI is a large-sized particle (~1 mm) and surface area is about 0.9 m^2/g whereas NZVI is about 24.4 to 33 m^2/g. Though ZVI is cheaper than NZVI, in terms of surface area, NZVI is about 100 times cheaper than micron ZVI. The NZVI is 10–1000 times more reactive than ZVI. Moreover NZVI can remove pollutants to end product whereas ZVI produces byproducts. For example NO_3^- can be removed to N_2 gas by NZVI, whereas ZVI produces NH_4^+ which is another pollutant to be removed. Table 68.6 shows the comparison of ZVI and NZVI in terms of cost and efficiency.

Table 68.5. Summary of As(V) removal by micron and nano ZVI (Kanel *et al.* 2006).

Fe^0 Type size	q (mg/g)	k_{sa} (ml/m^2/h)	S (m^2/g)	Dosages (g/l)	As (mg/l)	References
NZVI (nano)	10.0	1740	25.0	0.1	1 [As(V)]	Kanel *et al.* (2006)
NZVI (nano)	3.5	690	24.4	1.0	1 [As(III)]	Kanel *et al.* (2005)
Peerless (micro)	0.7	0.59	2.5	24.0	2 [As(V)]	Su and Puls (2001)
Master builder (micro)	0.3	–	1.0	–	As(III)/As(V)	Lackovic *et al.* (2000)
Iron fillings (micro)	7.2	–	1.0	1.0	As(V)	Richmond *et al.* (2004)

q: maximum sorption capacity (mg/g); k_{sa}: surface area normalized rate constant (ml/m^2/h); S: surface area (m^2/g).

Table 68.6.   Comparison of ZVI and NZVI for groundwater remediation.

Parameter	Micron ZVI (Fe^0)	Nano Fe^0	References
Diameter	~1 mm	~60 nm	Kanel *et al.* (2005)
Surface area	1.5 m^2/kg (~0.9 m^2/g)	15,000 m^2/kg (33 m^2/g)	Zhang (2004)
Cost/kg	~US$ 0.5/kga	<US$ 50/kga; ~200–300 $/kgb	Zhang (2004)
Iron surface area/dollar	3 m^2/US$	<300 m^2/US$	Zhang (2004)
Reactivity	x	10–1000*x	Kanel *et al.* (2005)
By-products	By-products remain	Remove pollutants to end product	Choe *et al.* (2000)
Mobility	Immobile, can be used in PRB	Mobile, can be used in CRB and MRB for *in situ* remediation	Kanel (2006b)

68.5   CHALLENGES OF INP AND INTRODUCTION OF S-INP

Nanoscale zerovalent iron (NZVI), also known as iron nanoparticle (INP), has been widely applied due to its large surface area and high reactivity (Kanel *et al.* 2005). Its natural occurrence in aggregated state prevents its movement through porous media such as sand and soil. Hence, it has remained a big challenge to make it mobile, since mobile NZVI is the most important prerequisite for *in situ* application, especially for subsurface remediation. In this perspective, Cantrell *et al.* (1997) and Cantrell and Kaplan (1997) reported mobilization of polymer stabilized micron zerovalent iron particles in coarse-grained porous media. Recently, Schrick *et al.* (2004) have introduced the delivery vehicle concept for the first time in environmental remediation, which holds promise for direct remediation of contaminated sites without excavation, which is not possible by conventional treatment and PRB. Most recently, we have synthesized surfactant stabilized INP (S-INP) and studied its transport in unsaturated porous media and its application to As remediation (Kanel *et al.* 2005a, 2005b, Kanel *et al.* 2007). Hence surface modification is a novel concept for mobilizing INP in the subsurface environment.

68.5.1   *Synthesis of S-INP*

Surfactant modified INP (S-INP) can be prepared by dispersing INP particles in aqueous 0.5% non-ionic surfactant followed by sonication with a VCX-400 Vibracell (Sonics and Materials, Inc., Danbury, CT) operating at 20 kHz. Recently, we reported synthesis of S-INP using sonication (Kanel *et al.* 2005b, Kanel *et al.* 2007).

68.5.2   *Characterization of INP and S-INP*

The morphological appearance of the pristine INP and S-INP are shown by atomic force microscope (AFM) (Fig. 68.7). The image of pristine INP (Fig. 68.5a) shows that more than 90% of the particles are less than 160 nm but they are in aggregated state. However, in case of S-INP (Fig. 68.7b) highly dispersed particles are observed. It is shown that S-INP is fully coated with the surfactant molecules (Kanel *et al.* 2005b).

68.5.3   *Stability of INP and S-INP*

The dramatic difference in stability of INP and S-INP is clearly illustrated in Figure 68.8a, b. It can be clearly observed that when pristine INP had been kept in a vial containing water and shaken vigorously, all INP precipitated within a minute at the bottom of the vial and was not dispersed. This means that pristine INP could not be mixed in water but formed aggregates due to its thermodynamic properties (Cushil *et al.* 2004). On the contrary, when S-INP was highly dispersed under same experimental condition, S-INP remained in a mixed form for long time.

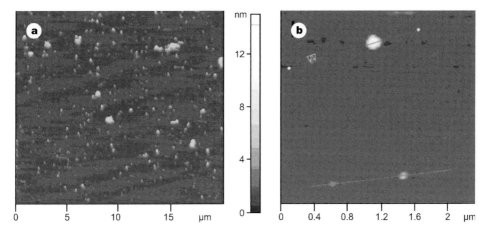

Figure 68.7. The morphological appearance of the pristine INP (a) and S-INP (b). The vertical color scale indicates size of the particle. The green and red lines are used to measure particle size.

Figure 68.8. Stability of INP (a) and S-INP (b). Comparison of effluents collected (c) without surfactant (INP) and (d) with surfactant (S-INP).

68.5.4 *Transport of INP and S-INP in porous media*

The large difference in transport of INP and S-INP through sand packed columns is clearly illustrated in Figure 68.8c, d. It can be clearly observed that when pristine INP passed through the column, all INP was clogged at the top and could not move even a centimeter. This prohibited water elution (Fig. 68.8c), which means that pristine INP would not be mobile in sandy subsurface environments (Kanel *et al.* 2005b). On the contrary, when S-INP circulated under the same experimental conditions, immediately S-INP started moving downwards. Figure 68.8d shows the elution of black solution of S-INP from the column. Therefore, S-INP mobility in porous media is associated with INPs that are fully coated with the surfactants.

The As(III) strongly sorbs on S-INP at neutral pH, which signifies its application in *in situ* groundwater treatment. Furthermore, to apply this adsorbent for *in situ* treatment long term, complexation of S-INP and As needs to be studied further in detail. Interestingly, a recent study by Ford (2002) confirmed that hydrospheric oxide could not desorb As(V) for more than 4 months after its adsorption. Similar observation was reported by Dixit and Herring (2003). This shows that S-INP can be applied for *in situ* remediation. More studies are needed for the application of S-INP to optimize the remediation efficiency. Further studies on dissolved organic and inorganic constituents in groundwater, and microorganisms for its *in situ* application are underway in our laboratory. Though results are promising, many fundamental questions remain unanswered about the transport of S-INP in porous media as well as the reactivity of this material with specific pollutants. Therefore more studies are needed to apply S-INP for *in situ* groundwater remediation.

68.5.5   *Column studies*

Pyrex glass columns (10 cm long and 2.5 cm internal diameter) are used for the experiments. A porous glass diffuser plate placed on each end cap provided an even distribution of S-INP in the soil column. Feed solutions containing 1 g/l suspended S-INP in 0.01 M NaCl at pH 7 is prepared freshly prior to each column experiment. Pretreatment of columns is performed by introduction of feed solution (S-INP) at a rate of 3 ml/min to the top of the column. When the initial concentration/final concentration is $\sim$same, the feed solution is stopped (Kanel *et al.* 2005b). Total Fe (particulate + dissolved) was measured by addition of 5 M HCl to effluent aliquots to dissolve particulate INP followed by reduction of Fe(III) by addition of hydroxylamine. Total dissolved Fe as Fe(II) was then determined using the phenanthroline colorimetric method (Cantrell and Kaplan 1997, Schrick *et al.* 2004).

68.5.6   *Influence of substrate type on S-INP transport*

Natural soil contains clay minerals, metal oxides (MO), and soil organic matter (SOM). In our experiments, different porous media such as glass beads, baked sand and unbaked sand were packed in columns (e.g., SiO_2, SiO_2 + MO, and SiO_2 + MO + SOM, respectively) providing approximate field conditions. It was observed that the S-INP particles were sorbed at porous media surfaces during transport through the column. Hence, experiments were conducted to investigate the effects of MO and SOM on transport of S-INP in sand. Figure 68.10 shows break through curves (BTC) of S-INP for different porous media. Transport of S-INP was most rapid through glass beads, followed by unbaked sand and baked sand. The glass beads, which only contain SiO_2 allowed S-INP to migrate with little surface interaction and its breakthrough at $\sim$25.2 pore volumes. Since sand contained small quantities of MO and 0.12% SOM S-INP breakthrough was delayed to $\sim$57 pore volumes (Kanel *et al.* 2007). The breakthrough time was directly related to the surface coverage of the collectors by the nanoparticles (Fig. 68.9). Once a monolayer coverage was attained, the C/C_0 approached 1.0.

68.6   PROSPECTIVE AND FUTURE CHALLENGES

We discussed that As(III) can be removed without any pretreatment by adsorption on NZVI in a very short time (minute scale). As(III) strongly sorbs on NZVI at neutral pH. Various co-precipitates on iron oxide corrosion products were involved. This study shows that NZVI is an efficient material for the treatment of As and may be used as a new material for CRB or MRB for *in situ* as well as *ex situ* treatment. The potential of removing As from water makes NZVI a particularly good candidate for treatment of groundwater.

Moreover, we have also presented evidence that As(V) can be removed by adsorption/ precipitation on NZVI (at neutral pH) in a very short time (minute scale). As(V) strongly sorbs on NZVI in a wide range of pH, which involves co-precipitation of various iron oxide corrosion products. The result shows that NZVI is an efficient material for the treatment of As(V), which holds promise for *in situ* as well as *ex situ* groundwater remediation.

We have found that NZVI can be used as efficient material for the removal of As at the time scale of minutes. After sorption of As on NZVI, the product changes into magnetite/maghemite

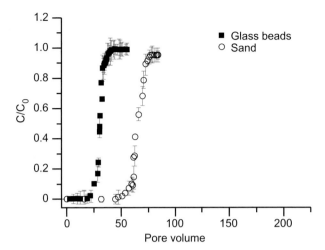

Figure 68.9. Breakthrough curves of S-INP in glass beads, sand baked sand and soil packed in column; INP concentration 1 g/l, surfactant: 0.5%, flow rate: 3.0 ml/min in downward direction (Kanel *et al.* 2006b).

and lepidocrocite. Further studies are required to investigate desorption after long-term aging of As on NZVI.

Surface modified INP has great potential to be used as a colloidal reactive barrier (CRB) as well as mobile reactive barrier (MRB). More work to study the application of nanoparticles to use them as MRB is warranted. Pollutants in the subsurface environment can be removed without excavation by this method; further studies are needed for its application in real field.

ACKNOWLEDGEMENTS

This work was supported by a grant from the National Research Laboratory Program by the Korea Science and Engineering Foundation.

REFERENCES

Agrawal, A. and Tratnyek, P.G.: Reduction of nitro aromatic compounds by zerovalent iron metal. *Environ. Sci. Tech.* 30:1 (1996), pp. 153–160.

Alowitz M.J. and Scherer, M.M.: Kinetics of nitrate, nitrite, and Cr (VI) reduction by iron metal. *Environ. Sci. Tech.* 36:3 (2002), pp. 299–306.

Baghai, A. and Bowen, H.J.M.: Analyst separation of rhodium and iridium using silicone rubber foam treated with tri-n-octylamine. *Analyst* 101 (1976), pp. 661–665.

Balaji, S., Ghosh, B., Das, M.C., Gangopadhyay, A.K., Singh, K., Lal, S., Das, A., Chatterjee, S.K. and Banerjee, N.N.: Removal kinetics of arsenics from aqueous media on modified alumina. *Indian Journal of Chemical Technology* 7 (2000), p. 30.

Bang, S., Johnson, M.D., Korfiatis, G.P. and Meng, X.: Chemical reactions between arsenic and zero-valent iron in water. *Water Res.* 39 (2005), pp. 763–770.

Bhumbla, D.K. and Keefer, R.F.: Arsenic mobilization and bioavailability in soils. In: J. O Nriagu (ed): *Arsenic in the environment.* Wiley, New York, 1994, pp. 51–82.

Bissen, M. and Frimmel, F.: Arsenic—a Review. Part I: Occurrence, toxicity, speciation, mobility. *Acta Hydrochim. Hydrobiol.* 31:1 (2003a), pp. 9–18.

Bissen, M. and Frimmel, F.: Arsenic—a Review. Part II: Oxidation of arsenic and its removal in water treatment. *Acta Hydrochim. Hydrobiol.* 31 (2003b), pp. 97–107.

Cantrell, K.J. and Kaplan, D.I.: Zero-valent iron colloid emplacement in sand columns. *J. Env. Eng.* 123 (1997), pp. 499–505.

Cantrell, K.J., Kaplan, D.I. and Gilmore, T.J.: Injection of colloidal FeO particles in sand with shear-thinning fluids. *J. Envir. Eng.* 123 (1997), pp. 786–791.

Chanda, M., O'Driscoll, K.F. and Rempel, G.L.: Ligand exchange sorption of arsenate and arsenite anions by chelating resins in ferric ion form I. Weak-base chelating resin dow XFS-4195. *React. Polym.* 7 (1988), pp. 251–261.

Charlet, L. and Manceau, A.: Structure, formation and reactivity of hydrous oxide particles; Insights from X-ray absorption spectroscopy. In: J. Buffle and H.P. Van Leeuwen (eds): *Environmental particles II. IUPAC environmental analytical and physical chemistry series.* Lewis Publ., Chelsea, Michigan, 1993, pp. 117–164.

Charlet, L., Silvester, E.J. and Liger, E.: N-compound reduction and actinide immobilisation in surficial fluids by Fe(II): the surface FeIIIOFeIIOH° species, as major reductant. *Chem. Geol.* 151 (1998), pp. 85–93.

Choe, S., Chang, Y., Hwang, K. and Khim, J.: Kinetics of reductive denitrification by nanoscale zero-valent iron. *Chemosphere* 41 (2000), pp. 1307–1311.

Chunming, S. and Robert, W.P.: Arsenate and arsenite removal by zerovalent iron; Kinetics, redox transformation, and implications for *in situ* groundwater remediation. *Environ. Sci. Tech.* 35 (2001), pp. 1487–1492.

Chiu, V.Q. and Hering, J.G.: Arsenic adsorption and oxidation at manganite surfaces. 1. Method for simultaneous determination of adsorbed and dissolved arsenic species. *Environ. Sci. Tech.* 34 (2000), pp. 2029–2034.

Clifford, D.A. and Zang, Z.: Arsenic chemistry and speciation. *Proc. Water Quality Technology Conference: 1955–1968.* AWWA, 1993, Denver, CO, 1994.

Cullen, W.R. and Reimer, K.J.: Arsenic speciation in the environment. *Chem. Rev.* 89 (1989), pp. 713–764.

Cushing, B.L., Kolesnichenko, V.L. and O'Connor, C.J.: Recent advances in the liquid-phase syntheses of inorganic nanoparticles. *Chem. Rev.* 104 (2004), pp. 3893–3946.

Davis, J.A. and Kent, D.B.: Surface complexation modeling in aqueous geochemistry. In: M.F. Hochella, and A.F. White (eds): Mineral-Water Interface Geochemistry. *Reviews in Mineralogy* 23 (1990), pp. 177–260, Mineralogical Society of America, Washington, DC.

Dixit, S. and Hering, J.G.: Comparison of arsenic (V) and arsenic (III) sorption onto iron oxide minerals: implications for arsenic mobility. *Environ. Sci. Tech.* 37 (2003), pp. 4182–4189.

Dzombak, D.A. and Morel, F.M.M: *Surface complexation modeling: hydrous ferric oxide.* John Wiley and Sons, New York, 1990.

Elliott, D.W. and Zhang, W.-X.: Field assessment of nanoscale bimetallic particles for groundwater treatment. *Environ. Sci. Tech.* 35:24 (2001), pp. 4922–4926.

EPA (US Environmental Protection Agency): Research plan for arsenic in drinking water. US Environmental Protection Agency, Office of research and development: EPA/68/R-98/042, Washington, DC, 1988.

Farrell, J., Wang, J., O'Day, P. and Coklin, M.: Electrochemical and spectroscopic study of arsenate removal from water using zero-valent iron media. *Environ. Sci. Tech.* 35 (2001), pp. 2026–2032.

Fendorf, S., Eick, M.J., Grossl, P. and Sparks, D.L.: Arsenate and chromate retention mechanisms on goethite. 1. Surface structure. *Environ. Sci. Technol.* 31 (1997), pp. 315–320.

Ferguson, J.F. and Gavis, J.: Review of the arsenic cycle in natural waters. *Water Res.* 6 (1972), pp. 1259–1274.

Ford, R.G.: Rates of hydrous ferric oxide crystallization and the influence on coprecipitated arsenate. *Environ. Sci. Tech.* 36 (2002), pp. 2459–2463.

Giasuddin, A.B.M., Kanel, S.R. and Choi, H.: Natural organic matter removal from groundwater using nano scale zero valent iron. *Environ. Sci. Tech.* 41 (2007), pp. 2022–2027.

Gillham, R.W. and O'Hannesin, S.F.: Enhanced degradation of halogenated aliphatics by zero-valent iron. *Ground Water* 32:6 (1994), pp. 958–967.

Glavee, G.N., Klabunde, K.J., Sorensen, C.M. and Hadjipanayis, G.C.: Chemistry of borohydride reduction of iron (II) and iron (III) ions in aqueous and nonaqueous media. Formation of nanoscale Fe, FeB, and Fe$_2$B powders. *Inorg. Chem.* 34 (1995), pp. 28–35.

Gulens, J., Champ, D.R. and Jackson, R.E. : In: A.J. Rubia (ed): *Chemistry of water supply treatment and distribution.* Ann Arbor Science Publishers, Ann Arbor, MI, 1973.

Hiemstra, T. and Van Riemsdijk, W.H.: Surface structural ion adsorption modeling of competitive binding of oxyanions by metal (hydr)oxides *J. Colloid Interface Sci.* 210 (1999), pp. 182–193.

Hozalski, R.M., Zhang, L. and Arnold, W.A.: Reduction of haloacetic acids by Fe0: implications for treatment and fate. *Environ. Sci. Tech.* 35 :11 (2001), pp. 2258–2263.

Huang, C.P. and FU, P.L.K. : Treatment of arsenic (V) containing water by activated carbon. *J. Water Pollut. Control Fed.* 56 (1984), pp. 233–242.

Jain, A., Raven, K.P. and Loeppert, R.H.: Arsenite and arsenate adsorption on ferrihydrite: surface charge reduction and net OH-release stoichiometry. *Environ. Sci. Tech.* 33 (1999), pp. 1179–1184.

Jiang, J.Q.: Removing As from groundwater for the developing world—a review. *Water Sci. Technol.* 44 (2001), pp. 89–98.

Johnson, P.R. and Elimelech, M.: Dynamics of colloid deposition in porous media: blocking based on random sequential adsorption. *Langmuir* 11 (1996), pp. 801–812.

Johnson, T.J., Scherer, M.M. and Tratnyek, P.G.: Kinetics of halogenated organic compound degradation by iron metal. *Environ. Sci. Tech.* 30 (1996), pp. 2634–2640.

Johnson, P.R., Sun, N. and Elimelech, M.: Colloid transport in geochemically heterogeneous porous media: modeling and measurements. *Environ. Sci. Tech.* 30 (1996), pp. 3284–3293.

Joshi, A. and Chaudhary, M.: ACSC removal of arsenic from groundwater by iron oxide-coated sand. *J. Environ. Eng.* 122 (1996), pp. 769–771.

Kanel, S.R., Manning, B., Charlet, L. and Choi, H.: Removal of arsenic (III) from groundwater by nano scale zero-valent iron. *Environ. Sci. Tech.* 39 (2005a), pp. 1291–1298.

Kanel, S.R., Greneche, J.M. and Choi, H.: Removal of arsenic (V) from groundwater by nano scale zero-valent iron. *Environ. Sci. Tech.* 40 (2006a), pp. 2045–2050.

Kanel, S.R., Kang, S., Jung, H. and Choi, H.: Transport characteristics of surfactant stabilized iron nano particle in unsaturated porous media. *The 230th ACS national Meeting*, in Washington DC, Aug. 28–Sept. 1, 2005, Volume 45, 2005b, pp. 683–687.

Kanel, S.R.: *Remediation and transport characteristics of arsenic in groundwater and its investigation by environmentally reactive nano particles.* PhD Thesis, Gwangju Institute of Science and Technology (GIST), South Korea, 2006b.

Kanel, S.R., Dhriti, N., Manning, B. and Choi, H.: Surface-modified iron nanoparticle transport in porous media and application to arsenic (III) remediation. *J. Nanoparticle Res.* (2007) DOI 10.1007/s11051-007-9225-7.

Ko, C. and Elimelech, M.: The "shadow effect" in colloid transport and deposition dynamics in granular porous media: measurements and mechanisms. *Environ. Sci. Technonol* 34 (2000), pp. 3681–3689.

Lackovic, J.A., Nilaidis, N.P. and Dobbs, G.M.: Innovative arsenic remediation technology (AsTR) for groundwater, drinking water, and waste streams. In: G.L. Christensen and R.P.S. Suri (eds): *Procceding of the 30th Mid Atlantic Industrial and Hazardous Waste.* Villanova, PA, 1998, pp. 604–613.

Lackovic, J.A., Nikolaids, N.P. and Dobbs, G.M.: Inorganic arsenic removal by zero-valent iron. *Environ. Eng. Sci.* 17 (2000), pp. 29–39.

Lafferty, B.J. and Loeppert, R.H.: Methyl arsenic adsorption and desorption behavior on iron oxides. *Environ. Sci. Tech.* 39 (2005), pp. 2120–2127.

Leupin O.X. and Hug, S.J.: Oxidation and removal of arsenic (III) from aerated groundwater by filtration through sand and zero-valent iron. *Water Res.* 39 (2005), pp. 1729–1740.

Lien, H.-L. and Zhang, W.: Transformation of chlorinated methanes by nanoscale iron particles. *J. Environ. Eng.* 125 (1999), pp. 1042–1047.

Liu, D., Johnson, P.R. and Elimelech, M.: Colloid deposition dynamics in flow-through porous media: role of electrolyte concentration. *Environ. Sci. Techn.* 29 (1995), pp. 2963–2973.

Lombi, E., Wenzel, W.W. and Sletten, R.S.: Arsenic adsorption by soils and iron-oxide-coated sand: kinetics and reversibility. *J. Plant. Nutr. Soil Sci.* 162 (1999), pp.451–456.

Lowry, G.V. and Johnson, K.M.: Congener-specific dechlorination of dissolved PCBs by microscale and nanoscale zerovalent iron in a water/methanol solution. *Environ. Sci. Tech.* 38 (2004), pp. 5208–5216.

Mandal, B.K. and Suzuki, K.T.: Arsenic round the world: a review. *Talanta* 58 (2002), pp. 201–235.

Manning, B.A., Hunt, M., Amrhein, C. and Yarmoff, J.A.: Arsenic (III) and arsenic (V) reactions with zerovalent iron corrosion products. *Environ. Sci. Tech.* 36 (2002), pp. 5455–5461.

Manning, B.A. and Martens, D.A.: Speciation of arsenic (III) and arsenic (V) in sediment extracts by high-performance liquid chromatography-hydride generation atomic absorption spectrophotometry. *Environ. Sci. Tech.* 31 (1997), pp. 171–177.

Manning, B.A., Kiser, J.R., Kwan, H. and Kanel, S.R.: Spectroscopic investigation of Cr (III) and Cr (VI) treated nanoscale zerovalent iron. *Environ. Sci. Tech.* 41 (2007), pp. 586–592.

Matis, K.A., Zouboulis, A.I., Malamas, F.B., Ramos Afonso, M.D. and Hudson, M.J.: Flotation removal of As(V) onto goethite. *Environ. Pollut.* 97 (1997), pp. 239–245.

Matsunaga, H., Yokoyama, T., Eldridge, R.J. and Bolto, B.A.: Adsorption characteristics of arsenic (III) and arsenic (V) on iron (III)-loaded chelating resin having lysine-Nα, Nα-diacetic acid moiety. *React. Funct. Polym.* 29 (1996), pp. 167–174.

Melitas, N., Wang, J., Conklin, M., O'Day, P. and Farrell, J.: Understanding soluble arsenate removal kinetics by zerovalent iron media. *Environ. Sci. Tech.* 36 (2002), pp. 2074–2081.

Munoz, J.A., Gonzalo, A. and Valiente, M.: Arsenic adsorption by Fe(III)-loaded open-celled cellulose sponge. Thermodynamic and selectivity aspects. *Environ. Sci. Tech.* 36 (2002), pp. 3405–3411.

Nam, S. and Tratnyek, P.G.: Reduction of azo dyes with zero-valent iron. *Water Res.* 34:6 (2000), pp. 1837–1845.

Nriagu, J.O.: *Arsenic in the environment, Part I: Cycling and characterization.* John Wiley and Sons Inc., New York, 1994.

Nurmi, J.T., Tratnyek, P.G., Sarathy, V., Baer, D.R., Amonette, J.E., Pecher, K., Wang, C., Linehan, J.C., Matson, D.W., Penn, R., L. and Driessen, M.D.: Characterization and properties of metallic iron nanoparticles: spectroscopy, electrochemistry, and kinetics, *Environ. Sci. Tech.* 39 (2005), pp. 1221–123.

Pierce M.L. and Moore, C.B.: Adsorption of arsenite and arsenate on amorphous Iron hydroxide. *Water Res.* 16 (1982), pp. 1247–1253.

Pichler T., Veizer, J. and Hall, G.E.M.: Natural input of arsenic into a coral-reef ecosystem by hydrothermal fluids and its removal. *Environ. Sci. Tech.* 33 (1999), pp. 1373–1378.

Ponder, S.M., Darab, J.C. and Mallouk, T.E.: Remediation of Cr (VI) and Pb (II) aqueous solutions using supported, nanoscale zero-valent iron. *Environ. Sci. Tech.* 34 (2000), pp. 2564–2569.

Ponder, S.M., Darab, J.G., Bucher, J., Caulder, D., Craig, I., Davis, L., Edelstein, N., Lukens, W., Nitsche, H., Rao, L., Shuh, D.K. and Mallouk, T.E.: Surface chemistry and electrochemistry of supported zerovalent iron nanoparticles in the remediation of aqueous metal contaminants. *Chem. Mater.* 13 (2001), pp. 479–486.

Rau, I., Gonzalo, A. and Valiente, M.: Arsenic (V) Removal from aqueous solutions by iron (III) loaded chelating resin. *J. Radioanal. Nucl. Chem.* 246 (2000), pp. 597–600.

Richmond, W.R., Loan, M., Morton, J. and Parkinson, G.M.: Arsenic removal from aqueous solution via ferrihydrite crystallization control. *Environ. Sci. Tech.* 38 (2004), pp. 2368–2372.

Roberts, A.L., Totten, L.A., Arnold W.A., Burris D.R. and Campell, T.J.: Reductive elimination of chlorinated ethylenes by zero-valent metals. *Environ. Sci. Tech.* 30 (1996), pp. 2654–2659.

Schrick, B., Blough, J.L., Jones, A.D. and Mallouk, T.E.: Hydrodechlorination of trichloroethylene to hydrocarbons using bimetallic nickel-iron nanoparticles. *Chem. Mater.* 14 (2002), pp. 5140–5147.

Schrick, B., Hydutsky, B.W., Blough, J.L. and Mallouk, T.E.: Delivery vehicles for zerovalent metal nanoparticles in soil and groundwater. *Chem. Mater.* 16 (2004), pp. 2187–2193.

Shiao-Shing, C., Hong-Der, H. and Chi-Wang Li.: A new method to produce nanoscale iron for nitrate removal. *J. Nanopart. Res.* 6 (2004), pp. 639–647.

Smith, A.H.: Cancer risks from As in drinking water. *Env. Health Persp.* 97 (1992), pp. 259–267.

Su, C. and Puls, R.W.: Arsenate and arsenite removal by zerovalent iron: kinetics, redox transformation, and implications for *in situ* groundwater remediation. *Environ. Sci. Tech.* 35 (2001), pp. 1487–1492.

Su, C. and Puls, R.W.: Arsenate and arsenite removal by zerovalent iron: effects of phosphate, silicate, carbonate, borate, sulfate, chromate, molybdate, and nitrate, relative to chloride. *Environ. Sci. Tech.* 35 (2001a), pp. 4562–4568.

Tseng, W.P., Chu, H.M., How, S.W., Fong, J.M., Lin, C.S. and Yeh, S.: Prevalence of skin cancer in an endemic area of chronic arsenism in Taiwan. *J. Natl. Cancer Inst.* 40 (1968), pp. 453–463.

VanStone, N., Przepiora, A. Vogan, J., Lacrampe-Couloume, G., Powers, B., Perez, E., Mabury, S. and Sherwood Lollar, B.: Monitoring trichloroethene remediation at an iron permeable reactive barrier using stable carbon isotopic analysis. *J. Cont. Hydrol.* 78:4 (2005), pp. 313–325.

Wang, C.B. and Zhang, W.: Synthesizing nanoscale iron particles for rapid and complete dechlorination of TCE and PCBs. *Environ. Sci. Tech.* 31 (1997a), pp. 2154–2156.

WHO: Guidelines for drinking water quality, Vol. 1: Recommendations. 2nd ed., World Health Organization, Geneva, Switzerland, 1993.

Zhang, Y., Charlet, L. and Schindler, P.W.: N-compound reduction and actinide immobilisation in surficial fluids by Fe(II): the surface FeIIIOFeIIOH° species, as major reductant. *Colloids and Surfaces* 63 (1992), pp. 259–268.

Zhang, W.X.: Nano scale iron particles for environmental remediation: an overview. *J. Nanoparticle Res.* 5 (2003), pp. 323–332.

Zhang W.X.: Environmental applications of iron nanoparticles—Fundamental issues. *International Symposium on Environmental Nanotechnology*. Taipei, Taiwan, 2004, pp. 13–18.

CHAPTER 69

Phytoremediation of arsenic by sorghum (*Sorghum biocolor*) under hydroponics

N. Haque
Environmental Science and Engineering PhD Program, The University of Texas at El Paso,
El Paso, TX, USA

N.S. Mokgalaka, J.R. Peralta-Videa & J.L. Gardea-Torresdey
Department of Chemistry, The University of Texas at El Paso, El Paso, TX, USA

ABSTRACT: Sorghum (*Sorghum biocolor*) was studied as a potential hyperaccumulator for arsenic (As) from aqueous solution. The plants were grown in hydroponics media containing different concentrations of As(III) (0.5 and 1 mg/l) and As(V) (1, 5, and 10 mg/l). The uptake of As(V) (2400, 810, and 300 mg/kg for roots, stems, and leaves, respectively, was 30 times higher than As(III) which is more toxic than As(V). As(V) concentrations of up to 10 mg/l had no effects on the morphology and growth of the plants, whereas 90% of the plants grown in 5 mg/l As(III) did not survive. The root and shoot elongation of plants grown in 1 mg/l As(III) were significantly smaller than the elongation of plants grown in 1 mg/l As(V). The significant amounts of As concentrated in the plants indicate that sorghum could be a potential hyperaccumulator of As for application in phytoremediation of As-contaminated soil and water.

69.1 INTRODUCTION

High levels of arsenic (As) in drinking water are a major concern in several developing regions (McLellan 2002, Sun 2004). The elevated levels are directly related to anthropogenic activities such as agriculture, manufacturing, mining, and smelting (Smedley and Kinniburgh 2002). Inorganic arsenic (i-As) is well recognized as a human poison and chronic exposure to elevated As concentrations in drinking water has caused vascular disorders, such as dermal pigments (black foot disease) and skin and lung cancer (Desesso *et al.* 1998).

A range of technologies, such as precipitation, ion exchange, solvent extraction, adsorption on activated carbon, and iron materials (Gupta *et al.* 2005, Vaishya and Gupta 2005, Hlavay and Polyak 2005) have been used for the removal of As from aqueous solutions. Many of these methods have high maintenance cost and require relatively expensive mineral adsorbents which offset performance and efficiency advantages.

A promising approach is phytoremediation technology, where living plants are used to remove trace metals from aqueous solution. Significant research has been conducted on phytoremediation for their metal-sorption capacity (USEPA 2000, Guadalupe *et al.* 2004), and because of their low cost and high efficiency for metal removal (Chaney *et al.* 1997 and 2000, Anastasios and Ioannis 2005).

The uptake kinetics of As by sorghum (*Sorghum biocolor*) plants under hydroponic media was studied in this work. In addition, the effects of As concentration as arsenate and arsenite on sorghum plants were investigated. Finally, the effects of plant growth and nutrient uptake were also demonstrated.

69.2   EXPERIMENTAL

69.2.1   *Medium preparation and seed planting*

Sorghum seeds were sown in an As-free seedbed with regular watering. After the germination of the seeds, approximately 20 plants were transferred into sterilized jars poured with media. A Hoagland modified nutrient solution previously described in literature (Peralta *et al.* 2001) was prepared as a medium. As(III) and As(V) were added to the nutrient solution to obtain different concentrations (0.5, 1, 5 and 10 mg/l), which were adjusted at pH 5.3. Plants were grown for 7 days and were examined for uptake after 3, 5, and 7 days. In addition, each treatment was replicated three times for statistical purposes.

69.2.2   *Kinetics of arsenic uptake*

To determine the kinetics of As uptake by sorghum, the experiments were carried out at different time intervals (3, 5, and 7 days). After 3, 5, and 7 days, plants were collected from the jars and washed for 5 minutes using a 5% HNO_3 solution to eliminate any external As and then rinsed with DI water. Later, they were separated in roots, stems, and leaves and oven dried at 64°C for 72 hours. The dried samples were digested in a microwave oven following the USEPA 3051 method (Kinston and Jassie 1988). Later, samples were diluted and analyzed by inductively coupled plasma/optical emission spectrometry (ICP/OES) to determine the As concentration as well as the micronutrients.

69.2.3   *Evaluation of the effect of As(III) and As(V) on plant growth*

To determine the effect of As species on plant growth (evaluated as plant elongation), 10 plants/replicate/treatment were randomly selected, and the size of the roots and shoots was measured. Each plant was measured from the main apex of the root to the crown and from the crown to the main apex of the shoot.

69.3   RESULTS AND DISCUSSION

69.3.1   *Kinetics of arsenic uptake*

In this experiment, As uptake kinetics was investigated to understand how fast sorghum could remove the As from the aqueous solution. Arsenic uptake rates as arsenate and arsenite over time are compared in Figure 69.1. The uptake rate of arsenate and arsenite by sorghum was increasing with time. The rate of uptake of As(V) (0.71 mg/l) was higher than of As(III) (0.41 mg/l) after 7 days. This might be due to the organic ligands such as thiols, induced probably by the exposure of the plant to As. Thiols should be able to complex As to avoid the damage of the plant cells by free As(III) (Zhang *et al.* 2002).

The uptake of As in sorghum plants after 7 days of treatment with either As(III) or As(V) is shown in Figure 69.2. It can be easily seen that more As was absorbed by the sorghum plants when the supplied solution was As(V). At a concentration of 10 mg/l, the roots of plants exposed to As(V) absorbed 30 times more As (ca. 2400 mg/kg dry wt.), than the roots of plants exposed to As(III) (80 mg/kg dry wt.). Also, the shoots of plants exposed to 1, and 5 mg/l of As(V) accumulated 8, and 22 times more As compared to the shoots of plants grown in the same concentrations of As(III). In addition, the shoots of plants exposed to 10 mg/l of As(V) accumulated 200 mg/kg dry wt of As whereas in case of As(III) grown media, not a single plant survived. These results show that the accumulation of As was moved from the roots to the shoots of the plants. Roots of sorghum might have greater ability to absorb As than shoots due to its physiological activity (Ma *et al.* 2004).

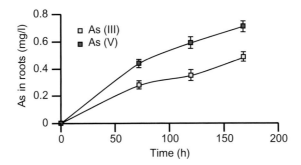

Figure 69.1.   Arsenate and arsenite concentration in roots of sorghum. The initial As concentration in the solution was 1 mg/l. The bars are standard error of the means from three replicates.

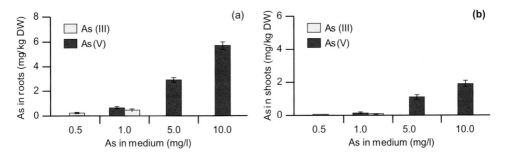

Figure 69.2.   Arsenic concentrations in (a) sorghum roots and (b) shoots after 7 days of growth in hydroponics containing either As(V) or As(III) at different concentrations in the media. Data represent average of 20 plants ± SE. The bars are standard error of the means from three replicates.

Therefore, it can be concluded that As(V) moves easier from roots to shoots than As(III). The amount of As accumulated in the sorghum plants indicated that this plant could be used for the phytoremediation of As-contaminated soils and waters.

In this research, the effect of As species on Ca, K, P, and Mg uptake in sorghum plants was also studied. According to the results (data not shown), the control roots and the plants grown in 1 mg/l of As(V) accumulated about 3200, and 3300 mg Ca/kg DW, respectively. On the other hand, both As species accumulated lesser amounts of K, P, and Mg in roots. In the case of stems, 1, 100 and 800 mg/kg DW were accumulated when the supplied solutions were 5 mg/l of As(V) and 1 mg/l of As(III), respectively.

69.3.2   *Effects of arsenic species on plant growth*

The elongation of the roots and shoots of the plants at different concentration as well as at different As species are shown in Figure 69.3. In general, Figure 69.3 clearly shows that the elongation of the roots at a concentration of 1, 5 and 10 mg/l As(V) were significantly larger than those grown in As(III). Similar results have been observed in the case of shoots. However, after 1 mg/l of either As(V) or As(III) concentration, the elongation of the shoots in As(III) media was dramatically decreased and most of the plants did not survive 5 mg/l As(III). The morphology of the plant changes with the high concentration of As that stops increasing growth of the roots (Ghaly *et al.* 2004).

A strong negative relationship between shoot elongation and As in tissues (when supplied as As(III)) was observed (Pearson correlation coefficients -0.801, $P < 0.01$). However, no significant relationship was found for these two variables in plants grown in As(V). The amount of root biomass

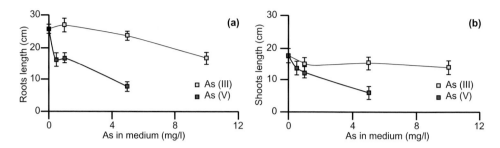

Figure 69.3.   Length of (a) sorghum roots and (b) shoots after 7 days of growth in hydroponics containing either As(V) or As(III) at different concentrations in the media. Data represent average of 20 plants ± SE. The bars are standard error of the means from three replicates.

accumulated from the plants grown in As(V) (data is not shown) was 15 times higher than those of As(III). These results suggest that As(V) might be less toxic than As(III) for sorghum plants.

69.4   CONCLUSIONS

The results of this study clearly demonstrate that the uptake of As by sorghum plants was completely dependent on As speciation. There were no significant changes in root elongation between plants grown in control and at a concentration of 5 mg/l of As(V). Approximately 30 times higher accumulation of As was found in sorghum plants when the supplied solution was As(V) rather than As(III). The movements of the nutrients in the plant is greatly affected by As(III). Finally, this research shows that sorghum (*Sorghum biocolor*) could be used as hyperaccumulating plants to uptake As from contaminated soils and water.

ACKNOWLEDGEMENTS

National Institutes of Health, Center for Environmental Resource Management (CERM) from the University of Texas at El Paso are greatly acknowledged for their financial support.

REFERENCES

Anastasios, I.Z. and Ioannis, A.K.: Recent advances in the bioremediation of arsenic-contaminated groundwaters. *Environ. Internat.* 31 (2005), pp. 213–219.

Chaney, R.L., Malik, M., Li, Y.-M., Brown, S.L., Brewer, E.P., Angle, J.S. and Baker, A.J.M.: Phytoremediation of soil metals. *Curr. Opin. Biotechnol.* 8 (1997), pp. 279–284.

Chaney, R.L., Li, Y.M., Angle, J.S., Baker, A.J.M., Reeves, R.D., Brown, S.L., Homer, F.A., Malik, M. and Chin, M.: Improving metal hyperaccumulation wild plants to develop commercial phytoextraction systems: approaches and progress. In: N. Terry and G.S. Bañuelos (eds): *Phytoremediation of contaminated soils and water*. CRC Press, Boca Raton, FL, 2000, pp. 131–160.

Desesso, J.M., Jacobson, C.F., Scialli, A.R., Farr, C.H. and Holson, J.F.: An assessment of the developmental toxicity of inorganic arsenic. *Reprod. Toxicol.* 12 (1998), pp. 385–433.

Ghaly, A.E., Kamal, M., Mahmoud, N. and Côté, R.: Phytoaccumulation of heavy metals by aquatic plants. *Environ. Internat.* 29 (2004), pp. 1029–1039.

Gupta, V.K., Saini, V.K. and Jain, N.: Adsorption of As(III) from aqueous solutions by iron oxide-coated sand. *J. Colloid Interface Sci.* 288 (2005), pp. 55–60.

Hlavay, J. and Polyak, K.: Determination of surface properties of iron hydroxide-coated alumina adsorbent prepared for removal of arsenic from drinking water. *J. Colloid Interface Sci.* 284 (2005), pp. 71–77.

Ma, L.Q., Tu, S., Fayiga, A.O. and Zillioux, E.J.: Phytoremediation of arsenic-contaminated groundwater by the arsenic hyperaccumulating fern *Pteris Vittata* L. *Int. J. Phytoremediation* 6 (2004), pp. 35–47.

McLellan, F.: Arsenic contamination affects millions in Bangladesh. *The Lancet* 359 (2002), p. 1127.

Peralta-Videa, J.R., Gardea-Torresdey, J.L., Gomez, E., Tiemann, K.J., Parsons, J.G., de la Rosa, G. and Carrillo, G.: Potential of alfalfa plant to phytoremediate individually contaminated montmorillonite-soils with Cd (II), Cr (VI), Cu (II), Ni (II), and Zn (II). *Bull. Environ. Contam. Toxicol.* 69 (2002), pp. 74–81.

Smedley, P.L. and Kinniburgh, D.G.: A review of the source, behaviour and distribution of arsenic in natural waters. *Appl. Geochem.* 17 (2002), pp. 517–568.

Sun, G.: Arsenic contamination and arsenicosis in China. *Toxicol. Appl. Pharmacol.* 198 (2004), pp. 268–271.

USEPA: Introduction to phytoremediation. US Environmental Protection Agency, Office of Research and Development, EPA/600/R-99/107, Washington DC, 2000.

Vaishya, R.C. and Gupta, S.K.: Arsenic removal from groundwater by iron impregnated sand. *J. Environ. Eng.* 129 (2003), pp. 89–92.

Zhang, W., Cai, Y, Tu, C. and Ma L.Q.: Arsenic speciation and distribution in an arsenic hyperaccumulating plant. *Sci. Total Environ.* 300 (2004), pp. 167–177.

CHAPTER 70

Potential use of sedges (Cyperaceae) in arsenic phytoremediation

M.T. Alarcón-Herrera
Centro de Investigación en Materiales Avanzados (CIMAV), Chihuahua, Chih., Mexico

O.G. Núñez-Montoya
Facultad de Zootecnia, Universidad Autónoma de Chihuahua (UACH), Chihuahua, Chih., Mexico

A. Melgoza-Castillo & M.H. Royo-Márquez
Instituto Nacional de Investigaciones Forestales, Agrícolas y Pecuarias (INIFAP), Universidad Autónoma de Chihuahua (UACH), Chihuahua, Chih., Mexico

F.A. Rodriguez Almeida
Facultad de Zootecnia, Universidad Autónoma de Chihuahua (UACH), Chihuahua, Chih., Mexico

ABSTRACT: Development of new technologies to achieve the reduction of arsenic (As) contamination in water and soil is a very important issue. Hyperaccumulator plants from the Pteridaceae family have been widely studied for purposes of As phytoextraction. Phytoremediation, however, can also be performed through phytostabilization using As-tolerant plant species. The objective of this study was to evaluate the tolerance of two plant species from the Cyperaceae family, *Schonoeplectous americanus* and *Eleocharis macrostachya*, to different levels of As and to quantify the accumulation of this element in the plants' biomass (roots and aerial parts). The study was conducted under greenhouse conditions over a period of five months. Plants were collected from native environments, propagated and transplanted into pots with coarse sand as substrate, and covered with 5 cm of water to simulate flood conditions. Initially, the used dosage levels were 1.5, 3.0, and 4.5 mg/l of As. Since these As levels had no visible effect on the plants, the doses were doubled at the end of the first month. The survival, height, number of tillers, and inflorescence of the plants were measured, and a variance analysis was applied to the data. Another measured variable was the As concentration in the plants' biomass. Based on these data, biological absorption coefficients (BAC) and translocation factors (TF) were calculated. The data were then adjusted to a regression model. Plant survival was above 97% for all treatments and species. Height and number of tillers were not affected by treatments. The As BAC of the *E. macrostachya* specimens were 3.45, 3.00 and 2.76; the TF were 0.18, 0.94 and 1.20, at 25, 50 and 75 mg of As, respectively. For *S. americanus*, the As BAC were 1.75, 1.09 and 1.72, and the TF were 1.16, 2.09 and 0.64 for the As concentrations used in the experiment. Results show that both species are tolerant to As and can be considered for use in As phytostabilization and rhizofiltration.

70.1 INTRODUCTION

Arsenic is a naturally occurring metalloid which has been used in pesticides, wood preservatives and many other products, leading to As-contaminated sites. High levels of As often occur naturally, in sources such as silicate or sulfidic ore deposits, tin mining activities, volcanic sites, and sea salt sprays. (Fitz and Wenzel 2002, Jankong *et al.* 2007). Water contamination is a local, regional and worldwide problem, whether it comes from anthropogenic or natural sources. High concentration of dissolved contaminants such as As can make water unfit for drinking. Arsenic contamination is widespread in the aquifers of the Indo-gangetic alluvium and the Bengal Delta (Bhattacharya *et al.* 2002, 2006, Smedley and Kinniburgh 2002, Mukherjee *et al.* 2006, Parijat Tripathi *et al.* 2009, this volume) leading to several health crisis in the region. Irrigation has dispersed As into surrounding soils, resulting in considerable bioaccumulation in the agricultural products and thereby increase the risks of As poisoning of humans and animals. Similar contamination is found worldwide in regions with As in their subsoils, including Mexico (Miller 1994 and 1996, Castro and Wong 1999, McArthur *et al.* 2001, Chao-Yang and Tong Bin 2006).

Phytoremediation is a term applied to a group of technologies that use plants to reduce, remove, degrade, or immobilize contaminants, with the aim of restoring area sites to a condition usable for different applications (Tu *et al.* 2004). To date, phytoremediation efforts have focused on the use of plants to accelerate degradation of organic contaminants or remove hazardous heavy metals and metalloids from soils or water. Phytoremediation of contaminated sites is appealing because it is less expensive and more aesthetically pleasing to the public than to alternate physicochemical remediation strategies (Kramer 2005, Chao-Yang and Tong Bin 2006).

Many different plant families are being tested both in laboratory and field conditions for decontaminating metalliferous substrates in the environment. About 400 plants that hyperaccumulate metals have been reported to this day, most of them are belonging to the Brassicaceae family (Prasad and Freitas 2003). Some fern species from the Pteridaceae family have been found to specifically hyperaccumulate As (Koeller *et al.* 2007). Plants that hyperaccumulate, accumulate and tolerate metals and metalloids have potential for application in the remediation of metal contamination in the environment. The water phytoremediation process involves raising plants hydroponically and transplanting them into metal or metalloid-polluted waters, where they absorb and concentrate the contaminants in their roots and shoots. As they become saturated with the metal pollutants, roots or the whole plants are harvested for treatment and safe disposal.

There are several species with promising results in As phytoremediation. For example, Ma *et al.* (2001) report 22,630 $\mu g/g$ of As in *Pteris vittata* leaves, as well as 8000 $\mu g/g$ in the leaves and 88 $\mu g/g$ in the roots of *Pityrogramma calomelanos*, both of which are hyperaccumulating plants. Fitz and Wenzel (2002) refer to *Agrostis capillaris*, with 3470 $\mu g/g$; and *Cynodon dactylon*, with 12,450 $\mu g/g$, as As-tolerating plants. Non-tolerating plants have a phytotoxic threshold at approximately 5–100 $\mu g/g$ of As in their dry weight (Prasad and Freitas 2003). Although hyperaccumulators are mostly considered for the phytoextraction of As from both soil and water of contaminated sites, accumulators and tolerators could also be useful in the phytostabilization and rhizofiltration of polluted sites.

Plant species found in metal-polluted soils and water bodies are expected to take up metals and eventually accumulate them. Therefore, this is a method to identify metal tolerant, accumulating or hyperaccumulating species. Bioconcentration and translocation factors have to be considered when evaluating whether a particular plant is metal-tolerant. The bioconcentration factor (BCF) is defined as the ratio of the element concentration in the shoots to the element concentration in the soil. Translocation factor (TF) depicts the effectiveness of a plant at translocation, and is defined as the ratio of the element concentration in the shoots to the concentration in the roots. Plants are classified as accumulators when both factors are greater than one, tolerant if one or both factors are less than one, and hyperaccumulators when they have the capacity to bioconcentrate the element in their tissues to a level over 100 times that of the surrounding soil (Fitz and Wenzel 2002, Zhang *et al.* 2002).

The selection of species for this study was based on the fact that, for natural reasons, these species live in flooded areas contaminated with As. In this study, both *S. americanus* and *E. macrostachya* were subjected to different concentrations of As. This was done with the objective of analyzing their capacity of adaptation to the metal, the effect of the metal on the plants' development, and the plants' bioconcentration and translocation factors.

70.2 MATERIALS AND METHODS

70.2.1 *Description of the species*

The Cyperaceae are grasslike, herbaceous plants comprising about 70 genera and 4000 species, commonly found in wet or saturated conditions. The stems are usually 3-angled and solid. The leaves are alternate, commonly in 3 ranks, usually with a closed sheathing base and a parallel-veined, strap-shaped blade. The flowers are very minute and are bisexual or unisexual. Each floret is in the axil of a chaffy bract and these are arranged spirally or distichously in spikelets (USDA 2002).

Eleocharis macrostachya Britt. Herbaceous plant, native, perennial of the rush family (Cyperaceae). It reaches a height of up to 60 cm, with reddish brown rhizomes of up to 30 cm in length and up to 2.5 mm in diameter.

Schoenoplectus americanus Pert. Herbaceous plant, native, perennial of the rush family (Cyperaceae). It reaches a height of up to 2 m, with long and strong rhizomes. Individual stalks grow in small, triangular shape (USDA 2002).

70.2.2 *Collection and adaptation of the plants to a greenhouse*

The species *E. macrostachya* and *S. americanus* were field-collected, the former from an artificial wetland near the town of Naica, Chihuahua, Mexico (27°50′81″N, 105°28′99″O), and the latter from the springs near the village of San Diego de Alcalá, Chihuahua, Mexico (28°35′27″N, 105°32′88″O). The specimens were transplanted and allowed to grow under greenhouse conditions before the experiment. The plants were separated in modules (from 2 to 5 stalks) and placed in pots inside the greenhouse. The pots had a height of 30 cm and a diameter of 25 cm, and were filled with 0.7 kg of river sand with a granulometric diameter between 1 and 6 mm, in which the plants were kept under flood conditions with a minimum water level 5 cm above the surface of the soil. Plants were preconditioned for 47 days, having 0.25 l of a nutritive solution used in hydroponic farming and containing macroelements (80%) and microelements (20%) applied twice to every pot.

70.2.3 *Experiment design*

After preconditioning, 4 treatments with 5 repetitions including the control plants were randomly assigned to each species. The As concentrations used to irrigate the plants during the experiments were 1.5, 3.0 and 4.5 mg/l of sodium arsenite ($NaAsO_2$). Since the plants tolerated these doses well, the concentrations were doubled at the end of the first month. At the end of the experiment, the accumulated As levels in the pots of the experimental units were 25, 50 and 76 mg. The variables measured weekly were: survival to transplant, number of sprouts, number of flowers, and the plants' height.

At the end of the experiment, two random pots were taken from each treatment and each species to evaluate the As concentrations in the plants' structure, both in the aerial part and the roots. Both sections were first washed with drinking water and then with tridistilled water. Next, they were placed in paper bags and dried at 40°C for 7 days. From each plant's dry material, 0.5 g samples were submitted to acid digestion with nitric acid (HNO_3) for 12 hours at laboratory ambient temperature (20°C). After that, digestion proceeded in a microwave oven, in accordance with the method recommended by CEM (2002). Analyses were performed with an atomic absorption spectrophotometer equipped with a BGC Avanta Sigma hydride generator.

The bioconcentration and translocation factors were determined according to the criteria established by Fitz and Wenzel (2002).

70.2.4 *Statistical analysis*

For the variables height, number of sprouts and number of flowers, data were analyzed for each species with a model for an experimental design of parcels divided in time, with PROC MIXED DE

SAS (SAS 1998). To determine the response to the applied As doses in terms of the accumulated concentrations of these elements in aerial and root of the plants, linear regression models were adjusted at the end of the tests.

70.3   RESULTS AND DISCUSSION

70.3.1   *Eleocharis macrostachya*

Concerning the plant height, number of sprouts and number of flowers, generally no effects ($P > 0.05$) were observed for the different As doses. The amount of As accumulated by the plants at the end of the experiment showed a linear behavior in the aerial part ($R^2 = 0.95$); for the roots, the behavior was different since the best correlation ($R^2 = 1$) for results was a third order polynomial equation (Fig. 70.1a). For the highest quantity of As added (75 mg), this species presented 10.8 $\mu g/g$ in the aerial part and 9.1 $\mu g/g$ in the roots. According to a correlation to a straight line of results from the aerial part, concentration in the stalk increased 0.15 $\mu g/g$ for the addition of each mg of As. In Table 70.1, As values are presented with their respective bioconcentration and translocation factors. These values indicate that *E. macrostachya* is tolerant to doses of 3 and 6 mg/l, and is As accumulating at 9 mg/l doses. This explains why no differences were observed between the treatments. The species accumulates As in the stalk proportionally to the applied quantity. For the highest As quantity in the medium, the translocation factor increases and the As concentration in roots decrease.

70.3.2   *Schoenoplectus americanus*

In the case of *S. americanus*, no effects were observed for the plant height and general growth with different As doses($P> 0.05$). The amount of As accumulated by the plants showed a linear behavior in the aerial part $R^2 = 0.98$. However, the behavior in the roots was different and more similar to that shown by *E. macrostachya*, since the best correlation obtained was also a third grade polynomial equation (Fig. 70.1b). For the highest As quantity, this species presented 7.6 $\mu g/g$ in the roots and 4.8 $\mu g/g$ in the aerial part. Considering that the behavior is a straight line for the aerial part of the plants, according to the equation obtained from

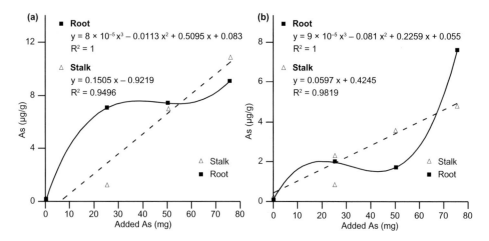

Figure 70.1.   Total arsenic concentration ($\mu g/g$) in the species *Eleocharis macrostachya* (a) and *Schoenoplectus americanus* (b) when exposed to different doses of sodium arsenite ($NaAsO_2$) for 5 months.

Table 70.1. Bioconcentration and translocation factors by species for different arsenic concentrations.

Species	As concentration (mg/l)	Bioconcentration factor (BCF)	Translocation factor (TF)
Eleocharis macrostachya	3	3.45	0.18
	6	3.00	0.94
	9	2.76	1.20
Schoenoplectus americanus	3	1.75	1.16
	6	1.09	2.09
	9	1.72	0.64
Other research works: (Ekkasit and Pornsawan 2004)			
Typha spp.	1		0.07
Canna spp.	1		0.21
Colocasia esculenta	1		0.02
Heliconia spittacorum	1		0.09
Thalia dealbato	1		0.06

results, the addition of each mg/l of As caused concentration to increase $x = 0.06$ μg/g in the aerial part. In contrast to the behavior found in *E. macrostachya*, a higher quantity of As in the medium causes this species of plant to tend to increase the concentration in the root tissue and decrease translocation to the aerial part. An increase in the value of the translocation factor implies that As has accumulated in the root system. The results of this experiment showed that *E. macrostachya* had the highest TF and BCF; therefore, it could be considered a better accumulator than *S. americanus*. Plants' physiology and genotype control their ability to accumulate the metal forms available to them; in consequence, plants differ in their extent of metal accumulation (Ray *et al.* 1995, Ekkasit and Pornsawan 2004). Thus, it is important to acquire knowledge of the ability of different families of wetland plant species to absorb toxic metals under different conditions.

Table 70.1 shows the As values with their respective bioconcentration and translocation factors. These values show that *S. americanus* is accumulating at doses of 3 and 6 mg/l, but becomes tolerant when concentration rises to 9 mg/l. Ekkasit and Pornsawan (2004) determined the As TF values for five emergent plants suitable for the removal of As from contaminated water (*Typha* spp., *Canna* spp., *Colocasia esculenta*, *Heliconia spittacorum* and *Thalia dealbato*); the TF values were in the range of 0.07 to 0.21 for each of the tested plants (Table 70.1).

The bioconcentration and translocation skills of hyperaccumulators are, by definition, much higher than those of accumulator and tolerant plants (Ma *et al.* 2001, Fitz and Wenzel 2002, Koller 2007). However, when hyperaccumulators are used in phytoremediation it will be necessary to consider that the improper disposal of As-rich plants may create further environmental problems (Rathinasabapathi *et al.* 2006). Accumulator and tolerant plants have an advantage over hyperaccumulators in this respect: even though they accumulate a high amount of As in their roots, they only translocate a small quantity to their aerial parts. In addition, many of these plant species are perennial; this includes *E. macrostachya* and *S. americanus*.

E. macrostachya and *S. americanus* are members of the Cyperaceae family; however, they haven't been considered as possible As phytoremediators. This study shows that they have the potential to tolerate and accumulate As, and therefore both plants can be useful in phytostabilization and rhizofiltration of polluted water. However, more laboratory and field studies are needed in order to determine other important characteristics such as As phytotoxic levels, the influence of the As oxidation state, and bioavailability in presence of other contaminants.

70.4   CONCLUSIONS

Eleocharis macrostachya and *Schoenoplectus americanus* have As phytoremediation potential, since no negative effects were observed during five months with the studied doses. *E. macrostachya* is As-tolerant at doses of 3 and 6 mg/l, and becomes an accumulator at a dose of 9 mg/l. *S. americanus* is an As accumulator at doses of 3 and 6 mg/l, and becomes tolerant at a dose of 9 mg/l. *E. macrostachya* accumulated the highest amount of As in their vegetal tissue. For both species, the behavior of the results for the aerial part of the plants was a straight line; for the roots, the behavior of results correlated to a third order polynomial equation. It is advisable to perform later experiments with higher doses than those used in this study, in order to determine each species' phytotoxic dose and speciation effect on bioconcentration and translocation factors. Both plants belong to the Cyperaceae family and have the potential to be considered for the phytostabilization and rhizofiltration of As-polluted sites.

REFERENCES

Bhattacharya, P., Ahmed, K.M., Hasan, M.A., Broms, S., Fogelström, J., Jacks, G., Sracek, O., von Brömssen, M. and Routh, J.: Mobility of arsenic in groundwater in part Brahmanbaria district, NE Bangladesh. In: Groundwater arsenic contamination in India: Extent and severity. In: R. Naidu, E. Smith, G. Owens, P. Bhattacharya and P. Nadebaum (eds): *Managing arsenic in the environment: from soil to human health.* CSIRO Publishing, Melbourne, Australia, 2006, pp. 95–115.

Bhattacharya, P., Jacks, G., Ahmed, K.M., Khan, A.A. and Routh, J.: Arsenic in groundwater of the Bengal Delta Plain aquifers in Bangladesh. *Bull. Env. Cont. Toxicol.* 69 (2002), pp. 538–545.

Castro, E.M.L. and Wong, M.: Remoción de arsénico a nivel domiciliario. Resumen Informe Técnico HDT–CEPIS N^O 74, OPS-OMS, Lima, Peru, 1999.

CEM Corporation: User's manual-digestion system. CEM innovators in microwave instrumentations, North Carolina, 2002, available from: http://www.cem.com/analytical/digestion.asp (accessed: April 2004).

Chao-Yang, W. and Tong-Bin, C.: Arsenic accumulation by two brake ferns growing on an arsenic mine and their potential in phytoremediation. *Chemosphere* 63 (2006), pp. 1048–1053.

Fitz, W.J. and Wenzel, W.W.: Arsenic transformations in the soil-rhizosphere-plant system: fundamentals and potential application to phytoremediation. *J. Biotecnol.* 99 (2002), pp. 259–278.

Jankong, P., Visoottiviseth, P. and Khokiattiwong, S.: Enhanced phytoremediation of arsenic contaminated land. *Chemosphere* 68 (2007), pp. 1906–1912.

Koller, C.E., Patrick, J.W., Rose, R.J., Offler, C.E. and MacFarlane, G.R.: *Pteris umbrosa* R. Br. as an arsenic hyperaccumulator: accumulation, partitioning and comparison with the established As hyperaccumulator Pteris vittata. *Chemosphere* 66 (2007), pp. 1256–1263.

Krämer, U.: Phytoremediation: novel approaches to cleaning up polluted soils. *Curr. Opin. Biotech.* 16 (2005), pp. 133–141.

Ma, L.Q., Komar, K.M., Tu, C., Zhang, W., Cai, Y. and Kennelley, E.: A fern that hyperaccumulates arsenic. *Nature* 409 (2001), p. 579.

McArthur, J.M., Ravenscroft, P., Safiullah, S. and Thirwall, M.F.: Arsenic in groundwater: testing pollution mechanisms for sedimentary aquifers in Bangladesh. *Water Resour. Res.* 37:1 (2001), pp. 109–117.

Miller, G.T.: Ecología and medio ambiente: Introducción a la ciencia ambiental, en desarrollo sustentable a la conciencia de conservación del planeta. Grupo Editorial Iberoamérica, Mexico City, Mexico, 1994.

Miller, R.R.: Groundwater remediation technologies analysis center. 1996, available from: http://www.gwrtac.org/pdf/Horiz_o.pdf (accessed: April 2004).

Mukherjee, A.B., Bhattacharya, P., Jacks, G., Banerjee, D.M., Ramanathan, A.L., Mahanta, C. Chandrashekharam, D., Chatterjee, D. and Naidu, R.: Groundwater arsenic contamination in India: Extent and severity. In: Groundwater arsenic contamination in India: Extent and severity. In: R. Naidu, E. Smith, G. Owens, P. Bhattacharya and P. Nadebaum (eds): *Managing arsenic in the environment: from soil to human health.* CSIRO Publishing, Melbourne, Australia, 2006, pp. 533–594.

Parijat Tripathi, Ramanathan, A.L., Pankaj Kumar, Anshumali Singh, Bhattacharya, P., Thunvik, R. and J. Bundschuh: Arsenic distribution in the groundwater in the central Gangetic plains of Uttar Pradesh, India. In: J. Bundschuh, M.A. Armienta, P. Birkle, P. Bhattacharya, J. Matschullat and A.B. Mukherjee (eds):

Natural arsenic in groundwater of Latin America. Taylor and Francis/Balkema. Leiden, The Netherlands, 2009 (this volume).

Prasad, M.N.V. and de Oliveira-Freitas, H.M.: Metal hyperaccumulation in plants—Biodiversity prospecting for phytoremediation technology. *Electronic J. Biotechnol.* [On line] 6:3 (2003), available from: http://www. ejbiotechnology.info/content/vol2/issue3/full/3/index.html (accessed: November 2005).

Rathinasabapathi, B., Ma L.Q. and Srivastava, M.: Arsenic hyperaccumulating ferns and their application to phytoremediation of arsenic contaminated sites. In: *Floriculture, Ornamental and Plant Biotechnology*, Volume III. Global Science Books, UK, 2006.

Ray, U.N., Sinha, S., Tripathi, R.D. and Chandra, P.: Wastewater treatability potential of some aquatic macrophytes: Removal of heavy metals. *Ecol. Eng.* 5 (1995), pp. 5–12.

SAS: User's Guide: Proc-Mixed. Ver. 8.2, SAS Institute Inc., Cary, NC, 1998.

Smedley, P.L. and Kinniburgh, D.G.: A review of the source, behavior and distribution of arsenic in natural waters. *Appl. Geochem.* 17 (2002), pp. 517–568.

Tu, S., Ma, L.Q., Fayiga, O.A. and Zillioux, E.J.: Phytoremediation of arsenic-contaminated groundwater by the arsenic hyperaccumulating fern *Pteris vittata* L. *Int. J. Phytorem.* 6:1 (2004), pp. 35–47.

USDA: *Schoenoplectus americanus.* US Department of Agriculture, 2002, available from: http://plants.usda. gov/plantguide/pdf/csscam6.pdf (accessed: June 2004).

Zhang, W., Cai, Y., Tu, C. and Ma, L.Q.: Arsenic speciation and distribution in an arsenic hyperaccumulating plant. *Sci. Total Environ.* 300 (2002), pp. 167–177.

CHAPTER 71

Filter development from low cost materials for arsenic removal from water

G. Muñiz, L.A. Manjarrez-Nevárez, J. Pardo-Rueda, A. Rueda-Ramírez,
V. Torres-Muñoz & M.L. Ballinas-Casarrubias
Facultad de Ciencias Químicas, Universidad Autónoma de Chihuahua, Chih., Mexico

G. González-Sánchez
Depto. de Medio Ambiente y Energía, Centro de Investigación en Materiales Avanzados, S.C., Chih., Mexico

ABSTRACT: Filters from low cost materials (cellulose and activated carbon from lignite) were developed and tested for arsenic (As) separation from the aqueous phase. Acetylated cellulose was synthesized from low-cost materials (cotton and Kraft cellulose) and combined with activated carbon. Both materials were characterized by different techniques such as scanning electron microscopy (SEM), transmission electron microscopy (TEM), X-ray diffraction spectrometry (XRD), laser diffractometry, Fourier transformed infrared spectroscopy (FTIR) and Brunauer Emett and Teller determination (BET). Pore distribution curves were shown, and the maximum was found for micropores of 5 Å. Laser diffractometry allows the calculation of particle media at 80 μm. A BET analysis reveals a 1500 m^2/g of carbon surface area. An elemental analysis reveals an iron concentration of 3% as an oxide. For the acetylated cellulose obtained, FTIR allowed the comparison of spectra that were in accordance with commercial triacetate cellulose. For film preparation, activated carbon and acetylated cellulose were put in contact independently, with methylene chloride for homogeneous dispersion. This was improved for low carbon loading (CL). So, CL was kept constant at 1% and 5% for filter preparation. Flat sheet composite films from cotton were obtained by the evaporation method, which occurred at a controlled temperature and relative humidity conditions. Microscopy techniques reveal a narrow carbon distribution and membrane structure, obtaining an open morphology for filters obtained with carbon, and a tight one for the one without carbon. Films were tested in a continuous flow cell at a pressure of 4 bar, with synthetic As solutions [200 and 400 μg/l As (V)]. There is variation in As removal due to As concentration and CL. The filters obtained allow separating As in the range of 16–63%. Fluxes are in the range of 47–49 l/m^2/h/bar, characteristic of the micro filtration membrane process.

71.1 INTRODUCTION

The presence of arsenic (As) in groundwater for human intake is of international concern due to its well known toxicity (Faust and Aly 1998). Natural geochemical pollution through mineral leaching is the primary source of dissolved As in groundwaters around the world. The recently promulgated As maximum contaminant level (MCL) in drinking water of 10 μg/l would require adjusting more than four thousand water supply systems serving 20 million people (Ng *et al.* 2003). A vast majority of these systems are fed by groundwaters. Therefore, there is a high demand to apply efficient methods for As removal from drinking water. Currently, a variety of methods have been developed for this purpose.

The conventional physical-chemical processes used for As removal can be classified on the basis of the principles involved: (1) precipitation, (2) adsorption, (3) ion exchange, and (4) membrane technology.

Membrane treatment is a technology theoretically capable of meeting the lowered As MCL (Kartinen *et al.* 1995) and of treating water with a high salt composition.

Pressure-driven processes can be divided into four overlapping categories of increasing selectivity: microfiltration (MF), ultrafiltration (UF), nanofiltration (NF) and hyperfiltration or reverse osmosis (RO). It should be noted that, in general, driving pressure increases as selectivity increases. Clearly, it is desirable to achieve the required degree of separation (rejection) at the maximum specific flux (membrane flux/driving pressure). In general, separation is accomplished by MF membranes and UF membranes via mechanical sieving, while capillary flow or solution diffusion is responsible for separation in NF membranes and RO membranes (AWWA 1992).

Operating conditions, such as source water characteristics, water contaminants, As species, and membrane characteristics affect the membrane process. RO and NF have an optimal removal efficiency of As, especially RO which can achieve a removal of over 95%. However, the percentage of product water that can be produced from the feed water for RO and NF is usually lower than the percentage of product water from other membrane operations. The small pore size makes RO and NF membranes use more energy to push water past the membrane than UF and MF membranes and are more prone to fouling than UF and MF membranes. The cost of membranes depends on membrane type, size and the degree of automation of the design. Operating conditions such as temperature, pressure, solution pH and chemical compatibility must therefore always be considered.

In recent years, a tremendous amount of research has been conducted to identify technologies for As removal that can be applied in rural areas. Membrane technologies could be the best choice because of all these advantages, provided operation reduces its cost (Shih 2005).

Vroenhoek *et al.* (2000) found that thin film composite-type membranes could improve As removal efficiency when compared to cellulose acetate in membrane operation. They found that the reason is the higher selectivity of the thin film composite-type membrane compared to the cellulose acetate membrane. Thin film composite membranes also present a higher permeated flow rate and need much lower driving pressures than those used by cellulose acetate membranes.

Considering the situation in developing countries, such as low annual incomes and low availability of electric power, traditional RO technology seems difficult to apply due to its high energy consumption. Therefore, a low operating cost of As removal technology through other types of membranes could be researched.

This work is a study of the synthesis of composite cellulose membranes with activated carbon. Cellulose fibers were treated for acetylation in order to obtain flat sheet composite filters. Synthesis was carried out by modifying a previously published procedure (Groggins 1958). Acetylated cellulose was analyzed by Fourier transformed infrared spectroscopy (FTIR). Activated carbon properties were measured by BET analysis, laser diffractometry, TEM and X ray diffraction spectrometry. Flat sheet composite filters were characterized by SEM and tested in a continuous flow cell with an As synthetic solution.

71.2   EXPERIMENT

71.2.1   *Acetylated cellulose synthesis and characterization*

Synthesis was carried out by modifying a previously published procedure (Groggins 1958). Cellulose (100 g either of Kraft cellulose, Cellulose of Mexico, or 100 g of commercial cotton) reacted with acetic acid (223 ml, 99.9% $\rho = 1.049$ g/cm^3, JT Baker) for 60 minutes at 37.8°C. Then, sulfuric acid (98.2%, $\rho = 1.834$ g/cm^3, JT Baker) was added drop wise (0.5 ml) with acetic acid (380 ml) and stirred for 45 minutes at 37°C. Afterwards, the solution was cooled down to 18°C. When the desired temperature was reached, acetic anhydride (250 ml 99.8%, $\rho = 1.08$ g/cm^3, JT Baker) was put in contact with the mixture, stirring for 2 hours. Acetylated cellulose was obtained after being purified with water until neutralization, precipitated with a solution of acetic acid (20%).

The acetylated cellulose was dried in an oven at 80°C and analyzed by FTIR, directly with the attenuated reflectance mode (ATR). Data acquisition was made over the entire wavelength rate and it was compared to cellulose triacetate (Sigma-Aldrich, 99%).

71.2.2   *Activated carbon characterization*

Activated carbon (LQ 1000 Carbochem) was used, triturated and sieved to obtain a narrow distribution in particle size. For the determination of surface carbon area and pore distribution, the BET equation was used for adsorption of N_2 in an Autosorb-1, Quantachrome device, at 77 K. Total volume was calculated by the adsorption volume at the maximum relative pressure reached. Particle size distribution was measured by laser diffractometry using a Mastersizer MS2000 (Malvern Instruments). Hexametaphosphate was used as a dispersant for measurement. A transmission electron microscope (Phillips CM 200 at 200 kVA) was used for carbon analysis.

71.2.3   *Composite film preparation*

Activated carbon (LQ1000, 1.8 μm < particle diameter < 8.7 μm) and acetylated cellulose were put in contact with methylene chloride for 24 hours, separately. Organic and activated carbon solutions were mixed up for composite filter synthesis for a final concentration of 0.01% of acetylated cellulose and 1% or 5% CL. The solution was poured into a glass plate for a 14 × 16 cm membrane. Evaporation occurred in a controlled temperature and relative humidity chamber (Shell Lab) for 10 hours, at a temperature of 40°C and relative humidity of 35%.

Carbon loading was kept constant (1% w/w). The prepared filters were dried at 110°C in an oven for 12 hours.

71.2.4   *Composite film characterization*

Films were analyzed by SEM (SEM JEOL JSM5800-LV). They were previously fractured in liquid nitrogen and then treated further in a covering system (Denton Desk-II Gatan) with gold. Micrographs were taken at 15 kV at top views. The structures were analyzed at a magnification of 5000×.

71.2.5   *Composite film performance*

Arsenic permeation flux (an As solution of 200 or 400 μg/l) was measured using a SEPA-Osmonics cell which is operated at constant pressure and temperature. The film area was 266 cm^2. Films were characterized in the module after pretreatment with pure water for 2 hours. The permeation flux was calculated as $F = V/AtP$, where V is the total volume of the water permeated during the experiment; A represents the membrane area, t is the operation time and P is the pressure. Operation time was 5 hours for each membrane after reaching a stationary state.

Hydride generation atomic absorption spectrometry was used for As analysis (Perkin Elmer 3100). The equipment settings were the following: 193.7 nm wavelength, 0.7 mm slit, 50 ml/min argon flow rate.

71.3   RESULTS AND DISCUSSION

71.3.1   *Acetylated cellulose synthesis and characterization*

Acetylated cellulose was obtained from two different raw materials. First, synthesis was carried out using Kraft cellulose and the method described above. The FTIR spectrum of the product obtained is shown in Figure 71.1. Cellulose triacetate spectra were acquired for comparison. Peaks that appear at 1730 1/cm (C=O ester group), 1381 1/cm [C-H bond and (–O(C=O)-CH$_3$)] and

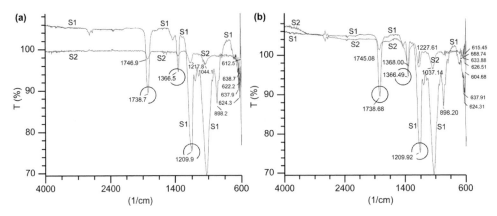

Figure 71.1.   (a) FTIR spectra of acetylated cellulose obtained using Kraft cellulose as a raw material; the spectrum S1 corresponds to commercial cellulose triacetate; the main peaks are surrounded by a circle; (b) FTIR spectra of acetylated cellulose obtained using cotton as a raw material. The spectrum S1 corresponds to commercial cellulose triacetate, the spectrum S2 corresponds to the material obtained.

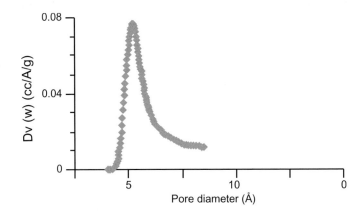

Figure 71.2.   Pore diameter distribution of activated carbon (LQ 1000) by BET analysis.

1200–1250 1/cm (-CO- stretching acetyl group) are characteristic of cellulose triacetate. These signals are surrounded by a circle in figures. The acetylated cellulose obtained from cotton appears to be more concentrated, showing the main signals of commercial triacetate cellulose. In Kraft cellulose, some of the peaks are not totally appreciated, due to concentration or to the non-existence of the groups associated with those signals.

71.3.2   *Activated carbon characterization*

Activated carbon was characterized by some techniques in order to obtain particle size distribution and pore diameter. These parameters are of fundamental importance due to the role of particles in the composite material. As the particle size decreases, polymer-particle interaction is better (Odergard *et al.* 2005). For composite materials, it is necessary to find a large surface area-to-volume ratio, which promotes a strong interaction and adhesion between the inorganic filler and the polymer (Frogley *et al.* 2003).

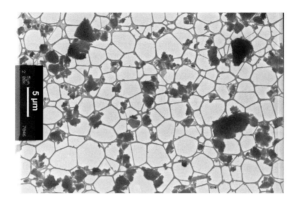

Figure 71.3.   TEM image of activated carbon size distribution.

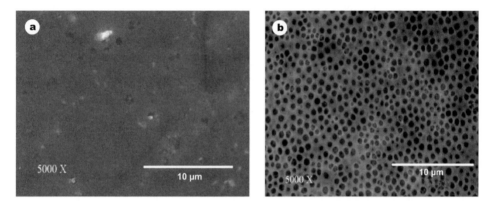

Figure 71.4.   Surface of triacetate cellulose membrane without carbon (a), and with 1% CL (b).

Figure 71.2 shows the pore distribution of activated carbon LQ 1000. Its average diameter pore was calculated at 5.57 Å, and its surface area at 1058 m^2/g. Anson *et al.* (2004) find optimal carbon to have an average diameter of 28.2 Å, 818 m^2/g surface area and mean particle size of 4.47 μm for ABS (acrylonitrite-butadiene-styrene) in a gas separation process.

Figure 71.3 is a TEM image of the particle size distribution of the carbon obtained. There are tiny particles of less than 1 μm but large particles predominate. The laser diffraction technique calculates 1.841 μm < particle diameter < 8.73 μm.

For elemental analysis, the following ratio (%) were found: C 88.09, oxygen 8.02, Al 0.88, Si 2.02, S 0.46, Fe 0.54. So there can be an effect on As adsorption due to metal content (as As tends to form insoluble arsenates) and from an interaction among activated carbon surfaces.

71.3.3   *Acetylated cellulose-activated carbon film characterization and performance*

Acetylated cellulose from cotton and activated carbon composite films was obtained at 40°C and 35% RH. The structure was observed by SEM, and micrographs are shown in Figure 71.4. There is a morphological change due to the presence of carbon. Even though carbon particles cannot be seen individually through SEM, it can be inferred that activated carbon is dispersed homogeneously throughout the material, interacting with the polymer to form a composite that is mechanically optimal for membrane operation.

Table 71.1.   Membrane fluxes and arsenic removal. Operating pressure: 4 bar; 5 h operation (n = 5).

Arsenic concentration	CL %	Arsenic removal	Flux ($l/m^2/h/bar$)
200 µg/l	1	23.49% ± 7.34	47.36 ± 1.11
400 µg/l	1	51.53% ± 7.45	47.96 ± 1.32
200 µg/l	5	30.92% ± 0.76	48.49 ± 2.20
400 µg/l	5	39.38% ± 7.37	49.06 ± 1.04

Arsenic solutions of 200 and 400 µg/l were treated with a membrane obtained at 1% and 5% CL in a continuous mode. Arsenic removal and fluxes are shown in Table 71.1. The fluxes obtained without carbon are two orders of magnitude lower than those obtained with carbon. This is explained with new characteristics of the composite material; carbon must act as a filler in the composite, optimizing its wetability and opening structure.

Arsenic is removed by membranes in the range of 16–63%. Fluxes are in the range of 47–49 $l/m^2/h/bar$, characteristic of the MF membrane process (Mulder 1996). Removal is a function of As concentration and carbon concentration into the film. It should be explained based on the interaction of As compounds with the film structure.

At typical pH values in natural waters (pH 5–8), arsenate exists as an anion while arsenite remains as a fully protonated neutral molecule. Redox conditions control the speciation of As between arsenate and arsenite. In natural waters, As is generally found in the oxidized arsenate form.

When a solution containing ions is brought in contact with membranes possessing a fixed surface charge, the passage of ions possessing the same charge as the membrane (co-ion) can be inhibited (Strathmann 1992). Activated carbon at the typical pH of water has a negative charge. So, membrane composition combined with solution characteristics can influence rejection via electrostatic double layer interactions. This rejection could increase with solute concentration (Bungay *et al.* 1986) as it is observed for 1% CL films.

More specifically, when a solution with anionic arsenate is brought in contact with a membrane possessing a fixed negative charge, the rejection of arsenate may be greater than if the membrane were uncharged. Hence, the selection of a membrane possessing a slight negative charge, such as using activated carbon at a pH higher than its pH_{ZPC} may be advantageous for the removal of As from drinking water. This is a particularly fortunate set of circumstances, because the speciation of As in natural waters is primarily in the anionic arsenate form.

71.4   CONCLUSIONS

Acetylation of cellulose is possible by making it reacts under controlled conditions by an acidic mixture. Raw material affects acetyl concentration during reaction (among cotton and Kraft cellulose) and it is comparable to triacetate cellulose. Activated carbon (LQ 1000) particles can be obtained in a narrow distribution range 1.841 µm < particle diameter < 8.73 µm. They also present an average diameter pore of 5.57 Å, and a surface area of 1058 m^2/g according to the BET method. Acetylated cellulose-activated carbon films are obtained and As could be removed by membranes in the range of 16–63%. Fluxes are in the range of 47–49 $l/m^2/h/bar$, which is characteristic of the MF membrane process.

REFERENCES

Anson, M., Marchese, J., Garis, E., Ochoa, N. and Pagliero, C.: ABS copolymer-activated carbon mixed matrix membranes for CO_2/CH_4 separation. *J. Membr. Sci.* 243 (2004), pp. 19–28.

AWWA: Membrane processes in potable water treatment, AWWA Membrane Technology Research Committee, Committee Report. *J. AWWA* 84 (1992), p. 59.

Bungay, P.M.: Transport principles—porous membranes. In: P.M. Bungay, H.K. Londsale and M.N. de Pinho (eds): Synthetic membranes: Science, engineering and applications. D. Reidel Publishing Co., Dordrecht, The Netherlands, 1986, pp. 57–107.

Faust, S.D. and Aly, O.M.: Chemistry of water treatment. 2nd ed., CRC Press, New York, USA, 1998.

Frogley, M.D., Ravich, D. and Wagner, H.D.: Mechanical properties of carbon nanoparticle-reinforced elastomers. *Comp. Sci. Technol.* 63 (2003), pp. 1647–1654.

Groggins, P.H.: *Unit processes in organic synthesis*. 4th ed. Mc Graw Hill, New York, 1958.

Kartinen, E.O. and Martin, C.J.: An overview of arsenic removal processes. *Desalination* 103 (1995), pp. 79–85.

Mulder, M.: Basic principles of membrane technology. 2nd ed., Kluwer Academic Publishers, Dordrecht, The Netherlands, 1996.

Ng, J.C., Wang, J. and Shraim, A.: A global health problem caused by arsenic from natural sources. *Chemosphere* 52 (2003), pp. 1353–1359.

Odegard, G.M., Clancy, T.C. and Gates, T.S.: Modeling of the mechanical properties of nanoparticle/polymer composites. *Polymer* 46 (2005), pp. 553–562.

Shih, M.: An overview of arsenic removal by pressure-driven membrane processes. *Desalination* 172 (2005), pp. 85–97.

Strathmann, H.: Ion-Exchange Membrane. In: W.S.W. Ho and K.K. Sirkar (eds): *Membrane Handbook*. Chapman and Hall, New York, 1992.

Van Nostrand, R., Vroenhoek, E. and Waypa, J.: Arsenic removal from drinking water by a loose nanofiltration membrane. *Desalination* 130 (2000), pp. 265–277.

CHAPTER 72

Arsenic removal from water of Huautla, Morelos, Mexico using capacitive deionization

S. Garrido, M. Aviles, A. Ramirez, C. Calderon & A. Ramirez-Orozco
Instituto Mexicano de Tecnologia del Agua (IMTA), Jiutepec, Mor., Mexico

A. Nieto, G. Shelp & L. Seed
Enpar Technologies Inc., Guelph, ON, Canada

M.E. Cebrian & E. Vera
Centro de Investigación y de Estudios Avanzados del Instituto Politécnico Nacional (CINVESTAV-IPN), Mexico City, Mexico

ABSTRACT: Concentrations of arsenic (As) that reach values of 2.35 mg/l have been identified in potable water sources in Mexico. Such concentrations by far surpass the maximum permissible limit of 0.025 mg/l set as the Mexican standard. It has been estimated that nearly 500,000 people living in rural areas are exposed through their water intake, to concentrations of As in excess of 0.05 mg/l. The objective of this study was to evaluate the removal of As present in the water for human consumption in the town of Huautla, Morelos, Mexico using electrochemical technology. This technology was developed in Canada and it is based on the principle of capacitive deionization, as an electrostatic charging system formed by carbon electrodes. The electrodes are supplied with direct current (~1 V, 0–375 A), which produces surfaces with positive and negative charges. The ionic compounds that contain As are thus electrostatically adsorbed onto the electrodes, and the water obtained reached concentrations below 0.005 mg/l of total As. The average percentage of As removal was 98.51% and the volume of reject water was 3%. The results of this study indicate that this technology is more efficient and potentially more economical than conventional technologies.

72.1 INTRODUCTION

Arsenic toxicity is widely known from poisoning cases and medical uses. Recent epidemiological reports of cancer and health problems such as "black foot" and non-cancerous skin diseases associated with arsenic (As) in Taiwan (Tseng 1977) and in other countries (Kumar and Suzuki 2002), have produced increasing interest on the public health effects related to the chronic exposure to As found in water for human use and consumption.

The presence of As in groundwater has been found to exceed the international standards in countries like India (Ahmed *et al.* 2004), Taiwán (Lui *et al.* 2003), the United States (Peters and Blum 2003), Argentina (Farias *et al.* 2003), as well as in the Mexican states Baja California Sur, Chihuahua, Coahuila, Durango, Guanajuato, Guerrero, Hidalgo, and Morelos, (Armienta *et al.* 1997, Cebrian *et al.* 1983, 1994, Cole *et al.* 2004, Rodríguez *et al.* 1996, 2004).

The natural occurrence of As in groundwater is linked to the lithology of the geological materials that comprise the aquifer and to the redox processes involving this element (Smedley and Kinniburgh 2002). The most common origin of As is the oxidation of minerals with high As content, such as arsenopyrite (FeAsS), scorodite ($FeAsO_4 \times 2H_2O$), and orpiment (As_2S_3) which can appear in different geological environments (Peters and Blum 2003, Sraceck *et al.* 2004). In these cases a close correlation can be generally observed between pH and the concentrations of As and Fe in water, which infer the hydrogeochemical characteristics of this element (Smedley and Kinniburgh 2002). Other origins of As in groundwater are linked to anthropogenic factors like lixiviation of mine residues (Armienta *et al.* 1997) or the use of pesticides (Liu *et al.* 2003), wood preservatives and pigments.

The World Health Organization (WHO) has established a maximum concentration of As in water for human consumption of 0.010 mg/l. The Mexican Standard, NOM-127-SSA1-1994 is used to define a maximum concentration of 0.050 mg/l. In year 2000, the maximum permissible limit was modified to 0.025 mg/l by 2005. However, a study by the National Academy of Sciences of the United States found in 2001 that even a concentration of 0.020 mg/l represents a risk of contracting cancer (USEPA 2001).

72.1.1 *Conventional technologies*

The following technologies are more effective in removing As(V) than in removing As(III):

- The removal of As through coagulation, followed by filtration is an effective process for the removal of As(V). Based on laboratory tests and pilot tests, it has been determined that the coagulant $FeCl_3$ has a greater As removal efficiency than $Al_2(SO_4)_3$ namely, 96% *vs.* 85% respectively. An important factor that must be considered in the use of coagulants is the pH of water (Pande *et al.* 1997, Petkova *et al.* 1997, Ngo *et al.* 2002).
- Lime softening [i.e., $CaCO_3$, $Mg(OH)_2$ and $Fe(OH)_3$] performed at pH above 10.5 can achieve high removal efficiencies of As(V). However, to reach a removal percentage above 80%, iron must be added to the water or further treatment such as ion exchange must be performed as a polishing stage (McNeil 1996, McNeil and Edwards 1997).
- Activated alumina is effective in treating water with high total dissolved solid (TDS) concentrations. It requires pH adjustment between 5.5 and 6.0 to favor the adsorption of As(V) over other anions present in the water, such as selenium, fluoride, chloride, sulfates and silicates. Activated alumina is highly selective towards As(V). Hence, regeneration problems can occur, where sorption capacity is gradually lost at a rate between 5% and 10% after each regeneration cycle (Gary *et al.* 1999).
- The process of ion exchange is recommended for water with low concentrations of TDS given that anions like sulfates (<120 mg/l), nitrates, fluoride, and selenium compete with As for the exchange sites, which can affect the length of the column. Suspended solids and precipitated iron can cause blockage of the treatment bed. If ion exchange treatment of water with these characteristics is required, pretreatment must be provided (Kartinen and Martín 1995).
- The process of adsorption using zeolite conditioned with iron oxides is effective for the removal of As from water. The main adsorption mechanism is through electrostatic interactions. The chemical affinity of As towards the iron coating surface is also involved in the removal process. The most important factor controlling the adsorption of As(V) onto the medium is pH. Acid pH values between 5.5 and 6.5 produce the highest removal efficiencies. Rivera and Piña (2000), determined that by the end of a 72 hour period, the As adsorption capacity of this medium was 39.81 mg As/190.5 g conditioned zeolite (0.2 mg As/g) without reaching saturation. They treated 212 l of water per l of packed bed, with a concentration of As below the maximum permissible limit of 0.03 mg/l in the first cycle. In the second cycle, the conditioned zeolite was capable of adsorbing 8.36 mg As/l, which is only 20% of the removal capacity during the first cycle. Information about this As removal system is limited with respect to the zeolite conditioning process and to the retention of the iron oxides onto the zeolite surface (Ngo *et al.* 2002).
- Membrane technologies like reverse osmosis provide removal efficiencies above 95% when the operating pressure is ideal. The reject water volume is a disadvantage of reverse osmosis in places where water is scarce (Kartinen and Martín 1995). Electrodialysis can provide As removal efficiencies up to 80%.
- The majority of these technologies are effective in removing As from water supplies. However, they present some problems. For example, most of them produce waste sludge contaminated with As, which requires proper disposal. Some are expensive given that they require sophisticated equipment, while others present difficulties when used in areas with low economical resources.

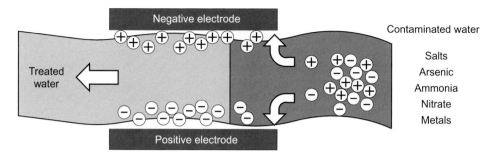

Figure 72.1. Principle of the first stage of the capacitive deionization process: Ionic adsorption.

- Biological treatment transforms, stabilizes and/or removes As by means of microorganisms. Microorganisms, primarily certain specific bacteria, accomplish this by oxidation/reduction, mineralization, detoxification or methylation. Critical factors include energy and carbon source; aerobic, anoxic or anaerobic conditions; temperature; pH.

72.1.2 *Unconventional technology*

The electrochemical process called capacitive deionization has been designed to remove ionic species from sea water, brackish water, or contaminated groundwater. The capacitive deionization system DesEL, the subject of this study, was developed by ENPAR Technologies Inc. from Guelph, Ontario, Canada. The main component of a DesEL plant is an electrostatic charging cell which behaves as a capacitor and is comprised of carbon electrodes. During the first stage of the process, the capacitor is energized using direct current, creating positive and negatively charged surfaces. Ionic compounds such as arsenic, ammonia, and nitrate are attracted to and electrostatically adsorbed onto the surface of the electrodes (Fig. 72.1).

To regenerate the electrode surfaces, the polarity of the cell is automatically reversed during stage two, causing the capacitor to release the contaminants into the cell channels. The third stage of the process is to remove the contaminants from the cell by flushing with a small quantity of liquid forming a concentrated solution. This mode of operation allows the DesEL system to combine high water recoveries with high ion removal efficiencies.

The operating potential is relatively low (approximately 1.2 V) so that no electrolysis reactions occur precluding breakdown of the capacitor material and the formation of secondary solid phases. In contrast to reverse osmosis or ion exchange systems, the DesEL system does not require softening or chloride removal as pretreatments. In addition, no chemical additives are needed for pretreatment or regeneration. The only pretreatment required consists of a 15 micron pre-filter, which is less costly than the 0.5 micron pre-filters required for reverse osmosis and ion exchange systems (note: the pre-treatment requirement for RO depends on the water source to be treated, often micro-filtration is required). The energy consumption of a DesEL plant decreases as the plant capacity increases. Typically, the energy consumption ranges from 1.5 to 1.0 kWh/m^3 for plants treating less than 10 m^3/day of brackish water and from 1.0 to 0.5 kWh/m^3 for larger plants. A DesEl-24 k plant treating 24 m^3/day of water with an initial TDS concentration of 1000 mg/l consumed 0.6 kWh/m^3 of treated water. These characteristics of the DesEL system result in low operational costs, high reliability and minimal maintenance requirements compared to conventional technologies (Nieto and Seed 2005).

72.2 GROUNDWATER SOURCE AREA

The groundwater used for this study was obtained from the community of Huautla, within the municipality of Tlalquitenango in the southern region of the state of Morelos (Fig. 72.2). This community

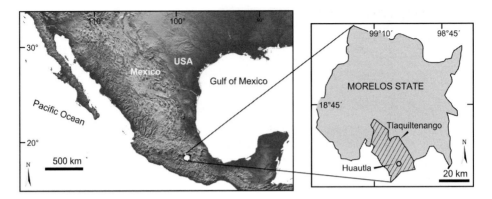

Figure 72.2.   Location map of Morelos state, study case.

has 1200 inhabitants and its water sources are the Cruz Pintada dam and the mine Pajaro Verde. It has a water distribution network and a storage tank with a capacity of 60 m^3 (INEGI 2000, CEAMA 2001).

72.3   METHODOLOGY

72.3.1   *Water quality at the source*

A water sample was obtained directly from the Pajaro Verde mine and its physico-chemical characteristics were determined according to Mexican standards and standard methods (APHA 1998). Total As and dissolved As were determined using an atomic absorption spectrophotometer (AAS) Perkin Elmer Model 2380, equipped with a hydride generator Perkin Elmer Model MHF-10. As(III) and As(V) were determined using an AAS Perkin Elmer Model 3100, equipped with an automatic injector Perkin Elmer Model FIAS 2000. The line source was an As electrodeless discharge lamp at 193.7 nm. Lead concentrations were determined using an AAS Varian Spectra Model AA220FS equipped with a graphite furnace and a hollow cathode lamp.

72.3.2   *Treatability tests*

The water used for the treatability tests was taken from the storage tank located in Huautla. The water in this tank was carried from the Pajaro Verde mine trough a 4 inch (10.16 cm) diameter pipe with an approximate length of 2 km. This water was disinfected using sodium hypochlorite at a concentration of 3 mg/l.

The treatability tests were performed using a capacitive deionization plant Model DesEL-4k (Fig. 72.3) at the Potable Water Systems laboratory of *Instituto Mexicano de Tecnologia del Agua* (IMTA). This plant has a nominal treatment capacity of 3.3 l/min for an initial TDS concentration of 1000 mg/l, it operates at a pressure of 1.0 kg/cm^2, and its electrical supply is rated for 127 V (ac), single phase, 15 A, 50/60 Hz. The two capacitive deionization cells of this plant are connected electrically in series and hydraulically in parallel (Fig. 72.4). Each cell is supplied with a peak voltage of 1.2 V (dc) and a maximum peak current of 375 A.

The DesEL process is configured by specifying five parameters through the human-machine-interface of the control panel. These parameters are: (1) adsorption time (stage 1) in seconds, (2) desorption time (stage 2) in seconds, (3) purge time (stage 3) in seconds, (4) pure water upper electrical conductivity (EC) limit in mS/cm, and (5) reject water lower EC limit in mS/cm.

The energy consumption of the DesEL plant was measured during the course of some experiments using a watt-hour meter Soar Model 2700.

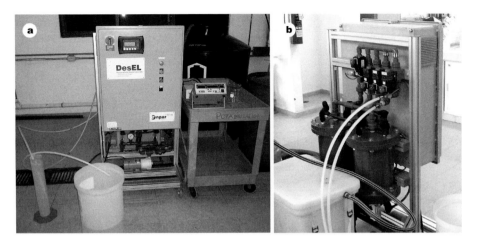

Figure 72.3. Front view (a) and rear view (b) of DesEL-4k plant showing pure and reject water outlets, deionization cells, and solenoid valves.

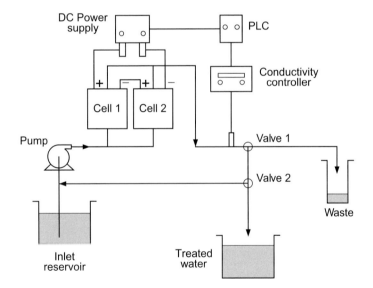

Figure 72.4. Typical schematic of a DesEL system.

72.3.3 *Design of experiments*

The experimental design involved 3 different configurations of the DesEL process, in which all parameters were kept constant except for the adsorption time (Table 72.1). In addition, raw waters with three different concentrations of As and Pb were used at the process inlet. The experiments 1 to 3, and 8 to 10 were performed with the raw water sample exactly as taken from the storage tank located in Huautla (Table 72.2). The same water was spiked with salts of As and Pb for all other experiments. Experiment 4 involved water with an As concentration of 0.8201 mg/l, while experiments 5 to 7 involved water with concentrations of As and Pb of 0.8201 mg/l and 0.0267 mg/l, respectively.

Table 72.1.   DesEL process configurations used in experiments.

Parameter	Units	DesEL configuration		
		A	B	C
Adsorption time	s	300	360	480
Desorption time	s	120	120	120
Purge time	s	20	20	20
Pure water upper EC limit	mS/cm	0.15	0.15	0.15
Reject water lower EC limit	mS/cm	3.00	3.00	3.00

Table 72.2.   Average characteristics of raw water and water treated with capacitive deionization.

Parameter	Raw water from Pajaro Verde mine	Water treated with capacitive deionization	Permissible limits (NOM 127-SSA1-1994)
Physical			
True color, UPt-Co	10	–	20
EC, μS/cm	492	25.21	–
Total suspended solids, mg/l	0.13	–	–
Total dissolved solids, mg/l	339.4	–	1000
Turbidity, UTN	0.1	0.02	5
Chemical			
pH	7.79	7.3	6.5–8.5
Arsenic$_{Total}$, mg/l	0.2098	<0.005	0.025
Arsenic$_{Dissolved}$, mg/l	0.2005	0.0048	
Lead, mg/l	0.0267	<0.01	0.01
Residual chlorine, mg/l	0.56	0.02	0.2–1.5
Total organic carbon, mg/l	2.99	–	–
Hardness$_{Total}$, mg/l	164	49.89	500

The salts used to spike the raw water samples were sodium arsenate ($Na_2HAsO_4 \times 7H_2O$), J.T. Baker and lead chloride ($PbCl_2$), CTR Scientific.

To distinguish between the energy consumed by the various parts of the DesEL process, incremental energy consumption measurements were performed with configuration C. In this way the energy consumed for controls, pump, and adsorption and desorption processes were determined during a particular experiment.

72.4   RESULTS AND DISCUSSION

72.4.1   *Water quality*

The average water quality results of the raw water sample taken from the Pajaro Verde mine and the samples of water treated with capacitive deionization are shown in Table 72.2. The working temperature of these measurements varied between 20.1 and 23.3°C.

The raw water from Pajaro Verde meets all the physico-chemical parameters and metal concentration limits of the Mexican Standard, NOM-127-SSA1-1994, except for total As, which is 8 times higher than the permissible limit. In Table 72.2 it can be seen that 95.57% of the total As is in dissolved form.

The removal of As from water using conventional technologies depends on its oxidation state, being As(V) removal more effective than As(III) removal. Arsenic can be pre-oxidized using oxidizers like chlorine, ferric chloride and permanganate. However, pre-oxidation with chlorine can generate undesirable chlorinated organic by-products in the presence of dissolved organic matter. Therefore, it is preferable to use ozone or hydrogen peroxide (USEPA 1998).

It has been proven under ideal conditions that the stoichiometric addition of sodium hypochlorite, NaOCl (free active chlorine concentration of 6%), required to oxidize 0.95 mg/l As(III) is 0.826 mg/l. In Table 72.3 and Figure 72.5 the inverse and direct relationships between the dose of NaOCl, and the concentrations of As(III) and As(V), respectively, are given. The correlation R^2 for respective sets of data were 0.9797 and 0.9855. In Figure 72.5a, the percentages of As(III) and As(V) are shown with respect to the NaOCl dose added. In the case of the water sample from Huautla, a dose of approximately 3 mg/l of NaOCl were added with a contact time of 60 minutes to ensure that the As(III) present in the water was oxidized to As(V).

72.4.2 *Treatability tests*

The treatability results of each experiment are presented in Table 72.4. The process configuration during the experiments 9 and 10 was changed after a few cycles to reach earlier a steady state of the bulk EC value of the treated water. As result of CO_2 solvation into the demineralized water, the pH of the treated water decreased by approximately 2 units with respect to the pH of 7.79 observed in the raw water. Aeration of the treated water increased the pH value back to 7.3. On average, the treated water showed reductions in EC of 94.88% and in hardness of 69.58%.

The change in total As concentration in the raw water from 0.2098 to 0.820 mg/l through the addition of sodium arsenate, as described in the methodology, did not have a major impact on the

Table 72.3. Oxidation of As(III) to As(V).

Dose of NaOCl mg/l	pH	Eh mV	As(III) mg/l	As(V) mg/l	As(III) %	As(V) %
0	6.14	324.3	0.95	0	100	0
0.048	–	325.5	0.72	0.229	75.87	24.13
0.144	6.08	359.7	0.57	0.379	60.06	39.94
0.288	5.99	396.1	0.43	0.519	45.31	54.69
0.432	5.96	436.0	0.28	0.669	29.50	70.50
0.576	5.76	492.2	0.20	0.749	21.07	78.93
0.720	5.66	553.8	0.11	0.839	11.59	88.41
0.792	5.68	606.2	0.05	0.899	5.27	94.73
0.826	–		0	0.949	0	100

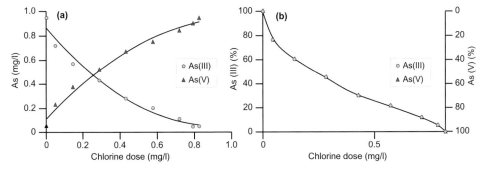

Figure 72.5. (a) Oxidation of As(III) to As(V) *vs.* chlorine dose; (b) Percentage of As(III) and As(V) species present during the oxidation of As(III).

Table 72.4.   Summary of treatability results.

Expt.	Config.	Raw water AsTotal mg/l	Raw water Pb mg/l	Treated water Volume 1	Treated water pH	Treated water EC μS/cm	Treated water HardnTot mg/l	Treated water AsTotal mg/l	Treated water Pb mg/l
1	A	0.2098	–	26	6.01	38.7	74	<0.005	–
2	B	0.2098	–	26.9	5.74	20.4	38	<0.005	–
3	C	0.2098	–	33.3	5.76	25.3	–	<0.005	–
4	C	0.8201	–	34.9	5.71	20.8	44	<0.005	–
5	C	0.8201	0.0267	35	5.46	17.4	59	<0.005	<0.01
6	C	0.8201	0.0267	35	5.55	12.5	54	<0.005	<0.01
7	C	0.8201	0.0267	35	5.75	18.7	54	<0.005	<0.01
8	C	0.2098	–	35	5.75	30.5	54		–
9	C: 6 cyc, A: 12cyc	0.2098	–	415	6.14	33.1	27	<0.005	–
10	A: 4 cyc, B: 16 cyc	0.2098	–	412	6.33	34.7	45	<0.005	–

As removal efficiencies of the respective experiments (97.62% and 99.39%). In all the experiments, the total As concentration in the treated water resulted in less than 0.005 mg/l, which is at least 5 times lower than the maximum permissible limit defined by NOM-127-SSA1-1994 and at most equal to the maximum contaminant level (MCL) proposed by USEPA (2001). The detection limit of the atomic absorption method used in the analysis of As is 0.005 mg/l, which is appropriate considering the maximum permissible limit of 0.025 mg/l set as Mexican standard. Moreover, the final concentration of As in the treated water produced from the various experiments did not exhibit any correlation with the various durations of the ion adsorption stage of the process. This would be expected given that the concentration of As in the raw water was very low with respect to the concentration of TDS, which has a strong correlation with the EC value used to control the process. Therefore, the removal efficiency of As with the DesEL process is expected to remain high regardless of the duration of the ion adsorption stage for As and TDS concentrations typically found in groundwater.

The adsorption rate of total As during the experiments 4 to 7 can be estimated as follows:

As adsorption rate $=$ Inlet $-$ Outlet
As adsorption rate $= C_i\, Q - C_0\, Q$
As adsorption rate $= Q(C_i - C_0)$
As adsorption rate $= 4.375$ l/min $(0.8201 - 0.005)$ mg/l
As adsorption rate $= 3.57$ mg/l

The reject water from the various experiments presented total As concentrations between 1.786 and 7.031 mg/l (Table 72.5). Moreover, the As adsorption rate was calculated for each experiment and multiplied by the respective adsorption times per cycle. The result is an estimate of the total milligrams of As adsorbed per cycle which, as illustrated in Figure 72.6, correlates ($R^2 = 0.9877$) to the concentrations of As found in the reject waters. Such correlation appears to indicate that the As ions are adsorbed onto the electrodes only as long as the electrical field promotes ionic adsorption and that they are not otherwise stored in the system.

An appropriate method for disposing of the As contained in the reject water will be subject of a future study.

With respect to the final Pb concentrations in the treated water of experiments 5, 6 and 7, they all were below 0.010 mg/l, which meets the maximum permissible limit of NOM-127-SSA1-1994.

72.4.3   Energy consumption

The specific energy consumption of the DesEL-4k plant was determined for experiments 6, 9 and 10. For each of these experiments, the average adsorption time was different, while the average

Table 72.5. Summary of reject water results.

Expt.	Volume %	pH	EC μS/cm	Arsenic$_{Total}$ mg/l	Pb mg/l
1	3.3	7.48	7310	1.80	–
2	3.4	7.53	7890	1.80	–
3	3.1	7.49	9320	2.77	–
4	2.9	7.55	9040	–	–
6	2.3	7.49	7850	7.03	–
7	2.9	7.96	7960	6.01	<0.01
8	2.9	7.09	8110	–	–
9	4.8	7.24	5600	–	–
10	4.1	7.39	5220	1.95	–

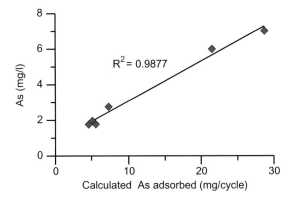

Figure 72.6. Concentration of As in reject water *vs.* calculated As adsorbed per cycle.

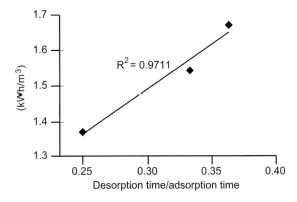

Figure 72.7. Specific energy consumption of plant DesEL-4k expressed in kWh per volume of treated water *vs.* the ratio of desorption time over adsorption time.

desorption time was kept equal. As indicated in Figure 72.7, a correlation exists ($R^2 = 0.9711$) between the specific energy consumption and the ratio of desorption to adsorption time. The values of specific energy consumption ranged from 1.371 to 1.67 kWh/m^3 the lowest for experiment 6, which had the longest adsorption time. Thus, each particular application of a DesEL plant requires finding optimum configuration values to attain a better economy for the process.

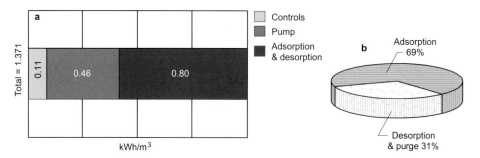

Figure 72.8.   (a) Specific energy consumption breakdown of controls, pump, and the adsorption and desorption parts of the capacitive deionization process using a DesEL-4k plant; (b) Comparison of the energy consumed during the desorption and purge stages and the adsorption stage. Both charts correspond to configuration C (Table 72.1).

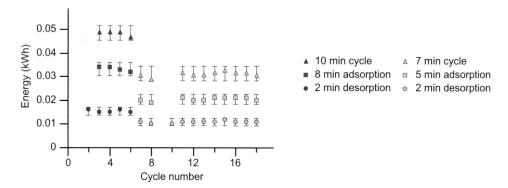

Figure 72.9.   Energy consumption per cycle and per process stage during experiment 9. Error bars are 6 sigma long.

The specific energy consumption of 1.371 kWh/m³ observed during experiment 6, was broken down into the energy consumed by the controls, the pump and the adsorption and desorption parts of the process. As presented in Figure 72.8a, the energy consumed by the controls and the pump is significant for the plant DesEL-4k, adding up to 0.57 kWh/m³. Typically, this amount is less significant in larger DesEL plants given that more water is treated with the same controls and with more efficient pumping. The specific energy dedicated by the DesEL-4k plant to the adsorption and desorption parts of the process was 0.8 kWh/m³, approximating the overall specific energy consumed by bigger DesEL plants.

Each stage of the DesEL process presented fairly stable energy consumptions from cycle to cycle (Fig. 72.9).

72.5   CONCLUSIONS

The As removal capability of the DesEL-4k plant proved to be sufficient in all the experiments, resulting in treated water with a total As concentration at least 5 times below the maximum permissible limit defined by NOM-127-SSA1-1994 and at least 2 times below the maximum contaminant level (MCL) proposed by USEPA.

The efficiency of As removal can be expected to be unaffected by changes in the duration of the ion adsorption stage for concentrations of As that occur naturally in groundwaters. Arsenic does not remain in the capacitive deionization cells, as indicated by the correlation between the concentration of As in the reject water and the calculated amount of As adsorbed onto the electrodes per cycle. The Pb removal capability of the DesEL-4k was sufficient to meet the Mexican Standard NOM-127-SSA1-1994.

The low values of specific energy consumption obtained are an attractive feature of the DesEL system considering that energy represents the sole input of the DesEL system. Along with very low maintenance requirements, low operational costs and free of chemical additives, the DesEL system appears to offer a manageable and, potentially, economically viable technology for the removal or As from groundwater, in comparison to the another techniques like as reverse osmosis, electrodialysis reversal, nanofiltration, coagulation/flocculation, activated alumina. The operating and maintenance cost was US$ 0.087/m³ for drinking water.

REFERENCES

Ahmed, K., Bhattacharya, P., Hasan, M., Akhter, S., Alam, S.M., Bhuyian, M.A., Imam, M.B., Khan, A. and Sracek, O.: Arsenic enrichment in groundwater of the alluvial aquifers in Bangladesh: an overview. *Appl. Geochem.* 19 (2004), pp. 181–200.

APHA, AWWA, WEF: Standard Methods for the examination of Water and Wastewater. 20th ed., American Public Health Association (APHA), the American Water Works Association (AWWA), and the Water Environment Federation (WEF), Washington, DC, 1998.

Armienta, M.A., Rodríguez, R., Aguayo, A., Ceniceros, N., Villaseñor, G. and Cruz, O.: Arsenic contamination of groundwater at Zimapán, Mexico. *Hydrogeol. J.* 5:2 (1997), pp. 39–46.

CEAMA: Diagnóstico e inventario de la infraestructura y los servicios hidráulicos en las localidades rurales de 16 municipios del Edo. de Morelos. Internal report, Comisión Estatal de Agua y Medio Ambiente de Morelos, Servicios de Ingeniería e Informática, Morelos, Mexico, 2001.

Cebrián, M., Albores, A., Aguilar, M. and Blakely, E.: Chronic arsenic poisoning in the north of Mexico. *Human Toxicol.* 2 (1983), pp. 121–133.

Cebrián, M.E., Albores, A., García-Vargas, G. and Del Razo, L.M.: Chronic arsenic poisoning in humans: The case of Mexico in the environment. Part II: Human health and ecosystem effects. V 27 in the Wiley Series in Advanced in Environmental Science and Technology, Wiley Interscience, New York, 1994, pp. 94–97.

Cole, J. M, Smith, S. and Bethune, D.: Arsenic source and fate at a village drinking water supply in Mexico and its relationship to sewage contamination. In: J. Bundschuh, P. Bhattacharya and D. Chandrasekharam (eds): *Natural arsenic in groundwater: occurrence, remediation and management.* Taylor and Francis/Balkema, Leiden, The Netherlands, 2005, pp. 67–75.

Farias, S., Casa, V., Vazquez, C., Ferpozzi, L., Pucci, G. and Cohen, I.: Natural contamination with arsenic and other trace elements in groundwaters of Argentina Pampean Plain. *Sci. Total Environ.* 309:1–3 (2003), pp. 187–199.

Gary, A., Edwards, M., Brandhuber, P., McNeill, L., Benjamin, M., Vagliasindi, F., Carlson, K. and Chwirka, J.: Arsenic treatibility options and evaluation of residuals management issues. AWWARF, Denver, CO, 1999.

INEGI: Anuario estadístico del Estado de Morelos, Mexico. Instituto Nacional de Estadística, Geografía e Informática, Mexico City, Mexico, 2000.

Kartinen, E.O. Jr. and Martín, C.J.: An overview of arsenic removal proceses. *Desalination* 103 (1995), pp. 79–85.

Kumar-Mandal, B. and Suzuki, T.: Arsenic round the world: a review. *Talanta* 58 (2002), pp. 201–235.

Liu, C., Lin, K. and Kuo, Y.: Application of factor analysis in the assessment of goundwater quality in a blackfoot disease area in Taiwan. *Sci. Total Environ.* 313 (2003), pp. 77–89.

McNeil, L.S.: *Understanding arsenic removal during conventional water treatment.* MSc Thesis, University of Colorado, Boulder, CO, 1996.

McNeil, L.S. and Edwards M.: Arsenic removal during precipitative softening. *J. Environ. Eng.* 123:5 (1997), p. 453.

Nieto, A. and Seed, L.: Test results of DesEL-24k unit. Internal report of Enpar Technologies Inc., Guelph, Canada, 2005.

Ngo, H.H., Vigneswaran, S., Hu, J.Y., Thirunavukkarasu, O. and T. Viraraghavan: A comparison of conventional treatment technologies on arsenic removal from water. *Water Supply* 2:5/6 (2002), pp. 119–125.

Norma Oficial Mexicana NOM-127-SSA1-1994, Salud ambiental. Agua para uso y consumo humano. Límites permisibles de calidad y tratamientos a que debe someterse el agua para su potabilización. Diario Oficial de la Federación 22 de noviembre de 2000, Mexico City, Mexico, 2000.

Pande, S.P., Deshpande, L.S. and Patni, P.M.: Arsenic removal in some ground waters of West Bengal, India. *J. Environ. Sci. Health* A 32:7 (1997), p. 1981.

Petkova, V., Rivera, L., Avilés, M., Piña, M. and Martín A.: Remoción de arsénico de agua para consumo humano. Final report, IMTA, Jiutepec, Mor., Mexico, 1997.

Peters, S.C. and Blum, J.D.: The source and transport of arsenic in a bedrock aquifer, New Hampshire, USA. *Appl. Geochem.* 18: (2003), pp. 1773–1787.

Rivera, L. and Piña, M.: Remoción de arsénico mediante zeolita recubierta con óxido de hierro. Proyecto TC-2009, IMTA, Jiutepec, Mor., Mexico, 2000.

Rodríguez, E.R., Gutierrez, P.A., Romero, G.J. and Velásquez, G.M.: Hidroarsenismo regional endémico en Acámbaro, Guanajuato. University of Guanajuato, Guanajuato, Gto., Mexico, 1996.

Rodríguez, R., Ramos, J.A. and Armienta A.: Groudwater arsenic variations: the role of local geology and rainfall. *Appl. Geochem.* 19:2 (2004), pp. 245–250.

Scareck, O., Bhattacharya, P., Jacks G., Gustafsson, J.P. and Bromssen, M.: Behavior of arsenic and geochemical modeling of arsenic enrichment in aqueous environments. *Appl. Geochem.* 19 (2004), pp. 169–180.

Smedley, P.L. and Kinninbeurg, D.G.: A review of the source, behavior and distribution of arsenic in natural waters. *Appl. Geochem.* 17 (2002), pp. 517–568.

Tseng, W.: Effects and dose-response relationships of skin cancer and blackfoot disease with arsenic. *Environ. Health Persp.* 19 (1977), pp. 109–119.

USEPA: Risk assessment guidance superfund. Volume I: Human health evaluation manual (Part A). EPA/540/1–89/002, Environmental Protection Agency, Washington, DC, 1989.

USEPA: National primary drinking water regulations; Arsenic and clarifications to compliance and new source contaminants monitoring; proposed rule. EPA 40 CFR Parts 141 and 142; Environmental Protection Agency, Washington, DC, 2001.

CHAPTER 73

Low-cost technologies for arsenic removal in the Chaco-Pampean plain, Argentina

M.E. Morgada de Boggio, I.K. Levy, M. Mateu & M.I. Litter
Gerencia Química, Comisión Nacional de Energía Atómica, San Martín, Prov. de Buenos Aires, Argentina

P. Bhattacharya
KTH-International Groundwater Arsenic Research Group, Department of Land and Water Resources Engineering, Royal Institute of Technology (KTH), Stockholm, Sweden

J. Bundschuh
International Technical Cooperation Program, CIM (GTZ/BA), Frankfurt, Germany
Instituto Costarricense de Electricidad (ICE), UEN, PySA, San José, Costa Rica

ABSTRACT: Groundwater of the Chaco-Pampean plain of Argentina contains levels of arsenic (As) exceeding drinking water standards. The situation is more serious in rural areas, causing a high incidence of CERHA (Chronic Endemic Regional Hydroarsenicism). Results of treatment experiments involving either heterogeneous photocatalysis (HP) or zerovalent iron (ZVI) in plastic bottles to remove As from groundwater are presented as low-cost technologies to remove As. For HP tests, synthetic or natural samples containing As placed in bottles impregnated with a TiO_2 layer and exposed to solar or artificial UV light followed by an addition of iron resulted in As concentration well below drinking water standards. For ZVI tests, iron wool was shown to be a better iron source than packing wire for As removal. Solar irradiation, in synthetic as well as in natural samples, improves As removal, avoiding the use of high amounts of iron.

73.1 INTRODUCTION

Groundwater of the Chaco-Pampean plain of Argentina (provinces of La Pampa, Buenos Aires, Córdoba, Santa Fe, Santiago del Estero, Chaco, Tucumán, Salta) contains levels of arsenic (As) well above the 10 μg/l limit established by WHO and the Argentine Food Code (CAA) (Código Alimentario Argentino 2007, Smedley *et al.* 1998, Nicolli *et al* 1989, 2001, Bundschuh *et al.* 2004, Bhattacharya *et al.* 2006). These groundwaters are characterized by high salinity, and the As concentrations are generally well correlated with other anions (fluoride, vanadium, bicarbonate, boron, molybdenum). Waters are predominantly oxic and have alkaline pH values (7.0–8.7). Hence, As is likely to be present as As(V) (Smedley *et al.* 1998). Metal oxides in the sediments (especially of Fe and Mn) are probably the main source of dissolved As (Smedley *et al.* 1998), although the dissolution of volcanic glass has also been cited as a potential source (Nicolli *et al.* 1989).

In many rural areas, the lack of a drinking water distribution network and the absence of important surface water bodies make it necessary to extract water from deep and shallow wells. Water is typically used without further treatment. In general, water quality is poor, especially in the phreatic zone, because of the bacteriological content and high levels of chemical contaminants (Nicolli *et al.* 2001). Exacerbated by poverty and malnutrition, the incidence of water-borne diseases, including Chronic Endemic Regional Hydroarsenicism (CERHA), is a concern.

Various methods have been used to remove As from drinking water, including anion exchange, precipitation, ion flotation, and adsorption (Edwards 1994, Hering *et al.* 1997, Harper *et al.* 1992). Attention has recently focused on advanced oxidation technologies or processes (AOTs, AOPs) such as heterogeneous photocatalysis (HP) (Dutta *et al.* 2005, Ferguson *et al.* 2005) and zerovalent iron (ZVI) (Bang *et al.* 2005, Su and Puls 2001) as simple and low-cost technologies to remove As from groundwater that could be applied in poor rural villages.

Heterogeneous photocatalysis is based on the use of a semiconductor, TiO_2, which under the action of UV light, originates chemical reactions leading to organic matter mineralization (Litter 2005, Quici *et al.* 2005), bacteria destruction (Ibáñez *et al.* 2003) and transformation of toxic metals (Litter 1999). The TiO_2 can be conveniently supported on the walls of plastic bottles. Exposure to solar irradiation causes oxidation of As(III) to As(V). In a further or simultaneous step, iron can be added to immobilize As(V) by adsorption or coprecipitation with iron oxides.

zerovalent iron has been extensively applied to remediation of contaminants such as chlorinated organic chemicals (Bremner *et al.* 2006) and metals (Rangsivek and Jekel 2005) in groundwater, and it appears promising for As remediation (Bang *et al.* 2005, Su and Puls 2001). The primary advantages for ZVI application include low cost, availability of iron in rural villages in the form of low-cost materials, and simplicity in handling and scalability. The mechanism of As removal by ZVI involves the formation of complexes of As(III) and As(V) on iron oxides formed *in situ* as a result of the Fe(0) corrosion reaction (Lackovic *et al.* 1999).

The objectives of this study are to evaluate the effectiveness of As removal (1) by HP in bottles with TiO_2 immobilized on the walls with iron addition, and (2) by using low-cost ZVI materials. This study was performed in natural waters from Los Pereyra, a small settlement of 1000 inhabitants located 70 km southeast of San Miguel de Tucumán city, and from Las Hermanas, a very poor area close to Santiago del Estero city. Waters in this region are mainly oxic and the main components are sodium, bicarbonate, sulfate, chloride and calcium, with bacteriological contamination and high levels of nitrate, boron, fluoride and arsenic. Both locations were selected because of unfavorable socioeconomic conditions and the detection of several cases of CERHA. Physiographically, the region belongs to the Dulce-Salí river hydrogeological basin.

73.2   EXPERIMENTAL

73.2.1   *Chemicals*

All chemicals were reagent grade and used without further purification. Arsenic (III) and As(V) stock solutions were prepared from $NaAsO_2$ and $NaH_2AsO_4 \times 7H_2O$, respectively, (Baker). Drinking water samples were obtained from wells of Las Hermanas and Los Pereyra. Synthetic samples of similar composition to that of Los Pereyra well waters were prepared with 1.3×10^{-4} M $MgSO_4$, 2.3×10^{-4} M $CaCl_2$, 4.0×10^{-5} M NH_4Cl, 9.0×10^{-6} M $FeCl_3$ and NaOH to adjust pH to 7.8.

Fe(III) was added as $FeCl_3$ (Mallinckrodt). zerovalent iron was introduced in the form of commercial packing wire or iron wool (Virulana®), both non-galvanized. XRD patterns showed that both materials were pure metallic Fe with only traces of Al and $CaSiO_3$ in the case of packing wire. TiO_2 (Degussa P-25) was provided by Degussa AG, Germany and used as received. For analytical determinations, $(NH_4)_6Mo_7O_{24} \times 4H_2O$ (Stanton), potassium and antimonil tartrate (Baker), L-ascorbic acid (Sigma-Aldrich) and $KMnO_4$ (Riedel-de-Häen) were used. Water was purified with a Millipore Milli-Q purification unit (resistivity $= 18$ M$\Omega \cdot$ cm).

73.2.2   *Arsenic removal experiments*

Experiments were performed in 600 ml plastic colorless polyethyleneterephtalate (PET) soft drink or mineral water bottles. For HP tests, the inner surface of the bottles was covered with a thin TiO_2 film, following an easy technique already reported by this group (Meichtry *et al.* 2007). Bottles were manually shaken in order to oxygenate solutions previous to the experiment and before each sampling.

Artificial light was provided by a black-light tubular UV lamp (Philips TLD/08, 15 W, maximum emission at 366 nm). The lamp intensity was 800 μW/cm^2. The intensity of the solar light used in the experiments corresponded to that of the Buenos Aires city spring (34°38' S, 58°28' W, October 2006) and ranged 1600–2000 μW/cm^2. Both light intensities were measured with a Spectroline DM-365 XA radiometer.

73.2.3   *Analytical determinations*

Arsenic (V) was determined colorimetrically through the formation of the arsenomolybdic complex (Lenoble *et al.* 2003). Total As was determined similarly following As(III) oxidation by adding 126 mg KMnO$_4$/mg As(contact time 120 minutes). The As(III) concentration was calculated by difference. In natural water samples, total As was determined by ICP-OES, using a Perkin-Elmer Optima 3100 XL apparatus or by total reflection X-ray fluorescence (TRXRF) using a PANalytical PW3830 X-ray generator. Before analysis, all samples were filtered through a 0.45 μm Millipore membrane.

73.3   RESULTS AND DISCUSSIONS

73.3.1   *HP experiments*

Artificial solutions with 1000 μg/l As(III) (250 ml, pure Milli-Q water) were placed in 600 ml bottles impregnated with TiO$_2$ and exposed to artificial UV light for 6 hours (h). After irradiation, 1.5 g packing wire were added in one or several pieces. In some experiments, packing wire was added at the beginning of the irradiation. Sampling was done always after 24 h in the dark to allow precipitates to settle, and remaining total As in solution was measured spectrophotometrically (Lenoble *et al.* 2003). Results (Table 73.1) indicate significant removal of As whether iron packing wire is added during or after irradiation. Also, As removal occurred whether it was added in one piece or as several fragments (although the latter procedure would be more complicated). Although the final As concentrations do not attain the limit of national regulations, it should be taken into account that the initial concentration is very high (representative of only a few wells), and that the exposure time could be prolonged to accomplish with the required values.

To analyze the reuse of bottles covered with TiO$_2$, three consecutive experiments were performed with the same bottle under the same conditions as sample 1 in Table 73.1. Arsenic removal was 90.5, 89.5 and 96.7% respectively, indicating that the efficiency of the catalyst is either not affected or improved by reuse.

Experiments with well-water samples from Las Hermanas were performed in bottles with TiO$_2$ exposed to sunlight for several hours. At the end of the irradiation, Fe(III) as FeCl$_3$ was added. The resulting precipitate was separated by filtration and the solution analyzed for As by ICP-OES. Results are listed in Table 73.2. In all cases, the attained concentrations are rather low, being in one case (W6) below the WHO standard. The initial Fe/As molar ratio used in the experiments ranged from 85 to 423. As this ratio is relatively high, future work will focus on the optimization of this parameter.

Table 73.1.   Arsenic removal from pure water by HP with 6.0 g/l iron packing wire added. Artificial UV light intensity = 800 μW/cm^2, 600 ml bottles, V$_{solution}$ = 250 ml.

Sample	As$_0$ (μg/l)	As$_f$ (μg/l)	t$_{irr}$ (h)	Form of packing wire addition	% As removal after 24 h
1	1000	140	6	one piece, after irradiation	86
2	1000	180	6	fragmented, after irradiation	82
3	1000	200	6	one piece, during irradiation	80

As$_0$: initial As concentration, As$_f$: final As concentration, t$_{irr}$: radiation time.

Table 73.2. As removal efficiency by HP and further Fe^{3+} addition in natural waters from Las Hermanas. Range of solar light intensity = 1600–2000 $\mu W/cm^2$, 600 ml bottles, $V_{solution}$ = 250 ml.

Sample	As_0 ($\mu g/l$)	As_f ($\mu g/l$)	t_{irr} (h)	Fe(III)/As	% As removal after 24 h
W1	961	31	10*	162	96.8
W2	1090	13	4.2	214	98.8
W3	551	30	4.9	423	94.5
W4	1530	14	3.8	152	99.1
W5	1830	25	10*	85	98.6
W6	1630	<10	5.1	143	>99.4

As_0: initial As concentration, As_f: final As concentration, t_{irr}: radiation time.

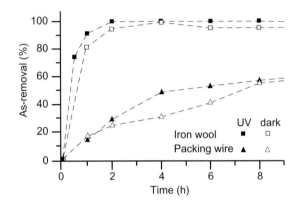

Figure 73.1. Arsenic removal from synthetic waters of similar composition to Los Pereyra well waters after addition of 6 g/l of either packing wire or iron wool. Initial As concentration $As(V)_0$ = 1000 $\mu g/l$, artificial UV light intensity = 800 $\mu W/cm^2$, 600 ml bottles, $V_{solution}$ = 250 ml.

73.3.2 ZVI experiments

Synthetic waters (250 ml) of similar composition to that of Los Pereyra well waters were spiked with 1000 $\mu g/l$ As(III) or As(V), and placed in 600 ml bottles. Iron was added as packing wire or iron wool, in both cases at 6.0 g/l, and the samples were irradiated with artificial light; similar experiments were performed in the dark. Previous tests had indicated that this is the optimal Fe concentration for As removal from Los Pereyra well waters (d'Hiriart *et al.* this volume). Arsenic concentrations were determined immediately after sampling (no settling in the dark). Trends in As removal with time are plotted in Figure 73.1. Under these experimental conditions, there is no significant effect of UV light. Arsenic removal is considerably faster with iron wool, reaching complete removal in 2 h, while with packing wire only 30% removal was obtained in the same timeframe. In addition, after settling for 24 h in the dark, 83 and 100% of the As was removed by packing wire and iron wool, respectively (not shown). This result can be attributed to a higher exposed oxidizable surface. After reaction, the remaining iron wool and packing wire were analyzed by XRD and patterns of maghemite, lepidocrocite and magnetite were observed in both cases.

In a second set of experiments, the effect of the amount of iron wool was studied. Three different iron concentrations were tested (viz., 0.6, 1.0 and 6.0 g/l). Water with 1000 $\mu g/l$ of both As(III) and As(V) were irradiated under artificial UV light for 8 h and in dark conditions. Figures 73.2a and b show that a higher initial rate is obtained at higher Fe(0) concentrations, although complete As removal is obtained for all samples after settling for 24 h in the dark (not shown). No significant differences could be observed when starting from As(III) or As(V).

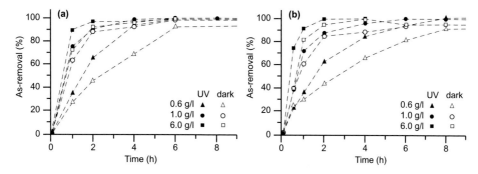

Figure 73.2.   Arsenic removal from synthetic waters similar in composition to Los Pereyra well waters for three different iron wool loadings whether in the dark or irradiated with artificial UV light (800 $\mu W/cm^2$) in 600 ml bottles, $V_{solution} = 250$ ml. Initial As concentration $As_0 = 1000\,\mu g/l$, starting with (a) As(III) or (b) As(V).

Table 73.3.   Arsenic removal efficiency achieved by adding Fe (0) as packing wire (6 g/l) to natural waters from Los Pereyra. Artificial UV light intensity $= 800\,\mu W/cm^2$, 600 ml bottles, $V_{solution} = 250$ ml.

Sample	As_0 ($\mu g/l$)	As_f ($\mu g/l$)	t_{irr} (h)	% As removal after 24 h
Irradiated	340	97	6	71.5
Dark	340	174	–	48.8

As_0: initial As concentration, As_f: final As concentration, t_{irr}: radiation time.

The effect of light is not evident at high Fe(0) concentrations, but it can be clearly observed in the experiment carried out with 0.6 g/l ZVI. This result indicates that prolonged solar irradiation could provide good As removal with lower amounts of iron, rendering the process more economical.

A well water sample (250 ml) from Los Pereyra was poured into a PET bottle containing 1.5 g of packing wire and exposed to sunlight for 6 h. A similar experiment was performed in the dark. After settling for 24 h in the dark, both samples were filtered and the remaining As was analyzed by TXRF. Table 73.3 shows As removal in both cases, but much better removal was obtained under irradiation. This result differs from the one obtained for synthetic samples where for the same Fe(0) concentration no difference is observed under irradiation (see Figure 73.1). Different experimental conditions might be responsible for this difference, such as the presence of species that accelerate Fe(0) oxidation under sunlight (for example organic matter) or that improve As adsorption on the Fe flocs.

73.4   CONCLUSIONS

Experimental results demonstrate that As can be effectively removed from As-spiked solutions and As-contaminated groundwater using low-cost technologies: (1) heterogeneous photocatalysis with TiO_2 supported on the walls of plastic PET bottles with Fe(III) or Fe(0) addition, and (2) zerovalent iron in PET bottles. In both cases, Fe can be added as very cheap materials such as iron wool or packing wire. In the case of ZVI, solar irradiation appears to improve As removal conditions. Although both methodologies yield similar results, use of HP could be superior to ZVI because simultaneous removal of As, organic matter, toxic metals and microbiological contamination can be achieved.

ACKNOWLEDGEMENTS

This work was performed as part of OAS/AE/141 AICD Project, *Comisión Nacional de Energía Atómica* P5-PID-36-4 Program and *Agencia Nacional de Promoción Científica y Tecnológica* PICT03-13-13261 Project. MIL is a member of CONICET. MEM thanks CONICET for a postdoctoral fellowship. IKL thanks OAS for a student fellowship.

REFERENCES

Bang, S., Johnson, M.D., Korfiatis, G.P. and Meng, X.: Chemical reactions between arsenic and zerovalent iron in water. *Water Res.* 39 (2005), pp. 763–770.

Bhattacharya, P., Claesson, M., Bundschuh, J., Sracek, O., Fagerberg, J., Jacks, G., Martin, R.A., Storniolo, A. and Thir, J.M.: Distribution and mobility of arsenic in the Rio Dulce alluvial aquifers in Santiago del Estero Province, Argentina. *Sci. Total Environ.* 358 (2006), pp. 97–120.

Bremner, D.H., Burgess, A.E., Houllemare, D. and Namkung, K.-C.: Phenol degradation using hydroxyl radicals generated from zerovalent iron and hydrogen peroxide. *App. Catal. B: Environ.* 63:1/2 (2006), pp. 15–19.

Bundschuh, J., Farías, B., Martin, R., Storniolo, A., Bhattacharya, P., Cortes, J., Bonorino, G. and Albouy, R.: Groundwater arsenic in the Chaco-Pampean Plain, Argentina: case study from Robles county, Santiago del Estero Province. *Appl. Geochem.* 19 (2004), pp. 231–243.

Código Alimentario Argentino, modification of articles 982 and 983, May 22, 2007, Buenos Aires, Argentina, 2007, http://www.anmat.gov.ar/normativa/normativa/Alimentos/Resolucion_Conj_68-2007_96-2007.pdf.

d'Hiriart, J., Hidalgo, M. del V., García, M.G., Litter, M.I. and Blesa, M.A.: Arsenic removal by solar oxidation in groundwater of Los Pereyra, Tucumán Province, Argentina. In: Bundschuh, J., Armienta, M.A., Birkle, P., Bhattacharya, P., Matschullat, J. and Mukherjee, A.B. (eds): *Natural arsenic in groundwater of Latin America.* Taylor and Francis/Balkema. Leiden, The Netherlands, 2009 (this volume).

Dutta, P.K., Pehkonen, S.O., Sharma, V. and Ray, A.: Photocatalytic oxidation of arsenic (III): evidence of hydroxyl radicals. *Environ. Sci. Technol.* 39:6 (2005), pp. 1827–1834.

Edwards, M.: Chemistry of arsenic removal during coagulation and Fe-Mn oxidation. *J. Am. Water Works Assoc.* 86:9 (1994), pp. 64–78.

Ferguson, M., Hoffmann, M.R. and Hering, J.G.: TiO₂-photocatalyzed As(III) oxidation in aqueous suspensions: reaction kinetics and effects of adsorption. *Environ. Sci. Technol.* 39:6 (2005), pp. 1880–1886.

Harper, T.R. and Kingham, N.W.: Removal of arsenic from wastewater using chemical precipitation methods. *Water. Environ. Res.* 64:3 (1992), pp. 200–203.

Hering, J.G., Chen, P.-Y., Wilkie, J.A. and Elimelech, M.: Arsenic removal from drinking water during coagulation. *J. Environ. Eng.* 123:8 (1997), pp. 800–807.

Ibáñez, J.A., Litter, M.I. and Pizarro, R.A.: Photocatalytic bactericidal effect of TiO₂ on Enterobacter cloacae: Comparative study with other Gram (−) bacteria. *J. Photochem. Photobiol. A.* 157:1 (2003), pp. 81–85.

Lackovic, J.A., Nikolaidis, N.P. and Dobbs, G.M.: Inorganic arsenic removal by zerovalent iron. *Environ. Eng. Sci.* 17:11 (1999), pp. 29–39.

Lenoble, V., Deluchat, V., Serpaud, B. and Bollinger, J.C.: Arsenite oxidation and arsenate determination by the molybdene blue method. *Talanta* 61:3 (2003), pp. 267–276.

Litter, M.I.: Environmental photochemistry, Part II. In: P. Boule, D.W. Bahnemann and P.K.J. Robertson (eds): *The handbook of environmental chemistry*, Vol. 2, Part M. Springer Verlag, Berlin, Heidelberg, Germany, 2005, pp. 325–366.

Litter, M.I.: Heterogeneous photocatalysis: Transition metal ions in photocatalytic systems. *Appl. Catal. B: Environ.* 23:2/3 (1999), pp. 89–114.

Meichtry, J.M., Lin, H., de la Fuente, L., Levy, I.K., Gautier, E.A., Blesa, M.A. and Litter, M.I.: Low-cost TiO₂ photocatalytic technology for water potabilization in plastic bottles for isolated regions. Photocatalyst fixation. *J. Solar Energy Eng.* 129:1 (2007), pp. 119–126.

Nicolli, H.B., Suriano, J.M., Gómez Peral, M.A., Ferpozzi, L.H. and Baleani, O.H.: Groundwater contamination with arsenic and other trace elements in an area of the Pampa, Province of Córdoba, Argentina. *Environ. Geol. Water Sci.* 14 (1989), pp. 3–16.

Nicolli, H.B., Tineo, A., García, J.W., Falcón, C.M. and Merino, M.H.: Trace-element quality problems in groundwater from Tucumán, Argentina. In R. Cidu (ed): *Water-Rock Interaction* 2. Balkema, Lisse, The Netherlands, 2001, pp. 993–996.

Quici, N., Morgada, M.E., Piperata, G., Babay, P.A., Gettar, R.T. and Litter, M.I.: Oxalic acid destruction at high concentrations by combined heterogeneous photocatalysis and photo-Fenton processes. *Catal. Today* 101:3/4 (2005), pp. 253–260.

Rangsivek, R. and Jekel, M.R.: Removal of dissolved metals by zerovalent iron (ZVI): Kinetics, equilibria, processes and implications for stormwater runoff treatment. *Water Res.* 39 (2005), pp. 4153–4163.

Smedley, P.L., Nicolli, H.B., Barros, A.J. and Tullio, J.O.: Origin and mobility of arsenic in groundwater from the Pampean Plain, Argentina. In: E.B. Arehart, and J.R. Hulston (eds): *Water-Rock Interaction.* Balkema, Rotterdam, 1998, pp. 275–278.

Su, C.M. and Puls, R.W.: Arsenate and arsenite removal by zerovalent iron: Kinetics, redox transformation, and implications for in situ groundwater remediation. *Environ. Sci. Technol.* 35:7 (2001), pp. 1487–1492.

Section VII
Innovative and sustainable options
for arsenic mitigation: Some experiences

CHAPTER 74

Arsenic-safe aquifers as a socially acceptable source of safe drinking water—What can rural Latin America learn from Bangladesh experiences?

J. Bundschuh
International Technical Cooperation Program, CIM (GTZ/BA), Frankfurt, Germany
Instituto Costarricense de Electricidad (ICE), San José, Costa Rica

P. Bhattacharya, M. von Brömssen, M. Jakariya, G. Jacks & R. Thunvik
KTH-International Groundwater Arsenic Research Group, Department of Land and Water Resources Engineering, Royal Institute of Technology (KTH), Stockholm, Sweden

M.I. Litter
Gerencia Química, Comisión Nacional de Energía Atómica, and Escuela de Posgrado, Universidad de Gral. San Martín, San Martín, Prov. de Buenos Aires, Argentina

ABSTRACT: Geogenic arsenic (As) are found at elevated levels in groundwaters resources in many regions of the world. Since 1998, As problem is "discovered" each year in groundwater in parts of the world, where the problem was not known. This calls for global approaches for sharing and exchange of information and experiences. Two pilot areas, Matlab Upazila, representative of the Bengal delta where 35 million of rural population is at risk, and Río Dulce alluvial cone (NW Argentina), representative of Latin America where >5 million people were at risk, were compared to identify mechanisms of As mobilization and sustainable mitigation options. In both areas, As mobilization is governed by site- or zone-specific hydrogeochemical conditions that result in extremely heterogeneous lateral and vertical distributions of As in the shallow groundwaters. In rural Bangladesh, different As removal techniques, e.g., sand filters, Alcan filters, Bishuddya filters, or rainwater harvesters have been provided by governmental and non-governmental organizations, but their social acceptance is low. In contrast, "Targeting As-safe aquifers", which calls for installation of safe wells, was identified as a socially accepted option. This was proven currently in our pilot project in Bangladesh, where sediment color could be used to identify As-safe aquifer zones for installation of tube wells. Initial results from our investigation in NW Argentina are complex because the conditions and sources of As to aquifers in Argentina are very different from those in Bangladesh.

74.1 INTRODUCTION

Arsenic (As) of geogenic origin contaminates drinking water resources in many parts of the world and every year such new regions are discovered (Fig. 74.1). This calls for common global approaches and free exchange of information and experiences from regions where the problem and its mitigation have previously been studied. In order to provide the rural population with microbe-free drinking water in Bangladesh, a campaign took place in the 1970s to drill shallow groundwater wells. Previously, the rural population depended on surface- and rainwater that were highly contaminated by bacteria and other microorganisms. However, the quality of the groundwater exploited from the shallow Holocene aquifers was not tested for As and caused, 10 years later, exposure to large parts of the population and toxicological effects for many years to come. This became the beginning of extended bilateral aid efforts to mitigate the As problem in Bangladesh and adjacent West-Bengal, India. Many treatment methods have been developed and tested for rural

Figure 74.1.   Regions of geogenic arsenic in the world.

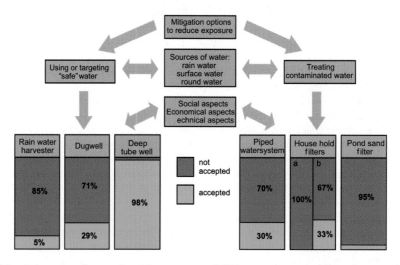

Figure 74.2.   Percentage of use and social acceptance of different options to mitigate the groundwater As problem of Bangladesh (data from Jakariya 2007, Jakariya *et al.* 2007).

areas in the regions. However, socioeconomic constraints were not addressed properly and thus many of the provided mitigation options such as pond sand filters, Alcan filters, Bishuddya filters and rainwater harvesters (Figs. 74.2 and 74.3) were not socially accepted and remained, in many cases, unused even after a short time of operation (Jakariya *et al.* 2005, 2007). In contrast, deep hand tube wells, which exploit As-safe pre-Holocene aquifers (>150 m, BGS 2001), are socially accepted.

These are, however, more expensive and cannot be afforded by the rural population especially if they are not going together for financing such option (Hoque *et al.* 2004, Jakariya *et al.* 2005, 2007, Jakariya 2007, Opar *et al.* 2007, Chen *et al.* 2007).

Similar problems have been observed in Latin America, where the mitigation measures have not been accepted by the population and/or where treatment devices installed and programs provided lasted only a short time and had no continuity. Like in Bangladesh, the same applies for deep tube wells to be widely accepted safe drinking water option in Latin American countries but economically not a very feasible option. As a consequence, socially accepted mitigation approaches remediation

Figure 74.3.   Mitigation techniques provided to the population of Matlab Upazila: (a) rainwater harvesters; (b) Bishuddya filter; (c) Alcan filter; (d) pond sand filter; (e) shallow dug well; (f) drilling of tube wells by local drillers at targeted depths; (g) deep tubewell drilling.

methods must be developed. The treatment of As-rich groundwater should be seen as one category of remediation methods. A different approach would be to target As-low and -safe aquifers, in contrast to treat As-rich groundwater (Bhattacharya *et al.* 2002a, van Geen *et al.* 2004, von Brömssen *et al.* 2005 and 2007, Jakariya *et al.* 2007). This would allow the rural population to use water from wells, which is a socially accepted way for water supply in both Bangladesh and Latin America. If targeting As-safe aquifers at relatively shallow depth is a sustainable option with respect to As contamination, such option would also be economically and technically feasible, making targeting shallow As-safe aquifers an ideal option. This method is currently studied in a pilot project in Bangladesh (van Geen *et al.* 2003, von Brömssen *et al.* 2007), and in an initial phase in Argentina (Chaco-Pampean plain; Río Dulce alluvial cone).

Dealing with geogenic As in groundwater in the developing world, gives rise to the question of what other regions, e.g., Latin American countries, can learn from the SE Asian experiences in order to (1) not to repeat the same errors, and (2) not to implement the same mitigation methods that have not been socioeconomically feasible in rural Bangladesh.

Consequences of neglect become evident when e.g., comparing Bangladesh with Nicaragua. In Nicaragua, the national water company drilled in 1994 a well in El Zapote in order to supply the population with clean drinking water. However, as in Bangladesh, the exploited water was not analyzed for As, resulting in a poisoning of the population for 2 years with As-rich (1320 μg/l) drinking water (Barragne 2004, Altamirano Espinoza and Bundschuh 2009). Similar negligence can be observed in many other countries, calling urgently for inclusion of As in the standard list of parameters for ensuring drinking water quality.

In Latin America, the As problem has been known for over a century; however the problem has not received enough attention by national authorities or international agencies to mitigate the problem. Furthermore, the rural people are often dependent on As-contaminated water as their only drinking water source. Thus, although suitable remediation methods were developed (e.g., solar oxidation methods, phytoremediation, use of natural materials as adsorbents for As removal) they were mostly tested only on laboratory scale, and only few pilot studies on field scale exist even today.

In many Latin American countries, groundwater from the shallow aquifers are the major sources of drinking water in rural areas, but they contain high levels of As. In the absence of any proper mitigation method and alternative drinking water sources, people are compelled to drink As-contaminated groundwater. Thus, to provide As-safe drinking water, it is crucial to develop tools to identify and delineate zones with As-safe shallow groundwater in order to achieve the UN Millennium Development Goals for ensuring safe drinking water.

This chapter comprises three sections. First, it addresses the areas of Latin America in which the problem is relevant, indicating the number of people exposed and affected—in order to show that the As problem in Latin America is of the same importance as in Bengal delta—and the experiences on treatment methods for As removal developed and tested so far in rural areas. The next part discusses the low-cost techniques for As removal in small communities as well as single household levels are addressed. In the third part, "targeting As-safe aquifers" is discussed as innovative mitigation methods and first results from Argentina are compared with experiences from Bangladesh. Finally, measures that must be taken in the future to mitigate the drinking water As problem of "Rural Latin America" are outlined.

74.2   THE LATIN AMERICA SITUATION

A detailed analysis of the As problem in different Latin American countries (Bundschuh and García 2008, Bundschuh *et al.* 2006 and 2009) shows that the contamination of ground- and surface water by As of geogenic origin (predominantly to sediments resulting from volcanic rocks, sulfide ore bodies and volcanic rocks) is a severe environmental problem in Latin America. In Argentina, Chile, Bolivia, Peru and Mexico, at least 4 million people depend on drinking water containing toxic As concentrations, which mostly originate from geogenic sources. For example, in Argentina (and until 1970 also in Chile, before the installation of treatment plants) over 1% of the population is exposed to unsafe levels of As, whereas in Bolivia, Brazil, Ecuador, Colombia, Costa Rica, El Salvador, and Guatemala, the numbers of people affected has not been assessed. The number of regions where the problem is being detected is increasing. Furthermore, due to the introduction of new national As limits for drinking water of 10 μg/l, several countries that until now had "safe" As levels will be classified as having unsafe concentrations. Thus, the population exposed to As above 10 µg/l will increase further. This new limit for As in drinking water was introduced recently in Nicaragua and Chile. Other Latin American countries are planning to adopt it in the near future.

74.3   ARSENIC REMEDIATION EXPERIENCES AND NEEDS IN LATIN AMERICA

The problem of As in drinking water has already been dealt with in most Latin American urban areas by installing treatment plants. However, many of them are not working properly or are too expensive. For example, the provincial capital Antofagasta in northern Chile, Calama, San Pedro and other cities treat their drinking water successfully using predominantly flocculation by $FeCl_3$ and subsequent filtration. In Peru, a treatment plant (using flocculation by $FeCl_3$) was constructed in the city of Ilo in 1982. However, the plant has high operation costs and is not working properly (Esparza 2004). In the cities and some big towns of Argentina, e.g., in the provinces of La Pampa, Santa Fe and Santiago del Estero, predominantly coagulation methods and reverse osmosis are applied to remove As from drinking water. All these treatment methods are expensive and are often not efficient for As removal due to the complex groundwater chemical characteristics (see viz. Sancha, 2000 and 2003). Therefore, the concerned countries, mainly Chile and Argentina, are permanently developing new and improving existing systems and methods of As remediation, looking at the same time on the reduction of the treatment costs and the decreasing maintenance needs.

In contrast to urban areas, practically no action has been performed by the authorities or international cooperation agencies to mitigate the As problem for Latin-American rural population. Lack of awareness and interest of authorities have hindered the mitigation of the problem for the affected rural population. This makes the dispersed rural population the most disadvantaged group. The rural areas should be an important target for further actions to reduce the As exposure.

Advanced treatment plants using $FeCl_3$, flocculation, etc., are in most cases not feasible for small communities, especially in the case of sparsely populated rural areas typical of the Latin American continent. In the Chaco-Pampean plain, Argentine, about 12% of the population lives in dispersed settlements with less than 50 inhabitants each, often the poorest of the region. The rural poor population requires low-cost remediation methods suitable for small communities from 50 inhabitants down to single household. Thus, low-cost and robust method that is simple to handle and maintain is required, provided such options are sustainable locally.

Many new techniques for As removal from drinking water have been developed and tested in laboratory. However, only a few have been applied in field, for example: (1) solar oxidation methods, (2) phytoremediation (e.g., using algae *Lessonia nigrescens*, Hansen (2004)), or lacustrine algae (Bundschuh *et al.* 2007), (3) the use of biomass (e.g., using sorghum biomass, Haque *et al.* 2005), and (4) natural or activated clay and lime as adsorbents.

One of the low-cost technologies is the so-called solar oxidation removal of As (SORAS). Countries like Bangladesh have used this technology to reduce As pollution in drinking water at rural community level, and some others like Chile and Argentina are testing to implement this method. It is based on the oxidation and precipitation of As assisted by light in the presence of citric acid (Hug *et al.* 2001). The practical procedure employed is to fill plastic bottles with the contaminated water, then to add some drops of lemon juice, and left the bottles under sunlight for a few hours. The capacity of As removal at the household level by this method is well proven, provided the respective adaptation of the technology to the geographic reality and chemical composition of the waters is made (García *et al.* 2004).

Another low-cost method uses TiO_2 as photocatalyst followed by iron addition. This option is suitable for small community and household-scale applications (Nieto *et al.* 2004, Morgada *et al.* 2006).

74.4   TARGETING ARSENIC-SAFE AQUIFERS AS INNOVATIVE SOCIALLY ACCEPTED REMEDIATION METHOD

The scientific community has been able to delineate the mechanisms of mobilization of As in many aquifers on the basis of the prevailing aquifer conditions such as certain prevailing geomorphological, geological, hydrogeological and hydrogeochemical conditions. In the case of Bangladesh, dissolved As is related to strongly reducing aquifers (As is mobilized through the reductive dissolution of non-crystalline Fe(III)-oxyhydroxides present as coatings on the sediments in reaction with organic matter), whereas the Argentine case study is related to As release in alkaline- and high pH, mostly oxidizing aquifers, which are encountered in many places of the Chaco-Pampean plain. In most of these regions, the distribution of As is extremely heterogeneous, both laterally and vertically. Consequently, the "patchy distribution" has often been explained in terms of "local variations in sedimentary characteristics, hydrogeological and hydrogeochemical conditions" both in Bangladesh (e.g., BGS 2001, Bhattacharya *et al.* 2002a, b and 2006a, McArthur *et al.* 2004, Ahmed *et al.* 2004, Horneman *et al.* 2004, Zheng *et al.* 2004 and 2005 Hasan *et al.* 2007), and in Argentina (Smedley *et al.* 2002 and 2005, Bundschuh *et al.* 2004, Bhattacharya *et al.* 2006b).

In Bangladesh, the aquifer sediments and groundwaters could be characterized to be used to distinguish As-safe and As-unsafe aquifers for installations of tube wells. In Matlab Upazila case study (representative for the Bengal delta), it has been found that the groundwater composition and redox conditions are strongly correlated to the color of the sediment (von Brömssen *et al.* 2005 and 2007) (Fig. 74.4). Groundwater extracted from the black sediments is most reduced, followed by white, off-white and red, which are less reduced. Consequently, neither Fe nor As were found at elevated levels in the groundwater extracted from the less reduced white, off-white and red sediments, since reduction of Fe(III)-oxyhydroxides is redox-buffered by Mn. If this feature is correlated with local geological conditions observed in field color, it could be used to target safe aquifers.

In Latin America including Argentina, targeting As-safe shallow aquifers has not been considered as a viable mitigation option so far. Similar studies to Bangladesh are taking place in the Río Dulce area. A pilot study indicates certain hydrogeochemical processes, which prevail in either areas with acceptable As concentrations, and non-acceptable As concentrations in groundwater. Since hydrogeochemical conditions cannot be visually derived from the sediment color as it is the case in Matlab Upazila, other indicators must be derived and used to distinguish between areas of the shallow aquifer with high and low As concentrations. Preliminary studies indicate that groundwater residence time and the way of groundwater recharge are the principal factors that drive the As concentration in groundwater. Thus, areas that are morphologically elevated by few meters (groundwater mountains), show often low concentrations of groundwater As. However, this parameter needs to be correlated with others and with local geological conditions in order to make it to a usable tool to delimitate As-safe parts of the shallow aquifer. Developing such tool for the Río Dulce project area is expected that it will be transferable to most of the Pampa-Chaco plain where people depend on the use of the shallow aquifer as the only available source of drinking water as well as to other regions of Latin America, where As occurrence is related to similar oxidizing aquifers.

In Argentina (as in Latin America in general), groundwater with elevated geogenic As is commonly encountered in the oxidizing aquifers with high alkalinity as well as high pH in many places. In the Chaco-Pampean plain of Argentina, at least 1.2 million people (about 3% of the total population) are exposed to elevated As mobilized primarily from volcanic ash interbedded or dispersed with the alluvial sediments (Bundschuh *et al.* 2004, Smedley *et al.* 2005). The matrix of the shallow aquifers shows that hydrogeochemical processes are the principal control to obtain zones with low and high concentrations of As in groundwater (Bundschuh *et al.* 2004, Bhattacharya *et al.* 2006b). The spatial distribution and variability of As concentrations in the shallow aquifers of Chaco-Pampean plain (covering about $1,000,000 \text{ km}^2$), in Argentina, is shown in Figure 74.5.

Common features of As-enriched hotspots in the Chaco-Pampean region of Argentina include: (1) prevalence of high As in Na-HCO$_3$ type of groundwater with high pH and EC values and with

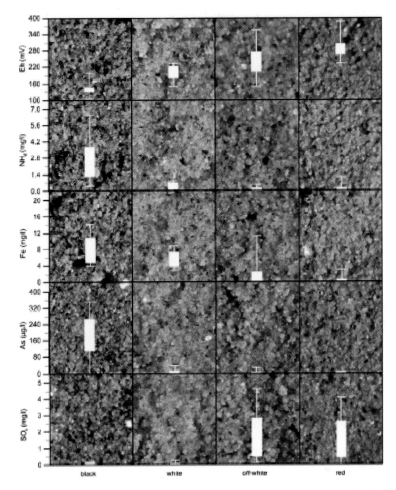

Figure 74.4. Relationship of redox-sensitive species in groundwater and sediment color in Matlab Upazila.

oxidizing or moderately reducing conditions, (2) aquifer sediments are young, of loessic origin, and volcanic ash/volcanic glass material is found in both discrete layers and dispersed in sediment, and (3) primary source of As is not known so far, but volcanic ash/volcanic glass is suggested on the basis of other dissolved species like Mo, V, F, B, etc. present in volcanic material (Smedley *et al.* 2005, Bundschuh *et al.* 2004, Bhattacharya *et al.* 2006b). The underlying mechanism for As mobilization is a result of desorption from oxyhydroxides of Al, Mn, and to less extent of Fe due to changes in pH versus pH_{ZPC} of the adsorbent and ion-competition of oxyanions of V, Mo, $PO_{4\,tot}^{3-}$ and HCO_3^- for adsorption sites. Low concentrations of dissolved As are found in zones with Ca-HCO_3 type of groundwater with about neutral pH (Fig. 74.5).

Studies have explored the mechanisms of As mobilization in the groundwater in Bangladesh and Latin America. The reasons for the absence of As in groundwater in these regions and its link to the site-specific geological and hydrogeological characteristics remain not fully resolved. On the basis of the prevailing geochemical conditions in the aquifers and the primary mechanisms of mobilization, many of the safe wells are expected to develop high As concentrations. However, in reality, we also find As-safe wells for the abstraction of safe water in an area with high percentage of wells with elevated As. Thus, there is an urgent need for a thorough investigation of the geological

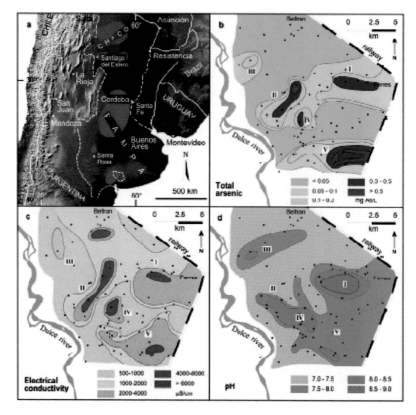

Figure 74.5.   (a) Map of Argentina showing the principal regions of groundwaters with elevated As and the location of Santiago del Estero province; (b) spatial distribution of As in the shallow aquifer of Río Dulce alluvial cone, which constitutes the only drinking water resource for most of the rural population and related spatial variability of (c) electrical conductivity; (d) pH value (modified after Bundschuh *et al.* 2004).

and hydrogeological characteristics of the areas to understand the hydrogeochemical processes that are responsible for As immobilization in the wells in these regions.

74.5   CONCLUSIONS

Two areas, Matlab Upazila in Bangladesh with predominantly reducing shallow aquifers and Río Dulce area in Argentina with oxidizing shallow aquifers were compared to identify mechanisms of As mobilization and sustainable mitigation options.

In both areas, the mobilization of As could be identified on the basis of zone-specific hydrogeochemical conditions, whose local variations result is extremely heterogeneous distributions of As concentrations in the shallow aquifers, both laterally and vertically.

In both areas, the treatment of As-rich drinking water is related to numerous problems. Especially in rural areas, the treatment options and equipments were often not accepted by the population or their provision by the authorities had no continuity.

Therefore, alternative remediation methods that are socially accepted are needed. Targeting As-safe shallow aquifers could be a socially accepted option in both Bangladesh and Latin America. This option would be economically and technically viable. This method was proved currently in a pilot project in Bangladesh, where sediment color could be used to identify aquifers for installations of As-safe tube wells and currently being practiced by the local drillers for As-safe tubewell installations. The preliminary results from ongoing investigation in Argentina are not encouraging, where the shallow aquifer is characterized by a more or less uniform matrix of loessic origin containing volcanic glass as principal source of As. Further research is needed to understand the geochemical parameters and their regional pattern of variations that would allow local drillers to identify safe aquifers.

ACKNOWLEDGEMENTS

The authors are thankful to the Strategic Environmental Research Foundation (Mistra) for the grant (dnr: 2005-035-137) for the research project "Targeting arsenic-safe aquifers in regions with high arsenic groundwater and its worldwide implications". PB gratefully acknowledges the Swedish Research Council for Environment, Agricultural Sciences and Spatial Planning (Formas) for the financial support for participation in the International Conference on Natural Arsenic in the groundwaters of Latin America in Mexico City during June 2006 (dnr: 214-2006-1619).

REFERENCES

Altamirano Espinoza, M. and Bundschuh, J.: Natural arsenic enrichment in the southwestern basin of the Sébaco Valley, Nicaragua. In: J. Bundschuh, M.A. Armienta, P. Birkle, P. Bhattacharya, J. Matschullat and A.B. Mukherjee (eds): *Natural arsenic in groundwater of Latin America*. Taylor and Francis/Balkema. Leiden, The Netherlands, 2009 (this volume).

Barragne, P.: Contribución al estudio de cinco zonas contaminadas naturalmente por arsénico en Nicaragua. UNICEF, Managua, Nicaragua, 2004.

BGS: Arsenic contamination of groundwater. Vol 2, Final Report, BGS Tech. Rep. WC/00/19. Edited by D.G. Kinniburgh and P.L. Smedley. British Geological Survey, Keyworth, UK, 2001, p. 267.

Bhattacharya, P., Frisbie, S.H., Smith, E., Naidu, R., Jacks, G. and Sarkar, B.: Arsenic in the environment: a global Perspective. In: B. Sarkar (ed): *Handbook of heavy metals in the environment*. Marcell Dekker, New York, 2002a, pp. 145–215.

Bhattacharya, P., Jacks, G., Ahmed, K.M., Khan, A.A. and Routh, J.: Arsenic in groundwater of the Bengal Delta Plain aquifers in Bangladesh. *Bull. Environ. Cont. Toxicol.* 69 (2002b), pp. 538–545.

Bhattacharya, P., Ahmed, K.M., Hasan, M.A., Broms, S., Fogelström, J., Jacks, G., Sracek, O., von Brömssen, M. and Routh, J.: Mobility of arsenic in groundwater in part Brahmanbaria district, NE Bangladesh. In: Groundwater arsenic contamination in India: extent and severity. In: R. Naidu, E. Smith, G. Owens, P. Bhattacharya and P. Nadebaum (eds): *Managing arsenic in the environment: from soil to human health*. CSIRO Publishing, Melbourne, Australia, 2006a, pp. 95–115.

Bhattacharya, P., Classon, M., Bundschuh, J., Sracek, O., Fagerberg, J., Jacks, G., Martin, R.A., Storniolo, A. del S. and Thir, J.M.: Distribution and mobility of arsenic in the río Dulce alluvial aquifers in Santiago del Estero Province, Argentina. *Sci. Total Environ.* 358 (2006b), pp. 97–120.

Bundschuh, J., Farías, B., Martin, R., Storniolo, A., Bhattacharya, P., Cortes, J., Bonorino, G. and Albouy. R.: Groundwater arsenic in the Chaco-Pampean Plain, Argentina: case study from Robles county, Santiago del Estero Province. *Appl. Geochem.* 19 (2004), pp. 231–243.

Bundschuh, J., García, M.E. and Bhattacharya, P.: Arsenic in groundwater of Latin America—A challenge of the 21st century. Geological Society of America Annual Meeting, Philadelphia, 22–25 Oct. 2006, *Geological Society of America Abstracts with Programs* 38:7 (2006), p. 320.

Bundschuh, J. and García, M.E.: Rural Latin America—A forgotten part of the global groundwater arsenic problem? In: P. Bhattacharya, AL. Ramanathan, J. Bundschuh, D. Chandrasekharam, A.K. Keshari and A.B. Mukherjee (eds): *Groundwater for sustainable development: problems, perspectives and challenges*. Taylor and Francis/Balkema, Leiden, The Netherlands, 2008.

Bundschuh, J., García, M.E., Birkle, P., Cumbal, L., Bhattacharya, P. and Matschullat, J.: Groundwater arsenic in rural Latin America—Occurrence, health effects and remediation experiences. In: J. Bundschuh, M.A. Armienta, P. Birkle, P. Bhattacharya, J. Matschullat and A.B. Mukherjee (eds): *Natural arsenic in groundwater of Latin America*. Taylor and Francis/Balkema, Leiden, The Netherlands, 2009 (this volume).

Bundschuh, J., García, M.E. and Alvarez, M.T.: Arsenic and heavy metal removal by phytofiltration and biogenic sulfide precipitation—A comparative study from Poopó Lake basin, Bolivia. *Abstract volume 3rd International Groundwater Conference IGC-2007, Water, Environment and Agriculture—Present Problems and Future Challenges*, February 7–10, 2007, Tamil Nadu Agricultural University, Coimbatore, India, 2007, p. 152.

Esparza. M.L.: Presencia de arsénico el el agua de bebida en América Latina y su efecto el la salud pública. In: A.M. Sancha: *Tercer Seminario Internacional sobre Evaluación y Manejo de las Fuentes de Agua de Bebida contaminadas con Arsénico* (proceedings available as CD), Universidad de Chile, November 08–11, 2004, Santiago de Chile, Chile, 2004.

García, M.G., Lin, H.J., Custo, G., d'Hiriart, J., Hidalgo, M. del V., Litter, M.I. and Blesa, M.A.: Advances in solar oxidation removal of arsenic in waters of Tucumán, Argentina. In: M.I. Litter and A. Jiménez González (eds): *Advances in low-cost technologies for disinfection, decontamination and arsenic removal in waters from rural communities of Latin America (HP and SORAS methods)*. AOS Project AE/141, Digital Grafic, La Plata, Argentina, 2004, pp. 43–63.

Hansen, H.K., Rojo, A., Oyarzun, C., Ottosen, A.R. and Mateus, E.: Biosorption of arsenic by *Lessonia nigrescens* in wastewater from cooper smelting. In: A.M. Sancha (ed): *Tercer Seminario Internacional sobre Evaluación y Manejo de las Fuentes de Agua de Bebida contaminadas con Arsénico* (proceedings available as CD), Universidad de Chile, November 08–11, 2004, Santiago de Chile, Chile, 2004.

Hasan, M.A., Ahmed, K.M., Sracek, O., Bhattacharya, P., von Brömssen, M., Broms, S., Fogelström, J., Mazumder, M.L. and Jacks, G.: Arsenic in shallow groundwater of Bangladesh: investigations from three different physiographic settings. *Hydrogeol. J.* 15:8 (2007), pp. 1507–1522.

Haque, N., Morrison, G., Perrusquía, G., Cano-Aguilera, I., Aguilera-Alvarado, A.F. and Gutiérrez-Valtierra, M.: Sorption of arsenic on sorghum biomass: A case study. In: J. Bundschuh, P. Bhattacharya and D. Chandrasekharam (eds): *Natural arsenic in groundwater: occurrence, remediation and management*. Taylor and Francis/Balkema, Leiden, The Netherlands, 2005, pp. 247–253.

Horneman, A., van Geen, A., Kent, D., Mathe, P.E., Zheng, Y., Dhar, R.K., O'Connell, S., Hoque, M., Aziz, Z., Shamsudduha, M., Seddique, A. and Ahmed, K.M.: Decoupling of As and Fe release to Bangladesh groundwater under reducing conditions. Part I: Evidence from sediment profiles. *Geochim. Cosmochim. Acta* 68 (2004). pp. 3459–3473.

Hoque, B.A., Hoque, M.M., Ahmed, T., Islam, S., Azad, A.K., Ali, N., Hossain, M. and Hossain, M.S.: Demand-based water options for arsenic mitigation: an experience from rural Bangladesh. *Public Health* 118:1 (2004), pp. 70–77.

Hug, S.J., Canonica, L., Wegelin, M., Gechter, D. and von Gunten, U.: Solar oxidation and removal of arsenic at circumneutral pH in iron containing waters. *Environ. Sci. Technol.* 35:10 (2001), pp. 2114–2121.

Jakariya, M.: *Arsenic in tubewell water of Bangladesh and approaches for sustainable mitigation*. TRITA LWR PhD Thesis 1033, Royal Institute of Technology (KTH), Stockholm, Sweden, 2007.

Jakariya, M., Rahman, M., Chowdhury, A.M.R., Rahman, M., Yunus, M., Bhiuya, M.A., Wahed, M.A., Bhattacharya, P., Jacks, G., Vahter, M. and Persson, L-Å.: Sustainable safe water options in Bangladesh: experiences from the arsenic project at Matlab (AsMat). In: J. Bundschuh, P. Bhattacharya and D. Chandrasekharam (eds): *Natural arsenic in groundwater: occurrence, remediation and management*. Taylor and Francis/Balkema, Leiden, The Netherlands, 2005, pp. 319–330.

Jakariya, M., von Brömssen, M., Jacks, G., Chowdhury, A.M.R., Ahmed, K.M. and Bhattacharya, P.: Searching for sustainable arsenic mitigation strategy in Bangladesh: experience from two upazilas. *Int. J. Environ. Poll.* 31:3–4 (2007), pp. 415–430.

McArthur, J.M., Banerjee, D.M., Hudson-Edwards, K.A., Mishra, R., Purohit, R., Ravenscroft, P., Cronin, A., Howarth, R.J., Chatterjee, A., Talukder, T. Lowry, D., Houghton, S. and Chadha, D.K.: Natural organic matter in sedimentary basins and its relation to arsenic in anoxic ground water: the example of West Bengal and its worldwide implications. *Appl. Geochem.* 19 (2004), pp. 1255–1293.

Morgada, M.E., Levy, I.K., Mateu, M. and Litter, M.I.: Low-cost technologies based on heterogeneous photocatalysis and zerovalent iron for arsenic removal in the Chaco-Pampean Plain, Argentina. In: M.I. Litter (ed): *Final results of the OAS/AE/141 Project: research, development, validation and application of solar technologies for water potabilization in isolated rural zones of Latin America and the Caribbean. Argentina-Brasil-Chile-México-Perú-Trinidad and Tobago*. OAS publication, Buenos Aires, Argentina, 2006, pp. 25–38, http://www.cnea.gov.ar/xxi/ambiental/iberoarsen/.

Nieto, J., Aguilar J., Ponce, S., Rodriguez, J., Solis, J. and Estrada, W.: TiO$_2$ thin films deposited by spray pyrolysis inside glass tubes for use in the HP technology for water potabilization. In: M.I. Litter and A. Jiménez González (eds): *Advances in low-cost technologies for disinfection, decontamination and arsenic removal in waters from rural communities of Latin America (HP and SORAS methods).* AOS Project AE/141, Digital Grafic, La Plata, Argentina, 2004, pp. 153–155.

Opar, A., Pfaff, A., Seddique, A.A., Ahmed, K.M., Graziano, J.H. and van Geen, A.: Responses of 6500 households to arsenic mitigation in Araihazar, Bangladesh. *Health and Place* 13 (2007), pp. 164–172.

Sancha A.M.: Removal of arsenic from drinking water supplies: Chile experience. *Water Supply* 18:1 (2000), pp. 621–625.

Sancha, A.M.: Removing arsenic from drinking water: A brief review of some lessons learned and gaps arised in Chilean water utilities. In: W.R. Chappell, C.O. Abernathy, R.L. Calderon and D.J. Thomas (eds): *Arsenic exposure and health effects V*. Elsevier Science, Amsterdam, The Netherlands, 2003, pp. 471–481.

Smedley, P.L., Nicolli, H.B., Macdonald, D.M.J., Barros, A.J. and Tullio, J.O.: Hydrogeochemistry of arsenic and other inorganic constituents in groundwaters from La Pampa, Argentina. *Appl. Geochem.* 17:3 (2002), pp. 259–284.

Smedley, P.L., Kinniburgh, D.G., Macdonald, D.M.J., Nicolli, H.B., Barros, A.J., Tullio, J.O., Pearce, J.M. and Alonso, M.S.: Arsenic associations in sediments from the loess aquifer of La Pampa, Argentina. *Appl. Geochem.* 20 (2005), pp. 989–1016

van Geen, A., Ahmed, K.M., Seddique, A.A. and Shamsudduha, M.: Community wells to mitigate the current arsenic crisis in Bangladesh. *Bull. World Health Organiz.* 82 (2003), pp. 632–638.

van Geen, A., Protus, T., Cheng, Z., Horneman, A., Seddique, A.A., Hoque, M.A. and Ahmed, K.M.: Testing groundwater for arsenic in Bangladesh before installing a well. *Env. Sci. Technol.* 38:24 (2004), pp. 6783–6789.

von Brömssen, M. Bhattacharya, P., Ahmed, K.M., Jakariya, M., Jonsson, L., Lundell, L. and Jacks, G.: Targeting safe aquifers in regions with elevated arsenic in groundwater of Matlab Upazila, Bangladesh. In: E. Lombi, S. Tyrell, A. Nolan, M. McLaughlin, G. Pierzynski, M. Gerzabek, N. Lepp, C. Leyval, M. Selim, F. Zhao, C. Grant and D. Parker (eds) *Book of Abstracts, 8th International Conference on the Biogeochemistry of Trace Elements*, Adelaide, Australia, 2005, pp. 190–191.

von Brömssen, M., Jakariya, M., Bhattacharya, P., Ahmed, K.M., Hasan, M.A., Sracek, O., Jonsson, L., Lundell, L. and Jacks, G.: Targeting low-arsenic aquifers in Matlab Upazila, Southeastern Bangladesh, *Sci. Total Environ.* 379 (2007), pp. 121–132.

Zheng. Y., Stute, M., van Geen. A., Gavrieli, I., Dhar, R., Simpson, H.J., Schlosser, P. and Ahmed, K.M.: Redox control of arsenic mobilization in Bangladesh groundwater. *Appl. Geochem.* 19:2 (2004), pp. 201–214.

Zheng, Y., van Geen, A., Stute, M., Dhar, R., Mo, Z., Cheng, Z., Horneman, A., Gavrieli, I., Simpson, H.J., Versteeg, R., Steckler, M., Grazioli-Venier, A., Goodbred, S., Shanewaz, M., Shamsudduha, M., Hoque, M. and Ahmed, K.M.: Geochemical and hydrogeological contrasts between shallow and deeper aquifers in two villages of Araihazar, Bangladesh: Implications for deeper aquifers as drinking water sources. *Geochim. Cosmochim. Acta* 69:22 (2005), pp. 5203–5218.

CHAPTER 75

Mitigation actions respond to arsenic exposure in Brazil

E. Deschamps & S.M. Oberdá
Fundação Estadual do Meio Ambiente (FEAM), Belo Horizonte, Minas Gerais, Brazil

J. Matschullat
Interdisciplinary Environmental Research Center, TU Bergakademie Freiberg, Freiberg, Germany

N.O.C. Silva
Laboratório de Contaminantes Metálicos, Divisão de Vigilância Sanitária, Instituto Octávio Magalhães, Fundação Ezequiel Dias (FUNED), Belo Horizonte, Brazil

O.R. Vasconcelos
Departamento de Engenharia Sanitária e Ambiental, Universidade Federal de Minas Gerais (UFMG), Belo Horizonte, Minas Gerais, Brazil

ABSTRACT: Globally, millions of people are at risk from adverse health effects related to both acute and chronic arsenic (As) exposure. While most of these persons are affected by contaminated drinking water, other important sources such as food, soil and air deserve attention. Known As anomalies in Brazil are mainly related to geological structures of hydrothermal gold mineralization, with dissipation due to centuries of gold mining and smelting activities. Most of the gold is associated with arsenopyrite and to a lesser extend with As-rich pyrite. Although not as severe and not exclusively water-related as in Bangladesh and West Bengal, As enrichments were recently detected in environmental and biological media in the Iron Quadrangle, Minas Gerais. This chapter presents the mitigation actions taken to improve the situation, starting with an environmental and health perception study which led to an environmental education program. Next, appropriate tailings deposits management was started to improve the control of tailings with very high total As concentrations (≤ 2 wt%), and to slow down As dissipation into the environment. Additionally, a low-cost water treatment plant was constructed with local materials to avoid ingestion via As-loaded particulates in drinking water.

75.1 INTRODUCTION

The Iron Quadrangle in Minas Gerais, Brazil, is one of the world's most important suppliers of mineral resources, including gold. The hydrothermal gold deposits, exploited since the early 1700's, contain arsenopyrite and pyrrhotite as major accompanying minerals. In the past, about 3.1 Mt (million metric tons) of tailings with a median value of 14,500 mg As/kg were deposited in nearby valleys. A single As-trioxide factory (Morro do Galo, Nova Lima district) produced about 100 t of As_2O_3 from 1962 to 1975. From 1983 to 1993, another 2.3 Mt of As were directly discarded into the drainage basin of Rio das Velhas (Borba *et al.* 2000). Those practices have contributed to an As release into surface and groundwaters, and via atmospheric emissions from ore dressing, smelters, and redistribution from tailings, onto the soils (Deschamps *et al.* 2002). Environmental awareness and related laws and regulations have been established since 1988. The well-known practice of mining and its long history gave rise to concern about possible health consequences in the area.

To answer the questions, whether low-level but chronic As exposure of people living in those areas, and whether up to 250 years of continual gold mining with As-bearing minerals have led to related health effects, a joint research project of several institutions and professionals was launched in 1998. This interdisciplinary project studied all environmental compartments (soils and tailings, ground- and surface waters, atmospheric deposition, garden vegetables and other plants) and performed human biomonitoring to assess the situation and deliver solutions in case of problems (Matschullat *et al.* 2007). Problems already became visible after the first series of field-work already. This contribution addresses the actions taken by the state government to minimize risks and work towards a sustainable future in the As-contaminated areas.

75.2   TRIGGERS FOR ACTION

From 1998 to 2004, more than 600 children and adults from two districts have been investigated regularly for As exposure. Persistent As enrichments in urine were encountered, but no direct evidence of arsenicosis was found. While children showed higher average values than adults, and a relevant percentage remained in a category with As values between 10 to 40 µg/l (Matschullat *et al.* 2000), early symptoms of As-related diseases (e.g., via UV-light scans on skin surfaces) were not found. Instead, learning difficulties, above average diarrhoea, bronchitis and similar anomalies prevailed—often related to people, and particularly children, living under conditions of malnutrition. Because of the typically retarded physiological reaction, direct As-related effects could not be determined, unless acutely toxic exposure was present (e.g., with the high death rate among the former As-trioxide factory workers).

Investigations of environmental compartments clearly showed the hot spots of As contamination. A total of 109 topsoil samples were taken from 1998 to 2003. Elevated to very high soil As concentrations (50–1000 mg As/kg) in living quarters, and extremely high (percentage range) As values on derelict industrial sites (tailings) were found (up to 2 wt% of total As, median 15,000 mg/kg). Elevated As concentrations occurred in fine (<63 µm) and fresh stream and river sediments (median 140 mg As/kg), equal to a 10-fold enrichment compared with the regional geogenic background in non-As-polluted soils and rocks (<20 mg/kg). Private garden vegetables, such as lettuce, onions, carrots, cabbage, and mustard were sampled, and presented elevated As concentrations in both edible and non-edible parts (>1 mg As/kg); the higher values were found close to the old mine tailings (Matschullat 2000 and 2002, Deschamps and Matschullat 2007).

The complex evaluation of all environmental compartments and their interactions delivered a thorough understanding of the dominating processes and As pathways. The results indicate that there are three major pathways to explain the increase of As values in the human body: (1) ingestion via As-loaded particulates in drinking water; (2) ingestion and inhalation via As-loaded dust, and (3) ingestion of As-contaminated foodstuff.

The risk assessment proposed As mitigation through rehabilitation of the old tailings as part of the actions to solve the environmental and human As-related health threat in the region. The results indicated that the full use of the locale results in an unacceptable risk.

75.3   ACTIONS TAKEN AND DISCUSSION

Next to these investigations, an environmental and health perception study was carried out to provide the responsible state institutions with information related to the values, faiths, attitudes and behavior of the communities directly involved and/or exposed to elevated arsenic (As) concentrations (Tostes de Macedo 2007). The instrument used for data collection was a semi-structured questionnaire (with closed subjects, pre-codified, and open subjects/spontaneous) containing general and specific questions with the main focus on As exposure. The results were an important tool for the definition of a program of environmental education. A total of 343 questionnaires were evaluated and the analyzed data allowed ascertaining not only the common subjects to all investigated communities but also their typical characteristics (Deschamps and Matschullat 2007).

To reduce the unacceptable risks related to the previous use of highly contaminated soils, such as gardening, the risk analysis suggested institutional actions as well as projects. Based on further field sampling and subsequent analysis, both the consumption of groundwater and the ingestion of meat and vegetables cultivated in this region were prohibited, and As was immobilized through coverage of the contaminated area, and re-vegetation. Three major tasks were considered of prime relevance:

(1) Set-up of an educational program for the people exposed to elevated As levels to generate awareness, adapt behavior and reduce personal risks and As uptake;
(2) Remediation and "neutralization" of the old tailings deposits as major sources for As dissipation;
(3) Development of low-cost water purification plants to help provide clean drinking water in more remote areas without access to the state distribution system.

75.3.1 *Educational program*

As a result of the environmental and health perception study, an environmental education program was launched with participation of the communities, local authorities, and schoolteachers (Fig. 75.1). It focused on the most important subjects in terms of As exposure reduction or elimination. The first topic "Have you heard about arsenic?" received much attention. Based on the results of As concentrations in topsoil and dust, parents and children received intensive pedagogic information on how to improve their personal hygiene. They were strongly encouraged to reduce the dust load in their houses and the direct hand-to-mouth contact mainly of younger children, since settling indoor dust can easily be inhaled through respiration and after remobilization during house-cleaning. The

Figure 75.1.   Leonardo Fittipaldi, biologist from FEAM, key person in the environmental awareness program, teaching school children and parents in personal hygiene.

feedback was straightforward; they learnt about the importance of changing habits and put it into practice.

Related to the As enrichments in food through home grown vegetables, two alternatives are being implemented: first to substitute normal gardens by suspended gardens using non-contaminated soil and second to exchange the topsoil material in their gardens. In addition, the local community is being educated to plant non-biomagnifying vegetables, but this seems more difficult to implement.

75.3.2   *Tailings remediation*

A risk assessment associated to the old tailings deposits was carried out in 2002 as part of the actions needed to solve the environmental and human As-related health threat in the region. It was based on the RESRAD-Chem methodology and considered residents and their exposition pathways like ingestion and dermal contact with soil and water, ingestion of meat, milk, fish, water, vapour inhalation, and the ingestion of particulates and groundwater (http://web. ead.anl.gov/resrad/home2/chem.cfm).

Hydrological studies suggested that in addition to the tailings coverage, a drainage system was to be implemented to bypass the maximum volume of rain water and derail the runoff from penetrating the deposits (Fig. 75.2). This project started in 2004 at the old tailing deposits. It preserved the silt layer lining from an initial mitigation project in late 1996. On top of that, three additional layers were assembled. These layers consist of (1) 20 cm of sand, (2) a high-density polyethylene (HDPE) sheet, and (3) uncontaminated topsoil. The drainage system was developed on the site. The system allows rainwater to be collected and diverted directly to the next creek. The layers were set in an angle to allow any water to flow towards the drainage system. The mining company established an effective monitoring program for surface and groundwater and also a maintenance project for the drainage system and erosion control. The total cost of rehabilitation of all old tailing deposits will amount to approximately 4 million Euro. This successful case study demonstrates how the mining industry handles this issue in partnership with Environmental Authorities and the local community.

75.3.3   *Low-cost water purification plant*

From 2001 to 2003, bench and pilot scale experiments were conducted to assess the efficacy of a natural enriched iron and manganese sorbent on the As removal from drinking water (Deschamps

Figure 75.2.   Left (a) Galo Novo tailings deposit in the Nova Lima district after the second mitigation phase, already planted. Right (b) concreted front yards and elevated house gardens.

Figure 75.3.   Left (a) site of the new local water purification plant in Santana do Morro, Santa Bárbara district. Right (b) Detailed view of the three purification stages (see text).

2003, Deschamps *et al.* 2003). Groundwater in the vicinity of the old tailings deposits was prohibited for human consumption and contaminated wells have been closed in 2005. A joint project with the Brazilian Federal Government/FNMA and the mayor of Santa Bárbara district led to a construction of a water treatment plant for the Santana do Morro community (Fig. 75.3).

Recently, the low-cost water treatment plant composed basically of a pre-filtration tank made from Fe-cement, followed by a coagulation tank, a disinfection unit and As adsorption filter started up. First results will be soon published.

75.4   CONCLUSIONS

The question whether As pollution poses a real threat for people living in areas like the Iron Quadrangle with elevated, but relatively low-level exposition, remains central—not just in Brazil. There is no doubt, that the results of this project and consequently the As enrichments encountered in biological media and other environmental compartments rectifies the implementation of those important mitigation measures as a contribution not only for today but for future generations. The educational program has resulted not only in increased awareness among the exposed population but also reduced individual As uptake.

ACKNOWLEDGEMENTS

The authors acknowledge the support from the National fund for the Environment—FNMA/MMA, the Millennium Science Initiative and CNPq, Brazil, and the German Federal Ministry for Research and Technology (*Bundesministerium für Forschung und Technologie*, BMBF), Germany.

REFERENCES

Borba, R.P., Figueiredo, B.R., Rawlins, B.G. and Matschullat, J.: Arsenic in water and sediment in the Iron Quadrangle, Minas Gerais state, Brazil. *Rev. Bras. Geoc.* 20 (2000), pp. 554–557.
Deschamps, E.: *Avaliação da contaminação humana e ambiental por arsênio e sua imobilisação em óxidos de ferro e manganês.* PhD Thesis in Metallurgical Engineering, Federal University of Minas Gerais (UFMG), Belo Horizonte, Brazil, 2003.
Deschamps, E. and Matschullat, J. (eds): *Arsênio antropogênico e natural. Um estudo em regiões do Quadrilátero Ferrífero.* Fundacao Estadual do Meio Ambiente, Belo Horizonte, Brazil, 2007.

Deschamps, E., Ciminelli, V.S.T., Frank, L.F.T., Matschullat, J., Raue, B. and Schmidt, H.: Soil and sediment geochemistry of the Iron Quadrangle, Brazil: the case of arsenic. *J. Soils Sedim.* 2:4 (2002), pp. 216–222.

Deschamps, E., Ciminelli, V.S.T., Weidler, P.G. and Ramos, A.Y.: Arsenic sorption onto soils enriched in Mn and Fe minerals. *Clays Clay Min.* 51:2 (2003), pp. 197–204.

Matschullat, J.: Arsenic in the geosphere—A review. *Sci. Total Environ.* 249 (2000), pp. 297–312.

Matschullat, J.: Biogeochemie des Arsens unter subtropischen Bedingungen am Beispiel Brasilien. *Umfeld Umwelt* 1-2002 (2002), pp. 36–38.

Matschullat, J., Borba, R.P., Deschamps, E., Figueiredo, B.R., Gabrio, T. and Schwenk, M.: Human and environmental contamination in the Iron Quadrangle, Brazil. *Appl. Geochem.* 15 (2000), pp. 181–190.

Matschullat, J., Birmann, K., Borba, R.P., Ciminelli, V.S.T., Deschamps, E.M., Figueiredo, B.R., Gabrio, T., Haßler, S., Hilscher, A., Junghänel, I., de Oliveira, N., Schmidt, H., Schwenk, M., de Oliveira Vilhena, M.J. and Weidner, U.: Long-term environmental impact of As-dispersion in Minas Gerais, Brazil. In: P. Bhattacharya, A.B. Mukherjee, J. Bundschuh, R. Zevenhoven and R.H. Loeppert (eds): *Arsenic in soil and groundwater environments: biogeochemical interactions.* In: *Trace metals and other contaminants in the environment* 9. Elsevier, Amsterdam, 2007, pp. 355–372.

Tostes de Macedo, A.: Estudo de percepção ambiental e em saúde. In: E. Deschamps and J. Matschullat (eds): *Arsênio antropogênico e natural. Um estudo em regiões do Quadrilátero Ferrífero*; Chapter 3.1. Fundacao Estadual do Meio Ambiente, Belo Horizonte, Brazil, 2007, pp. 125–174.

Index I: Subject index

Localities, stratigraphic units, tectonic and structural elements are included in Index II. Note that individual analytical methods and techniques are listed below "analytical methods". Arsenic in individual plant or animal species are listed in the rubrics "arsenic in plants" and "arsenic in animals", respectively.

A

2-oxoglutarate dehydrogenase 19
absorption (*see arsenic absorption*)
accumulation of arsenic (*see arsenic accumulation*)
acid drainage (*see arsenic sources*)
adenosine 5-triphosphate 19
adsorption of arsenic (*see arsenic adsorption*)
adsorbed arsenic 201, 205, 387
adsorbents of arsenic (*see arsenic sorbents*)
advanced light source (ALS) synchrotron 257, 263
advanced oxidation technologies (*see arsenic removal technologies*)
advanced treatment plants (*see arsenic removal, water*)
aeolic processes 398
Al$_2$O$_3$ (*see also arsenic sorbents*) 41, 55, 130, 198, 296, 312, 313, 507, 512, 513, 519, 522, 566
albite 53, 61, 63, 64, 66, 151, 513
Alcan filters (*see arsenic removal, water*)
alkali feldspar 41
alkalinity 18, 35, 37, 53, 71, 597, 616, 617, 692
aluminium oxides 42, 43, 47, 57, 522
  arsenate sorption 43
  surface 47, 57
alluvial
  aquifer (*see aquifer*)
  deposit 124, 409
  fan 48, 61, 66
alpha tocopherol (α-TOC) 481–486
alterations in locomotor activity (*see arsenic toxicity*)
altered gene expression (*see arsenic toxicity*)
altered hematopoiesis (*see arsenic toxicity*)
amphibole 41, 55, 151
analytical methods 4, 26, 114, 166, 313, 338, 398, 400, 401, 423, 496, 539

atomic absorption spectrometer (AAS) 84, 114, 133, 166, 218, 306, 328, 448, 561, 567, 574, 617, 651, 666
atomic absorption spectroscopy graphite furnace (AAS-GF) 173, 246, 247, 561, 583
atomic absorption spectroscopy hydride generation (AAS-HG) 173, 247, 289, 540
atomic emission spectrometry (AES) 18, 311, 313, 530, 699
atomic fluorescence spectrometry (AFS) 311
colorimetric methods 114, 247
electrochemical techniques 265, 266, 270
electron microprobe 61, 63–66, 195, 198, 199, 274
energy dispersed X-ray fluorescence (EDXRF) 265, 266, 270
flame photometer 114, 468
flow injection-hydride generation-atomic absorption spectrometry (FI-HG-AAS) 235, 237, 242, 328, 338, 540
hydride generation-atomic fluorescence spectroscopy (HG-AFS) 275, 311, 313, 314, 316, 365–367
hydride generator 84, 114, 389, 497, 651, 668
inductively coupled plasma atomic emission spectroscopy (ICP-AES) 125, 247, 275, 311, 313, 314, 530
inductively coupled plasma mass spectrometer (ICP-MS) 125, 133, 146, 195, 196, 245, 247, 251–253, 289, 296, 360, 365–367, 374, 610
inductively coupled plasma spectroscopy (ICP) 125, 133, 146, 182, 195, 196, 207, 245, 247, 251–253, 275, 289, 296, 297, 311, 313, 314, 360, 365–367, 374, 400, 401, 421, 530, 540, 542, 610, 679

inductively coupled plasma-optical emission spectrometry (ICP-OES) 182, 400, 401, 421, 540, 679
ion chromatograph 71, 114, 125, 207, 574, 610
liquid chromatography coupled to hydride generation atomic fluorescence (LC-HG-AFS) 275
low-cost assessment of inorganic and total arsenic in food 235
microprobe 61, 63–66, 195, 198, 199, 274, 276
near infrared spectroscopy (NIRS) 235–238, 240, 242
optical microscopy 63
polarography 173
scanning electron microscopy (SEM) 173, 263, 657
stripping voltammetry 247, 265, 266, 270
   anodic (ASV) 247, 251–253, 266
   cathodic (CSV) 265, 266, 270
UV-Vis spectrophotometry 173
X-ray absorption fine structure spectroscopy 295
X-ray diffraction analysis 55, 63, 85, 633
X-ray fluorescence spectrometry (XRF) 103, 195, 198, 255–257, 265–267, 270, 312, 506, 513, 522, 679, 681
μ-X-ray fluorescence spectrometry (μ-XRF) 103, 195, 255–257, 266, 269, 296, 312, 313, 549, 679
andesine 85
andesite 56, 113, 118, 119, 132, 147
andesitic lava 92, 103
animal models 28, 320, 321, 436, 442, 459, 474
anion exchange 84, 270, 288, 388, 390, 391, 539, 571, 572, 574–578, 580, 616, 629, 630, 677
anthropogenic arsenic (*see arsenic sources*)
antimony 56, 95, 349, 499
   trioxide 95
antioxidant 430, 460, 463, 467, 468, 470, 478, 482, 486
   enzymes 463, 486
   mechanisms 460
   proteins 430
apatite 55, 64, 66, 139
aquifer (*see also arsenic in aquifer*)
   alluvial 8, 109, 110, 113–115, 118, 119, 132, 163, 222, 226
   anoxic conditions 216
   arsenic hot spot 5, 62, 66, 109, 226, 692, 700
   arsenic-safe 122, 687, 689, 691–693, 695
   artesian 50–54, 57

characteristics 227
confined 50, 69, 71, 157, 164, 206, 217, 218, 398, 419–422
deep 50, 72, 75, 77, 218, 219, 618
fracture system 118, 156, 180, 211
hardrock 118, 119
heterogeneity 201
high salinity 47, 48, 397, 677
hydraulic conductivity 115, 208, 218, 521, 522, 524, 540
hydraulic gradient 83, 115, 541
hydrodynamic regime 207
hydrogeological conditions 207
hydrogeological environment 48
lateral variations of As concentration 118
loess 35, 36, 39, 40, 41, 43, 57
micro-fissuring 211
multilayered 71
oxidizing 7, 38, 50, 56, 201, 692
oxidizing conditions 7, 38, 50, 56, 201, 692
phreatic 6
redox changes 88
sedimentary 67, 109, 111, 113, 115, 117, 119, 121, 127, 193, 195, 197, 199, 201, 203, 216, 223, 357, 524
sediments (*see also arsenic in aquifer sediments*) 43, 61, 63, 65, 85, 118, 120, 194, 196, 219, 225, 398, 692, 693
unconfined 50, 71, 157, 164, 217, 398, 419–422
aragonite 51, 57
arsenate (*see arsenic species*)
   esters 19
arsenian pyrite 77, 193, 196, 199, 200, 201, 511, 547
arseniasis (*see arsenic health effects*)
arsenate-phosphate competition for sorption sites 8
arsenic absorption 65, 84, 114, 133, 166, 173, 218, 225, 226, 236, 237, 241, 247, 255–258, 262, 268, 295, 296, 306, 311, 319, 320, 322, 323, 328, 338, 343, 346, 353, 360, 361, 371, 374, 375, 421, 436, 447, 448, 454, 461, 482, 506, 522, 540, 561, 567, 574, 583, 596, 617, 627, 631, 649, 651, 659, 668, 672
   by gastrointestinal epithelium 322
arsenic accumulation (*see also arsenic in, arsenic metabolites and arsenic biomarkers*)
   humans
      blood (*see arsenic biomarkers*)
      bones 19
      breast milk 25
      creatinine 410, 447–449, 454

hair (*see arsenic biomarkers*)
nails (*see arsenic biomarkers*)
skin (*see also arsenic health effects, skin*)
   19
urine (*see arsenic biomarkers*)
plants (*see also phytoremediation*) 360, 365,
   380, 387, 646, 650, 653
ferns 359, 360, 361, 362
vegetables 343, 345, 355, 359, 361, 363
arsenic (ad)sorption (*see also arsenic
   sorbents*) 61, 62, 66, 69, 77, 132, 146,
   169, 183, 197, 200, 207, 255, 274, 285,
   292, 341, 374, 375, 511, 512, 516, 518,
   521–523, 527, 528, 530, 531, 533, 538,
   539, 543, 544, 547, 551, 555, 556, 557,
   562, 578, 582, 589, 595, 596, 607–609,
   612, 613, 616, 625–627, 629–635, 637,
   638, 643, 657, 659, 661, 664–670,
   672–681, 691, 693, 701, 703
capacity 61, 66, 521–523, 539, 578, 613,
   635, 664, 666
competition for adsorption sites 8, 62
of As oxyanions 77
of As(V) onto goethite 527
on ferric oxide and hydroxides 62, 65
sites for As 61, 62, 66, 200, 528, 607, 608,
   612, 693
arsenic anomalous map 101, 104–106
arsenic bioaccessibility 320–323, 389
after cooking 321
in rice 322
of arsenosugars 322
of inorganic arsenic 321, 322
of total arsenic 321, 323
arsenic bioaccumulation 340, 341, 343, 360,
   361, 498, 650
factor (BF) 340, 341, 360, 361
for vegetables 343
arsenic bioavailability 265, 319, 320–324, 336,
   339, 341–343, 345, 348, 360, 361, 383,
   384, 387, 495, 653
for human 319, 323
from soil 338, 341–343
in food 319, 323
in raw and cooked fish 323
in rice 322
in vitro studies 319, 320
in vivo studies 319
to plants 360
arsenic biomarkers (*see also arsenic in,
   arsenic metabolites and arsenic
   accumulation*) 20, 443
blood 19, 20, 22, 182, 410, 435–437, 439,
   440, 453–457, 461, 462, 468, 474, 475,
   477, 478

hair 19, 20, 23, 155, 415, 436
nails 19, 20, 617, 619, 623
urinary creatinine 410, 447–450, 454
urine 10, 19–25, 187, 320–322, 359, 410,
   435, 439, 440, 442, 443, 447–450, 453,
   456, 457, 462, 571, 700
arsenic excreted in urine 19, 24 320, 322
association between F and As and IQ
   scores 456
F and As in urine 453, 454, 456, 457
of girls and boys 439
samples 20, 25, 359, 410, 436, 447, 454
children 447
arsenic biotransfer factor (BTF) 419–421,
   423–424
value for As from water to bovine milk 423
arsenic biotransformation 463
in brain tissue 464
arsenic carcinogenesis (*see arsenic health
   effects*)
arsenic coagulants (/flocculants) (*see also
   arsenic removal technologies*) 537,
   555–560, 562, 563, 582, 583, 590, 596,
   607, 664
aluminum salts 555, 556, 589
aluminum sulfate 589
polyaluminum chloride (PACl) 589–591,
   593, 594
dose 555, 558, 560, 562, 563, 582–584, 590,
   591
iron salts 556, 596
ferric 511, 512, 528
chloride (FeCl$_3$) 11, 527–529, 532–534,
   538, 542, 545, 555–562, 572, 583–586,
   589, 617, 619, 628, 666, 678, 679, 691
dose 584
sulfate 527, 529–534
ferrous 511, 512
arsenic concentrations (*see arsenic in*)
arsenic contaminated site 61, 295, 496, 649
arsenic contamination (*see also arsenic in*) 3,
   4, 7–10, 12, 13, 18, 23–25, 82, 91, 92,
   95, 97, 109, 120, 132, 141, 161, 179,
   182, 187, 201, 206, 215, 216, 225, 226,
   228, 246, 295, 313, 335, 340, 341, 346,
   355, 399, 405, 409, 427, 450, 498, 505,
   521, 606, 649, 650, 690, 700
arsenic desorption (*see arsenic mobilization*)
arsenic detoxification in cells 430
arsenic effects in liver (*see arsenic health
   effects*)
arsenic enrichment in organic matter 285
arsenic exposure (*see also arsenic in and
   arsenic health effects*) 3, 4, 6, 9–12, 18,
   19, 20, 23–25, 35, 123, 155, 156, 161,

179, 182, 185, 187, 230, 245, 265, 323,
    327, 343, 345, 351, 381, 387, 392, 397,
    409, 410, 412, 413, 415, 419, 427–429,
    435, 436, 439, 442, 443, 447, 449, 450,
    455, 457, 459, 460, 463, 467, 468, 471,
    473, 474, 477, 478, 482, 486, 489–493,
    606, 629, 665, 691, 699–701
acute (short-term)
  by dust or fumes 245
assessment 343
chronic (long-term) 3, 11, 12, 18, 24, 35,
    155, 182, 265, 327, 435, 442, 443, 453,
    459, 467, 471, 473, 474, 485, 489, 643,
    665, 699
  by inhalation of i-As 18, 23, 132, 245,
    606, 698, 700, 702
  dusts 10, 359
  by ingestion (*see also arsenic in plants and
    arsenic in*) 10, 11, 17, 18, 20, 23, 25, 26,
    123, 187, 327, 414, 473, 595, 699
  food chain 10, 92, 336, 341, 343, 351,
    371, 380
doses 179, 185–187, 459, 474, 489
  calculation 186
high 4, 10, 11, 412
levels
  daily intake 185, 186, 404, 405
  by diet 185, 355, 356, 398
  maximum admissible arsenic level 205
  maximum contaminant level (MCL) 181,
    555, 565, 581, 584, 595, 605, 606, 672
  provisional tolerable daily intake (PTDI)
    186
  provisional tolerable weekly intake
    (PTWI) 186, 398, 404, 405
  tolerable daily intake (TDI) 186, 321–323
  WHO guideline 22, 37, 38, 54, 56, 92, 97,
    123, 179, 225, 453, 521
of arsenic-trioxide factory workers 700
of children 405, 435, 436
of human 9, 10, 18, 161, 179, 182, 187, 245,
    343, 428, 436, 442, 443, 459, 467, 489,
    606, 625, 665, 703
  assessment 319
  by soil 359
  via breast milk 25
of mine workers 4
of people living in mining areas 4
pathways 18, 429, 492
populations 4, 9, 24, 26, 112, 427, 436, 443,
    688, 703
  estimates 9
risk 9, 22, 23
  analysis study 23
  assessment tool 201

of arsenicism 22
of cancer 9, 20, 22–24
of death 23
to the population 11
subchronic 473, 474, 476–478
to airborne dust 10
to arsenate 428
under co-exposure of F 453, 457
under co-exposure of Se 474
arsenic fractions in rhizospheric soils 392
arsenic freshwater food chain models 336
arsenic from drinking water to the milk 420
arsenic from water (*see also arsenic in*) 397,
    404, 421, 528, 556, 582, 596, 606, 609,
    638, 666
arsenic geochemical anomalies 104
arsenic hot spot sites 5, 62, 66, 109, 226, 692,
    700
arsenic health effects (*see also arsenic toxicity*)
    3–6, 10, 11, 26, 91, 163, 179, 186, 187,
    265, 379, 473, 479, 555, 581, 605, 665,
    699, 700
arseniasis 409, 410, 415
arsenicism 17, 20, 22–26, 409, 616
arsenicosis 10, 123, 124, 230, 355, 595, 700
cancer 9–11, 17, 19–24, 26, 35, 48, 132,
    155, 182, 186, 187, 193, 246, 327,
    379, 397, 409, 413, 416, 419, 427, 428,
    435, 442, 453, 459, 467, 473, 481, 489,
    492, 527, 555, 606, 625, 643, 663,
    665, 666
  bladder 17, 20, 22, 23, 26, 48, 397, 427,
    453, 555, 606
  kidney 22, 427, 428
  liver 22, 186, 246, 427, 467
    cytokeratin 18 (CK18) expression 467
  lung 22, 23, 35, 48, 397, 416, 427, 643
  skin 10, 20–22, 187, 397, 413, 419, 427,
    435, 443, 489, 527, 606, 643
    p53 DNA binding in human keratinocytes
    489
carcinogenesis 427–430, 443
  molecular mechanisms of 489, 492
    altered gene expression 427, 428, 430
    cell proliferation 427, 428, 430
    chromosomal instability 427, 430
    inhibition of DNA repair 427–430
    oxidative stress 427–430
    p53 DNA binding 489–493
diabetogenic action 473
  in selenium-deficient rats 473
Endemic Chronic Regional Hydroarsenism
    (ERCH) (= *Hidroarsenicismo Crónico
    Regional Endémico, HACRE*) 21, 22,
    419, 616

hydroarsenicism 409
hyperinsulinaemia 473, 478
 in selenium-deficient rats 473
immunodepression 435, 442, 443
 CD4/CD8 ratio 435
 cytokine secretion by mononucleated cells 435
 GM-CSF secretion by mononucleated cells 435
 in Argentina 419
 in Brazil 447
 in Central America 409
 in children 435, 447
 co-exposed to fluoride
 in Nicaragua 409
 in Mexico 304, 454, 511, 665, 695
 neurological 453, 457
 IQ 453
 toxic effects in sural nerves 481, 482
 protection against by α-tocopherol (α-TOC) 481, 482
 respiratory system 409
 rodent models for study of 459, 467, 473, 481
 susceptibility to opportunistic infections 435
 skin lesions 10, 17, 21, 23–26, 193, 246, 410, 413, 415, 442, 443, 489, 555, 595
 hyperkeratosis 9, 10, 17, 19, 21, 26, 112, 156, 182, 379, 409, 411, 489, 527, 595, 606
 hyperpigmentation 17, 21, 25, 26, 123, 156, 246, 379, 409, 410, 413–415, 489, 595, 625
arsenic hydrogeochemistry 47, 71
arsenic hyperaccumulating plants (*see also phytoremediation*) 360, 646, 650, 653
arsenic immobilization and remobilization in aquifer 537, 542
arsenic in animal fodder 345
arsenic in aquifer (*see also arsenic in groundwater*) 6–8, 11, 24, 31, 35, 36, 39–41, 43, 44, 47, 48, 50–54, 56, 57, 61, 62, 64, 66, 69–72, 76, 77, 81, 85, 98, 109, 110, 113–115, 118–121, 123–125, 132, 141, 146, 147, 149, 156, 157, 163, 164, 166, 167, 179–182, 184, 185, 193–201, 205–209, 211–213, 215–219, 222, 223, 225–228, 230, 255, 285, 289, 292, 398, 402, 419–422, 521, 527–529, 537, 538, 543, 544, 547, 606, 618, 627, 629, 631, 650, 665, 687–690, 692–695
arsenic in drinking water 3, 7, 9, 17, 19, 20, 24–26, 37, 48, 72, 109, 110, 124, 155, 181, 185, 246, 265, 379, 397, 401–407,

419, 421, 423, 436, 442, 443, 454–456, 473, 527, 565, 599, 606, 625, 643, 691
arsenic in fish and seafood (including products) (*see also arsenic in food*)
 hake fillets 327, 328
 jack mackerel 327, 328, 330, 331
 Orcorhynchus mykiss (freshwater trout) 335, 336, 341
 Procambarus clarkii (crayfish) 239–241
 salmon pulp 327, 328, 330
 shellfish 327, 328, 330, 331
 Uca tangeri (fiddler crab) 235, 238, 239
arsenic in food (human diet) (*see also arsenic exposure and arsenic in*) 10, 23, 92, 185–187, 235, 236, 265, 281, 304, 319, 320, 327, 336, 341, 343, 351, 371, 380, 404
 fish (*for individual species see arsenic in fish and seafood*) 92, 132, 141, 235–238, 240, 241, 245, 319, 320, 323, 327, 328, 330, 331, 336, 338, 340, 343, 355, 527, 702
 influence of cooking 319, 321–323
 meat 185, 345, 355, 701, 702
 milk (bovine) 419–424
 plants (without vegetables) (*see arsenic in plants*)
 seafood (*for individual species see arsenic in fish and seafood*) 185, 327–329, 331
 vegetables (*for individual species see arsenic in plants*) 308, 320, 335–337, 340–344, 351–353, 355, 356, 359–363, 376, 379, 380, 384–385, 447, 700–702
 cultivated by indigenous people 346
 from private gardens 359, 360, 362
 water used during food preparation 10
arsenic in food chain (*see also arsenic exposure*) 10, 336, 343, 351, 371, 380
arsenic in geothermal fluids 70, 75, 77, 81–83, 85, 87–89, 119, 129, 141, 145–151, 181–183, 185
arsenic in humans (*see arsenic accumulation, arsenic exposure and arsenic biomarkers*)
arsenic in milk 25, 419–424, 702
arsenic in oil reservoir 145, 148, 149, 151
arsenic in oilfield fluids 148
arsenic in plants (edible and non-edible)
 Abelmoschus esculentus (okra) 351, 353, 355
 alga 12, 138, 322, 335–340, 691
 Gracilaria sp. (freshwater alga) 335, 336, 339, 340
 lacustrine 12, 691

Lessonia nigrescens 12, 691
Amaranthus gangeticus (red amaranth) 351, 353, 355
apple 348
asparagus 335, 336, 341, 342
Basella alba (indian spinach) 351, 353, 355
bean
　broad beans 346, 347
　Lablab niger (hyacinth bean) 351, 353, 355
　Vigna sesquipedalis (string bean) 351, 353, 355
beets 345, 347
broccoli 379–385
　Brassica oleracea L. var. *italica* Plenck 379, 380
　"Chevalier" 380, 383, 384
　"Romanesco" 380, 383, 384
Buxus sempervirens L. (boxtree) 367
cabbage 346, 347, 700
cauliflower 379, 380, 381, 383–385
　Brassica oleracea L. var. *botrytis* L. 379, 380
　"Premia" 380, 383, 384
　"Thasca" 380, 383, 384
Capsicum sp. (chilli) 351, 353, 355
carrot 335, 336, 341, 342, 346, 347, 360, 700
　Daucus carota L. 342, 360
cereals 185, 308, 336, 342, 343, 373
corn 10, 304, 335, 336, 341, 342, 346, 347
crops (edible) 360, 373, 380
Cyperaceae (sedges) 649, 651, 653, 654
downy oak 367
fern 359–362, 650
　Dryopteris filix-max (L.) Schott. 367
　Pityrogramma calomelanos 359–362, 650
　Pteris vittata 359–362, 650
fungus 495–498
　Eupenicillium cinnamopurpureum 495–498
　Neosartorya fischeri 495–498
　Talaromyces flavus 495–498
　Talaromyces wortmannii 495–498
garlic 245, 346, 347, 495
gourd
　Cucurbita maxima (sweet gourd) 351, 353, 355
　Lagenaria siceraria (bottle gourd) 351, 353, 355
　Momordica charantia (bitter gourd) 351, 353, 355
grape 348
grass 286
guavapple 348

Hydnum cupressiforme Hedw. (moss) 367
Juncus inflexus L. (hard rush) 367
kale 360
kiwi 348
lemon 12, 348, 616–623, 691
lettuce 335, 336, 341–343, 346, 347, 360, 362, 382, 700
　Lactuca sativa 342
Lycopersicon esculentum (tomato) 351, 353, 355
mango 348
Medicago sativa L. (alfalfa) 303, 304, 306, 308, 342, 423
melon 348
Mentha viridis (mint) 351, 353, 355
mustard 360, 362, 380, 700
　Sinapis alba L. (white mustard) 384
nopal 303, 304, 306–309
onions 347, 700
Opuntia robusta L. (prickly pear) 303, 306, 308
orange 348
papaya 351, 353, 355
　Carica papaya (green papaya) 351, 353, 355
parsley 346, 347
peach 348, 371, 372, 374, 376
peas 347
pepino 348
potatoes 342, 345–347
radish 347, 379–385
　"ICEBERG" 380, 383, 384
　"Middle East Giant" 380, 383, 384
　Raphanus raphanistrum L. 384
　Raphanus sativus L. 379, 380
rice 222, 235–239, 241, 242, 319–323, 351, 355
　cooked 321–323
　Oriza sativa 236
Rubus ulmifolius (L.) Schott. (blackberry) 367
seaweed 235, 319–322, 339
　cooked 321, 322
　edible 320, 322
　Enteromorpha sp. 321
　Porphyra sp. 321, 322
Solanum melongena (brinjal) 351, 353, 355
spinach 346, 347, 351, 353, 355
Stellaria halostea (Stitchwort) 367
sugar beets 345
sweet potato 360, 362
swiss chard 346, 347
terrestrial 360, 365, 366, 368
tomato 304, 351, 353, 355, 380
Trichosanthes dioica (palwal) 351, 353, 355

vegetables (*see arsenic in food*)
Viguiera dentata 387, 388, 392, 393
wheat 345, 388, 389, 615
arsenic in porewater 39, 171–175
arsenic in residential dust 95, 96
arsenic in sediments 81, 85, 87–89, 140, 171, 175
arsenic in soil 4, 95, 118, 120, 255, 265, 285, 295, 299, 300, 306–308, 311, 315, 335, 341–343, 345, 360, 361, 371, 372, 375–377, 384, 387, 547
  andosol 296
  As(III)-contaminated 295, 298–300
  As(V)-contaminated 295, 298–300
  calcic gleysol 287
  cultivated 351
  gleysols 286, 291
  histosols 285, 286, 291
  rhizospheric 387–389, 392
arsenic in sural nerves 481, 482, 486
arsenic in water (freshwater) 4, 13, 17–19, 22, 26, 81, 82, 84, 86, 91, 137, 138, 140, 181, 182, 185, 206, 210, 228, 247, 251, 266, 270, 288, 339, 341, 342, 347, 359, 361, 399, 402, 404, 405, 421, 423, 424, 453–455, 467, 506, 528, 533, 555, 581, 595, 599, 606, 607, 649, 665
  groundwater (*see also arsenic in aquifers*) 4, 6–11, 18, 22, 35, 43, 61, 62, 86, 97, 98, 109–111, 113, 116, 118, 123–125, 127, 155, 163, 167, 171, 172, 181, 183, 185–187, 193, 196, 206, 215, 216, 220–223, 225–227, 229, 230, 245–247, 251, 253, 285, 351, 354–356, 401, 419, 422, 427, 511, 521, 537, 541, 543, 555, 605, 606, 629, 657, 665, 674, 675, 677, 678, 681, 687–695, 699
  arsenic correlation with
  B 135
  Cl$^-$ 74, 77
  F$^-$ 75
  Fe 53, 135, 200, 206, 211, 215, 222, 290
  high-pH groundwater 53
  Li 137, 139, 140, 211
  Mn 53, 210, 290
  Na-HCO$_3$ water type 61, 62, 66, 692
  organic carbon 290
  organic matter 135, 137, 140
  SO$_4^{2-}$ 74, 77
  temperature 185
  well age 228
  well depth 88, 227, 355
  temporal variability of arsenic concentrations 88, 225, 226
  irrigation water 10, 341, 345, 346, 356

wastewater 306
arsenic ingestion/intake (*see also arsenic in and arsenic exposure*) 7, 10, 11, 20, 123, 124, 186, 187, 226, 319, 320, 323, 327, 356, 398, 402, 404, 405, 412–415, 419
  by food 7, 185, 419
  by hot beverages 185
  by water 7, 404
arsenic ion exchangers (*see also arsenic removal technologies*)
  hybrid anion exchangers (HAIX) 574, 576
    durability 578
    macroporous (HAIX-M) 571, 572, 574–578
    breakthrough curve of arsenic 574
    regenerant 576
  hybrid cation exchangers (HCIX) 572, 574–576
    gel-type (HCIX-G) 574–576, 578
  hybrid sorbents
    durability 578
    materials 572
    regeneration efficiency 576
    selective removal of As 571, 572
  polymeric 573
  resins 505, 571, 608, 629
arsenic metabolism 19, 22, 322, 371, 427, 428, 429, 474, 478, 481
  interactions between Se and As 474
arsenic metabolites 24, 436, 438, 439, 459, 462, 471, 474, 481, 483, 485, 486
  dimethyl sulfoxide (DMSO) 430
  dimethylarsenate (DMA) 19, 311, 320, 322, 323, 335, 336, 427, 428, 430, 437, 438, 439, 448, 459–463, 471, 474, 481–483, 486
  dimethylarsine 495
  dimethylarsinic acid [DMA(V)] 320, 323, 335, 427, 471, 495
  dimethylarsinous acid [DMA(III)] 471, 495
  in brain (mice) 461
  in nerves 483
  in urine (*see also arsenic biomarkers*) 436–439
    between boys and girls 439
  methylarsonic acid 427, 459, 462
  monomethylarsenate (MMA) 19, 320, 323, 335, 336, 362, 427, 428, 436, 438, 439, 448, 450, 459, 460–463, 471, 527, 703
  monomethylarsonic acid [MMA(V)] 330, 335, 495, 527
  monomethylarsonous acid [MMA(III)] 471, 527
  of inorganic arsenic 471, 474

tetramethylarsonium ion (TMA$^+$) 320, 323, 335
trimethylarsine oxide (TMAO) 320, 335, 342, 460, 462, 471, 481
arsenic methylation 19, 320, 427, 429, 430, 460, 461, 463, 467, 474, 489, 667
  of arsenate 430
  of arsenite 430
  oxidative 427, 460
    of trivalent arsenicals 427
arsenic mine (*see also arsenic sources*) 159, 161, 256, 262, 312, 359, 366
arsenic mitigation (*see also arsenic remediation*) 226, 230, 700
  innovative methods 690
  innovative strategies 156, 226
    local drillers to identify safe aquifers 695
    targeting arsenic-safe aquifers 66, 687, 690, 692, 695
    targeting arsenic-safe tubewell 695
    tool to delimitate As-safe parts 692
  program 230, 702
  risks and benefits analysis 230
  social acceptance 687, 688, 690, 692, 694
    increasing awareness 186, 193, 285, 703
  socioeconomic constraints 688
    dispersed settlements 12, 123, 691
    education campaign 25
    educational program 701, 703
    educational status 92
    environmental awareness 699, 701
      increased 186, 193, 703
    lack of local community participation 92
    low-income communities 505
    pedagogic information 701
    poor rural villages 678
    poverty 193, 615, 677
    small communities 3, 4, 12, 92, 156, 690, 691
    sparsely populated rural areas 691
    urban areas 3, 11, 44, 163, 179, 627, 691
arsenic mobility 7, 35, 41, 43, 75, 77, 109, 118, 172, 200, 256, 290, 295, 374, 375, 387, 495, 547
  between sediment water 172
  control 7, 35, 41, 75, 77, 118
  in groundwater 57, 113, 118, 511
  in rhizosphere 387
  in soils 255, 290, 295
arsenic mobilization 8, 66, 82, 109, 110, 113, 119, 120, 124, 171, 174–176, 183, 193, 199–201, 205, 211, 213, 216, 222, 306, 354, 387, 390–392, 419, 505, 547, 687, 692–694
  by anionic exchange 387, 392

by desorption
  desorption of As from mineral 212
  from As(III)-contaminated soil 245, 298, 299, 300
  from As(V)-contaminated soil 300
  from contaminated Kuroboku soil 296, 299
  from Fe- and Mn-oxyhydroxides 85
  in soil 295
by dissolution
  dissolution/flow velocity relationship 56
  from the solid phase 207
  leaching of mine tailings 94
  of As-containing minerals 81
  of carbonates 62, 66, 127
  of fluorite 75, 77
  of silicates 62, 127
  of sulfides 175, 176
  oxidation of pyrite 201, 205, 207
  rates of FeAsS 279
by reductive dissolution 66, 85, 87, 127, 172, 193, 201, 606, 692
  of ferric iron minerals 66, 127
  of iron and manganese minerals 87
  of iron oxyhydroxides 193, 201
controls 199
  by changes in Eh 206
  by changes in pH 206, 691
  factors 226
  influx of oxidized rainwater 201
  oxygenated recharge water 88
  pH-influenced desorption 201
conceptual model 62, 66
from rocks 109
from sediments 82, 175
  extractable Fe and Mn concentrations 43
from sediments to water 174
in aquatic sediments 171
in aquifer 201
in rhizospheric soil 387, 392
in sulfide production zone 175
into groundwater 44, 113, 119, 222
mechanisms 69, 110, 124, 171, 687, 692–694
processes 175, 200, 201, 207, 255
arsenic neurotoxicity (*see arsenic toxicity*) 460, 464
arsenic occurrence (*see arsenic sources*)
arsenic oxidation state (*see also arsenic species*) 559, 607–609, 612, 653
arsenic oxyanions (*see also arsenic species*) 77, 511, 512, 547–549, 606
arsenic phytotoxicity 336
arsenic poisoning 9, 20, 21, 24, 26, 215, 246, 379, 409, 416, 467, 547, 565, 595, 606, 612, 650

arsenic pollution 12, 155, 206, 306, 380, 447, 505, 691, 703
arsenic pyrite 206
arsenic release (*see arsenic mobilization*)
arsenic remediation (*see also arsenic mitigation*) 11, 13, 625, 626, 678, 691
  approaches 226, 627
  arsenic removal from water (*see arsenic removal*)
  bioremediation 496, 498
    arsenic volatilization by fungi 497, 498
  experiences 691
  methods 3, 4, 12, 226, 247, 495, 626, 627, 651, 687–689, 692
  of mine tailings 94, 158
    monitoring program 702
    redistribution from tailings 699
    rehabilitation of the old tailings 700
  phytoremediation 3, 12, 360, 380, 606, 643, 645, 649–651, 653, 654, 690, 691
    of arsenic-contaminated soils 360, 645
    of water 650
    by As hyperaccumulating plants 360, 646
    Agrostis capillaris 650
    Brassicaceae 380, 650
    Cynodon dactylon 650
    Cyperaceae 649
    Eleocharis macrostachya Britt. 649–654
    fern 360–362, 650
      Pityrogramma calomelanos 359–362, 650
      Pteris vittata 359, 362, 650
    Pteridaceae 649, 650
    Schoenoplectus americanus Pert. 651–654
    Sorghum biocolor 643, 646, 689
      uptake kinetics 643, 644
  techniques 626
    low-cost 3, 4, 12, 678, 690, 691, 701, 702
    removal of As(III) and/or As(V) 512
    techniques 565, 689
    technologies 12, 678, 681, 691
    water purification plant 701, 702
    water treatment plant 703
  in situ (groundwater/aquifer) 625, 627, 629, 635, 637, 638
    by micron size ZVI and NZVI 635, 636
    colloidal reactive barrier (CRB) 639
    mobile reactive barrier (MRB) 627, 636, 638, 639
    permeable reactive barriers 627
  in situ immobilization of arsenic 538
  water treatment (*see arsenic removal, water*)
arsenic remobilization in aquifer 539

arsenic sinks 196
arsenic soil-crop-food transfer (*see also arsenic transfer factor*) 356
arsenic soil-plant transfer 372, 373
arsenic(-rich) minerals (*see arsenic sources*)
arsenic-producing industries (*see also arsenic sources*) 11
arsenic removal (water) (*see also arsenic removal technologies*) 13, 505, 528, 555, 565, 595, 596, 599, 600, 611, 657, 658, 690, 691, 702
  break-through 560
    characteristics 524
    curves 568, 574
  efficiency 495, 507, 533, 556, 589, 595, 597, 679
    from mining water 553
    of As(III) species 77, 314, 514
    of As(V) species 528
    of As(V) species by NZVI 503
    pH-dependency 521
    plant 555, 557, 559, 584, 586, 589, 690
    process 556, 557, 589
  from mining water 553
  in rural areas 528
  low-cost household-scale applications 505, 565, 690, 691
    Alcan filters 687, 688
    Bishuddya filters 687, 688
    bottles with TiO_2 678, 679
    household filter systems 524
    pond sand filters 688
    rainwater harvesters 687–689
  plant 6, 10, 226, 360, 365, 606, 629
    advanced 689
    backwash 559–561, 563, 581, 583, 584
    conditions 560, 563
    frequency 560
    interval 581, 584
    clogging processes 538, 544
    DesEL (capacitive deionization system) 667–670, 672–674
    DesEL-4k 669, 672–675
    pilot scale 555, 556, 584, 630, 702
    pilot scale experiments 555, 702
    pre-filtration tank 703
    pump-and-treat 537, 627
    real scale 589
arsenic removal technologies (water) 230, 521, 556, 626
  (ad)sorption (*for specific sorbents, see arsenic sorbents*) 3, 595, 596, 625, 626, 627, 629, 630, 690, 691
    and filtration by manganese greensand 521
    As(III) and As(V) sorption 511, 514

As(III)/As(V) sorption efficiency 515
As(V) adsorption 518, 616, 634, 635
As(V) absorption onto coagulated flocs
    555, 556
As(V)-Fe(III) complexes 511, 547
batch experiments 507
blast furnace slag 626
of organic As compounds 538
on low-cost media 595
onto coagulated flocs 555, 556
physical adsorption 627
temperature dependence 528
best available technology (BAT) 582
capacitive deionization 665, 667, 668, 670,
    674, 675
coagulation (*see also arsenic coagulant*)
    626, 666, 675
    assisted microfiltration 606
    coagulation followed by direct filtration
        (CF) 581, 582, 584, 585
    conventional coagulation 596
    electrocoagulation (EC) 528, 595–600
    followed by filtration 666
    followed by flocculation and filtration 582
    followed by microfiltration 557
    iron coagulation 505, 584, 626
    modified coagulation-filtration 582
    process 555, 584, 597, 600, 607
coagulation-adsorption 589
coagulation-filtration 521
coagulation-flocculation 556, 557, 563,
    581–582, 584
coagulation-rapid sand filtration 557
conventional 555, 625, 626, 665, 667
    adsorption/co-precipitation 533
    decantation and/or filtration 547, 553
    direct filtration 555–557, 559, 563, 581,
        582, 584
    experiments 563
    dissolved air flotation 596
    ion flotation 511, 616, 677
    lime softening 505, 582, 606, 607, 626,
        630, 666
    modified 582
    lime-soda softening 607
    pressure-driven processes 658
    rapid filter 591, 593
    rapid sand filtration 557
    simplified coagulation 582–585
    simplified coagulation-filtration process
        582
    split lime treatment 607
    subsequent filtration 11, 691
filter bed design 563
filter rate 563

filtration 521, 538, 541, 543, 544, 547, 553,
    555–557, 559, 560, 563, 565, 576, 581,
    584, 589, 591, 594, 600, 605, 608, 619,
    649, 657, 666, 679, 691, 703
    column studies 559, 563
    novel As filtration method 605
    process 557, 559, 563, 581, 582, 584, 585,
        589
    roughing filter 591–593
    step 557
    up flow roughing filtration 589, 594
flocculation (*see also arsenic coagulant*) 11,
    691
    conditions 557, 558, 563
    flocs 555–557, 559, 563, 582, 589, 591,
        594, 681
    microflocs 584
    separation system 557
    tapered flocculation 558, 563
innovative (new technologies) 582, 626
ion exchange (IE) (*see also arsenic ion
    exchangers*) 516, 521, 608, 629
    exchangeable Fe^{3+} 516
    process 608
    systems 608
membrane 511, 540, 556, 558, 559, 563,
    576, 584, 608, 610, 617, 619, 621, 626,
    657–659, 661, 679
    cellulose acetate 71, 195, 617, 658
    cellulose triacetate 659
    composite film 659, 671
    characterization 658, 659, 661
    performance 659, 661
    preparation 659
    filtration techniques 556
    flux/driving pressure 658
    processes 511, 608
    separation 626
    technologies 658, 665
    thin film composite-type membranes 658
    treatment 658
membrane applications (*see also membrane*)
    electrodialysis 505, 582, 666, 675
    electrodialysis reversal 582, 675
    high pressure processes 608
    hyperfiltration 658
    low pressure processes (MF and UF) 608
    microfiltration (MF) 557, 563, 584, 606,
        608, 658, 667
    membrane 584, 657, 659
    nanofiltration (NF) 576, 608, 658, 675
    membrane 608, 610, 658
    ultrafiltration (UF) 608, 658
    reverse osmosis (RO) 44, 207, 505, 528,
        565, 582, 608, 630, 658, 666, 675, 691

membrane 658
traditional RO technology 658
oxidation 505, 507, 518, 521, 528, 538, 539,
    549, 551, 555, 556, 559, 571, 582–585,
    589, 605–609, 612–616, 626, 631, 653,
    665, 667, 677–679, 681, 690, 691
  advanced oxidation technologies (AOTs)
    677
  air oxidation 626
  by heterogeneous photocatalysis (HP)
    677–681
    effectiveness 678
    experiment 677
    materials 678, 681
    with TiO$_2$ 678, 679, 681
      iron wool 677, 678, 680, 681
      packing wire 617, 619, 621–623,
        677–681
  chemical 595, 616
  Donnan co-ion exclusion effect 572, 576,
    578
  Fe/citrate containing systems 616
  Fenton reaction 616
  of As(III) to As(V) 267, 299, 528, 556,
    571, 595, 616, 671, 678
  previous oxidation of As(III) 556
  solar oxidation (SORAS) 12, 615–623,
    691
  tests 618–623
  with Cl$_2$ 584
oxidation-coagulation-flocculation-filtration
    582
oxidation-filtration 582
pre-oxidation 556, 585, 607, 671
pre-oxidation-coagulation-filtration 584
precipitation (*for specific*
    coagulants/flocculants, see arsenic
    coagulants/flocculants) 511, 528, 529,
    531, 533, 538, 542, 543, 547–549,
    551–553, 556, 565, 596, 606, 607, 615,
    616, 619, 625–631, 633, 638, 643, 657,
    677, 691
  Fe/As molar ratio 529–534, 679
  precipitates 65, 66, 87, 131, 132, 255, 257,
    262, 528, 551, 552, 638, 679
  sludges 380, 589, 593, 626
    dewatered 593
    disposal 582
    iron 538
    disposal options 509
sieving
  mechanical 658
  physical 608
arsenic sorbents (*see also arsenic removal*
    technologies and arsenic sorption) 3, 12,

61, 66, 85, 197, 375, 511–513, 516, 518,
    565, 572, 574, 576, 578, 600, 608, 609,
    612
activated
  alumina 505, 512, 521,556, 565, 582, 608,
    627
  carbon 565, 574, 582, 584, 627, 629, 630,
    643, 657–662, 665, 667, 670, 675
  charcoal 505
aluminium oxide
  Al$_2$O$_3$ 507, 512, 513, 522, 566
  α-Al$_2$O$_3$ 512
aluminum-based sorbents 512
aluminum-loaded materials 565
arsenic-selective sorbent 571
ARTI-64^{tm} 605, 609, 610, 612, 613
  capacity 613
  effectiveness 612
  particles 605, 609, 610, 612, 613
  treatment 610
bentonite (pre-treated) 511–518
  spectra 516, 518
  surface 515
biomass 691
  sorghum 691
carbon filtering bed 584
ferric oxi-hydroxide (*see hydrous ferric*
    oxides)
ferrihydrite (Fe$_5$HO$_8$ × 4H$_2$O) 516, 518,
    547, 548, 551, 553, 596, 629, 633
iron oxide 4, 521, 522, 547, 577, 578, 596,
    617, 619, 626, 627, 629
  amorphous 596
  Fe$_2$O$_3$ 507, 513, 522, 547, 633
  ferric 521
  α-Fe$_2$O$_3$ 547
granular ferric hydroxide (GFH) 572, 573,
    578
hydrous ferric oxides (HFO) 43, 509,
    571–574, 576–578, 596, 599
  particles 571, 573, 574, 576–578
  α-FeOOH 295, 547, 617
iron coated materials 565, 626
  sand 505, 522
iron hydroxide [Fe(OH)$_3$] 53, 57, 169, 176,
    291, 537, 538, 540, 542, 543, 551, 595,
    599
  ferric 521
  precipitation 538, 542, 543
iron nanoparticles (INP) 625, 627–629, 631,
    636–639
  characterization 636
  stability 636
  surface modified 627
  surface stabilized 625

transport in porous media 637
iron pre-treatment 518
  Fe(III)-treatment 515, 518
    of bentonite 515–518
    of metakaoline 516, 517
    process 514
    solution 514
  Fe(II)-treatment 513–518
    of bentonite 514, 515, 517, 518
    of metakaoline 517
    process 513
  mechanism of Fe-loading 511, 518
    of bentonite 517
    of metakaoline 516, 517
iron oxide coated materials 626
iron-coated light expanded clay aggregates
    (Fe-LECA) 565–569
  capacity 566
  for As(V) removal 565, 567–569
  in batch experiments 566, 567
  in column experiments 566, 568
light expanded clay aggregates (LECA)
    565–569
  metakaolin (pre-treated) 514–517
nanoscale hydrated ferric oxide 571, 572
  amorphous 573, 577
  for As(V) removal 578
nanoscale zerovalent iron (NZVI) 625, 626,
    628–631, 634–636
  comparison of ZVI and NZVI 635, 636
  for As(III) removal 626
  for As(V) removal 626, 635
  material 628, 631
  surface-modified NZVI (*see surface
    stabilized iron nanoparticles*)
  surface stabilized iron nanoparticles
    (S-INP) 625, 627, 636
  synthesis 628, 636
  characterization 636
  stability 636, 637
  transport in porous media 637, 638
natural materials 3, 522, 527, 528, 555–557,
    565, 581, 605, 606, 649, 663, 665, 674,
    690, 695
  bentonite 511
  capacity 521
  goethite (α-FeOOH) 518, 527, 529–534,
    547, 551–553, 577, 578, 617, 619–623,
    629
  hematite (α-Fe₂O₃) 53, 57, 151, 206, 512,
    522, 538, 547, 577, 578, 617, 622, 629
  iron ores rich in hematite 512
  iron oxide minerals 199, 522
  limestones 156, 506, 509
    for domestic water-treatment 509

  retention capacity 509
  rock particles 505–509
  metakaolin 512
  natural red earth (NRE) 521–524
    Al sites available 523
    As desorption behavior 522
    capacity 522, 523
    Fe sites available 523
    for As(III) removal 522
    for As(V) removal 522
    usable in household filter 522
  red mud 512
  zeolites 505, 512
natural or activated clay and lime
    12, 691
pre-treated carrier materials 512
  clays treated with Fe ions 518
zerovalent iron (ZVI) 537, 538, 541, 625,
    626, 628–631, 633–636, 677, 678, 681
  experiments 680
  low-cost materials 678
arsenic sorption
  competitive sorption
    ions competing for adsorption/exchange
      sites 582
  models/modeling
    chemisorption of arsenic 528
    diffuse double-layer model 43, 629
    diffuse layer surface complexation model
      522
    electric repulsion 608
    inner-sphere complexes 547, 553, 576, 578
    inner-sphere Fe(III)-As(III/V) complexes
      516
    surface complexation 522, 557, 635
  of As(III) by Kuroboku soil 295–297
  of As(III) by soil 298
  of As(V) by Kuroboku soil 295–297
  of As(V) by soil 298
  processes 57, 118, 174, 375, 518,
      589, 670
  stability of sorption/immobilization
    arsenate desorption 523
    arsenite desorption 523, 524
    desorption experiments 523
    reductive dissolution of As-rich iron
      oxyhydroxide 606
    remobilization of As 538
    remobilization test 539
    remobilized 538
arsenic sorption/desorption reactions 35, 43
arsenic source 4, 6–9, 41, 42, 43, 64, 65, 81,
      94, 109, 119, 124, 141, 155, 166, 182,
      193, 196, 201, 206, 392, 402, 447, 547

anthropogenic 8, 11, 17, 81, 94, 95, 115,
    155, 196, 182, 200, 201, 205, 215, 255,
    285, 311, 379, 419, 427, 521, 547, 565,
    595, 605, 625, 643, 650, 665
agriculture
  activities 179
  pollutants 50, 51
arsenicals 295, 380, 415, 427, 460, 462,
    471, 474, 477, 478, 481, 486, 495, 497,
    498, 625
  organoarsenical species 319, 322
  pesticides 17, 18, 24, 565
arsenic-producing industries 11
arsenic-trioxide factory 699, 700
atmospheric deposition 18, 306,
    373, 700
burning of fossil fuels 138, 182
chemical weapons 537
combustion of municipal waste 379
electrolytic processes producing metals 17
electrolytic production of zinc 18
fossil fuels 138, 182, 379
industrial effluents 304, 565, 629
industrial waste 547, 555
manufacturing zinc laminates 95
metallurgical processes 95
mineral processing 255, 256
mining 4, 5, 7–12, 17, 18, 25, 26, 84,
    91–95, 97, 98, 155, 156, 163, 166, 171,
    172, 176, 206, 255, 256, 265, 273, 311,
    345, 359, 365, 368, 379, 387–390, 392,
    406, 447, 495, 547–549, 551, 553, 565,
    595, 625, 643, 649, 699, 702
  acid drainage water 91, 94
  foundry 95
  historical 92, 548
  impacted rhizospheric soil 392
  lead-zinc-silver mining 156
  of arsenic 4, 91, 97, 547
  of gold 91, 97, 359, 699
    As/Au ratios 359
    refinement of gold 18
    gold ore 359, 447
  of pyrite 355, 380
  of silver 91
  ore melters 8
  ore processing plants 97
  smelting activities 25, 447, 699
  sulfide production 175, 176
  tailings 8, 94, 155, 158, 159, 161, 171,
    172, 174, 255, 256, 262, 273, 312,
    359, 365, 366, 509, 699–703
  toasting process 95
  water 547, 548, 551, 553
  waste mineral deposit 159–161

natural
  geogenic sources 3–5, 8, 9, 18, 25, 81, 91,
    94, 97, 109, 182, 196, 201, 246, 273,
    285, 405, 406, 547, 555, 650, 690
  geothermal activities (*see geothermal*)
  minerals (arsenic-containing) 81, 245,
    565, 625
    amphibole 41, 55, 151
    apatite 55, 64, 66, 139
    arsenian pyrite 77, 193, 196, 199–201,
    511, 547
    arsenic pyrite 206
    arsenopyrite (*see arsenopyrite*)
    biotite 41, 55, 61, 65, 66, 149, 151, 208
    galena 206
    muscovite 65
    orpiment 81, 132, 176, 245, 258, 260, 665
    pyrite (*see pyrite*)
    pyroxen 41, 55
    pyrrholite 8
    realgar 81, 245
    scorodite 505
    sulfide minerals/ores 8, 91, 98, 132, 273,
    391
    sulfur minerals 206
  oxidation of pyrite 201, 205, 207
  primary source 61, 62, 66, 246, 657, 693
  sediment 42, 43
    aquifer 61, 196
  sediment-hosted Au deposits 273
  volcanic manifestations (*see also*
    volcanic) 8
    volcanic gases 139
    volcanic ash/glass (*see volcanic ash and*
    glass)
    volcanic rocks (*see volcanic rocks*)
    crater lake 138
  oil spill 81
  poultry 419
  raw wolfram concentrate 95
  semiconductor manufacturing 182
  wood preservatives 182, 342, 379, 419,
    649, 666
arsenic speciation 19, 22, 71, 97, 175, 183,
    184, 255, 265, 270, 273, 275, 314, 342,
    365, 366, 556, 558, 581, 646, 662
analysis 270, 311, 314, 482
as a function of pH 539
calculations 71
in water 86
modeling 54, 97
of inorganic arsenic 84
studies 266
techniques 265, 266

arsenic species 84, 97, 256, 265, 273, 311,
　314, 319, 356, 360, 361, 365, 367, 428,
　438, 460, 461, 482, 483, 487, 495, 521,
　522, 527, 528, 533, 539, 540, 547, 558,
　567, 568, 612, 629, 635, 644, 645
　bioavailability for humans 323
　calculation 71
　determination 314, 316
　in environmental samples 342, 343, 365
　in food 319, 360
　in rat sural nerves 483
　in terrestrial plants 360
　in urine 438
　inorganic 8, 11, 19, 20, 22, 24, 123, 132,
　　181, 235, 237, 241, 266, 273, 311, 314,
　　319, 321, 327–331, 335, 338, 340, 341,
　　356, 360, 380, 405, 409, 427, 428, 436,
　　459, 460, 462, 473, 481, 489, 495, 497,
　　528, 537, 538, 540, 543, 544, 595, 629
　　arsenate [As(V)] 4, 7, 8, 19, 22, 35, 37, 43,
　　　56, 57, 65, 77, 84, 87, 88, 110, 114, 123,
　　　125, 127, 132, 155, 167, 175, 183, 184,
　　　235, 237, 245, 251, 255–260, 262, 263,
　　　273, 277, 279, 295–300, 311, 314–316,
　　　319, 321, 322, 328–330, 332, 335, 336,
　　　338, 340, 342, 361, 369, 375, 383, 391,
　　　392, 419, 428, 430, 479, 495, 511–519,
　　　521–524, 527–529, 531, 533, 538, 539,
　　　541, 542, 547, 548, 553–558, 561–563,
　　　566–568, 571, 574–579, 581, 585,
　　　595–597, 600, 605–610, 612, 616, 622,
　　　623, 625, 626, 630–632, 634, 635, 637,
　　　638, 640, 641, 643–646, 659, 660, 664,
　　　666, 668, 669, 675, 676, 678, 679
　　arsenilic acid 245
　　arsenite [As(III)] 4, 19, 22, 65, 71, 75, 77,
　　　84, 86, 87, 114, 125, 132, 173, 175, 183,
　　　184, 209, 235, 237, 245, 249–251,
　　　265–270, 273, 277, 279, 295–300, 311,
　　　313–315, 319, 321, 322, 328–330, 332,
　　　335, 336, 338, 340, 342, 367, 369,
　　　428–430, 435, 436, 442, 459, 460–464,
　　　467, 468, 470, 471, 473, 474, 476, 477,
　　　481–490, 492, 493, 495, 511–519,
　　　521–524, 527, 528, 533, 538, 539, 540,
　　　541, 542, 547, 548, 555–558, 561–563,
　　　571, 576, 577, 579–581, 589, 595–597,
　　　600, 601, 605–610, 612, 614, 616, 617,
　　　623, 625, 626, 629–635, 637, 638, 640,
　　　643–646, 651, 652, 660, 664, 666, 668,
　　　669, 676–681
　　arsine 18, 19, 114, 245, 247, 269, 460,
　　　462, 482, 567

　oxyanions 35, 37, 39, 40, 42, 43, 53, 57,
　　66, 75, 77, 183, 511, 512, 518, 547–549,
　　557, 571, 606, 693
　organic (*see also arsenic metabolites*)
　　arsenobetaine 245, 320, 321, 335, 340, 527
　　arsenocholine 320, 335
　　arsenosugars 319, 320, 322, 336, 340
　　arsphenamine 245
　　organic As(III) 19
　　organic As(V) 19
　　organoarsenical species 319, 322
arsenic sulfides 173, 175–177
　amorphous 175, 176, 177
arsenic toxicity (effects) (*see also arsenic
　　health effects*) 9, 11, 19, 21, 92, 132,
　　184, 186, 187, 205, 235, 245, 247, 251,
　　265, 279, 295, 319, 320, 322, 323, 330,
　　336, 360, 365, 371, 379, 409, 428, 430,
　　433, 435, 436, 442, 459, 461, 463, 466,
　　468, 469, 471–474, 480–482, 486, 489,
　　495, 497, 527, 543, 555, 565, 581, 585,
　　595, 600, 606, 625, 629, 639, 646
　abnormal cytosine methylation of DNA 429
　alterations in locomotor activity 460
　altered gene expression 427, 428, 430
　altered hematopoiesis 19
　apoptosis 460, 471
　　in cerebellar neurons 460
　bilateral Babinsky signs 464
　carcinogen 3, 11, 12, 20, 21, 132, 235, 246,
　　295, 319, 327, 397, 419, 427, 430, 435,
　　443, 473, 489, 492, 521, 555, 565, 595,
　　606, 625
　carcinogenesis 419, 427, 428
　　in humans 21, 428, 430
　carcinogenic
　　dose in humans 21
　　effects 3, 12, 132, 565, 606
　　in humans 435
　changes in gene expression 429
　changes in methylation 429
　changes in p53 levels 490
　chromosomal abnormalities 429
　chromosomal damage 429, 430
　chromosome abnormality 19
　chronic 9, 215, 409, 442
　co-carcinogen with UV light exposure 435
　co-mutagen 155, 429
　cytotoxic effects 473
　cytotoxicity 428, 471
　damage to pancreatic beta cells 478
　damage to respiratory system 11
　decrease Trx activity 478
　degenerative effects on circulatory system 9
　degree 187, 235

dermal carcinogenesis 443
DNA damage 429, 430, 460, 489, 492
  oxidative 460
DNA methylation changes 429
DNA mutation 430
DNA repair inhibition 427–430
enhance the mutagenesis 429
F and As in urine and scores of IQ 457
fetal growth 20
fetal, neonatal and postneonatal mortality
    10, 23
fetuses are heavily exposed to As 25
for central nervous system 453, 481, 595
genotoxic
  effects 430
  risks 24
  stress 489
    responses 489
genotoxicity 430, 435, 474, 489
  reduction by co-exposure to Se 474
hepatocellular carcinoma 467
hepatoportal sclerosis 467
hepatotoxicity 9
hyperglycemia 477, 478
immunodepression 435, 442, 443
immunosuppressive effect 416
immunotoxic effects 436
impact on visuospatial organization 453
indicator of oxidative stress 478
induce chromosomal abnormalities 429
inducing damage to immune cells 435
induction of chromosomal abnormalities
    428
induction of chromosomal instability 427,
    430
inhibition of
  glutathione reductase 430
  IL-2 secretion by human lymphocytes
    436
  lymphocyte proliferation 436, 442, 443
  protein tyrosine phosphatases 478
  pyruvate dehydrogenase 428
  T lymphocyte proliferation 442
  TrxR activity 478
intellectual development of children 17, 20
intelligence quotient 453, 459
interactions between Se and As 474
IQ affected by arsenic 173, 303, 453–457,
    459, 572
irritation of the respiratory organs 19
mechanism in cancer development 489
mechanism of arsenic induced
    carcinogenesis 428
mental deterioration 459, 464
mutagen induced tumors 429

mutagenesis 429
mutagenicity of arsenic 429
nervous system disorders 17, 20, 26
neurological and neurobehavioral disorders
    428
neurological effects of F co-exposure 457
neurological functions 453, 481
neurological symptoms 25
neuropathic symptoms 481
neurotoxic alterations 460, 482
neurotoxicity 9, 459, 460, 464, 481
on peripheral nervous system 482
oxidative damage 430, 468, 471, 473, 474,
    477, 478, 481, 482, 485, 486
oxidative stress 427–430, 467, 468, 471,
    474, 478, 482
p53 activation 489, 492
p53 response after As-exposure 489
p53-DNA binding activity after As exposure
    490
pancreatic oxidative damage 474
protective role of CK 471
protein damage in cells 428
reproduction in animals 20
species dependent 24, 319, 342
stress and oxidative damage in pancreas
    477
stress proteins 467
toxicological interactions between
    Se and As 474
tumor 20, 416, 428, 429, 435, 443, 489
  tumor suppressor protein 489
type 2 diabetes mellitus 478
under co-exposure of fluoride 187
arsenic transfer
  factor (TF) 371–373, 375, 377
    soil-to-leaf in peach 371
  into food chain (*see also arsenic exposure*)
    10, 92, 336, 341, 343, 351, 371, 380
  models 371, 373, 376
    freshwater food chain 336
  soil-crop-food transfer 356, 371
  soil-plant 372, 373
arsenic transport through the epithelium 320
arsenic treatment plant (*see arsenic removal*)
arsenic uptake (*see arsenic exposure and
    arsenic in*)
arsenic volatilization 391, 495, 497, 498
arsenicals (*see arsenic sources*)
arsenic-affected area 380, 597, 606
arsenic-associated cancers (*see arsenic health
    effects*)
arsenic-based pesticides (*see also arsenic
    sources*) 17, 24
arsenic-contaminated soils (*see arsenic in*)

arsenic-contaminated water (*see arsenic in*)
arsenic-endemic areas 322, 323, 343, 348, 453, 478
arsenic-free water 26
arsenic-induced carcinogenesis in humans (*see arsenic health effects*)
arsenicism (*see arsenic health effects*)
arsenicosis (*see arsenic health effects*)
arsenic-safe aquifers 685, 687, 689, 690, 692, 695
arsenic-waste stabilization 626
arsenic-trioxide 245, 596, 699, 700
arsenilic acid (*see arsenic species*)
arsenite (*see arsenic species*)
arsenobetaine (*see arsenic species*)
arsenocholine (*see arsenic species*)
arsenosugars (*see arsenic species*)
arsenopyrite (FeAsS) 8, 74, 77, 81, 103, 104, 150, 174, 206, 246, 258, 260, 262, 273, 274–279, 313, 316, 447, 505, 511, 547, 665, 699
    dissolution 273, 274, 277–279
        rate 273, 274, 278, 279
    oxidative dissolution rate 279
    surface 273, 274, 278, 279
arsine (*see arsenic species*)
arsine gas (AsH_3) (*see arsenic species*)
arsphenamine (*see arsenic species*)
ARTI-64^{tm} (*see arsenic sorbents*)
As(III) methyltransferase (*see AS3MT*)
As_2S_3 (orpiment) 81, 132, 176–178, 245, 258, 280, 663
AS3MT 427, 474
    activity 474
$AsCl_3$ 245
ash (*see volcanic ash*)
AsH_3 (*see arsenic species*)
AsS (realgar) 81, 174, 176, 245, 258, 273, 279, 665
assimilation of arsenic into edible plants 351
atmospheric deposition (*see arsenic source*)
ATP (adenosine 5-triphosphate) 19, 428
    synthesis 19, 428
awareness of arsenic problem 186, 193, 285, 404, 605, 691, 700

B
back-arc 103
basalt 49, 71, 92, 103, 113, 119, 146, 147
batch experiments 507–509, 543, 544, 566–568, 609, 635
behavioral change communication (BCC) 230

bench studies 557, 561
beta cells 477, 478
    function 478
bicarbonate 47, 51–53, 57, 88, 149, 167, 168, 195, 196, 207, 209, 218, 219, 557, 597, 607, 616–618, 677, 678
bioaccessibility of arsenic (*see arsenic bioaccessibility*)
bioaccumulation of arsenic (*see arsenic bioaccumulation*)
bioavailable arsenic (*see arsenic bioavailability*)
biogeochemical arsenic cycle 497
biomarker (*see arsenic biomarkers*)
bioremediation (*see arsenic remediation*)
biotite 41, 55, 61, 65, 66, 149, 151, 208
biotransfer factor of arsenic (*see arsenic biotransfer factor*)
biotransformation of arsenic (*see arsenic biotransformation*)
Bishuddya filters (*see arsenic removal, water*)
bladder cancer (*see arsenic health effects*)
blood (*see arsenic biomarkers*)
    glucose 477, 478
boron 18, 56, 94, 129, 132–135, 137, 138, 140, 166, 207, 304, 397, 615, 618, 677, 678
bottles with TiO_2 (*see arsenic removal, water*)
brain tissue 463, 464
BTF (*see arsenic biotransfer factor*)

C
C33-A cells 491, 492
    exposed to arsenic 491
caco-2 cells 322–324
calc-aluminum silicate zone 151
calcite 51, 53, 55, 57, 64, 66, 74–77, 149, 151, 167–169, 176, 200, 206, 219, 257–259, 262, 263, 507, 509
calcium, 52, 53, 114, 149, 167–169, 207, 209, 255, 256, 259, 262, 382, 392, 528, 557, 561, 607, 618, 678
Ca-montmorillonite 53, 57
cancer (*see arsenic health effects*)
capacitive deionization (*see arsenic removal technologies*)
    system DesEL (*see arsenic removal, water*)
capillary fringe 62, 290
caprock seal 146
carbon dioxide (CO_2) 52–54, 131, 138, 139, 167–169, 219, 289, 437, 461, 468, 490, 522, 548, 549, 616, 671
    anomalies 131
    dominated lake 138

carbonate 51, 52, 57, 62, 64, 66, 70, 88, 115, 127, 145, 146, 149, 151, 159, 160, 167, 173, 219, 262, 288, 289, 296, 382, 387, 388, 390–392, 557, 607
   equilibrium 51
   reactions 51, 57
carboxylic acids 388
carcinogenesis of arsenic (*see arsenic toxicity*)
caspase activation 460
catalase 430
cation exchange 4, 51, 62, 66, 74, 127, 296, 372, 388, 390–392, 571–575, 578
   capacity 4, 296, 372
cDNA 469
cell carcinomas 428
cell proliferation (*see also arsenic toxicity*) 427, 428, 430, 436, 437, 471, 489–491
central nervous system (*see also arsenic toxicity*) 453, 481, 595
cerebellar neurons 460
chalcedony 53, 55, 57
chelation 20
chemical contaminants 235, 677
chemical weathering (*see weathering*)
chloride 11, 40, 51, 52, 71, 77, 88, 97, 133–135, 138, 139, 140, 149, 150, 166, 167, 169, 196, 207, 218, 219, 527–529, 532, 533, 534, 537, 538, 541–544, 555–559, 562, 574, 589, 594, 617, 618, 628, 630, 655, 657, 666, 667, 670, 671, 676, 678
chromosomal damage (*see arsenic toxicity*)
chronic exposure to arsenic (*see arsenic exposure*)
chronic human arsenic intoxication 215
CK (cytokeratin) 467
   accumulation 467, 468
   CK18 467–471
      CK18 mRNA 468–470
      CK18 synthesis 469
      overexpression of CK18 471
   expression 471
   protective role 471
   synthesis in liver 471
clay 12, 41, 48–50, 55, 57, 70–72, 113, 149, 151, 164, 168, 194, 196–200, 217, 341, 342, 345, 348, 360, 374, 380, 508, 511, 512, 515, 518, 522, 551, 565, 566, 617, 627, 638, 691
clogging processes (*see arsenic removal, water*)
co-exposure to As and Se 474
co-exposure to F and As 457
coagulant for arsenic removal (*see arsenic coagulants*)

coagulation (*see arsenic removal technologies*)
coagulation-filtration (*see arsenic removal technologies*)
coagulation-flocculation (*see arsenic removal technologies*)
COD (chemical oxygen demand) 52
colloidal reactive barrier (CRB) (*see also arsenic remediation*) 639
column experiments 506–508, 509, 543, 560, 566
confined aquifer (*see aquifer*)
conglomerate 70, 103, 208
contaminated soils (*see arsenic in soil*)
control mobility of arsenic (*see arsenic mobility*)
controls on arsenic mobilization (*see arsenic mobilization*)
conventional treatment technology (*see arsenic removal technologies*)
conventional water treatment processes (*see arsenic removal technologies*)
copper deposits 18
cow's milk (*see arsenic in*)
creatinine 410, 447–450, 454
cutaneous manifestations (*see arsenic health effects*)
cytokeratin (*see CK*)
cytokines
   secretion 442
cytotoxicity of arsenic (*see arsenic toxicity*)

D
dacite 36, 55, 56, 118, 130, 146, 419
dacitic composition 130, 398
daily arsenic intake (*see also arsenic exposure*) 186, 404, 405
   adults 186
   babies 186
   by diet 185, 355, 356, 398
   children 186
deaths associated with arsenicism 22, 23
defluoridation of water 596
degree of arsenic toxicity (*see arsenic toxicity*)
deiodinases 473
deoxyribonucleic acid (*see DNA*)
dermal carcinogenesis (*see arsenic toxicity*)
dermatosis (*see also arsenic toxicity*) 23
DesEL plant (*see arsenic removal, water*)
developing countries 600, 626, 629, 630, 658
development projects 92
diabetes 17, 20, 26, 379, 435, 467, 473, 474, 478
diabetes mellitus 17, 20, 26, 379, 435, 473, 478

diet (*see arsenic in & dietary intake of arsenic*)

dietary intake of arsenic (*see also arsenic in plants, arsenic exposure and arsenic in*) 320, 351, 355, 356, 381, 404, 420, 473–478, 591

digestive enzymes (*see enzymes*)

dimethyl sulfoxide (DMSO) (*see arsenic metabolites*)

dimethylarsenate (DMA) (*see arsenic metabolites*)

dimethylarsine (*see arsenic metabolites*)

dimethylarsinic acid [DMA(V)] (*see arsenic metabolites*)

dimethylarsinoylribose 322

dimethylarsinous acid [DMA(III)] (*see arsenic metabolites*)

dissolution (*see also arsenic mobilization*)
  iron oxyhydroxides 193, 201, 205
  of As-containing minerals 81
  of carbonates 62, 66, 127
  of ferric minerals 66, 127
  of fluorite 75, 77
  of iron and manganese minerals 87
  of silicates 62, 127
  rates of FeAsS 279

dissolved oxygen 4, 37, 57, 71, 84, 114, 118, 121, 273, 274, 279, 296, 617, 622

dissolved silica 71, 218

distribution of As between sediments and water 172

DMA (*see arsenic metabolites*)

DMSO (*see arsenic metabolites*)

DNA (deoxyribonucleic acid) (*see also arsenic toxicity*) 19, 155, 427–430, 460, 469, 489, 490–493
  binding ability 490, 492
  damage 429, 430, 460, 489, 492
    oxidative 460
  methylation 429
    changes in As-induced tumors 429
  mutation 430
  p53-DNA binding activity 489–492
  repair 155, 427, 428–430, 489
    altered 489
    inhibition 427–430
    processes 489

DNA-dependent protein kinase 489

DNA-PK 489, 492

DNA-protein interactions 491

drinking water (*see also arsenic in drinking water*) 3, 4, 6, 7, 9–13, 17–26, 35, 37, 44, 47, 48, 70, 72, 81, 85, 92, 94, 109, 110, 112, 113, 115, 118, 120, 123, 124, 127, 132, 155, 161, 163, 166, 172, 179–182, 185, 186, 187, 193, 215, 217, 220, 222, 223, 227, 230, 245, 246, 251, 265, 270, 285, 319, 320, 322, 343, 359, 379, 397, 401–405, 411, 415, 419–424, 427, 435, 436, 442, 443, 453–456, 459, 473, 491, 505, 506, 512, 521, 527, 528, 538, 539, 555, 556, 565, 581, 582, 589, 595, 599, 600, 605–607, 615, 616, 625, 630, 643, 651, 657, 662, 675, 677, 687, 690–692, 694, 699–702
  sources 7, 17, 18, 109, 120, 123, 230, 402, 690

dolomite 51, 57, 147, 149, 206

dusts (*see also arsenic sources*) 10, 359

E

echelon blocks 113

economic development 47, 91

Eh (*see redox potential*)

electrical conductivity 71, 77, 84, 85, 114, 125, 160, 195, 207, 210, 218, 288, 421, 668, 694

electrolytic processes (*see arsenic sources*)

electrolytic production of zinc (*see also arsenic sources*) 18

elemental arsenic 19, 266, 269

Endemic Regional Chronic Hydroarcenicism (ERCH) (*see arsenic health effects*)

enrichment of As during pedogenesis 292

enstatite ($Mg_2Si_2O_6$) 55, 85, 88

environmental
  concern 6, 109, 171, 511
  problems 6, 92, 94, 205, 265, 653

enzyme 19, 20, 319, 321, 323, 428, 429, 460–464, 467, 473, 474, 486
  digestive 319, 321, 323
  inhibitors 460

epidemiological studies 17, 21, 22, 26, 187, 320, 327, 453

epidote 55, 149, 151

epithelioms (*see arsenic health effects*)

evaporation
  seasonal 40

evaporative concentration (*see also groundwater*) 40

evaporitic facies 398

F

Fe(OH)$_3$ (*see iron hydroxide*)

Fe$_2$O$_3$ (*see also arsenic sorbents*) 41, 55, 130, 198, 200, 296, 312, 313, 316, 507, 513, 522, 547, 633

FeAsS (*see arsenopyrite*)

feed additive for livestock 265, 419

feldspar 41, 53, 55, 57, 71, 151, 194, 208, 551

ferric chloride (FeCl$_3$) (*see arsenic coagulants*)
ferric sulfate (*see arsenic coagulants*)
ferrihydrite [Fe$_5$HO$_8$ × 4H$_2$O] 516, 518, 547,
    548, 551, 553, 596, 629, 633
fertilizers 24, 50, 133, 304, 547, 631
fish farming 92
    trout 83
flocculation (*see arsenic removal technologies*)
flood plain sediments 223
fluoride 69, 73, 77, 114, 132, 166, 168, 179,
    183, 185, 187, 207, 397, 421, 453, 605,
    615, 618, 666, 676–678
fluorine 18, 56, 589–593, 615
    removal 589
flushing of groundwater 229
fluvial
    processes 35, 55, 419
    sediments 6, 286
food (*see arsenic in*)
    chain (*see also arsenic exposure*) 10, 92,
        281, 336, 341, 343, 351, 371, 380
    safety 235, 322
        concern 235
fossil fuels (*see also arsenic sources*) 138, 182,
    379
fracture system (*see aquifer*)
freshwater food chain models (*see also arsenic
    transfer*) 336
fumarole 8, 18, 94, 148

G
galena 206
garnet 55, 61, 65, 66, 149
gastrointestinal
    digestion (human) 319–323
        simulation 320
    effects 179, 182, 187, 245, 246
    manifestation 416
genomic stability 489
genotoxicity of arsenic (*see arsenic toxicity*)
geochemical
    geothermometer 70
    modeling 163, 166, 629
geochemistry of arsenic 72, 181
geogenic arsenic (*see arsenic sources*)
    enrichment in histosols 285, 286
geographic information system 101, 103, 155,
    183
geothermal (*see also arsenic sources*) 7, 18,
    50, 69–71, 74, 77, 81–83, 85–89, 94,
    145–151, 179, 181–183, 185, 345, 547
    activities 94
    anomalies 104
    area 50, 71
    arsenic 89, 179, 181

field 81, 85, 146, 148
fluids 145, 146, 148, 149, 150, 151, 182
gradient 71
phenomena 18
sources 82, 85
system 77, 146
water 7, 74, 81, 82, 85, 87–89, 181–183, 185
well 146
geothermometer 70
gibbsite 53, 57, 257, 258, 260, 565
glass (*see volcanic glass*)
gleysol (*see soil*)
global positioning system 84, 114, 125, 218
glucagon 475, 476, 478
    level 475, 476, 478
glucose
    from blood 478
    level 410, 475–478
    tolerance 474
glutathione (GSH) 19, 430, 459, 460, 462, 463,
    467, 473, 486, 495
    in brain 463
    peroxidase (GPx) 463, 473, 480, 486
    reductase (GR) 19, 430, 459, 460–463, 486
        activity 459, 462
GM-CSF (granulocyte-macrophage
    colony-stimulating factor)
    secretion 435, 441, 443
        by macrophages 443
gneiss 103
goethite (α-FeOOH) 53, 57, 206, 257–260,
    262, 295, 375, 527, 529–534, 547,
    551–553, 577, 578, 617, 619, 620, 622,
    623, 629
gold mines (*see also arsenic sources*) 91, 273,
    359, 699
    As/Au ratios 359
    extraction and refinement of gold 18
    gold ore 359, 447
GPx (*see glutathione peroxidase*)
GR (*see glutathione reductase*)
granite 71
granitoid 103, 195, 197
granular ferric hydroxide (GFH) (*see arsenic
    sorbents*)
graywacke 103
groundwater (*see also arsenic in*) 3, 4, 6–11,
    13, 17, 18, 22, 35–44, 47–54, 56, 57,
    61–64, 66, 67, 69–77, 86, 88, 91, 92, 94,
    97, 98, 109–111, 113–121, 123–127,
    132, 145, 155, 156, 163, 164, 166–169,
    171, 172, 180–183, 185–187, 193–196,
    199–201, 205–209, 215–223, 225–227,
    229, 230, 245–247, 251–253, 255, 273,
    285, 286, 290, 292, 307–309, 351,

353–356, 359, 397–404, 419, 421–424,
427, 501, 511, 521, 524, 529, 537–544,
547, 555, 556, 559, 565, 569, 571, 581,
582, 584, 585, 589, 595, 596, 605–607,
609–612, 615, 618, 625–627, 629–632,
635–638, 657, 665, 667, 672, 674, 675,
677, 678, 681, 687–695, 699, 701–703
arsenic (*see arsenic in*)
  hot spot 62, 66, 109, 226, 692, 700
chemistry 36, 51, 56, 64, 66, 115, 124, 218
  composition 40, 118, 692
  ionic strength 61, 66
  major ions 37, 51, 71–73, 118, 129, 145,
    195, 207, 219, 618
  spatial variability 56
classification 119
  Ca-Cl type 219, 220
  Ca-HCO₃ type 62, 219, 220, 693
  Ca-Mg-Cl type 219, 220
  Na-Cl type 36, 220
  Na-HCO₃ type 6, 36, 40, 43, 61, 62, 66,
    116, 117, 692
  Piper diagram 115, 118, 119, 167, 209,
    211, 218
deep 40, 51, 53, 54, 71, 76, 119, 183, 185,
  618
evaporative concentration 40
flow 35, 40, 56, 71, 72, 114, 115, 118, 156,
  200, 218, 225, 229, 398, 627
  pattern 40, 71, 72, 114, 115, 225, 229
flushing 229
  since sediment deposition 40
high salinity 47, 48, 397, 677
hydrogeochemical studies 7, 110
in faults and fractures 119
in fissure network 209
increased residence time 40, 43
level fluctuations 206, 229
oxidizing conditions 8, 50, 56, 155, 201,
  206, 209, 273, 391, 522, 692
oxygenated recharge water 88
periodic discharge 43
piezometric levels 209
recharge 115, 118, 692
regional flow path 71
residence time 6, 40, 43, 118, 205, 690,
  692
resources management 92
saline 36, 50, 51, 151, 422, 437, 461, 468,
  475, 605
sample 6, 22, 36, 37, 43, 44, 50, 54, 56, 115,
  118, 125, 195, 196, 218, 219, 221, 222,
  251, 253, 292, 354, 397, 422, 423, 542,
  609, 610
seasonal discharge 35

shallow 4, 22, 47, 50, 51, 53, 54, 56, 57, 69,
  72, 88, 110, 118, 125, 422, 424, 618,
  626, 627, 630, 687, 690
table 56, 62, 114, 115, 164–166, 206, 207,
  251, 285, 287, 288, 292, 354, 544, 606,
  617
  seasonal fluctuations 229
  temperature 69, 70, 77, 183
    anomalies 70, 77
  trapped 119
GSH (*see glutathione*)
gypsum 49, 51, 66, 73, 167, 169, 176, 209,
  552, 553

H
HACRE (*Hidroarsenicismo Crónico Regional
  Endémico*) (*see arsenic health effects,
  Endemic Regional Chronic
  Hydroarcenicism*)
hair (*see arsenic in*)
haploxeralf 380
hardrock aquifer (*see aquifer*)
hardrocks outcroping 119
HCO₃⁻ (*see bicarbonate*)
HCys (*see homocysteine*)
health (*see also arsenic health effects*) 3, 4, 6,
  10, 11, 26, 91, 163, 179, 186, 187, 265,
  379, 473, 479, 555, 581, 605, 665, 699,
  700
  authorities 26, 186, 581
  effects 3, 4, 6, 10, 11, 26, 91, 163, 179, 186,
    187, 265, 379, 473, 479, 555, 581, 605,
    699, 700
  hazards 25
  impact assessment 23
  problems 26, 35, 47, 91, 182, 331, 511, 665
  risk 3, 4, 9, 13, 141, 155, 180, 185, 187, 273,
    319, 356, 404, 405, 420, 612
heavy metal contamination 91, 92, 94
hematite (*α*-Fe₂O₂) (*see also arsenic sorbents*)
  53, 57, 151, 206, 512, 522, 538, 547,
  577, 578, 617, 622, 629
hepatotoxicity of arsenic (*see arsenic
  toxicity*)
heterogeneity of aquifer (*see aquifer*)
heterogeneous photocatalysis (HP) (*see
  arsenic removal technologies*)
Hidroarsenicismo Crónico Regional Endémico
  (Endemic Regional Chronic
  Hydroarcenicism) (*see arsenic health
  effects*)
high arsenic ingestion (*see arsenic
  exposure*)
home garden 351, 353, 355
  soils 351

homocysteine (HCys) 478
  concentrations 478
  metabolism 478
histosols (*see soil*)
hornblende 55, 147
hot spot (*see arsenic hot spot*)
hot spring (*see also geothermal*) 8, 83, 85–87, 148
household filter systems (*see arsenic removal, water*)
Hsp70 expression in human 471
human
  exposure to As (*see arsenic exposure*)
  health (*see also arsenic health effects*) 81, 92, 182, 185, 304, 319, 360, 379, 380, 419, 447, 454, 473, 521, 547, 555, 597
  urine (*see arsenic metabolites and arsenic biomarkers*)
HVO_4^{2-} (*see also vanadium*) 54, 57
hydraulic
  conductivity (*see aquifer*)
  hydraulic gradient (*see aquifer*)
hydroarsenicism (*see arsenic health effects*)
hydrodynamic regime (*see aquifer*)
hydrogeochemical studies (*see also aquifer*) 7, 110
hydrogeological conditions (*see aquifer*)
hydrothermal
  alteration 8, 109, 111, 119, 140
  deposits 85
  fluid 119, 129, 141
  mineral deposit 273
  system 85, 88, 132
  veins 208
  zones 145, 149, 151
hyperglycemia (*see arsenic toxicity*)
hyperkeratosis (*see arsenic health effects*)
hyperpigmentation (*see arsenic health effects*)
hyperstene 55

I
igneous rocks 132, 156, 179
ignimbrite 113, 147, 398
IL-2 secretion 435, 436, 439, 441, 442, 443
illite 41, 55, 57, 151, 197–199
illite-chlorite zone 151
ilmenite 41, 61, 65, 66, 374, 522
immune system (*see arsenic toxicity*)
immunotoxic effects of As (*see arsenic toxicity*)
in situ remediation (*see remediation*)
  of groundwater (*see arsenic remediation*)
in vitro

bioavailability methods (*see also arsenic bioavailability*) 320
  digestion 320–323
    methods 320
  gastrointestinal digestion 319, 320, 321
    methods 323
increased cell proliferation 428, 430
increased residence time (*see groundwater*)
indicator of oxidative stress (*see arsenic toxicity*)
indigenous population 345, 346
induction of chromosomal abnormalities (*see arsenic toxicity*)
industrial waste (*see arsenic sources*)
ingestion of arsenic (*see arsenic exposure*)
inhalation of arsenic (*see arsenic exposure*)
inorganic arsenic 84, 132, 235, 241, 311, 314, 319, 321, 327, 328–331, 335, 338, 340, 341, 356, 360, 380, 409, 427, 436, 460, 462, 481, 489, 495, 537, 540, 542, 643
  bioaccessibility (*see also arsenic bioaccessibility*) 319–323
  in food (*see also arsenic in plants, fish and seafood*) 235
    Procambarus clarkii (crayfish) 235, 236, 240
    rice 235, 239, 241, 321
    sea food products 328
    shellfish 331
    Uca tangeri (fiddler crab) 235, 239
insulin 473–478
  level 476, 477
  resistance 473–478
  secretion 478
  sensitivity 475–478
insulin-like effects of inorganic Se 478
insulin-stimulated glucose metabolism 478
insulinomimetic properties of inorganic Se 478
intake of arsenic (*see arsenic ingestion and arsenic exposure*)
ion
  exchange 4, 51, 62, 66, 74, 84, 118, 127, 168, 169, 288, 296, 372, 388, 390–392, 505, 511, 516, 521, 539, 565, 571–578, 582, 608, 616, 626, 629, 630, 643, 657, 666, 667, 677
    reactions 51
  exchangers for arsenic removal (*see arsenic ion exchangers*)
  transport across root membranes 372
ionic activity product (IAP) 53, 175, 176
IQ affected by arsenic (*see arsenic toxicity*)
iron
  chlorosis (plants) 371, 372, 374
  content in soil 361

deficiency (plants) 371, 372
hydroxide [Fe(OH)$_3$] (*see also arsenic
    sorbents*) 53, 57, 87, 169, 176, 291,
    537–540, 542, 543, 547, 551, 595, 599,
    623, 630, 631, 664, 666
monosulfides 175
nanoparticles (INP) (*see arsenic sorbents*)
oxidation 87
oxide 4, 75, 197, 199–201, 205–208, 255,
    262, 288, 290, 292, 314, 511, 516, 518,
    521, 522, 547, 549, 577, 578, 596, 617,
    619, 626, 627, 629, 638, 666, 678
    arsenate sorption 43
    coating sediment grains 201
    content 209, 345, 361, 565, 567, 569, 575,
        576
    ferric 44, 132, 571, 572, 576, 578, 629,
        631, 633
    minerals 75, 199, 375, 522
    surface 47, 57
oxihydroxides 47, 57, 61, 65, 66, 85, 193,
    200, 201, 206, 216, 222, 285, 290, 511,
    528, 533, 547, 551, 692, 693
    adsorption on 69
    precipitation 61, 66, 87
    reduction 216
iron-coated adsorbents (*see arsenic sorbents*)
iron-coated light expanded clay aggregates
    (Fe-LECA) (*see arsenic sorbents*)
irrigation 9, 10, 12, 48, 50, 69, 91, 97, 113,
    119, 125, 163, 164, 182, 193, 194, 217,
    222, 223, 226, 229, 245, 303, 304, 306,
    308, 335, 336, 339, 341, 343, 345, 346,
    348, 351, 354, 356, 374, 375, 398, 401,
    402, 423, 511
    by contaminated water 351
    by raw wastewater 303–305, 306–308
    increasing demand 222
    water (*see also arsenic in*) 10, 304, 336, 339,
        341, 345, 346, 356, 402, 511
    samples 346
irritation of the respiratory organs (*see arsenic
    toxicity*)

J
jar test 555, 557–559, 583, 590, 591
jar test apparatus 558
jar test experiment 555, 557, 558

K
kaolinite (*see also arsenic sorbents*) 53, 55, 57,
    197, 512, 617
keratosis (*see arsenic health effects*)
kidney cancer (*see arsenic health effects*)

kinetic study on the As(V)-contaminated
    soil 300
KSi$_3$AlO$_8$ (*see sanidine*)

L
lack of local community participation 92
lacustrine deposit 131, 164
lagoonal deposit 166
lake water 8, 81, 85, 98, 131, 134, 135, 139,
    141
lamprobolite 55
lava 18, 82, 89, 92, 103, 146, 179, 398
    andesitic 92, 103
    cones 92
    flow 18, 82, 146, 179, 398
leaching tests of sediments (*see also sequential
    extraction*) 82
light expanded clay aggregates (LECA) (*see
    arsenic sorbents*)
limestone (*see also arsenic sorbents*) 92, 147,
    156, 206, 372, 505, 506, 509, 522
linear arsenic transfer models 373
lipid peroxidation 475, 477, 478, 481, 482,
    484, 487
    in pancreas 477
lithic fragments 55
livestock (*see also arsenic in*) 163, 185, 343,
    419–423
    health 420
loess 6, 35, 36, 39–44, 47, 48, 51, 53, 55–57,
    61, 64, 69–71, 419, 695
    aquifer (*see also aquifer*) 35, 36, 39–41, 43,
        57
    chemical composition 56
    content of volcanic ash 40
    deposit 35, 41, 42, 44, 47, 48, 55, 56
    sediment 36, 41, 47, 51, 53, 55, 56
long-term exposure (*see arsenic exposure*)
low-cost remediation (*see remediation*)
LPS/IFN-γ-activated macrophages 437, 443
lymphocyte proliferation (*see arsenic toxicity*)

M
mafic minerals 66
magmatic rocks 6
magnesium 53, 168, 169, 207, 209, 528, 607
magnetite 41, 43, 61, 64–66, 522, 571, 577,
    578, 631, 633, 635, 638, 680
major ions (*see groundwater*)
malnutrition 20, 26, 474, 615, 677, 700
manganese oxides 206, 285, 387, 388,
    390–392
manufacturing zinc laminates (*see also arsenic
    sources*) 95
marcasite 132, 206

mass poisoning 246, 351
maximum admissible arsenic level 205
maximum contaminant level (MCL) 181, 555, 565, 581, 584, 605, 606, 657, 672, 674
MCL (*see maximum contaminant level*)
MDA231 cells 471
mechanisms of As mobilization (*see arsenic mobilization*)
mercury 131
metabolic homeostasis 467
metabolism 19, 22, 322, 371, 395, 427–429, 460, 474, 478, 481
   of arsenic (*see arsenic metabolism*)
   of ATP synthesis 19
   of sulfur aminoacids 478
metabolites (*see arsenic metabolites*)
metallothionein 430
metallurgical area (*see arsenic sources*)
metal-rich acid effluent (*see arsenic sources*)
metamorphic basement 48
metasediment 103
methionine 19, 186, 427, 460, 463, 495
   methylarsonic acid (*see arsenic metabolites*)
methylation (*see arsenic methylation*)
Mg$_2$Si$_2$O$_6$ (*see enstatite*)
mice 130, 135, 147, 436, 442, 443, 460, 461, 468, 469, 478
micro-fissuring of aquifer (*see aquifer*)
microbial
   activity in rhizospheric soil 387
   volatilization of arsenic 495
migmatite 103
military site (*see also arsenic source*) 537, 539, 543
milk 25, 419–421, 423, 424, 702
   bovine 423
mine workers (*see arsenic exposure*) 4
mineral equilibria 118
mineralized bodies (*see also arsenic sources*) 119
mineralogical
   analyses 65, 85, 507
   characteristics 522
mineralogy 39, 40, 205, 256, 509
mining (*see arsenic sources, anthropogenic*)
   impacted rhizospheric soil 392
mitochondrial pathways 430
MMA (*see arsenic metabolites*)
mobile reactive barrier (*see arsenic remediation*)
mobility of arsenic (*see arsenic mobility*)
molybdenum 54, 195, 345, 397, 513, 597, 677
monomethylarsenate (MMA) (*see arsenic metabolites*)

monomethylarsonic acid [MMA(V)] (*see arsenic metabolites*)
monomethylarsonous acid [MMA(III)] (*see arsenic metabolites*)
montmorillonite 53, 55, 57, 513
multilayered aquifer (*see aquifer*)
muscovite 55, 65, 208, 513
mutagen induced tumors (*see arsenic toxicity*)
mutagenicity of arsenic (*see arsenic toxicity*)

N
NAC 430, 468, 470, 471
NAC diminished CK18 induction 470
N-acetylcysteine 430
N-acetyl-L-cysteine 468
nails (*see arsenic biomarkers*)
natural arsenic contamination (*see arsenic sources*)
natural materials as arsenic adsorbents (*see arsenic sorbents*)
nerves
   peripheral 482, 486
neurological effects of F exposure 457
neurotoxicity of arsenic (*see arsenic toxicity*)
nitrate 37, 50, 52, 133, 134, 200, 207, 218, 267, 596, 615, 618, 626, 630, 665–667, 678
nitrite 51, 52, 114, 659

O
oil
   reservoir (*see arsenic sources*)
   spill (*see also arsenic sources*) 81
oilfield fluids 148
olivine basalt 147
opal 55
organic
   arsenic 11, 17–20, 181, 182, 235, 335, 428, 430, 537, 538, 542–544, 581
   carbon content 4, 288, 291
   matter 64, 84, 85, 88, 129, 133, 135–138, 140, 199–201, 217, 285–288, 292, 304, 341, 342, 345, 348, 360, 372, 380, 388, 390, 391, 497, 557, 606, 608, 626, 638, 671, 678, 679, 681, 692
organoarsenical species (*see also arsenic species*) 319, 322
orogenic cycle 92
orpiment 81, 132, 176, 245, 258, 260, 665
oxidation 3, 12, 19, 87, 98, 132, 160, 171, 173, 183–185, 187, 196, 201, 205–207, 209, 218, 246, 248, 255, 256, 262, 266, 267, 273, 277, 279, 296, 298, 299, 311, 314, 367, 368, 387, 391, 427, 475, 477, 478, 481, 505, 507, 518, 521, 528, 538, 539,

549, 551, 555, 556, 559, 571, 582–585,
589, 595, 598, 605–609, 612, 615, 616,
626, 631, 653, 665, 667–669, 677–681,
691
oxidizing aquifer (*see aquifer*)
oxidizing condition of groundwaters (*see
groundwater*)

P
p53 (*see also arsenic toxicity*) 429, 489–493
    activation 489, 492
    as gene activator 489
    phosphorylated 489
    promoter 429
    response after As-exposure 489
p53-DNA 489–492
    binding 489–492
        activity 490–492
            after As exposure 490
palaeosols (*see soil*)
paleochannel 48
pancreatic
    beta cells 477, 478
    oxidative damage 474
    TBARS level 477
    TrxR activity, 475, 477, 478
pathway of arsenic intake (*see arsenic
    exposure*)
pattern of groundwater flow (*see groundwater*)
PBMC (peripheral blood mononuclear cells)
    435, 437, 439, 441, 443
    activation 443
pedogenesis 292
    enrichment of arsenic 292
pegmatite 71, 273
people exposed (*see arsenic exposure*)
people living in mining areas (*see arsenic
    exposure*)
periodic groundwater discharge (*see also
    groundwater*) 43
peripheral blood mononuclear cells (*see
    PBMC*)
permeable reactive barriers (*see also
    remediation*) 627
petroleum (*see also arsenic sources*)
    exploitation 147
    refining 182
    reservoir 145–149
PHA (*see phytohemagglutinin*)
phosphate 8, 19, 132–135, 138, 218, 288, 291,
    322, 361, 383, 388, 391, 413, 428, 437,
    461, 468, 475, 482, 528, 538, 557, 574,
    597, 600, 607–609, 612, 659
    competition with As absorption sites (soil)
        361

minerals 132, 133
phosphate-exchangeable As 291
phosphorylation 428
    oxidative 428
photocatalytic method (*see arsenic removal
    technologies*)
phreatic aquifer (*see aquifer*)
phytoremediation (*see arsenic remediation*)
phytostabilization and rhizofiltration of
    polluted sites 650
phytotoxic threshold of arsenic (*see also
    arsenic phytotoxicity*) 650
phytotoxicity of arsenic (*see arsenic
    phytotoxicity*)
phytohemagglutinin (PHA)
    stimulus 435, 436, 441, 442
piedmont 48, 50, 398
    fanglomerate 398
piezometric levels (*see groundwater*)
Piper classification (*see groundwater*)
placenta 25, 26, 415
plagioclase 41, 51, 55, 64, 70
plants irrigated with As-contaminated water
    226
polyaluminum chloride (PACl) (*see arsenic
    coagulants*)
polyaluminum sulfates (*see arsenic
    coagulants*)
pond sand filters (*see arsenic removal, water*)
porewater 35, 39, 40, 42, 43, 171–176
postglacial fluvial sediments 286
potassium 51, 151, 168, 169, 196, 207, 208,
    304, 678
    feldspar 53, 55, 57, 71, 151
    mica 53, 57
precipitation of
    arsenian pyrite 193, 201
    arsenic 12, 89, 175, 177, 201, 266, 533, 548,
        631, 633, 691
        into iron sulfides 201
    arsenic sulfides 175, 177
    iron hydroxides 87
    sulfur 131
pre-oxidation (*see arsenic removal
    technologies*)
production of ROS 430, 486
proliferation of activated T cells 442
proliferation of PHA-stimulated lymphocytes
    442
protein-DNA interactions 491
provisional tolerable daily intake (PTDI) 186
provisional tolerable weekly intake (PTWI)
    186, 398, 404, 405

protein (*see also arsenic toxicity*)19, 20, 240, 321, 340, 428, 430, 461, 467–469, 470, 471, 473, 475, 477, 478, 489–492
    damage in cells 428
PTDI (*see provisional tolerable daily intake*)
PTWI (*see provisional tolerable weekly intake*)
public
    drinking water supply 6, 10, 48, 50, 265
    health 3, 17, 22, 24, 26, 155, 225, 226, 351, 427, 527, 581, 665
        problem 17, 22, 24, 26
pumice deposit 130
pyrite 8, 74, 77, 81, 103, 104, 132, 149–151, 160, 175, 193, 196, 199, 200, 201, 205–207, 246, 258, 260, 262, 273, 279, 313, 316, 335, 380, 447, 505, 511, 547, 606, 665, 699
    mine 380
    oxidation 200, 201, 205, 207
pyroclastic deposit 130, 140
pyroxen 41, 55
pyrrholite 8
pyrrhotite 160, 699
pyruvate dehydrogenase 19, 428

Q
quartz 41, 48, 50, 53, 55, 57, 61, 63, 64, 66, 70, 71, 85, 88, 92, 103, 149, 151, 176, 194, 197–199, 206, 208, 287, 289, 400, 461, 507, 513, 522, 540, 596, 617
quartzite 92, 208
quiescent lakes 138

R
radon 129, 131, 138 140
    soil gas 139
rainwater harvesters (*see arsenic removal, water*)
Raman spectra 549, 551–553
Raman spectroscopy 549, 551
reactive oxygen species (ROS) 429, 430, 468, 470, 471, 473, 478, 481, 482, 486, 616, 623
realgar (AsS) 81, 245, 258, 665
red earth (*see arsenic sorbents*)
red mud (*see arsenic sorbents*)
redox
    changes 88
    conditions 87, 109, 125, 127, 193, 199, 201, 209, 547, 629, 631, 662, 692
        reducing 4, 61, 71, 81, 132, 184, 185, 200, 201, 205, 209, 211, 390, 538, 547, 548, 551, 606, 693

potential 50, 52, 57, 69, 75, 77, 84, 87, 114, 115, 121, 125, 183, 200, 209, 267, 345, 372, 387, 511, 539, 630
    reactions of cells 19
reducing
    conditions (*see redox*)
    environment (*see also redox*) 201, 205, 538, 548, 595
reductive dissolution (*see arsenic mobilization*)
refinement of gold 18
refining of metals 17
regenerative proliferation 428
regional flow path (*see groundwater*)
remediation of Fe-deficiency (plants) 372
removal of arsenic (*see arsenic removal*)
residential dust (*see also arsenic in*) 95
respiratory effects of arsenic 124, 415
reverse osmosis (*see arsenic removal technologies*)
rhizospheric soil (*see soil*)
rhyodacite 130, 146
rhyolite 56, 132, 140, 146
risk of As exposure (*see arsenic exposure*)
rock
    outcrop 48, 113, 115, 119
    samples 118, 506, 509
rock-groundwater interactions 114
ROS (*see reactive oxygen species*)
rural
    areas 3, 4, 7, 9, 12, 13, 17, 20, 26, 123, 132, 193, 455, 456, 528, 616, 658, 665, 677, 690, 691, 694
    community 10, 12, 123, 409, 691
        wells 123
    dwellers 26
    population 3, 11, 12, 47, 50, 92, 109, 397, 687, 689, 694
    settlements 527
rutile 55

S
S-adenosylmethionine (SAM) 19, 427, 460, 464
    homeostasis 464
salar 93
salinity 18, 36, 38, 47, 48, 50, 145, 149, 151, 200, 335, 397, 677
salt
    deposits 94
    pan 91–93
SAM (*see S-adenosylmethionine*)
sandstone 49, 72, 147–149, 205–209, 211, 212, 285, 292, 617, 619, 620, 623
sanidine (KSi₃AlO₈) 85

Sb 41, 51, 52, 54–56, 196, 311, 312, 365, 366, 380

scanning electron microscopy (*see analytical methods*)

schist 92, 103, 208

scorodite (FeAsO₄ × 2H₂O) 505

seafood products (*see arsenic in fish and seafood*)

sediment (*see also arsenic in sediment*) 6, 7, 8, 10, 31, 35, 36, 40—43, 47, 48, 50, 51, 53, 55–58, 61, 63–66, 69–71, 75, 77, 81–85, 87–89, 97, 98, 101–103, 106, 109, 110, 113, 114, 118–120, 125, 129–133, 135–141, 145, 147, 148, 150, 151, 156, 164, 166, 171, 177, 182, 183, 193–198, 201, 206, 208, 216–219, 222, 223, 225, 228, 236, 246, 273, 286, 292, 311, 359, 375, 380, 398, 402, 419, 447, 496, 511, 547, 557, 563, 590, 593, 606, 615, 617, 631, 677, 687, 691–696, 700, 703

  alluvial 48, 109, 113, 217, 692

  chemistry 40, 41, 130, 135, 196

  extract 36, 42, 43

  flood plain 223

  fluvial 6, 286

  lake 8, 81, 82, 98, 129, 132, 136, 138, 139, 236

  leaching tests (*see also sequential extraction*) 82

  postglacial fluvial 286

  pyroclastic deposit 130, 140

  river 7, 359, 700

  sample 41, 85, 103, 133, 138–140, 141, 173, 195

  volcaniclastic 398

sedimentary aquifer (*see aquifer*)

sediment-groundwater interactions 114

sediment-hosted Au deposits 273

sediment-water interaction 201

sediment-water interface 171, 174, 175, 176

selenate [Se(VI)] 474, 478

selenite [Se(IV)] 265–267, 269, 474

  supranutritional intake (rats) 478

selenium (Se)

  deficiency 473

  deficient diet (rats) 473, 474, 476–478

  influence on i-As-exposed populations 479

  inorganic 474, 478

  insulin-like effects (i-Se) 478

  insulinomimetic properties (i-Se) 478

  intake (rats) 474, 477, 478

    from food 474

  sufficient diet (rats) 474, 476

  sufficient rats 474

semiconductor manufacturing (*see arsenic source*)

sequential extraction 65, 85, 288, 289, 387, 388

Ser15 489, 492

Ser15 p53 phosphorylation 489, 492

serum glucagon 476, 478

shallow groundwater (*see groundwater*)

silica (SiO₂) 41, 51–53, 55, 56, 57, 61, 62, 65, 66, 71, 72, 85, 114, 115, 125–127, 130, 140, 151, 196, 198, 207, 208, 210, 218, 219, 257, 296, 312, 313, 391, 392, 507, 513, 522, 528, 538, 557, 566, 583, 600, 609, 612, 638, 649, 666

silicate minerals 51, 53, 61, 65, 66, 85, 391, 392

silicic volcanic centers 146

silicosis 416

silt 35, 41, 48–50, 55, 62, 70–72, 113, 164, 217, 360, 380, 419, 702

siltstone 72

slate 92, 259, 303, 582

smectite 41, 55, 512

smelting activities (*see arsenic sources*)

socioeconomic development 47

sodium 47, 51, 53, 71, 77, 97, 166–169, 207, 265, 267, 268, 270, 400, 421, 436, 437, 442, 443, 448, 460–463, 467, 473–475, 481–483, 489, 490, 492, 493, 540, 542, 567, 576, 590, 591, 609, 612, 618, 651, 652, 668, 670, 671, 678

soil (*see also arsenic in soil*) 4, 7, 8, 10, 12, 18, 48, 61, 64, 66, 85, 86, 91, 92, 95, 109, 111, 113–115, 118–120, 124, 125, 127, 129, 131, 132, 138–141, 155, 158, 161, 171, 172, 183, 217, 219, 236, 247, 251, 255–258, 262, 265, 281, 283, 285–290, 292, 295–300, 303–309, 311–316, 320, 321, 327, 335–343, 345–348, 351, 355, 356, 359–362, 365, 371–377, 379–385, 387–392, 402, 423, 427, 447, 473, 495–498, 511, 537, 540, 541, 543, 544, 547, 596, 606, 629, 631, 636, 638, 639, 643, 645, 646, 649–651, 700–702

  andosol 296

  calcic gleysol 287

  diffuse gas 138–140

  gas 129, 138–140

  gleyic horizon 287, 291, 292

  gleysols 286, 287, 291

  haploxeralf 380

  histosols histosols 285, 286, 291

  palaeosols 36

  polluted with copper 342

  radon 131, 138, 139

  rhizospheric soil 387–389, 392

microbial activity 387, 392
sample 114, 115, 120, 255–258, 297, 303, 311–316, 335, 337, 338, 342, 343, 360, 361, 374, 380, 388, 390, 496, 700
saprolite 113
soil-crop-food transfer 356
soil-irrigation water system 341
soil-plant route 371
soil-plant transfer 371, 372, 373
soil-to-leaf transfer factor for arsenic in peach 371
SORAS (*see arsenic removal technologies*)
sorbents for arsenic removal (*see arsenic sorbents*)
sorption for arsenic removal (*see arsenic sorbents*)
sorption of arsenic (*see arsenic sorption*)
speciation of As (*see arsenic speciation*)
spinel-type minerals 64
squamous epithelium 428
stripping voltammetry (*see analytical methods*)
subchronic exposure to i-As (*see arsenic exposure*)
subsurface treatment of arsenic in groundwater (*see arsenic remediation*)
sulfate (SO_4^{2-}) 51, 52, 66, 71, 77, 97, 125, 132–135, 138, 139, 155, 159, 160, 161, 166, 167, 172, 174–176, 200, 201, 206–211, 215, 218, 220, 237, 258, 265–268, 270, 273–275, 322, 329, 338, 374, 516, 527–534, 538, 542, 543, 547, 549, 551, 553, 556, 557, 559, 574, 583, 589, 607–609, 618, 666, 678
sulfate/sulfide redox process 174–176
sulfhydryl groups 240, 428
sulfide
    dissolution 174, 175, 274
    minerals 98, 132, 391
    ores 8, 273
    production 175, 176
sulfur
    minerals 206
    salts 206, 207
superoxide dismutase 430
sural nerves (*see also arsenic in*) 481—486
surface water 4, 5, 7, 8, 12, 18, 35, 44, 50, 88, 91, 92, 95–98, 149, 163, 166, 167, 169, 180, 217, 230, 246, 337, 339, 401–404, 505, 521, 582, 595, 605, 615, 677, 688, 690, 700

T
T cells 439, 442, 443
T lymphocytes 443
tailings (*see arsenic sources*)

targeting arsenic-safe aquifers 66, 223, 687, 690, 692, 694, 695
TBARS 475, 477, 478, 482–486
    concentration 478, 483–486
tetramethylarsonium ion (TMA$^+$) (*see arsenic metabolites*)
TF (*see arsenic transfer factor*)
Th/Tc
    cells 440, 442
    lymphocyte ratio 439
    ratio 437, 438, 442, 443
thermal fluids (*see arsenic sources*)
thiobarbituric acid reactive substances (*see TBARS*)
thiol groups 19, 471
thioredoxin (Trx) (*see also arsenic toxicity*) 460, 467, 473, 475, 477–480
    system 478
    reductase 473
    TrxR activity 473, 475, 477, 478
    reduced 477
TiO$_2$ 12, 41, 55, 130, 507, 513, 677–679, 681, 691, 694
titano-magnetite 61, 64–66
TMA$^+$ (*see arsenic metabolites*)
TMAO (*see arsenic metabolites*)
toba 113
total
    arsenic (t-As) (*see also arsenic in*) 8, 22, 41, 75, 81, 84–87, 98, 123, 125, 126, 159, 171–173, 175, 176, 179, 181–183, 185, 186, 218, 235, 242, 255, 265–268, 270, 277, 289, 299, 307, 311, 313, 314, 316, 319, 321, 323, 327–331, 335–343, 345, 345–348, 353, 355, 356, 360, 365, 366, 368, 369, 375, 376, 387–392, 405, 441–443, 458, 461, 462, 482–486, 495–498, 527, 533, 540, 542–544, 559, 561, 567
    diet 351
    dissolved solids (TDS) 36, 50–52, 84–86, 88, 126, 149, 151, 166, 219, 421, 422, 608, 618, 666–668, 672
    inorganic carbon 61, 63, 64, 66, 574
    organic carbon 61, 63, 64, 66, 288–291
tourmaline 55, 208
toxicological
    effects of arsenic (*see arsenic toxicity*)
    interactions between Se and As (*see arsenic toxicity*)
    problems (*see arsenic toxicity*)
    risk of arsenic (*see also arsenic toxicity*) 343, 398, 406
trace-element 38, 43, 47, 48, 53, 55–57, 306, 374, 398

transfer factor of arsenic (*see arsenic transfer factor*)
transformation of primary and secondary minerals 62
trimethylarsine oxide (TMAO) (*see arsenic metabolites*)
Trx 475, 477, 478
TrxR (*see also arsenic toxicity*) 475, 477, 478
tube well (*see well*)
tuff 72, 103, 147, 179, 398
tumor (*see arsenic toxicity*)

U
ulvöspinel 64
unconfined aquifer (*see aquifer*)
unsaturated zone 109, 119, 120
$UO_2(CO_2)_3^{2-}$ (*see also uranium*) 54
$UO_2(CO_3)_3^{4-}$ (*see also uranium*) 54
uptake of As (*see arsenic exposure and arsenic in*)
uranium 54, 56, 397
    dissolved species 54
urinary
    arsenic (*see arsenic biomarkers*)
    creatinine (*see arsenic biomarkers*)
    tract 21, 23
urine (*see arsenic biomarkers*)

V
vanadate 35, 43, 54
vanadium 18, 397, 677
vegetables (*see arsenic in*)
vitamin A 23, 186
vitamin C 186
vitamin E 430, 482
vitamin supplements 20, 26
volatile arsenicals 495, 497, 498, 625
volatile elements 94
volatilization of arsenic (*see arsenic volatilization*)
volatilizing microbes 495
volcanic
    activity 94, 129, 138, 140, 141, 166, 379, 398
    arsenic 129
    ash 6, 18, 36, 40, 41, 56, 61, 73, 75, 125, 132, 201, 296, 419, 692, 693
    emissions 215, 265, 625
    gases 129, 132, 138, 139, 140, 182
    glass 6, 36, 41, 53, 55, 56, 61, 64–66, 70, 125, 127, 132, 419, 677, 693
        composition 56

dissolution 56
particles 61, 64, 66
shards 36, 55, 419
rocks 6–9, 41, 74, 86, 97, 103, 109, 110, 115, 118–120, 130, 132, 139, 140, 156, 185, 398, 409, 690
volcaniclastic sediments 398
volcanism (*see also arsenic sources*) 4, 9, 18, 71
    caldera 8, 129–131, 138–140
        collapse 130
    calderic eruption 8, 141
    crater lake 129, 138
    fumarole 8, 18, 94, 148
    pyroclastic deposit 130, 140

W
waste mineral deposit (*see also arsenic source*) 159–161
water
    quality (*see also arsenic in water*) 44, 48, 123, 137, 138, 141, 164, 215, 216, 218, 230, 285, 342, 402, 409, 557, 563, 607, 612, 615, 619, 623, 668, 670, 677, 690
    resources 35, 47, 91, 109, 120, 201, 209, 581, 687
    samples 6–8, 22, 36, 37, 43, 50, 53, 54, 56, 71, 82, 84, 88, 109, 114, 115, 118, 123, 125, 131, 133, 138, 139, 141, 167, 179, 182, 184, 195, 196, 210, 215, 218–222, 226, 229, 230, 247, 251, 253, 267, 270, 292, 337, 346, 353, 354, 397, 402, 405, 421, 422, 454, 529, 530, 539, 540, 542, 597, 668, 670, 678, 679
    source for the rural population 109
    supply 6, 7, 10, 12, 22, 24, 36, 48, 50, 109, 110, 112, 113, 127, 155, 179, 205, 207, 212, 222, 223, 230, 265, 285, 354, 400, 450, 505, 506, 527, 538, 590, 657, 690
        problem 110
        program 230
water borne contaminants 230
water-rock interaction 145, 149–151, 205, 207, 212, 289, 547
water-sediment interaction 201
water-soluble and exchangeable As 291
well
    shallow 11, 52, 161, 220, 398, 400, 615, 677
    tube (well) 7, 10, 25, 109, 113, 114, 123, 125, 215, 217, 225–230, 351, 353–355, 409, 687, 688, 692, 693, 695

weathering 4, 6–8, 12, 42, 61, 65, 66, 73, 75, 77, 103, 109, 113, 138, 215, 219, 245, 246, 379, 398, 565, 625

    chemical 113, 219

WHO guideline for arsenic 22, 37, 38, 54, 56, 92, 97, 123, 179, 225, 453, 521

wood preservation (*see also arsenic sources*) 182, 342, 379, 419, 649, 666

X

Y

Z

zircon 55, 61, 65, 66, 208

zerovalent iron (*see arsenic sorbents*)

Zn-deficiency (plant) 371

zone of sulfide production 175, 176

αFeOOH 295, 547, 617

Index II: Localities, stratigraphic units, tectonic and structural elements

Note: Many of the names refer to different expressions. So, e.g. "Pazña" may refer to Pazña town, Pazña river, etc. Only in special cases, where it is essential, the full expressions are given.

Abkatún 146–148
Acatic 181, 183, 184, 186
Achichilco 307
Aconquija 70, 71
Actopan 307, 308
Agua Suja stream 360
Aguafría 126
Aguanaval river 163, 164, 166
Aguas Calientes 24
Aguas Frias 113, 115, 116, 118, 120
Alfarcito 403
Altiplano 91–94, 172, 345–347, 397, 399
Alto Comedero 404
Alto Lima II 25, 95
Alvaro Obregón 156
Amajac 307
Andean mountain range 6
Antequera river 92
Antofagasta 6, 9–11, 18, 23, 24, 81, 345, 415, 586, 689
Antonio Pereira 360
Arandas 181, 184, 186
Árboles Grandes 72, 75
Argentina 3, 4, 6, 9–12, 17–19, 21, 22, 24, 26, 35, 36, 38, 40, 47–49, 57, 61, 66, 70, 81, 85, 91, 118, 127, 132, 155, 181, 193, 201, 245, 397, 400, 419, 420, 427, 511, 521, 547, 565, 589, 590, 595, 615, 618, 663, 675, 685, 688–690, 692, 693
Argentine Puna 397, 398
Arica 23, 345, 351, 353
Aricota lake 7, 18
Aruwakkalu 522
Atacama desert 6, 23
Atitalaquia 307
Atotonilco de Tula 307
Avaroa 93
Avicaya 94
Azapa 341, 345
Aznalcóllar 380, 382–384

Badenian sands 285
Bahia state 101

Baja California 8, 9, 146, 148, 663
Bajo Guadalquivir 236
Ballia 215–223
Banda 22
Bangladesh 12, 17, 20, 61, 64, 65, 87, 118, 127, 155, 181, 185, 186, 193, 201, 205, 216, 222, 225–230, 245–247, 251, 253, 307, 308, 351, 352, 354–356, 397, 427, 511, 521, 547, 565, 595, 600, 605, 606, 685, 686, 688–692, 697
Barra Feliz 362, 447, 449
Batab 147, 148
Bell Ville 21, 35, 420
Benito Juárez 156
Bihar 215
Bohemia 512, 548
Bolivia 3, 4, 6, 10, 17, 18, 25, 91, 92, 97, 688
Bolivian Altiplano 91, 397
Brahmanbaria 65
Brazil 3, 4, 6, 10, 17, 18, 21, 25, 91, 101, 253, 361, 362, 448, 688, 697, 701
Brumal 447, 449
Buenos Aires 6, 21, 400, 675, 677
Buntsandstein 205–208
Burruyacú 40, 47–52, 54–57

Caan 147, 148
Cactus-Sitio Grande 146–149, 151
Calama 11, 23, 336, 345, 689
Callazas river 18
Camarones river 12
Camarones valley 341, 342
Cañadas de Obregón 181
Caquiza 96
Cara Pujio 62, 64
Caracilla 96
Cardonal 307
Cartaya 374
Caspana 23
Castellón 207
Catalonia 311, 316, 365
Central America 10, 124, 129, 130, 141, 409
Central American graben 109, 130

Central depression of Tucumán 70
Cercado 93
Cerdanya region 274
Cerro La Chorrea 113
Cerro La Mina 113, 117, 118
Cerro Los Patos island 131
Cerro Mina de Agua 7, 25, 110
Cerro Prieto 145, 146, 148, 151
Cerros Quemados domes 129
Cerros Quemados islands 129, 133,
    138
Chaco 6, 10, 21, 61, 615, 675,
    690
Chaco-Pampean plain 6, 35, 36, 44, 47, 48, 56,
    69, 70, 72, 132, 193, 201, 397, 419, 675,
    688–690
Chaguite river 8, 130
Chhagalnaiya 352, 353, 355
Chhoti Saryu 217
Chihuahua 8, 24, 651, 663
Chile 3, 4, 6, 9–12, 17, 18, 21, 23, 61, 81, 85,
    91, 92, 155, 181, 185, 245, 246, 327,
    328, 335–337, 341–343, 345–349, 397,
    415, 427, 581–583, 586, 595, 605, 688,
    689
Chilean Puna de Atacama 398
China 17, 21, 155, 181, 187, 245, 521,
    595
Chittagong 354
Chiu-Chiu 23, 335, 336, 342
Chuquiña 96
Cintra 420
Ciudad del Carmen 147
Coahuila 8, 11, 24, 163, 181, 527, 529, 530,
    533, 663
Coatepeque lake 8
Cochinoca 398, 399
Colina 341
Collada de Tosses 312, 366
Collazas 7
Comalcalco 147
Comalcalco basin 147
Comarca Lagunera 164, 187, 423,
    527–529
Copiapó 23
Coquimbo 345
Córdoba 10, 21, 22, 35, 36, 48, 56, 385, 419,
    420, 677
Cortadera river 97
Costa Rica 3, 91, 688
Cotopaxi 7
Coyol group 109, 113, 409
Cruz de la India 7, 25, 110
Cruz Pintada dam 666
Cuesta de Lipán 403

Cuiabá 359
Cumbaya 7
Cuquío 180, 181
Czech Republic 355, 511, 512
Daganbhuiyan 352–355
Dendhó 307
Desaguadero 93, 94, 96
Desaguadero river 92–96, 98
Desaque river 130
Detzani 156
Doñana national park 380
Dorsal de Tacanas 48
Doxey 307
Durango 8, 11, 24, 163, 181, 454, 527, 529,
    530, 663

Ecuador 3, 6, 7, 9, 81, 82, 86, 89, 688
Eduardo Castex 36, 39
Ejido de Salamanca 529, 530
El Alto 25, 95, 181, 183, 184, 186
El Angel river 7
El Arbolito 72, 75
El Cacao 113, 116, 118, 120
El Carchi 7
El Charco 7, 110
El Derrumbado 113, 117, 118
El Espinal 72, 75
El Mayrán 533
El Mojon 25
El Morcillo 126
El Moreno 402, 403
El Palmar 126
El Quinche 7
El Salvador 3, 4, 6, 8, 9, 17, 129–131, 141,
    688
El Tambo hot springs 83
El Tatio 81
El Vicario 380
El Zapote 7, 10, 25, 109, 110, 112, 113, 115,
    117–120, 123, 409–412, 414–416, 688
Embalse La Virgen 114
Encarnación de Díaz 181, 183, 184, 186
Enseño Cove 18
Español bridge 92, 96
Estelí 113, 123–125

Fco I Madero 164
Feni 351, 352, 354–356
Feni Sadar 352–355
First Region (Chile) 341
Francisco I. Madero 156
Francisco Zarco 163
Franconian sandstone 285
Frontera fault 147

Ganga 215, 217, 222, 223
Ganges delta 216
Gangetic plain 215–217, 222
Ghagra 215, 217, 222, 223
Ghazipur 215, 216, 218–223
Gómez Palacio 163
Gran Porvenir 22
Guadalquivir 236
Guadalupe 156
Guadiamar river 380
Guadiana valley 181
Guanajuato 8, 24, 569, 663
Guasayán 70–72
Guatemala 3, 129, 688
Guayllabamba 7
Guerrero 8, 11, 663

Hajek 511, 512
Hermosillo 181, 187
Hidalgo 8, 155–159, 161, 171, 181, 184, 186,
    303–308, 436, 663
Huancané river 96, 97
Huancar 402, 403
Huancavelica 24
Huanuni 94, 96
Huari 96, 97
Huari river 96
Huautla 663, 665–667, 669
Huaya Pajchi river 97
Huaytara 24
Huelva 374
Huitel 307

Iberian Peninsula 205
Ilo 7, 11, 18, 24, 689
Ilopango 8, 129, 130, 132, 137, 140, 141
Ilopango caldera 129–131, 140
Ilopango lake 8, 129–141
Imbabura 7
India 17, 20, 21, 23, 61, 115, 155, 181, 185,
    186, 215, 216, 218, 220, 223, 245, 246,
    251, 253, 415, 416, 427, 496, 511, 521,
    547, 595, 596, 600, 605, 606, 663, 685
Indo-Gangetic plain 215–217
Iquique 23
Iron Quadrangle 359, 361, 697, 701
Itapicuru greenstone 101, 102
Itos 94
Iturbe 307
Ixmiquilpan 303, 307, 308

Jalostotitlán 181, 184, 186
Jamanco hot springs 83, 85
Jessore 226
Jesús María 181, 184, 186

Jharkhand 215
Jhikargachha 225–227, 229
Jinotega 7, 110
Jocote Renco 126
Juchusuma 96
Juchusuma river 96
Jujo-Tecominoacán 146–149
Jujuy 397–399, 402, 404
Julian Villagrán 307

Kaňk 548
Kinuma 7, 25, 110
Kutna Hora 548

La Banda 22
La Ceiba 113, 118, 120
La Cruz de la India 25
La Grecia 126
La Libertad 7, 110
La Montañita 113
La Pampa 11, 21, 35–41, 43, 44, 56, 58, 61,
    65, 127, 675, 689
La Paz 18, 25, 92, 95, 98
La Poma 398, 399
La Puna 397–406
La Ramada range 48
La Sabaneta 113
La Serena 23, 345
La Soledad 72, 75
La Union 113, 116, 118
Ladislao Cabrera 93
Lagos de Moreno 181, 183, 184, 186
Lagunera 11, 24, 163, 164, 166, 169, 171, 181,
    187, 423, 527–529
Lagunilla 307, 308, 403
Lamadrid 71, 72, 75
Las Cañas 72
Las Mangas 109, 110, 113, 117, 118, 120
Las Pilas 25
Las Trancas formation 156, 505
Lasana 335, 336, 342
Latin America 3–5, 9–13, 17, 18, 21, 26, 67,
    91, 129, 193, 201, 685, 686, 688–693
Laxmipur 65
Lázaro Cárdenas 163
Lerdo 163, 529, 530
Llallagua 94
Llano La Tejera 7, 110
Llapallapani 94
Lluta 341
Loa 85, 335–337, 339–341, 346
Locumba river 24, 25
Los Altos de Jalisco 179–187
Los Andes 398, 399
Los Azufres 146, 148, 150, 151

Los Cercos 72
Los Humeros 146, 148, 149, 151
Los Humeros geothermal field 146, 148
Los Pereyra 36, 39, 53, 55, 56, 615, 616–619, 621, 623, 676, 678, 679
Lules 617, 620
Luna-Sen 146–149

M. Juarez 22
Macleay Arm estuary 194, 200
Macleay river 195, 196
Macuspana basin 147
Manuel Belgrano 398
Mariana 360, 443
Márquez 93
Márquez river 93, 94
Matehuala 172, 174
Matlab 222, 685, 687, 690–692
Mauri 93
Mayrán 164, 166, 167, 169
Mayrán Viesca lagoon
Merkur quarry 511, 512
Mexico 3, 4, 8–11, 17–19, 21, 24, 26, 67, 91, 145, 147–151, 155, 156, 163, 169, 171, 172, 174, 176, 179, 181, 185, 187, 223, 245, 246, 253, 255, 256, 303–305, 307, 308, 387, 397, 423, 427, 442, 443, 454, 455, 479, 487, 505, 511, 527, 529, 530, 533, 547, 565, 569, 595, 650, 651, 656, 663, 688, 693
Mexico City 67, 147, 156, 304, 693
Mexticacán 179, 181, 183, 184, 186
Mezquital valley 304
Michoacán 8
Middle America 4, 9, 130, 147
Middle America trench 147
Mili 22
Milluni 18
Minas Gerais 18, 25, 359, 360, 362, 697
Mixquiahuala 307
Moctezuma 156, 454–456
Monsenhor Horta 360
Monterrey 255, 256
Moquegua 18, 24
Morelos 24
Morrison 420
Morro Velho 359
Mt Yarrahapinni 194, 195
Munich 285, 286, 289, 291
Murillo 25

Naica 651
Napo province 82
Nazas river 163, 164, 167, 169
Nepal 216, 511, 547

New Zealand 74, 81, 85, 235, 321
Nicaragua 4, 6–10, 17, 25, 91, 109–114, 116, 118, 120, 121, 123–125, 409, 410, 416, 688, 689
Nicaraguan depression 109, 113
Nicolás Flores 156
Nispero 149
Nova Lima 18, 25, 697, 700
Nuevo León 8, 24, 255, 256
Nuevo Libano 62, 64

Ojuelos de Jalisco 181, 183, 184, 186
Olomega 8
Oruro 25, 92–96
Ouro Preto-Mariana 359

Pachac 24
Pacific Ring of Fire 7, 18
Pacula 156
Pajaro Verde 666, 668
Pajaro Verde mine 666, 668
Palpalá 398, 399, 404
Palwal 351, 353, 355
Pampa Aullagas 97
Pampean hills 47
Pampean plain 6, 419
Pampean region 61, 66
Papallacta 83
Papallacta lake 7, 81–85, 88, 89
Parcila 126
Parshuram 352–355
Paso Julian 93
Passagem de Mariana 359
Pazña 95, 96
Pedro Ma. Anaya 307
Perico 399, 404
Peru 3, 4, 7, 9–11, 17, 18, 21, 24, 91, 92, 245, 427, 688, 689
Phulgazi 352–355
Pichincha 7
Pichincha province 7
Pifo 7
Planoles 312–314, 366
Platanares 126
Pol-Chuc 146–149
Poopó 93, 94, 96
Poopó basin 7, 93, 97, 98
Poopó lake 7, 91–98
Poopó river 92, 96
Potosí 8, 24, 92, 255, 256, 387, 454
Potrero de las Tablas 617
Pozo Colorado 403
Presa Requena 307
Puebla 8, 24
Puelche aquifer 6

Puembo 7
Puesto Sey 402, 403, 405
Puno 7, 18, 24, 25
Purmamarca 403
Pyrenees 274, 311, 312, 316, 365, 366

Queralbs 312, 313, 366
Querétaro 156
Quijos 82

Raigón aquifer 6
Ramos Arizpe 529, 530
Rancagua 23
Raposos 359
Real de Ángeles 255, 256
Real de la Cruz 113, 117, 118, 120
Represo 307
Ribes de Freser 312, 316, 366
Rimac river 24
Rinconadilla 403
Río Coco 124, 125
Río Cuarto 22
Rio das Velhas 697
Río Dulce 132, 685, 688, 690, 692
Río Itapicuru greenstone belt 101, 102
Río Primero 22
Ripollés 312, 366
Roberto Centeno 113
Robles 22
Rosario 303, 307, 308, 403, 589
Rosario de Lerma 22
Rua do Carrapato 447, 449

Sabana Larga 113, 120
Salado river 7, 18
Salamanca 8, 11, 530
Salar del Carmen 342
Salí river 36, 40, 48, 615, 676
Salitral de Carrera 454, 455
Salta 21, 398, 399, 402, 675
Salton Sea 146
Samaria-Sitio Grande reservoir 146
Samta village 355
San Antonio 160, 398, 399
San Antonio de Litín 420
San Antonio de los Cobres 22, 399, 402, 403
San Antonio-El Triunfo 9
San Diego de Alcalá 651
San Diego de Alejandría 181, 184, 186
San José department 6
San Juan de Limay 123–125
San Juan de los Lagos 181, 183, 184, 186
San Juan de Sora Sora 95
San Juan de Sora-Sora river 92

San Juan Sólis 307
San Juan Tepa 303, 307
San Julián 181, 184, 186
San Justo 22
San Lorenzo 126
San Luis 21, 36
San Luis Potosí 8, 24, 255, 256, 387, 454, 457
San Marcos 307, 308
San Miguel El Alto 181, 183, 184, 186
San Pedro de Atacama 23
San Pedro de las Colonias 533
San Ramon de la Uva 113, 117
San Salvador de Jujuy 399
Santa Barbara 18, 25, 359, 360, 362, 701
Santa Fe 6, 11, 21, 22, 36, 419, 589, 675, 689
Santa Inés formation 164
Santa Rosa del Peñón 7, 25, 110
Santana do Morro 447, 449, 450, 701
Santiago del Estero 11, 21, 22, 36, 56, 61, 62, 65, 66, 675, 676, 689, 692
Santuario de Quillacas 97
São Bento 359
São Francisco craton 101
Sathkira 66
Saucari 93
Sébaco valley 8, 25, 109–116, 118–121, 409
Sebastián Pagador 93
Second Region (chile) 335–337, 342, 343
Seville (Sevilla) 236, 380
Sierra Daxi 156
Sierra de Espadán 207
Sierra del Monte 156
Sierra El Mechudo 9
Sitio Grande 149
Socaire 342
Soledad de Graciano Sanchez 454, 455
Sonagazi 352–355
Sonora 8, 24, 181
South Carangas 93
South German Molasse Trough 286
Soyatal 506, 507, 509
Soyatal formation 156, 505
Soyatal limestone 506, 509
Spain 185, 205, 212, 236, 274, 311, 312, 316, 327, 343, 355, 365, 366, 374, 380
Sri Lanka 522
Stuarts Point aquifer 193–201
Sub Andine valleys 397–402, 404, 405
Subandean hills 48
Sucus
Sucus lake 82
Sucus river 82, 83, 85, 86, 89

Sumidouro 447, 449
Susques 398, 399, 402, 403

Tacagua 96
Tacna 18
Taco Pozo 10, 22
Taco Ralo 71
Taiwan 17, 20, 21, 23, 61, 155, 181, 187,
    193, 246, 397, 427, 565, 595, 605,
    663
Talabre 342
Talleres Norte 36, 39, 40, 42, 43
Taltal 23, 581, 584, 586
Tamagnoni 36, 39, 40, 42, 43
Tambo 86
Tambo river 7, 81, 83, 85–89
Taratunich 147, 148
Tatazcame 109, 113, 116, 118
Tecozautla 156
Temuthe 156
Teocalco 307
Teocaltiche 181, 183, 184, 186
Tepatepec 307
Tepatitlán de Morelos 181, 183, 184, 186
Terai 216
Termas de Río Hondo 70, 77
Tezontepec 307
Tierra Blanca Joven deposit 130
Titicaca 94
Titicaca lake 92, 93
Tlahualilo 164, 166, 167
Tlalquitenango 665
Tlamacazapa 8, 11
Toconce river 10
Tocopilla 23
Toledo bridge 96
Torreón 8, 24, 163
Transecuatorian pipeline 82
Transmexican volcanic belt 145–148, 179–181
Tres Morros 402, 403
Tumbaco 7

Tumbaya 398, 399, 403
Tungurahua 7

Unión 22, 116, 216, 303, 306, 373, 397
Unión de San Antonio 181, 184, 186
Uru Uru 91, 93, 94, 97, 98
Urueña river 48
Uruguay 6
Uttar Pradesh 215, 216, 220, 222
Uyuni salar 93

Valencia 207, 236
Vall d'Uixo 207
Vall de Ribes 311, 365
Valle de Guadalupe 181, 183, 184, 186
Valparaíso 10, 23
Venustiano Carranza 156
Viesca 164, 166
Vilavella 209, 210
Villa Cañás 590, 593, 594
Villa de la Paz 171, 172, 176, 178, 387
Villa Hidalgo 181, 184, 186
Villanueva 7, 110
Vinto 25, 95, 96
Vº Río Hondo 72, 75
Volcán 403

Wairakei geothermal field 81
West Bohemia 512

Xibi-Xibi river 404
Xochitlán 307, 308

Yahualica de González Gallo 180, 181
Yaruqui 7
Yucamane volcano 7, 18

Zacatecas 8, 24, 163, 255, 256
Zimapán 8, 11, 24, 155–161, 171, 181, 187,
    436, 442, 443, 505–509
Zimapán Centro 156

Author index

Acero, P. 273
Acosta, L. 435
Aguilera-Alvarado, A.F. 565
Aguirre, V. 81
Ahmed, K.M. 225
Alam, M.G.M. 295
Alarcón-Herrera, M.T. 649
Alfaro-De la Torre, M.C. 171
Alonso, M.S. 47
Altamirano Espinoza, M. 109
Alvarez-Silva, M. 527
Alves, T.V. 447
Andrade, G. 327
Armienta, M.A. 155, 505
Asta, M.P. 273
Ávila Carrera, M.E. 397
Aviles, M. 665

Ballinas-Casarrubias, M.L. 657
Barahona, F. 129
Barberá, R. 319
Bastías, J.M. 327
Bastida, M. 435
Beltrán-Hernández, R.I. 303
Bhattacharya, P. 3, 61, 91, 215,
    225, 677, 687
Bianco de Salas, G. 397
Billib, M. 537
Birkle, P. 3, 145
Blesa, M.A. 615
Boochs, P.-W. 537
Bovi Mitre, G. 397
Brůha, T. 547
Briones-Gallardo, R. 387
Bundschuh, J. 3, 61, 81, 91,
    109, 129, 145, 215, 225,
    677, 687

Calderon, C. 665
Calderón, E. 435
Calderón, J. 453
Cama, J. 273
Cano-Aguilera, I. 565
Canyellas, C. 123
Carrizales, L. 453
Castro de Esparza, M.L. 17
Castro-Larragoitia, J. 171

Cebrián, M. 435
Cebrian, M.E. 303, 665
Čerňanský, S. 495
Chávez, C. 81
Chandrajith, R. 521
Chaudhari, S. 595
Choi, H. 625
Conde, P. 435
Cortina, J.L. 123
Cruz-Jiménez, G. 565
Cumbal, L.H. 3, 81, 571

Díaz, O. 335
De Giudicci, G. 273
De Haro-Bailón, A. 235, 379
De Haro-Bravo, M.I. 379
de la Rosa, G. 565
Del Río-Celestino, M. 235, 379
Del Razo, L.M. 303, 459, 467,
    473, 481
Deschamps, E. 359, 447, 699
d'Hiriart, J. 615
Doušová, B. 511, 547

Escalante, J. 397
Estévez, J.R. 265
Esteller, M.V. 205
Etchichury, M.C. 47

Falcón, C.M. 47
Farías, S.S. 397
Farré, R. 319
Fernández, D.S. 69
Fernández, R.G. 555, 589
Fernández-Cirelli, A. 419
Flores-Valverde, E. 505
Font, R. 235, 379
Fuentealba, C. 581
Fuentes-Ramírez, R. 565
Fuitová, L. 511
Furet, N.R. 371

Gómez, A. 409
Galindo, M.C. 69
García, J.W. 47
García, M.E. 3, 91
García, M.G. 69, 615
García-Chávez, E. 481

Gardea-Torresdey, J.L. 179, 643
Garrido, S. 665
Giménez, E. 205
Giordano, M. 459
Gonsebatt, M.E. 459, 467
González-Sánchez, G. 657
Grygar, T. 511
Gutiérrez-Ojeda, C. 163
Gutiérrez-Ospina, G. 459
Gutiérrez-Ruiz, M. 255
Gutiérrez-Valtierra, M. 565

Haque, N. 565, 643
Harazim, B. 537
Hasan, M.A. 225
Hasan, M.T. 351
Hassan, M.M. 225
Hernández, J.C. 371
Hidalgo, M. del V. 69, 615
Holländer, H.M. 537
Hossain, M.M. 351
Hurtado-Jiménez, R. 179

Ingallinella, A.M. 589
Izquierdo-Vega, J.A. 473

Jacks, G. 687
Jakariya, M. 225, 687
Jankowski, J. 193
Jiménez, I. 481

Königskötter, H. 285
Kanel, S.R. 625
Khilar, K.C. 595
Kinniburgh, D.G. 35
Koloušek, D. 511, 547
Krüger, T. 537
Kumar, Pankaj 215

López, D.L. 129
López, L. 435
López-Bayghen, E. 489
López-Bellido, R.J. 371
López-Sánchez, J.F. 311, 365
López-Zepeda, J.L. 255
Laparra, J.M. 319
Lara-Castro, R.H. 171
Levy, I.K. 677

Limón-Pacheco, J. 459
Litter, M.I. 615, 677, 687
Lucho-Constantino, C.A. 303
Luna, A. 435

Macdonald, D.M.J. 35
Machovič, V. 511, 547
Mahatantila, K. 521
Manjarrez-Nevárez, L.A. 657
Marchetti, N. 345
Marcus, M. 255
Marijuán, L. 371
Martaus, A. 511, 547
Martin, R. 61
Mateu, M. 677
Matschullat, J. 3, 447, 699
Medina, H. 91
Melgoza-Castillo, A. 649
Menezes, M.A.B.C. 359
Merchant, H. 481
Merino, M.H. 47
Micete, S. 505
Mokgalaka, N.S. 643
Monroy-Fernández, M.G. 387
Montero, A. 265
Monterrosa, J. 129
Montoro, R. 235, 319, 335
Morales, L. 123
Morales, M. 489
Morales, R. 453
Morell, I. 205
Moreno, C. 69
Morgada de Boggio, M.E. 677
Moscuzza, C. 419
Muñiz, G. 657
Muñoz, O. 327
Mugica, V. 155
Murgueitio, E. 81

Núñez, N. 335
Núñez-Montoya, O.G. 649
Nahar, S. 225
Nalini, H.A., Jr. 359
Nava-Alonso, F. 527
Navarro, M.E. 453
Nazim Uddin, M. 351
Nemade, P.D. 595
Nicolli, H.B. 35, 47
Nieto, A. 665
Novák, M. 61

Oberdá, S.M. 447, 699
Olmos, R. 129

Orihuela, D.L. 371
Ormachea, M. 91
Ortega, A. 489
Ortiz, E. 155
O'Shea, B. 193
Ostrosky-Wegman, P. 489
Pérez-Carrera, A. 419
Pérez-Garibay, R. 527
Pérez-Mohedano, S. 371
Pažout, R. 547
Palmieri, H.E.L. 359
Pardo-Rueda, J. 657
Pastene, R. 335
Pelallo-Martinez, N.A. 171
Peralta-Videa, J.R. 643
Petrusevski, B. 555
Poggi-Varaldo, H.M. 303
Ponce, R.I. 397
Prieto-García, F.R. 303
Puigdomènech, C. 123
Puntí, A. 123
Pupo, I. 265

Quintanilla, J. 91

Rüde, T.R. 285
Rahman, I.M.M. 351
Ramírez, E. 155
Ramírez, P. 467
Ramírez-Ramírez, M.L. 565
Ramanathan, AL. 215
Ramirez, A. 665
Ramirez-Orozco, A. 665
Ramos, O. 91
Ransom, L. 129
Recabarren G., E. 335
Reddy, K.J. 605
Reséndiz, R. 155
Rocha, C.A. 447
Rocha-Amador, D.O. 453
Rodríguez, V.M. 459
Rodriguez Almeida, F.A. 649
Romero, F. 255
Roth, T.R. 605
Royo-Márquez, M.H. 649
Rubio, R. 311, 365
Rueda-Ramírez, A. 657
Ruiz-Chancho, M.J. 311, 365

Sánchez-Peña, L.C. 459, 473
Sancha, A.M. 123, 345, 581
Sandoval, M. 489
Sastre-Conde, I. 303

Schippers, J. 555
Seed, L. 665
Segura, B. 481
SenGupta, A.K. 571
Servant, R.E. 397
Ševc, J. 495
Sharma, S. 555
Shelp, G. 665
Shroff, M. 427
Silva, A.B. 101
Silva, N.O.C. 447, 697
Singh, Anshumali 215
Singh, K.K. 427
Smedley, P.L. 35
Soriano, T. 129
Soto, C. 473
Soto, G. 435
Sposito, G. 255
Sracek, O. 61, 69
Stecca, L.M. 589
Stummeyer, J. 537
Sulovský, P. 61

Thunvik, R. 91, 215, 687
Tineo, A. 47
Tipán, I. 81
Tofalo, O.R. 47
Tokunaga, S. 295
Torres, E. 123
Torres-Muñoz, V. 657
Tripathi, P. 215

Urík, M. 495
Uribe-Querol, E. 459
Uribe-Salas, A. 527

Vázquez-Rodríguez, G. 387
Vélez, D. 235, 319, 335
Valcárcel, L.A. 265
Vasconcelos, O.R. 359, 699
Vega, L. 435
Vera, E. 665
Vilches, S. 327
Villalobos, M. 255
Villatoro-Pulido, M.M. 379
Vithanage, M. 521
von Brömssen, M. 687
Vujcic, M. 427

Wajrak, M. 245
Weerasooriya, R. 521